NINTH EDITION

BIOSTATISTICS

Basic Concepts and Methodology for the Health Sciences

WAYNE W. DANIEL
Professor Emeritus
Georgia State University

INTERNATIONAL STUDENT VERSION

WILEY
JOHN WILEY & SONS, INC.

Cover image from © Laurence Dutton/Stone/Getty Images

ISBN: 978-0-470-41333-3

Printed in Asia

10 9 8 7 6 5 4 3 2 1

To my wife, Mary,
and my children, Jean,
Carolyn, and John

Special thanks are extended to Professor Chad Cross, of the School of Public Health at the University of Nevada, Las Vegas, for his many contributions to this edition. He has introduced numerous new topics and explanatory comments that have broadened greatly the scope of the book.

PREFACE

This ninth edition of *Biostatistics: Basic Concepts and Methodology for the Health Sciences* should appeal to the same audience for which the first eight editions were written: advanced undergraduate students, beginning graduate students, and health professionals in need of a reference book on statistical methodology.

Like its predecessors, this edition requires few mathematical prerequisites. Only reasonable proficiency in algebra is required for an understanding of the concepts and methods underlying the calculations. The emphasis continues to be on an intuitive understanding of principles rather than an understanding based on mathematical sophistication.

For most of the statistical techniques covered in this edition, we discuss the capabilities of one or more software packages (MINITAB, SAS, SPSS, and NCSS) that may be used to perform the calculations needed for their application. Resulting screen displays are also shown.

NEW TO THIS EDITION

Chapter Overviews. In this edition, we introduce each chapter with a brief chapter overview that alerts students to the concepts that they will encounter as they read and study the chapter. The chapter overviews use non-technical language in order to provide students with a general understanding of the chapter contents without having to be confronted with unfamiliar terminology.

Leaning Outcomes. Before they begin reading each chapter, students are provided with a list of learning outcomes that inform them of what they will be expected to know after having read and studied the chapter. Instructors may also use the learning outcomes as guides when preparing chapter-by-chapter syllabi.

Summaries of Equations. Where appropriate, students will find at the ends of chapters a summary of the equations that were used in the chapter. This feature will provide students with a quick reference source when working on homework assignments. Instructors who wish to do so may allow students to bring copies of the equation summaries to the classroom for use during tests.

New Topics. Following is a chapter-by-chapter summary of the topics that are new to this edition of *Biostatistics*.

Chapter 1 Getting Acquainted With Biostatistics. The sampling topic is expanded to include a discussion of systematic sampling, including comments on stratified systematic sampling and stratified sampling proportional to size. A new section is devoted to the scientific method and the design of experiments.

Chapter 2 Strategies for Understanding the Meanings of Data. Skewness and kurtosis are discussed in considerable detail and illustrated with computer-generated graphs.

Chapter 3 Probability: The Basis of Statistical Inference. The discussion of Bayesian probability is expanded to enhance students' understanding of the application of Bayes's theorem.

Chapter 4 Probabilistic Features of Certain Data Distributions. The discussion of probability distributions is expanded.

Chapter 5 Probabilistic Features of the Distributions of Certain Sample Statistics. This chapter is essentially unchanged.

Chapter 6 Using Sample Data to Make Estimates about Population Parameters. Several brief comments and new computer output are added for the purpose of clarifying certain topics.

Chapter 7 Using Sample Statistics to Test Hypotheses About Population Parameters. This chapter contains several new computer printouts.

Chapter 8 Statistical Inference and the Analysis of Data Variability. Additional comments and new computer printouts are added to help clarify several topics covered in this chapter.

Chapter 9 Statistical Inference and the Relationship Between Two Variables. Several explanatory comments, a section on testing the regression assumptions, and several computer printouts are new to this chapter.

Chapter 10 Statistical Inference and the Relationships Among Three or More Variables. New to this chapter are several computer printouts and comments for added clarification.

Chapter 11 Additional Techniques for the Analysis of Relationships Among Variables. The discussion of regression assumptions is expanded to include the following topics: non-normal data, unequal error variances, and correlated independent variables. The discussion of variable selection procedures is expanded to include forward selection and backward elimination strategies. Discussions of multiple logistic regression and polytomous logistic regression have been added to the logistic regression section.

Chapter 12 Analysis of Frequency Data: An Introduction to the Chi-Square Distribution. This chapter contains several new explanatory paragraphs, new examples, and new computer printouts. The section on survival analysis has been expanded and augmented with new computer output.

Chapter 13 Special Techniques for Use When Population Parameters and/or Population Distributions Are Unknown. New explanatory comments and computer printouts have been added in this chapter.

Chapter 14 Evaluating the Health of Human Groups: Vital Statistics. The introduction now includes a paragraph on epidemiology.

SUPPLEMENTS

Instructor's Solutions Manual. Prepared by Chad Cross, University of Nevada, Las Vegas. Includes solutions to all problems found in the text. Available only to instructors who have adopted the text.

Data Sets. More than 250 data sets of varying sizes have been integrated throughout the exposition, evidencing this edition's focus on currency and relevance to modern students. All examples, section exercises, and review exercises of more then 20 entries are available at the Wiley Web site below. The large data sets are designed for analysis by the following techniques: probability (Chapter 3), interval estimation (Chapter 6), hypothesis testing (Chapter 7), analysis of variance (Chapter 8), simple linear regression (Chapter 9), multiple regression (Chapter 10), advanced regression analysis (Chapter 11), and chi-square (Chapter 12). Exercises at the end of these chapters instruct students on how to use the large data sets. The data sets are available to both instructor and student for download from the Wiley Web site at

www.wiley.com/go/global/daniel

If you do not have Internet access, please contact your Wiley representative.

ACKNOWLEDGMENTS

For their many helpful suggestions on how to make this edition of *Biostatistics* better, I wish to express my gratitude to the many readers of the previous editions and to the instructors who have used the book in their classrooms. In particular, I thank the following people who made valuable recommendations for this revision:

Emilia Bagiella	Columbia University
Yi Cheng	Indiana University, South Bend
Casey Cremins	University of Maryland
Kate Crespi	University of California, Los Angeles
Edward Danial	Morgan State University
David Elashoff	University of California, Los Angeles

Stephen Ganocy	Case Western Reserve University
Bhisham Gupta	University of Southern Maine
James Halavin	Rochester Institute of Technology
Jean Lambert	University of Montreal
Mia Markey	University of Texas, Austin
Nutan Mishra	University of Southern Alabama
Amal Mitra	University of Southern Mississippi
Thomas Jay Morrison	Ohio Northern University
Theophile Niyonsenga	Florida International University
Leonard Onyiah	St. Cloud State University
Robert Serfling	University of Texas, Dallas
Richard Shiavi	Vanderbilt University
Pramil Singh	Loma Linda University
James Stanley	Baylor University
Gary Stanek	Youngstown State University
Paul Weiss	Emory University

I wish to acknowledge the cooperation of Minitab, Inc. for making available to me over the years the latest versions of the MINITAB software package for illustrating the use of the microcomputer in statistical analysis.

I gratefully acknowledge a major contribution to this edition of *Biostatistics* by Professor John P. Holcomb of the Department of Mathematics at Cleveland State University, Cleveland, Ohio.

Thanks go to Professor Chad L. Cross of the School of Public Health at the University of Nevada, Las Vegas, for the new material that he provided for this ninth edition.

Thanks are due to my colleagues at Georgia State University—Professors Geoffrey Churchill and Brian Schott, who wrote computer programs for generating some of the Appendix tables—and Professor Lillian Lin, who read the section on logistic regression and made valuable suggestions for its improvement.

I gratefully acknowledge the help of Dr. James T. Wassell for his help with the section on survival analysis.

I am grateful to the many researchers in the health sciences field who so generously made available raw data from their research projects. These data appear in the examples and exercises and are acknowledged individually wherever they appear. I would also like to thank the editors and publishers of the various journals who gave permission to reprint data from their publications for use in many of the examples and exercises.

Wayne W. Daniel

CONTENTS

CHAPTER 1

GETTING ACQUAINTED WITH BIOSTATISTICS

CHAPTER OVERVIEW

This chapter is intended to provide an overview of the basic statistical concepts used throughout the textbook. A course in statistics requires the student to learn many new terms and concepts. This chapter lays the foundation necessary for the understanding of basic statistical terms and concepts and the role that statisticians play in promoting scientific discovery and wisdom.

TOPICS

LEARNING OUTCOMES

After studying this chapter, the student will

1. understand the basic concepts and terminology of biostatistics, including the various kinds of variables, measurement, and measurement scales.
2. be able to select a simple random sample and other scientific samples from a population of subjects.
3. understand the processes involved in the scientific method and the design of experiments.
4. appreciate the advantages of using computers in the statistical analysis of data generated by studies and experiments conducted by researchers in the health sciences.

1.1 INTRODUCTION

We are frequently reminded of the fact that we are living in the information age. Appropriately, then, this book is about information—how it is obtained, how it is analyzed, and how it is interpreted. The information about which we are concerned we call data, and the data are available to us in the form of numbers.

The objectives of this book are twofold: (1) to teach the student to organize and summarize data, and (2) to teach the student how to reach decisions about a large body of data by examining only a small part of the data. The concepts and methods necessary for achieving the first objective are presented under the heading of *descriptive statistics,* and the second objective is reached through the study of what is called *inferential statistics*. This chapter discusses descriptive statistics. Chapters 2 through 5 discuss topics that form the foundation of statistical inference, and most of the remainder of the book deals with inferential statistics.

Because this volume is designed for persons preparing for or already pursuing a career in the health field, the illustrative material and exercises reflect the problems and activities that these persons are likely to encounter in the performance of their duties.

1.2 SOME BASIC CONCEPTS

Like all fields of learning, statistics has its own vocabulary. Some of the words and phrases encountered in the study of statistics will be new to those not previously exposed to the subject. Other terms, though appearing to be familiar, may have specialized meanings that are different from the meanings that we are accustomed to associating with these terms. The following are some terms that we will use extensively in this book.

Data The raw material of statistics is *data*. For our purposes we may define data as *numbers*. The two kinds of numbers that we use in statistics are numbers that result from the taking—in the usual sense of the term—of a *measurement,* and those that result from the process of *counting*. For example, when a nurse weighs a patient or takes a patient's temperature, a measurement, consisting of a number such as 150 pounds or 100 degrees Fahrenheit, is obtained. Quite a different type of number is obtained when a hospital administrator counts the number of patients—perhaps 20—discharged from the hospital on a given day. Each of the three numbers is a *datum,* and the three taken together are data.

Statistics The meaning of *statistics* is implicit in the previous section. More concretely, however, we may say that *statistics is a field of study concerned with* (1) *the collection, organization, summarization, and analysis of data; and* (2) *the drawing of inferences about a body of data when only a part of the data is observed.*

The person who performs these statistical activities must be prepared to *interpret* and to *communicate* the results to someone else as the situation demands. Simply put, we may say that data are numbers, numbers contain information, and the purpose of statistics is to investigate and evaluate the nature and meaning of this information.

Sources of Data The performance of statistical activities is motivated by the need to answer a question. For example, clinicians may want answers to questions regarding the relative merits of competing treatment procedures. Administrators may want answers to questions regarding such areas of concern as employee morale or facility utilization. When we determine that the appropriate approach to seeking an answer to a question will require the use of statistics, we begin to search for suitable data to serve as the raw material for our investigation. Such data are usually available from one or more of the following sources:

1. **Routinely kept records.** It is difficult to imagine any type of organization that does not keep records of day-to-day transactions of its activities. Hospital medical records, for example, contain immense amounts of information on patients, while hospital accounting records contain a wealth of data on the facility's business activities. When the need for data arises, we should look for them first among routinely kept records.
2. **Surveys.** If the data needed to answer a question are not available from routinely kept records, the logical source may be a survey. Suppose, for example, that the administrator of a clinic wishes to obtain information regarding the mode of transportation used by patients to visit the clinic. If admission forms do not contain a question on mode of transportation, we may conduct a survey among patients to obtain this information.
3. **Experiments.** Frequently the data needed to answer a question are available only as the result of an experiment. A nurse may wish to know which of several strategies is best for maximizing patient compliance. The nurse might conduct an experiment in which the different strategies of motivating compliance are tried with different patients. Subsequent evaluation of the responses to the different strategies might enable the nurse to decide which is most effective.
4. **External sources.** The data needed to answer a question may already exist in the form of published reports, commercially available data banks, or the research literature. In other words, we may find that someone else has already asked the same question, and the answer obtained may be applicable to our present situation.

Biostatistics The tools of statistics are employed in many fields—business, education, psychology, agriculture, and economics, to mention only a few. When the data analyzed are derived from the biological sciences and medicine, we use the term *biostatistics* to distinguish this particular application of statistical tools and concepts. This area of application is the concern of this book.

Variable If, as we observe a characteristic, we find that it takes on different values in different persons, places, or things, we label the characteristic a *variable*. We do this for the simple reason that the characteristic is not the same when observed in different possessors of it. Some examples of variables include diastolic blood pressure, heart rate, the heights of adult males, the weights of preschool children, and the ages of patients seen in a dental clinic.

Quantitative Variables A *quantitative variable* is one that can be measured in the usual sense. We can, for example, obtain measurements on the heights of adult males, the weights of preschool children, and the ages of patients seen in a dental clinic. These are examples of *quantitative variables*. Measurements made on quantitative variables convey information regarding amount.

Qualitative Variables Some characteristics are not capable of being measured in the sense that height, weight, and age are measured. Many characteristics can be categorized only, as, for example, when an ill person is given a medical diagnosis, a person is designated as belonging to an ethnic group, or a person, place, or object is said to possess or not to possess some characteristic of interest. In such cases measuring consists of categorizing. We refer to variables of this kind as *qualitative variables*. Measurements made on qualitative variables convey information regarding attribute.

Although, in the case of qualitative variables, measurement in the usual sense of the word is not achieved, we can count the number of persons, places, or things belonging to various categories. A hospital administrator, for example, can count the number of patients admitted during a day under each of the various admitting diagnoses. These counts, or *frequencies* as they are called, are the numbers that we manipulate when our analysis involves qualitative variables.

Random Variable Whenever we determine the height, weight, or age of an individual, the result is frequently referred to as a *value* of the respective variable. When the values obtained arise as a result of chance factors, so that they cannot be exactly predicted in advance, the variable is called a *random variable*. An example of a random variable is adult height. When a child is born, we cannot predict exactly his or her height at maturity. Attained adult height is the result of numerous genetic and environmental factors. Values resulting from measurement procedures are often referred to as *observations* or *measurements*.

Discrete Random Variable Variables may be characterized further as to whether they are *discrete* or *continuous*. Since mathematically rigorous definitions of discrete and continuous variables are beyond the level of this book, we offer, instead, nonrigorous definitions and give an example of each.

A discrete variable is characterized by gaps or interruptions in the values that it can assume. These gaps or interruptions indicate the absence of values between particular values that the variable can assume. Some examples illustrate the point. The number of daily admissions to a general hospital is a discrete random variable since the number of admissions each day must be represented by a whole number, such as 0, 1, 2, or 3. The number of admissions on a given day cannot be a number such as 1.5, 2.997, or 3.333. The number of decayed, missing, or filled teeth per child in an elementary school is another example of a discrete variable.

Continuous Random Variable *A continuous random variable does not possess the gaps or interruptions characteristic of a discrete random variable*. A continuous random variable can assume any value within a specified relevant interval of values assumed by the variable. Examples of continuous variables include the various measurements that can be made on individuals such as height, weight, and skull circumference. No matter how close together the observed heights of two people, for example, we can, theoretically, find another person whose height falls somewhere in between.

Because of the limitations of available measuring instruments, however, observations on variables that are inherently continuous are recorded as if they were discrete. Height, for example, is usually recorded to the nearest one-quarter, one-half, or whole inch, whereas, with a perfect measuring device, such a measurement could be made as precise as desired.

Population The average person thinks of a population as a collection of entities, usually people. A population or collection of entities may, however, consist of animals, machines, places, or cells. For our purposes, we define a *population of entities as the largest collection of entities for which we have an interest at a particular time*. If we take a measurement of some variable on each of the entities in a population, we generate a population of values of that variable. We may, therefore, define a *population of values as the largest collection of values of a random variable for which we have an interest at a particular time*. If, for example, we are interested in the weights of all the children enrolled in a certain county elementary school system, our population consists of all these weights. If our interest lies only in the weights of first-grade students in the system, we have a different population—weights of first-grade students enrolled in the school system. Hence, populations are determined or defined by our sphere of interest. Populations may be *finite* or *infinite*. If a population of values consists of a fixed number of these values, the population is said to be *finite*. If, on the other hand, a population consists of an endless succession of values, the population is an *infinite* one.

Sample A sample may be defined simply as *a part of a population*. Suppose our population consists of the weights of all the elementary school children enrolled in a certain county school system. If we collect for analysis the weights of only a fraction of these children, we have only a part of our population of weights, that is, we have a *sample*.

1.3 MEASUREMENT AND MEASUREMENT SCALES

In the preceding discussion we used the word *measurement* several times in its usual sense, and presumably the reader clearly understood the intended meaning. The word *measurement*, however, may be given a more scientific definition. In fact, there is a whole body of scientific literature devoted to the subject of measurement. Part of this literature is concerned also with the nature of the numbers that result from measurements. Authorities on the subject of measurement speak of measurement scales that result in the categorization of measurements according to their nature. In this section we define measurement and the four resulting measurement scales. A more detailed discussion of the subject is to be found in the writings of Stevens (1, 2).

Measurement This may be defined as the assignment of numbers to objects or events according to a set of rules. The various measurement scales result from the fact that measurement may be carried out under different sets of rules.

The Nominal Scale The lowest measurement scale is the *nominal scale*. As the name implies it consists of "naming" observations or classifying them into various mutually exclusive and collectively exhaustive categories. The practice of using numbers to distinguish among the various medical diagnoses constitutes measurement on a nominal scale. Other examples include such dichotomies as male–female, well–sick, under 65 years of age–65 and over, child–adult, and married–not married.

The Ordinal Scale Whenever observations are not only different from category to category but can be ranked according to some criterion, they are said to be measured on an ordinal scale. Convalescing patients may be characterized as unimproved, improved, and much improved. Individuals may be classified according to socioeconomic status as low, medium, or high. The intelligence of children may be above average, average, or below average. In each of these examples the members of any one category are all considered equal, but the members of one category are considered lower, worse, or smaller than those in another category, which in turn bears a similar relationship to another category. For example, a much improved patient is in better health than one classified as improved, while a patient who has improved is in better condition than one who has not improved. It is usually impossible to infer that the difference between members of one category and the next adjacent category is equal to the difference between members of that category and the members of the next category adjacent to it. The degree of improvement between unimproved and improved is probably not the same as that between improved and much improved. The implication is that if a finer breakdown were made resulting in more categories, these, too, could be ordered in a similar manner. The function of numbers assigned to ordinal data is to order (or rank) the observations from lowest to highest and, hence, the term *ordinal*.

The Interval Scale The *interval scale* is a more sophisticated scale than the nominal or ordinal in that with this scale not only is it possible to order measurements, but also the distance between any two measurements is known. We know, say, that the difference between a measurement of 20 and a measurement of 30 is equal to the difference between measurements of 30 and 40. The ability to do this implies the use of a unit distance and a zero point, both of which are arbitrary. The selected zero point is not necessarily a true zero in that it does not have to indicate a total absence of the quantity being measured. Perhaps the best example of an interval scale is provided by the way in which temperature is usually measured (degrees Fahrenheit or Celsius). The unit of measurement is the degree, and the point of comparison is the arbitrarily chosen "zero degrees," which does not indicate a lack of heat. The interval scale unlike the nominal and ordinal scales is a truly quantitative scale.

The Ratio Scale The highest level of measurement is the *ratio scale*. This scale is characterized by the fact that equality of ratios as well as equality of intervals may be determined. Fundamental to the ratio scale is a true zero point. The measurement of such familiar traits as height, weight, and length makes use of the ratio scale.

1.4 SAMPLING AND STATISTICAL INFERENCE

As noted earlier, one of the purposes of this book is to teach the concepts of statistical inference, which we may define as follows:

DEFINITION

Statistical inference is the procedure by which we reach a conclusion about a population on the basis of the information contained in a sample that has been drawn from that population.

There are many kinds of samples that may be drawn from a population. Not every kind of sample, however, can be used as a basis for making valid inferences about a population. In general, in order to make a valid inference about a population, we need a scientific sample from the population. There are also many kinds of scientific samples that may be drawn from a population. The simplest of these is the *simple random sample*. In this section we define a simple random sample and show you how to draw one from a population.

If we use the letter N to designate the size of a finite population and the letter n to designate the size of a sample, we may define a simple random sample as follows:

DEFINITION

If a sample of size n is drawn from a population of size N in such a way that every possible sample of size n has the same chance of being selected, the sample is called a simple random sample.

The mechanics of drawing a sample to satisfy the definition of a simple random sample is called *simple random sampling*.

We will demonstrate the procedure of simple random sampling shortly, but first let us consider the problem of whether to sample *with replacement* or *without replacement*. When sampling with replacement is employed, every member of the population is available at each draw. For example, suppose that we are drawing a sample from a population of former hospital patients as part of a study of length of stay. Let us assume that the sampling involves selecting from the shelves in the medical records department a sample of charts of discharged patients. In sampling with replacement we would proceed as follows: select a chart to be in the sample, record the length of stay, and return the chart to the shelf. The chart is back in the "population" and may be drawn again on some subsequent draw, in which case the length of stay will again be recorded. In sampling without replacement, we would not return a drawn chart to the shelf after recording the length of stay, but would lay it aside until the entire sample is drawn. Following this procedure, a given chart could appear in the sample only once. As a rule, in practice, sampling is always done without replacement. The significance and consequences of this will be explained later, but first let us see how one goes about selecting a simple random sample. To ensure true randomness of selection, we will need to follow some objective procedure. We certainly will want to avoid using our own judgment to decide which members of the population constitute a random sample. The following example illustrates one method of selecting a simple random sample from a population.

EXAMPLE 1.4.1

Gold et al. (A-1) studied the effectiveness on smoking cessation of bupropion SR, a nicotine patch, or both, when co-administered with cognitive-behavioral therapy. Consecutive consenting patients assigned themselves to one of the three treatments. For illustrative purposes, let us consider all these subjects to be a population of size $N = 189$. We wish to select a simple random sample of size 10 from this population whose ages are shown in Table 1.4.1.

TABLE 1.4.1 Ages of 189 Subjects Who Participated in a Study on Smoking Cessation

Subject No.	Age	Subject No.	Age	Subject No.	Age	Subject No.	Age
1	48	49	38	97	51	145	52
2	35	50	44	98	50	146	53
3	46	51	43	99	50	147	61
4	44	52	47	100	55	148	60
5	43	53	46	101	63	149	53
6	42	54	57	102	50	150	53
7	39	55	52	103	59	151	50
8	44	56	54	104	54	152	53
9	49	57	56	105	60	153	54
10	49	58	53	106	50	154	61
11	44	59	64	107	56	155	61
12	39	60	53	108	68	156	61
13	38	61	58	109	66	157	64
14	49	62	54	110	71	158	53
15	49	63	59	111	82	159	53
16	53	64	56	112	68	160	54
17	56	65	62	113	78	161	61
18	57	66	50	114	66	162	60
19	51	67	64	115	70	163	51
20	61	68	53	116	66	164	50
21	53	69	61	117	78	165	53
22	66	70	53	118	69	166	64
23	71	71	62	119	71	167	64
24	75	72	57	120	69	168	53

(*Continued*)

Subject No.	Age	Subject No.	Age	Subject No.	Age	Subject No.	Age
25	72	73	52	121	78	169	60
26	65	74	54	122	66	170	54
27	67	75	61	123	68	171	55
28	38	76	59	124	71	172	58
29	37	77	57	125	69	173	62
30	46	78	52	126	77	174	62
31	44	79	54	127	76	175	54
32	44	80	53	128	71	176	53
33	48	81	62	129	43	177	61
34	49	82	52	130	47	178	54
35	30	83	62	131	48	179	51
36	45	84	57	132	37	180	62
37	47	85	59	133	40	181	57
38	45	86	59	134	42	182	50
39	48	87	56	135	38	183	64
40	47	88	57	136	49	184	63
41	47	89	53	137	43	185	65
42	44	90	59	138	46	186	71
43	48	91	61	139	34	187	71
44	43	92	55	140	46	188	73
45	45	93	61	141	46	189	66
46	40	94	56	142	48		
47	48	95	52	143	47		
48	49	96	54	144	43		

Source: Paul B. Gold, Ph.D. Used with permission.

Solution: One way of selecting a simple random sample is to use a table of random numbers like that shown in the Appendix, Table A. As the first step, we locate a random starting point in the table. This can be done in a number of ways, one of which is to look away from the page while touching it with the point of a pencil. The random starting point is the digit closest to where the pencil touched the page. Let us assume that following this procedure led to a random starting point in Table A at the intersection of row 21 and column 28. The digit at this point is 5. Since we have 189 values to choose from, we can use only the random numbers 1 through 189. It will be convenient to pick three-digit numbers so that the numbers 001 through 189 will be the only eligible numbers. The first three-digit number, beginning at our random starting point is 532, a number we cannot use. The next number (going down) is 196, which again we cannot use. Let us move down past 196, 372, 654, and 928 until we come to 137, a number we can use. The age of the 137th subject from Table 1.4.1 is 43, the first value in our sample. We record the random number and the corresponding age in Table 1.4.2. We record the random number to keep track of the random numbers selected. Since we want to sample without replacement, we do not want to include the same individual's age twice. Proceeding in the manner just

TABLE 1.4.2 Sample of 10 Ages Drawn from the Ages in Table 1.4.1

Random Number	Sample Subject Number	Age
137	1	43
114	2	66
155	3	61
183	4	64
185	5	65
028	6	38
085	7	59
181	8	57
018	9	57
164	10	50

described leads us to the remaining nine random numbers and their corresponding ages shown in Table 1.4.2. Notice that when we get to the end of the column, we simply move over three digits to 028 and proceed up the column. We could have started at the top with the number 369.

Thus we have drawn a simple random sample of size 10 from a population of size 189. In future discussions, whenever the term simple random sample is used, it will be understood that the sample has been drawn in this or an equivalent manner. ■

The preceding discussion of random sampling is presented because of the important role that the sampling process plays in designing *research studies* and *experiments*. The methodology and concepts employed in sampling processes will be described in more detail in Section 1.5.

DEFINITION

A research study is a scientific study of a phenomenon of interest. Research studies involve designing sampling protocols, collecting and analyzing data, and providing valid conclusions based on the results of the analyses.

DEFINITION

Experiments are a special type of research study in which observations are made after specific manipulations of conditions have been carried out; they provide the foundation for scientific research.

Despite the tremendous importance of random sampling in the design of research studies and experiments, there are some occasions when random sampling may not be the most appropriate method to use. Consequently, other sampling methods must be considered. The intention here is not to provide a comprehensive review of sampling methods, but rather to acquaint the student with two additional sampling methods that are employed in the health sciences, *systematic sampling* and *stratified random sampling*. Interested readers are referred to the books by Thompson (3) and Levy and Lemeshow (4) for detailed overviews of various sampling methods and explanations of how sample statistics are calculated when these methods are applied in research studies and experiments.

Systematic Sampling A sampling method that is widely used in healthcare research is the systematic sample. Medical records, which contain raw data used in healthcare research, are generally stored in a file system or on a computer and hence are easy to select in a systematic way. Using systematic sampling methodology, a researcher calculates the total number of records needed for the study or experiment at hand. A random numbers table is then employed to select a starting point in the file system. The record located at this starting point is called record x. A second number, determined by the number of records desired, is selected to define the sampling interval (call this interval k). Consequently, the data set would consist of records x, $x + k$, $x + 2k$, $x + 3k$, and so on, until the necessary number of records are obtained.

EXAMPLE 1.4.2

Continuing with the study of Gold et al. (A-1) illustrated in the previous example, imagine that we wanted a systematic sample of 10 subjects from those listed in Table 1.4.1.

Solution: To obtain a starting point, we will again use Appendix Table A. For purposes of illustration, let us assume that the random starting point in Table A was the intersection of row 10 and column 25. The digit is a 4 and will serve as our starting point, x. Since we are starting at subject 4, this leaves 185 remaining subjects from which to choose. Since we wish to select 10 subjects, one method to define the sample interval, k, would be to take $185/10 = 18.5$. To ensure that there will be enough subjects, it is customary to round this quotient down, and hence we will round the result to 18. The resulting sample is shown in Table 1.4.3.

TABLE 1.4.3 Sample of 10 Ages Selected Using a Systematic Sample from the Ages in Table 1.4.1

Systematically Selected Subject Number	Age
4	44
22	66
40	47
58	53
76	59
94	56
112	68
130	47
148	60
166	64

■

Stratified Random Sampling A common situation that may be encountered in a population under study is one in which the sample units occur together in a grouped fashion. On occasion, when the sample units are not inherently grouped, it may be possible and desirable to group them for sampling purposes. In other words, it may be desirable to partition a population of interest into groups, or *strata*, in which the sample units within a particular stratum are more similar to each other than they are to the sample units that compose the other strata. After the population is stratified, it is customary to take a random sample independently from each stratum. This technique is called *stratified random sampling*. The resulting sample is called a *stratified random sample*. Although the benefits of stratified random sampling may not be readily observable, it is most often the case that random samples taken within a stratum will have much less variability than a random sample taken across all strata. This is true because sample units within each stratum tend to have characteristics that are similar.

EXAMPLE 1.4.3

Hospital trauma centers are given ratings depending on their capabilities to treat various traumas. In this system, a level 1 trauma center is the highest level of available trauma care and a level 4 trauma center is the lowest level of available trauma care. Imagine that we are interested in estimating the survival rate of trauma victims treated at hospitals within a large metropolitan area. Suppose that the metropolitan area has a level 1, a level 2, and a level 3 trauma center. We wish to take samples of patients from these trauma centers in such a way that the total sample size is 30.

Solution: We assume that the survival rates of patients may depend quite significantly on the trauma that they experienced and therefore on the level of care that they receive. As a result, a simple random sample of all trauma patients, without regard to the center at which they were treated, may not represent true survival rates, since patients receive different care at the various trauma centers. One way to better estimate the survival rate is to treat each trauma center as a stratum and then randomly select 10 patient files from each of the three centers. This procedure is based on the fact that we suspect that the survival rates within the trauma centers are less variable than the survival rates across trauma centers. Therefore, we believe that the stratified random sample provides a better representation of survival than would a sample taken without regard to differences within strata. ■

It should be noted that two slight modifications of the stratified sampling technique are frequently employed. To illustrate, consider again the trauma center example. In the first place, a systematic sample of patient files could have been selected from each trauma center (stratum). Such a sample is called a *stratified systematic sample*.

The second modification of stratified sampling involves selecting the sample from a given stratum in such a way that the number of sample units selected from that stratum is proportional to the size of the population of that stratum. Suppose, in our trauma center example that the level 1 trauma center treated 100 patients and the level 2 and level 3 trauma centers treated only 10 each. In that case, selecting a random sample of 10 from each trauma center overrepresents the trauma centers with smaller patient loads. To avoid this problem, we adjust the size of the sample taken from a stratum so that it is proportional to the size of the stratum's population. This type of sampling is called *stratified sampling proportional to size*. The within-stratum samples can be either random or systematic as described above.

EXERCISES

1.4.1 Using the table of random numbers, select a new random starting point, and draw another simple random sample of size 10 from the data in Table 1.4.1. Record the ages of the subjects in this new sample. Save your data for future use. What is the variable of interest in this exercise? What measurement scale was used to obtain the measurements?

1.4.2 Select another simple random sample of size 10 from the population represented in Table 1.4.1. Compare the subjects in this sample with those in the sample drawn in Exercise 1.4.1. Are there any subjects who showed up in both samples? How many? Compare the ages of the subjects in the two samples. How many ages in the first sample were duplicated in the second sample?

1.4.3 Using the table of random numbers, select a random sample and a systematic sample, each of size 15, from the data in Table 1.4.1. Visually compare the distributions of the two samples. Do they appear similar? Which appears to be the best representation of the data?

1.4.4 Construct an example where it would be appropriate to use stratified sampling. Discuss how you would use stratified random sampling and stratified sampling proportional to size with this example. Which do you think would best represent the population that you described in your example? Why?

1.5 THE SCIENTIFIC METHOD AND THE DESIGN OF EXPERIMENTS

Data analyses by statistical methods play a significant role in scientific studies. The previous section highlighted the importance of obtaining samples in a scientific manner. Appropriate sampling techniques enhance the likelihood that the results of statistical analyses of a data set will provide valid and scientifically defensible results. Because of the importance of the proper collection of data to support scientific discovery, it is necessary to consider the foundation of such discovery—the *scientific method*—and to explore the role of statistics in the context of this method.

DEFINITION

The scientific method is a process by which scientific information is collected, analyzed, and reported in order to produce unbiased and replicable results in an effort to provide an accurate representation of observable phenomena.

The scientific method is recognized universally as the only truly acceptable way to produce new scientific understanding of the world around us. It is based on an *empirical approach*, in that decisions and outcomes are based on data. There are several key elements associated with the scientific method, and the concepts and techniques of statistics play a prominent role in all these elements.

Making an Observation First, an *observation* is made of a phenomenon or a group of phenomena. This observation leads to the formulation of questions or uncertainties that can be answered in a scientifically rigorous way. For example, it is readily observable that regular exercise reduces body weight in many people. It is also readily observable that changing diet may have a similar effect. In this case there are two observable phenomena, regular exercise and diet change, that have the same endpoint. The nature of this endpoint can be determined by use of the scientific method.

Formulating a Hypothesis In the second step of the scientific method a *hypothesis* is formulated to explain the observation and to make quantitative *predictions* of new observations. Often hypotheses are generated as a result of extensive background research and literature reviews. The objective is to produce hypotheses that are scientifically sound. Hypotheses may be stated as either *research hypotheses* or *statistical hypotheses*. Explicit definitions of these terms are given in Chapter 7, which discusses the science of testing hypotheses. Suffice it to say for now that a research hypothesis from the weight-loss example would be a statement such as, "Exercise appears to reduce body weight." There is certainly nothing incorrect about this conjecture, but it lacks a truly quantitative basis for testing. A statistical hypothesis may be stated using quantitative terminology as follows: "The average (mean) loss of body weight of people who exercise is greater than the average (mean) loss of body weight of people who do not exercise." In this statement a quantitative measure, the "average" or "mean" value, is hypothesized to be greater in the sample of patients who exercise. The role of the statistician in this step of the scientific method is to state the hypothesis in a way that valid conclusions may be drawn and to interpret correctly the results of such conclusions.

Designing an Experiment The third step of the scientific method involves *designing an experiment* that will yield the data necessary to validly test an appropriate statistical hypothesis. This step of the scientific method, like that of data analysis, requires the expertise of a statistician. Improperly designed experiments are the leading cause of invalid results and unjustified conclusions. Further, most studies that are challenged by experts are challenged on the basis of the appropriateness or inappropriateness of the study's research design.

Those who properly design research experiments make every effort to ensure that the measurement of the phenomenon of interest is both accurate and precise. *Accuracy* refers to the correctness of a measurement. *Precision*, on the other hand, refers to the consistency of a measurement. It should be noted that in the social sciences, the term *validity* is sometimes used to mean accuracy and that *reliability* is sometimes used to mean precision. In the context of the weight-loss example given earlier, the scale used to measure the weight of study participants would be accurate if the measurement is validated using a scale that is properly calibrated. If, however, the scale is off by +3 pounds, then each participant's weight would be 3 pounds heavy; the measurements would be precise in that each would be wrong by +3 pounds, but the measurements would not be accurate. Measurements that are inaccurate or imprecise may invalidate research findings.

The design of an experiment depends on the type of data that need to be collected to test a specific hypothesis. As discussed in Section 1.2, data may be collected or made available through a variety of means. For much scientific research, however, the standard for data collection is experimentation. A true *experimental design* is one in which study subjects are randomly assigned to an *experimental group* (or *treatment group*) and a *control group* that is not directly exposed to a treatment. Continuing the weight-loss example, a sample of 100 participants could be randomly assigned to two conditions using the methods of Section 1.4. A sample of 50 of the participants would be assigned to a specific exercise program and the remaining 50 would be monitored, but asked not to exercise for a specific period of time. At the end of this experiment the average (mean) weight losses of the two groups could be compared. The reason that experimental designs are desirable is that if all other potential factors are controlled, a *cause–effect relationship* may be tested; that is, all else being equal, we would be able to conclude or fail to conclude that the experimental group lost weight as a result of exercising.

The potential complexity of research designs requires statistical expertise, and Chapter 8 highlights some commonly used experimental designs. For a more in-depth discussion of research designs, the interested reader may wish to refer to texts by Kuehl (5), Keppel and Wickens (6), and Tabachnick and Fidell (7).

Conclusion In the execution of a research study or experiment, one would hope to have collected the data necessary to draw conclusions, with some degree of confidence, about the hypotheses that were posed as part of the design. It is often the case that hypotheses need to be modified and retested with new data and a different design. Whatever the conclusions of the scientific process, however, results are rarely considered to be conclusive. That is, results need to be *replicated*, often a large number of times, before scientific credence is granted them.

EXERCISES

1.5.1 Using the example of weight loss as an endpoint, discuss how you would use the scientific method to test the observation that change in diet is related to weight loss. Include all of the steps, including the hypothesis to be tested and the design of your experiment.

1.5.2 Continuing with Exercise 1.5.1, consider how you would use the scientific method to test the observation that both exercise and change in diet are related to weight loss. Include all of the steps, paying particular attention to how you might design the experiment and which hypotheses would be testable given your design.

1.6 COMPUTERS AND BIOSTATISTICAL ANALYSIS

The widespread use of computers has had a tremendous impact on health sciences research in general and biostatistical analysis in particular. The necessity to perform long and tedious arithmetic computations as part of the statistical analysis of data lives only in the memory of those researchers and practitioners whose careers antedate the so-called computer revolution. Computers can perform more calculations faster and far more accurately than can human technicians. The use of computers makes it possible for investigators to devote more time to the improvement of the quality of raw data and the interpretation of the results.

The current prevalence of microcomputers and the abundance of available statistical software programs have further revolutionized statistical computing. The reader in search of a statistical software package may wish to consult *The American Statistician,* a quarterly publication of the American Statistical Association. Statistical software packages are regularly reviewed and advertised in the periodical.

Many of the computers currently on the market are equipped with random number generating capabilities. As an alternative to using printed tables of random numbers, investigators may use computers to generate the random numbers they need. Actually, the "random" numbers generated by most computers are in reality *pseudorandom numbers* because they are the result of a deterministic formula. However, as Fishman (8) points out, the numbers appear to serve satisfactorily for many practical purposes.

The usefulness of the computer in the health sciences is not limited to statistical analysis. The reader interested in learning more about the use of computers in the health sciences will find the books by Hersh (4), Johns (5), Miller et al. (6), and Saba and McCormick (7) helpful. Those who wish to derive maximum benefit from the Internet may wish to consult the books *Physicians' Guide to the Internet* (13) and *Computers in Nursing's Nurses' Guide to the Internet* (14). Current developments in the use of computers in biology, medicine, and related fields are reported in several periodicals devoted to the subject. A few such periodicals are *Computers in Biology and Medicine*, *Computers and Biomedical Research, International Journal of Bio-Medical Computing*, *Computer Methods and Programs in Biomedicine*, *Computer Applications in the Biosciences*, and *Computers in Nursing*.

Computer printouts are used throughout this book to illustrate the use of computers in biostatistical analysis. The MINITAB, SPSS, and SAS® statistical software packages for the personal computer have been used for this purpose.

1.7 SUMMARY

In this chapter we introduced the reader to the basic concepts of statistics. We defined statistics as an area of study concerned with collecting and describing data and with making statistical inferences. We defined statistical inference as the procedure by which we reach a conclusion about a population on the basis of information contained in a sample drawn from that population. We learned that a basic type of sample that will allow us to make valid inferences is the simple random sample. We learned how to use a table of random numbers to draw a simple random sample from a population.

The reader is provided with the definitions of some basic terms, such as variable and sample, that are used in the study of statistics. We also discussed measurement and defined four measurement scales—nominal, ordinal, interval, and ratio. The reader is also introduced to the scientific method and the role of statistics and the statistician in this process.

Finally, we discussed the importance of computers in the performance of the activities involved in statistics.

REVIEW QUESTIONS AND EXERCISES

1. Explain what is meant by descriptive statistics.
2. Explain what is meant by inferential statistics.
3. Define:
   (a) Statistics
   (b) Biostatistics
   (c) Variable
   (d) Quantitative variable
   (e) Qualitative variable
   (f) Random variable
   (g) Population
   (h) Finite population
   (i) Infinite population
   (j) Sample
   (k) Discrete variable
   (l) Continuous variable
   (m) Simple random sample
   (n) Sampling with replacement
   (o) Sampling without replacement
4. Define the word *measurement*.
5. List, describe, and compare the four measurement scales.
6. For each of the following variables, indicate whether it is quantitative or qualitative and specify the measurement scale that is employed when taking measurements on each:
   (a) Class standing of the members of this class relative to each other
   (b) Admitting diagnosis of patients admitted to a mental health clinic
   (c) Weights of babies born in a hospital during a year
   (d) Gender of babies born in a hospital during a year
   (e) Range of motion of elbow joint of students enrolled in a university health sciences curriculum
   (f) Under-arm temperature of day-old infants born in a hospital
7. For each of the following situations, answer questions a through e:
   (a) What is the sample in the study?
   (b) What is the population?
   (c) What is the variable of interest?
   (d) How many measurements were used in calculating the reported results?
   (e) What measurement scale was used?

   Situation A. A study of 300 households in a small southern town revealed that 20 percent had at least one school-age child present.

   Situation B. A study of 250 patients admitted to a hospital during the past year revealed that, on the average, the patients lived 15 miles from the hospital.
8. Consider the two situations given in Exercise 7. For Situation A describe how you would use a stratified random sample to collect the data. For Situation B describe how you would use systematic sampling of patient records to collect the data.

REFERENCES

Methodology References

1. S. S. Stevens, "On the Theory of Scales of Measurement," *Science, 103* (1946), 677–680.
2. S. S. Stevens, "Mathematics, Measurement and Psychophysics," in S. S. Stevens (ed.), *Handbook of Experimental Psychology,* Wiley, New York, 1951.
3. Steven K. Thompson, *Sampling* (2nd ed.), Wiley, New York, 2002.
4. Paul S. Levy and Stanley Lemeshow, *Sampling of Populations: Methods and Applications* (3rd ed.), Wiley, New York, 1999.
5. Robert O. Kuehl, *Statistical Principles of Research Design and Analysis* (2nd ed.), Duxbury Press, Belmont, CA, 1999.
6. Geoffrey Keppel and Thomas D. Wickens, *Design and Analysis: A Researcher's Handbook* (4th ed.), Prentice Hall, Upper Saddle River, NJ, 2004.
7. Barbara G. Tabachnick and Linda S. Fidell, *Experimental Designs using ANOVA*, Thomson, Belmont, CA, 2007.
8. George S. Fishman, *Concepts and Methods in Discrete Event Digital Simulation,* Wiley, New York, 1973.
9. William R. Hersh, *Information Retrieval: A Health Care Perspective,* Springer, New York, 1996.
10. Merida L. Johns, *Information Management for Health Professions,* Delmar Publishers, Albany, NY, 1997.
11. Marvin J. Miller, Kenric W. Hammond, and Matthew G. Hile (eds.), *Mental Health Computing,* Springer, New York, 1996.
12. Virginia K. Saba and Kathleen A. McCormick, *Essentials of Computers for Nurses,* McGraw-Hill, New York, 1996.
13. Lee Hancock, *Physicians' Guide to the Internet,* Lippincott Williams & Wilkins Publishers, Philadelphia, 1996.
14. Leslie H. Nicoll and Teena H. Ouellette, *Computers in Nursing's Nurses' Guide to the Internet, 3rd ed.,* Lippincott Williams & Wilkins Publishers, Philadelphia, 2001.

Applications References

A-1. Paul B. Gold, Robert N. Rubey, and Richard T. Harvey, "Naturalistic, Self-Assignment Comparative Trial of Bupropion SR, a Nicotine Patch, or Both for Smoking Cessation Treatment in Primary Care," *American Journal on Addictions, 11* (2002), 315–331.

CHAPTER **2**

STRATEGIES FOR UNDERSTANDING THE MEANINGS OF DATA

CHAPTER OVERVIEW

This chapter introduces a set of basic procedures and statistical measures for describing data. Data generally consist of an extensive number of measurements or observations that are too numerous or complicated to be understood through simple observation. Therefore, this chapter introduces several techniques including the construction of tables, graphical displays, and basic statistical computations that provide ways to condense and organize information into a set of descriptive measures and visual devices that enhance the understanding of complex data.

TOPICS

LEARNING OUTCOMES

After studying this chapter, the student will

1. understand how data can be appropriately organized and displayed.
2. understand how to reduce data sets into a few useful, descriptive measures.
3. be able to calculate and interpret measures of central tendency, such as the mean, median, and mode.
4. be able to calculate and interpret measures of dispersion, such as the range, variance, and standard deviation.

2.1 INTRODUCTION

In Chapter 1 we stated that the taking of a measurement and the process of counting yield numbers that contain information. The objective of the person applying the tools of statistics to these numbers is to determine the nature of this information. This task is made much easier if the numbers are organized and summarized. When measurements of a random variable are taken on the entities of a population or sample, the resulting values are made available to the researcher or statistician as a mass of unordered data. Measurements that have not been organized, summarized, or otherwise manipulated are called *raw data*. Unless the number of observations is extremely small, it will be unlikely that these raw data will impart much information until they have been put into some kind of order.

In this chapter we learn several techniques for organizing and summarizing data so that we may more easily determine what information they contain. The ultimate in summarization of data is the calculation of a single number that in some way conveys important information about the data from which it was calculated. Such single numbers that are used to describe data are called *descriptive measures*. After studying this chapter you will be able to compute several descriptive measures for both populations and samples of data.

The purpose of this chapter is to equip you with skills that will enable you to manipulate the information—in the form of numbers—that you encounter as a health sciences professional. The better able you are to manipulate such information, the better understanding you will have of the environment and forces that generate the information.

2.2 THE ORDERED ARRAY

A first step in organizing data is the preparation of an ordered array. An *ordered array* is a listing of the values of a collection (either population or sample) in order of magnitude from the smallest value to the largest value. If the number of measurements to be ordered is of any appreciable size, the use of a computer to prepare the ordered array is highly desirable.

An ordered array enables one to determine quickly the value of the smallest measurement, the value of the largest measurement, and other facts about the arrayed data that might be needed in a hurry. We illustrate the construction of an ordered array with the data discussed in Example 1.4.1.

EXAMPLE 2.2.1

Table 1.4.1 contains a list of the ages of subjects who participated in the study on smoking cessation discussed in Example 1.4.1. As can be seen, this unordered table requires considerable searching for us to ascertain such elementary information as the age of the youngest and oldest subjects.

Solution: Table 2.2.1 presents the data of Table 1.4.1 in the form of an ordered array. By referring to Table 2.2.1 we are able to determine quickly the age of the youngest subject (30) and the age of the oldest subject (82). We also readily note that about one-third of the subjects are 50 years of age or younger.

TABLE 2.2.1 Ordered Array of Ages of Subjects from Table 1.4.1

30	34	35	37	37	38	38	38	38	39	39	40	40	42	42
43	43	43	43	43	43	44	44	44	44	44	44	44	45	45
45	46	46	46	46	46	46	47	47	47	47	47	47	48	48
48	48	48	48	48	49	49	49	49	49	49	49	50	50	50
50	50	50	50	50	51	51	51	51	52	52	52	52	52	52
53	53	53	53	53	53	53	53	53	53	53	53	53	53	53
53	53	54	54	54	54	54	54	54	54	54	54	54	55	55
55	56	56	56	56	56	56	57	57	57	57	57	57	57	58
58	59	59	59	59	59	59	60	60	60	60	61	61	61	61
61	61	61	61	61	61	61	62	62	62	62	62	62	62	63
63	64	64	64	64	64	64	65	65	66	66	66	66	66	66
67	68	68	68	69	69	69	70	71	71	71	71	71	71	71
72	73	75	76	77	78	78	78	82						

■

Computer Analysis If additional computations and organization of a data set have to be done by hand, the work may be facilitated by working from an ordered array. If the data are to be analyzed by a computer, it may be undesirable to prepare an ordered array, unless one is needed for reference purposes or for some other use. A computer does not need its user to first construct an ordered array before entering data for the construction of frequency distributions and the performance of other analyses. However, almost all computer statistical packages and spreadsheet programs contain a routine for sorting data in either an ascending or descending order. See Figure 2.2.1, for example.

Dialog box:

Data ➤ Sort

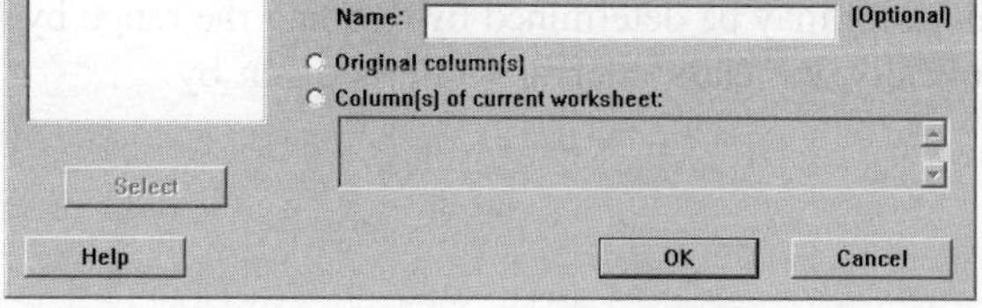

Session command:

```
MTB > Sort C1  C2;
SUBC>   By C1.
```

FIGURE 2.2.1 MINITAB dialog box for Example 2.2.1.

2.3 GROUPED DATA: THE FREQUENCY DISTRIBUTION

Although a set of observations can be made more comprehensible and meaningful by means of an ordered array, further useful summarization may be achieved by grouping the data. Before the days of computers one of the main objectives in grouping large data sets was to facilitate the calculation of various descriptive measures such as percentages and averages. Because computers can perform these calculations on large data sets without first grouping the data, the main purpose in grouping data now is summarization. One must bear in mind that data contain information and that summarization is a way of making it easier to determine the nature of this information.

To group a set of observations we select a set of contiguous, nonoverlapping intervals such that each value in the set of observations can be placed in one, and only one, of the intervals. These intervals are usually referred to as *class intervals.*

One of the first considerations when data are to be grouped is how many intervals to include. Too few intervals are undesirable because of the resulting loss of information. On the other hand, if too many intervals are used, the objective of summarization will not be met. The best guide to this, as well as to other decisions to be made in grouping data, is your knowledge of the data. It may be that class intervals have been determined by precedent, as in the case of annual tabulations, when the class intervals of previous years are maintained for comparative purposes. A commonly followed rule of thumb states that there should be no fewer than five intervals and no more than 15. If there are fewer than five intervals, the data have been summarized too much and the information they contain has been lost. If there are more than 15 intervals, the data have not been summarized enough.

Those who need more specific guidance in the matter of deciding how many class intervals to employ may use a formula given by Sturges (1). This formula gives $k = 1 + 3.322(\log_{10} n)$, where k stands for the number of class intervals and n is the number of values in the data set under consideration. The answer obtained by applying *Sturges's rule* should not be regarded as final, but should be considered as a guide only. The number of class intervals specified by the rule should be increased or decreased for convenience and clear presentation.

Suppose, for example, that we have a sample of 275 observations that we want to group. The logarithm to the base 10 of 275 is 2.4393. Applying Sturges's formula gives $k = 1 + 3.322(2.4393) \simeq 9$. In practice, other considerations might cause us to use eight or fewer or perhaps 10 or more class intervals.

Another question that must be decided regards the width of the class intervals. Class intervals generally should be of the same width, although this is sometimes impossible to accomplish. This width may be determined by dividing the range by k, the number of class intervals. Symbolically, the class interval width is given by

$$w = \frac{R}{k} \tag{2.3.1}$$

where R (the range) is the difference between the smallest and the largest observation in the data set. As a rule this procedure yields a width that is inconvenient for use. Again, we may exercise our good judgment and select a width (usually close to one given by Equation 2.3.1) that is more convenient.

There are other rules of thumb that are helpful in setting up useful class intervals. When the nature of the data makes them appropriate, class interval widths of 5 units, 10 units, and widths that are multiples of 10 tend to make the summarization more comprehensible. When these widths are employed it is generally good practice to have the lower limit of each interval end in a zero or 5. Usually class intervals are ordered from smallest to largest; that is, the first class interval contains the smaller measurements and the last class interval contains the larger measurements. When this is the case, the lower limit of the first class interval should be equal to or smaller than the smallest measurement in the data set, and the upper limit of the last class interval should be equal to or greater than the largest measurement.

Most statistical packages allow users to interactively change the number of class intervals and/or the class widths, so that several visualizations of the data can be obtained quickly. This feature allows users to exercise their judgment in deciding which data display is most appropriate for a given purpose. Let us use the 189 ages shown in Table 1.4.1 and arrayed in Table 2.2.1 to illustrate the construction of a frequency distribution.

EXAMPLE 2.3.1

We wish to know how many class intervals to have in the frequency distribution of the data. We also want to know how wide the intervals should be.

Solution: To get an idea as to the number of class intervals to use, we can apply Sturges's rule to obtain

$$\begin{aligned} k &= 1 + 3.322(\log 189) \\ &= 1 + 3.322(2.2764618) \\ &\approx 9 \end{aligned}$$

Now let us divide the range by 9 to get some idea about the class interval width. We have

$$\frac{R}{k} = \frac{82 - 30}{9} = \frac{52}{9} = 5.778$$

It is apparent that a class interval width of 5 or 10 will be more convenient to use, as well as more meaningful to the reader. Suppose we decide on 10. We may now construct our intervals. Since the smallest value in Table 2.2.1 is 30 and the largest value is 82, we may begin our intervals with 30 and end with 89. This gives the following intervals:

30–39
40–49
50–59
60–69

70–79

80–89

We see that there are six of these intervals, three fewer than the number suggested by Sturges's rule.

It is sometimes useful to refer to the center, called the *midpoint*, of a class interval. The midpoint of a class interval is determined by obtaining the sum of the upper and lower limits of the class interval and dividing by 2. Thus, for example, the midpoint of the class interval 30–39 is found to be $(30 + 39)/2 = 34.5$. ■

When we group data manually, determining the number of values falling into each class interval is merely a matter of looking at the ordered array and counting the number of observations falling in the various intervals. When we do this for our example, we have Table 2.3.1.

A table such as Table 2.3.1 is called a *frequency distribution*. This table shows the way in which the values of the variable are distributed among the specified class intervals. By consulting it, we can determine the frequency of occurrence of values within any one of the class intervals shown.

Relative Frequencies It may be useful at times to know the proportion, rather than the number, of values falling within a particular class interval. We obtain this information by dividing the number of values in the particular class interval by the total number of values. If, in our example, we wish to know the proportion of values between 50 and 59, inclusive, we divide 70 by 189, obtaining .3704. Thus we say that 70 out of 189, or 70/189ths, or .3704, of the values are between 50 and 59. Multiplying .3704 by 100 gives us the percentage of values between 50 and 59. We can say, then, that 37.04 percent of the subjects are between 50 and 59 years of age. We may refer to the proportion of values falling within a class interval as the *relative frequency of occurrence* of values in that interval. In Section 3.2 we shall see that a relative frequency may be interpreted also as the probability of occurrence within the given interval. This probability of occurrence is also called the *experimental probability* or the *empirical probability*.

TABLE 2.3.1 Frequency Distribution of Ages of 189 Subjects Shown in Tables 1.4.1 and 2.2.1

Class Interval	Frequency
30–39	11
40–49	46
50–59	70
60–69	45
70–79	16
80–89	1
Total	189

TABLE 2.3.2 Frequency, Cumulative Frequency, Relative Frequency, and Cumulative Relative Frequency Distributions of the Ages of Subjects Described in Example 1.4.1

Class Interval	Frequency	Cumulative Frequency	Relative Frequency	Cumulative Relative Frequency
30–39	11	11	.0582	.0582
40–49	46	57	.2434	.3016
50–59	70	127	.3704	.6720
60–69	45	172	.2381	.9101
70–79	16	188	.0847	.9948
80–89	1	189	.0053	1.0001
Total	189		1.0001	

Note: Frequencies do not add to 1.0000 exactly because of rounding.

In determining the frequency of values falling within two or more class intervals, we obtain the sum of the number of values falling within the class intervals of interest. Similarly, if we want to know the relative frequency of occurrence of values falling within two or more class intervals, we add the respective relative frequencies. We may sum, or *cumulate,* the frequencies and relative frequencies to facilitate obtaining information regarding the frequency or relative frequency of values within two or more contiguous class intervals. Table 2.3.2 shows the data of Table 2.3.1 along with the *cumulative frequencies,* the *relative frequencies,* and *cumulative relative frequencies.*

Suppose that we are interested in the relative frequency of values between 50 and 79. We use the cumulative relative frequency column of Table 2.3.2 and subtract .3016 from .9948, obtaining .6932.

We may use a statistical package to obtain a table similar to that shown in Table 2.3.2. Tables obtained from both MINITAB and SPSS software are shown in Figure 2.3.1.

The Histogram We may display a frequency distribution (or a relative frequency distribution) graphically in the form of a *histogram,* which is a special type of bar graph.

When we construct a histogram the values of the variable under consideration are represented by the horizontal axis, while the vertical axis has as its scale the frequency (or relative frequency if desired) of occurrence. Above each class interval on the horizontal axis a rectangular bar, or cell, as it is sometimes called, is erected so that the height corresponds to the respective frequency when the class intervals are of equal width. The cells of a histogram must be joined and, to accomplish this, we must take into account the true boundaries of the class intervals to prevent gaps from occurring between the cells of our graph.

The level of precision observed in reported data that are measured on a continuous scale indicates some order of rounding. The order of rounding reflects either the reporter's personal preference or the limitations of the measuring instrument employed. When a frequency distribution is constructed from the data, the class interval limits usually reflect the degree of precision of the raw data. This has been done in our illustrative example.

Dialog box:

Stat ➤ Tables ➤ Tally Individual Variables

Type *C2* in **Variables.** Check **Counts, Percents, Cumulative counts,** and **Cumulative percents** in **Display.** Click **OK.**

Session command:

```
MTB > Tally C2;
SUBC>   Counts;
SUBC>   CumCounts;
SUBC>   Percents;
SUBC>   CumPercents;
```

Output:

Tally for Discrete Variables: C2

MINITAB Output

```
C2  Count  CumCnt  Percent  CumPct
 0     11      11     5.82    5.82
 1     46      57    24.34   30.16
 2     70     127    37.04   67.20
 3     45     172    23.81   91.01
 4     16     188     8.47   99.47
 5      1     189     0.53  100.00
N=    189
```

SPSS Output

		Frequency	Percent	Valid Percent	Cumulative Percent
Valid	30-39	11	5.8	5.8	5.8
	40-49	46	24.3	24.3	30.2
	50-59	70	37.0	37.0	67.2
	60-69	45	23.8	23.8	91.0
	70-79	16	8.5	8.5	99.5
	80-89	1	.5	.5	100.0
	Total	189	100.0	100.0	

FIGURE 2.3.1 Frequency, cumulative frequencies, percent, and cumulative percent distribution of the ages of subjects described in Example 1.4.1 as constructed by MINITAB and SPSS.

We know, however, that some of the values falling in the second class interval, for example, when measured precisely, would probably be a little less than 40 and some would be a little greater than 49. Considering the underlying continuity of our variable, and assuming that the data were rounded to the nearest whole number, we find it convenient to think of 39.5 and 49.5 as the true limits of this second interval. The true limits for each of the class intervals, then, we take to be as shown in Table 2.3.3.

If we construct a graph using these class limits as the base of our rectangles, no gaps will result, and we will have the histogram shown in Figure 2.3.2. We used MINITAB to construct this histogram, as shown in Figure 2.3.3.

We refer to the space enclosed by the boundaries of the histogram as the *area* of the histogram. Each observation is allotted one unit of this area. Since we have 189 observations, the histogram consists of a total of 189 units. Each cell contains a certain proportion of the total area, depending on the frequency. The second cell, for example, contains 46/189 of the area. This, as we have learned, is the relative frequency of occurrence of values between 39.5 and 49.5. From this we see that subareas of the histogram defined by the cells correspond to the frequencies of occurrence of values between the horizontal scale boundaries of the areas. The ratio of a particular subarea to the total area of the histogram is equal to the relative frequency of occurrence of values between the corresponding points on the horizontal axis.

TABLE 2.3.3 The Data of Table 2.3.1 Showing True Class Limits

True Class Limits	Frequency
29.5–39.5	11
39.5–49.5	46
49.5–59.5	70
59.5–69.5	45
69.5–79.5	16
79.5–89.5	1
Total	189

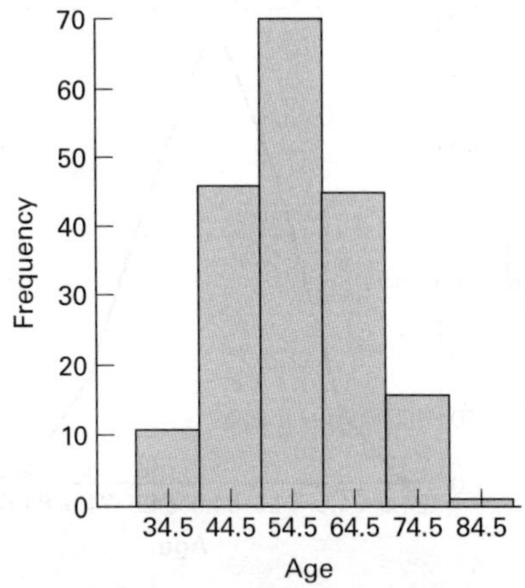

FIGURE 2.3.2 Histogram of ages of 189 subjects from Table 2.3.1.

The Frequency Polygon A frequency distribution can be portrayed graphically in yet another way by means of a *frequency polygon,* which is a special kind of line graph. To draw a frequency polygon we first place a dot above the midpoint of each class interval represented on the horizontal axis of a graph like the one shown in Figure 2.3.2. The height of a given dot above the horizontal axis corresponds to the frequency of the relevant class interval. Connecting the dots by straight lines produces the frequency polygon. Figure 2.3.4 is the frequency polygon for the age data in Table 2.2.1.

Note that the polygon is brought down to the horizontal axis at the ends at points that would be the midpoints if there were an additional cell at each end of the corresponding histogram. This allows for the total area to be enclosed. The total area under the frequency polygon is equal to the area under the histogram. Figure 2.3.5 shows the frequency polygon of Figure 2.3.4 superimposed on the histogram of Figure 2.3.2. This figure allows you to see, for the same set of data, the relationship between the two graphic forms.

Dialog box:	Session command:
Graph ➤ Histogram ➤ Simple ➤ OK	`MTB > Histogram 'Age';`
	`SUBC> MidPoint 34.5:84.5/10;`
Type *Age* in **Graph Variables:** Click **OK.**	`SUBC> Bar.`

Now double click the histogram and click **Binning** Tab.
Type 34.5:84.5/10 in **MidPoint/CutPoint positions:**
Click **OK.**

FIGURE 2.3.3 MINITAB dialog box and session command for constructing histogram from data on ages in Example 1.4.1.

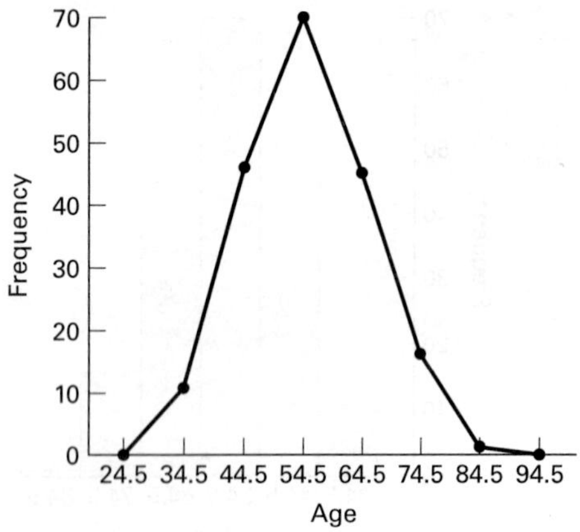

FIGURE 2.3.4 Frequency polygon for the ages of 189 subjects shown in Table 2.2.1.

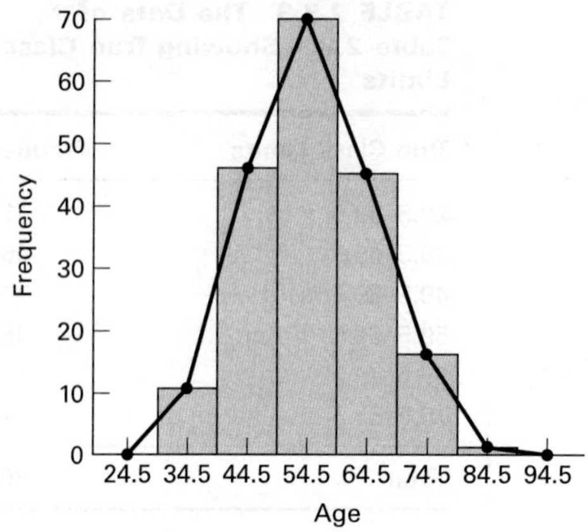

FIGURE 2.3.5 Histogram and frequency polygon for the ages of 189 subjects shown in Table 2.2.1.

Stem-and-Leaf Displays Another graphical device that is useful for representing quantitative data sets is the *stem-and-leaf display*. A stem-and-leaf display bears a strong resemblance to a histogram and serves the same purpose. A properly constructed stem-and-leaf display, like a histogram, provides information regarding the range of the data set, shows the location of the highest concentration of measurements, and reveals the presence or absence of symmetry. An advantage of the stem-and-leaf display over the histogram is the fact that it preserves the information contained in the individual measurements. Such information is lost when measurements are assigned to the class intervals of a histogram. As will become apparent, another advantage of stem-and-leaf displays is the fact that they can be constructed during the tallying process, so the intermediate step of preparing an ordered array is eliminated.

To construct a stem-and-leaf display we partition each measurement into two parts. The first part is called the *stem,* and the second part is called the *leaf.* The stem consists of one or more of the initial digits of the measurement, and the leaf is composed of one or more of the remaining digits. All partitioned numbers are shown together in a single display; the stems form an ordered column with the smallest stem at the top and the largest at the bottom. We include in the stem column all stems within the range of the data even when a measurement with that stem is not in the data set. The rows of the display contain the leaves, ordered and listed to the right of their respective stems. When leaves consist of more than one digit, all digits after the first may be deleted. Decimals when present in the original data are omitted in the stem-and-leaf display. The stems are separated from their leaves by a vertical line. Thus we see that a stem-and-leaf display is also an ordered array of the data.

Stem-and-leaf displays are most effective with relatively small data sets. As a rule they are not suitable for use in annual reports or other communications aimed at the general public. They are primarily of value in helping researchers and decision makers understand the nature of their data. Histograms are more appropriate for externally circulated publications. The following example illustrates the construction of a stem-and-leaf display.

Stem	Leaf
3	04577888899
4	0022333333444444455566666667777778888888999999
5	0000000111222222233333333333333334444444444555666666677777788999999
6	0000111111111111222222233444444556666667888999
7	011111123567888
8	2

FIGURE 2.3.6 Stem-and-leaf display of ages of 189 subjects shown in Table 2.2.1 (stem unit = 10, leaf unit = 1).

EXAMPLE 2.3.2

Let us use the age data shown in Table 2.2.1 to construct a stem-and-leaf display.

Solution: Since the measurements are all two-digit numbers, we will have one-digit stems and one-digit leaves. For example, the measurement 30 has a stem of 3 and a leaf of 0. Figure 2.3.6 shows the stem-and-leaf display for the data.

The MINITAB statistical software package may be used to construct stem-and-leaf displays. The MINITAB procedure and output are as shown in Figure 2.3.7. The increment subcommand specifies the distance from one stem to the next. The numbers in the leftmost output column of Figure 2.3.7

Dialog box:

Graph ➤ Stem-and-Leaf

Type *Age* in **Graph Variables.** Type *10* in **Increment.** Click **OK.**

Session command:

```
MTB > Stem-and-Leaf 'Age';
SUBC>   Increment 10.
```

Output:

Stem-and-Leaf Display: Age

```
Stem-and-leaf of Age    N  =  189
Leaf Unit = 1.0

 11  3 04577888899
 57  4 0022333333444444455566666667777778888888999999
(70) 5 00000001112222222333333333333333344444444445556666666777777889+
 62  6 0000111111111111222222233444444556666667888999
 17  7 011111123567888
  1  8 2
```

FIGURE 2.3.7 Stem-and-leaf display prepared by MINITAB from the data on subjects' ages shown in Table 2.2.1.

```
Stem-and-leaf of Age              N = 189
Leaf Unit = 1.0
    2    3 04
   11    3 577888899
   28    4 00223333334444444
   57    4 55566666677777788888889999999
 (46)    5 0000000011112222222333333333333333344444444444
   86    5 555666666777777788999999
   62    6 000011111111111222222233444444
   32    6 556666667888999
   17    7 0111111123
    7    7 567888
    1    8 2
```

FIGURE 2.3.8 Stem-and-leaf display prepared by MINITAB from the data on subjects' ages shown in Table 2.2.1; class interval width = 5.

provide information regarding the number of observations (leaves) on a given line and above or the number of observations on a given line and below. For example, the number 57 on the second line shows that there are 57 observations (or leaves) on that line and the one above it. The number 62 on the fourth line from the top tells us that there are 62 observations on that line and all the ones below. The number in parentheses tells us that there are 70 observations on that line. The parentheses mark the line containing the middle observation if the total number of observations is odd or the two middle observations if the total number of observations is even.

The + at the end of the third line in Figure 2.3.7 indicates that the frequency for that line (age group 50 through 59) exceeds the line capacity, and that there is at least one additional leaf that is not shown. In this case, the frequency for the 50–59 age group was 70. The line contains only 65 leaves, so the + indicates that there are five more leaves, the number 9, that are not shown. ■

One way to avoid exceeding the capacity of a line is to have more lines. This is accomplished by making the distance between lines shorter, that is, by decreasing the widths of the class intervals. For the present example, we may use class interval widths of 5, so that the distance between lines is 5. Figure 2.3.8 shows the result when MINITAB is used to produce the stem-and-leaf display.

EXERCISES

2.3.1 In a study of the oral home care practice and reasons for seeking dental care among individuals on renal dialysis, Atassi (A-1) studied 90 subjects on renal dialysis. The oral hygiene status of all subjects was examined using a plaque index with a range of 0 to 3 (0 = no soft plaque deposits,

3 = an abundance of soft plaque deposits). The following table shows the plaque index scores for all 90 subjects.

1.17	2.50	2.00	2.33	1.67	1.33
1.17	2.17	2.17	1.33	2.17	2.00
2.17	1.17	2.50	2.00	1.50	1.50
1.00	2.17	2.17	1.67	2.00	2.00
1.33	2.17	2.83	1.50	2.50	2.33
0.33	2.17	1.83	2.00	2.17	2.00
1.00	2.17	2.17	1.33	2.17	2.50
0.83	1.17	2.17	2.50	2.00	2.50
0.50	1.50	2.00	2.00	2.00	2.00
1.17	1.33	1.67	2.17	1.50	2.00
1.67	0.33	1.50	2.17	2.33	2.33
1.17	0.00	1.50	2.33	1.83	2.67
0.83	1.17	1.50	2.17	2.67	1.50
2.00	2.17	1.33	2.00	2.33	2.00
2.17	2.17	2.00	2.17	2.00	2.17

Source: Farhad Atassi, DDS, MSc, FICOI.
Used with permission.

(a) Use these data to prepare:

A frequency distribution
A relative frequency distribution
A cumulative frequency distribution
A cumulative relative frequency distribution
A histogram
A frequency polygon

(b) What percentage of the measurements are less than 2.00?

(c) What proportion of the subjects have measurements greater than or equal to 1.50?

(d) What percentage of the measurements are between 1.50 and 1.99 inclusive?

(e) How many of the measurements are greater than 2.49?

(f) What proportion of the measurements are either less than 1.0 or greater than 2.49?

(g) Someone picks a measurement at random from this data set and asks you to guess the value. What would be your answer? Why?

(h) Frequency distributions and their histograms may be described in a number of ways depending on their shape. For example, they may be symmetric (the left half is at least approximately a mirror image of the right half), skewed to the left (the frequencies tend to increase as the measurements increase in size), skewed to the right (the frequencies tend to decrease as the measurements increase in size), or U-shaped (the frequencies are high at each end of the distribution and small in the center). How would you describe the present distribution?

2.3.2 Janardhan et al. (A-2) conducted a study in which they measured incidental intracranial aneurysms (IIAs) in 125 patients. The researchers examined postprocedural complications and concluded that IIAs can be safely treated without causing mortality and with a lower complications rate than previously reported. The following are the sizes (in millimeters) of the 159 IIAs in the sample.

8.1	10.0	5.0	7.0	10.0	3.0
20.0	4.0	4.0	6.0	6.0	7.0

(Continued)

10.0	4.0	3.0	5.0	6.0	6.0
6.0	6.0	6.0	5.0	4.0	5.0
6.0	25.0	10.0	14.0	6.0	6.0
4.0	15.0	5.0	5.0	8.0	19.0
21.0	8.3	7.0	8.0	5.0	8.0
5.0	7.5	7.0	10.0	15.0	8.0
10.0	3.0	15.0	6.0	10.0	8.0
7.0	5.0	10.0	3.0	7.0	3.3
15.0	5.0	5.0	3.0	7.0	8.0
3.0	6.0	6.0	10.0	15.0	6.0
3.0	3.0	7.0	5.0	4.0	9.2
16.0	7.0	8.0	5.0	10.0	10.0
9.0	5.0	5.0	4.0	8.0	4.0
3.0	4.0	5.0	8.0	30.0	14.0
15.0	2.0	8.0	7.0	12.0	4.0
3.8	10.0	25.0	8.0	9.0	14.0
30.0	2.0	10.0	5.0	5.0	10.0
22.0	5.0	5.0	3.0	4.0	8.0
7.5	5.0	8.0	3.0	5.0	7.0
8.0	5.0	9.0	11.0	2.0	10.0
6.0	5.0	5.0	12.0	9.0	8.0
15.0	18.0	10.0	9.0	5.0	6.0
6.0	8.0	12.0	10.0	5.0	
5.0	16.0	8.0	5.0	8.0	
4.0	16.0	3.0	7.0	13.0	

Source: Vallabh Janardhan, M.D. Used with permission.

(a) Use these data to prepare:

A frequency distribution
A relative frequency distribution
A cumulative frequency distribution
A cumulative relative frequency distribution
A histogram
A frequency polygon

(b) What percentage of the measurements are between 10 and 14.9 inclusive?

(c) How many observations are less than 20?

(d) What proportion of the measurements are greater than or equal to 25?

(e) What percentage of the measurements are either less than 10.0 or greater than 19.95?

(f) Refer to Exercise 2.3.1, part h. Describe the distribution of the size of the aneurysms in this sample.

2.3.3 Hoekema et al. (A-3) studied the craniofacial morphology of patients diagnosed with obstructive sleep apnea syndrome (OSAS) in healthy male subjects. One of the demographic variables the researchers collected for all subjects was the Body Mass Index (calculated by dividing weight in kg by the square of the patient's height in cm). The following are the BMI values of 29 OSAS subjects.

33.57	27.78	40.81
38.34	29.01	47.78
26.86	54.33	28.99

(Continued)

25.21	30.49	27.38
36.42	41.50	29.39
24.54	41.75	44.68
24.49	33.23	47.09
29.07	28.21	42.10
26.54	27.74	33.48
31.44	30.08	

Source: A. Hoekema, D.D.S.
Used with permission.

(a) Use these data to construct:

A frequency distribution
A relative frequency distribution
A cumulative frequency distribution
A cumulative relative frequency distribution
A histogram
A frequency polygon

(b) What percentage of the measurements are less than 30?

(c) What percentage of the measurements are between 40.0 and 49.99 inclusive?

(d) What percentage of the measurements are greater than 34.99?

(e) Describe these data with respect to symmetry and skewness as discussed in Exercise 2.3.1, part h.

(f) How many of the measurements are less than 40?

2.3.4 David Holben (A-4) studied selenium levels in beef raised in a low selenium region of the United States. The goal of the study was to compare selenium levels in the region-raised beef to selenium levels in cooked venison, squirrel, and beef from other regions of the United States. The data below are the selenium levels calculated on a dry weight basis in $\mu g/100$ g for a sample of 53 region-raised cattle.

11.23	15.82
29.63	27.74
20.42	22.35
10.12	34.78
39.91	35.09
32.66	32.60
38.38	37.03
36.21	27.00
16.39	44.20
27.44	13.09
17.29	33.03
56.20	9.69
28.94	32.45
20.11	37.38
25.35	34.91
21.77	27.99
31.62	22.36
32.63	22.68
30.31	26.52
46.16	46.01

(Continued)

56.61	38.04
24.47	30.88
29.39	30.04
40.71	25.91
18.52	18.54
27.80	25.51
19.49	

Source: David Holben, Ph.D.
Used with permission.

(a) Use these data to construct:

A frequency distribution
A relative frequency distribution
A cumulative frequency distribution
A cumulative relative frequency distribution
A histogram
A frequency polygon

(b) Describe these data with respect to symmetry and skewness as discussed in Exercise 2.3.1, part h.

(c) How many of the measurements are greater than 40?

(d) What percentage of the measurements are less than 25?

2.3.5 The following table shows the number of hours 45 hospital patients slept following the administration of a certain anesthetic.

7	10	12	4	8	7	3	8	5
12	11	3	8	1	1	13	10	4
4	5	5	8	7	7	3	2	3
8	13	1	7	17	3	4	5	5
3	1	17	10	4	7	7	11	8

(a) From these data construct:

A frequency distribution
A relative frequency distribution
A histogram
A frequency polygon

(b) Describe these data relative to symmetry and skewness as discussed in Exercise 2.3.1, part h.

2.3.6 The following are the number of babies born during a year in 60 community hospitals.

30	55	27	45	56	48	45	49	32	57	47	56
37	55	52	34	54	42	32	59	35	46	24	57
32	26	40	28	53	54	29	42	42	54	53	59
39	56	59	58	49	53	30	53	21	34	28	50
52	57	43	46	54	31	22	31	24	24	57	29

(a) From these data construct:

A frequency distribution
A relative frequency distribution
A frequency polygon

(b) Describe these data relative to symmetry and skewness as discussed in Exercise 2.3.1, part h.

2.3.7 In a study of physical endurance levels of male college freshman, the following composite endurance scores based on several exercise routines were collected.

254	281	192	260	212	179	225	179	181	149
182	210	235	239	258	166	159	223	186	190
180	188	135	233	220	204	219	211	245	151
198	190	151	157	204	238	205	229	191	200
222	187	134	193	264	312	214	227	190	212
165	194	206	193	218	198	241	149	164	225
265	222	264	249	175	205	252	210	178	159
220	201	203	172	234	198	173	187	189	237
272	195	227	230	168	232	217	249	196	223
232	191	175	236	152	258	155	215	197	210
214	278	252	283	205	184	172	228	193	130
218	213	172	159	203	212	117	197	206	198
169	187	204	180	261	236	217	205	212	218
191	124	199	235	139	231	116	182	243	217
251	206	173	236	215	228	183	204	186	134
188	195	240	163	208					

(a) From these data construct:
A frequency distribution
A relative frequency distribution
A frequency polygon
A histogram

(b) Describe these data relative to symmetry and skewness as discussed in Exercise 2.3.1, part h.

2.3.8 The following are the ages of 30 patients seen in the emergency room of a hospital on a Friday night. Construct a stem-and-leaf display from these data. Describe these data relative to symmetry and skewness as discussed in Exercise 2.3.1, part h.

35	32	21	43	39	60
36	12	54	45	37	53
45	23	64	10	34	22
36	45	55	44	55	46
22	38	35	56	45	57

2.3.9 The following are the emergency room charges made to a sample of 25 patients at two city hospitals. Construct a stem-and-leaf display for each set of data. What does a comparison of the two displays suggest regarding the two hospitals? Describe the two sets of data with respect to symmetry and skewness as discussed in Exercise 2.3.1, part h.

Hospital A

249.10	202.50	222.20	214.40	205.90
214.30	195.10	213.30	225.50	191.40
201.20	239.80	245.70	213.00	238.80
171.10	222.00	212.50	201.70	184.90
248.30	209.70	233.90	229.80	217.90

Hospital B				
199.50	184.00	173.20	186.00	214.10
125.50	143.50	190.40	152.00	165.70
154.70	145.30	154.60	190.30	135.40
167.70	203.40	186.70	155.30	195.90
168.90	166.70	178.60	150.20	212.40

2.3.10 Refer to the ages of patients discussed in Example 1.4.1 and displayed in Table 1.4.1.

(a) Use class interval widths of 5 and construct:

A frequency distribution
A relative frequency distribution
A cumulative frequency distribution
A cumulative relative frequency distribution
A histogram
A frequency polygon

(b) Describe these data with respect to symmetry and skewness as discussed in Exercise 2.3.1, part h.

2.3.11 The objectives of a study by Skjelbo et al. (A-5) were to examine (a) the relationship between chloroguanide metabolism and efficacy in malaria prophylaxis and (b) the mephenytoin metabolism and its relationship to chloroguanide metabolism among Tanzanians. From information provided by urine specimens from the 216 subjects, the investigators computed the ratio of unchanged *S*-mephenytoin to *R*-mephenytoin (*S/R* ratio). The results were as follows:

0.0269	0.0400	0.0550	0.0550	0.0650	0.0670	0.0700	0.0720
0.0760	0.0850	0.0870	0.0870	0.0880	0.0900	0.0900	0.0990
0.0990	0.0990	0.0990	0.0990	0.0990	0.0990	0.0990	0.0990
0.0990	0.0990	0.0990	0.0990	0.0990	0.0990	0.0990	0.0990
0.0990	0.0990	0.0990	0.0990	0.0990	0.0990	0.0990	0.0990
0.0990	0.0990	0.0990	0.0990	0.0990	0.1000	0.1020	0.1040
0.1050	0.1050	0.1080	0.1080	0.1090	0.1090	0.1090	0.1160
0.1190	0.1200	0.1230	0.1240	0.1340	0.1340	0.1370	0.1390
0.1460	0.1480	0.1490	0.1490	0.1500	0.1500	0.1500	0.1540
0.1550	0.1570	0.1600	0.1650	0.1650	0.1670	0.1670	0.1677
0.1690	0.1710	0.1720	0.1740	0.1780	0.1780	0.1790	0.1790
0.1810	0.1880	0.1890	0.1890	0.1920	0.1950	0.1970	0.2010
0.2070	0.2100	0.2100	0.2140	0.2150	0.2160	0.2260	0.2290
0.2390	0.2400	0.2420	0.2430	0.2450	0.2450	0.2460	0.2460
0.2470	0.2540	0.2570	0.2600	0.2620	0.2650	0.2650	0.2680
0.2710	0.2800	0.2800	0.2870	0.2880	0.2940	0.2970	0.2980
0.2990	0.3000	0.3070	0.3100	0.3110	0.3140	0.3190	0.3210
0.3400	0.3440	0.3480	0.3490	0.3520	0.3530	0.3570	0.3630
0.3630	0.3660	0.3830	0.3900	0.3960	0.3990	0.4080	0.4080
0.4090	0.4090	0.4100	0.4160	0.4210	0.4260	0.4290	0.4290
0.4300	0.4360	0.4370	0.4390	0.4410	0.4410	0.4430	0.4540
0.4680	0.4810	0.4870	0.4910	0.4980	0.5030	0.5060	0.5220
0.5340	0.5340	0.5460	0.5480	0.5480	0.5490	0.5550	0.5920
0.5930	0.6010	0.6240	0.6280	0.6380	0.6600	0.6720	0.6820

(Continued)

0.6870	0.6900	0.6910	0.6940	0.7040	0.7120	0.7200	0.7280
0.7860	0.7950	0.8040	0.8200	0.8350	0.8770	0.9090	0.9520
0.9530	0.9830	0.9890	1.0120	1.0260	1.0320	1.0620	1.1600

Source: Erik Skjelbo, M.D. Used with permission.

(a) From these data construct the following distributions: frequency, relative frequency, cumulative frequency, and cumulative relative frequency; and the following graphs: histogram, frequency polygon, and stem-and-leaf plot.

(b) Describe these data with respect to symmetry and skewness as discussed in Exercise 2.3.1, part h.

(c) The investigators defined as poor metabolizers of mephenytoin any subject with an *S/R* mephenytoin ratio greater than .9. How many and what percentage of the subjects were poor metabolizers?

(d) How many and what percentage of the subjects had ratios less than .7? Between .3 and .6999 inclusive? Greater than .4999?

2.3.12 Schmidt et al. (A-6) conducted a study to investigate whether autotransfusion of shed mediastinal blood could reduce the number of patients needing homologous blood transfusion and reduce the amount of transfused homologous blood if fixed transfusion criteria were used. The following table shows the heights in centimeters of the 109 subjects of whom 97 were males.

1.720	1.710	1.700	1.655	1.800	1.700
1.730	1.700	1.820	1.810	1.720	1.800
1.800	1.800	1.790	1.820	1.800	1.650
1.680	1.730	1.820	1.720	1.710	1.850
1.760	1.780	1.760	1.820	1.840	1.690
1.770	1.920	1.690	1.690	1.780	1.720
1.750	1.710	1.690	1.520	1.805	1.780
1.820	1.790	1.760	1.830	1.760	1.800
1.700	1.760	1.750	1.630	1.760	1.770
1.840	1.690	1.640	1.760	1.850	1.820
1.760	1.700	1.720	1.780	1.630	1.650
1.660	1.880	1.740	1.900	1.830	
1.600	1.800	1.670	1.780	1.800	
1.750	1.610	1.840	1.740	1.750	
1.960	1.760	1.730	1.730	1.810	
1.810	1.775	1.710	1.730	1.740	
1.790	1.880	1.730	1.560	1.820	
1.780	1.630	1.640	1.600	1.800	
1.800	1.780	1.840	1.830		
1.770	1.690	1.800	1.620		

Source: Erik Skjelbo, M.D. Used with permission.

(a) For these data construct the following distributions: frequency, relative frequency, cumulative frequency, and cumulative relative frequency; and the following graphs: histogram, frequency polygon, and stem-and-leaf plot.

(b) Describe these data with respect to symmetry and skewness as discussed in Exercise 2.3.1, part h.

(c) How do you account for the shape of the distribution of these data?

(d) How tall were the tallest 6.42 percent of the subjects?

(e) How tall were the shortest 10.09 percent of the subjects?

2.4 DESCRIPTIVE STATISTICS: MEASURES OF CENTRAL TENDENCY

Although frequency distributions serve useful purposes, there are many situations that require other types of data summarization. What we need in many instances is the ability to summarize the data by means of a single number called a *descriptive measure.* Descriptive measures may be computed from the data of a sample or the data of a population. To distinguish between them we have the following definitions:

DEFINITIONS

1. **A descriptive measure computed from the data of a sample is called a *statistic*.**
2. **A descriptive measure computed from the data of a population is called a *parameter*.**

Several types of descriptive measures can be computed from a set of data. In this chapter, however, we limit discussion to *measures of central tendency* and *measures of dispersion.* We consider measures of central tendency in this section and measures of dispersion in the following one.

In each of the measures of central tendency, of which we discuss three, we have a single value that is considered to be typical of the set of data as a whole. Measures of central tendency convey information regarding the average value of a set of values. As we will see, the word *average* can be defined in different ways.

The three most commonly used measures of central tendency are the *mean,* the *median,* and the *mode.*

Arithmetic Mean The most familiar measure of central tendency is the arithmetic mean. It is the descriptive measure most people have in mind when they speak of the "average." The adjective *arithmetic* distinguishes this mean from other means that can be computed. Since we are not covering these other means in this book, we shall refer to the arithmetic mean simply as the *mean.* The mean is obtained by adding all the values in a population or sample and dividing by the number of values that are added.

EXAMPLE 2.4.1

We wish to obtain the mean age of the population of 189 subjects represented in Table 1.4.1.

Solution: We proceed as follows:

$$\text{mean age} = \frac{48 + 35 + 46 + \cdots + 73 + 66}{189} = 55.032$$

■

The three dots in the numerator represent the values we did not show in order to save space.

General Formula for the Mean It will be convenient if we can generalize the procedure for obtaining the mean and, also, represent the procedure in a more compact notational form. Let us begin by designating the random variable of interest by the capital letter X. In our present illustration we let X represent the random variable, age. Specific values of a random variable will be designated by the lowercase letter x. To distinguish one value from another, we attach a subscript to the x and let the subscript refer to the first, the second, the third value, and so on. For example, from Table 1.4.1 we have

$$x_1 = 48, x_2 = 35, \quad \ldots, \quad x_{189} = 66$$

In general, a typical value of a random variable will be designated by x_i and the final value, in a finite population of values, by x_N, where N is the number of values in the population. Finally, we will use the Greek letter μ to stand for the population mean. We may now write the general formula for a finite population mean as follows:

$$\mu = \frac{\sum_{i=1}^{N} x_i}{N} \tag{2.4.1}$$

The symbol $\sum_{i=1}^{N}$ instructs us to add all values of the variable from the first to the last. This symbol Σ, called the *summation sign,* will be used extensively in this book. When from the context it is obvious which values are to be added, the symbols above and below Σ will be omitted.

The Sample Mean When we compute the mean for a sample of values, the procedure just outlined is followed with some modifications in notation. We use $\bar{x}$ to designate the sample mean and n to indicate the number of values in the sample. The sample mean then is expressed as

$$\bar{x} = \frac{\sum_{i=1}^{n} x_i}{n} \tag{2.4.2}$$

EXAMPLE 2.4.2

In Chapter 1 we selected a simple random sample of 10 subjects from the population of subjects represented in Table 1.4.1. Let us now compute the mean age of the 10 subjects in our sample.

Solution: We recall (see Table 1.4.2) that the ages of the 10 subjects in our sample were $x_1 = 43, x_2 = 66, x_3 = 61, x_4 = 64, x_5 = 65, x_6 = 38, x_7 = 59, x_8 = 57, x_9 = 57, x_{10} = 50$. Substitution of our sample data into Equation 2.4.2 gives

$$\bar{x} = \frac{\sum_{i=1}^{n} x_i}{n} = \frac{43 + 66 + \cdots + 50}{10} = 56$$

■

Properties of the Mean The arithmetic mean possesses certain properties, some desirable and some not so desirable. These properties include the following:

1. Uniqueness. For a given set of data there is one and only one arithmetic mean.
2. Simplicity. The arithmetic mean is easily understood and easy to compute.
3. Since each and every value in a set of data enters into the computation of the mean, it is affected by each value. Extreme values, therefore, have an influence on the mean and, in some cases, can so distort it that it becomes undesirable as a measure of central tendency.

As an example of how extreme values may affect the mean, consider the following situation. Suppose the five physicians who practice in an area are surveyed to determine their charges for a certain procedure. Assume that they report these charges: \$75, \$75, \$80, \$80, and \$280. The mean charge for the five physicians is found to be \$118, a value that is not very representative of the set of data as a whole. The single atypical value had the effect of inflating the mean.

Median The median of a finite set of values is that value which divides the set into two equal parts such that the number of values equal to or greater than the median is equal to the number of values equal to or less than the median. If the number of values is odd, the median will be the middle value when all values have been arranged in order of magnitude. When the number of values is even, there is no single middle value. Instead there are two middle values. In this case the median is taken to be the mean of these two middle values, when all values have been arranged in the order of their magnitudes. In other words, the median observation of a data set is the $(n + 1)/2$th one when the observation have been ordered. If, for example, we have 11 observations, the median is the $(11 + 1)/2 = 6$th ordered observation. If we have 12 observations the median is the $(12 + 1)/2 = 6.5$th ordered observation and is a value halfway between the 6th and 7th ordered observations.

EXAMPLE 2.4.3

Let us illustrate by finding the median of the data in Table 2.2.1.

Solution: The values are already ordered so we need only to find the two middle values. The middle value is the $(n + 1)/2 = (189 + 1)/2 = 190/2 =$ 95th one. Counting from the smallest up to the 95th value we see that it is 54. Thus the median age of the 189 subjects is 54 years. ■

EXAMPLE 2.4.4

We wish to find the median age of the subjects represented in the sample described in Example 2.4.2.

Solution: Arraying the 10 ages in order of magnitude from smallest to largest gives 38, 43, 50, 57, 57, 59, 61, 64, 65, 66. Since we have an even number of ages, there is no middle value. The two middle values, however, are 57 and 59. The median, then, is $(57 + 59)/2 = 58$. ■

Properties of the Median Properties of the median include the following:

1. Uniqueness. As is true with the mean, there is only one median for a given set of data.
2. Simplicity. The median is easy to calculate.
3. It is not as drastically affected by extreme values as is the mean.

The Mode The mode of a set of values is that value which occurs most frequently. If all the values are different there is no mode; on the other hand, a set of values may have more than one mode.

EXAMPLE 2.4.5

Find the modal age of the subjects whose ages are given in Table 2.2.1.

Solution: A count of the ages in Table 2.2.1 reveals that the age 53 occurs most frequently (17 times). The mode for this population of ages is 53. ■

For an example of a set of values that has more than one mode, let us consider a laboratory with 10 employees whose ages are 20, 21, 20, 20, 34, 22, 24, 27, 27, and 27. We could say that these data have two modes, 20 and 27. The sample consisting of the values 10, 21, 33, 53, and 54 has no mode since all the values are different.

The mode may be used for describing qualitative data. For example, suppose the patients seen in a mental health clinic during a given year received one of the following diagnoses: mental retardation, organic brain syndrome, psychosis, neurosis, and personality disorder. The diagnosis occurring most frequently in the group of patients would be called the modal diagnosis.

An attractive property of a data distribution occurs when the mean, median, and mode are all equal. The well-known "bell-shaped curve" is a graphical representation of a distribution for which the mean, median, and mode are all equal. Much statistical inference is based on this distribution, the most common of which is the normal distribution. The normal distribution is introduced in Section 4.6 and discussed further in subsequent chapters. Another common distribution of this type is the t-distribution, which is introduced in Section 6.3.

Skewness Data distributions may be classified on the basis of whether they are symmetric or asymmetric. If a distribution is symmetric, the left half of its graph (histogram or frequency polygon) will be a mirror image of its right half. When the left half and right half of the graph of a distribution are not mirror images of each other, the distribution is asymmetric.

DEFINITION

If the graph (histogram or frequency polygon) of a distribution is asymmetric, the distribution is said to be *skewed*. If a distribution is not symmetric because its graph extends further to the right than to the left, that is, if it has a long tail to the right, we say that the distribution is *skewed to the right* or is *positively skewed*. If a distribution is not symmetric because its graph extends further to the left than to the right, that is, if it has a long tail to the left, we say that the distribution is *skewed to the left* or is *negatively skewed*.

A distribution will be skewed to the right, or positively skewed, if its mean is greater than its mode. A distribution will be skewed to the left, or negatively skewed, if its mean is less than its mode. Skewness can be expressed as follows:

$$\text{Skewness} = \frac{\sqrt{n}\sum_{i=1}^{n}(x_i - \bar{x})^3}{\left(\sum_{i=1}^{n}(x_i - \bar{x})^2\right)^{3/2}} = \frac{\sqrt{n}\sum_{i=1}^{n}(x_i - \bar{x})^3}{(n-1)\sqrt{n-1}\,s^3} \tag{2.4.3}$$

In Equation 2.4.3, s is the standard deviation of a sample as defined in Equation 2.5.4. Most computer statistical packages include this statistic as part of a standard printout. A value of skewness > 0 indicates positive skewness and a value of skewness < 0 indicates negative skewness. An illustration of skewness is shown in Figure 2.4.1.

EXAMPLE 2.4.6

Consider the three distributions shown in Figure 2.4.1. Given that the histograms represent frequency counts, the data can be easily re-created and entered into a statistical package. For example, observation of the "No Skew" distribution would yield the following data: 5, 5, 6, 6, 6, 7, 7, 7, 7, 8, 8, 8, 8, 8, 9, 9, 9, 9, 10, 10, 10, 11, 11. Values can be

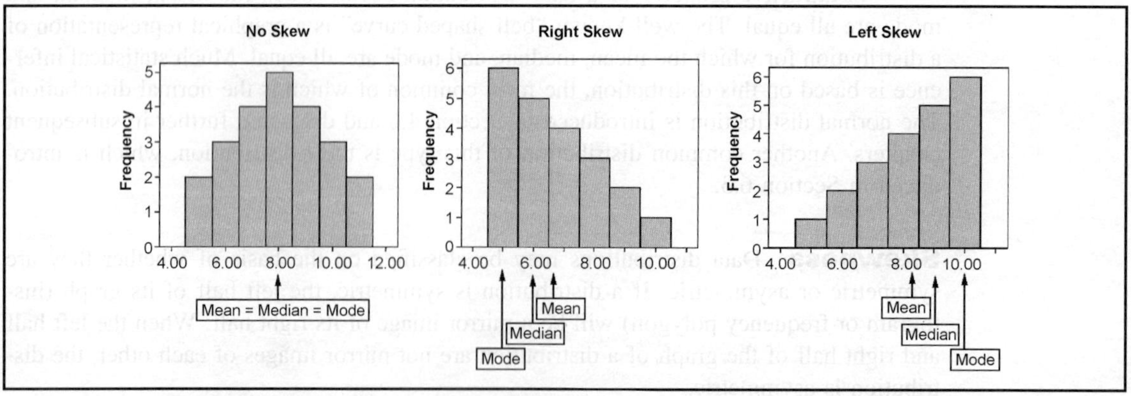

FIGURE 2.4.1 Three histograms illustrating skewness.

obtained from the skewed distributions in a similar fashion. Using SPSS software, the following descriptive statistics were obtained for these three distributions

	No Skew	Right Skew	Left Skew
Mean	8.0000	6.6667	8.3333
Median	8.0000	6.0000	9.0000
Mode	8.00	5.00	10.00
Skewness	.000	.627	−.627

■

2.5 DESCRIPTIVE STATISTICS: MEASURES OF DISPERSION

The *dispersion* of a set of observations refers to the variety that they exhibit. A measure of dispersion conveys information regarding the amount of variability present in a set of data. If all the values are the same, there is no dispersion; if they are not all the same, dispersion is present in the data. The amount of dispersion may be small when the values, though different, are close together. Figure 2.5.1 shows the frequency polygons for two populations that have equal means but different amounts of variability. Population B, which is more variable than population A, is more spread out. If the values are widely scattered, the dispersion is greater. Other terms used synonymously with dispersion include *variation, spread,* and *scatter*.

The Range One way to measure the variation in a set of values is to compute the *range*. The range is the difference between the largest and smallest value in a set of observations. If we denote the range by R, the largest value by x_L, and the smallest value by x_s, we compute the range as follows:

$$R = x_L - x_S \tag{2.5.1}$$

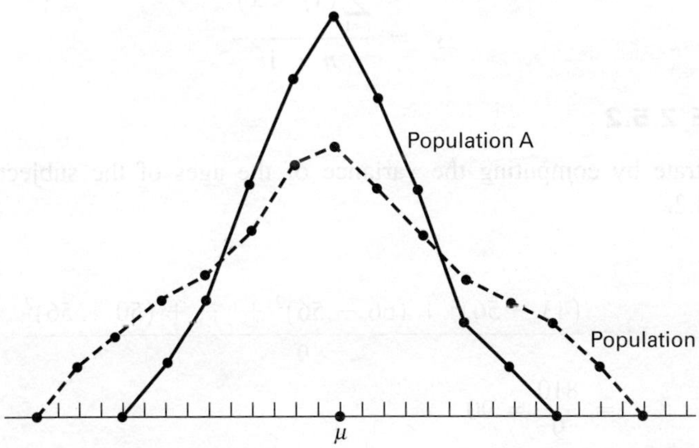

FIGURE 2.5.1 Two frequency distributions with equal means but different amounts of dispersion.

EXAMPLE 2.5.1

We wish to compute the range of the ages of the sample subjects discussed in Example 2.4.2.

Solution: Since the youngest subject in the sample is 30 years old and the oldest is 82, we compute the range to be

$$R = 82 - 30 = 52$$

■

The usefulness of the range is limited. The fact that it takes into account only two values causes it to be a poor measure of dispersion. The main advantage in using the range is the simplicity of its computation. Since the range, expressed as a single measure, imparts minimal information about a data set and therefore, is of limited use, it is often preferable to express the range as a number pair, $[x_S, x_L]$, in which x_S and x_L are the smallest and largest values in the data set, respectively. For the data in Example 2.5.1, we may express the range as the number pair [30, 82]. Although this is not the traditional expression for the range, it is intuitive to imagine that knowledge of the minimum and maximum values in this data set would convey more information than knowing only that the range is equal to 52. An infinite number of distributions, each with quite different minimum and maximum values, may have a range of 52.

The Variance When the values of a set of observations lie close to their mean, the dispersion is less than when they are scattered over a wide range. Since this is true, it would be intuitively appealing if we could measure dispersion relative to the scatter of the values about their mean. Such a measure is realized in what is known as the *variance*. In computing the variance of a sample of values, for example, we subtract the mean from each of the values, square the resulting differences, and then add up the squared differences. This sum of the squared deviations of the values from their mean is divided by the sample size, minus 1, to obtain the sample variance. Letting s^2 stand for the sample variance, the procedure may be written in notational form as follows:

$$s^2 = \frac{\sum_{i=1}^{n}(x_i - \bar{x})^2}{n - 1} \tag{2.5.2}$$

EXAMPLE 2.5.2

Let us illustrate by computing the variance of the ages of the subjects discussed in Example 2.4.2.

Solution:

$$s^2 = \frac{(43 - 56)^2 + (66 - 56)^2 + \cdots + (50 - 56)^2}{9}$$

$$= \frac{810}{9} = 90$$

■

Degrees of Freedom The reason for dividing by $n - 1$ rather than n, as we might have expected, is the theoretical consideration referred to as *degrees of freedom*.

In computing the variance, we say that we have $n - 1$ *degrees of freedom*. We reason as follows. The sum of the deviations of the values from their mean is equal to zero, as can be shown. If, then, we know the values of $n - 1$ of the deviations from the mean, we know the nth one, since it is automatically determined because of the necessity for all n values to add to zero. From a practical point of view, dividing the squared differences by $n - 1$ rather than n is necessary in order to use the sample variance in the inference procedures discussed later. The concept of degrees of freedom will be revisited in a later Chapter. Students interested in pursuing the matter further at this time should refer to the article by Walker (2).

When we compute the variance from a finite population of N values, the procedures outlined above are followed except that we subtract μ from each x and divide by N rather than $N - 1$. If we let σ^2 stand for the finite population variance, the formula is as follows:

$$\sigma^2 = \frac{\sum_{i=1}^{N}(x_i - \mu)^2}{N} \tag{2.5.3}$$

Standard Deviation The variance represents squared units and, therefore, is not an appropriate measure of dispersion when we wish to express this concept in terms of the original units. To obtain a measure of dispersion in original units, we merely take the square root of the variance. The result is called the *standard deviation*. In general, the standard deviation of a sample is given by

$$s = \sqrt{s^2} = \sqrt{\frac{\sum_{i=1}^{n}(x_i - \bar{x})^2}{n - 1}} \tag{2.5.4}$$

The standard deviation of a finite population is obtained by taking the square root of the quantity obtained by Equation 2.5.3.

The Coefficient of Variation The standard deviation is useful as a measure of variation within a given set of data. When one desires to compare the dispersion in two sets of data, however, comparing the two standard deviations may lead to fallacious results. It may be that the two variables involved are measured in different units. For example, we may wish to know, for a certain population, whether serum cholesterol levels, measured in milligrams per 100 ml, are more variable than body weight, measured in pounds.

Furthermore, although the same unit of measurement is used, the two means may be quite different. If we compare the standard deviation of weights of first-grade children with the standard deviation of weights of high school freshmen, we may find that the latter standard deviation is numerically larger than the former, because the weights themselves are larger, not because the dispersion is greater.

What is needed in situations like these is a measure of relative variation rather than absolute variation. Such a measure is found in the *coefficient of variation*, which expresses the standard deviation as a percentage of the mean. The formula is given by

$$\text{C.V.} = \frac{s}{\bar{x}}(100)\% \tag{2.5.5}$$

We see that, since the mean and standard deviations are expressed in the same unit of measurement, the unit of measurement cancels out in computing the coefficient of variation. What we have, then, is a measure that is independent of the unit of measurement.

EXAMPLE 2.5.3

Suppose two samples of human males yield the following results:

	Sample 1	Sample 2
Age	25 years	11 years
Mean weight	145 pounds	80 pounds
Standard deviation	10 pounds	10 pounds

We wish to know which is more variable, the weights of the 25-year-olds or the weights of the 11-year-olds.

Solution: A comparison of the standard deviations might lead one to conclude that the two samples possess equal variability. If we compute the coefficients of variation, however, we have for the 25-year-olds

$$\text{C.V.} = \frac{10}{145}(100) = 6.9\%$$

and for the 11-year-olds

$$\text{C.V.} = \frac{10}{80}(100) = 12.5\%$$

If we compare these results, we get quite a different impression. It is clear from this example that variation is much higher in the sample of 11-year-olds than in the sample of 25-year-olds. ■

The coefficient of variation is also useful in comparing the results obtained by different persons who are conducting investigations involving the same variable. Since the coefficient of variation is independent of the scale of measurement, it is a useful statistic for comparing the variability of two or more variables measured on different scales. We could, for example, use the coefficient of variation to compare the variability in weights of one sample of subjects whose weights are expressed in pounds with the variability in weights of another sample of subjects whose weights are expressed in kilograms.

Computer Analysis Computer software packages provide a variety of possibilities in the calculation of descriptive measures. Figure 2.5.2 shows a printout of the descriptive measures available from the MINITAB package. The data consist of the ages from Example 2.4.2.

In the printout Q_1 and Q_3 are the first and third quartiles, respectively. These measures are described later in this chapter. N stands for the number of data observations, and N* stands for the number of missing values. The term SEMEAN stands for *standard*

```
Variable   N N*   Mean SE Mean StDev Minimum     Q1 Median    Q3 Maximum
C1        10  0 56.00     3.00  9.49   38.00 48.25  58.00 64.25   66.00
```

FIGURE 2.5.2 Printout of descriptive measures computed from the sample of ages in Example 2.4.2, MINITAB software package.

error of the mean. This measure will be discussed in detail in a later chapter. Figure 2.5.3 shows, for the same data, the SAS® printout obtained by using the PROC MEANS statement.

Percentiles and Quartiles The mean and median are special cases of a family of parameters known as *location parameters.* These descriptive measures are called location parameters because they can be used to designate certain positions on the horizontal axis when the distribution of a variable is graphed. In that sense the so-called location parameters "locate" the distribution on the horizontal axis. For example, a distribution with a median of 100 is located to the right of a distribution with a median of 50 when the two distributions are graphed. Other location parameters include percentiles and quartiles. We may define a percentile as follows:

DEFINITION

Given a set of n observations $x_1, x_2, \ldots x_n$, the pth percentile P is the value of X such that p percent or less of the observations are less than P and $(100 - p)$ percent or less of the observations are greater than P.

Subscripts on P serve to distinguish one percentile from another. The 10th percentile, for example, is designated P_{10}, the 70th is designated P_{70}, and so on. The 50th percentile is the median and is designated P_{50}. The 25th percentile is often referred to as the *first quartile* and denoted Q_1. The 50th percentile (the median) is referred to as the second or *middle quartile* and written Q_2, and the 75th percentile is referred to as the *third quartile*, Q_3.

```
                    The MEANS Procedure
                 Analysis Variable: Age
     N           Mean      Std Dev      Minimum        Maximum
    10     56.0000000    9.4868330   38.0000000     66.0000000

                                        Coeff of
Std Error          Sum     Variance    Variation
3.0000000  560.0000000   90.0000000   16.9407732
```

FIGURE 2.5.3 Printout of descriptive measures computed from the sample of ages in Example 2.4.2, SAS® software package.

When we wish to find the quartiles for a set of data, the following formulas are used:

$$\left.\begin{aligned} Q_1 &= \frac{n+1}{4}\text{th ordered observation} \\ Q_2 &= \frac{2(n+1)}{4} = \frac{n+1}{2}\text{th ordered observation} \\ Q_3 &= \frac{3(n+1)}{4}\text{th ordered observation} \end{aligned}\right\} \tag{2.5.6}$$

Interquartile Range As we have seen, the range provides a crude measure of the variability present in a set of data. A disadvantage of the range is the fact that it is computed from only two values, the largest and the smallest. A similar measure that reflects the variability among the middle 50 percent of the observations in a data set is the *interquartile range.*

DEFINITION

The interquartile range (IQR) is the difference between the third and first quartiles: that is,

$$\text{IQR} = Q_3 - Q_1 \tag{2.5.7}$$

A large IQR indicates a large amount of variability among the middle 50 percent of the relevant observations, and a small IQR indicates a small amount of variability among the relevant observations. Since such statements are rather vague, it is more informative to compare the interquartile range with the range for the entire data set. A comparison may be made by forming the ratio of the IQR to the range (R) and multiplying by 100. That is, $100(\text{IQR}/R)$ tells us what percent the IQR is of the overall range.

Kurtosis Just as we may describe a distribution in terms of skewness, we may describe a distribution in terms of kurtosis.

DEFINITION

***Kurtosis* is a measure of the degree to which a distribution is "peaked" or flat in comparison to a normal distribution whose graph is characterized by a bell-shaped appearance.**

A distribution, in comparison to a normal distribution, may possess an excessive proportion of observations in its tails, so that its graph exhibits a flattened appearance. Such a distribution is said to be *platykurtic*. Conversely, a distribution, in comparison to a normal distribution, may possess a smaller proportion of observations in its tails, so that its graph exhibits a more peaked appearance. Such a distribution is said to be *leptokurtic*. A normal, or bell-shaped distribution, is said to be *mesokurtic*.

Kurtosis can be expressed as

$$Kurtosis = \frac{n\sum_{i=1}^{n}(x_i - \bar{x})^4}{\left(\sum_{i=1}^{n}(x_i - \bar{x})^2\right)^2} - 3 = \frac{n\sum_{i=1}^{n}(x_i - \bar{x})^4}{(n-1)^2 s^4} - 3 \qquad (2.5.8)$$

Manual calculation using Equation 2.5.8 is usually not necessary, since most statistical packages calculate and report information regarding kurtosis as part of the descriptive statistics for a data set. Note that each of the two parts of Equation 2.5.8 has been reduced by 3. A perfectly mesokurtic distribution has a kurtosis measure of 3 based on the equation. Most computer algorithms reduce the measure by 3, as is done in Equation 2.5.8, so that the kurtosis measure of a mesokurtic distribution will be equal to 0. A leptokurtic distribution, then, will have a kurtosis measure > 0, and a platykurtic distribution will have a kurtosis measure < 0. Be aware that not all computer packages make this adjustment. In such cases, comparisons with a mesokurtic distribution are made against 3 instead of against 0. Graphs of distributions representing the three types of kurtosis are shown in Figure 2.5.4.

EXAMPLE 2.5.4

Consider the three distributions shown in Figure 2.5.4. Given that the histograms represent frequency counts, the data can be easily re-created and entered into a statistical package. For example, observation of the "mesokurtic" distribution would yield the following data: 1, 2, 2, 3, 3, 3, 3, 3, … , 9, 9, 9, 9, 9, 10, 10, 11. Values can be obtained from the other distributions in a similar fashion. Using SPSS software, the following descriptive statistics were obtained for these three distributions:

	Mesokurtic	Leptokurtic	Platykurtic
Mean	6.0000	6.0000	6.0000
Median	6.0000	6.0000	6.0000
Mode	6.00	6.00	6.00
Skewness	.000	.608	−1.158

■

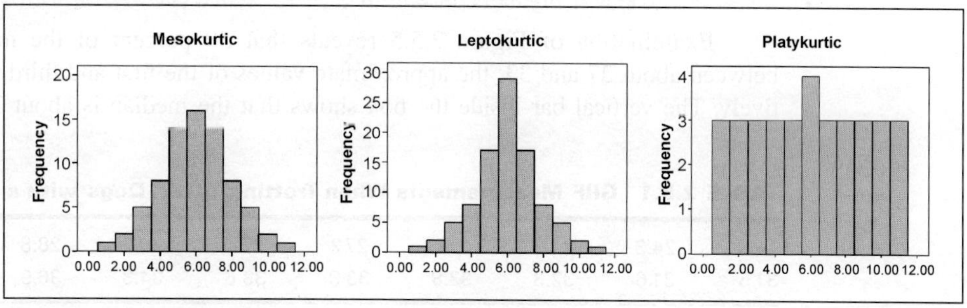

FIGURE 2.5.4 Three histograms representing kurtosis.

Box-and-Whisker Plots A useful visual device for communicating the information contained in a data set is the *box-and-whisker plot.* The construction of a box-and-whisker plot (sometimes called, simply, a *boxplot*) makes use of the quartiles of a data set and may be accomplished by following these five steps:

1. Represent the variable of interest on the horizontal axis.
2. Draw a box in the space above the horizontal axis in such a way that the left end of the box aligns with the first quartile Q_1 and the right end of the box aligns with the third quartile Q_3.
3. Divide the box into two parts by a vertical line that aligns with the median Q_2.
4. Draw a horizontal line called a *whisker* from the left end of the box to a point that aligns with the smallest measurement in the data set.
5. Draw another horizontal line, or whisker, from the right end of the box to a point that aligns with the largest measurement in the data set.

Examination of a box-and-whisker plot for a set of data reveals information regarding the amount of spread, location of concentration, and symmetry of the data.

The following example illustrates the construction of a box-and-whisker plot.

EXAMPLE 2.5.5

Evans et al. (A-7) examined the effect of velocity on ground reaction forces (GRF) in dogs with lameness from a torn cranial cruciate ligament. The dogs were walked and trotted over a force platform, and the GRF was recorded during a certain phase of their performance. Table 2.5.1 contains 20 measurements of force where each value shown is the mean of five force measurements per dog when trotting.

Solution: The smallest and largest measurements are 14.6 and 44, respectively. The first quartile is the $Q_1 = (20 + 1)/4 = 5.25$th measurement, which is $27.2 + (.25)(27.4 - 27.2) = 27.25$. The median is the $Q_2 + (20 + 1)/2 = 10.5$th measurement or $30.7 + (.5)(31.5 - 30.7) = 31.1$; and the third quartile is the $Q_3 + 3(20 + 1)/4 = 15.75$th measurement, which is equal to $33.3 + (.75)(33.6 - 33.3) = 33.525$. The interquartile range is $\text{IQR} = 33.525 - 27.25 = 6.275$. The range is 29.4, and the IQR is $100(6.275/29.4) = 21$ percent of the range. The resulting box-and-whisker plot is shown in Figure 2.5.5. ■

Examination of Figure 2.5.5 reveals that 50 percent of the measurements are between about 27 and 33, the approximate values of the first and third quartiles, respectively. The vertical bar inside the box shows that the median is about 31.

TABLE 2.5.1 GRF Measurements When Trotting of 20 Dogs with a Lame Ligament

14.6	24.3	24.9	27.0	27.2	27.4	28.2	28.8	29.9	30.7
31.5	31.6	32.3	32.8	33.3	33.6	34.3	36.9	38.3	44.0

Source: Richard Evans, Ph.D. Used with permission.

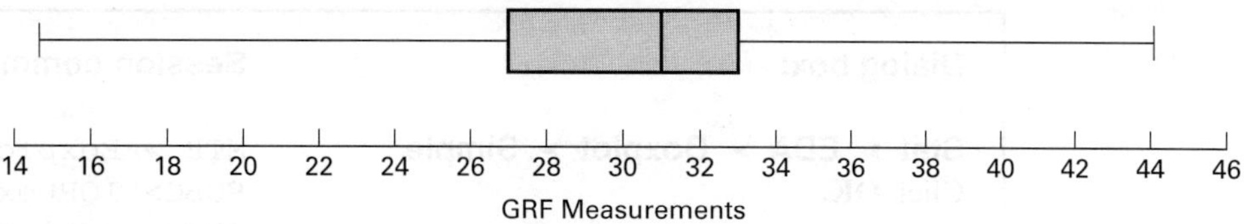

FIGURE 2.5.5 Box-and-whisker plot for Example 2.5.5.

Many statistical software packages have the capability of constructing box-and-whisker plots. Figure 2.5.6 shows one constructed by MINITAB and one constructed by NCSS from the data of Table 2.5.1. The procedure to produce the MINITAB plot is shown in Figure 2.5.7. The asterisks in Figure 2.5.6 alert us to the fact that the data set contains one unusually large and one unusually small value, called *outliers*. The outliers are the dogs that generated forces of 14.6 and 44. Figure 2.5.6 illustrates the fact that box-and-whisker plots may be displayed vertically as well as horizontally.

An outlier, or a typical observation, may be defined as follows.

DEFINITION

An *outlier* is an observation whose value, x, either exceeds the value of the third quartile by a magnitude greater than 1.5(IQR) or is less than the value of the first quartile by a magnitude greater than 1.5(IQR). That is, an observation of $x > Q_3 + 1.5(\text{IQR})$ or an observation of $x < Q_1 - 1.5(\text{IQR})$ is called an outlier.

For the data in Table 2.5.1 we may use the previously computed values of Q_1, Q_3, and IQR to determine how large or how small a value would have to be in order to be considered an outlier. The calculations are as follows:

$$x > 27.25 - 1.5(6.275) = 17.8375 \quad \text{and} \quad x > 33.525 + 1.5(6.275) = 42.9375$$

For the data in Table 2.5.1, then, an observed value smaller than 17.8375 or larger than 42.9375 would be considered an outlier.

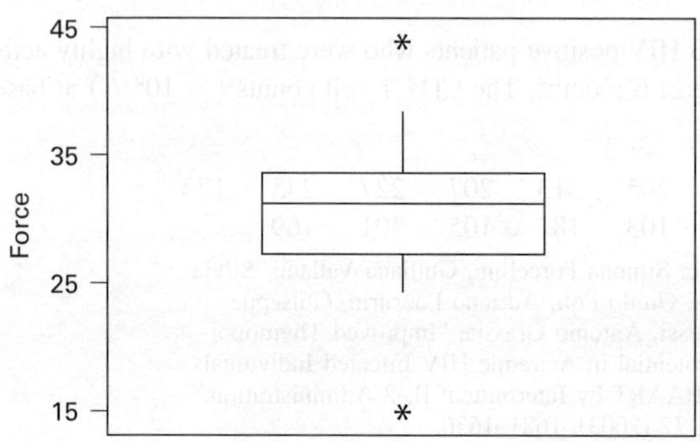

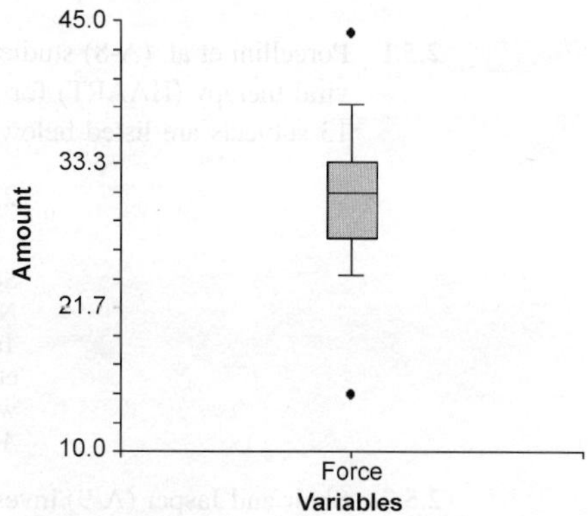

FIGURE 2.5.6 Box-and-whisker plot constructed by MINITAB (left) and by NCSS (right) from the data of Table 2.5.1.

Dialog box:	**Session command:**

Stat ➤ EDA ➤ Boxplot ➤ Simple
Click **OK.**
 	
Type *Force* **Graph Variables.**
Click **OK.**
	`MTB > Boxplot 'Force';`
	`SUBC> IQRbox;`
	`SUBC> Outlier.`

FIGURE 2.5.7 MINITAB procedure to produce Figure 2.5.6.

The SAS® statement PROC UNIVARIATE may be used to obtain a box-and-whisker plot. The statement also produces other descriptive measures and displays, including stem-and-leaf plots, means, variances, and quartiles.

Exploratory Data Analysis Box-and-whisker plots and stem-and-leaf displays are examples of what are known as *exploratory data analysis* techniques. These techniques, made popular as a result of the work of Tukey (3), allow the investigator to examine data in ways that reveal trends and relationships, identify unique features of data sets, and facilitate their description and summarization.

EXERCISES

For each of the data sets in the following exercises compute (a) the mean, (b) the median, (c) the mode, (d) the range, (e) the variance, (f) the standard deviation, (g) the coefficient of variation, and (h) the interquartile range. Treat each data set as a sample. For those exercises for which you think it would be appropriate, construct a box-and-whisker plot and discuss the usefulness in understanding the nature of the data that this device provides. For each exercise select the measure of central tendency that you think would be most appropriate for describing the data. Give reasons to justify your choice.

2.5.1 Porcellini et al. (A-8) studied 13 HIV-positive patients who were treated with highly active antiretroviral therapy (HAART) for at least 6 months. The CD4 T cell counts ($\times 10^6/L$) at baseline for the 13 subjects are listed below.

230	205	313	207	227	245	173
58	103	181	105	301	169	

Source: Simona Porcellini, Guiliana Vallanti, Silvia Nozza, Guido Poli, Adriano Lazzarin, Guiseppe Tambussi, Antonio Grassia, "Improved Thymopoietic Potential in Aviremic HIV Infected Individuals with HAART by Intermittent IL-2 Administration," *AIDS, 17* (2003), 1621–1630.

2.5.2 Shair and Jasper (A-9) investigated whether decreasing the venous return in young rats would affect ultrasonic vocalizations (USVs). Their research showed no significant change in the number of ultrasonic vocalizations when blood was removed from either the superior vena cava or the carotid artery. Another important variable measured was the heart rate (bmp) during the withdrawal of blood. The table below presents the heart rate of seven rat pups from the experiment involving the carotid artery.

500	570	560	570	450	560	570

Source: Harry N. Shair and Anna Jasper, "Decreased Venous Return Is Neither Sufficient nor Necessary to Elicit Ultrasonic Vocalization of Infant Rat Pups," *Behavioral Neuroscience, 117* (2003), 840–853.

2.5.3 Butz et al. (A-10) evaluated the duration of benefit derived from the use of noninvasive positive-pressure ventilation by patients with amyotrophic lateral sclerosis on symptoms, quality of life, and survival. One of the variables of interest is partial pressure of arterial carbon dioxide ($PaCO_2$). The values below (mm Hg) reflect the result of baseline testing on 30 subjects as established by arterial blood gas analyses.

40.0	47.0	34.0	42.0	54.0	48.0	53.6	56.9	58.0	45.0
54.5	54.0	43.0	44.3	53.9	41.8	33.0	43.1	52.4	37.9
34.5	40.1	33.0	59.9	62.6	54.1	45.7	40.6	56.6	59.0

Source: M. Butz, K. H. Wollinsky, U. Widemuth-Catrinescu, A. Sperfeld, S. Winter, H. H. Mehrkens, A. C. Ludolph, and H. Schreiber, "Longitudinal Effects of Noninvasive Positive-Pressure Ventilation in Patients with Amyotrophic Lateral Sclerosis," *American Journal of Medical Rehabilitation, 82* (2003), 597–604.

2.5.4 According to Starch et al. (A-11), hamstring tendon grafts have been the "weak link" in anterior cruciate ligament reconstruction. In a controlled laboratory study, they compared two techniques for reconstruction: either an interference screw or a central sleeve and screw on the tibial side. For eight cadaveric knees, the measurements below represent the required force (in newtons) at which initial failure of graft strands occurred for the central sleeve and screw technique.

172.5	216.63	212.62	98.97	66.95	239.76	19.57	195.72

Source: David W. Starch, Jerry W. Alexander, Philip C. Noble, Suraj Reddy, and David M. Lintner, "Multistranded Hamstring Tendon Graft Fixation with a Central Four-Quadrant or a Standard Tibial Interference Screw for Anterior Cruciate Ligament Reconstruction," *American Journal of Sports Medicine, 31* (2003), 338–344.

2.5.5 Cardosi et al. (A-12) performed a 4-year retrospective review of 102 women undergoing radical hysterectomy for cervical or endometrial cancer. Catheter-associated urinary tract infection was observed in 12 of the subjects. Below are the numbers of postoperative days until diagnosis of the infection for each subject experiencing an infection.

16	10	49	15	6	15
8	19	11	22	13	17

Source: Richard J. Cardosi, Rosemary Cardosi, Edward C. Grendys Jr., James V. Fiorica, and Mitchel S. Hoffman, "Infectious Urinary Tract Morbidity with Prolonged Bladder Catheterization After Radical Hysterectomy," *American Journal of Obstetrics and Gynecology, 189* (2003), 380–384.

2.5.6 The purpose of a study by Nozawa et al. (A-13) was to evaluate the outcome of surgical repair of a pars interarticularis defect by segmental wire fixation in young adults with lumbar spondylolysis. The authors found that segmental wire fixation historically has been successful in the treatment of nonathletes with spondylolysis, but no information existed on the results of this type of surgery in athletes. In a retrospective study, the authors found 20 subjects who had the surgery between 1993 and 2000. For these subjects, the data below represent the duration in months of follow-up care after the operation.

103	68	62	60	60	54	49	44	42	41
38	36	34	30	19	19	19	19	17	16

Source: Satoshi Nozawa, Katsuji Shimizu, Kei Miyamoto, and Mizuo Tanaka, "Repair of Pars Interarticularis Defect by Segmental Wire Fixation in Young Athletes with Spondylolysis," *American Journal of Sports Medicine, 31* (2003), 359–364.

2.5.7 See Exercise 2.3.1.

2.5.8 See Exercise 2.3.2.

2.5.9 See Exercise 2.3.3.

2.5.10 See Exercise 2.3.4.

2.5.11 See Exercise 2.3.5.

2.5.12 See Exercise 2.3.6.

2.5.13 See Exercise 2.3.7.

2.5.14 In a pilot study, Huizinga et al. (A-14) wanted to gain more insight into the psychosocial consequences for children of a parent with cancer. For the study, 14 families participated in semistructured interviews and completed standardized questionnaires. Below is the age of the sick parent with cancer (in years) for the 14 families.

37	48	53	46	42	49	44
38	32	32	51	51	48	41

Source: Gea A. Huizinga, Winette T.A. van der Graaf, Annemike Visser, Jos S. Dijkstra, and Josette E. H. M. Hoekstra-Weebers, "Psychosocial Consequences for Children of a Parent with Cancer," *Cancer Nursing, 26* (2003), 195–202.

2.6 SUMMARY

In this chapter various descriptive statistical procedures are explained. These include the organization of data by means of the ordered array, the frequency distribution, the relative frequency distribution, the histogram, and the frequency polygon. The concepts of central tendency and variation are described, along with methods for computing their more common measures: the mean, median, mode, range, variance, and standard deviation. The reader is also introduced to the concepts of skewness and kurtosis, and to exploratory data analysis through a description of stem-and-leaf displays and box-and-whisker plots.

We emphasize the use of the computer as a tool for calculating descriptive measures and constructing various distributions from large data sets.

SUMMARY OF FORMULAS FOR CHAPTER 2

Formula Number	Name	Formula
2.3.1	Class interval width using Sturges's rule	$w = \frac{R}{k}$
2.4.1	Mean of a population	$\mu = \frac{\sum_{i=1}^{N} x_i}{N}$
2.4.2	Skewness	$Skewness = \frac{\sqrt{n}\sum_{i=1}^{n}(x_i - \bar{x})^3}{\left(\sum_{i=1}^{n}(x_i - \bar{x})^2\right)^{3/2}} = \frac{\sqrt{n}\sum_{i=1}^{n}(x_i - \bar{x})^3}{(n-1)\sqrt{n-1}\, s^3}$
2.4.2	Mean of a sample	$\bar{x} = \frac{\sum_{i=1}^{n} x_i}{n}$
2.5.1	Range	$R = x_L - x_S$
2.5.2	Sample variance	$s^2 = \frac{\sum_{i=1}^{n}(x_i - \bar{x})^2}{n-1}$
2.5.3	Population variance	$\sigma^2 = \frac{\sum_{i=1}^{n}(x_i - \mu)^2}{N}$
2.5.4	Standard deviation	$s = \sqrt{s^2} = \sqrt{\frac{\sum_{i=1}^{n}(x_i - \bar{x})^2}{n-1}}$
2.5.5	Coefficient of variation	$C.V. = \frac{S}{\bar{x}}(100)\%$
2.5.6	Quartile location in ordered array	$Q_1 = \frac{1}{4}(n+1)$ $Q_2 = \frac{1}{2}(n+1)$ $Q_3 = \frac{3}{4}(n+1)$
2.5.7	Interquartile range	$IQR = Q_3 - Q_1$

(Continued)

2.5.8	Kurtosis	$$Kurtosis = \frac{\sum_{i=1}^{n}(x_i - \bar{x})^4}{\left(\sum_{i=1}^{n}(x_i - \bar{x})^2\right)^2} - 3$$ $$= \frac{n\sum_{i=1}^{n}(x_i - \bar{x})^4}{(n-1)^2 s^4} - 3$$
Symbol Key	• C.V. = coefficient of variation • IQR = Interquartile range • k = number of class intervals • μ = population mean • N = population size • n = sample size • $(n - 1)$ = degrees of freedom • Q_1 = first quartile • Q_2 = second quartile = median • Q_3 = third quartile • R = range • s = standard deviation • s^2 = sample variance • σ^2 = population variance • x_i = i^{th} data observation • x_L = largest data point • x_S = smallest data point • $\bar{x}$ = sample mean • w = class width	

REVIEW QUESTIONS AND EXERCISES

1. Define:

   (a) Stem-and-leaf display **(b)** Box-and-whisker plot
   (c) Percentile **(d)** Quartile
   (e) Location parameter **(f)** Exploratory data analysis
   (g) Ordered array **(h)** Frequency distribution
   (i) Relative frequency distribution **(j)** Statistic
   (k) Parameter **(l)** Frequency polygon
   (m) True class limits **(n)** Histogram

2. Define and compare the characteristics of the mean, the median, and the mode.

3. What are the advantages and limitations of the range as a measure of dispersion?

4. Explain the rationale for using $n - 1$ to compute the sample variance.

5. What is the purpose of the coefficient of variation?

6. What is the purpose of Sturges's rule?

7. What is another name for the 50th percentile (second or middle quartile)?

8. Describe from your field of study a population of data where knowledge of the central tendency and dispersion would be useful. Obtain real or realistic synthetic values from this population and compute the mean, median, mode, variance, and standard deviation.

9. Collect a set of real, or realistic, data from your field of study and construct a frequency distribution, a relative frequency distribution, a histogram, and a frequency polygon.

10. Compute the mean, median, mode, variance, and standard deviation for the data in Exercise 9.

11. Find an article in a journal from your field of study in which some measure of central tendency and dispersion have been computed.

12. The purpose of a study by Tam et al. (A-15) was to investigate the wheelchair maneuvering in individuals with lower-level spinal cord injury (SCI) and healthy controls. Subjects used a modified wheelchair to incorporate a rigid seat surface to facilitate the specified experimental measurements. Interface pressure measurement was recorded by using a high-resolution pressure-sensitive mat with a spatial resolution of 4 sensors per square centimeter taped on the rigid seat support. During static sitting conditions, average pressures were recorded under the ischial tuberosities. The data for measurements of the left ischial tuberosity (in mm Hg) for the SCI and control groups are shown below.

Control	131	115	124	131	122	117	88	114	150	169
SCI	60	150	130	180	163	130	121	119	130	148

Source: Eric W. Tam, Arthur F. Mak, Wai Nga Lam, John H. Evans, and York Y. Chow, "Pelvic Movement and Interface Pressure Distribution During Manual Wheelchair Propulsion," *Archives of Physical Medicine and Rehabilitation, 84* (2003), 1466–1472.

(a) Find the mean, median, variance, and standard deviation for the controls.

(b) Find the mean, median variance, and standard deviation for the SCI group.

(c) Construct a box-and-whisker plot for the controls.

(d) Construct a box-and-whisker plot for the SCI group.

(e) Do you believe there is a difference in pressure readings for controls and SCI subjects in this study?

13. Johnson et al. (A-16) performed a retrospective review of 50 fetuses that underwent open fetal myelomeningocele closure. The data below show the gestational age in weeks of the 50 fetuses undergoing the procedure.

25	25	26	27	29	29	29	30	30	31
32	32	32	33	33	33	33	34	34	34
35	35	35	35	35	35	35	35	35	36
36	36	36	36	36	36	36	36	36	36
36	36	36	36	36	36	36	36	37	37

Source: Mark P. Johnson, Leslie N. Sutton, Natalie Rintoul, Timothy M. Crombleholme, Alan W. Flake, Lori J. Howell, Holly L. Hedrick, R. Douglas Wilson, and N. Scott Adzick, "Fetal Myelomeningocele Repair: Short-Term Clinical Outcomes," *American Journal of Obstetrics and Gynecology, 189* (2003), 482–487.

(a) Construct a stem-and-leaf plot for these gestational ages.

(b) Based on the stem-and-leaf plot, what one word would you use to describe the nature of the data?

(c) Why do you think the stem-and-leaf plot looks the way it does?

(d) Compute the mean, median, variance, and standard deviation.

14. The following table gives the age distribution for the number of deaths in New York State due to accidents for residents age 25 and older.

Age (Years)	Number of Deaths Due to Accidents
25–34	393
35–44	514
45–54	460
55–64	341
65–74	365
75–84	616
85–94*	618

Source: New York State Department of Health, Vital Statistics of New York State, 2000, Table 32: *Death Summary Information by Age.*

*May include deaths due to accident for adults over age 94.

For these data construct a cumulative frequency distribution, a relative frequency distribution, and a cumulative relative frequency distribution.

15. Krieser et al. (A-17) examined glomerular filtration rate (GFR) in pediatric renal transplant recipients. GFR is an important parameter of renal function assessed in renal transplant recipients. The following are measurements from 19 subjects of GFR measured with diethylenetriamine penta-acetic acid. (Note: some subjects were measured more than once.)

18	42
21	43
21	43
23	48
27	48
27	51
30	55
32	58
32	60
32	62
36	67
37	68
41	88
42	63

Source: D. M. Z. Krieser, M.D. Used with permission.

(a) Compute mean, median, variance, standard deviation, and coefficient of variation.

(b) Construct a stem-and-leaf display.

(c) Construct a box-and-whisker plot.

(d) What percentage of the measurements is within one standard deviation of the mean? Two standard deviations? Three standard deviations?

16. The following are the cystatin C levels (mg/L) for the patients described in Exercise 15 (A-17). Cystatin C is a cationic basic protein that was investigated for its relationship to GFR levels. In addition, creatinine levels are also given. (Note: Some subjects were measured more than once.)

Cystatin C (mg/L)		Creatinine (mmol/L)	
1.78	4.69	0.35	0.14
2.16	3.78	0.30	0.11
1.82	2.24	0.20	0.09
1.86	4.93	0.17	0.12
1.75	2.71	0.15	0.07
1.83	1.76	0.13	0.12
2.49	2.62	0.14	0.11
1.69	2.61	0.12	0.07
1.85	3.65	0.24	0.10
1.76	2.36	0.16	0.13
1.25	3.25	0.17	0.09
1.50	2.01	0.11	0.12
2.06	2.51	0.12	0.06
2.34			

Source: D. M. Z. Krieser, M.D. Used with permission.

(a) For each variable, compute the mean, median, variance, standard deviation, and coefficient of variation.

(b) For each variable, construct a stem-and-leaf display and a box-and-whisker plot.

(c) Which set of measurements is more variable, cystatin C or creatinine? On what do you base your answer?

17. Give three synonyms for variation (variability).

18. The following table shows the age distribution of live births in Albany County, New York, for 2000.

Mother's Age	Number of Live Births
10–14	7
15–19	258
20–24	585
25–29	841
30–34	981
35–39	526
40–44	99
45–49*	4

Source: New York State Department of Health, Annual Vital Statistics 2000, Table 7, *Live Births by Resident County and Mother's Age.*
*May include live births to mothers over age 49.

For these data construct a cumulative frequency distribution, a relative frequency distribution, and a cumulative relative frequency distribution.

19. Spivack (A-18) investigated the severity of disease associated with *C. difficile* in pediatric inpatients. One of the variables they examined was number of days patients experienced diarrhea. The data for the 22 subjects in the study appear below. Compute the mean, median, variance, and standard deviation.

3	11	3	4	14	2	4	5	3	11	2
2	3	2	1	1	7	2	1	1	3	2

Source: Jordan G. Spivack, Stephen C. Eppes, and Joel D. Klien, "*Clostridium Difficile*–Associated Diarrhea in a Pediatric Hospital," *Clinical Pediatrics, 42* (2003), 347–352.

20. Express in words the following properties of the sample mean:

(a) $\Sigma(x - \bar{x})^2 = \text{a minimum}$

(b) $n\bar{x} = \Sigma x$

(c) $\Sigma(x - \bar{x}) = 0$

21. Your statistics instructor tells you on the first day of class that there will be five tests during the term. From the scores on these tests for each student, the instructor will compute a measure of central tendency that will serve as the student's final course grade. Before taking the first test, you must choose whether you want your final grade to be the mean or the median of the five test scores. Which would you choose? Why?

22. Consider the following possible class intervals for use in constructing a frequency distribution of serum cholesterol levels of subjects who participated in a mass screening:

(a)	(b)	(c)
50–74	50–74	50–75
75–99	75–99	75–100
100–149	100–124	100–125
150–174	125–149	125–150
175–199	150–174	150–175
200–249	175–199	175–200
250–274	200–224	200–225
etc.	225–249	225–250
	etc.	etc.

Which set of class intervals do you think is most appropriate for the purpose? Why? State specifically for each one why you think the other two are less desirable.

23. On a statistics test students were asked to construct a frequency distribution of the blood creatine levels (units/liter) for a sample of 300 healthy subjects. The mean was 95, and the standard deviation was 40. The following class interval widths were used by the students:

(a) 1 **(d)** 15

(b) 5 **(e)** 20

(c) 10 **(f)** 25

Comment on the appropriateness of these choices of widths.

24. Give a health sciences–related example of a population of measurements for which the mean would be a better measure of central tendency than the median.

25. Give a health sciences–related example of a population of measurements for which the median would be a better measure of central tendency than the mean.

26. Indicate for the following variables which you think would be a better measure of central tendency, the mean, the median, or mode, and justify your choice:

(a) Annual incomes of licensed practical nurses in the Southeast.
(b) Diagnoses of patients seen in the emergency department of a large city hospital.
(c) Weights of high-school male basketball players.

27. Refer to Exercise 2.3.11. Compute the mean, median, variance, standard deviation, first quartile, third quartile, and interquartile range. Construct a boxplot of the data. Are the mode, median, and mean equal? If not, explain why. Discuss the data in terms of variability. Compare the IQR with the range. What does the comparison tell you about the variability of the observations?

28. Refer to Exercise 2.3.12. Compute the mean, median, variance, standard deviation, first quartile, third quartile, and interquartile range. Construct a boxplot of the data. Are the mode, median, and mean equal? If not, explain why. Discuss the data in terms of variability. Compare the IQR with the range. What does the comparison tell you about the variability of the observations?

29. Thilothammal et al. (A-19) designed a study to determine the efficacy of BCG (bacillus Calmette-Guérin) vaccine in preventing tuberculous meningitis. Among the data collected on each subject was a measure of nutritional status (actual weight expressed as a percentage of expected weight for actual height). The following table shows the nutritional status values of the 107 cases studied.

73.3	54.6	82.4	76.5	72.2	73.6	74.0
80.5	71.0	56.8	80.6	100.0	79.6	67.3
50.4	66.0	83.0	72.3	55.7	64.1	66.3
50.9	71.0	76.5	99.6	79.3	76.9	96.0
64.8	74.0	72.6	80.7	109.0	68.6	73.8
74.0	72.7	65.9	73.3	84.4	73.2	70.0
72.8	73.6	70.0	77.4	76.4	66.3	50.5
72.0	97.5	130.0	68.1	86.4	70.0	73.0
59.7	89.6	76.9	74.6	67.7	91.9	55.0
90.9	70.5	88.2	70.5	74.0	55.5	80.0
76.9	78.1	63.4	58.8	92.3	100.0	84.0
71.4	84.6	123.7	93.7	76.9	79.6	
45.6	92.5	65.6	61.3	64.5	72.7	
77.5	76.9	80.2	76.9	88.7	78.1	
60.6	59.0	84.7	78.2	72.4	68.3	
67.5	76.9	82.6	85.4	65.7	65.9	

Source: Dr. N. Thilothammal. Used with permission.

(a) For these data compute the following descriptive measures: mean, median, mode, variance, standard deviation, range, first quartile, third quartile, and IQR.

(b) Construct the following graphs for the data: histogram, frequency polygon, stem-and-leaf plot, and boxplot.

(c) Discuss the data in terms of variability. Compare the IQR with the range. What does the comparison tell you about the variability of the observations?

(d) What proportion of the measurements are within one standard deviation of the mean? Two standard deviations of the mean? Three standard deviations of the mean?

(e) What proportion of the measurements are less than 100?

(f) What proportion of the measurements are less than 50?

Exercises for Use with Large Data Sets Available on the Following Website: www.wiley.com/college/daniel

Refer to the dataset NCBIRTH800. The North Carolina State Center for Health Statistics and Howard W. Odum Institute for Research in Social Science at the University of North Carolina at Chapel Hill (A-20) make publicly available birth and infant death data for all children born in the state of North Carolina. These data can be accessed at www.irss.unc.edu/ncvital/bfd1down.html. Records on birth data go back to 1968. This comprehensive data set for the births in 2001 contains 120,300 records. The data represents a random sample of 800 of those births and selected variables. The variables are as follows:

Variable Label	Description
PLURALITY	Number of children born of the pregnancy
SEX	Sex of child (1 = male, 2 = female)
MAGE	Age of mother (years)
WEEKS	Completed weeks of gestation (weeks)
MARITAL	Marital status (1 = married, 2 = not married)
RACEMOM	Race of mother (0 = other non-White, 1 = White, 2 = Black 3 = American Indian, 4 = Chinese, 5 = Japanese, 6 = Hawaiian, 7 = Filipino, 8 = Other Asian or Pacific Islander)
HISPMOM	Mother of Hispanic origin (C = Cuban, M = Mexican, N = Non-Hispanic, O = other and unknown Hispanic, P = Puerto Rican, S = Central/South American, U = not classifiable)
GAINED	Weight gained during pregnancy (pounds)
SMOKE	0 = mother did not smoke during pregnancy 1 = mother did smoke during pregnancy
DRINK	0 = mother did not consume alcohol during pregnancy 1 = mother did consume alcohol during pregnancy
TOUNCES	Weight of child (ounces)
TGRAMS	Weight of child (grams)
LOW	0 = infant was not low birth weight 1 = infant was low birth weight
PREMIE	0 = infant was not premature 1 = infant was premature Premature defined at 36 weeks or sooner

For the variables of MAGE, WEEKS, GAINED, TOUNCES, and TGRAMS:

1. Calculate the mean, median, standard deviation, IQR, and range.
2. For each, construct a histogram and comment on the shape of the distribution.
3. Do the histograms for TOUNCES and TGRAMS look strikingly similar? Why?
4. Construct box-and-whisker plots for all four variables.
5. Construct side-by-side box-and-whisker plots for the variable of TOUNCES for women who admitted to smoking and women who did not admit to smoking. Do you see a difference in birth weight in the two groups? Which group has more variability?
6. Construct side-by-side box-and-whisker plots for the variable of MAGE for women who are and are not married. Do you see a difference in ages in the two groups? Which group has more variability? Are the results surprising?
7. Calculate the skewness and kurtosis of the data set. What do they indicate?

REFERENCES

Methodology References

1. H. A. STURGES, "The Choice of a Class Interval," *Journal of the American Statistical Association, 21* (1926), 65–66.
2. HELEN M. WALKER, "Degrees of Freedom," *Journal of Educational Psychology, 31* (1940), 253–269.
3. JOHN W. TUKEY, *Exploratory Data Analysis,* Addison-Wesley, Reading, MA, 1977.

Applications References

A-1. FARHAD ATASSI, "Oral Home Care and the Reasons for Seeking Dental Care by Individuals on Renal Dialysis," *Journal of Contemporary Dental Practice, 3* (2002), 31–41.

A-2. VALLABH JANARDHAN, ROBERT FRIEDLANDER, HOWARD RIINA, and PHILIP EDWIN STIEG, "Identifying Patients at Risk for Postprocedural Morbidity after Treatment of Incidental Intracranial Aneurysms: The Role of Aneurysm Size and Location," *Neurosurgical Focus, 13* (2002), 1–8.

A-3. A. HOEKEMA, B. HOVINGA, B. STEGENGA, and L. G. M. De BONT, "Craniofacial Morphology and Obstructive Sleep Apnoea: A Cephalometric Analysis," *Journal of Oral Rehabilitation, 30* (2003), 690–696.

A-4. DAVID H. HOLBEN, "Selenium Content of Venison, Squirrel, and Beef Purchased or Produced in Ohio, a Low Selenium Region of the United States," *Journal of Food Science, 67* (2002), 431–433.

A-5. ERIK SKJELBO, THEONEST K. MUTABINGWA, IB BYGBJERG, KARIN K. NIELSEN, LARS F. GRAM, and KIM BRØSEN, "Chloroguanide Metabolism in Relation to the Efficacy in Malaria Prophylaxis and the *S*-Mephenytoin Oxidation in Tanzanians," *Clinical Pharmacology & Therapeutics, 59* (1996), 304–311.

A-6. HENRIK SCHMIDT, POUL ERIK MORTENSEN, SÀREN LARS FÀLSGAARD, and ESTHER A. JENSEN, "Autotransfusion After Coronary Artery Bypass Grafting Halves the Number of Patients Needing Blood Transfusion," *Annals of Thoracic Surgery, 61* (1996), 1178–1181.

A-7. RICHARD EVANS, WANDA GORDON, and MIKE CONZEMIUS, "Effect of Velocity on Ground Reaction Forces in Dogs with Lameness Attributable to Tearing of the Cranial Cruciate Ligament," *American Journal of Veterinary Research, 64* (2003), 1479–1481.

A-8. SIMONA PORCELLINI, GUILIANA VALLANTI, SILVIA NOZZA, GUIDO POLI, ADRIANO LAZZARIN, GUISEPPE TAMBUSSI, and ANTONIO GRASSIA, "Improved Thymopoietic Potential in Aviremic HIV Infected Individuals with HAART by Intermittent IL-2 Administration," *AIDS, 17* (2003), 1621–1630.

A-9. HARRY N. SHAIR and ANNA JASPER, "Decreased Venous Return Is Neither Sufficient nor Necessary to Elicit Ultrasonic Vocalization of Infant Rat Pups," *Behavioral Neuroscience, 117* (2003), 840–853.

A-10. M. BUTZ, K. H. WOLLINSKY, U. WIDEMUTH-CATRINESCU, A. SPERFELD, S. WINTER, H. H. MEHRKENS, A. C. LUDOLPH, and H. SCHREIBER, "Longitudinal Effects of Noninvasive Positive-Pressure Ventilation in Patients with Amyotophic Lateral Sclerosis," *American Journal of Medical Rehabilitation, 82* (2003), 597–604.

A-11. DAVID W. STARCH, JERRY W. ALEXANDER, PHILIP C. NOBLE, SURAJ REDDY, and DAVID M. LINTNER, "Multistranded Hamstring Tendon Graft Fixation with a Central Four-Quadrant or a Standard Tibial Interference Screw for Anterior Cruciate Ligament Reconstruction," *American Journal of Sports Medicine, 31* (2003), 338–344.

A-12. RICHARD J. CARDOSI, ROSEMARY CARDOSI, EDWARD C. GRENDYS Jr., JAMES V. FIORICA, and MITCHEL S. HOFFMAN, "Infectious Urinary Tract Morbidity with Prolonged Bladder Catheterization After Radical Hysterectomy," *American Journal of Obstetrics and Gynecology, 189* (2003), 380–384.

A-13. SATOSHI NOZAWA, KATSUJI SHIMIZU, KEI MIYAMOTO, and MIZUO TANAKA, "Repair of Pars Interarticularis Defect by Segmental Wire Fixation in Young Athletes with Spondylolysis," *American Journal of Sports Medicine, 31* (2003), 359–364.

A-14. GEA A. HUIZINGA, WINETTE T. A. van der GRAAF, ANNEMIKE VISSER, JOS S. DIJKSTRA, and JOSETTE E. H. M. HOEKSTRA-WEEBERS, "Psychosocial Consequences for Children of a Parent with Cancer," *Cancer Nursing, 26* (2003), 195–202.

A-15. ERIC W. TAM, ARTHUR F. MAK, WAI NGA LAM, JOHN H. EVANS, and YORK Y. CHOW, "Pelvic Movement and Interface Pressure Distribution During Manual Wheelchair Propulsion," *Archives of Physical Medicine and Rehabilitation, 84* (2003), 1466–1472.

A-16. MARK P. JOHNSON, LESLIE N. SUTTON, NATALIE RINTOUL, TIMOTHY M. CROMBLEHOLME, ALAN W. FLAKE, LORI J. HOWELL, HOLLY L. HEDRICK, R. DOUGLAS WILSON, and N. SCOTT ADZICK, "Fetal Myelomeningocele Repair: Short-Term Clinical Outcomes," *American Journal of Obstetrics and Gynecology, 189* (2003), 482–487.

A-17. D. M. Z. Krieser, A. R. Rosenberg, G. Kainer, and D. Naidoo, "The Relationship between Serum Creatinine, Serum Cystatin C, and Glomerular Filtration Rate in Pediatric Renal Transplant Recipients: A Pilot Study," *Pediatric Transplantation, 6* (2002), 392–395.

A-18. Jordan G. Spivack, Stephen C. Eppes, and Joel D. Klien, "*Clostridium Difficile*–Associated Diarrhea in a Pediatric Hospital," *Clinical Pediatrics, 42* (2003), 347–352.

A-19. N. Thilothammal, P. V. Krishnamurthy, Desmond K. Runyan, and K. Banu, "Does BCG Vaccine Prevent Tuberculous Meningitis?" *Archives of Disease in Childhood, 74* (1996), 144–147.

A-20. North Carolina State Center for Health Statistics and Howard W. Odum Institute for Research in Social Science at the University of North Carolina at Chapel Hill. Birth data set for 2001 found at www.irss.unc.edu/ncvital/bfd1down.html. All calculations were performed by John Holcomb and do not represent the findings of the Center or Institute.

CHAPTER **3**

PROBABILITY: THE BASIS OF STATISTICAL INFERENCE

CHAPTER OVERVIEW

Probability lays the foundation for statistical inference. This chapter provides a brief overview of the probability concepts necessary for the understanding of topics covered in the chapters that follow. It also provides a context for understanding the probability distributions used in statistical inference, and introduces the student to several measures commonly found in the medical literature (e.g., the sensitivity and specificity of a test).

TOPICS

LEARNING OUTCOMES

After studying this chapter, the student will

1. understand classical, relative frequency, and subjective probability.
2. understand the properties of probability and selected probability rules.
3. be able to calculate the probability of an event.
4. be able to apply Bayes' theorem when calculating screening test results.

3.1 INTRODUCTION

The theory of probability provides the foundation for statistical inference. However, this theory, which is a branch of mathematics, is not the main concern of this book, and, consequently, only its fundamental concepts are discussed here. Students who desire to pursue this subject should refer to the many books on probability available in most college and university libraries. The books by Gut (1), Isaac (2), and Larson (3) are recommended. The objectives of this chapter are to help students gain some mathematical ability in the area of probability and to assist them in developing an understanding of the more important concepts. Progress along these lines will contribute immensely to their success in understanding the statistical inference procedures presented later in this book.

The concept of probability is not foreign to health workers and is frequently encountered in everyday communication. For example, we may hear a physician say that a patient has a 50–50 chance of surviving a certain operation. Another physician may say that she is 95 percent certain that a patient has a particular disease. A public health nurse may say that nine times out of ten a certain client will break an appointment. As these examples suggest, most people express probabilities in terms of percentages. In dealing with probabilities mathematically, it is more convenient to express probabilities as fractions. (Percentages result from multiplying the fractions by 100.) Thus, we measure the probability of the occurrence of some event by a number between zero and one. The more likely the event, the closer the number is to one; and the more unlikely the event, the closer the number is to zero. An event that cannot occur has a probability of zero, and an event that is certain to occur has a probability of one.

Health sciences researchers continually ask themselves if the results of their efforts could have occurred by chance alone or if some other force was operating to produce the observed effects. For example, suppose six out of ten patients suffering from some disease are cured after receiving a certain treatment. Is such a cure rate likely to have occurred if the patients had not received the treatment, or is it evidence of a true curative effect on the part of the treatment? We shall see that questions such as these can be answered through the application of the concepts and laws of probability.

3.2 TWO VIEWS OF PROBABILITY: OBJECTIVE AND SUBJECTIVE

Until fairly recently, probability was thought of by statisticians and mathematicians only as an *objective* phenomenon derived from objective processes.

The concept of *objective probability* may be categorized further under the headings of (1) *classical,* or *a priori, probability;* and (2) the *relative frequency,* or *a posteriori,* concept of probability.

Classical Probability The classical treatment of probability dates back to the 17th century and the work of two mathematicians, Pascal and Fermat. Much of this theory developed out of attempts to solve problems related to games of chance, such as those involving the rolling of dice. Examples from games of chance illustrate very well the principles involved in classical probability. For example, if a fair six-sided die is rolled, the probability that a 1 will be observed is equal to $1/6$ and is the same for the other five faces. If a card is picked at random from a well-shuffled deck of ordinary playing cards, the probability of picking a heart is $13/52$. Probabilities such as these are calculated by the processes of abstract reasoning. It is not necessary to roll a die or draw a card to compute these probabilities. In the rolling of the die, we say that each of the six sides is *equally likely* to be observed if there is no reason to favor any one of the six sides. Similarly, if there is no reason to favor the drawing of a particular card from a deck of cards, we say that each of the 52 cards is equally likely to be drawn. We may define probability in the classical sense as follows:

DEFINITION

If an event can occur in N mutually exclusive and equally likely ways, and if m of these possess a trait E, the probability of the occurrence of E is equal to m/N.

If we read $P(E)$ as "the probability of E," we may express this definition as

$$P(E) = \frac{m}{N} \tag{3.2.1}$$

Relative Frequency Probability The relative frequency approach to probability depends on the repeatability of some process and the ability to count the number of repetitions, as well as the number of times that some event of interest occurs. In this context we may define the probability of observing some characteristic, E, of an event as follows:

DEFINITION

If some process is repeated a large number of times, n, and if some resulting event with the characteristic E occurs m times, the relative frequency of occurrence of E, m/n, will be approximately equal to the probability of E.

To express this definition in compact form, we write

$$P(E) = \frac{m}{n} \tag{3.2.2}$$

We must keep in mind, however, that, strictly speaking, m/n is only an estimate of $P(E)$.

Subjective Probability In the early 1950s, L. J. Savage (4) gave considerable impetus to what is called the "personalistic" or subjective concept of probability. This view holds that probability measures the confidence that a particular individual has in the truth of a particular proposition. This concept does not rely on the repeatability of any process. In fact, by applying this concept of probability, one may evaluate the probability of an event that can only happen once, for example, the probability that a cure for cancer will be discovered within the next 10 years.

Although the subjective view of probability has enjoyed increased attention over the years, it has not been fully accepted by statisticians who have traditional orientations.

Bayesian Methods Bayesian methods are named in honor of the Reverend Thomas Bayes (1702–1761), an English clergyman who had an interest in mathematics. Bayesian methods are an example of subjective probability, since it takes into consideration the degree of belief that one has in the chance that an event will occur. While probabilities based on classical or relative frequency concepts are designed to allow for decisions to be made solely on the basis of collected data, Bayesian methods make use of what are known as *prior probabilities* and *posterior probabilities.*

DEFINITION

The *prior probability* of an event is a probability based on prior knowledge, prior experience, or results derived from prior data collection activity.

DEFINITION

The *posterior probability* of an event is a probability obtained by using new information to update or revise a prior probability.

As more data are gathered, the more is likely to be known about the "true" probability of the event under consideration. Although the idea of updating probabilities based on new information is in direct contrast to the philosophy behind frequency-of-occurrence probability, Bayesian concepts are widely used. For example, Bayesian techniques have found recent application in the construction of e-mail spam filters. Typically, the application of Bayesian concepts makes use of a mathematical formula called *Bayes' theorem*. In Section 3.5 we employ Bayes' theorem in the evaluation of diagnostic screening test data.

3.3 ELEMENTARY PROPERTIES OF PROBABILITY

In 1933 the axiomatic approach to probability was formalized by the Russian mathematician A. N. Kolmogorov (5). The basis of this approach is embodied in three properties from which a whole system of probability theory is constructed through the use of mathematical logic. The three properties are as follows.

1. Given some process (or experiment) with n mutually exclusive outcomes (called events), $E_1, E_2, \ldots, E_n$, the probability of any event E_i is assigned a nonnegative number. That is,

$$P(E_i) \geq 0 \tag{3.3.1}$$

In other words, all events must have a probability greater than or equal to zero, a reasonable requirement in view of the difficulty of conceiving of negative probability. A key concept in the statement of this property is the concept of *mutually exclusive* outcomes. Two events are said to be mutually exclusive if they cannot occur simultaneously.

2. The sum of the probabilities of the mutually exclusive outcomes is equal to 1.

$$P(E_1) + P(E_2) + \cdots + P(E_n) = 1 \tag{3.3.2}$$

This is the property of *exhaustiveness* and refers to the fact that the observer of a probabilistic process must allow for all possible events, and when all are taken together, their total probability is 1. The requirement that the events be mutually exclusive is specifying that the events $E_1, E_2, \ldots, E_n$ do not overlap; that is, no two of them can occur at the same time.

3. Consider any two mutually exclusive events, E_i and E_j. The probability of the occurrence of either E_i or E_j is equal to the sum of their individual probabilities.

$$P(E_i + E_j) = P(E_i) + P(E_j) \tag{3.3.3}$$

Suppose the two events were not mutually exclusive; that is, suppose they could occur at the same time. In attempting to compute the probability of the occurrence of either E_i or E_j the problem of overlapping would be discovered, and the procedure could become quite complicated. This concept will be discussed further in the next section.

3.4 CALCULATING THE PROBABILITY OF AN EVENT

We now make use of the concepts and techniques of the previous sections in calculating the probabilities of specific events. Additional ideas will be introduced as needed.

EXAMPLE 3.4.1

The primary aim of a study by Carter et al. (A-1) was to investigate the effect of the age at onset of bipolar disorder on the course of the illness. One of the variables investigated was family history of mood disorders. Table 3.4.1 shows the frequency of a family history of mood disorders in the two groups of interest (Early age at onset defined to be

TABLE 3.4.1 Frequency of Family History of Mood Disorder by Age Group Among Bipolar Subjects

Family History of Mood Disorders	Early = 18(*E*)	Later > 18(*L*)	Total
Negative (*A*)	28	35	63
Bipolar disorder (*B*)	19	38	57
Unipolar (*C*)	41	44	85
Unipolar and bipolar (*D*)	53	60	113
Total	141	177	318

Source: Tasha D. Carter, Emanuela Mundo, Sagar V. Parkh, and James L. Kennedy, "Early Age at Onset as a Risk Factor for Poor Outcome of Bipolar Disorder," *Journal of Psychiatric Research, 37* (2003), 297–303.

18 years or younger and Later age at onset defined to be later than 18 years). Suppose we pick a person at random from this sample. What is the probability that this person will be 18 years old or younger?

Solution: For purposes of illustrating the calculation of probabilities, we consider this group of 318 subjects to be the largest group for which we have an interest. In other words, for this example, we consider the 318 subjects as a population. We assume that Early and Later are mutually exclusive categories and that the likelihood of selecting any one person is equal to the likelihood of selecting any other person. We define the desired probability as the number of subjects with the characteristic of interest (Early) divided by the total number of subjects. We may write the result in probability notation as follows:

$$
\begin{aligned}
P(E) &= \text{number of Early subjects/total number of subjects} \\
&= 141/318 = .4434
\end{aligned}
$$

■

Conditional Probability On occasion, the set of "all possible outcomes" may constitute a subset of the total group. In other words, the size of the group of interest may be reduced by conditions not applicable to the total group. When probabilities are calculated with a subset of the total group as the denominator, the result is a *conditional probability.*

The probability computed in Example 3.4.1, for example, may be thought of as an unconditional probability, since the size of the total group served as the denominator. No conditions were imposed to restrict the size of the denominator. We may also think of this probability as a *marginal probability* since one of the marginal totals was used as the numerator.

We may illustrate the concept of conditional probability by referring again to Table 3.4.1.

EXAMPLE 3.4.2

Suppose we pick a subject at random from the 318 subjects and find that he is 18 years or younger (E). What is the probability that this subject will be one who has no family history of mood disorders (A)?

Solution: The total number of subjects is no longer of interest, since, with the selection of an Early subject, the Later subjects are eliminated. We may define the desired probability, then, as follows: What is the probability that a subject has no family history of mood disorders (A), given that the selected subject is Early (E)? This is a conditional probability and is written as $P(A \mid E)$ in which the vertical line is read "given." The 141 Early subjects become the denominator of this conditional probability, and 28, the number of Early subjects with no family history of mood disorders, becomes the numerator. Our desired probability, then, is

$$P(A \mid E) = 28/141 = .1986$$ ■

Joint Probability Sometimes we want to find the probability that a subject picked at random from a group of subjects possesses two characteristics at the same time. Such a probability is referred to as a *joint probability*. We illustrate the calculation of a joint probability with the following example.

EXAMPLE 3.4.3

Let us refer again to Table 3.4.1. What is the probability that a person picked at random from the 318 subjects will be Early (E) *and* will be a person who has no family history of mood disorders (A)?

Solution: The probability we are seeking may be written in symbolic notation as $P(E \cap A)$ in which the symbol $\cap$ is read either as "intersection" or "and." The statement $E \cap A$ indicates the joint occurrence of conditions E and A. The number of subjects satisfying both of the desired conditions is found in Table 3.4.1 at the intersection of the column labeled E and the row labeled A and is seen to be 28. Since the selection will be made from the total set of subjects, the denominator is 318. Thus, we may write the joint probability as

$$P(E \cap A) = 28/318 = .0881$$ ■

The Multiplication Rule A probability may be computed from other probabilities. For example, a joint probability may be computed as the product of an appropriate marginal probability and an appropriate conditional probability. This relationship is known as the *multiplication rule* of probability. We illustrate with the following example.

EXAMPLE 3.4.4

We wish to compute the joint probability of Early age at onset (E) and a negative family history of mood disorders (A) from knowledge of an appropriate marginal probability and an appropriate conditional probability.

Solution: The probability we seek is $P(E \cap A)$. We have already computed a marginal probability, $P(E) = 141/318 = .4434$, and a conditional probability, $P(A \mid E) = 28/141 = .1986$. It so happens that these are appropriate marginal and conditional probabilities for computing the desired joint probability. We may now compute $P(E \cap A) = P(E)P(A \mid E) = (.4434)(.1986) = .0881$. This, we note, is, as expected, the same result we obtained earlier for $P(E \cap A)$. ■

We may state the multiplication rule in general terms as follows: For any two events A and B,

$$P(A \cap B) = P(B)P(A \mid B), \qquad \text{if } P(B) \neq 0 \tag{3.4.1}$$

For the same two events A and B, the multiplication rule may also be written as $P(A \cap B) = P(A)P(B \mid A)$, if $P(A) \neq 0$.

We see that through algebraic manipulation the multiplication rule as stated in Equation 3.4.1 may be used to find any one of the three probabilities in its statement if the other two are known. We may, for example, find the conditional probability $P(A \mid B)$ by dividing $P(A \cap B)$ by $P(B)$. This relationship allows us to formally define conditional probability as follows.

DEFINITION

The *conditional probability* of *A* given *B* is equal to the probability of A ∩ B divided by the probability of *B*, provided the probability of *B* is not zero.

That is,

$$P(A \mid B) = \frac{P(A \cap B)}{P(B)}, \qquad P(B) \neq 0 \tag{3.4.2}$$

We illustrate the use of the multiplication rule to compute a conditional probability with the following example.

EXAMPLE 3.4.5

We wish to use Equation 3.4.2 and the data in Table 3.4.1 to find the conditional probability, $P(A \mid E)$.

Solution: According to Equation 3.4.2,

$$P(A \mid E) = P(A \cap E)/P(E)$$

■

Earlier we found $P(E \cap A) = P(A \cap E) = 28/318 = .0881$. We have also determined that $P(E) = 141/318 = .4434$. Using these results we are able to compute $P(A \mid E) = .0881/.4434 = .1987$, which, as expected, is the same result we obtained by using the frequencies directly from Table 3.4.1. (The slight discrepancy is due to rounding.)

The Addition Rule The third property of probability given previously states that the probability of the occurrence of either one or the other of two mutually exclusive events is equal to the sum of their individual probabilities. Suppose, for example, that we pick a person at random from the 318 represented in Table 3.4.1. What is the probability that this person will be Early age at onset (E) or Later age at onset (L)? We state this probability in symbols as $P(E \cup L)$, where the symbol $\cup$ is read either as "union" or "or." Since the two age conditions are mutually exclusive, $P(E \cap L) = (141/318) + (177/318) = .4434 + .5566 = 1$.

What if two events are not mutually exclusive? This case is covered by what is known as the *addition rule,* which may be stated as follows:

DEFINITION

Given two events *A* and *B*, the probability that event *A*, or event *B*, or both occur is equal to the probability that event *A* occurs, plus the probability that event *B* occurs, minus the probability that the events occur simultaneously.

The addition rule may be written

$$P(A \cup B) = P(A) + P(B) - P(A \cap B) \tag{3.4.3}$$

When events A and B cannot occur simultaneously, $P(A \cap B)$ is sometimes called "exclusive or," and $P(A \cup B) = 0$. When events A and B can occur simultaneously, $P(A \cup B)$ is sometimes called "inclusive or," and we use the addition rule to calculate $P(A \cup B)$. Let us illustrate the use of the addition rule by means of an example.

EXAMPLE 3.4.6

If we select a person at random from the 318 subjects represented in Table 3.4.1, what is the probability that this person will be an Early age of onset subject (E) or will have no family history of mood disorders (A) or both?

Solution: The probability we seek is $P(E \cup A)$. By the addition rule as expressed by Equation 3.4.3, this probability may be written as $P(E \cup A) = P(E) + P(A) - P(E \cap A)$. We have already found that $P(E) = 141/318 = .4434$ and $P(E \cap A) = 28/318 = .0881$. From the information in Table 3.4.1 we calculate $P(A) = 63/318 = .1981$. Substituting these results into the equation for $P(E \cup A)$ we have $P(E \cup A) = .4434 + .1981 - .0881 = .5534$. ■

Note that the 28 subjects who are *both* Early *and* have no family history of mood disorders are included in the 141 who are Early as well as in the 63 who have no family history of mood disorders. Since, in computing the probability, these 28 have been added into the numerator twice, they have to be subtracted out once to overcome the effect of duplication, or overlapping.

Independent Events Suppose that, in Equation 3.4.2, we are told that event B has occurred, but that this fact has no effect on the probability of A. That is, suppose that the probability of event A is the same regardless of whether or not B occurs. In this situation, $P(A \mid B) = P(A)$. In such cases we say that A and B are *independent events.* The multiplication rule for two independent events, then, may be written as

$$P(A \cap B) = P(A)P(B); \quad P(A) \neq 0, \quad P(B) \neq 0 \tag{3.4.4}$$

Thus, we see that if two events are independent, the probability of their joint occurrence is equal to the product of the probabilities of their individual occurrences.

Note that when two events with nonzero probabilities are independent, each of the following statements is true:

$$P(A \mid B) = P(A), \quad P(B \mid A) = P(B), \quad P(A \cap B) = P(A)P(B)$$

Two events are not independent unless all these statements are true. It is important to be aware that the terms *independent* and *mutually exclusive* do not mean the same thing.

Let us illustrate the concept of independence by means of the following example.

EXAMPLE 3.4.7

In a certain high school class, consisting of 60 girls and 40 boys, it is observed that 24 girls and 16 boys wear eyeglasses. If a student is picked at random from this class, the probability that the student wears eyeglasses, $P(E)$, is 40/100, or .4.

(a) What is the probability that a student picked at random wears eyeglasses, given that the student is a boy?

Solution: By using the formula for computing a conditional probability, we find this to be

$$P(E \mid B) = \frac{P(E \cap B)}{P(B)} = \frac{16/100}{40/100} = .4$$

Thus the additional information that a student is a boy does not alter the probability that the student wears eyeglasses, and $P(E) = P(E \mid B)$. We say that the events being a boy and wearing eyeglasses for this group are independent. We may also show that the event of wearing eyeglasses, E, and *not* being a boy, $\overline{B}$ are also independent as follows:

$$P(E \mid \overline{B}) = \frac{P(E \cap \overline{B})}{P(\overline{B})} = \frac{24/100}{60/100} = \frac{24}{60} = .4$$

(b) What is the probability of the joint occurrence of the events of wearing eyeglasses and being a boy?

Solution: Using the rule given in Equation 3.4.1, we have

$$P(E \cap B) = P(B)P(E \mid B)$$

but, since we have shown that events E and B are independent we may replace $P(E \mid B)$ by $P(E)$ to obtain, by Equation 3.4.4,

$$\begin{aligned} P(E \cap B) &= P(B)P(E) \\ &= \left(\frac{40}{100}\right)\left(\frac{40}{100}\right) \\ &= .16 \end{aligned}$$

■

Events Earlier, using the data in Table 3.4.1, we computed the
picked at random from the 318 subjects will be an Early age of onset
/318 = .4434. We found the probability of a Later age at onset to
.5566. The sum of these two probabilities we found to be equal to
he events being Early age at onset and being Later age at onset are
n general, we may make the following statement about complemen-
lity of an event A is equal to 1 minus the probability of its comple-
, and

$$P(\overline{A}) = 1 - P(A) \tag{3.4.5}$$

the third property of probability since the event, A, and its com-
exclusive.

nissions to a general hospital during a certain period of time,
. If we designate these as set A, then $\overline{A}$ is equal to 1200 minus
pute

$$P(A) = 750/1200 = .625$$

$$P(\overline{A}) = 450/1200 = .375$$

$$\begin{aligned} P(\overline{A}) &= 1 - P(A) \\ .375 &= 1 - .625 \\ .375 &= .375 \end{aligned}$$

■

Marginal Probability Earlier we used the term *marginal probability* to refer to a probability in which the numerator of the probability is a marginal total from a table such as Table 3.4.1. For example, when we compute the probability that a person picked at random from the 318 persons represented in Table 3.4.1 is an Early age of onset subject, the numerator of the probability is the total number of Early subjects, 141. Thus, $P(E) = 141/318 = .4434$. We may define marginal probability more generally as follows:

DEFINITION

Given some variable that can be broken down into m categories designated by $A_1, A_2, \ldots, A_i, \ldots, A_m$ and another jointly occurring variable that is broken down into n categories designated by $B_1, B_2, \ldots, B_j, \ldots, B_n$, the marginal probability of A_i, $P(A_i)$, is equal to the sum of the joint probabilities of A_i with all the categories of B. That is,

$$P(A_i) = \sum P(A_i \cap B_j), \qquad \text{for all values of } j \qquad (3.4.6)$$

The following example illustrates the use of Equation 3.4.6 in the calculation of a marginal probability.

EXAMPLE 3.4.9

We wish to use Equation 3.4.6 and the data in Table 3.4.1 to compute the marginal probability $P(E)$.

Solution: The variable age at onset is broken down into two categories, Early for onset 18 years or younger (E) and Later for onset occurring at an age over 18 years (L). The variable family history of mood disorders is broken down into four categories: negative family history (A), bipolar disorder only (B), unipolar disorder only (C), and subjects with a history of both unipolar and bipolar disorder (D). The category Early occurs jointly with all four categories of the variable family history of mood disorders. The four joint probabilities that may be computed are

$$P(E \cap A) = 28/318 = .0881$$
$$P(E \cap B) = 19/318 = .0597$$
$$P(E \cap C) = 41/318 = .1289$$
$$P(E \cap D) = 53/318 = .1667$$

We obtain the marginal probability $P(E)$ by adding these four joint probabilities as follows:

$$\begin{aligned} P(E) &= P(E \cap A) + P(E \cap B) + P(E \cap C) + P(E \cap D) \\ &= .0881 + .0597 + .1289 + .1667 \\ &= .4434 \end{aligned}$$

■

The result, as expected, is the same as the one obtained by using the marginal total for Early as the numerator and the total number of subjects as the denominator.

EXERCISES

3.4.1 In a study of violent victimization of women and men, Porcerelli et al. (A-2) collected information from 679 women and 345 men aged 18 to 64 years at several family practice centers in the metropolitan Detroit area. Patients filled out a health history questionnaire that included a question about victimization. The following table shows the sample subjects cross-classified by sex and the type of violent victimization reported. The victimization categories are defined as no victimization, partner victimization (and not by others), victimization by persons other than partners (friends, family members, or strangers), and those who reported multiple victimization.

	No Victimization	Partners	Nonpartners	Multiple Victimization	Total
Women	611	34	16	18	679
Men	308	10	17	10	345
Total	919	44	33	28	1024

Source: John H. Porcerelli, Ph.D., Rosemary Cogan, Ph.D. Used with permission.

(a) Suppose we pick a subject at random from this group. What is the probability that this subject will be a woman?

(b) What do we call the probability calculated in part a?

(c) Show how to calculate the probability asked for in part a by two additional methods.

(d) If we pick a subject at random, what is the probability that the subject will be a woman and have experienced partner abuse?

(e) What do we call the probability calculated in part d?

(f) Suppose we picked a man at random. Knowing this information, what is the probability that he experienced abuse from nonpartners?

(g) What do we call the probability calculated in part f?

(h) Suppose we pick a subject at random. What is the probability that it is a man or someone who experienced abuse from a partner?

(i) What do we call the method by which you obtained the probability in part h?

3.4.2 Fernando et al. (A-3) studied drug-sharing among injection drug users in the South Bronx in New York City. Drug users in New York City use the term "split a bag" or "get down on a bag" to refer to the practice of dividing a bag of heroin or other injectable substances. A common practice includes splitting drugs after they are dissolved in a common cooker, a procedure with considerable HIV risk. Although this practice is common, little is known about the prevalence of such practices. The researchers asked injection drug users in four neighborhoods in the South Bronx if they ever "got down on" drugs in bags or shots. The results classified by gender and splitting practice are given below:

Gender	Split Drugs	Never Split Drugs	Total
Male	349	324	673
Female	220	128	348
Total	569	452	1021

Source: Daniel Fernando, Robert F. Schilling, Jorge Fontdevila, and Nabila El-Bassel, "Predictors of Sharing Drugs Among Injection Drug Users in the South Bronx: Implications for HIV Transmission," *Journal of Psychoactive Drugs, 35* (2003), 227–236.

(a) How many marginal probabilities can be calculated from these data? State each in probability notation and do the calculations.

(b) How many joint probabilities can be calculated? State each in probability notation and do the calculations.

(c) How many conditional probabilities can be calculated? State each in probability notation and do the calculations.

(d) Use the multiplication rule to find the probability that a person picked at random never split drugs and is female.

(e) What do we call the probability calculated in part d?

(f) Use the multiplication rule to find the probability that a person picked at random is male, given that he admits to splitting drugs.

(g) What do we call the probability calculated in part f?

3.4.3 Refer to the data in Exercise 3.4.2. State the following probabilities in words and calculate:

(a) $P(\text{Male} \cap \text{Split Drugs})$

(b) $P(\text{Male} \cup \text{Split Drugs})$

(c) $P(\text{Male} \mid \text{Split Drugs})$

(d) $P(\text{Male})$

3.4.4 Laveist and Nuru-Jeter (A-4) conducted a study to determine if doctor–patient race concordance was associated with greater satisfaction with care. Toward that end, they collected a national sample of African-American, Caucasian, Hispanic, and Asian-American respondents. The following table classifies the race of the subjects as well as the race of their physician:

	Patient's Race				
Physician's Race	**Caucasian**	**African-American**	**Hispanic**	**Asian-American**	**Total**
White	779	436	406	175	1796
African-American	14	162	15	5	196
Hispanic	19	17	128	2	166
Asian/Pacific-Islander	68	75	71	203	417
Other	30	55	56	4	145
Total	910	745	676	389	2720

Source: Thomas A. Laveist and Amani Nuru-Jeter, "Is Doctor-Patient Race Concordance Associated with Greater Satisfaction with Care?" *Journal of Health and Social Behavior, 43* (2002), 296–306.

(a) What is the probability that a randomly selected subject will have an Asian/Pacific-Islander physician?

(b) What is the probability that an African-American subject will have an African-American physician?

(c) What is the probability that a randomly selected subject in the study will be Asian-American and have an Asian/Pacific-Islander physician?

(d) What is the probability that a subject chosen at random will be Hispanic or have a Hispanic physician?

(e) Use the concept of complementary events to find the probability that a subject chosen at random in the study does not have a white physician.

3.4.5 If the probability of left-handedness in a certain group of people is .05, what is the probability of right-handedness (assuming no ambidexterity)?

3.4.6 The probability is .6 that a patient selected at random from the current residents of a certain hospital will be a male. The probability that the patient will be a male who is in for surgery is .2. A patient randomly selected from current residents is found to be a male; what is the probability that the patient is in the hospital for surgery?

3.4.7 In a certain population of hospital patients the probability is .35 that a randomly selected patient will have heart disease. The probability is .86 that a patient with heart disease is a smoker. What is the probability that a patient randomly selected from the population will be a smoker *and* have heart disease?

3.5 BAYES' THEOREM, SCREENING TESTS, SENSITIVITY, SPECIFICITY, AND PREDICTIVE VALUE POSITIVE AND NEGATIVE

In the health sciences field a widely used application of probability laws and concepts is found in the evaluation of screening tests and diagnostic criteria. Of interest to clinicians is an enhanced ability to correctly predict the presence or absence of a particular disease from knowledge of test results (positive or negative) and/or the status of presenting symptoms (present or absent). Also of interest is information regarding the likelihood of positive and negative test results and the likelihood of the presence or absence of a particular symptom in patients with and without a particular disease.

In our consideration of screening tests, we must be aware of the fact that they are not always infallible. That is, a testing procedure may yield a *false positive* or a *false negative.*

DEFINITIONS

1. **A false positive results when a test indicates a positive status when the true status is negative.**
2. **A false negative results when a test indicates a negative status when the true status is positive.**

In summary, the following questions must be answered in order to evaluate the usefulness of test results and symptom status in determining whether or not a subject has some disease:

1. Given that a subject has the disease, what is the probability of a positive test result (or the presence of a symptom)?
2. Given that a subject does not have the disease, what is the probability of a negative test result (or the absence of a symptom)?
3. Given a positive screening test (or the presence of a symptom), what is the probability that the subject has the disease?
4. Given a negative screening test result (or the absence of a symptom), what is the probability that the subject does not have the disease?

Table 3.5.1 Sample of n Subjects (Where n Is Large) Cross-Classified According to Disease Status and Screening Test Result

	Disease		
Test Result	**Present (D)**	**Absent ($\overline{D}$)**	**Total**
Positive (T)	a	b	$a + b$
Negative ($\overline{T}$)	c	d	$c + d$
Total	$a + c$	$b + d$	n

Suppose we have for a sample of n subjects (where n is a large number) the information shown in Table 3.5.1. The table shows for these n subjects their status with regard to a disease and results from a screening test designed to identify subjects with the disease. The cell entries represent the number of subjects falling into the categories defined by the row and column headings. For example, a is the number of subjects who have the disease and whose screening test result was positive.

As we have learned, a variety of probability estimates may be computed from the information displayed in a two-way table such as Table 3.5.1. For example, we may compute the conditional probability estimate $P(T \mid D) = a/(a + c)$. This ratio is an estimate of the *sensitivity* of the screening test.

DEFINITION

The sensitivity of a test (or symptom) is the probability of a positive test result (or presence of the symptom) given the presence of the disease.

We may also compute the conditional probability estimate $P(\overline{T} \mid \overline{D}) = d/(b + d)$. This ratio is an estimate of the *specificity* of the screening test.

DEFINITION

The specificity of a test (or symptom) is the probability of a negative test result (or absence of the symptom) given the absence of the disease.

From the data in Table 3.5.1 we answer Question 3 by computing the conditional probability estimate $P(D \mid T)$. This ratio is an estimate of a probability called the *predictive value positive* of a screening test (or symptom).

DEFINITION

The predictive value positive of a screening test (or symptom) is the probability that a subject has the disease given that the subject has a positive screening test result (or has the symptom).

Similarly, the ratio $P(\overline{D} \mid \overline{T})$ is an estimate of the conditional probability that a subject does not have the disease given that the subject has a negative screening test result (or does not have the symptom). The probability estimated by this ratio is called the *predictive value negative* of the screening test or symptom.

DEFINITION

The predictive value negative of a screening test (or symptom) is the probability that a subject does not have the disease, given that the subject has a negative screening test result (or does not have the symptom).

Estimates of the predictive value positive and predictive value negative of a test (or symptom) may be obtained from knowledge of a test's (or symptom's) sensitivity and specificity and the probability of the relevant disease in the general population. To obtain these predictive value estimates, we make use of Bayes's theorem. The following statement of Bayes's theorem, employing the notation established in Table 3.5.1, gives the predictive value positive of a screening test (or symptom):

$$P(D \mid T) = \frac{P(T \mid D)\,P(D)}{P(T \mid D)\,P(D) + P(T \mid \overline{D})\,P(\overline{D})} \tag{3.5.1}$$

It is instructive to examine the composition of Equation 3.5.1. We recall from Equation 3.4.2 that the conditional probability $P(D \mid T)$ is equal to $P(D \cap T)/P(T)$. To understand the logic of Bayes's theorem, we must recognize that the numerator of Equation 3.5.1 represents $P(D \cap T)$ and that the denominator represents $P(T)$. We know from the multiplication rule of probability given in Equation 3.4.1 that the numerator of Equation 3.5.1, $P(T \mid D)\,P(D)$, is equal to $P(D \cap T)$.

Now let us show that the denominator of Equation 3.5.1 is equal to $P(T)$. We know that event T is the result of a subject's being classified as positive with respect to a screening test (or classified as having the symptom). A subject classified as positive may have the disease or may not have the disease. Therefore, the occurrence of T is the result of a subject having the disease and being positive $[P(D \cap T)]$ or not having the disease and being positive $[P(\overline{D} \cap T)]$. These two events are mutually exclusive (their intersection is zero), and consequently, by the addition rule given by Equation 3.4.3, we may write

$$P(T) = P(D \cap T) + P(\overline{D} \cap T) \tag{3.5.2}$$

Since, by the multiplication rule, $P(D \cap T) = P(T \mid D)\,P(D)$ and $P(\overline{D} \cap T) = P(T \mid \overline{D})\,P(\overline{D})$, we may rewrite Equation 3.5.2 as

$$P(T) = P(T \mid D)\,P(D) + P(T \mid \overline{D})\,P(\overline{D}) \tag{3.5.3}$$

which is the denominator of Equation 3.5.1.

Note, also, that the numerator of Equation 3.5.1 is equal to the sensitivity times the rate (prevalence) of the disease and the denominator is equal to the sensitivity times the rate of the disease plus the term *1 minus the sensitivity* times the term *1 minus the rate of the disease.* Thus, we see that the predictive value positive can be calculated from knowledge of the sensitivity, specificity, and the rate of the disease.

Evaluation of Equation 3.5.1 answers Question 3. To answer Question 4 we follow a now familiar line of reasoning to arrive at the following statement of Bayes's theorem:

$$P(\overline{D} \mid \overline{T}) = \frac{P(\overline{T} \mid \overline{D})\, P(\overline{D})}{P(\overline{T} \mid \overline{D})\, P(\overline{D}) + P(\overline{T} \mid D)\, P(D)} \tag{3.5.4}$$

Equation 3.5.4 allows us to compute an estimate of the probability that a subject who is negative on the test (or has no symptom) does not have the disease, which is the predictive value negative of a screening test or symptom.

We illustrate the use of Bayes' theorem for calculating a predictive value positive with the following example.

EXAMPLE 3.5.1

A medical research team wished to evaluate a proposed screening test for Alzheimer's disease. The test was given to a random sample of 450 patients with Alzheimer's disease and an independent random sample of 500 patients without symptoms of the disease. The two samples were drawn from populations of subjects who were 65 years of age or older. The results are as follows:

	Alzheimer's Diagnosis?		
Test Result	**Yes (D)**	**No ($\overline{D}$)**	**Total**
Positive (T)	436	5	441
Negative ($\overline{T}$)	14	495	509
Total	450	500	950

Using these data we estimate the sensitivity of the test to be $P(T \mid D) = 436/450 = .97$. The specificity of the test is estimated to be $P(\overline{T} \mid \overline{D}) = 495/500 = .99$. We now use the results of the study to compute the predictive value positive of the test. That is, we wish to estimate the probability that a subject who is positive on the test has Alzheimer's disease. From the tabulated data we compute $P(T \mid D) = 436/450 = .9689$ and $P(T \mid \overline{D}) = 5/500 = .01$. Substitution of these results into Equation 3.5.1 gives

$$P(D \mid T) = \frac{(.9689)\, P(D)}{(.9689)\, P(D) + (.01)\, P(\overline{D})} \tag{3.5.5}$$

We see that the predictive value positive of the test depends on the rate of the disease in the relevant population in general. In this case the relevant population consists of subjects who are 65 years of age or older. We emphasize that the rate of disease in the relevant general population, $P(D)$, cannot be computed from the sample data, since two independent samples were drawn from two different populations. We must look elsewhere for an estimate of $P(D)$. Evans et al. (A-5) estimated that 11.3 percent of the U.S. population aged 65 and over have Alzheimer's disease. When we substitute this estimate of $P(D)$ into Equation 3.5.5 we obtain

$$P(D \mid T) = \frac{(.9689)(.113)}{(.9689)(.113) + (.01)(1 - .113)} = .93$$

As we see, in this case, the predictive value of the test is very high.

Similarly, let us now consider the predictive value negative of the test. We have already calculated all entries necessary except for $P(\overline{T} \mid \overline{D}) = 14/450 = .0311$. Using the values previously obtained and our new value, we find

$$P(\overline{D} \mid \overline{T}) = \frac{(.99)(1 - .113)}{(.99)(1 - .113) + (.0311)(.113)} = .996$$

As we see, the predictive value negative is also quite high. ■

EXERCISES

3.5.1 A medical research team wishes to assess the usefulness of a certain symptom (call it S) in the diagnosis of a particular disease. In a random sample of 775 patients with the disease, 744 reported having the symptom. In an independent random sample of 1380 subjects without the disease, 21 reported that they had the symptom.

(a) In the context of this exercise, what is a false positive?

(b) What is a false negative?

(c) Compute the sensitivity of the symptom.

(d) Compute the specificity of the symptom.

(e) Suppose it is known that the rate of the disease in the general population is .001. What is the predictive value positive of the symptom?

(f) What is the predictive value negative of the symptom?

(g) Find the predictive value positive and the predictive value negative for the symptom for the following hypothetical disease rates: .0001, .01, and .10.

(h) What do you conclude about the predictive value of the symptom on the basis of the results obtained in part g?

3.5.2 In an article entitled "Bucket-Handle Meniscal Tears of the Knee: Sensitivity and Specificity of MRI signs," Dorsay and Helms (A-6) performed a retrospective study of 71 knees scanned by MRI. One of the indicators they examined was the absence of the "bow-tie sign" in the MRI as evidence of a bucket-handle or "bucket-handle type" tear of the meniscus. In the study, surgery confirmed that 43 of the 71 cases were bucket-handle tears. The cases may be cross-classified by "bow-tie sign" status and surgical results as follows:

	Tear Surgically Confirmed (D)	Tear Surgically Confirmed As Not Present ($\overline{D}$)	Total
Positive Test (absent bow-tie sign) (T)	38	10	48
Negative Test (bow-tie sign present) ($\overline{T}$)	5	18	23
Total	43	28	71

Source: Theodore A. Dorsay and Clyde A. Helms, "Bucket-Handle Meniscal Tears of the Knee: Sensitivity and Specificity of MRI Signs," *Skeletal Radiology, 32* (2003), 266–272.

(a) What is the sensitivity of testing to see if the absent bow-tie sign indicates a meniscal tear?

(b) What is the specificity of testing to see if the absent bow-tie sign indicates a meniscal tear?

(c) What additional information would you need to determine the predictive value of the test?

3.5.3 Oexle et al. (A-7) calculated the negative predictive value of a test for carriers of X-linked ornithine transcarbamylase deficiency (OTCD—a disorder of the urea cycle). A test known as the "allopurinol test" is often used as a screening device of potential carriers whose relatives are OTCD patients. They cited a study by Brusilow and Horwich (A-8) that estimated the sensitivity of the allopurinol test as .927. Oexle et al. themselves estimated the specificity of the allopurinol test as .997. Also they estimated the prevalence in the population of individuals with OTCD as $1/32000$. Use this information and Bayes's theorem to calculate the predictive value negative of the allopurinol screening test.

3.6 SUMMARY

In this chapter some of the basic ideas and concepts of probability were presented. The objective has been to provide enough of a "feel" for the subject so that the probabilistic aspects of statistical inference can be more readily understood and appreciated when this topic is presented later.

We defined probability as a number between 0 and 1 that measures the likelihood of the occurrence of some event. We distinguished between subjective probability and objective probability. Objective probability can be categorized further as classical or relative frequency probability. After stating the three properties of probability, we defined and illustrated the calculation of the following kinds of probabilities: marginal, joint, and conditional. We also learned how to apply the addition and multiplication rules to find certain probabilities. We learned the meaning of independent, mutually exclusive, and complementary events. We learned the meaning of specificity, sensitivity, predictive value positive, and predictive value negative as applied to a screening test or disease symptom. Finally, we learned how to use Bayes's theorem to calculate the probability that a subject has a disease, given that the subject has a positive screening test result (or has the symptom of interest).

SUMMARY OF FORMULAS FOR CHAPTER 3

Formula Number	Name	Formula
3.2.1	Classical probability	$P(E) = \frac{m}{N}$
3.2.2	Relative frequency probability	$P(E) = \frac{m}{n}$
3.3.1–3.3.3	Properties of probability	$P(E_i) \geq 0$ $P(E_1) + P(E_2) + \ldots + P(E_n) = 1$ $P(E_i + E_j) = P(E_i) + P(E_j)$
3.4.1	Multiplication rule	$P(A \cap B) = P(B)P(A \mid B) = P(A)P(B \mid A)$
3.4.2	Conditional probability	$P(A \mid B) = \frac{P(A \cap B)}{P(B)}$
3.4.3	Addition rule	$P(A \cup B) = P(A) + P(B) - P(A \cap B)$
3.4.4	Independent events	$P(A \cap B) = P(A)P(B)$
3.4.5	Complementary events	$P(\overline{A}) = 1 - P(A)$
3.4.6	Marginal Probability	$P(A_i) = \sum P(A_i \cap B_j)$
	Sensitivity of a screening test	$P(T \mid D) = \frac{a}{(a + c)}$
	Specificity of a screening test	$P(\overline{T} \mid \overline{D}) = \frac{d}{(b + d)}$
3.5.1	Predictive value positive of a screening test	$P(D \mid T) = \frac{P(T \mid D)P(D)}{P(T \mid D)P(D) + P(T \mid \overline{D})P(\overline{D})}$
3.5.2	Predictive value negative of a screening test	$P(\overline{D} \mid \overline{T}) = \frac{P(\overline{T} \mid \overline{D})P(\overline{D})}{P(\overline{T} \mid \overline{D})P(\overline{D}) + P(\overline{T} \mid D)P(D)}$
Symbol Key	• D = disease • m = the number of times an event E_i occurs • n = sample size or the total number of time a process occurs • N = population size or the total number of mutually exclusive and equally likely events	

(*Continued*)

	• $P(\overline{A})$ = a complementary event; the probability of an event A, *not* occurring • $P(E_i)$ = probability of some event E_i occurring • $P(A \cap B)$ = an "intersection" or "and" statement; the probability of an event A *and* an event B occurring • $P(A \cup B)$ = a "union" or "or" statement; the probability of an event A *or* an event B or both occurring • $P(A \mid B)$ = a conditional statement; the probability of an event A occurring *given that* an event B has already occurred • T = test results

REVIEW QUESTIONS AND EXERCISES

1. Define the following:

(a) Probability
(b) Objective probability
(c) Subjective probability
(d) Classical probability
(e) The relative frequency concept of probability
(f) Mutually exclusive events
(g) Independence
(h) Marginal probability
(i) Joint probability
(j) Conditional probability
(k) The addition rule
(l) The multiplication rule
(m) Complementary events
(n) False positive
(o) False negative
(p) Sensitivity
(q) Specificity
(r) Predictive value positive
(s) Predictive value negative
(t) Bayes's theorem

2. Name and explain the three properties of probability.

3. Coughlin et al. (A-9) examined the breast and cervical screening practices of Hispanic and non-Hispanic women in counties that approximate the U.S. southern border region. The study used data from the Behavioral Risk Factor Surveillance System surveys of adults age 18 years or older conducted in 1999 and 2000. The table below reports the number of observations of Hispanic and non-Hispanic women who had received a mammogram in the past 2 years cross-classified with marital status.

Marital Status	Hispanic	Non-Hispanic	Total
Currently Married	319	738	1057
Divorced or Separated	130	329	459
Widowed	88	402	490
Never Married or Living As an Unmarried Couple	41	95	136
Total	578	1564	2142

Source: Steven S. Coughlin, Robert J. Uhler, Thomas Richards, and Katherine M. Wilson, "Breast and Cervical Cancer Screening Practices Among Hispanic and Non-Hispanic Women Residing Near the United States–Mexico Border, 1999–2000," *Family and Community Health, 26* (2003), 130–139.

(a) We select at random a subject who had a mammogram. What is the probability that she is divorced or separated?

(b) We select at random a subject who had a mammogram and learn that she is Hispanic. With that information, what is the probability that she is married?

(c) We select at random a subject who had a mammogram. What is the probability that she is non-Hispanic and divorced or separated?

(d) We select at random a subject who had a mammogram. What is the probability that she is Hispanic or she is widowed?

(e) We select at random a subject who had a mammogram. What is the probability that she is not married?

4. Swor et al. (A-10) looked at the effectiveness of cardiopulmonary resuscitation (CPR) training in people over 55 years old. They compared the skill retention rates of subjects in this age group who completed a course in traditional CPR instruction with those who received chest-compression only cardiopulmonary resuscitation (CC-CPR). Independent groups were tested 3 months after training. The table below shows the skill retention numbers in regard to overall competence as assessed by video ratings done by two video evaluators.

Rated Overall Competent	CPR	CC-CPR	Total
Yes	12	15	27
No	15	14	29
Total	27	29	56

Source: Robert Swor, Scott Compton, Fern Vining, Lynn Ososky Farr, Sue Kokko, Rebecca Pascual, and Raymond E. Jackson, "A Randomized Controlled Trial of Chest Compression Only CPR for Older Adults—a Pilot Study," *Resuscitation, 58* (2003), 177–185.

(a) Find the following probabilities and explain their meaning:

1. A randomly selected subject was enrolled in the CC-CPR class.
2. A randomly selected subject was rated competent.
3. A randomly selected subject was rated competent and was enrolled in the CPR course.
4. A randomly selected subject was rated competent or was enrolled in CC-CPR.
5. A randomly selected subject was rated competent given that he or she enrolled in the CC-CPR course.

(b) We define the following events to be

A = a subject enrolled in the CPR course
B = a subject enrolled in the CC-CPR course
C = a subject was evaluated as competent
D = a subject was evaluated as not competent

Then explain why each of the following equations is or is not a true statement:

1. $P(A \cap C) = P(C \cap A)$ **2.** $P(A \cup B) = P(B \cup A)$
3. $P(A) = P(A \cup C) + P(A \cup D)$ **4.** $P(B \cup C) = P(B) + P(C)$
5. $P(D \mid A) = P(D)$ **6.** $P(C \cap B) = P(C)P(B)$
7. $P(A \cap B) = 0$ **8.** $P(C \cap B) = P(B)\,P(C \mid B)$
9. $P(A \cap D) = P(A)\,P(A \mid D)$

5. Pillmann et al. (A-11) studied patients with acute brief episodes of psychoses. The researchers classified subjects into four personality types: obsessoid, asthenic/low self-confident, asthenic/high self-confident, nervous/tense, and undeterminable. The table below cross-classifies these personality types with three groups of subjects—those with acute and transient psychotic disorders (ATPD), those with "positive" schizophrenia (PS), and those with bipolar schizo-affective disorder (BSAD):

Personality Type	ATPD (1)	PS (2)	BSAD (3)	Total
Obsessoid (*O*)	9	2	6	17
Asthenic/low Self-confident (*A*)	20	17	15	52
Asthenic/high Self-confident (*S*)	5	3	8	16
Nervous/tense (*N*)	4	7	4	15
Undeterminable (*U*)	4	13	9	26
Total	42	42	42	126

Source: Frank Pillmann, Raffaela Blōink, Sabine Balzuweit, Annette Haring, and Andreas Marneros, "Personality and Social Interactions in Patients with Acute Brief Psychoses," *Journal of Nervous and Mental Disease, 191* (2003), 503–508.

Find the following probabilities if a subject in this study is chosen at random:

(a) $P(O)$ **(b)** $P(A \cup 2)$ **(c)** $P(1)$ **(d)** $P(\overline{A})$
(e) $P(A \mid 3)$ **(f)** $P(\overline{3})$ **(g)** $P(2 \cap 3)$ **(h)** $P(2 \mid A)$

6. A certain county health department has received 25 applications for an opening that exists for a public health nurse. Of these applicants 10 are over 30 and 15 are under 30. Seventeen hold bachelor's degrees only, and eight have master's degrees. Of those under 30, six have master's degrees. If a selection from among these 25 applicants is made at random, what is the probability that a person over 30 *or* a person with a master's degree will be selected?

7. The following table shows 1000 nursing school applicants classified according to scores made on a college entrance examination and the quality of the high school from which they graduated, as rated by a group of educators:

	Quality of High Schools			
Score	Poor (*P*)	Average (*A*)	Superior (*S*)	Total
Low (*L*)	105	60	55	220
Medium (*M*)	70	175	145	390
High (*H*)	25	65	300	390
Total	200	300	500	1000

(a) Calculate the probability that an applicant picked at random from this group:

1. Made a low score on the examination.
2. Graduated from a superior high school.
3. Made a low score on the examination and graduated from a superior high school.
4. Made a low score on the examination given that he or she graduated from a superior high school.
5. Made a high score or graduated from a superior high school.

(b) Calculate the following probabilities:
1. $P(A)$ **2.** $P(H)$ **3.** $P(M)$
4. $P(A \mid H)$ **5.** $P(M \cap P)$ **6.** $(H \mid S)$

8. If the probability that a public health nurse will find a client at home is .7, what is the probability (assuming independence) that on two home visits made in a day both clients will be home?

9. For a variety of reasons, self-reported disease outcomes are frequently used without verification in epidemiologic research. In a study by Parikh-Patel et al. (A-12), researchers looked at the relationship between self-reported cancer cases and actual cases. They used the self-reported cancer data from a California Teachers Study and validated the cancer cases by using the California Cancer Registry data. The following table reports their findings for breast cancer:

Cancer Reported (*A*)	**Cancer in Registry (*B*)**	**Cancer Not in Registry**	**Total**
Yes	2991	2244	5235
No	112	115849	115961
Total	3103	118093	121196

Source: Arti Parikh-Patel, Mark Allen, William E. Wright, and the California Teachers Study Steering Committee, "Validation of Self-Reported Cancers in the California Teachers Study," *American Journal of Epidemiology, 157* (2003), 539–545.

(a) Let *A* be the event of reporting breast cancer in the California Teachers Study. Find the probability of *A* in this study.

(b) Let *B* be the event of having breast cancer confirmed in the California Cancer Registry. Find the probability of *B* in this study.

(c) Find $P(A \cap B)$

(d) Find $P(A \mid B)$

(e) Find $P(B \mid A)$

(f) Find the sensitivity of using self-reported breast cancer as a predictor of actual breast cancer in the California registry.

(g) Find the specificity of using self-reported breast cancer as a predictor of actual breast cancer in the California registry.

10. In a certain population the probability that a randomly selected subject will have been exposed to a certain allergen and experience a reaction to the allergen is .60. The probability is .8 that a subject exposed to the allergen will experience an allergic reaction. If a subject is selected at random from this population, what is the probability that he or she will have been exposed to the allergen?

11. Suppose that 3 percent of the people in a population of adults have attempted suicide. It is also known that 20 percent of the population are living below the poverty level. If these two events are independent, what is the probability that a person selected at random from the population will have attempted suicide *and* be living below the poverty level?

12. In a certain population of women 4 percent have had breast cancer, 20 percent are smokers, and 3 percent are smokers and have had breast cancer. A woman is selected at random from the population. What is the probability that she has had breast cancer or smokes or both?

13. The probability that a person selected at random from a population will exhibit the classic symptom of a certain disease is .2, and the probability that a person selected at random has the disease is .23. The probability that a person who has the symptom also has the disease is .18. A person selected at random from the population does not have the symptom. What is the probability that the person has the disease?

14. For a certain population we define the following events for mother's age at time of giving birth: A = under 20 years; B = 20–24 years; C = 25–29 years; D = 30–44 years. Are the events A, B, C, and D pairwise mutually exclusive?

15. Refer to Exercise 14. State in words the event $E = (A \cup B)$.

16. Refer to Exercise 14. State in words the event $F = (B \cup C)$.

17. Refer to Exercise 14. Comment on the event $G = (A \cap B)$.

18. For a certain population we define the following events with respect to plasma lipoprotein levels (mg/dl): $A = (10\text{–}15)$; $B = (\geq 30)$; $C = (\leq 20)$. Are the events A and B mutually exclusive? A and C? B and C? Explain your answer to each question.

19. Refer to Exercise 18. State in words the meaning of the following events:

(a) $A \cup B$ **(b)** $A \cap B$ **(c)** $A \cap C$ **(d)** $A \cup C$

20. Refer to Exercise 18. State in words the meaning of the following events:

(a) $\overline{A}$ **(b)** $\overline{B}$ **(c)** $\overline{C}$

21. Rothenberg et al. (A-13) investigated the effectiveness of using the Hologic Sahara Sonometer, a portable device that measures bone mineral density (BMD) in the ankle, in predicting a fracture. They used a Hologic estimated bone mineral density value of .57 as a cutoff. The results of the investigation yielded the following data:

	Confirmed Fracture		
	Present (D)	**Not Present ($\overline{D}$)**	**Total**
BMD = .57 (T)	214	670	884
BMD > .57 ($\overline{T}$)	73	330	403
Total	287	1000	1287

Source: Ralph J. Rothenberg, M.D., Joan L. Boyd, Ph.D., and John P. Holcomb, Ph.D. Used with permission.

(a) Calculate the sensitivity of using a BMD value of .57 as a cutoff value for predicting fracture and interpret your results.

(b) Calculate the specificity of using a BMD value of .57 as a cutoff value for predicting fracture and interpret your results.

22. Verma et al. (A-14) examined the use of heparin-PF4 ELISA screening for heparin-induced thrombocytopenia (HIT) in critically ill patients. Using C-serotonin release assay (SRA) as the way of validating HIT, the authors found that in 31 patients tested negative by SRA, 22 also tested negative by heparin-PF4 ELISA.

(a) Calculate the specificity of the heparin-PF4 ELISA testing for HIT.

(b) Using a "literature-derived sensitivity" of 95 percent and a prior probability of HIT occurrence as 3.1 percent, find the positive predictive value.

(c) Using the same information as part (b), find the negative predictive value.

23. The sensitivity of a screening test is .95, and its specificity is .85. The rate of the disease for which the test is used is .002. What is the predictive value positive of the test?

Exercises for Use with Large Data Sets Available on the Following Website: www.wiley.com/college/daniel

Refer to the random sample of 800 subjects from the North Carolina birth registry we investigated in the Chapter 2 review exercises.

1. Create a table that cross-tabulates the counts of mothers in the classifications of whether the baby was premature or not (PREMIE) and whether the mother admitted to smoking during pregnancy (SMOKE) or not.

(a) Find the probability that a mother in this sample admitted to smoking.

(b) Find the probability that a mother in this sample had a premature baby.

(c) Find the probability that a mother in the sample had a premature baby given that the mother admitted to smoking.

(d) Find the probability that a mother in the sample had a premature baby given that the mother did not admit to smoking.

(e) Find the probability that a mother in the sample had a premature baby or that the mother did not admit to smoking.

2. Create a table that cross-tabulates the counts of each mother's marital status (MARITAL) and whether she had a low birth weight baby (LOW).

(a) Find the probability a mother selected at random in this sample had a low birth weight baby.

(b) Find the probability a mother selected at random in this sample was married.

(c) Find the probability a mother selected at random in this sample had a low birth weight child given that she was married.

(d) Find the probability a mother selected at random in this sample had a low birth weight child given that she was not married.

(e) Find the probability a mother selected at random in this sample had a low birth weight child and the mother was married.

REFERENCES

Methodology References

1. Allan Gut, *An Intermediate Course in Probability,* Springer-Verlag, New York, 1995.
2. Richard Isaac, *The Pleasures of Probability,* Springer-Verlag, New York, 1995.
3. Harold J. Larson, *Introduction to Probability,* Addison-Wesley, Reading, MA, 1995.
4. L. J. Savage, *Foundations of Statistics,* Second Revised Edition, Dover, New York, 1972.
5. A. N. Kolmogorov, *Foundations of the Theory of Probability,* Chelsea, New York, 1964 (original German edition published in 1933).

Applications References

A-1. Tasha D. Carter, Emanuela Mundo, Sagar V. Parkh, and James L. Kennedy, "Early Age at Onset as a Risk Factor for Poor Outcome of Bipolar Disorder," *Journal of Psychiatric Research, 37* (2003), 297–303.

A-2. John H. Porcerelli, Rosemary Cogan, Patricia P. West, Edward A. Rose, Dawn Lambrecht, Karen E. Wilson, Richard K. Severson, and Dunia Karana,"Violent Victimization of Women and Men: Physical and Psychiatric Symptoms," *Journal of the American Board of Family Practice, 16* (2003), 32–39.

A-3. DANIEL FERNANDO, ROBERT F. SCHILLING, JORGE FONTDEVILA, and NABILA EL-BASSEL, "Predictors of Sharing Drugs Among Injection Drug Users in the South Bronx: Implications for HIV Transmission," *Journal of Psychoactive Drugs, 35* (2003), 227–236.

A-4. THOMAS A. LAVEIST and AMANI NURU-JETER, "Is Doctor–Patient Race Concordance Associated with Greater Satisfaction with Care?" *Journal of Health and Social Behavior, 43* (2002), 296–306.

A-5. D. A. EVANS, P. A. SCHERR, N. R. COOK, M. S. ALBERT, H. H. FUNKENSTEIN, L. A. SMITH, L. E. HEBERT, T. T. WETLE, L. G. BRANCH, M. CHOWN, C. H. HENNEKENS, and J. O. TAYLOR, "Estimated Prevalence of Alzheimer's Disease in the United States," *Milbank Quarterly, 68* (1990), 267–289.

A-6. THEODORE A. DORSAY and CLYDE A. HELMS,"Bucket-Handle Meniscal Tears of the Knee: Sensitivity and Specificity of MRI Signs," *Skeletal Radiology, 32* (2003), 266–272.

A-7. KONRAD OEXLE, LUISA BONAFE, and BEAT STENMANN, "Remark on Utility and Error Rates of the Allopurinol Test in Detecting Mild Ornithine Transcarbamylase Deficiency," *Molecular Genetics and Metabolism, 76* (2002), 71–75.

A-8. S. W. BRUSILOW, A. L. HORWICH, "Urea Cycle Enzymes," in: C. R. SCRIVER, A. L. BEAUDET, W. S. SLY, D. VALLE (Eds.), *The Metabolic and Molecular Bases of Inherited Disease,* 8th ed., McGraw-Hill, New York, 2001, pp. 1909–1963.

A-9. STEVEN S. COUGHLIN, ROBERT J. UHLER, THOMAS RICHARDS, and KATHERINE M. WILSON, "Breast and Cervical Cancer Screening Practices Among Hispanic and Non-Hispanic Women Residing Near the United States–Mexico Border, 1999–2000," *Family and Community Health, 26* (2003), 130–139.

A-10. ROBERT SWOR, SCOTT COMPTON, FERN VINING, LYNN OSOSKY FARR, SUE KOKKO, REBECCA PASCUAL, and RAYMOND E. JACKSON, "A Randomized Controlled Trial of Chest Compression Only CPR for Older Adults—a Pilot Study," *Resuscitation, 58* (2003), 177–185.

A-11. FRANK PILLMANN, RAFFAELA BLÕINK, SABINE BALZUWEIT, ANNETTE HARING, and ANDREAS MARNEROS, "Personality and Social Interactions in Patients with Acute Brief Psychoses," *Journal of Nervous and Mental Disease, 191* (2003), 503–508.

A-12. ARTI PARIKH-PATEL, MARK ALLEN, WILLIAM E. WRIGHT, and the California Teachers Study Steering Committee, "Validation of Self-Reported Cancers in the California Teachers Study," *American Journal of Epidemiology, 157* (2003), 539–545.

A-13. RALPH J. ROTHENBERG, JOAN L. BOYD, and JOHN P. HOLCOMB, "Quantitative Ultrasound of the Calcaneus as a Screening Tool to Detect Osteoporosis: Different Reference Ranges for Caucasian Women, African-American Women, and Caucasian Men," *Journal of Clinical Densitometry, 7* (2004), 101–110.

A-14. ARUN K. VERMA, MARC LEVINE, STEPHEN J. SHALANSKY, CEDRIC J. CARTER, and JOHN G. KELTON, "Frequency of Heparin-Induced Thrombocytopenia in Critical Care Patients," *Pharmacotheray, 23* (2003), 645–753.

CHAPTER **4**

PROBABILISTIC FEATURES OF CERTAIN DATA DISTRIBUTIONS

CHAPTER OVERVIEW

Probability distributions of random variables assume powerful roles in statistical analyses. Since they show all possible values of a random variable and the probabilities associated with these values, probability distributions may be summarized in ways that enable researchers to easily make objective decisions based on samples drawn from the populations that the distributions represent. This chapter introduces frequently used discrete and continuous probability distributions that are used in later chapters to make statistical inferences.

TOPICS

LEARNING OUTCOMES

After studying this chapter, the student will

1. understand selected discrete distributions and how to use them to calculate probabilities in real-world problems.
2. understand selected continuous distributions and how to use them to calculate probabilities in real-world problems.
3. be able to explain the similarities and differences between distributions of the discrete type and the continuous type and when the use of each is appropriate.

4.1 INTRODUCTION

In the preceding chapter we introduced the basic concepts of probability as well as methods for calculating the probability of an event. We build on these concepts in the present chapter and explore ways of calculating the probability of an event under somewhat more complex conditions. In this chapter we shall see that the relationship between the values of a random variable and the probabilities of their occurrence may be summarized by means of a device called a *probability distribution.* A probability distribution may be expressed in the form of a table, graph, or formula. Knowledge of the probability distribution of a random variable provides the clinician and researcher with a powerful tool for summarizing and describing a set of data and for reaching conclusions about a population of data on the basis of a sample of data drawn from the population.

4.2 PROBABILITY DISTRIBUTIONS OF DISCRETE VARIABLES

Let us begin our discussion of probability distributions by considering the probability distribution of a discrete variable, which we shall define as follows:

DEFINITION

The *probability distribution* of a discrete random variable is a table, graph, formula, or other device used to specify all possible values of a discrete random variable along with their respective probabilities.

If we let the discrete probability distribution be represented by $p(x)$, then $p(x) = P(X = x)$ is the probability of the discrete random variable X to assume a value x.

EXAMPLE 4.2.1

In an article appearing in the *Journal of the American Dietetic Association,* Holben et al. (A-1) looked at food security status in families in the Appalachian region of southern Ohio. The purpose of the study was to examine hunger rates of families with children in a local Head Start program in Athens, Ohio. The survey instrument included the 18-question U.S. Household Food Security Survey Module for measuring hunger and food security. In addition, participants were asked how many food assistance programs they had used in the last 12 months. Table 4.2.1 shows the number of food assistance programs used by subjects in this sample.

We wish to construct the probability distribution of the discrete variable X, where $X =$ number of food assistance programs used by the study subjects.

Solution: The values of X are $x_1 = 1, x_2 = 2, \ldots, x_7 = 7$, and $x_8 = 8$. We compute the probabilities for these values by dividing their respective frequencies by the total, 297. Thus, for example, $p(x_1) = P(X = x_1) = 62/297 = .2088$.

TABLE 4.2.1 Number of Assistance Programs Utilized by Families with Children in Head Start Programs in Southern Ohio

Number of Programs	Frequency
1	62
2	47
3	39
4	39
5	58
6	37
7	4
8	11
Total	297

Source: David H. Holben, Ph.D. and John P. Holcomb, Ph.D. Used with permission.

TABLE 4.2.2 Probability Distribution of Programs Utilized by Families Among the Subjects Described in Example 4.2.1

Number of Programs (x)	$P(X = x)$
1	.2088
2	.1582
3	.1313
4	.1313
5	.1953
6	.1246
7	.0135
8	.0370
Total	1.0000

We display the results in Table 4.2.2, which is the desired probability distribution. ■

Alternatively, we can present this probability distribution in the form of a graph, as in Figure 4.2.1. In Figure 4.2.1 the length of each vertical bar indicates the probability for the corresponding value of x.

It will be observed in Table 4.2.2 that the values of $p(x) = P(X = x)$ are all positive, they are all less than 1, and their sum is equal to 1. These are not phenomena peculiar to this particular example, but are characteristics of all probability distributions of discrete variables. If $x_1, x_2, x_3, \ldots, x_k$ are all possible values of the discrete random

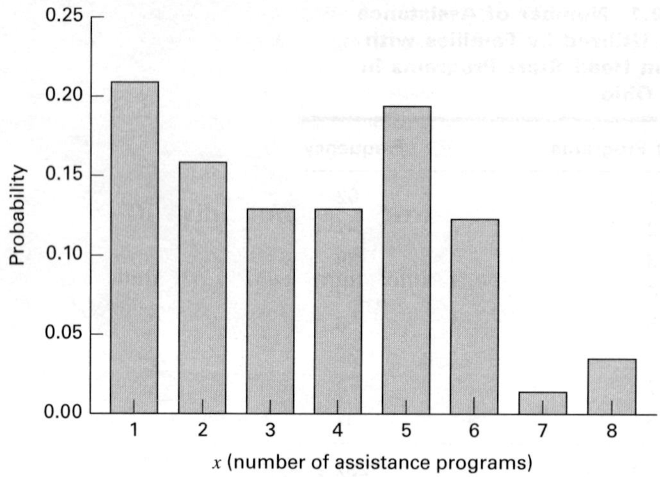

FIGURE 4.2.1 Graphical representation of the probability distribution shown in Table 4.2.1.

variable X, then we may then give the following two essential properties of a probability distribution of a discrete variable:

$$(1) \quad 0 \leq P(X = x) \leq 1$$

$$(2) \quad \sum P(X = x) = 1, \quad \text{for all } x$$

The reader will also note that each of the probabilities in Table 4.2.2 is the *relative frequency of occurrence* of the corresponding value of X.

With its probability distribution available to us, we can make probability statements regarding the random variable X. We illustrate with some examples.

EXAMPLE 4.2.2

What is the probability that a randomly selected family used three assistance programs?

Solution: We may write the desired probability as $p(3) = P(X = 3)$. We see in Table 4.2.2 that the answer is .1313. ■

EXAMPLE 4.2.3

What is the probability that a randomly selected family used either one or two programs?

Solution: To answer this question, we use the addition rule for mutually exclusive events. Using probability notation and the results in Table 4.2.2, we write the answer as $P(1 \cup 2) = P(1) + P(2) = .2088 + .1582 = .3670$. ■

TABLE 4.2.3 Cumulative Probability Distribution of Number of Programs Utilized by Families Among the Subjects Described in Example 4.2.1

Number of Programs (x)	Cumulative Frequency $P(X \leq x)$
1	.2088
2	.3670
3	.4983
4	.6296
5	.8249
6	.9495
7	.9630
8	1.0000

Cumulative Distributions Sometimes it will be more convenient to work with the *cumulative probability distribution* of a random variable. The cumulative probability distribution for the discrete variable whose probability distribution is given in Table 4.2.2 may be obtained by successively adding the probabilities, $P(X = x_i)$, given in the last column. The cumulative probability for x_i is written as $F(x_i) = P(X \leq x_i)$. It gives the probability that X is less than or equal to a specified value, x_i.

The resulting cumulative probability distribution is shown in Table 4.2.3. The graph of the cumulative probability distribution is shown in Figure 4.2.2. The graph of a cumulative probability distribution is called an *ogive*. In Figure 4.2.2 the graph of $F(x)$ consists solely of the horizontal lines. The vertical lines only give the graph a connected appearance. The length of each vertical line represents the same probability as that of the corresponding line in Figure 4.2.1. For example, the length of the vertical line at $X = 3$

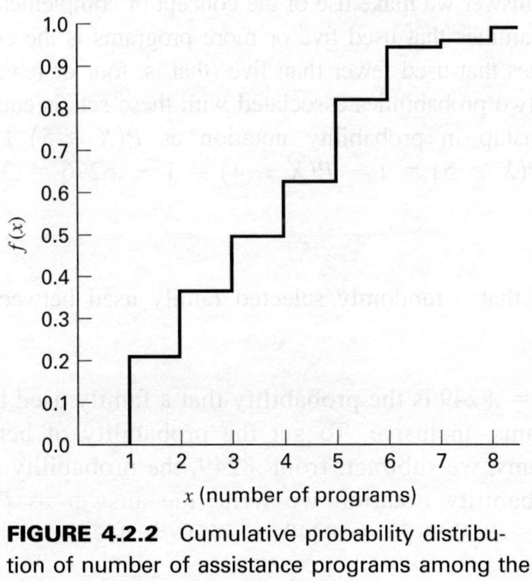

FIGURE 4.2.2 Cumulative probability distribution of number of assistance programs among the subjects described in Example 4.2.1.

in Figure 4.2.2 represents the same probability as the length of the line erected at $X = 3$ in Figure 4.2.1, or .1313 on the vertical scale.

By consulting the cumulative probability distribution we may answer quickly questions like those in the following examples.

EXAMPLE 4.2.4

What is the probability that a family picked at random used two or fewer assistance programs?

Solution: The probability we seek may be found directly in Table 4.2.3 by reading the cumulative probability opposite $x = 2$, and we see that it is .3670. That is, $P(X \leq 2) = .3670$. We also may find the answer by inspecting Figure 4.2.2 and determining the height of the graph (as measured on the vertical axis) above the value $X = 2$. ■

EXAMPLE 4.2.5

What is the probability that a randomly selected family used fewer than four programs?

Solution: Since a family that used fewer than four programs used either one, two, or three programs, the answer is the cumulative probability for 3. That is, $P(X < 4) = P(X \leq 3) = .4983$. ■

EXAMPLE 4.2.6

What is the probability that a randomly selected family used five or more programs?

Solution: To find the answer we make use of the concept of complementary probabilities. The set of families that used five or more programs is the complement of the set of families that used fewer than five (that is, four or fewer) programs. The sum of the two probabilities associated with these sets is equal to 1. We write this relationship in probability notation as $P(X \geq 5) + P(X \leq 4) = 1$. Therefore, $P(X \geq 5) = 1 - P(X \leq 4) = 1 - .6296 = .3704$. ■

EXAMPLE 4.2.7

What is the probability that a randomly selected family used between three and five programs, inclusive?

Solution: $P(X \leq 5) = .8249$ is the probability that a family used between one and five programs, inclusive. To get the probability of between three and five programs, we subtract, from .8249, the probability of two or fewer. Using probability notation we write the answer as $P(3 \leq X \leq 5) = P(X \leq 5) - P(X \leq 2) = .8249 - .3670 = .4579$. ■

The probability distribution given in Table 4.2.1 was developed out of actual experience, so to find another variable following this distribution would be coincidental. The probability distributions of many variables of interest, however, can be determined or assumed on the basis of theoretical considerations. In later sections, we study in detail three of these theoretical probability distributions: the *binomial,* the *Poisson,* and the *normal.*

Mean and Variance of Discrete Probability Distributions The mean and variance of a discrete probability distribution can easily be found using the formulae below.

$$\mu = \sum xp(x) \tag{4.2.1}$$

$$\sigma^2 = \sum (x - \mu)^2 p(x) = \sum x^2 p(x) - \mu^2 \tag{4.2.2}$$

where $p(x)$ is the relative frequency of a given random variable X. The standard deviation is simply the positive square root of the variance.

EXAMPLE 4.2.8

What are the mean, variance, and standard deviation of the distribution from Example 4.2.1?

Solution:

$$\mu = (1)(.2088) + (2)(.1582) + (3)(.1313) + \cdots + (8)(.0370) = 3.5589$$

$$\sigma^2 = (1 - 3.5589)^2(.2088) + (2 - 3.5589)^2(.1582) + (3 - 3.5589)^2(.1313) + \cdots + (8 - 3.5589)^2(.0370) = 3.8559$$

We therefore can conclude that the mean number of programs utilized was 3.5589 with a variance of 3.8559. The standard deviation is therefore $\sqrt{3.5589} = 1.9637$ programs. ■

EXERCISES

4.2.1 In a study by Cross et al. (A-2), patients who were involved in problem gambling treatment were asked about co-occurring drug and alcohol addictions. Let the discrete random variable X represent the number of co-occurring addictive substances used by the subjects. Table 4.2.4 summarizes the frequency distribution for this random variable.

(a) Construct a table of the relative frequency and the cumulative frequency for this discrete distribution.

(b) Construct a graph of the probability distribution and a graph representing the cumulative probability distribution for these data.

4.2.2 Refer to Exercise 4.2.1.

(a) What is probability that an individual selected at random used five addictive substances?

(b) What is the probability that an individual selected at random used fewer than three addictive substances?

(c) What is the probability that an individual selected at random used more than six addictive substances?

(d) What is the probability that an individual selected at random used between two and five addictive substances, inclusive?

4.2.3 Refer to Exercise 4.2.1. Find the mean, variance, and standard deviation of this frequency distribution.

Table 4.2.4 Number of Co-occurring Addictive Substances Used by Patients in Selected Gambling Treatment Programs

Number of Substances Used	Frequency
0	144
1	342
2	142
3	72
4	39
5	20
6	6
7	9
8	2
9	1
Total	777

4.3 THE BINOMIAL DISTRIBUTION

The *binomial distribution* is one of the most widely encountered probability distributions in applied statistics. The distribution is derived from a process known as a *Bernoulli trial,* named in honor of the Swiss mathematician James Bernoulli (1654–1705), who made significant contributions in the field of probability, including, in particular, the binomial distribution. When a random process or experiment, called a trial, can result in only one of two mutually exclusive outcomes, such as dead or alive, sick or well, full-term or premature, the trial is called a Bernoulli trial.

The Bernoulli Process A sequence of Bernoulli trials forms a *Bernoulli process* under the following conditions.

1. Each trial results in one of two possible, mutually exclusive, outcomes. One of the possible outcomes is denoted (arbitrarily) as a success, and the other is denoted a failure.
2. The probability of a success, denoted by p, remains constant from trial to trial. The probability of a failure, $1 - p$, is denoted by q.
3. The trials are independent; that is, the outcome of any particular trial is not affected by the outcome of any other trial.

EXAMPLE 4.3.1

We are interested in being able to compute the probability of x successes in n Bernoulli trials. For example, if we examine all birth records from the North Carolina State Center for Health Statistics (A-3) for the calendar year 2001, we find that 85.8 percent of the pregnancies had delivery in week 37 or later. We will refer to this as a full-term birth. With that percentage, we can interpret the probability of a recorded birth in week 37 or later as .858. If we randomly select five birth records from this population, what is the probability that exactly three of the records will be for full-term births?

Solution: Let us designate the occurrence of a record for a full-term birth (F) as a "success," and hasten to add that a premature birth (P) is not a failure, but medical research indicates that children born in week 36 or sooner are at risk for medical complications. If we are looking for birth records of premature deliveries, these would be designated successes, and birth records of full-term would be designated failures.

It will also be convenient to assign the number 1 to a success (record for a full-term birth) and the number 0 to a failure (record of a premature birth).

The process that eventually results in a birth record we consider to be a Bernoulli process.

Suppose the five birth records selected resulted in this sequence of full-term births:

FPFFP

In coded form we would write this as

10110

Since the probability of a success is denoted by p and the probability of a failure is denoted by q, the probability of the above sequence of outcomes is found by means of the multiplication rule to be

$$P(1, 0, 1, 1, 0) = pqppq = q^2p^3$$

The multiplication rule is appropriate for computing this probability since we are seeking the probability of a full-term, and a premature, and a full-term, and a full-term, and a premature, in that order or, in other words, the joint probability of the five events. For simplicity, commas, rather than intersection notation, have been used to separate the outcomes of the events in the probability statement.

The resulting probability is that of obtaining the specific sequence of outcomes in the order shown. We are not, however, interested in the order of occurrence of records for full-term and premature births but, instead, as has been stated already, the probability of the occurrence of exactly three records of full-term births out of five randomly selected records. Instead of occurring in the sequence shown above (call it sequence number 1), three successes and two failures could occur in any one of the following additional sequences as well:

Number	Sequence
2	11100
3	10011
4	11010
5	11001
6	10101
7	01110
8	00111
9	01011
10	01101

Each of these sequences has the same probability of occurring, and this probability is equal to q^2p^3, the probability computed for the first sequence mentioned.

When we draw a single sample of size five from the population specified, we obtain only one sequence of successes and failures. The question now becomes, What is the probability of getting sequence number 1 or sequence number 2 . . . or sequence number 10? From the addition rule we know that this probability is equal to the sum of the individual probabilities. In the present example we need to sum the $10q^2p^3$'s or, equivalently, multiply q^2p^3 by 10. We may now answer our original question: What is the probability, in a random sample of size 5, drawn from the specified population, of observing three successes (record of a full-term birth) and two failures (record of a premature birth)? Since in the population, $p = .858$, $q = (1 - p) = (1 - .858) = .142$ the answer to the question is

$$10(.142)^2(.858)^3 = 10(.0202)(.6316) = .1276$$ ■

Large Sample Procedure: Use of Combinations We can easily anticipate that, as the size of the sample increases, listing the number of sequences becomes more and more difficult and tedious. What is needed is an easy method of counting the number of sequences. Such a method is provided by means of a counting formula that allows us to determine quickly how many subsets of objects can be formed when we use in the subsets different numbers of the objects that make up the set from which the objects are selected. When the order of the objects in a subset is immaterial, the subset is called a combination of objects. When the order of objects in a subset does matter, we refer to the subset as a permutation of objects. Though permutations of objects are often used in probability theory, they will not be used in our current discussion. If a set consists of n objects, and we wish to form a subset of x objects from these n objects, without regard to the order of the objects in the subset, the result is called a *combination.* For examples, we define a combination as follows when the combination is formed by taking x objects from a set of n objects.

DEFINITION

A combination of n objects taken x at a time is an unordered subset of x of the n objects.

The number of combinations of n objects that can be formed by taking x of them at a time is given by

$$_nC_x = \frac{n!}{x!(n-x)!} \tag{4.3.1}$$

where $x!$, read x factorial, is the product of all the whole numbers from x down to 1. That is, $x! = x(x-1)(x-2)\ldots(1)$. We note that, by definition, $0! = 1$.

Let us return to our example in which we have a sample of $n = 5$ birth records and we are interested in finding the probability that three of them will be for full-term births.

TABLE 4.3.1 The Binomial Distribution

Number of Successes, x	Probability, $f(x)$
0	${}_nC_0 q^{n-0} p^0$
1	${}_nC_1 q^{n-1} p^1$
2	${}_nC_2 q^{n-2} p^2$
⋮	⋮
x	${}_nC_x q^{n-x} p^x$
⋮	⋮
n	${}_nC_n q^{n-n} p^n$
Total	1

The number of sequences in our example is found by Equation 4.3.1 to be

$$_5C_3 = \frac{5 \cdot 4 \cdot 3 \cdot 2 \cdot 1}{3 \cdot 2 \cdot 1 \cdot 2 \cdot 1} = \frac{120}{12} = 10$$

In our example we let $x = 3$, the number of successes, so that $n - x = 2$, the number of failures. We then may write the probability of obtaining exactly x successes in n trials as

$$\begin{aligned} f(x) &= {}_nC_x q^{n-x} p^x = {}_nC_x\, p^x q^{n-x} \quad \text{for } x = 0, 1, 2, \ldots, n \\ &= 0, \quad \text{elsewhere} \end{aligned} \tag{4.3.2}$$

This expression is called the binomial distribution. In Equation 4.3.2 $f(x) = P(X = x)$, where X is the random variable, the number of successes in n trials. We use $f(x)$ rather than $P(X = x)$ because of its compactness and because of its almost universal use.

We may present the binomial distribution in tabular form as in Table 4.3.1.

We establish the fact that Equation 4.3.2 is a probability distribution by showing the following:

1. $f(x) \geq 0$ for all real values of x. This follows from the fact that n and p are both nonnegative and, hence, ${}_nC_x$, p^x, and $(1 - p)^{n-x}$ are all nonnegative and, therefore, their product is greater than or equal to zero.
2. $\Sigma f(x) = 1$. This is seen to be true if we recognize that $\Sigma\, {}_nC_x q^{n-x} p^x$ is equal to $[(1 - p) + p]^n = 1^n = 1$, the familiar binomial expansion. If the binomial $(q + p)^n$ is expanded, we have

$$(q + p)^n = q^n + nq^{n-1}p^1 + \frac{n(n-1)}{2} q^{n-2}p^2 + \cdots + nq^1 p^{n-1} + p^n$$

If we compare the terms in the expansion, term for term, with the $f(x)$ in Table 4.3.1 we see that they are, term for term, equivalent, since

$$f(0) = {}_nC_0 q^{n-0} p^0 = q^n$$
$$f(1) = {}_nC_1 q^{n-1} p^1 = nq^{n-1}p$$

$$f(2) = {}_nC_2 q^{n-2}p^2 = \frac{n(n-1)}{2}q^{n-2}p^2$$
$$\vdots \qquad \vdots \qquad \vdots$$
$$f(n) = {}_nC_n q^{n-n}p^n = p^n$$

EXAMPLE 4.3.2

As another example of the use of the binomial distribution, the data from the North Carolina State Center for Health Statistics (A-3) show that 14 percent of mothers admitted to smoking one or more cigarettes per day during pregnancy. If a random sample of size 10 is selected from this population, what is the probability that it will contain exactly four mothers who admitted to smoking during pregnancy?

Solution: We take the probability of a mother admitting to smoking to be .14. Using Equation 4.3.2 we find

$$\begin{aligned} f(4) &= {}_{10}C_4(.86)^6(.14)^4 \\ &= \frac{10!}{4!6!}(.4045672)(.0003842) \\ &= .0326 \end{aligned}$$

■

Binomial Table The calculation of a probability using Equation 4.3.2 can be a tedious undertaking if the sample size is large. Fortunately, probabilities for different values of n, p, and x have been tabulated, so that we need only to consult an appropriate table to obtain the desired probability. Table B of the Appendix is one of many such tables available. It gives the probability that X is less than or equal to some specified value. That is, the table gives the cumulative probabilities from $x = 0$ up through some specified positive number of successes.

Let us illustrate the use of the table by using Example 4.3.2, where it was desired to find the probability that $x = 4$ when $n = 10$ and $p = .14$. Drawing on our knowledge of cumulative probability distributions from the previous section, we know that $P(x = 4)$ may be found by subtracting $P(X \leq 3)$ from $P(X \leq 4)$. If in Table B we locate $p = .14$ for $n = 10$, we find that $P(X \leq 4) = .9927$ and $P(X \leq 3) = .9600$. Subtracting the latter from the former gives $.9927 - .9600 = .0327$, which nearly agrees with our hand calculation (discrepancy due to rounding).

Frequently we are interested in determining probabilities, not for specific values of X, but for intervals such as the probability that X is between, say, 5 and 10. Let us illustrate with an example.

EXAMPLE 4.3.3

Suppose it is known that 10 percent of a certain population is color blind. If a random sample of 25 people is drawn from this population, use Table B in the Appendix to find the probability that:

(a) Five or fewer will be color blind.

Solution: This probability is an entry in the table. No addition or subtraction is necessary. $P(X \leq 5) = .9666$.

(b) Six or more will be color blind.

Solution: We cannot find this probability directly in the table. To find the answer, we use the concept of complementary probabilities. The probability that six or more are color blind is the complement of the probability that five or fewer are color blind. That is, this set is the complement of the set specified in part a; therefore,

$$P(X \geq 6) = 1 - P(X \leq 5) = 1 - .9666 = .0334$$

(c) Between six and nine inclusive will be color blind.

Solution: We find this by subtracting the probability that X is less than or equal to 5 from the probability that X is less than or equal to 9. That is,

$$P(6 \leq X \leq 9) = P(X \leq 9) - P(X \leq 5) = .9999 - .9666 = .0333$$

(d) Two, three, or four will be color blind.

Solution: This is the probability that X is between 2 and 4 inclusive.

$$P(2 \leq X \leq 4) = P(X \leq 4) - P(X \leq 1) = .9020 - .2712 = .6308$$

■

Using Table B When $p > .5$ Table B does not give probabilities for values of p greater than .5. We may obtain probabilities from Table B, however, by restating the problem in terms of the probability of a failure, $1 - p$, rather than in terms of the probability of a success, p. As part of the restatement, we must also think in terms of the number of failures, $n - x$, rather than the number of successes, x. We may summarize this idea as follows:

$$P(X = x \mid n, p > .50) = P(X = n - x \mid n, 1 - p) \tag{4.3.3}$$

In words, Equation 4.3.3 says, "The probability that X is equal to some specified value given the sample size and a probability of success greater than .5 is equal to the probability that X is equal to $n - x$ given the sample size and the probability of a failure of $1 - p$." For purposes of using the binomial table we treat the probability of a failure as though it were the probability of a success. When p is greater than .5, we may obtain cumulative probabilities from Table B by using the following relationship:

$$P(X \leq x \mid n, p > .50) = P(X \geq n - x \mid n, 1 - p) \tag{4.3.4}$$

Finally, to use Table B to find the probability that X is greater than or equal to some x when $P > .5$, we use the following relationship:

$$P(X \geq x \mid n, p > .50) = P(X \leq n - x \mid n, 1 - p) \tag{4.3.5}$$

EXAMPLE 4.3.4

According to a June 2003 poll conducted by the Massachusetts Health Benchmarks project (A-4), approximately 55 percent of residents answered "serious problem" to the question, "Some people think that childhood obesity is a national health problem. What do you think? Is it a very serious problem, somewhat of a problem, not much of a problem, or not a problem at all?" Assuming that the probability of giving this answer to the question is .55 for any Massachusetts resident, use Table B to find the probability that if 12 residents are chosen at random:

(a) Exactly seven will answer "serious problem."

Solution: We restate the problem as follows: What is the probability that a randomly selected resident gives an answer other than "serious problem" from exactly five residents out of 12, if 45 percent of residents give an answer other than "serious problem." We find the answer as follows:

$$\begin{aligned} P(X = 5 \mid n = 12, p = .45) &= P(X \leq 5) - P(X \leq 4) \\ &= .5269 - .3044 = .2225 \end{aligned}$$

(b) Five or fewer households will answer "serious problem."

Solution: The probability we want is

$$\begin{aligned} P(X \leq 5 \mid n = 12, p = .55) &= P(X \geq 12 - 5 \mid n = 12, p = .45) \\ &= P(X \geq 7 \mid n = 12, p = .45) \\ &= 1 - P(X \leq 6 \mid n = 12, p = .45) \\ &= 1 - .7393 = .2607 \end{aligned}$$

(c) Eight or more households will answer "serious problem."

Solution: The probability we want is

$$P(X \geq 8 \mid n = 12, p = .55) = P(X \leq 4 \mid n = 12, p = .45) = .3044$$ ■

Figure 4.3.1 provides a visual representation of the solution to the three parts of Example 4.3.4.

The Binomial Parameters The binomial distribution has two parameters, n and p. They are parameters in the sense that they are sufficient to specify a binomial distribution. The binomial distribution is really a family of distributions with each possible value of n and p designating a different member of the family. The mean and variance of the binomial distribution are $\mu = np$ and $\sigma^2 = np(1 - p)$, respectively.

Strictly speaking, the binomial distribution is applicable in situations where sampling is from an infinite population or from a finite population with replacement. Since in actual practice samples are usually drawn without replacement from finite populations, the question arises as to the appropriateness of the binomial distribution under these circumstances. Whether or not the binomial is appropriate depends on how drastic the effect of these conditions is on the constancy of p from trial to trial. It is generally agreed

	Possible number of successes (serious) = x P(SUCCESS) = .55	Probability statement	Possible number of failures (not serious) = $n - x$ P(FAILURE) = .45	Probability statement
Part b	0, 1, 2, 3, 4, 5	$P(X \leq 5 \mid 12, .55)$	12, 11, 10, 9, 8, 7	$P(X \geq 7 \mid 12, .45)$
	6		6	
Part a	7	$P(X = 7 \mid 12, .55)$	5	$P(X = 5 \mid 12, .45)$
Part c	8, 9, 10, 11, 12	$P(X \geq 8 \mid 12, .55)$	4, 3, 2, 1, 0	$P(X \leq 4 \mid 12, .45)$

FIGURE 4.3.1 Schematic representation of solutions to Example 4.3.4 (the relevant numbers of successes and failures in each case are circled).

that when n is small relative to N, the binomial model is appropriate. Some writers say that n is small relative to N if N is at least 10 times as large as n.

Most statistical software programs allow for the calculation of binomial probabilities with a personal computer. EXCEL, for example, can be used to calculate individual or cumulative probabilities for specified values of x, n, and p. Suppose we wish to find the individual probabilities for $x = 0$ through $x = 6$ when $n = 6$ and $p = .3$. We enter the numbers 0 through 6 in Column 1 and proceed as shown in Figure 4.3.2. We may follow a similar procedure to find the cumulative probabilities. For this illustration, we use MINITAB and place the numbers 1 through 6 in Column 1. We proceed as shown in Figure 4.3.3.

Using the following cell command:

BINOMDIST(A*, 6, .3, false), ***where A* is the appropriate cell reference***

We obtain the following output:

0	0.117649
1	0.302526
2	0.324135
3	0.185220
4	0.059535
5	0.010206
6	0.000729

FIGURE 4.3.2 Excel calculation of individual binomial probabilities for $x = 0$ through $x = 6$ when $n = 6$ and $p = 3$.

Data:

```
C1:  0 1 2 3 4 5 6
```

Dialog box:

Calc ➤ Probability Distributions ➤ Binomial

Choose **Cumulative probability.** Type *6* in **Number of trials.** Type *0.3* in **Probability of success.** Choose **Input column** and type *C1.* Click **OK.**

Session command:

```
MTB > CDF C1;
SUBC>    BINOMIAL 6 0.3.
```

Output:

Cumulative Distribution Function

```
Binomial with n = 6 and p = 0.300000

        x      P( X <= x)
     0.00       0.1176
     1.00       0.4202
     2.00       0.7443
     3.00       0.9295
     4.00       0.9891
     5.00       0.9993
     6.00       1.0000
```

FIGURE 4.3.3 MINITAB calculation of cumulative binomial probabilities for $x = 0$ through $x = 6$ when $n = 6$ and $p = 3$.

EXERCISES

In each of the following exercises, assume that N is sufficiently large relative to n that the binomial distribution may be used to find the desired probabilities.

4.3.1 Based on data collected by the National Center for Health Statistics and made available to the public in the Sample Adult database (A-5), an estimate of the percentage of adults who have at some point in their life been told they have hypertension is 23.53 percent. If we select a simple random sample of 20 U.S. adults and assume that the probability that each has been told that he or she has hypertension is .24, find the probability that the number of people in the sample who have been told that they have hypertension will be:

(a) Exactly three **(b)** Three or more

(c) Fewer than three **(d)** Between three and seven, inclusive

4.3.2 Refer to Exercise 4.3.1. How many adults who have been told that they have hypertension would you expect to find in a sample of 20?

4.3.3 Refer to Exercise 4.3.1. Suppose we select a simple random sample of five adults. Use Equation 4.3.2 to find the probability that, in the sample, the number of people who have been told that they have hypertension will be:

(a) Zero **(b)** More than one
(c) Between one and three, inclusive **(d)** Two or fewer
(e) Five

4.3.4 The same survey database cited in exercise 4.3.1 (A-5) shows that 32 percent of U.S. adults indicated that they have been tested for HIV at some point in their life. Consider a simple random sample of 15 adults selected at that time. Find the probability that the number of adults who have been tested for HIV in the sample would be:

(a) Three **(b)** Less than five
(c) Between five and nine, inclusive **(d)** More than five, but less than 10
(e) Six or more

4.3.5 Refer to Exercise 4.3.4. Find the mean and variance of the number of people tested for HIV in samples of size 15.

4.3.6 Refer to Exercise 4.3.4. Suppose we were to take a simple random sample of 25 adults today and find that two have been tested for HIV at some point in their life. Would these results be surprising? Why or why not?

4.3.7 Coughlin et al. (A-6) estimated the percentage of women living in border counties along the southern United States with Mexico (designated counties in California, Arizona, New Mexico, and Texas) who have less than a high school education to be 18.7. Assume the corresponding probability is .19. Suppose we select three women at random. Find the probability that the number with less than a high-school education is:

(a) Exactly zero **(b)** Exactly one
(c) More than one **(d)** Two or fewer
(e) Two or three **(f)** Exactly three

4.3.8 In a survey of nursing students pursuing a master's degree, 75 percent stated that they expect to be promoted to a higher position within one month after receiving the degree. If this percentage holds for the entire population, find, for a sample of 15, the probability that the number expecting a promotion within a month after receiving their degree is:

(a) Six **(b)** At least seven
(c) No more than five **(d)** Between six and nine, inclusive

4.3.9 Given the binomial parameters $p = .8$ and $n = 3$, show by means of the binomial expansion given in Table 4.3.1 that $\sum f(x) = 1$.

4.4 THE POISSON DISTRIBUTION

The next discrete distribution that we consider is the *Poisson distribution,* named for the French mathematician Simeon Denis Poisson (1781–1840), who is generally credited for publishing its derivation in 1837. This distribution has been used extensively as a probability model in biology and medicine. Haight (1) presents a fairly extensive catalog of such applications in Chapter 7 of his book.

If x is the number of occurrences of some random event in an interval of time or space (or some volume of matter), the probability that x will occur is given by

$$f(x) = \frac{e^{-\lambda}\lambda^x}{x!}, \qquad x = 0, 1, 2, \ldots \tag{4.4.1}$$

The Greek letter λ (lambda) is called the parameter of the distribution and is the average number of occurrences of the random event in the interval (or volume). The symbol e is the constant (to four decimals) 2.7183.

It can be shown that $f(x) \geq 0$ for every x and that $\sum_x f(x) = 1$ so that the distribution satisfies the requirements for a probability distribution.

The Poisson Process We have seen that the binomial distribution results from a set of assumptions about an underlying process yielding a set of numerical observations. Such, also, is the case with the Poisson distribution. The following statements describe what is known as the *Poisson process.*

1. The occurrences of the events are independent. The occurrence of an event in an interval[1] of space or time has no effect on the probability of a second occurrence of the event in the same, or any other, interval.
2. Theoretically, an infinite number of occurrences of the event must be possible in the interval.
3. The probability of the single occurrence of the event in a given interval is proportional to the length of the interval.
4. In any infinitesimally small portion of the interval, the probability of more than one occurrence of the event is negligible.

An interesting feature of the Poisson distribution is the fact that the mean and variance are equal.

When to Use the Poisson Model The Poisson distribution is employed as a model when counts are made of events or entities that are distributed at random in space or time. One may suspect that a certain process obeys the Poisson law, and under this assumption probabilities of the occurrence of events or entities within some unit of space or time may be calculated. For example, under the assumptions that the distribution of some parasite among individual host members follows the Poisson law, one may, with knowledge of the parameter λ, calculate the probability that a randomly selected individual host will yield x number of parasites. In a later chapter we will learn how to decide whether the assumption that a specified process obeys the Poisson law is plausible. An additional use of the Poisson distribution in practice occurs

[1] For simplicity, the Poisson distribution is discussed in terms of intervals, but other units, such as a volume of matter, are implied.

when n is large and p is small. In this case, the Poisson distribution can be used to approximate the binomial distribution. In other words,

$$_{n}C_{x}p^{x}q^{n-x} \approx \frac{e^{-\lambda}\lambda^{x}}{x!}, \quad x = 0, 1, 2, \ldots$$

where $\lambda = np$.

To illustrate the use of the Poisson distribution for computing probabilities, let us consider the following examples.

EXAMPLE 4.4.1

In a study of drug-induced anaphylaxis among patients taking rocuronium bromide as part of their anesthesia, Laake and Røttingen (A-7) found that the occurrence of anaphylaxis followed a Poisson model with $\lambda = 12$ incidents per year in Norway. Find the probability that in the next year, among patients receiving rocuronium, exactly three will experience anaphylaxis.

Solution: By Equation 4.4.1, we find the answer to be

$$P(X = 3) = \frac{e^{-12}12^{3}}{3!} = .00177$$

■

EXAMPLE 4.4.2

Refer to Example 4.4.1. What is the probability that at least three patients in the next year will experience anaphylaxis if rocuronium is administered with anesthesia?

Solution: We can use the concept of complementary events in this case. Since $P(X \leq 2)$ is the complement of $P(X \geq 3)$, we have

$$\begin{aligned} P(X \geq 3) &= 1 - P(X \leq 2) = 1 - [P(X = 0) + P(X = 1) + P(X = 2)] \\ &= 1 - \left[\frac{e^{-12}12^{0}}{0!} + \frac{e^{-12}12^{1}}{1!} + \frac{e^{-12}12^{2}}{2!}\right] \\ &= 1 - [.00000614 + .00007373 + .00044238] \\ &= 1 - .00052225 \\ &= .99947775 \end{aligned}$$

■

In the foregoing examples the probabilities were evaluated directly from the equation. We may, however, use Appendix Table C, which gives cumulative probabilities for various values of λ and X.

EXAMPLE 4.4.3

In the study of a certain aquatic organism, a large number of samples were taken from a pond, and the number of organisms in each sample was counted. The average number of organisms per sample was found to be two. Assuming that the number of organisms follows a Poisson distribution, find the probability that the next sample taken will contain one or fewer organisms.

Solution: In Table C we see that when $\lambda = 2$, the probability that $X \leq 1$ is .406. That is, $P(X \leq 1 \mid 2) = .406$. ■

EXAMPLE 4.4.4

Refer to Example 4.4.3. Find the probability that the next sample taken will contain exactly three organisms.

Solution:

$$P(X = 3 \mid 2) = P(X \leq 3) - P(X \leq 2) = .857 - .677 = .180$$ ■

Data:

```
C1: 0 1 2 3 4 5 6
```

Dialog box:

Calc ➤ Probability Distributions ➤ Poisson

Choose **Probability.** Type *.70* in **Mean.** Choose **Input column** and type *C1.* Click **OK.**

Session command:

```
MTB > PDF C1;
SUBC>   Poisson .70.
```

Output:

Probability Density Function

```
Poisson with mu = 0.700000

        x          P( X = x)
     0.00           0.4966
     1.00           0.3476
     2.00           0.1217
     3.00           0.0284
     4.00           0.0050
     5.00           0.0007
     6.00           0.0001
```

FIGURE 4.4.1 MINITAB calculation of individual Poisson probabilities for $x = 0$ through $x = 6$ and $\lambda = .7$.

Using commands found in:

Analysis ➤ Other ➤ Probability Calculator

We obtain the following output:

0 <= X	Prob(x <= X)
0	0.4966
1	0.8442
2	0.9659
3	0.9942
4	0.9992
5	0.9999
6	1.0000

FIGURE 4.4.2 MINITAB calculation of cumulative Poisson probabilities for $x = 0$ through $x = 6$ and $\lambda = .7$.

EXAMPLE 4.4.5

Refer to Example 4.4.3. Find the probability that the next sample taken will contain more than five organisms.

Solution: Since the set of more than five organisms does not include five, we are asking for the probability that six or more organisms will be observed. This is obtained by subtracting the probability of observing five or fewer from one. That is,

$$P(X > 5 \mid 2) = 1 - P(X \leq 5) = 1 - .983 = .017$$ ■

Poisson probabilities are obtainable from most statistical software packages. To illustrate the use of MINITAB for this purpose, suppose we wish to find the individual probabilities for $x = 0$ through $x = 6$ when $\lambda = .7$. We enter the values of x in Column 1 and proceed as shown in Figure 4.4.1. We obtain the cumulative probabilities for the same values of x and λ as shown in Figure 4.4.2.

EXERCISES

4.4.1 Singh et al. (A-8) looked at the occurrence of retinal capillary hemangioma (RCH) in patients with von Hippel–Lindau (VHL) disease. RCH is a benign vascular tumor of the retina. Using a retrospective consecutive case series review, the researchers found that the number of RCH tumor incidents followed a Poisson distribution with $\lambda = 4$ tumors per eye for patients with VHL. Using this model, find the probability that in a randomly selected patient with VHL:

(a) There are exactly five occurrences of tumors per eye.

(b) There are more than five occurrences of tumors per eye.

(c) There are fewer than five occurrences of tumors per eye.

(d) There are between five and seven occurrences of tumors per eye, inclusive.

4.4.2 Tubert-Bitter et al. (A-9) found that the number of serious gastrointestinal reactions reported to the British Committee on Safety of Medicine was 538 for 9,160,000 prescriptions of the anti-inflammatory drug piroxicam. This corresponds to a rate of .058 gastrointestinal reactions per 1000 prescriptions written. Using a Poisson model for probability, with $\lambda = .06$, find the probability of

(a) Exactly one gastrointestinal reaction in 1000 prescriptions

(b) Exactly two gastrointestinal reactions in 1000 prescriptions

(c) No gastrointestinal reactions in 1000 prescriptions

(d) At least one gastrointestinal reaction in 1000 prescriptions

4.4.3 If the mean number of serious accidents per year in a large factory (where the number of employees remains constant) is five, find the probability that in the current year there will be:

(a) Exactly seven accidents **(b)** Ten or more accidents

(c) No accidents **(d)** Fewer than five accidents

4.4.4 In a study of the effectiveness of an insecticide against a certain insect, a large area of land was sprayed. Later the area was examined for live insects by randomly selecting squares and counting the number of live insects per square. Past experience has shown the average number of live insects per square after spraying to be .5. If the number of live insects per square follows a Poisson distribution, find the probability that a selected square will contain:

(a) Exactly one live insect **(b)** No live insects

(c) Exactly four live insects **(d)** One or more live insects

4.4.5 In a certain population an average of 13 new cases of esophageal cancer are diagnosed each year. If the annual incidence of esophageal cancer follows a Poisson distribution, find the probability that in a given year the number of newly diagnosed cases of esophageal cancer will be:

(a) Exactly 10 **(b)** At least eight

(c) No more than 12 **(d)** Between nine and 15, inclusive

(e) Fewer than seven

4.5 CONTINUOUS PROBABILITY DISTRIBUTIONS

The probability distributions considered thus far, the binomial and the Poisson, are distributions of discrete variables. Let us now consider distributions of continuous random variables. In Chapter 1 we stated that a continuous variable is one that can assume any value within a specified interval of values assumed by the variable. Consequently, between any two values assumed by a continuous variable, there exist an infinite number of values.

To help us understand the nature of the distribution of a continuous random variable, let us consider the data presented in Table 1.4.1 and Figure 2.3.2. In the table we have 189 values of the random variable, age. The histogram of Figure 2.3.2 was constructed by locating specified points on a line representing the measurement of interest and erecting a series of rectangles, whose widths were the distances between two specified points on the line, and whose heights represented the number of values of the variable falling between the two specified points. The intervals defined by any two consecutive specified points we called class intervals. As was noted in Chapter 2, subareas of the histogram correspond to the frequencies of occurrence of values of the variable between the horizontal scale boundaries of these subareas. This provides a way whereby the relative frequency of occurrence of values between any two specified points can be calculated: merely determine the proportion of the histogram's total area falling between the specified points. This can be done more conveniently by consulting the relative frequency or cumulative relative frequency columns of Table 2.3.2.

Imagine now the situation where the number of values of our random variable is very large and the width of our class intervals is made very small. The resulting histogram could look like that shown in Figure 4.5.1.

If we were to connect the midpoints of the cells of the histogram in Figure 4.5.1 to form a frequency polygon, clearly we would have a much smoother figure than the frequency polygon of Figure 2.3.4.

In general, as the number of observations, n, approaches infinity, and the width of the class intervals approaches zero, the frequency polygon approaches a smooth curve such as is shown in Figure 4.5.2. Such smooth curves are used to represent graphically the distributions of continuous random variables. This has some important consequences when we deal with probability distributions. First, the total area under the curve is equal to one, as was true with the histogram, and the relative frequency of occurrence of values between any two points on the x-axis is equal to the total area bounded by the curve, the x-axis, and perpendicular lines erected at the two points on the x-axis. See

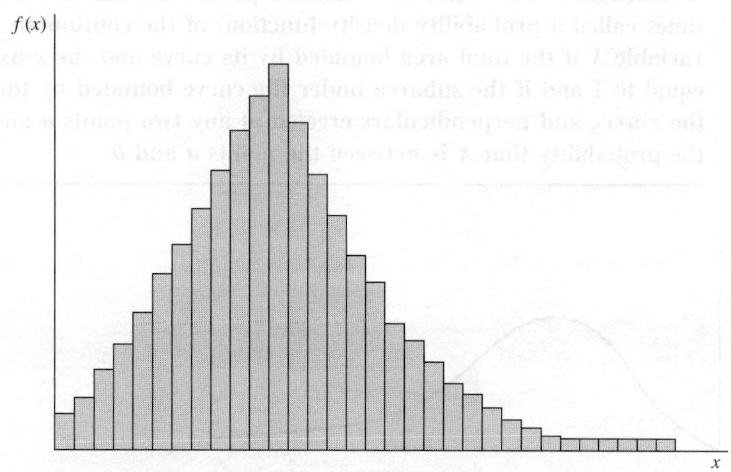

FIGURE 4.5.1 A histogram resulting from a large number of values and small class intervals.

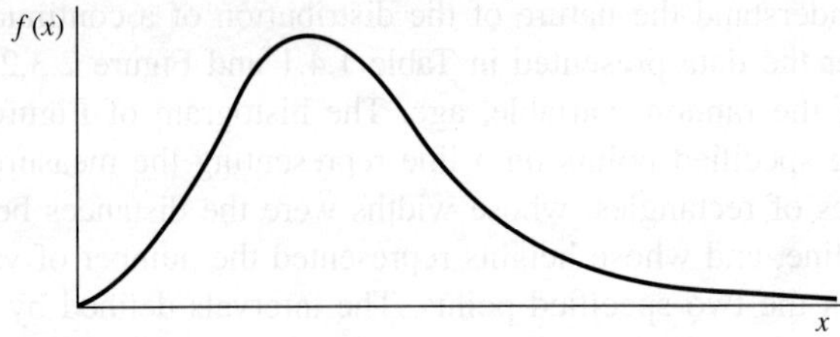

FIGURE 4.5.2 Graphical representation of a continuous distribution.

Figure 4.5.3. The probability of *any specific value* of the random variable is *zero.* This seems logical, since a specific value is represented by a point on the x-axis and the area above a point is zero.

Finding Area Under a Smooth Curve With a histogram, as we have seen, subareas of interest can be found by adding areas represented by the cells. We have no cells in the case of a smooth curve, so we must seek an alternate method of finding subareas. Such a method is provided by the integral calculus. To find the area under a smooth curve between any two points a and b, the *density function* is integrated from a to b. A *density function* is a formula used to represent the distribution of a continuous random variable. Integration is the limiting case of summation, but we will not perform any integrations, since the level of mathematics involved is beyond the scope of this book. As we will see later, for all the continuous distributions we will consider, there will be an easier way to find areas under their curves.

Although the definition of a probability distribution for a continuous random variable has been implied in the foregoing discussion, by way of summary, we present it in a more compact form as follows.

DEFINITION

A nonnegative function $f(x)$ is called a probability distribution (sometimes called a probability density function) of the continuous random variable X if the total area bounded by its curve and the x-axis is equal to 1 and if the subarea under the curve bounded by the curve, the x-axis, and perpendiculars erected at any two points a and b give the probability that X is between the points a and b.

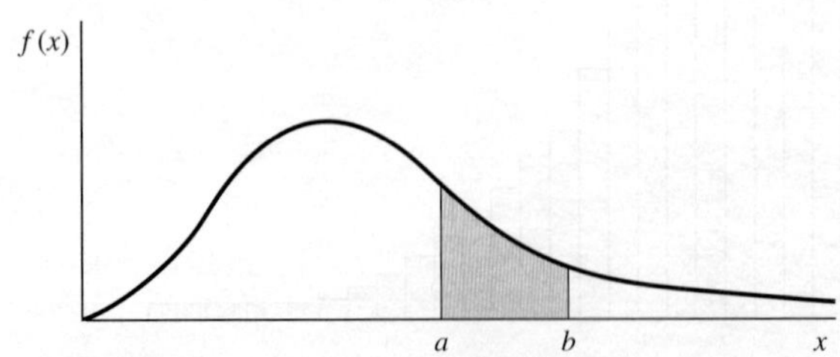

FIGURE 4.5.3 Graph of a continuous distribution showing area between a and b.

Thus, the probability of a continuous random variable to assume values between a and b is denoted by $P(a < X < b)$

4.6 THE NORMAL DISTRIBUTION

We come now to the most important distribution in all of statistics—the *normal distribution.* The formula for this distribution was first published by Abraham De Moivre (1667–1754) on November 12, 1733. Many other mathematicians figure prominently in the history of the normal distribution, including Carl Friedrich Gauss (1777–1855). The distribution is frequently called the *Gaussian distribution* in recognition of his contributions.

The normal density is given by

$$f(x) = \frac{1}{\sqrt{2\pi}\sigma} e^{-(x-\mu)^2/2\sigma^2}, \qquad -\infty < x < \infty \tag{4.6.1}$$

In Equation 4.6.1, π and e are the familiar constants, 3.14159 . . . and 2.71828 . . . , respectively, which are frequently encountered in mathematics. The two parameters of the distribution are μ, the mean, and σ, the standard deviation. For our purposes we may think of μ and σ of a normal distribution, respectively, as measures of central tendency and dispersion as discussed in Chapter 2. Since, however, a normally distributed random variable is continuous and takes on values between $-\infty$ and $+\infty$, its mean and standard deviation may be more rigorously defined; but such definitions cannot be given without using calculus. The graph of the normal distribution produces the familiar bell-shaped curve shown in Figure 4.6.1.

Characteristics of the Normal Distribution The following are some important characteristics of the normal distribution.

1. It is symmetrical about its mean, μ. As is shown in Figure 4.6.1, the curve on either side of μ is a mirror image of the other side.
2. The mean, the median, and the mode are all equal.
3. The total area under the curve above the x-axis is one square unit. This characteristic follows from the fact that the normal distribution is a probability distribution. Because of the symmetry already mentioned, 50 percent of the area is to the right of a perpendicular erected at the mean, and 50 percent is to the left.

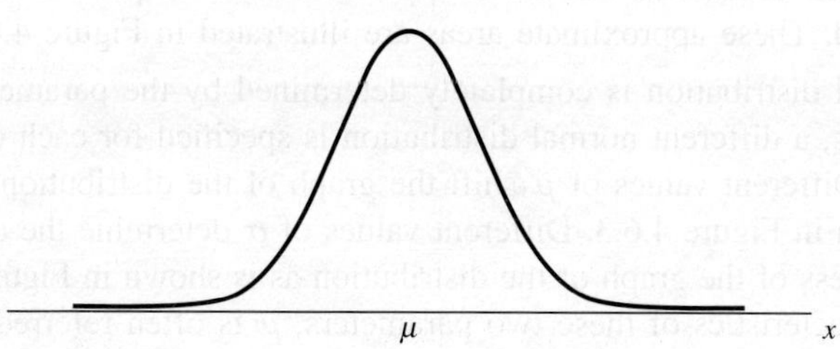

FIGURE 4.6.1 Graph of a normal distribution.

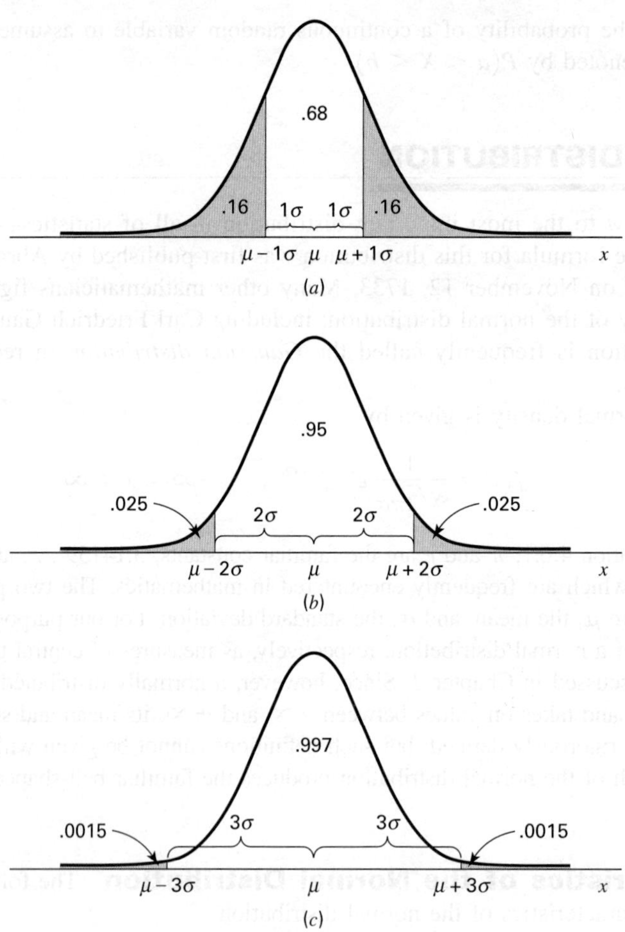

FIGURE 4.6.2 Subdivision of the area under the normal curve (areas are approximate).

4. If we erect perpendiculars a distance of 1 standard deviation from the mean in both directions, the area enclosed by these perpendiculars, the x-axis, and the curve will be approximately 68 percent of the total area. If we extend these lateral boundaries a distance of two standard deviations on either side of the mean, approximately 95 percent of the area will be enclosed, and extending them a distance of three standard deviations will cause approximately 99.7 percent of the total area to be enclosed. These approximate areas are illustrated in Figure 4.6.2.

5. The normal distribution is completely determined by the parameters μ and σ. In other words, a different normal distribution is specified for each different value of μ and σ. Different values of μ shift the graph of the distribution along the x-axis as is shown in Figure 4.6.3. Different values of σ determine the degree of flatness or peakedness of the graph of the distribution as is shown in Figure 4.6.4. Because of the characteristics of these two parameters, μ is often referred to as a *location parameter* and σ is often referred to as a *shape parameter*.

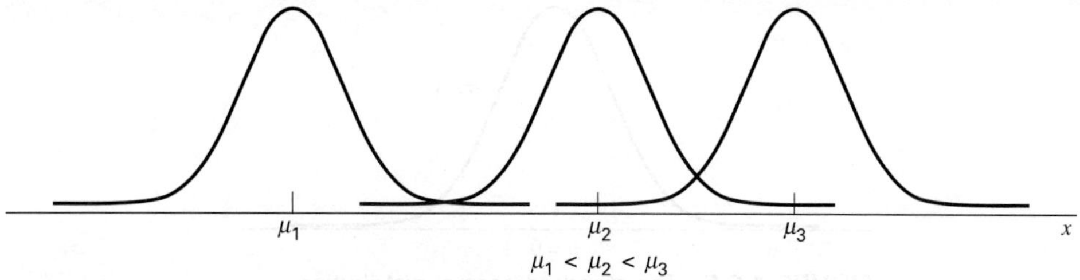

FIGURE 4.6.3 Three normal distributions with different means but the same amount of variability.

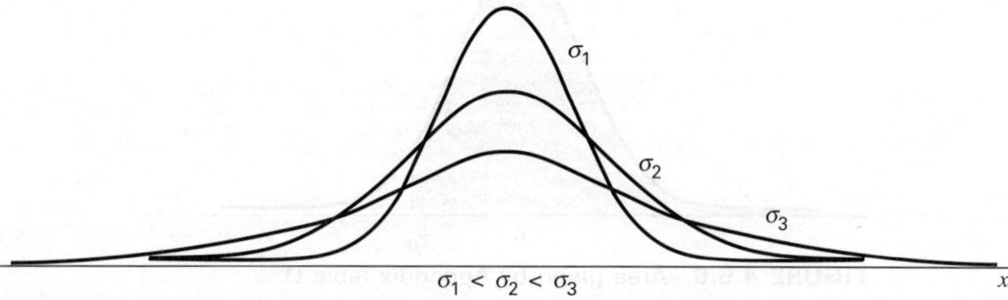

FIGURE 4.6.4 Three normal distributions with different standard deviations but the same mean.

The Standard Normal Distribution The last-mentioned characteristic of the normal distribution implies that the normal distribution is really a family of distributions in which one member is distinguished from another on the basis of the values of μ and σ. The most important member of this family is the *standard normal distribution* or *unit normal distribution,* as it is sometimes called, because it has a mean of 0 and a standard deviation of 1. It may be obtained from Equation 4.6.1 by creating a random variable.

$$z = (x - \mu)/\sigma \tag{4.6.2}$$

The equation for the standard normal distribution is written

$$f(z) = \frac{1}{\sqrt{2\pi}} e^{-z^2/2}, \quad -\infty < z < \infty \tag{4.6.3}$$

The graph of the standard normal distribution is shown in Figure 4.6.5.

The z-transformation will prove to be useful in the examples and applications that follow. This value of z denotes, for a value of a random variable, the number of standard deviations that value falls above $(+z)$ or below $(-z)$ the mean, which in this case is 0. For example, a z-transformation that yields a value of $z = 1$ indicates that the value of x used in the transformation is 1 standard deviation above 0. A value of $z = -1$ indicates that the value of x used in the transformation is 1 standard deviation below 0. This property is illustrated in the examples of Section 4.7.

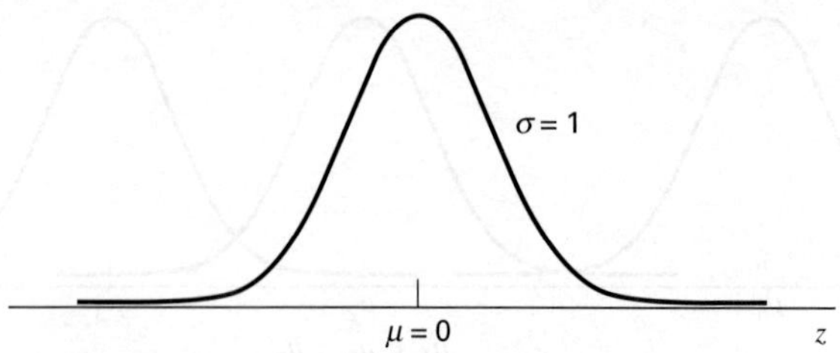

FIGURE 4.6.5 The standard normal distribution.

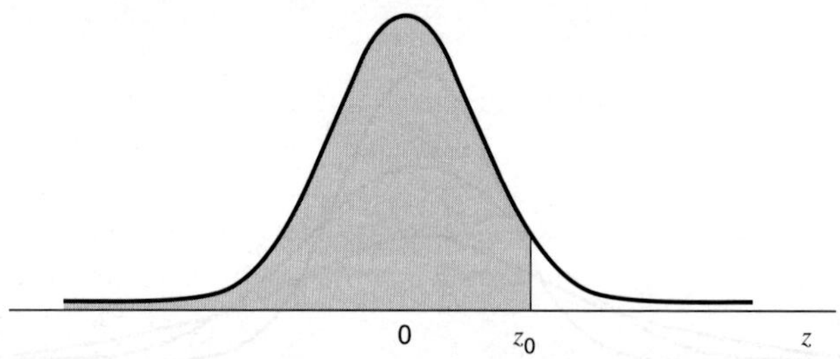

FIGURE 4.6.6 Area given by Appendix Table D.

To find the probability that z takes on a value between any two points on the z-axis, say, z_0 and z_1, we must find the area bounded by perpendiculars erected at these points, the curve, and the horizontal axis. As we mentioned previously, areas under the curve of a continuous distribution are found by integrating the function between two values of the variable. In the case of the standard normal, then, to find the area between z_0 and z_1 directly, we would need to evaluate the following integral:

$$\int_{z_0}^{z_1} \frac{1}{\sqrt{2\pi}} e^{-z^2/2}\, dz$$

Although a closed-form solution for the integral does not exist, we can use numerical methods of calculus to approximate the desired areas beneath the curve to a desired accuracy. Fortunately, we do not have to concern ourselves with such matters, since there are tables available that provide the results of any integration in which we might be interested. Table D in the Appendix is an example of these tables. In the body of Table D are found the areas under the curve between $-\infty$ and the values of z shown in the leftmost column of the table. The shaded area of Figure 4.6.6 represents the area listed in the table as being between $-\infty$ and z_0, where z_0 is the specified value of z.

We now illustrate the use of Table D by several examples.

EXAMPLE 4.6.1

Given the standard normal distribution, find the area under the curve, above the z-axis between $z = -\infty$ and $z = 2$.

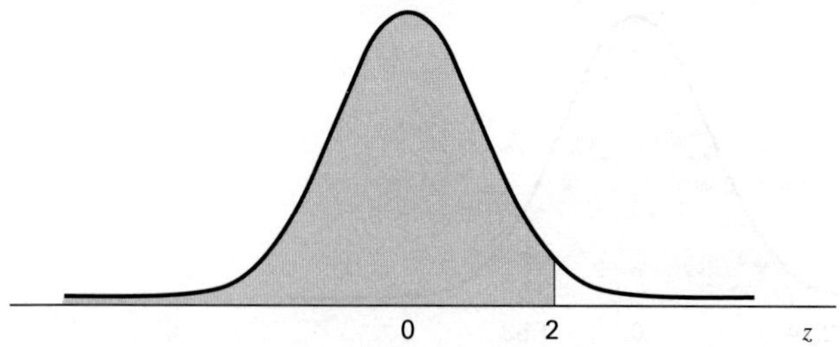

FIGURE 4.6.7 The standard normal distribution showing area between $z = -\infty$ and $z = 2$.

Solution: It will be helpful to draw a picture of the standard normal distribution and shade the desired area, as in Figure 4.6.7. If we locate $z = 2$ in Table D and read the corresponding entry in the body of the table, we find the desired area to be .9772. We may interpret this area in several ways. We may interpret it as the probability that a z picked at random from the population of z's will have a value between $-\infty$ and 2. We may also interpret it as the relative frequency of occurrence (or proportion) of values of z between $-\infty$ and 2, or we may say that 97.72 percent of the z's have a value between $-\infty$ and 2. ■

EXAMPLE 4.6.2

What is the probability that a z picked at random from the population of z's will have a value between -2.55 and $+2.55$?

Solution: Figure 4.6.8 shows the area desired. Table D gives us the area between $-\infty$ and 2.55, which is found by locating 2.5 in the leftmost column of the table and then moving across until we come to the entry in the column headed by 0.05. We find this area to be .9946. If we look at the picture we draw, we see that this is more area than is desired. We need to subtract from .9946 the area to the left of -2.55. Reference to Table D shows that the area to the left of -2.55 is .0054. Thus the desired probability is

$$P(-2.55 < z < 2.55) = .9946 - .0054 = .9892$$

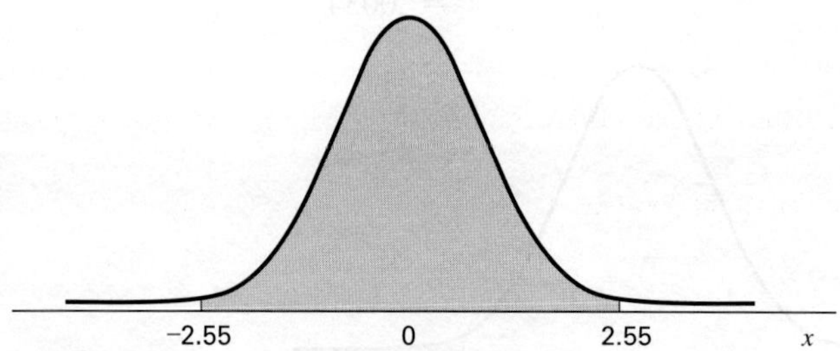

FIGURE 4.6.8 Standard normal curve showing $P(-2.55 < z < 2.55)$.

■

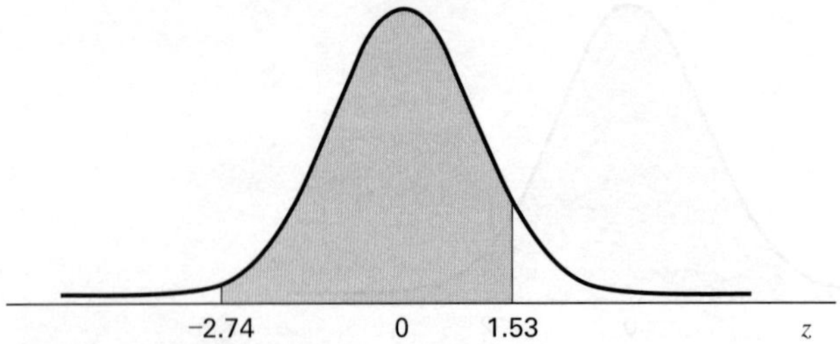

FIGURE 4.6.9 Standard normal curve showing proportion of z values between $z = -2.74$ and $z = 1.53$.

Suppose we had been asked to find the probability that z is between -2.55 and 2.55 inclusive. The desired probability is expressed as $P(-2.55 \leq z \leq 2.55)$. Since, as we noted in Section 4.5, $P(z = z_0) = 0, P(-2.55 \leq z \leq 2.55) = P(-2.55 < z < 2.55) = .9892$.

EXAMPLE 4.6.3

What proportion of z values are between -2.74 and 1.53?

Solution: Figure 4.6.9 shows the area desired. We find in Table D that the area between $-\infty$ and 1.53 is .9370, and the area between $-\infty$ and -2.74 is .0031. To obtain the desired probability we subtract .0031 from .9370. That is,

$$P(-2.74 \leq z \leq 1.53) = .9370 - .0031 = .9339$$

■

EXAMPLE 4.6.4

Given the standard normal distribution, find $P(z \geq 2.71)$.

Solution: The area desired is shown in Figure 4.6.10. We obtain the area to the right of $z = 2.71$ by subtracting the area between $-\infty$ and 2.71 from 1. Thus,

$$\begin{aligned} P(z \geq 2.71) &= 1 - P(z \leq 2.71) \\ &= 1 - .9966 \\ &= .0034 \end{aligned}$$

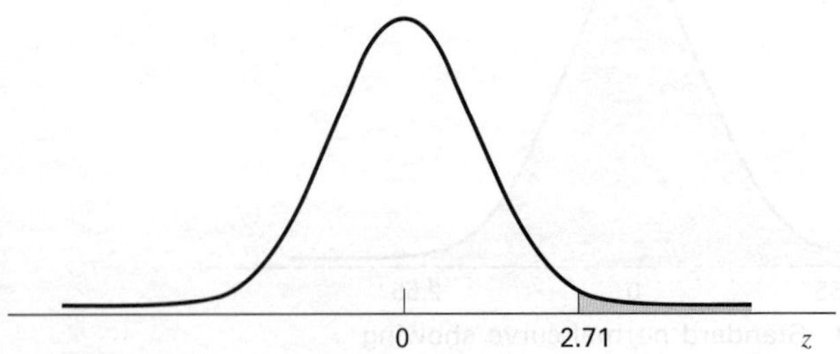

FIGURE 4.6.10 Standard normal distribution showing $P(z \geq 2.71)$.

■

EXAMPLE 4.6.5

Given the standard normal distribution, find $P(.84 \leq z \leq 2.45)$.

Solution: The area we are looking for is shown in Figure 4.6.11. We first obtain the area between $-\infty$ and 2.45 and from that subtract the area between $-\infty$ and .84. In other words,

$$\begin{aligned} P(.84 \leq z \leq 2.45) &= P(z \leq 2.45) - P(z \leq .84) \\ &= .9929 - .7995 \\ &= .1934 \end{aligned}$$

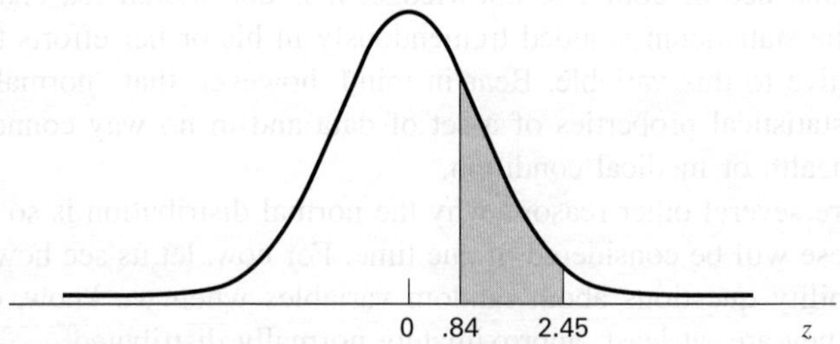

FIGURE 4.6.11 Standard normal curve showing $P(.84 \leq z \leq 2.45)$.

■

EXERCISES

Given the standard normal distribution find:

4.6.1 The area under the curve between $z = 0$ and $z = 1.43$

4.6.2 The probability that a z picked at random will have a value between $z = -2.87$ and $z = 2.64$

4.6.3 $P(z \geq .55)$

4.6.4 $P(z \geq -.55)$

4.6.5 $P(z < -2.33)$

4.6.6 $P(z < 2.33)$

4.6.7 $P(-1.96 \leq z \leq 1.96)$

4.6.8 $P(-2.58 \leq z \leq 2.58)$

4.6.9 $P(-1.65 \leq z \leq 1.65)$

4.6.10 $P(z = .74)$

Given the following probabilities, find z_1:

4.6.11 $P(z \leq z_1) = .0055$

4.6.12 $P(-2.67 \leq z \leq z_1) = .9718$

4.6.13 $P(z > z_1) = .0384$

4.6.14 $P(z_1 \leq z \leq 2.98) = .1117$

4.6.15 $P(-z_1 \leq z \leq z_1) = .8132$

4.7 NORMAL DISTRIBUTION APPLICATIONS

Although its importance in the field of statistics is indisputable, one should realize that the normal distribution is not a law that is adhered to by all measurable characteristics occurring in nature. It is true, however, that many of these characteristics are approximately normally distributed. Consequently, even though no variable encountered in practice is precisely normally distributed, the normal distribution can be used to model the distribution of many variables that are of interest. Using the normal distribution as a model allows us to make useful probability statements about some variables much more conveniently than would be the case if some more complicated model had to be used.

Human stature and human intelligence are frequently cited as examples of variables that are approximately normally distributed. On the other hand, many distributions relevant to the health field cannot be described adequately by a normal distribution. Whenever it is known that a random variable is approximately normally distributed, or when, in the absence of complete knowledge, it is considered reasonable to make this assumption, the statistician is aided tremendously in his or her efforts to solve practical problems relative to this variable. Bear in mind, however, that "normal" in this context refers to the statistical properties of a set of data and in no way connotes normality in the sense of health or medical condition.

There are several other reasons why the normal distribution is so important in statistics, and these will be considered in due time. For now, let us see how we may answer simple probability questions about random variables when we know, or are willing to assume, that they are, at least, approximately normally distributed.

EXAMPLE 4.7.1

The Uptimer is a custom-made lightweight battery-operated activity monitor that records the amount of time an individual spends in the upright position. In a study of children ages 8 to 15 years, Eldridge et al. (A-10) studied 529 normally developing children who each wore the Uptimer continuously for a 24-hour period that included a typical school day. The researchers found that the amount of time children spent in the upright position followed a normal distribution with a mean of 5.4 hours and standard deviation of 1.3 hours. Assume that this finding applies to all children 8 to 15 years of age. Find the probability that a child selected at random spends less than 3 hours in the upright position in a 24-hour period.

Solution: First let us draw a picture of the distribution and shade the area corresponding to the probability of interest. This has been done in Figure 4.7.1.

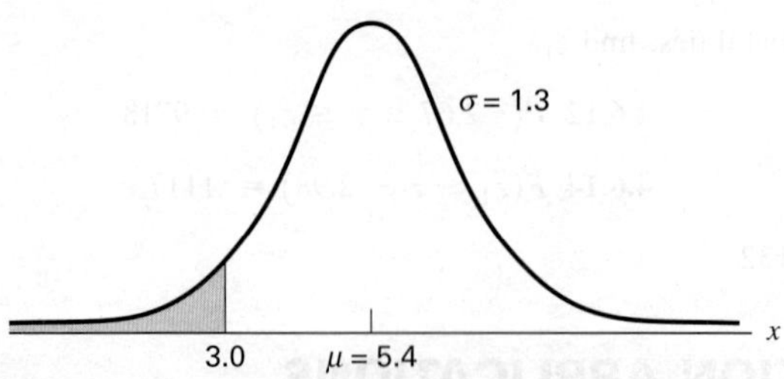

FIGURE 4.7.1 Normal distribution to approximate distribution of amount of time children spent in upright position (mean and standard deviation estimated).

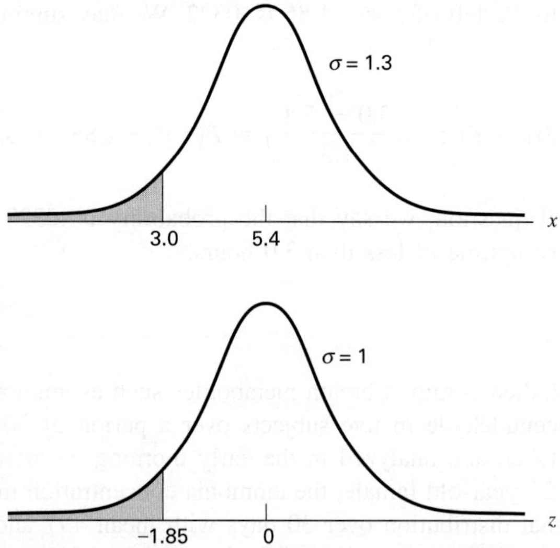

FIGURE 4.7.2 Normal distribution of time spent upright (x) and the standard normal distribution (z).

If our distribution were the standard normal distribution with a mean of 0 and a standard deviation of 1, we could make use of Table D and find the probability with little effort. Fortunately, it is possible for any normal distribution to be transformed easily to the standard normal. What we do is transform all values of X to corresponding values of z. This means that the mean of X must become 0, the mean of z. In Figure 4.7.2 both distributions are shown. We must determine what value of z, say, z_0, corresponds to an x of 3.0. This is done using formula 4.6.2, $z = (x - \mu)/\sigma$, which transforms any value of x in any normal distribution to the corresponding value of z in the standard normal distribution. For the present example we have

$$z = \frac{3.0 - 5.4}{1.3} = -1.85$$

The value of z_0 we seek, then, is -1.85. ■

Let us examine these relationships more closely. It is seen that the distance from the mean, 5.4, to the x-value of interest, 3.0, is $3.0 - 5.4 = -2.4$, which is a distance of 1.85 standard deviations. When we transform x values to z values, the distance of the z value of interest from its mean, 0, is equal to the distance of the corresponding x value from its mean, 5.4, in standard deviation units. We have seen that this latter distance is 1.85 standard deviations. In the z distribution a standard deviation is equal to 1, and consequently the point on the z scale located a distance of 1.85 standard deviations below 0 is $z = -1.85$, the result obtained by employing the formula. By consulting Table D, we find that the area to the left of $z = -1.85$ is .0322. We may summarize this discussion as follows:

$$P(x < 3.0) = P\left(z < \frac{3.0 - 5.4}{1.3}\right) = P(z < -1.85) = .0322$$

To answer the original question, we say that the probability is .0322 that a randomly selected child will have uptime of less than 3.0 hours.

EXAMPLE 4.7.2

Diskin et al. (A-11) studied common breath metabolites such as ammonia, acetone, isoprene, ethanol, and acetaldehyde in five subjects over a period of 30 days. Each day, breath samples were taken and analyzed in the early morning on arrival at the laboratory. For subject A, a 27-year-old female, the ammonia concentration in parts per billion (ppb) followed a normal distribution over 30 days with mean 491 and standard deviation 119. What is the probability that on a random day, the subject's ammonia concentration is between 292 and 649 ppb?

Solution: In Figure 4.7.3 are shown the distribution of ammonia concentrations and the z distribution to which we transform the original values to determine the desired probabilities. We find the z value corresponding to an x of 292 by

$$z = \frac{292 - 491}{119} = -1.67$$

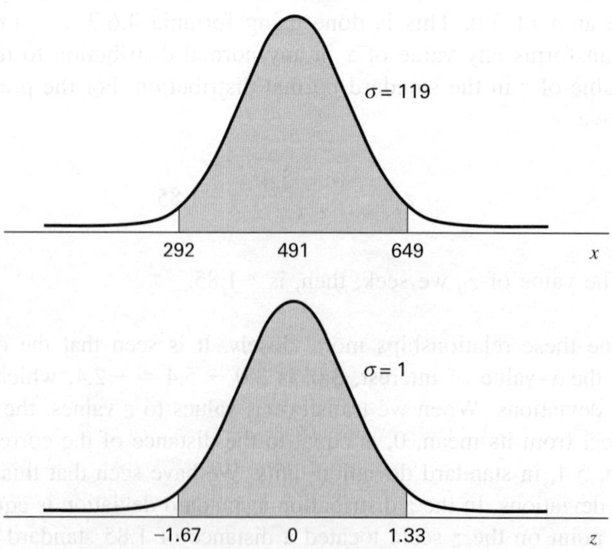

FIGURE 4.7.3 Distribution of ammonia concentration (x) and the corresponding standard normal distribution (z).

Similarly, for $x = 649$ we have

$$z = \frac{649 - 491}{119} = 1.33$$

From Table D we find the area between $-\infty$ and -1.67 to be .0475 and the area between $-\infty$ and 1.33 to be .9082. The area desired is the difference between these, $.9082 - .0475 = .8607$. To summarize,

$$\begin{aligned} P(292 \leq x \leq 649) &= P\left(\frac{292 - 491}{119} \leq z \leq \frac{649 - 491}{119}\right) \\ &= P(-1.67 \leq z \leq 1.33) \\ &= P(-\infty \leq z \leq 1.33) - P(-\infty \leq z \leq -1.67) \\ &= .9082 - .0475 \\ &= .8607 \end{aligned}$$

The probability asked for in our original question, then, is .8607. ■

EXAMPLE 4.7.3

In a population of 10,000 of the children described in Example 4.7.1, how many would you expect to be upright more than 8.5 hours?

Solution: We first find the probability that one child selected at random from the population would be upright more than 8.5 hours. That is,

$$P(x \geq 8.5) = P\left(z \geq \frac{8.5 - 5.4}{1.3}\right) = P(z \geq 2.38) = 1 - .9913 = .0087$$

Out of 10,000 people we would expect $10{,}000(.0087) = 87$ to spend more than 8.5 hours upright. ■

We may use MINITAB to calculate cumulative standard normal probabilities. Suppose we wish to find the cumulative probabilities for the following values of z: $-3, -2, -1, 0, 1, 2$, and 3. We enter the values of z into Column 1 and proceed as shown in Figure 4.7.4.

The preceding two sections focused extensively on the normal distribution, the most important and most frequently used continuous probability distribution. Though much of what will be covered in the next several chapters uses this distribution, it is not the only important continuous probability distribution. We will be introducing several other continuous distributions later in the text, namely the *t-distribution,* the *chi-square distribution,* and the *F-distribution.* The details of these distributions will be discussed in the chapters in which we need them for inferential tests.

Data:

```
C1: -3 -2 -1 0 1 2 3
```

Dialog box:

Calc ➤ Probability Distributions ➤ Normal

Choose **Cumulative probability.** Choose **Input column** and type *C1.* Click **OK.**

Session command:

```
MTB > CDF C1;
SUBC>   Normal 0 1.
```

Output:

Cumulative Distribution Function

```
Normal with mean = 0 and standard
deviation = 1.00000

        x          P( X <= x)
  -3.0000          0.0013
  -2.0000          0.0228
  -1.0000          0.1587
   0.0000          0.5000
   1.0000          0.8413
   2.0000          0.9772
   3.0000          0.9987
```

FIGURE 4.7.4 MINITAB calculation of cumulative standard normal probabilities.

EXERCISES

4.7.1 For another subject (a 29-year-old male) in the study by Diskin et al. (A-11), acetone levels were normally distributed with a mean of 870 and a standard deviation of 211 ppb. Find the probability that on a given day the subject's acetone level is:

(a) Between 600 and 1000 ppb
(b) Over 900 ppb
(c) Under 500 ppb
(d) Between 900 and 1100 ppb

4.7.2 In the study of fingerprints, an important quantitative characteristic is the total ridge count for the 10 fingers of an individual. Suppose that the total ridge counts of individuals in a certain population are approximately normally distributed with a mean of 140 and a standard deviation of 50. Find the probability that an individual picked at random from this population will have a ridge count of:

(a) 200 or more
(b) Less than 100

(c) Between 100 and 200

(d) Between 200 and 250

(e) In a population of 10,000 people how many would you expect to have a ridge count of 200 or more?

4.7.3 One of the variables collected in the North Carolina Birth Registry data (A-3) is pounds gained during pregnancy. According to data from the entire registry for 2001, the number of pounds gained during pregnancy was approximately normally distributed with a mean of 30.23 pounds and a standard deviation of 13.84 pounds. Calculate the probability that a randomly selected mother in North Carolina in 2001 gained:

(a) Less than 15 pounds during pregnancy **(b)** More than 40 pounds
(c) Between 14 and 40 pounds **(d)** Less than 10 pounds
(e) Between 10 and 20 pounds

4.7.4 Suppose the average length of stay in a chronic disease hospital of a certain type of patient is 60 days with a standard deviation of 15. If it is reasonable to assume an approximately normal distribution of lengths of stay, find the probability that a randomly selected patient from this group will have a length of stay:

(a) Greater than 50 days **(b)** Less than 30 days
(c) Between 30 and 60 days **(d)** Greater than 90 days

4.7.5 If the total cholesterol values for a certain population are approximately normally distributed with a mean of 200 mg/100 ml and a standard deviation of 20 mg/100 ml, find the probability that an individual picked at random from this population will have a cholesterol value:

(a) Between 180 and 200 mg/100 ml **(b)** Greater than 225 mg/100 ml
(c) Less than 150 mg/100 ml **(d)** Between 190 and 210 mg/100 ml

4.7.6 Given a normally distributed population with a mean of 75 and a variance of 625, find:

(a) $P(50 \leq x \leq 100)$ **(b)** $P(x > 90)$
(c) $P(x < 60)$ **(d)** $P(x \geq 85)$
(e) $P(30 \leq x \leq 110)$

4.7.7 The weights of a certain population of young adult females are approximately normally distributed with a mean of 132 pounds and a standard deviation of 15. Find the probability that a subject selected at random from this population will weigh:

(a) More than 155 pounds **(b)** 100 pounds or less
(c) Between 105 and 145 pounds

4.8 SUMMARY

In this chapter the concepts of probability described in the preceding chapter are further developed. The concepts of discrete and continuous random variables and their probability distributions are discussed. In particular, two discrete probability distributions, the binomial and the Poisson, and one continuous probability distribution, the normal, are examined in considerable detail. We have seen how these theoretical distributions allow us to make probability statements about certain random variables that are of interest to the health professional.

SUMMARY OF FORMULAS FOR CHAPTER 4

Formula Number	Name	Formula
4.2.1	Mean of a frequency distribution	$\mu = \sum xp(x)$
4.2.2	Variance of a frequency distribution	$\sigma^2 = \sum (x - \mu)^2 p(x)$ or $\sigma^2 = \sum x^2 p(x) - \mu^2$
4.3.1	Combination of objects	${}_nC_x = \dfrac{n!}{x!(n-1)!}$
4.3.2	Binomial distribution function	$f(x) = {}_nC_x p^x q^{n-x}, x = 0, 1, 2, \ldots$
4.3.3–4.3.5	Tabled binomial probability equalities	$P(X = x \mid n, p > 50) = P(X = n - x \mid n, 1 - p)$ $P(X \leq x \mid n, p > 50) = P(X \geq n - x \mid n, 1 - p)$ $P(X \geq x \mid n, p > 50) = P(X \leq n - x \mid n, 1 - p)$
4.4.1	Poisson distribution function	$f(x) = \dfrac{e^{-\lambda}\lambda^x}{x!}, x = 0, 1, 2, \ldots$
4.6.1	Normal distribution function	$f(x) = \dfrac{1}{\sqrt{2\pi}\sigma} e^{-(x-\mu)^2/2\sigma^2}, \quad -\infty < x < \infty$ $-\infty < \mu < \infty$ $\sigma > 0$
4.6.2	*z*-transformation	$Z = \dfrac{X - \mu}{\sigma}$
4.6.3	Standard normal distribution function	$f(z) = \dfrac{1}{\sqrt{2\pi}} e^{-z^2/2}, \quad -\infty < z < \infty$
Symbol Key		• ${}_nC_x$ = a *combination* of n events taken x at a time • e = Euler's constant = 2.71828… • $f(x)$ = function of x • λ = the parameter of the Poisson distribution • n = sample size or the total number of time a process occurs • p = binomial "success" probability • $p(x)$ = discrete probability of random variable X • $q = 1 - p$ = binomial "failure" probability • π = pi = constant = 3.14159… • σ = population standard deviation • σ^2 = population variance • μ = population mean • x^2 = a quantity of individual value of X • X = random variable • z = standard normal transformation

REVIEW QUESTIONS AND EXERCISES

1. What is a discrete random variable? Give three examples that are of interest to the health professional.

2. What is a continuous random variable? Give three examples of interest to the health professional.

3. Define the probability distribution of a discrete random variable.

4. Define the probability distribution of a continuous random variable.

5. What is a cumulative probability distribution?

6. What is a Bernoulli trial?

7. Describe the binomial distribution.

8. Give an example of a random variable that you think follows a binomial distribution.

9. Describe the Poisson distribution.

10. Give an example of a random variable that you think is distributed according to the Poisson law.

11. Describe the normal distribution.

12. Describe the standard normal distribution and tell how it is used in statistics.

13. Give an example of a random variable that you think is, at least approximately, normally distributed.

14. Using the data of your answer to Question 13, demonstrate the use of the standard normal distribution in answering probability questions related to the variable selected.

15. Kanjanarat et al. (A-12) estimate the rate of preventable adverse drug events (ADEs) in hospitals to be 35.2 percent. Preventable ADEs typically result from inappropriate care or medication errors, which include errors of commission and errors of omission. Suppose that 10 hospital patients experiencing an ADE are chosen at random. Let $p = .35$, and calculate the probability that:

    (a) Exactly seven of those drug events were preventable
    (b) More than half of those drug events were preventable
    (c) None of those drug events were preventable
    (d) Between three and six inclusive were preventable

16. In a poll conducted by the Pew Research Center in 2003 (A-13), a national sample of adults answered the following question, "All in all, do you strongly favor, favor, oppose, or strongly oppose . . . making it legal for doctors to give terminally ill patients the means to end their lives?" The results showed that 43 percent of the sample subjects answered "strongly favor" or "favor" to this question. If 12 subjects represented by this sample are chosen at random, calculate the probability that:

    (a) Exactly two of the respondents answer "strongly favor" or "favor"
    (b) No more than two of the respondents answer "strongly favor" or "favor"
    (c) Between five and nine inclusive answer "strongly favor" or "favor"

17. In a study by Thomas et al. (A-14) the Poisson distribution was used to model the number of patients per month referred to an oncologist. The researchers use a rate of 15.8 patients per month that are referred to the oncologist. Use Table C in the Appendix and a rate of 16 patients per month to calculate the probability that in a month:

    (a) Exactly 10 patients are referred to an oncologist
    (b) Between five and 15 inclusive are referred to an oncologist

(c) More than 10 are referred to an oncologist

(d) Less than eight are referred to an oncologist

(e) Less than 12, but more than eight are referred to an oncologist

18. On the average, two students per hour report for treatment to the first-aid room of a large elementary school.

(a) What is the probability that during a given hour three students come to the first-aid room for treatment?

(b) What is the probability that during a given hour two or fewer students will report to the first-aid room?

(c) What is the probability that during a given hour between three and five students, inclusive, will report to the first-aid room?

19. A Harris Interactive poll conducted in Fall, 2002 (A-15) via a national telephone survey of adults asked, "Do you think adults should be allowed to legally use marijuana for medical purposes if their doctor prescribes it, or do you think that marijuana should remain illegal even for medical purposes." The results showed that 80 percent of respondents answered "Yes" to the above question. Assuming 80 percent of Americans would say "Yes" to the above question, find the probability when eight Americans are chosen at random that:

(a) Six or fewer said "Yes"

(b) Seven or more said "Yes"

(c) All eight said "Yes"

(d) Fewer than four said "Yes"

(e) Between four and seven inclusive said "Yes"

20. In a study of the relationship between measles vaccination and Guillain-Barré syndrome (GBS), Silveira et al. (A-16) used a Poisson model in the examination of the occurrence of GBS during latent periods after vaccinations. They conducted their study in Argentina, Brazil, Chile, and Colombia. They found that during the latent period, the rate of GBS was 1.28 cases per day. Using this estimate rounded to 1.3, find the probability on a given day of:

(a) No cases of GBS

(b) At least one case of GBS

(c) Fewer than five cases of GBS

21. The IQs of individuals admitted to a state school for the mentally retarded are approximately normally distributed with a mean of 60 and a standard deviation of 10.

(a) Find the proportion of individuals with IQs greater than 75.

(b) What is the probability that an individual picked at random will have an IQ between 55 and 75?

(c) Find $P(50 \leq X \leq 70)$.

22. A nurse supervisor has found that staff nurses, on the average, complete a certain task in 10 minutes. If the times required to complete the task are approximately normally distributed with a standard deviation of 3 minutes, find:

(a) The proportion of nurses completing the task in less than 4 minutes

(b) The proportion of nurses requiring more than 5 minutes to complete the task

(c) The probability that a nurse who has just been assigned the task will complete it within 3 minutes

23. Scores made on a certain aptitude test by nursing students are approximately normally distributed with a mean of 500 and a variance of 10,000.

(a) What proportion of those taking the test score below 200?

(b) A person is about to take the test. What is the probability that he or she will make a score of 650 or more?

(c) What proportion of scores fall between 350 and 675?

24. Given a binomial variable with a mean of 20 and a variance of 16, find n and p.

25. Suppose a variable X is normally distributed with a standard deviation of 10. Given that .0985 of the values of X are greater than 70, what is the mean value of X?

26. Given the normally distributed random variable X, find the numerical value of k such that $P(\mu - k\sigma \leq X \leq \mu + k\sigma) = .754$.

27. Given the normally distributed random variable X with mean 100 and standard deviation 15, find the numerical value of k such that:

(a) $P(X \leq k) = .0094$
(b) $P(X \geq k) = .1093$
(c) $P(100 \leq X \leq k) = .4778$
(d) $P(k' \leq X \leq k) = .9660$, where k' and k are equidistant from μ

28. Given the normally distributed random variable X with $\sigma = 10$ and $P(X \leq 40) = .0080$, find μ.

29. Given the normally distributed random variable X with $\sigma = 15$ and $P(X \leq 50) = .9904$, find μ.

30. Given the normally distributed random variable X with $\sigma = 5$ and $P(X \geq 25) = .0526$, find μ.

31. Given the normally distributed random variable X with $\mu = 25$ and $P(X \leq 10) = .0778$, find σ.

32. Given the normally distributed random variable X with $\mu = 30$ and $P(X \leq 50) = .9772$, find σ.

33. Explain why each of the following measurements is or is not the result of a Bernoulli trial:

(a) The gender of a newborn child
(b) The classification of a hospital patient's condition as stable, critical, fair, good, or poor
(c) The weight in grams of a newborn child

34. Explain why each of the following measurements is or is not the result of a Bernoulli trial:

(a) The number of surgical procedures performed in a hospital in a week
(b) A hospital patient's temperature in degrees Celsius
(c) A hospital patient's vital signs recorded as normal or not normal

35. Explain why each of the following distributions is or is not a probability distribution:

(a)

x	$P(X = x)$
0	0.15
1	0.25
2	0.10
3	0.25
4	0.30

(b)

x	$P(X = x)$
0	0.15
1	0.20
2	0.30
3	0.10

(c)

x	$P(X = x)$
0	0.15
1	−0.20
2	0.30
3	0.20
4	0.15

(d)

x	$P(X = x)$
−1	0.15
0	0.30
1	0.20
2	0.15
3	0.10
4	0.10

REFERENCES

Methodology References

1. FRANK A. HAIGHT, *Handbook of the Poisson Distribution,* Wiley, New York, 1967.

Applications References

A-1. DAVID H. HOLBEN, MEGAN C. MCCLINCY, and JOHN P. HOLCOMB, "Food Security Status of Households in Appalachian Ohio with Children in Head Start," *Journal of American Dietetic Association, 104* (2004), 238–241.

A-2. CHAD L. CROSS, BO BERNHARD, ANNE ROTHWEILER, MELANIE MULLIN, and ANNABELLE JANAIRO, "Research and Evaluation of Problem Gambling: Nevada Annual Report, April 2006–April 2007," Final Grant Report to the Nevada Division of Health and Human Services.

A-3. North Carolina State Center for Health Statistics and Howard W. Odum Institute for Research in Social Science at the University of North Carolina at Chapel Hill. Birth Data set for 2001 found at www.irss.unc.edu/ncvital/bfd1down.html. All calculations were performed by John Holcomb and do not represent the findings of the Center or Institute.

A-4. The University of Massachusetts Poll, Conducted May 29–June 2, 2003. Data provided by Massachusetts Health Benchmarks. www.healthbenchmarks.org/Poll/June2003Survey.cfm.

A-5. National Center for Health Statistics (2001) Data File Documentation, National Health Interview Survey, CD Series 10, No. 16A, National Center for Health Statistics, Hyattsville, Maryland. All calculations are the responsibility of John Holcomb and do not reflect efforts by NCHS.

A-6. STEVEN S. COUGHLIN, ROBERT J. UHLER, THOMAS RICHARDS, and KATHERINE M. WILSON, "Breast and Cervical Cancer Screening Practices Among Hispanic and Non-Hispanic Women Residing Near the United States–Mexico Border, 1999–2000," *Family Community Health, 26* (2003), 130–139.

A-7. J. H. LAAKE and J. A. RØTTINGEN, "Rocuronium and Anaphylaxis—a Statistical Challenge," *Acta Anasthesiologica Scandinavica, 45* (2001), 1196–1203.

A-8. ARUN D. SINGH, MAHNAZ NOURI, CAROL L. SHIELDS, JERRY A. SHIELDS, and ANDREW F. SMITH, "Retinal Capillary Hemangioma," *Ophthalmology, 10* (2001), 1907–1911.

A-9. PASCALE TUBERT-BITTER, BERNARD BEGAUD, YOLA MORIDE, ANICET CHASLERIE, and FRANCOISE HARAMBURU, "Comparing the Toxicity of Two Drugs in the Framework of Spontaneous Reporting: A Confidence Interval Approach," *Journal of Clinical Epidemiology, 49* (1996), 121–123.

A-10. B. ELDRIDGE, M. GALEA, A. MCCOY, R. WOLFE, H., and K. GRAHAM, "Uptime Normative Values in Children Aged 8 to 15 Years," *Developmental Medicine and Child Neurology, 45* (2003), 189–193.

A-11. ANN M. DISKIN, PATRIK ŠPANEL, and DAVID SMITH, "Time Variation of Ammonia, Acetone, Isoprene, and Ethanol in Breath: A Quantitative SIFT-MS study over 30 days," *Physiological Measurement, 24* (2003), 107–119.

A-12. PENKARN KANJANARAT, ALMUT G. WINTERSTEIN, THOMAS E. JOHNS, RANDY C. HATTON, RICARDO GONZALEZ-ROTHI, and RICHARD SEGAL, "Nature of Preventable Adverse Drug Events in Hospitals: A Literature Review," *American Journal of Health-System Pharmacy, 60* (2003), 1750–1759.

A-13. Pew Research Center survey conducted by Princeton Survey Research Associates, June 24–July 8, 2003. Data provided by the Roper Center for Public Opinion Research. www.kaisernetwork.org/health_poll/hpoll_index.cfm.

A-14. S. J. THOMAS, M. V. WILLIAMS, N. G. BURNET, and C. R. BAKER, "How Much Surplus Capacity Is Required to Maintain Low Waiting Times?," *Clinical Oncology, 13* (2001), 24–28.

A-15. Time, Cable News Network survey conducted by Harris Associates, October 22–23, 2002. Data provided by the Roper Center for Public Opinion Research. www.kaisernetwork.org/health_poll/hpoll_index.cfm.

A-16. CLAUDIO M. DA SILVEIRA, DAVID M. SALISBURY, and CIRO A DE QUADROS, "Measles Vaccination and Guillain-Barré Syndrome," *The Lancet*, 349 (1997), 14–16.

CHAPTER 5

PROBABILISTIC FEATURES OF THE DISTRIBUTIONS OF CERTAIN SAMPLE STATISTICS

CHAPTER OVERVIEW

This chapter ties together the foundations of applied statistics: descriptive measures, basic probability, and inferential procedures. This chapter also includes a discussion of one of the most important theorems in statistics, the central limit theorem. Students may find it helpful to revisit this chapter from time to time as they study the remaining chapters of the book.

TOPICS

LEARNING OUTCOMES

After studying this chapter, the student will

1. be able to construct a sampling distribution of a statistic.
2. understand how to use a sampling distribution to calculate basic probabilities.
3. understand the central limit theorem and when to apply it.
4. understand the basic concepts of sampling with replacement and without replacement.

5.1 INTRODUCTION

Before we examine the subject matter of this chapter, let us review the high points of what we have covered thus far. Chapter 1 introduces some basic and useful statistical vocabulary and discusses the basic concepts of data collection. In Chapter 2, the organization and summarization of data are emphasized. It is here that we encounter the concepts of central tendency and dispersion and learn how to compute their descriptive measures. In Chapter 3, we are introduced to the fundamental ideas of probability, and in Chapter 4 we consider the concept of a probability distribution. These concepts are fundamental to an understanding of statistical inference, the topic that comprises the major portion of this book.

This chapter serves as a bridge between the preceding material, which is essentially descriptive in nature, and most of the remaining topics, which have been selected from the area of statistical inference.

5.2 SAMPLING DISTRIBUTIONS

The topic of this chapter is *sampling distributions.* The importance of a clear understanding of sampling distributions cannot be overemphasized, as this concept is the very key to the understanding of statistical inference. Sampling distributions serve two purposes: (1) they allow us to answer probability questions about sample statistics, and (2) they provide the necessary theory for making statistical inference procedures valid. In this chapter we use sampling distributions to answer probability questions about sample statistics. We recall from Chapter 2 that a sample statistic is a descriptive measure, such as the mean, median, variance, or standard deviation, that is computed from the data of a sample. In the chapters that follow, we will see how sampling distributions make statistical inferences valid.

We begin with the following definition.

DEFINITION

The distribution of all possible values that can be assumed by some statistic, computed from samples of the same size randomly drawn from the same population, is called the *sampling distribution* of that statistic.

Sampling Distributions: Construction Sampling distributions may be constructed empirically when sampling from a discrete, finite population. To construct a sampling distribution we proceed as follows:

1. From a finite population of size N, randomly draw all possible samples of size n.
2. Compute the statistic of interest for each sample.
3. List in one column the different distinct observed values of the statistic, and in another column list the corresponding frequency of occurrence of each distinct observed value of the statistic.

The actual construction of a sampling distribution is a formidable undertaking if the population is of any appreciable size and is an impossible task if the population is infinite. In such cases, sampling distributions may be approximated by taking a large number of samples of a given size.

Sampling Distributions: Important Characteristics We usually are interested in knowing three things about a given sampling distribution: its *mean,* its *variance,* and its *functional form* (how it looks when graphed).

We can recognize the difficulty of constructing a sampling distribution according to the steps given above when the population is large. We also run into a problem when considering the construction of a sampling distribution when the population is infinite. The best we can do experimentally in this case is to approximate the sampling distribution of a statistic.

Both these problems may be obviated by means of mathematics. Although the procedures involved are not compatible with the mathematical level of this text, sampling distributions can be derived mathematically. The interested reader can consult one of many mathematical statistics textbooks, for example, Larsen and Marx (1) or Rice (2).

In the sections that follow, some of the more frequently encountered sampling distributions are discussed.

5.3 DISTRIBUTION OF THE SAMPLE MEAN

An important sampling distribution is the distribution of the sample mean. Let us see how we might construct the sampling distribution by following the steps outlined in the previous section.

EXAMPLE 5.3.1

Suppose we have a population of size $N = 5$, consisting of the ages of five children who are outpatients in a community mental health center. The ages are as follows: $x_1 = 6, x_2 = 8, x_3 = 10, x_4 = 12$, and $x_5 = 14$. The mean, μ, of this population is equal to $\sum x_i/N = 10$ and the variance is

$$\sigma^2 = \frac{\sum(x_i - \mu)^2}{N} = \frac{40}{5} = 8$$

Let us compute another measure of dispersion and designate it by capital S as follows:

$$S^2 = \frac{\sum(x_i - \mu)^2}{N - 1} = \frac{40}{4} = 10$$

We will refer to this quantity again in the next chapter. We wish to construct the sampling distribution of the sample mean, $\bar{x}$, based on samples of size $n = 2$ drawn from this population.

Solution: Let us draw all possible samples of size $n = 2$ from this population. These samples, along with their means, are shown in Table 5.3.1.

TABLE 5.3.1 All Possible Samples of Size $n = 2$ from a Population of Size $N = 5$. Samples Above or Below the Principal Diagonal Result When Sampling Is Without Replacement. Sample Means Are in Parentheses

		Second Draw				
		6	8	10	12	14
First Draw	6	6, 6 (6)	6, 8 (7)	6, 10 (8)	6, 12 (9)	6, 14 (10)
	8	8, 6 (7)	8, 8 (8)	8, 10 (9)	8, 12 (10)	8, 14 (11)
	10	10, 6 (8)	10, 8 (9)	10, 10 (10)	10, 12 (11)	10, 14 (12)
	12	12, 6 (9)	12, 8 (10)	12, 10 (11)	12, 12 (12)	12, 14 (13)
	14	14, 6 (10)	14, 8 (11)	14, 10 (12)	14, 12 (13)	14, 14 (14)

TABLE 5.3.2 Sampling Distribution of $\bar{x}$ Computed from Samples in Table 5.3.1

$\bar{x}$	Frequency	Relative Frequency
6	1	1/25
7	2	2/25
8	3	3/25
9	4	4/25
10	5	5/25
11	4	4/25
12	3	3/25
13	2	2/25
14	1	1/25
Total	25	25/25

We see in this example that, when sampling is with replacement, there are 25 possible samples. In general, when sampling is with replacement, the number of possible samples is equal to N^n.

We may construct the sampling distribution of $\bar{x}$ by listing the different values of $\bar{x}$ in one column and their frequency of occurrence in another, as in Table 5.3.2. ■

We see that the data of Table 5.3.2 satisfy the requirements for a probability distribution. The individual probabilities are all greater than 0, and their sum is equal to 1.

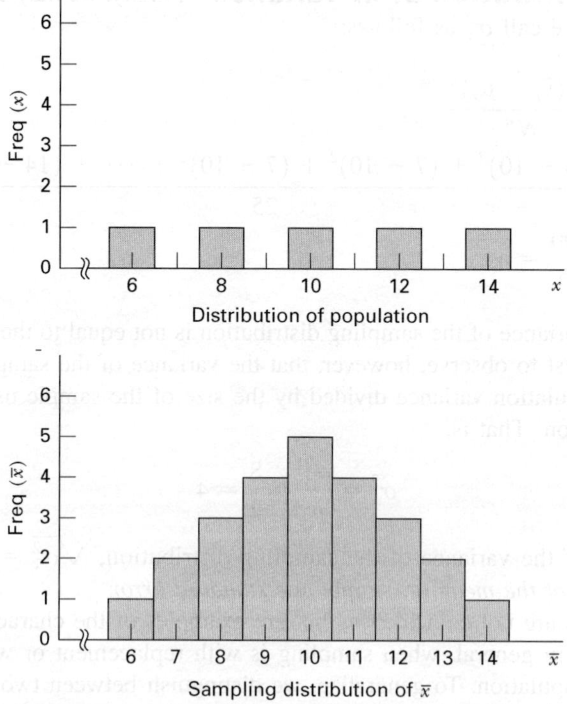

FIGURE 5.3.1 Distribution of population and sampling distribution of $\bar{x}$.

It was stated earlier that we are usually interested in the functional form of a sampling distribution, its mean, and its variance. We now consider these characteristics for the sampling distribution of the sample mean, $\bar{x}$.

Sampling Distribution of $\bar{x}$: Functional Form Let us look at the distribution of $\bar{x}$ plotted as a histogram, along with the distribution of the population, both of which are shown in Figure 5.3.1. We note the radical difference in appearance between the histogram of the population and the histogram of the sampling distribution of $\bar{x}$. Whereas the former is uniformly distributed, the latter gradually rises to a peak and then drops off with perfect symmetry.

Sampling Distribution of $\bar{x}$: Mean Now let us compute the mean, which we will call $\mu_{\bar{x}}$, of our sampling distribution. To do this we add the 25 sample means and divide by 25. Thus,

$$\mu_{\bar{x}} = \frac{\Sigma \bar{x}_i}{N^n} = \frac{6 + 7 + 7 + 8 + \cdots + 14}{25} = \frac{250}{25} = 10$$

We note with interest that the mean of the sampling distribution of $\bar{x}$ has the same value as the mean of the original population.

Sampling Distribution of $\bar{x}$: Variance Finally, we may compute the variance of $\bar{x}$, which we call $\sigma_{\bar{x}}^2$ as follows:

$$\begin{aligned}\sigma_{\bar{x}}^2 &= \frac{\Sigma(\bar{x}_i - \mu_{\bar{x}})^2}{N^n} \\ &= \frac{(6-10)^2 + (7-10)^2 + (7-10)^2 + \cdots + (14-10)^2}{25} \\ &= \frac{100}{25} = 4\end{aligned}$$

We note that the variance of the sampling distribution is not equal to the population variance. It is of interest to observe, however, that the variance of the sampling distribution is equal to the population variance divided by the size of the sample used to obtain the sampling distribution. That is,

$$\sigma_{\bar{x}}^2 = \frac{\sigma^2}{n} = \frac{8}{2} = 4$$

The square root of the variance of the sampling distribution, $\sqrt{\sigma_{\bar{x}}^2} = \sigma/\sqrt{n}$ is called the *standard error of the mean* or, simply, the *standard error.*

These results are not coincidences but are examples of the characteristics of sampling distributions in general, when sampling is with replacement or when sampling is from an infinite population. To generalize, we distinguish between two situations: sampling from a normally distributed population and sampling from a nonnormally distributed population.

Sampling Distribution of $\bar{x}$: Sampling from Normally Distributed Populations When sampling is from a normally distributed population, the distribution of the sample mean will possess the following properties:

1. The distribution of $\bar{x}$ will be normal.
2. The mean, $\mu_{\bar{x}}$, of the distribution of $\bar{x}$ will be equal to the mean of the population from which the samples were drawn.
3. The variance, $\sigma_{\bar{x}}^2$ of the distribution of $\bar{x}$ will be equal to the variance of the population divided by the sample size.

Sampling from Nonnormally Distributed Populations For the case where sampling is from a nonnormally distributed population, we refer to an important mathematical theorem known as the *central limit theorem.* The importance of this theorem in statistical inference may be summarized in the following statement.

The Central Limit Theorem

Given a population of any nonnormal functional form with a mean μ and finite variance σ^2, the sampling distribution of $\bar{x}$, computed from samples of size n from this population, will have mean μ and variance σ^2/n and will be approximately normally distributed when the sample size is large.

A mathematical formulation of the central limit theorem is that the distribution of

$$\frac{\bar{x} - \mu}{\sigma/\sqrt{n}}$$

approaches a normal distribution with mean 0 and variance 1 as $n \rightarrow \infty$. Note that the central limit theorem allows us to sample from nonnormally distributed populations with a guarantee of approximately the same results as would be obtained if the populations were normally distributed provided that we take a large sample.

The importance of this will become evident later when we learn that a normally distributed sampling distribution is a powerful tool in statistical inference. In the case of the sample mean, we are assured of at least an approximately normally distributed sampling distribution under three conditions: (1) when sampling is from a normally distributed population; (2) when sampling is from a nonnormally distributed population and our sample is large; and (3) when sampling is from a population whose functional form is unknown to us as long as our sample size is large.

The logical question that arises at this point is, How large does the sample have to be in order for the central limit theorem to apply? There is no one answer, since the size of the sample needed depends on the extent of nonnormality present in the population. One rule of thumb states that, in most practical situations, a sample of size 30 is satisfactory. In general, the approximation to normality of the sampling distribution of $\bar{x}$ becomes better and better as the sample size increases.

Sampling Without Replacement The foregoing results have been given on the assumption that sampling is either with replacement or that the samples are drawn from infinite populations. In general, we do not sample with replacement, and in most practical situations it is necessary to sample from a finite population; hence, we need to become familiar with the behavior of the sampling distribution of the sample mean under these conditions. Before making any general statements, let us again look at the data in Table 5.3.1. The sample means that result when sampling is without replacement are those above the principal diagonal, which are the same as those below the principal diagonal, if we ignore the order in which the observations were drawn. We see that there are 10 possible samples. In general, when drawing samples of size n from a finite population of size N without replacement, and ignoring the order in which the sample values are drawn, the number of possible samples is given by the combination of N things taken n at a time. In our present example we have

$${}_NC_n = \frac{N!}{n!(N-n)!} = \frac{5!}{2!3!} = \frac{5 \cdot 4 \cdot 3!}{2!3!} = 10 \text{ possible samples.}$$

The mean of the 10 sample means is

$$\mu_{\bar{x}} = \frac{\sum \bar{x}_i}{{}_NC_n} = \frac{7 + 8 + 9 + \cdots + 13}{10} = \frac{100}{10} = 10$$

We see that once again the mean of the sampling distribution is equal to the population mean.

The variance of this sampling distribution is found to be

$$\sigma_{\bar{x}}^2 = \frac{\Sigma(\bar{x}_i - \mu_{\bar{x}})^2}{{}_N C_n} = \frac{30}{10} = 3$$

and we note that this time the variance of the sampling distribution is not equal to the population variance divided by the sample size, since $\sigma_{\bar{x}}^2 = 3 \neq 8/2 = 4$. There is, however, an interesting relationship that we discover by multiplying σ^2/n by $(N - n)/(N - 1)$. That is,

$$\frac{\sigma^2}{n} \cdot \frac{N - n}{N - 1} = \frac{8}{2} \cdot \frac{5 - 2}{4} = 3$$

This result tells us that if we multiply the variance of the sampling distribution that would be obtained if sampling were with replacement, by the factor $(N - n)/(N - 1)$, we obtain the value of the variance of the sampling distribution that results when sampling is without replacement. We may generalize these results with the following statement.

When sampling is without replacement from a finite population, the sampling distribution of $\bar{x}$ will have mean μ and variance

$$\sigma_{\bar{x}}^2 = \frac{\sigma^2}{n} \cdot \frac{N - n}{N - 1}$$

If the sample size is large, the central limit theorem applies and the sampling distribution of $\bar{x}$ will be approximately normally distributed.

The Finite Population Correction The factor $(N - n)/(N - 1)$ is called the finite population correction and can be ignored when the sample size is small in comparison with the population size. When the population is much larger than the sample, the difference between σ^2/n and $(\sigma^2/n)[(N - n)/(N - 1)]$ will be negligible. Imagine a population of size 10,000 and a sample from this population of size 25; the finite population correction would be equal to $(10{,}000 - 25)/(9999) = .9976$. To multiply σ^2/n by .9976 is almost equivalent to multiplying it by 1. Most practicing statisticians do not use the finite population correction unless the sample is more than 5 percent of the size of the population. That is, the finite population correction is usually ignored when $n/N \leq .05$.

The Sampling Distribution of $\bar{x}$: A Summary Let us summarize the characteristics of the sampling distribution of $\bar{x}$ under two conditions.

1. Sampling is from a normally distributed population with a known population variance:
   (a) $\mu_{\bar{x}} = \mu$
   (b) $\sigma_{\bar{x}} = \sigma/\sqrt{n}$
   (c) The sampling distribution of $\bar{x}$ is normal.

2. Sampling is from a nonnormally distributed population with a known population variance:

(a) $\mu_{\bar{x}} = \mu$

(b) $\sigma_{\bar{x}} = \sigma/\sqrt{n}, \quad$ when $\quad n/N \leq .05$

$$\sigma_{\bar{x}} = (\sigma/\sqrt{n})\sqrt{\frac{N-n}{N-1}}, \quad \text{otherwise}$$

(c) The sampling distribution of $\bar{x}$ is approximately normal.

Applications As we will see in succeeding chapters, knowledge and understanding of sampling distributions will be necessary for understanding the concepts of statistical inference. The simplest application of our knowledge of the sampling distribution of the sample mean is in computing the probability of obtaining a sample with a mean of some specified magnitude. Let us illustrate with some examples.

EXAMPLE 5.3.2

Suppose it is known that in a certain large human population cranial length is approximately normally distributed with a mean of 185.6 mm and a standard deviation of 12.7 mm. What is the probability that a random sample of size 10 from this population will have a mean greater than 190?

Solution: We know that the single sample under consideration is one of all possible samples of size 10 that can be drawn from the population, so that the mean that it yields is one of the $\bar{x}$'s constituting the sampling distribution of $\bar{x}$ that, theoretically, could be derived from this population.

When we say that the population is approximately normally distributed, we assume that the sampling distribution of $\bar{x}$ will be, for all practical purposes, normally distributed. We also know that the mean and standard deviation of the sampling distribution are equal to 185.6 and $\sqrt{(12.7)^2/10} = 12.7/\sqrt{10} = 4.0161$, respectively. We assume that the population is large relative to the sample so that the finite population correction can be ignored.

We learn in Chapter 4 that whenever we have a random variable that is normally distributed, we may very easily transform it to the standard normal distribution. Our random variable now is $\bar{x}$, the mean of its distribution is $\mu_{\bar{x}}$, and its standard deviation is $\sigma_{\bar{x}} = \sigma/\sqrt{n}$. By appropriately modifying the formula given previously, we arrive at the following formula for transforming the normal distribution of $\bar{x}$ to the standard normal distribution:

$$z = \frac{\bar{x} - \mu_{\bar{x}}}{\sigma/\sqrt{n}} \tag{5.3.1}$$

■

The probability that answers our question is represented by the area to the right of $\bar{x} = 190$ under the curve of the sampling distribution. This area is equal to the area to the right of

$$z = \frac{190 - 185.6}{4.0161} = \frac{4.4}{4.0161} = 1.10$$

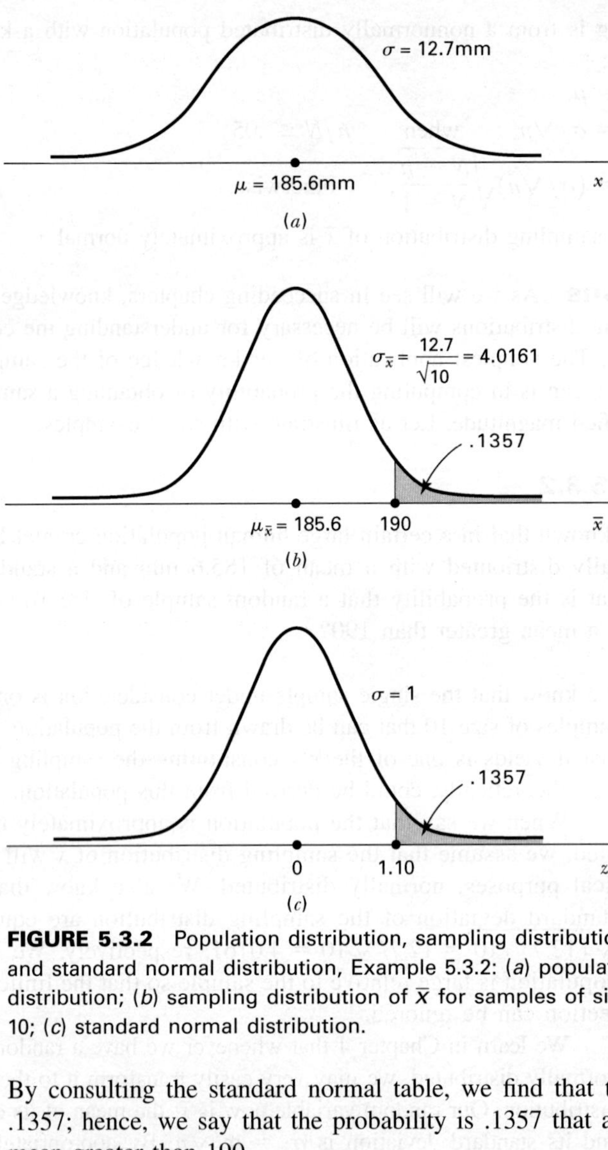

FIGURE 5.3.2 Population distribution, sampling distribution, and standard normal distribution, Example 5.3.2: (*a*) population distribution; (*b*) sampling distribution of $\bar{x}$ for samples of size 10; (*c*) standard normal distribution.

By consulting the standard normal table, we find that the area to the right of 1.10 is .1357; hence, we say that the probability is .1357 that a sample of size 10 will have a mean greater than 190.

Figure 5.3.2 shows the relationship between the original population, the sampling distribution of $\bar{x}$ and the standard normal distribution.

EXAMPLE 5.3.3

If the mean and standard deviation of serum iron values for healthy men are 120 and 15 micrograms per 100 ml, respectively, what is the probability that a random sample of 50 normal men will yield a mean between 115 and 125 micrograms per 100 ml?

Solution: The functional form of the population of serum iron values is not specified, but since we have a sample size greater than 30, we make use of the central limit theorem and transform the resulting approximately normal sampling distribution of $\bar{x}$ (which has a mean of 120 and a standard deviation of $15/\sqrt{50} = 2.1213$) to the standard normal. The probability we seek is

$$\begin{aligned} P(115 \le \bar{x} \le 125) &= P\left(\frac{115 - 120}{2.12} \le z \le \frac{125 - 120}{2.12}\right) \\ &= P(-2.36 \le z \le 2.36) \\ &= .9909 - .0091 \\ &= .9818 \end{aligned}$$

■

EXERCISES

5.3.1 The National Health and Nutrition Examination Survey of 1988–1994 (NHANES III, A-1) estimated the mean serum cholesterol level for U.S. females aged 20–74 years to be 204 mg/dl. The estimate of the standard deviation was approximately 44. Using these estimates as the mean μ and standard deviation σ for the U.S. population, consider the sampling distribution of the sample mean based on samples of size 50 drawn from women in this age group. What is the mean of the sampling distribution? The standard error?

5.3.2 The study cited in Exercise 5.3.1 reported an estimated mean serum cholesterol level of 183 for women aged 20–29 years. The estimated standard deviation was approximately 37. Use these estimates as the mean μ and standard deviation σ for the U.S. population. If a simple random sample of size 60 is drawn from this population, find the probability that the sample mean serum cholesterol level will be:

(a) Between 170 and 195 **(b)** Below 175
(c) Greater than 190

5.3.3 If the uric acid values in normal adult males are approximately normally distributed with a mean and standard deviation of 5.7 and 1 mg percent, respectively, find the probability that a sample of size 9 will yield a mean:

(a) Greater than 6 **(b)** Between 5 and 6
(c) Less than 5.2

5.3.4 Wright et al. (A-2) used the 1999–2000 National Health and Nutrition Examination Survey (NHANES) to estimate dietary intake of 10 key nutrients. One of those nutrients was calcium (mg). They found in all adults 60 years or older a mean daily calcium intake of 721 mg with a standard deviation of 454. Using these values for the mean and standard deviation for the U.S. population, find the probability that a random sample of size 50 will have a mean:

(a) Greater than 800 mg **(b)** Less than 700 mg
(c) Between 700 and 850 mg

5.3.5 In the study cited in Exercise 5.3.4, researchers found the mean sodium intake in men and women 60 years or older to be 2940 mg with a standard deviation of 1476 mg. Use these values for the mean and standard deviation of the U.S. population and find the probability that a random sample of 75 people from the population will have a mean:

(a) Less than 2450 mg **(b)** Over 3100 mg
(c) Between 2500 and 3300 mg **(d)** Between 2500 and 2900 mg

5.3.6 Given a normally distributed population with a mean of 100 and a standard deviation of 20, find the following probabilities based on a sample of size 16:
(a) $P(\bar{x} \geq 100)$ **(b)** $P(\bar{x} \leq 110)$
(c) $P(96 \leq \bar{x} \leq 108)$

5.3.7 Given $\mu = 50$, $\sigma = 16$, and $n = 64$, find:
(a) $P(45 \leq \bar{x} \leq 55)$ **(b)** $P(\bar{x} > 53)$
(c) $P(\bar{x} < 47)$ **(d)** $P(49 \leq \bar{x} \leq 56)$

5.3.8 Suppose a population consists of the following values: 1, 3, 5, 7, 9. Construct the sampling distribution of $\bar{x}$ based on samples of size 2 selected without replacement. Find the mean and variance of the sampling distribution.

5.3.9 Use the data of Example 5.3.1 to construct the sampling distribution of $\bar{x}$ based on samples of size 3 selected without replacement. Find the mean and variance of the sampling distribution.

5.3.10 Use the data cited in Exercise 5.3.1. Imagine we take samples of size 5, 25, 50, 100, and 500 from the women in this age group.
(a) Calculate the standard error for each of these sampling scenarios.
(b) Discuss how sample size affects the standard error estimates calculated in part (a) and the potential implications this may have in statistical practice.

5.4 DISTRIBUTION OF THE DIFFERENCE BETWEEN TWO SAMPLE MEANS

Frequently the interest in an investigation is focused on two populations. Specifically, an investigator may wish to know something about the difference between two population means. In one investigation, for example, a researcher may wish to know if it is reasonable to conclude that two population means are different. In another situation, the researcher may desire knowledge about the magnitude of the difference between two population means. A medical research team, for example, may want to know whether or not the mean serum cholesterol level is higher in a population of sedentary office workers than in a population of laborers. If the researchers are able to conclude that the population means are different, they may wish to know by how much they differ. A knowledge of the sampling distribution of the difference between two means is useful in investigations of this type.

Sampling from Normally Distributed Populations The following example illustrates the construction of and the characteristics of the sampling distribution of the difference between sample means when sampling is from two normally distributed populations.

EXAMPLE 5.4.1

Suppose we have two populations of individuals—one population (population 1) has experienced some condition thought to be associated with mental retardation, and the other population (population 2) has not experienced the condition. The distribution of intelligence scores in each of the two populations is believed to be approximately normally distributed with a standard deviation of 20.

Suppose, further, that we take a sample of 15 individuals from each population and compute for each sample the mean intelligence score with the following results: $\bar{x}_1 = 92$ and $\bar{x}_2 = 105$. If there is no difference between the two populations, with respect to their true mean intelligence scores, what is the probability of observing a difference this large or larger $(\bar{x}_1 - \bar{x}_2)$ between sample means?

Solution: To answer this question we need to know the nature of the sampling distribution of the relevant statistic, the *difference between two sample means*, $\bar{x}_1 - \bar{x}_2$. Notice that we seek a probability associated with the difference between two sample means rather than a single mean. ■

Sampling Distribution of $\bar{x}_1 - \bar{x}_2$: Construction Although, in practice, we would not attempt to construct the desired sampling distribution, we can conceptualize the manner in which it could be done when sampling is from finite populations. We would begin by selecting from population 1 all possible samples of size 15 and computing the mean for each sample. We know that there would be ${}_{N_1}C_{n_1}$ such samples where N_1 is the population size and $n_1 = 15$. Similarly, we would select all possible samples of size 15 from population 2 and compute the mean for each of these samples. We would then take all possible pairs of sample means, one from population 1 and one from population 2, and take the difference. Table 5.4.1 shows the results of following this procedure. Note that the 1's and 2's in the last line of this table are not exponents, but indicators of population 1 and 2, respectively.

Sampling Distribution of $\bar{x}_1 - \bar{x}_2$: Characteristics It is the distribution of the differences between sample means that we seek. If we plotted the sample differences against their frequency of occurrence, we would obtain a normal distribution with a mean equal to $\mu_1 - \mu_2$, the difference between the two population means, and a variance equal to $(\sigma_1^2/n_1) + (\sigma_2^2/n_2)$. That is, the standard error of the difference between

TABLE 5.4.1 Working Table for Constructing the Distribution of the Difference Between Two Sample Means

Samples from Population 1	Samples from Population 2	Sample Means Population 1	Sample Means Population 2	All Possible Differences Between Means
n_{11}	n_{12}	$\bar{x}_{11}$	$\bar{x}_{12}$	$\bar{x}_{11} - \bar{x}_{12}$
n_{21}	n_{22}	$\bar{x}_{21}$	$\bar{x}_{22}$	$\bar{x}_{11} - \bar{x}_{22}$
n_{31}	n_{32}	$\bar{x}_{31}$	$\bar{x}_{32}$	$\bar{x}_{11} - \bar{x}_{32}$
.	.	.	.	.
.	.	.	.	.
.	.	.	.	.
$n_{{}_{N_1}C_{n_1}1}$	$n_{{}_{N_2}C_{n_2}2}$	$\bar{x}_{{}_{N_1}C_{n_1}1}$	$\bar{x}_{{}_{N_2}C_{n_2}2}$	$\bar{x}_{{}_{N_1}C_{n_1}1} - \bar{x}_{{}_{N_2}C_{n_2}2}$

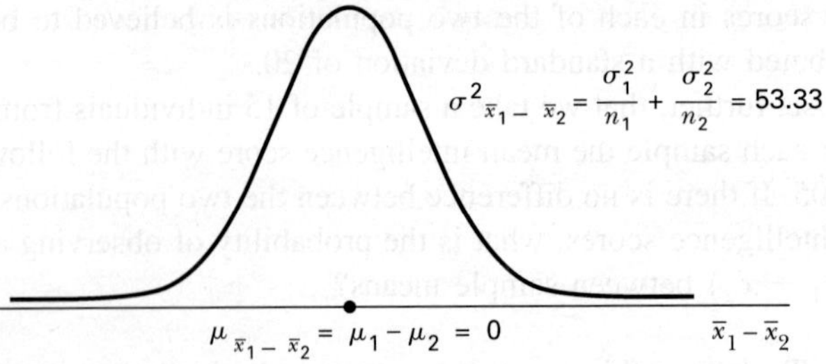

FIGURE 5.4.1 Graph of the sampling distribution of $\bar{x}_1 - \bar{x}_2$ when there is no difference between population means, Example 5.4.1.

sample means would be equal to $\sqrt{(\sigma_1^2/n_1) + (\sigma_2^2/n_2)}$. It should be noted that these properties convey two important points. First, the means of two distributions can be subtracted from one another, or summed together, using standard arithmetic operations. Second, since the overall variance of the sampling distribution will be affected by both contributing distributions, the variances will always be summed even if we are interested in the difference of the means. This last fact assumes that the two distributions are independent of one another.

For our present example we would have a normal distribution with a mean of 0 (if there is no difference between the two population means) and a variance of $[(20)^2/15] + [(20)^2/15] = 53.3333$. The graph of the sampling distribution is shown in Figure 5.4.1.

Converting to *z* We know that the normal distribution described in Example 5.4.1 can be transformed to the standard normal distribution by means of a modification of a previously learned formula. The new formula is as follows:

$$z = \frac{(\bar{x}_1 - \bar{x}_2) - (\mu_1 - \mu_2)}{\sqrt{\dfrac{\sigma_1^2}{n_1} + \dfrac{\sigma_2^2}{n_2}}} \tag{5.4.1}$$

The area under the curve of $\bar{x}_1 - \bar{x}_2$ corresponding to the probability we seek is the area to the left of $\bar{x}_1 - \bar{x}_2 = 92 - 105 = -13$. The z value corresponding to -13, assuming that there is no difference between population means, is

$$z = \frac{-13 - 0}{\sqrt{\dfrac{(20)^2}{15} + \dfrac{(20)^2}{15}}} = \frac{-13}{\sqrt{53.3}} = \frac{-13}{7.3} = -1.78$$

By consulting Table D, we find that the area under the standard normal curve to the left of -1.78 is equal to .0375. In answer to our original question, we say that if there is no difference between population means, the probability of obtaining a difference between sample means as large as or larger than 13 is .0375.

Sampling from Normal Populations The procedure we have just followed is valid even when the sample sizes, n_1 and n_2, are different and when the population variances, σ_1^2 and σ_2^2 have different values. The theoretical results on which this procedure is based may be summarized as follows.

Given two normally distributed populations with means μ_1 and μ_2 and variances σ_1^2 and σ_2^2, respectively, the sampling distribution of the difference, $\bar{x}_1 - \bar{x}_2$, between the means of independent samples of size n_1 and n_2 drawn from these populations is normally distributed with mean $\mu_1 - \mu_2$ and variance $\sqrt{(\sigma_1^2/n_1) + (\sigma_2^2/n_2)}$.

Sampling from Nonnormal Populations Many times a researcher is faced with one or the other of the following problems: the necessity of (1) sampling from nonnormally distributed populations, or (2) sampling from populations whose functional forms are not known. A solution to these problems is to take large samples, since when the sample sizes are large the central limit theorem applies and the distribution of the difference between two sample means is at least approximately normally distributed with a mean equal to $\mu_1 - \mu_2$ and a variance of $(\sigma_1^2/n_1) + (\sigma_2^2/n_2)$. To find probabilities associated with specific values of the statistic, then, our procedure would be the same as that given when sampling is from normally distributed populations.

EXAMPLE 5.4.2

Suppose it has been established that for a certain type of client the average length of a home visit by a public health nurse is 45 minutes with a standard deviation of 15 minutes, and that for a second type of client the average home visit is 30 minutes long with a standard deviation of 20 minutes. If a nurse randomly visits 35 clients from the first and 40 from the second population, what is the probability that the average length of home visit will differ between the two groups by 20 or more minutes?

Solution: No mention is made of the functional form of the two populations, so let us assume that this characteristic is unknown, or that the populations are not normally distributed. Since the sample sizes are large (greater than 30) in both cases, we draw on the results of the central limit theorem to answer the question posed. We know that the difference between sample means is at least approximately normally distributed with the following mean and variance:

$$\mu_{\bar{x}_1 - \bar{x}_2} = \mu_1 - \mu_2 = 45 - 30 = 15$$

$$\sigma^2_{\bar{x}_1 - \bar{x}_2} = \frac{\sigma_1^2}{n_1} + \frac{\sigma_2^2}{n_2} = \frac{(15)^2}{35} + \frac{(20)^2}{40} = 16.4286$$

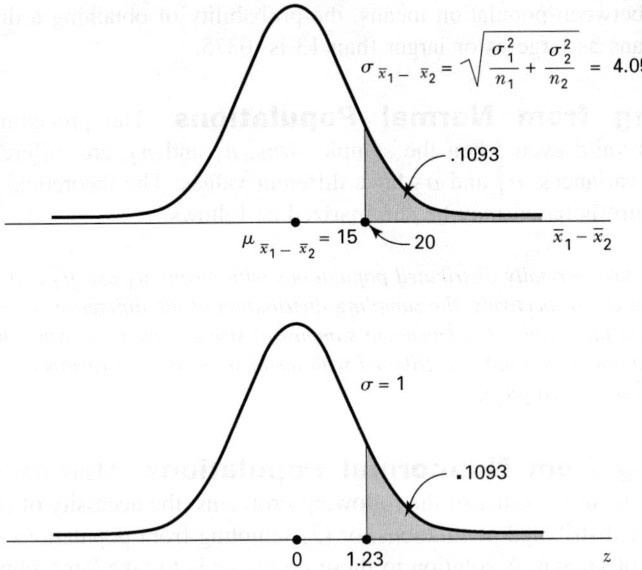

FIGURE 5.4.2 Sampling distribution of $\bar{x}_1 - \bar{x}_2$ and the corresponding standard normal distribution, home visit example.

The area under the curve of $\bar{x}_1 - \bar{x}_2$ that we seek is that area to the right of 20. The corresponding value of z in the standard normal is

$$z = \frac{(\bar{x}_1 - \bar{x}_2) - (\mu_1 - \mu_2)}{\sqrt{\dfrac{\sigma_1^2}{n_1} + \dfrac{\sigma_2^2}{n_2}}} = \frac{20 - 15}{\sqrt{16.4286}} = \frac{5}{4.0532} = 1.23$$

In Table D we find that the area to the right of $z = 1.23$ is $1 - .8907 = .1093$. We say, then, that the probability of the nurse's random visits resulting in a difference between the two means as great as or greater than 20 minutes is .1093. The curve of $\bar{x}_1 - \bar{x}_2$ and the corresponding standard normal curve are shown in Figure 5.4.2. ■

EXERCISES

5.4.1 The study cited in Exercises 5.3.1 and 5.3.2 gives the following data on serum cholesterol levels in U.S. females:

Population	Age	Mean	Standard Deviation
A	20–29	183	37.2
B	30–39	189	34.7

Use these estimates as the mean μ and standard deviation σ for the respective U.S. populations. Suppose we select a simple random sample of size 50 independently from each population. What is the probability that the difference between sample means $\bar{x}_B - \bar{x}_A$ will be more than 8?

5.4.2 In the study cited in Exercises 5.3.4 and 5.3.5, the calcium levels in men and women ages 60 years or older are summarized in the following table:

	Mean	Standard Deviation
Men	797	482
Women	660	414

Use these estimates as the mean μ and standard deviation σ for the U.S. populations for these age groups. If we take a random sample of 40 men and 35 women, what is the probability of obtaining a difference between sample means of 100 mg or more?

5.4.3 Given two normally distributed populations with equal means and variances of $\sigma_1^2 = 100$ and $\sigma_2^2 = 80$, what is the probability that samples of size $n_1 = 25$ and $n_2 = 16$ will yield a value of $\bar{x}_1 - \bar{x}_2$ greater than or equal to 8?

5.4.4 Given two normally distributed populations with equal means and variances of $\sigma_1^2 = 240$ and $\sigma_2^2 = 350$, what is the probability that samples of size $n_1 = 40$ and $n_2 = 35$ will yield a value of $\bar{x}_1 - \bar{x}_2$ as large as or larger than 12?

5.4.5 For a population of 17-year-old boys and 17-year-old girls, the means and standard deviations, respectively, of their subscapular skinfold thickness values are as follows: boys, 9.7 and 6.0; girls, 15.6 and 9.5. Simple random samples of 40 boys and 35 girls are selected from the populations. What is the probability that the difference between sample means $\bar{x}_{\text{girls}} - \bar{x}_{\text{boys}}$ will be greater than 10?

5.5 DISTRIBUTION OF THE SAMPLE PROPORTION

In the previous sections we have dealt with the sampling distributions of statistics computed from measured variables. We are frequently interested, however, in the sampling distribution of a statistic, such as a sample proportion, that results from counts or frequency data.

EXAMPLE 5.5.1

Results (A-3) from the 1999–2000 National Health and Nutrition Examination Survey (NHANES), show that 31 percent of U.S. adults ages 20–74 are obese (obese as defined with body mass index greater than or equal to 30.0). We designate this population proportion as $p = .31$. If we randomly select 150 individuals from this population, what is the probability that the proportion in the sample who are obese will be as great as .40?

Solution: To answer this question, we need to know the properties of the sampling distribution of the sample proportion. We will designate the sample proportion by the symbol $\hat{p}$.

You will recognize the similarity between this example and those presented in Section 4.3, which dealt with the binomial distribution. The variable obesity is a *dichotomous variable,* since an individual can be classified into one or the other of two mutually exclusive categories obese or not obese. In Section 4.3, we were given similar information and were asked to find the number with the characteristic of interest, whereas here we are seeking the proportion in the sample possessing the characteristic of interest. We could with a sufficiently large table of binomial probabilities, such as Table B, determine the probability associated with the number corresponding to the proportion of interest. As we will see, this will not be necessary, since there is available an alternative procedure, when sample sizes are large, that is generally more convenient. ■

Sampling Distribution of $\hat{p}$: Construction The sampling distribution of a sample proportion would be constructed experimentally in exactly the same manner as was suggested in the case of the arithmetic mean and the difference between two means. From the population, which we assume to be finite, we would take all possible samples of a given size and for each sample compute the sample proportion, $\hat{p}$. We would then prepare a frequency distribution of $\hat{p}$ by listing the different distinct values of $\hat{p}$ along with their frequencies of occurrence. This frequency distribution (as well as the corresponding relative frequency distribution) would constitute the sampling distribution of $\hat{p}$.

Sampling Distribution of $\hat{p}$: Characteristics When the sample size is large, the distribution of sample proportions is approximately normally distributed by virtue of the central limit theorem. The mean of the distribution, $\mu_{\hat{p}}$, that is, the average of all the possible sample proportions, will be equal to the true population proportion, p, and the variance of the distribution, $\sigma^2_{\hat{p}}$, will be equal to $p(1-p)/n$ or pq/n, where $q = 1 - p$. To answer probability questions about p, then, we use the following formula:

$$z = \frac{\hat{p} - p}{\sqrt{\dfrac{p(1-p)}{n}}} \tag{5.5.1}$$

The question that now arises is, How large does the sample size have to be for the use of the normal approximation to be valid? A widely used criterion is that both np and $n(1-p)$ must be greater than 5, and we will abide by that rule in this text.

We are now in a position to answer the question regarding obesity in the sample of 150 individuals from a population in which 31 percent are obese. Since both np and $n(1-p)$ are greater than $5(150 \times .31 = 46.5$ and $150 \times .69 = 103.5)$, we can say that, in this case, $\hat{p}$ is approximately normally distributed with a mean $\mu_{\hat{p}}, = p = .31$ and $\sigma^2_{\hat{p}} = p(1-p)/n = (.31)(.69)/150 = .001426$. The probability we seek is the area under the curve of $\hat{p}$ that is to the right of .40. This area is equal to the area under the standard normal curve to the right of

$$z = \frac{\hat{p} - p}{\sqrt{\frac{p(1-p)}{n}}} = \frac{.40 - .31}{\sqrt{.001426}} = 2.38$$

The transformation to the standard normal distribution has been accomplished in the usual manner: z is found by dividing the difference between a value of a statistic and its mean by the standard error of the statistic. Using Table D we find that the area to the right of $z = 2.38$ is $1 - .9913 = .0087$. We may say, then, that the probability of observing $\hat{p} \geq .40$ in a random sample of size $n = 150$ from a population in which $p = .31$ is .0087. If we should, in fact, draw such a sample, most people would consider it a rare event.

Correction for Continuity The normal approximation may be improved by the *correction for continuity,* a device that makes an adjustment for the fact that a discrete distribution is being approximated by a continuous distribution. Suppose we let $x = n\hat{p}$, the number in the sample with the characteristic of interest when the proportion is $\hat{p}$. To apply the correction for continuity, we compute

$$z_c = \frac{\frac{x + .5}{n} - p}{\sqrt{pq/n}}, \quad \text{for } x < np \tag{5.5.2}$$

or

$$z_c = \frac{\frac{x - .5}{n} - p}{\sqrt{pq/n}}, \quad \text{for } x > np \tag{5.5.3}$$

where $q = 1 - p$. The correction for continuity will not make a great deal of difference when n is large. In the above example $n\hat{p} = 150(.4) = 60$, and

$$z_c = \frac{\frac{60 - .5}{150} - .31}{\sqrt{(.31)(.69)/150}} = 2.30$$

and $P(\hat{p} \geq .40) = 1 - .9893 = .0107$, a result not greatly different from that obtained without the correction for continuity. This adjustment is not often done by hand, since most statistical computer programs automatically apply the appropriate continuity correction when necessary.

EXAMPLE 5.5.2

Blanche Mikhail (A-4) studied the use of prenatal care among low-income African-American women. She found that only 51 percent of these women had adequate prenatal care. Let us assume that for a population of similar low-income African-American women, 51 percent had adequate prenatal care. If 200 women from this population are drawn at random, what is the probability that less than 45 percent will have received adequate prenatal care?

Solution: We can assume that the sampling distribution of $\hat{p}$ is approximately normally distributed with $\mu_{\hat{p}} = .51$ and $\sigma^2_{\hat{p}} = (.51)(.49)/200 = .00125$. We compute

$$z = \frac{.45 - .51}{\sqrt{.00125}} = \frac{-.06}{.0353} = -1.70$$

The area to the left of -1.70 under the standard normal curve is .0446. Therefore, $P(\hat{p} \leq .45) = P(z \leq -1.70) = .0446$. ■

EXERCISES

5.5.1 Smith et al. (A-5) performed a retrospective analysis of data on 782 eligible patients admitted with myocardial infarction to a 46-bed cardiac service facility. Of these patients, 248 (32 percent) reported a past myocardial infarction. Use .32 as the population proportion. Suppose 50 subjects are chosen at random from the population. What is the probability that over 40 percent would report previous myocardial infarctions?

5.5.2 In the study cited in Exercise 5.5.1, 13 percent of the patients in the study reported previous episodes of stroke or transient ischemic attack. Use 13 percent as the estimate of the prevalence of stroke or transient ischemic attack within the population. If 70 subjects are chosen at random from the population, what is the probability that 10 percent or less would report an incidence of stroke or transient ischemic attack?

5.5.3 In the same 1999–2000 NHANES (A-3) report cited in Example 5.5.1, researchers estimated that 64 percent of U.S. adults ages 20–74 were overweight or obese (overweight: BMI 25-29, obese: BMI 30 or greater). Use this estimate as the population proportion for U.S. adults ages 20–74. If 125 subjects are selected at random from the population, what is the probability that 70 percent or more would be found to be overweight or obese?

5.5.4 Gallagher et al. (A-6) reported on a study to identify factors that influence women's attendance at cardiac rehabilitation programs. They found that by 12 weeks post-discharge, only 64 percent of eligible women attended such programs. Using 64 percent as an estimate of the attendance percentage of all eligible women, find the probability that in a sample of 45 women selected at random from the population of eligible women less than 50 percent would attend programs.

5.5.5 Given a population in which $p = .6$ and a random sample from this population of size 100, find:

(a) $P(\hat{p} \geq .65)$ **(b)** $P(\hat{p} \leq .58)$

(c) $P(.56 \leq \hat{p} \leq .63)$

5.5.6 It is known that 35 percent of the members of a certain population suffer from one or more chronic diseases. What is the probability that in a sample of 200 subjects drawn at random from this population 80 or more will have at least one chronic disease?

5.6 DISTRIBUTION OF THE DIFFERENCE BETWEEN TWO SAMPLE PROPORTIONS

Often there are two population proportions in which we are interested and we desire to assess the probability associated with a difference in proportions computed from samples drawn from each of these populations. The relevant sampling distribution is the distribution of the difference between the two sample proportions.

Sampling Distribution of $\hat{p}_1 - \hat{p}_2$: Characteristics The characteristics of this sampling distribution may be summarized as follows:

If independent random samples of size n_1 and n_2 are drawn from two populations of dichotomous variables where the proportions of observations with the characteristic of interest in the two populations are p_1 and p_2, respectively, the distribution of the difference between sample proportions, $\hat{p}_1 - \hat{p}_2$, is approximately normal with mean

$$\mu_{\hat{p}_1 - \hat{p}_2} = p_1 - p_2$$

and variance

$$\sigma^2_{\hat{p}_1 - \hat{p}_2} = \frac{p_1(1-p_1)}{n_1} + \frac{p_2(1-p_2)}{n_2}$$

when n_1 and n_2 are large.

We consider n_1 and n_2 sufficiently large when $n_1 p_1, n_2 p_2, n_1(1-p_1)$, and $n_2(1-p_2)$ are all greater than 5.

Sampling Distribution of $\hat{p}_1 - \hat{p}_2$: Construction To physically construct the sampling distribution of the difference between two sample proportions, we would proceed in the manner described in Section 5.4 for constructing the sampling distribution of the difference between two means.

Given two sufficiently small populations, one would draw, from population 1, all possible simple random samples of size n_1 and compute, from each set of sample data, the sample proportion $\hat{p}_1$. From population 2, one would draw independently all possible simple random samples of size n_2 and compute, for each set of sample data, the sample proportion $\hat{p}_2$. One would compute the differences between all possible pairs of sample proportions, where one number of each pair was a value of $\hat{p}_1$ and the other a value of $\hat{p}_2$. The sampling distribution of the difference between sample proportions, then, would consist of all such distinct differences, accompanied by their frequencies (or relative frequencies) of occurrence. For large finite or infinite populations, one could approximate the sampling distribution of the difference between sample proportions by drawing a large number of independent simple random samples and proceeding in the manner just described.

To answer probability questions about the difference between two sample proportions, then, we use the following formula:

$$z = \frac{(\hat{p}_1 - \hat{p}_2) - (p_1 - p_2)}{\sqrt{\frac{p_1(1-p_1)}{n_1} + \frac{p_2(1-p_2)}{n_2}}} \tag{5.6.1}$$

EXAMPLE 5.6.1

The 1999 National Health Interview Survey, released in 2003 (A-7), reported that 28 percent of the subjects self-identifying as white said they had experienced lower back pain during the three months prior to the survey. Among subjects of Hispanic origin, 21 percent reported lower back pain. Let us assume that .28 and .21 are the proportions for the respective races reporting lower back pain in the United States. What is the probability that independent random samples of size 100 drawn from each of the populations will yield a value of $\hat{p}_1 - \hat{p}_2$ as large as .10?

Solution: We assume that the sampling distribution of $\hat{p}_1 - \hat{p}_2$ is approximately normal with mean

$$\mu_{\hat{p}_1 - \hat{p}_2} = .28 - .21 = .07$$

and variance

$$\begin{aligned} \sigma^2_{\hat{p}_1 - \hat{p}_2} &= \frac{(.28)(.72)}{100} + \frac{(.21)(.79)}{100} \\ &= .003675 \end{aligned}$$

The area corresponding to the probability we seek is the area under the curve of $\hat{p}_1 - \hat{p}_2$ to the right of .10. Transforming to the standard normal distribution gives

$$z = \frac{(\hat{p}_1 - \hat{p}_2) - (p_1 - p_2)}{\sqrt{\frac{p_1(1-p_1)}{n_1} + \frac{p_2(1-p_2)}{n_2}}} = \frac{.10 - .07}{\sqrt{.003675}} = .49$$

Consulting Table D, we find that the area under the standard normal curve that lies to the right of $z = .49$ is $1 - .6879 = .3121$. The probability of observing a difference as large as .10 is, then, .3121. ■

EXAMPLE 5.6.2

In the 1999 National Health Interview Survey (A-7), researchers found that among U.S. adults ages 75 or older, 34 percent had lost all their natural teeth and for U.S. adults ages 65–74, 26 percent had lost all their natural teeth. Assume that these proportions are the parameters for the United States in those age groups. If a random sample of 250 adults ages 65–74 and an independent random sample of 200 adults ages 45–64 years old are drawn from these populations, find the probability that the difference in percent of total natural teeth loss is less than 5 percent between the two populations.

Solution: We assume that the sampling distribution $\hat{p}_1 - \hat{p}_2$ is approximately normal. The mean difference in proportions of those losing all their teeth is

$$\mu_{\hat{p}_1 - \hat{p}_2} = .34 - .26 = .08$$

and the variance is

$$\sigma^2_{\hat{p}_1 - \hat{p}_2} = \frac{p_1(1-p_1)}{n_1} + \frac{p_2(1-p_2)}{n_2} = \frac{(.34)(.66)}{250} + \frac{(.26)(.74)}{200} = .00186$$

The area of interest under the curve of $\hat{p}_1 - \hat{p}_2$ is that to the left of .05. The corresponding z value is

$$z = \frac{.05 - (.08)}{\sqrt{.00186}} = -.70$$

Consulting Table D, we find that the area to the left of $z = -.70$ is .2420. ■

EXERCISES

5.6.1 According to the 2000 U.S. Census Bureau (A-8), in 2000, 9.5 percent of children in the state of Ohio were not covered by private or government health insurance. In the neighboring state of Pennsylvania, 4.9 percent of children were not covered by health insurance. Assume that these proportions are parameters for the child populations of the respective states. If a random sample of size 100 children is drawn from the Ohio population, and an independent random sample of size 120 is drawn from the Pennsylvania population, what is the probability that the samples would yield a difference, $\hat{p}_1 - \hat{p}_2$ of .09 or more?

5.6.2 In the report cited in Exercise 5.6.1 (A-8), the Census Bureau stated that for Americans in the age group 18–24 years, 64.8 percent had private health insurance. In the age group 25–34 years, the percentage was 72.1. Assume that these percentages are the population parameters in those age groups for the United States. Suppose we select a random sample of 250 Americans from the 18–24 age group and an independent random sample of 200 Americans from the age group 25–34; find the probability that $\hat{p}_2 - \hat{p}_1$ is less than 6 percent.

5.6.3 From the results of a survey conducted by the U.S. Bureau of Labor Statistics (A-9), it was estimated that 21 percent of workers employed in the Northeast participated in health care benefits programs that included vision care. The percentage in the South was 13 percent. Assume these percentages are population parameters for the respective U.S. regions. Suppose we select a simple random sample of size 120 northeastern workers and an independent simple random sample of 130 southern workers. What is the probability that the difference between sample proportions, $\hat{p}_1 - \hat{p}_2$, will be between .04 and .20?

5.7 SUMMARY

This chapter is concerned with sampling distributions. The concept of a sampling distribution is introduced, and the following important sampling distributions are covered:

1. The distribution of a single sample mean.
2. The distribution of the difference between two sample means.
3. The distribution of a sample proportion.
4. The distribution of the difference between two sample proportions.

We emphasize the importance of this material and urge readers to make sure that they understand it before proceeding to the next chapter.

SUMMARY OF FORMULAS FOR CHAPTER 5

Formula Number	Name	Formula
5.3.1	z-transformation for sample mean	$Z = \dfrac{\bar{X} - \mu_{\bar{x}}}{\sigma/\sqrt{n}}$
5.4.1	z-transformation for difference between two means	$Z = \dfrac{(\bar{X}_1 - \bar{X}_2) - (\mu_1 - \mu_2)}{\sqrt{\dfrac{\sigma_1^2}{n_1} + \dfrac{\sigma_2^2}{n_2}}}$
5.5.1	z-transformation for sample proportion	$Z = \dfrac{\hat{p} - p}{\sqrt{\dfrac{p(1-p)}{n}}}$
5.5.2	Continuity correction when $x < np$	$Z_c = \dfrac{\dfrac{x + .5}{n} - p}{\sqrt{pq/n}}$
5.5.3	Continuity correction when $x > np$	$Z_c = \dfrac{\dfrac{X + .5}{n} + p}{\sqrt{pq/n}}$
5.6.1	z-transformation for difference between two proportions	$Z = \dfrac{(\hat{p}_1 - \hat{p}_2) - (p_1 - p_2)}{\sqrt{\dfrac{p_1(1-p_1)}{n_1} + \dfrac{p_2(1-p_2)}{n_2}}}$
Symbol Key	• μ_i = mean of population i • $\mu_{\bar{x}}$ = mean of sampling distribution if $\bar{x}$ • n_i = sample size for sample i from population i • p_i = proportion for population i • $\hat{p}_i$ = proportion for sample i from population i • σ_i^2 = variance for population i • $\bar{X}_i$ = mean of sample i from population i • z = standard normal random variable	

REVIEW QUESTIONS AND EXERCISES

1. What is a sampling distribution?

2. Explain how a sampling distribution may be constructed from a finite population.

3. Describe the sampling distribution of the sample mean when sampling is with replacement from a normally distributed population.

4. Explain the central limit theorem.

5. How does the sampling distribution of the sample mean, when sampling is without replacement, differ from the sampling distribution obtained when sampling is with replacement?

6. Describe the sampling distribution of the difference between two sample means.

7. Describe the sampling distribution of the sample proportion when large samples are drawn.

8. Describe the sampling distribution of the difference between two sample means when large samples are drawn.

9. Explain the procedure you would follow in constructing the sampling distribution of the difference between sample proportions based on large samples from finite populations.

10. Suppose it is known that the response time of healthy subjects to a particular stimulus is a normally distributed random variable with a mean of 15 seconds and a variance of 16. What is the probability that a random sample of 16 subjects will have a mean response time of 12 seconds or more?

11. Janssen et al. (A-10) studied Americans ages 60 and over. They estimated the mean body mass index of women over age 60 with normal skeletal muscle to be 23.1 with a standard deviation of 3.7. Using these values as the population mean and standard deviation for women over age 60 with normal skeletal muscle index, find the probability that 45 randomly selected women in this age range with normal skeletal muscle index will have a mean BMI greater than 25.

12. In the study cited in Review Exercise 11, the researchers reported the mean BMI for men ages 60 and older with normal skeletal muscle index to be 24.7 with a standard deviation of 3.3. Using these values as the population mean and standard deviation, find the probability that 50 randomly selected men in this age range with normal skeletal muscle index will have a mean BMI less than 24.

13. Using the information in Review Exercises 11 and 12, find the probability that the difference in mean BMI for 45 women and 50 men selected independently and at random from the respective populations will exceed 3.

14. In the results published by Wright et al. (A-2) based on data from the 1999–2000 NHANES study referred to in Exercises 5.4.1 and 5.4.2, investigators reported on their examination of iron levels. The mean iron level for women ages 20–39 years was 13.7 mg with an estimated standard deviation of 8.9 mg. Using these as population values for women ages 20–39, find the probability that a random sample of 100 women will have a mean iron level less than 12 mg.

15. Refer to Review Exercise 14. The mean iron level for men between the ages of 20 and 39 years is 17.9 mg with an estimated standard deviation of 10.9 mg. Using 17.9 and 10.9 as population parameters, find the probability that a random sample of 120 men will have a mean iron level higher than 19 mg.

16. Using the information in Review Exercises 14 and 15, and assuming independent random samples of size 100 and 120 for women and men, respectively, find the probability that the difference in sample mean iron levels is greater than 5 mg.

17. The results of the 1999 National Health Interview Survey released in 2003 (A-7) showed that among U.S. adults ages 60 and older, 19 percent had been told by a doctor or other health care provider that they had some form of cancer. If we use this as the percentage for all adults 65 years old and older living in the United States, what is the probability that among 65 adults chosen at random more than 25 percent will have been told by their doctor or some other health care provider that they have cancer?

18. Refer to Review Exercise 17. The reported cancer rate for women subjects ages 65 and older is 17 percent. Using this estimate as the true percentage of all females ages 65 and over who have been told by a health care provider that they have cancer, find the probability that if 220 women are selected at random from the population, more than 20 percent will have been told they have cancer.

19. Refer to Review Exercise 17. The cancer rate for men ages 65 and older is 23 percent. Use this estimate as the percentage of all men ages 65 and older who have been told by a health care provider that they have cancer. Find the probability that among 250 men selected at random that fewer than 20 percent will have been told they have cancer.

20. Use the information in Review Exercises 18 and 19 to find the probability that the difference in the cancer percentages between men and women will be less than 5 percent when 220 women and 250 men aged 65 and older are selected at random.

21. How many simple random samples (without replacement) of size 5 can be selected from a population of size 10?

22. It is estimated by the 1999–2000 NHANES (A-7) that among adults 18 years old or older 53 percent have never smoked. Assume the proportion of U.S. adults who have never smoked to be .53. Consider the sampling distribution of the sample proportion based on simple random samples of size 110 drawn from this population. What is the functional form of the sampling distribution?

23. Refer to Exercise 22. Compute the mean and variance of the sampling distribution.

24. Refer to Exercise 22. What is the probability that a single simple random sample of size 110 drawn from this population will yield a sample proportion smaller than .50?

25. In a population of subjects who died from lung cancer following exposure to asbestos, it was found that the mean number of years elapsing between exposure and death was 25. The standard deviation was 7 years. Consider the sampling distribution of sample means based on samples of size 35 drawn from this population. What will be the shape of the sampling distribution?

26. Refer to Exercise 25. What will be the mean and variance of the sampling distribution?

27. Refer to Exercise 25. What is the probability that a single simple random sample of size 35 drawn from this population will yield a mean between 22 and 29?

28. For each of the following populations of measurements, state whether the sampling distribution of the sample mean is normally distributed, approximately normally distributed, or not approximately normally distributed when computed from samples of size (A) 10, (B) 50, and (C) 200.
   (a) The logarithm of metabolic ratios. The population is normally distributed.
   (b) Resting vagal tone in healthy adults. The population is normally distributed.
   (c) Insulin action in obese subjects. The population is not normally distributed.

29. For each of the following sampling situations indicate whether the sampling distribution of the sample proportion can be approximated by a normal distribution and explain why or why not.

(a) $p = .50, n = 8$ **(b)** $p = .40, n = 30$
(c) $p = .10, n = 30$ **(d)** $p = .01, n = 1000$
(e) $p = .90, n = 100$ **(f)** $p = .05, n = 150$

REFERENCES

Methodology References

1. Richard J. Larsen and Morris L. Marx, *An Introduction to Mathematical Statistics and Its Applications,* 2nd ed., Prentice-Hall, Englewood Cliffs, NJ, 1986.
2. John A. Rice, *Mathematical Statistics and Data Analysis,* 2nd ed., Duxbury, Belmont, CA, 1995.

Applications References

A-1. The Third National Health and Nutrition Examination Survey, NHANES III (1988–94), Table 2. National Center for Health Statistics, Division of Data Services, Hyattsville, MD. Available at http://www.cdc.gov/nchs/about/major/nhanes/datatblelink.htm.

A-2. Jacqueline D. Wright, Chia-Yih Wang, Jocelyn Kennedy-Stephenson, and R. Bethene Ervin, "Dietary Intake of Ten Key Nutrients for Public Health, United States: 1999–2000," National Center for Health Statistics. Advance Data from Vital and Health Statistics, No. 334 (2003).

A-3. Health, United States, 2002, Table 70: Healthy weight, overweight, and obesity among persons 20 years of age and over, according to sex, age, race, and Hispanic origin: United States, 1960–62, 1971–74, 1976–80, 1988–1994, and 1999–2000. National Center for Health Statistics, Division of Data Services, Hyattsville, MD, 20782. http://www.cdc.gov/nchs/products/pubs/pubd/hestats/obese/obse99.htm.

A-4. Blanche Mikhail, "Prenatal Care Utilization among Low-Income African American Women," *Journal of Community Health Nursing, 17* (2000), 235–246.

A-5. James P. Smith, Rajendra H. Mehta, Sugata K. Das, Thomas Tsai, Dean J. Karavite, Pamela L. Rusman, David Bruckman, and Kim A. Eagle, "Effects of End-of-Month Admission on Length of Stay and Quality of Care Among Inpatients with Myocardial Infarction," *American Journal of Medicine, 113* (2002), 288–293.

A-6. Robyn Gallagher, Sharon McKinley, and Kathleen Dracup, "Predictor's of Women's Attendance at Cardiac Rehabilitation Programs," *Progress in Cardiovascular Nursing, 18* (2003), 121–126.

A-7. J. R. Pleis and R. Coles, "Summary Health Statistics for U.S. Adults: National Health Interview Survey, 1999," National Center for Health Statistics. *Vital and Health Statistics, 10* (212), (2003).

A-8. U.S. Census Bureau, *Current Population Reports,* P60–215, as reported in Statistical Abstract of the United States: 2002 (118th edition), U.S. Bureau of the Census, Washington, DC, 2002, Table Nos. 137–138.

A-9. U.S. Bureau of Labor Statistics, *News,* USDL 01–473, as reported in Statistical Abstract of the United States: 2002 (118th edition), U.S. Bureau of the Census, Washington, DC, 2002, Table No. 139.

A-10. Ian Janssen, Steven B. Heymsfield, and Robert Ross, "Low Relative Skeletal Muscle Mass (Sacopenia) in Older Persons Is Associated with Functional Impairment and Physical Disability," *Journal of the American Geriatrics Society, 50* (2002), 889–896.

CHAPTER **6**

USING SAMPLE DATA TO MAKE ESTIMATES ABOUT POPULATION PARAMETERS

CHAPTER OVERVIEW

This chapter covers estimation, one of the two types of statistical inference. As discussed in earlier chapters, statistics, such as means and variances, can be calculated from samples drawn from populations. These statistics serve as estimates of the corresponding population parameters. We expect these estimates to differ by some amount from the parameters they estimate. This chapter introduces estimation procedures that take these differences into account, thereby providing a foundation for statistical inference procedures discussed in the remaining chapters of the book.

TOPICS

LEARNING OUTCOMES

After studying this chapter, the student will

1. understand the importance and basic principles of estimation.
2. be able to calculate interval estimates for a variety of parameters.
3. be able to interpret a confidence interval from both a practical and a probabilistic viewpoint.
4. understand the basic properties and uses of the *t* distribution, chi-square distribution, and *F* distribution.

6.1 INTRODUCTION

We come now to a consideration of *estimation,* the first of the two general areas of statistical inference. The second general area, *hypothesis testing,* is examined in the next chapter.

We learned in Chapter 1 that inferential statistics is defined as follows.

DEFINITION

***Statistical inference* is the procedure by which we reach a conclusion about a population on the basis of the information contained in a sample drawn from that population.**

The process of estimation entails calculating, from the data of a sample, some statistic that is offered as an approximation of the corresponding parameter of the population from which the sample was drawn.

The rationale behind estimation in the health sciences field rests on the assumption that workers in this field have an interest in the parameters, such as means and proportions, of various populations. If this is the case, there is a good reason why one must rely on estimating procedures to obtain information regarding these parameters. Many populations of interest, although finite, are so large that a 100 percent examination would be prohibitive from the standpoint of cost.

Suppose the administrator of a large hospital is interested in the mean age of patients admitted to his hospital during a given year. He may consider it too expensive to go through the records of all patients admitted during that particular year and, consequently, elect to examine a sample of the records from which he can compute an estimate of the mean age of patients admitted that year.

A physician in general practice may be interested in knowing what proportion of a certain type of individual, treated with a particular drug, suffers undesirable side effects. No doubt, her concept of the population consists of all those persons who ever have been or ever will be treated with this drug. Deferring a conclusion until the entire population has been observed could have an adverse effect on her practice.

These two examples have implied an interest in estimating, respectively, a population mean and a population proportion. Other parameters, the estimation of which we will cover in this chapter, are the difference between two means, the difference between two proportions, the population variance, and the ratio of two variances.

We will find that for each of the parameters we discuss, we can compute two types of estimate: a point estimate and an interval estimate.

DEFINITION

A *point estimate* is a single numerical value used to estimate the corresponding population parameter.

DEFINITION

An *interval estimate* consists of two numerical values defining a range of values that, with a specified degree of confidence, most likely includes the parameter being estimated.

These concepts will be elaborated on in the succeeding sections.

Choosing an Appropriate Estimator Note that a single computed value has been referred to as an *estimate.* The rule that tells us how to compute this value, or estimate, is referred to as an *estimator.* Estimators are usually presented as formulas. For example,

$$\bar{x} = \frac{\sum x_i}{n}$$

is an estimator of the population mean, μ. The single numerical value that results from evaluating this formula is called an estimate of the parameter μ.

In many cases, a parameter may be estimated by more than one estimator. For example, we could use the sample median to estimate the population mean. How then do we decide which estimator to use for estimating a given parameter? The decision is based on an objective measure or set of criteria that reflect some desired property of a particular estimator. When measured against these criteria, some estimators are better than others. One of these criteria is the property of *unbiasedness.*

DEFINITION

An estimator, say, T, of the parameter θ is said to be an unbiased *estimator* of θ if $E(T) = \theta$.

$E(T)$ is read, "the expected value of T." For a finite population, $E(T)$ is obtained by taking the average value of T computed from all possible samples of a given size that may be drawn from the population. That is, $E(T) = \mu_T$. For an infinite population, $E(T)$ is defined in terms of calculus.

In the previous chapter we have seen that the sample mean, the sample proportion, the difference between two sample means, and the difference between two sample proportions are each unbiased estimates of their corresponding parameters. This property was implied when the parameters were said to be the means of the respective sampling distributions. For example, since the mean of the sampling distribution of $\bar{x}$ is equal to μ, we know that $\bar{x}$ is an unbiased estimator of μ. The other criteria of good estimators will not be discussed in this book. The interested reader will find them covered in detail in most mathematical statistics texts.

Sampled Populations and Target Populations The health researcher who uses statistical inference procedures must be aware of the difference between two kinds of population—the sampled population and the target population.

DEFINITION

The *sampled population* is the population from which one actually draws a sample.

DEFINITION

The *target population* is the population about which one wishes to make an inference.

These two populations may or may not be the same. Statistical inference procedures allow one to make inferences about sampled populations (provided proper sampling methods have been employed). Only when the target population and the sampled population are the same is it possible for one to use statistical inference procedures to reach conclusions about the target population. If the sampled population and the target population are different, the researcher can reach conclusions about the target population only on the basis of nonstatistical considerations.

Suppose, for example, that a researcher wishes to assess the effectiveness of some method for treating rheumatoid arthritis. The target population consists of all patients suffering from the disease. It is not practical to draw a sample from this population. The researcher may, however, select a sample from all rheumatoid arthritis patients seen in some specific clinic. These patients constitute the sampled population, and, if proper sampling methods are used, inferences about this sampled population may be drawn on the basis of the information in the sample. If the researcher wishes to make inferences about all rheumatoid arthritis sufferers, he or she must rely on nonstatistical means to do so. Perhaps the researcher knows that the sampled population is similar, with respect to all important characteristics, to the target population. That is, the researcher may know that the age, sex, severity of illness, duration of illness, and so on are similar in both populations. And on the strength of this knowledge, the researcher may be willing to extrapolate his or her findings to the target population.

In many situations the sampled population and the target population are identical; when this is the case, inferences about the target population are straightforward. The researcher, however, should be aware that this is not always the case and not fall into the trap of drawing unwarranted inferences about a population that is different from the one that is sampled.

Random and Nonrandom Samples In the examples and exercises of this book, we assume that the data available for analysis have come from random samples. The strict validity of the statistical procedures discussed depends on this assumption. In many instances in real-world applications it is impossible or impractical to use truly random samples. In animal experiments, for example, researchers usually use whatever animals are available from suppliers or their own breeding stock. If the researchers had to depend on randomly selected material, very little research of this type would be conducted. Again, nonstatistical considerations must play a part in the generalization process. Researchers may contend that the samples actually used are equivalent to simple random samples, since there is no reason to believe that the material actually used is not representative of the population about which inferences are desired.

In many health research projects, samples of convenience, rather than random samples, are employed. Researchers may have to rely on volunteer subjects or on readily available subjects such as students in their classes. Samples obtained from such sources are examples of *convenience samples*. Again, generalizations must be made on the basis of nonstatistical considerations. The consequences of such generalizations, however, may be useful or they may range from misleading to disastrous.

In some situations it is possible to introduce randomization into an experiment even though available subjects are not randomly selected from some well-defined population. In comparing two treatments, for example, each subject may be randomly assigned to one or the other of the treatments. Inferences in such cases apply to the treatments and not the subjects, and hence the inferences are valid.

6.2 CONFIDENCE INTERVAL FOR A POPULATION MEAN

Suppose researchers wish to estimate the mean of some normally distributed population. They draw a random sample of size n from the population and compute $\bar{x}$, which they use as a point estimate of μ. Although this estimator of μ possesses all the qualities of a good estimator, we know that because random sampling inherently involves chance, $\bar{x}$ cannot be expected to be equal to μ.

It would be much more meaningful, therefore, to estimate μ by an interval that somehow communicates information regarding the probable magnitude of μ.

Sampling Distributions and Estimation To obtain an interval estimate, we must draw on our knowledge of sampling distributions. In the present case, because we are concerned with the sample mean as an estimator of a population mean, we must recall what we know about the sampling distribution of the sample mean.

In the previous chapter we learned that if sampling is from a normally distributed population, the sampling distribution of the sample mean will be normally distributed with a mean $\mu_{\bar{x}}$ equal to the population mean μ, and a variance $\sigma_{\bar{x}}^2$ equal to σ^2/n. We could plot the sampling distribution if we only knew where to locate it on the $\bar{x}$-axis. From our knowledge of normal distributions, in general, we know even more about the distribution of $\bar{x}$ in this case. We know, for example, that regardless of where the distribution of $\bar{x}$ is located, approximately 95 percent of the possible values of $\bar{x}$ constituting the distribution are within two standard deviations of the mean. The two points that are two standard deviations from the mean are $\mu - 2\sigma_{\bar{x}}$ and $\mu + 2\sigma_{\bar{x}}$, so that the interval $\mu \pm 2\sigma_{\bar{x}}$ will contain approximately 95 percent of the possible values of $\bar{x}$. We know that μ and, hence $\mu_{\bar{x}}$, are unknown, but we may arbitrarily place the sampling distribution of $\bar{x}$ on the $\bar{x}$-axis.

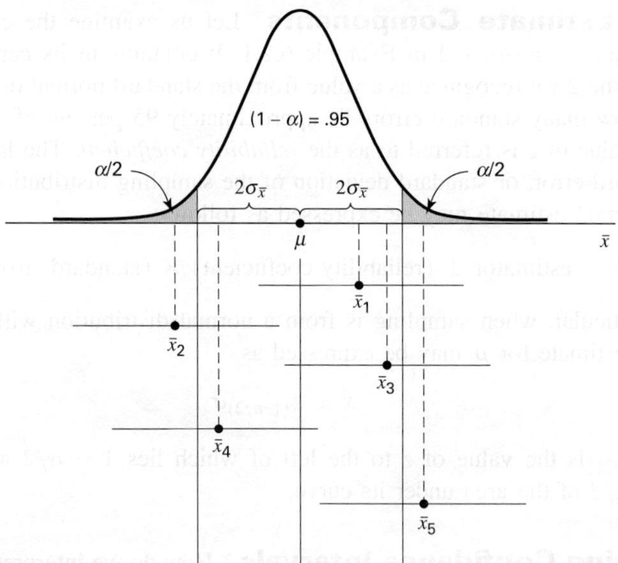

FIGURE 6.2.1 The 95 percent confidence interval for μ.

Since we do not know the value of μ, not a great deal is accomplished by the expression $\mu \pm 2\sigma_{\bar{x}}$. We do, however, have a point estimate of μ, which is $\bar{x}$. Would it be useful to construct an interval about this point estimate of μ? The answer is yes. Suppose we constructed intervals about every possible value of $\bar{x}$ computed from all possible samples of size n from the population of interest. We would have a large number of intervals of the form $\bar{x} \pm 2\sigma_{\bar{x}}$ with widths all equal to the width of the interval about the unknown μ. Approximately 95 percent of these intervals would have centers falling within the $\pm 2\sigma_{\bar{x}}$ interval about μ. Each of the intervals whose centers fall within $2\sigma_{\bar{x}}$ of μ would contain μ. These concepts are illustrated in Figure 6.2.1, in which we see that $\bar{x}_1, \bar{x}_3$, and $\bar{x}_4$ all fall within the interval about μ, and, consequently, the $2\sigma_{\bar{x}}$ intervals about these sample means include the value of μ. The sample means $\bar{x}_2$ and $\bar{x}_5$ do not fall within the $2\sigma_{\bar{x}}$ interval about μ, and the $2\sigma_{\bar{x}}$ intervals about them do not include μ.

EXAMPLE 6.2.1

Suppose a researcher, interested in obtaining an estimate of the average level of some enzyme in a certain human population, takes a sample of 10 individuals, determines the level of the enzyme in each, and computes a sample mean of $\bar{x} = 22$. Suppose further it is known that the variable of interest is approximately normally distributed with a variance of 45. We wish to estimate μ.

Solution: An approximate 95 percent confidence interval for μ is given by

$$\bar{x} \pm 2\sigma_{\bar{x}}$$
$$22 \pm 2\sqrt{\frac{45}{10}}$$
$$22 \pm 2(2.1213)$$
$$17.76, 26.24$$

■

Interval Estimate Components Let us examine the composition of the interval estimate constructed in Example 6.2.1. It contains in its center the point estimate of μ. The 2 we recognize as a value from the standard normal distribution that tells us within how many standard errors lie approximately 95 percent of the possible values of $\bar{x}$. This value of z is referred to as the *reliability coefficient.* The last component, $\sigma_{\bar{x}}$, is the standard error, or standard deviation of the sampling distribution of $\bar{x}$. In general, then, an interval estimate may be expressed as follows:

$$\text{estimator} \pm (\text{reliability coefficient}) \times (\text{standard error}) \tag{6.2.1}$$

In particular, when sampling is from a normal distribution with known variance, an interval estimate for μ may be expressed as

$$\bar{x} \pm z_{(1-\alpha/2)}\sigma_{\bar{x}} \tag{6.2.2}$$

where $z_{(1-\alpha/2)}$ is the value of z to the left of which lies $1 - \alpha/2$ and to the right of which lies $\alpha/2$ of the area under its curve.

Interpreting Confidence Intervals How do we interpret the interval given by Expression 6.2.2? In the present example, where the reliability coefficient is equal to 2, we say that in repeated sampling approximately 95 percent of the intervals constructed by Expression 6.2.2 will include the population mean. This interpretation is based on the probability of occurrence of different values of $\bar{x}$. We may generalize this interpretation if we designate the total area under the curve of $\bar{x}$ that is outside the interval $\mu \pm 2\sigma_{\bar{x}}$ as α and the area within the interval as $1 - \alpha$ and give the following *probabilistic interpretation* of Expression 6.2.2.

Probabilistic Interpretation

In repeated sampling, from a normally distributed population with a known standard deviation, $100(1 - \alpha)$ percent of all intervals of the form $\bar{x} \pm z_{(1-\alpha/2)}\sigma_{\bar{x}}$ will in the long run include the population mean μ.

The quantity $1 - \alpha$, in this case .95, is called the *confidence coefficient* (or confidence level), and the interval $\bar{x} \pm z_{(1-\alpha/2)}\sigma_{\bar{x}}$ is called a *confidence interval* for μ. When $(1 - \alpha) = .95$, the interval is called the 95 percent confidence interval for μ. In the present example we say that we are 95 percent confident that the population mean is between 17.76 and 26.24. This is called the *practical interpretation* of Expression 6.2.2. In general, it may be expressed as follows.

Practical Interpretation

When sampling is from a normally distributed population with known standard deviation, we are $100(1 - \alpha)$ percent confident that the single computed interval, $\bar{x} \pm z_{(1-\alpha/2)}\sigma_{\bar{x}}$, contains the population mean μ.

In the example given here we might prefer, rather than 2, the more exact value of z, 1.96, corresponding to a confidence coefficient of .95. Researchers may use any confidence coefficient they wish; the most frequently used values are .90, .95, and .99, which have associated reliability factors, respectively, of 1.645, 1.96, and 2.58.

Precision The quantity obtained by multiplying the reliability factor by the standard error of the mean is called the *precision* of the estimate. This quantity is also called the *margin of error.*

EXAMPLE 6.2.2

A physical therapist wished to estimate, with 99 percent confidence, the mean maximal strength of a particular muscle in a certain group of individuals. He is willing to assume that strength scores are approximately normally distributed with a variance of 144. A sample of 15 subjects who participated in the experiment yielded a mean of 84.3.

Solution: The z value corresponding to a confidence coefficient of .99 is found in Appendix Table D to be 2.58. This is our reliability coefficient. The standard error is $\sigma_{\bar{x}} = 12/\sqrt{15} = 3.0984$. Our 99 percent confidence interval for μ, then, is

$$84.3 \pm 2.58(3.0984)$$
$$84.3 \pm 8.0$$
$$76.3, 92.3$$

We say we are 99 percent confident that the population mean is between 76.3 and 92.3 since, in repeated sampling, 99 percent of all intervals that could be constructed in the manner just described would include the population mean. ■

Situations in which the variable of interest is approximately normally distributed with a known variance are so rare as to be almost nonexistent. The purpose of the preceding examples, which assumed that these ideal conditions existed, was to establish the theoretical background for constructing confidence intervals for population means. In most practical situations either the variables are not approximately normally distributed or the population variances are not known or both. Example 6.2.3 and Section 6.3 explain the procedures that are available for use in the less than ideal, but more common, situations.

Sampling from Nonnormal Populations As noted, it will not always be possible or prudent to assume that the population of interest is normally distributed. Thanks to the central limit theorem, this will not deter us if we are able to select a large enough sample. We have learned that for large samples, the sampling distribution of $\bar{x}$ is approximately normally distributed regardless of how the parent population is distributed.

EXAMPLE 6.2.3

Punctuality of patients in keeping appointments is of interest to a research team. In a study of patient flow through the offices of general practitioners, it was found that a sample of 35 patients were 17.2 minutes late for appointments, on the average. Previous research had shown the standard deviation to be about 8 minutes. The population distribution was felt to be nonnormal. What is the 90 percent confidence interval for μ, the true mean amount of time late for appointments?

Solution: Since the sample size is fairly large (greater than 30), and since the population standard deviation is known, we draw on the central limit theorem and assume the sampling distribution of $\bar{x}$ to be approximately normally distributed. From Appendix Table D we find the reliability coefficient corresponding to a confidence coefficient of .90 to be about 1.645, if we interpolate. The standard error is $\sigma_{\bar{x}} = 8/\sqrt{35} = 1.3522$, so that our 90 percent confidence interval for μ is

$$17.2 \pm 1.645(1.3522)$$
$$17.2 \pm 2.2$$
$$15.0, 19.4$$

■

Frequently, when the sample is large enough for the application of the central limit theorem, the population variance is unknown. In that case we use the sample variance as a replacement for the unknown population variance in the formula for constructing a confidence interval for the population mean.

Computer Analysis When confidence intervals are desired, a great deal of time can be saved if one uses a computer, which can be programmed to construct intervals from raw data.

EXAMPLE 6.2.4

The following are the activity values (micromoles per minute per gram of tissue) of a certain enzyme measured in normal gastric tissue of 35 patients with gastric carcinoma.

.360	1.189	.614	.788	.273	2.464	.571
1.827	.537	.374	.449	.262	.448	.971
.372	.898	.411	.348	1.925	.550	.622
.610	.319	.406	.413	.767	.385	.674
.521	.603	.533	.662	1.177	.307	1.499

We wish to use the MINITAB computer software package to construct a 95 percent confidence interval for the population mean. Suppose we know that the population variance is .36. It is not necessary to assume that the sampled population of values is normally distributed since the sample size is sufficiently large for application of the central limit theorem.

Solution: We enter the data into Column 1 and proceed as shown in Figure 6.2.2. These instructions tell the computer that the reliability factor is z, that a 95 percent confidence interval is desired, that the population standard deviation is .6, and that the data are in Column 1. The output tells us that the sample mean is .718, the sample standard deviation is .511, and the standard error of the mean, $\sigma/\sqrt{n}$ is $.6/\sqrt{35} = .101$.

■

We are 95 percent confident that the population mean is somewhere between .519 and .917. Confidence intervals may be obtained through the use of many other software packages. Users of SAS®, for example, may wish to use the output from PROC MEANS or PROC UNIVARIATE to construct confidence intervals.

Dialog box:

Stat ➤ Basic Statistics ➤ 1-Sample z

Type *C1* in **Samples in Columns.**
Type *.6* in **Standard deviation.** Click **OK.**

Session command:

```
MTB > ZINTERVAL 95 .6 C1
```

Output:

One-Sample Z: C1

```
The assumed standard deviaion = 0.600

Variable     N    Mean    StDev    SE Mean      95.0 % C.I.
MicMoles    35   0.718    0.511      0.101    ( 0.519, 0.917)
```

FIGURE 6.2.2 MINITAB procedure for constructing 95 percent confidence interval for a population mean, Example 6.2.4.

Alternative Estimates of Central Tendency As noted previously, the mean is sensitive to extreme values—those values that deviate appreciably from most of the measurements in a data set. They are sometimes referred to as *outliers.* We also noted earlier that the median, because it is not so sensitive to extreme measurements, is sometimes preferred over the mean as a measure of central tendency when outliers are present. For the same reason, we may prefer to use the sample median as an estimator of the population median when we wish to make an inference about the central tendency of a population. Not only may we use the sample median as a point estimate of the population median, we also may construct a confidence interval for the population median. The formula is not given here but may be found in the book by Rice (1).

Trimmed Mean Estimators that are insensitive to outliers are called *robust estimators.* Another robust measure and estimator of central tendency is the *trimmed mean.* For a set of sample data containing n measurements we calculate the 100α percent trimmed mean as follows:

1. Order the measurements.
2. Discard the smallest 100α percent and the largest 100α percent of the measurements. The recommended value of α is something between .1 and .2.
3. Compute the arithmetic mean of the remaining measurements.

Note that the median may be regarded as a 50 percent trimmed mean.

EXERCISES

For each of the following exercises construct 90, 95, and 99 percent confidence intervals for the population mean, and state the practical and probabilistic interpretations of each. Indicate which interpretation you think would be more appropriate to use when discussing confidence intervals with someone who has not had a course in statistics, and state the reason for your choice. Explain why the three intervals that you construct are not of equal width. Indicate which of the three intervals you would prefer to use as an estimate of the population mean, and state the reason for your choice.

6.2.1 We wish to estimate the average number of heartbeats per minute for a certain population. The average number of heartbeats per minute for a sample of 49 subjects was found to be 90. Assume that these 49 patients constitute a random sample, and that the population is normally distributed with a standard deviation of 10.

6.2.2 We wish to estimate the mean serum indirect bilirubin level of 4-day-old infants. The mean for a sample of 16 infants was found to be 5.98 mg/100 cc. Assume that bilirubin levels in 4-day-old infants are approximately normally distributed with a standard deviation of 3.5 mg/100 cc.

6.2.3 In a length of hospitalization study conducted by several cooperating hospitals, a random sample of 64 peptic ulcer patients was drawn from a list of all peptic ulcer patients ever admitted to the participating hospitals and the length of hospitalization per admission was determined for each. The mean length of hospitalization was found to be 8.25 days. The population standard deviation is known to be 3 days.

6.2.4 A sample of 100 apparently normal adult males, 25 years old, had a mean systolic blood pressure of 125. It is believed that the population standard deviation is 15.

6.2.5 Some studies of Alzheimer's disease (AD) have shown an increase in $^{14}CO_2$ production in patients with the disease. In one such study the following $^{14}CO_2$ values were obtained from 16 neocortical biopsy samples from AD patients.

1009	1280	1180	1255	1547	2352	1956	1080
1776	1767	1680	2050	1452	2857	3100	1621

Assume that the population of such values is normally distributed with a standard deviation of 350.

6.3 THE *t* DISTRIBUTION

In Section 6.2, a procedure was outlined for constructing a confidence interval for a population mean. The procedure requires knowledge of the variance of the population from which the sample is drawn. It may seem somewhat strange that one can have knowledge of the population variance and not know the value of the population mean. Indeed, it is the usual case, in situations such as have been presented, that the population variance, as well as the population mean, is unknown. This condition presents a problem with respect to constructing confidence intervals. Although, for example, the statistic

$$z = \frac{\bar{x} - \mu}{\sigma/\sqrt{n}}$$

is normally distributed when the population is normally distributed and is at least approximately normally distributed when n is large, regardless of the functional form of the population, we cannot make use of this fact because σ is unknown. However, all is not lost, and the most logical solution to the problem is the one followed. We use the sample standard deviation

$$s = \sqrt{\Sigma(x_i - \bar{x})^2/(n - 1)}$$

to replace σ. When the sample size is large, say, greater than 30, our faith in s as an approximation of σ is usually substantial, and we may be appropriately justified in using normal distribution theory to construct a confidence interval for the population mean. In that event, we proceed as instructed in Section 6.2.

It is when we have small samples that it becomes mandatory for us to find an alternative procedure for constructing confidence intervals.

As a result of the work of Gosset (2), writing under the pseudonym of "Student," an alternative, known as *Student's t distribution,* usually shortened to *t distribution,* is available to us.

The quantity

$$t = \frac{\bar{x} - \mu}{s/\sqrt{n}} \tag{6.3.1}$$

follows this distribution.

Properties of the *t* Distribution The t distribution has the following properties.

1. It has a mean of 0.
2. It is symmetrical about the mean.
3. In general, it has a variance greater than 1, but the variance approaches 1 as the sample size becomes large. For $df > 2$, the variance of the t distribution is $df/(df - 2)$, where df is the degrees of freedom. Alternatively, since here $df = n - 1$ for $n > 3$, we may write the variance of the t distribution as $(n - 1)/(n - 3)$.
4. The variable t ranges from $-\infty$ to $+\infty$.
5. The t distribution is really a family of distributions, since there is a different distribution for each sample value of $n - 1$, the divisor used in computing s^2. We recall that $n - 1$ is referred to as degrees of freedom. Figure 6.3.1 shows t distributions corresponding to several degrees-of-freedom values.

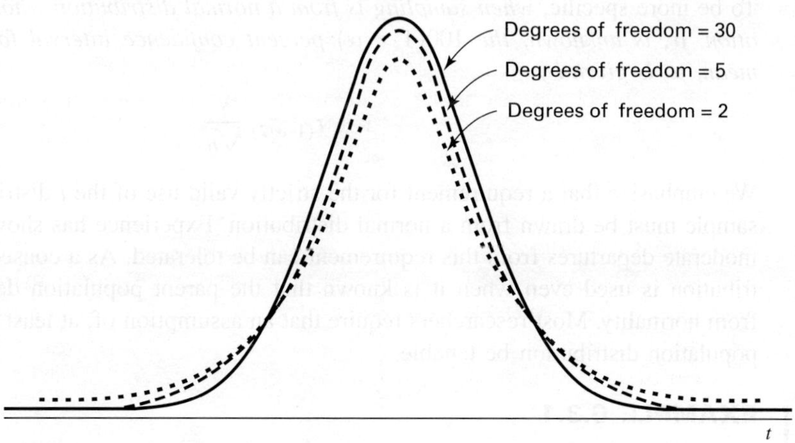

FIGURE 6.3.1 The t distribution for different degrees-of-freedom values.

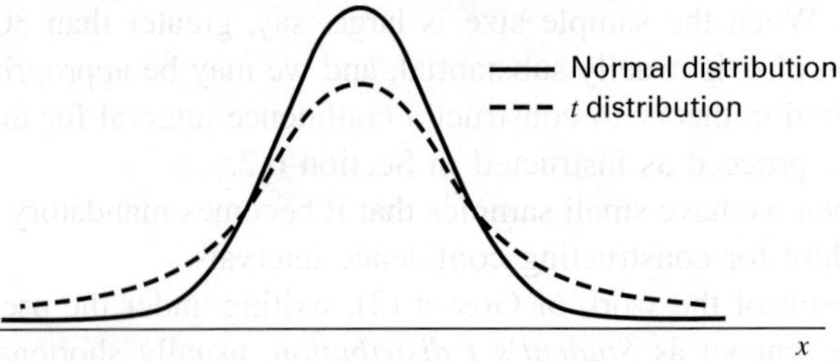

FIGURE 6.3.2 Comparison of normal distribution and t distribution.

6. Compared to the normal distribution, the t distribution is less peaked in the center and has thicker tails. Figure 6.3.2 compares the t distribution with the normal.
7. The t distribution approaches the normal distribution as $n - 1$ approaches infinity.

The t distribution, like the standard normal, has been extensively tabulated. One such table is given as Table E in the Appendix. As we will see, we must take both the confidence coefficient and degrees of freedom into account when using the table of the t distribution.

You may use MINITAB to graph the t distribution (for specified degrees-of-freedom values) and other distributions. After designating the horizontal axis by following directions in the Set Patterned Data box, choose menu path Calc and then Probability Distributions. Finally, click on the distribution desired and follow instructions. Use the Plot dialog box to plot the graph.

Confidence Intervals Using *t* The general procedure for constructing confidence intervals is not affected by our having to use the t distribution rather than the standard normal distribution. We still make use of the relationship expressed by

$$\text{estimator} \pm (\text{reliability coefficient}) \times (\text{standard error of the estimator})$$

What is different is the source of the reliability coefficient. It is now obtained from the table of the t distribution rather than from the table of the standard normal distribution. To be more specific, *when sampling is from a normal distribution whose standard deviation, σ, is unknown, the* $100(1 - \alpha)$ *percent confidence interval for the population mean, μ, is given by*

$$\bar{x} \pm t_{(1-\alpha/2)} \frac{s}{\sqrt{n}} \tag{6.3.2}$$

We emphasize that a requirement for the strictly valid use of the t distribution is that the sample must be drawn from a normal distribution. Experience has shown, however, that moderate departures from this requirement can be tolerated. As a consequence, the t distribution is used even when it is known that the parent population deviates somewhat from normality. Most researchers require that an assumption of, at least, a mound-shaped population distribution be tenable.

EXAMPLE 6.3.1

Maffulli et al. (A-1) studied the effectiveness of early weightbearing and ankle mobilization therapies following acute repair of a ruptured Achilles tendon. One of the variables they measured following treatment was the isometric gastrocsoleus muscle strength. In 19 subjects, the mean isometric strength for the operated limb (in newtons) was 250.8 with a standard deviation of 130.9. We assume that these 19 patients constitute a random sample from a population of similar subjects. We wish to use these sample data to estimate for the population the mean isometric strength after surgery.

Solution: We may use the sample mean, 250.8, as a point estimate of the population mean but, because the population standard deviation is unknown, we must assume the population of values to be at least approximately normally distributed before constructing a confidence interval for μ. Let us assume that such an assumption is reasonable and that a 95 percent confidence interval is desired. We have our estimator, $\bar{x}$, and our standard error is $s/\sqrt{n} = 130.9/\sqrt{19} = 30.0305$. We need now to find the reliability coefficient, the value of t associated with a confidence coefficient of .95 and $n - 1 = 18$ degrees of freedom. Since a 95 percent confidence interval leaves .05 of the area under the curve of t to be equally divided between the two tails, we need the value of t to the right of which lies .025 of the area. We locate in Appendix Table E the column headed $t_{.975}$. This is the value of t to the left of which lies .975 of the area under the curve. The area to the right of this value is equal to the desired .025. We now locate the number 18 in the degrees-of-freedom column. The value at the intersection of the row labeled 18 and the column labeled $t_{.975}$ is the t we seek. This value of t, which is our reliability coefficient, is found to be 2.1009. We now construct our 95 percent confidence interval as follows:

$$250.8 \pm 2.1009(30.0305)$$
$$250.8 \pm 63.1$$
$$187.7, 313.9$$

■

This interval may be interpreted from both the probabilistic and practical points of view. We are 95 percent confident that the true population mean, μ, is somewhere between 187.7 and 313.9 because, in repeated sampling, 95 percent of intervals constructed in like manner will include μ.

Deciding Between z and t When we construct a confidence interval for a population mean, we must decide whether to use a value of z or a value of t as the reliability factor. To make an appropriate choice we must consider sample size, whether the sampled population is normally distributed, and whether the population variance is known. Figure 6.3.3 provides a flowchart that one can use to decide quickly whether the reliability factor should be z or t.

Computer Analysis If you wish to have MINITAB construct a confidence interval for a population mean when the t statistic is the appropriate reliability factor, the command is TINTERVAL. In Windows choose 1-Sample t from the Basic Statistics menu.

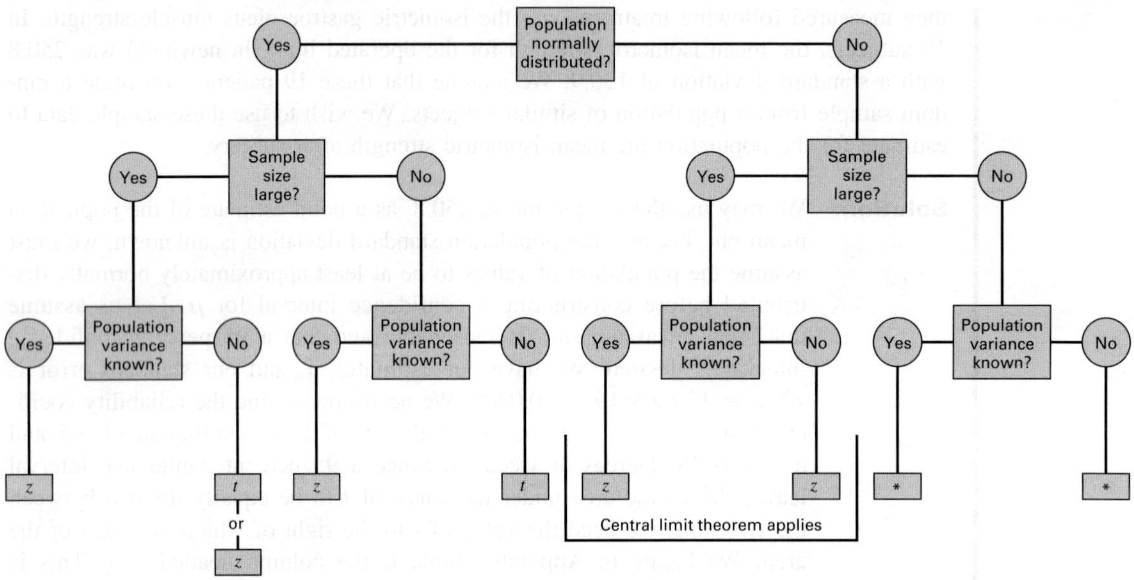

FIGURE 6.3.3 Flowchart for use in deciding between *z* and *t* when making inferences about population means. (*Use a nonparametric procedure. See Chapter 13.)

EXERCISES

6.3.1 Use the *t* distribution to find the reliability factor for a confidence interval based on the following confidence coefficients and sample sizes:

	a	*b*	*c*	*d*
Confidence coefficient	.95	.99	.90	.95
Sample size	15	24	8	30

6.3.2 In a study of the effects of early Alzheimer's disease on nondeclarative memory, Reber et al. (A-2) used the Category Fluency Test to establish baseline persistence and semantic memory and language abilities. The eight subjects in the sample had Category Fluency Test scores of 11, 10, 6, 3, 11, 10, 9, 11. Assume that the eight subjects constitute a simple random sample from a normally distributed population of similar subjects with early Alzheimer's disease.

(a) What is the point estimate of the population mean?

(b) What is the standard deviation of the sample?

(c) What is the estimated standard error of the sample mean?

(d) Construct a 95 percent confidence interval for the population mean category fluency test score.

(e) What is the precision of the estimate?

(f) State the probabilistic interpretation of the confidence interval you constructed.

(g) State the practical interpretation of the confidence interval you constructed.

6.3.3 Pedroletti et al. (A-3) reported the maximal nitric oxide diffusion rate in a sample of 15 asthmatic schoolchildren and 15 controls as mean ± standard error of the mean. For asthmatic children, they reported 3.5 ± 0.4 nL/s (nanoliters per second) and for control subjects they reported $0.7 \pm .1$ nL/s. For each group, determine the following:

(a) What was the sample standard deviation?

(b) What is the 95 percent confidence interval for the mean maximal nitric oxide diffusion rate of the population?

(c) What assumptions are necessary for the validity of the confidence interval you constructed?

(d) What are the practical and probabilistic interpretations of the interval you constructed?

(e) Which interpretation would be more appropriate to use when discussing confidence intervals with someone who has not had a course in statistics? State the reasons for your choice.

(f) If you were to construct a 90 percent confidence interval for the population mean from the information given here, would the interval be wider or narrower than the 95 percent confidence interval? Explain your answer without actually constructing the interval.

(g) If you were to construct a 99 percent confidence interval for the population mean from the information given here, would the interval be wider or narrower than the 95 percent confidence interval? Explain your answer without actually constructing the interval.

6.3.4 The concern of a study by Beynnon et al. (A-4) were nine subjects with chronic anterior cruciate ligament (ACL) tears. One of the variables of interest was the laxity of the anteroposterior, where higher values indicate more knee instability. The researchers found that among subjects with ACL-deficient knees, the mean laxity value was 17.4 mm with a standard deviation of 4.3 mm.

(a) What is the estimated standard error of the mean?

(b) Construct the 99 percent confidence interval for the mean of the population from which the nine subjects may be presumed to be a random sample.

(c) What is the precision of the estimate?

(d) What assumptions are necessary for the validity of the confidence interval you constructed?

6.3.5 A sample of 16 ten-year-old girls had a mean weight of 71.5 and a standard deviation of 12 pounds, respectively. Assuming normality, find the 90, 95, and 99 percent confidence intervals for μ.

6.3.6 The subjects of a study by Dugoff et al. (A-5) were 10 obstetrics and gynecology interns at the University of Colorado Health Sciences Center. The researchers wanted to assess competence in performing clinical breast examinations. One of the baseline measurements was the number of such examinations performed. The following data give the number of breast examinations performed for this sample of 10 interns.

Intern Number	No. of Breast Exams Performed
1	30
2	40
3	8
4	20
5	26
6	35
7	35
8	20
9	25
10	20

Source: Lorraine Dugoff, Mauritha R. Everett, Louis Vontver, and Gwyn E. Barley, "Evaluation of Pelvic and Breast Examination Skills of Interns in Obstetrics and Gynecology and Internal Medicine," *American Journal of Obstetrics and Gynecology, 189* (2003), 655–658.

Construct a 95 percent confidence interval for the mean of the population from which the study subjects may be presumed to have been drawn.

6.4 CONFIDENCE INTERVAL FOR THE DIFFERENCE BETWEEN TWO POPULATION MEANS

Sometimes there arise cases in which we are interested in estimating the difference between two population means. From each of the populations an independent random sample is drawn and, from the data of each, the sample means $\bar{x}_1$ and $\bar{x}_2$, respectively, are computed. We learned in the previous chapter that the estimator $\bar{x}_1 - \bar{x}_2$ yields an unbiased estimate of $\mu_1 - \mu_2$, the difference between the population means. The variance of the estimator is $(\sigma_1^2/n_1) + (\sigma_2^2/n_2)$. We also know from Chapter 5 that, depending on the conditions, the sampling distribution of $\bar{x}_1 - \bar{x}_2$ may be, at least, approximately normally distributed, so that in many cases we make use of the theory relevant to normal distributions to compute a confidence interval for $\mu_1 - \mu_2$. When the population variances are known, the $100(1 - \alpha)$ percent confidence interval for $\mu_1 - \mu_2$ is given by

$$(\bar{x}_1 - \bar{x}_2) \pm z_{1-\alpha/2}\sqrt{\frac{\sigma_1^2}{n_1} + \frac{\sigma_2^2}{n_2}} \qquad (6.4.1)$$

An examination of a confidence interval for the difference between population means provides information that is helpful in deciding whether or not it is likely that the two population means are equal. When the constructed interval does not include zero, we say that the interval provides evidence that the two population means are not equal. When the interval includes zero, we say that the population means may be equal.

Let us illustrate for the case where sampling is from normal distributions.

EXAMPLE 6.4.1

A research team is interested in the difference between serum uric acid levels in patients with and without Down's syndrome. In a large hospital for the treatment of the mentally retarded, a sample of 12 individuals with Down's syndrome yielded a mean of $\bar{x}_1 = 4.5$ mg/100 ml. In a general hospital a sample of 15 normal individuals of the same age and sex were found to have a mean value of $\bar{x}_2 = 3.4$. If it is reasonable to assume that the two populations of values are normally distributed with variances equal to 1 and 1.5, find the 95 percent confidence interval for $\mu_1 - \mu_2$.

Solution: For a point estimate of $\mu_1 - \mu_2$, we use $\bar{x}_1 - \bar{x}_2 = 4.5 - 3.4 = 1.1$. The reliability coefficient corresponding to .95 is found in Appendix Table D to be 1.96. The standard error is

$$\sigma_{\bar{x}_1 - \bar{x}_2} = \sqrt{\frac{\sigma_1^2}{n_1} + \frac{\sigma_2^2}{n_2}} = \sqrt{\frac{1}{12} + \frac{1.5}{15}} = .4282$$

The 95 percent confidence interval, then, is

$$1.1 \pm 1.96(.4282)$$
$$1.1 \pm .84$$
$$.26, 1.94$$

We say that we are 95 percent confident that the true difference, $\mu_1 - \mu_2$, is somewhere between .26 and 1.94 because, in repeated sampling, 95 percent of the intervals constructed in this manner would include the difference between the true means.

Since the interval does not include zero, we conclude that the two population means are not equal. ■

Sampling from Nonnormal Populations The construction of a confidence interval for the difference between two population means when sampling is from nonnormal populations proceeds in the same manner as in Example 6.4.1 if the sample sizes n_1 and n_2 are large. Again, this is a result of the central limit theorem. If the population variances are unknown, we use the sample variances to estimate them.

EXAMPLE 6.4.2

Despite common knowledge of the adverse effects of doing so, many women continue to smoke while pregnant. Mayhew et al. (A-6) examined the effectiveness of a smoking cessation program for pregnant women. The mean number of cigarettes smoked daily at the close of the program by the 328 women who completed the program was 4.3 with a standard deviation of 5.22. Among 64 women who did not complete the program, the mean number of cigarettes smoked per day at the close of the program was 13 with a standard deviation of 8.97. We wish to construct a 99 percent confidence interval for the difference between the means of the populations from which the samples may be presumed to have been selected.

Solution: No information is given regarding the shape of the distribution of cigarettes smoked per day. Since our sample sizes are large, however, the central limit theorem assures us that the sampling distribution of the difference between sample means will be approximately normally distributed even if the distribution of the variable in the populations is not normally distributed. We may use this fact as justification for using the z statistic as the reliability factor in the construction of our confidence interval. Also, since the population standard deviations are not given, we will use the sample standard deviations to estimate them. The point estimate for the difference between population means is the difference between sample means, $4.3 - 13.0 = -8.7$. In Appendix Table D we find the reliability factor to be 2.58. The estimated standard error is

$$s_{\bar{x}_1 - \bar{x}_2} = \sqrt{\frac{5.22^2}{328} + \frac{8.97^2}{64}} = 1.1577$$

By Equation 6.4.1, our 99 percent confidence interval for the difference between population means is

$$-8.7 \pm 2.58(1.1577)$$
$$-11.7, -5.7$$

We are 99 percent confident that the mean number of cigarettes smoked per day for women who complete the program is between 5.7 and 11.7 lower than the mean for women who do not complete the program. ■

The *t* Distribution and the Difference Between Means When population variances are unknown, and we wish to estimate the difference between two population means with a confidence interval, we can use the *t* distribution as a source of the reliability factor if certain assumptions are met. We must know, or be willing to assume, that the two sampled populations are normally distributed. With regard to the population variances, we distinguish between two situations: (1) the situation in which the population variances are equal, and (2) the situation in which they are not equal. Let us consider each situation separately.

Population Variances Equal If the assumption of equal population variances is justified, the two sample variances that we compute from our two independent samples may be considered as estimates of the same quantity, the common variance. It seems logical, then, that we should somehow capitalize on this in our analysis. We do just that and obtain a *pooled estimate* of the common variance. This pooled estimate is obtained by computing the weighted average of the two sample variances. Each sample variance is weighted by its degrees of freedom. If the sample sizes are equal, this weighted average is the arithmetic mean of the two sample variances. If the two sample sizes are unequal, the weighted average takes advantage of the additional information provided by the larger sample. The pooled estimate is given by the formula

$$s_p^2 = \frac{(n_1 - 1)s_1^2 + (n_2 - 1)s_2^2}{n_1 + n_2 - 2} \tag{6.4.2}$$

The standard error of the estimate, then, is given by

$$s_{\bar{x}_1 - \bar{x}_2} = \sqrt{\frac{s_p^2}{n_1} + \frac{s_p^2}{n_2}} \tag{6.4.3}$$

and the $100(1 - \alpha)$ percent confidence interval for $\mu_1 - \mu_2$ is given by

$$(\bar{x}_1 - \bar{x}_2) \pm t_{(1-\alpha/2)} \sqrt{\frac{s_p^2}{n_1} + \frac{s_p^2}{n_2}} \tag{6.4.4}$$

The number of degrees of freedom used in determining the value of t to use in constructing the interval is $n_1 + n_2 - 2$, the denominator of Equation 6.4.2. We interpret this interval in the usual manner.

Methods that may be used in reaching a decision about the equality of population variances are discussed in Sections 6.10 and 7.8.

EXAMPLE 6.4.3

The purpose of a study by Granholm et al. (A-7) was to determine the effectiveness of an integrated outpatient dual-diagnosis treatment program for mentally ill subjects. The authors were addressing the problem of substance abuse issues among people with severe mental disorders. A retrospective chart review was carried out on 50 consecutive patient referrals to the Substance Abuse/Mental Illness program at the VA San Diego Healthcare System. One of the outcome variables examined was the number of inpatient treatment days for psychiatric disorder during the year following the end of the program. Among 18 subjects with schizophrenia, the mean number of treatment days was 4.7 with a standard deviation of 9.3. For 10 subjects with bipolar disorder, the mean number of psychiatric disorder treatment days was 8.8 with a standard deviation of 11.5. We wish to construct a 95 percent confidence interval for the difference between the means of the populations represented by these two samples.

Solution: First we use Equation 6.4.2 to compute the pooled estimate of the common population variance.

$$s_p^2 = \frac{(18 - 1)(9.3^2) + (10 - 1)(11.5)^2}{18 + 10 - 2} = 102.33$$

When we enter Appendix Table E with $18 + 10 - 2 = 26$ degrees of freedom and a desired confidence level of .95, we find that the reliability factor is 2.0555. By Expression 6.4.4 we compute the 95 percent confidence interval for the difference between population means as follows:

$$(4.7 - 8.8) \pm 2.0555\sqrt{\frac{102.33}{18} + \frac{102.33}{10}}$$

$$-4.1 \pm 8.20$$

$$-12.3, 4.10$$

We are 95 percent confident that the difference between population means is somewhere between -12.3 and 4.10. We can say this because we know that if we were to repeat the study many, many times, and compute confidence intervals in the same way, about 95 percent of the intervals would include the difference between the population means.

Since the interval includes zero, we conclude that the population means may be equal. ■

Population Variances Not Equal When one is unable to conclude that the variances of two populations of interest are equal, even though the two populations may be assumed to be normally distributed, it is not proper to use the t distribution as just outlined in constructing confidence intervals.

A solution to the problem of unequal variances was proposed by Behrens (3) and later was verified and generalized by Fisher (4, 5). Solutions have also been proposed by Neyman (6), Scheffé (7, 8), and Welch (9, 10). The problem is discussed in detail by Cochran (11).

The problem revolves around the fact that the quantity

$$\frac{(\bar{x}_1 - \bar{x}_2) - (\mu_1 - \mu_2)}{\sqrt{\frac{s_p^2}{n_1} + \frac{s_p^2}{n_2}}}$$

does not follow a t distribution with $n_1 + n_2 - 2$ degrees of freedom when the population variances are not equal. Consequently, the t distribution cannot be used in the usual way to obtain the reliability factor for the confidence interval for the difference between the means of two populations that have unequal variances. The solution proposed by Cochran consists of computing the reliability factor, $t'_{1-\alpha/2}$, by the following formula:

$$t'_{1-\alpha/2} = \frac{w_1 t_1 + w_2 t_2}{w_1 + w_2} \tag{6.4.5}$$

where $w_1 = s_1^2/n_1$, $w_2 = s_2^2/n_2$, $t_1 = t_{1-\alpha/2}$ for $n_1 - 1$ degrees of freedom, and $t_2 = t_{1-\alpha/2}$ for $n_2 - 1$ degrees of freedom. An approximate $100(1 - \alpha)$ percent confidence interval for $\mu_1 - \mu_2$ is given by

$$(\bar{x}_1 - \bar{x}_2) \pm t'_{(1-\alpha/2)}\sqrt{\frac{s_1^2}{n_1} + \frac{s_2^2}{n_2}} \tag{6.4.6}$$

Adjustments to the reliability coefficient may also be made by reducing the number of degrees of freedom instead of modifying t in the manner just demonstrated. Many computer programs calculate an adjusted reliability coefficient in this way.

EXAMPLE 6.4.4

Let us reexamine the data presented in Example 6.4.3 from the study by Granholm et al. (A-7). Recall that among the 18 subjects with schizophrenia, the mean number of treatment days was 4.7 with a standard deviation of 9.3. In the bipolar disorder treatment group of 10 subjects, the mean number of psychiatric disorder treatment days was 8.8 with a standard deviation of 11.5. We assume that the two populations of number of psychiatric disorder days are approximately normally distributed. Now let us assume, however, that the two population variances are not equal. We wish to construct a 95 percent confidence interval for the difference between the means of the two populations represented by the samples.

Solution: We will use t' as found in Equation 6.4.5 for the reliability factor. Reference to Appendix Table E shows that with 17 degrees of freedom and $1 - .05/2 = .975$, $t_1 = 2.1098$. Similarly, with 9 degrees of freedom and $1 - .05/2 = .975$, $t_2 = 2.2622$. We now compute

$$t' = \frac{(9.3^2/18)(2.1098) + (11.5^2/10)(2.2622)}{(9.3^2/18) + (11.5^2/10)} = 2.2216$$

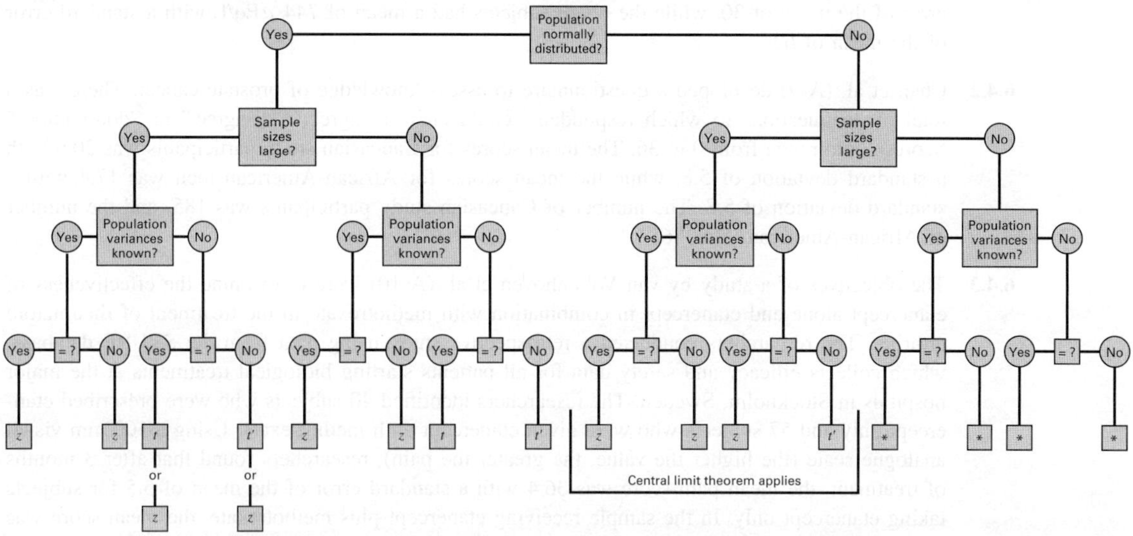

FIGURE 6.4.1 Flowchart for use in deciding whether the reliability factor should be z, t, or t' when making inferences about the difference between two population means. (*Use a nonparametric procedure. See Chapter 13.)

By Expression 6.4.6 we now construct the 95 percent confidence interval for the difference between the two population means.

$$(4.7 - 8.8) \pm 2.2216\sqrt{\frac{9.3^2}{18} + \frac{11.5^2}{10}}$$

$$(4.7 - 8.8) \pm 2.2216(4.246175)$$

$$-13.5, 5.3$$

Since the interval does include zero, we conclude that the two population means may be equal. ■

When constructing a confidence interval for the difference between two population means one may use Figure 6.4.1 to decide quickly whether the reliability factor should be z, t, or t'.

EXERCISES

For each of the following exercises construct 90, 95, and 99 percent confidence intervals for the difference between population means. Where appropriate, state the assumptions that make your method valid. State the practical and probabilistic interpretations of each interval that you construct. Consider the variables under consideration in each exercise, and state what use you think researchers might make of your results.

6.4.1 Iannelo et al. (A-8) performed a study that examined free fatty acid concentrations in 18 lean subjects and 11 obese subjects. The lean subjects had a mean level of 299 μEq/L with a standard error of the mean of 30, while the obese subjects had a mean of 744 μEq/L with a standard error of the mean of 62.

6.4.2 Chan et al. (A-9) developed a questionnaire to assess knowledge of prostate cancer. There was a total of 36 questions to which respondents could answer "agree," "disagree," or "don't know." Scores could range from 0 to 36. The mean scores for Caucasian study participants was 20.6 with a standard deviation of 5.8, while the mean scores for African-American men was 17.4 with a standard deviation of 5.8. The number of Caucasian study participants was 185, and the number of African-Americans was 86.

6.4.3 The objectives of a study by van Vollenhoven et al. (A-10) were to examine the effectiveness of etanercept alone and etanercept in combination with methotrexate in the treatment of rheumatoid arthritis. The researchers conducted a retrospective study using data from the STURE database, which collects efficacy and safety data for all patients starting biological treatments at the major hospitals in Stockholm, Sweden. The researchers identified 40 subjects who were prescribed etanercept only and 57 subjects who were given etanercept with methotrexate. Using a 100-mm visual analogue scale (the higher the value, the greater the pain), researchers found that after 3 months of treatment, the mean pain score was 36.4 with a standard error of the mean of 5.5 for subjects taking etanercept only. In the sample receiving etanercept plus methotrexate, the mean score was 30.5 with a standard error of the mean of 4.6.

6.4.4 The purpose of a study by Nozawa et al. (A-11) was to determine the effectiveness of segmental wire fixation in athletes with spondylolysis. Between 1993 and 2000, 20 athletes (6 women and 14 men) with lumbar spondylolysis were treated surgically with the technique. The following table gives the Japanese Orthopaedic Association (JOA) evaluation score for lower back pain syndrome for men and women prior to the surgery. The lower score indicates less pain.

Gender	JOA scores
Female	14, 13, 24, 21, 20, 21
Male	21, 26, 24, 24, 22, 23, 18, 24, 13, 22, 25, 23, 21, 25

Source: Satoshi Nozawa, Katsuji Shimizu, Kei Miyamoto, and Mizuo Tanaka, "Repair of Pars Interarticularis Defect by Segmental Wire Fixation in Young Athletes with Spondylolysis," *American Journal of Sports Medicine, 31* (2003), 359–364.

6.4.5 Krantz et al. (A-12) investigated dose-related effects of methadone in subjects with torsade de pointes, a polymorphic ventricular tachycardia. In the study of 17 subjects, nine were being treated with methadone for opiate dependency and eight for chronic pain. The mean daily dose of methadone in the opiate dependency group was 541 mg/day with a standard deviation of 156, while the chronic pain group received a mean dose of 269 mg/day with a standard deviation of 316.

6.4.6 Transverse diameter measurements on the hearts of adult males and females gave the following results:

Group	Sample Size	$\bar{x}$ (cm)	s (cm)
Males	12	13.21	1.05
Females	9	11.00	1.01

Assume normally distributed populations with equal variances.

6.4.7 Twenty-four experimental animals with vitamin D deficiency were divided equally into two groups. Group 1 received treatment consisting of a diet that provided vitamin D. The second group was not treated. At the end of the experimental period, serum calcium determinations were made with the following results:

Treated group: $\bar{x} = 11.1$ mg/100 ml, $s = 1.5$

Untreated group: $\bar{x} = 7.8$ mg/100 ml, $s = 2.0$

Assume normally distributed populations with equal variances.

6.4.8 Two groups of children were given visual acuity tests. Group 1 was composed of 11 children who receive their health care from private physicians. The mean score for this group was 26 with a standard deviation of 5. Group 2 was composed of 14 children who receive their health care from the health department, and had an average score of 21 with a standard deviation of 6. Assume normally distributed populations with equal variances.

6.4.9 The average length of stay of a sample of 20 patients discharged from a general hospital was 7 days with a standard deviation of 2 days. A sample of 24 patients discharged from a chronic disease hospital had an average length of stay of 36 days with a standard deviation of 10 days. Assume normally distributed populations with unequal variances.

6.4.10 In a study of factors thought to be responsible for the adverse effects of smoking on human reproduction, cadmium level determinations (nanograms per gram) were made on placenta tissue of a sample of 14 mothers who were smokers and an independent random sample of 18 nonsmoking mothers. The results were as follows:

Nonsmokers: 10.0, 8.4, 12.8, 25.0, 11.8, 9.8, 12.5, 15.4, 23.5, 9.4, 25.1, 19.5, 25.5, 9.8, 7.5, 11.8, 12.2, 15.0

Smokers: 30.0, 30.1, 15.0, 24.1, 30.5, 17.8, 16.8, 14.8, 13.4, 28.5, 17.5, 14.4, 12.5, 20.4

Does it appear likely that the mean cadmium level is higher among smokers than nonsmokers? Why do you reach this conclusion?

6.5 CONFIDENCE INTERVAL FOR A POPULATION PROPORTION

Many questions of interest to the health worker relate to population proportions. What proportion of patients who receive a particular type of treatment recover? What proportion of some population has a certain disease? What proportion of a population is immune to a certain disease?

To estimate a population proportion we proceed in the same manner as when estimating a population mean. A sample is drawn from the population of interest, and the sample proportion, $\hat{p}$, is computed. This sample proportion is used as the point estimator of the population proportion. A confidence interval is obtained by the general formula

$$\text{estimator} \pm (\text{reliability coefficient}) \times (\text{standard error of the estimator})$$

In the previous chapter we saw that when both np and $n(1 - p)$ are greater than 5, we may consider the sampling distribution of $\hat{p}$ to be quite close to the normal distribution. When this condition is met, our reliability coefficient is some value of z from the standard normal distribution. The standard error, we have seen, is equal to $\sigma_{\hat{p}} = \sqrt{p(1-p)/n}$. Since p, the parameter we are trying to estimate, is unknown, we must use $\hat{p}$ as an estimate. Thus, we estimate $\sigma_{\hat{p}}$ by $\sqrt{\hat{p}(1-\hat{p})/n}$, and our $100(1-\alpha)$ percent confidence interval for p is given by

$$\hat{p} \pm z_{1-\alpha/2}\sqrt{\hat{p}(1-\hat{p})/n} \tag{6.5.1}$$

We give this interval both the probabilistic and practical interpretations.

EXAMPLE 6.5.1

The Pew Internet and American Life Project (A-13) reported in 2003 that 18 percent of Internet users have used it to search for information regarding experimental treatments or medicines. The sample consisted of 1220 adult Internet users, and information was collected from telephone interviews. We wish to construct a 95 percent confidence interval for the proportion of Internet users in the sampled population who have searched for information on experimental treatments or medicines.

Solution: We shall assume that the 1220 subjects were sampled in random fashion. The best point estimate of the population proportion is $\hat{p} = .18$. The size of the sample and our estimate of p are of sufficient magnitude to justify use of the standard normal distribution in constructing a confidence interval. The reliability coefficient corresponding to a confidence level of .95 is 1.96, and our estimate of the standard error $\sigma_{\hat{p}}$ is $\sqrt{\hat{p}(1-\hat{p})/n} = \sqrt{(.18)(.82)/1220} = .0110$. The 95 percent confidence interval for p, based on these data, is

$$.18 \pm 1.96(.0110)$$
$$.18 \pm .022$$
$$.158, .202$$

We are 95 percent confident that the population proportion p is between .158 and .202 because, in repeated sampling, about 95 percent of the intervals constructed in the manner of the present single interval would include the true p. On the basis of these results we would expect, with 95 percent confidence, to find somewhere between 15.8 percent and 20.2 percent of adult Internet users to have used it for information on medicine or experimental treatments. ■

EXERCISES

For each of the following exercises state the practical and probabilistic interpretations of the interval that you construct. Identify each component of the interval: point estimate, reliability coefficient, and standard error. Explain why the reliability coefficients are not the same for all exercises.

6.5.1 Luna et al. (A-14) studied patients who were mechanically ventilated in the intensive care unit of six hospitals in Buenos Aires, Argentina. The researchers found that of 472 mechanically ventilated patients, 63 had clinical evidence of ventilator-associated pneumonia (VAP). Construct a 95 percent confidence interval for the proportion of all mechanically ventilated patients at these hospitals who may be expected to develop VAP.

6.5.2 Q waves on the electrocardiogram, according to Schinkel et al. (A-15), are often considered to be reflective of irreversibly scarred myocardium. These researchers assert, however, that there are some indications that residual viable tissue may be present in Q-wave-infarcted regions. Their study of 150 patients with chronic electrocardiographic Q-wave infarction found 202 dysfunctional Q-wave regions. With dobutamine stress echocardiography (DSE), they noted that 118 of these 202 regions were viable with information from the DSE testing. Construct a 90 percent confidence interval for the proportion of viable regions that one might expect to find a population of dysfunctional Q-wave regions.

6.5.3 In a study by von zur Muhlen et al. (A-16), 136 subjects with syncope or near syncope were studied. Syncope is the temporary loss of consciousness due to a sudden decline in blood flow to the brain. Of these subjects, 75 also reported having cardiovascular disease. Construct a 99 percent confidence interval for the population proportion of subjects with syncope or near syncope who also have cardiovascular disease.

6.5.4 In a simple random sample of 125 unemployed male high-school dropouts between the ages of 16 and 21, inclusive, 88 stated that they were regular consumers of alcoholic beverages. Construct a 95 percent confidence interval for the population proportion.

6.6 CONFIDENCE INTERVAL FOR THE DIFFERENCE BETWEEN TWO POPULATION PROPORTIONS

The magnitude of the difference between two population proportions is often of interest. We may want to compare, for example, men and women, two age groups, two socioeconomic groups, or two diagnostic groups with respect to the proportion possessing some characteristic of interest. An unbiased point estimator of the difference between two population proportions is provided by the difference between sample proportions, $\hat{p}_1 - \hat{p}_2$. As we have seen, when n_1 and n_2 are large and the population proportions are not too close to 0 or 1, the central limit theorem applies and normal distribution theory may be employed to obtain confidence intervals. The standard error of the estimate usually must be estimated by

$$\hat{\sigma}_{\hat{p}_1 - \hat{p}_2} = \sqrt{\frac{\hat{p}_1(1-\hat{p}_1)}{n_1} + \frac{\hat{p}_2(1-\hat{p}_2)}{n_2}}$$

because, as a rule, the population proportions are unknown. A $100(1-\alpha)$ percent confidence interval for $p_1 - p_2$ is given by

$$(\hat{p}_1 - \hat{p}_2) \pm z_{1-\alpha/2}\sqrt{\frac{\hat{p}_1(1-\hat{p}_1)}{n_1} + \frac{\hat{p}_2(1-\hat{p}_2)}{n_2}} \tag{6.6.1}$$

We may interpret this interval from both the probabilistic and practical points of view.

EXAMPLE 6.6.1

Connor et al. (A-17) investigated gender differences in proactive and reactive aggression in a sample of 323 children and adolescents (68 females and 255 males). The subjects were from unsolicited consecutive referrals to a residential treatment center and a pediatric psychopharmacology clinic serving a tertiary hospital and medical school. In the sample, 31 of the females and 53 of the males reported sexual abuse. We wish to construct a 99 percent confidence interval for the difference between the proportions of sexual abuse in the two sampled populations.

Solution: The sample proportions for the females and males are, respectively, $\hat{p}_F = 31/68 = .4559$ and $\hat{p}_M = 53/255 = .2078$. The difference between sample proportions is $\hat{p}_F - \hat{p}_M = .4559 - .2078 = .2481$. The estimated standard error of the difference between sample proportions is

$$\hat{\sigma}_{\hat{p}_F - \hat{p}_M} = \sqrt{\frac{(.4559)(.5441)}{68} + \frac{(.2078)(.7922)}{255}}$$
$$= .0655$$

The reliability factor from Appendix Table D is 2.58, so that our confidence interval, by Expression 6.6.1, is

$$.2481 \pm 2.58(.0655)$$
$$.0791, .4171$$

We are 99 percent confident that for the sampled populations, the proportion of cases of reported sexual abuse among females exceeds the proportion of cases of reported sexual abuse among males by somewhere between .0791 and .4171.

Since the interval does not include zero, we conclude that the two population proportions are not equal. ■

EXERCISES

For each of the following exercises state the practical and probabilistic interpretations of the interval that you construct. Identify each component of the interval: point estimate, reliability coefficient, and standard error. Explain why the reliability coefficients are not the same for all exercises.

6.6.1 Horwitz et al. (A-18) studied 637 persons who were identified by court records from 1967 to 1971 as having experienced abuse or neglect. For a control group, they located 510 subjects who as children attended the same elementary school and lived within a five-block radius of those in the abused/neglected group. In the abused/neglected group, and control group, 114 and 57 subjects, respectively, had developed antisocial personality disorders over their lifetimes. Construct a 95 percent confidence interval for the difference between the proportions of subjects developing antisocial personality disorders one might expect to find in the populations of subjects from which the subjects of this study may be presumed to have been drawn.

6.6.2 The objective of a randomized controlled trial by Adab et al. (A-19) was to determine whether providing women with additional information on the pros and cons of screening for cervical cancer would increase the willingness to be screened. A treatment group of 138 women received a leaflet on screening that contained more information (average individual risk for cervical cancer, likelihood of positive finding, the possibility of false positive/negative results, etc.) than the standard leaflet developed by the British National Health Service that 136 women in a control group received. In the treatment group, 109 women indicated they wanted to have the screening test for cervical cancer while in the control group, 120 indicated they wanted the screening test. Construct a 95 percent confidence interval for the difference in proportions for the two populations represented by these samples.

6.6.3 Spertus et al. (A-20) performed a randomized single blind study for subjects with stable coronary artery disease. They randomized subjects into two treatment groups. The first group had current angina medications optimized, and the second group was tapered off existing medications and then started on long-acting diltiazem at 180 mg/day. The researchers performed several tests to determine if there were significant differences in the two treatment groups at baseline. One of the characteristics of interest was the difference in the percentages of subjects who had reported a history of congestive heart failure. In the group where current medications were optimized, 16 of 49 subjects reported a history of congestive heart failure. In the subjects placed on the diltiazem, 12 of the 51 subjects reported a history of congestive heart failure. State the assumptions that you think are necessary and construct a 95 percent confidence interval for the difference between the proportions of those reporting congestive heart failure within the two populations from which we presume these treatment groups to have been selected.

6.6.4 To study the difference in drug therapy adherence among subjects with depression who received usual care and those who received care in a collaborative care model was the goal of a study conducted by Finley et al. (A-21). The collaborative care model emphasized the role of clinical pharmacists in providing drug therapy management and treatment follow-up. Of the 50 subjects receiving usual care, 24 adhered to the prescribed drug regimen, while 50 out of 75 subjects in the collaborative care model adhered to the drug regimen. Construct a 90 percent confidence interval for the difference in adherence proportions for the populations of subjects represented by these two samples.

6.7 DETERMINATION OF SAMPLE SIZE FOR ESTIMATING MEANS

The question of how large a sample to take arises early in the planning of any survey or experiment. This is an important question that should not be treated lightly. To take a larger sample than is needed to achieve the desired results is wasteful of resources, whereas very small samples often lead to results that are of no practical use. Let us consider, then, how one may go about determining the sample size that is needed in a given situation. In this section, we present a method for determining the sample size required for estimating a population mean, and in the next section we apply this method to the case of sample size determination when the parameter to be estimated is a population proportion. By straightforward extensions of these methods, sample sizes required for more complicated situations can be determined.

Objectives The objectives in interval estimation are to obtain narrow intervals with high reliability. If we look at the components of a confidence interval, we see that the width of the interval is determined by the magnitude of the quantity

$$(\text{reliability coefficient}) \times (\text{standard error of the estimator})$$

since the total width of the interval is twice this amount. We have learned that this quantity is usually called the precision of the estimate or the margin of error. For a given standard error, increasing reliability means a larger reliability coefficient. But a larger reliability coefficient for a fixed standard error makes for a wider interval.

On the other hand, if we fix the reliability coefficient, the only way to reduce the width of the interval is to reduce the standard error. Since the standard error is equal to $\sigma/\sqrt{n}$, and since σ is a constant, the only way to obtain a small standard error is to take a large sample. How large a sample? That depends on the size of σ, the population standard deviation, the desired degree of reliability, and the desired interval width.

Let us suppose we want an interval that extends d units on either side of the estimator. We can write

$$d = (\text{reliability coefficient}) \times (\text{standard error of the estimator}) \tag{6.7.1}$$

If sampling is to be with replacement, from an infinite population, or from a population that is sufficiently large to warrant our ignoring the finite population correction, Equation 6.7.1 becomes

$$d = z\frac{\sigma}{\sqrt{n}} \tag{6.7.2}$$

which, when solved for n, gives

$$n = \frac{z^2\sigma^2}{d^2} \tag{6.7.3}$$

When sampling is without replacement from a small finite population, the finite population correction is required and Equation 6.7.1 becomes

$$d = z\frac{\sigma}{\sqrt{n}}\sqrt{\frac{N-n}{N-1}} \tag{6.7.4}$$

which, when solved for n, gives

$$n = \frac{Nz^2\sigma^2}{d^2(N-1) + z^2\sigma^2} \tag{6.7.5}$$

If the finite population correction can be ignored, Equation 6.7.5 reduces to Equation 6.7.3.

Estimating σ^2 The formulas for sample size require knowledge of σ^2 but, as has been pointed out, the population variance is, as a rule, unknown. As a result, σ^2 has to be estimated. The most frequently used sources of estimates for σ^2 are the following:

1. A *pilot* or preliminary sample may be drawn from the population, and the variance computed from this sample may be used as an estimate of σ^2. Observations used in the pilot sample may be counted as part of the final sample, so that n (the computed sample size) $- n_1$ (the pilot sample size) $= n_2$ (the number of observations needed to satisfy the total sample size requirement).

2. Estimates of σ^2 may be available from previous or similar studies.
3. If it is thought that the population from which the sample is to be drawn is approximately normally distributed, one may use the fact that the range is approximately equal to six standard deviations and compute $\sigma \approx R/6$. This method requires some knowledge of the smallest and largest value of the variable in the population.

EXAMPLE 6.7.1

A health department nutritionist, wishing to conduct a survey among a population of teenage girls to determine their average daily protein intake (measured in grams), is seeking the advice of a biostatistician relative to the sample size that should be taken.

What procedure does the biostatistician follow in providing assistance to the nutritionist? Before the statistician can be of help to the nutritionist, the latter must provide three items of information: (1) the desired width of the confidence interval, (2) the level of confidence desired, and (3) the magnitude of the population variance.

Solution: Let us assume that the nutritionist would like an interval about 10 grams wide; that is, the estimate should be within about 5 grams of the population mean in either direction. In other words, a margin of error of 5 grams is desired. Let us also assume that a confidence coefficient of .95 is decided on and that, from past experience, the nutritionist feels that the population standard deviation is probably about 20 grams. The statistician now has the necessary information to compute the sample size: $z = 1.96$, $\sigma = 20$, and $d = 5$. Let us assume that the population of interest is large so that the statistician may ignore the finite population correction and use Equation 6.7.3. On making proper substitutions, the value of n is found to be

$$n = \frac{(1.96)^2 (20)^2}{(5)^2} = 61.47$$

The nutritionist is advised to take a sample of size 62. When calculating a sample size by Equation 6.7.3 or Equation 6.7.5, we round up to the next-largest whole number if the calculations yield a number that is not itself an integer. ■

EXERCISES

6.7.1 A hospital administrator wishes to estimate the mean weight of babies born in her hospital. How large a sample of birth records should be taken if she wants a 99 percent confidence interval that is 1 pound wide? Assume that a reasonable estimate of σ is 1 pound. What sample size is required if the confidence coefficient is lowered to .95?

6.7.2 The director of the rabies control section in a city health department wishes to draw a sample from the department's records of dog bites reported during the past year in order to estimate the mean age of persons bitten. He wants a 95 percent confidence interval, he will be satisfied to let $d = 2.5$, and from previous studies he estimates the population standard deviation to be about 15 years. How large a sample should be drawn?

6.7.3 A physician would like to know the mean fasting blood glucose value (milligrams per 100 ml) of patients seen in a diabetes clinic over the past 10 years. Determine the number of records the physician should examine in order to obtain a 90 percent confidence interval for μ if the desired width of the interval is 6 units and a pilot sample yields a variance of 60.

6.7.4 For multiple sclerosis patients we wish to estimate the mean age at which the disease was first diagnosed. We want a 95 percent confidence interval that is 10 years wide. If the population variance is 90, how large should our sample be?

6.8 DETERMINATION OF SAMPLE SIZE FOR ESTIMATING PROPORTIONS

The method of sample size determination when a population proportion is to be estimated is essentially the same as that described for estimating a population mean. We make use of the fact that one-half the desired interval, d, may be set equal to the product of the reliability coefficient and the standard error.

Assuming that random sampling and conditions warranting approximate normality of the distribution of $\hat{p}$ leads to the following formula for n when sampling is with replacement, when sampling is from an infinite population, or when the sampled population is large enough to make use of the finite population correction unnecessary,

$$n = \frac{z^2pq}{d^2} \tag{6.8.1}$$

where $q = 1 - p$.

If the finite population correction cannot be disregarded, the proper formula for n is

$$n = \frac{Nz^2pq}{d^2(N-1) + z^2pq} \tag{6.8.2}$$

When N is large in comparison to n (that is, $n/N \leq .05$) the finite population correction may be ignored, and Equation 6.8.2 reduces to Equation 6.8.1.

Estimating p As we see, both formulas require knowledge of p, the proportion in the population possessing the characteristic of interest. Since this is the parameter we are trying to estimate, it, obviously, will be unknown. One solution to this problem is to take a pilot sample and compute an estimate to be used in place of p in the formula for n. Sometimes an investigator will have some notion of an upper bound for p that can be used in the formula. For example, if it is desired to estimate the proportion of some population who have a certain disability, we may feel that the true proportion cannot be greater than, say, .30. We then substitute .30 for p in the formula for n. If it is impossible to come up with a better estimate, one may set p equal to .5 and solve for n. Since $p = .5$ in the formula yields the maximum value of n, this procedure will give a large enough sample for the desired reliability and interval width. It may, however, be larger than needed and result in a more expensive sample than if a better estimate of p had been available. This procedure should be used only if one is unable to arrive at a better estimate of p.

EXAMPLE 6.8.1

A survey is being planned to determine what proportion of families in a certain area are medically indigent. It is believed that the proportion cannot be greater than .35. A 95 percent confidence interval is desired with $d = .05$. What size sample of families should be selected?

Solution: If the finite population correction can be ignored, we have

$$n = \frac{(1.96)^2(.35)(.65)}{(.05)^2} = 349.59$$

The necessary sample size, then, is 350. ■

EXERCISES

6.8.1 An epidemiologist wishes to know what proportion of adults living in a large metropolitan area have subtype ayr hepatitis B virus. Determine the sample size that would be required to estimate the true proportion to within .03 with 95 percent confidence. In a similar metropolitan area the proportion of adults with the characteristic is reported to be .20. If data from another metropolitan area were not available and a pilot sample could not be drawn, what sample size would be required?

6.8.2 A survey is planned to determine what proportion of the high-school students in a metropolitan school system have regularly smoked marijuana. If no estimate of p is available from previous studies, a pilot sample cannot be drawn, a confidence coefficient of .95 is desired, and $d = .04$ is to be used, determine the appropriate sample size. What sample size would be required if 99 percent confidence were desired?

6.8.3 A hospital administrator wishes to know what proportion of discharged patients is unhappy with the care received during hospitalization. How large a sample should be drawn if we let $d = .05$, the confidence coefficient is .95, and no other information is available? How large should the sample be if p is approximated by .25?

6.8.4 A health planning agency wishes to know, for a certain geographic region, what proportion of patients admitted to hospitals for the treatment of trauma die in the hospital. A 95 percent confidence interval is desired, the width of the interval must be .06, and the population proportion, from other evidence, is estimated to be .20. How large a sample is needed?

6.9 CONFIDENCE INTERVAL FOR THE VARIANCE OF A NORMALLY DISTRIBUTED POPULATION

Point Estimation of the Population Variance In previous sections it has been suggested that when a population variance is unknown, the sample variance may be used as an estimator. You may have wondered about the quality of this estimator. We have discussed only one criterion of quality—unbiasedness—so let us see if the sample variance is an unbiased estimator of the population variance. To be unbiased, the average value of the sample variance over all possible samples must be equal to the population variance. That is, the expression $E(s^2) = \sigma^2$ must hold. To see if this condition holds for a particular situation, let us refer to the example of constructing a sampling distribution given in Section 5.3. In Table 5.3.1 we have all possible samples of size 2 from the population consisting of the values 6, 8, 10, 12, and 14. It will be recalled that two measures of dispersion for this population were computed as follows:

$$\sigma^2 = \frac{\Sigma(x_i - \mu)^2}{N} = 8 \quad \text{and} \quad S^2 = \frac{\Sigma(x_i - \mu)^2}{N - 1} = 10$$

If we compute the sample variance $s^2 = \Sigma(x_i - \bar{x})^2/(n - 1)$ for each of the possible samples shown in Table 5.3.1, we obtain the sample variances shown in Table 6.9.1.

Sampling with Replacement If sampling is with replacement, the expected value of s^2 is obtained by taking the mean of all sample variances in Table 6.9.1. When we do this, we have

$$E(s^2) = \frac{\Sigma s_i^2}{N^n} = \frac{0 + 2 + \cdots + 2 + 0}{25} = \frac{200}{25} = 8$$

and we see, for example, that when sampling is with replacement $E(s^2) = \sigma^2$, where $s^2 = \Sigma(x_i - \bar{x})^2/(n - 1)$ and $\sigma^2 = \Sigma(x_i - \mu)^2/N$.

TABLE 6.9.1 Variances Computed from Samples Shown in Table 5.3.1

		Second Draw				
		6	8	10	12	14
	6	0	2	8	18	32
	8	2	0	2	8	18
First Draw	10	8	2	0	2	8
	12	18	8	2	0	2
	14	32	18	8	2	0

Sampling Without Replacement If we consider the case where sampling is without replacement, the expected value of s^2 is obtained by taking the mean of all variances above (or below) the principal diagonal. That is,

$$E(s^2) = \frac{\Sigma s_i^2}{_N C_n} = \frac{2 + 8 + \cdots + 2}{10} = \frac{100}{10} = 10$$

which, we see, is not equal to σ^2, but is equal to $S^2 = \Sigma(x_i - \mu)^2/(N - 1)$.

These results are examples of general principles, as it can be shown that, in general,

$$E(s^2) = \sigma^2 \quad \textit{when sampling is with replacement}$$
$$E(s^2) = S^2 \quad \textit{when sampling is without replacement}$$

When N is large, $N - 1$ and N will be approximately equal and, consequently, σ^2 and S^2 will be approximately equal.

These results justify our use of $s^2 = \Sigma(x_i - \bar{x})^2/(n - 1)$ when computing the sample variance. In passing, let us note that although s^2 is an unbiased estimator of σ^2, s is not an unbiased estimator of σ. The bias, however, diminishes rapidly as n increases.

Interval Estimation of a Population Variance With a point estimate available, it is logical to inquire about the construction of a confidence interval for a population variance. Whether we are successful in constructing a confidence interval for σ^2 will depend on our ability to find an appropriate sampling distribution.

The Chi-Square Distribution Confidence intervals for σ^2 are usually based on the sampling distribution of $(n - 1)s^2/\sigma^2$. If samples of size n are drawn from a normally distributed population, this quantity has a distribution known as the *chi-square* (χ^2) *distribution* with $n - 1$ degrees of freedom. As we will say more about this distribution in chapter 12, we only say here that it is the distribution that the quantity $(n - 1)s^2/\sigma^2$ follows and that it is useful in finding confidence intervals for σ^2 when the assumption that the population is normally distributed holds true.

Figure 6.9.1 shows chi-square distributions for several values of degrees of freedom. Percentiles of the chi-square distribution are given in Appendix Table F. The column headings give the values of χ^2 to the left of which lies a proportion of the total area under the curve equal to the subscript of χ^2. The row labels are the degrees of freedom.

To obtain a $100(1 - \alpha)$ percent confidence interval for σ^2, we first obtain the $100(1 - \alpha)$ percent confidence interval for $(n - 1)s^2/\sigma^2$. To do this, we select the values of χ^2 from Appendix Table F in such a way that $\alpha/2$ is to the left of the smaller value and $\alpha/2$ is to the right of the larger value. In other words, the two values of χ^2 are selected in such a way that α is divided equally between the two tails of the distribution. We may

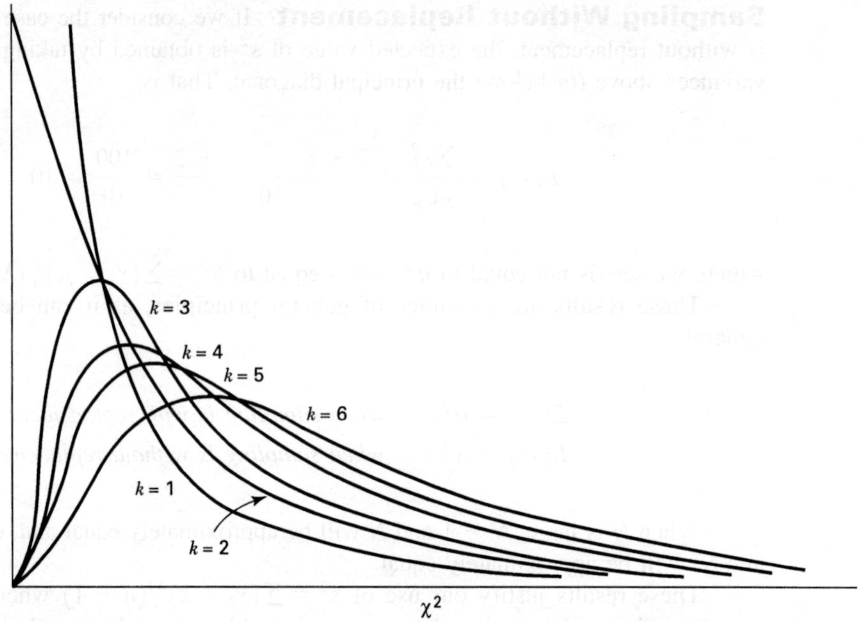

FIGURE 6.9.1 Chi-square distributions for several values of degrees of freedom *k*. (*Source:* Paul G. Hoel and Raymond J. Jessen, *Basic Statistics for Business and Economics,* Wiley, 1971. Used with permission.)

designate these two values of χ^2 as $\chi^2_{\alpha/2}$ and $\chi^2_{1-(\alpha/2)}$, respectively. The $100(1-\alpha)$ percent confidence interval for $(n-1)s^2/\sigma^2$, then, is given by

$$\chi^2_{\alpha/2} < \frac{(n-1)\,s^2}{\sigma^2} < \chi^2_{1-(\alpha/2)}$$

We now manipulate this expression in such a way that we obtain an expression with σ^2 alone as the middle term. First, let us divide each term by $(n-1)s^2$ to get

$$\frac{\chi^2_{\alpha/2}}{(n-1)s^2} < \frac{1}{\sigma^2} < \frac{\chi^2_{1-(\alpha/2)}}{(n-1)s^2}$$

If we take the reciprocal of this expression, we have

$$\frac{(n-1)s^2}{\chi^2_{\alpha/2}} > \sigma^2 > \frac{(n-1)s^2}{\chi^2_{1-(\alpha/2)}}$$

Note that the direction of the inequalities changed when we took the reciprocals. If we reverse the order of the terms, we have

$$\frac{(n-1)s^2}{\chi^2_{1-(\alpha/2)}} < \sigma^2 < \frac{(n-1)s^2}{\chi^2_{\alpha/2}} \qquad (6.9.1)$$

which is the $100(1-\alpha)$ percent confidence interval for σ^2. If we take the square root of each term in Expression 6.9.1, we have the following $100(1-\alpha)$ percent confidence interval for σ, the population standard deviation:

$$\sqrt{\frac{(n-1)s^2}{\chi^2_{1-(\alpha/2)}}} < \sigma < \sqrt{\frac{(n-1)s^2}{\chi^2_{\alpha/2}}} \qquad (6.9.2)$$

EXAMPLE 6.9.1

In a study of the effectiveness of a gluten-free diet in first-degree relatives of patients with type I diabetics, Hummel et al. (A-22) placed seven subjects on a gluten-free diet for 12 months. Prior to the diet, they took baseline measurements of several antibodies and autoantibodies, one of which was the diabetes related insulin autoantibody (IAA). The IAA levels were measured by radiobinding assay. The seven subjects had IAA units of

9.7, 12.3, 11.2, 5.1, 24.8, 14.8, 17.7

We wish to estimate from the data in this sample the variance of the IAA units in the population from which the sample was drawn and construct a 95 percent confidence interval for this estimate.

Solution: The sample yielded a value of $s^2 = 39.763$. The degrees of freedom are $n - 1 = 6$. The appropriate values of χ^2 from Appendix Table F are $\chi^2_{1-(\alpha/2)} = 14.449$ and $\chi^2_{\alpha/2} = 1.237$. Our 95 percent confidence interval for σ^2 is

$$\frac{6(39.763)}{14.449} < \sigma^2 < \frac{6(39.763)}{1.237}$$

$$16.512 < \sigma^2 < 192.868$$

The 95 percent confidence interval for σ is

$$4.063 < \sigma < 13.888$$

We are 95 percent confident that the parameters being estimated are within the specified limits, because we know that in the long run, in repeated sampling, 95 percent of intervals constructed as illustrated would include the respective parameters. ■

Some Precautions Although this method of constructing confidence intervals for σ^2 is widely used, it is not without its drawbacks. First, the assumption of the normality of the population from which the sample is drawn is crucial, and results may be misleading if the assumption is ignored.

Another difficulty with these intervals results from the fact that the estimator is not in the center of the confidence interval, as is the case with the confidence interval for μ. This is because the chi-square distribution, unlike the normal, is not symmetric. The practical implication of this is that the method for the construction of confidence intervals for σ^2, which has just been described, does not yield the shortest possible confidence intervals. Tate and Klett (12) give tables that may be used to overcome this difficulty.

EXERCISES

6.9.1 A study by Aizenberg et al. (A-23) examined the efficacy of sildenafil, a potent phosphodiesterase inhibitor, in the treatment of elderly men with erectile dysfunction induced by antidepressant treatment for major depressive disorder. The ages of the 10 enrollees in the study were

74, 81, 70, 70, 74, 77, 76, 70, 71, 72

Assume that the subjects in this sample constitute a simple random sample drawn from a population of similar subjects. Construct a 95 percent confidence interval for the variance of the ages of subjects in the population.

6.9.2 Borden et al. (A-24) performed experiments on cadaveric knees to test the effectiveness of several meniscal repair techniques. Specimens were loaded into a servohydraulic device and tension-loaded to failure. The biomechanical testing was performed by using a slow loading rate to simulate the stresses that the medial meniscus might be subjected to during early rehabilitation exercises and activities of daily living. One of the measures is the amount of displacement that occurs. Of the 12 specimens receiving the vertical mattress suture and the FasT-FIX method, the displacement values measured in millimeters are 16.9, 20.2, 20.1, 15.7, 13.9, 14.9, 18.0, 18.5, 9.2, 18.8, 22.8, 17.5. Construct a 90 percent confidence interval for the variance of the displacement in millimeters for a population of subjects receiving these repair techniques.

6.9.3 Forced vital capacity determinations were made on 20 healthy adult males. The sample variance was 1,000,000. Construct 90 percent confidence intervals for σ^2 and σ.

6.9.4 In a study of myocardial transit times, appearance transit times were obtained on a sample of 30 patients with coronary artery disease. The sample variance was found to be 1.03. Construct 99 percent confidence intervals for σ^2 and σ.

6.9.5 A sample of 25 physically and mentally healthy males participated in a sleep experiment in which the percentage of each participant's total sleeping time spent in a certain stage of sleep was recorded. The variance computed from the sample data was 2.25. Construct 95 percent confidence intervals for σ^2 and σ.

6.9.6 Hemoglobin determinations were made on 16 animals exposed to a harmful chemical. The following observations were recorded: 15.6, 14.8, 14.4, 16.6, 13.8, 14.0, 17.3, 17.4, 18.6, 16.2, 14.7, 15.7, 16.4, 13.9, 14.8, 17.5. Construct 95 percent confidence intervals for σ^2 and σ.

6.9.7 Twenty air samples taken at the same site over a period of 6 months showed the following amounts of suspended particulate matter (micrograms per cubic meter of air):

68	22	36	32
42	24	28	38
30	44	28	27
28	43	45	50
79	74	57	21

Consider these measurements to be a random sample from a population of normally distributed measurements, and construct a 95 percent confidence interval for the population variance.

6.10 CONFIDENCE INTERVAL FOR THE RATIO OF THE VARIANCES OF TWO NORMALLY DISTRIBUTED POPULATIONS

It is frequently of interest to compare two variances, and one way to do this is to form their ratio, σ_1^2/σ_2^2. If two variances are equal, their ratio will be equal to 1. We usually will not know the variances of populations of interest, and, consequently, any comparisons we make will be based on sample variances. In other words, we may wish to estimate the ratio of two population variances. We learned in Section 6.4 that the valid use of the t distribution to construct a confidence interval for the difference between two population means requires that the population variances be equal. The use of the ratio of two population variances for determining equality of variances has been formalized into a statistical test. The distribution of this test provides test values for determining if the ratio exceeds the value 1 to a large enough extent that we may conclude that the variances are not equal. The test is referred to as the F-max Test by Hartley (13) or the Variance Ratio Test by Zar (14). Many computer programs provide some formalized test of the equality of variances so that the assumption of equality of variances associated with many of the tests in the following chapters can be examined. If the confidence interval for the ratio of two population variances includes 1, we conclude that the two population variances may, in fact, be equal. Again, since this is a form of inference, we must rely on some sampling distribution, and this time the distribution of $(s_1^2/\sigma_1^2)/(s_2^2/\sigma_2^2)$ is utilized provided certain assumptions are met. The assumptions are that s_1^2 and s_2^2 are computed from independent samples of size n_1 and n_2, respectively, drawn from two normally distributed populations. We use s_1^2 to designate the larger of the two sample variances.

The *F* Distribution If the assumptions are met, $(s_1^2/\sigma_1^2)/(s_2^2/\sigma_2^2)$ follows a distribution known as the *F distribution.* We defer a more complete discussion of this distribution until chapter 8, but note that this distribution depends on two-degrees-of-freedom values, one corresponding to the value of $n_1 - 1$ used in computing s_1^2 and the other corresponding to the value of $n_2 - 1$ used in computing s_2^2. These are usually referred to as the *numerator degrees of freedom* and the *denominator degrees of freedom.* Figure 6.10.1 shows some F distributions for several numerator and denominator degrees-of-freedom combinations. Appendix Table G contains, for specified combinations of degrees of freedom and values of α, F values to the right of which lies $\alpha/2$ of the area under the curve of F.

A Confidence Interval for σ_1^2/σ_2^2 To find the $100(1-\alpha)$ percent confidence interval for σ_1^2/σ_2^2, we begin with the expression

$$F_{\alpha/2} < \frac{s_1^2/\sigma_1^2}{s_2^2/\sigma_2^2} < F_{1-(\alpha/2)}$$

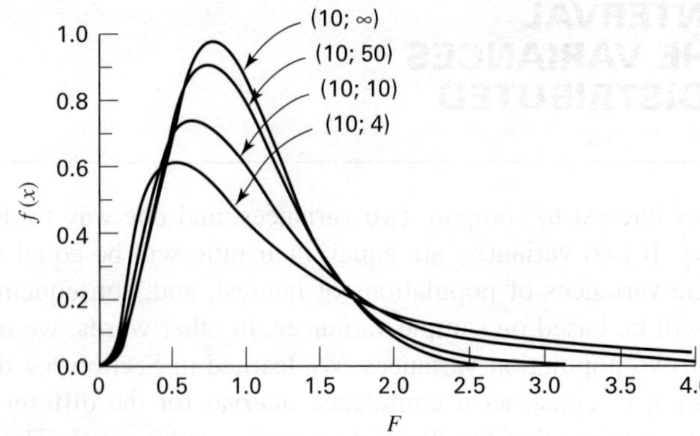

FIGURE 6.10.1 The F distribution for various degrees of freedom. (From *Documenta Geigy, Scientific Tables,* Seventh Edition, 1970. Courtesy of Ciba-Geigy Limited, Basel, Switzerland.)

where $F_{\alpha/2}$ and $F_{1-(\alpha/2)}$ are the values from the F table to the left and right of which, respectively, lies $\alpha/2$ of the area under the curve. The middle term of this expression may be rewritten so that the entire expression is

$$F_{\alpha/2} < \frac{s_1^2}{s_2^2} \cdot \frac{\sigma_2^2}{\sigma_1^2} < F_{1-(\alpha/2)}$$

If we divide through by s_1^2/s_2^2, we have

$$\frac{F_{\alpha/2}}{s_1^2/s_2^2} < \frac{\sigma_2^2}{\sigma_1^2} < \frac{F_{1-(\alpha/2)}}{s_1^2/s_2^2}$$

Taking the reciprocals of the three terms gives

$$\frac{s_1^2/s_2^2}{F_{\alpha/2}} > \frac{\sigma_1^2}{\sigma_2^2} > \frac{s_1^2/s_2^2}{F_{1-(\alpha/2)}}$$

and if we reverse the order, we have the following $100(1-\alpha)$ percent confidence interval for σ_1^2/σ_2^2:

$$\frac{s_1^2/s_2^2}{F_{1-(\alpha/2)}} < \frac{\sigma_1^2}{\sigma_2^2} < \frac{s_1^2/s_2^2}{F_{\alpha/2}} \qquad (6.10.1)$$

EXAMPLE 6.10.1

Allen and Gross (A-25) examine toe flexors strength in subjects with plantar fasciitis (pain from heel spurs, or general heel pain), a common condition in patients with musculoskeletal problems. Inflammation of the plantar fascia is often costly to treat and frustrating for both the patient and the clinician. One of the baseline measurements was the body mass index (BMI). For the 16 women in the study, the standard deviation for BMI was 8.1 and for four men in the study, the standard deviation was 5.9. We wish to construct a 95 percent confidence interval for the ratio of the variances of the two populations from which we presume these samples were drawn.

Solution: We have the following information:

$$n_1 = 16 \qquad n_2 = 4$$
$$s_1^2 = (8.1)^2 = 65.61 \qquad s_2^2 = (5.9)^2 = 34.81$$
$$df_1 = \text{numerator degrees of freedom} = n_1 - 1 = 15$$
$$df_2 = \text{denominator degrees of freedom} = n_2 - 1 = 3$$
$$\alpha = .05$$
$$F_{.025} = .24096 \qquad F_{.975} = 14.25$$

We are now ready to obtain our 95 percent confidence interval for σ_1^2/σ_2^2 by substituting appropriate values into Expression 6.10.1:

$$\frac{65.61/34.81}{14.25} < \frac{\sigma_1^2}{\sigma_2^2} < \frac{65.61/34.81}{.24096}$$

$$.1323 < \frac{\sigma_1^2}{\sigma_2^2} < 7.8221$$

We give this interval the appropriate probabilistic and practical interpretations.

Since the interval .1323 to 7.8221 includes 1, we are able to conclude that the two population variances may be equal. ■

Finding $F_{1-(\alpha/2)}$ and $F_{\alpha/2}$ At this point we must make a cumbersome, but unavoidable, digression and explain how the values $F_{.975} = 14.25$ and $F_{.025} = .24096$ were obtained. The value of $F_{.975}$ at the intersection of the column headed $df_1 = 15$ and the row labeled $df_2 = 3$ is 14.25. If we had a more extensive table of the F distribution, finding $F_{.025}$ would be no trouble; we would simply find $F_{.025}$ as we found $F_{.975}$. We would take the value at the intersection of the column headed 15 and the row headed 3. To include every possible percentile of F would make for a very lengthy table. Fortunately, however, there exists a relationship that enables us to compute the lower percentile values from our limited table. The relationship is as follows:

$$F_{\alpha,df_1,df_2} = \frac{1}{F_{1-\alpha,df_2,df_1}} \tag{6.10.2}$$

We proceed as follows.

Interchange the numerator and denominator degrees of freedom and locate the appropriate value of F. For the problem at hand we locate 4.15, which is at the intersection of the column headed 3 and the row labeled 15. We now take the reciprocal of this value, $1/4.15 = .24096$. In summary, the lower confidence limit (LCL) and upper confidence limit (UCL) σ_1^2/σ_2^2 are as follows:

$$LCL = \frac{s_1^2}{s_2^2} \frac{1}{F_{(1-\alpha/2),df_1,df_2}}$$

$$UCL = \frac{s_1^2}{s_2^2} F_{1-(\alpha/2),df_2,df_1}$$

Alternative procedures for making inferences about the equality of two variances when the sampled populations are not normally distributed may be found in the book by Daniel (15).

Some Precautions Similar to the discussion in the previous section of constructing confidence intervals for σ^2, the assumption of normality of the populations from which the samples are drawn is crucial to obtaining correct intervals for the ratio of variances discussed in this section. Fortunately, most statistical computer programs provide alternatives to the F-ratio, such as Levene's test, when the underlying distributions cannot be assumed to be normally distributed. Computationally, Levene's test uses a measure of distance from a sample median instead of a sample mean, hence removing the assumption of normality.

EXERCISES

6.10.1 The purpose of a study by Moneim et al. (A-26) was to examine thumb amputations from team roping at rodeos. The researchers reviewed 16 cases of thumb amputations. Of these, 11 were complete amputations while five were incomplete. The ischemia time is the length of time that insufficient oxygen is supplied to the amputated thumb. The ischemia times (hours) for 11 subjects experiencing complete amputations were

4.67, 10.5, 2.0, 3.18, 4.00, 3.5, 3.33, 5.32, 2.0, 4.25, 6.0

For five victims of incomplete thumb amputation, the ischemia times were

3.0, 10.25, 1.5, 5.22, 5.0

Treat the two reported sets of data as sample data from the two populations as described. Construct a 95 percent confidence interval for the ratio of the two unknown population variances.

6.10.2 The objective of a study by Horesh et al. (A-27) was to explore the hypothesis that some forms of suicidal behavior among adolescents are related to anger and impulsivity. The sample consisted of 65 adolescents admitted to a university-affiliated adolescent psychiatric unit. The researchers used the Impulsiveness-Control Scale (ICS, A-28) where higher numbers indicate higher degrees of impulsiveness and scores can range from 0 to 45. The 33 subjects classified as suicidal had an ICS score standard deviation of 8.4 while the 32 nonsuicidal subjects had a standard deviation of 6.0. Assume that these two groups constitute independent simple random samples from two populations of similar subjects. Assume also that the ICS scores in these two populations are normally distributed. Find the 99 percent confidence interval for the ratio of the two population variances of scores on the ICS.

6.10.3 Stroke index values were statistically analyzed for two samples of patients suffering from myocardial infarction. The sample variances were 12 and 10. There were 21 patients in each sample. Construct the 95 percent confidence interval for the ratio of the two population variances.

6.10.4 Thirty-two adult asphasics seeking speech therapy were divided equally into two groups. Group 1 received treatment 1, and group 2 received treatment 2. Statistical analysis of the treatment effectiveness scores yielded the following variances: $s_1^2 = 8$, $s_2^2 = 15$. Construct the 90 percent confidence interval for σ_2^2/σ_1^2.

6.10.5 Sample variances were computed for the tidal volumes (milliliters) of two groups of patients suffering from atrial septal defect. The results and sample sizes were as follows:

$$n_1 = 31, \quad s_1^2 = 35{,}000$$

$$n_2 = 41, \quad s_2^2 = 20{,}000$$

Construct the 95 percent confidence interval for the ratio of the two population variances.

6.10.6 Glucose responses to oral glucose were recorded for 11 patients with Huntington's disease (group 1) and 13 control subjects (group 2). Statistical analysis of the results yielded the following sample variances: $s_1^2 = 105$, $s_2^2 = 148$. Construct the 95 percent confidence interval for the ratio of the two population variances.

6.10.7 Measurements of gastric secretion of hydrochloric acid (milliequivalents per hour) in 16 normal subjects and 10 subjects with duodenal ulcer yielded the following results:

Normal subjects:	6.3, 2.0, 2.3, 0.5, 1.9, 3.2, 4.1, 4.0, 6.2, 6.1, 3.5, 1.3, 1.7, 4.5, 6.3, 6.2
Ulcer subjects:	13.7, 20.6, 15.9, 28.4, 29.4, 18.4, 21.1, 3.0, 26.2, 13.0

Construct a 95 percent confidence interval for the ratio of the two population variances. What assumptions must be met for this procedure to be valid?

6.11 SUMMARY

This chapter is concerned with one of the major areas of statistical inference—estimation. Both point estimation and interval estimation are covered. The concepts and methods involved in the construction of confidence intervals are illustrated for the following parameters: means, the difference between two means, proportions, the difference between two proportions, variances, and the ratio of two variances. In addition, we learned in this chapter how to determine the sample size needed to estimate a population mean and a population proportion at specified levels of precision.

We learned, also, in this chapter that interval estimates of population parameters are more desirable than point estimates because statements of confidence can be attached to interval estimates.

SUMMARY OF FORMULAS FOR CHAPTER 6

Formula Number	Name	Formula
6.2.1	Expression of an interval estimate	estimator ± (reliability coefficient) × (standard error of the estimator)
6.2.2	Interval estimate for μ when σ is known	$\bar{x} \pm z_{(1-\alpha/2)}\sigma_{\bar{x}}$
6.3.1	t-transformation	$t = \frac{\bar{x} - \mu}{s/\sqrt{n}}$
6.3.2	Interval estimate for μ when σ is unknown	$\bar{x} \pm t_{(1-\alpha/2)} = \frac{s}{\sqrt{n}}$
6.4.1	Interval estimate for the difference between two population means when σ_1 and σ_2 are known	$(\bar{x}_1 - \bar{x}_2) \pm z_{(1-\alpha/2)}\sqrt{\frac{\sigma_1^2}{n_1} + \frac{\sigma_2^2}{n_2}}$
6.4.2	Pooled variance estimate	$s_p^2 = \frac{(n_1 - 1)s_1^2 + (n_2 - 1)s_2^2}{n_1 + n_2 - 2}$
6.4.3	Standard error of estimate	$s_{(\bar{x}_1 - \bar{x}_2)} = \sqrt{\frac{s_p^2}{n_1} + \frac{s_p^2}{n_2}}$
6.4.4	Interval estimate for the difference between two population means when σ is unknown	$(\bar{x}_1 - \bar{x}_2) \pm t_{(1-\alpha/2)}\sqrt{\frac{s_p^2}{n_1} + \frac{s_p^2}{n_2}}$
6.4.5	Cochran's correction for reliability coefficient when variances are not equal	$t'_{(1-\alpha/2)} = \frac{w_1 t_1 + w_2 t_2}{w_1 + w_2}$
6.4.6	Interval estimate using Cochran's correction for t	$(\bar{x}_1 - \bar{x}_2) \pm t'_{(1-\alpha/2)}\sqrt{\frac{s_1^2}{n_1} + \frac{s_2^2}{n_2}}$
6.5.1	Interval estimate for a population proportion	$\hat{p} \pm z_{(1-\alpha/2)}\sqrt{\hat{p}(1 - \hat{p})/n}$

6.6.1	Interval estimate for the difference between two population proportions	$(\hat{p}_1 - \hat{p}_2) \pm z_{(1-\alpha/2)}\sqrt{\frac{\hat{p}_1(1-\hat{p}_1)}{n_1} + \frac{\hat{p}_2(1-\hat{p}_2)}{n_2}}$
6.7.1–6.7.3	Sample size determination when sampling with replacement	$d = \text{(reliability coefficient)} \times \text{(standard error)}$ $d = z\frac{\sigma}{\sqrt{n}}$ $\therefore$ $n = \frac{z^2\sigma^2}{d^2}$
6.7.4–6.7.5	Sample size determination when sampling without replacement	$d = z\frac{\sigma}{\sqrt{n}}\sqrt{\frac{N-n}{N-1}}$ $\therefore$ $n = \frac{Nz^2\sigma^2}{d^2(N-1) + z^2\sigma^2}$
6.8.1	Sample size determination for proportions when sampling with replacement	$n = \frac{z^2pq}{d^2}$
6.8.2	Sample size determination for proportions when sampling without replacement	$n = \frac{Nz^2\sigma^2}{d^2(N-1) + z^2\sigma^2}$
6.9.1	Interval estimate for σ^2	$\frac{(n-1)s^2}{\chi^2_{1-(\alpha/2)}} < \sigma^2 < \frac{(n-1)s^2}{\chi^2_{(\alpha/2)}}$
6.9.2	Interval estimate for σ	$\sqrt{\frac{(n-1)s^2}{\chi^2_{1-(\alpha/2)}}} < \sigma < \sqrt{\frac{(n-1)s^2}{\chi^2_{(\alpha/2)}}}$
6.10.1	Interval estimate for the ratio of two variances	$\frac{s_1^2/s_2^2}{F_{1-(\alpha/2)}} < \frac{\sigma_1^2}{\sigma_2^2} < \frac{s_1^2/s_2^2}{F_{\alpha/2}}$
6.10.2	Relationship among F ratios	$F_{\alpha,df_1,df_2} = \frac{1}{F_{1-\alpha,df_2,df_1}}$
Symbol Key		• α = Type 1 error rate • χ^2 = Chi-square distribution • d = error component of interval estimate

(Continued)

	• df = degrees of freedom • F = F-distribution • μ = mean of population • n = sample size • p = proportion for population • $q = (1 - p)$ • $\hat{p}$ = estimated proportion for sample • σ^2 = population variance • σ = population standard deviation • $\sigma_{\bar{x}}$ = standard error • s = standard deviation of sample • s_p = pooled standard deviation • t = Student's t-transformation • t' = Cochran's correction to t • $\bar{x}$ = mean of sample • z = standard normal distribution

REVIEW QUESTIONS AND EXERCISES

1. What is statistical inference?
2. Why is estimation an important type of inference?
3. What is a point estimate?
4. Explain the meaning of unbiasedness.
5. Define the following:
   (a) Reliability coefficient
   (b) Confidence coefficient
   (c) Precision
   (d) Standard error
   (e) Estimator
   (f) Margin of error
6. Give the general formula for a confidence interval.
7. State the probabilistic and practical interpretations of a confidence interval.
8. Of what use is the central limit theorem in estimation?
9. Describe the t distribution.
10. What are the assumptions underlying the use of the t distribution in estimating a single population mean?
11. What is the finite population correction? When can it be ignored?
12. What are the assumptions underlying the use of the t distribution in estimating the difference between two population means?
13. Arterial blood gas analyses performed on a sample of 15 physically active adult males yielded the following resting PaO_2 values:

    75, 80, 80, 74, 84, 78, 89, 72, 83, 76, 75, 87, 78, 79, 88

Compute the 95 percent confidence interval for the mean of the population.

14. What proportion of asthma patients are allergic to house dust? In a sample of 140, 35 percent had positive skin reactions. Construct the 95 percent confidence interval for the population proportion.

15. An industrial hygiene survey was conducted in a large metropolitan area. Of 70 manufacturing plants of a certain type visited, 21 received a "poor" rating with respect to absence of safety hazards. Construct a 95 percent confidence interval for the population proportion deserving a "poor" rating.

16. Refer to the previous problem. How large a sample would be required to estimate the population proportion to within .05 with 95 percent confidence (.30 is the best available estimate of p):
(a) If the finite population correction can be ignored?
(b) If the finite population correction is not ignored and $N = 1500$?

17. In a dental survey conducted by a county dental health team, 500 adults were asked to give the reason for their last visit to a dentist. Of the 220 who had less than a high-school education, 44 said they went for preventative reasons. Of the remaining 280, who had a high-school education or better, 150 stated that they went for preventative reasons. Construct a 95 percent confidence interval for the difference between the two population proportions.

18. A breast cancer research team collected the following data on tumor size:

Type of Tumor	n	$\bar{x}$	s
A	21	3.85 cm	1.95 cm
B	16	2.80 cm	1.70 cm

Construct a 95 percent confidence interval for the difference between population means.

19. A certain drug was found to be effective in the treatment of pulmonary disease in 180 of 200 cases treated. Construct the 90 percent confidence interval for the population proportion.

20. Seventy patients with stasis ulcers of the leg were randomly divided into two equal groups. Each group received a different treatment for edema. At the end of the experiment, treatment effectiveness was measured in terms of reduction in leg volume as determined by water displacement. The means and standard deviations for the two groups were as follows:

Group (Treatment)	$\bar{x}$	s
A	95 cc	25
B	125 cc	30

Construct a 95 percent confidence interval for the difference in population means.

21. What is the average serum bilirubin level of patients admitted to a hospital for treatment of hepatitis? A sample of 10 patients yielded the following results:

20.5, 14.8, 21.3, 12.7, 15.2, 26.6, 23.4, 22.9, 15.7, 19.2

Construct a 95 percent confidence interval for the population mean.

22. Determinations of saliva pH levels were made in two independent random samples of seventh-grade schoolchildren. Sample A children were caries-free while sample B children had a high incidence of caries. The results were as follows:

A: 7.14, 7.11, 7.61, 7.98, 7.21, 7.16, 7.89
7.24, 7.86, 7.47, 7.82, 7.37, 7.66, 7.62, 7.65

B: 7.36, 7.04, 7.19, 7.41, 7.10, 7.15, 7.36,
7.57, 7.64, 7.00, 7.25, 7.19

Construct a 90 percent confidence interval for the difference between the population means. Assume that the population variances are equal.

23. Drug A was prescribed for a random sample of 12 patients complaining of insomnia. An independent random sample of 16 patients with the same complaint received drug B. The number of hours of sleep experienced during the second night after treatment began were as follows:

A: 3.5, 5.7, 3.4, 6.9, 17.8, 3.8, 3.0, 6.4, 6.8, 3.6, 6.9, 5.7

B: 4.5, 11.7, 10.8, 4.5, 6.3, 3.8, 6.2, 6.6, 7.1, 6.4, 4.5,
5.1, 3.2, 4.7, 4.5, 3.0

Construct a 95 percent confidence interval for the difference between the population means. Assume that the population variances are equal.

24. The objective of a study by Crane et al. (A-29) was to examine the efficacy, safety, and maternal satisfaction of (a) oral misoprostol and (b) intravenous oxytocin for labor induction in women with premature rupture of membranes at term. Researchers randomly assigned women to the two treatments. For the 52 women who received oral misoprostol, the mean time in minutes to active labor was 358 minutes with a standard deviation of 308 minutes. For the 53 women taking oxytocin, the mean time was 483 minutes with a standard deviation of 144 minutes. Construct a 99 percent confidence interval for the difference in mean time to active labor for these two different medications. What assumptions must be made about the reported data? Describe the population about which an inference can be made.

25. Over a 2-year period, 34 European women with previous gestational diabetes were retrospectively recruited from West London antenatal databases for a study conducted by Kousta et al. (A-30). One of the measurements for these women was the fasting nonesterified fatty acids concentration (NEFA) measured in μmol/L. In the sample of 34 women, the mean NEFA level was 435 with a sample standard deviation of 215.0. Construct a 95 percent confidence interval for the mean fasting NEFA level for a population of women with gestational diabetes. State all necessary assumptions about the reported data and subjects.

26. Scheid et al. (A-31) questioned 387 women receiving free bone mineral density screening. The questions focused on past smoking history. Subjects undergoing hormone replacement therapy (HRT), and subjects not undergoing HRT, were asked if they had ever been a regular smoker. In the HRT group, 29.3 percent of 220 women stated that they were at some point in their life a regular smoker. In the non–HRT group, 17.3 percent of 106 women responded positively to being at some point in their life a regular smoker. (Sixty-one women chose not to answer the question.) Construct a 95 percent confidence interval for the difference in smoking percentages for the two populations of women represented by the subjects in the study. What assumptions about the data are necessary?

27. The purpose of a study by Elliott et al. (A-32) was to assess the prevalence of vitamin D deficiency in women living in nursing homes. The sample consisted of 39 women in a 120-bed skilled nursing facility. Women older than 65 years of age who were long-term residents were invited to participate if they had no diagnosis of terminal cancer or metastatic disease. In the sample,

23 women had 25-hydroxyvitamin D levels of 20 ng/ml or less. Construct a 95 percent confidence interval for the percent of women with vitamin D deficiency in the population presumed to be represented by this sample.

28. In a study of the role of dietary fats in the etiology of ischemic heart disease the subjects were 60 males between 40 and 60 years of age who had recently had a myocardial infarction and 50 apparently healthy males from the same age group and social class. One variable of interest in the study was the proportion of linoleic acid (L.A.) in the subjects' plasma triglyceride fatty acids. The data on this variable were as follows:

Subjects with Myocardial Infarction

Subject	L.A.	Subject	L.A.	Subject	L.A.	Subject	L.A.
1	18.0	2	17.6	3	9.6	4	5.5
5	16.8	6	12.9	7	14.0	8	8.0
9	8.9	10	15.0	11	9.3	12	5.8
13	8.3	14	4.8	15	6.9	16	18.3
17	24.0	18	16.8	19	12.1	20	12.9
21	16.9	22	15.1	23	6.1	24	16.6
25	8.7	26	15.6	27	12.3	28	14.9
29	16.9	30	5.7	31	14.3	32	14.1
33	14.1	34	15.1	35	10.6	36	13.6
37	16.4	38	10.7	39	18.1	40	14.3
41	6.9	42	6.5	43	17.7	44	13.4
45	15.6	46	10.9	47	13.0	48	10.6
49	7.9	50	2.8	51	15.2	52	22.3
53	9.7	54	15.2	55	10.1	56	11.5
57	15.4	58	17.8	59	12.6	60	7.2

Healthy Subjects

Subject	L.A.	Subject	L.A.	Subject	L.A.	Subject	L.A.
1	17.1	2	22.9	3	10.4	4	30.9
5	32.7	6	9.1	7	20.1	8	19.2
9	18.9	10	20.3	11	35.6	12	17.2
13	5.8	14	15.2	15	22.2	16	21.2
17	19.3	18	25.6	19	42.4	20	5.9
21	29.6	22	18.2	23	21.7	24	29.7
25	12.4	26	15.4	27	21.7	28	19.3
29	16.4	30	23.1	31	19.0	32	12.9
33	18.5	34	27.6	35	25.0	36	20.0
37	51.7	38	20.5	39	25.9	40	24.6
41	22.4	42	27.1	43	11.1	44	32.7
45	13.2	46	22.1	47	13.5	48	5.3
49	29.0	50	20.2				

Construct the 95 percent confidence interval for the difference between population means. What do these data suggest about the levels of linoleic acid in the two sampled populations?

29. The purpose of a study by Tahmassebi and Curzon (A-33) was to compare the mean salivary flow rate among subjects with cerebral palsy and among subjects in a control group. Each group had 10 subjects. The following table gives the mean flow rate in ml/minute as well as the standard error.

Group	Sample Size	Mean ml/minute	Standard Error
Cerebral palsy	10	0.220	0.0582
Control	10	0.334	0.1641

Source: J. F. Tahmassebi and M. E. J. Curzon, "The Cause of Drooling in Children with Cerebral Palsy—Hypersalivation or Swallowing Defect?" *International Journal of Paediatric Dentistry, 13* (2003), 106–111.

Construct the 90 percent confidence interval for the difference in mean salivary flow rate for the two populations of subjects represented by the sample data. State the assumptions necessary for this to be a valid confidence interval.

30. Culligan et al. (A-34) compared the long-term results of two treatments: (a) a modified Burch procedure, and (b) a sling procedure for stress incontinence with a low-pressure urethra. Thirty-six women took part in the study with 19 in the Burch treatment group and 17 in the sling procedure treatment group. One of the outcome measures at three months post-surgery was maximum urethral closure pressure (cm H_2O). In the Burch group the mean and standard deviation were 16.4 and 8.2 cm, respectively. In the sling group, the mean and standard deviation were 39.8 and 23.0, respectively. Construct the 99 percent confidence interval for the difference in mean maximum urethral closure pressure for the two populations represented by these subjects. State all necessary assumptions.

31. In general, narrow confidence intervals are preferred over wide ones. We can make an interval narrow by using a small confidence coefficient. For a given set of other conditions, what happens to the level of confidence when we use a small confidence coefficient? What would happen to the interval width and the level of confidence if we were to use a confidence coefficient of zero?

32. In general, a high level of confidence is preferred over a low level of confidence. For a given set of other conditions, suppose we set our level of confidence at 100 percent. What would be the effect of such a choice on the width of the interval?

33. The subjects of a study by Borland et al. (A-35) were children in acute pain. Thirty-two children who presented at an emergency room were enrolled in the study. Each child used the visual analogue scale to rate pain on a scale from 0 to 100 mm. The mean pain score was 61.3 mm with a 95 percent confidence interval of 53.2 mm–69.4 mm. Which would be the appropriate reliability factor for the interval, z or t? Justify your choice. What is the precision of the estimate? The margin of error?

34. Does delirium increase hospital stay? That was the research question investigated by McCusker et al. (A-36). The researchers sampled 204 patients with prevalent delirium and 118 without delirium. The conclusion of the study was that patients with prevalent delirium did not have a higher mean length of stay compared to those without delirium. What was the target population? The sampled population?

35. Assessing driving self-restriction in relation to vision performance was the objective of a study by West et al. (A-37). The researchers studied 629 current drivers ages 55 and older for 2 years. The variables of interest were driving behavior, health, physical function, and vision function. The subjects were part of a larger vision study at the Smith-Kettlewell Eye Research Institute. A conclusion of the study was that older adults with early changes in spatial vision function and depth perception appear to recognize their limitations and restrict their driving. What was the target population? The sampled population?

36. In a pilot study conducted by Ayouba et al. (A-38), researchers studied 123 children born of HIV-1-infected mothers in Yaoundé, Cameroon. Counseled and consenting pregnant women were given a single dose of nevirapine at the onset of labor. Babies were given a syrup containing nevirapine within the first 72 hours of life. The researchers found that 87 percent of the children were considered not infected at 6–8 weeks of age. What is the target population? What is the sampled population?

37. Refer to Exercise 2.3.11. Construct a 95 percent confidence interval for the population mean *S/R* ratio. Should you use t or z as the reliability coefficient? Why? Describe the population about which inferences based on this study may be made.

38. Refer to Exercise 2.3.12. Construct a 90 percent confidence interval for the population mean height. Should you use t or z as the reliability coefficient? Why? Describe the population about which inferences based on this study may be made.

Exercises for Use with Large Data Sets Available on the Following Website: www.wiley.com/college/daniel

1. Refer to North Carolina Birth Registry Data NCBIRTH800 with 800 observations (see Large Data Exercise 1 in Chapter 2). Calculate 95 percent confidence intervals for the following:

(a) the percentage of male children

(b) the mean age of a mother giving birth

(c) the mean weight gained during pregnancy

(d) the percentage of mothers admitting to smoking during pregnancy

(e) the difference in the average weight gained between smoking and nonsmoking mothers

(f) the difference in the average birth weight in grams between married and nonmarried mothers

(g) the difference in the percentage of low birth weight babies between married and nonmarried mothers

2. Refer to the serum cholesterol levels for 1000 subjects (CHOLEST). Select a simple random sample of size 15 from this population and construct a 95 percent confidence interval for the population mean. Compare your results with those of your classmates. What assumptions are necessary for your estimation procedure to be valid?

3. Refer to the serum cholesterol levels for 1000 subjects (CHOLEST). Select a simple random sample of size 50 from the population and construct a 95 percent confidence interval for the proportion of subjects in the population who have readings greater than 225. Compare your results with those of your classmates.

4. Refer to the weights of 1200 babies born in a community hospital (BABY WGTS). Draw a simple random sample of size 20 from this population and construct a 95 percent confidence interval for the population mean. Compare your results with those of your classmates. What assumptions are necessary for your estimation procedure to be valid?

5. Refer to the weights of 1200 babies born in a community hospital (BABY WGTS). Draw a simple random sample of size 35 from the population and construct a 95 percent confidence interval for the population mean. Compare this interval with the one constructed in Exercise 4.

6. Refer to the heights of 1000 twelve-year-old boys (BOY HGTS). Select a simple random sample of size 15 from this population and construct a 99 percent confidence interval for the population mean. What assumptions are necessary for this procedure to be valid?

7. Refer to the heights of 1000 twelve-year-old boys (BOY HGTS). Select a simple random sample of size 35 from the population and construct a 99 percent confidence interval for the population mean. Compare this interval with the one constructed in Exercise 5.

REFERENCES

Methodology References

1. JOHN A. RICE, *Mathematical Statistics and Data Analysis,* 2nd ed., Duxbury, Belmont, CA, 1988.
2. W. S. GOSSET ("Student"), "The Probable Error of a Mean," *Biometrika, 6* (1908), 1–25.
3. W. V. BEHRENS, "Ein Beitrag zu Fehlerberechnung bei wenige Beobachtungen," *Landwirtsschaftliche Jahrbücher, 68* (1929), 807–837.
4. R. A. FISHER, "The Comparison of Samples with Possibly Unequal Variances," *Annals of Eugenics, 9* (1939), 174–180.
5. R. A. FISHER, "The Asymptotic Approach to Behrens' Integral with Further Tables for the *d* Test of Significance," *Annals of Eugenics, 11* (1941), 141–172.
6. J. NEYMAN, "Fiducial Argument and the Theory of Confidence Intervals," *Biometrika, 32* (1941), 128–150.
7. H. SCHEFFÉ, "On Solutions of the Behrens-Fisher Problem Based on the *t*-Distribution," *Annals of Mathematical Statistics, 14* (1943), 35–44.
8. H. SCHEFFÉ, "A Note on the Behrens-Fisher Problem," *Annals of Mathematical Statistics, 15* (1944), 430–432.
9. B. L. WELCH, "The Significance of the Difference Between Two Means When the Population Variances Are Unequal," *Biometrika, 29* (1937), 350–361.
10. B. L. WELCH, "The Generalization of 'Student's Problem When Several Different Population Variances Are Involved," *Biometrika, 34* (1947), 28–35.
11. WILLIAM G. COCHRAN, "Approximate Significance Levels of the Behrens-Fisher Test," *Biometrics, 20* (1964), 191–195.
12. R. F. TATE and G. W. KLETT, "Optimal Confidence Intervals for the Variance of a Normal Distribution," *Journal of the American Statistical Association, 54* (1959), 674–682.
13. H. O. HARTLEY, "The Maximum F-Ratio as a Short Cut Test for Heterogeneity of Variances," *Biometrika, 37* (1950), 308–312.
14. J. H. ZAR, *Biostatistical Analysis*, 4th ed., Prentice-Hall, Upper Saddle River, NJ, 1999.
15. WAYNE W. DANIEL, *Applied Nonparametric Statistics,* 2nd ed., PWS-KENT, Boston, 1989.

Applications References

A-1. NICOLA MAFFULLI, CHERLY TALLON, JASON WONG, KIM PENG, and ROBERT BLEAKNEY, "Early Weightbearing and Ankle Mobilization after Open Repair of Acute Midsubstance Tears of the Achilles Tendon," *American Journal of Sports Medicine, 31* (2003), 692–700.

A-2. PAUL J. REBER, LUCY A. MARTINEZ, and SANDRA WEINTRAUB, "Artificial Grammar Learning in Alzheimer's Disease," *Cognitive, Affective, and Behavioral Neuroscience, 3* (2003), 145–153.

A-3. CHRISTOPHE PEDROLETTI, MARIEANN HÖGMAN, PEKKA MERILÄINEN, LENNART S. NORDVALL, GUNILLA HEDLIN, and KJELL ALVING, "Nitric Oxide Airway Diffusing Capacity and Mucosal Concentration in Asthmatic Schoolchildren," *Pediatric Research, 54* (2003), 496–501.

A-4. BRUCE D. BEYNNON, BRADEN C. FLEMING, DAVID L. CHURCHILL, and DANIEL BROWN, "The Effect of Anterior Cruciate Ligament Deficiency and Functional Bracing on Translation of the Tibia Relative to the Femur during Nonweightbearing and Weightbearing," *American Journal of Sports Medicine, 31* (2003), 99–105.

A-5. LORRAINE DUGOFF, MAURITHA R. EVERETT, LOUIS VONTVER, and GWYN E. BARLEY, "Evaluation of Pelvic and Breast Examination Skills of Interns in Obstetrics and Gynecology and Internal Medicine," *American Journal of Obstetrics and Gynecology, 189* (2003), 655–658.

A-6. DIONNE MAYHEW, KAREN M. PERRIN, and WENDY STRUCHEN, "An Analysis of a Healthy Start Smoking Cessation Program," *American Journal of Health Studies, 17* (2002), 186–190.

A-7. ERIC GRANHOLM, ROBERT ANTHENELLI, RITA MONTEIRO, JOHN SEVCIK, and MARILYN STOLER, "Brief Integrated Outpatient Dual-Diagnosis Treatment Reduces Psychiatric Hospitalizations," *American Journal on Addictions, 12* (2003), 306–313.

A-8. Silvia Iannelo, Antonina Cavaleri, Paolina Milazzo, Santi Cantarella, and Francesco Belfiore, "Low Fasting Serum Triglyceride Level as a Precocious Marker of Autoimmune Disorders," *Medscape General Medicine, 5*(3)(2003).

A-9. Evelyn C. Y. Chan, Sally W. Vernon, Frederick T. O'Donnell, Chul Ahn, Anthony Gresinger, and Donnie W. Aga, "Informed Consent for Cancer Screening with Prostate-Specific Antigen: How Well Are Men Getting the Message?" *American Journal of Public Health, 93* (2003), 779–785.

A-10. Ronald F. van Vollenhoven, Sofia Ernestam, Anders Harju, Johan Bratt, and Lars Klareskog, "Etanercept Versus Etanercept Plus Methotrexate: A Registry-Based Study Suggesting that the Combination Is Clinically More Effacious," *Arthritis Research and Therapy, 5* (2003), R347–R351.

A-11. Satoshi Nozawa, Katsuji Shimizu, Kei Miyamoto, and Mizuo Tanaka, "Repair of Pars Interarticularis Defect by Segmental Wire Fixation in Young Athletes with Spondylolysis," *American Journal of Sports Medicine, 31* (2003), 359–364.

A-12. Mori J. Krantz, Ilana B. Kutinsky, Alastair D. Robertson, and Philip S. Mehler, "Dose-Related Effects of Methadone on QT Prolongation in a Series of Patients with Torsade de Pointes," *Pharmacotherapy, 23* (2003), 802–805.

A-13. Susannah Fox and Deborah Fellows, "Health Searches and Email Have Become More Commonplace, But There Is Room for Improvement in Searches and Overall Internet Access," *Pew Internet and American Life Project, www.pewinternet.org/PPF/r/95/report_display.asp.*

A-14. Carlos M. Luna, Daniel Blanzaco, Michael S. Niderman, Walter Matarucco, Natalio C. Baredes, Pablo Desmery, Fernando Palizas, Guillermo Menga, Fernando Rios, and Carlos Apezteguia, "Resolution of Ventilator-Associated Pneumonia: Prospective Evaluation of the Clinical Pulmonary Infection Score as an Early Clinical Predictor of Outcome," *Critical Care Medicine, 31*(3)(2003), 676–682.

A-15. Arend F. L. Schinkel, Jeroen J. Bax, Eric Boersma, Abdou Elhendy, Eleni C. Vourvouri, Jos R. T. C. Roelandt, and Don Poldermans, "Assessment of Residual Myocardial Viability in Regions with Chronic Electrocardiographic Q-Wave Infarction," *American Heart Journal, 144* (2002), 865–869.

A-16. Friederike von zur Muhlen, Weilum Quan, David J. D'Agate, and Todd J. Cohen, "A Study of Carotid Sinus Massage and Head-Up Tilt Table Testing in Patients with Syncope and Near-Syncope," *Journal of Invasive Cardiology, 14* (2002), 477–482.

A-17. Daniel F. Connor, Ronald J. Steingard, Jennifer J. Anderson, and Richard H. Melloni, Jr., "Gender Differences in Reactive and Proactive Aggression," *Child Psychiatry and Human Development, 33* (2003), 279–294.

A-18. Allan V. Horwitz, Cathy Spatz Widom, Julie McLaughlin, and Helene Raskin White, "The Impact of Childhood Abuse and Neglect on Adult Mental Health: A Prospective Study," *Journal of Health and Social Behavior, 42* (2001), 184–201.

A-19. P. Adab, T. Marshall, A. Rouse, B. Randhawa, H. Sangha, and N. Bhangoo, "Randomised Controlled Trial of the Effect of Evidence Based Information on Women's Willingness to Participate in Cervical Cancer Screening," *Journal of Epidemiology and Community Health, 57* (2003), 589–593.

A-20. John A. Spertus, Tim Kewhurst, Cynthia M. Dougherty, and Paul Nichol, "Testing the Effectiveness of Converting Patients to Long-Acting Antianginal Medications: The Quality of Life in Angina Research," *American Heart Journal, 141* (2001), 550–558.

A-21. Patrick R. Finley, Heidi R. Rens, Joan T. Pont, Susan L. Gess, Clifton Louie, Scott A. Bull, Janelle Y. Lee, and Lisa A. Bero, "Impact of a Collaborative Care Model on Depression in a Primary Care Setting: A Randomized, Controlled Trial," *Pharmacotherapy, 23*(9)(2003), 1175–1185.

A-22. Michael Hummel, Ezio Bonifacio, Heike E. Naserke, and Anette G. Ziegler, "Elimination of Dietary Gluten Does Not Reduce Titers of Type 1 Diabetes-Associated Autoantibodies in High-Risk Subjects," *Diabetes Care, 25* (2002), 1111–1116.

A-23. Dov Aizenberg, Abraham Weizman, and Yoram Barak, "Sildenafil for Selective Serotonin Reuptake Inhibitor-Induced Erectile Dysfunction in Elderly Male Depressed Patients," *Journal of Sex and Marital Therapy, 29* (2003), 297–303.

A-24. Peter Borden, John Nyland, David N. M. Caborn, and David Pienkowski, "Biomechanical Comparison of the FasT-Fix Meniscal Repair Suture System with Vertical Mattress Sutures and Meniscus Arrows," *American Journal of Sports Medicine,* 31 (2003), 374–378.

A-25. Rachel H. Allen and Michael T. Gross, "Toe Flexors Strength and Passive Extension Range of Motion of the First Metatarsophalangeal Joint in Individuals with Plantar Fasciitis," *Journal of Orthopaedic and Sports Physical Therapy, 33* (2003), 468–477.

A-26. MOHEB S. MONEIM, KEIKHOSROW FIROOZBAKHSH, DOMINIC GROSS, STEVEN D. YOUNG, and GEORGE OMER, "Thumb Amputations from Team Roping," *American Journal of Sports Medicine, 31* (2003), 728–735.

A-27. NETTA HORESH, ISRAEL ORBACH, DORON GOTHELF, MEIR EFRATI, and ALAN APTER, "Comparison of the Suicidal Behavior of Adolescent Inpatients with Borderline Personality Disorder and Major Depression," *Journal of Nervous and Mental Disease, 191* (2003), 582–588.

A-28. R. PLUTCHIK, and H. VAN PRAAG, "The Measurement of Suicidality, Aggressivity and Impulsivity," *Clinical Neuropharmacology, 9*(suppl)(1989), 380–382.

A-29. JOAN M. G. CRANE, TINA DELANEY, and DONNA HUTCHENS, "Oral Misoprostol for Premature Rupture of Membranes at Term," *American Journal of Obstetrics and Gynecology, 189* (2003), 720–724.

A-30. ELENI KOUSTA, NATASHA J. LAWRENCE, IAN F. GODSLAND, ANNA PENNY, VICTOR ANYAOKU, BARBARA A. MILLAUER, ESTER CELA, DESMOND G. JOHNSTON, STEPHEN ROBINSON, and MARK I. MCCARTHY, "Insulin Resistance and Beta-Cell Dysfunction in Normoglycaemic European Women with a History of Gestational Diabetes," *Clinical Endocrinology, 59* (2003), 289–297.

A-31. DEWEY C. SCHEID, MARIO T. COLEMAN, and ROBERT M. HAMM, "Do Perceptions of Risk and Quality of Life Affect Use of Hormone Replacement Therapy by Postmenopausal Women?" *Journal of the American Board of Family Practice, 16* (2003), 270–277.

A-32. MARY E. ELLIOTT, NEIL C. BINKLEY, MOLLY CARNES, DAVID R. ZIMMERMAN, KIM PETERSEN, KATHY KNAPP, JESSICA J. BEHLKE, NANCY AHMANN, and MARY A. KIESER, "Fracture Risks for Women in Long-Term Care: High Prevalence of Calcaneal Osteoporosis and Hypovitaminosis D," *Pharmacotherapy, 23* (2003), 702–710.

A-33. J. F. TAHMASSEBI and M. E. J. CURZON, "The Cause of Drooling in Children with Cerebral Palsy—Hypersalivation or Swallowing Defect?" *International Journal of Paediatric Dentistry, 13* (2003), 106–111.

A-34. PATRICK J. CULLIGAN, ROGER P. GOLDBERG, and PETER K. SAND, "A Randomized Controlled Trial Comparing a Modified Burch Procedure and a Suburethral Sling: Long-Term Follow-Up," *International Urogynecology Journal, 14* (2003), 229–233.

A-35. MEREDITH L. BORLAND, IAN JACOBS, and GARY GEELHOE, "Intranasal Fentanyl Reduces Acute Pain in Children in the Emergency Department: A Safety and Efficacy Study," *Emergency Medicine, 14* (2002), 275–280.

A-36. JANE MCCUSKER, MARTIN G. COLE, NANDINI DENDUKURI, and ERIC BELZILE, "Does Delirium Increase Hospital Stay?" *Journal of the American Geriatrics Society, 51* (2003), 1539–1546.

A-37. CATHERINE G. WEST, GINNY GILDENGORIN, GUNILLA HAEGERSTROM PORTNOY, LORI A. LOTT, MARILYN E. SCHNECK, and JOHN A. BRABYN, "Vision and Driving Self-Restriction in Older Adults," *Journal of the American Geriatrics Society, 51* (2003), 1348–1355.

A-38. AHIDJO AYOUBA, GILBERT TENE, PATRICK CUNIN, YACOUBA FOUPOUAPOUOGNIGNI, ELISABETH MENU, ANFUMBOM KFUTWAH, JOCELYN THONNON, GABRIELLA SCARLATTI, MARCEL MONNY-LOBÉ, NICOLE ETEKI, CHARLES KOUANFACK, MICHÈLE TARDY, ROBERT LEKE, MAURICE NKAM, ANNE E. NLEND, FRANÇOISE BARRÉ-SINOUSSI, PAUL M. V. MARTIN, and ERIC NERRIENET, "Low Rate of Mother-to-Child Transmission of HIV-1 After Nevirapine Intervention in a Pilot Public Health Program in Yaoundé, Cameroon," *Journal of Acquired Immune Deficiency Syndrome, 34* (2003), 274–280.

CHAPTER 7

USING SAMPLE STATISTICS TO TEST HYPOTHESES ABOUT POPULATION PARAMETERS

CHAPTER OVERVIEW

This chapter covers hypothesis testing, the second of two general areas of statistical inference. Hypothesis testing is a topic with which you as a student are likely to have some familiarity. Interval estimation, discussed in the preceding chapter, and hypothesis testing are based on similar concepts. In fact, confidence intervals may be used to arrive at the same conclusions that are reached through the use of hypothesis tests. This chapter provides a format, followed throughout the remainder of this book, for conducting a hypothesis test.

TOPICS

LEARNING OUTCOMES

After studying this chapter, the student will

1. understand how to correctly state a null and alternative hypothesis and carry out a structured hypothesis test.

2. understand the concepts of type I error, type II error, and the power of a test.
3. be able to calculate and interpret *z*, *t*, F, and chi-square test statistics for making statistical inferences.
4. understand how to calculate and interpret *p* values.

7.1 INTRODUCTION

One type of statistical inference, estimation, is discussed in the preceding chapter. The other type, hypothesis testing, is the subject of this chapter. As is true with estimation, the *purpose of hypothesis testing is to aid the clinician, researcher, or administrator in reaching a conclusion concerning a population by examining a sample from that population*. Estimation and hypothesis testing are not as different as they are made to appear by the fact that most textbooks devote a separate chapter to each. As we will explain later, one may use confidence intervals to arrive at the same conclusions that are reached by using the hypothesis testing procedures discussed in this chapter.

Basic Concepts In this section some of the basic concepts essential to an understanding of hypothesis testing are presented. The specific details of particular tests will be given in succeeding sections.

DEFINITION

A *hypothesis* may be defined simply as *a statement about one or more populations*.

The hypothesis is frequently concerned with the parameters of the populations about which the statement is made. A hospital administrator may hypothesize that the average length of stay of patients admitted to the hospital is 5 days; a public health nurse may hypothesize that a particular educational program will result in improved communication between nurse and patient; a physician may hypothesize that a certain drug will be effective in 90 percent of the cases for which it is used. By means of hypothesis testing one determines whether or not such statements are compatible with the available data.

Types of Hypotheses Researchers are concerned with two types of hypotheses—research hypotheses and statistical hypotheses.

DEFINITION

The *research hypothesis* is the conjecture or supposition that motivates the research.

It may be the result of years of observation on the part of the researcher. A public health nurse, for example, may have noted that certain clients responded more readily to a particular type of health education program. A physician may recall numerous instances in which certain combinations of therapeutic measures were more effective than any one of them alone. Research projects often result from the desire of such health practitioners to determine whether or not their theories or suspicions can be supported when subjected to the rigors of scientific investigation.

Research hypotheses lead directly to statistical hypotheses.

DEFINITION

***Statistical hypotheses* are hypotheses that are stated in such a way that they may be evaluated by appropriate statistical techniques.**

In this book the hypotheses that we will focus on are statistical hypotheses. We will assume that the research hypotheses for the examples and exercises have already been considered.

Hypothesis Testing Steps For convenience, hypothesis testing will be presented as a ten-step procedure. There is nothing magical or sacred about this particular format. It merely breaks the process down into a logical sequence of actions and decisions.

1. **Data.** The nature of the data that form the basis of the testing procedures must be understood, since this determines the particular test to be employed. Whether the data consist of counts or measurements, for example, must be determined.
2. **Assumptions.** As we learned in the chapter on estimation, different assumptions lead to modifications of confidence intervals. The same is true in hypothesis testing: A general procedure is modified depending on the assumptions. In fact, the same assumptions that are of importance in estimation are important in hypothesis testing. We have seen that these include assumptions about the normality of the population distribution, equality of variances, and independence of samples.
3. **Hypotheses.** There are two statistical hypotheses involved in hypothesis testing, and these should be stated explicitly. The *null hypothesis* is the *hypothesis to be tested.* It is designated by the symbol H_0. The null hypothesis is sometimes referred to as a *hypothesis of no difference,* since it is a statement of agreement with (or no difference from) conditions presumed to be true in the population of interest. In general, the null hypothesis is set up for the express purpose of being discredited. Consequently, the complement of the conclusion that the researcher is seeking to reach becomes the statement of the null hypothesis. In the testing process the null hypothesis either is rejected or is not rejected. If the null hypothesis is not rejected, we will say that the data on which the test is based do not provide sufficient evidence to cause rejection. If the testing procedure leads to rejection, we will say that the data at hand are not compatible with the null hypothesis, but are supportive of some other hypothesis. The *alternative hypothesis* is a statement of what we will believe is true if our sample data cause us to reject the null hypothesis. Usually the alternative hypothesis and the research hypothesis are the same, and in fact the two terms are used interchangeably. We shall designate the alternative hypothesis by the symbol H_A.

Rules for Stating Statistical Hypotheses When hypotheses are of the type considered in this chapter an indication of equality (either $=$, $\leq$, or $\geq$) must appear in the null hypothesis. Suppose, for example, that we want to answer the question: Can we conclude that a certain population mean is not 50? The null hypothesis is

$$H_0: \mu = 50$$

and the alternative is

$$H_A: \mu \neq 50$$

Suppose we want to know if we can conclude that the population mean is greater than 50. Our hypotheses are

$$H_0: \mu \leq 50 \qquad H_A: \mu > 50$$

If we want to know if we can conclude that the population mean is less than 50, the hypotheses are

$$H_0: \mu \geq 50 \qquad H_A: \mu < 50$$

In summary, we may state the following rules of thumb for deciding what statement goes in the null hypothesis and what statement goes in the alternative hypothesis:

(a) What you hope or expect to be able to conclude as a result of the test usually should be placed in the alternative hypothesis.

(b) The null hypothesis should contain a statement of equality, either $=$, $\leq$, or $\geq$.

(c) The null hypothesis is the hypothesis that is tested.

(d) The null and alternative hypotheses are complementary. That is, the two together exhaust all possibilities regarding the value that the hypothesized parameter can assume.

A Precaution It should be pointed out that neither hypothesis testing nor statistical inference, in general, leads to the proof of a hypothesis; it merely indicates whether the hypothesis is supported or is not supported by the available data. When we fail to reject a null hypothesis, therefore, we do not say that it is true, but that it may be true. When we speak of accepting a null hypothesis, we have this limitation in mind and do not wish to convey the idea that accepting implies proof.

4. **Test statistic.** The test statistic is some statistic that may be computed from the data of the sample. As a rule, there are many possible values that the test statistic may assume, the particular value observed depending on the particular sample drawn. As we will see, the test statistic serves as a decision maker, since the decision

to reject or not to reject the null hypothesis depends on the magnitude of the test statistic. An example of a test statistic is the quantity

$$z = \frac{\bar{x} - \mu_0}{\sigma/\sqrt{n}} \tag{7.1.1}$$

where μ_0 is a hypothesized value of a population mean. This test statistic is related to the statistic

$$z = \frac{\bar{x} - \mu}{\sigma/\sqrt{n}} \tag{7.1.2}$$

with which we are already familiar.

General Formula for Test Statistic The following is a general formula for a test statistic that will be applicable in many of the hypothesis tests discussed in this book:

$$\text{test statistic} = \frac{\text{relevant statistic} - \text{hypothesized parameter}}{\text{standard error of the relevant statistic}}$$

In Equation 7.1.1, $\bar{x}$ is the relevant statistic, μ_0 is the hypothesized parameter, and $\sigma/\sqrt{n}$ is the standard error of $\bar{x}$, the relevant statistic.

5. **Distribution of test statistic.** It has been pointed out that the key to statistical inference is the sampling distribution. We are reminded of this again when it becomes necessary to specify the probability distribution of the test statistic. The distribution of the test statistic

$$z = \frac{\bar{x} - \mu_0}{\sigma/\sqrt{n}}$$

for example, follows the standard normal distribution if the null hypothesis is true and the assumptions are met.

6. **Decision rule.** All possible values that the test statistic can assume are points on the horizontal axis of the graph of the distribution of the test statistic and are divided into two groups; one group constitutes what is known as the *rejection region* and the other group makes up the *nonrejection region*. The values of the test statistic forming the rejection region are those values that are less likely to occur if the null hypothesis is true, while the values making up the acceptance region are more likely to occur if the null hypothesis is true. *The decision rule tells us to reject the null hypothesis if the value of the test statistic that we compute from our sample is one of the values in the rejection region and to not reject the null hypothesis if the computed value of the test statistic is one of the values in the nonrejection region.*

Significance Level The decision as to which values go into the rejection region and which ones go into the nonrejection region is made on the basis of the desired *level of significance,* designated by α. The term *level of significance* reflects the fact that hypothesis tests are sometimes called significance tests, and a computed value of the test statistic that falls in the rejection region is said to be *significant*. The level of significance, α, specifies the area under the curve of the distribution of the test statistic that is above the values on the horizontal axis constituting the rejection region.

DEFINITION

The *level of significance* α is a probability and, in fact, is the *probability of rejecting a true null hypothesis.*

Since to reject a true null hypothesis would constitute an error, it seems only reasonable that we should make the probability of rejecting a true null hypothesis small and, in fact, that is what is done. We select a small value of α in order to make the probability of rejecting a true null hypothesis small. The more frequently encountered values of α are .01, .05, and .10.

Types of Errors The error committed when a true null hypothesis is rejected is called the *type I error*. The *type II error* is the error committed when a false null hypothesis is not rejected. The probability of committing a type II error is designated by β.

Whenever we reject a null hypothesis there is always the concomitant risk of committing a type I error, rejecting a true null hypothesis. Whenever we fail to reject a null hypothesis the risk of failing to reject a false null hypothesis is always present. We make α small, but we generally exercise no control over β, although we know that in most practical situations it is larger than α.

We never know whether we have committed one of these errors when we reject or fail to reject a null hypothesis, since the true state of affairs is unknown. If the testing procedure leads to rejection of the null hypothesis, we can take comfort from the fact that we made α small and, therefore, the probability of committing a type I error was small. If we fail to reject the null hypothesis, we do not know the concurrent risk of committing a type II error, since β is usually unknown but, as has been pointed out, we do know that, in most practical situations, it is larger than α.

Figure 7.1.1 shows for various conditions of a hypothesis test the possible actions that an investigator may take and the conditions under which each of the two types of error will be made. The table shown in this figure is an example of what is generally referred to as a *confusion matrix*.

7. **Calculation of test statistic.** From the data contained in the sample we compute a value of the test statistic and compare it with the rejection and nonrejection regions that have already been specified.

		Condition of Null Hypothesis	
		True	False
Possible Action	Fail to reject H_0	Correct action	Type II error
	Reject H_0	Type I error	Correct action

FIGURE 7.1.1 Conditions under which type I and type II errors may be committed.

8. **Statistical decision.** The statistical decision consists of rejecting or of not rejecting the null hypothesis. It is rejected if the computed value of the test statistic falls in the rejection region, and it is not rejected if the computed value of the test statistic falls in the nonrejection region.
9. **Conclusion.** If H_0 is rejected, we conclude that H_A is true. If H_0 is not rejected, we conclude that H_0 may be true.
10. ***p* values.** The p value is a number that tells us how unusual our sample results are, given that the null hypothesis is true. A p value indicating that the sample results are not likely to have occurred, if the null hypothesis is true, provides justification for doubting the truth of the null hypothesis.

DEFINITION

A *p value* is the probability that the computed value of a test statistic is at least as extreme as a specified value of the test statistic when the null hypothesis is true. Thus, the *p value* is the smallest value of α for which we can reject a null hypothesis.

We emphasize that when the null hypothesis is not rejected one should not say that the null hypothesis is accepted. We should say that the null hypothesis is "not rejected." We avoid using the word "accept" in this case because we may have committed a type II error. Since, frequently, the probability of committing a type II error can be quite high, we do not wish to commit ourselves to accepting the null hypothesis.

Figure 7.1.2 is a flowchart of the steps that we follow when we perform a hypothesis test.

Purpose of Hypothesis Testing The purpose of hypothesis testing is to assist administrators and clinicians in making decisions. The administrative or clinical decision usually depends on the statistical decision. If the null hypothesis is rejected, the administrative or clinical decision usually reflects this, in that the decision is compatible with the alternative hypothesis. The reverse is usually true if the null hypothesis is not rejected. The administrative or clinical decision, however, may take other forms, such as a decision to gather more data.

We also emphasize that the hypothesis testing procedures highlighted in the remainder of this chapter generally examine the case of normally distributed data or cases where the procedures are appropriate because the central limit theorem applies. In practice, it is not uncommon for samples to be small relative to the size of the population, or to have samples that are highly skewed, and hence the assumption of normality is violated. Methods to handle this situation, that is *distribution-free* or *nonparametric methods,* are examined in detail in Chapter 13. Most computer packages include an analytical procedure (for example, the Shapiro-Wilk or Anderson-Darling test) for testing normality. It is important that such tests are carried out prior to analysis of data. Further, when testing two samples, there is an implicit assumption that the variances are equal. Tests for this assumption are provided in Section 7.8. Finally, it should be noted that hypothesis

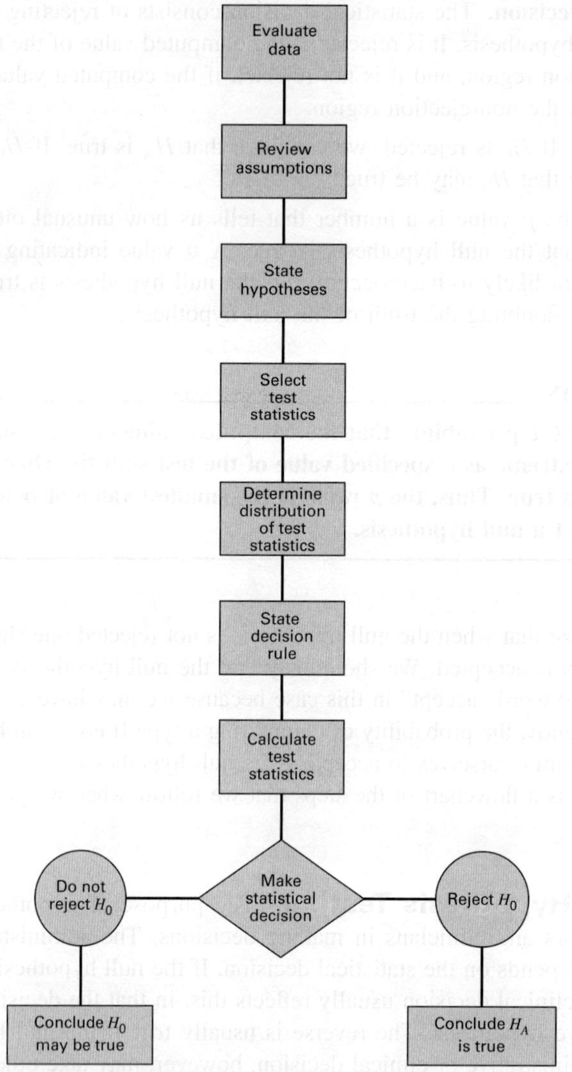

FIGURE 7.1.2 Steps in the hypothesis testing procedure.

tests, just like confidence intervals, are relatively sensitive to the size of the samples being tested, and caution should be taken when interpreting results involving very small sample sizes.

We must emphasize at this point, however, that the outcome of the statistical test is only one piece of evidence that influences the administrative or clinical decision. The statistical decision should not be interpreted as definitive but should be considered along with all the other relevant information available to the experimenter.

With these general comments as background, we now discuss specific hypothesis tests.

7.2 HYPOTHESIS TESTING: A SINGLE POPULATION MEAN

In this section we consider the testing of a hypothesis about a population mean under three different conditions: (1) when sampling is from a normally distributed population of values with known variance; (2) when sampling is from a normally distributed population with unknown variance, and (3) when sampling is from a population that is not normally distributed. Although the theory for conditions 1 and 2 depends on normally distributed populations, it is common practice to make use of the theory when relevant populations are only approximately normally distributed. This is satisfactory as long as the departure from normality is not drastic. When sampling is from a normally distributed population and the population variance is known, the test statistic for testing H_0: $\mu = \mu_0$ is

$$z = \frac{\bar{x} - \mu}{\sigma/\sqrt{n}} \tag{7.2.1}$$

which, when H_0 is true, is distributed as the standard normal. Examples 7.2.1 and 7.2.2 illustrate hypothesis testing under these conditions.

Sampling from Normally Distributed Populations: Population Variances Known As we did in Chapter 6, we again emphasize that situations in which the variable of interest is normally distributed with a known variance are rare. The following example, however, will serve to illustrate the procedure.

EXAMPLE 7.2.1

Researchers are interested in the mean age of a certain population. Let us say that they are asking the following question: Can we conclude that the mean age of this population is different from 30 years?

Solution: Based on our knowledge of hypothesis testing, we reply that they can conclude that the mean age is different from 30 if they can reject the null hypothesis that the mean is equal to 30. Let us use the ten-step hypothesis testing procedure given in the previous section to help the researchers reach a conclusion.

1. **Data.** The data available to the researchers are the ages of a simple random sample of 10 individuals drawn from the population of interest. From this sample a mean of $\bar{x} = 27$ has been computed.
2. **Assumptions.** It is assumed that the sample comes from a population whose ages are approximately normally distributed. Let us also assume that the population has a known variance of $\sigma^2 = 20$.
3. **Hypotheses.** The hypothesis to be tested, or null hypothesis, is that the mean age of the population is equal to 30. The alternative hypothesis is

that the mean age of the population is not equal to 30. Note that we are identifying with the alternative hypothesis the conclusion the researchers wish to reach, so that if the data permit rejection of the null hypothesis, the researchers' conclusion will carry more weight, since the accompanying probability of rejecting a true null hypothesis will be small. We will make sure of this by assigning a small value to α, the probability of committing a type I error. We may present the relevant hypotheses in compact form as follows:

$$H_0: \mu = 30$$
$$H_A: \mu \neq 30$$

4. **Test statistic.** Since we are testing a hypothesis about a population mean, since we assume that the population is normally distributed, and since the population variance is known, our test statistic is given by Equation 7.2.1.

5. **Distribution of test statistic.** Based on our knowledge of sampling distributions and the normal distribution, we know that the test statistic is normally distributed with a mean of 0 and a variance of 1, if H_0 is true. There are many possible values of the test statistic that the present situation can generate; one for every possible sample of size 10 that can be drawn from the population. Since we draw only one sample, we have only one of these possible values on which to base a decision.

6. **Decision rule.** The decision rule tells us to reject H_0 if the computed value of the test statistic falls in the rejection region and to fail to reject H_0 if it falls in the nonrejection region. We must now specify the rejection and nonrejection regions. We can begin by asking ourselves what magnitude of values of the test statistic will cause rejection of H_0. If the null hypothesis is false, it may be so either because the population mean is less than 30 or because the population mean is greater than 30. Therefore, either sufficiently small values or sufficiently large values of the test statistic will cause rejection of the null hypothesis. We want these extreme values to constitute the rejection region. How extreme must a possible value of the test statistic be to qualify for the rejection region? The answer depends on the significance level we choose, that is, the size of the probability of committing a type I error. Let us say that we want the probability of rejecting a true null hypothesis to be $\alpha = .05$. Since our rejection region is to consist of two parts, sufficiently small values and sufficiently large values of the test statistic, part of α will have to be associated with the large values and part with the small values. It seems reasonable that we should divide α equally and let $\alpha/2 = .025$ be associated with small values and $\alpha/2 = .025$ be associated with large values.

Critical Value of Test Statistic

What value of the test statistic is so large that, when the null hypothesis is true, the probability of obtaining a value this large or larger is .025? In other words, what is the value of z to the right of which lies .025 of the area under the standard normal distribution? The value of z to the right of which lies .025 of the area is the same value that has .975 of the area between it and $-\infty$. We look in the body of Appendix Table D until we find .975 or its closest value and read the corresponding marginal entries to obtain our z value. In the present example the value of z is 1.96. Similar reasoning will lead us to find -1.96 as the value of the test statistic so small that when the null hypothesis is true, the probability of obtaining a value this small or smaller is .025. Our rejection region, then, consists of all values of the test statistic equal to or greater than 1.96 and less than or equal to -1.96. The nonrejection region consists of all values in between. We may state the decision rule for this test as follows: *reject* H_0 *if the computed value of the test statistic is either* ≥ 1.96 *or* ≤ -1.96. Otherwise, do not reject H_0. The rejection and nonrejection regions are shown in Figure 7.2.1. The values of the test statistic that separate the rejection and nonrejection regions are called *critical values* of the test statistic, and the rejection region is sometimes referred to as the *critical region*.

The decision rule tells us to compute a value of the test statistic from the data of our sample and to reject H_0 if we get a value that is either equal to or greater than 1.96 or equal to or less than -1.96 and to fail to reject H_0 if we get any other value. The value of α and, hence, the decision rule should be decided on before gathering the data. This prevents our being accused of allowing the sample results to influence our choice of α. This condition of objectivity is highly desirable and should be preserved in all tests.

7. Calculation of test statistic. From our sample we compute

$$z = \frac{27 - 30}{\sqrt{20/10}} = \frac{-3}{1.4142} = -2.12$$

8. Statistical decision. Abiding by the decision rule, we are able to reject the null hypothesis since -2.12 is in the rejection region. We

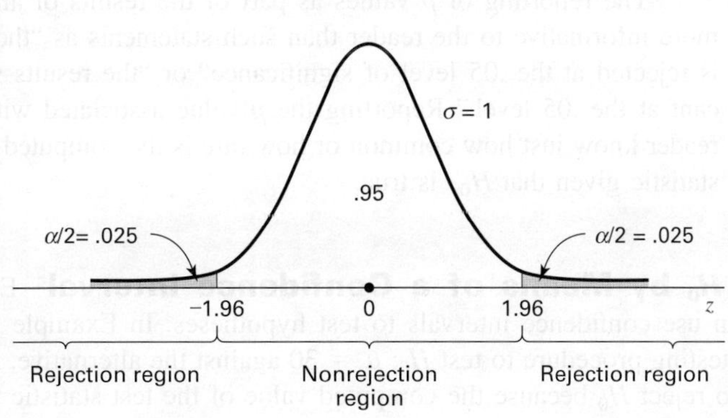

FIGURE 7.2.1 Rejection and nonrejection regions for Example 7.2.1.

can say that the computed value of the test statistic is significant at the .05 level.

9. Conclusion. We conclude that μ is not equal to 30 and let our administrative or clinical actions be in accordance with this conclusion.

10. *p* values. Instead of saying that an observed value of the test statistic is significant or is not significant, most writers in the research literature prefer to report the exact probability of getting a value as extreme as or more extreme than that observed if the null hypothesis is true. In the present instance these writers would give the computed value of the test statistic along with the statement $p = .0340$. The statement $p = .0340$ means that the probability of getting a value as extreme as 2.12 in either direction, when the null hypothesis is true, is .0340. The value .0340 is obtained from Appendix Table D and is the probability of observing a $z \geq 2.12$ or a $z \leq -2.12$ when the null hypothesis is true. That is, when H_0 is true, the probability of obtaining a value of z as large as or larger than 2.12 is .0170, and the probability of observing a value of z as small as or smaller than -2.12 is .0170. The probability of one or the other of these events occurring, when H_0 is true, is equal to the sum of the two individual probabilities, and hence, in the present example, we say that $p = .0170 + .0170 = .0340$.

Recall that the p value for a test may be defined also as the smallest value of α for which the null hypothesis can be rejected. Since, in Example 7.2.1, our p value is .0340, we know that we could have chosen an α value as small as .0340 and still have rejected the null hypothesis. If we had chosen an α smaller than .0340, we would not have been able to reject the null hypothesis. A general rule worth remembering, then, is this: *if the p value is less than or equal to α, we reject the null hypothesis; if the p value is greater than α, we do not reject the null hypothesis.*

The reporting of p values as part of the results of an investigation is more informative to the reader than such statements as "the null hypothesis is rejected at the .05 level of significance" or "the results were not significant at the .05 level." Reporting the p value associated with a test lets the reader know just how common or how rare is the computed value of the test statistic given that H_0 is true. ■

Testing H_0 by Means of a Confidence Interval Earlier, we stated that one can use confidence intervals to test hypotheses. In Example 7.2.1 we used a hypothesis testing procedure to test H_0: $\mu = 30$ against the alternative, H_A: $\mu \neq 30$. We were able to reject H_0 because the computed value of the test statistic fell in the rejection region.

Let us see how we might have arrived at this same conclusion by using a 100 $(1 - \alpha)$ percent confidence interval. The 95 percent confidence interval for μ is

$$27 \pm 1.96\sqrt{20/10}$$
$$27 \pm 1.96(1.414)$$
$$27 \pm 2.7714$$
$$24.2286, 29.7714$$

Since this interval does not include 30, we say 30 is not a candidate for the mean we are estimating and, therefore, μ is not equal to 30 and H_0 is rejected. This is the same conclusion reached by means of the hypothesis testing procedure.

If the hypothesized parameter, 30, had been within the 95 percent confidence interval, we would have said that H_0 is not rejected at the .05 level of significance. In general, *when testing a null hypothesis by means of a two-sided confidence interval, we reject H_0 at the α level of significance if the hypothesized parameter is not contained within the $100(1 - \alpha)$ percent confidence interval. If the hypothesized parameter is contained within the interval, H_0 cannot be rejected at the α level of significance.*

One-Sided Hypothesis Tests The hypothesis test illustrated by Example 7.2.1 is an example of a *two-sided test,* so called because the rejection region is split between the two sides or tails of the distribution of the test statistic. A hypothesis test may be *one-sided,* in which case all the rejection region is in one or the other tail of the distribution. Whether a one-sided or a two-sided test is used depends on the nature of the question being asked by the researcher.

If both large and small values will cause rejection of the null hypothesis, a two-sided test is indicated. When either sufficiently "small" values only or sufficiently "large" values only will cause rejection of the null hypothesis, a one-sided test is indicated.

EXAMPLE 7.2.2

Refer to Example 7.2.1. Suppose, instead of asking if they could conclude that $\mu \neq 30$, the researchers had asked: Can we conclude that $\mu < 30$? To this question we would reply that they can so conclude if they can reject the null hypothesis that $\mu \geq 30$.

Solution: Let us go through the ten-step procedure to reach a decision based on a one-sided test.

1. **Data.** See the previous example.
2. **Assumptions.** See the previous example.
3. **Hypotheses.**

$$H_0: \mu \geq 30$$
$$H_A: \mu < 30$$

The inequality in the null hypothesis implies that the null hypothesis consists of an infinite number of hypotheses. The test will be made only at the point of equality, since it can be shown that if H_0 is rejected when the test is made at the point of equality it would be rejected if the test were done for any other value of μ indicated in the null hypothesis.

4. **Test statistic.**

$$z = \frac{\bar{x} - \mu_0}{\sigma/\sqrt{n}}$$

5. **Distribution of test statistic.** See the previous example.
6. **Decision rule.** Let us again use $\alpha = .05$. To determine where to place the rejection region, let us ask ourselves what magnitude of values would cause rejection of the null hypothesis. If we look at the hypotheses, we see that sufficiently small values would cause rejection and that large values would tend to reinforce the null hypothesis. We will want our rejection region to be where the small values are—at the lower tail of the distribution. This time, since we have a one-sided test, all of α will go in the one tail of the distribution. By consulting Appendix Table D, we find that the value of z to the left of which lies .05 of the area under the standard normal curve is -1.645 after interpolating. Our rejection and nonrejection regions are now specified and are shown in Figure 7.2.2.

   Our decision rule tells us to reject H_0 if the computed value of the test statistic is less than or equal to -1.645.
7. **Calculation of test statistic.** From our data we compute

$$z = \frac{27 - 30}{\sqrt{20/10}} = -2.12$$

8. **Statistical decision.** We are able to reject the null hypothesis since $-2.12 < -1.645$.
9. **Conclusion.** We conclude that the population mean is smaller than 30 and act accordingly.
10. ***p* value.** The p value for this test is .0170, since $P(z \leq -2.12)$, when H_0 is true, is .0170 as given by Appendix Table D when we determine

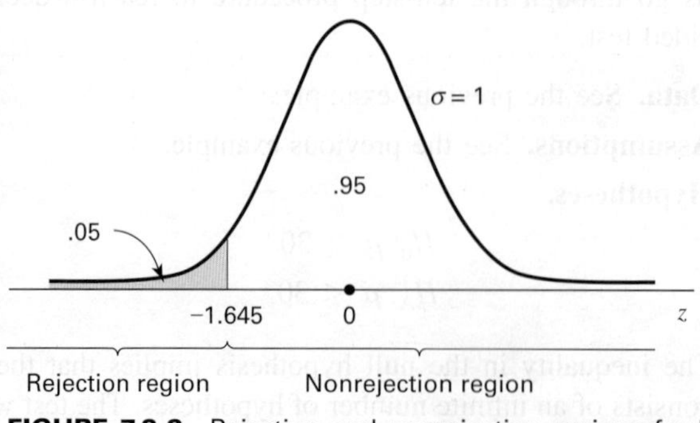

FIGURE 7.2.2 Rejection and nonrejection regions for Example 7.2.2.

the magnitude of the area to the left of −2.12 under the standard normal curve. One can test a one-sided null hypothesis by means of a one-sided confidence interval. However, we will not cover the construction and interpretation of this type of confidence interval in this book.

If the researcher's question had been, "Can we conclude that the mean is greater than 30?," following the above ten-step procedure would have led to a one-sided test with all the rejection region at the upper tail of the distribution of the test statistic and a critical value of +1.645. ■

Sampling from a Normally Distributed Population: Population Variance Unknown As we have already noted, the population variance is usually unknown in actual situations involving statistical inference about a population mean. When sampling is from an approximately normal population with an unknown variance, the test statistic for testing H_0: $\mu = \mu_0$ is

$$t = \frac{\bar{x} - \mu_0}{s/\sqrt{n}} \tag{7.2.2}$$

which, when H_0 is true, is distributed as Student's t with $n - 1$ degrees of freedom. The following example illustrates the hypothesis testing procedure when the population is assumed to be normally distributed and its variance is unknown. This is the usual situation encountered in practice.

EXAMPLE 7.2.3

Nakamura et al. (A-1) studied subjects with medial collateral ligament (MCL) and anterior cruciate ligament (ACL) tears. Between February 1995 and December 1997, 17 consecutive patients with combined acute ACL and grade III MCL injuries were treated by the same physician at the research center. One of the variables of interest was the length of time in days between the occurrence of the injury and the first magnetic resonance imaging (MRI). The data are shown in Table 7.2.1. We wish to know if we can conclude that the mean number of days between injury and initial MRI is not 15 days in a population presumed to be represented by these sample data.

TABLE 7.2.1 Number of Days Until MRI for Subjects with MCL and ACL Tears

Subject	Days	Subject	Days	Subject	Days	Subject	Days
1	14	6	0	11	28	16	14
2	9	7	10	12	24	17	9
3	18	8	4	13	24		
4	26	9	8	14	2		
5	12	10	21	15	3		

Source: Norimasa Nakamura, Shuji Horibe, Yukyoshi Toritsuka, Tomoki Mitsuoka, Hideki Yoshikawa, and Konsei Shino, "Acute Grade III Medial Collateral Ligament Injury of the Knee Associated with Anterior Cruciate Ligament Tear," *American Journal of Sports Medicine, 31* (2003), 261–267.

Solution: We will be able to conclude that the mean number of days for the population is not 15 if we can reject the null hypothesis that the population mean is equal to 15.

1. **Data.** The data consist of number of days until MRI on 17 subjects as previously described.

2. **Assumptions.** The 17 subjects constitute a simple random sample from a population of similar subjects. We assume that the number of days until MRI in this population is approximately normally distributed.

3. **Hypotheses.**

$$H_0: \mu = 15$$
$$H_A: \mu \neq 15$$

4. **Test statistic.** Since the population variance is unknown, our test statistic is given by Equation 7.2.2.

5. **Distribution of test statistic.** Our test statistic is distributed as Student's t with $n - 1 = 17 - 1 = 16$ degrees of freedom if H_0 is true.

6. **Decision rule.** Let $\alpha = .05$. Since we have a two-sided test, we put $\alpha/2 = .025$ in each tail of the distribution of our test statistic. The t values to the right and left of which .025 of the area lies are 2.1199 and -2.1199. These values are obtained from Appendix Table E. The rejection and nonrejection regions are shown in Figure 7.2.3.

   The decision rule tells us to compute a value of the test statistic and reject H_0 if the computed t is either greater than or equal to 2.1199 or less than or equal to -2.1199.

7. **Calculation of test statistic.** From our sample data we compute a sample mean of 13.2941 and a sample standard deviation of 8.88654. Substituting these statistics into Equation 7.2.2 gives

$$t = \frac{13.2941 - 15}{8.88654/\sqrt{17}} = \frac{-1.7059}{2.1553} = -.791$$

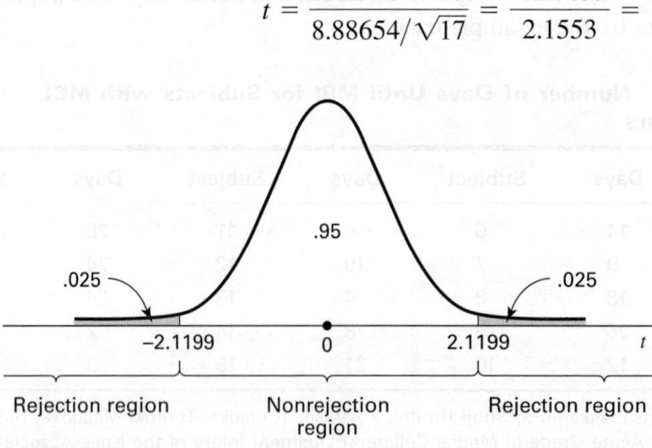

FIGURE 7.2.3 Rejection and nonrejection regions for Example 7.2.3.

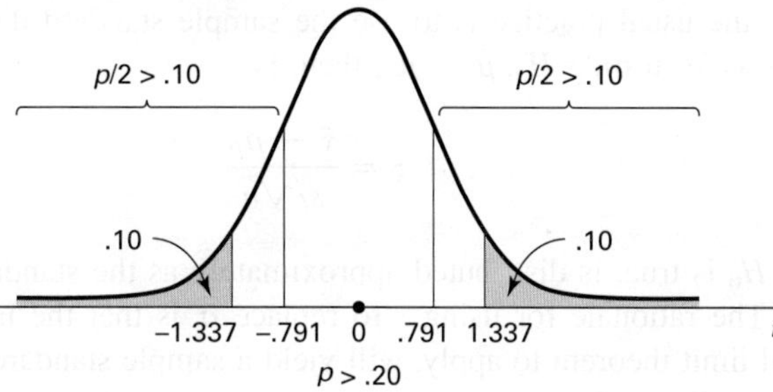

FIGURE 7.2.4 Determination of p value for Example 7.2.3.

8. **Statistical decision.** Do not reject H_0, since $-.791$ falls in the nonrejection region.
9. **Conclusion.** Our conclusion, based on these data, is that the mean of the population from which the sample came may be 15.
10. **p value.** The exact p value for this test cannot be obtained from Appendix Table E since it gives t values only for selected percentiles. The p value can be stated as an interval, however. We find that $-.791$ is less than -1.337, the value of t to the left of which lies .10 of the area under the t with 16 degrees of freedom. Consequently, when H_0 is true, the probability of obtaining a value of t as small as or smaller than $-.791$ is greater than .10. That is $P(t \leq -.791) > .10$. Since the test was two-sided, we must allow for the possibility of a computed value of the test statistic as large in the opposite direction as that observed. Appendix Table E reveals that $P(t \geq .791) > .10$ also. The p value, then, is $p > .20$. Figure 7.2.4 shows the p value for this example.

If in the previous example the hypotheses had been

$$H_0: \mu \geq 15$$
$$H_A: \mu < 15$$

the testing procedure would have led to a one-sided test with all the rejection region at the lower tail of the distribution, and if the hypotheses had been

$$H_0: \mu \leq 15$$
$$H_A: \mu > 15$$

we would have had a one-sided test with all the rejection region at the upper tail of the distribution. ■

Sampling from a Population That Is Not Normally Distributed

If, as is frequently the case, the sample on which we base our hypothesis test about a population mean comes from a population that is not normally distributed, we may, if our sample is large (greater than or equal to 30), take advantage of the central limit theorem and use $z = (\bar{x} - \mu_0)/(\sigma/\sqrt{n})$ as the test statistic. If the population standard deviation is not known, the usual practice is to use the sample standard deviation as an estimate. The test statistic for testing H_0: $\mu = \mu_0$, then, is

$$z = \frac{\bar{x} - \mu_0}{s/\sqrt{n}} \tag{7.2.3}$$

which, when H_0 is true, is distributed approximately as the standard normal distribution if n is large. The rationale for using s to replace σ is that the large sample, necessary for the central limit theorem to apply, will yield a sample standard deviation that closely approximates σ.

EXAMPLE 7.2.4

The goal of a study by Klingler et al. (A-2) was to determine how symptom recognition and perception influence clinical presentation as a function of race. They characterized symptoms and care-seeking behavior in African-American patients with chest pain seen in the emergency department. One of the presenting vital signs was systolic blood pressure. Among 157 African-American men, the mean systolic blood pressure was 146 mm Hg with a standard deviation of 27. We wish to know if, on the basis of these data, we may conclude that the mean systolic blood pressure for a population of African-American men is greater than 140.

Solution: We will say that the data do provide sufficient evidence to conclude that the population mean is greater than 140 if we can reject the null hypothesis that the mean is less than or equal to 140. The following test may be carried out:

1. **Data.** The data consist of systolic blood pressure scores for 157 African-American men with $\bar{x} = 146$ and $s = 27$.
2. **Assumptions.** The data constitute a simple random sample from a population of African-American men who report to an emergency department with symptoms similar to those in the sample. We are unwilling to assume that systolic blood pressure values are normally distributed in such a population.
3. **Hypotheses.**

$$H_0: \mu \leq 140$$
$$H_A: \mu > 140$$

4. **Test statistic.** The test statistic is given by Equation 7.2.3, since s is unknown.
5. **Distribution of test statistic.** Because of the central limit theorem, the test statistic is at worst approximately normally distributed with $\mu = 0$ if H_0 is true.
6. **Decision rule.** Let $\alpha = .05$. The critical value of the test statistic is 1.645. The rejection and nonrejection regions are shown in Figure 7.2.5. Reject H_0 if computed $z \geq 1.645$.

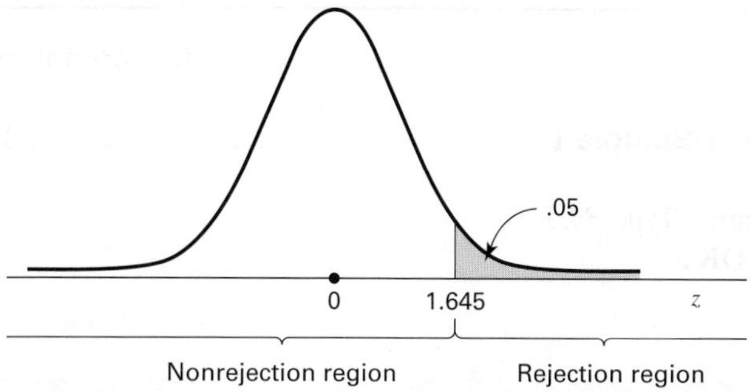

FIGURE 7.2.5 Rejection and nonrejection regions for Example 7.2.4.

7. **Calculation of test statistic.**

$$z = \frac{146 - 140}{27/\sqrt{157}} = \frac{6}{2.1548} = 2.78$$

8. **Statistical decision.** Reject H_0 since $2.78 > 1.645$.
9. **Conclusion.** Conclude that the mean systolic blood pressure for the sampled population is greater than 140.
10. ***p* value.** The p value for this test is $1 - .9973 = .0027$, since as shown in Appendix Table D, the area (.0027) to the right of 2.78 is less than .05, the area to the right of 1.645. ■

Procedures for Other Conditions If the population variance had been known, the procedure would have been identical to the above except that the known value of σ, instead of the sample value s, would have been used in the denominator of the computed test statistic.

Depending on what the investigators wished to conclude, either a two-sided test or a one-sided test, with the rejection region at the lower tail of the distribution, could have been made using the above data.

When testing a hypothesis about a single population mean, we may use Figure 6.3.3 to decide quickly whether the test statistic is z or t.

Computer Analysis To illustrate the use of computers in testing hypotheses we consider the following example.

EXAMPLE 7.2.5

The following are the head circumferences (centimeters) at birth of 15 infants:

33.38	32.15	33.99	34.10	33.97
34.34	33.95	33.85	34.23	32.73
33.46	34.13	34.45	34.19	34.05

We wish to test H_0: $\mu = 34.5$ against H_A: $\mu \neq 34.5$.

Dialog box:

Stat ➤ Basic Statistics ➤ 1-Sample t

Type *C1* in **Samples in columns.** Type *34.5* in the **Test mean** box. Click **OK**.

Session command:

```
MTB > TTEST 34.5 C1
```

Output:

One-Sample T: C1

```
TEST OF MU = 34.5 VS NOT = 34.5
VARIABLE   N     MEAN   STDEV  SE MEAN          95% CI              T      P
C1        15  33.7980  0.6303   0.1627  (33.4490, 34.1470)  -4.31  0.001
```

FIGURE 7.2.6 MINITAB procedure and output for Example 7.2.5.

Solution: We assume that the assumptions for use of the t statistic are met. We enter the data into Column 1 and proceed as shown in Figure 7.2.6.

To indicate that a test is one-sided when in Windows, click on the **Options** button and then choose "less than" or "greater than" as appropriate in the **Alternative** box. If z is the appropriate test statistic, we choose 1-Sample z from the Basic Statistics menu. The remainder of the commands are the same as for the t test.

We learn from the printout that the computed value of the test statistic is -4.31 and the p value for the test is .0007. SAS® users may use the output from PROC MEANS or PROC UNIVARIATE to perform hypothesis tests.

When both the z statistic and the t statistic are inappropriate test statistics for use with the available data, one may wish to use a nonparametric technique to test a hypothesis about a single population measure of central tendency. One such procedure, the sign test, is discussed in Chapter 13. ■

EXERCISES

For each of the following exercises carry out the ten-step hypothesis testing procedure for the given significance level. For each exercise, as appropriate, explain why you chose a one-sided test or a two-sided test. Discuss how you think researchers and/or clinicians might use the results of your hypothesis test. What clinical and/or research decisions and/or actions do you think would be appropriate in light of the results of your test?

7.2.1 Escobar et al. (A-3) performed a study to validate a translated version of the Western Ontario and McMaster Universities Osteoarthritis Index (WOMAC) questionnaire used with Spanish-speaking patients with hip or knee osteoarthritis. For the 76 women classified with severe hip pain, the

WOMAC mean function score (on a scale from 0 to 100 with a higher number indicating less function) was 70.7 with a standard deviation of 14.6. We wish to know if we may conclude that the mean function score for a population of similar women subjects with severe hip pain is less than 75. Let $\alpha = .01$.

7.2.2 A study by Thienprasiddhi et al. (A-4) examined a sample of 16 subjects with open-angle glaucoma and unilateral hemifield defects. The ages (years) of the subjects were:

62	62	68	48	51	60	51	57
57	41	62	50	53	34	62	61

Source: Phamornsak Thienprasiddhi, Vivienne C. Greenstein, Candice S. Chen, Jeffrey M. Liebmann, Robert Ritch, and Donald C. Hood, "Multifocal Visual Evoked Potential Responses in Glaucoma Patients with Unilateral Hemifield Defects," *American Journal of Ophthalmology, 136* (2003), 34–40.

Can we conclude that the mean age of the population from which the sample may be presumed to have been drawn is less than 60 years? Let $\alpha = .05$.

7.2.3 The purpose of a study by Luglié et al. (A-5) was to investigate the oral status of a group of patients diagnosed with thalassemia major (TM). One of the outcome measures was the decayed, missing, and filled teeth index (DMFT). In a sample of 18 patients the mean DMFT index value was 10.3 with a standard deviation of 7.3. Is this sufficient evidence to allow us to conclude that the mean DMFT index is greater than 9.0 in a population of similar subjects? Let $\alpha = .10$.

7.2.4 A study was made of a sample of 25 records of patients seen at a chronic disease hospital on an outpatient basis. The mean number of outpatient visits per patient was 4.8, and the sample standard deviation was 2. Can it be concluded from these data that the population mean is greater than four visits per patient? Let the probability of committing a type I error be .05. What assumptions are necessary?

7.2.5 In a sample of 49 adolescents who served as the subjects in an immunologic study, one variable of interest was the diameter of skin test reaction to an antigen. The sample mean and standard deviation were 21 and 11 mm erythema, respectively. Can it be concluded from these data that the population mean is less than 30? Let $\alpha = .05$.

7.2.6 Nine laboratory animals were infected with a certain bacterium and then immunosuppressed. The mean number of organisms later recovered from tissue specimens was 6.5 (coded data) with a standard deviation of .6. Can one conclude from these data that the population mean is greater than 6? Let $\alpha = .05$. What assumptions are necessary?

7.2.7 A sample of 25 freshman nursing students made a mean score of 77 on a test designed to measure attitude toward the dying patient. The sample standard deviation was 10. Do these data provide sufficient evidence to indicate, at the .05 level of significance, that the population mean is less than 80? What assumptions are necessary?

7.2.8 We wish to know if we can conclude that the mean daily caloric intake in the adult rural population of a developing country is less than 2000. A sample of 500 had a mean of 1985 and a standard deviation of 210. Let $\alpha = .05$.

7.2.9 A survey of 100 similar-sized hospitals revealed a mean daily census in the pediatrics service of 27 with a standard deviation of 6.5. Do these data provide sufficient evidence to indicate that the population mean is greater than 25? Let $\alpha = .05$.

7.2.10 Following a week-long hospital supervisory training program, 16 assistant hospital administrators made a mean score of 74 on a test administered as part of the evaluation of the training program. The sample standard deviation was 12. Can it be concluded from these data that the population mean is greater than 70? Let $\alpha = .05$. What assumptions are necessary?

7.2.11 A random sample of 16 emergency reports was selected from the files of an ambulance service. The mean time (computed from the sample data) required for ambulances to reach their destinations was 13 minutes. Assume that the population of times is normally distributed with a variance of 9. Can we conclude at the .05 level of significance that the population mean is greater than 10 minutes?

7.2.12 The following data are the oxygen uptakes (milliliters) during incubation of a random sample of 15 cell suspensions:

14.0, 14.1, 14.5, 13.2, 11.2, 14.0, 14.1, 12.2,
11.1, 13.7, 13.2, 16.0, 12.8, 14.4, 12.9

Do these data provide sufficient evidence at the .05 level of significance that the population mean is not 12 ml? What assumptions are necessary?

7.2.13 Can we conclude that the mean maximum voluntary ventilation value for apparently healthy college seniors is not 110 liters per minute? A sample of 20 yielded the following values:

132, 33, 91, 108, 67, 169, 54, 203, 190, 133,
96, 30, 187, 21, 63, 166, 84, 110, 157, 138

Let $\alpha = .01$. What assumptions are necessary?

7.2.14 The following are the systolic blood pressures (mm Hg) of 12 patients undergoing drug therapy for hypertension:

183, 152, 178, 157, 194, 163, 144, 114, 178, 152, 118, 158

Can we conclude on the basis of these data that the population mean is less than 165? Let $\alpha = .05$. What assumptions are necessary?

7.2.15 Can we conclude that the mean age at death of patients with homozygous sickle-cell disease is less than 30 years? A sample of 50 patients yielded the following ages in years:

15.5	2.0	45.1	1.7	.8	1.1	18.2	9.7	28.1	18.2
27.6	45.0	1.0	66.4	2.0	67.4	2.5	61.7	16.2	31.7
6.9	13.5	1.9	31.2	9.0	2.6	29.7	13.5	2.6	14.4
20.7	30.9	36.6	1.1	23.6	.9	7.6	23.5	6.3	40.2
23.7	4.8	33.2	27.1	36.7	3.2	38.0	3.5	21.8	2.4

Let $\alpha = .05$. What assumptions are necessary?

7.2.16 The following are intraocular pressure (mm Hg) values recorded for a sample of 21 elderly subjects:

14.5	12.9	14.0	16.1	12.0	17.5	14.1	12.9	17.9	12.0
16.4	24.2	12.2	14.4	17.0	10.0	18.5	20.8	16.2	14.9
19.6									

Can we conclude from these data that the mean of the population from which the sample was drawn is greater than 14? Let $\alpha = .05$. What assumptions are necessary?

7.2.17 Suppose it is known that the IQ scores of a certain population of adults are approximately normally distributed with a standard deviation of 15. A simple random sample of 25 adults drawn from this population had a mean IQ score of 105. On the basis of these data can we conclude that the mean IQ score for the population is not 100? Let the probability of committing a type I error be .05.

7.2.18 A research team is willing to assume that systolic blood pressures in a certain population of males are approximately normally distributed with a standard deviation of 16. A simple random sample of 64 males from the population had a mean systolic blood pressure reading of 133. At the .05 level of significance, do these data provide sufficient evidence for us to conclude that the population mean is greater than 130?

7.2.19 A simple random sample of 16 adults drawn from a certain population of adults yielded a mean weight of 63 kg. Assume that weights in the population are approximately normally distributed with a variance of 49. Do the sample data provide sufficient evidence for us to conclude that the mean weight for the population is less than 70 kg? Let the probability of committing a type I error be .01.

7.3 HYPOTHESIS TESTING: THE DIFFERENCE BETWEEN TWO POPULATION MEANS

Hypothesis testing involving the difference between two population means is most frequently employed to determine whether or not it is reasonable to conclude that the two population means are unequal. In such cases, one or the other of the following hypotheses may be formulated:

1. $H_0: \mu_1 - \mu_2 = 0, \quad H_A: \mu_1 - \mu_2 \neq 0$
2. $H_0: \mu_1 - \mu_2 \geq 0, \quad H_A: \mu_1 - \mu_2 < 0$
3. $H_0: \mu_1 - \mu_2 \leq 0, \quad H_A: \mu_1 - \mu_2 > 0$

It is possible, however, to test the hypothesis that the difference is equal to, greater than or equal to, or less than or equal to some value other than zero.

As was done in the previous section, hypothesis testing involving the difference between two population means will be discussed in three different contexts: (1) when sampling is from normally distributed populations with known population variances, (2) when sampling is from normally distributed populations with unknown population variances, and (3) when sampling is from populations that are not normally distributed.

Sampling from Normally Distributed Populations: Population Variances Known When each of two independent simple random samples has been drawn from a normally distributed population with a known variance, the test statistic for testing the null hypothesis of equal population means is

$$z = \frac{(\bar{x}_1 - \bar{x}_2) - (\mu_1 - \mu_2)_0}{\sqrt{\dfrac{\sigma_1^2}{n_1} + \dfrac{\sigma_2^2}{n_2}}} \tag{7.3.1}$$

where the subscript 0 indicates that the difference is a hypothesized parameter. When H_0 is true the test statistic of Equation 7.3.1 is distributed as the standard normal.

EXAMPLE 7.3.1

Researchers wish to know if the data they have collected provide sufficient evidence to indicate a difference in mean serum uric acid levels between normal individuals and individuals with Down's syndrome. The data consist of serum uric acid readings on 12 individuals with Down's syndrome and 15 normal individuals. The means are $\bar{x}_1 = 4.5$ mg/100 ml and $\bar{x}_2 = 3.4$ mg/100 ml.

Solution: We will say that the sample data do provide evidence that the population means are not equal if we can reject the null hypothesis that the population means are equal. Let us reach a conclusion by means of the ten-step hypothesis testing procedure.

1. **Data.** See problem statement.
2. **Assumptions.** The data constitute two independent simple random samples each drawn from a normally distributed population with a variance equal to 1 for the Down's syndrome population and 1.5 for the normal population.
3. **Hypotheses.**

$$H_0: \mu_1 - \mu_2 = 0$$
$$H_A: \mu_1 - \mu_2 \neq 0$$

An alternative way of stating the hypotheses is as follows:

$$H_0: \mu_1 = \mu_2$$
$$H_A: \mu_1 \neq \mu_2$$

4. **Test statistic.** The test statistic is given by Equation 7.3.1.
5. **Distribution of test statistic.** When the null hypothesis is true, the test statistic follows the standard normal distribution.
6. **Decision rule.** Let $\alpha = .05$. The critical values of z are ± 1.96. Reject H_0 unless $-1.96 < z_{\text{computed}} < 1.96$. The rejection and nonrejection regions are shown in Figure 7.3.1.
7. **Calculation of test statistic.**

$$z = \frac{(4.5 - 3.4) - 0}{\sqrt{1/12 + 1.5/15}} = \frac{1.1}{.4282} = 2.57$$

8. **Statistical decision.** Reject H_0, since $2.57 > 1.96$.
9. **Conclusion.** Conclude that, on the basis of these data, there is an indication that the two population means are not equal.
10. ***p* value.** For this test, $p = .0102$.

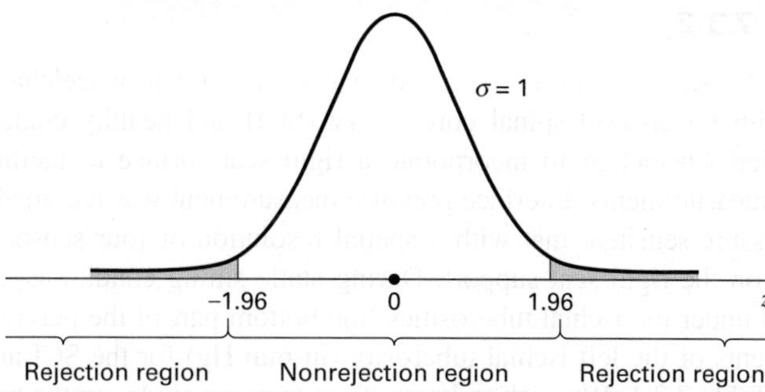

FIGURE 7.3.1 Rejection and nonrejection regions for Example 7.3.1.

■

A 95 Percent Confidence Interval for $\mu_1 - \mu_2$ In the previous chapter the 95 percent confidence interval for $\mu_1 - \mu_2$, computed from the same data, was found to be .26 to 1.94. Since this interval does not include 0, we say that 0 is not a candidate for the difference between population means, and we conclude that the difference is not zero. Thus we arrive at the same conclusion by means of a confidence interval.

Sampling from Normally Distributed Populations: Population Variances Unknown As we have learned, when the population variances are unknown, two possibilities exist. The two population variances may be equal or they may be unequal. We consider first the case where it is known, or it is reasonable to assume, that they are equal. A test of the hypothesis that two population variances are equal is described in Section 7.8.

Population Variances Equal When the population variances are unknown, but assumed to be equal, we recall from Chapter 6 that it is appropriate to pool the sample variances by means of the following formula:

$$s_p^2 = \frac{(n_1 - 1)s_1^2 + (n_2 - 1)s_2^2}{n_1 + n_2 - 2}$$

When each of two independent simple random samples has been drawn from a normally distributed population and the two populations have equal but unknown variances, the test statistic for testing $H_0: \mu_1 = \mu_2$ is given by

$$t = \frac{(\bar{x}_1 - \bar{x}_2) - (\mu_1 - \mu_2)_0}{\sqrt{\dfrac{s_p^2}{n_1} + \dfrac{s_p^2}{n_2}}} \tag{7.3.2}$$

which, when H_0 is true, is distributed as Student's t with $n_1 + n_2 - 2$ degrees of freedom.

EXAMPLE 7.3.2

The purpose of a study by Tam et al. (A-6) was to investigate wheelchair maneuvering in individuals with lower-level spinal cord injury (SCI) and healthy controls (C). Subjects used a modified wheelchair to incorporate a rigid seat surface to facilitate the specified experimental measurements. Interface pressure measurement was recorded by using a high-resolution pressure-sensitive mat with a spatial resolution of four sensors per square centimeter taped on the rigid seat support. During static sitting conditions, average pressures were recorded under the ischial tuberosities (the bottom part of the pelvic bones). The data for measurements of the left ischial tuberosity (in mm Hg) for the SCI and control groups are shown in Table 7.3.1. We wish to know if we may conclude, on the basis of these data, that, in general, healthy subjects exhibit lower pressure than SCI subjects.

Solution:

1. **Data.** See statement of problem.
2. **Assumptions.** The data constitute two independent simple random samples of pressure measurements, one sample from a population of control subjects and the other sample from a population with lower-level spinal cord injury. We shall assume that the pressure measurements in both populations are approximately normally distributed. The population variances are unknown but are assumed to be equal.
3. **Hypotheses.** H_0: $\mu_C \geq \mu_{SCI}$, H_A: $\mu_C < \mu_{SCI}$.
4. **Test statistic.** The test statistic is given by Equation 7.3.2.
5. **Distribution of test statistic.** When the null hypothesis is true, the test statistic follows Student's t distribution with $n_1 + n_2 - 2$ degrees of freedom.
6. **Decision rule.** Let $\alpha = .05$. The critical value of t is -1.7341. Reject H_0 unless $t_{\text{computed}} > -1.7341$.
7. **Calculation of test statistic.** From the sample data we compute

$$\bar{x}_C = 126.1, \qquad s_C = 21.8, \qquad \bar{x}_{SCI} = 133.1, \qquad s_{SCI} = 32.2$$

Next, we pool the sample variances to obtain

$$s_p^2 = \frac{9(21.8)^2 + 9(32.2)^2}{9 + 9} = 756.04$$

TABLE 7.3.1 Pressures (mm Hg) Under the Pelvis during Static Conditions for Example 7.3.2

Control	131	115	124	131	122	117	88	114	150	169
SCI	60	150	130	180	163	130	121	119	130	148

Source: Eric W. Tam, Arthur F. Mak, Wai Nga Lam, John H. Evans, and York Y. Chow, "Pelvic Movement and Interface Pressure Distribution During Manual Wheelchair Propulsion," *Archives of Physical Medicine and Rehabilitation, 84* (2003), 1466–1472.

We now compute

$$t = \frac{(126.1 - 133.1) - 0}{\sqrt{\dfrac{756.04}{10} + \dfrac{756.04}{10}}} = -.569$$

8. **Statistical decision.** We fail to reject H_0, since $-1.7341 < -.569$; that is, $-.569$ falls in the nonrejection region.
9. **Conclusion.** On the basis of these data, we cannot conclude that the population mean pressure is less for healthy subjects than for SCI subjects.
10. ***p* value.** For this test, $p > .10$ since $-1.330 < -.569$. ■

Population Variances Unequal When two independent simple random samples have been drawn from normally distributed populations with unknown and unequal variances, the test statistic for testing $H_0\colon \mu_1 = \mu_2$ is

$$t' = \frac{(\bar{x}_1 - \bar{x}_2) - (\mu_1 - \mu_2)_0}{\sqrt{\dfrac{s_1^2}{n_1} + \dfrac{s_2^2}{n_2}}} \tag{7.3.3}$$

The critical value of t' for an α level of significance and a two-sided test is approximately

$$t'_{1-(\alpha/2)} = \frac{w_1 t_1 + w_2 t_2}{w_1 + w_2} \tag{7.3.4}$$

where $w_1 = s_1^2/n_1$, $w_2 = s_2^2/n_2$, $t_1 = t_{1-(\alpha/2)}$ for $n_1 - 1$ degrees of freedom, and $t_2 = t_{1-(\alpha/2)}$ for $n_2 - 1$ degrees of freedom. The critical value of t' for a one-sided test is found by computing $t'_{1-\alpha}$ by Equation 7.3.4, using $t_1 = t_{1-\alpha}$ for $n_1 - 1$ degrees of freedom and $t_2 = t_{1-\alpha}$ for $n_2 - 1$ degrees of freedom.

For a two-sided test, reject H_0 if the computed value of t' is either greater than or equal to the critical value given by Equation 7.3.4 or less than or equal to the negative of that value.

For a one-sided test with the rejection region in the right tail of the sampling distribution, reject H_0 if the computed t' is equal to or greater than the critical t'. For a one-sided test with a left-tail rejection region, reject H_0 if the computed value of t' is equal to or smaller than the negative of the critical t' computed by the indicated adaptation of Equation 7.3.4.

EXAMPLE 7.3.3

Dernellis and Panaretou (A-7) examined subjects with hypertension and healthy control subjects. One of the variables of interest was the aortic stiffness index. Measures of this variable were calculated from the aortic diameter evaluated by M-mode echocardiography and blood pressure measured by a sphygmomanometer. Generally, physicians wish to reduce aortic stiffness. In the 15 patients with hypertension (group 1), the mean aortic stiffness index was 19.16 with a standard deviation of 5.29. In the 30 control subjects (group 2), the mean aortic stiffness index was 9.53 with a standard deviation of 2.69. We wish to determine if the two populations represented by these samples differ with respect to mean aortic stiffness index.

Solution:

1. **Data.** The sample sizes, means, and sample standard deviations are:

$$n_1 = 15, \quad \bar{x}_1 = 19.16, \quad s_1 = 5.29$$
$$n_2 = 30, \quad \bar{x}_2 = 9.53, \quad s_2 = 2.69$$

2. **Assumptions.** The data constitute two independent random samples, one from a population of subjects with hypertension and the other from a control population. We assume that aortic stiffness values are approximately normally distributed in both populations. The population variances are unknown and unequal.

3. **Hypotheses.**

$$H_0: \mu_1 - \mu_2 = 0$$
$$H_A: \mu_1 - \mu_2 \neq 0$$

4. **Test statistic.** The test statistic is given by Equation 7.3.3.

5. **Distribution of test statistic.** The statistic given by Equation 7.3.3 does not follow Student's t distribution. We, therefore, obtain its critical values by Equation 7.3.4.

6. **Decision rule.** Let $\alpha = .05$. Before computing t' we calculate $w_1 = (5.29)^2/15 = 1.8656$ and $w_2 = (2.69)^2/30 = .2412$. In Appendix Table E we find that $t_1 = 2.1448$ and $t_2 = 2.0452$. By Equation 7.3.4 we compute

$$t' = \frac{1.8656(2.1448) + .2412(2.0452)}{1.8656 + .2412} = 2.133$$

Our decision rule, then, is reject H_0 if the computed t is either ≥ 2.133 or ≤ -2.133.

7. **Calculation of test statistic.** By Equation 7.3.3 we compute

$$t' = \frac{(19.16 - 9.53) - 0}{\sqrt{\dfrac{(5.29)^2}{15} + \dfrac{(2.69)^2}{30}}} = \frac{9.63}{1.4515} = 6.63$$

8. **Statistical decision.** Since $6.63 > 2.133$, we reject H_0.
9. **Conclusion.** On the basis of these results we conclude that the two population means are different.
10. ***p* value.** For this test $p < .05$. ■

Sampling from Populations That Are Not Normally Distributed

When sampling is from populations that are not normally distributed, the results of the central limit theorem may be employed if sample sizes are large (say, ≥ 30). This will allow the use of normal theory since the distribution of the difference between sample means will be approximately normal. When each of two large independent simple random samples has been drawn from a population that is not normally distributed, the test statistic for testing H_0: $\mu_1 = \mu_2$ is

$$z = \frac{(\bar{x}_1 - \bar{x}_2) - (\mu_1 - \mu_2)_0}{\sqrt{\dfrac{\sigma_1^2}{n_1} + \dfrac{\sigma_2^2}{n_2}}} \tag{7.3.5}$$

which, when H_0 is true, follows the standard normal distribution. If the population variances are known, they are used; but if they are unknown, as is the usual case, the sample variances, which are necessarily based on large samples, are used as estimates. Sample variances are not pooled, since equality of population variances is not a necessary assumption when the z statistic is used.

EXAMPLE 7.3.4

The objective of a study by Sairam et al. (A-8) was to identify the role of various disease states and additional risk factors in the development of thrombosis. One focus of the study was to determine if there were differing levels of the anticardiolipin antibody IgG in subjects with and without thrombosis. Table 7.3.2 summarizes the researchers' findings:

TABLE 7.3.2 IgG Levels for Subjects with and Without Thrombosis for Example 7.3.4

Group	Mean IgG Level (ml/unit)	Sample Size	Standard Deviation
Thrombosis	59.01	53	44.89
No thrombosis	46.61	54	34.85

Source: S. Sairam, B. A. Baethge and T. McNearney, "Analysis of Risk Factors and Comorbid Diseases in the Development of Thrombosis in Patients with Anticardiolipin Antibodies," *Clinical Rheumatology, 22* (2003), 24–29.

We wish to know if we may conclude, on the basis of these results, that, in general, persons with thrombosis have, on the average, higher IgG levels than persons without thrombosis.

Solution:

1. **Data.** See statement of example.
2. **Assumptions.** The statistics were computed from two independent samples that behave as simple random samples from a population of persons with thrombosis and a population of persons who do not have thrombosis. Since the population variances are unknown, we will use the sample variances in the calculation of the test statistic.
3. **Hypotheses.**

$$H_0: \mu_T - \mu_{NT} \leq 0$$
$$H_A: \mu_T - \mu_{NT} > 0$$

or, alternatively,

$$H_0: \mu_T \leq \mu_{NT}$$
$$H_A: \mu_T > \mu_{NT}$$

4. **Test statistic.** Since we have large samples, the central limit theorem allows us to use Equation 7.3.5 as the test statistic.
5. **Distribution of test statistic.** When the null hypothesis is true, the test statistic is distributed approximately as the standard normal.
6. **Decision rule.** Let $\alpha = .01$. This is a one-sided test with a critical value of z equal to 2.33. Reject H_0 if $z_{\text{computed}} \geq 2.33$.
7. **Calculation of test statistic.**

$$z = \frac{59.01 - 46.61}{\sqrt{\frac{44.89^2}{53} + \frac{34.85^2}{54}}} = 1.59$$

8. **Statistical decision.** Fail to reject H_0, since $z = 1.59$ is in the nonrejection region.
9. **Conclusion.** These data indicate that on the average, persons with thrombosis and persons without thrombosis may not have differing IgG levels.
10. **p value.** For this test, $p = .0559$. When testing a hypothesis about the difference between two populations means, we may use Figure 6.4.1 to decide quickly whether the test statistic should be z or t. ■

We may use MINITAB to perform two-sample t tests. To illustrate, let us refer to the data in Table 7.3.1. We put the data for control subjects and spinal cord

Dialog box:

Stat ➤ Basic Statistics ➤ 2-Sample t

Choose **Samples in different columns.** Type *C1* in **First** and *C2* in **Second.** Click the **Options** box and select "less than" in the **Alternatives** box. Check **Assume equal variances.** Click **OK.**

Session command:

```
MTB > TwoSample 95.0 C1 C2
SUBC> Alternative -1,
SUBC> Pooled.
```

Output:

```
Two-Sample T-Test and CI: C, SCI

Two-sample T for C vs SCI
       N    Mean   StDev   SE Mean
C     10   126.1    21.8       6.9
SCI   10   133.1    32.2        10

Difference = mu C - mu SCI
Estimate for difference: -7.0
95% upper bound for difference: 14.3
T-Test of difference = 0 (vs <): T-Value = -0.57 P-Value = 0.288
DF = 18
Both use Pooled StDev = 27.5
```

FIGURE 7.3.2 MINITAB procedure and output for two-sample t test, Example 7.3.2 (data in Table 7.3.1).

injury subjects in Column 1 and Column 2, respectively, and proceed as shown in Figure 7.3.2.

The SAS® statistical package performs the t test for equality of population means under both assumptions regarding population variances: that they are equal and that they are not equal. Note that SAS® designates the p value as $\Pr > |t|$. The default output is a p value for a two-sided test. The researcher using SAS® must divide this quantity in half when the hypothesis test is one-sided. The SAS® package also tests for equality of population variances as described in Section 7.8. Figure 7.3.3 shows the SAS® output for Example 7.3.2.

Alternatives to z and t Sometimes neither the z statistic nor the t statistic is an appropriate test statistic for use with the available data. When such is the case, one may wish to use a nonparametric technique for testing a hypothesis about the difference between two population measures of central tendency. The Mann-Whitney test statistic and the median test, discussed in Chapter 13, are frequently used alternatives to the z and t statistics.

The SAS System
The TTEST Procedure

Statistics

Variable	group	N	Lower CL Mean	Mean	Upper CL Mean	Lower CL Std Dev	Std Dev	Upper CL Std Dev	Std Err
pressure	C	10	110.49	126.1	141.71	15.008	21.82	39.834	6.9
pressure	SCI	10	110.08	133.1	156.12	22.133	32.178	58.745	10.176
pressure	Diff (1-2)		−32.83	−7	18.83	20.773	27.491	40.655	12.294

T-Tests

Variable	Method	Variances	DF	t Value	Pr > |t|
pressure	Pooled	Equal	18	−0.57	0.5761
pressure	Satterthwaite	Unequal	15.8	−0.57	0.5771

Equality of Variances

Variable	Method	Num DF	Den DF	F Value	Pr > F
pressure	Folded F	9	9	2.17	0.2626

FIGURE 7.3.3 SAS® output for Example 7.3.2 (data in Table 7.3.1).

EXERCISES

In each of the following exercises, complete the ten-step hypothesis testing procedure. State the assumptions that are necessary for your procedure to be valid. For each exercise, as appropriate, explain why you chose a one-sided test or a two-sided test. Discuss how you think researchers or clinicians might use the results of your hypothesis test. What clinical or research decisions or actions do you think would be appropriate in light of the results of your test?

7.3.1 Subjects in a study by Dabonneville et al. (A-9) included a sample of 40 men who claimed to engage in a variety of sports activities (multisport). The mean body mass index (BMI) for these men was 22.41 with a standard deviation of 1.27. A sample of 24 male rugby players had a mean BMI of 27.75 with a standard deviation of 2.64. Is there sufficient evidence for one to claim that, in general, rugby players have a higher BMI than the multisport men? Let $\alpha = .01$.

7.3.2 The purpose of a study by Ingle and Eastell (A-10) was to examine the bone mineral density (BMD) and ultrasound properties of women with ankle fractures. The investigators recruited 31 postmenopausal women with ankle fractures and 31 healthy postmenopausal women to serve as controls. One of the baseline measurements was the stiffness index of the lunar Achilles. The mean stiffness index for the ankle fracture group was 76.9 with a standard deviation of 12.6. In the control group, the mean was 90.9 with a standard deviation of 12.5. Do these data provide sufficient evidence to allow you to conclude that, in general, the mean stiffness index is higher in healthy postmenopausal women than in postmenopausal women with ankle fractures? Let $\alpha = .05$.

7.3.3 Hoekema et al. (A-11) studied the craniofacial morphology of 26 male patients with obstructive sleep apnea syndrome (OSAS) and 37 healthy male subjects (non–OSAS). One of the variables of interest was the length from the most superoanterior point of the body of the hyoid bone to the Frankfort horizontal (measured in millimeters).

Length (mm) Non–OSAS				Length (mm) OSAS		
96.80	97.00	101.00	88.95	105.95	114.90	113.70
100.70	97.70	88.25	101.05	114.90	114.35	116.30
94.55	97.00	92.60	92.60	110.35	112.25	108.75
99.65	94.55	98.25	97.00	123.10	106.15	113.30
109.15	106.45	90.85	91.95	119.30	102.60	106.00
102.75	94.55	95.25	88.95	110.00	102.40	101.75
97.70	94.05	88.80	95.75	98.95	105.05	
92.10	89.45	101.40		114.20	112.65	
91.90	89.85	90.55		108.95	128.95	
89.50	98.20	109.80		105.05	117.70	

Source: A. Hoekema, D.D.S. Used with permission.

Do these data provide sufficient evidence to allow us to conclude that the two sampled populations differ with respect to length from the hyoid bone to the Frankfort horizontal? Let $\alpha = .01$.

7.3.4 Can we conclude that patients with primary hypertension (PH), on the average, have higher total cholesterol levels than normotensive (NT) patients? This was one of the inquiries of interest for Rossi et al. (A-12). In the following table are total cholesterol measurements (mg/dl) for 133 PH patients and 41 NT patients. Can we conclude that PH patients have, on average, higher total cholesterol levels than NT patients? Let $\alpha = .05$.

Total Cholesterol (mg/dl)						
Primary Hypertensive Patients					Normotensive Patients	
207	221	212	220	190	286	189
172	223	260	214	245	226	196
191	181	210	215	171	187	142
221	217	265	206	261	204	179
203	208	206	247	182	203	212
241	202	198	221	162	206	163
208	218	210	199	182	196	196
199	216	211	196	225	168	189
185	168	274	239	203	229	142
235	168	223	199	195	184	168
214	214	175	244	178	186	121
134	203	203	214	240	281	
226	280	168	236	222	203	

(Continued)

Total Cholesterol (mg/dl)

Primary Hypertensive Patients					Normotensive Patients	
222	203	178	249	117	177	135
213	225	217	212	252	179	161
272	227	200	259	203	194	
185	239	226	189	245	206	
181	265	207	235	218	219	
238	228	232	239	152	173	
141	226	182	239	231	189	
203	236	215	210	237	194	
222	195	239	203		196	
221	284	210	188		212	
180	183	207	237		168	
276	266	224	231		188	
226	258	251	222		232	
224	214	212	174		242	
206	260	201	219		200	

Source: Gian Paolo Rossi, M.D., F.A.C.C., F.A.H.A. Used with permission.

7.3.5 Garção and Cabrita (A-13) wanted to evaluate the community pharmacist's capacity to positively influence the results of antihypertensive drug therapy through a pharmaceutical care program in Portugal. Eighty-two subjects with essential hypertension were randomly assigned to an intervention or a control group. The intervention group received monthly monitoring by a research pharmacist to monitor blood pressure, assess adherence to treatment, prevent, detect, and resolve drug-related problems, and encourage nonpharmacologic measures for blood pressure control. The changes after 6 months in diastolic blood pressure (pre − post, mm Hg) are given in the following table for patients in each of the two groups.

Intervention Group				Control Group			
20	4	12	16	0	4	12	0
2	24	6	10	12	2	2	8
36	6	24	16	18	2	0	10
26	−2	42	10	0	8	0	14
2	8	20	6	8	10	−4	8
20	8	14	6	10	0	12	0
2	16	−2	2	8	6	4	2
14	14	10	8	14	10	28	−8
30	8	2	16	4	−2	−18	16
18	20	18	−12	−2	2	12	12
6				−6			

Source: José Garção, M.S., Pharm.D. Used with permission.

On the basis of these data, what should the researcher conclude? Let $\alpha = .05$.

7.3.6 A test designed to measure mothers' attitudes toward their labor and delivery experiences was given to two groups of new mothers. Sample 1 (attenders) had attended prenatal classes held at the local health department. Sample 2 (nonattenders) did not attend the classes. The sample sizes and means and standard deviations of the test scores were as follows:

Sample	n	$\bar{x}$	s
1	15	4.75	1.0
2	22	3.00	1.5

Do these data provide sufficient evidence to indicate that attenders, on the average, score higher than nonattenders? Let $\alpha = .05$.

7.3.7 Cortisol level determinations were made on two samples of women at childbirth. Group 1 subjects underwent emergency cesarean section following induced labor. Group 2 subjects delivered by either cesarean section or the vaginal route following spontaneous labor. The sample sizes, mean cortisol levels, and standard deviations were as follows:

Sample	n	$\bar{x}$	s
1	10	435	65
2	12	645	80

Do these data provide sufficient evidence to indicate a difference in the mean cortisol levels in the populations represented? Let $\alpha = .05$.

7.3.8 Protoporphyrin levels were measured in two samples of subjects. Sample 1 consisted of 50 adult male alcoholics with ring sideroblasts in the bone marrow. Sample 2 consisted of 40 apparently healthy adult nonalcoholic males. The mean protoporphyrin levels and standard deviations for the two samples were as follows:

Sample	$\bar{x}$	s
1	340	250
2	45	25

Can one conclude on the basis of these data that protoporphyrin levels are higher in the represented alcoholic population than in the nonalcoholic population? Let $\alpha = .01$.

7.3.9 A researcher was interested in knowing if preterm infants with late metabolic acidosis and preterm infants without the condition differ with respect to urine levels of a certain chemical. The mean levels, standard deviations, and sample sizes for the two samples studied were as follows:

Sample	n	$\bar{x}$	s
With condition	35	8.5	5.5
Without condition	40	4.8	3.6

What should the researcher conclude on the basis of these results? Let $\alpha = .05$.

7.3.10 Researchers wished to know if they could conclude that two populations of infants differ with respect to mean age at which they walked alone. The following data (ages in months) were collected:

Sample from population A: 9.5, 10.5, 9.0, 9.75, 10.0, 13.0, 10.0, 13.5, 10.0, 9.5, 10.0, 9.75

Sample from population B: 12.5, 9.5, 13.5, 13.75, 12.0, 13.75, 12.5, 9.5, 12.0, 13.5, 12.0, 12.0

What should the researchers conclude? Let $\alpha = .05$.

7.3.11 Does sensory deprivation have an effect on a person's alpha-wave frequency? Twenty volunteer subjects were randomly divided into two groups. Subjects in group A were subjected to a 10-day period of sensory deprivation, while subjects in group B served as controls. At the end of the experimental period, the alpha-wave frequency component of subjects' electroencephalograms was measured. The results were as follows:

Group A: 10.2, 9.5, 10.1, 10.0, 9.8, 10.9, 11.4, 10.8, 9.7, 10.4

Group B: 11.0, 11.2, 10.1, 11.4, 11.7, 11.2, 10.8, 11.6, 10.9, 10.9

Let $\alpha = .05$.

7.3.12 Can we conclude that, on the average, lymphocytes and tumor cells differ in size? The following are the cell diameters (μm) of 40 lymphocytes and 50 tumor cells obtained from biopsies of tissue from patients with melanoma:

Lymphocytes

9.0	9.4	4.7	4.8	8.9	4.9	8.4	5.9
6.3	5.7	5.0	3.5	7.8	10.4	8.0	8.0
8.6	7.0	6.8	7.1	5.7	7.6	6.2	7.1
7.4	8.7	4.9	7.4	6.4	7.1	6.3	8.8
8.8	5.2	7.1	5.3	4.7	8.4	6.4	8.3

Tumor Cells

12.6	14.6	16.2	23.9	23.3	17.1	20.0	21.0	19.1	19.4
16.7	15.9	15.8	16.0	17.9	13.4	19.1	16.6	18.9	18.7
20.0	17.8	13.9	22.1	13.9	18.3	22.8	13.0	17.9	15.2
17.7	15.1	16.9	16.4	22.8	19.4	19.6	18.4	18.2	20.7
16.3	17.7	18.1	24.3	11.2	19.5	18.6	16.4	16.1	21.5

Let $\alpha = .05$.

7.4 PAIRED COMPARISONS

In our previous discussion involving the difference between two population means, it was assumed that the samples were independent. A method frequently employed for assessing the effectiveness of a treatment or experimental procedure is one that makes use of related observations resulting from nonindependent samples. A hypothesis test based on this type of data is known as a *paired comparisons* test.

Reasons for Pairing It frequently happens that true differences do not exist between two populations with respect to the variable of interest, but the presence of extraneous sources of variation may cause rejection of the null hypothesis of no difference. On the other hand, true differences also may be masked by the presence of extraneous factors.

Suppose, for example, that we wish to compare two sunscreens. There are at least two ways in which the experiment may be carried out. One method would be to select a simple random sample of subjects to receive sunscreen A and an independent simple random sample of subjects to receive sunscreen B. We send the subjects out into the sunshine for a specified length of time, after which we will measure the amount of damage from the rays of the sun. Suppose we employ this method, but inadvertently, most of the subjects receiving sunscreen A have darker complexions that are naturally less sensitive to sunlight. Let us say that after the experiment has been completed we find that subjects receiving sunscreen A had less sun damage. We would not know if they had less sun damage because sunscreen A was more protective than sunscreen B or because the subjects were naturally less sensitive to the sun.

A better way to design the experiment would be to select just one simple random sample of subjects and let each member of the sample receive both sunscreens. We could, for example, randomly assign the sunscreens to the left or the right side of each subject's back with each subject receiving both sunscreens. After a specified length of exposure to the sun, we would measure the amount of sun damage to each half of the back. If the half of the back receiving sunscreen A tended to be less damaged, we could more confidently attribute the result to the sunscreen, since in each instance both sunscreens were applied to equally pigmented skin.

The objective in paired comparisons tests is to eliminate a maximum number of sources of extraneous variation by making the pairs similar with respect to as many variables as possible.

Related or paired observations may be obtained in a number of ways. The same subjects may be measured before and after receiving some treatment. Litter mates of the same sex may be assigned randomly to receive either a treatment or a placebo. Pairs of twins or siblings may be assigned randomly to two treatments in such a way that members of a single pair receive different treatments. In comparing two methods of analysis, the material to be analyzed may be divided equally so that one-half is analyzed by one method and one-half is analyzed by the other. Or pairs may be formed by matching individuals on some characteristic, for example, digital dexterity, which is closely related to the measurement of interest, say, posttreatment scores on some test requiring digital manipulation.

Instead of performing the analysis with individual observations, we use d_i, the difference between pairs of observations, as the variable of interest.

When the n sample differences computed from the n pairs of measurements constitute a simple random sample from a normally distributed population of differences, the test statistic for testing hypotheses about the population mean difference μ_d is

$$t = \frac{\bar{d} - \mu_{d_0}}{s_{\bar{d}}} \tag{7.4.1}$$

where $\bar{d}$ is the sample mean difference, μ_{d_0} is the hypothesized population mean difference, $s_{\bar{d}} = s_d/\sqrt{n}$, n is the number of sample differences, and s_d is the standard deviation of the sample differences. When H_0 is true, the test statistic is distributed as Student's t with $n - 1$ degrees of freedom.

Although to begin with we have two samples—say, before levels and after levels—we do not have to worry about equality of variances, as with independent samples, since our variable is the difference between readings in the same individual, or matched individuals, and, hence, only one variable is involved. The arithmetic involved in performing a paired comparisons test, therefore, is the same as for performing a test involving a single sample as described in Section 7.2.

The following example illustrates the procedures involved in a paired comparisons test.

EXAMPLE 7.4.1

John M. Morton et al. (A-14) examined gallbladder function before and after fundoplication—a surgery used to stop stomach contents from flowing back into the esophagus (reflux)—in patients with gastroesophageal reflux disease. The authors measured gallbladder functionality by calculating the gallbladder ejection fraction (GBEF) before and after fundoplication. The goal of fundoplication is to increase GBEF, which is measured as a percent. The data are shown in Table 7.4.1. We wish to know if these data provide sufficient evidence to allow us to conclude that fundoplication increases GBEF functioning.

Solution: We will say that sufficient evidence is provided for us to conclude that the fundoplication is effective if we can reject the null hypothesis that the population mean change μ_d is different from zero in the appropriate direction. We may reach a conclusion by means of the ten-step hypothesis testing procedure.

1. **Data.** The data consist of the GBEF for 12 individuals, before and after fundoplication. We shall perform the statistical analysis on the differences in preop and postop GBEF. We may obtain the differences in one of two ways: by subtracting the preop percents from the postop percents or by subtracting the postop percents from the preop percents. Let us obtain the differences by subtracting the preop

TABLE 7.4.1 Gallbladder Function in Patients with Presentations of Gastroesophageal Reflux Disease Before and After Treatment

Preop (%)	22	63.3	96	9.2	3.1	50	33	69	64	18.8	0	34
Postop (%)	63.5	91.5	59	37.8	10.1	19.6	41	87.8	86	55	88	40

Source: John M. Morton, Steven P. Bowers, Tananchai A. Lucktong, Samer Mattar, W. Alan Bradshaw, Kevin E. Behrns, Mark J. Koruda, Charles A. Herbst, William McCartney, Raghuveer K. Halkar, C. Daniel Smith, and Timothy M. Farrell, "Gallbladder Function Before and After Fundoplication," *Journal of Gastrointestinal Surgery*, 6 (2002), 806–811.

percents from the postop percents. The d_i = postop − preop differences are:

41.5, 28.2, −37.0, 28.6, 7.0, −30.4, 8.0, 18.8, 22.0, 36.2, 88.0, 6.0

2. **Assumptions.** The observed differences constitute a simple random sample from a normally distributed population of differences that could be generated under the same circumstances.

3. **Hypotheses.** The way we state our null and alternative hypotheses must be consistent with the way in which we subtract measurements to obtain the differences. In the present example, we want to know if we can conclude that the fundoplication is useful in increasing GBEF percentage. If it is effective in improving GBEF, we would expect the postop percents to tend to be higher than the preop percents. If, therefore, we subtract the preop percents from the postop percents (postop − preop), we would expect the differences to tend to be positive. Furthermore, we would expect the mean of a population of such differences to be positive. So, under these conditions, asking if we can conclude that the fundoplication is effective is the same as asking if we can conclude that the population mean difference is positive (greater than zero).

   The null and alternative hypotheses are as follows:

$$H_0: \mu_d \leq 0$$
$$H_A: \mu_d > 0$$

   If we had obtained the differences by subtracting the postop percents from the preop weights (preop − postop), our hypotheses would have been

$$H_0: \mu_d \geq 0$$
$$H_A: \mu_d < 0$$

   If the question had been such that a two-sided test was indicated, the hypotheses would have been

$$H_0: \mu_d = 0$$
$$H_A: \mu_d \neq 0$$

   regardless of the way we subtracted to obtain the differences.

4. **Test statistic.** The appropriate test statistic is given by Equation 7.4.1.

5. **Distribution of test statistic.** If the null hypothesis is true, the test statistic is distributed as Student's t with $n - 1$ degrees of freedom.

6. **Decision rule.** Let $\alpha = .05$. The critical value of t is 1.7959. Reject H_0 if computed t is greater than or equal to the critical value. The rejection and nonrejection regions are shown in Figure 7.4.1.

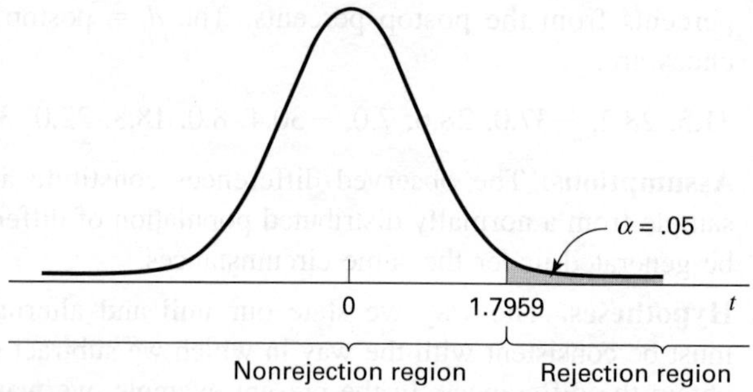

FIGURE 7.4.1 Rejection and nonrejection regions for Example 7.4.1.

7. **Calculation of test statistic.** From the $n = 12$ differences d_i, we compute the following descriptive measures:

$$\bar{d} = \frac{\sum d_i}{n} = \frac{(41.5) + (28.2) + (-37.0) + \cdots + (6.0)}{12} = \frac{216.9}{12} = 18.075$$

$$s_{\bar{d}}^2 = \frac{\sum(d_i - \bar{d})^2}{n-1} = \frac{n\sum d_i^2 - (\sum d_i)^2}{n(n-1)} = \frac{12(15669.49) - (216.9)^2}{(12)(11)} = 1068.0930$$

$$t = \frac{18.075 - 0}{\sqrt{1068.0930/12}} = \frac{18.075}{9.4344} = 1.9159$$

8. **Statistical decision.** Reject H_0, since 1.9159 is in the rejection region.
9. **Conclusion.** We may conclude that the fundoplication procedure increases GBEF functioning.
10. **p value.** For this test, $.025 < p < .05$, since $1.7959 < 1.9159 < 2.2010$. ■

A Confidence Interval for μ_d A 95 percent confidence interval for μ_d may be obtained as follows:

$$\bar{d} \pm t_{1-(\alpha/2)} s_{\bar{d}}$$

$$18.075 \pm 2.2010\sqrt{1068.0930/12}$$

$$18.075 \pm 20.765$$

$$-2.690, 38.840$$

The Use of z If, in the analysis of paired data, the population variance of the differences is known, the appropriate test statistic is

$$z = \frac{\bar{d} - \mu_d}{\sigma_d/\sqrt{n}} \tag{7.4.2}$$

It is unlikely that σ_d will be known in practice.

```
Paired T-Test and CI: C2, C1

Paired T for C2 - C1

                N     Mean    StDev   SE Mean
C2             12  56.6083  27.8001    8.0252
C1             12  38.5333  30.0587    8.6772
Difference     12  18.0750  32.6817    9.4344


95% lower bound for mean difference: 1.1319
T-Test of mean difference = 0 (vs > 0): T-Value = 1.92 P-Value =
0.041
```

FIGURE 7.4.2 MINITAB procedure and output for paired comparisons test, Example 7.4.1 (data in Table 7.4.1).

If the assumption of normally distributed d_i's cannot be made, the central limit theorem may be employed if n is large. In such cases, the test statistic is Equation 7.4.2, with s_d used to estimate σ_d when, as is generally the case, the latter is unknown.

We may use MINITAB to perform a paired t-test. The output from this procedure is given in Figure 7.4.2.

Disadvantages The use of the paired comparisons test is not without its problems. If different subjects are used and randomly assigned to two treatments, considerable time and expense may be involved in our trying to match individuals on one or more relevant variables. A further price we pay for using paired comparisons is a loss of degrees of freedom. If we do not use paired observations, we have $2n - 2$ degrees of freedom available as compared to $n - 1$ when we use the paired comparisons procedure.

In general, in deciding whether or not to use the paired comparisons procedure, one should be guided by the economics involved as well as by a consideration of the gains to be realized in terms of controlling extraneous variation.

Alternatives If neither z nor t is an appropriate test statistic for use with available data, one may wish to consider using some nonparametric technique to test a hypothesis about a median difference. The sign test, discussed in Chapter 13, is a candidate for use in such cases.

EXERCISES

In the following exercises, carry out the ten-step hypothesis testing procedure at the specified significance level. For each exercise, as appropriate, explain why you chose a one-sided test or a two-sided test. Discuss how you think researchers or clinicians might use the results of your hypothesis test.

What clinical or research decisions or actions do you think would be appropriate in light of the results of your test?

7.4.1 Ellen Davis Jones (A-15) studied the effects of reminiscence therapy for older women with depression. She studied 15 women 60 years or older residing for 3 months or longer in an assisted living long-term care facility. For this study, depression was measured by the Geriatric Depression Scale (GDS). Higher scores indicate more severe depression symptoms. The participants received reminiscence therapy for long-term care, which uses family photographs, scrapbooks, and personal memorabilia to stimulate memory and conversation among group members. Pre-treatment and post-treatment depression scores are given in the following table. Can we conclude, based on these data, that subjects who participate in reminiscence therapy experience, on average, a decline in GDS depression scores? Let $\alpha = .01$.

Pre–GDS:	12	10	16	2	12	18	11	16	16	10	14	21	9	19	20
Post–GDS:	11	10	11	3	9	13	8	14	16	10	12	22	9	16	18

Source: Ellen Davis Jones, N.D., R.N., FNP-C. Used with permission.

7.4.2 Beney et al. (A-16) evaluated the effect of telephone follow-up on the physical well-being dimension of health-related quality of life in patients with cancer. One of the main outcome variables was measured by the physical well-being subscale of the Functional Assessment of Cancer Therapy Scale–General (FACT-G). A higher score indicates higher physical well-being. The following table shows the baseline FACT-G score and the follow-up score to evaluate the physical well-being during the 7 days after discharge from hospital to home for 66 patients who received a phone call 48–72 hours after discharge that gave patients the opportunity to discuss medications, problems, and advice. Is there sufficient evidence to indicate that quality of physical well-being significantly decreases in the first week of discharge among patients who receive a phone call? Let $\alpha = .05$.

Subject	Baseline FACT-G	Follow-up FACT-G	Subject	Baseline FACT-G	Follow-up FACT-G
1	16	19	34	25	14
2	26	19	35	21	17
3	13	9	36	14	22
4	20	23	37	23	22
5	22	25	38	19	16
6	21	20	39	19	15
7	20	10	40	18	23
8	15	20	41	20	21
9	25	22	42	18	11
10	20	18	43	22	22
11	11	6	44	7	17
12	22	21	45	23	9
13	18	17	46	19	16
14	21	13	47	17	16
15	25	25	48	22	20
16	17	21	49	19	23
17	26	22	50	5	17
18	18	22	51	22	17
19	7	9	52	12	6

(Continued)

Subject	Baseline FACT-G	Follow-up FACT-G	Subject	Baseline FACT-G	Follow-up FACT-G
20	25	24	53	19	19
21	22	15	54	17	20
22	15	9	55	7	6
23	19	7	56	27	10
24	23	20	57	22	16
25	19	19	58	16	14
26	21	24	59	26	24
27	24	23	60	17	19
28	21	15	61	23	22
29	28	27	62	23	23
30	18	26	63	13	3
31	25	26	64	24	22
32	25	26	65	17	21
33	28	28	66	22	21

Source: Johnny Beney, Ph.D. and E. Beth Devine, Pharm.D., M.B.A. et al. Used with permission.

7.4.3 The purpose of an investigation by Morley et al. (A-17) was to evaluate the analgesic effectiveness of a daily dose of oral methadone in patients with chronic neuropathic pain syndromes. The researchers used a visual analogue scale (0–100 mm, higher number indicates higher pain) ratings for maximum pain intensity over the course of the day. Each subject took either 20 mg of methadone or a placebo each day for 5 days. Subjects did not know which treatment they were taking. The following table gives the mean maximum pain intensity scores for the 5 days on methadone and the 5 days on placebo. Do these data provide sufficient evidence, at the .05 level of significance, to indicate that in general the maximum pain intensity is lower on days when methadone is taken?

Subject	Methadone	Placebo
1	29.8	57.2
2	73.0	69.8
3	98.6	98.2
4	58.8	62.4
5	60.6	67.2
6	57.2	70.6
7	57.2	67.8
8	89.2	95.6
9	97.0	98.4
10	49.8	63.2
11	37.0	63.6

Source: John S. Morley, John Bridson, Tim P. Nash, John B. Miles, Sarah White, and Matthew K. Makin, "Low-Dose Methadone Has an Analgesic Effect in Neuropathic Pain: A Double-Blind Randomized Controlled Crossover Trial," *Palliative Medicine, 17* (2003), 576–587.

7.4.4 Woo and McKenna (A-18) investigated the effect of broadband ultraviolet B (UVB) therapy and topical calcipotriol cream used together on areas of psoriasis. One of the outcome variables is the Psoriasis Area and Severity Index (PASI). The following table gives the PASI scores for 20 subjects measured at baseline and after eight treatments. Do these data provide sufficient evidence, at the .01 level of significance, to indicate that the combination therapy reduces PASI scores?

Subject	Baseline	After 8 Treatments
1	5.9	5.2
2	7.6	12.2
3	12.8	4.6
4	16.5	4.0
5	6.1	0.4
6	14.4	3.8
7	6.6	1.2
8	5.4	3.1
9	9.6	3.5
10	11.6	4.9
11	11.1	11.1
12	15.6	8.4
13	6.9	5.8
14	15.2	5.0
15	21.0	6.4
16	5.9	0.0
17	10.0	2.7
18	12.2	5.1
19	20.2	4.8
20	6.2	4.2

Source: W. K. Woo, M.D. Used with permission.

7.4.5 One of the purposes of an investigation by Porcellini et al. (A-19) was to investigate the effect on CD4 T cell count of administration of intermittent interleukin (IL-2) in addition to highly active antiretroviral therapy (HAART). The following table shows the CD4 T cell count at baseline and then again after 12 months of HAART therapy with IL-2. Do the data show, at the .05 level, a significant change in CD4 T cell count?

Subject	1	2	3	4	5	6	7
CD4 T cell count at entry ($\times$ 10⁶/L)	173	58	103	181	105	301	169
CD4 T cell count at end of follow-up ($\times$ 10⁶/L)	257	108	315	362	141	549	369

Source: Simona Procellini, Giuliana Vallanti, Silvia Nozza, Guido Poli, Adraino Lazzarin, Guiseppe Tabussi, and Antonio Grassia, "Improved Thymopoietic Potential in Aviremic HIV-Infected Individuals with HAART by Intermittent IL-2 Administration," *AIDS, 17* (2003), 1621–1630.

7.5 HYPOTHESIS TESTING: A SINGLE POPULATION PROPORTION

Testing hypotheses about population proportions is carried out in much the same way as for means when the conditions necessary for using the normal curve are met. One-sided or two-sided tests may be made, depending on the question being asked. When a sample sufficiently large for application of the central limit theorem as discussed in Section 5.5 is available for analysis, the test statistic is

$$z = \frac{\hat{p} - p_0}{\sqrt{\frac{p_0 q_0}{n}}} \tag{7.5.1}$$

which, when H_0 is true, is distributed approximately as the standard normal.

EXAMPLE 7.5.1

Wagenknecht et al. (A-20) collected data on a sample of 301 Hispanic women living in San Antonio, Texas. One variable of interest was the percentage of subjects with impaired fasting glucose (IFG). IFG refers to a metabolic stage intermediate between normal glucose homeostasis and diabetes. In the study, 24 women were classified in the IFG stage. The article cites population estimates for IFG among Hispanic women in Texas as 6.3 percent. Is there sufficient evidence to indicate that the population of Hispanic women in San Antonio has a prevalence of IFG higher than 6.3 percent?

Solution:

1. **Data.** The data are obtained from the responses of 301 individuals of which 24 possessed the characteristic of interest; that is, $\hat{p} = 24/301 = .080$.
2. **Assumptions.** The study subjects may be treated as a simple random sample from a population of similar subjects, and the sampling distribution of $\hat{p}$ is approximately normally distributed in accordance with the central limit theorem.
3. **Hypotheses.**

$$H_0: p \leq .063$$
$$H_A: p > .063$$

We conduct the test at the point of equality. The conclusion we reach will be the same as we would reach if we conducted the test using any other hypothesized value of p greater than .063. If H_0 is true, $p = .063$ and the standard error $\sigma_{\hat{p}} = \sqrt{(.063)(.937)/301}$. Note that we use the hypothesized value of p in computing $\sigma_{\hat{p}}$. We do this because the entire test is based on the assumption that the null hypothesis is true. To use the sample proportion, $\hat{p}$, in computing $\sigma_{\hat{p}}$ would not be consistent with this concept.

4. **Test statistic.** The test statistic is given by Equation 7.5.1.
5. **Distribution of test statistic.** If the null hypothesis is true, the test statistic is approximately normally distributed with a mean of zero.
6. **Decision rule.** Let $\alpha = .05$. The critical value of z is 1.645. Reject H_0 if the computed z is ≥ 1.645.

7. **Calculation of test statistic.**

$$z = \frac{.080 - .063}{\sqrt{\frac{(.063)(.937)}{301}}} = 1.21$$

8. **Statistical decision.** Do not reject H_0 since $1.21 < 1.645$.
9. **Conclusion.** We cannot conclude that in the sampled population the proportion who are IFG is higher than 6.3 percent.
10. ***p* value.** $p = .1131$. ■

Tests involving a single proportion can be carried out using a variety of computer programs. Outputs from MINITAB and NCSS, using the data from Example 7.5.1, are shown in Figure 7.5.1. It should be noted that the results will vary slightly, because of rounding errors, if calculations are done by hand. It should also be noted that some programs, such as NCSS, use a continuity correction in calculating the z-value, and therefore the test statistic values and corresponding p values differ slightly from the MINITAB output.

MINITAB Output

Test and CI for One Proportion

```
Test of p = 0.063 vs p > 0.063

                              95% Lower
Sample     X    N   Sample p     Bound      Z-Value    P-Value
1         24  301   0.079734  0.054053       1.19      0.116

Using the normal approximation.
```

NCSS Output

Normal Approximation using (P0)

Alternative Hypothesis	Z-Value	Prob Level	Decision (5%)
P<>P0	1.0763	0.281780	Accept H0
P<P0	1.0763	0.859110	Accept H0
P>P0	1.0763	0.140890	Accept H0

FIGURE 7.5.1 MINITAB and partial NCSS output for the data in Example 7.5.1

EXERCISES

For each of the following exercises, carry out the ten-step hypothesis testing procedure at the designated level of significance. For each exercise, as appropriate, explain why you chose a one-sided test or a two-sided test. Discuss how you think researchers or clinicians might use the results of your hypothesis test. What clinical or research decisions or actions do you think would be appropriate in light of the results of your test?

7.5.1 Jacquemyn et al. (A-21) conducted a survey among gynecologists-obstetricians in the Flanders region and obtained 295 responses. Of those responding, 90 indicated that they had performed at least one cesarean section on demand every year. Does this study provide sufficient evidence for us to conclude that less than 35 percent of the gynecologists-obstetricians in the Flanders region perform at least one cesarean section on demand each year? Let $\alpha = .05$.

7.5.2 In an article in the journal *Health and Place,* Hui and Bell (A-22) found that among 2428 boys ages 7 to 12 years, 461 were overweight or obese. On the basis of this study, can we conclude that more than 15 percent of the boys ages 7 to 12 in the sampled population are obese or overweight? Let $\alpha = .05$.

7.5.3 Becker et al. (A-23) conducted a study using a sample of 50 ethnic Fijian women. The women completed a self-report questionnaire on dieting and attitudes toward body shape and change. The researchers found that five of the respondents reported at least weekly episodes of binge eating during the previous 6 months. Is this sufficient evidence to conclude that less than 20 percent of the population of Fijian women engage in at least weekly episodes of binge eating? Let $\alpha = .05$.

7.5.4 The following questionnaire was completed by a simple random sample of 250 gynecologists. The number checking each response is shown in the appropriate box.

1. When you have a choice, which procedure do you prefer for obtaining samples of endometrium?
   (a) Dilation and curettage [175]
   (b) Vobra aspiration [75]

2. Have you seen one or more pregnant women during the past year whom you knew to have elevated blood lead levels?
   (a) Yes [25]
   (b) No [225]

3. Do you routinely acquaint your pregnant patients who smoke with the suspected hazards of smoking to the fetus?
   (a) Yes [238]
   (b) No [12]

Can we conclude from these data that in the sampled population more than 60 percent prefer dilation and curettage for obtaining samples of endometrium? Let $\alpha = .01$.

7.5.5 Refer to Exercise 7.5.4. Can we conclude from these data that in the sampled population fewer than 15 percent have seen (during the past year) one or more pregnant women with elevated blood lead levels? Let $\alpha = .05$.

7.5.6 Refer to Exercise 7.5.4. Can we conclude from these data that more than 90 percent acquaint their pregnant patients who smoke with the suspected hazards of smoking to the fetus? Let $\alpha = .05$.

7.6 HYPOTHESIS TESTING: THE DIFFERENCE BETWEEN TWO POPULATION PROPORTIONS

The most frequent test employed relative to the difference between two population proportions is that their difference is zero. It is possible, however, to test that the difference is equal to some other value. Both one-sided and two-sided tests may be made.

When the null hypothesis to be tested is $p_1 - p_2 = 0$, we are hypothesizing that the two population proportions are equal. We use this as justification for combining the results of the two samples to come up with a pooled estimate of the hypothesized common proportion. If this procedure is adopted, one computes

$$\overline{p} = \frac{x_1 + x_2}{n_1 + n_2}, \text{ and } \overline{q} = 1 - \overline{p}$$

where x_1 and x_2 are the numbers in the first and second samples, respectively, possessing the characteristic of interest. This pooled estimate of $p = p_1 = p_2$ is used in computing $\hat{\sigma}_{\hat{p}_1 - \hat{p}_2}$, the estimated standard error of the estimator, as follows:

$$\hat{\sigma}_{\hat{p}_1 - \hat{p}_2} = \sqrt{\frac{\overline{p}(1 - \overline{p})}{n_1} + \frac{\overline{p}(1 - \overline{p})}{n_2}} \tag{7.6.1}$$

The test statistic becomes

$$z = \frac{(\hat{p}_1 - \hat{p}_2) - (p_1 - p_2)_0}{\hat{\sigma}_{\hat{p}_1 - \hat{p}_2}} \tag{7.6.2}$$

which is distributed approximately as the standard normal if the null hypothesis is true.

EXAMPLE 7.6.1

Noonan syndrome is a genetic condition that can affect the heart, growth, blood clotting, and mental and physical development. Noonan et al. (A-24) examined the stature of men and women with Noonan syndrome. The study contained 29 male and 44 female adults. One of the cut-off values used to assess stature was the third percentile of adult height. Eleven of the males fell below the third percentile of adult male height, while 24 of the females fell below the third percentile of female adult height. Does this study provide sufficient evidence for us to conclude that among subjects with Noonan syndrome, females are more likely than males to fall below the respective third percentile of adult height? Let $\alpha = .05$.

Solution:

1. **Data.** The data consist of information regarding the height status of Noonan syndrome males and females as described in the statement of the example.
2. **Assumptions.** We assume that the patients in the study constitute independent simple random samples from populations of males and females with Noonan syndrome.
3. **Hypotheses.**

$$H_0: p_F \leq p_M \quad \text{or} \quad p_F - p_M \leq 0$$
$$H_A: p_F > p_M \quad \text{or} \quad p_F - p_M > 0$$

where p_F is the proportion of females below the third percentile of female adult height and p_M is the proportion of males below the third percentile of male adult height.

4. **Test statistic.** The test statistic is given by Equation 7.6.2.
5. **Distribution of test statistic.** If the null hypothesis is true, the test statistic is distributed approximately as the standard normal.
6. **Decision rule.** Let $\alpha = .05$. The critical value of z is 1.645. Reject H_0 if computed z is greater than 1.645.
7. **Calculation of test statistic.** From the sample data we compute $\hat{p}_F = 24/44 = .545$, $\hat{p}_M = 11/29 = .379$, and $\bar{p} = (24 + 11)/(44 + 29) = .479$. The computed value of the test statistic, then, is

$$z = \frac{(.545 - .379)}{\sqrt{\frac{(.479)(.521)}{44} + \frac{(.479)(.521)}{29}}} = 1.39$$

8. **Statistical decision.** Fail to reject H_0 since $1.39 < 1.645$.
9. **Conclusion.** In the general population of adults with Noonan syndrome there may be no difference in the proportion of males and females who have heights below the third percentile of adult height.
10. ***p* value.** For this test $p = .0823$. ■

Tests involving two proportions, using the data from Example 7.6.1, can be carried out with a variety of computer programs. Outputs from MINITAB and NCSS are shown in Figure 7.6.1. Again, it should be noted that, because of rounding errors, the results will vary slightly if calculations are done by hand.

MINITAB Output

Test and CI for Two Proportions

```
Sample    X    N   Sample p
1        24   44   0.545455
2        11   29   0.379310

Difference = p (1) - p (2)
Estimate for difference: 0.166144
95% lower bound for difference: -0.0267550
Test for difference = 0 (vs > 0):  Z = 1.39  P-Value = 0.082
```

NCSS Output

Test Name	Test Statistic's Distribution	Test Statistic Value	Prob Level	Conclude H1 at 5% Significance?
Z-Test	Normal	1.390	0.0822	No

FIGURE 7.6.1 MINITAB and partial NCSS output for the data in Example 7.6.1

EXERCISES

In each of the following exercises use the ten-step hypothesis testing procedure. For each exercise, as appropriate, explain why you chose a one-sided test or a two-sided test. Discuss how you think researchers or clinicians might use the results of your hypothesis test. What clinical or research decisions or actions do you think would be appropriate in light of the results of your test?

7.6.1 Ho et al. (A-25) used telephone interviews of randomly selected respondents in Hong Kong to obtain information regarding individuals' perceptions of health and smoking history. Among 1222 current male smokers, 72 reported that they had "poor" or "very poor" health, while 30 among 282 former male smokers reported that they had "poor" or "very poor" health. Is this sufficient evidence to allow one to conclude that among Hong Kong men there is a difference between current and former smokers with respect to the proportion who perceive themselves as having "poor" and "very poor" health? Let $\alpha = .01$.

7.6.2 Landolt et al. (A-26) examined rates of posttraumatic stress disorder (PTSD) in mothers and fathers. Parents were interviewed 5 to 6 weeks after an accident or a new diagnosis of cancer or diabetes mellitus type I for their child. Twenty-eight of the 175 fathers interviewed and 43 of the

180 mothers interviewed met the criteria for current PTSD. Is there sufficient evidence for us to conclude that fathers are less likely to develop PTSD than mothers when a child is traumatized by an accident, cancer diagnosis, or diabetes diagnosis? Let $\alpha = .05$.

7.6.3 In a *Kidney International* article, Avram et al. (A-27) reported on a study involving 529 hemodialysis patients and 326 peritoneal dialysis patients. They found that at baseline 249 subjects in the hemodialysis treatment group were diabetic, while at baseline 134 of the subjects in the peritoneal dialysis group were diabetic. Is there a significant difference in diabetes prevalence at baseline between the two groups of this study? Let $\alpha = .05$. What does your finding regarding sample significance imply about the populations of subjects?

7.6.4 In a study of obesity the following results were obtained from samples of males and females between the ages of 20 and 75:

	n	**Number Overweight**
Males	150	21
Females	200	48

Can we conclude from these data that in the sampled populations there is a difference in the proportions who are overweight? Let $\alpha = .05$.

7.7 HYPOTHESIS TESTING: A SINGLE POPULATION VARIANCE

In Section 6.9 we examined how it is possible to construct a confidence interval for the variance of a normally distributed population. The general principles presented in that section may be employed to test a hypothesis about a population variance. When the data available for analysis consist of a simple random sample drawn from a normally distributed population, the test statistic for testing hypotheses about a population variance is

$$\chi^2 = (n - 1)s^2/\sigma^2 \tag{7.7.1}$$

which, when H_0 is true, is distributed as χ^2 with $n - 1$ degrees of freedom.

EXAMPLE 7.7.1

The purpose of a study by Wilkins et al. (A-28) was to measure the effectiveness of recombinant human growth hormone (rhGH) on children with total body surface area burns > 40 percent. In this study, 16 subjects received daily injections at home of rhGH. At baseline, the researchers wanted to know the current levels of insulin-like growth factor (IGF-I) prior to administration of rhGH. The sample variance of IGF-I levels (in ng/ml) was 670.81. We wish to know if we may conclude from these data that the population variance is not 600.

Solution:

1. **Data.** See statement in the example.
2. **Assumptions.** The study sample constitutes a simple random sample from a population of similar children. The IGF-I levels are normally distributed.
3. **Hypotheses.**

$$H_0: \sigma^2 = 600$$
$$H_A: \sigma^2 \neq 600$$

4. **Test statistic.** The test statistic is given by Equation 7.7.1.
5. **Distribution of test statistic.** When the null hypothesis is true, the test statistic is distributed as χ^2 with $n - 1$ degrees of freedom.
6. **Decision rule.** Let $\alpha = .05$. Critical values of χ^2 are 6.262 and 27.488. Reject H_0 unless the computed value of the test statistic is between 6.262 and 27.488. The rejection and nonrejection regions are shown in Figure 7.7.1.
7. **Calculation of test statistic.**

$$\chi^2 = \frac{15(670.81)}{600} = 16.77$$

8. **Statistical decision.** Do not reject H_0 since $6.262 < 16.77 < 27.488$.
9. **Conclusion.** Based on these data we are unable to conclude that the population variance is not 600.
10. ***p* value.** The determination of the p value for this test is complicated by the fact that we have a two-sided test and an asymmetric sampling distribution. When we have a two-sided test and a symmetric sampling distribution such as the standard normal or t, we may, as we have seen, double the one-sided p value. Problems arise when we attempt to do this with an asymmetric sampling distribution such as the

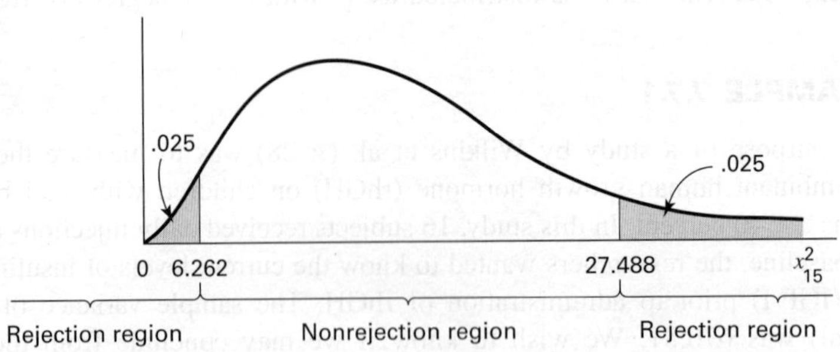

FIGURE 7.7.1 Rejection and nonrejection regions for Example 7.7.1.

chi-square distribution. In this situation the one-sided p value is reported along with the direction of the observed departure from the null hypothesis. In fact, this procedure may be followed in the case of symmetric sampling distributions. Precedent, however, seems to favor doubling the one-sided p value when the test is two-sided and involves a symmetric sampling distribution.

For the present example, then, we may report the p value as follows: $p > .05$ (two-sided test). A population variance greater than 600 is suggested by the sample data, but this hypothesis is not strongly supported by the test.

If the problem is stated in terms of the population standard deviation, one may square the sample standard deviation and perform the test as indicated above. ■

One-Sided Tests Although this was an example of a two-sided test, one-sided tests may also be made by logical modification of the procedure given here.

For $H_A: \sigma^2 > \sigma_0^2$, reject H_0 if computed $\chi^2 \geq x_{1-\alpha}^2$
For $H_A: \sigma^2 < \sigma_0^2$, reject H_0 if computed $\chi^2 \leq x_{\alpha}^2$

Tests involving a single population variance can be carried out using MINITAB software. Most other statistical computer programs lack procedures for carrying out these tests directly. The output from MINITAB, using the data from Example 7.7.1, is shown in Figure 7.7.2.

```
Test and CI for One Variance

Statistics

 N  StDev  Variance
16   25.9       671

95% Confidence Intervals

                CI for       CI for
Method          StDev        Variance
Standard   (19.1, 40.1)   (366, 1607)

Tests

Method     Chi-Square    DF    P-Value
Standard        16.77    15      0.666
```

FIGURE 7.7.2 MINITAB output for the data in Example 7.7.1.

EXERCISES

In each of the following exercises, carry out the ten-step testing procedure. For each exercise, as appropriate, explain why you chose a one-sided test or a two-sided test. Discuss how you think researchers or clinicians might use the results of your hypothesis test. What clinical or research decisions or actions do you think would be appropriate in light of the results of your test?

7.7.1 Recall Example 7.2.3, where Nakamura et al. (A-1) studied subjects with acute medial collateral ligament injury (MCL) with anterior cruciate ligament tear (ACL). The ages of the 17 subjects were:

31, 26, 21, 15, 26, 16, 19, 21, 28, 27, 22, 20, 25, 31, 20, 25, 15

Use these data to determine if there is sufficient evidence for us to conclude that in a population of similar subjects, the variance of the ages of the subjects is not 20 years. Let $\alpha = .01$.

7.7.2 Robinson et al. (A-29) studied nine subjects who underwent baffle procedure for transposition of the great arteries (TGA). At baseline, the systemic vascular resistance (SVR) (measured in WU $\times$ m^2) values at rest yielded a standard deviation of 28. Can we conclude from these data that the SVR variance of a population of similar subjects with TGA is not 700? Let $\alpha = .10$.

7.7.3 Vital capacity values were recorded for a sample of 10 patients with severe chronic airway obstruction. The variance of the 10 observations was .75. Test the null hypothesis that the population variance is 1.00. Let $\alpha = .05$.

7.7.4 Hemoglobin (g percent) values were recorded for a sample of 20 children who were part of a study of acute leukemia. The variance of the observations was 5. Do these data provide sufficient evidence to indicate that the population variance is greater than 4? Let $\alpha = .05$.

7.7.5 A sample of 25 administrators of large hospitals participated in a study to investigate the nature and extent of frustration and emotional tension associated with the job. Each participant was given a test designed to measure the extent of emotional tension he or she experienced as a result of the duties and responsibilities associated with the job. The variance of the scores was 30. Can it be concluded from these data that the population variance is greater than 25? Let $\alpha = .05$.

7.7.6 In a study in which the subjects were 15 patients suffering from pulmonary sarcoid disease, blood gas determinations were made. The variance of the Pa_{O_2} (mm Hg) values was 450. Test the null hypothesis that the population variance is greater than 250. Let $\alpha = .05$.

7.7.7 Analysis of the amniotic fluid from a simple random sample of 15 pregnant women yielded the following measurements on total protein (grams per 100 ml) present:

.69, 1.04, .39, .37, .64, .73, .69, 1.04,
.83, 1.00, .19, .61, .42, .20, .79

Do these data provide sufficient evidence to indicate that the population variance is greater than .05? Let $\alpha = .05$. What assumptions are necessary?

7.8 HYPOTHESIS TESTING: THE RATIO OF TWO POPULATION VARIANCES

As we have seen, the use of the t distribution in constructing confidence intervals and in testing hypotheses for the difference between two population means assumes that the population variances are equal. As a rule, the only hints available about the magnitudes of the respective variances are the variances computed from samples taken from the populations. We would like to know if the difference that, undoubtedly, will exist between the sample variances is indicative of a real difference in population variances, or if the difference is of such magnitude that it could have come about as a result of chance alone when the population variances are equal.

Two methods of chemical analysis may give the same results on the average. It may be, however, that the results produced by one method are more variable than the results of the other. We would like some method of determining whether this is likely to be true.

Variance Ratio Test Decisions regarding the comparability of two population variances are usually based *on the variance ratio test,* which is a test of the null hypothesis that two population variances are equal. When we test the hypothesis that two population variances are equal, we are, in effect, testing the hypothesis that their ratio is equal to 1.

We learned in the preceding chapter that, when certain assumptions are met, the quantity $(s_1^2/\sigma_1^2)/(s_2^2/\sigma_2^2)$ is distributed as F with $n_1 - 1$ numerator degrees of freedom and $n_2 - 1$ denominator degrees of freedom. If we are hypothesizing that $\sigma_1^2 = \sigma_2^2$, we assume that the hypothesis is true, and the two variances cancel out in the above expression leaving s_1^2/s_2^2, which follows the same F distribution. The ratio s_1^2/s_2^2 will be designated V.R. for variance ratio.

For a two-sided test, we follow the convention of placing the larger sample variance in the numerator and obtaining the critical value of F for $\alpha/2$ and the appropriate degrees of freedom. However, for a one-sided test, which of the two sample variances is to be placed in the numerator is predetermined by the statement of the null hypothesis. For example, for the null hypothesis that σ_1^2/σ_2^2, the appropriate test statistic is V.R. $= s_1^2/s_2^2$. The critical value of F is obtained for α (not $\alpha/2$) and the appropriate degrees of freedom. In like manner, if the null hypothesis is that $\sigma_1^2 \geq \sigma_2^2$, the appropriate test statistic is V.R. $= s_2^2/s_1^2$. In all cases, the decision rule is to reject the null hypothesis if the computed V.R. is equal to or greater than the critical value of F.

EXAMPLE 7.8.1

Borden et al. (A-30) compared meniscal repair techniques using cadaveric knee specimens. One of the variables of interest was the load at failure (in newtons) for knees fixed with the FasT-FIX technique (group 1) and the vertical suture method (group 2). Each technique was applied to six specimens. The standard deviation for the FasT-FIX method was 30.62, and the standard deviation for the vertical suture method was 11.37. Can we conclude that, in general, the variance of load at failure is higher for the FasT-FIX technique than the vertical suture method?

Solution:

1. **Data.** See the statement of the example.
2. **Assumptions.** Each sample constitutes a simple random sample of a population of similar subjects. The samples are independent. We assume the loads at failure in both populations are approximately normally distributed.

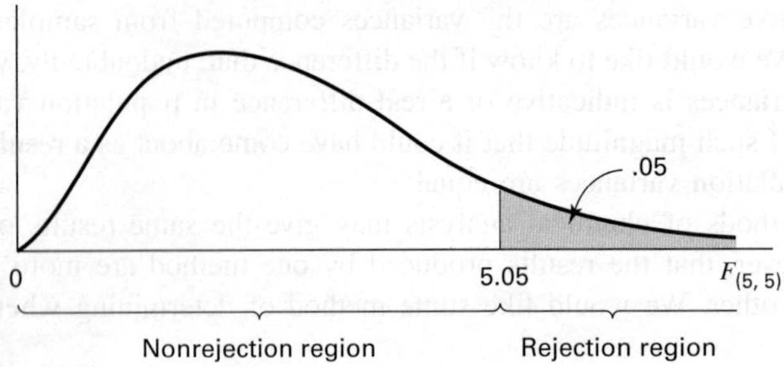

FIGURE 7.8.1 Rejection and nonrejection regions, Example 7.8.1.

3. Hypotheses.

$$H_0: \sigma_1^2 \leq \sigma_2^2$$
$$H_A: \sigma_1^2 > \sigma_2^2$$

4. Test statistic.

$$\text{V.R.} = \frac{s_1^2}{s_2^2} \tag{7.8.1}$$

5. Distribution of test statistic. When the null hypothesis is true, the test statistic is distributed as F with $n_1 - 1$ numerator and $n_2 - 1$ denominator degrees of freedom.

6. Decision rule. Let $\alpha = .05$. The critical value of F, from Appendix Table G, is 5.05. Note that if Table G does not contain an entry for the given numerator degrees of freedom, we use the column closest in value to the given numerator degrees of freedom. Reject H_0 if V.R. ≥ 5.05. The rejection and nonrejection regions are shown in Figure 7.8.1.

7. Calculation of test statistic.

$$\text{V.R.} = \frac{(30.62)^2}{(11.37)^2} = 7.25$$

8. Statistical decision. We reject H_0, since $7.25 > 5.05$; that is, the computed ratio falls in the rejection region.

9. Conclusion. The failure load variability is higher when using the FasT-FIX method than the vertical suture method.

10. *p* value. Because the computed V.R. of 7.25 is greater than 5.05, the p value for this test is less than 0.05. ■

Several computer programs can be used to test the equality of two variances. Outputs from these programs will differ depending on the test that is used. We saw in Figure 7.3.3, for example, that the SAS system uses a folded F-test procedure. MINITAB uses two different tests. The first is an F-test under the assumption of normality, and the other is a modified Levene's test (1) that is used when normality cannot be assumed. SPSS uses an unmodified Levene's test (2). Regardless of the options, these tests are generally considered superior to the variance ratio test that is presented in Example 7.8.1. Discussion of the mathematics behind these tests is beyond the scope of this book, but an example is given to illustrate these procedures, since results from these tests are often provided automatically as outputs when a computer program is used to carry out a *t*-test.

EXAMPLE 7.8.2

Using the data from Example 7.3.2, we are interested in testing whether the assumption of the equality of variances can be assumed prior to performing a *t*-test. For ease of discussion, the data are reproduced below (Table 7.8.1):

TABLE 7.8.1 Pressures (mm Hg) Under the Pelvis During Static Conditions for Example 7.3.2

Control	131	115	124	131	122	117	88	114	150	169
SCI	60	150	130	180	163	130	121	119	130	148

Partial outputs for MINITAB, SAS, and SPSS are shown in Figure 7.8.2. Regardless of the test or program that is used, we fail to reject the null hypothesis of equal variances (H_0: $\sigma_1^2 = \sigma_2^2$) because all *p* values > 0.05. We may now proceed with a *t*-test under the assumption of equal variances. ■

MINITAB Output

F-Test	
Test Statistic	0.46
P-Value	0.263
Levene's Test	
Test Statistic	0.49
P-Value	0.495

SPSS Output

Levene's Test for Equality of Variances	
F	Sig.
.664	.482

SAS Output

Equality of Variances

Variable	Method	Num DF	Den DF	F Value	Pr > F
pressure	Folded F	9	9	2.17	0.2626

FIGURE 7.8.2 Partial MINITAB, SPSS, and SAS outputs for testing the equality of two population variances.

EXERCISES

In the following exercises perform the ten-step test. For each exercise, as appropriate, explain why you chose a one-sided test or a two-sided test. Discuss how you think researchers or clinicians might use the results of your hypothesis test. What clinical or research decisions or actions do you think would be appropriate in light of the results of your test?

7.8.1 Dora et al. (A-31) investigated spinal canal dimensions in 30 subjects symptomatic with disc herniation selected for a discectomy and 45 asymptomatic individuals. The researchers wanted to know if spinal canal dimensions are a significant risk factor for the development of sciatica. Toward that end, they measured the spinal canal dimension between vertebrae L3 and L4 and obtained a mean of 17.8 mm in the discectomy group with a standard deviation of 3.1. In the control group, the mean was 18.5 mm with a standard deviation of 2.8 mm. Is there sufficient evidence to indicate that in relevant populations the variance for subjects symptomatic with disc herniation is larger than the variance for control subjects? Let $\alpha = .05$.

7.8.2 Nagy et al. (A-32) studied 50 stable patients who were admitted for a gunshot wound that traversed the mediastinum. Of these, eight were deemed to have a mediastinal injury and 42 did not. The standard deviation for the ages of the eight subjects with mediastinal injury was 4.7 years, and the standard deviation of ages for the 42 without injury was 11.6 years. Can we conclude from these data that the variance of age is larger for a population of similar subjects without injury compared to a population with mediastinal injury? Let $\alpha = .05$.

7.8.3 A test designed to measure level of anxiety was administered to a sample of male and a sample of female patients just prior to undergoing the same surgical procedure. The sample sizes and the variances computed from the scores were as follows:

$$\text{Males:} \quad n = 16, s^2 = 150$$

$$\text{Females:} \quad n = 21, s^2 = 275$$

Do these data provide sufficient evidence to indicate that in the represented populations the scores made by females are more variable than those made by males? Let $\alpha = .05$.

7.8.4 In an experiment to assess the effects on rats of exposure to cigarette smoke, 11 animals were exposed and 11 control animals were not exposed to smoke from unfiltered cigarettes. At the end of the experiment, measurements were made of the frequency of the ciliary beat (beats/min at 20°C) in each animal. The variance for the exposed group was 3400 and 1200 for the unexposed group. Do these data indicate that in the populations represented the variances are different? Let $\alpha = .05$.

7.8.5 Two pain-relieving drugs were compared for effectiveness on the basis of length of time elapsing between administration of the drug and cessation of pain. Thirteen patients received drug 1, and 13 received drug 2. The sample variances were $s_1^2 = 64$ and $s_2^2 = 16$. Test the null hypothesis that the two populations variances are equal. Let $\alpha = .05$.

7.8.6 Packed cell volume determinations were made on two groups of children with cyanotic congenital heart disease. The sample sizes and variances were as follows:

Group	n	s^2
1	10	40
2	16	84

Do these data provide sufficient evidence to indicate that the variance of population 2 is larger than the variance of population 1? Let $\alpha = .05$.

7.8.7 Independent simple random samples from two strains of mice used in an experiment yielded the following measurements on plasma glucose levels following a traumatic experience:

Strain A:	54, 99, 105, 46, 70, 87, 55, 58, 139, 91
Strain B:	93, 91, 93, 150, 80, 104, 128, 83, 88, 95, 94, 97

Do these data provide sufficient evidence to indicate that the variance is larger in the population of strain A mice than in the population of strain B mice? Let $\alpha = .05$. What assumptions are necessary?

7.9 THE TYPE II ERROR AND THE POWER OF A TEST

In our discussion of hypothesis testing our focus has been on α, the probability of committing a type I error (rejecting a true null hypothesis). We have paid scant attention to β, the probability of committing a type II error (failing to reject a false null hypothesis). There is a reason for this difference in emphasis. For a given test, α is a single number assigned by the investigator in advance of performing the test. It is a measure of the acceptable risk of rejecting a true null hypothesis. On the other hand, β may assume one of many values. Suppose we wish to test the null hypothesis that some population parameter is equal to some specified value. If H_0 is false and we fail to reject it, we commit a type II error. If the hypothesized value of the parameter is not the true value, the value of β (the probability of committing a type II error) depends on several factors: (1) the true value of the parameter of interest, (2) the hypothesized value of the parameter, (3) the value of α, and (4) the sample size, n. For fixed α and n, then, we may, before performing a hypothesis test, compute many values of β by postulating many values for the parameter of interest given that the hypothesized value is false.

For a given hypothesis test it is of interest to know how well the test controls type II errors. If H_0 is in fact false, we would like to know the probability that we will reject it. The *power* of a test, designated $1 - \beta$, provides this desired information. The quantity $1 - \beta$ is the probability that we will reject a false null hypothesis; it may be computed for any alternative value of the parameter about which we are testing a hypothesis. Therefore, $1 - \beta$ is the probability that we will take the correct action when H_0 is false because the true parameter value is equal to the one for which we computed $1 - \beta$. For a given test we may specify any number of possible values of the parameter of interest and for each compute the value of $1 - \beta$. The result is called a *power function.* The graph of a power function, called a *power curve,* is a helpful device for quickly assessing the nature of the power of a given test. The following example illustrates the procedures we use to analyze the power of a test.

EXAMPLE 7.9.1

Suppose we have a variable whose values yield a population standard deviation of 3.6. From the population we select a simple random sample of size $n = 100$. We select a value of $\alpha = .05$ for the following hypotheses:

$$H_0: \mu = 17.5, \quad H_A: \mu \neq 17.5$$

Solution: When we study the power of a test, we locate the rejection and nonrejection regions on the $\bar{x}$ scale rather than the z scale. We find the critical values of $\bar{x}$ for a two-sided test using the following formulas:

$$\bar{x}_{\text{U}} = \mu_0 + z\frac{\sigma}{\sqrt{n}} \tag{7.9.1}$$

and

$$\bar{x}_{\text{L}} = \mu_0 - z\frac{\sigma}{\sqrt{n}} \tag{7.9.2}$$

where $\bar{x}_{\text{U}}$ and $\bar{x}_{\text{L}}$ are the upper and lower critical values, respectively, of $\bar{x}$; $+z$ and $-z$ are the critical values of z; and μ_0 is the hypothesized value of μ. For our example, we have

$$\begin{aligned}\bar{x}_{\text{U}} &= 17.50 + 1.96\frac{(3.6)}{(10)} = 17.50 + 1.96(.36) \\ &= 17.50 + .7056 = 18.21\end{aligned}$$

and

$$\bar{x}_{\text{L}} = 17.50 - 1.96(.36) = 17.50 - .7056 = 16.79$$

Suppose that H_0 is false, that is, that μ is not equal to 17.5. In that case, μ is equal to some value other than 17.5. We do not know the actual value of μ. But if H_0 is false, μ is one of the many values that are greater than or smaller than 17.5. Suppose that the true population mean is $\mu_1 = 16.5$. Then the sampling distribution of $\bar{x}_1$ is also approximately normal, with $\mu_{\bar{x}} = \mu = 16.5$. We call this sampling distribution $f(\bar{x}_1)$, and we call the sampling distribution under the null hypothesis $f(\bar{x}_0)$.

β, the probability of the type II error of failing to reject a false null hypothesis, is the area under the curve of $f(\bar{x}_1)$ that overlaps the nonrejection region specified under H_0. To determine the value of β, we find the area under $f(\bar{x}_1)$, above the $\bar{x}$ axis, and between $\bar{x} = 16.79$ and $\bar{x} = 18.21$. The value of β is equal to $P(16.79 \leq \bar{x} \leq 18.21)$ when $\mu = 16.5$. This is the same as

$$\begin{aligned}P\left(\frac{16.79 - 16.5}{.36} \leq z \leq \frac{18.21 - 16.5}{.36}\right) &= P\left(\frac{.29}{.36} \leq z \leq \frac{1.71}{.36}\right) \\ &= P(.81 \leq z \leq 4.75) \\ &\approx 1 - .7910 = .2090\end{aligned}$$

Thus, the probability of taking an appropriate action (that is, rejecting H_0) when the null hypothesis states that $\mu = 17.5$, but in fact $\mu = 16.5$, is

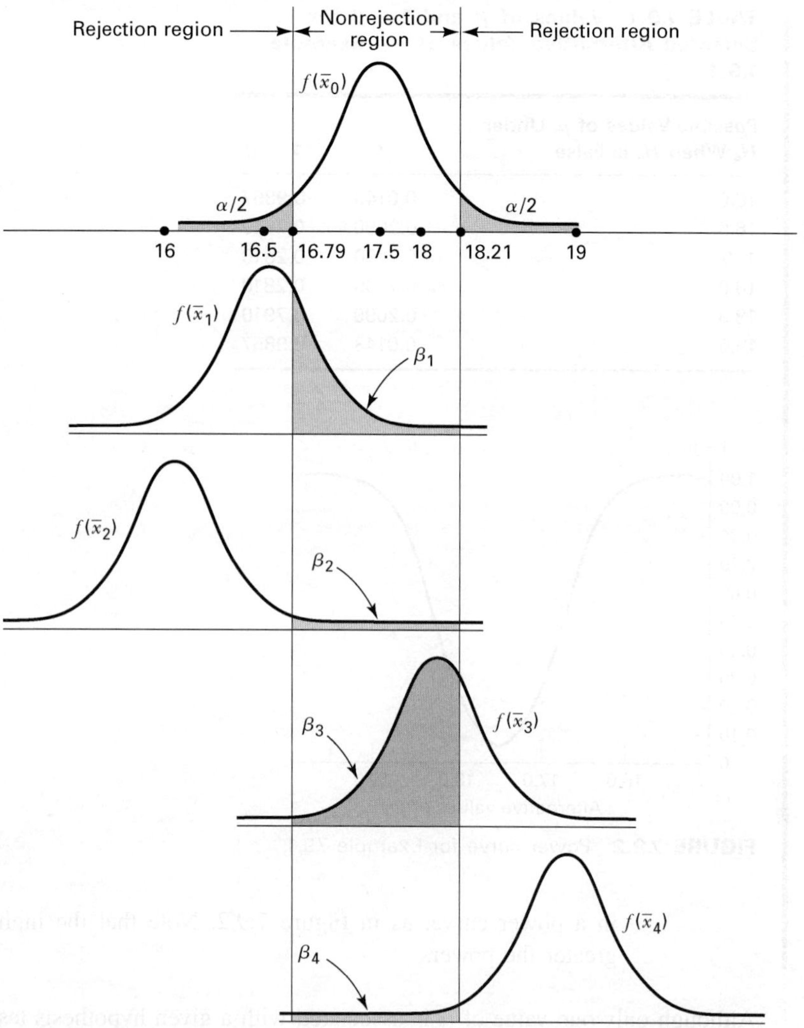

FIGURE 7.9.1 Size of β for selected values for H_1 for Example 7.9.1.

$1 - .2090 = .7910$. As we noted, μ may be one of a large number of possible values when H_0 is false. Figure 7.9.1 shows a graph of several such possibilities. Table 7.9.1 shows the corresponding values of β and $1 - \beta$ (which are approximate), along with the values of β for some additional alternatives.

Note that in Figure 7.9.1 and Table 7.9.1 those values of μ under the alternative hypothesis that are closer to the value of μ specified by H_0 have larger associated β values. For example, when $\mu = 18$ under the alternative hypothesis, $\beta = .7190$; and when $\mu = 19.0$ under H_A, $\beta = .0143$. The power of the test for these two alternatives, then, is $1 - .7190 = .2810$ and $1 - .0143 = .9857$, respectively. We show the power of the test graphically

TABLE 7.9.1 Values of β and $1 - \beta$ for Selected Alternative Values of μ_1, Example 7.9.1

Possible Values of μ Under H_A When H_0 is False	β	$1 - \beta$
16.0	0.0143	0.9857
16.5	0.2090	0.7910
17.0	0.7190	0.2810
18.0	0.7190	0.2810
18.5	0.2090	0.7910
19.0	0.0143	0.9857

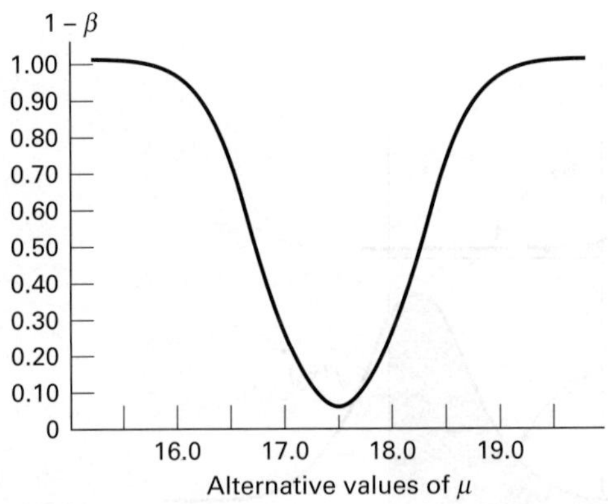

FIGURE 7.9.2 Power curve for Example 7.9.1.

in a power curve, as in Figure 7.9.2. Note that the higher the curve, the greater the power. ■

Although only one value of α is associated with a given hypothesis test, there are many values of β, one for each possible value of μ if μ_0 is not the true value of μ as hypothesized. Unless alternative values of μ are much larger or smaller than μ_0, β is relatively large compared with α. Typically, we use hypothesis-testing procedures more often in those cases in which, when H_0 is false, the true value of the parameter is fairly close to the hypothesized value. In most cases, β, the computed probability of failing to reject a false null hypothesis, is larger than α, the probability of rejecting a true null hypothesis. These facts are compatible with our statement that a decision based on a rejected null hypothesis is more conclusive than a decision based on a null hypothesis that is not rejected. The probability of being wrong in the latter case is generally larger than the probability of being wrong in the former case.

Figure 7.9.2 shows the V-shaped appearance of a power curve for a two-sided test. In general, a two-sided test that discriminates well between the value of the parameter in H_0 and values in H_1 results in a narrow V-shaped power curve. A wide V-shaped curve indicates that the test discriminates poorly over a relatively wide interval of alternative values of the parameter.

Power Curves for One-Sided Tests The shape of a power curve for a one-sided test with the rejection region in the upper tail is an elongated S. If the rejection region of a one-sided test is located in the lower tail of the distribution, the power curve takes the form of a reverse elongated S. The following example shows the nature of the power curve for a one-sided test.

EXAMPLE 7.9.2

The mean time laboratory employees now take to do a certain task on a machine is 65 seconds, with a standard deviation of 15 seconds. The times are approximately normally distributed. The manufacturers of a new machine claim that their machine will reduce the mean time required to perform the task. The quality-control supervisor designs a test to determine whether or not she should believe the claim of the makers of the new machine. She chooses a significance level of $\alpha = 0.01$ and randomly selects 20 employees to perform the task on the new machine. The hypotheses are

$$H_0: \mu \geq 65, \quad H_A: \mu < 65$$

The quality-control supervisor also wishes to construct a power curve for the test.

Solution: The quality-control supervisor computes, for example, the following value of $1 - \beta$ for the alternative $\mu = 55$. The critical value of $1 - \beta$ for the test is

$$65 - 2.33\left(\frac{15}{\sqrt{20}}\right) = 57$$

We find β as follows:

$$\beta = P(\bar{x} > 57 \mid \mu = 55) = P\left(z > \frac{57 - 55}{15/\sqrt{20}}\right) = P(z > .60)$$
$$= 1 - .7257 = .2743$$

Consequently, $1 - \beta = 1 - .2743 = .7257$. Figure 7.9.3 shows the calculation of β. Similar calculations for other alternative values of μ also yield

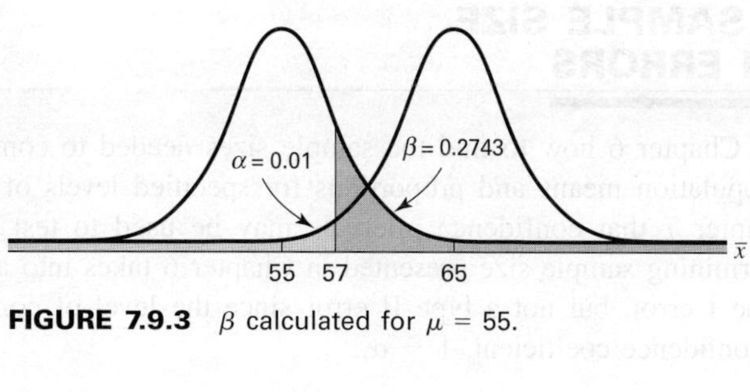

FIGURE 7.9.3 β calculated for $\mu = 55$.

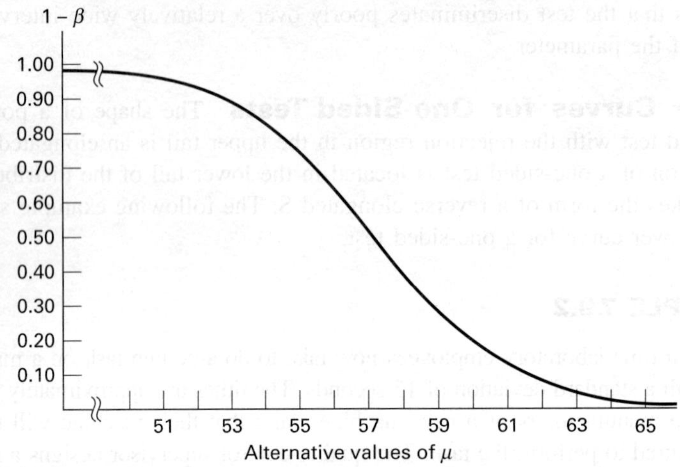

FIGURE 7.9.4 Power curve for Example 7.9.2.

values of $1 - \beta$. When plotted against the values of μ, these give the power curve shown in Figure 7.9.4. ■

Operating Characteristic Curves Another way of evaluating a test is to look at its *operating characteristic* (OC) *curve*. To construct an OC curve, we plot values of β, rather than $1 - \beta$, along the vertical axis. Thus, an OC curve is the complement of the corresponding power curve.

EXERCISES

Construct and graph the power function for each of the following situations.

7.9.1 H_0: $\mu \leq 516$, H_A: $\mu > 516$, $n = 16, \sigma = 32, \alpha = 0.05$.

7.9.2 H_0: $\mu = 3$, H_A: $\mu \neq 3$, $n = 100, \sigma = 1, \alpha = 0.05$.

7.9.3 H_0: $\mu \leq 4.25$, H_A: $\mu > 4.25$, $n = 81, \sigma = 1.8, \alpha = 0.01$.

7.10 DETERMINING SAMPLE SIZE TO CONTROL TYPE II ERRORS

You learned in Chapter 6 how to find the sample sizes needed to construct confidence intervals for population means and proportions for specified levels of confidence. You learned in Chapter 7 that confidence intervals may be used to test hypotheses. The method of determining sample size presented in Chapter 6 takes into account the probability of a type I error, but not a type II error since the level of confidence is determined by the confidence coefficient, $1 - \alpha$.

In many statistical inference procedures, the investigator wishes to consider the type II error as well as the type I error when determining the sample size. To illustrate the procedure, we refer again to Example 7.9.2.

EXAMPLE 7.10.1

In Example 7.9.2, the hypotheses are

$$H_0: \mu \geq 65, \quad H_A: \mu < 65$$

The population standard deviation is 15, and the probability of a type I error is set at .01. Suppose that we want the probability of failing to reject $H_0(\beta)$ to be .05 if H_0 is false because the true mean is 55 rather than the hypothesized 65. How large a sample do we need in order to realize, simultaneously, the desired levels of α and β?

Solution: For $\alpha = .01$ and $n = 20$, β is equal to .2743. The critical value is 57. Under the new conditions, the critical value is unknown. Let us call this new critical value C. Let μ_0 be the hypothesized mean and μ_1 the mean under the alternative hypothesis. We can transform each of the relevant sampling distributions of $\bar{x}$, the one with a mean of μ_0 and the one with a mean of μ_1 to a z distribution. Therefore, we can convert C to a z value on the horizontal scale of each of the two standard normal distributions. When we transform the sampling distribution of $\bar{x}$ that has a mean of μ_0 to the standard normal distribution, we call the z that results z_0. When we transform the sampling distribution $\bar{x}$ that has a mean of μ_1 to the standard normal distribution, we call the z that results z_1. Figure 7.10.1 represents the situation described so far.

We can express the critical value C as a function of z_0 and μ_0 and also as a function of z_1 and μ_1. This gives the following equations:

$$C = \mu_0 - z_0 \frac{\sigma}{\sqrt{n}} \tag{7.10.1}$$

$$C = \mu_1 + z_1 \frac{\sigma}{\sqrt{n}} \tag{7.10.2}$$

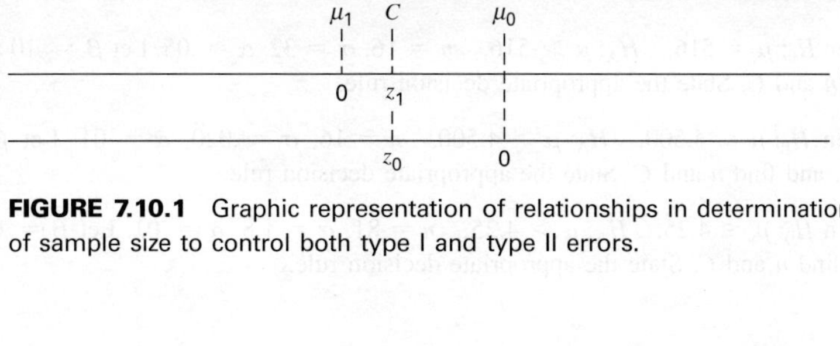

FIGURE 7.10.1 Graphic representation of relationships in determination of sample size to control both type I and type II errors.

We set the right-hand sides of these equations equal to each other and solve for n, to obtain

$$n = \left[\frac{(z_0 + z_1)\sigma}{(\mu_0 - \mu_1)}\right]^2 \tag{7.10.3}$$

To find n for our illustrative example, we substitute appropriate quantities into Equation 7.10.3. We have $\mu_0 = 65$, $\mu_1 = 55$, and $\sigma = 15$. From Appendix Table D, the value of z that has .01 of the area to its left is -2.33. The value of z that has .05 of the area to its right is 1.645. Both z_0 and z_1 are taken as positive. We determine whether C lies above or below either μ_0 or μ_1 when we substitute into Equations 7.10.1 and 7.10.2. Thus, we compute

$$n = \left[\frac{(2.33 + 1.645)(15)}{(65 - 55)}\right]^2 = 35.55$$

We would need a sample of size 36 to achieve the desired levels of α and β when we choose $\mu_1 = 55$ as the alternative value of μ.

We now compute C, the critical value for the test, and state an appropriate decision rule. To find C, we may substitute known numerical values into either Equation 7.10.1 or Equation 7.10.2. For illustrative purposes, we solve both equations for C. First we have

$$C = 65 - 2.33\left(\frac{15}{\sqrt{36}}\right) = 59.175$$

From Equation 7.10.2, we have

$$C = 55 - 1.645\left(\frac{15}{\sqrt{36}}\right) = 59.1125$$

The difference between the two results is due to rounding error.

The decision rule, when we use the first value of C, is as follows:

Select a sample of size 36 and compute $\bar{x}$, if $\bar{x} \leq 59.175$, reject H_0. If $\bar{x} > 59.175$, do not reject H_0.

We have limited our discussion of the type II error and the power of a test to the case involving a population mean. The concepts extend to cases involving other parameters. ■

EXERCISES

7.10.1 Given H_0: $\mu = 516$, H_A: $\mu > 516$, $n = 16$, $\sigma = 32$, $\alpha = .05$. Let $\beta = .10$ and $\mu_1 = 520$, and find n and C. State the appropriate decision rule.

7.10.2 Given H_0: $\mu \leq 4.500$, H_A: $\mu > 4.500$, $n = 16$, $\sigma = .020$, $\alpha = .01$. Let $\beta = .05$ and $\mu_1 = 4.52$, and find n and C. State the appropriate decision rule.

7.10.3 Given H_0: $\mu \leq 4.25$, H_A: $\mu > 4.25$, $n = 81$, $\sigma = 1.8$, $\alpha = .01$. Let $\beta = .03$ and $\mu_1 = 5.00$, and find n and C. State the appropriate decision rule.

7.11 SUMMARY

In this chapter the general concepts of hypothesis testing are discussed. A general procedure for carrying out a hypothesis test consisting of the following ten steps is suggested.

1. Description of data.
2. Statement of necessary assumptions.
3. Statement of null and alternative hypotheses.
4. Specification of the test statistic.
5. Specification of the distribution of the test statistic.
6. Statement of the decision rule.
7. Calculation of test statistic from sample data.
8. The statistical decision based on sample results.
9. Conclusion.
10. Determination of p value.

A number of specific hypothesis tests are described in detail and illustrated with appropriate examples. These include tests concerning population means, the difference between two population means, paired comparisons, population proportions, the difference between two population proportions, a population variance, and the ratio of two population variances. In addition we discuss the power of a test and the determination of sample size for controlling both type I and type II errors.

SUMMARY OF FORMULAS FOR CHAPTER 7

Formula Number	Name	Formula
7.1.1, 7.1.2, 7.2.1	z-transformation (using either μ or μ_0)	$z = \dfrac{\bar{x} - \mu_0}{\sigma/\sqrt{n}}$
7.2.2	t-transformation	$t = \dfrac{\bar{x} - \mu_0}{s/\sqrt{n}}$
7.2.3	Test statistic when sampling from a population that is not normally distributed	$z = \dfrac{\bar{x} - \mu_0}{s/\sqrt{n}}$
7.3.1	Test statistic when sampling from normally distributed populations: population variances known	$z = \dfrac{(\bar{x}_1 - \bar{x}_2) - (\mu_1 - \mu_2)_0}{\sqrt{\dfrac{\sigma_1^2}{n_1} + \dfrac{\sigma_2^2}{n_2}}}$

(*Continued*)

7.3.2	Test statistic when sampling from normally distributed populations: population variances unknown and equal	$t = \frac{(\bar{x}_1 - \bar{x}_2) - (\mu_1 - \mu_2)_0}{\sqrt{\frac{s_p^2}{n_1} + \frac{s_p^2}{n_2}}}$, where $s_p^2 = \frac{(n_1 - 1)s_1^2 + (n_2 - 1)s_2^2}{n_1 + n_2 - 2}$
7.3.3, 7.3.4	Test statistic when sampling from normally distributed populations: population variances unknown and unequal	$t' = \frac{(\bar{x}_1 - \bar{x}_2) - (\mu_1 - \mu_2)_0}{\sqrt{\frac{s_1^2}{n_1} + \frac{s_2^2}{n_2}}}$, where $t'_{(1-\alpha/2)} = \frac{w_1 t_1 + w_2 t_2}{w_1 + w_2}$
7.3.5	Sampling from populations that are not normally distributed	$z = \frac{(\bar{x}_1 - \bar{x}_2) - (\mu_1 - \mu_2)_0}{\sqrt{\frac{\sigma_1^2}{n_1} + \frac{\sigma_2^2}{n_2}}}$
7.4.1	Test statistic for paired differences when the population variance is unknown	$t = \frac{\bar{d} - \mu_{d_0}}{s_{\bar{d}}}$
7.4.2	Test statistic for paired differences when the population variance is known	$z = \frac{\bar{d} - \mu_{d_0}}{\sigma_d / \sqrt{n}}$
7.5.1	Test statistic for a single population proportion	$z = \frac{\hat{p} - p_0}{\sqrt{\frac{p_0 q_0}{n}}}$
7.6.1, 7.6.2	Test statistic for the difference between two population proportions	$z = \frac{(\hat{p}_1 - \hat{p}_2) - (p_1 - p_2)_0}{\hat{\sigma}_{\hat{p}_1 - \hat{p}_2}}$, where $\bar{p} = \frac{x_1 + x_2}{n_1 + n_2}$, and $\hat{\sigma}_{\hat{p}_1 - \hat{p}_2} = \sqrt{\frac{\bar{p}(1 - \bar{p})}{n_1} + \frac{\bar{p}(1 - \bar{p})}{n_2}}$
7.7.1	Test statistic for a single population variance	$\chi^2 = \frac{(n - 1)s^2}{\sigma^2}$
7.8.1	Variance ratio	$V.R. = \frac{s_1^2}{s_2^2}$

7.9.1, 7.9.2	Upper and lower critical values for $\bar{x}$	$\bar{X}_U = \mu_0 + z\frac{\sigma}{\sqrt{n}}$ $\bar{X}_L = \mu_0 - z\frac{\sigma}{\sqrt{n}}$
7.10.1, 7.10.2	Critical value for determining sample size to control type II errors	$C = \mu_0 - z_0\frac{\sigma}{\sqrt{n}} = \mu_1 + z_1\frac{\sigma}{\sqrt{n}}$
7.10.3	Sample size to control type II errors	$n = \left[\frac{(z_0 + z_1)\sigma}{(\mu_0 - \mu_1)}\right]^2$
Symbol Key	• α = type 1 error rate • C = critical value • χ^2 = chi-square distribution • $\bar{d}$ = average difference • μ = mean of population • μ_0 = hypothesized mean • n = sample size • p = proportion for population • $\bar{p}$ = average proportion • $q = (1-p)$ • $\hat{p}$ = estimated proportion for sample • σ^2 = population variance • σ = population standard deviation • $\sigma_{\bar{d}}$ = standard error of difference • $\sigma_{\bar{x}}$ = standard error • s = standard deviation of sample • $s_{\bar{d}}$ = standard deviation of the difference • s_p = pooled standard deviation • t = Student's t-transformation • t' = Cochran's correction to t • $\bar{x}$ = mean of sample • $\bar{x}_L$ = lower limit of critical value for $\bar{x}$ • $\bar{x}_U$ = upper limit of critical value for $\bar{x}$ • z = standard normal transformation	

REVIEW QUESTIONS AND EXERCISES

1. What is the purpose of hypothesis testing?
2. What is a hypothesis?
3. List and explain each step in the ten-step hypothesis testing procedure.

4. Define:

(a) Type I error **(b)** Type II error

(c) The power of a test **(d)** Power function

(e) Power curve **(f)** Operating characteristic curve

5. Explain the difference between the power curves for one-sided tests and two-sided tests.

6. Explain how one decides what statement goes into the null hypothesis and what statement goes into the alternative hypothesis.

7. What are the assumptions underlying the use of the t statistic in testing hypotheses about a single mean? The difference between two means?

8. When may the z statistic be used in testing hypotheses about

(a) a single population mean?

(b) the difference between two population means?

(c) a single population proportion?

(d) the difference between two population proportions?

9. In testing a hypothesis about the difference between two population means, what is the rationale behind pooling the sample variances?

10. Explain the rationale behind the use of the paired comparisons test.

11. Give an example from your field of interest where a paired comparisons test would be appropriate. Use real or realistic data and perform an appropriate hypothesis test.

12. Give an example from your field of interest where it would be appropriate to test a hypothesis about the difference between two population means. Use real or realistic data and carry out the ten-step hypothesis testing procedure.

13. Do Exercise 12 for a single population mean.

14. Do Exercise 12 for a single population proportion.

15. Do Exercise 12 for the difference between two population proportions.

16. Do Exercise 12 for a population variance.

17. Do Exercise 12 for the ratio of two population variances.

18. Ochsenkühn et al. (A-33) studied birth as a result of in vitro fertilization (IVF) and birth from spontaneous conception. In the sample, there were 163 singleton births resulting from IVF with a mean birth weight of 3071 g and sample standard deviation of 761 g. Among the 321 singleton births resulting from spontaneous conception, the mean birth weight was 3172 g with a standard deviation of 702 g. Determine if these data provide sufficient evidence for us to conclude that the mean birth weight in grams of singleton births resulting from IVF is lower, in general, than the mean birth weight of singleton births resulting from spontaneous conception. Let $\alpha = .10$.

19. William Tindall (A-34) performed a retrospective study of the records of patients receiving care for hypercholesterolemia. The following table gives measurements of total cholesterol for patients before and 6 weeks after taking a statin drug. Is there sufficient evidence at the $\alpha = .01$ level of significance for us to conclude that the drug would result in reduction in total cholesterol in a population of similar hypercholesterolemia patients?

Id. No.	Before	After	Id. No.	Before	After	Id. No.	Before	After
1	195	125	37	221	191	73	205	151
2	208	164	38	245	164	74	298	163
3	254	152	39	250	162	75	305	171
4	226	144	40	266	180	76	262	129
5	290	212	41	240	161	77	320	191
6	239	171	42	218	168	78	271	167
7	216	164	43	278	200	79	195	158
8	286	200	44	185	139	80	345	192
9	243	190	45	280	207	81	223	117
10	217	130	46	278	200	82	220	114
11	245	170	47	223	134	83	279	181
12	257	182	48	205	133	84	252	167
13	199	153	49	285	161	85	246	158
14	277	204	50	314	203	86	304	190
15	249	174	51	235	152	87	292	177
16	197	160	52	248	198	88	276	148
17	279	205	53	291	193	89	250	169
18	226	159	54	231	158	90	236	185
19	262	170	55	208	148	91	256	172
20	231	180	56	263	203	92	269	188
21	234	161	57	205	156	93	235	172
22	170	139	58	230	161	94	184	151
23	242	159	59	250	150	95	253	156
24	186	114	60	209	181	96	352	219
25	223	134	61	269	186	97	266	186
26	220	166	62	261	164	98	321	206
27	277	170	63	255	164	99	233	173
28	235	136	64	275	195	100	224	109
29	216	134	65	239	169	101	274	109
30	197	138	66	298	177	102	222	136
31	253	181	67	265	217	103	194	131
32	209	147	68	220	191	104	293	228
33	245	164	69	196	129	105	262	211
34	217	159	70	177	142	106	306	192
35	187	139	71	211	138	107	239	174
36	265	171	72	244	166			

Source: William Tindall, Ph.D. and the Wright State University Consulting Center. Used with permission.

20. The objective of a study by van Vollenhoven et al. (A-35) was to examine the effectiveness of Etanercept alone and Etanercept in combination with methotrexate in the treatment of rheumatoid arthritis. They performed a retrospective study using data from the STURE database, which collects efficacy and safety data for all patients starting biological treatments at the major hospitals in Stockholm, Sweden. The researchers identified 40 subjects who were prescribed Etanercept only and 57 who were given Etanercept with methotrexate. One of the outcome measures was the number of swollen joints. The following table gives the mean number of swollen joints in the two groups as well as the standard error of the mean. Is there sufficient evidence at the $\alpha = .05$ level of significance for us to conclude that there is a difference in mean swollen joint counts in the relevant populations?

Treatment	Mean	Standard Error of Mean
Etanercept	5.56	0.84
Etanercept plus methotrexate	4.40	0.57

21. Miyazaki et al. (A-36) examined the recurrence-free rates of stripping with varicectomy and stripping with sclerotherapy for the treatment of primary varicose veins. The varicectomy group consisted of 122 limbs for which the procedure was done, and the sclerotherapy group consisted of 98 limbs for which that procedure was done. After 3 years, 115 limbs of the varicectomy group and 87 limbs of the sclerotherapy group were recurrence-free. Is this sufficient evidence for us to conclude there is no difference, in general, in the recurrence-free rate between the two procedures for treating varicose veins? Let $\alpha = .05$.

22. Recall the study, reported in Exercise 7.8.1, in which Dora et al. (A-37) investigated spinal canal dimensions in 30 subjects symptomatic with disc herniation selected for a discectomy and 45 asymptomatic individuals (control group). One of the areas of interest was determining if there is a difference between the two groups in the spinal canal cross-sectional area (cm^2) between vertebrae L5/S1. The data in the following table are simulated to be consistent with the results reported in the paper. Do these simulated data provide evidence for us to conclude that a difference in the spinal canal cross-sectional area exists between a population of subjects with disc herniations and a population of those who do not have disc herniations? Let $\alpha = .05$.

Herniated Disc Group					Control Group					
2.62	2.57	1.98	3.21	3.59	3.72	4.30	2.87	3.87	2.73	5.28
1.60	1.80	3.91	2.56	1.53	1.33	2.36	3.67	1.64	3.54	3.63
2.39	2.67	3.53	2.26	2.82	4.26	3.08	3.32	4.00	2.76	3.58
2.05	1.19	3.01	2.39	3.61	3.11	3.94	4.39	3.73	2.22	2.73
2.09	3.79	2.45	2.55	2.10	5.02	3.62	3.02	3.15	3.57	2.37
2.28	2.33	2.81	3.70	2.61	5.42	3.35	2.62	3.72	4.37	5.28
					4.97	2.58	2.25	3.12	3.43	
					3.95	2.98	4.11	3.08	2.22	

Source: Simulated data.

23. Iannelo et al. (A-38) investigated differences between triglyceride levels in healthy obese (control) subjects and obese subjects with chronic active B or C hepatitis. Triglyceride levels of 208 obese controls had a mean value of 1.81 with a standard error of the mean of .07 mmol/L. The 19 obese hepatitis subjects had a mean of .71 with a standard error of the mean of .05. Is this sufficient evidence for us to conclude that, in general, a difference exists in average triglyceride levels between obese healthy subjects and obese subjects with hepatitis B or C? Let $\alpha = .01$.

24. Kindergarten students were the participants in a study conducted by Susan Bazyk et al. (A-39). The researchers studied the fine motor skills of 37 children receiving occupational therapy. They used an index of fine motor skills that measured hand use, eye–hand coordination, and manual dexterity before and after 7 months of occupational therapy. Higher values indicate stronger fine motor skills. The scores appear in the following table.

Subject	Pre	Post	Subject	Pre	Post
1	91	94	20	76	112
2	61	94	21	79	91
3	85	103	22	97	100
4	88	112	23	109	112
5	94	91	24	70	70
6	112	112	25	58	76
7	109	112	26	97	97
8	79	97	27	112	112
9	109	100	28	97	112
10	115	106	29	112	106
11	46	46	30	85	112
12	45	41	31	112	112
13	106	112	32	103	106
14	112	112	33	100	100
15	91	94	34	88	88
16	115	112	35	109	112
17	59	94	36	85	112
18	85	109	37	88	97
19	112	112			

Source: Susan Bazyk, M.H.S. Used with permission.

Can one conclude on the basis of these data that after 7 months, the fine motor skills in a population of similar subjects would be stronger? Let $\alpha = .05$. Determine the p value.

25. A survey of 90 recently delivered women on the rolls of a county welfare department revealed that 27 had a history of intrapartum or postpartum infection. Test the null hypothesis that the population proportion with a history of intrapartum or postpartum infection is less than or equal to .25. Let $\alpha = .05$. Determine the p value.

26. In a sample of 150 hospital emergency admissions with a certain diagnosis, 128 listed vomiting as a presenting symptom. Do these data provide sufficient evidence to indicate, at the .01 level of significance, that the population proportion is less than .92? Determine the p value.

27. A research team measured tidal volume in 15 experimental animals. The mean and standard deviation were 45 and 5 cc, respectively. Do these data provide sufficient evidence to indicate that the population mean is greater than 40 cc? Let $\alpha = .05$.

28. A sample of eight patients admitted to a hospital with a diagnosis of biliary cirrhosis had a mean IgM level of 160.55 units per milliliter. The sample standard deviation was 50. Do these data provide sufficient evidence to indicate that the population mean is greater than 150? Let $\alpha = .05$. Determine the p value.

29. Some researchers have observed a greater airway resistance in smokers than in nonsmokers. Suppose a study, conducted to compare the percent of tracheobronchial retention of particles in smoking-discordant monozygotic twins, yielded the following results:

Percent Retention		Percent Retention	
Smoking Twin	**Nonsmoking Twin**	**Smoking Twin**	**Nonsmoking Twin**
60.6	47.5	57.2	54.3
12.0	13.3	62.7	13.9
56.0	33.0	28.7	8.9
75.2	55.2	66.0	46.1
12.5	21.9	25.2	29.8
29.7	27.9	40.1	36.2

Do these data support the hypothesis that tracheobronchial clearance is slower in smokers? Let $\alpha = .05$. Determine the p value for this test.

30. Circulating levels of estrone were measured in a sample of 25 postmenopausal women following estrogen treatment. The sample mean and standard deviation were 73 and 16, respectively. At the .05 significance level can one conclude on the basis of these data that the population mean is higher than 70?

31. Systemic vascular resistance determinations were made on a sample of 16 patients with chronic, congestive heart failure while receiving a particular treatment. The sample mean and standard deviation were 1600 and 700, respectively. At the .05 level of significance do these data provide sufficient evidence to indicate that the population mean is less than 2000?

32. The mean length at birth of 14 male infants was 53 cm with a standard deviation of 9 cm. Can one conclude on the basis of these data that the population mean is not 50 cm? Let the probability of committing a type I error be .10.

For each of the studies described in Exercises 33 through 38, answer as many of the following questions as possible: (a) What is the variable of interest? (b) Is the parameter of interest a mean, the difference between two means (independent samples), a mean difference (paired data), a proportion, or the difference between two proportions (independent samples)? (c) What is the sampled population? (d) What is the target population? (e) What are the null and alternative hypotheses? (f) Is the alternative one-sided (left tail), one-sided (right tail), or two-sided? (g) What type I and type II errors are possible? (h) Do you think the null hypothesis was rejected? Explain why or why not.

33. During a one-year period, Hong et al. (A-40) studied all patients who presented to the surgical service with possible appendicitis. One hundred eighty-two patients with possible appendicitis were randomized to either clinical assessment (CA) alone or clinical evaluation and abdominal/pelvic CT. A true-positive case resulted in a laparotomy that revealed a lesion requiring operation. A true-negative case did not require an operation at one-week follow-up evaluation. At the close of the study, they found no significant difference in the hospital length of stay for the two treatment groups.

34. Recall the study reported in Exercise 7.8.2 in which Nagy et al. (A-32) studied 50 stable patients admitted for a gunshot wound that traversed the mediastinum. They found that eight of the subjects had a mediastinal injury, while 42 did not have such an injury. They performed a student's t test to determine if there was a difference in mean age (years) between the two groups. The reported p value was .59.

35. Dykstra et al. (A-41) studied 15 female patients with urinary frequency with or without incontinence. The women were treated with botulinum toxin type B (BTX-B). A t test of the pre/post-difference in frequency indicated that these 15 patients experienced an average of 5.27 fewer frequency episodes per day after treatment with BTX-B. The p value for the test was less than 0.001.

36. Recall the study reported in Exercise 6.10.2 in which Horesh et al. (A-42) investigated suicidal behavior among adolescents. In addition to impulsivity, the researchers studied hopelessness among the 33 subjects in the suicidal group and the 32 subjects in the nonsuicidal group. The means for the two groups on the Beck Hopelessness Scale were 11.6 and 5.2, respectively, and the t value for the test was 5.13.

37. Mauksch et Al. (A-43) surveyed 500 consecutive patients (ages 18 to 64 years) in a primary care clinic serving only uninsured, low-income patients. They used self-report questions about why patients were coming to the clinic, and other tools to classify subjects as either having or not having major mental illness. Compared with patients without current major mental illness, patients with a current major mental illness reported significantly ($p < .001$) more concerns, chronic illnesses, stressors, forms of maltreatment, and physical symptoms.

38. A study by Hosking et al. (A-44) was designed to compare the effects of alendronate and risedronate on bone mineral density (BMD). One of the outcome measures was the percent increase in BMD at 12 months. Alendronate produced a significantly higher percent change (4.8 percent) in BMD than risedronate (2.8 percent) with a p value $< .001$.

39. For each of the following situations, identify the type I and type II errors and the correct actions.

(a) H_0: A new treatment is not more effective than the traditional one.
- **(1)** Adopt the new treatment when the new one is more effective.
- **(2)** Continue with the traditional treatment when the new one is more effective.
- **(3)** Continue with the traditional treatment when the new one is not more effective.
- **(4)** Adopt the new treatment when the new one is not more effective.

(b) H_0: A new physical therapy procedure is satisfactory.
- **(1)** Employ a new procedure when it is unsatisfactory.
- **(2)** Do not employ a new procedure when it is unsatisfactory.
- **(3)** Do not employ a new procedure when it is satisfactory.
- **(4)** Employ a new procedure when it is satisfactory.

(c) H_0: A production run of a drug is of satisfactory quality.
- **(1)** Reject a run of satisfactory quality.
- **(2)** Accept a run of satisfactory quality.
- **(3)** Reject a run of unsatisfactory quality.
- **(4)** Accept a run of unsatisfactory quality.

For each of the studies described in Exercises 40 through 55, do the following:

(a) Perform a statistical analysis of the data (including hypothesis testing and confidence interval construction) that you think would yield useful information for the researchers.

(b) State all assumptions that are necessary to validate your analysis.

(c) Find p values for all computed test statistics.

(d) Describe the population(s) about which you think inferences based on your analysis would be applicable.

40. A study by Bell (A-45) investigated the hypothesis that alteration of the vitamin D–endocrine system in blacks results from reduction in serum 25-hydroxyvitamin D and that the alteration is reversed by oral treatment with 25-hydroxyvitamin D_3. The eight subjects (three men and five women) were studied while on no treatment (control) and after having been given 25-hydroxyvitamin D_3 for 7 days

($25\text{-}OHD_3$). The following are the urinary calcium (mg/d) determinations for the eight subjects under the two conditions.

Subject	Control	$25\text{-}OHD_3$
A	66	98
B	115	142
C	54	78
D	88	101
E	82	134
F	115	158
G	176	219
H	46	60

Source: Dr. Norman H. Bell. Used with permission.

41. Montner et al. (A-46) conducted studies to test the effects of glycerol-enhanced hyperhydration (GEH) on endurance in cycling performance. The 11 subjects, ages 22–40 years, regularly cycled at least 75 miles per week. The following are the pre-exercise urine output volumes (ml) following ingestion of glycerol and water:

Subject #	Experimental, ml (Glycerol)	Control, ml (Placebo)
1	1410	2375
2	610	1610
3	1170	1608
4	1140	1490
5	515	1475
6	580	1445
7	430	885
8	1140	1187
9	720	1445
10	275	890
11	875	1785

Source: Dr. Paul Montner. Used with permission.

42. D'Alessandro et al. (A-47) wished to know if preexisting airway hyperresponsiveness (HR) predisposes subjects to a more severe outcome following exposure to chlorine. Subjects were healthy volunteers between the ages of 18 and 50 years who were classified as with and without HR. The following are the FEV_1 and specific airway resistance (Sraw) measurements taken on the subjects before and after exposure to appropriately diluted chlorine gas:

Hyperreactive Subjects

Subject	Pre-Exposure FEV_1	Pre-Exposure Sraw	Post-Exposure FEV_1	Post-Exposure Sraw
1	3.0	5.80	1.8	21.4
2	4.1	9.56	3.7	12.5
3	3.4	7.84	3.0	14.3
4	3.3	6.41	3.0	10.9
5	3.3	9.12	3.0	17.1

Normal Subjects				
	Pre-Exposure		Post-Exposure	
Subject	FEV_1	Sraw	FEV_1	Sraw
1	4.3	5.52	4.2	8.70
2	3.9	6.43	3.7	6.94
3	3.6	5.67	3.3	10.00
4	3.6	3.77	3.5	4.54
5	5.1	5.53	4.9	7.37

Source: Dr. Paul Blanc. Used with permission.

43. Noting the paucity of information on the effect of estrogen on platelet membrane fatty acid composition, Ranganath et al. (A-48) conducted a study to examine the possibility that changes may be present in postmenopausal women and that these may be reversible with estrogen treatment. The 31 women recruited for the study had not menstruated for at least 3 months or had symptoms of the menopause. No woman was on any form of hormone replacement therapy (HRT) at the time she was recruited. The following are the platelet membrane linoleic acid values before and after a period of HRT:

Subject	Before	After	Subject	Before	After	Subject	Before	After
1	6.06	5.34	12	7.65	5.55	23	5.04	4.74
2	6.68	6.11	13	4.57	4.25	24	7.89	7.48
3	5.22	5.79	14	5.97	5.66	25	7.98	6.24
4	5.79	5.97	15	6.07	5.66	26	6.35	5.66
5	6.26	5.93	16	6.32	5.97	27	4.85	4.26
6	6.41	6.73	17	6.12	6.52	28	6.94	5.15
7	4.23	4.39	18	6.05	5.70	29	6.54	5.30
8	4.61	4.20	19	6.31	3.58	30	4.83	5.58
9	6.79	5.97	20	4.44	4.52	31	4.71	4.10
10	6.16	6.00	21	5.51	4.93			
11	6.41	5.35	22	8.48	8.80			

Source: Dr. L. Ranganath. Used with permission.

44. The purpose of a study by Goran et al. (A-49) was to examine the accuracy of some widely used body-composition techniques for children through the use of the dual-energy X-ray absorptiometry (DXA) technique. Subjects were children between the ages of 4 and 10 years. The following are fat mass measurements taken on the children by three techniques—DXA, skinfold thickness (ST), and bioelectrical resistance (BR):

DXA	ST	BR	Sex (1 = Male, 0 = Female)
3.6483	4.5525	4.2636	1
2.9174	2.8234	6.0888	0
7.5302	3.8888	5.1175	0
6.2417	5.4915	8.0412	0
10.5891	10.4554	14.1576	0
9.5756	11.1779	12.4004	0

(*Continued*)

DXA	ST	BR	Sex (1 = Male, 0 = Female)
2.4424	3.5168	3.7389	1
3.5639	5.8266	4.3359	1
1.2270	2.2467	2.7144	1
2.2632	2.4499	2.4912	1
2.4607	3.1578	1.2400	1
4.0867	5.5272	6.8943	0
4.1850	4.0018	3.0936	1
2.7739	5.1745	*	1
4.4748	3.6897	4.2761	0
4.2329	4.6807	5.2242	0
2.9496	4.4187	4.9795	0
2.9027	3.8341	4.9630	0
5.4831	4.8781	5.4468	0
3.6152	4.1334	4.1018	1
5.3343	3.6211	4.3097	0
3.2341	2.0924	2.5711	1
5.4779	5.3890	5.8418	0
4.6087	4.1792	3.9818	0
2.8191	2.1216	1.5406	1
4.1659	4.5373	5.1724	1
3.7384	2.5182	4.6520	1
4.8984	4.8076	6.5432	1
3.9136	3.0082	3.2363	1
12.1196	13.9266	16.3243	1
15.4519	15.9078	18.0300	0
20.0434	19.5560	21.7365	0
9.5300	8.5864	4.7322	1
2.7244	2.8653	2.7251	1
3.8981	5.1352	5.2420	0
4.9271	8.0535	6.0338	0
3.5753	4.6209	5.6038	1
6.7783	6.5755	6.6942	1
3.2663	4.0034	3.2876	0
1.5457	2.4742	3.6931	0
2.1423	2.1845	2.4433	1
4.1894	3.0594	3.0203	1
1.9863	2.5045	3.2229	1
3.3916	3.1226	3.3839	1
2.3143	2.7677	3.7693	1
1.9062	3.1355	12.4938	1
3.7744	4.0693	5.9229	1
2.3502	2.7872	4.3192	0
4.6797	4.4804	6.2469	0
4.7260	5.4851	7.2809	0
4.2749	4.4954	6.6952	0
2.6462	3.2102	3.8791	0

(*Continued*)

DXA	ST	BR	Sex (1 = Male, 0 = Female)
2.7043	3.0178	5.6841	0
4.6148	4.0118	5.1399	0
3.0896	3.2852	4.4280	0
5.0533	5.6011	4.3556	0
6.8461	7.4328	8.6565	1
11.0554	13.0693	11.7701	1
4.4630	4.0056	7.0398	0
2.4846	3.5805	3.6149	0
7.4703	5.5016	9.5402	0
8.5020	6.3584	9.6492	0
6.6542	6.8948	9.3396	1
4.3528	4.1296	6.9323	0
3.6312	3.8990	4.2405	1
4.5863	5.1113	4.0359	1
2.2948	2.6349	3.8080	1
3.6204	3.7307	4.1255	1
2.3042	3.5027	3.4347	1
4.3425	3.7523	4.3001	1
4.0726	3.0877	5.2256	0
1.7928	2.8417	3.8734	1
4.1428	3.6814	2.9502	1
5.5146	5.2222	6.0072	0
3.2124	2.7632	3.4809	1
5.1687	5.0174	3.7219	1
3.9615	4.5117	2.7698	1
3.6698	4.9751	1.8274	1
4.3493	7.3525	4.8862	0
2.9417	3.6390	3.4951	1
5.0380	4.9351	5.6038	0
7.9095	9.5907	8.5024	0
1.7822	3.0487	3.0028	1
3.4623	3.3281	2.8628	1
11.4204	14.9164	10.7378	1
1.2216	2.2942	2.6263	1
2.9375	3.3124	3.3728	1
4.6931	5.4706	5.1432	0
8.1227	7.7552	7.7401	0
10.0142	8.9838	11.2360	0
2.5598	2.8520	4.5943	0
3.7669	3.7342	4.7384	0
4.2059	2.6356	4.0405	0
6.7340	6.6878	8.1053	0
3.5071	3.4947	4.4126	1
2.2483	2.8100	3.6705	0
7.1891	5.4414	6.6332	0
6.4390	3.9532	5.1693	0

* Missing data.

Source: Dr. Michael I. Goran. Used with permission.

45. Hartard et al. (A-50) conducted a study to determine whether a certain training regimen can counteract bone density loss in women with postmenopausal osteopenia. The following are strength measurements for five muscle groups taken on 15 subjects before (B) and after (A) 6 months of training:

	Leg Press		Hip Flexor		Hip Extensor	
Subject	(B)	(A)	(B)	(A)	(B)	(A)
1	100	180	8	15	10	20
2	155	195	10	20	12	25
3	115	150	8	13	12	19
4	130	170	10	14	12	20
5	120	150	7	12	12	15
6	60	140	5	12	8	16
7	60	100	4	6	6	9
8	140	215	12	18	14	24
9	110	150	10	13	12	19
10	95	120	6	8	8	14
11	110	130	10	12	10	14
12	150	220	10	13	15	29
13	120	140	9	20	14	25
14	100	150	9	10	15	29
15	110	130	6	9	8	12

	Arm Abductor		Arm Adductor	
Subject	(B)	(A)	(B)	(A)
1	10	12	12	19
2	7	20	10	20
3	8	14	8	14
4	8	15	6	16
5	8	13	9	13
6	5	13	6	13
7	4	8	4	8
8	12	15	14	19
9	10	14	8	14
10	6	9	6	10
11	8	11	8	12
12	8	14	13	15
13	8	19	11	18
14	4	7	10	22
15	4	8	8	12

Source: Dr. Manfred Hartard. Used with permission.

46. Vitacca et al. (A-51) conducted a study to determine whether the supine position or sitting position worsens static, forced expiratory flows and measurements of lung mechanics. Subjects were aged persons living in a nursing home who were clinically stable and without clinical evidence of cardiorespiratory diseases. Among the data collected were the following FEV_1 percent values for subjects in sitting and supine postures:

Sitting	Supine	Sitting	Supine
64	56	103	94
44	37	109	92
44	39	−99	−99
40	43	169	165
32	32	73	66
70	61	95	94
82	58	−99	−99
74	48	73	58
91	63		

Source: Dr. M. Vitacca. Used with permission.

47. The purpose of an investigation by Young et al. (A-52) was to examine the efficacy and safety of a particular suburethral sling. Subjects were women experiencing stress incontinence who also met other criteria. Among the data collected were the following pre- and postoperative cystometric capacity (ml) values:

Pre	Post	Pre	Post	Pre	Post	Pre	Post
350	321	340	320	595	557	475	344
700	483	310	336	315	221	427	277
356	336	361	333	363	291	405	514
362	447	339	280	305	310	312	402
361	214	527	492	200	220	385	282
304	285	245	330	270	315	274	317
675	480	313	310	300	230	340	323
367	330	241	230	792	575	524	383
387	325	313	298	275	140	301	279
535	325	323	349	307	192	411	383
328	250	438	345	312	217	250	285
557	410	497	300	375	462	600	618
569	603	302	335	440	414	393	355
260	178	471	630	300	250	232	252
320	362	540	400	379	335	332	331
405	235	275	278	682	339	451	400
351	310	557	381				

Source: Dr. Stephen B. Young. Used with permission.

48. Diamond et al. (A-53) wished to know if cognitive screening should be used to help select appropriate candidates for comprehensive inpatient rehabilitation. They studied a sample of geriatric rehabilitation patients using standardized measurement strategies. Among the data collected were the following admission and discharge scores made by the subjects on the Mini Mental State Examination (MMSE):

Admission	Discharge	Admission	Discharge
9	10	24	26
11	11	24	30
14	19	24	28
15	15	25	26
16	17	25	22
16	15	26	26
16	17	26	28
16	17	26	26
17	14	27	28
17	18	27	28
17	21	27	27
18	21	27	27
18	21	27	27
19	21	28	28
19	25	28	29
19	21	28	29
19	22	28	29
19	19	29	28
20	22	29	28
21	23	29	30
22	22	29	30
22	19	29	30
22	26	29	30
23	21	29	30
24	21	30	30
24	20		

Source: Dr. Stephen N. Macciocchi. Used with permission.

49. In a study to explore the possibility of hormonal alteration in asthma, Weinstein et al. (A-54) collected data on 22 postmenopausal women with asthma and 22 age-matched, postmenopausal, women without asthma. The following are the dehydroepiandrosterone sulfate (DHEAS) values collected by the investigators:

Without Asthma	With Asthma	Without Asthma	With Asthma
20.59	87.50	15.90	166.02
37.81	111.52	49.77	129.01
76.95	143.75	25.86	31.02
77.54	25.16	55.27	47.66
19.30	68.16	33.83	171.88
35.00	136.13	56.45	241.88
146.09	89.26	19.91	235.16
166.02	96.88	24.92	25.16
96.58	144.34	76.37	78.71
24.57	97.46	6.64	111.52
53.52	82.81	115.04	54.69

Source: Dr. Robert E. Weinstein. Used with permission.

50. The motivation for a study by Gruber et al. (A-55) was a desire to find a potentially useful serum marker in rheumatoid arthritis (RA) that reflects underlying pathogenic mechanisms. They measured, among other variables, the circulating levels of gelatinase B in the serum and synovial fluid (SF) of patients with RA and of control subjects. The results were as follows:

Serum		Synovial Fluid		Serum		Synovial Fluid	
RA	Control	RA	Control	RA	Control	RA	Control
26.8	23.4	71.8	3.0	36.7			
19.1	30.5	29.4	4.0	57.2			
249.6	10.3	185.0	3.9	71.3			
53.6	8.0	114.0	6.9	25.2			
66.1	7.3	69.6	9.6	46.7			
52.6	10.1	52.3	22.1	30.9			
14.5	17.3	113.1	13.4	27.5			
22.7	24.4	104.7	13.3	17.2			
43.5	19.7	60.7		10.3			
25.4	8.4	116.8		7.5			
29.8	20.4	84.9		31.6			
27.6	16.3	215.4		30.0			
106.1	16.5	33.6		42.0			
76.5	22.2	158.3		20.3			

Source: Dr. Darius Sorbi. Used with permission.

51. Benini et al. (A-56) conducted a study to evaluate the severity of esophageal acidification in achalasia following successful dilatation of the cardias and to determine which factors are associated with pathological esophageal acidification in such patients. Twenty-two subjects, of whom seven were males; ranged in ages from 28 to 78 years. On the basis of established criteria they were classified as refluxers or nonrefluxers. The following are the acid clearance values (min/reflux) for the 22 subjects:

Refluxers	Nonrefluxers
8.9	2.3
30.0	0.2
23.0	0.9
6.2	8.3
11.5	0.0
	0.9
	0.4
	2.0
	0.7
	3.6
	0.5
	1.4
	0.2
	0.7
	17.9
	2.1
	0.0

Source: Dr. Luigi Benini. Used with permission.

52. The objective of a study by Baker et al. (A-57) was to determine whether medical deformation alters in vitro effects of plasma from patients with preeclampsia on endothelial cell function to produce a paradigm similar to the in vivo disease state. Subjects were 24 nulliparous pregnant women before delivery, of whom 12 had preeclampsia and 12 were normal pregnant patients. Among the data collected were the following gestational ages (weeks) at delivery:

Preeclampsia	Normal Pregnant
38	40
32	41
42	38
30	40
38	40
35	39
32	39
38	41
39	41
29	40
29	40
32	40

Source: Dr. James M. Roberts. Used with permission.

53. Zisselman et al. (A-58) conducted a study to assess benzodiazepine use and the treatment of depression before admission to an inpatient geriatric psychiatry unit in a sample of elderly patients. Among the data collected were the following behavior disorder scores on 27 patients treated with benzodiazepines (W) and 28 who were not (WO).

W		WO	
.00	1.00	.00	.00
.00	1.00	.00	10.00
.00	.00	.00	.00
.00	.00	.00	18.00
.00	10.00	.00	.00
.00	2.00	.00	2.00
.00		.00	5.00
.00		.00	
.00		4.00	
.00		1.00	
4.00		2.00	
3.00		.00	
2.00		6.00	
.00		.00	
10.00		.00	
2.00		1.00	
.00		2.00	
9.00		1.00	
.00		22.00	
1.00		.00	
16.00		.00	

Source: Dr. Yochi Shmuely. Used with permission.

54. The objective of a study by Reinecke et al. (A-59) was to investigate the functional activity and expression of the sarcolemmal Na^+/Ca^{2+} exchange in the failing human heart. The researchers obtained left ventricular samples from failing human hearts of 11 male patients (mean age 51 years) undergoing cardiac transplantation. Nonfailing control hearts were obtained from organ donors (four females, two males, mean age 41 years) whose hearts could not be transplanted for noncardiac reasons. The following are the Na^+/Ca^{2+} exchanger activity measurements for the patients with end-stage heart failure (CHF) and nonfailing controls (NF).

NF	CHF
0.075	0.221
0.073	0.231
0.167	0.145
0.085	0.112
0.110	0.170
0.083	0.207
	0.112
	0.291
	0.164
	0.195
	0.185

Source: Dr. Hans Reinecke. Used with permission.

55. Reichman et al. (A-60) conducted a study with the purpose of demonstrating that negative symptoms are prominent in patients with Alzheimer's disease and are distinct from depression. The following are scores made on the Scale for the Assessment of Negative Symptoms in Alzheimer's Disease by patients with Alzheimer's disease (PT) and normal elderly, cognitively intact, comparison subjects (C).

PT	C
19	6
5	5
36	10
22	1
1	1
18	0
24	5
17	5
7	4
19	6
5	6
2	7
14	5
9	3
34	5
13	12

(Continued)

PT	C
0	0
21	5
30	1
43	2
19	3
31	19
21	3
41	5
	24
	3

Source: Dr. Andrew C. Coyne. Used with permission.

Exercises for Use with Large Data Sets Available on the Following Website: www.wiley.com/college/daniel

1. Refer to the creatine phosphokinase data on 1005 subjects (PCKDATA). Researchers would like to know if psychologically stressful situations cause an increase in serum creatine phosphokinase (CPK) levels among apparently healthy individuals. To help the researchers reach a decision, select a simple random sample from this population, perform an appropriate analysis of the sample data, and give a narrative report of your findings and conclusions. Compare your results with those of your classmates.

2. Refer to the prothrombin time data on 1000 infants (PROTHROM). Select a simple random sample of size 16 from each of these populations and conduct an appropriate hypothesis test to determine whether one should conclude that the two populations differ with respect to mean prothrombin time. Let $\alpha = .05$. Compare your results with those of your classmates. What assumptions are necessary for the validity of the test?

3. Refer to the head circumference data of 1000 matched subjects (HEADCIRC). Select a simple random sample of size 20 from the population and perform an appropriate hypothesis test to determine if one can conclude that subjects with the sex chromosome abnormality tend to have smaller heads than normal subjects. Let $\alpha = .05$. Construct a 95 percent confidence interval for the population mean difference. What assumptions are necessary? Compare your results with those of your classmates.

4. Refer to the hemoglobin data on 500 children with iron deficiency anemia and 500 apparently healthy children (HEMOGLOB). Select a simple random sample of size 16 from population A and an independent simple random sample of size 16 from population B. Does your sample data provide sufficient evidence to indicate that the two populations differ with respect to mean Hb value? Let $\alpha = .05$. What assumptions are necessary for your procedure to be valid? Compare your results with those of your classmates.

5. Refer to the manual dexterity scores of 500 children with learning disabilities and 500 children with no known learning disabilities (MANDEXT). Select a simple random sample of size 10 from population A and an independent simple random sample of size 15 from population B. Do your samples provide sufficient evidence for you to conclude that learning-disabled children, on the average, have lower manual dexterity scores than children without a learning disability? Let $\alpha = .05$. What assumptions are necessary in order for your procedure to be valid? Compare your results with those of your classmates.

REFERENCES

Methodology References

1. M. B. Brown and A. B. Forsythe, "Robust Tests for the Equality of Variances," *Journal of the American Statistical Association, 69* (1974), 364–367.
2. H. Levene, "Robust Tests for Equality of Variances," in I. Olkin, ed., *Contributions to Probability and Statistics*, Stanford University Press, Palo Alto, CA, 1960, 278–292.

Applications References

A-1. Norimasa Nakamura, Shuji Horibe, Yukyoshi Toritsuka, Tomoki Mitsuoka, Hideki Yoshikawa, and Konsei Shino, "Acute Grade III Medial Collateral Ligament Injury of the Knee Associated with Anterior Cruciate Ligament Tear,"*American Journal of Sports Medicine, 31* (2003), 261–267.

A-2. Diane Klingler, Robbya Green-Weir, David Nerenz, Suzanne Havstad, Howard S. Rosman, Leonard Cetner, Samir Shah, Frances Wimbush, and Steven Borzak, "Perceptions of Chest Pain Differ by Race," *American Heart Journal, 144* (2002), 51–59.

A-3. A. Escobar, J. M. Quintana, A. Bilbao, J. Azkárate, and J. I. Güenaga, "Validation of the Spanish Version of the WOMAC Questionnaire for Patients with Hip or Knee Osteoarthritis," *Clinical Rheumatology, 21* (2002), 466–471.

A-4. Phamornsak Thienprasiddhi, Vivienne C. Greenstein, Candice S. Chen, Jeffrey M. Liebmann, Robert Ritch, and Donald C. Hood, "Multifocal Visual Evoked Potential Responses in Glaucoma Patients with Unilateral Hemifield Defects," *American Journal of Opthalmology, 136* (2003), 34–40.

A-5. P. F. Lugliè, Guglielmo Campus, C. Deiola, M. G. Mela, and D. Gallisai, "Oral Condition, Chemistry of Saliva, and Salivary Levels of *Streptococcus Mutans* in Thalassemic Patients," *Clinical Oral Investigations, 6* (2002), 223–226.

A-6. Eric W. Tam, Arthur F. Mak, Wai Nga Lam, John H. Evans, and York Y. Chow, "Pelvic Movement and Interface Pressure Distribution During Manual Wheelchair Propulsion," *Archives of Physical Medicine and Rehabilitation, 84* (2003), 1466–1472.

A-7. John Dernellis and Maria Panaretou, "Effects of Thyroid Replacement Therapy on Arterial Blood Pressure in Patients with Hypertension and Hypothyroidism," *American Heart Journal, 143* (2002), 718–724.

A-8. S. Sairam, B. A. Baethge, and T. McNearney, "Analysis of Risk Factors and Comorbid Diseases in the Development of Thrombosis in Patients with Anticardiolipin Antibodies," *Clinical Rheumatology, 22* (2003), 24–29.

A-9. Michel Dabonneville, Paul Berthon, Philippe Vaslin, and Nicole Fellmann, "The 5 M in Running Field Test: Test and Retest Reliability on Trained Men and Women," *European Journal of Applied Physiology, 88* (2003), 353–360.

A-10. B. M. Ingle and R. Eastell, "Site-Specific Bone Measurements in Patients with Ankle Fracture," *Osteoporosis International, 13* (2002), 342–347.

A-11. A. Hoekema, B. Hovinga, B. Stegenga, and L. G. M. De Bont, "Craniofacial Morphology and Obstructive Steep Apnoea: A Cephalometric Analysis," *Journal of Oral Rehabilitation, 30* (2003), 690–696.

A-12. Gian Paolo Rossi, Stefano Taddei, Agostino Virdis, Martina Cavallin, Lorenzo Ghiadoni, Stefania Favilla, Daniele Versari, Isabella Sudano, Achille C. Pessina, and Antonio Salvetti, "The $T^{786}C$ and Glu298Asp Polymorphisms of the Endothelial Nitric Oxide Gene Affect the Forearm Blood Flow Responses of Caucasian Hypertensive Patients," *Journal of the American College of Cardiology, 41* (2003), 938–945.

A-13. José Garção and José Cabrita, "Evaluation of a Pharmaceutical Care Program for Hypertensive Patients in Rural Portugal," *Journal of the American Pharmaceutical Association, 42* (2002), 858–864.

A-14. John M. Morton, Steven P. Bowers, Tananchai A. Lucktong, Samer Mattar, W. Alan Bradshaw, Kevin E. Behrns, Mark J. Koruda, Charles A. Herbst, William McCartney, Raghuveer K. Halkar, C. Daniel Smith, and Timothy M. Farrell, "Gallbladder Function Before and After Fundoplication," *Journal of Gastrointestinal Surgery, 6* (2002), 806–811.

A-15. Ellen Davis Jones, "Reminiscence Therapy for Older Women with Depression: Effects of Nursing Intervention Classification in Assisted-Living Long-Term Care," *Journal of Gerontological Nursing, 29* (2003), 26–33.

A-16. Johnny Beney, E. Beth Devine, Valby Chow, Robert J. Ignoffo, Lisa Mitsunaga, Mina Shahkarami, Alex McMillan, and Lisa A. Bero. "Effect of Telephone Follow-Up on the Physical Well-Being Dimension of Quality of Life in Patients with Cancer," *Pharmacotherapy, 22* (2002), 1301–1311.

A-17. John S. Morley, John Bridson, Tim P. Nash, John B. Miles, Sarah White, and Matthew K. Makin, "Low-Dose Methadone Has an Analgesic Effect in Neuropathic Pain: A Double-Blind Randomized Controlled Crossover Trial," *Palliative Medicine, 17* (2003), 576–587.

A-18. W. K. Woo and K. E. McKenna, "Combination TL01 Ultraviolet B Phototherapy and Topical Calcipotriol for Psoriasis: A Prospective Randomized Placebo-Controlled Clinical Trial," *British Journal of Dermatology, 149* (2003), 146–150.

A-19. Simona Porcellini, Giuliana Vallanti, Silvia Nozza, Guido Poli, Adraino Lazzarin, Guiseppe Tabussi, and Antonio Grassia, "Improved Thymopoietic Potential in Aviremic HIV-Infected Individuals with HAART by Intermittent IL-2 Administration," *AIDS, 17* (2003), 1621–1630.

A-20. Lynne E. Wagenknecht, Carl D. Langefeld, Ann L. Scherzinger, Jill M. Norris, Steven M. Haffner, Mohammed F. Saad, and Richard N. Bergman, "Insulin Sensitivity, Insulin Secretion, and Abdominal Fat," *Diabetes, 52* (2003), 2490–2496.

A-21. Yves Jacquemyn, Fatima Ahankour, and Guy Martens, "Flemish Obstetricians' Personal Preference Regarding Mode of Delivery and Attitude Towards Caesarean Section on Demand," *European Journal of Obstetrics and Gynecology and Reproductive Biology, 111* (2003), 164–166.

A-22. Li Hui and A. Colin Bell, "Overweight and Obesity in Children from Shenzhen, People's Republic of China," *Health and Place, 9* (2003), 371–376.

A-23. Anne E. Becker, Rebecca A. Burwell, Kesaia Navara, and Stephen E. Gilman, "Binge Eating and Binge Eating Disorder in a Small-Scale, Indigenous Society: The View from Fiji," *International Journal of Eating Disorders, 34* (2003), 423–431.

A-24. Jacqueline A. Noonan, Renske Raaijmakers, and Bryan D. Hall, "Adult Height in Noonan Syndrome," *American Journal of Medical Genetics, 123A* (2003), 68–71.

A-25. Sai Yin Ho, Tai Hing Lam, Richard Fielding, and Edward Denis Janus, "Smoking and Perceived Health in Hong Kong Chinese," *Social Science and Medicine, 57* (2003), 1761–1770.

A-26. Markus A. Landolt, Margarete Vollrath, Karin Ribi, Hanspeter E. Gnehm, and Felix H. Sennhauser, "Incidence and Associations of Parental and Child Posttraumatic Stress Symptoms in Pediatric Patients," *Journal of Child Psychology and Psychiatry, 44* (2003), 1199–1207.

A-27. Morell Michael Avram, Daniel Blaustein, Paul A. Fein, Naveen Goel, Jyotiprakas Chattopadyay, and Neal Mittman, "Hemoglobin Predicts Long-Term Survival in Dialysis Patients: A 15-Year Single-Center Longitudinal Study and a Correlation Trend Between Prealbumin and Hemoglobin," *Kidney International, 64* (2003), 6–11.

A-28. Judy P. Wilkins, Oscar E. Suman, Deb A. Benjamin, and David N. Hemdon, "Comparison of Self-Reported and Monitored Compliance of Daily Injection of Human Growth Hormone in Burned Children," *Burns, 29* (2003), 697–701.

A-29. B. Robinson, C. T. Heise, J. W. Moore, J. Anella, M. Sokoloski, and E. Eshaghpour, "Afterload Reduction Therapy in Patients Following Intraatrial Baffle Operation for Transposition of the Great Arteries," *Pediatric Cardiology, 23* (2003), 618–623.

A-30. Peter Borden, John Nyland, David N. M. Caborn, and David Pienowski, "Biomechanical Comparison of the FasT-FIX Meniscal Repair Suture System with Vertical Mattress and Meniscus Arrows," *American Journal of Sports Medicine, 31* (2003), 374–378.

A-31. Claudio Dora, Beat Wälchli, Achim Elfering, Imre Gal, Dominik Weishaupt, and Norbert Boos, "The Significance of Spinal Canal Dimensions in Discriminating Symptomatic from Asymptomatic Disc Herniations," *European Spine Journal, 11* (2002), 575–581.

A-32. Kimberly K. Nagy, Roxanne R. Roberts, Robert F. Smith, Kimberly T. Joseph, Gary, C. An, Faran Bokhari, and John Barrett, "Trans-Mediastinal Gunshot Wounds: Are 'Stable' Patients Really Stable?" *World Journal of Surgery, 26* (2002), 1247–1250.

A-33. Robert Ochsenkühn, Thomas Strowitzki, Monika Gurtner, Alexander Strauss, Andreas Schulze, Hermann Hepp, and Peter Hillemanns, "Pregnancy Complications, Obstetric Risks, and Neonatal Outcome in Singleton and Twin Pregnancies after GIFT and IVF," *Archives of Gynecology and Obstetrics, 268* (2003), 256–261.

A-34. William Tindall and the Wright State University Consulting Center (2003).

A-35. Ronald F. van Vollenhoven, Sofia Ernestam, Anders Harju, Johan Bratt, and Lars Klareskog, "Etanercept Versus Etanercept Plus Methotrexate: A Registry-Based Study Suggesting that the Combination Is Clinically More Effacious," *Arthritis Research & Therapy, 5* (2003), R347–R351.

A-36. Kicko Miyazaki, Toshiya Nishibe, Fumihiro Sata, Takahiro Imai, Favio A. Kudo, Jorge Flores, Yasumasa J. Miyazaki, and Keishu Yasuda, "Stripping Operation with Sclerotherapy for Primary Varicose

Veins Due to Greater Saphenous Vein Reflux: Three Year Results," *World Journal of Surgery, 27* (2003), 551–553.

A-37. Claudio Dora, Beat Wälchli, Achim Elfering, Imre Gal, Dominik Weishaupt, and Norbert Boos, "The Significance of Spinal Canal Dimensions in Discriminating Symptomatic from Asymptomatic Disc Herniations," *European Spine Journal, 11* (2002), 575–581.

A-38. Silvia Iannelo, Antonina Cavaleri, Paolina Milazzo, Santi Cantarella, and Francesco Belfiore, "Low Fasting Serum Triglyceride Level as a Precocious Marker of Autoimmune Disorders," *Medscape General Medicine, 5*(3) (2003).

A-39. Susan Bazyk, Paula Michaud, Glenn Goodman, and Paula Papp, "Integrating Occupational Therapy Services in a Kindergarten Curriculum: A Look at the Outcomes." Submitted to *American Journal of Occupational Therapy,* June 30, 2004.

A-40. John J. Hong, Stephen M. Cohn, A. Peter Ekeh, Martin Newman, Moises Salama, and Suzanne D. Leblang, "A Prospective Randomized Study of Clinical Assessment Versus Computed Tomography for the Diagnosis of Acute Appendicitis," *Surgical Infections, 4* (2003), 231–239.

A-41. Dennis Dykstra, Al Enriquez, and Michael Valley, "Treatment of Overactive Bladder with Botulinum Toxin Type B: A Pilot Study," *International Urogynecological Journal, 14* (2003), 424–426.

A-42. Netta Horesh, Israel Orbach, Doron Gothelf, Meir Efrati, and Alan Apter, "Comparison of the Suicidal Behavior of Adolescent Inpatients with Borderline Personality Disorder and Major Depression," *Journal of Nervous and Mental Disease, 191* (2003), 582–588.

A-43. Larry B. Mauksch, Wayne J. Katon, Joan Russo, Suzanne M. Tucker, Edward Walker, and Janet Cameron, "The Content of a Low-Income, Uninsured Primary Care Population: Including the Patient Agenda," *Journal of the American Board of Family Practice, 16* (2003), 278–289.

A-44. David Hosking, Silvano Adami, Dieter Felsenberg, Jorge Cannata Andia, Matti Välimäki, Laurent Benbamou, Jean-Yves Reginster, Carol Yacik, Andrea Rybak-Feglin, Richard A. Petruschke, Luna Zaru, and Arthur C. Santora, "Comparison of Change in Bone Resorption and Bone Mineral Density with Once-Weekly Alendronate and Daily Risedronate: A Randomised, Placebo-Controlled Study," *Current Medical Research Opinion, 19* (2003), 383–394.

A-45. Norman H. Bell, "25-Hydroxyvitamin D3[r] Reverses Alteration of the Vitamin D–Endocrine System in Blacks," *The American Journal of Medicine, 99* (December 1995), 597–599.

A-46. P. Montner, D. M. Stark, M. L. Riedesel, G. Murata, R. Robergs, M. Timms, and T. W. Chick, "Pre-exercise Glycerol Hydration Improves Cycling Endurance Time," *International Journal of Sports Medicine, 17* (1996), 27–33.

A-47. Alessandra D'Alessandro, Ware Kushner, Hofer Wong, Homer A. Boushey, and Paul D. Blanc, "Exaggerated Responses to Chlorine Inhalation Among Persons with Nonspecific Airway Hyperreactivity," *Chest, 109* (1996), 331–337.

A-48. L. R. Ranganath, J. A. Christofides, J. W. Wright, and V. Marks, "Effect of Hormone Replacement Therapy on Platelet Membrane Fatty Acid Composition," *Journal of Endocrinology, 148* (1996), 207–212.

A-49. Michael I. Goran, Patricia Driscoll, Rachel Johnson, Tim R. Nagy, and Gary Hunter, "Cross-Calibration of Body-Composition Techniques Against Dual-Energy X-Ray Absorptiometry in Young Children," *American Journal of Clinical Nutrition, 63* (1996), 299–305.

A-50. Manfred Hartard, Paul Haber, Diana Ilieva, Elisabeth Preisinger, Georg Seidl, and Johannes Huber, "Systematic Strength Training as a Model of Therapeutic Intervention," *American Journal of Physical Medicine & Rehabilitation, 75* (1996), 21–28.

A-51. M. Vitacca, E. Clini, W. Spassini, L. Scaglia, P. Negrini, and A. Quadri, "Does the Supine Position Worsen Respiratoy Function in Elderly Subjects?" *Gerontology, 42* (1996), 46–53.

A-52. Stephen B. Young, Peter L. Rosenblatt, Diane M. Pingeton, Allison E. Howard, and Stephen P. Baker, "The Mersilene Mesh Suburethral Sling: A Clinical and Urodynamic Evaluation," *American Journal of Obstetrics and Gynecology, 173* (1995), 1719–1726.

A-53. Paul T. Diamond, Gerald Felsenthal, Stephen N. Macciocchi, Donna H. Butler, and Denise Lally-Cassady, "Effect of Cognitive Impairment on Rehabilitation Outcome," *American Journal of Physical Medicine & Rehabilitation, 75* (1996), 40–43.

A-54. Robert E. Weinstein, Catherine A. Lobocki, Sharon Gravett, Hillman Hum, Ronald Negrich, Jonathan Herbst, Dana Greenberg, and David R. Pieper, "Decreased Adrenal Sex Steroid Levels in the Absence of Glucocorticoid Suppression in Postmenopausal Asthmatic Women," *Journal of Allergy and Clinical Immunology, 97* (1996), 1–8.

A-55. Barry L. Gruber, Darius Sorbi, Deborah L. French, Mary J. Marchese, Gerard J. Nuovo, Richard R. Kew, and Leonard A. Arbeit, "Markedly Elevated Serum MMP-9 (Gelatinase B) Levels in

Rheumatoid Arthritis: A Potentially Useful Laboratory Marker," *Clinical Immunology and Immunopathology, 78* (1996), 161–171.

A-56. LUIGI BENINI, CARLO SEMBENINI, GRETA CASTELLANI, ENRICO BARDELLI, MARIA TERESA BRENTEGANI, PIERGIORGIO GIORGETTI, and ITALO VANTINI, "Pathological Esophageal Acidification and Pneumatic Dilatation in Achalasic Patients: Too Much or Not Enough?" *Digestive Diseases and Sciences, 41* (1996), 365–371.

A-57. PHILIP N. BAKER, CHRISTOPHER P. STRANKO, SANDRA T. DAVIDGE, PAUL S. DAVIES, and JAMES M. ROBERTS, "Mechanical Stress Eliminates the Effects of Plasma from Patients with Preeclampsia on Endothelial Cells," *American Journal of Obstetrics and Gynecology, 174* (1996), 703–706.

A-58. MARC H. ZISSELMAN, BARRY W. ROVNER, and YOCHI SHMUELY, "Benzodiazepine Use in the Elderly Prior to Psychiatric Hospitalization," *Psychosomatics, 37* (1996), 38–42.

A-59. HANS REINECKE, ROLAND STRUDER, ROLAND VETTER, JURGEN HOLTZ, and HELMUT DREXLER, "Cardiac Na^+/Ca^{2+} Exchange Activity in Patients with End-Stage Heart Failure," *Cardiovascular Research, 31* (1996), 48–54.

A-60. WILLIAM E. REICHMAN, ANDREW C. COYNE, SATISH AMIRNENI, BRUNO MOLINO, and SANDRA EGAN, "Negative Symptoms in Alzheimer's Disease," *American Journal of Psychiatry, 153* (1996), 424–426.

CHAPTER 8

STATISTICAL INFERENCE AND THE ANALYSIS OF DATA VARIABILITY

CHAPTER OVERVIEW

The topic of this chapter, analysis of variance, provides a methodology for partitioning the total variance computed from a data set into components, each of which represents the amount of the total variance that can be attributed to a specific source of variation. The results of this partitioning can then be used to estimate and test hypotheses about population variances and means. In this chapter we focus our attention on hypothesis testing of means. Specifically, we discuss the testing of differences among means when there is interest in more than two populations or two or more variables. The techniques discussed in this chapter are widely used in the health sciences.

TOPICS

LEARNING OUTCOMES

After studying this chapter, the student will

1. understand how the total variation in a data set can be partitioned into different components.
2. be able to compare the means of more than two samples simultaneously.
3. understand multiple comparison tests and when their use is appropriate.
4. understand commonly used experimental designs.

8.1 INTRODUCTION

In the preceding chapters the basic concepts of statistics have been examined, and they provide a foundation for the present and succeeding chapters.

This chapter is concerned with *analysis of variance,* which may be defined as *a technique whereby the total variation present in a set of data is partitioned into two or more components. Associated with each of these components is a specific source of variation, so that in the analysis it is possible to ascertain the magnitude of the contributions of each of these sources to the total variation.*

The development of analysis of variance (ANOVA) is due mainly to the work of R. A. Fisher (1), whose contributions to statistics, spanning the years 1912 to 1962, have had a tremendous influence on modern statistical thought (2, 3).

Applications Analysis of variance finds its widest application in the analysis of data derived from experiments. The principles of the design of experiments are well covered in many books, including those by Hinkelmann and Kempthorne (4), Montgomery (5), and Myers and Well (6). We do not study this topic in detail, since to do it justice would require a minimum of an additional chapter. Some of the important concepts in experimental design, however, will become apparent as we discuss analysis of variance.

Analysis of variance is used for two different purposes: (1) to estimate and test hypotheses about population variances, and (2) to estimate and test hypotheses about population means. We are concerned here with the latter use. However, as we will see, our conclusions regarding the means will depend on the magnitudes of the observed variances.

As we shall see, the concepts and techniques that we cover under the heading of analysis of variance are extensions of the concepts and techniques covered in Chapter 7. In Chapter 7 we learned to test the null hypothesis that two means are equal. In this chapter we learn to test the null hypothesis that three or more means are equal. Whereas, for example, what we learned in Chapter 7 enables us to determine if we can conclude that two treatments differ in effectiveness, what we learn in this chapter enables us to determine if we can conclude that three or more treatments differ in effectiveness. The following example illustrates some basic ideas involved in the application of analysis of variance. These will be extended and elaborated on later in this chapter.

EXAMPLE 8.1.1

Suppose we wish to know if three drugs differ in their effectiveness in lowering serum cholesterol in human subjects. Some subjects receive drug A, some drug B, and some drug C. After a specified period of time measurements are taken to determine the extent to which serum cholesterol was reduced in each subject. We find that the amount by which serum cholesterol was lowered is not the same in all subjects. In other words, there is *variability* among the measurements. Why, we ask ourselves, are the measurements not all the same? Presumably, one reason they are not the same is that the subjects received different drugs. We now look at the measurements of those subjects who received drug A. We find that the amount by which serum cholesterol was lowered is not the same among these subjects. We find this to be the case when we look at the measurements for subjects who received drug B and those subjects who received drug C. We see that there is *variability* among the measurements within the treatment groups. Why, we ask ourselves again, are these measurements not the same? Among the reasons that come to mind are differences in the genetic makeup of the subjects and differences in their diets. Through an analysis of the *variability* that we have observed, we will be able to reach a conclusion regarding the equality of the effectiveness of the three drugs. To do this we employ the techniques and concepts of analysis of variance. ■

Variables In our example we allude to three kinds of variables. We find these variables to be present in all situations in which the use of analysis of variance is appropriate. First we have the *treatment variable,* which in our example was "drug." We had three "values" of this variable, drug A, drug B, and drug C. The second kind of variable we refer to is the *response variable.* In the example it is change in serum cholesterol. The response variable is the variable that we expect to exhibit different values when different "values" of the treatment variable are employed. Finally, we have the other variables that we mention—genetic composition and diet. These are called *extraneous variables.* These variables may have an effect on the response variable, but they are not the focus of our attention in the experiment. The treatment variable is the variable of primary concern, and the question to be answered is: Do the different "values" of the treatment variable result in differences, on the average, in the response variable?

Assumptions Underlying the valid use of analysis of variance as a tool of statistical inference are a set of fundamental assumptions. Although an experimenter must not expect to find all the assumptions met to perfection, it is important that the user of analysis of variance techniques be aware of the underlying assumptions and be able to recognize when they are substantially unsatisfied. Because experiments in which all the assumptions are perfectly met are rare, analysis of variance results should be considered as approximate rather than exact. These assumptions are pointed out at appropriate points in the following sections.

We discuss analysis of variance as it is used to analyze the results of two different experimental designs, the completely randomized and the randomized complete block designs. In addition to these, the concept of a factorial experiment is given through its use in a completely randomized design. These do not exhaust the possibilities. A discussion of additional designs may be found in the references (4–6).

The ANOVA Procedure In our presentation of the analysis of variance for the different designs, we follow the ten-step procedure presented in Chapter 7. The following is a restatement of the steps of the procedure, including some new concepts necessary for its adaptation to analysis of variance.

1. **Description of data.** In addition to describing the data in the usual way, we display the sample data in tabular form.
2. **Assumptions.** Along with the assumptions underlying the analysis, we present the model for each design we discuss. The model consists of a symbolic representation of a typical value from the data being analyzed.
3. **Hypotheses.**

4. **Test statistic.**
5. **Distribution of test statistic.**
6. **Decision rule.**
7. **Calculation of test statistic.** The results of the arithmetic calculations will be summarized in a table called the analysis of variance (ANOVA) table. The entries in the table make it easy to evaluate the results of the analysis.
8. **Statistical decision.**
9. **Conclusion.**
10. **Determination of *p* value.**

We discuss these steps in greater detail in Section 8.2.

The Use of Computers The calculations required by analysis of variance are lengthier and more complicated than those we have encountered in preceding chapters. For this reason the computer assumes an important role in analysis of variance. All the exercises appearing in this chapter are suitable for computer analysis and may be used with the statistical packages mentioned in Chapter 1. The output of the statistical packages may vary slightly from that presented in this chapter, but this should pose no major problem to those who use a computer to analyze the data of the exercises. The basic concepts of analysis of variance that we present here should provide the necessary background for understanding the description of the programs and their output in any of the statistical packages.

8.2 THE COMPLETELY RANDOMIZED DESIGN

We saw in Chapter 7 how it is possible to test the null hypothesis of no difference between two population means. It is not unusual for the investigator to be interested in testing the null hypothesis of no difference among several population means. The student first encountering this problem might be inclined to suggest that all possible pairs of sample means be tested separately by means of the Student t test. Suppose there are five populations involved. The number of possible pairs of sample means is ${}_5C_2 = 10$. As the amount of work involved in carrying out this many t tests is substantial, it would be worthwhile if a more efficient alternative for analysis were available. A more important consequence of performing all possible t tests, however, is that it is very likely to lead to a false conclusion.

Suppose we draw five samples from populations having equal means. As we have seen, there would be 10 tests if we were to do each of the possible tests separately. If we select a significance level of $\alpha = .05$ for each test, the probability of failing to reject a hypothesis of no difference in each case would be .95. By the multiplication rule of probability, if the tests were independent of one another, the probability of failing to reject a hypothesis of no difference in all 10 cases would be $(.95)^{10} = .5987$. The probability of rejecting at least one hypothesis of no difference, then, would be $1 - .5987 = .4013$. Since we know that the null hypothesis is true in every case in this illustrative example, rejecting the null hypothesis constitutes the committing of a type I error. In the long run, then, in testing all possible pairs of means from five samples, we would commit a type I error 40 percent of the time. The problem becomes even more complicated in practice, since three or more t tests based on the same data would not be independent of one another.

It becomes clear, then, that some other method for testing for a significant difference among several means is needed. Analysis of variance provides such a method.

One-Way ANOVA The simplest type of analysis of variance is that known as *one-way analysis of variance*, in which only one source of variation, or *factor*, is investigated. It is an extension to three or more samples of the t test procedure (discussed in Chapter 7) for use with two independent samples. Stated another way, we can say that the t test for use with two independent samples is a special case of one-way analysis of variance.

In a typical situation we want to use one-way analysis of variance to test the null hypothesis that three or more treatments are equally effective. The necessary experiment is designed in such a way that the treatments of interest are assigned completely at random to the subjects or objects on which the measurements to determine treatment effectiveness are to be made. For this reason the design is called the *completely randomized experimental design.*

We may randomly allocate subjects to treatments as follows. Suppose we have 16 subjects available to participate in an experiment in which we wish to compare four drugs. We number the subjects from 01 through 16. We then go to a table of random numbers and select 16 consecutive, unduplicated numbers between 01 and 16. To illustrate, let us use Appendix Table A and a random starting point that, say, is at the intersection of Row 4 and Columns 11 and 12. The two-digit number at this intersection is 98. The succeeding (moving downward) 16 consecutive two-digit numbers between 01 and 16 are 16, 09, 06, 15, 14, 11, 02, 04, 10, 07, 05, 13, 03, 12, 01, and 08. We allocate subjects 16, 09, 06, and 15 to drug A; subjects 14, 11, 02, and 04 to drug B; subjects 10, 07, 05, and 13 to drug C; and subjects 03, 12, 01, and 08 to drug D. We emphasize that the number of subjects in each treatment group does not have to be the same. Figure 8.2.1 illustrates the scheme of random allocation.

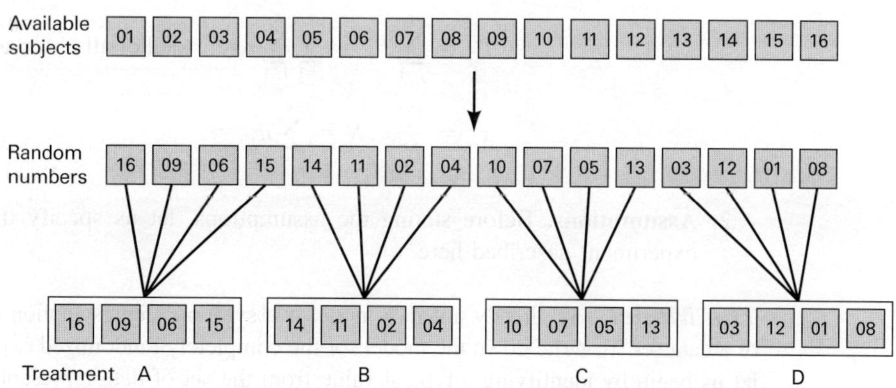

FIGURE 8.2.1 Allocation of subjects to treatments, completely randomized design.

TABLE 8.2.1 Table of Sample Values for the Completely Randomized Design

	Treatment					
	1	**2**	**3**	⋯	***k***	
	x_{11}	x_{12}	x_{13}	⋯	x_{1k}	
	x_{21}	x_{22}	x_{23}	⋯	x_{2k}	
	x_{31}	x_{32}	x_{33}	⋯	x_{3k}	
	⋮	⋮	⋮	⋮	⋮	
	$x_{n_1 1}$	$x_{n_2 2}$	$x_{n_3 3}$	⋯	$x_{n_k k}$	
Total	$T_{.1}$	$T_{.2}$	$T_{.3}$	⋯	$T_{.k}$	$T_{..}$
Mean	$\bar{x}_{.1}$	$\bar{x}_{.2}$	$\bar{x}_{.3}$	⋯	$\bar{x}_{.k}$	$\bar{x}_{..}$

Hypothesis Testing Steps Once we decide that the completely randomized design is the appropriate design, we may proceed with the hypothesis testing steps. We discuss these in detail first, and follow with an example.

1. **Description of data.** The measurements (or observations) resulting from a completely randomized experimental design, along with the means and totals that can be computed from them, may be displayed for convenience as in Table 8.2.1. The symbols used in Table 8.2.1 are defined as follows:

$$x_{ij} = \text{the } i\text{th observation resulting from the } j\text{th treatment (there are a total of } k \text{ treatments)}$$

$$i = 1, 2, \ldots, n_j, \qquad j = 1, 2, \ldots, k$$

$$T_{.j} = \sum_{i=1}^{n_j} x_{ij} = \text{total of the } j\text{th treatment}$$

$$\bar{x}_{.j} = \frac{T_{.j}}{n_j} = \text{mean of the } j\text{th treatment}$$

$$T_{..} = \sum_{j=1}^{k} T_{.j} = \sum_{j=1}^{k}\sum_{i=1}^{n_j} x_{ij} = \text{total of all observations}$$

$$\bar{x}_{..} = \frac{T_{..}}{N}, \quad N = \sum_{j=1}^{k} n_j$$

2. **Assumptions.** Before stating the assumptions, let us specify the model for the experiment described here.

The Model As already noted, a model is a symbolic representation of a typical value of a data set. To write down the model for the completely randomized experimental design, let us begin by identifying a typical value from the set of data represented by the sample displayed in Table 8.2.1. We use the symbol x_{ij} to represent this typical value.

The one-way analysis of variance model may be written as follows:

$$x_{ij} = \mu + \tau_j + \epsilon_{ij}; \qquad i = 1, 2, \ldots, n_j, \qquad j = 1, 2, \ldots, k \qquad (8.2.1)$$

The terms in this model are defined as follows:

1. μ represents the mean of all the k population means and is called the *grand mean.*
2. τ_j represents the difference between the mean of the jth population and the grand mean and is called the *treatment effect.*
3. ϵ_{ij} represents the amount by which an individual measurement differs from the mean of the population to which it belongs and is called the *error term.*

Components of the Model By looking at our model we can see that a typical observation from the total set of data under study is composed of (1) the grand mean, (2) a treatment effect, and (3) an error term representing the deviation of the observation from its group mean.

In most situations we are interested only in the k treatments represented in our experiment. Any inferences that we make apply only to these treatments. We do not wish to extend our inference to any larger collection of treatments. When we place such a restriction on our inference goals, we refer to our model as the *fixed-effects model*, or *model 1.* The discussion in this book is limited to this model.

Assumptions of the Model The assumptions for the fixed-effects model are as follows:

(a) The k sets of observed data constitute k independent random samples from the respective populations.

(b) Each of the populations from which the samples come is normally distributed with mean μ_j and variance σ_j^2.

(c) Each of the populations has the same variance. That is, $\sigma_1^2 = \sigma_2^2 = \cdots = \sigma_k^2 = \sigma^2$ the common variance.

(d) The τ_j are unknown constants and $\sum \tau_j = 0$ since the sum of all deviations of the μ_j from their mean, μ, is zero.

(e) The ϵ_{ij} have a mean of 0, since the mean of x_{ij} is μ_j.

(f) The ϵ_{ij} have a variance equal to the variance of the x_{ij}, since the ϵ_{ij} and x_{ij} differ only by a constant; that is, the error variance is equal to σ^2, the common variance specified in assumption c.

(g) The ϵ_{ij} are normally (and independently) distributed.

3. **Hypotheses.** We test the null hypothesis that all population or treatment means are equal against the alternative that the members of at least one pair are not equal. We may state the hypotheses formally as follows:

$$H_0: \mu_1 = \mu_2 = \cdots = \mu_k$$
$$H_A: \text{not all } \mu_j \text{ are equal}$$

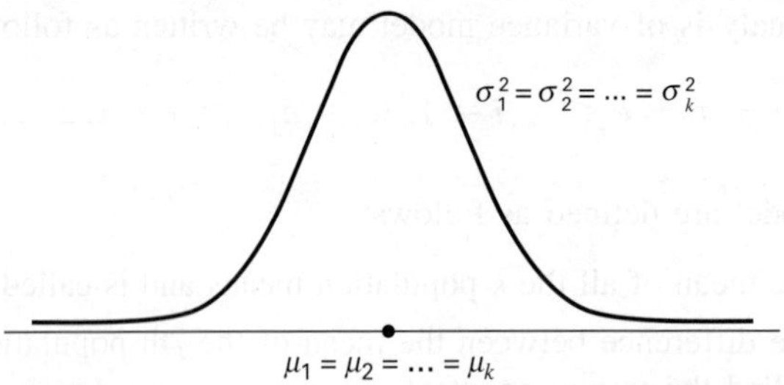

FIGURE 8.2.2 Picture of the populations represented in a completely randomized design when H_0 is true and the assumptions are met.

If the population means are equal, each treatment effect is equal to zero, so that, alternatively, the hypotheses may be stated as

$$H_0\colon \tau_j = 0, \qquad j = 1, 2, \ldots, k$$
$$H_A\colon \text{not all } \tau_j = 0$$

If H_0 is true and the assumptions of equal variances and normally distributed populations are met, a picture of the populations will look like Figure 8.2.2. When H_0 is true the population means are all equal, and the populations are centered at the same point (the common mean) on the horizontal axis. If the populations are all normally distributed with equal variances the distributions will be identical, so that in drawing their pictures each is superimposed on each of the others, and a single picture sufficiently represents them all.

When H_0 is false it may be false because one of the population means is different from the others, which are all equal. Or, perhaps, all the population means are different. These are only two of the possibilities when H_0 is false. There are many other possible combinations of equal and unequal means. Figure 8.2.3 shows a picture of the populations when the assumptions are met, but H_0 is false because no two population means are equal.

4. **Test statistic.** The test statistic for one-way analysis of variance is a computed variance ratio, which we designate by V.R. as we did in Chapter 7. The two variances

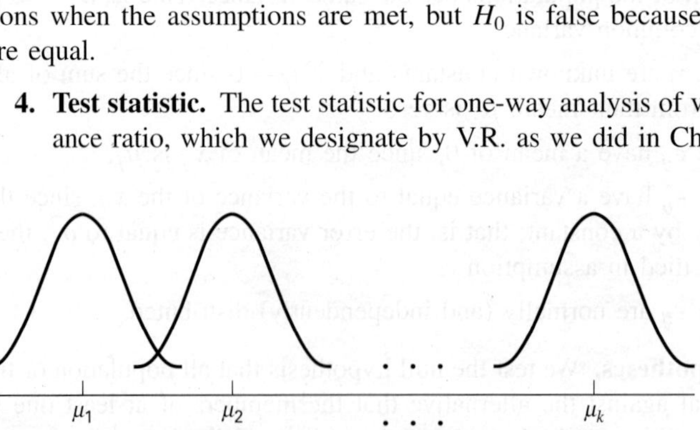

FIGURE 8.2.3 Picture of the populations represented in a completely randomized design when the assumptions of equal variances and normally distributed populations are met, but H_0 is false because none of the population means are equal.

from which V.R. is calculated are themselves computed from the sample data. The methods by which they are calculated will be given in the discussion that follows.

5. **Distribution of test statistic.** As discussed in Section 7.8, V.R. is distributed as the F distribution when H_0 is true and the assumptions are met.
6. **Decision rule.** In general, the decision rule is: reject the null hypothesis if the computed value of V.R. is equal to or greater than the critical value of F for the chosen α level.
7. **Calculation of test statistic.** We have defined analysis of variance as a process whereby the total variation present in a set of data is partitioned into components that are attributable to different sources. The term *variation* used in this context refers to the *sum of squared deviations of observations from their mean*, or *sum of squares* for short.

The initial computations performed in one-way ANOVA consist of the partitioning of the total variation present in the observed data into its basic components, each of which is attributable to an identifiable source.

Those who use a computer for calculations may wish to skip the following discussion of the computations involved in obtaining the test statistic.

The Total Sum of Squares Before we can do any partitioning, we must first obtain the total sum of squares. The total sum of squares is the sum of the squares of the deviations of individual observations from the mean of all the observations taken together. This *total sum of squares* is defined as

$$SST = \sum_{j=1}^{k}\sum_{i=1}^{n_j}(x_{ij} - \bar{x}_{..})^2 \tag{8.2.2}$$

where $\sum_{i=1}^{n_j}$ tells us to sum the squared deviations for each treatment group, and $\sum_{j=1}^{k}$ tells us to add the k group totals obtained by applying $\sum_{i=1}^{n_j}$. The reader will recognize Equation 8.2.2 as the numerator of the variance that may be computed from the complete set of observations taken together.

The Within Groups Sum of Squares Now let us show how to compute the first of the two components of the total sum of squares.

The first step in the computation calls for performing certain calculations *within* each group. These calculations involve computing within each group the sum of the squared deviations of the individual observations from their mean. When these calculations have been performed within each group, we obtain the sum of the individual group results. This component of variation is called the *within groups sum of squares* and may be designated *SSW*. This quantity is sometimes referred to as the *residual* or *error* sum of squares. The expression for these calculations is written as follows:

$$SSW = \sum_{j=1}^{k}\sum_{i=1}^{n_j}(x_{ij} - \bar{x}_{.j})^2 \tag{8.2.3}$$

The Among Groups Sum of Squares To obtain the second component of the total sum of squares, we compute for each group the squared deviation of the group mean from the grand mean and multiply the result by the size of the group. Finally, we add these results over all groups. This quantity is a measure of the variation among groups and is referred to as the *sum of squares among groups* or *SSA*. The formula for calculating this quantity is as follows:

$$SSA = \sum_{j=1}^{k} n_j(\bar{x}_{.j} - \bar{x}_{..})^2 \tag{8.2.4}$$

In summary, then, we have found that the total sum of squares is equal to the sum of the among and the within sum of squares. We express this relationship as follows:

$$SST = SSA + SSW$$

From the sums of squares that we have now learned to compute, it is possible to obtain two estimates of the common population variance, σ^2. It can be shown that when the assumptions are met and the population means are all equal, both the among sum of squares and the within sum of squares, when divided by their respective degrees of freedom, yield independent and unbiased estimates of σ^2.

The First Estimate of σ^2 Within any sample,

$$\frac{\sum_{i=1}^{n_j}(x_{ij} - \bar{x}_{.j})^2}{n_j - 1}$$

provides an unbiased estimate of the true variance of the population from which the sample came. Under the assumption that the population variances are all equal, we may pool the k estimates to obtain

$$MSW = \frac{\sum_{j=1}^{k}\sum_{i=1}^{n_j}(x_{ij} - \bar{x}_{.j})^2}{\sum_{j=1}^{k}(n_j - 1)} \tag{8.2.5}$$

This is our first estimate of σ^2 and may be called the *within groups variance,* since it is the within groups sum of squares of Equation 8.2.3 divided by the appropriate degrees of freedom. The student will recognize this as an extension to k samples of the pooling of variances procedure encountered in Chapters 6 and 7 when the variances from two samples were pooled in order to use the t distribution. The quantity in Equation 8.2.5 is customarily referred to as the within groups *mean square* rather than the within groups variance.

The within groups mean square is a valid estimate of σ^2 only if the population variances are equal. It is not necessary, however, for H_0 to be true in order for the within groups mean square to be a valid estimate of σ^2; that is, the within groups mean square estimates σ^2 regardless of whether H_0 is true or false, as long as the population variances are equal.

The Second Estimate of σ^2 The second estimate of σ^2 may be obtained from the familiar formula for the variance of sample means, $\sigma_{\bar{x}}^2 = \sigma^2/n$. If we solve this equation for σ^2, the variance of the population from which the samples were drawn, we have

$$\sigma^2 = n\sigma_{\bar{x}}^2 \tag{8.2.6}$$

An unbiased estimate of $\sigma_{\bar{x}}^2$ computed from sample data is provided by

$$\frac{\sum_{j=1}^{k}(\bar{x}_{.j} - \bar{x}_{..})^2}{k-1}$$

If we substitute this quantity into Equation 8.2.6, we obtain the desired estimate of σ^2,

$$MSA = \frac{n\sum_{j=1}^{k}(\bar{x}_{.j} - \bar{x}_{..})^2}{k-1} \tag{8.2.7}$$

The reader will recognize the numerator of Equation 8.2.7 as the among groups sum of squares for the special case when all sample sizes are equal. This sum of squares when divided by the associated degrees of freedom $k - 1$ is referred to as the *among groups mean square.*

When the sample sizes are not all equal, an estimate of σ^2 based on the variability among sample means is provided by

$$MSA = \frac{\sum_{j=1}^{k}n_j(\bar{x}_{.j} - \bar{x}_{..})^2}{k-1} \tag{8.2.8}$$

If, indeed, the null hypothesis is true we would expect these two estimates of σ^2 to be fairly close in magnitude. If the null hypothesis is false, that is, if all population means are not equal, we would expect the among groups mean square, which is computed by using the squared deviations of the sample means from the overall mean, to be larger than the within groups mean square.

In order to understand analysis of variance we must realize that the among groups mean square provides a valid estimate of σ^2 when the assumption of equal population variances is met *and when* H_0 is true. Both conditions, a true null hypothesis and equal population variances, must be met in order for the among groups mean square to be a valid estimate of σ^2.

The Variance Ratio What we need to do now is to compare these two estimates of σ^2, and we do this by computing the following variance ratio, which is the desired test statistic:

$$\text{V.R.} = \frac{\text{among groups mean square}}{\text{within groups means square}}$$

If the two estimates are about equal, V.R. will be close to 1. A ratio close to 1 tends to support the hypothesis of equal population means. If, on the other hand, the among groups mean square is considerably larger than the within groups mean square, V.R. will be considerably greater than 1. A value of V.R. sufficiently greater than 1 will cast doubt on the hypothesis of equal population means.

We know that because of the vagaries of sampling, even when the null hypothesis is true, it is unlikely that the among and within groups mean squares will be equal. We must decide, then, how big the observed difference has to be before we can conclude that the difference is due to something other than sampling fluctuation. In other words, how large a value of V.R. is required for us to be willing to conclude that the observed difference between our two estimates of σ^2 is not the result of chance alone?

The *F* Test To answer the question just posed, we must consider the sampling distribution of the ratio of two sample variances. In Chapter 6 we learned that the quantity $(s_1^2/\sigma_1^2)/(s_2^2/\sigma_2^2)$ follows a distribution known as the F distribution when the sample variances are computed from random and independently drawn samples from normal populations. The F distribution, introduced by R. A. Fisher in the early 1920s, has become one of the most widely used distributions in modern statistics. We have already become acquainted with its use in constructing confidence intervals for, and testing hypotheses about, population variances. In this chapter, we will see that it is the distribution fundamental to analysis of variance. For this reason the ratio that we designate V.R. is frequently referred to as F, and the testing procedure is frequently called the F test. It is of interest to note that the F distribution is the ratio of two Chi-square distributions.

In Chapter 7 we learned that when the population variances are the same, they cancel in the expression $(s_1^2/\sigma_1^2)/(s_2^2/\sigma_2^2)$, leaving s_1^2/s_2^2, which is itself distributed as F. The F distribution is really a family of distributions, and the particular F distribution we use in a given situation depends on the number of degrees of freedom associated with the sample variance in the numerator (*numerator degrees of freedom*) and the number of degrees of freedom associated with the sample variance in the denominator (*denominator degrees of freedom*).

Once the appropriate F distribution has been determined, the size of the observed V.R. that will cause rejection of the hypothesis of equal population variances depends on the significance level chosen. The significance level chosen determines the critical value of F, the value that separates the nonrejection region from the rejection region.

TABLE 8.2.2 Analysis of Variance Table for the Completely Randomized Design

Source of Variation	Sum of Squares	Degrees of Freedom	Mean Square	Variance Ratio
Among samples	$SSA = \sum_{j=1}^{k} n_j(\bar{x}_{.j} - \bar{x}_{..})^2$	$k - 1$	$MSA = SSA/(k-1)$	$\text{V.R.} = \frac{MSA}{MSW}$
Within samples	$SSW = \sum_{j=1}^{k}\sum_{i=1}^{n_j}(x_{ij} - \bar{x}_{.j})^2$	$N - k$	$MSW = SSW/(N-k)$	
Total	$SST = \sum_{j=1}^{k}\sum_{i=1}^{n_j}(x_{ij} - \bar{x}_{..})^2$	$N - 1$		

As we have seen, we compute V.R. in situations of this type by placing the among groups mean square in the numerator and the within groups mean square in the denominator, so that the numerator degrees of freedom is equal to $(k - 1)$, the number of groups minus 1, and the denominator degrees of freedom value is equal to

$$\sum_{j=1}^{k}(n_j - 1) = \left(\sum_{j=1}^{k} n_j\right) - k = N - k$$

The ANOVA Table The calculations that we perform may be summarized and displayed in a table such as Table 8.2.2, which is called the ANOVA table.

8. **Statistical decision.** To reach a decision we must compare our computed V.R. with the critical value of F, which we obtain by entering Appendix Table G with $k - 1$ numerator degrees of freedom and $N - k$ denominator degrees of freedom.

If the computed V.R. is equal to or greater than the critical value of F, we reject the null hypothesis. If the computed value of V.R. is smaller than the critical value of F, we do not reject the null hypothesis.

Explaining a Rejected Null Hypothesis There are two possible explanations for a rejected null hypothesis. If the null hypothesis is true, that is, if the two sample variances are estimates of a common variance, we know that the probability of getting a value of V.R. as large as or larger than the critical F is equal to our chosen level of significance. When we reject H_0 we may, if we wish, conclude that the null hypothesis is true and assume that because of chance we got a set of data that gave rise to a rare event. On the other hand, we may prefer to take the position that our large computed V.R. value does not represent a rare event brought about by chance but, instead, reflects the fact that something other than chance is operative. This other something we conclude to be a false null hypothesis.

It is this latter explanation that we usually give for computed values of V.R. that exceed the critical value of F. In other words, if the computed value of V.R. is greater than the critical value of F, we reject the null hypothesis.

It will be recalled that the original hypothesis we set out to test was

$$H_0: \mu_1 = \mu_2 = \cdots = \mu_k$$

Does rejection of the hypothesis about variances imply a rejection of the hypothesis of equal population means? The answer is yes. A large value of V.R. resulted from the fact that the among groups mean square was considerably larger than the within groups mean square. Since the among groups mean square is based on the dispersion of the sample means about their mean (called the grand mean), this quantity will be large when there is a large discrepancy among the sizes of the sample means. Because of this, then, a significant value of V.R. tells us to reject the null hypothesis that all population means are equal.

9. **Conclusion.** When we reject H_0, we conclude that not all population means are equal. When we fail to reject H_0, we conclude that the population means may all be equal.
10. **Determination of *p* value.**

EXAMPLE 8.2.1

Game meats, including those from white-tailed deer and eastern gray squirrels, are used as food by families, hunters, and other individuals for health, cultural, or personal reasons. A study by David Holben (A-1) assessed the selenium content of meat from free-roaming white-tailed deer (venison) and gray squirrel (squirrel) obtained from a low selenium region of the United States. These selenium content values were also compared to those of beef produced within and outside the same region. We want to know if the selenium levels are different in the four meat groups.

Solution:

1. **Description of data.** Selenium content of raw venison (VEN), squirrel meat (SQU), region-raised beef (RRB), and nonregion-raised beef (NRB), in μg/100 g of dry weight, are shown in Table 8.2.3. A graph of the data in the form of a *dotplot* is shown in Figure 8.2.4. Such a graph highlights

TABLE 8.2.3 Selenium Content, in μg/100 g, of Four Different Meat Types

Meat Type							
VEN		SQU		RRB		NRB	
26.72	14.86	37.42	37.57	11.23	15.82	44.33	
28.58	16.47	56.46	25.71	29.63	27.74	76.86	
29.71	25.19	51.91	23.97	20.42	22.35	4.45	
26.95	37.45	62.73	13.82	10.12	34.78	55.01	
10.97	45.08	4.55	42.21	39.91	35.09	58.21	
21.97	25.22	39.17	35.88	32.66	32.60	74.72	
14.35	22.11	38.44	10.54	38.38	37.03	11.84	

(Continued)

Meat Type						
VEN		SQU		RRB		NRB
32.21	33.01	40.92	27.97	36.21	27.00	139.09
19.19	31.20	58.93	41.89	16.39	44.20	69.01
30.92	26.50	61.88	23.94	27.44	13.09	94.61
10.42	32.77	49.54	49.81	17.29	33.03	48.35
35.49	8.70	64.35	30.71	56.20	9.69	37.65
36.84	25.90	82.49	50.00	28.94	32.45	66.36
25.03	29.80	38.54	87.50	20.11	37.38	72.48
33.59	37.63	39.53	68.99	25.35	34.91	87.09
33.74	21.69			21.77	27.99	26.34
18.02	21.49			31.62	22.36	71.24
22.27	18.11			32.63	22.68	90.38
26.10	31.50			30.31	26.52	50.86
20.89	27.36			46.16	46.01	
29.44	21.33			56.61	38.04	
				24.47	30.88	
				29.39	30.04	
				40.71	25.91	
				18.52	18.54	
				27.80	25.51	
				19.49		

Source: David H. Holben, Ph.D. Used with permission.

the main features of the data and brings into clear focus differences in selenium levels among the different meats.

2. **Assumptions.** We assume that the four sets of data constitute independent simple random samples from the four indicated populations. We assume that the four populations of measurements are normally distributed with equal variances.

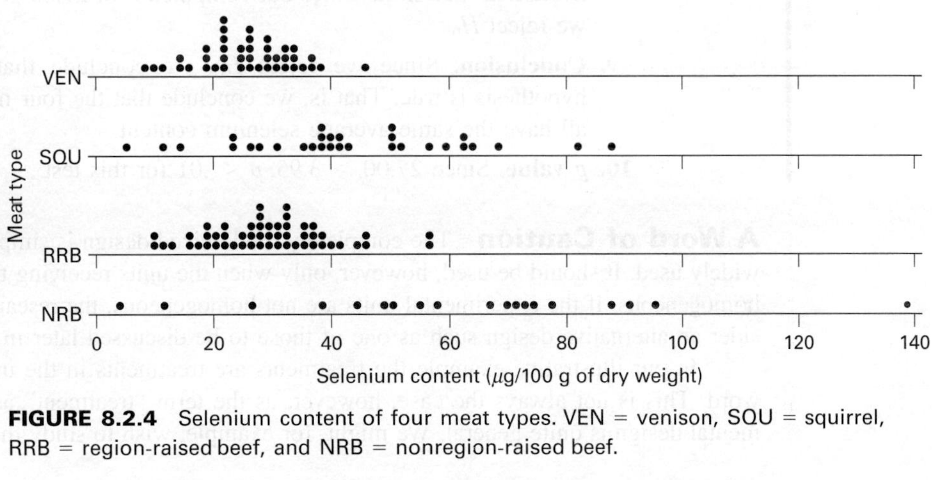

FIGURE 8.2.4 Selenium content of four meat types. VEN = venison, SQU = squirrel, RRB = region-raised beef, and NRB = nonregion-raised beef.

TABLE 8.2.4 ANOVA Table for Example 8.2.1

Source	SS	df	MS	F
Among samples	21261.82886	3	7087.27629	27.00
Within samples	36747.22674	140	262.48019	
Total	58009.05560	143		

3. **Hypotheses.**
   H_0: $\mu_1 = \mu_2 = \mu_3 = \mu_4$ (On average the four meats have the same selenium content.)

   H_A: Not all μ's are equal (At least one meat yields an average selenium content different from the average selenium content of at least one other meat.)

4. **Test statistic.** The test statistic is V.R. $= MSA/MSW$.
5. **Distribution of test statistic.** If H_0 is true and the assumptions are met, the V.R. follows the F distribution with $4 - 1 = 3$ numerator degrees of freedom and $144 - 4 = 140$ denominator degrees of freedom.
6. **Decision rule.** Suppose we let $\alpha = .01$. The critical value of F from Appendix Table G is < 3.95. The decision rule, then, is reject H_0 if the computed V.R. statistic is equal to or greater than 3.95.
7. **Calculation of test statistic.** By Equation 8.2.2 we compute

$$SST = 58009.05560$$

   By Equation 8.2.4 we compute

$$SSA = 21261.82886$$

$$SSW = 58009.05560 - 21261.82886 = 36747.22674$$

   The results of our calculations are displayed in Table 8.2.4.

8. **Statistical decision.** Since our computed F of 27.00 is greater than 3.95 we reject H_0.
9. **Conclusion.** Since we reject H_0, we conclude that the alternative hypothesis is true. That is, we conclude that the four meat types do not all have the same average selenium content.
10. ***p* value.** Since $27.00 > 3.95$, $p < .01$ for this test. ■

A Word of Caution The completely randomized design is simple and, therefore, widely used. It should be used, however, only when the units receiving the treatments are homogeneous. If the experimental units are not homogeneous, the researcher should consider an alternative design such as one of those to be discussed later in this chapter.

In our illustrative example the treatments are treatments in the usual sense of the word. This is not always the case, however, as the term "treatment" as used in experimental design is quite general. We might, for example, wish to study the response to the

Dialog box:

Stat ➤ ANOVA ➤ Oneway (Unstacked)

Type *C1-C4* in **responses (in separate columns)**
Click **OK.**

Session command:

```
MTB>AOVONEWAY C1-C4
```

Output:

One-way ANOVA: NRB, RRB, SQU, VEN

```
Analysis of Variance for Selenium

Source      DF       SS      MS       F        P
Meat Typ     3    21262    7087   27.00    0.000
Error      140    36747     262
Total      143    58009

Individual 95% CIs For Mean

                                   Based on Pooled StDev
Level    N      Mean    StDev   ------+---------+-------+----------
NRB     19     62.05    31.15                           (----*----)
RRB     53     29.08    10.38        (--*--)
SQU     30     43.25    19.51                  (---*---)
VEN     42     25.88     8.03     (--*---)
                                -------+---------+--------+----------
Pooled StDeV = 16.20                   30        45       60
```

FIGURE 8.2.5 MINITAB procedure and output for Example 8.2.1.

same treatment (in the usual sense of the word) of several breeds of animals. We would, however, refer to the breed of animal as the "treatment."

We must also point out that, although the techniques of analysis of variance are more often applied to data resulting from controlled experiments, the techniques also may be used to analyze data collected by a survey, provided that the underlying assumptions are reasonably well met.

Computer Analysis Figure 8.2.5 shows the computer procedure and output for Example 8.2.1 provided by a one-way analysis of variance program found in the MINITAB package. The data were entered into Columns 1 through 4. When you compare the ANOVA table on this printout with the one given in Table 8.2.4, you see that the printout uses the label "factor" instead of "among samples." The different treatments are referred to on the printout as levels. Thus level 1 = treatment 1, level 2 = treatment 2, and so on. The printout gives the four sample means and standard deviations as well as

The SAS System

Analysis of Variance Procedure

```
Dependent Variable: selen

                           Sum of
Source          DF        Squares    Mean Square    F Value    Pr > F
Model            3    21261.82886     7087.27629      27.00    <.0001
Error          140    36747.22674      262.48019
Corrected Total 143   58009.05560
          R-Square   Coeff Var    Root MSE     selen Mean
          0.366526    45.70507    16.20124       35.44736
```

FIGURE 8.2.6 Partial SAS® printout for Example 8.2.1.

the pooled standard deviation. This last quantity is equal to the square root of the error mean square shown in the ANOVA table. Finally, the computer output gives graphic representations of the 95 percent confidence intervals for the mean of each of the four populations represented by the sample data.

Figure 8.2.6 contains a partial SAS® printout resulting from analysis of the data of Example 8.2.1 through use of the SAS® statement PROC ANOVA. SAS® computes some additional quantities as shown in the output. R-Square = SSA/SST. This quantity tells us what proportion of the total variability present in the observations is accounted for by differences in response to the treatments. C.V. = 100 (root MSE/selen mean). Root MSE is the square root of MSW, and selen mean is the mean of the 18 observations.

Note that the test statistic V.R. is labeled differently by different statistical software programs. MINITAB, for example, uses F rather than V.R. SAS® uses the label F Value.

A useful device for displaying important characteristics of a set of data analyzed by one-way analysis of variance is a graph consisting of side-by-side boxplots. For each sample a boxplot is constructed using the method described in Chapter 2. Figure 8.2.7 shows the side-by-side boxplots for Example 8.2.1. Note that in Figure 8.2.7 the variable of interest is represented by the vertical axis rather than the horizontal axis.

Alternatives If the data available for analysis do not meet the assumptions for one-way analysis of variance as discussed here, one may wish to consider the use of the Kruskal-Wallis procedure, a nonparametric technique discussed in Chapter 13.

Testing for Significant Differences Between Individual Pairs of Means When the analysis of variance leads to a rejection of the null hypothesis of no difference among population means, the question naturally arises regarding just which pairs of means are different. In fact, the desire, more often than not, is to carry

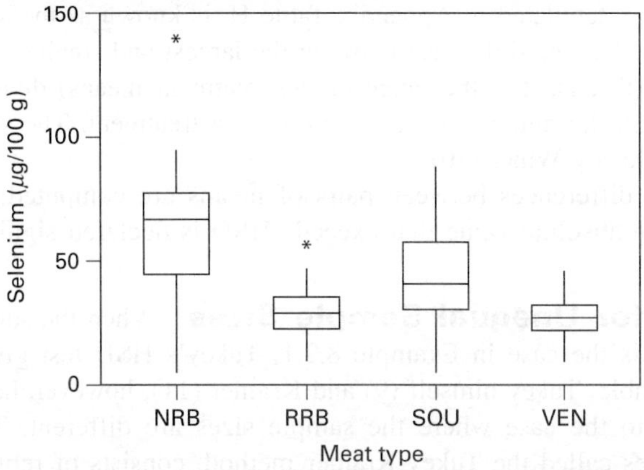

FIGURE 8.2.7 Side-by-side boxplots for Example 8.2.1.

out a significance test on each and every pair of treatment means. For instance, in Example 8.2.1, where there are four treatments, we may wish to know, after rejecting H_0: $\mu_1 = \mu_2 = \mu_3 = \mu_4$, which of the six possible individual hypotheses should be rejected. The experimenter, however, must exercise caution in testing for significant differences between individual means and must always make certain that the procedure is valid. The critical issue in the procedure is the level of significance. Although the probability, α, of rejecting a true null hypothesis for the test as a whole is made small, the probability of rejecting at least one true hypothesis when several pairs of means are tested is, as we have seen, greater than α. There are several multiple comparison procedures commonly used in practice. Below we illustrate two popular procedures, namely Tukey's HSD test and Bonferroni's method. The interested student is referred to the books by Hsu (7) and Westfall et al. (8) for additional techniques.

Tukey's HSD Test Over the years several procedures for making multiple comparisons have been suggested. A multiple comparison procedure developed by Tukey (9) is frequently used for testing the null hypothesis that all possible pairs of treatment means are equal when the samples are all of the same size. When this test is employed we select an overall significance level of α. The probability is α, then, that one or more of the null hypotheses is false.

Tukey's test, which is usually referred to as the HSD (*honestly significant difference*) test, makes use of a single value against which all differences are compared. This value, called the HSD, is given by

$$\text{HSD} = q_{\alpha,k,N-k}\sqrt{\frac{MSE}{n}} \tag{8.2.9}$$

where α is the chosen level of significance, k is the number of means in the experiment, N is the total number of observations in the experiment, n is the number of observations in a treatment, MSE is the error or within mean square from the ANOVA table, and q is obtained by entering Appendix Table H with α, k, and $N - k$.

The statistic q, tabulated in Appendix Table H, is known as the studentized range statistic. It is defined as the difference between the largest and smallest treatment means from an ANOVA (that is, it is the range of the treatment means) divided by the error mean square over n, the number of observations in a treatment. The studentized range is discussed in detail by Winer (10).

All possible differences between pairs of means are computed, and any difference that yields an absolute value that exceeds HSD is declared significant.

Tukey's Test for Unequal Sample Sizes When the samples are not all the same size, as is the case in Example 8.2.1, Tukey's HSD test given by Equation 8.2.9 is not applicable. Tukey himself (9) and Kramer (11), however, have extended the Tukey procedure to the case where the sample sizes are different. Their procedure, which is sometimes called the Tukey-Kramer method, consists of replacing *MSE/n* in Equation 8.2.9 with $(MSE/2)(1/n_i + 1/n_j)$, where n_i and n_j are the sample sizes of the two groups to be compared. If we designate the new quantity by HSD*, we have as the new test criterion

$$\text{HSD}^* = q_{\alpha,k,N-k}\sqrt{\frac{MSE}{2}\left(\frac{1}{n_i} + \frac{1}{n_j}\right)} \tag{8.2.10}$$

Any absolute value of the difference between two sample means that exceeds HSD* is declared significant.

Bonferroni's Method Another very commonly used multiple comparison test is based on a method developed by C. E. Bonferroni. As with Tukey's method, we desire to maintain an overall significance level of α for the total of all pair-wise tests. In the Bonferroni method, we simply divide the desired significance level by the number of individual pairs that we are testing. That is, instead of testing at a significance level of α, we test at a significance level of α/k, where k is the number of paired comparisons. The sum of all α/k terms cannot, then, possibly exceed our stated level of α. For example, if one has three samples, A, B, and C, then there are $k = 3$ pair-wise comparisons. These are $\mu_A = \mu_B$, $\mu_A = \mu_C$, and $\mu_B = \mu_C$. If we choose a significance level of $\alpha = .05$, then we would proceed with the comparisons and use a Bonferroni-corrected significance level of $\alpha/3 = .017$ Therefore, our p value must be no greater then .017 in order to reject the null hypothesis and conclude that two means differ.

Most computer packages compute values using the Bonferroni method and produce an output similar to the Tukey's HSD or other multiple comparison procedures. In general, these outputs report the actual corrected p value using the Bonferroni method. Given the basic relationship that $p = \alpha/k$, then algebraically we can multiply both sides of the equation by k to obtain $\alpha = pk$. In other words, the total α is simply the sum of all of the pk values, and the actual corrected p value is simply the calculated p value multiplied by the number of tests that were performed.

EXAMPLE 8.2.2

Let us illustrate the use of the HSD test with the data from Example 8.2.1.

Solution: The first step is to prepare a table of all possible (ordered) differences between means. The results of this step for the present example are displayed in Table 8.2.5.

Suppose we let $\alpha = .05$. Entering Table H with $\alpha = .05, k = 4$, and $N - k = 140$, we find that $q < 3.68$. The actual value is $q = 3.667$, which can be obtained from SAS®. In Table 8.2.4 we have MSE = 262.4802.

The hypotheses that can be tested, the value of HSD*, and the statistical decision for each test are shown in Table 8.2.6.

SAS® uses Tukey's procedure to test the hypothesis of no difference between population means for all possible pairs of sample means. The output also contains

TABLE 8.2.5 Differences Between Sample Means (Absolute Value) for Example 8.2.2

	VEN	RRB	SQR	NRB
VEN	—	3.208	17.37	36.171
RRB		—	14.163	32.963
SQU			—	18.801
NRB				—

Table 8.2.6 Multiple Comparison Tests Using Data of Example 8.2.1 and HSD*

Hypotheses	HSD*	Statistical Decision
$H_0\colon \mu_{VEN} = \mu_{RRB}$	$\text{HSD}^* = 3.677\sqrt{\frac{262.4802}{2}\left(\frac{1}{42}+\frac{1}{53}\right)} = 8.68$	Do not reject H_0 since $3.208 < 8.68$
$H_0\colon \mu_{VEN} = \mu_{SQU}$	$\text{HSD}^* = 3.677\sqrt{\frac{262.4802}{2}\left(\frac{1}{42}+\frac{1}{30}\right)} = 10.04$	Reject H_0 since $17.37 > 10.04$
$H_0\colon \mu_{VEN} = \mu_{NRB}$	$\text{HSD}^* = 3.677\sqrt{\frac{262.4802}{2}\left(\frac{1}{42}+\frac{1}{19}\right)} = 11.61$	Reject H_0 since $36.171 > 11.61$
$H_0\colon \mu_{RRB} = \mu_{SQU}$	$\text{HSD}^* = 3.677\sqrt{\frac{262.4802}{2}\left(\frac{1}{53}+\frac{1}{30}\right)} = 9.60$	Reject H_0 since $14.163 > 9.60$
$H_0\colon \mu_{RRB} = \mu_{NRB}$	$\text{HSD}^* = 3.677\sqrt{\frac{262.4802}{2}\left(\frac{1}{53}+\frac{1}{19}\right)} = 11.23$	Reject H_0 since $32.963 > 11.23$
$H_0\colon \mu_{SQU} = \mu_{NRB}$	$\text{HSD}^* = 3.677\sqrt{\frac{262.4802}{2}\left(\frac{1}{30}+\frac{1}{19}\right)} = 12.32$	Reject H_0 since $18.801 > 12.32$

The SAS System

Analysis of Variance Procedure

```
          Tukey's Studentized Range (HSD) Test for selen

NOTE: This test controls the Type I experimentwise error rate.

          Alpha                                          0.05
          Error Degrees of Freedom                        140
          Error Mean Square                          262.4802
          Critical Value of Studentized Range         3.67719

Comparisons significant at the 0.05 level are indicated by ***.

                 Difference
   type            Between       Simultaneous 95%
Comparison           Means       Confidence Limits

NRB  -  SQU         18.801        6.449       31.152   ***
NRB  -  RRB         32.963       21.699       44.228   ***
NRB  -  VEN         36.171       24.524       47.818   ***
SQU  -  NRB        -18.801      -31.152       -6.449   ***
SQU  -  RRB         14.163        4.538       23.787   ***
SQU  -  VEN         17.370        7.300       27.440   ***
RRB  -  NRB        -32.963      -44.228      -21.699   ***
RRB  -  SQU        -14.163      -23.787       -4.538   ***
RRB  -  VEN          3.208       -5.495       11.910
VEN  -  NRB        -36.171      -47.818      -24.524   ***
VEN  -  SQU        -17.370      -27.440       -7.300   ***
VEN  -  RRB         -3.208      -11.910        5.495
```

FIGURE 8.2.8 SAS® multiple comparisons for Example 8.2.1.

confidence intervals for the difference between all possible pairs of population means. This SAS output for Example 8.2.1 is displayed in Figure 8.2.8.

One may also use SPSS to perform multiple comparisons by a variety of methods, including Tukey's. The SPSS outputs for Tukey's HSD and Bonferroni's method for the data for Example 8.2.1 are shown in Figures 8.2.9 and 8.2.10. The outputs contain an exhaustive comparison of sample means, along with the associated standard errors, p values, and 95% confidence intervals. ■

Multiple Comparisons

Dependent Variable: Selenium
Tukey HSD

(I) Meat_type	(J) Meat_type	Mean Difference (I–J)	Std. Error	Sig.	95% Confidence Interval	
					Lower Bound	Upper Bound
VEN	SQU	−17.370190*	3.872837210	.000	−27.44017302	−7.30020793
	RRB	−3.2075427	3.346936628	.773	−11.91010145	5.49501609
	NRB	−36.170840*	4.479316382	.000	−47.81776286	−24.52391634
SQU	VEN	17.370190*	3.872837210	.000	7.30020793	27.44017302
	RRB	14.162648*	3.701593729	.001	4.53792509	23.78737051
	NRB	−18.800649*	4.750167007	.001	−31.15182638	−6.44947187
RRB	VEN	3.2075427	3.346936628	.773	−5.49501609	11.91010145
	SQU	−14.162648*	3.701593729	.001	−23.78737051	−4.53792509
	NRB	−32.963297*	4.332113033	.000	−44.22746845	−21.69912540
NRB	VEN	36.170840*	4.479316382	.000	24.52391634	47.81776286
	SQU	18.800649*	4.750167007	.001	6.44947187	31.15182638
	RRB	32.963297*	4.332113033	.000	21.69912540	44.22746845

*The mean difference is significant at the .05 level.

FIGURE 8.2.9 SPSS output for Tukey's HSD using data from Example 8.2.1.

Multiple Comparisons

Dependent Variable: Selenium
Bonferroni

(I) Meat_type	(J) Meat_type	Mean Difference (I–J)	Std. Error	Sig.	95% Confidence Interval	
					Lower Bound	Upper Bound
VEN	RRB	−3.20754	3.34694	1.000	−12.1648	5.7497
	SQU	−17.37019*	3.87284	.000	−27.7349	−7.0055
	NRB	−36.17084*	4.47932	.000	−48.1587	−24.1830
RRB	VEN	3.20754	3.34694	1.000	−5.7497	12.1648
	SQU	−14.16265*	3.70159	.001	−24.0691	−4.2562
	NRB	−32.96330*	4.33211	.000	−44.5572	−21.3694
SQU	VEN	17.37019*	3.87284	.000	7.0055	27.7349
	RRB	14.16265*	3.70159	.001	4.2562	24.0691
	NRB	−18.80065*	4.75017	.001	−31.5134	−6.0879
NRB	VEN	36.17084*	4.47932	.000	24.1830	48.1587
	RRB	32.96330*	4.33211	.000	21.3694	44.5572
	SQU	18.80065*	4.75017	.001	6.0879	31.5134

*The mean difference is significant at the .05 level.

FIGURE 8.2.10 SPSS output for Bonferroni's method using data from Example 8.2.1.

EXERCISES

In Exercises 8.2.1 to 8.2.7, go through the ten steps of analysis of variance hypothesis testing to see if you can conclude that there is a difference among population means. Let $\alpha = .05$ for each test. Use Tukey's HSD procedure to test for significant differences among individual pairs of means (if appropriate). Use the same α value for the F test. Construct a dot plot and side-by-side boxplots of the data.

8.2.1 Researchers at Case Western Reserve University (A-2) wanted to develop and implement a transducer, manageable in a clinical setting, for quantifying isometric moments produced at the elbow joint by individuals with tetraplegia (paralysis or paresis of all four limbs). The apparatus, called an elbow moment transducer (EMT), measures the force the elbow can exert when flexing. The output variable is voltage. The machine was tested at four different elbow extension angles, 30, 60, 90, and 120 degrees, on a mock elbow consisting of two hinged aluminum beams. The data are shown in the following table.

Elbow Angle (Degrees)							
30		60	90				120
−0.003	1.094	0.000	−0.001	0.000	−0.007	0.558	0.003
0.050	1.061	0.053	0.010	0.006	0.012	0.529	0.062
0.272	1.040	0.269	0.028	0.026	−0.039	0.524	0.287
0.552	1.097	0.555	0.055	0.053	−0.080	0.555	0.555
1.116	1.080	1.103	0.105	0.108	−0.118	0.539	1.118
2.733	1.051	2.727	0.272	0.278	−0.291	0.536	2.763
0.000	1.094	−0.002	0.553	0.555	−0.602	0.557	0.006
0.056	1.075	0.052	0.840	0.834	−0.884	0.544	0.050
0.275	1.035	0.271	1.100	1.106	−1.176	0.539	0.277
0.556	1.096	0.550	1.647	1.650	−1.725	1.109	0.557
1.100	1.100	1.097	2.728	2.729	0.003	1.085	1.113
2.723	1.096	2.725	−0.001	0.005	0.003	1.070	2.759
−0.003	1.108	0.003	0.014	−0.023	−0.011	1.110	0.010
0.055	1.099	0.052	0.027	−0.037	−0.060	1.069	0.060
0.273	1.089	0.270	0.057	−0.046	−0.097	1.045	0.286
0.553	1.107	0.553	0.111	−0.134	−0.320	1.110	0.564
1.100	1.094	1.100	0.276	−0.297	−0.593	1.066	1.104
2.713	1.092	2.727	0.555	−0.589	−0.840	1.037	2.760
0.007	1.092	0.022	0.832	−0.876	−1.168	2.728	−0.003
−0.066	1.104	−0.075	1.099	−1.157	−1.760	2.694	−0.060
−0.258	1.121	−0.298	1.651	−1.755	0.004	2.663	−0.289
−0.581	1.106	−0.585	2.736	−2.862	0.566	2.724	−0.585
−1.162	1.135	−1.168	0.564	0.000	1.116	2.693	−1.180
0.008	1.143	0.017	0.556	0.245	2.762	2.670	0.000
−0.045	1.106	−0.052	0.555	0.497	0.563	2.720	−0.034
−0.274	1.135	−0.258	0.567	0.001	0.551	2.688	−0.295
−0.604	1.156	−0.548	0.559	0.248	0.551	2.660	−0.579
−1.143	1.112	−1.187	0.551	0.498	0.561	0.556	−1.165
−0.004	1.104	0.019	1.107	0.001	0.555	0.560	−0.019

(Continued)

Elbow Angle (Degrees)							
30		60	90				120
−0.050	1.107	−0.044	1.104	0.246	0.558	0.557	−0.056
−0.290	1.107	−0.292	1.102	0.491	0.551	0.551	−0.270
−0.607	1.104	−0.542	1.112	0.001	0.566	0.564	−0.579
−1.164	1.117	−1.189	1.103	0.262	0.560	0.555	−1.162
1.105	1.101		1.104	0.527	1.107	0.551	
	1.103		1.114	0.001	1.104	0.563	
			1.095	0.260	1.109	0.559	
			1.100	0.523	1.108	1.113	
			2.739	−0.005	1.106	1.114	
			2.721	0.261	1.102	1.101	
			2.687	0.523	1.111	1.113	
			2.732	2.696	1.102	1.113	
			2.702	2.664	1.107	1.097	
			2.660	2.722	2.735	1.116	
			2.743	2.686	2.733	1.112	
			2.687	2.661	2.659	1.098	
			2.656	0.548	2.727	2.732	
			2.733	2.739	0.542	2.722	
			2.731	2.742	0.556	2.734	
					2.728	2.747	

Source: S. A. Snyder, M.S. Used with permission.

8.2.2 Patients suffering from rheumatic diseases or osteoporosis often suffer critical losses in bone mineral density (BMD). Alendronate is one medication prescribed to build or prevent further loss of BMD. Holcomb and Rothenberg (A-3) looked at 96 women taking alendronate to determine if a difference existed in the mean percent change in BMD among five different primary diagnosis classifications. Group 1 patients were diagnosed with rheumatoid arthritis (RA). Group 2 patients were a mixed collection of patients with diseases including lupus, Wegener's granulomatosis and polyarteritis, and other vasculitic diseases (LUPUS). Group 3 patients had polymyalgia rheumatica or temporal arthritis (PMRTA). Group 4 patients had osteoarthritis (OA) and group 5 patients had osteoporosis (O) with no other rheumatic diseases identified in the medical record. Changes in BMD are shown in the following table.

Diagnosis					
RA		LUPUS	PMRTA	OA	O
11.091	7.412	2.961	−3.669	11.146	2.937
24.414	5.559	0.293	−7.816	−0.838	15.968
10.025	4.761	8.394	4.563	4.082	5.349
−3.156	−3.527	2.832	−0.093	6.645	1.719
6.835	4.839	−1.369	−0.185	4.329	6.445
3.321	1.850	11.288	1.302	1.234	20.243
1.493	−3.933	3.997	5.299	−2.817	3.290

(Continued)

Diagnosis					
RA		LUPUS	PMRTA	OA	O
−1.864	9.669	7.260	10.734	3.544	8.992
5.386	4.659	5.546	1.399	4.160	6.120
3.868	1.137		0.497	1.160	25.655
6.209	7.521		0.592	−0.247	
−5.640	0.073		3.950	5.372	
3.514	−8.684		0.674	6.721	
−2.308	−0.372		9.354	9.950	
15.981	21.311		2.610	10.820	
−9.646	10.831		5.682	7.280	
5.188	3.351			6.605	
−1.892	9.557			7.507	
16.553				5.075	
				0.163	
				12.767	
				3.481	
				0.917	
				15.853	

Source: John P. Holcomb, Ph.D. and Ralph J. Rothenberg, M.D. Used with permission.

8.2.3 Ilich-Ernst et al. (A-4) investigated dietary intake of calcium among a cross section of 113 healthy women ages 20–88. The researchers formed four age groupings as follows: Group A, 20.0–45.9 years; group B, 46.0–55.9 years; group C, 56.0–65.9 years; and group D, over 66 years. Calcium from food intake was measured in mg/day. The data below are consistent with summary statistics given in the paper.

Age Groups (Years)				Age Groups (Years)			
A	B	C	D	A	B	C	D
1820	191	724	1652			1020	775
2588	1098	613	1309			805	1393
2670	644	918	1002			631	533
1022	136	949	966			641	734
1555	1605	877	788			760	485
222	1247	1368	472				449
1197	1529	1692	471				236
1249	1422	697	771				831
1520	445	849	869				698
489	990	1199	513				167
2575	489	429	731				824
1426	2408	798	1130				448
1846	1064	631	1034				991
1088	629	1016	1261				590
912		1025	42				994
1383		948	767				1781

(Continued)

Age Groups (Years)				Age Groups (Years)			
A	B	C	D	A	B	C	D
1483		1085	752				937
1723		775	804				1022
727		1307	1182				1073
1463		344	1243				948
1777		961	985				222
1129		239	1295				721
		944	1676				375
		1096	754				1187

8.2.4 Gold et al. (A-5) investigated the effectiveness on smoking cessation of a nicotine patch, bupropion SR, or both, when co-administered with cognitive-behavioral therapy. Consecutive consenting patients ($n = 164$) assigned themselves to one of three treatments according to personal preference: nicotine patch (NTP, $n = 13$), bupropion SR (B; $n = 92$), and bupropion SR plus nicotine patch (BNTP, $n = 59$). At their first smoking cessation class, patients estimated the number of packs of cigarettes they currently smoked per day and the numbers of years they smoked. The "pack years" is the average number of packs the subject smoked per day multiplied by the number of years the subject had smoked. The results are shown in the following table.

	Pack Years				
NTP	B			BNTP	
15	8	60	90	8	80
17	10	60	90	15	80
18	15	60	90	25	82
20	20	60	95	25	86
20	22	60	96	25	87
20	24	60	98	26	90
30	25	60	98	30	90
37	26	66	99	34	90
43	27	66	100	35	90
48	29	67	100	36	90
60	30	68	100	40	95
100	30	68	100	45	99
100	35	70	100	45	100
	35	70	100	45	102
	39	70	105	45	105
	40	75	110	48	105
	40	75	110	48	105
	40	75	120	49	111
	40	75	120	52	113
	40	76	123	60	120
	40	80	125	60	120
	45	80	125	60	125

(Continued)

	Pack Years				
NTP	B			BNTP	
	45	80	126	64	125
	45	80	130	64	129
	50	80	130	70	130
	51	80	132	70	133
	52	80	132	70	135
	55	84	142	75	140
	58	84	157	75	154
	60	84	180	76	
	60	90			

Source: Paul B. Gold, Ph.D. Used with Permission.

8.2.5 In a study by Wang et al. (A-6), researchers examined bone strength. They collected 10 cadaveric femurs from subjects in three age groups: young (19–49 years), middle-aged (50–69 years), and elderly (70 years or older) [Note: one value was missing in the middle-aged group]. One of the outcome measures (W) was the force in newtons required to fracture the bone. The following table shows the data for the three age groups.

Young (Y)	Middle-aged (MA)	Elderly (E)
193.6	125.4	59.0
137.5	126.5	87.2
122.0	115.9	84.4
145.4	98.8	78.1
117.0	94.3	51.9
105.4	99.9	57.1
99.9	83.3	54.7
74.0	72.8	78.6
74.4	83.5	53.7
112.8		96.0

Source: Xiaodu Wang, Ph.D. Used with permission.

8.2.6 In a study of 90 patients on renal dialysis, Farhad Atassi (A-7) assessed oral home care practices. He collected data from 30 subjects who were in (1) dialysis for less than 1 year, (2) dialysis for 1 to 3 years, and (3) dialysis for more than 3 years. The following table shows plaque index scores for these subjects. A higher score indicates a greater amount of plaque.

Group 1		Group 2		Group 3	
2.00	2.67	2.83	2.83	1.83	1.83
1.00	2.17	2.00	1.83	2.00	2.67
2.00	1.00	2.67	2.00	1.83	1.33
1.50	2.00	2.00	1.83	1.83	2.17
2.00	2.00	2.83	2.00	2.83	3.00

(Continued)

Group 1		Group 2		Group 3	
1.00	2.00	2.17	2.17	2.17	2.33
1.00	2.33	2.17	1.67	2.83	2.50
1.00	1.50	2.00	2.33	2.50	2.83
1.00	1.00	2.00	2.00	2.17	2.83
1.67	2.00	1.67	2.00	1.67	2.33
1.83	.83	2.33	2.17	2.17	2.33
2.17	.50	2.00	3.00	1.83	2.67
1.00	2.17	1.83	2.50	2.83	2.00
2.17	2.33	1.67	2.17	2.33	2.00
2.83	2.83	2.17	2.00	2.00	2.00

Source: Farhad Atassi, DDS, MSC, FICOI. Used with permission.

8.2.7 Thrombocytopaenia is a condition of abnormally low platelets that often occurs during necrotizing enterocolitis (NEC)—a serious illness in infants that can cause tissue damage to the intestines. Ragazzi et al. (A-8) investigated differences in the $\log_{10}$ of platelet counts in 178 infants with NEC. Patients were grouped into four categories of NEC status. Group 0 referred to infants with no gangrene, group 1 referred to subjects in whom gangrene was limited to a single intestinal segment, group 2 referred to patients with two or more intestinal segments of gangrene, and group 3 referred to patients with the majority of small and large bowel involved. The following table gives the $\log_{10}$ platelet counts for these subjects.

Gangrene Grouping							
0			1		2		3
1.97	2.33	2.48	1.38	2.45	1.87	2.37	1.77
0.85	2.60	2.23	1.86	2.60	1.90	1.75	1.68
1.79	1.88	2.51	2.26	1.83	2.43	2.57	1.46
2.30	2.33	2.38	1.99	2.47	1.32	1.51	1.53
1.71	2.48	2.31	1.32	1.92	2.06	1.08	1.36
2.66	2.15	2.08	2.11	2.51	1.04	2.36	1.65
2.49	1.41	2.49	2.54	1.79	1.99	1.58	2.12
2.37	2.03	2.21	2.06	2.17	1.52	1.83	1.73
1.81	2.59	2.45	2.41	2.18	1.99	2.55	1.91
2.51	2.23	1.96	2.23	2.53	2.52	1.80	1.57
2.38	1.61	2.29	2.00	1.98	1.93	2.44	2.27
2.58	1.86	2.54	2.74	1.93	2.29	2.81	1.00
2.58	2.33	2.23	2.00	2.42	1.75	2.17	1.81
2.84	2.34	2.78	2.51	0.79	2.16	2.72	2.27
2.55	1.38	2.36	2.08	1.38	1.81	2.44	2.43
1.90	2.52	1.89			2.46	1.98	1.74
2.28	2.35	2.26			1.66	1.57	1.60
2.33	2.63	1.79			2.51	2.05	2.08
1.77	2.03	1.87			1.76	2.30	2.34
1.83	1.08	2.51			1.72	1.36	1.89
1.67	2.40	2.29			2.57	2.48	1.75

(Continued)

Gangrene Grouping						
0			1	2		3
2.67	1.77	2.38		2.30	1.40	1.69
1.80				0.70	1.75	2.49
				2.16	2.67	1.75
				2.17	2.37	1.86
				2.12	1.46	1.26
				2.27	1.91	2.36

Source: Simon Eaton, M.D. Used with permission.

8.2.8 The objective of a study by Romito et al. (A-9) was to determine whether there is a different response to different calcium channel blockers. Two hundred and fifty patients with mild-to-moderate hypertension were randomly assigned to 4 weeks of treatment with once-daily doses of (1) lercanidipine, (2) felodipine, or (3) nifedipine. Prior to treatment and at the end of 4 weeks, each of the subjects had his or her systolic blood pressure measured. Researchers then calculated the change in systolic blood pressure. What is the treatment variable in this study? The response variable? What extraneous variables can you think of whose effects would be included in the error term? What are the "values" of the treatment variable? Construct an analysis of variance table in which you specify for this study the sources of variation and the degrees of freedom.

8.2.9 Kosmiski et al. (A-10) conducted a study to examine body fat distributions of men infected and not infected with HIV, taking and not taking protease inhibitors (PI), and having been diagnosed and not diagnosed with lipodystrophy. Lipodystrophy is a syndrome associated with HIV/PI treatment that remains controversial. Generally, it refers to fat accumulation in the abdomen or viscera accompanied by insulin resistance, glucose intolerance, and dyslipidemia. In the study, 14 subjects were taking protease inhibitors and were diagnosed with lipodystrophy, 12 were taking protease inhibitors, but were not diagnosed with lipodystrophy, five were HIV positive, not taking protease inhibitors, nor had diagnosed lypodystrophy, and 43 subjects were HIV negative and not diagnosed with lipodystrophy. Each of the subjects underwent body composition and fat distribution analyses by dual-energy X-ray absorptiometry and computed tomography. Researchers were able to then examine the percent of body fat in the trunk. What is the treatment variable? The response variable? What are the "values" of the treatment variable? Who are the subjects? What extraneous variables can you think of whose effects would be included in the error term? What was the purpose of including HIV-negative men in the study? Construct an ANOVA table in which you specify the sources of variation and the degrees of freedom for each. The authors reported a computed V.R. of 11.79. What is the p value for the test?

8.3 THE RANDOMIZED COMPLETE BLOCK DESIGN

The *randomized complete block* design was developed about 1925 by R. A. Fisher, who was seeking methods of improving agricultural field experiments. The randomized complete block design is a design in which the units (called *experimental units*) to which the treatments are applied are subdivided into homogeneous groups called *blocks*, so that the number of experimental units in a block is equal to the number (or some multiple of the number) of treatments being studied. The treatments are then assigned at random to the experimental units within each block. It should be emphasized that each treatment appears in every block, and each block receives every treatment.

Objective The objective in using the randomized complete block design is to isolate and remove from the error term the variation attributable to the blocks, while assuring that treatment means will be free of block effects. The effectiveness of the design depends on the ability to achieve homogeneous blocks of experimental units. The ability to form homogeneous blocks depends on the researcher's knowledge of the experimental material. When blocking is used effectively, the error mean square in the ANOVA table will be reduced, the V.R. will be increased, and the chance of rejecting the null hypothesis will be improved.

In animal experiments, the breed of animal may be used as a blocking factor. Litters may also be used as blocks, in which case an animal from each litter receives a treatment. In experiments involving human beings, if it is desired that differences resulting from age be eliminated, then subjects may be grouped according to age so that one person of each age receives each treatment. The randomized complete block design also may be employed effectively when an experiment must be carried out in more than one laboratory (block) or when several days (blocks) are required for completion.

The random allocation of treatments to subjects is restricted in the randomized complete block design. That is, each treatment must be represented an equal number of times (one or more times) within each blocking unit. In practice this is generally accomplished by assigning a random permutation of the order of treatments to subjects within each block. For example, if there are four treatments representing three drugs and a placebo (drug A, drug B, drug C, and placebo [p]), then there are $4! = 24$ possible permutations of the four treatments: (A, B, C, P) or (A, C, B, P) or (C, A, P, B), and so on. One permutation is then randomly assigned to each block.

Advantages One of the advantages of the randomized complete block design is that it is easily understood. Furthermore, certain complications that may arise in the course of an experiment are easily handled when this design is employed.

It is instructive here to point out that the paired comparisons analysis presented in Chapter 7 is a special case of the randomized complete block design. Example 7.4.1, for example, may be treated as a randomized complete block design in which the two points in time (Pre-op and Post-op) are the treatments and the individuals on whom the measurements were taken are the blocks.

Data Display In general, the data from an experiment utilizing the randomized complete block design may be displayed in a table such as Table 8.3.1. The following new notation in this table should be observed:

$$\text{total of the } i\text{th block} = T_{i.} = \sum_{j=1}^{k} x_{ij}$$

$$\text{mean of the } i\text{th block} = \bar{x}_{i.} = \frac{\sum_{j=1}^{k} x_{ij}}{k} = \frac{T_{i.}}{k}$$

$$\text{grand total} = T_{..} = \sum_{j=1}^{k} T_{.j} = \sum_{i=1}^{n} T_{i.}$$

TABLE 8.3.1 Table of Sample Values for the Randomized Complete Block Design

	Treatments						
Blocks	**1**	**2**	**3**	$\cdots$	$\boldsymbol{k}$	**Total**	**Mean**
1	x_{11}	x_{12}	x_{13}	$\cdots$	x_{1k}	$T_{1.}$	$\bar{x}_{1.}$
2	x_{21}	x_{22}	x_{23}	$\cdots$	x_{2k}	$T_{2.}$	$\bar{x}_{2.}$
3	x_{31}	x_{32}	x_{33}	$\cdots$	x_{3k}	$T_{3.}$	$\bar{x}_{3.}$
$\vdots$	$\vdots$	$\vdots$	$\vdots$	$\vdots$	$\vdots$	$\vdots$	$\vdots$
n	x_{n1}	x_{n2}	x_{n3}	$\cdots$	x_{nk}	$T_{n.}$	$\bar{x}_{n.}$
Total	$T_{.1}$	$T_{.2}$	$T_{.3}$	$\cdots$	$T_{.k}$	$T_{..}$	
Mean	$\bar{x}_{.1}$	$\bar{x}_{.2}$	$\bar{x}_{.3}$	$\cdots$	$\bar{x}_{.k}$		$\bar{x}_{..}$

indicating that the grand total may be obtained either by adding row totals or by adding column totals.

Two-Way ANOVA The technique for analyzing the data from a randomized complete block design is called *two-way analysis of variance* since an observation is categorized on the basis of two criteria—the block to which it belongs as well as the treatment group to which it belongs.

The steps for hypothesis testing when the randomized complete block design is used are as follows:

1. **Data.** After identifying the treatments, the blocks, and the experimental units, the data, for convenience, may be displayed as in Table 8.3.1.
2. **Assumptions.** The model for the randomized complete block design and its underlying assumptions are as follows:

The Model

$$x_{ij} = \mu + \beta_i + \tau_j + \epsilon_{ij} \qquad (8.3.1)$$
$$i = 1, 2, \ldots, n; \quad j = 1, 2, \ldots, k$$

In this model

x_{ij} is a typical value from the overall population.

μ is an unknown constant.

β_i represents a block effect reflecting the fact that the experimental unit fell in the ith block.

τ_j represents a treatment effect, reflecting the fact that the experimental unit received the jth treatment.

ϵ_{ij} is a residual component representing all sources of variation other than treatments and blocks.

Assumptions of the Model

(a) Each x_{ij} that is observed constitutes a random independent sample of size 1 from one of the kn populations represented.

(b) Each of these kn populations is normally distributed with mean μ_{ij} and the same variance σ^2. This implies that the ϵ_{ij} are independently and normally distributed with mean 0 and variance σ^2.

(c) The block and treatment effects are additive. This assumption may be interpreted to mean that there is no *interaction* between treatments and blocks. In other words, a particular block-treatment combination does not produce an effect that is greater or less than the sum of their individual effects. It can be shown that when this assumption is met,

$$\sum_{j=1}^{k} \tau_j = \sum_{i=1}^{n} \beta_i = 0$$

The consequences of a violation of this assumption are misleading results. One need not become concerned with the violation of the additivity assumption unless the largest mean is more than 50 percent greater than the smallest.

When these assumptions hold true, the τ_j and β_i are a set of fixed constants, and we have a situation that fits the fixed-effects model.

3. Hypotheses. We may test

$$H_0\colon \tau_j = 0, \qquad j = 1, 2, \ldots, k$$

against the alternative

$$H_A\colon \text{not all } \tau_j = 0$$

A hypothesis test regarding block effects is not usually carried out under the assumptions of the fixed-effects model for two reasons. First, the primary interest is in treatment effects, the usual purpose of the blocks being to provide a means of eliminating an extraneous source of variation. Second, although the experimental units are randomly assigned to the treatments, the blocks are obtained in a nonrandom manner.

4. Test statistic. The test statistic is V.R.

5. Distribution of test statistic. When H_0 is true and the assumptions are met, V.R. follows an F distribution.

6. Decision rule. Reject the null hypothesis if the computed value of the test statistic V.R. is equal to or greater than the critical value of F.

7. Calculation of test statistic. It can be shown that the total sum of squares for the randomized complete block design can be partitioned into three components, one each attributable to treatments ($SSTr$), blocks ($SSBl$), and error (SSE). That is,

$$SST = SSBl + SSTr + SSE \tag{8.3.2}$$

The formulas for the quantities in Equation 8.3.2 are as follows:

$$SST = \sum_{j=1}^{k}\sum_{i=1}^{n}(x_{ij} - \bar{x}_{..})^2 \tag{8.3.3}$$

$$SSBl = \sum_{j=1}^{k}\sum_{i=1}^{n}(\bar{x}_{i.} - \bar{x}_{..})^2 \tag{8.3.4}$$

$$SSTr = \sum_{j=1}^{k}\sum_{i=1}^{n}(\bar{x}_{.j} - \bar{x}_{..})^2 \tag{8.3.5}$$

$$SSE = SST - SSBl - SSTr \tag{8.3.6}$$

The appropriate degrees of freedom for each component of Equation 8.3.2 are

$$\begin{array}{ccccccc} \text{total} & & \text{blocks} & & \text{treatments} & & \text{residual (error)} \\ kn - 1 & = & (n-1) & + & (k-1) & + & (n-1)(k-1) \end{array}$$

The residual degrees of freedom, like the residual sum of squares, may be obtained by subtraction as follows:

$$\begin{aligned}(kn - 1) - (n - 1) - (k - 1) &= kn - 1 - n + 1 - k + 1 \\ &= n(k - 1) - 1(k - 1) = (n - 1)(k - 1)\end{aligned}$$

The ANOVA Table The results of the calculations for the randomized complete block design may be displayed in an ANOVA table such as Table 8.3.2.

8. **Statistical decision.** It can be shown that when the fixed-effects model applies and the null hypothesis of no treatment effects (all $\tau_i = 0$) is true, both the error, or residual, mean square and the treatments mean square are estimates of the common variance σ^2. When the null hypothesis is true, therefore, the quantity

$$MSTr/MSE$$

is distributed as F with $k - 1$ numerator degrees of freedom and $(n - 1) \times (k - 1)$ denominator degrees of freedom. The computed variance ratio, therefore, is compared with the critical value of F.

TABLE 8.3.2 ANOVA Table for the Randomized Complete Block Design

Source	SS	d.f.	MS	V.R.
Treatments	$SSTr$	$(k-1)$	$MSTr = SSTr/(k-1)$	$MSTr/MSE$
Blocks	$SSBl$	$(n-1)$	$MSBl = SSBl/(n-1)$	
Residual	SSE	$(n-1)(k-1)$	$MSE = SSE/(n-1)(k-1)$	
Total	SST	$kn-1$		

9. **Conclusion.** If we reject H_0, we conclude that the alternative hypothesis is true. If we fail to reject H_0, we conclude that H_0 may be true.

10. ***p* value.**

The following example illustrates the use of the randomized complete block design.

EXAMPLE 8.3.1

A physical therapist wished to compare three methods for teaching patients to use a certain prosthetic device. He felt that the rate of learning would be different for patients of different ages and wished to design an experiment in which the influence of age could be taken into account.

Solution: The randomized complete block design is the appropriate design for this physical therapist.

1. **Data.** Three patients in each of five age groups were selected to participate in the experiment, and one patient in each age group was randomly assigned to each of the teaching methods. The methods of instruction constitute our three treatments, and the five age groups are the blocks. The data shown in Table 8.3.3 were obtained.
2. **Assumptions.** We assume that each of the 15 observations constitutes a simple random of size 1 from one of the 15 populations defined by a block-treatment combination. For example, we assume that the number 7 in the table constitutes a randomly selected response from a population of responses that would result if a population of subjects under the age of 20 received teaching method A. We assume that the responses in the 15 represented populations are normally distributed with equal variances.

TABLE 8.3.3 Time (in Days) Required to Learn the Use of a Certain Prosthetic Device

	Teaching Method				
Age Group	**A**	**B**	**C**	**Total**	**Mean**
Under 20	7	9	10	26	8.67
20 to 29	8	9	10	27	9.00
30 to 39	9	9	12	30	10.00
40 to 49	10	9	12	31	10.33
50 and over	11	12	14	37	12.33
Total	45	48	58	151	
Mean	9.0	9.6	11.6		10.07

3. **Hypotheses.**

$$H_0: \tau_j = 0 \qquad j = 1, 2, 3$$
$$H_A: \text{not all } \tau_j = 0$$

Let $\alpha = .05$.

4. **Test statistic.** The test statistic is V.R. = $MSTr/MSE$.
5. **Distribution of test statistic.** When H_0 is true and the assumptions are met, V.R. follows an F distribution with 2 and 8 degrees of freedom.
6. **Decision rule.** Reject the null hypothesis if the computed V.R. is equal to or greater than the critical F, which we find in Appendix Table G to be 4.46.
7. **Calculation of test statistic.** We compute the following sums of squares:

$$SST = (7 - 10.07)^2 + (8 - 10.07)^2 + \cdots + (14 - 10.07)^2 = 46.9335$$
$$SSBl = 3\left[(8.67 - 10.07)^2 + (9.00 - 10.07)^2 + \cdots + (12.33 - 10.07)^2\right] = 24.855$$
$$SSTr = 5\left[(9 - 10.07)^2 + (9.6 - 10.07)^2 + (11.6 - 10.07)^2\right] = 18.5335$$
$$SSE = 46.9335 - 24.855 - 18.5335 = 3.545$$

The degrees of freedom are total = $(3)(5) - 1 = 14$, blocks = $5 - 1 = 4$, treatments = $3 - 1 = 2$, and residual = $(5 - 1)(3 - 1) = 8$. The results of the calculations may be displayed in an ANOVA table as in Table 8.3.4.

8. **Statistical decision.** Since our computed variance ratio, 20.91, is greater than 4.46, we reject the null hypothesis of no treatment effects on the assumption that such a large V.R. reflects the fact that the two sample mean squares are not estimating the same quantity. The only other explanation for this large V.R. would be that the null hypothesis is really true, and we have just observed an unusual set of results. We rule out the second explanation in favor of the first.

TABLE 8.3.4 ANOVA Table for Example 8.3.1

Source	SS	*d.f.*	MS	V.R.
Treatments	18.5335	2	9.26675	20.91
Blocks	24.855	4	6.21375	
Residual	3.545	8	.443125	
Total	46.9335	14		

9. **Conclusion.** We conclude that not all treatment effects are equal to zero, or equivalently, that not all treatment means are equal.

10. ***p* value.** For this test $p < .005$. ■

Computer Analysis Most statistics software packages will analyze data from a randomized complete block design. We illustrate the input and output for MINITAB. We use the data from the experiment to set up a MINITAB worksheet consisting of three columns. Column 1 contains the observations, Column 2 contains numbers that identify the block to which each observation belongs, and Column 3 contains numbers that identify the treatment to which each observation belongs. Figure 8.3.1 shows the MINITAB worksheet for Example 8.3.1. Figure 8.3.2 contains the MINITAB dialog box that initiates the analysis and the resulting ANOVA table.

The ANOVA table from the SAS® output for the analysis of Example 8.3.1 is shown in Figure 8.3.3. Note that in this output the model SS is equal to the sum of *SSBl* and *SSTr*.

Alternatives When the data available for analysis do not meet the assumptions of the randomized complete block design as discussed here, the Friedman procedure discussed in Chapter 13 may prove to be a suitable nonparametric alternative.

ROW	C1	C2	C3
1	7	1	1
2	9	1	2
3	10	1	3
4	8	2	1
5	9	2	2
6	10	2	3
7	9	3	1
8	9	3	2
9	12	3	3
10	10	4	1
11	9	4	2
12	12	4	3
13	11	5	1
14	12	5	2
15	14	5	3

FIGURE 8.3.1 MINITAB worksheet for the data in Figure 8.3.2.

Dialog box:

Stat ➤ ANOVA ➤ Twoway

Type *C1* in **Response.** Type *C2* in **Row factor** and check **Display means.** Type *C3* in **Column factor** and check **Display means.** Click **OK.**

Session command:

```
MTB > TWOWAY C1 C2 C3;
SUBC > MEANS C2 C3.
```

Output:

Two-Way ANOVA: C1 versus C2, C3

```
Analysis of Variance for C1
Source    DF       SS       MS       F        P
C2         4   24.933    6.233   14.38    0.001
C3         2   18.533    9.267   21.38    0.001
Error      8    3.467    0.433
Total     14   46.933

                         Individual 95% CI
  C2     Mean            ---+---------+----------+---------+--
   1     8.67        (-----*-----)
   2     9.00          (-----*-----)
   3    10.00                  (-----*-----)
   4    10.33                    (-----*-----)
   5    12.33                               (-----*-----)
                         ---+---------+----------+---------+--
                         9.00     10.50      12.00     13.50

                         Individual 95% CI
  C3     Mean            ---+---------+----------+---------+--
   1     9.00        (-----*-----)
   2     9.60            (-----*-----)
   3    11.60                                  (----*----)
                         ---+---------+----------+---------+--
                         9.00     10.00      11.00     12.00
```

FIGURE 8.3.2 MINITAB dialog box and output for two-way analysis of variance, Example 8.3.1.

EXERCISES

For Exercises 8.3.1 to 8.3.5 perform the ten-step hypothesis testing procedure for analysis of variance.

8.3.1 The objective of a study by Brooks et al. (A-11) was to evaluate the efficacy of using a virtual kitchen for vocational training of people with learning disabilities. Twenty-four students participated

The SAS System

Analysis of Variance Procedure

```
Dependent Variable: DAYS

Source              DF  Sum of Squares  Mean Square   F Value       Pr > F

Model                6     43.46666667   7.24444444     16.72       0.0004

Error                8      3.46666667   0.43333333

Corrected Total     14     46.93333333

              R-Square            C.V.     Root MSE             DAYS Mean

              0.926136        6.539211   0.65828059           10.06666667

Source              DF        Anova SS  Mean Square   F Value       Pr > F

GROUP                2     18.53333333   9.26666667     21.38       0.0006
AGE                  4     24.93333333   6.23333333     14.38       0.0010
```

FIGURE 8.3.3 Partial SAS® output for analysis of Example 8.3.1.

in the study. Each participant performed four food preparation tasks and they were scored on the quality of the preparation. Then each participant received regular vocational training in food preparation (real training), virtual training using a TV and computer screen of a typical kitchen, workbook training with specialized reading materials, and no training (to serve as a control). After each of these trainings, the subjects were tested on food preparation. Improvement scores for each of the four training methods are shown in the following table.

Subject No.	Real Training	Virtual Training	Workbook Training	No Training
1	2	10	2	−4
2	4	3	2	1
3	4	13	0	1
4	6	11	2	1
5	5	13	5	1
6	2	0	1	4
7	10	17	2	6
8	5	5	2	2
9	10	4	5	2
10	3	6	9	3
11	11	9	8	7
12	10	9	6	10
13	5	8	4	1

(*Continued*)

Subject No.	Real Training	Virtual Training	Workbook Training	No Training
14	8	11	1	1
15	4	8	5	2
16	11	8	10	2
17	6	11	1	3
18	2	5	1	2
19	3	1	0	−3
20	7	5	0	−6
21	7	10	4	4
22	8	7	−2	8
23	4	9	3	0
24	9	6	3	5

Source: B. M. Brooks, Ph.D. Used with permission.

After eliminating subject effects, can we conclude that the improvement scores differ among methods of training? Let $\alpha = .05$.

8.3.2 McConville et al. (A-12) report the effects of chewing one piece of nicotine gum (containing 2 mg nicotine) on tic frequency in patients whose Tourette's disorder was inadequately controlled by haloperidol. The following are the tic frequencies under four conditions:

	Number of Tics During 30-Minute Period			
			After End of Chewing	
Patient	Baseline	Gum Chewing	0–30 Minutes	30–60 Minutes
1	249	108	93	59
2	1095	593	600	861
3	83	27	32	61
4	569	363	342	312
5	368	141	167	180
6	326	134	144	158
7	324	126	312	260
8	95	41	63	71
9	413	365	282	321
10	332	293	525	455

Source: Brian J. McConville, M. Harold Fogelson, Andrew B. Norman, William M. Klykylo, Pat Z. Manderscheid, Karen W. Parker, and Paul R. Sanberg. "Nicotine Potentiation of Haloperidol in Reducing Tic Frequency in Tourette's Disorder," *American Journal of Psychiatry, 148* (1991), 793–794. Copyright© 1991, American Psychiatric Association. Reprinted by permission.

After eliminating patient effects, can we conclude that the mean number of tics differs among the four conditions? Let $\alpha = .01$.

8.3.3 A remotivation team in a psychiatric hospital conducted an experiment to compare five methods for remotivating patients. Patients were grouped according to level of initial motivation. Patients in each group were randomly assigned to the five methods. At the end of the experimental period the patients were evaluated by a team composed of a psychiatrist, a psychologist, a nurse, and a social worker, none of whom was aware of the method to which patients had been assigned. The team assigned each patient a composite score as a measure of his or her level of motivation. The results were as follows:

Level of Initial Motivation	Remotivation Method				
	A	B	C	D	E
Nil	58	68	60	68	64
Very low	62	70	65	80	69
Low	67	78	68	81	70
Average	70	81	70	89	74

Do these data provide sufficient evidence to indicate a difference in mean scores among methods? Let $\alpha = .05$.

8.3.4 The nursing supervisor in a local health department wished to study the influence of time of day on length of home visits by the nursing staff. It was thought that individual differences among nurses might be large, so the nurse was used as a blocking factor. The nursing supervisor collected the following data:

	Length of Home Visit by Time of Day			
Nurse	Early Morning	Late Morning	Early Afternoon	Late Afternoon
A	27	28	30	23
B	31	30	27	20
C	35	38	34	30
D	20	18	20	14

Do these data provide sufficient evidence to indicate a difference in length of home visit among the different times of day? Let $\alpha = .05$.

8.3.5 Four subjects participated in an experiment to compare three methods of relieving stress. Each subject was placed in a stressful situation on three different occasions. Each time a different method for reducing stress was used with the subject. The response variable is the amount of decrease in stress level as measured before and after treatment application. The results were as follows:

	Treatment		
Subject	A	B	C
1	16	26	22
2	16	20	23
3	17	21	22
4	28	29	36

Can we conclude from these data that the three methods differ in effectiveness? Let $\alpha = .05$.

8.3.6 In a study by Valencia et al. (A-13), the effects of environmental temperature and humidity on 24-hour energy expenditure were measured using whole-body indirect calorimetry in eight normal-weight young men who wore standardized light clothing and followed a controlled activity regimen. Temperature effects were assessed by measurements at 20, 23, 26, and 30 degrees Celsius at ambient humidity and at 20 and 30 degrees Celsius with high humidity. What is the blocking variable? The treatment variable? How many blocks are there? How many treatments? Construct an ANOVA table in which you specify the sources of variability and the degrees of freedom for each. What are the experimental units? What extraneous variables can you think of whose effects would be included in the error term?

8.3.7 Hodgson et al. (A-14) conducted a study in which they induced gastric dilatation in six anesthetized dogs maintained with constant-dose isoflurane in oxygen. Cardiopulmonary measurements prior to stomach distension (baseline) were compared with measurements taken during .1, .5, 1.0, 1.5, 2.5, and 3.5 hours of stomach distension by analyzing the change from baseline. After distending the stomach, cardiac index increased from 1.5 to 3.5 hours. Stroke volume did not change. During inflation, increases were observed in systemic arterial, pulmonary arterial, and right atrial pressure. Respiratory frequency was unchanged. Pa_{O_2} tended to decrease during gastric dilatation. What are the experimental units? The blocks? Treatment variable? Response variable(s)? Can you think of any extraneous variable whose effect would contribute to the error term? Construct an ANOVA table for this study in which you identify the sources of variability and specify the degrees of freedom.

8.4 THE REPEATED MEASURES DESIGN

One of the most frequently used experimental designs in the health sciences field is the repeated measures design.

DEFINITION

A *repeated measures* design is one in which measurements of the same variable are made on each subject on two or more different occasions.

The different occasions during which measurements are taken may be either points in time or different conditions such as different treatments.

When to Use Repeated Measures The usual motivation for using a repeated measures design is a desire to control for variability among subjects. In such a design each subject serves as its own control. When measurements are taken on only two occasions, we have the paired comparisons design that we discussed in Chapter 7. One of the most frequently encountered situations in which the repeated measures design is used is the situation in which the investigator is concerned with responses over time.

Advantages The major advantage of the repeated measures design is, as previously mentioned, its ability to control for extraneous variation among subjects. An additional advantage is the fact that fewer subjects are needed for the repeated measures design than for a design in which different subjects are used for each occasion on which measurements are made. Suppose, for example, that we have four treatments (in the usual sense) or four points in time on each of which we would like to have 10 measurements. If a different sample of subjects is used for each of the four treatments or points in time, 40 subjects would be required. If we are able to take measurements on the same subject for each treatment or point in time—that is, if we can use a repeated measures design—only 10 subjects would be required. This can be a very attractive advantage if subjects are scarce or expensive to recruit.

Disadvantages A major potential problem to be on the alert for is what is known as the *carry-over effect*. When two or more treatments are being evaluated, the investigator should make sure that a subject's response to one treatment does not reflect a residual effect from previous treatments. This problem can frequently be solved by allowing a sufficient length of time between treatments.

Another possible problem is the *position effect*. A subject's response to a treatment experienced last in a sequence may be different from the response that would have occurred if the treatment had been first in the sequence. In certain studies, such as those involving physical participation on the part of the subjects, enthusiasm that is high at the beginning of the study may give way to boredom toward the end. A way around this problem is to randomize the sequence of treatments independently for each subject.

Single-Factor Repeated Measures Design The simplest repeated measures design is the one in which, in addition to the treatment variable, one additional variable is considered. The reason for introducing this additional variable is to measure and isolate its contribution to the total variability among the observations. We refer to this additional variable as a *factor*.

DEFINITION

The repeated measures design in which one additional factor is introduced into the experiment is called a *single-factor repeated measures design*.

We refer to the additional factor as subjects. In the single-factor repeated measures design, each subject receives each of the treatments. The order in which the subjects are exposed to the treatments, when possible, is random, and the randomization is carried out independently for each subject.

Assumptions The following are the assumptions of the single-factor repeated measures design that we consider in this text. A design in which these assumptions are met is called a *fixed-effects additive* design.

1. The subjects under study constitute a simple random sample from a population of similar subjects.
2. Each observation is an independent simple random sample of size 1 from each of kn populations, where n is the number of subjects and k is the number of treatments to which each subject is exposed.

3. The kn populations have potentially different means, but they all have the same variance.
4. The k treatments are fixed; that is, they are the only treatments about which we have an interest in the current situation. We do not wish to make inferences to some larger collection of treatments.
5. There is no interaction between treatments and subjects; that is, the treatment and subject effects are additive.

Experimenters may find frequently that their data do not conform to the assumptions of fixed treatments and/or additive treatment and subject effects. For such cases the references at the end of this chapter may be consulted for guidance.

In addition to the assumptions just listed, it should be noted that in a repeated-measures experiment there is a presumption that correlations should exist among the repeated measures. That is, measurements at time 1 and 2 are likely correlated, as are measurements at time 1 and 3, 2 and 3, and so on. This is expected because the measurements are taken on the same individuals through time.

An underlying assumption of the repeated-measures ANOVA design is that all of these correlations are the same, a condition referred to as *compound symmetry*. This assumption, coupled with assumption 3 concerning equal variances, is referred to as *sphericity*. Violations of the sphericity assumption can result in an inflated type I error. Most computer programs provide a formal test for the sphericity assumption along with alternative estimation methods if the sphericity assumption is violated.

The Model The model for the fixed-effects additive single-factor repeated measures design is

$$x_{ij} = \mu + \beta_i + \tau_j + \epsilon_{ij}$$
$$i = 1, 2, \ldots, n; \qquad j = 1, 2, \ldots, k \tag{8.4.1}$$

The reader will recognize this model as the model for the randomized complete block design discussed in Section 8.3. The subjects are the blocks. Consequently, the notation, data display, and hypothesis testing procedure are the same as for the randomized complete block design as presented earlier. The following is an example of a repeated measures design.

EXAMPLE 8.4.1

Licciardone et al. (A-15) examined subjects with chronic, nonspecific low back pain. In this study, 18 of the subjects completed a survey questionnaire assessing physical functioning at baseline, and after 1, 3, and 6 months. Table 8.4.1 shows the data for these subjects who received a sham treatment that appeared to be genuine osteopathic manipulation. Higher values indicate better physical functioning. The goal of the experiment was to determine if subjects would report improvement over time even though the treatment they received would provide minimal improvement. We wish to know if there is a difference in the mean survey values among the four points in time.

TABLE 8.4.1 SF-36 Health Scores at Four Different Points in Time

Subject	Baseline	Month 1	Month 3	Month 6
1	80	60	95	100
2	95	90	95	95
3	65	55	50	45
4	50	45	70	70
5	60	75	80	85
6	70	70	75	70
7	80	80	85	80
8	70	60	75	65
9	80	80	70	65
10	65	30	45	60
11	60	70	95	80
12	50	50	70	60
13	50	65	80	65
14	85	45	85	80
15	50	65	90	70
16	15	30	20	25
17	10	15	55	75
18	80	85	90	70

Source: John C. Licciardone. Used with permission.

Solution:

1. **Data.** See Table 8.4.1.
2. **Assumptions.** We assume that the assumptions for the fixed-effects, additive single-factor repeated measures design are met.
3. **Hypotheses.**

$$H_0: \mu_B = \mu_{M1} = \mu_{M3} = \mu_{M6}$$
$$H_A: \text{not all } \mu\text{'s are equal}$$

4. **Test statistic.** V.R. = treatment MS/error MS.
5. **Distribution of test statistic.** F with $4 - 1 = 3$ numerator degrees of freedom and $71 - 3 - 17 = 51$ denominator degrees of freedom.
6. **Decision rule.** Let $\alpha = .05$. The critical value of F is 2.80 (obtained by interpolation). Reject H_0 if computed V.R. is equal to or greater than 2.80.
7. **Calculation of test statistic.** We use MINITAB to perform the calculations. We first enter the measurements in Column 1, the row (subject) codes in Column 2, the treatment (time period) codes in Column 3, and proceed as shown in Figure 8.4.1.

Dialog box:

Stat ➤ ANOVA ➤ Twoway

Type *C1* in **Response.** Type *C2* in **Row factor** and Check **Display means.** Type *C3* in **Column factor** and Check **Display means.** Click **OK.**

Session command:

```
MTB > TWOWAY C1 C2 C3;
SUBC> MEANS C2 C3.
```

Output:

Two-way ANOVA: C1 versus C2, C3

```
Analysis of Variance for C1
Source      DF        SS       MS       F       P
C2          17     20238     1190    8.20   0.000
C3           3      2396      799    5.50   0.002
Error       51      7404      145
Total       71     30038
```

FIGURE 8.4.1 MINITAB procedure and output (ANOVA table) for Example 8.4.1.

8. **Statistical decision.** Since V.R. = 5.50 is greater than 2.80, we are able to reject the null hypothesis.

9. **Conclusion.** We conclude that there is a difference in the four population means.

10. ***p* value.** Since 5.50 is greater than 4.98, the F value for $\alpha = .005$ and $df = 40$, the p value is less than .005.

Figure 8.4.2. shows the SAS® output for the analysis of Example 8.4.1 and Figure 8.4.3 shows the SPSS output for the same example. Note that SPSS provides four potential tests. The first test is used under an assumption of sphericity and matches the outputs in Figures 8.4.1 and 8.4.2. The next three tests are modifications if the assumption of sphericity is violated. Note that SPSS modifies the degrees of freedom for these three tests, which changes the mean squares and the p values, but not the V. R. Note that the assumption of sphericity was violated for these data, but that the decision rule did not change, since all of the p values were less than $\alpha = .05$. ■

EXERCISES

For Exercises 8.4.1 to 8.4.3 perform the ten-step hypothesis testing procedure. Let $\alpha = .05$.

8.4.1 One of the purposes of a study by Liu et al. (A-16) was to determine the effects of MRZ 2/579 on neurological deficit in Sprague-Dawley rats. In this study, 10 rats were measured at four time

```
The ANOVA Procedure

Dependent Variable: sf36

Source                 DF   Sum of Squares   Mean Square   F Value    Pr > F

Model                  20      22633.33333    1131.66667      7.79    <.0001

Error                  51       7404.16667     145.17974

Corrected Total        71      30037.50000

          R-Square   Coeff Var   Root MSE   sf36 Mean

          0.753503    18.18725   12.04906    66.25000

Source                 DF      Anova SS   Mean Square   F Value    Pr > F

subj                   17   20237.50000    1190.44118      8.20    <.0001
time                    3    2395.83333     798.61111      5.50    0.0024
```

FIGURE 8.4.2 SAS® output for analysis of Example 8.4.1.

Tests of Within-Subjects Effects

Measure: MEASURE_1

Source		Type III Sum of Squares	df	Mean Square	F	Sig.
factor 1	Sphericity Assumed	2395.833	3	798.611	5.501	.002
	Greenhouse-Geisser	2395.833	2.216	1080.998	5.501	.006
	Huynh-Feldt	2395.833	2.563	934.701	5.501	.004
	Lower-bound	2395.833	1.000	2395.833	5.501	.031
Error (factor 1)	Sphericity Assumed	7404.167	51	145.180		
	Greenhouse-Geisser	7404.167	37.677	196.515		
	Huynh-Feldt	7404.167	43.575	169.919		
	Lower-bound	7404.167	17.000	435.539		

FIGURE 8.4.3 SPSS output for the analysis of Example 8.4.1.

periods following occlusion of the middle carotid artery and subsequent treatment with the uncompetitive *N*-methly-D-aspartate antagonist MRZ 2/579, which previous studies had suggested provides neuroprotective activity. The outcome variable was a neurological function variable measured on a scale of 0–12. A higher number indicates a higher degree of neurological impairment.

Rat	60 Minutes	24 Hours	48 Hours	72 Hours
1	11	9	8	4
2	11	7	5	3
3	11	10	8	6
4	11	4	3	2
5	11	10	9	9
6	11	6	5	5
7	11	6	6	6
8	11	7	6	5
9	11	7	5	5
10	11	9	7	7

Source: Ludmila Belayev, M.D. Used with permission.

8.4.2 Starch et al. (A-17) wanted to show the effectiveness of a central four-quadrant sleeve and screw in anterior cruciate ligament reconstruction. The researchers performed a series of reconstructions on eight cadaveric knees. The following table shows the loads (in newtons) required to achieve different graft laxities (mm) for seven specimens (data not available for one specimen) using five different load weights. Graft laxity is the separation (in mm) of the femur and the tibia at the points of graft fixation. Is there sufficient evidence to conclude that different loads are required to produce different levels of graft laxity? Let $\alpha = .05$.

	Graft Laxity (mm)				
	Loads				
Specimen	**1**	**2**	**3**	**4**	**5**
1	297.1	297.1	297.1	297.1	297.1
2	264.4	304.6	336.4	358.2	379.3
3	188.8	188.8	188.8	188.8	188.8
4	159.3	194.7	211.4	222.4	228.1
5	228.2	282.1	282.1	334.8	334.8
6	100.3	105.0	106.3	107.7	108.7
7	116.9	140.6	182.4	209.7	215.4

Source: David W. Starch, Jerry W. Alexander, Philip C. Noble, Suraj Reddy, and David M. Lintner, "Multistranded Hamstring Tendon Graft Fixation with a Central Four-Quadrant or a Standard Tibial Interference Screw for Anterior Cruciate Ligament Reconstruction," *American Journal of Sports Medicine, 31* (2003), 338–344.

8.4.3 Holben et al. (A-18) designed a study to evaluate selenium intake in young women in the years of puberty. The researchers studied a cohort of 16 women for three consecutive summers. One of the outcome variables was the selenium intake per day. The researchers examined dietary journals of the subjects over the course of 2 weeks and then computed the average daily selenium intake. The following table shows the average daily selenium intake values (in μg/d) for the 16 women in years 1, 2, and 3 of the study.

Subject	Year 1	Year 2	Year 3	Subject	Year 1	Year 2	Year 3
1	112.51	121.28	94.99	9	95.05	93.89	73.26
2	106.20	121.14	145.69	10	112.65	100.47	145.69
3	102.00	121.14	130.37	11	103.74	121.14	123.97
4	103.74	90.21	135.91	12	103.74	121.14	135.91
5	103.17	121.14	145.69	13	112.67	104.66	136.87
6	112.65	98.11	145.69	14	106.20	121.14	126.42
7	106.20	121.14	136.43	15	103.74	121.14	136.43
8	83.57	102.87	144.35	16	106.20	100.47	135.91

Source: David H. Holben, Ph.D. and John P. Holcomb, Ph.D. Used with permission.

8.4.4 Linke et al. (A-19) studied seven male mongrel dogs. They induced diabetes by injecting the animals with alloxan monohydrate. The researchers measured the arterial glucose (mg/gl), arterial lactate (mmol/L), arterial free fatty acid concentration, and arterial β-hydroxybutyric acid concentration prior to the alloxan injection, and again in weeks 1, 2, 3, and 4 post-injection. What is the response variable(s)? Comment on carryover effect and position effect as they may or may not be of concern in this study. Construct an ANOVA table for this study in which you identify the sources of variability and specify the degrees of freedom for each.

8.4.5 Werther et al. (A-20) examined the vascular endothelial growth factor (VEGF) concentration in blood from colon cancer patients. Research suggests that inhibiting VEGF may disrupt tumor growth. The researchers measured VEGF concentration (ng/L) for 10 subjects and found an upward trend in VEGF concentrations during the clotting time measured at baseline, and hours 1 and 2. What is the response variable? What is the treatment variable? Construct an ANOVA table for this study in which you identify the sources of variability and specify the degrees of freedom for each.

8.5 THE FACTORIAL EXPERIMENT

In the experimental designs that we have considered up to this point, we have been interested in the effects of only one variable—the treatments. Frequently, however, we may be interested in studying, simultaneously, the effects of two or more variables. We refer to the variables in which we are interested as *factors*. The experiment in which two or more factors are investigated simultaneously is called a *factorial experiment*.

The different designated categories of the factors are called *levels*. Suppose, for example, that we are studying the effect on reaction time of three dosages of some drug. The drug factor, then, is said to occur at three levels. Suppose the second factor of interest in the study is age, and it is thought that two age groups, under 65 years and 65 years and older, should be included. We then have two levels of the age factor. In general, we say that factor A occurs at a levels and factor B occurs at b levels.

In a factorial experiment we may study not only the effects of individual factors but also, if the experiment is properly conducted, the interaction between factors. To illustrate the concept of interaction let us consider the following example.

EXAMPLE 8.5.1

Suppose, in terms of effect on reaction time, that the true relationship between three dosage levels of some drug and the age of human subjects taking the drug is known. Suppose further that age occurs at two levels—"young" (under 65) and "old" (65 and older). If the true relationship between the two factors is known, we will know, for the three dosage levels, the mean effect on reaction time of subjects in the two age groups. Let us assume that effect is measured in terms of reduction in reaction time to some stimulus. Suppose these means are as shown in Table 8.5.1.

The following important features of the data in Table 8.5.1 should be noted.

1. For both levels of factor A the difference between the means for any two levels of factor B is the same. That is, for both levels of factor A, the difference between means for levels 1 and 2 is 5, for levels 2 and 3 the difference is 10, and for levels 1 and 3 the difference is 15.
2. For all levels of factor B the difference between means for the two levels of factor A is the same. In the present case the difference is 5 at all three levels of factor B.
3. A third characteristic is revealed when the data are plotted as in Figure 8.5.1. We note that the curves corresponding to the different levels of a factor are all parallel.

When population data possess the three characteristics listed above, we say that there is no interaction present.

The presence of interaction between two factors can affect the characteristics of the data in a variety of ways depending on the nature of the interaction. We illustrate

TABLE 8.5.1 Mean Reduction in Reaction Time (milliseconds) of Subjects in Two Age Groups at Three Drug Dosage Levels

	Factor B—Drug Dosage		
Factor A—Age	$j = 1$	$j = 2$	$j = 3$
Young ($i = 1$)	$\mu_{11} = 5$	$\mu_{12} = 10$	$\mu_{13} = 20$
Old ($i = 2$)	$\mu_{21} = 10$	$\mu_{22} = 15$	$\mu_{23} = 25$

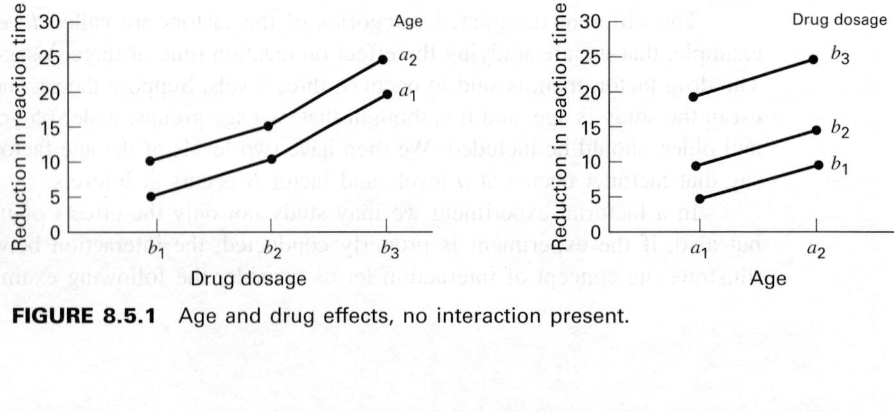

FIGURE 8.5.1 Age and drug effects, no interaction present.

TABLE 8.5.2 Data of Table 8.5.1 Altered to Show the Effect of One Type of Interaction

	Factor *B*—Drug Dosage		
Factor *A*—Age	$j = 1$	$j = 2$	$j = 3$
Young ($i = 1$)	$\mu_{11} = 5$	$\mu_{12} = 10$	$\mu_{13} = 20$
Old ($i = 2$)	$\mu_{21} = 15$	$\mu_{22} = 10$	$\mu_{23} = 5$

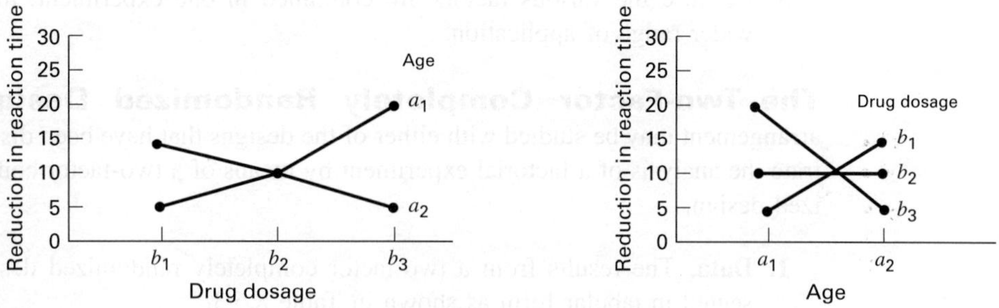

FIGURE 8.5.2 Age and drug effects, interaction present.

the effect of one type of interaction by altering the data of Table 8.5.1 as shown in Table 8.5.2.

The important characteristics of the data in Table 8.5.2 are as follows.

1. The difference between means for any two levels of factor *B* is not the same for both levels of factor *A*. We note in Table 8.5.2, for example, that the difference between levels 1 and 2 of factor *B* is -5 for the young age group and $+5$ for the old age group.
2. The difference between means for both levels of factor *A* is not the same at all levels of factor *B*. The differences between factor *A* means are -10, 0, and 15 for levels 1, 2, and 3, respectively, of factor *B*.
3. The factor level curves are not parallel, as shown in Figure 8.5.2.

When population data exhibit the characteristics illustrated in Table 8.5.2 and Figure 8.5.2, we say that there is interaction between the two factors. We emphasize that the kind of interaction illustrated by the present example is only one of many types of interaction that may occur between two factors. ■

In summary, then, we can say that *there is interaction between two factors if a change in one of the factors produces a change in response at one level of the other factor different from that produced at other levels of this factor.*

Advantages The advantages of the factorial experiment include the following.

1. The interaction of the factors may be studied.
2. There is a saving of time and effort.

In the factorial experiment all the observations may be used to study the effects of each of the factors under investigation. The alternative, when two factors are being investigated, would be to conduct two different experiments, one to study each of the two factors. If this were done, some of the observations would yield information only on one of the factors, and the remainder would yield information only on the other factor. To achieve the level of accuracy of the factorial experiment, more experimental units would be needed if the factors were studied through two experiments. It is seen, then, that 1 two-factor experiment is more economical than 2 one-factor experiments.

3. Because the various factors are combined in one experiment, the results have a wider range of application.

The Two-Factor Completely Randomized Design A factorial arrangement may be studied with either of the designs that have been discussed. We illustrate the analysis of a factorial experiment by means of a two-factor completely randomized design.

1. **Data.** The results from a two-factor completely randomized design may be presented in tabular form as shown in Table 8.5.3.

   Here we have a levels of factor A, b levels of factor B, and n observations for each combination of levels. Each of the ab combinations of levels of factor A with levels of factor B is a treatment. In addition to the totals and means shown in Table 8.5.3, we note that the total and mean of the ijth cell are

$$T_{ij.} = \sum_{k=1}^{n} x_{ijk} \quad \text{and} \quad \bar{x}_{ij.} = T_{ij.}/n$$

TABLE 8.5.3 Table of Sample Data from a Two-Factor Completely Randomized Experiment

	Factor B					
Factor A	**1**	**2**	$\cdots$	**b**	**Totals**	**Means**
1	x_{111}	x_{121}	$\cdots$	x_{1b1}		
	$\vdots$	$\vdots$	$\vdots$	$\vdots$	$T_{1..}$	$\bar{x}_{1..}$
	x_{11n}	x_{12n}	$\cdots$	x_{1bn}		
2	x_{211}	x_{221}	$\cdots$	x_{2b1}		
	$\vdots$	$\vdots$	$\vdots$	$\vdots$	$T_{2..}$	$\bar{x}_{2..}$
	x_{21n}	x_{22n}	$\cdots$	x_{2bn}		
$\vdots$	$\vdots$	$\vdots$	$\vdots$	$\vdots$	$\vdots$	$\vdots$
a	x_{a11}	x_{a21}	$\cdots$	x_{ab1}		
	$\vdots$	$\vdots$	$\vdots$	$\vdots$	$T_{a..}$	$\bar{x}_{a..}$
	x_{a1n}	x_{a2n}	$\cdots$	x_{abn}		
Totals	$T_{.1.}$	$T_{.2.}$	$\cdots$	$T_{.b.}$	$T_{...}$	
Means	$\bar{x}_{.1.}$	$\bar{x}_{.2.}$	$\cdots$	$\bar{x}_{.b.}$		$\bar{x}_{...}$

respectively. The subscript i runs from 1 to a and j runs from 1 to b. The total number of observations is nab.

To show that Table 8.5.3 represents data from a completely randomized design, we consider that each combination of factor levels is a treatment and that we have n observations for each treatment. An alternative arrangement of the data would be obtained by listing the observations of each treatment in a separate column. Table 8.5.3 may also be used to display data from a two-factor randomized block design if we consider the first observation in each cell as belonging to block 1, the second observation in each cell as belonging to block 2, and so on to the nth observation in each cell, which may be considered as belonging to block n.

Note the similarity of the data display for the factorial experiment as shown in Table 8.5.3 to the randomized complete block data display of Table 8.3.1. The factorial experiment, in order that the experimenter may test for interaction, requires at least two observations per cell, whereas the randomized complete block design requires only one observation per cell. We use two-way analysis of variance to analyze the data from a factorial experiment of the type presented here.

2. **Assumptions.** We assume a fixed-effects model and a two-factor completely randomized design. For a discussion of other designs, consult the references at the end of this chapter.

The Model The fixed-effects model for the two-factor completely randomized design may be written as

$$x_{ijk} = \mu + \alpha_i + \beta_j + (\alpha\beta)_{ij} + \epsilon_{ijk}$$
$$i = 1, 2, \ldots, a; \quad j = 1, 2, \ldots, b; \quad k = 1, 2, \ldots, n \tag{8.5.1}$$

where x_{ijk} is a typical observation, μ is a constant, α represents an effect due to factor A, β represents an effect due to factor B, $(\alpha\beta)$ represents an effect due to the interaction of factors A and B, and ϵ_{ijk} represents the experimental error.

Assumptions of the Model

a. The observations in each of the ab cells constitute a random independent sample of size n drawn from the population defined by the particular combination of the levels of the two factors.

b. Each of the ab populations is normally distributed.

c. The populations all have the same variance.

3. **Hypotheses.** The following hypotheses may be tested:

a. H_0: $\alpha_i = 0$ $\quad i = 1, 2, \ldots, a$
H_A: not all $\alpha_i = 0$

b. H_0: $\beta_j = 0$ $\quad j = 1, 2, \ldots, b$
H_A: not all $\beta_j = 0$

c. H_0: $(\alpha\beta)_{ij} = 0$ $\quad i = 1, 2, \ldots, a; j = 1, 2, \ldots, b$
H_A: not all $(\alpha\beta)_{ij} = 0$

Before collecting data, the researchers may decide to test only one of the possible hypotheses. In this case they select the hypothesis they wish to test, choose a significance level α, and proceed in the familiar, straightforward fashion. This procedure is free of the complications that arise if the researchers wish to test all three hypotheses.

When all three hypotheses are tested, the situation is complicated by the fact that the three tests are not independent in the probability sense. If we let α be the significance level associated with the test as a whole, and α', α'', and α''' the significance levels associated with hypotheses 1, 2, and 3, respectively, we find

$$\alpha < 1 - (1 - \alpha')(1 - \alpha'')(1 - \alpha''') \tag{8.5.2}$$

If $\alpha' = \alpha'' = \alpha''' = .05$, then $\alpha < 1 - (.95)^3$, or $\alpha < .143$. This means that the probability of rejecting one or more of the three hypotheses is something less than .143 when a significance level of .05 has been chosen for the hypotheses and all are true. To demonstrate the hypothesis testing procedure for each case, we perform all three tests. The reader, however, should be aware of the problem involved in interpreting the results.

4. **Test statistic.** The test statistic for each hypothesis set is V.R.
5. **Distribution of test statistic.** When H_0 is true and the assumptions are met, each of the test statistics is distributed as F.
6. **Decision rule.** Reject H_0 if the computed value of the test statistic is equal to or greater than the critical value of F.
7. **Calculation of test statistic.** By an adaptation of the procedure used in partitioning the total sum of squares for the completely randomized design, it can be shown that the total sum of squares under the present model can be partitioned into two parts as follows:

$$\sum_{i=1}^{a}\sum_{j=1}^{b}\sum_{k=1}^{n}(x_{ijk} - \bar{x}_{...})^2 = \sum_{i=1}^{a}\sum_{j=1}^{b}\sum_{k=1}^{n}(\bar{x}_{ij.} - \bar{x}_{...})^2 + \sum_{i=1}^{a}\sum_{j=1}^{b}\sum_{k=1}^{n}(x_{ijk} - \bar{x}_{ij.})^2 \tag{8.5.3}$$

or

$$SST = SSTr + SSE \tag{8.5.4}$$

The sum of squares for treatments can be partitioned into three parts as follows:

$$\begin{aligned}\sum_{i=1}^{a}\sum_{j=1}^{b}\sum_{k=1}^{n}(\bar{x}_{ij.} - \bar{x}_{...})^2 &= \sum_{i=1}^{a}\sum_{j=1}^{b}\sum_{k=1}^{n}(\bar{x}_{i..} - \bar{x}_{...})^2 \\ &+ \sum_{i=1}^{a}\sum_{j=1}^{b}\sum_{k=1}^{n}(\bar{x}_{.j.} - \bar{x}_{...})^2 \\ &+ \sum_{i=1}^{a}\sum_{j=1}^{b}\sum_{k=1}^{n}(\bar{x}_{ij.} - \bar{x}_{i..} - \bar{x}_{.j.} + \bar{x}_{...})^2\end{aligned} \tag{8.5.5}$$

TABLE 8.5.4 Analysis of Variance Table for a Two-Factor Completely Randomized Experiment (Fixed-Effects Model)

Source	SS	d.f.	MS	V.R.
A	SSA	$a - 1$	$MSA = SSA/(a - 1)$	MSA/MSE
B	SSB	$b - 1$	$MSB = SSB/(b - 1)$	MSB/MSE
AB	$SSAB$	$(a - 1)(b - 1)$	$MSAB = SSAB/(a - 1)(b - 1)$	$MSAB/MSE$
Treatments	$SSTr$	$ab - 1$		
Residual	SSE	$ab(n - 1)$	$MSE = SSE/ab(n - 1)$	
Total	SST	$abn - 1$		

or

$$SSTr = SSA + SSB + SSAB$$

The ANOVA Table The results of the calculations for the fixed-effects model for a two-factor completely randomized experiment may, in general, be displayed as shown in Table 8.5.4.

8. Statistical decision. If the assumptions stated earlier hold true, and if each hypothesis is true, it can be shown that each of the variance ratios shown in Table 8.5.4 follows an F distribution with the indicated degrees of freedom. We reject H_0 if the computed V.R. values are equal to or greater than the corresponding critical values as determined by the degrees of freedom and the chosen significance levels.

9. Conclusion. If we reject H_0, we conclude that H_A is true. If we fail to reject H_0, we conclude that H_0 may be true.

10. *p* value.

EXAMPLE 8.5.2

In a study of length of time spent on individual home visits by public health nurses, data were reported on length of home visit, in minutes, by a sample of 80 nurses. A record was made also of each nurse's age and the type of illness of each patient visited. The researchers wished to obtain from their investigation answers to the following questions:

1. Does the mean length of home visit differ among different age groups of nurses?
2. Does the type of patient affect the mean length of home visit?
3. Is there interaction between nurse's age and type of patient?

Solution:

1. Data. The data on length of home visit that were obtained during the study are shown in Table 8.5.5.

TABLE 8.5.5 Length of Home Visit in Minutes by Public Health Nurses by Nurse's Age Group and Type of Patient

Factor *A* (Type of Patient) Levels	Factor *B* (Nurse's Age Group) Levels 1 (20 to 29)	2 (30 to 39)	3 (40 to 49)	4 (50 and Over)
1 (Cardiac)	20	25	24	28
	25	30	28	31
	22	29	24	26
	27	28	25	29
	21	30	30	32
2 (Cancer)	30	30	39	40
	45	29	42	45
	30	31	36	50
	35	30	42	45
	36	30	40	60
3 (C.V.A.)	31	32	41	42
	30	35	45	50
	40	30	40	40
	35	40	40	55
	30	30	35	45
4 (Tuberculosis)	20	23	24	29
	21	25	25	30
	20	28	30	28
	20	30	26	27
	19	31	23	30

2. Assumptions. To analyze these data, we assume a fixed-effects model and a two-factor completely randomized design.

For our illustrative example we may test the following hypotheses subject to the conditions mentioned above.

a. H_0: $\alpha_1 = \alpha_2 = \alpha_3 = \alpha_4 = 0$ $\quad$ H_A: not all $\alpha_i = 0$
b. H_0: $\beta_1 = \beta_2 = \beta_3 = \beta_4 = 0$ $\quad$ H_A: not all $\beta_j = 0$
c. H_0: all $(\alpha\beta)_{ij} = 0$ $\quad$ H_A: not all $(\alpha\beta)_{ij} = 0$
Let $\alpha = .05$

3. Test statistic. The test statistic for each hypothesis set is V.R.

4. Distribution of test statistic. When H_0 is true and the assumptions are met, each of the test statistics is distributed as *F.*

5. **Decision rule.** Reject H_0 if the computed value of the test statistic is equal to or greater than the critical value of F. The critical values of F for testing the three hypotheses of our illustrative example are 2.76, 2.76, and 2.04, respectively. Since denominator degrees of freedom equal to 64 are not shown in Appendix Table G, 60 was used as the denominator degrees of freedom.

6. **Calculation of test statistic.** We use MINITAB to perform the calculations. We put the measurements in Column 1, the row (factor A) codes in Column 2, and the column (factor B) codes in Column 3. The resulting column contents are shown in Table 8.5.6. The MINITAB output is shown in Figure 8.5.3.

TABLE 8.5.6 Column Contents for MINITAB Calculations, Example 8.5.2

Row	C1	C2	C3	Row	C1	C2	C3
1	20	1	1	41	31	3	1
2	25	1	1	42	30	3	1
3	22	1	1	43	40	3	1
4	27	1	1	44	35	3	1
5	21	1	1	45	30	3	1
6	25	1	2	46	32	3	2
7	30	1	2	47	35	3	2
8	29	1	2	48	30	3	2
9	28	1	2	49	40	3	2
10	30	1	2	50	30	3	2
11	24	1	3	51	41	3	3
12	28	1	3	52	45	3	3
13	24	1	3	53	40	3	3
14	25	1	3	54	40	3	3
15	30	1	3	55	35	3	3
16	28	1	4	56	42	3	4
17	31	1	4	57	50	3	4
18	26	1	4	58	40	3	4
19	29	1	4	59	55	3	4
20	32	1	4	60	45	3	4
21	30	2	1	61	20	4	1
22	45	2	1	62	21	4	1
23	30	2	1	63	20	4	1
24	35	2	1	64	20	4	1
25	36	2	1	65	19	4	1
26	30	2	2	66	23	4	2
27	29	2	2	67	25	4	2
28	31	2	2	68	28	4	2
29	30	2	2	69	30	4	2

(*Continued*)

Row	C1	C2	C3	Row	C1	C2	C3
30	30	2	2	70	31	4	2
31	39	2	3	71	24	4	3
32	42	2	3	72	25	4	3
33	36	2	3	73	30	4	3
34	42	2	3	74	26	4	3
35	40	2	3	75	23	4	3
36	40	2	4	76	29	4	4
37	45	2	4	77	30	4	4
38	50	2	4	78	28	4	4
39	45	2	4	79	27	4	4
40	60	2	4	80	30	4	4

7. **Statistical decision.** The variance ratios are V.R. $(A) = 997.5/14.7 = 67.86$, V.R. $(B) = 400.4/14.7 = 27.24$, and V.R. $(AB) = 67.6/14.7 = 4.60$. Since the three computed values of V.R. are all greater than the corresponding critical values, we reject all three null hypotheses.
8. **Conclusion.** When H_0: $\alpha_1 = \alpha_2 = \alpha_3 = \alpha_4$ is rejected, we conclude that there are differences among the levels of A, that is, differences in the average amount of time spent in home visits with different types of patients. Similarly, when H_0: $\beta_1 = \beta_2 = \beta_3 = \beta_4$ is rejected, we conclude that there are differences among the levels of B, or differences in the average amount of time spent on home visits among the different nurses when grouped by age. When H_0: $(\alpha\beta)_{ij} = 0$ is rejected, we conclude that factors A and B interact; that is, different combinations of levels of the two factors produce different effects.
9. ***p* value.** Since 67.86, 27.24, and 4.60 are all greater than the critical values of $F_{.995}$ for the appropriate degrees of freedom, the p value for each of the tests is less than .005. When the hypothesis of no interaction is rejected, interest in the levels of factors A and B usually become subordinate to interest in the interaction effects. In other words, we are more interested in learning what combinations of levels are significantly different.

Figure 8.5.4 shows the SAS® output for the analysis of Example 8.5.2. ■

We have treated only the case where the number of observations in each cell is the same. When the number of observations per cell is not the same for every cell, the analysis becomes more complex.

In such cases the design is said to be unbalanced. To analyze these designs with MINITAB we use the general linear (GLM) procedure. Other software packages such as SAS® also will accommodate unequal cell sizes.

Dialog box:

Stat ➤ ANOVA ➤ Twoway

Type *C1* in **Response.** Type *C2* in **Row factor** and check **Display means.** Type *C3* in **Column factor** and check **Display means.** Click **OK.**

Session command:

```
MTB > TWOWAY C1 C2 C3;
SUBC > MEANS C2 C3.
```

Output:

Two-Way ANOVA: C1 versus C2, C3

```
Analysis of Variance for C1
Source          DF       SS          MS        F         P
C2               3   2992.4     997.483    67.94     0.000
C3               3   1201.1     400.350    27.27     0.000
Interaction      9    608.5      67.606     4.60     0.000
Error           64    939.6      14.681
Total           79   5741.5

                             Individual 95% CI
   C2       Mean             -+---------+---------+---------+---------+
    1      26.70                  (----*---)
    2      38.25                                                 (----*---)
    3      38.30                                                 (----*---)
    4      25.45             (----*---)
                             -+---------+---------+---------+---------+
                           24.00     28.00     32.00     36.00     40.00

                             Individual 95% CI
   C3       Mean             ------+---------+---------+---------+-----
    1      27.85                 (----*----)
    2      29.80                     (----*----)
    3      32.95                         (----*---)
    4      38.10                                            (----*---)
                             ------+---------+---------+---------+-----
                                 28.00     31.50     35.00     38.50
```

FIGURE 8.5.3 MINITAB procedure and ANOVA table for Example 8.5.2.

The SAS System

Analysis of Variance Procedure

```
Dependent Variable: TIME

Source                 DF   Sum of Squares   Mean Square   F Value      Pr > F

Model                  15    4801.95000000  320.13000000     21.81      0.0001

Error                  64     939.60000000   14.68125000

Corrected Total        79    5741.55000000

                 R-Square             C.V.      Root MSE              TIME Mean

                 0.836351         11.90866    3.83161193            32.17500000

Source                 DF         Anova SS   Mean Square   F Value      Pr > F

FACTORB                 3    1201.05000000  400.35000000     27.27      0.0001
FACTORA                 3    2992.45000000  997.48333333     67.94      0.0001
FACTORB*FACTORA         9    608.450000000   67.60555556      4.60      0.0001
```

FIGURE 8.5.4 SAS® output for analysis of Example 8.5.2.

EXERCISES

For Exercises 8.5.1 to 8.5.4, perform the analysis of variance, test appropriate hypotheses at the .05 level of significance, and determine the *p* value associated with each test.

8.5.1 Uryu et al. (A-21) studied the effect of three different doses of troglitazone (μM) on neuro cell death. Cell death caused by stroke partially results from the accumulation of high concentrations of glutamate. The researchers wanted to determine if different doses of troglitazone (1.3, 4.5, and 13.5 μM) and different ion forms ($-$ and $+$) of LY294002, a PI3-kinase inhibitor, would give different levels of neuroprotection. Four rats were studied at each dose and ion level, and the measured variable is the percent of cell death as compared to glutamate. Therefore, a higher value implies less neuroprotection. The results are in the table below.

Percent Compared to Glutamate	$-$LY294002 vs $+$LY294002	Troglitazone Dose (μM)
73.61	Negative	1.3
130.69	Negative	1.3
118.01	Negative	1.3
140.20	Negative	1.3

(*Continued*)

Percent Compared to Glutamate	−LY294002 vs +LY294002	Troglitazone Dose (μM)
97.11	Positive	1.3
114.26	Positive	1.3
120.26	Positive	1.3
92.39	Positive	1.3
26.95	Negative	4.5
53.23	Negative	4.5
59.57	Negative	4.5
53.23	Negative	4.5
28.51	Positive	4.5
30.65	Positive	4.5
44.37	Positive	4.5
36.23	Positive	4.5
−8.83	Negative	13.5
25.14	Negative	13.5
20.16	Negative	13.5
34.65	Negative	13.5
−35.80	Positive	13.5
−7.93	Positive	13.5
−19.08	Positive	13.5
5.36	Positive	13.5

Source: Shigeko Uryu. Used with permission.

8.5.2 Researchers at a trauma center wished to develop a program to help brain-damaged trauma victims regain an acceptable level of independence. An experiment involving 72 subjects with the same degree of brain damage was conducted. The objective was to compare different combinations of psychiatric treatment and physical therapy. Each subject was assigned to one of 24 different combinations of four types of psychiatric treatment and six physical therapy programs. There were three subjects in each combination. The response variable is the number of months elapsing between initiation of therapy and time at which the patient was able to function independently. The results were as follows:

Physical Therapy Program	Psychiatric Treatment			
	A	*B*	*C*	*D*
	11.0	9.4	12.5	13.2
I	9.6	9.6	11.5	13.2
	10.8	9.6	10.5	13.5
	10.5	10.8	10.5	15.0
II	11.5	10.5	11.8	14.6
	12.0	10.5	11.5	14.0
	12.0	11.5	11.8	12.8
III	11.5	11.5	11.8	13.7
	11.8	12.3	12.3	13.1

(Continued)

Physical Therapy Program	Psychiatric Treatment			
	A	*B*	*C*	*D*
	11.5	9.4	13.7	14.0
IV	11.8	9.1	13.5	15.0
	10.5	10.8	12.5	14.0
	11.0	11.2	14.4	13.0
V	11.2	11.8	14.2	14.2
	10.0	10.2	13.5	13.7
	11.2	10.8	11.5	11.8
VI	10.8	11.5	10.2	12.8
	11.8	10.2	11.5	12.0

Can one conclude on the basis of these data that the different psychiatric treatment programs have different effects? Can one conclude that the physical therapy programs differ in effectiveness? Can one conclude that there is interaction between psychiatric treatment programs and physical therapy programs? Let $\alpha = .05$ for each test.

Exercises 8.5.3 and 8.5.4 are optional since they have unequal cell sizes. It is recommended that the data for these be analyzed using SAS® or some other software package that will accept unequal cell sizes.

8.5.3 Main et al. (A-22) state, "Primary headache is a very common condition and one that nurses encounter in many different care settings. Yet there is a lack of evidence as to whether advice given to sufferers is effective and what improvements may be expected in the conditions." The researchers assessed frequency of headaches at the beginning and end of the study for 19 subjects in an intervention group (treatment 1) and 25 subjects in a control group (treatment 2). Subjects in the intervention group received health education from a nurse, while the control group did not receive education. In the 6 months between pre- and post-evaluation, the subjects kept a headache diary. The following table gives as the response variable the difference (pre − post) in frequency of headaches over the 6 months for two factors: (1) treatment with two levels (intervention and control), and (2) migraine status with two levels (migraine sufferer and nonmigraine sufferer).

Change in Frequency of Headaches	Migraine Sufferer (1 = No, 2 = Yes)	Treatment	Change in Frequency of Headaches	Migraine Sufferer (1 = No, 2 = Yes)	Treatment
−2	1	1	−3	2	2
2	2	1	−6	2	2
33	1	1	11	1	2
−6	2	1	64	1	2
6	2	1	65	1	2
98	1	1	14	1	2
2	2	1	8	1	2
6	2	1	6	2	2

(*Continued*)

Change in Frequency of Headaches	Migraine Sufferer (1 = No, 2 = Yes)	Treatment	Change in Frequency of Headaches	Migraine Sufferer (1 = No, 2 = Yes)	Treatment
33	1	1	14	1	2
−7	2	1	−11	2	2
−1	2	1	53	1	2
−12	2	1	26	2	2
12	1	1	3	1	2
64	1	1	15	1	2
36	2	1	3	1	2
6	2	1	41	1	2
4	2	1	16	1	2
11	2	1	−4	2	2
0	2	1	−6	2	2
			9	1	2
			9	2	2
			−3	2	2
			9	2	2
			3	1	2
			4	2	2

Source: A. Main, H. Abu-Saad, R. Salt, I. Vlachonikolis, and A. Dowson, "Management by Nurses of Primary Headache: A Pilot Study," *Current Medical Research Opinion, 18* (2002), 471–478. Used with Permission.

Can one conclude on the basis of these data that there is a difference in the reduction of headache frequency between the control and treatment groups? Can one conclude that there is a difference in the reduction of headache frequency between migraine and non-migraine sufferers? Can one conclude that there is interaction between treatments and migraine status? Let $\alpha = .05$ for each test.

8.5.4 The purpose of a study by Porcellini et al. (A-23) was to study the difference in CD4 cell response in patients taking highly active antiretroviral therapy (HAART, treatment 1) and patients taking HAART plus intermittent interleukin (IL-2, treatment 2). Another factor of interest was the HIV-RNA plasma count at baseline of study. Subjects were classified as having fewer than 50 copies/ml (plasma 1) or having 50 or more copies/ml (plasma 2). The outcome variable is the percent change in CD4 T cell count from baseline to 12 months of treatment. Can one conclude that there is a difference in the percent change in CD4 T cell count between the two treatments? The results are shown in the following table. Can one conclude that there is a difference in the percent change in CD4 T cell count between those who have fewer than 50/ml plasma copies of HIV-RNA and those who do not? Can one conclude that there is interaction between treatments and plasma levels? Let $\alpha = .05$ for each test.

Percent Change in CD4 T Cell	Treatment	Plasma
−12.60	1	1
−14.60	2	1
28.10	2	1

(Continued)

Percent Change in CD4 T Cell	Treatment	Plasma
77.30	1	1
−0.44	1	1
50.20	1	1
48.60	2	2
86.20	2	2
205.80	1	2
100.00	1	2
34.30	1	2
82.40	1	2
118.30	1	2

Source: Simona Porcellini, Guiliana Vallanti, Silvia Nozza, Guido Poli, Adriano Lazzarin, Guiseppe Tabussi, and Antonio Grassia, "Improved Thymopoietic Potential in Aviremic HIV Infected Individuals with HAART by Intermittent IL-2 Administration," *AIDS, 17* (2003), 1621–1630. Used with permission.

8.5.5 A study by Górecka et al. (A-24) assessed the manner in which among middle-aged smokers the diagnosis of airflow limitation (AL) combined with advice to stop smoking influences the smoking cessation rate. Their concerns were whether having AL, whether the subject successfully quit smoking, and whether interaction between AL and smoking status were significant factors in regard to baseline variables and lung capacity variables at the end of the study. Some of the variables of interest were previous years of smoking (pack years), age at which subject first began smoking, forced expiratory volume in one second (FEV_1), and forced vital capacity (FVC). There were 368 subjects in the study. What are the factors in this study? At how many levels does each occur? Who are the subjects? What is (are) the response variable(s)? Can you think of any extraneous variables whose effects are included in the error term?

8.5.6 A study by Meltzer et al. (A-25) examined the response to 5 mg desloratadine, an H1-receptor antagonist, in patients with seasonal allergies. During the fall allergy season, 172 subjects were randomly assigned to receive treatments of desloratadine and 172 were randomly assigned to receive a placebo. Subjects took the medication for 2 weeks after which changes in the nasal symptom score were calculated. A significant reduction was noticed in the treatment group compared to the placebo group, but gender was not a significant factor. What are the factors in the study? At how many levels does each occur? What is the response variable?

8.6 SUMMARY

The purpose of this chapter is to introduce the student to the basic ideas and techniques of analysis of variance. Two experimental designs, the completely randomized and the randomized complete block, are discussed in considerable detail. In addition, the concept of repeated measures designs and a factorial experiment as used with the completely randomized design are introduced. Individuals who wish to pursue further any aspect of analysis of variance will find the methodology references at the end of the chapter most helpful.

SUMMARY OF FORMULAS FOR CHAPTER 8

Formula Number	Name	Formula
8.2.1	One-way ANOVA model	$X_{ij} = \mu + \tau_i + \epsilon_{ij}$
8.2.2	Total sum-of-squares	$SST = \sum_{j=1}^{k}\sum_{i=1}^{n_j}(x_{ij} - \bar{x}_{..})^2$
8.2.3	Within-group sum-of-squares	$SSW = \sum_{j=1}^{k}\sum_{i=1}^{n_j}(x_{ij} - \bar{x}_{.j})^2$
8.2.4	Among-group sum-of-squares	$SSA = \sum_{j=1}^{k} n_j(\bar{x}_{.j} - \bar{x}_{..})^2$
8.2.5	Within-group variance	$MSW = \dfrac{\sum_{j=1}^{k}\sum_{i=1}^{n_j}(\bar{x}_{ij} - \bar{x}_{.j})^2}{\sum_{j=1}^{k}(n_j - 1)}$
8.2.6	Among-group variance I	$\sigma^2 = n\sigma_{\bar{x}}^2$
8.2.7	Among-group variance II (equal sample sizes)	$MSA = \dfrac{n\sum_{j=1}^{k}(\bar{x}_{.j} - \bar{x}_{..})^2}{(k-1)}$
8.2.8	Among-group variance III (unequal sample sizes)	$MSA = \dfrac{\sum_{j=1}^{k} n_j(\bar{x}_{.j} - \bar{x}_{..})^2}{(k-1)}$
8.2.9	Tukey's HSD (equal sample sizes)	$HSD = q_{a,k,N-k}\sqrt{\dfrac{MSE}{n}}$
8.2.10	Tukey's HSD (unequal sample sizes)	$HSD^* = q_{a,k,N-k}\sqrt{\dfrac{MSE}{n}\left(\dfrac{1}{n_i} + \dfrac{1}{n_j}\right)}$
8.3.1	Two-way ANOVA model	$X_{ij} = \mu + \beta_i + \tau_i + \epsilon_{ij}$
8.3.2	Sum-of-squares representation	SST = SSBl + SSTr + SSE

(*Continued*)

8.3.3	Sum-of-squares total	$SST = \sum_{j=1}^{k}\sum_{i=1}^{n}(\bar{x}_{ij} - \bar{x}_{..})^2$
8.3.4	Sum-of-squares block	$SSBl = \sum_{j=1}^{k}\sum_{i=1}^{n}(\bar{x}_{i.} - \bar{x}_{..})^2$
8.3.5	Sum-of-squares treatments	$SSTr = \sum_{j=1}^{k}\sum_{i=1}^{n}(\bar{x}_{.j} - \bar{x}_{..})^2$
8.3.6	Sum-of-squares error	SSE = SST − SSBl − SSTr
8.4.1	Fixed-effects, additive single-factor, repeated-measures ANOVA model	$x_{ij} = \mu + \beta_i + \tau_i + \epsilon_{ij}$
8.5.1	Two-factor completely randomized fixed-effects factorial model	$X_{ijk} = \mu + \alpha_i + \beta_j + (\alpha\beta)_{ij} + \epsilon_{ijk}$
8.5.2	Probabilistic representation of α	$\alpha < 1 - (1 - \alpha')(1 - \alpha'')(1 - \alpha''')$
8.5.3	Sum-of-squares total I	$\sum_{i=1}^{a}\sum_{j=1}^{b}\sum_{k=1}^{n}(x_{ijk} - \bar{x}_{...}) = \sum_{i=1}^{a}\sum_{j=1}^{b}\sum_{k=1}^{n}(\bar{x}_{ij.} - \bar{x}_{...}) + \sum_{i=1}^{a}\sum_{j=1}^{b}\sum_{k=1}^{n}(x_{ijk} - \bar{x}_{ij.})$
8.5.4	Sum-of-squares total II	SST = SSTr + SSE
8.5.5	Sum-of-squares treatment partition	$\sum_{i=1}^{a}\sum_{j=1}^{b}\sum_{k=1}^{n}(\bar{x}_{ij.} - \bar{x}_{...})^2 = \sum_{i=1}^{a}\sum_{j=1}^{b}\sum_{k=1}^{n}(\bar{x}_{i..} - \bar{x}_{...})^2 + \sum_{i=1}^{a}\sum_{j=1}^{b}\sum_{k=1}^{n}(\bar{x}_{.j.} - \bar{x}_{...})^2 + \sum_{i=1}^{a}\sum_{j=1}^{b}\sum_{k=1}^{n}(\bar{x}_{ij.} - \bar{x}_{i..} - \bar{x}_{.j.} + \bar{x}_{...})^2$
Symbol Key	• α = probability of type I error • α_i = treatment A effect • β_j = treatment B effect • β_i = block effect • $(\alpha\beta)_{ij}$ = interaction effect • ϵ_{ij} = error term • HSD = honestly significant difference • k = number of treatments	

	• μ = mean of population (or the grand mean) • n = number of blocks • n_x = sample size • σ^2 = variance • SSX = sum-of-squares (where X: A = among, Bl = block, T = total, Tr = treatment, W = within) • τ_i = treatment effect • X_{xxx} = measurement

REVIEW QUESTIONS AND EXERCISES

1. Define analysis of variance.
2. Describe the completely randomized design.
3. Describe the randomized block design.
4. Describe the repeated measures design.
5. Describe the factorial experiment as used in the completely randomized design.
6. What is the purpose of Tukey's HSD test?
7. What is an experimental unit?
8. What is the objective of the randomized complete block design?
9. What is interaction?
10. What is a mean square?
11. What is an ANOVA table?
12. For each of the following designs describe a situation in your particular field of interest where the design would be an appropriate experimental design. Use real or realistic data and do the appropriate analysis of variance for each one:
    (a) Completely randomized design
    (b) Randomized complete block design
    (c) Completely randomized design with a factorial experiment
    (d) Repeated measures designs
13. Werther et al. (A-26) examined the β-leucocyte count ($\times 10^9$/L) in 51 subjects with colorectal cancer and 19 healthy controls. The cancer patients were also classified into Dukes's classification (A, B, C) for colorectal cancer that gives doctors a guide to the risk, following surgery, of the cancer coming back or spreading to other parts of the body. An additional category (D) identified patients with disease that had not been completely resected. The results are displayed in the following table. Perform an analysis of these data in which you identify the sources of variability and specify the degrees of freedom for each. Do these data provide sufficient evidence to indicate that, on the average, leucocyte counts differ among the five categories? Let $\alpha = .01$ and find the p value. Use Tukey's procedure to test for significant differences between individual pairs of sample means.

Healthy	A	B	C	D
6.0	7.7	10.4	8.0	9.5
6.3	7.8	5.6	6.7	7.8
5.1	6.1	7.0	9.3	5.7
6.2	9.6	8.2	6.6	8.0
10.4	5.5	9.0	9.3	9.6
4.4	5.8	8.4	7.2	13.7
7.4	4.0	8.1	5.2	6.3
7.0	5.4	8.0	9.8	7.3
5.6		6.5	6.2	6.2
5.3		9.1	10.1	
2.6		11.0	9.3	
6.3		10.9	9.4	
6.1		10.6	6.5	
5.3		5.2	5.4	
5.4		7.9	7.6	
5.2		7.6	9.2	
4.3		5.8		
4.9		7.0		
7.3				
4.9				
6.9				
4.3				
5.6				
5.1				

Source: Kim Werther, M.D., Ph.D. Used with permission.

14. In Example 8.4.1, we examined data from a study by Licciardone et al. (A-15) on osteopathic manipulation as a treatment for chronic back pain. At the beginning of that study, there were actually 91 subjects randomly assigned to one of three treatments: osteopathic manipulative treatment (OMT), sham manipulation (SHAM), or non-intervention (CONTROL). One important outcome variable was the rating of back pain at the beginning of the study. The researchers wanted to know if the treatment had essentially the same mean pain level at the start of the trial. The results are displayed in the following table. The researchers used a visual analog scale from 0 to 10 cm where 10 indicated "worst pain possible." Can we conclude, on the basis of these data, that, on the average, pain levels differ in the three treatment groups? Let $\alpha = .05$ and find the p value. If warranted, use Tukey's procedure to test for differences between individual pairs of sample means.

CONTROL	SHAM	OMT	
2.6	5.8	7.8	3.5
5.6	1.3	4.1	3.4
3.3	2.4	1.7	1.1
4.6	1.0	3.3	0.5
8.4	3.2	4.3	5.1
0.0	0.4	6.5	1.9
2.5	5.4	5.4	2.0
5.0	4.5	4.0	2.8

(Continued)

CONTROL	SHAM	OMT	
1.7	1.5	4.1	3.7
3.8	0.0	2.6	1.6
2.4	0.6	3.2	0.0
1.1	0.0	2.8	0.2
0.7	7.6	3.4	7.3
2.4	3.5	6.7	1.7
3.3	3.9	7.3	7.5
6.6	7.0	2.1	1.6
0.4	7.4	3.7	3.0
0.4	6.5	2.3	6.5
0.9	1.6	4.4	3.0
6.0	1.3	2.8	3.3
	6.6	0.4	7.3
	6.3	0.7	4.6
	7.0	7.9	4.8
		1.3	4.9

Source: J. C. Licciardone, D.O. Used with permission.

15. The goal of a study conducted by Meshack and Norman (A-27) was to evaluate the effects of weights on postural hand tremor related to self-feeding in subjects with Parkinson's disease (PD). Each of the 16 subjects had the tremor amplitude measured (in mm) under three conditions: holding a built-up spoon (108 grams), holding a weighted spoon (248 grams), and holding the built-up spoon while wearing a weighted wrist cuff (470 grams). The data are displayed in the following table.

	Tremor Amplitude (mm)		
Subject	Built-Up Spoon	Weighted Spoon	Built-Up Spoon + Wrist Cuff
1	.77	1.63	1.02
2	.78	.88	1.11
3	.17	.14	.14
4	.30	.27	.26
5	.29	.27	.28
6	1.60	1.49	1.73
7	.38	.39	.37
8	.24	.24	.24
9	.17	.17	.16
10	.38	.29	.27
11	.93	1.21	.90
12	.63	.52	.66
13	.49	.73	.76
14	.42	.60	.29
15	.19	.21	.21
16	.19	.20	.16

Source: Rubia P. Meshack and Kathleen E. Norman, "A Randomized Controlled Trial of the Effects of Weights on Amplitude and Frequency of Postural Hand Tremor in People with Parkinson's Disease," *Clinical Rehabilitation, 16* (2003), 481–492.

Can one conclude on the basis of these data that the three experimental conditions, on the average, have different effects on tremor amplitude? Let $\alpha = .05$. Determine the p value.

16. In a study of pulmonary effects on guinea pigs, Lacroix et al. (A-28) exposed 18 ovalbumin-sensitized guinea pigs and 18 nonsensitized guinea pigs to regular air, benzaldehyde, and acetaldehyde. At the end of exposure, the guinea pigs were anesthetized and allergic responses were assessed in bronchoalveolar lavage (BAL). The following table shows the alveolar cell count ($\times 10^6$) by treatment group for the ovalbumin-sensitized and nonsensitized guinea pigs.

Ovalbumin-Sensitized	Treatment	Alveolar Count $\times 10^6$
no	acetaldehyde	49.90
no	acetaldehyde	50.60
no	acetaldehyde	50.35
no	acetaldehyde	44.10
no	acetaldehyde	36.30
no	acetaldehyde	39.15
no	air	24.15
no	air	24.60
no	air	22.55
no	air	25.10
no	air	22.65
no	air	26.85
no	benzaldehyde	31.10
no	benzaldehyde	18.30
no	benzaldehyde	19.35
no	benzaldehyde	15.40
no	benzaldehyde	27.10
no	benzaldehyde	21.90
yes	acetaldehyde	90.30
yes	acetaldehyde	72.95
yes	acetaldehyde	138.60
yes	acetaldehyde	80.05
yes	acetaldehyde	69.25
yes	acetaldehyde	31.70
yes	air	40.20
yes	air	63.20
yes	air	59.10
yes	air	79.60
yes	air	102.45
yes	air	64.60
yes	benzaldehyde	22.15
yes	benzaldehyde	22.75
yes	benzaldehyde	22.15
yes	benzaldehyde	37.85
yes	benzaldehyde	19.35
yes	benzaldehyde	66.70

Source: G. Lacroix, Docteur en Toxicologie. Used with permission.

Test for differences (a) between ovalbumin-sensitized and nonsensitized outcomes, (b) among the three different exposures, and (c) interaction. Let $\alpha = .05$ for all tests.

17. Watanabe et al. (A-29) studied 52 healthy middle-aged male workers. The researchers used the Masstricht Vital Exhaustion Questionnaire to assess vital exhaustion. Based on the resultant scores, they assigned subjects into three groups: VE1, VE2, and VE3. VE1 indicates the fewest signs of exhaustion, and VE3 indicates the most signs of exhaustion. The researchers also asked subjects about their smoking habits. Smoking status was categorized as follows: SMOKE1 are nonsmokers, SMOKE2 are light smokers (20 cigarettes or fewer per day), SMOKE3 are heavy smokers (more than 20 cigarettes per day). One of the outcome variables of interest was the amplitude of the high-frequency spectral analysis of heart rate variability observed during an annual health checkup. This variable, HF-amplitude, was used as an index of parasympathetic nervous function. The data are summarized in the following table:

	HF-Amplitude			
	Smoking Status			
Vital Exhaustion Group	**SMOKE1**		**SMOKE2**	**SMOKE3**
VE1	23.33	13.37	16.14	16.83
	31.82	9.76	20.80	29.40
	10.61	22.24	15.44	6.50
	42.59	8.77	13.73	10.18
	23.15	20.28		13.86
	17.29			
VE2	20.69	11.67	44.92	27.91
	16.21	30.17		36.89
	28.49	29.20		16.80
	25.67	8.73		17.08
	15.29	9.08		18.77
	7.51	22.53		18.33
	22.03	17.19		
	10.27			
VE3	9.44		17.59	5.57
	19.16		18.90	13.51
	14.46			17.37
				10.63
				13.83

Source: Takemasa Watanabe, M.D., Ph.D. Used with permission.

Perform an analysis of variance on these data and test the three possible hypotheses. Let $\alpha' = \alpha'' = \alpha''' = .05$. Determine the p values.

18. The effects of thermal pollution on *Corbicula fluminea* (Asiatic clams) at three different geographical locations were analyzed by John Brooker (A-30). Sample data on clam shell length, width, and height are displayed in the following table. Determine if there is a significant difference in mean length, height, or width (measured in mm) of the clam shell at the three different locations by performing three analyses. What inferences can be made from your results? What are the assumptions underlying your inferences? What are the target populations?

Location 1			Location 2			Location 3		
Length	Width	Height	Length	Width	Height	Length	Width	Height
7.20	6.10	4.45	7.25	6.25	4.65	5.95	4.75	3.20
7.50	5.90	4.65	7.23	5.99	4.20	7.60	6.45	4.56
6.89	5.45	4.00	6.85	5.61	4.01	6.15	5.05	3.50
6.95	5.76	4.02	7.07	5.91	4.31	7.00	5.80	4.30
6.73	5.36	3.90	6.55	5.30	3.95	6.81	5.61	4.22
7.25	5.84	4.40	7.43	6.10	4.60	7.10	5.75	4.10
7.20	5.83	4.19	7.30	5.95	4.29	6.85	5.55	3.89
6.85	5.75	3.95	6.90	5.80	4.33	6.68	5.50	3.90
7.52	6.27	4.60	7.10	5.81	4.26	5.51	4.52	2.70
7.01	5.65	4.20	6.95	5.65	4.31	6.85	5.53	4.00
6.65	5.55	4.10	7.39	6.04	4.50	7.10	5.80	4.45
7.55	6.25	4.72	6.54	5.89	3.65	6.81	5.45	3.51
7.14	5.65	4.26	6.39	5.00	3.72	7.30	6.00	4.31
7.45	6.05	4.85	6.08	4.80	3.51	7.05	6.25	4.71
7.24	5.73	4.29	6.30	5.05	3.69	6.75	5.65	4.00
7.75	6.35	4.85	6.35	5.10	3.73	6.75	5.57	4.06
6.85	6.05	4.50	7.34	6.45	4.55	7.35	6.21	4.29
6.50	5.30	3.73	6.70	5.51	3.89	6.22	5.11	3.35
6.64	5.36	3.99	7.08	5.81	4.34	6.80	5.81	4.50
7.19	5.85	4.05	7.09	5.95	4.39	6.29	4.95	3.69
7.15	6.30	4.55	7.40	6.25	4.85	7.55	5.93	4.55
7.21	6.12	4.37	6.00	4.75	3.37	7.45	6.19	4.70
7.15	6.20	4.36	6.94	5.63	4.09	6.70	5.55	4.00
7.30	6.15	4.65				7.51	6.20	4.74
6.35	5.25	3.75				6.95	5.69	4.29
						7.50	6.20	4.65

Source: John Brooker, M.S. and the Wright State University Statistical Consulting Center. Used with permission.

19. Eleftherios Kellis (A-31) conducted an experiment on 18 pubertal males. He recorded the electromyographic (EMG) activity at nine different angular positions of the biceps femoris muscle. The EMG values are expressed as a percent (0–100 percent) of the maximal effort exerted with the muscle and represent an average in a range of flexion angles. The nine positions correspond to testing knee flexion angles of 1–10°, 11–20°, 21–30°, 31–40°, 41–50°, 51–60°, 61–70°, 71–80°, and 81–90°. The results are displayed in the following table. For subject 1, for example, the value of 30.96 percent represents the average maximal percent of effort in angular positions from 1 to 10 degrees.

Subject	1–10°	11–20°	21–30°	31–40°	41–50°	51–60°	61–70°	71–80°	81–90°
1	30.96	11.32	4.34	5.99	8.43	10.50	4.49	10.93	33.26
2	3.61	1.47	3.50	10.25	3.30	3.62	10.14	11.05	8.78
3	8.46	2.94	1.83	5.80	11.59	15.17	13.04	10.57	8.22
4	0.69	1.06	1.39	1.08	0.96	2.52	2.90	3.27	5.52
5	4.40	3.02	3.74	3.83	3.73	10.16	9.31	12.70	11.45
6	4.59	9.80	10.71	11.64	9.78	6.91	8.53	8.30	11.75
7	3.31	3.31	4.12	12.56	4.60	1.88	2.42	2.46	2.19
8	1.98	6.49	2.61	3.28	10.29	7.56	16.68	14.52	13.49
9	10.43	4.96	12.37	24.32	17.16	34.71	35.30	37.03	45.65
10	20.91	20.72	12.70	15.06	12.03	11.31	28.47	26.81	25.08
11	5.59	3.13	2.83	4.31	6.37	13.95	13.48	11.15	30.97
12	8.67	4.32	2.29	6.20	13.01	19.30	9.33	12.30	12.20
13	2.11	1.59	2.40	2.56	2.83	2.55	5.84	5.23	8.84
14	3.82	5.04	6.81	10.74	10.10	13.14	19.39	13.31	12.02
15	39.51	62.34	70.46	20.48	17.38	54.04	25.76	50.32	46.84
16	3.31	4.95	12.49	9.18	14.00	16.17	25.75	11.82	13.17
17	11.42	7.53	4.65	4.70	7.57	9.86	5.30	4.47	3.99
18	2.97	2.18	2.36	4.61	7.83	17.49	42.55	61.84	39.70

Source: Eleftherios Kellis, Ph.D. Used with permission.

Can we conclude on the basis of these data that the average EMG values differ among the nine angular locations? Let $\alpha = .05$.

20. In a study of Marfan syndrome, Pyeritz et al. (A-32) reported the following severity scores of patients with no, mild, and marked dural ectasia. May we conclude, on the basis of these data, that mean severity scores differ among the three populations represented in the study? Let $\alpha = .05$ and find the p value. Use Tukey's procedure to test for significant differences among individual pairs of sample means.

No dural ectasia: 18, 18, 20, 21, 23, 23, 24, 26, 26, 27, 28, 29, 29, 29, 30, 30, 30, 30, 32, 34, 34, 38

Mild dural ectasia: 10, 16, 22, 22, 23, 26, 28, 28, 28, 29, 29, 30, 31, 32, 32, 33, 33, 38, 39, 40, 47

Marked dural ectasia: 17, 24, 26, 27, 29, 30, 30, 33, 34, 35, 35, 36, 39

Source: Reed E. Pyeritz, M.D., Ph.D. Used with permission.

21. The following table shows the arterial plasma epinephrine concentrations (nanograms per milliliter) found in 10 laboratory animals during three types of anesthesia:

	Animal									
Anesthesia	**1**	**2**	**3**	**4**	**5**	**6**	**7**	**8**	**9**	**10**
A	.28	.50	.68	.27	.31	.99	.26	.35	.38	.34
B	.20	.38	.50	.29	.38	.62	.42	.87	.37	.43
C	1.23	1.34	.55	1.06	.48	.68	1.12	1.52	.27	.35

Can we conclude from these data that the three types of anesthesia, on the average, have different effects? Let $\alpha = .05$.

22. The aim of a study by Hartman-Maeir et al. (A-33) was to evaluate the awareness of deficit profiles among stroke patients undergoing rehabilitation. She studied 35 patients with a stroke lesion in the right hemisphere and 19 patients with a lesion on the left hemisphere. She also grouped lesion size as

$$2 = 1\text{–}3 \text{ cm},\ 3 = 3 - 5 \text{ cm, and } 4 = 5 \text{ cm or greater.}$$

One of the outcome variables was a measure of each patient's total unawareness of their own limitations. Scores ranged from 8 to 24, with higher scores indicating more unawareness.

	Unawareness Score		
Lesion Size Group	**Left Hemisphere**	**Right Hemisphere**	
2	11	10	8
	13	11	10
	10	13	9
	11	10	9
	9	13	9
	10	10	
	9	10	
	8	9	
	10	8	
3	13	11	10
	8	10	11
	10	10	12
	10	14	11
	10	8	
4	11	10	11
	13	13	9
	14	10	19
	13	10	10
	14	15	9
		8	10

Source: Adina Hartman-Maeir, Ph.D., O.T.R. Used with permission.

Test for a difference in lesion size, hemisphere, and interaction. Let $\alpha = .05$ for all tests.

23. A random sample of the records of single births was selected from each of four populations. The weights (grams) of the babies at birth were as follows:

Sample			
A	**B**	**C**	**D**
2946	3186	2300	2286
2913	2857	2903	2938
2280	3099	2572	2952
3685	2761	2584	2348
2310	3290	2675	2691

(*Continued*)

Sample			
A	B	C	D
2582	2937	2571	2858
3002	3347		2414
2408			2008
			2850
			2762

Do these data provide sufficient evidence to indicate, at the .05 level of significance, that the four populations differ with respect to mean birth weight? Test for a significant difference between all possible pairs of means.

24. The following table shows the aggression scores of 30 laboratory animals reared under three different conditions. One animal from each of 10 litters was randomly assigned to each of the three rearing conditions.

	Rearing Condition		
Litter	Extremely Crowded	Moderately Crowded	Not Crowded
1	30	20	10
2	30	10	20
3	30	20	10
4	25	15	10
5	35	25	20
6	30	20	10
7	20	20	10
8	30	30	10
9	25	25	10
10	30	20	20

Do these data provide sufficient evidence to indicate that level of crowding has an effect on aggression? Let $\alpha = .05$.

25. The following table shows the vital capacity measurements of 60 adult males classified by occupation and age group:

	Occupation			
Age Group	A	B	C	D
1	4.31	4.68	4.17	5.75
	4.89	6.18	3.77	5.70
	4.05	4.48	5.20	5.53
	4.44	4.23	5.28	5.97
	4.59	5.92	4.44	5.52

(Continued)

	Occupation			
Age Group	A	B	C	D
2	4.13	3.41	3.89	4.58
	4.61	3.64	3.64	5.21
	3.91	3.32	4.18	5.50
	4.52	3.51	4.48	5.18
	4.43	3.75	4.27	4.15
3	3.79	4.63	5.81	6.89
	4.17	4.59	5.20	6.18
	4.47	4.90	5.34	6.21
	4.35	5.31	5.94	7.56
	3.59	4.81	5.56	6.73

Test for differences among occupations, for differences among age groups, and for interaction. Let $\alpha = .05$ for all tests.

26. Complete the following ANOVA table and state which design was used.

Source	SS	*d.f.*	MS	V.R.	*p*
Treatments	154.9199	4			
Error					
Total	200.4773	39			

27. Complete the following ANOVA table and state which design was used.

Source	SS	*d.f.*	MS	V.R.	*p*
Treatments		3			
Blocks	183.5	3			
Error	26.0				
Total	709.0	15			

28. Consider the following ANOVA table.

Source	SS	*d.f.*	MS	V.R.	*p*
A	12.3152	2	6.15759	29.4021	<.005
B	19.7844	3	6.59481	31.4898	<.005
AB	8.94165	6	1.49027	7.11596	<.005
Treatments	41.0413	11			
Error	10.0525	48	0.209427		
Total	51.0938	59			

(a) What sort of analysis was employed?

(b) What can one conclude from the analysis? Let $\alpha = .05$.

29. Consider the following ANOVA table.

Source	SS	*d.f.*	MS	V.R.
Treatments	5.05835	2	2.52917	1.0438
Error	65.42090	27	2.4230	

(a) What design was employed?

(b) How many treatments were compared?

(c) How many observations were analyzed?

(d) At the .05 level of significance, can one conclude that there is a difference among treatments? Why?

30. Consider the following ANOVA table.

Source	SS	*d.f.*	MS	V.R.
Treatments	231.5054	2	115.7527	2.824
Blocks	98.5000	7	14.0714	
Error	573.7500	14	40.9821	

(a) What design was employed?

(b) How many treatments were compared?

(c) How many observations were analyzed?

(d) At the .05 level of significance, can one conclude that the treatments have different effects? Why?

31. In a study of the relationship between smoking and serum concentrations of high-density lipoprotein cholesterol (HDL-C), the following data (coded for ease of calculation) were collected from samples of adult males who were nonsmokers, light smokers, moderate smokers, and heavy smokers. We wish to know if these data provide sufficient evidence to indicate that the four populations differ with respect to mean serum concentration of HDL-C. Let the probability of committing a type I error be .05. If an overall significant difference is found, determine which pairs of individual sample means are significantly different.

Smoking Status			
Nonsmokers	**Light**	**Moderate**	**Heavy**
12	9	5	3
10	8	4	2
11	5	7	1
13	9	9	5
9	9	5	4
9	10	7	6
12	8	6	2

32. Polyzogopoulou et al. (A-34) report the effects of bariatric surgery on fasting glucose levels (mmol/L) on 12 obese subjects with type 2 diabetes at four points in time: pre-operation, at 3 months, 6 months, and 12 months. Can we conclude, after eliminating subject effects, that fasting glucose levels differ over time after surgery? Let $\alpha = .05$.

Subject No.	Pre-op	3 Months	6 Months	12 Months
1	108.0	200.0	94.3	92.0
2	96.7	119.0	84.0	93.0
3	77.0	130.0	76.0	74.0
4	92.0	181.0	82.5	80.5
5	97.0	134.0	81.0	76.0
6	94.0	163.0	96.0	71.0
7	76.0	125.0	74.0	75.5
8	100.0	189.0	97.0	88.5
9	82.0	282.0	91.0	93.0
10	103.5	226.0	86.0	80.5
11	85.5	145.0	83.5	83.0
12	74.5	156.0	71.0	87.0

Source: Theodore K. Alexandrides, M.D. Used with permission.

33. Refer to Review Exercise 32. In addition to studying the 12 type 2 diabetes subjects (group 1), Polyzogopoulou et al. (A-34) studied five subjects with impaired glucose tolerance (group 2), and eight subjects with normal glucose tolerance (group 3). The following data are the 12-month post-surgery fasting glucose levels for the three groups.

Group	
1.0	92.0
1.0	93.0
1.0	74.0
1.0	80.5
1.0	76.0
1.0	71.0
1.0	75.5
1.0	88.5
1.0	93.0
1.0	80.5
1.0	83.0
1.0	87.0
2.0	79.0
2.0	78.0
2.0	100.0
2.0	76.5
2.0	68.0
3.0	81.5
3.0	75.0
3.0	76.5

(*Continued*)

Group	
3.0	70.5
3.0	69.0
3.0	73.8
3.0	74.0
3.0	80.0

Source: Theodore K. Alexandrides, M.D. Used with permission.

Can we conclude that there is a difference among the means of the three groups? If so, which pairs of means differ? Let $\alpha = .05$ for all tests.

For exercises 34 to 38 do the following:

(a) Indicate which technique studied in this chapter (the completely randomized design, the randomized block design, the repeated measures design, or the factorial experiment) is appropriate.

(b) Identify the response variable and treatment variables.

(c) As appropriate, identify the factors and the number of levels of each, the blocking variables, and the subjects.

(d) List any extraneous variables whose effects you think might be included in the error term.

(e) As appropriate, comment on carry-over and position effects.

(f) Construct an ANOVA table in which you indicate the sources of variability and the number of degrees of freedom for each.

34. Johnston and Bowling (A-35) studied the ascorbic acid content (vitamin C) in several orange juice products. One of the products examined was ready-to-drink juice packaged in a re-sealable, screw-top container. One analysis analyzed the juice for reduced and oxidized vitamin C content at time of purchase and reanalyzed three times weekly for 4 to 5 weeks.

35. A study by Pittini et al. (A-36) assessed the effectiveness of a simulator-based curriculum on 30 trainees learning the basic practice of amniocentesis. Pre- and post-training performance were evaluated with the same instrument. The outcome variable was the post-training score − pretraining score. Trainees were grouped by years of postgraduate experience: PGY 0–2, PGY 3–5, Fellows, and Faculty.

36. Anim-Nyame et al. (A-37) studied three sets of women in an effort to understand factors related to pre-eclampsia. Enrolled in the study were 18 women with pre-eclampsia, 18 normal pregnant women, and 18 nonpregnant female matched controls. Blood samples were obtained to measure plasma levels of vascular endothelial growth factor, leptin, TNF-α plasma protein concentrations, and full blood count.

37. In a study by Iwamoto et al. (A-38) 26 women were randomly assigned to the medication alfacalcidol for treatment of lumbar bone mineral density (BMD). BMD of the lumbar spine was measured at baseline and every year for 5 years.

38. Inoue et al. (A-39) studied donor cell type and genotype on the efficiency of mouse somatic cell cloning. They performed a factorial experiment with two donor cell types (Sertoli cells or cumulus) and six genotypes. Outcome variables were the cleavage rate and the birth rate of pups in each treatment combination.

For the studies described in Exercises 39 through 66, do the following:

(a) Perform a statistical analysis of the data (including hypothesis testing and confidence interval construction) that you think would yield useful information for the researchers.

(b) Determine p values for each computed test statistic.

(c) State all assumptions that are necessary to validate your analysis.

(d) Describe the population(s) about which you think inferences based on your analysis would be applicable.

39. Shirakami et al. (A-40) investigated the clinical significance of endothelin (ET), natriuretic peptides, and the renin-angiotensin-aldosterone system in pediatric liver transplantation. Subjects were children ages 6 months to 12 years undergoing living-related liver transplantation due to congenital biliary atresia and severe liver cirrhosis. Among the data collected were the following serum total bilirubin (mg/dl) levels after transplantation (h–hours, d–days):

			Time After Reperfusion of Donor Liver						
Preoperative	**Liver Transection**	**Anhepatic Phase**	**1 h**	**2 h**	**4 h**	**8 h**	**1 d**	**2 d**	**3 d**
6.2	1.2	0.9	0.8	1.1	1.5	2	1.4	1.6	1.3
17.6	11.9	9.3	3.5	3	6.1	9	6.3	6.4	6.2
13.2	10.2	7.9	5.3	4.9	3.3	3.6	2.8	1.9	1.9
3.9	3.3	3	2.9	2.3	1.4	1.2	0.8	0.8	0.9
20.8	19.4	*	9.4	8.4	6.8	7.1	3.7	3.8	3.2
1.8	1.8	1.6	1.4	1.4	1.1	1.9	0.7	0.8	0.7
8.6	6.5	4.8	3.1	2.1	1	1.3	1.5	1.6	3.2
13.4	12	10.1	5.8	5.6	4.5	4.1	3	3.1	3.6
16.8	13.9	8.3	3.7	3.7	2.2	2.1	1.9	3.1	4.1
20.4	17.8	17	10.8	9.3	8.9	7	2.8	3.8	4.8
25	21.5	13.8	7.6	7	5	11.5	12.3	10.1	11.4
9.2	6.3	6.8	5.3	4.8	0.2	4	4.2	3.7	3.5
8	6.5	6.4	4.1	3.8	3.8	3.5	3.1	2.9	2.8
2.9	3	4.1	3.4	3.4	3.7	4.2	3.3	2	1.9
21.3	17.3	13.6	9.2	7.9	7.9	9.8	8.6	4.7	5.5
25	25	24	20.1	19.3	18.6	23.6	25	14.4	20.6
23.3	23.7	15.7	13.2	11	9.6	9.3	7.2	6.3	6.3
17.5	16.2	14.4	12.6	12.7	11.5	10	7.8	5.5	4.9

*Missing observation.
Source: Dr. Gotaro Shirakami. Used with permission.

Note that there is a missing observation in the data set. You may handle this problem in at least three ways.

(a) Omit the subject whose datum is missing, and analyze the data of the remaining 17 subjects.

(b) Use a computer package that automatically deals with missing data.

(c) Analyze the data using a missing data procedure. For such a procedure, see Jerome L. Myers and Arnold D. Well, *Research Design and Statistical Analysis,* Erlbaum Associates, Hillsdale, NJ, 1995, pp. 256–258.

40. The purpose of a study by Sakakibara and Hayano (A-41) was to examine the effect of voluntarily slowed respiration on the cardiac parasympathetic response to a threat (the anticipation of an electric shock). Subjects were 30 healthy college students whose mean age was 23 years with a standard deviation of 1.5 years. An equal number of subjects were randomly assigned to slow (six males, four females), fast (seven males, three females), and nonpaced (five males, five females) breathing groups. Subjects in the slow- and fast-paced breathing groups regulated their breathing rate to 8 and 30 cpm, respectively. The nonpaced group breathed spontaneously. The following are the subjects' scores on the State Anxiety Score of State-Trait Anxiety Inventory after baseline and period of threat:

Slow Paced		Fast Paced		Nonpaced	
Baseline	**Threat**	**Baseline**	**Threat**	**Baseline**	**Threat**
39	59	37	49	36	51
44	47	40	42	34	71
48	51	39	48	50	37
50	61	47	57	49	53
34	48	45	49	38	52
54	69	43	44	39	56
34	43	32	45	66	67
38	52	27	54	39	49
44	48	44	44	45	65
39	65	41	61	42	57

Source: Dr. Masahito Sakakibara. Used with permission.

41. Takahashi et al. (A-42) investigated the correlation of magnetic resonance signal intensity with spinal cord evoked potentials and spinal cord morphology after 5 hours of spinal cord compression in cats. Twenty-four adult cats were divided into four groups on the basis of a measure of spinal cord function plus a control group that did not undergo spinal compression. Among the data collected were the following compression ratio [(sagittal diameter/transverse diameter) × 100] values after 5 hours of compression:

Control	80.542986	Group III	36.923077
	79.111111		31.304348
	70.535714		53.333333
	87.323944		55.276382
	80.000000		40.725806
	82.222222	Group IV	66.666667
Group I	83.928571		29.565217
	84.183673		12.096774
	48.181818		34.274194
	98.461538		24.000000
Group II	30.263158		
	34.865900		
	43.775100		
	82.439024		

Source: Dr. Toshiaki Takahashi. Used with permission.

42. The objective of a study by Yamashita et al. (A-43) was to investigate whether pentoxifylline administered in the flush solution or during reperfusion would reduce ischemia-reperfusion lung injury in preserved canine lung allografts. Three groups of animals were studied. Pentoxifylline was not administered to animals in group 1 (C), was administered only during the reperfusion period (P) to animals in group 2, and was administered only in the flush solution to animals in group 3 (F). A total of 14 left lung allotransplantations were performed. The following are the aortic pressure readings for each animal during the 6-hour assessment period:

Group	0 min	60 min	120 min	180 min	240 min	300 min	360 min
C	85.0	100.0	120.0	80.0	72.0	75.0	*
C	85.0	82.0	80.0	80.0	85.0	80.0	80.0
C	100.0	75.0	85.0	98.0	85.0	80.0	82.0
C	57.0	57.0	57.0	30.0	*	*	*
C	57.0	75.0	52.0	56.0	65.0	95.0	75.0
P	112.0	67.0	73.0	90.0	71.0	70.0	66.0
P	92.0	70.0	90.0	80.0	75.0	80.0	*
P	105.0	62.0	73.0	75.0	70.0	55.0	50.0
P	80.0	73.0	50.0	35.0	*	*	*
F	70.0	95.0	105.0	115.0	110.0	105.0	100.0
F	60.0	63.0	140.0	135.0	125.0	130.0	120.0
F	67.0	65.0	75.0	75.0	80.0	80.0	80.0
F	115.0	107.0	90.0	103.0	110.0	112.0	95.0
F	90.0	99.0	102.0	110.0	117.0	118.0	103.0

*Missing observation.
Source: Dr. Motohiro Yamashita. Used with permission.

43. In a study investigating the relative bioavailability of beta-carotene (BC) and alpha-carotene (AC) from different sources of carrots, Zhou et al. (A-44) used ferrets as experimental animals. Among the data collected were the following concentrations of BC, AC, and AC/BC molar ratios in the sera of 24 ferrets provided with different sources of carotenoids for 3 days in their drinking water:

BC (μmol/g)	AC (μmol/g)	AC/BC (mol/mol)
	Unheated Juice	
0.637	0.506	0.795
0.354	0.297	0.840
0.287	0.249	0.869
0.533	0.433	0.813
0.228	0.190	0.833
0.632	0.484	0.767
	Heated Juice	
0.303	0.266	0.878
0.194	0.180	0.927

(Continued)

BC (μmol/g)	AC (μmol/g)	AC/BC (mol/mol)
	Heated Juice	
0.293	0.253	0.864
0.276	0.238	0.859
0.226	0.207	0.915
0.395	0.333	0.843
	Unheated Chromoplast	
0.994	0.775	0.780
0.890	0.729	0.819
0.809	0.661	0.817
0.321	0.283	0.882
0.712	0.544	0.763
0.949	0.668	0.704
	Heated Chromoplast	
0.933	0.789	0.845
0.280	0.289	1.031
0.336	0.307	0.916
0.678	0.568	0.837
0.714	0.676	0.947
0.757	0.653	0.862

Source: Dr. Jin-R. Zhou. Used with permission.

44. Potteiger et al. (A-45) wished to determine if sodium citrate ingestion would improve cycling performance and facilitate favorable metabolic conditions during the cycling ride. Subjects were eight trained male competitive cyclists whose mean age was 25.4 years with a standard deviation of 6.5. Each participant completed a 30-km cycling time trial under two conditions, following ingestion of sodium citrate and following ingestion of a placebo. Blood samples were collected prior to treatment ingestion (PRE-ING); prior to exercising (PRE-EX); during the cycling ride at completion of 10, 20, and 30 km; and 15 minutes after cessation of exercise (POST-EX). The following are the values of partial pressures of oxygen (Po_2) and carbon dioxide (Pco_2) for each subject, under each condition, at each measurement time:

		Po_2 (mm Hg)					
		Measurement Times					
Subject	**Treatment**[a]	**PRE-ING**	**PRE-EX**	**10-km**	**20-km**	**30-km**	**15-POST-EX**
1	1	42.00	20.00	53.00	51.00	56.00	41.00
1	2	43.00	29.00	58.00	49.00	55.00	56.00
2	1	44.00	38.00	66.00	66.00	76.00	58.00
2	2	40.00	26.00	57.00	47.00	46.00	45.00

(*Continued*)

P_{O_2} (mm Hg)

		Measurement Times					
Subject	Treatment[a]	PRE-ING	PRE-EX	10-km	20-km	30-km	15-POST-EX
3	1	37.00	22.00	59.00	58.00	56.00	52.00
3	2	36.00	30.00	52.00	65.00	65.00	36.00
4	1	34.00	21.00	65.00	62.00	62.00	59.00
4	2	46.00	36.00	65.00	72.00	72.00	66.00
5	1	36.00	24.00	41.00	43.00	50.00	46.00
5	2	41.00	25.00	52.00	60.00	67.00	54.00
6	1	28.00	31.00	52.00	60.00	53.00	46.00
6	2	34.00	21.00	57.00	58.00	57.00	41.00
7	1	39.00	28.00	72.00	69.00	65.00	72.00
7	2	40.00	27.00	64.00	61.00	57.00	60.00
8	1	49.00	27.00	67.00	61.00	51.00	49.00
8	2	27.00	22.00	56.00	64.00	49.00	34.00

P_{CO_2} (mm Hg)

		Measurement Times					
Subject	Treatment[a]	PRE-ING	PRE-EX	10-km	20-km	30-km	15-POST-EX
1	1	31.70	30.20	28.20	29.80	28.20	30.10
1	2	24.60	24.40	34.40	35.20	30.90	34.00
2	1	27.10	35.90	31.30	35.40	34.10	42.00
2	2	21.70	37.90	31.90	39.90	45.10	48.00
3	1	37.40	49.60	39.90	39.70	39.80	42.80
3	2	38.40	42.10	40.90	37.70	37.70	45.60
4	1	36.60	45.50	34.80	33.90	34.00	40.50
4	2	39.20	40.20	31.90	32.30	33.70	45.90
5	1	33.70	39.50	32.90	30.50	28.50	37.20
5	2	31.50	37.30	32.40	31.90	30.20	31.70
6	1	35.00	41.00	38.70	37.10	35.80	40.00
6	2	27.20	36.10	34.70	36.30	34.10	40.60
7	1	28.00	36.50	30.70	34.60	34.30	38.60
7	2	28.40	31.30	48.10	43.70	35.10	34.70
8	1	22.90	28.40	25.70	28.20	32.30	34.80
8	2	41.40	41.80	29.50	29.90	31.30	39.00

[a]1 = Sodium citrate; 2 = placebe.
Source: Dr. Jeffrey A. Potteiger. Used with permission.

45. Teitge et al. (A-46) describe a radiographic method to demonstrate patellar instability. The 90 subjects ranged in age from 13 to 52 years and were divided into the following four groups on the basis of clinical findings regarding the nature of instability of the knee: normal (no symptoms or signs related to the knee), lateral, medial, and multidirectional instability. Among the data collected were the following radiographic measurements of the congruence angle (degrees):

Normal	Lateral		Medial		Multidirectional
−8	4	12	−16	10	15
−16	18	−8	−25	−5	−26
−22	5	−8	20	−10	−8
−26	−6	−20	−8	−12	−12
−8	32	−5	8	−14	−40
12	30	−10	−14		−20
−8	−10		−18		−16
12	28		−4		−34
−20	6		−20		−14
−20	9		−20		−6
−5	10		−20		−35
10	20		−22		−24
−4	−9		−15		−25
−2	−10		−10		10
6	12		5		16
−7	0		−5		−30
0	35		−6		−30
−2	−1		−15		
−15	5		−25		
−5	22		−10		
	22		−20		

Source: Dr. Robert A. Teitge. Used with permission.

46. A study by Ikeda et al. (A-47) was designed to determine the dose of ipratropium bromide aerosol that improves exercise performance using progressive cycle ergometry in patients with stable chronic obstructive pulmonary disease. The mean age of the 20 male subjects was 69.2 years with a standard deviation of 4.6 years. Among the data collected were the following maximum ventilation (VE_{max}, L/min) values at maximum achieved exercise for different ipratropium bromide dosage levels (μg):

Placebo	40	80	160	240
26	24	23	25	28
38	39	43	43	37
49	46	54	57	52
37	39	39	38	38
34	33	37	37	41
42	38	44	44	42
23	26	28	27	22
38	41	44	37	40
37	37	36	38	39
33	35	34	38	36
40	37	40	46	40
52	58	48	58	63
45	48	47	51	38

(Continued)

Placebo	40	80	160	240
24	30	23	27	30
41	37	39	46	42
56	54	51	58	58
35	51	49	51	46
28	41	37	33	38
28	34	34	35	35
38	40	43	39	45

Source: Dr. Akihiko Ikeda. Used with permission.

47. Pertovaara et al. (A-48) compared the effect of skin temperature on the critical threshold temperature eliciting heat pain with the effect of skin temperature on the response latency to the first heat pain sensation. Subjects were healthy adults between the ages of 23 and 54 years. Among the data collected were the following latencies (seconds) to the first pain response induced by radiant heat stimulation at three different skin temperatures:

Subject	25°C	30°C	35°C
1	6.4	4.5	3.6
2	8.1	5.7	6.3
3	9.4	6.8	3.2
4	6.75	4.6	3.9
5	10	6.2	6.2
6	4.5	4.2	3.4

Source: Dr. Antti Pertovaara. Used with permission.

48. A study for the development and validation of a sensitive and specific method for quantifying total activin-A concentrations has been reported on by Knight et al. (A-49). As part of the study they collected the following peripheral serum concentrations of activin-A in human subjects of differing reproductive status: normal follicular phase (FP), normal luteal phase (LP), pregnant (PREG), ovarian hyperstimulated for in vivo fertilization (HYP), postmenopausal (PM), and normal adult males. Hint: Convert responses to logarithms before performing analysis.

FP	LP	PREG	HYP	PM	Male
134.5	78.0	2674.0	253.1	793.1	196.7
159.2	130.4	945.6	294.3	385.1	190.6
133.2	128.3	5507.6	170.2	270.9	185.3
225.0	166.4	7796.5	219.8	640.3	335.4
146.4	115.2	5077.5	165.8	459.8	214.6
180.5	148.9	4541.9	159.0		

Source: Dr. Philip G. Knight. Used with permission.

49. The purpose of a study by Maheux et al. (A-50) was to evaluate the effect of labor on glucose production and glucose utilization. Subjects were six normal pregnant women. Among the data collected were the following glucose concentrations during four stages of labor: latent (A1) and active (A2) phases of cervical dilatation, fetal expulsion (B), and placental expulsion (C).

A1	A2	B	C
3.60	4.40	5.30	6.20
3.53	3.70	4.10	3.80
4.02	4.80	5.40	5.27
4.90	5.33	6.30	6.20
4.06	4.65	6.10	6.90
3.97	5.20	4.90	4.60

Source: Dr. Pierre C. Maheux. Used with permission.

50. Trachtman et al. (A-51) conducted studies (1) to assess the effect of recombinant human (rh) IGF-I on chronic puromycin aminonucleoside (PAN) nephropathy and (2) to compare the results of rhIGF-I versus rhGH treatment in a model of focal segmental glomerulosclerosis. As part of the studies, male Sprague-Dawley rats were divided into four groups: PAN (IA), PAN + rhIGF-I (IB), normal (IIA), and normal + rhIGF-I (IIB). The animals yielded the following data on creatinine levels before (pre) and after 4, 8, and 12 weeks of treatment:

Group			
IA	IB	IIA	IIB
	Pre		
44	44	44	35
44	44	44	44
44	44	44	44
53	44	44	35
44	44		
44	53		
	4 Weeks		
97	44	53	44
88	35	44	53
62	44	44	53
53	35	53	44
62	62		
53	53		
	8 Weeks		
53	53	62	44
53	53	53	62

(Continued)

Group			
IA	IB	IIA	IIB
8 Weeks			
44	53	62	44
53	44	53	44
62	53		
70	62		
12 Weeks			
88	79	53	53
70	79	62	62
53	79	53	53
70	62	62	53
88	79		
88	70		

Source: Dr. Howard Trachtman. Used with permission.

51. Twelve healthy men, ages 22 through 35 years, yielded the following serum T_3 (nmol/L) levels at 0800 hours after 8 (day 1), 32 (day 2), and 56 (day 3) hours of fasting, respectively. Subjects were participants in a study of fasting-induced alterations in pulsatile glycoprotein secretion conducted by Samuels and Kramer (A-52).

Subject	T_3	Day	Subject	T_3	Day	Subject	T_3	Day	Subject	T_3	Day
1	88	1	2	115	1	3	119	1	4	164	1
1	73	2	2	77	2	3	93	2	4	120	2
1	59	3	2	75	3	3	65	3	4	86	3
Subject	**T_3**	**Day**	**Subject**	**T_3**	**Day**	**Subject**	**T_3**	**Day**	**Subject**	**T_3**	**Day**
5	93	1	6	119	1	7	152	1	8	121	1
5	91	2	6	57	2	7	70	2	8	107	2
5	113	3	6	44	3	7	74	3	8	133	3
Subject	**T_3**	**Day**	**Subject**	**T_3**	**Day**	**Subject**	**T_3**	**Day**	**Subject**	**T_3**	**Day**
9	108	1	10	124	1	11	102	1	12	131	1
9	93	2	10	97	2	11	56	2	12	83	2
9	75	3	10	74	3	11	58	3	12	66	3

Source: Dr. Mary H. Samuels. Used with permission.

52. To determine the nature and extent to which neurobehavioral changes occur in association with the toxicity resulting from exposure to excess dietary iron (Fe), Sobotka et al. (A-53) used weanling male Sprague-Dawley rats as experimental subjects. The researchers randomly assigned the animals, according to ranked body weights, to one of five diet groups differentiated on the basis of amount of Fe present: Control—35 (1), 350 (2), 3500 (3), 4 (iron deficient) (4), and 20,000 (5) ppm, respectively. The following are the body weights of the animals (grams) at the end of 10 weeks.

Diet	Weight	Diet	Weight	Diet	Weight
1	396	1	335	1	373
2	368	2	349	4	292
3	319	3	302	5	116
4	241	4	220	4	291
5	138	5	118	5	154
1	331	1	394	4	281
2	325	2	300	5	118
3	331	3	285	4	250
4	232	4	237	5	119
5	116	5	113	4	242
1	349	1	377	5	118
2	364	2	366	4	277
3	392	3	269	5	104
4	310	4	344	5	120
5	131	5	Dead	5	102
1	341	1	336		
2	399	2	379		
3	274	3	195		
4	319	4	277		
5	131	5	148		
1	419	1	301		
2	373	2	368		
3	Dead	3	308		
4	220	4	299		
5	146	5	Dead		

Source: Dr. Thomas J. Sobotka. Used with permission.

53. Hansen (A-54) notes that brain bilirubin concentrations are increased by hyperosmolality and hypercarbia, and that previous studies have not addressed the question of whether increased brain bilirubin under different conditions is due to effects on the entry into or clearance of bilirubin from brain. In a study, he hypothesized that the kinetics of increased brain bilirubin concentration would differ in respiratory acidosis (hypercarbia) and hyperosmolality. Forty-four young adult male Sprague-Dawley rats were sacrificed at various time periods following infusion with bilirubin. The following are the blood bilirubin levels (μmol/L) of 11 animals just prior to sacrifice 60 minutes after the start of bilirubin infusion:

Controls	Hypercarbia	Hyperosmolality
30	48	102
94	20	118
78	58	74
52	74	

Source: Dr. Thor Willy Ruud Hansen. Used with permission.

54. Johansson et al. (A-55) compared the effects of short-term treatments with growth hormone (GH) and insulin-like growth factor I (IGF-I) on biochemical markers of bone metabolism in men with idiopathic osteoporosis. Subjects ranged in age from 32 to 57 years. Among the data collected were the following serum concentrations of IGF binding protein-3 at 0 and 7 days after first injection and 1, 4, 8, and 12 weeks after last injection with GH and IGF-I.

Patient No.	Treatment	0 Day	7 Days	1 Week	4 Weeks	8 Weeks	12 Weeks
1	GH	4507	4072	3036	2484	3540	3480
1	IGF-I	3480	3515	4003	3667	4263	4797
2	GH	2055	4095	2315	1840	2483	2354
2	IGF-I	2354	3570	3630	3666	2700	2782
3	GH	3178	3574	3196	2365	4136	3088
3	IGF-I	3088	3405	3309	3444	2357	3831
4	IGF-I	2905	2888	2797	3083	3376	3464
4	GH	3464	5874	2929	3903	3367	2938
5	GH	4142	4465	3967	4213	4321	4990
5	IGF-I	4990	4590	2989	4081	4806	4435
6	IGF-I	3504	3529	4093	4114	4445	3622
6	GH	3622	6800	6185	4247	4450	4199
7	IGF-I	5130	4784	4093	4852	4943	5390
7	GH	5390	5188	4788	4602	4926	5793
8	IGF-I	3074	2691	2614	3003	3145	3161
8	GH	3161	4942	3222	2699	3514	2963
9	GH	3228	5995	3315	2919	3235	4379
9	IGF-I	4379	3548	3339	2379	2783	3000
10	IGF-I	5838	5025	4137	5777	5659	5628
10	GH	5628	6152	4415	5251	3334	3910
11	GH	2304	4721	3700	3228	2440	2698
11	IGF-I	2698	2621	3072	2383	3075	2822

Source: Dr. Anna G. Johansson. Used with permission.

55. The objective of a study by Strijbos et al. (A-56) was to compare the results of a 12-week hospital-based outpatient rehabilitation program (group 1) with those of a 12-week home-care rehabilitation program (group 2) in chronic obstructive pulmonary disease with moderate to severe airflow limitation. A control group (group 3) did not receive rehabilitation therapy. Among the data collected were the following breathing frequency scores of subjects 18 months after rehabilitation:

Group			Group		
1	2	3	1	2	3
12	16	24	12	16	24
16	14	16	12	12	14
16	12	18	14	12	15
14	12	18	16	12	16

(*Continued*)

Group			Group		
1	2	3	1	2	3
12	18	24	12	12	16
12	12	24	12	15	18
12	10	18		20	16

Source: Dr. Jaap H. Strijbos. Used with permission.

56. Seven healthy males (mean age 27.4 years with a standard deviation of 4.4) participated in a study by Lambert et al. (A-57), who measured intestinal absorption following oral ingestion and intestinal perfusion of a fluid. As part of the study the researchers recorded the following percent changes in plasma volume at six points during 85 minutes of cycle exercise in the drinking and infusion experiments:

	Subject	1	2	3
Drinking	1	−8.4151514	−7.4902674	−8.02277330
	2	−12.1966790	−5.1496679	−10.46486300
	3	−9.7418719	−5.9062747	−7.06516950
	4	−15.0291920	−14.4165470	−16.61268200
	5	−5.8845683	−5.8845683	−3.57781750
	6	−9.7100000	−7.5700000	−3.52995560
	7	−6.9787024	−6.5752716	−5.07020210
Infusion	8	−13.5391010	−11.7186910	−10.77312900
	9	−8.8259516	−8.9029745	−6.38160030
	10	−4.2410016	−1.3448910	−2.49740390
	11	−10.7192870	−9.7651132	−11.12140900
	12	−6.9487760	−2.9830660	1.77828157
	13	−7.1160660	−5.4111706	−7.07086340
	14	−7.0497788	−5.7725485	−5.18045500
	Subject	**4**	**5**	**6**
Drinking	1	−7.35202650	−7.89172340	−7.84726700
	2	−8.40517240	−9.02789810	5.13333985
	3	−4.19974130	−3.33795970	−5.65380700
	4	−15.36239700	−17.63314100	−14.43982000
	5	−5.50433470	−5.12242600	−6.26313790
	6	−4.22938570	−7.86923080	−7.51168220
	7	−5.94416340	−5.21535350	−6.34285620

(Continued)

	Subject	4	5	6
	8	−11.64145400	−12.40814000	−8.26411320
	9	−5.69396590	−6.38160030	−7.37350920
	10	−1.01234570	−5.58572150	−2.81811090
Infusion	11	−12.13053100	−15.98360700	−12.64667500
	12	2.28844839	2.59034233	1.56622058
	13	−8.35430040	−10.60663700	−9.45689580
	14	−7.92841880	−8.38462720	−8.44542770

Source: Dr. C. V. Gisolfi. Used with permission.

57. Roemer et al. (A-58) developed a self-report measure of generalized anxiety disorder (GAD) for use with undergraduate populations. In reliability studies the undergraduate subjects completed the GAD questionnaire (GAD-Q) as well as the Penn State Worry Questionnaire (PSWQ). The following are the PSWQ scores made by four groups of subjects determined by their GAD status: GAD by questionnaire, Study II (group 1); non-GAD by questionnaire, Study II (group 2); GAD by questionnaire, Study I (group 3); and clinical GAD (group 4).

		Group		
1	**2**	**3**	**4**	
59.0	50.0	46.0	65.0	65.0
51.0	28.0	77.0	62.0	66.0
58.0	43.0	80.0	76.0	69.0
61.0	36.0	60.0	66.0	73.0
64.0	36.0	59.0	78.0	67.0
68.0	30.0	56.0	76.0	78.0
64.0	24.0	44.0	74.0	76.0
67.0	39.0	71.0	73.0	66.0
56.0	29.0	54.0	61.0	55.0
78.0	48.0	64.0	63.0	59.0
48.0	36.0	66.0	75.0	44.0
62.0	38.0	59.0	63.0	68.0
77.0	42.0	68.0	55.0	64.0
72.0	26.0	59.0	67.5	41.0
59.0	35.0	61.0	70.0	54.0
	32.0	78.0	70.0	72.0
	43.0	70.0	55.0	74.0
	55.0	74.0	73.0	59.0
	42.0	73.0	80.0	63.0
	37.0		79.0	51.0
	36.0		79.0	72.0
	41.0		61.0	63.0
	36.0		61.0	58.0
	34.0		72.0	71.0

(*Continued*)

Group			
1	2	3	4
	42.0		67.0
	35.0		74.0
	51.0		65.0
	37.0		68.0
	50.0		72.0
	39.0		75.0
			56.0

Source: Dr. T. D. Borkovec. Used with permission.

58. Noting that non-Hodgkin's lymphomas (NHL) represent a heterogeneous group of diseases in which prognosis is difficult to predict, Christiansen et al. (A-59) report on the prognostic aspects of soluble intercellular adhesion molecule-1 (sICAM-1) in NHL. Among the data collected were the following serum sICAM-1 (ng/ml) levels in four groups of subjects: healthy controls (C), high-grade NHL (hNHL), low-grade NHL (lNHL), and patients with hairy cell leukemia (HCL).

C	hNHL			lNHL		HCL
309	460	844	824	961	581	382
329	222	503	496	1097	601	975
314	663	764	656	1099	572	663
254	1235	1088	1038	625	439	429
304	500	470	1050	473	1135	1902
335	739	806	446	654	590	1842
381	1847	482	1218	508	404	314
456	477	734	511	454	382	430
294	818	616	317	889	692	645
450	585	836	334	805	484	637
422	1837	1187	1026	541	438	712
528	362	581	534	655	787	581
461	671	381	292	654	77	860
286	375	699	782	1859	478	448
309	543	1854	1136	619	602	735
226	352	769	476	1837	802	
388	443	510		534	568	
377	359	571		424	665	
310	383	1248		571		
261	587	784		420		
350	648	514		408		
405	782	678		391		
319	472	1264		493		
289	506	618		1162		
310	663	1123		460		
227	873	912		1113		

(*Continued*)

C	hNHL		lNHL	HCL
206	987	520	572	
226	859	1867	653	
309	1193	485	1340	
382	1836	287	656	
325	691	455		
		522		

Source: Dr. Ilse Christiansen. Used with permission.

59. Cossette et al. (A-60) examined gender and kinship with regard to caregivers' use of informal and formal support and to two models of support. Among the data collected were the following ages of three groups of caregivers of a demented relative living at home: husbands, wives, and adult daughters.

Husband			Wife		Daughter	
64	66	73	59	67	40	50
70	58	71	66	67	47	58
55	81	70	80	57	46	46
67	77	71	76	53	45	47
79	76	56	68	50	69	50
67	64	68	53	70	48	53
77	82	76	78	70	53	57
68	85	67	75		50	65
72	63	66	74		47	50
67	72	67	86		62	43
77	77	72	63		55	59
70	79	72	52		49	44
65	63	70	55		43	45
65	80	66	71		44	41
74	70	73	67		47	50
86	85	78	78		57	58
72	76	64	70		49	35
71	67	78	68			50
78	72	59	78			59
71	60	71	59			45
88	74	70	72			50
77	65	67	73			48
75	53	78	75			51
66	70	67	54			46
80	72	55	65			62
76	74	64	67			55
67	79	69	83			50
65	63	59	70			43
62	77	55	72			39
82	78	75	71			50

(Continued)

Husband		Wife		Daughter
75	69	68	76	50
80		65	74	43
74		81	68	28
70		79	69	
		75	72	

Source: Sylvie Cossette, M.Sc., R.N. Used with permission.

60. Tasaka et al. (A-61) note that *Corynebacterium parvum* (CP) increases susceptibility to endotoxin, which is associated with increased production of tumor necrosis factor (TNF). They investigated the effect of CP-priming on the pathogenesis of acute lung injury caused by intratracheal *Escherichia coli* endotoxin (lipopolysaccharide [LPS]). Experimental animals consisted of female guinea pigs divided into four groups. Animals in two groups received a 4-mg/kg treatment of CP 7 days before the study. Subsequently, nonpretreated animals received either saline alone (Control) or endotoxin (LPS-alone). The pretreated groups received either saline (CP-alone) or LPS (CP + LPS). Among the data collected were the following values of lung tissue-to-plasma ratio of radio-iodized serum albumin assay:

Control	CP-alone	LPS-alone	CP + LPS
0.12503532	0.18191647	0.17669093	0.3651166
0.10862729	0.30887462	0.25344761	0.64062964
0.10552931	0.25011885	0.17372285	0.39208734
0.15587316	0.23858085	0.1786867	0.49942059
0.13672624	0.26558231	0.22209666	0.85718475
0.11290446	0.32298454	0.27064831	0.93030465

Source: Dr. Sadatomo Tasaka. Used with permission.

61. According to Takahashi et al. (A-62) research indicates that there is an association between alterations in calcium metabolism and various bone diseases in patients with other disabilities. Using subjects with severe mental retardation (mean age 16 years) who had been living in institutions for most of their lives, Takahashi et al. examined the relationship between bone change and other variables. Subjects were divided into groups on the basis of severity of bone change. Among the data collected were the following serum alkaline phosphatase (IU/L) values:

Grade I: 109, 86, 79, 103, 47, 105, 188, 96, 249

Grade II: 86, 106, 164, 146, 111, 263, 162, 111

Grade III: 283, 201, 208, 301, 135, 192, 135, 83, 193, 175, 174, 193, 224, 192, 233

Source: Dr. Mitsugi Takahashi. Used with permission.

62. Research indicates that dietary copper deficiency reduces growth rate in rats. In a related study, Allen (A-63) assigned weanling male Sprague-Dawley rats to one of three food groups: copper-deficient (CuD), copper-adequate (CuA), and pair-fed (PF). Rats in the PF group were initially weight-matched to rats of the CuD group and then fed the same weight of the CuA diet as that consumed by their CuD counterparts. After 20 weeks, the rats were anesthetized, blood samples were drawn, and organs were harvested. As part of the study the following data were collected:

Rat	Diet	Body Weight (BW)(g)	Heart Weight (HW)(g)	Liver Weight (LW)(g)	Kidney Weight (KW)(g)	Spleen Weight (SW)(g)
1		253.66	0.89	2.82	1.49	0.41
2		400.93	1.41	3.98	2.15	0.76
3	CuD	355.89	1.24	5.15	2.27	0.69
4		404.70	2.18	4.77	2.99	0.76
6		397.28	0.99	2.34	1.84	0.50
7		421.88	1.20	3.26	2.32	0.79
8	PF	386.87	0.88	3.05	1.86	0.84
9		401.74	1.02	2.80	2.06	0.76
10		437.56	1.22	3.94	2.25	0.75
11		490.56	1.21	4.51	2.30	0.78
12		528.51	1.34	4.38	2.75	0.76
13	CuA	485.51	1.36	4.40	2.46	0.82
14		509.50	1.27	4.67	2.50	0.79
15		489.62	1.31	5.83	2.74	0.81

Rat	Diet	HW/BW (g/100 g)	LW/BW (g/100 g)	KW/BW (g/100 g)	SW/BW (g/100 g)	Ceruloplasmin (mg/dl)
1		0.00351	0.01112	0.00587	0.00162	nd
2		0.00352	0.00993	0.00536	0.00190	5.27
3	CuD	0.00348	0.01447	0.00638	0.00194	4.80
4		0.00539	0.01179	0.00739	0.00188	4.97
6		0.00249	0.00589	0.00463	0.00126	35.30
7		0.00284	0.00773	0.00550	0.00187	39.00
8	PF	0.00227	0.00788	0.00481	0.00217	28.00
9		0.00254	0.00697	0.00513	0.00189	34.20
10		0.00279	0.00900	0.00514	0.00171	45.20
11		0.00247	0.00919	0.00469	0.00159	34.60
12		0.00254	0.00829	0.00520	0.00144	39.00
13	CuA	0.00280	0.00906	0.00507	0.00169	37.10
14		0.00249	0.00917	0.00491	0.00155	33.40
15		0.00268	0.01191	0.00560	0.00165	37.30

nd, no data.
Source: Corrie B. Allen. Used with permission.

63. Hughes et al. (A-64) point out that systemic complications in acute pancreatitis are largely responsible for mortality associated with the disease. They note further that proinflammatory cytokines, particularly TNFα, may play a central role in acute pancreatitis by mediating the systemic sequelae. In their research they used a bile-infusion model of acute pancreatitis to show amelioration of disease severity as well as an improvement in overall survival by TNFα inhibition. Experimental material consisted of adult male Sprague-Dawley rats weighing between 250 and 300 grams divided into three groups: untreated (bile solution infused without treatment); treated (bile solution infused preceded by treatment with polyclonal anti-TNFα antibody); and sham (saline infused). Among the data collected were the following hematocrit (%) values for animals surviving more than 48 hours:

Sham	Untreated	Treated
38	56	40
40	60	42
32	50	38
36	50	46
40	50	36
40		35
38		40
40		40
38		55
40		35
		36
		40
		40
		35
		45

Source: Dr. A. Osama Gaber. Used with permission.

64. A study by Smárason et al. (A-65) was motivated by the observations of other researchers that sera from pre-eclamptic women damaged cultured human endothelial cells. Subjects for the present study were women with pre-eclampsia, matched control women with normal pregnancies, and nonpregnant women of childbearing age. Among the data collected were the following observations on a relevant variable measured on subjects in the three groups.

Pre-Eclampsia	Pregnant Controls	Nonpregnant Controls
113.5	91.4	94.5
106.6	95.6	115.9
39.1	113.1	107.2
95.5	100.8	103.2
43.5	88.2	104.7
49.2	92.2	94.9
99.5	78.6	93.0
102.9	96.9	100.4
101.2	91.6	107.1
104.9	108.6	105.5
75.4	77.3	119.3
71.1	100.0	88.2
73.9	61.7	82.2

(Continued)

Pre-Eclampsia	Pregnant Controls	Nonpregnant Controls
76.0	83.3	125.0
81.3	103.6	126.1
72.7	92.3	129.1
75.3	98.6	106.9
55.2	85.0	110.0
90.5	128.2	127.3
55.8	88.3	128.6

Source: Dr. Alexander Smárason. Used with permission.

65. The objective of a study by LeRoith et al. (A-66) was to evaluate the effect of a 7-week administration of recombinant human GH (rhGH) and recombinant human insulin-like growth factor (rhIGF-I) separately and in combination on immune function in elderly female rhesus monkeys. The assay for the in vivo function of the immune system relied on the response to an immunization with tetanus toxoid. The following are the responses for the three treatment groups and a control group:

Saline	rhIGF-I	rhGH	rhIGF-I + rhGH
11.2	12.2	12.15	11.5
9.0	9.4	11.20	12.4
10.8	10.7	10.60	10.8
10.0	10.8	11.30	11.9
	9.1	11.00	11.0
	12.6		

Source: Dr. Jack A. Yanowski. Used with permission.

66. Hampl et al. (A-67) note that inhaled nitric oxide (NO) is a selective pulmonary vasodilator. They hypothesized that a nebulized diethylenetriamine/NO (DETA/NO) would stay in the lower airways and continuously supply sufficient NO to achieve sustained vasodilation in chronic pulmonary hypertension. Experimental material consisted of adult, male, specific pathogen-free Sprague-Dawley rats randomly divided into four groups: untreated, pulmonary normotensive controls; monocrotaline-injected (to induce hypertension) with no treatment (MCT); monocrotaline-injected treated with either a 5-μmol dose or a 50-μmol dose of DETA/NO. Nineteen days after inducing pulmonary hypertension in the two groups of rats, the researchers began the treatment procedure, which lasted for 4 days. They collected, among other data, the following measurements on cardiac output for the animals in the four groups:

		MCT + DETA/NO	
Control	MCT	5 μmol	50 μmol
71.8	42.8	72.5	47.1
66.1	53.2	62.9	86.6
67.6	56.1	58.9	56.0
66.4	56.5	69.3	

Source: Dr. Stephen L. Archer. Used with permission.

Exercises for Use with Large Data Sets Available on the Following Website: www.wiley.com/college/daniel

1. In Kreiter et al. (A-68) medical school exams were delivered via computer format. Because there were not enough computer stations to test the entire class simultaneously, the exams were administered over 2 days. Both students and faculty wondered if students testing on day 2 might have an advantage due to extra study time or a breach in test security. Thus, the researchers examined a large medical class ($n = 193$) tested over 2 days with three 2-hour 80-item multiple-choice exams. Students were assigned testing days via pseudorandom assignment. Of interest was whether taking a particular exam on day 1 or day 2 had a significant impact on scores. Use the data set MEDSCORES to determine if test, day, or interaction has significant impact on test scores. Let $\alpha = .05$.

2. Refer to the serum lipid-bound sialic acid data on 1400 subjects (LSADATA). We wish to conduct a study to determine if the measurement of serum lipid-bound sialic acid (LSA) might be of use in the detection of breast cancer. The LSA measurements (mg/dl) are for four populations of subjects: normal controls, A; patients with benign breast disease, B; patients with primary breast cancer, C; and patients with recurrent metastatic breast cancer, D. Select a simple random sample of size 10 from each population and perform an appropriate analysis to determine if we may conclude that the four population means are different. Let $\alpha = .05$ and determine the p value. Test all possible pairs of sample means for significance. What conclusions can one draw from the analysis? Prepare a verbal report of the findings. Compare your results with those of your classmates.

3. Refer to the serum angiotensin-converting enzyme data on 1600 subjects (SACEDATA). Sarcoidosis, found throughout the world, is a systemic granulomatous disease of unknown cause. The assay of serum angiotensin-converting enzyme (SACE) is helpful in the diagnosis of active sarcoidosis. The activity of SACE is usually increased in patients with the disease, while normal levels occur in subjects who have not had the disease, those who have recovered, and patients with other granulomatous disorders. The data are the SACE values for four populations of subjects classified according to status regarding sarcoidosis: never had, A; active, B; stable, C; recovered, D. Select a simple random sample of 15 subjects from each population and perform an analysis to determine if you can conclude that the population means are different. Let $\alpha = .05$. Use Tukey's test to test for significant differences among individual pairs of means. Prepare a written report on your findings. Compare your results with those of your classmates.

4. Refer to the urinary colony-stimulating factor data on 1500 subjects (CSFDATA). The data are the urinary colony-stimulating factor (CSF) levels in five populations: normal subjects and subjects with four different diseases. Each observation represents the mean colony count of four plates from a single urine specimen from a given subject. Select a simple random sample of size 15 from each of the five populations and perform an analysis of variance to determine if one may conclude that the population means are different. Let $\alpha = .05$. Use Tukey's HSD statistic to test for significant differences among all possible pairs of sample means. Prepare a narrative report on the results of your analysis. Compare your results with those of your classmates.

5. Refer to the red blood cell data on 1050 subjects (RBCDATA). Suppose that you are a statistical consultant to a medical researcher who is interested in learning something about the relationship between blood folate concentrations in adult females and the quality of their diet. The researcher has available three populations of subjects: those whose diet quality is rated as good, those whose diets are fair, and those with poor diets. For each subject there is also available her red blood cell (RBC) folate value (in μg/liter of red cells). Draw a simple random sample of size 10 from each population and determine whether the researcher can conclude that the three populations differ

with respect to mean RBC folate value. Use Tukey's test to make all possible comparisons. Let $\alpha = .05$ and find the p value for each test. Compare your results with those of your classmates.

6. Refer to the serum cholesterol data on 350 subjects under three diet regimens (SERUMCHO). Three-hundred-fifty adult males between the ages of 30 and 65 participated in a study to investigate the relationship between the consumption of meat and serum cholesterol levels. Each subject ate beef as his only meat for a period of 20 weeks, pork as his only meat for another period of 20 weeks, and chicken or fish as his only meat for another 20-week period. At the end of each period serum cholesterol determinations (mg/100 ml) were made on each subject. Select a simple random sample of 10 subjects from the population of 350. Use two-way analysis of variance to determine whether one should conclude that there is a difference in population mean serum cholesterol levels among the three diets. Let $\alpha = .05$. Compare your results with those of your classmates.

REFERENCES

Methodology References

1. R. A. FISHER, *The Design of Experiments*, Eighth Edition, Oliver and Boyd, Edinburgh, 1966.
2. R. A. FISHER, *Contributions to Mathematical Statistics*, Wiley, New York, 1950.
3. R. A. FISHER, *Statistical Methods for Research Workers*, Fourteenth Edition, Hafner, New York, 1973.
4. KLAUS HINKELMANN and OSCAR KEMPTHORNE, *Design and Analysis of Experiments,* Revised Edition, Wiley, New York, 1994.
5. DOUGLAS C. MONTGOMERY, *Design and Analysis of Experiments,* Fourth Edition, Wiley, New York, 1997.
6. JEROME L. MYERS and ARNOLD D. WELL, *Research Design and Statistical Analysis,* Earlbaum Associates, Hillsdale, NJ, 1995.
7. JASON C. HSU, *Multiple Comparisons: Theory and Methods*, Chapman & Hall, New York, 1996.
8. PETER H. WESTFALL, RANDALL D. TOBIAS, DROR ROM, RUSSELL D. WOLFINGER, and YOSEF HOCHBERG, *Multiple Comparisons and Multiple Tests*, SAS Publishing, Cary, NC. 2001.
9. J. W. TUKEY, "The Problem of Multiple Comparisons," Mimeographed Monograph, Princeton University, 1953; cited in Yosef Hochberg and Ajit C. Tamhane, *Multiple Comparison Procedures,* John Wiley and Sons, New York, 1987.
10. B. J. WINER, *Statistical Principles in Experimental Design,* Second Edition, McGraw-Hill, New York, 1971.
11. C. Y. KRAMER, "Extension of Multiple Range Tests to Group Means with Unequal Sample Sizes," *Biometrics, 63* (1956), 307–310.

Applications References

A-1. DAVID H. HOLBEN, "Selenium Content of Venison, Squirrel, and Beef Purchased or Produced in Ohio, a Low Selenium Region of the United States," *Journal of Food Science, 67* (2002), 431–433.

A-2. W. D. MEMBER, W. M. MURRY, S. I. RINGLEB, K. L. KILGORE, and S. A. SNYDER, "A Transducer to Measure Isometric Elbow Moments," *Clinical Biomechanics, 16* (2001), 918–920.

A-3. JOHN P. HOLCOMB and RALPH J. ROTHENBERG, "The Effectiveness of Alendronate and Etidronate in Clinical Practice," (unpublished manuscript).

A-4. JASMINKA ILICH-ERNST, RHONDA A. BROWNBILL, MARTHA A. LUDEMANN, and RONGWEI FU, "Critical Factors for Bone Health in Women across the Age Span: How Important Is Muscle Mass?" *Medscape Women's Health eJournal, 7* (2002).

A-5. PAUL B. GOLD, ROBERT N. RUBEY, and RICHARD T. HARVEY, "Naturalistic, Self-Assignment Comparative Trial of Bupropion SR, a Nicotine Patch, or Both for Smoking Cessation Treatment in Primary Care," *American Journal on Addictions, 11* (2002), 315–331.

A-6. X. WANG, X. SHEN, X. LI, and C. MAULI AGRAWAL, "Age-Related Changes in the Collagen Network and Toughness of Bone," *Bone, 31* (2002), 1–7.

A-7. Farhad Atassi, "Oral Home Care and the Reasons for Seeking Dental Care by Individuals on Renal Dialysis," *Journal of Contemporary Dental Practice, 3* (2002), 031–041.

A-8. Simone Ragazzi, Agostino Pierro, Mark Peters, Lorella Fasoli, and Simon Eaton, "Early Full Blood Count and Severity of Disease in Neonates with Necrotizing Enterocolitis," *Pediatric Surgery International, 19* (2003), 376–379.

A-9. Roberta Romito, Maria Ida Pansini, Francesco Perticone, Gianfranco Antonelli, Mariavittoria Pitzalis, and Paolo Rizzon, "Comparative Effect of Lercanidipine, Felodipine, and Nifedipine FITS on Blood Pressure and Heart Rate in Patients with Mild to Moderate Arterial Hypertension: The Lercanidipine in Adults (LEAD) Study," *Journal of Clinical Hypertension, 5* (2003), 249–253.

A-10. L. Kosmiski, D. Kuritzkes, J. Hamilton, T. Sharp, K. Lichtenstien, J. Hill, and R. Eckel, "Fat Distribution Is Altered in HIV-Infected Men Without Clinical Evidence of the HIV Lipodystrophy Syndrome," *HIV Medicine, 4* (2003), 235–240.

A-11. B. M. Brooks, F. D. Rose, E. A. Attree, and A. Elliot-Square, "An Evaluation of the Efficacy of Training People with Learning Disabilities in a Virtual Environment," *Disability and Rehabilitation, 24* (2002), 622–626.

A-12. Brian J. McConville, M. Harold Fogelson, Andrew B. Norman, William M. Klykylo, Pat Z. Manderscheid, Karen W. Parker, and Paul R. Sanberg, "Nicotine Potentiation of Haloperidol in Reducing Tic Frequency in Tourette's Disorder," *American Journal of Psychiatry, 148* (1991), 793–794.

A-13. M. E. Valencia, G. McNeill, J. M. Brockway, and J. S. Smith, "The Effect of Environmental Temperature and Humidity on 24h Energy Expenditure in Men," *British Journal of Nutrition, 68* (September 1992), 319–327.

A-14. D. S. Hodgson, C. I. Dunlop, P. L. Chapman, and J. L. Grandy, "Cardiopulmonary Responses to Experimentally Induced Gastric Dilatation in Isoflurane-Anesthetized Dogs," *American Journal of Veterinary Research, 53* (June 1992), 938–943.

A-15. John C. Licciardone, Scott T. Stoll, Kimberly G. Fulda, David P. Russo, Jeff Siu, William Winn, and John Swift, Jr., "Osteopathic Manipulative Treatment for Chronic Low Back Pain: A Randomized Controlled Trial," *Spine*, 28 (2003), 1355–1362.

A-16. Yitao Liu, Ludmila Belayev, Weizhao Zhao, Raul Busto, and Myron D. Ginsburg, "MRZ 2/579, a Novel Uncompetitive *N*-Methyl-D-Aspartate Antagonist, Reduces Infarct Volume and Brain Swelling and Improves Neurological Deficit After Focal Cerebral Ischemia in Rats," *Brain Research, 862* (2000), 111–119.

A-17. David W. Starch, Jerry W. Alexander, Philip C. Noble, Suraj Reddy, and David M. Lintner, "Multistranded Hamstring Tendon Graft Fixation with a Central Four-Quadrant or a Standard Tibial Interference Screw for Anterior Cruciate Ligament Reconstruction," *American Journal of Sports Medicine, 31* (2003), 338–344.

A-18. D. Holben, A. Smith, J. Ilich, J. Landoll, J. Holcomb, and V. Matkovik, "Puberty Does Not Negatively Affect Selenium Status in Young Females," *Journal of the American Dietetic Association, 102* (2002), 1082–1087.

A-19. Axel Linke, Gong Zhao, Fabio A. Recchia, Jeffrey Williams, Xiaobin Xu, and Thomas H. Hintze, "Shift in Metabolic Substrate Uptake by the Heart During Development of Alloxan-Induced Diabetes," *American Journal of Physiology—Heart and Circulatory Physiology, 285* (2003), H1007–H1014.

A-20. K. Werther, I. J. Christensen, and H. J. Nielsen, "Determination of Vascular Endothelial Growth Factor (VEGF) in Circulating Blood: Significance of VEGF in Various Leucocytes and Platelets," *Scandinavian Journal of Clinical and Laboratory Investigation, 62* (2002), 343–350.

A-21. Shigeko Uryu, Jun Harada, Marie Hisamoto, and Tomiichiro Oda, "Troglitazone Inhibits Both Postglutamate Neurotoxicity and Low-Potassium-Induced Apoptosis in Cerebellar Granule Neurons," *Brain Research, 924* (2002), 229–236.

A-22. A. Main, H. Abu-Saad, R. Salt, I. Vlachonikolis, and A. Dowson, "Management by Nurses of Primary Headache: A Pilot Study," *Current Medical Research Opinion, 18* (2002), 471–478.

A-23. Simona Porcellini, Guiliana Vallanti, Silvia Nozza, Guido Poli, Adriano Lazzarin, Guiseppe Tabussi, and Antonio Grassia, "Improved Thymopoietic Potential in Aviremic HIV Infected Individuals with HAART by Intermittent IL-2 Administration," *AIDS, 17* (2003), 1621–1630.

A-24. Dorota Górecka, Michal Benarek, Adam Nowinski, Elzbieta Puscinska, Anna Goljan-Geremek, and Jan Zielinski, "Diagnosis of Airflow Limitation Combined with Smoking Cessation Advice Increases Stop-Smoking Rate," *Chest, 123* (2003), 1916–1923.

A-25. Eli Meltzer, Bruce M. Prenner, and Anjuli Nayak, "Efficacy and Tolerability of Once-Daily 5 mg of Desloratadine, an H1-Receptor Antagonist, in Patients with Seasonal Allergic Rhinitis Assessment During the Spring and Fall Allergy Seasons," *Clinical Drug Investigations, 21* (2001), 25–32.

A-26. K. Werther, I. J. Christensen, and H. J. Nielsen, "Determination of Vascular Endothelial Growth Factor (VEGF) in Circulating Blood: Significance of VEGF in Various Leucocytes and Platelets," *Scandinavian Journal of Clinical and Laboratory Investigation, 62* (2002), 343–350.

A-27. Rubia P. Meshack and Kathleen E. Norman, "A Randomized Controlled Trial of the Effects of Weights on Amplitude and Frequency of Postural Hand Tremor in People with Parkinson's Disease," *Clinical Rehabilitation, 16* (2003), 481–492.

A-28. G. Lacroix, S. Tissot, F. Rogerieux, R. Beaulieu, L. Cornu, C. Gillet, F. R. Obidel, J. P. Lefevre, and F. Y. Bois, "Decrease in Ovalbumin-Induced Pulmonary Allergic Response by Benzaldehyde but Not Acetaldehyde Exposure in a Guinea Pig Model," *Journal of Toxicology and Environmental Health, Part A* (2002), 995–1012.

A-29. Takemasa Watanabe, Yoshiki Sugiyama, Yoshiko Sumi, Misuzu Watanabe, Kiyomi Takeuchi, Fumio Kobayashi, and Koichi Kono, "Effects of Vital Exhaustion on Cardiac Autonomic Nervous Functions Assessed by Heart Rate Variability at Rest in Middle-Aged Male Workers," *International Journal of Behavioral Medicine, 9* (2002), 68–75.

A-30. John Brooker, Data analyzed by the Wright State University Statistical Consulting Center, Wright State University, Dayton, OH (1997).

A-31. Eleftherios Kellis, Ph.D., "Pubertal Children Are Unable to Relax During the Passive Gravity Correction Procedure on the Iskinetic Dynamometer," *Isokinetics and Exercise Science, 10* (2002), 97–105.

A-32. Reed E. Pyeritz, Elliot K. Fishman, Barbara A. Bernhardt, and Stanley S. Siegelman, "Dural Ectasia Is a Common Feature of the Marfan Syndrome," *American Journal of Human Genetics, 43* (1988), 726–732.

A-33. Adina Hartman-Maeir, Nachum Soroker, Haim Ring, and Noomi Katz, "Awareness of Deficits in Stroke Rehabilitation," *Journal of Rehabilitation Medicine, 34* (2002), 158–164.

A-34. Eftihia V. Polyzogopoulou, Fotios Kalfarentzos, Apostolos G. Vagenakis, and Theodore K. Alexandrides, "Restoration of Euglycemia and Normal Acute Insulin Response to Glucose in Obese Subjects with Type 2 Diabetes Following Bariatric Surgery," *Diabetes, 53* (2003), 1098–1103.

A-35. Carol S. Johnston and D. L. Bowling, "Stability of Ascorbic Acid in Commercially Available Orange Juices," *Journal of the American Dietetic Association, 102* (2002), 525–529.

A-36. R. Pittini, D. Oepkess, K. Macrury, R. Reznick, J. Beyene, and R. Windrim, "Teaching Invasive Perinatal Procedures: Assessment of a High Fidelity Simulator-Based Curriculum," *Ultrasound Obstetrics and Gynecology, 19* (2002), 478–483.

A-37. Nick Anim-Nyame, John Gamble, Suren R. Sooranna, Mark R. Johnson, and Philip J. Steer, "Microvascular Permeability Is Related to Circulating Levels of Tumour Necrosis Factor-a in Pre-Eclampsia," *Cardiovascular Research, 58* (2003), 162–169.

A-38. Jun Iwamoto, Tsuyoshi Takeda, Shoichi Ichimura, and Mitsuyoshi Uzawa, "Effects of 5-Year Treatment with Elcatonin and Alfacalcidol on Lumbar Bone Mineral Density and the Incidence of Vertebral Fractures in Postmenopausal Women with Osteoporosis: A Retrospective Study," *Journal of Orthopaedic Science, 7* (2002), 637–643.

A-39. Kimiko Inoue, Narumi Ogonuki, Keiji Mochida, Yoshie Yamamoto, Kaoru Takano, Fumitoshi Ishino, and Atsuo Ogura, "Effects of Donor Cell Type and Genotype on the Efficiency of Mouse Somatic Cell Cloning," *Biology of Reproduction, 69* (2003), 1394–1400.

A-40. Gotaro Shirakami, Masahiro Murakawa, Koh Shingu, Susumu Mashima, Shin-ichi Suga, Tatsuo Magaribuchi, Toyohiko O'higashi, Tsutomu Shichino, Kazuwa Nakao, and Kenjiro Mori, "Perioperative Plasma Concentrations of Endothelin and Natriuretic Peptides in Children Undergoing Living-Related Liver Transplantation," *Anesthesia and Analgesia, 82* (1996), 235–240.

A-41. Masahito Sakakibara and Junichiro Hayano, "Effect of Slowed Respiration on Cardiac Parasympathetic Response to Threat," *Psychosomatic Medicine, 58* (1996), 32–37.

A-42. Toshiaki Takahashi, Yuji Suto, Shinsuke Kato, and Eisaku Ohama, "Experimental Acute Dorsal Compression of Cat Spinal Cord," *Spine, 21* (1996), 166–173.

A-43. Motohiro Yamashita, Ralph A. Schmid, Kan Okabayashi, Koei Ando, Jun Kobayashi, Joel D. Cooper, and G. Alexander Patterson, "Pentoxifylline in Flush Solution Improves Early Lung Allograft Function," *Annals of Thoracic Surgery, 61* (1996), 1055–1061.

A-44. Jin-R. Zhou, Eric T. Gugger, and John W. Erdman, Jr., "The Crystalline Form of Carotenes and the Food Matrix in Carrot Root Decrease the Relative Bioavailability of Beta- and Alpha-Carotene in the Ferret Model," *Journal of the American College of Nutrition, 15* (1996), 84–91.

A-45. J. A. Potteiger, G. L. Nickel, M. J. Webster, M. D. Haub, and R. J. Palmer, "Sodium Citrate Ingestion Enhances 30 km Cycling Performance," *International Journal of Sports Medicine, 17* (1996), 7–11.

A-46. Robert A. Teitge, Wade Faerber, Patricia Des Madryl, and Thomas M. Matelic, "Stress Radiographs of the Patellofemoral Joint," *Journal of Bone and Joint Surgery, 78-A* (February 1996), 193–203.

A-47. Akihiko Ikeda, Koichi Nishimura, Hiroshi Koyama, Mitsuhiro Tsukino, Michiaki Mishima, and Takateru Izumi, "Dose Response Study of Ipratropium Bromide Aerosol on Maximum Exercise Performance in Stable Patients with Chronic Obstructive Pulmonary Disease," *Thorax, 51* (1996), 48–53.

A-48. Antti Pertovaara, Timo Kauppila, and Minna M. Hämäläinen, "Influence of Skin Temperature on Heat Pain Threshold in Humans," *Experimental Brain Research, 107* (1996), 497–503.

A-49. P. G. Knight, S. Muttukrishna, and N. P. Groome, "Development and Application of a Two-Site Enzyme Immunoassay for the Determination of 'Total' Activin-A Concentrations in Serum and Follicular Fluid," *Journal of Endocrinology, 148* (1996), 267–279.

A-50. Pierre C. Maheux, Brigitte Bonin, Anne Dizazo, Pierre Guimond, Danielle Monier, Josée Bourque, and Jean-Louis Chiasson, "Glucose Homeostasis During Spontaneous Labor in Normal Human Pregnancy," *Journal of Clinical Endocrinology and Metabolism, 81* (1996), 209–215.

A-51. Howard Trachtman, John Maesaka, Stephen Futterweit, Chan Ma, Elsa Valderrama, James C. M. Chan, and Winnie Chan, "The Effect of Recombinant Human Insulin-Like Growth Factor-I on Chronic Puromycin Aminonucleoside Nephropathy in Rats," *Pediatric Research, 39* (1996), 13–19.

A-52. M. H. Samuels and P. Kramer, "Differential Effects of Short-Term Fasting on Pulsatile Thyrotropin, Gonadotropin, and a-Subunit Secretion in Healthy Men—A Clinical Research Center Study," *Journal of Clinical Endocrinology and Metabolism, 81* (1996), 32–36.

A-53. T. J. Sobotka, P. Whittaker, J. M. Sobotka, R. E. Brodie, D. Y. Quander, M. Robl, M. Bryant, and C. N. Barton, "Neurobehavioral Dysfunctions Associated with Dietary Iron Overload," *Physiology and Behavior, 58* (1996), 213–219.

A-54. Thor Willy Ruud Hansen, "Bilirubin Entry into and Clearance from Rat Brain During Hypercarbia and Hyperosmolality," *Pediatric Research, 39* (1996), 72–76.

A-55. Anna G. Johansson, Erik Lindh, Werner F. Blum, Gina Kollerup, Ole H. Sørensen, and Sverker Ljunghall, "Effects of Growth Hormone and Insulin-Like Growth Factor I in Men with Idiopathic Osteoporosis," *Journal of Clinical Endocrinology and Metabolism, 81* (1996), 44–48.

A-56. Jaap H. Strijbos, Dirkje S. Postma, Richard van Altena, Fernando Gimeno, and Gerard H. Koëter, "A Comparison Between an Outpatient Hospital-Based Pulmonary Rehabilitation Program and a Home-Care Pulmonary Rehabilitation Program in Patients with COPD," *Chest, 109* (1996), 366–372.

A-57. G. P. Lambert, R. T. Chang, D. Joensen, X. Shi, R. W. Summers, H. P. Schedl, and C. V. Gisolfi, "Simultaneous Determination of Gastric Emptying and Intestinal Absorption During Cycle Exercise in Humans," *International Journal of Sports Medicine, 17* (1996), 48–55.

A-58. Lizabeth Roemer, Mary Borkovec, Sharon Posa, and T. D. Borkovec, "A Self-Diagnostic Measure of Generalized Anxiety Disorder," *Journal of Behavior Therapy & Experimental Psychiatry, 26* (1995), 345–350.

A-59. Ilse Christiansen, Cecilia Gidlöf, Karl-Michael Kälkner, Hans Hagberg, Helge Bennmarker, and Thomas Tötterman, "Elevated Serum Levels of Soluble ICAM-1 in Non-Hodgkin's Lymphomas Correlate with Tumour Burden, Disease Activity and Other Prognostic Markers," *British Journal of Haematology, 92* (1996), 639–646.

A-60. Sylvie Cossette, Louise Lévesque, and Liane Laurin, "Informal and Formal Support for Caregivers of a Demented Relative: Do Gender and Kinship Make a Difference?" *Research in Nursing & Health, 18* (1995), 437–451.

A-61. Sadatomo Tasaka, Akitoshi Ishizaka, Koichi Sayama, Fumio Sakamaki, Hidetoshi Nakamura, Takeshi Terashima, Yasuhiro Waki, Kenzo Soejima, Morio Nakamura, Hiroaki Matsubara, Seitaro Fujishima, and Minoru Kanazawa, "Heat-Killed *Corynebacterium parvum* Enhances Endotoxin Lung Injury with Increased TNF Production in Guinea Pigs," *American Journal of Respiratory and Critical Care Medicine, 153* (1996), 1047–1055.

A-62. Mitsugi Takahashi, Hideo Nagao, and Hirosi Matsuda, "Estimating Bone Change in Patients with Severe Disability," *Development Medicine and Child Neurology, 37* (1995), 1085–1093.

A-63. Corrie B. Allen, "Effects of Dietary Copper Deficiency on Relative Food Intake and Growth Efficiency in Rats," *Physiology & Behavior, 59* (1996), 247–253.

A-64. Christopher B. Hughes, Lillian W. Gaber, Abou Bakr Mohey El-Din, Hani P. Grewal, Malak Kotb, Linda Mann, and A. Osama Gaber, "Inhibition of TNFa Improves Survival in an Experimental Model of Acute Pancreatitis," *American Surgeon, 62* (1996), 8–13.

A-65. Alexander K. Smárason, Ian L. Sargent, and Christopher W. G. Redman, "Endothelial Cell Proliferation Is Suppressed by Plasma but Not Serum from Women with Preeclampsia," *American Journal of Obstetrics and Gynecology, 174* (1996), 787–793.

A-66. Derek LeRoith, Jack Yanowski, Eric P. Kaldjian, Elaine S. Jaffe, Tanya LeRoith, Kathy Purdue, Barbara D. Cooper, Robert Pyle, and William Adler, "The Effects of Growth Hormone and Insulin-Like Growth Factor I on the Immune System of Aged Female Monkeys," *Endocrinology, 137* (1996), 1071–1079.

A-67. Václav Hampl, Martin Tristani-Firouzi, Thomas C. Hutsell, and Stephen L. Archer, "Nebulized Nitric Oxide/Nucleophile Adduct Reduces Chronic Pulmonary Hypertension," *Cardiovascular Research, 31* (1996), 55–62.

A-68. C. Kreiter, M. W. Peterson, K. Ferguson, and S. Elliott, "The Effects of Testing in Shifts on a Clinical In-Course Computerized Exam," *Medical Education, 37* (2003), 202–204.

CHAPTER **9**

STATISTICAL INFERENCE AND THE RELATIONSHIP BETWEEN TWO VARIABLES

CHAPTER OVERVIEW

This chapter provides an introduction and overview of two common techniques for exploring the strength of the relationship between two variables. The first technique, linear regression, will help us find an objective way to predict or estimate the value of one variable given a value of another variable. The second technique, correlation, will help us find an objective measure of the strength of the relationship between two variables.

TOPICS

LEARNING OUTCOMES

After studying this chapter, the student will

1. be able to obtain a simple linear regression model and use it to make predictions.
2. be able to calculate the coefficient of determination and to interpret tests of regression coefficients.
3. be able to calculate correlations among variables.
4. understand how regression and correlation differ and when the use of each is appropriate.

9.1 INTRODUCTION

In analyzing data for the health sciences disciplines, we find that it is frequently desirable to learn something about the relationship between two numeric variables. We may, for example, be interested in studying the relationship between blood pressure and age, height and weight, the concentration of an injected drug and heart rate, the consumption level of some nutrient and weight gain, the intensity of a stimulus and reaction time, or total family income and medical care expenditures. The nature and strength of the relationships between variables such as these may be examined by *regression* and *correlation* analysis, two statistical techniques that, although related, serve different purposes.

Regression Regression analysis is helpful in assessing specific forms of the relationship between variables, and the ultimate objective when this method of analysis is employed usually is to *predict* or *estimate* the value of one variable corresponding to a given value of another variable. The ideas of regression were first elucidated by the English scientist Sir Francis Galton (1822–1911) in reports of his research on heredity—first in sweet peas and later in human stature. He described a tendency of adult offspring, having either short or tall parents, to revert back toward the average height of the general population. He first used the word *reversion,* and later *regression,* to refer to this phenomenon.

Correlation Correlation analysis, on the other hand, is concerned with measuring the strength of the relationship between variables. When we compute measures of correlation from a set of data, we are interested in the degree of the *correlation* between variables. Again, the concepts and terminology of correlation analysis originated with Galton, who first used the word *correlation* in 1888.

In this chapter our discussion is limited to the exploration of the linear relationship between two variables. The concepts and methods of regression are covered first, beginning in the next section. In Section 9.6 the ideas and techniques of correlation are introduced. In the next chapter we consider the case where there is an interest in the relationships among three or more variables.

Regression and correlation analysis are areas in which the speed and accuracy of a computer are most appreciated. The data for the exercises of this chapter, therefore, are presented in a way that makes them suitable for computer processing. As is always the case, the input requirements and output features of the particular programs and software packages to be used should be studied carefully.

9.2 THE REGRESSION MODEL

In the typical regression problem, as in most problems in applied statistics, researchers have available for analysis a sample of observations from some real or hypothetical population. Based on the results of their analysis of the sample data, they are interested in reaching decisions about the population from which the sample is presumed to have been drawn. It is important, therefore, that the researchers understand the nature of the population in which they are interested. They should know enough about the population to be able either to construct a mathematical model for its representation or to determine if it reasonably fits some established model. A researcher about to analyze a set of data by the methods of simple linear regression, for example, should be secure in the knowledge that the simple linear regression model is, at least, an approximate representation of the population. It is unlikely that the model will be a perfect portrait of the real situation, since this characteristic is seldom found in models of practical value. A model constructed so that it corresponds precisely with the details of the situation is usually too complicated to yield any information of value. On the other hand, the results obtained from the analysis of data that have been forced into a model that does not fit are also worthless. Fortunately, however, a perfectly fitting model is not a requirement for obtaining useful results. Researchers, then, should be able to distinguish between the occasion when their chosen models and the data are sufficiently compatible for them to proceed and the case where their chosen model must be abandoned.

Assumptions Underlying Simple Linear Regression In the simple linear regression model two variables, usually labeled X and Y, are of interest. The letter X is usually used to designate a variable referred to as the *independent variable,* since frequently it is controlled by the investigator; that is, values of X may be selected by the investigator and, corresponding to each preselected value of X, one or more values of another variable, labeled Y, are obtained. The variable, Y, accordingly, is called the *dependent variable,* and we speak of the regression of Y on X. The following are the assumptions underlying the simple linear regression model.

1. Values of the independent variable X are said to be "fixed." This means that the values of X are preselected by the investigator so that in the collection of the data they are not allowed to vary from these preselected values. In this model, X is referred to by some writers as a *nonrandom* variable and by others as a *mathematical* variable. It should be pointed out at this time that the statement of this assumption classifies our model as the *classical regression model.* Regression analysis also can be carried out on data in which X is a random variable.
2. The variable X is measured without error. Since no measuring procedure is perfect, this means that the magnitude of the measurement error in X is negligible.
3. For each value of X there is a subpopulation of Y values. For the usual inferential procedures of estimation and hypothesis testing to be valid, these subpopulations must be normally distributed. In order that these procedures may be presented it will be assumed that the Y values are normally distributed in the examples and exercises that follow.
4. The variances of the subpopulations of Y are all equal and denoted by σ^2.
5. The means of the subpopulations of Y all lie on the same straight line. This is known as the *assumption of linearity.* This assumption may be expressed symbolically as

$$\mu_{y|x} = \beta_0 + \beta_1 x \tag{9.2.1}$$

where $\mu_{y|x}$ is the mean of the subpopulation of Y values for a particular value of X, and β_0 and β_1 are called the population regression coefficients. Geometrically, β_0 and β_1 represent the y-intercept and slope, respectively, of the line on which all of the means are assumed to lie.

6. The Y values are statistically independent. In other words, in drawing the sample, it is assumed that the values of Y chosen at one value of X in no way depend on the values of Y chosen at another value of X.

These assumptions may be summarized by means of the following equation, which is called the regression model:

$$y = \beta_0 + \beta_1 x + \epsilon \tag{9.2.2}$$

where y is a typical value from one of the subpopulations of Y, β_0 and β_1 are as defined for Equation 9.2.1, and ϵ is called the error term. If we solve 9.2.2 for ϵ, we have

$$\begin{aligned} \epsilon &= y - (\beta_0 + \beta_1 x) \\ &= y - \mu_{y|x} \end{aligned} \tag{9.2.3}$$

and we see that ϵ shows the amount by which y deviates from the mean of the subpopulation of Y values from which it is drawn. As a consequence of the assumption that the subpopulations of Y values are normally distributed with equal variances, the ϵ's for each subpopulation are normally distributed with a variance equal to the common variance of the subpopulations of Y values.

The following acronym will help the reader remember most of the assumptions necessary for inference in linear regression analysis:

LINE [Linear (assumption 5), Independent (assumption 6), Normal (assumption 3), Equal variances (assumption 4)]

A graphical representation of the regression model is given in Figure 9.2.1.

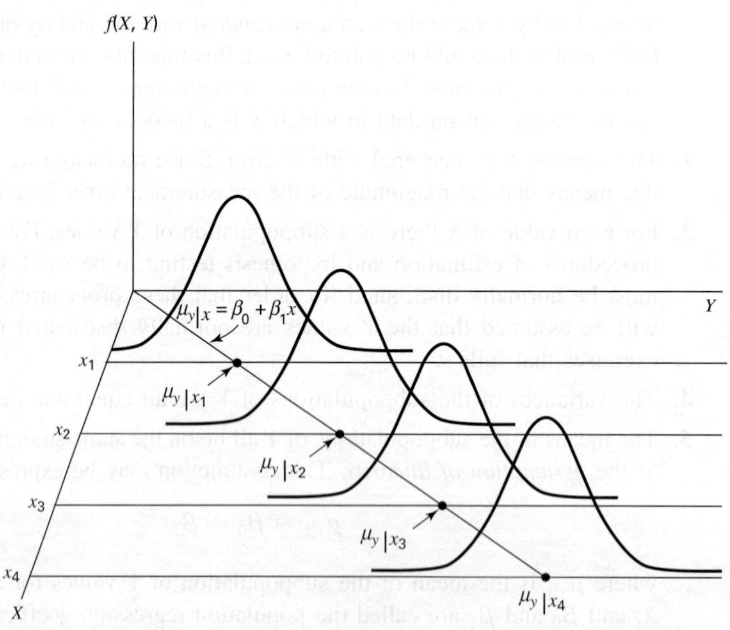

FIGURE 9.2.1 Representation of the simple linear regression model.

9.3 THE SAMPLE REGRESSION EQUATION

In simple linear regression the object of the researcher's interest is the population regression equation—the equation that describes the true relationship between the dependent variable Y and the independent variable X. The variable designated by Y is sometimes called the *response variable* and X is sometimes called the *predictor variable.*

In an effort to reach a decision regarding the likely form of this relationship, the researcher draws a sample from the population of interest and using the resulting data, computes a sample regression equation that forms the basis for reaching conclusions regarding the unknown population regression equation.

Steps in Regression Analysis In the absence of extensive information regarding the nature of the variables of interest, a frequently employed strategy is to assume initially that they are linearly related. Subsequent analysis, then, involves the following steps.

1. Determine whether or not the assumptions underlying a linear relationship are met in the data available for analysis.
2. Obtain the equation for the line that best fits the sample data.
3. Evaluate the equation to obtain some idea of the strength of the relationship and the usefulness of the equation for predicting and estimating.
4. If the data appear to conform satisfactorily to the linear model, use the equation obtained from the sample data to predict and to estimate.

When we use the regression equation to *predict,* we will be predicting the value Y is likely to have when X has a given value. When we use the equation to *estimate,* we will be estimating the mean of the subpopulation of Y values assumed to exist at a given value of X. Note that the sample data used to obtain the regression equation consist of known values of both X and Y. When the equation is used to predict and to estimate Y, only the corresponding values of X will be known. We illustrate the steps involved in simpler linear regression analysis by means of the following example.

EXAMPLE 9.3.1

Després et al. (A-1) point out that the topography of adipose tissue (AT) is associated with metabolic complications considered as risk factors for cardiovascular disease. It is important, they state, to measure the amount of intraabdominal AT as part of the evaluation of the cardiovascular-disease risk of an individual. Computed tomography (CT), the only available technique that precisely and reliably measures the amount of deep abdominal AT, however, is costly and requires irradiation of the subject. In addition, the technique is not available to many physicians. Després and his colleagues conducted a study to develop equations to predict the amount of deep abdominal AT from simple anthropometric measurements. Their subjects were men between the ages of 18 and 42 years who were free from metabolic disease that would require treatment.

Among the measurements taken on each subject were deep abdominal AT obtained by CT and waist circumference as shown in Table 9.3.1. A question of interest is how well one can predict and estimate deep abdominal AT from knowledge of the waist circumference. This question is typical of those that can be answered by means of regression analysis. Since deep abdominal AT is the variable about which we wish to make predictions and estimations, it is the dependent variable. The variable waist measurement, knowledge of which will be used to make the predictions and estimations, is the independent variable.

TABLE 9.3.1 Waist Circumference (cm), *X*, and Deep Abdominal AT, *Y*, of 109 Men

Subject	*X*	*Y*	Subject	*X*	*Y*	Subject	*X*	*Y*
1	74.75	25.72	38	103.00	129.00	75	108.00	217.00
2	72.60	25.89	39	80.00	74.02	76	100.00	140.00
3	81.80	42.60	40	79.00	55.48	77	103.00	109.00
4	83.95	42.80	41	83.50	73.13	78	104.00	127.00
5	74.65	29.84	42	76.00	50.50	79	106.00	112.00
6	71.85	21.68	43	80.50	50.88	80	109.00	192.00
7	80.90	29.08	44	86.50	140.00	81	103.50	132.00
8	83.40	32.98	45	83.00	96.54	82	110.00	126.00
9	63.50	11.44	46	107.10	118.00	83	110.00	153.00
10	73.20	32.22	47	94.30	107.00	84	112.00	158.00
11	71.90	28.32	48	94.50	123.00	85	108.50	183.00
12	75.00	43.86	49	79.70	65.92	86	104.00	184.00
13	73.10	38.21	50	79.30	81.29	87	111.00	121.00
14	79.00	42.48	51	89.80	111.00	88	108.50	159.00
15	77.00	30.96	52	83.80	90.73	89	121.00	245.00
16	68.85	55.78	53	85.20	133.00	90	109.00	137.00
17	75.95	43.78	54	75.50	41.90	91	97.50	165.00
18	74.15	33.41	55	78.40	41.71	92	105.50	152.00
19	73.80	43.35	56	78.60	58.16	93	98.00	181.00
20	75.90	29.31	57	87.80	88.85	94	94.50	80.95
21	76.85	36.60	58	86.30	155.00	95	97.00	137.00
22	80.90	40.25	59	85.50	70.77	96	105.00	125.00
23	79.90	35.43	60	83.70	75.08	97	106.00	241.00
24	89.20	60.09	61	77.60	57.05	98	99.00	134.00
25	82.00	45.84	62	84.90	99.73	99	91.00	150.00
26	92.00	70.40	63	79.80	27.96	100	102.50	198.00
27	86.60	83.45	64	108.30	123.00	101	106.00	151.00
28	80.50	84.30	65	119.60	90.41	102	109.10	229.00
29	86.00	78.89	66	119.90	106.00	103	115.00	253.00
30	82.50	64.75	67	96.50	144.00	104	101.00	188.00
31	83.50	72.56	68	105.50	121.00	105	100.10	124.00
32	88.10	89.31	69	105.00	97.13	106	93.30	62.20
33	90.80	78.94	70	107.00	166.00	107	101.80	133.00
34	89.40	83.55	71	107.00	87.99	108	107.90	208.00
35	102.00	127.00	72	101.00	154.00	109	108.50	208.00
36	94.50	121.00	73	97.00	100.00			
37	91.00	107.00	74	100.00	123.00			

Source: Jean-Pierre Després, Ph.D. Used with permission.

The Scatter Diagram

A first step that is usually useful in studying the relationship between two variables is to prepare a *scatter diagram* of the data such as is shown in Figure 9.3.1. The points are plotted by assigning values of the independent variable X to the horizontal axis and values of the dependent variable Y to the vertical axis.

The pattern made by the points plotted on the scatter diagram usually suggests the basic nature and strength of the relationship between two variables. As we look at Figure 9.3.1, for example, the points seem to be scattered around an invisible straight line. The scatter diagram also shows that, in general, subjects with large waist circumferences also have larger amounts of deep abdominal AT. These impressions suggest that the relationship between the two variables may be described by a straight line crossing the Y-axis below the origin and making approximately a 45-degree angle with the X-axis. It looks as if it would be simple to draw, freehand, through the data points the line that describes the relationship between X and Y. It is highly unlikely, however, that the lines drawn by any two people would be exactly the same. In other words, for every person drawing such a line by eye, or freehand, we would expect a slightly different line. The question then arises as to which line best describes the relationship between the two variables. We cannot obtain an answer to this question by inspecting the lines. In fact, it is not likely that any freehand line drawn

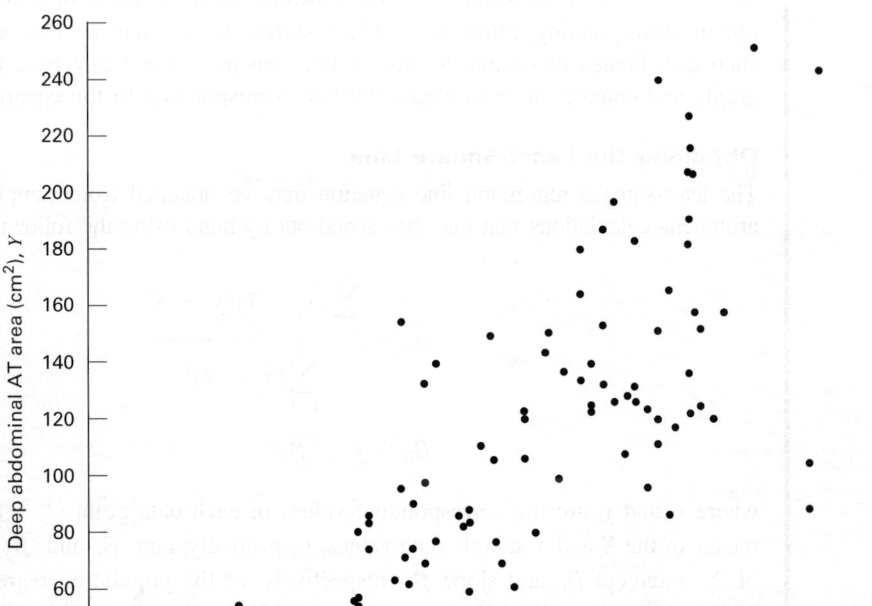

FIGURE 9.3.1 Scatter diagram of data shown in Table 9.3.1.

through the data will be the line that best describes the relationship between X and Y, since freehand lines will reflect any defects of vision or judgment of the person drawing the line. Similarly, when judging which of two lines best describes the relationship, subjective evaluation is liable to the same deficiencies.

What is needed for obtaining the desired line is some method that is not fraught with these difficulties.

The Least-Squares Line

The method usually employed for obtaining the desired line is known as the *method of least squares,* and the resulting line is called the *least-squares line.* The reason for calling the method by this name will be explained in the discussion that follows.

We recall from algebra that the general equation for a straight line may be written as

$$y = a + bx \tag{9.3.1}$$

where y is a value on the vertical axis, x is a value on the horizontal axis, a is the point where the line crosses the vertical axis, and b shows the amount by which y changes for each unit change in x. We refer to a as the y-intercept and b as the *slope* of the line. To draw a line based on Equation 9.3.1, we need the numerical values of the constants a and b. Given these constants, we may substitute various values of x into the equation to obtain corresponding values of y. The resulting points may be plotted. Since any two such coordinates determine a straight line, we may select any two, locate them on a graph, and connect them to obtain the line corresponding to the equation.

Obtaining the Least-Square Line

The least-squares regression line equation may be obtained from sample data by simple arithmetic calculations that may be carried out by hand using the following equations

$$\hat{\beta}_1 = \frac{\sum_{i=1}^{n}(x_i - \bar{x})(y_1 - \bar{y})}{\sum_{i=1}^{n}(x_i - \bar{x})^2} \tag{9.3.2}$$

$$\hat{\beta}_0 = \bar{y} - \hat{\beta}_1\bar{x} \tag{9.3.3}$$

where x_i and y_i are the corresponding values of each data point (X, Y), $\bar{x}$ and $\bar{y}$ are the means of the X and Y sample data values, respectively, and $\hat{\beta}_0$ and $\hat{\beta}_1$ are the estimates of the intercept β_0 and slope β_1, respectively, of the population regression line. Since the necessary hand calculations are time consuming, tedious, and subject to error, the regression line equation is best obtained through the use of a computer software package. Although the typical researcher need not be concerned with the arithmetic involved, the interested reader will find them discussed in references listed at the end of this chapter.

For the data in Table 9.3.1 we obtain the least-squares regression equation by means of MINITAB. After entering the X values in Column 1 and the Y values in Column 2 we proceed as shown in Figure 9.3.2.

For now, the only information from the output in Figure 9.3.2 that we are interested in is the regression equation. Other information in the output will be discussed later.

Dialog box:	**Session command:**
Stat ➤ Regression ➤ Regression	MTB > Name C3 = 'FITS1' C4 = 'RESI1'
Type *y* in **Response** and *x* in **Predictors.**	MTB > Regress 'y' 1 'x';
Click **Storage.** Check **Residuals** and **Fits.**	SUBC> Fits 'FITS1';
Click **OK.**	SUBC> Constant;
	SUBC> Residuals 'RESI1'.

Output:

Regression Analysis: y versus x

```
The regression equation is
y = -216 + 3.46  x

Predictor          Coef      Stdev    t-ratio         p
Constant        -215.98      21.80      -9.91     0.000
x                3.4589     0.2347      14.74     0.000

s = 33.06   R-sq = 67.0%   R-sq(adj) = 66.7%

Analysis  of  Variance

SOURCE            DF         SS         MS          F          p
Regression         1     237549     237549     217.28      0.000
Error            107     116982       1093
Total            108     354531

Unusual  Observations
Obs.       x          y        Fit     Stdev.Fit      Residual     St.Resid
  58      86     155.00      82.52          3.43         72.48        2.20R
  65     120      90.41     197.70          7.23       -107.29       -3.33R
  66     120     106.00     198.74          7.29        -92.74       -2.88R
  71     107      87.99     154.12          4.75        -66.13       -2.02R
  97     106     241.00     150.66          4.58         90.34        2.76R
 102     109     229.00     161.38          5.13         67.62        2.07R
 103     115     253.00     181.79          6.28         71.21        2.19R

R denotes an obs. with a large st. resid.
```

FIGURE 9.3.2 MINITAB procedure and output for obtaining the least-squares regression equation from the data in Table 9.3.1.

From Figure 9.3.2 we see that the linear equation for the least-squares line that describes the relationship between waist circumference and deep abdominal AT may be written, then, as

$$\hat{y} = -216 + 3.46x$$

This equation tells us that since $\hat{\beta}_0$ is negative, the line crosses the Y-axis below the origin, and that since $\hat{\beta}_1$ the slope, is positive, the line extends from the lower left-hand corner of the graph to the upper right-hand corner. We see further that for each unit increase in x, y increases by an amount equal to 3.46. The symbol y denotes a value of y computed from the equation, rather than an observed value of Y.

By substituting two convenient values of X into Equation 9.3.2, we may obtain the necessary coordinates for drawing the line. Suppose, first, we let $X = 70$ and obtain

$$\hat{y} = -216 + 3.46(70) = 26.2$$

If we let $X = 110$ we obtain

$$\hat{y} = -216 + 3.46(110) = 164$$

The line, along with the original data, is shown in Figure 9.3.3.

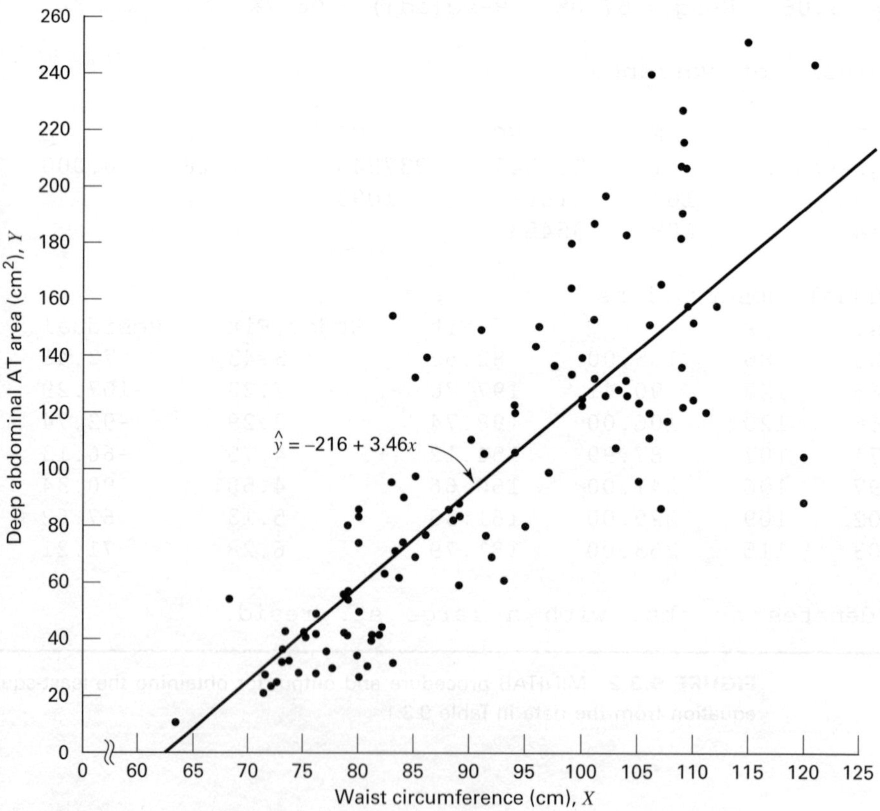

FIGURE 9.3.3 Original data and least-squares line for Example 9.3.1. ■

The Least-Squares Criterion Now that we have obtained what we call the "best fit" line for describing the relationship between our two variables, we need to determine by what criterion it is considered best. Before the criterion is stated, let us examine Figure 9.3.3. We note that generally the least-squares line does not pass through the observed points that are plotted on the scatter diagram. In other words, most of the observed points *deviate* from the line by varying amounts.

The line that we have drawn through the points is best in this sense:

The sum of the squared vertical deviations of the observed data points (y_i) from the least-squares line is smaller than the sum of the squared vertical deviations of the data points from any other line.

In other words, if we square the vertical distance from each observed point (y_i) to the least-squares line and add these squared values for all points, the resulting total will be smaller than the similarly computed total for any other line that can be drawn through the points. For this reason the line we have drawn is called the least-squares line.

EXERCISES

9.3.1 Plot each of the following regression equations on graph paper and state whether X and Y are directly or inversely related.

(a) $\hat{y} = -3 + 2x$

(b) $\hat{y} = 3 + 0.5x$

(c) $\hat{y} = 10 - 0.75x$

9.3.2 The following scores represent a nurse's assessment (X) and a physician's assessment (Y) of the condition of 10 patients at time of admission to a trauma center.

X:	18	13	18	15	10	12	8	4	7	3
Y:	23	20	18	16	14	11	10	7	6	4

(a) Construct a scatter diagram for these data.

(b) Plot the following regression equations on the scatter diagram and indicate which one you think best fits the data. State the reason for your choice.

(1) $\hat{y} = 8 + 0.5x$

(2) $\hat{y} = -10 + 2x$

(3) $\hat{y} = 1 + 1x$

For each of the following exercises (a) draw a scatter diagram and (b) obtain the regression equation and plot it on the scatter diagram.

9.3.3 Methadone is often prescribed in the treatment of opioid addiction and chronic pain. Krantz et al. (A-2) studied the relationship between dose of methadone and the corrected QT (QTc) interval for 17 subjects who developed *torsade de pointes* (ventricular tachycardia nearly always due to medications). QTc is calculated from an electrocardiogram and is measured in mm/sec. A higher QTc value indicates a higher risk of cardiovascular mortality. A question of interest is how well one can predict and estimate the QTc value from a knowledge of methadone dose. This question is typical of those that can be answered by means of regression analysis. Since QTc is the variable about which we wish to make predictions and estimations, it is the dependent variable. The variable methadone dose, knowledge of which will be used to make the predictions and estimations, is the independent variable.

Methadone Dose (mg/day)	QTc (mm/sec)	Methadone Dose (mg/day)	QTc (mm/sec)
1000	600	650	785
550	625	600	765
97	560	660	611
90	585	270	600
85	590	680	625
126	500	540	650
300	700	600	635
110	570	330	522
65	540		

Source: Mori J. Krantz, Ilana B. Kutinsky, Alastair D. Roberston, and Philip S. Mehler, "Dose-Related Effects of Methadone on QT Prolongation in a Series of Patients with Torsade de Pointes," *Pharmacotherapy, 23* (2003), 802–805.

9.3.4 Reiss et al. (A-3) compared point-of-care and standard hospital laboratory assays for monitoring patients receiving a single anticoagulant or a regimen consisting of a combination of anticoagulants. It is quite common when comparing two measuring techniques, to use regression analysis in which one variable is used to predict another. In the present study, the researchers obtained measures of international normalized ratio (INR) by assay of capillary and venous blood samples collected from 90 subjects taking warfarin. INR, used especially when patients are receiving warfarin, measures the clotting ability of the blood. Point-of-care testing for INR was conducted with the CoaguChek assay product. Hospital testing was done with standard hospital laboratory assays. The authors used the hospital assay INR level to predict the CoaguChek INR level. The measurements are given in the following table.

CoaguChek (Y)	Hospital (X)	CoaguChek (Y)	Hospital (X)	CoaguChek (Y)	Hospital (X)
1.8	1.6	2.4	1.2	3.1	2.4
1.6	1.9	2.3	2.3	1.7	1.8
2.5	2.8	2.0	1.6	1.8	1.6
1.9	2.4	3.3	3.8	1.9	1.7
1.3	1.5	1.9	1.6	5.3	4.2
2.3	1.8	1.8	1.5	1.6	1.6
1.2	1.3	2.8	1.8	1.6	1.4
2.3	2.4	2.5	1.5	3.3	3.3
2.0	2.1	0.8	1.0	1.5	1.5
1.5	1.5	1.3	1.2	2.2	2.8
2.1	2.4	3.7	1.4	1.1	1.6
1.5	1.5	2.4	1.6	2.6	2.6

(Continued)

CoaguChek (Y)	Hospital (X)	CoaguChek (Y)	Hospital (X)	CoaguChek (Y)	Hospital (X)
1.5	1.7	4.1	3.2	6.4	5.0
1.8	2.1	2.4	1.2	1.5	1.4
1.0	1.2	2.3	2.3	3.0	2.8
2.1	1.9	3.1	1.6	2.6	2.3
1.6	1.6	1.5	1.4	1.2	1.2
1.7	1.6	3.6	2.1	2.1	1.9
2.0	1.9	2.5	1.7	1.1	1.1
1.8	1.6	2.1	1.7	1.0	1.0
1.3	4.1	1.8	1.2	1.4	1.5
1.5	1.9	1.5	1.3	1.7	1.3
3.6	2.1	2.5	1.1	1.2	1.1
2.4	2.2	1.5	1.2	2.5	2.4
2.2	2.3	1.5	1.1	1.2	1.3
2.7	2.2	1.6	1.2	2.5	2.9
2.9	3.1	1.4	1.4	1.9	1.7
2.0	2.2	4.0	2.3	1.8	1.7
1.0	1.2	2.0	1.2	1.2	1.1
2.4	2.6	2.5	1.5	1.3	1.1

Source: Curtis E. Haas, Pharm.D. Used with permission.

9.3.5 Digoxin is a drug often prescribed to treat heart ailments. The purpose of a study by Parker et al. (A-4) was to examine the interactions of digoxin with common grapefruit juice. In one experiment, subjects took digoxin with water for 2 weeks, followed by a 2-week period during which digoxin was withheld. During the next 2 weeks subjects took digoxin with grapefruit juice. For seven subjects, the average peak plasma digoxin concentration (Cmax) when taking water is given in the first column of the following table. The second column contains the percent change in Cmax concentration when subjects were taking the digoxin with grapefruit juice [GFJ (%) change]. Use the Cmax level when taking digoxin with water to predict the percent change in Cmax concentration when taking digoxin with grapefruit juice.

Cmax (ngl/ml) with Water	Change in Cmax with GFJ (%)
2.34	29.5
2.46	40.7
1.87	5.3
3.09	23.3
5.59	−45.1
4.05	−35.3
6.21	−44.6
2.34	29.5

Source: Robert B. Parker, Pharm.D. Used with permission.

9.3.6 Evans et al. (A-5) examined the effect of velocity on ground reaction forces (GRF) in dogs with lameness from a torn cranial cruciate ligament. The dogs were walked and trotted over a force platform and the GRF recorded (in newtons) during the stance phase. The following table contains 22 measurements of force expressed as the mean of five force measurements per dog when walking and the mean of five force measurements per dog when trotting. Use the GRF value when walking to predict the GRF value when trotting.

GRF-Walk	GRF-Trot	GRF-Walk	GRF-Trot
31.5	50.8	24.9	30.2
33.3	43.2	33.6	46.3
32.3	44.8	30.7	41.8
28.8	39.5	27.2	32.4
38.3	44.0	44.0	65.8
36.9	60.1	28.2	32.2
14.6	11.1	24.3	29.5
27.0	32.3	31.6	38.7
32.8	41.3	29.9	42.0
27.4	38.2	34.3	37.6
31.5	50.8	24.9	30.2

Source: Richard Evans, Ph.D. Used with permission.

9.3.7 Glomerular filtration rate (GFR) is the most important parameter of renal function assessed in renal transplant recipients. Although inulin clearance is regarded as the gold standard measure of GFR, its use in clinical practice is limited. Krieser et al. (A-6) examined the relationship between the inverse of Cystatin C (a cationic basic protein measured in mg/L) and inulin GFR as measured by technetium radionuclide labeled diethylenetriamine penta-acetic acid) (DTPA GFR) clearance (ml/min/1.73 m^2). The results of 27 tests are shown in the following table. Use DTPA GFR as the predictor of inverse Cystatin C.

DTPA GFR	1/Cystatin C	DTPA GFR	1/Cystatin C
18	0.213	42	0.485
21	0.265	42	0.427
21	0.446	43	0.562
23	0.203	43	0.463
27	0.369	48	0.549
27	0.568	48	0.538
30	0.382	51	0.571
32	0.383	55	0.546
32	0.274	58	0.402
32	0.424	60	0.592
36	0.308	62	0.541
37	0.498	67	0.568
41	0.398	68	0.800
		88	0.667

Source: David Krieser, M.D. Used with permission.

9.4 EVALUATING THE REGRESSION EQUATION

Once the regression equation has been obtained it must be evaluated to determine whether it adequately describes the relationship between the two variables and whether it can be used effectively for prediction and estimation purposes.

When H_0: $\beta_1 = 0$ Is Not Rejected If in the population the relationship between X and Y is linear, β_1, the slope of the line that describes this relationship, will be either positive, negative, or zero. If β_1 is zero, sample data drawn from the population will, in the long run, yield regression equations that are of little or no value for prediction and estimation purposes. Furthermore, even though we assume that the relationship between X and Y is linear, it may be that the relationship could be described better by some nonlinear model. When this is the case, sample data when fitted to a linear model will tend to yield results compatible with a population slope of zero. Thus, following a test in which the null hypothesis that β_1 equals zero is not rejected, we may conclude (assuming that we have not made a type II error by accepting a false null hypothesis) either (1) that although the relationship between X and Y may be linear it is not strong enough for X to be of much value in predicting and estimating Y, or (2) that the relationship between X and Y is not linear; that is, some curvilinear model provides a better fit to the data. Figure 9.4.1 shows the kinds of relationships between X and Y in a population that may prevent rejection of the null hypothesis that $\beta_1 = 0$.

When H_0: $\beta_1 = 0$ Is Rejected Now let us consider the situations in a population that may lead to rejection of the null hypothesis that $\beta_1 = 0$. Assuming that we do not commit a type I error, rejection of the null hypothesis that $\beta_1 = 0$ may be attributed to one of the following conditions in the population: (1) the relationship is linear and of sufficient strength to justify the use of sample regression equations to predict and estimate Y for given values of X; and (2) there is a good fit of the data to a linear model, but some curvilinear model might provide an even better fit. Figure 9.4.2 illustrates the two population conditions that may lead to rejection of H_0: $\beta_1 = 0$.

Thus, we see that before using a sample regression equation to predict and estimate, it is desirable to test H_0: $\beta_1 = 0$. We may do this either by using analysis of variance and the F statistic or by using the t statistic. We will illustrate both methods. Before we do this, however, let us see how we may investigate the strength of the relationship between X and Y.

The Coefficient of Determination One way to evaluate the strength of the regression equation is to compare the scatter of the points about the regression line with the scatter about $\bar{y}$, the mean of the sample values of Y. If we take the scatter diagram for Example 9.3.1 and draw through the points a line that intersects the Y-axis at $\bar{y}$ and is parallel to the X-axis, we may obtain a visual impression of the relative magnitudes of the scatter of the points about this line and the regression line. This has been done in Figure 9.4.3.

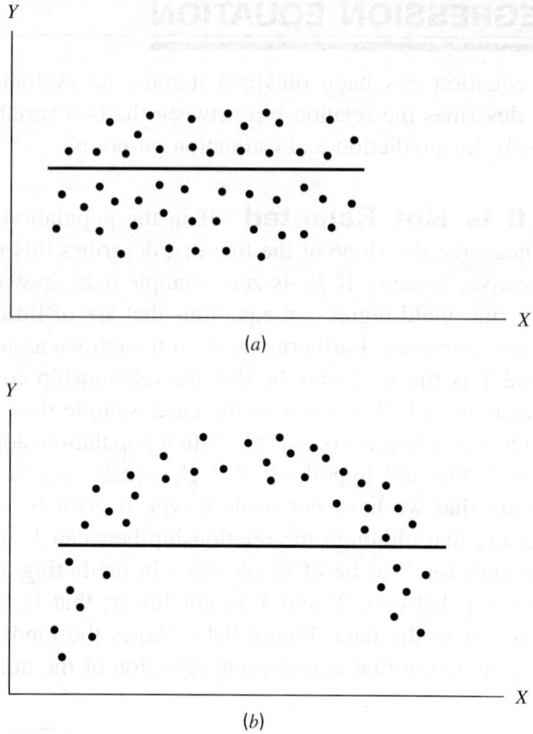

FIGURE 9.4.1 Conditions in a population that may prevent rejection of the null hypothesis that $\beta_1 = 0$. (*a*) The relationship between *X* and *Y* is linear, but β_1 is so close to zero that sample data are not likely to yield equations that are useful for predicting *Y* when *X* is given. (*b*) The relationship between *X* and *Y* is not linear; a curvilinear model provides a better fit to the data; sample data are not likely to yield equations that are useful for predicting *Y* when *X* is given.

It appears rather obvious from Figure 9.4.3 that the scatter of the points about the regression line is much less than the scatter about the $\bar{y}$ line. We would not wish, however, to decide on this basis alone that the equation is a useful one. The situation may not be always this clear-cut, so that an objective measure of some sort would be much more desirable. Such an objective measure, called the *coefficient of determination,* is available.

The Total Deviation Before defining the coefficient of determination, let us justify its use by examining the logic behind its computation. We begin by considering the point corresponding to any observed value, y_i, and by measuring its vertical distance from the $\bar{y}$ line. We call this the *total deviation* and designate it $(y_i - \bar{y})$.

The Explained Deviation If we measure the vertical distance from the regression line to the $\bar{y}$ line, we obtain $(\hat{y}_i - \bar{y})$, which is called the *explained deviation,* since it shows by how much the total deviation is reduced when the regression line is fitted to the points.

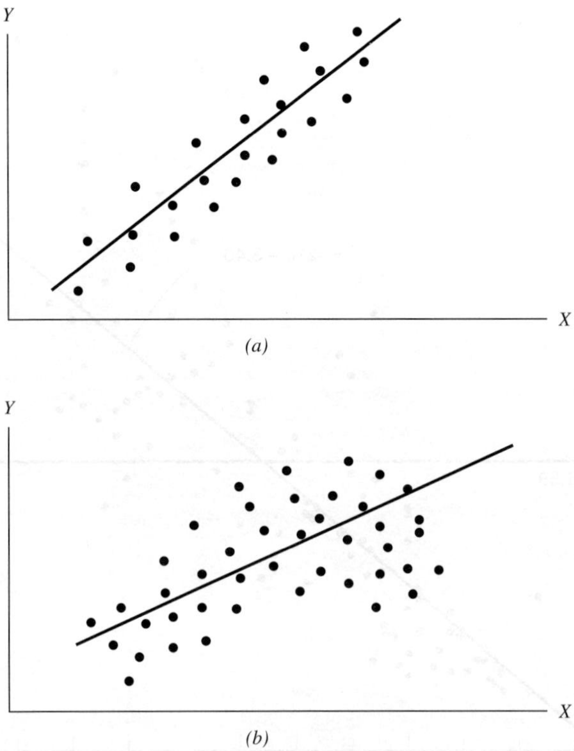

FIGURE 9.4.2 Population conditions relative to X and Y that may cause rejection of the null hypothesis that $\beta_1 = 0$. (*a*) The relationship between X and Y is linear and of sufficient strength to justify the use of a sample regression equation to predict and estimate Y for given values of X. (*b*) A linear model provides a good fit to the data, but some curvilinear model would provide an even better fit.

Unexplained Deviation Finally, we measure the vertical distance of the observed point from the regression line to obtain $(y_i - \hat{y}_i)$, which is called the *unexplained deviation,* since it represents the portion of the total deviation not "explained" or accounted for by the introduction of the regression line. These three quantities are shown for a typical value of Y in Figure 9.4.4. The difference between the observed value of Y and the predicted value of Y, $(y_i - \hat{y}_i)$, is also referred to as a *residual*. The set of residuals can be used to test the underlying linearity and equal-variances assumptions of the regression model described in Section 9.2. This procedure is illustrated at the end of this section.

It is seen, then, that the total deviation for a particular y_i is equal to the sum of the explained and unexplained deviations. We may write this symbolically as

$$\underset{\substack{\text{total}\\\text{deviation}}}{(y_i - \bar{y})} = \underset{\substack{\text{explained}\\\text{deviation}}}{(\hat{y}_i - \bar{y})} + \underset{\substack{\text{unexplained}\\\text{deviation}}}{(y_i - \hat{y}_i)} \tag{9.4.1}$$

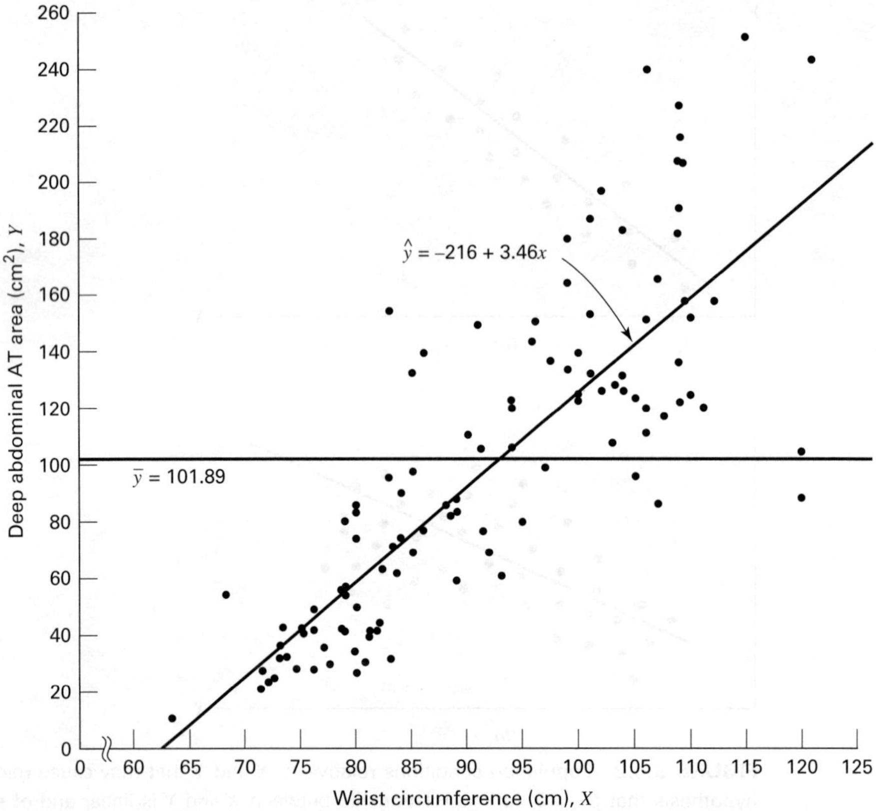

FIGURE 9.4.3 Scatter diagram, sample regression line, and $\bar{y}$ line for Example 9.3.1.

If we measure these deviations for each value of y_i and $\hat{y}_i$, square each deviation, and add up the squared deviations, we have

$$\underset{\substack{\text{total}\\ \text{sum}\\ \text{of squares}}}{\sum (y_i - \bar{y})^2} = \underset{\substack{\text{explained}\\ \text{sum}\\ \text{of squares}}}{\sum (\hat{y}_i - \bar{y})^2} + \underset{\substack{\text{unexplained}\\ \text{sum}\\ \text{of squares}}}{\sum (y_i - \hat{y}_i)^2} \tag{9.4.2}$$

These quantities may be considered measures of dispersion or variability.

Total Sum of Squares The *total sum of squares* (*SST*), for example, is a measure of the dispersion of the observed values of Y about their mean $\bar{y}$; that is, this term is a measure of the total variation in the observed values of Y. The reader will recognize this term as the numerator of the familiar formula for the sample variance.

Explained Sum of Squares The *explained sum of squares* measures the amount of the total variability in the observed values of Y that is accounted for by the linear relationship between the observed values of X and Y. This quantity is referred to also as the *sum of squares due to linear regression* (*SSR*).

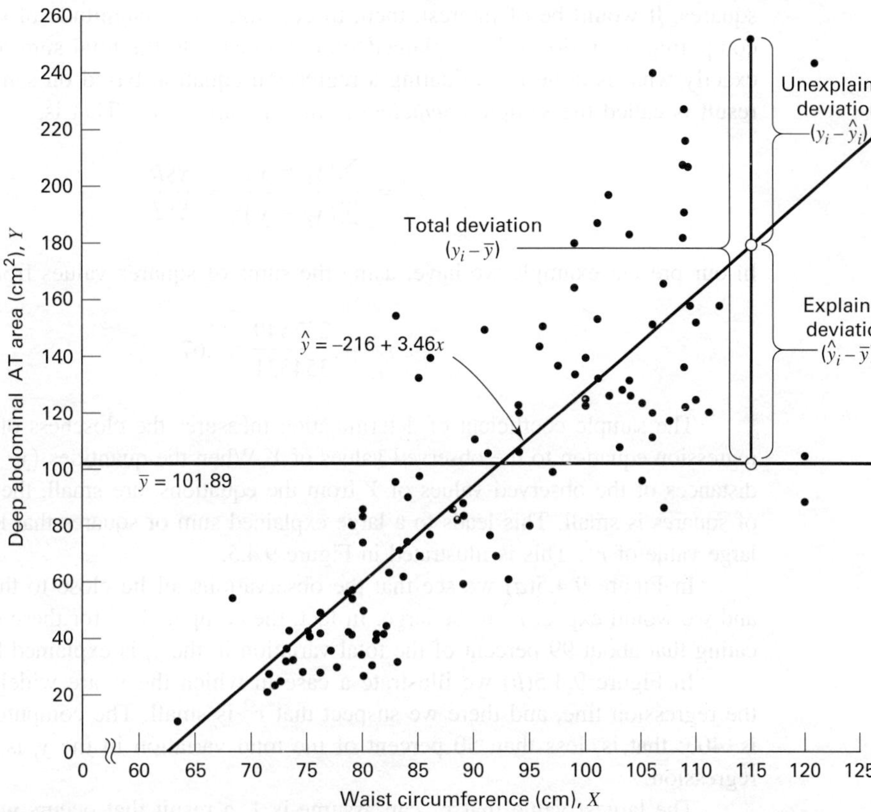

FIGURE 9.4.4 Scatter diagram showing the total, explained, and unexplained deviations for a selected value of *Y*, Example 9.3.1.

Unexplained Sum of Squares The *unexplained sum of squares* is a measure of the dispersion of the observed *Y* values about the regression line and is sometimes called the *error sum of squares*, or the *residual sum of squares* (*SSE*). It is this quantity that is minimized when the least-squares line is obtained.

We may express the relationship among the three sums of squares values as

$$SST = SSR + SSE$$

The numerical values of these sums of squares for our illustrative example appear in the analysis of variance table in Figure 9.3.2. Thus, we see that $SST = 354531$, $SSR = 237549$, $SSE = 116982$, and

$$354531 = 237549 + 116982$$
$$354531 = 354531$$

Calculating r^2 It is intuitively appealing to speculate that if a regression equation does a good job of describing the relationship between two variables, the explained or regression sum of squares should constitute a large proportion of the total sum of squares. It would be of interest, then, to determine the magnitude of this proportion by computing the ratio of the explained sum of squares to the total sum of squares. This is exactly what is done in evaluating a regression equation based on sample data, and the result is called the sample *coefficient of determination,* r^2. That is,

$$r^2 = \frac{\Sigma(\hat{y}_i - \bar{y})^2}{\Sigma(y_i - \bar{y})^2} = \frac{SSR}{SST}$$

In our present example we have, using the sums of squares values from Figure 9.3.2,

$$r^2 = \frac{237549}{354531} = .67$$

The sample coefficient of determination measures the closeness of fit of the sample regression equation to the observed values of Y. When the quantities $(y_i - \hat{y}_i)$, the vertical distances of the observed values of Y from the equations, are small, the unexplained sum of squares is small. This leads to a large explained sum of squares that leads, in turn, to a large value of r^2. This is illustrated in Figure 9.4.5.

In Figure 9.4.5(a) we see that the observations all lie close to the regression line, and we would expect r^2 to be large. In fact, the computed r^2 for these data is .986, indicating that about 99 percent of the total variation in the y_i is explained by the regression.

In Figure 9.4.5(b) we illustrate a case in which the y_i are widely scattered about the regression line, and there we suspect that r^2 is small. The computed r^2 for the data is .403; that is, less than 50 percent of the total variation in the y_i is explained by the regression.

The largest value that r^2 can assume is 1, a result that occurs when all the variation in the y_i is explained by the regression. When $r^2 = 1$ all the observations fall on the regression line. This situation is shown in Figure 9.4.5(c).

The lower limit of r^2 is 0. This result is obtained when the regression line and the line drawn through $\bar{y}$ coincide. In this situation none of the variation in the y_i is explained by the regression. Figure 9.4.5(d) illustrates a situation in which r^2 is close to zero.

When r^2 is large, then, the regression has accounted for a large proportion of the total variability in the observed values of Y, and we look with favor on the regression equation. On the other hand, a small r^2 which indicates a failure of the regression to account for a large proportion of the total variation in the observed values of Y, tends to cast doubt on the usefulness of the regression equation for predicting and estimating purposes. We do not, however, pass final judgment on the equation until it has been subjected to an objective statistical test.

Testing $H_0: \beta_1 = 0$ with the F Statistic The following example illustrates one method for reaching a conclusion regarding the relationship between X and Y.

EXAMPLE 9.4.1

Refer to Example 9.3.1. We wish to know if we can conclude that, in the population from which our sample was drawn, X and Y are linearly related.

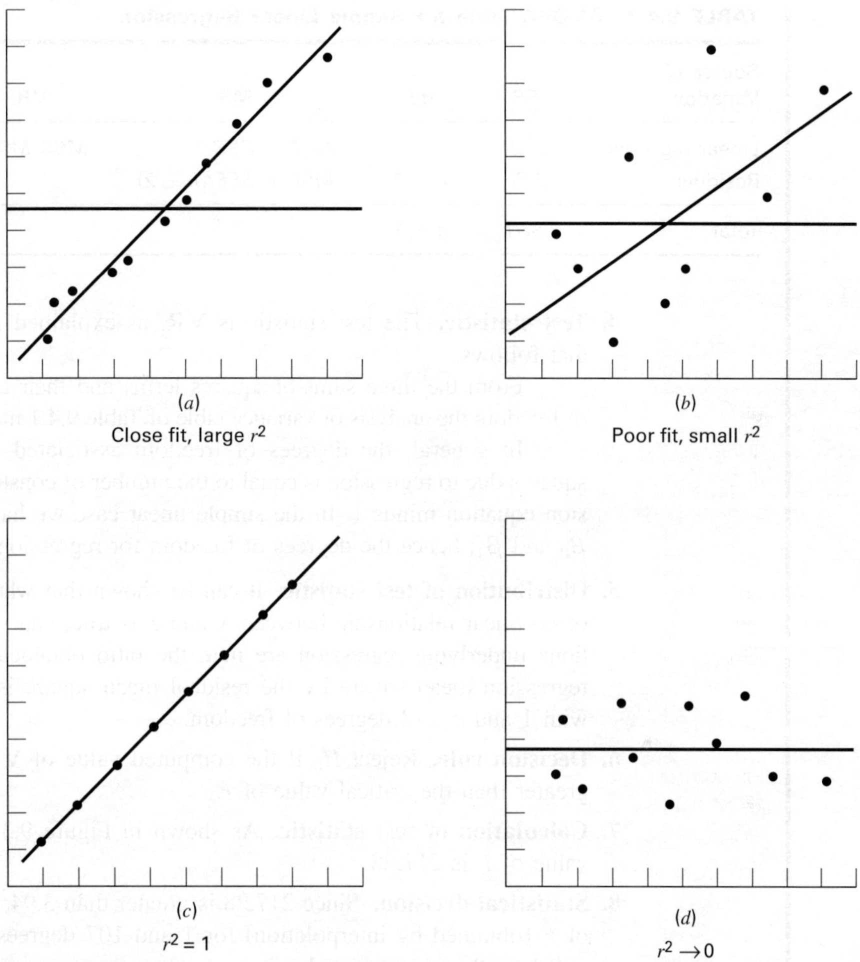

FIGURE 9.4.5 r^2 as a measure of closeness-of-fit of the sample regression line to the sample observations.

Solution: The steps in the hypothesis testing procedure are as follows:

1. **Data.** The data were described in the opening statement of Example 9.3.1.

2. **Assumptions.** We presume that the simple linear regression model and its underlying assumptions as given in Section 9.2 are applicable.

3. **Hypotheses.**

$$H_0\colon \beta_1 = 0$$
$$H_A\colon \beta_1 \neq 0$$
$$\alpha = .05$$

TABLE 9.4.1 ANOVA Table for Simple Linear Regression

Source of Variation	SS	d.f.	MS	V.R.
Linear regression	*SSR*	1	$MSR = SSR/1$	*MSR/MSE*
Residual	*SSE*	$n - 2$	$MSE = SSE/(n-2)$	
Total	*SST*	$n - 1$		

4. **Test statistic.** The test statistic is V.R. as explained in the discussion that follows.

   From the three sums-of-squares terms and their associated degrees of freedom the analysis of variance table of Table 9.4.1 may be constructed.

   In general, the degrees of freedom associated with the sum of squares due to regression is equal to the number of constants in the regression equation minus 1. In the simple linear case we have two estimates, β_0 and β_1; hence the degrees of freedom for regression are $2 - 1 = 1$.

5. **Distribution of test statistic.** It can be shown that when the hypothesis of no linear relationship between X and Y is true, and when the assumptions underlying regression are met, the ratio obtained by dividing the regression mean square by the residual mean square is distributed as F with 1 and $n - 2$ degrees of freedom.

6. **Decision rule.** Reject H_0 if the computed value of V.R. is equal to or greater than the critical value of F.

7. **Calculation of test statistic.** As shown in Figure 9.3.2, the computed value of F is 217.28.

8. **Statistical decision.** Since 217.28 is greater than 3.94, the critical value of F (obtained by interpolation) for 1 and 107 degrees of freedom, the null hypothesis is rejected.

9. **Conclusion.** We conclude that the linear model provides a good fit to the data.

10. ***p* value.** For this test, since $217.28 > 8.25$, we have $p < .005$. ■

Estimating the Population Coefficient of Determination The sample coefficient of determination provides a point estimate of ρ^2 the *population coefficient of determination.* The population coefficient of determination, ρ^2 has the same function relative to the population as r^2 has to the sample. It shows what proportion of the total population variation in Y is explained by the regression of Y on X. When the number of degrees of freedom is small, r^2 is positively biased. That is, r^2 tends to be large. An unbiased estimator of ρ^2 is provided by

$$\tilde{r}^2 = 1 - \frac{\Sigma(y_i - \hat{y}_i)^2/(n-2)}{\Sigma(y_i - \bar{y})^2/(n-1)} \tag{9.4.3}$$

Observe that the numerator of the fraction in Equation 9.4.3 is the unexplained mean square and the denominator is the total mean square. These quantities appear in the analysis of variance table. For our illustrative example we have, using the data from Figure 9.3.2,

$$\tilde{r}^2 = 1 - \frac{116982/107}{354531/108} = .66695$$

This quantity is labeled R-sq(adj) in Figure 9.3.2 and is reported as 66.7 percent. We see that this value is less than

$$r^2 = 1 - \frac{116982}{354531} = .67004$$

We see that the difference in r^2 and $\tilde{r}^2$ is due to the factor $(n - 1)/(n - 2)$. When n is large, this factor will approach 1 and the difference between r^2 and $\tilde{r}^2$ will approach zero.

Testing $H_0: \beta_1 = 0$ with the t Statistic When the assumptions stated in Section 9.2 are met, $\hat{\beta}_0$ and $\hat{\beta}_1$ are unbiased point estimators of the corresponding parameters β_0 and β_1. Since, under these assumptions, the subpopulations of Y values are normally distributed, we may construct confidence intervals for and test hypotheses about β_0 and β_1. When the assumptions of Section 9.2 hold true, the sampling distributions of $\hat{\beta}_0$ and $\hat{\beta}_1$ are each normally distributed with means and variances as follows:

$$\mu_{\hat{\beta}_0} = \beta_0 \tag{9.4.4}$$

$$\sigma^2_{\hat{\beta}_0} = \frac{\sigma^2_{y/x} \sum x_i^2}{n \sum (x_i - \bar{x})^2} \tag{9.4.5}$$

$$\mu_{\hat{\beta}_1} = \beta_1 \tag{9.4.6}$$

and

$$\sigma^2_{\hat{\beta}_1} = \frac{\sigma^2_{y/x}}{\sum (x_i - \bar{x})^2} \tag{9.4.7}$$

In Equations 9.4.5 and 9.4.7 $\sigma^2_{y/x}$ is the unexplained variance of the subpopulations of Y values.

With knowledge of the sampling distributions of $\hat{\beta}_0$ and $\hat{\beta}_1$ we may construct confidence intervals and test hypotheses relative to β_0 and β_1 in the usual manner. Inferences regarding α are usually not of interest. On the other hand, as we have seen, a great deal of interest centers on inferential procedures with respect to $\hat{\beta}_1$. The reason for this is the fact that β_1 tells us so much the form of the relationship between X and Y. When X and Y are linearly related a positive $\hat{\beta}_1$ indicates that, in general, Y increases as X increases, and we say that there is a *direct linear relationship* between X and Y. A negative $\hat{\beta}_1$ indicates that values of Y tend to decrease as values of X increase, and we say that there is an *inverse linear relationship* between X and

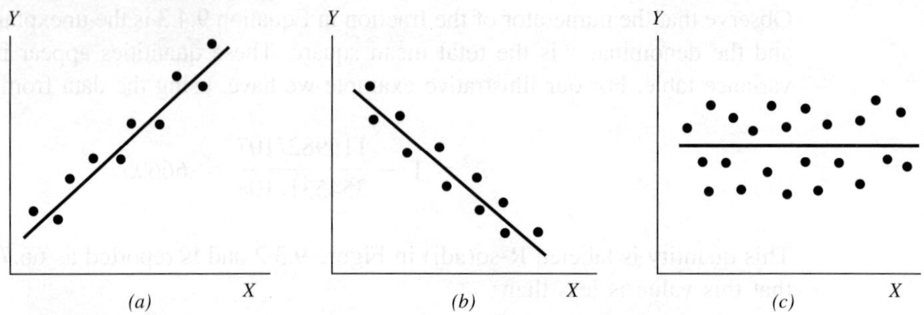

FIGURE 9.4.6 Scatter diagrams showing (*a*) direct linear relationship, (*b*) inverse linear relationship, and (*c*) no linear relationship between *X* and *Y*.

Y. When there is no linear relationship between *X* and *Y*, $\hat{\beta}_1$ is equal to zero. These three situations are illustrated in Figure 9.4.6.

The Test Statistic For testing hypotheses about β_1 the test statistic when $\sigma^2_{y/x}$ is known is

$$z = \frac{\hat{\beta}_1 - (\beta_1)_0}{\sigma_{\hat{\beta}_1}} \tag{9.4.8}$$

where $(\beta_1)_0$ is the hypothesized value of β_1. The hypothesized value of β_1 does not have to be zero, but in practice, more often than not, the null hypothesis of interest is that $\beta_1 = 0$.

As a rule $\sigma^2_{y|x}$ is unknown. When this is the case, the test statistic is

$$t = \frac{\hat{\beta}_1 - (\beta_1)_0}{s_{\hat{\beta}_1}} \tag{9.4.9}$$

where $s_{\hat{\beta}_1}$ is an estimate of $\sigma_{\hat{\beta}_1}$ and t is distributed as Student's t with $n - 2$ degrees of freedom.

If the probability of observing a value as extreme as the value of the test statistic computed by Equation 9.4.9 when the null hypothesis is true is less than $\alpha/2$ (since we have a two-sided test), the null hypothesis is rejected.

EXAMPLE 9.4.2

Refer to Example 9.3.1. We wish to know if we can conclude that the slope of the population regression line describing the relationship between X and Y is zero.

Solution:

1. **Data.** See Example 9.3.1.
2. **Assumptions.** We presume that the simple linear regression model and its underlying assumptions are applicable.

3. **Hypotheses.**

$$H_0: \beta_1 = 0$$
$$H_A: \beta_1 \neq 0$$
$$\alpha = .05$$

4. **Test statistic.** The test statistic is given by Equation 9.4.9.
5. **Distribution of test statistic.** When the assumptions are met and H_0 is true, the test statistic is distributed as Student's t with $n - 2$ degrees of freedom.
6. **Decision rule.** Reject H_0 if the computed value of t is either greater than or equal to 1.9826 or less than or equal to -1.9826.
7. **Calculation of statistic.** The output in Figure 9.3.2 shows that $\hat{\beta}_1 = 3.4589$, $s_{\hat{\beta}_1} = .2347$, and

$$t = \frac{3.4589 - 0}{.2347} = 14.74$$

8. **Statistical decision.** Reject H_0 because $14.74 > 1.9826$.
9. **Conclusion.** We conclude that the slope of the true regression line is not zero.
10. ***p* value.** The p value for this test is less than .01, since, when H_0 is true, the probability of getting a value of t as large as or larger than 2.6230 (obtained by interpolation) is .005, and the probability of getting a value of t as small as or smaller than -2.6230 is also .005. Since 14.74 is greater than 2.6230, the probability of observing a value of t as large as or larger than 14.74 (when the null hypothesis is true) is less than .005. We double this value to obtain $2(.005) = .01$.

Either the F statistic or the t statistic may be used for testing $H_0: \beta_1 = 0$. The value of the variance ratio is equal to the square of the value of the t statistic (i.e., $t^2 = F$) and, therefore, both statistics lead to the same conclusion. For the current example, we see that $(14.74)^2 = 217.27$, the value obtained by using the F statistic in Example 9.4.1.

The practical implication of our results is that we can expect to get better predictions and estimates of Y if we use the sample regression equation than we would get if we ignore the relationship between X and Y. The fact that b is positive leads us to believe that β_1 is positive and that the relationship between X and Y is a direct linear relationship. ■

As has already been pointed out, Equation 9.4.9 may be used to test the null hypothesis that β_1 is equal to some value other than 0. The hypothesized value for β_1, $(\beta_1)_0$ is substituted into Equation 9.4.9. All other quantities, as well as the computations, are the same as in the illustrative example. The degrees of freedom and the method of determining significance are also the same.

A Confidence Interval for β_1 Once we determine that it is unlikely, in light of sample evidence, that β_1 is zero, we may be interested in obtaining an interval estimate of β_1. The general formula for a confidence interval,

$$\text{estimator} \pm (\text{reliability factor})(\text{standard error of the estimate})$$

may be used. When obtaining a confidence interval for β_1, the estimator is $\hat{\beta}_1$, the reliability factor is some value of z or t (depending on whether or not $\sigma^2_{y|x}$ is known), and the standard error of the estimator is

$$\sigma_{\hat{\beta}_1} = \sqrt{\frac{\sigma^2_{y|x}}{\Sigma(x_i - \bar{x})^2}}$$

When $\sigma^2_{y|x}$ is unknown, σ_b is estimated by

$$s_{\hat{\beta}_1} = \sqrt{\frac{s^2_{y|x}}{\Sigma(x_i - \bar{x})^2}}$$

where $s^2_{y|x} = MSE$

In most practical situations our $100(1 - \alpha)$ percent confidence interval for β is

$$\hat{\beta}_1 \pm t_{(1-\alpha/2)} s_{\hat{\beta}_1} \tag{9.4.10}$$

For our illustrative example we construct the following 95 percent confidence interval for β:

$$3.4589 \pm 1.9826(.2347)$$
$$2.99, 3.92$$

We interpret this interval in the usual manner. From the probabilistic point of view we say that in repeated sampling 95 percent of the intervals constructed in this way will include β_1. The practical interpretation is that we are 95 percent confident that the single interval constructed includes β_1.

Using the Confidence Interval to Test H_0: $\beta_1 = 0$ It is instructive to note that the confidence interval we constructed does not include zero, so that zero is not a candidate for the parameter being estimated. We feel, then, that it is unlikely that $\beta_1 = 0$. This is compatible with the results of our hypothesis test in which we rejected the null hypothesis that $\beta_1 = 0$. Actually, we can always test H_0: $\beta_1 = 0$ at the α significance level by constructing the $100(1 - \alpha)$ percent confidence interval for β_1, and we can reject or fail to reject the hypothesis on the basis of whether or not the interval includes zero. If the interval contains zero, the null hypothesis is not rejected; and if zero is not contained in the interval, we reject the null hypothesis.

Interpreting the Results It must be emphasized that failure to reject the null hypothesis that $\beta_1 = 0$ does not mean that X and Y are not related. Not only is it possible that a type II error may have been committed but it may be true that X and Y are related in some nonlinear manner. On the other hand, when we reject the null hypothesis that $\beta_1 = 0$, we cannot conclude that the *true* relationship between X and Y is linear. Again, it may be that although the data fit the linear regression model fairly well (as evidenced by the fact that the null hypothesis that $\beta_1 = 0$ is rejected), some nonlinear model would provide an even better fit. Consequently, when we reject H_0 that $\beta_1 = 0$, the best we can say is that more useful results (discussed below) may be obtained by taking into account the regression of Y on X than in ignoring it.

Testing the Regression Assumptions The values of the set of residuals, $(y_i - \hat{y}_i)$, for a data set are often used to test the linearity and equal-variances assumptions (assumptions 4 and 5 of Section 9.2) underlying the regression model. This is done by plotting the values of the residuals on the y-axis and the predicted values of y on the x-axis. If these plots show a relatively random scatter of points above and below a horizontal line at $(y_i - \hat{y}_i) = 0$, these assumptions are assumed to have been met for a given set of data. A non-random pattern of points can indicate violation of the linearity assumption, and a funnel-shaped pattern of the points can indicate violation of the equal-variances assumption. Examples of these patterns are shown in Figure 9.4.7. Many

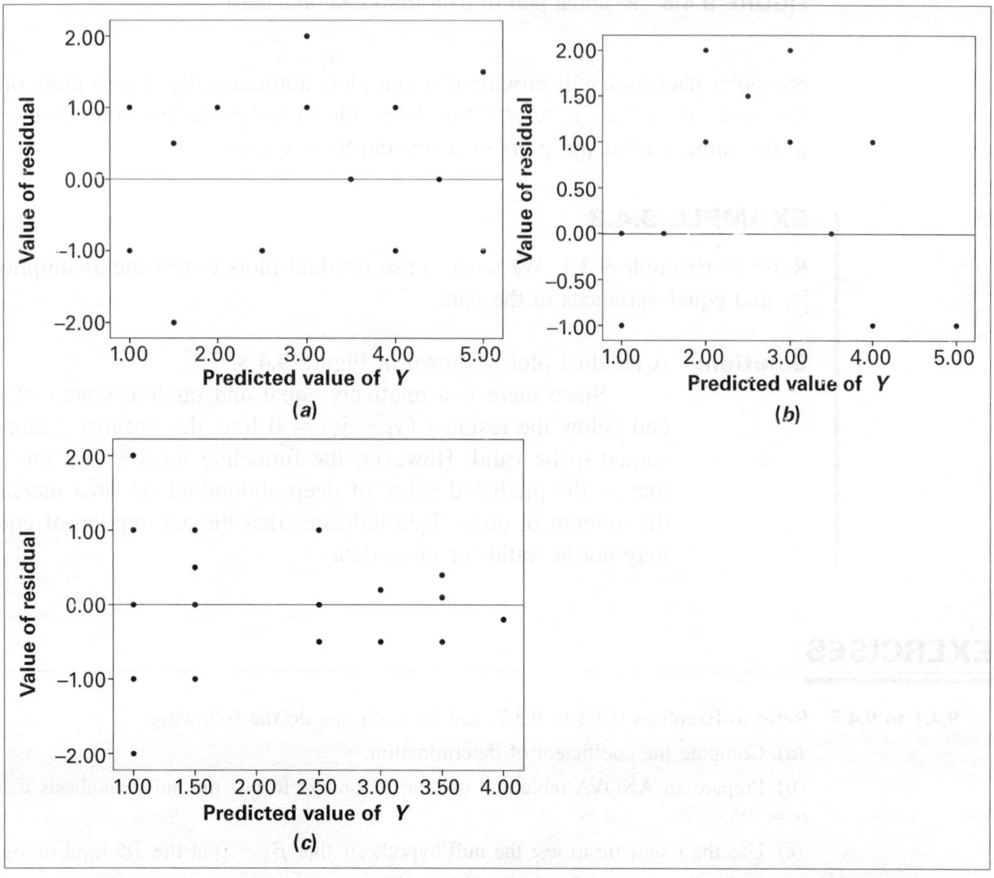

FIGURE 9.4.7 Residual plots useful for testing the linearity and equal-variances assumptions of the regression model. (*a*) A random pattern of points illustrating non-violation of the assumptions. (*b*) A non-random pattern illustrating a likely violation of the linearity assumption. (*c*) A funneling pattern illustrating a likely violation of the equal-variances assumption.

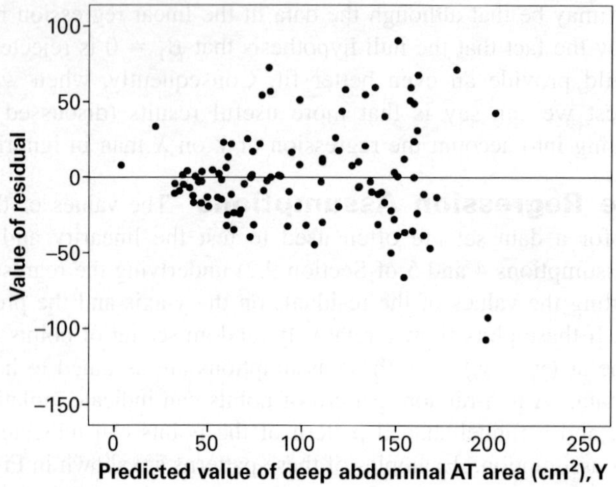

FIGURE 9.4.8 Residual plot of data from Example 9.3.1.

computer packages will provide residual plots automatically. These plots often use standardized values (i.e., $e_i/\sqrt{MSE}$) of the residuals and predicted values, but are interpreted in the same way as are plots of unstandardized values.

EXAMPLE 9.4.3

Refer to Example 9.3.1. We wish to use residual plots to test the assumptions of linearity and equal variances in the data.

Solution: A residual plot is shown in Figure 9.4.8.

Since there is a relatively equal and random scatter of points above and below the residual $(y_i - \hat{y}_i) = 0$ line, the linearity assumption is presumed to be valid. However, the funneling tendency of the plot suggests that as the predicted value of deep abdominal AT area increases, so does the amount of error. This indicates that the assumption of equal variances may not be valid for these data. ■

EXERCISES

9.4.1 to 9.4.5 Refer to Exercises 9.3.3 to 9.3.7, and for each one do the following:

(a) Compute the coefficient of determination.

(b) Prepare an ANOVA table and use the F statistic to test the null hypothesis that $\beta_1 = 0$. Let $\alpha = .05$.

(c) Use the t statistic to test the null hypothesis that $\beta_1 = 0$ at the .05 level of significance.

(d) Determine the p value for each hypothesis test.

(e) State your conclusions in terms of the problem.

(f) Construct the 95 percent confidence interval for β_1.

9.5 USING THE REGRESSION EQUATION

If the results of the evaluation of the sample regression equation indicate that there is a relationship between the two variables of interest, we can put the regression equation to practical use. There are two ways in which the equation can be used. It can be used to *predict* what value Y is likely to assume given a particular value of X. When the normality assumption of Section 9.2 is met, a *prediction interval* for this predicted value of Y may be constructed.

We may also use the regression equation to *estimate* the mean of the subpopulation of Y values assumed to exist at any particular value of X. Again, if the assumption of normally distributed populations holds, a confidence interval for this parameter may be constructed. The predicted value of Y and the point estimate of the mean of the subpopulation of Y will be numerically equivalent for any particular value of X but, as we will see, the prediction interval will be wider than the confidence interval.

Predicting Y for a Given X If it is known, or if we are willing to assume that the assumptions of Section 9.2 are met, and when $\sigma^2_{y|x}$ is unknown, then the $100(1 - \alpha)$ percent prediction interval for Y is given by

$$\hat{y} \pm t_{(1-\alpha/2)} s_{y|x} \sqrt{1 + \frac{1}{n} + \frac{(x_p - \bar{x})^2}{\sum (x_i - \bar{x})^2}} \tag{9.5.1}$$

where x_p is the particular value of x at which we wish to obtain a prediction interval for Y and the degrees of freedom used in selecting t are $n - 2$.

Estimating the Mean of Y for a Given X The $100(1 - \alpha)$ percent confidence interval for $\mu_{y|x}$, when $\sigma^2_{y|x}$ is unknown, is given by

$$\hat{y} \pm t_{(1-\alpha/2)} s_{y|x} \sqrt{\frac{1}{n} + \frac{(x_p - \bar{x})^2}{\sum (x_i - \bar{x})^2}} \tag{9.5.2}$$

We use MINITAB to illustrate, for a specified value of X, the calculation of a 95 percent confidence interval for the mean of Y and a 95 percent prediction interval for an individual Y measurement.

Suppose, for our present example, we wish to make predictions and estimates about AT for a waist circumference of 100 cm. In the regression dialog box click on "Options." Enter 100 in the "Prediction interval for new observations" box. Click on "Confidence limits," and click on "Prediction limits."

We obtain the following output:

Fit	Stdev.Fit	95.0% C.I.	95.0% P.I.
129.90	3.69	(122.58, 137.23)	(63.93, 195.87)

We interpret the 95 percent confidence interval (C.I.) as follows.

If we repeatedly drew samples from our population of men, performed a regression analysis, and estimated $\mu_{y|x=100}$ with a similarly constructed confidence interval, about 95 percent of such intervals would include the mean amount of deep abdominal AT for the population. For this reason we are 95 percent confident that the single interval constructed contains the population mean and that it is somewhere between 122.58 and 137.23.

Our interpretation of a prediction interval (P.I.) is similar to the interpretation of a confidence interval. If we repeatedly draw samples, do a regression analysis, and construct prediction intervals for men who have a waist circumference of 100 cm, about 95 percent of them will include the man's deep abdominal AT value. This is the probabilistic interpretation. The practical interpretation is that we are 95 percent confident that a man who has a waist circumference of 100 cm will have a deep abdominal AT area of somewhere between 63.93 and 195.87 square centimeters.

Simultaneous confidence intervals and prediction intervals can be calculated for all possible points along a fitted regression line. Plotting lines through these points will then provide a graphical representation of these intervals. Since the mean data point $(\bar{X}, \bar{Y})$ is always included in the regression equation, as illustrated by equations 9.3.2 and 9.3.3, plots of the simultaneous intervals will always provide the best estimates at the middle of the line and the error will increase toward the ends of the line. This illustrates the fact that estimation within the bounds of the data set, called *interpolation,* is acceptable, but that estimation outside of the bounds of the data set, called *extrapolation,* is not advisable since the pridiction error can be quite large. See Figure 9.5.1.

Figure 9.5.2 contains a partial printout of the SAS® simple linear regression analysis of the data of Example 9.3.1.

Resistant Line Frequently, data sets available for analysis by linear regression techniques contain one or more "unusual" observations; that is, values of x or y, or both, may be either considerably larger or considerably smaller than most of the other measurements. In the output of Figure 9.3.2, we see that the computer detected seven

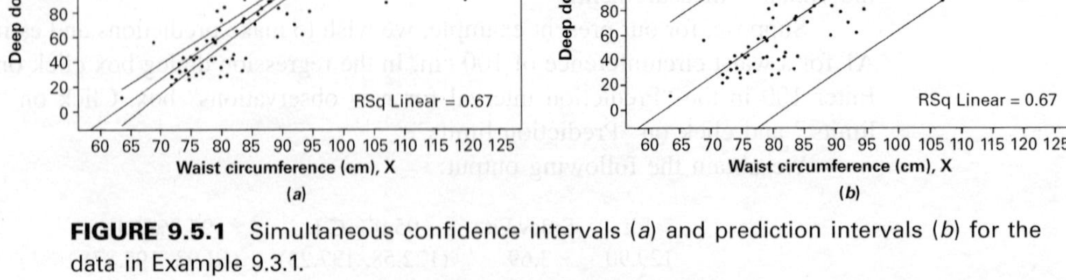

FIGURE 9.5.1 Simultaneous confidence intervals (*a*) and prediction intervals (*b*) for the data in Example 9.3.1.

```
The SAS System

Model: MODEL1
Dependent Variable: Y

Analysis of Variance

                       Sum of              Mean
Source      DF        Squares            Square     F Value    Prob>F

Model        1   237548.51620      237548.51620     217.279    0.0001
Error      107   116981.98602        1093.28959
C Total    108   354530.50222

    Root MSE      33.06493     R-square      0.6700
    Dep Mean     101.89404     Adj R-sq      0.6670
    C.V.          32.45031

Parameter Estimates

                 Parameter     Standard     T for H0:
Variable DF       Estimate        Error   Parameter=0   Prob > |T|

INTERCEP  1    -215.981488  21.79627076        -9.909       0.0001
X         1       3.458859   0.23465205        14.740       0.0001
```

FIGURE 9.5.2 Partial printout of the computer analysis of the data given in Example 9.3.1, using the SAS® software package.

unusual observations in the waist circumference and deep abdominal AT data shown in Table 9.3.1.

The least-squares method of fitting a straight line to data is sensitive to unusual observations, and the location of the fitted line can be affected substantially by them. Because of this characteristic of the least-squares method, the resulting least-squares line is said to lack *resistance* to the influence of unusual observations. Several methods have been devised for dealing with this problem, including one developed by John W. Tukey. The resulting line is variously referred to as *Tukey's line* and the *resistant line*.

Based on medians, which, as we have seen, are descriptive measures that are themselves resistant to extreme values, the resistant line methodology is an exploratory data analysis tool that enables the researcher to quickly fit a straight line to a set of data consisting of paired x, y measurements. The technique involves partitioning, on the basis of the independent variable, the sample measurements into three groups of as near equal size as possible: the smallest measurements, the largest measurements, and those in between. The resistant line is the line fitted in such a way that there are

Dialog box:	**Session command:**

Stat ➤ EDA ➤ Resistant Line	`MTB > Name C3 = 'RESI1' C4 = 'FITS1'`
	`MTB > RLine C2 C1 'RESI1' 'FITS1';`
	`SUBC>   MaxIterations 10.`

Type *C2* in **Response** and *C1* in **Predictors.**
Check **Residuals** and **Fits.** Click **OK.**

Output:

Resistant Line Fit: C2 versus C1

```
Slope =        3.2869  Level =    -203.7868  Half-slope ratio =    0.690
```

FIGURE 9.5.3 MINITAB resistant line procedure and output for the data of Table 9.3.1.

an equal number of values above and below it in both the smaller group and the larger group. The resulting slope and y-intercept estimates are resistant to the effects of either extreme y values, extreme x values, or both. To illustrate the fitting of a resistant line, we use the data of Table 9.3.1 and MINITAB. The procedure and output are shown in Figure 9.5.3.

We see from the output in Figure 9.5.3 that the resistant line has a slope of 3.2869 and a y-intercept of -203.7868. The *half-slope ratio,* shown in the output as equal to .690, is an indicator of the degree of linearity between x and y. A slope, called a half-slope, is computed for each half of the sample data. The ratio of the right half-slope, b_R, and the left half-slope, b_L, is equal to b_R/b_L. If the relationship between x and y is straight, the half-slopes will be equal, and their ratio will be 1. A half-slope ratio that is not close to 1 indicates a lack of linearity between x and y.

The resistant line methodology is discussed in more detail by Hartwig and Dearing (1), Johnstone and Velleman (2), McNeil (3), and Velleman and Hoaglin (4).

EXERCISES

In each exercise refer to the appropriate previous exercise and, for the value of X indicated, (a) construct the 95 percent confidence interval for $\mu_{y|x}$ and (b) construct the 95 percent prediction interval for Y.

9.5.1 Refer to Exercise 9.3.3 and let $X = 400$.

9.5.2 Refer to Exercise 9.3.4 and let $X = 1.6$.

9.5.3 Refer to Exercise 9.3.5 and let $X = 4.16$.

9.5.4 Refer to Exercise 9.3.6 and let $X = 29.4$.

9.5.5 Refer to Exercise 9.3.7 and let $X = 35$.

9.6 THE CORRELATION MODEL

In the classic regression model, which has been the underlying model in our discussion up to this point, only Y, which has been called the dependent variable, is required to be random. The variable X is defined as a fixed (nonrandom or mathematical) variable and is referred to as the independent variable. Recall, also, that under this model observations are frequently obtained by preselecting values of X and determining corresponding values of Y.

When both Y and X are random variables, we have what is called the *correlation model*. Typically, under the correlation model, sample observations are obtained by selecting a random sample of the *units of association* (which may be persons, places, animals, points in time, or any other element on which the two measurements are taken) and taking on each a measurement of X and a measurement of Y. In this procedure, values of X are not preselected but occur at random, depending on the unit of association selected in the sample.

Although correlation analysis cannot be carried out meaningfully under the classic regression model, regression analysis can be carried out under the correlation model. Correlation involving two variables implies a co-relationship between variables that puts them on an equal footing and does not distinguish between them by referring to one as the dependent and the other as the independent variable. In fact, in the basic computational procedures, which are the same as for the regression model, we may fit a straight line to the data either by minimizing $\sum(y_i - \hat{y}_i)^2$ or by minimizing $\sum(x_i - \hat{x}_i)^2$. In other words, we may do a regression of X on Y as well as a regression of Y on X. The fitted line in the two cases in general will be different, and a logical question arises as to which line to fit.

If the objective is solely to obtain a measure of the strength of the relationship between the two variables, it does not matter which line is fitted, since the measure usually computed will be the same in either case. If, however, it is desired to use the equation describing the relationship between the two variables for the purposes discussed in the preceding sections, it does matter which line is fitted. The variable for which we wish to estimate means or to make predictions should be treated as the dependent variable; that is, this variable should be regressed on the other variable.

The Bivariate Normal Distribution Under the correlation model, X and Y are assumed to vary together in what is called a *joint distribution*. If this joint distribution is a normal distribution, it is referred to as a *bivariate normal distribution*. Inferences regarding this population may be made based on the results of samples properly drawn from it. If, on the other hand, the form of the joint distribution is known to be nonnormal, or if the form is unknown and there is no justification for assuming normality, inferential procedures are invalid, although descriptive measures may be computed.

Correlation Assumptions The following assumptions must hold for inferences about the population to be valid when sampling is from a bivariate distribution.

1. For each value of X there is a normally distributed subpopulation of Y values.
2. For each value of Y there is a normally distributed subpopulation of X values.
3. The joint distribution of X and Y is a normal distribution called the *bivariate normal distribution*.

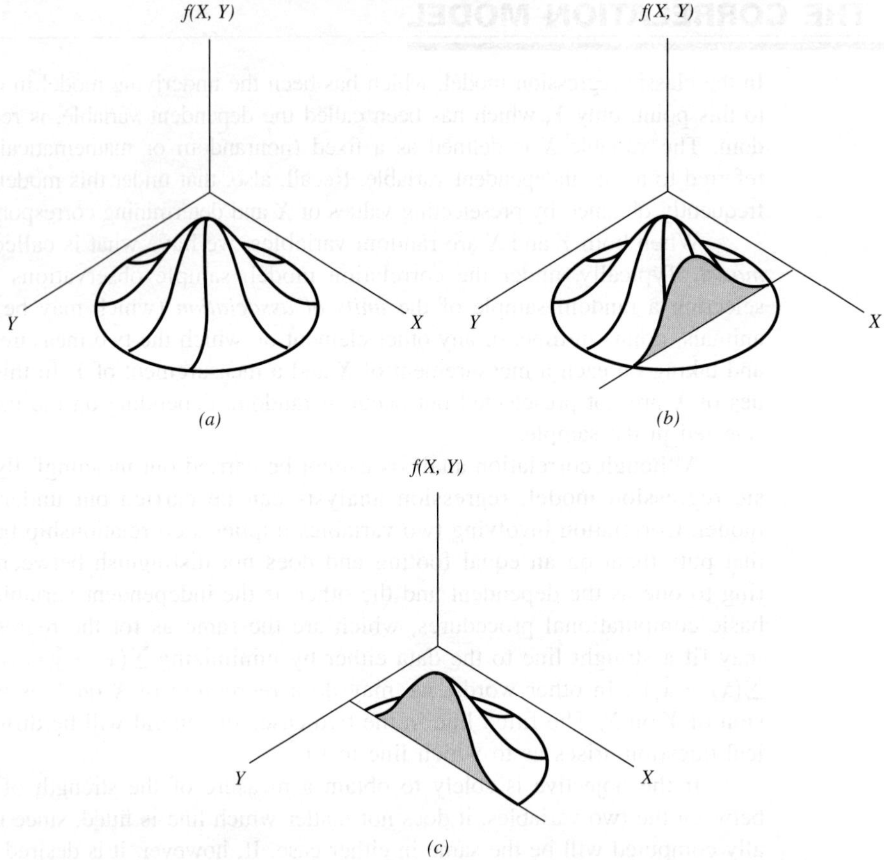

FIGURE 9.6.1 A bivariate normal distribution. (*a*) A bivariate normal distribution. (*b*) A cutaway showing normally distributed subpopulation of *Y* for given *X*. (*c*) A cutaway showing normally distributed subpopulation of *X* for given *Y*.

4. The subpopulations of *Y* values all have the same variance.
5. The subpopulations of *X* values all have the same variance.

The bivariate normal distribution is represented graphically in Figure 9.6.1. In this illustration we see that if we slice the mound parallel to *Y* at some value of *X*, the cutaway reveals the corresponding normal distribution of *Y*. Similarly, a slice through the mound parallel to *X* at some value of *Y* reveals the corresponding normally distributed subpopulation of *X*.

9.7 THE CORRELATION COEFFICIENT

The bivariate normal distribution discussed in Section 9.6 has five parameters, σ_x, σ_y, μ_x, μ_y, and ρ. The first four are, respectively, the standard deviations and means associated with the individual distributions. The other parameter, ρ, is called the population

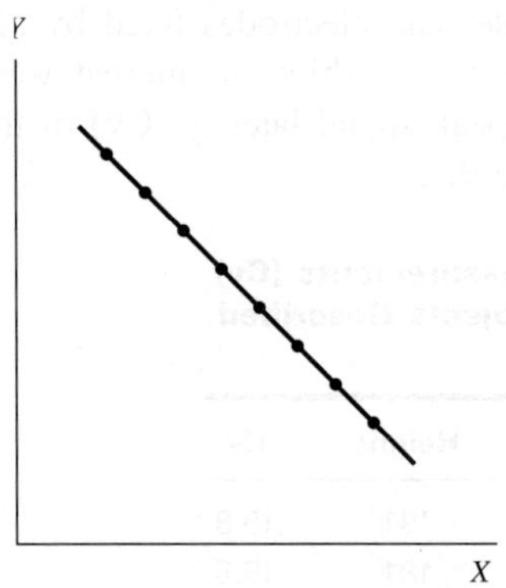

FIGURE 9.7.1 Scatter diagram for $r = -1$.

correlation coefficient and measures the strength of the linear relationship between X and Y.

The population correlation coefficient is the positive or negative square root of ρ^2, the population coefficient of determination previously discussed, and since the coefficient of determination takes on values between 0 and 1 inclusive, ρ may assume any value between -1 and $+1$. If $\rho = 1$ there is a perfect direct linear correlation between the two variables, while $\rho = -1$ indicates perfect inverse linear correlation. If $\rho = 0$ the two variables are not linearly correlated. The sign of ρ will always be the same as the sign of β_1, the slope of the population regression line for X and Y.

The sample correlation coefficient, r, describes the linear relationship between the sample observations on two variables in the same way that ρ describes the relationship in a population. The sample correlation coefficient is the square root of the sample coefficient of determination that was defined earlier.

Figures 9.4.5(*d*) and 9.4.5(*c*), respectively, show typical scatter diagrams where $r \to 0$ $(r^2 \to 0)$ and $r = +1$ $(r^2 = 1)$. Figure 9.7.1 shows a typical scatter diagram where $r = -1$.

We are usually interested in knowing if we may conclude that $\rho \neq 0$, that is, that X and Y are linearly correlated. Since ρ is usually unknown, we draw a random sample from the population of interest, compute r, the estimate of ρ, and test $H_0\colon \rho = 0$ against the alternative $\rho \neq 0$. The procedure will be illustrated in the following example.

EXAMPLE 9.7.1

The purpose of a study by Kwast-Rabben et al. (A-7) was to analyze somatosensory evoked potentials (SEPs) and their interrelations following stimulation of digits I, III, and V in the hand. The researchers wanted to establish reference criteria in a control population. Thus, healthy volunteers were recruited for the study. In the future this information could be quite valuable as SEPs may provide a method to demonstrate functional disturbances in patients with suspected cervical root lesion who have pain and sensory symptoms. In the study, stimulation below-pain-level intensity was applied to the fingers.

Recordings of spinal responses were made with electrodes fixed by adhesive electrode cream to the subject's skin. One of the relationships of interest was the correlation between a subject's height (cm) and the peak spinal latency (Cv) of the SEP. The data for 155 measurements are shown in Table 9.7.1.

TABLE 9.7.1 Height and Spine SEP Measurements (Cv) from Stimulation of Digit I for 155 Subjects Described in Example 9.7.1

Height	Cv	Height	Cv	Height	Cv
149	14.4	168	16.3	181	15.8
149	13.4	168	15.3	181	18.8
155	13.5	168	16.0	181	18.6
155	13.5	168	16.6	182	18.0
156	13.0	168	15.7	182	17.9
156	13.6	168	16.3	182	17.5
157	14.3	168	16.6	182	17.4
157	14.9	168	15.4	182	17.0
158	14.0	170	16.6	182	17.5
158	14.0	170	16.0	182	17.8
160	15.4	170	17.0	184	18.4
160	14.7	170	16.4	184	18.5
161	15.5	171	16.5	184	17.7
161	15.7	171	16.3	184	17.7
161	15.8	171	16.4	184	17.4
161	16.0	171	16.5	184	18.4
161	14.6	172	17.6	185	19.0
161	15.2	172	16.8	185	19.6
162	15.2	172	17.0	187	19.1
162	16.5	172	17.6	187	19.2
162	17.0	173	17.3	187	17.8
162	14.7	173	16.8	187	19.3
163	16.0	174	15.5	188	17.5
163	15.8	174	15.5	188	18.0
163	17.0	175	17.0	189	18.0
163	15.1	175	15.6	189	18.8
163	14.6	175	16.8	190	18.3
163	15.6	175	17.4	190	18.6
163	14.6	175	17.6	190	18.8
164	17.0	175	16.5	190	19.2
164	16.3	175	16.6	191	18.5
164	16.0	175	17.0	191	18.5
164	16.0	176	18.0	191	19.0
165	15.7	176	17.0	191	18.5
165	16.3	176	17.4	194	19.8

(*Continued*)

Height	Cv	Height	Cv	Height	Cv
165	17.4	176	18.2	194	18.8
165	17.0	176	17.3	194	18.4
165	16.3	177	17.2	194	19.0
166	14.1	177	18.3	195	18.0
166	14.2	179	16.4	195	18.2
166	14.7	179	16.1	196	17.6
166	13.9	179	17.6	196	18.3
166	17.2	179	17.8	197	18.9
167	16.7	179	16.1	197	19.2
167	16.5	179	16.0	200	21.0
167	14.7	179	16.0	200	19.2
167	14.3	179	17.5	202	18.6
167	14.8	179	17.5	202	18.6
167	15.0	180	18.0	182	20.0
167	15.5	180	17.9	190	20.0
167	15.4	181	18.4	190	19.5
168	17.3	181	16.4		

Source: Olga Kwast-Rabben, Ph.D. Used with permission.

Solution: The scatter diagram and least-squares regression line are shown in Figure 9.7.2.

Let us assume that the investigator wishes to obtain a regression equation to use for estimating and predicting purposes. In that case the sample correlation coefficient will be obtained by the methods discussed under the regression model.

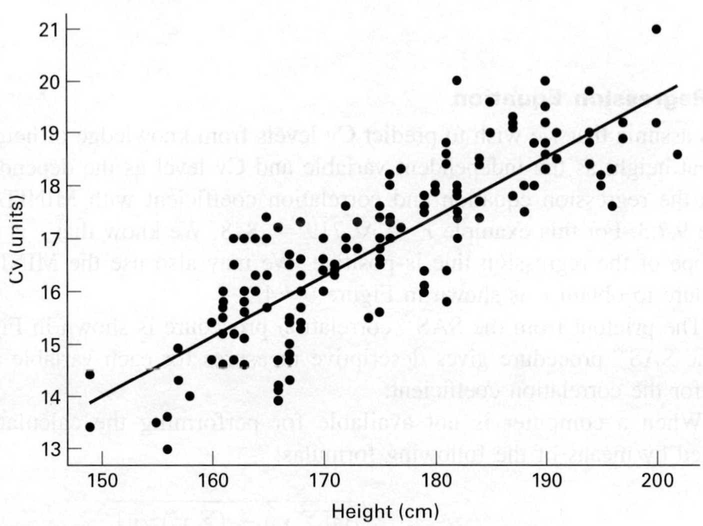

FIGURE 9.7.2 Height and cervical (spine) potentials in digit I stimulation for the data described in Example 9.7.1.

```
The regression equation is
Cv = -3.20 + 0.115 Height

Predictor        Coef     SE Coef       T       P
Constant       -3.198       1.016   -3.15   0.002
Height       0.114567    0.005792   19.78   0.000

S = 0.8573      R-Sq = 71.9%     R-Sq(adj) = 71.7%

Analysis of Variance

Source            DF        SS        MS         F       P
Regression         1    287.56    287.56    391.30   0.000
Residual Error   153    112.44      0.73
Total            154    400.00

Unusual Observations
Obs    Height        Cv       Fit    SE Fit   Residual   St Resid
 39       166   14.1000   15.8199    0.0865    -1.7199    -2.02R
 42       166   13.9000   15.8199    0.0865    -1.9199    -2.25R
105       181   15.8000   17.5384    0.0770    -1.7384    -2.04R
151       202   18.6000   19.9443    0.1706    -1.3443    -1.60 X
152       202   18.6000   19.9443    0.1706    -1.3443    -1.60 X
153       182   20.0000   17.6529    0.0798     2.3471     2.75R

R denotes an observation with a large standardized residual
X denotes an observation whose X value gives it large influence.
```

FIGURE 9.7.3 MINITAB output for Example 9.7.1 using the simple regression procedure.

The Regression Equation

Let us assume that we wish to predict Cv levels from knowledge of heights. In that case we treat height as the independent variable and Cv level as the dependent variable and obtain the regression equation and correlation coefficient with MINITAB as shown in Figure 9.7.3. For this example $r = \sqrt{.719} = .848$. We know that r is positive because the slope of the regression line is positive. We may also use the MINITAB correlation procedure to obtain r as shown in Figure 9.7.4.

The printout from the SAS® correlation procedure is shown in Figure 9.7.5. Note that the SAS® procedure gives descriptive measures for each variable as well as the p value for the correlation coefficient.

When a computer is not available for performing the calculations, r may be obtained by means of the following formulas:

$$r = \sqrt{\frac{\hat{\beta}_1^2\left[\sum x_i^2 - \left(\sum x_i\right)^2/n\right]}{\sum y_i^2 - \left(\sum y_i\right)^2/n}} \tag{9.7.1}$$

Data:

```
C1: Height
C2: Cv
```

Dialog Box:

Stat ➤ Basic Statistics ➤ Correlation

Type *C1 C2* in **Variables.** Click **OK.**

Session command:

```
MTB > Correlation C1 C2.
```

OUTPUT:

Correlations: Height, Cv

```
Pearson correlation of Height and Cv = 0.848
P-Value = 0.000
```

FIGURE 9.7.4 MINITAB procedure for Example 9.7.1 using the correlation command.

```
                     The CORR Procedure
                  2 Variables: HEIGHT CV

                     Simple Statistics

Variable     N        Mean    Std Dev      Sum     Minimum      Maximum
HEIGHT     155   175.04516   11.92745    27132   149.00000    202.00000
CV         155    16.85613    1.61165     2613    13.00000     21.00000

           Pearson Correlation Coefficients, N = 155
                 Prob > |r| under H0: Rho=0

                            HEIGHT           CV
           HEIGHT          1.00000      0.84788
                                         <.0001

           CV              0.84788      1.00000
                            <.0001
```

FIGURE 9.7.5 SAS® printout for Example 9.7.1.

An alternative formula for computing r is given by

$$r = \frac{n\sum x_i y_i - (\sum x_i)(\sum y_i)}{\sqrt{n\sum x_i^2 - (\sum x_i)^2}\sqrt{n\sum y_i^2 - (\sum y_i)^2}} \tag{9.7.2}$$

An advantage of this formula is that r may be computed without first computing b. This is the desirable procedure when it is not anticipated that the regression equation will be used.

Remember that the sample correlation coefficient, r, will always have the same sign as the sample slope, b. ■

EXAMPLE 9.7.2

Refer to Example 9.7.1. We wish to see if the sample value of $r = .848$ is of sufficient magnitude to indicate that, in the population, height and Cv SEP levels are correlated.

Solution: We conduct a hypothesis test as follows.

1. **Data.** See the initial discussion of Example 9.7.1.
2. **Assumptions.** We presume that the assumptions given in Section 9.6 are applicable.
3. **Hypotheses.**

$$H_0: \rho = 0$$
$$H_A: \rho \neq 0$$

4. **Test statistic.** When $\rho = 0$, it can be shown that the appropriate test statistic is

$$t = r\sqrt{\frac{n-2}{1-r^2}} \tag{9.7.3}$$

5. **Distribution of test statistic.** When H_0 is true and the assumptions are met, the test statistic is distributed as Student's t distribution with $n - 2$ degrees of freedom.
6. **Decision rule.** If we let $\alpha = .05$, the critical values of t in the present example are $\pm$ 1.9754 (by interpolation). If, from our data, we compute a value of t that is either greater than or equal to $+1.9754$ or less than or equal to -1.9754, we will reject the null hypothesis.
7. **Calculation of test statistic.** Our calculated value of t is

$$t = .848\sqrt{\frac{153}{1-.719}} = 19.787$$

8. **Statistical decision.** Since the computed value of the test statistic does exceed the critical value of t, we reject the null hypothesis.

9. **Conclusion.** We conclude that, in the population, height and SEP levels in the spine are linearly correlated.

10. ***p* value.** Since $t = 19.787 > 2.6085$ (interpolated value of t for 153, .995), we have for this test, $p < .005$. ■

A Test for Use When the Hypothesized ρ Is a Nonzero Value

The use of the t statistic computed in the above test is appropriate only for testing $H_0: \rho = 0$. If it is desired to test $H_0: \rho = \rho_0$, where ρ_0 is some value other than zero, we must use another approach. Fisher (5) suggests that r be transformed to z_r as follows:

$$z_r = \frac{1}{2} \ln \frac{1 + r}{1 - r} \tag{9.7.4}$$

where ln is a natural logarithm. It can be shown that z_r is approximately normally distributed with a mean of $z_\rho = \frac{1}{2}\ln\{(1 + \rho)/(1 - \rho)\}$ and estimated standard deviation of

$$\sigma_{z_\rho} = \frac{1}{\sqrt{n - 3}} \tag{9.7.5}$$

To test the null hypothesis that ρ is equal to some value other than zero, the test statistic is

$$Z = \frac{z_r - z_\rho}{1/\sqrt{n - 3}} \tag{9.7.6}$$

which follows approximately the standard normal distribution.

To determine z_r for an observed r and z_ρ for a hypothesized ρ, we consult Table I, thereby avoiding the direct use of natural logarithms.

Suppose in our present example we wish to test

$$H_0: \rho = .80$$

against the alternative

$$H_A: \rho \neq .80$$

at the .05 level of significance. By consulting Table I (and interpolating), we find that for

$$r = .848, \quad z_r = 1.24726$$

and for

$$\rho = .80, \quad z_\rho = 1.09861$$

Our test statistic, then, is

$$Z = \frac{1.24726 - 1.09861}{1/\sqrt{155 - 3}} = 1.83$$

Since 1.83 is less than the critical value of $z = 1.96$, we are unable to reject H_0. We conclude that the population correlation coefficient may be .80.

For sample sizes less than 25, Fisher's Z transformation should be used with caution, if at all. An alternative procedure from Hotelling (6) may be used for sample sizes equal to or greater than 10. In this procedure the following transformation of r is employed:

$$z^* = z_r - \frac{3z_r + r}{4n} \tag{9.7.7}$$

The standard deviation of z^* is

$$\sigma_{z^*} = \frac{1}{\sqrt{n-1}} \tag{9.7.8}$$

The test statistic is

$$Z^* = \frac{z^* - \zeta^*}{1/\sqrt{n-1}} = (z^* - \zeta^*)\sqrt{n-1} \tag{9.7.9}$$

where

$$\zeta^* \text{ (pronounced zeta)} = z_\rho - \frac{(3z_\rho + \rho)}{4n}$$

Critical values for comparison purposes are obtained from the standard normal distribution.

In our present example, to test $H_0: \rho = .80$ against $H_A: \rho \neq .80$ using the Hotelling transformation and $\alpha = .05$, we have

$$z^* = 1.24726 - \frac{3(1.24726) + .848}{4(155)} = 1.2339$$

$$\zeta^* = 1.09861 - \frac{3(1.09861) + .8}{4(155)} = 1.0920$$

$$Z^* = (1.2339 - 1.0920)\sqrt{155 - 1} = 1.7609$$

Since 1.7609 is less than 1.96, the null hypothesis is not rejected, and the same conclusion is reached as when the Fisher transformation is used.

Alternatives In some situations the data available for analysis do not meet the assumptions necessary for the valid use of the procedures discussed here for testing hypotheses about a population correlation coefficient. In such cases it may be more appropriate to use the Spearman rank correlation technique discussed in Chapter 13.

Confidence Interval for ρ Fisher's transformation may be used to construct $100(1 - \alpha)$ percent confidence intervals for ρ. The general formula for a confidence interval

$$\text{estimator} \pm (\text{reliability factor})(\text{standard error})$$

is employed. We first convert our estimator, r, to z_r, construct a confidence interval about z_ρ, and then reconvert the limits to obtain a $100(1 - \alpha)$ percent confidence interval about ρ. The general formula then becomes

$$z_r \pm z(1/\sqrt{n-3}) \tag{9.7.10}$$

For our present example the 95 percent confidence interval for z_ρ is given by

$$1.24726 \pm 1.96(1/\sqrt{155-3})$$
$$1.08828, 1.40624$$

Converting these limits (by interpolation in Appendix Table I), which are values of z_r, into values of r gives

z_r	r
1.08828	.7962
1.40624	.8866

We are 95 percent confident, then, that ρ is contained in the interval .7962 to .88866. Because of the limited entries in the table, these limits must be considered as only approximate.

EXERCISES

In each of the following exercises:

(a) Prepare a scatter diagram.

(b) Compute the sample correlation coefficient.

(c) Test $H_0: \rho = 0$ at the .05 level of significance and state your conclusions.

(d) Determine the p value for the test.

(e) Construct the 95 percent confidence interval for ρ.

9.7.1 The purpose of a study by Brown and Persley (A-8) was to characterize acute hepatitis A in patients more than 40 years old. They performed a retrospective chart review of 20 subjects who were diagnosed with acute hepatitis A, but were not hospitalized. Of interest was the use of age (years) to predict bilirubin levels (mg/dl). The following data were collected.

Age (Years)	Bilirubin (mg/dl)	Age (Years)	Bilirubin (mg/dl)
78	7.5	44	7.0
72	12.9	42	1.8
81	14.3	45	.8
59	8.0	78	3.8
64	14.1	47	3.5
48	10.9	50	5.1
46	12.3	57	16.5

(Continued)

Age (Years)	Bilirubin (mg/dl)	Age (Years)	Bilirubin (mg/dl)
42	1.0	52	3.5
58	5.2	58	5.6
52	5.1	45	1.9

Source: Geri R. Brown, M.D. Used with permission.

9.7.2 Another variable of interest in the study by Reiss et al. (A-3) (see Exercise 9.3.4) was partial thromboplastin (aPTT), the standard test used to monitor heparin anticoagulation. Use the data in the following table to examine the correlation between aPTT levels as measured by the CoaguCheck point-of-care assay and standard laboratory hospital assay in 90 subjects receiving heparin alone, heparin with warfarin, and warfarin and exoenoxaparin.

Heparin		Warfarin		Warfarin and Exoenoxaparin	
CoaguCheck aPTT	Hospital aPTT	CoaguCheck aPTT	Hospital aPTT	CoaguCheck aPTT	Hospital aPTT
49.3	71.4	18.0	77.0	56.5	46.5
57.9	86.4	31.2	62.2	50.7	34.9
59.0	75.6	58.7	53.2	37.3	28.0
77.3	54.5	75.2	53.0	64.8	52.3
42.3	57.7	18.0	45.7	41.2	37.5
44.3	59.5	82.6	81.1	90.1	47.1
90.0	77.2	29.6	40.9	23.1	27.1
55.4	63.3	82.9	75.4	53.2	40.6
20.3	27.6	58.7	55.7	27.3	37.8
28.7	52.6	64.8	54.0	67.5	50.4
64.3	101.6	37.9	79.4	33.6	34.2
90.4	89.4	81.2	62.5	45.1	34.8
64.3	66.2	18.0	36.5	56.2	44.2
89.8	69.8	38.8	32.8	26.0	28.2
74.7	91.3	95.4	68.9	67.8	46.3
150.0	118.8	53.7	71.3	40.7	41.0
32.4	30.9	128.3	111.1	36.2	35.7
20.9	65.2	60.5	80.5	60.8	47.2
89.5	77.9	150.0	150.0	30.2	39.7
44.7	91.5	38.5	46.5	18.0	31.3
61.0	90.5	58.9	89.1	55.6	53.0
36.4	33.6	112.8	66.7	18.0	27.4
52.9	88.0	26.7	29.5	18.0	35.7
57.5	69.9	49.7	47.8	78.3	62.0
39.1	41.0	85.6	63.3	75.3	36.7
74.8	81.7	68.8	43.5	73.2	85.3
32.5	33.3	18.0	54.0	42.0	38.3
125.7	142.9	92.6	100.5	49.3	39.8
77.1	98.2	46.2	52.4	22.8	42.3
143.8	108.3	60.5	93.7	35.8	36.0

Source: Curtis E. Haas, Pharm. D. Used with permission.

9.7.3 In the study by Parker et al. (A-4) (see Exercise 9.3.5), the authors also looked at the change in AUC (area under the curve of plasma concentration of digoxin) when comparing digoxin levels taken with and without grapefruit juice. The following table gives the AUC when digoxin was consumed with water (ng · hr/ml) and the change in AUC compared to the change in AUC when digoxin is taken with grapefruit juice (GFJ, %).

Water AUC Level (ng · hr/ml)	Change in AUC with GFJ (%)
6.96	17.4
5.59	24.5
5.31	8.5
8.22	20.8
11.91	−26.7
9.50	−29.3
11.28	−16.8

Source: Robert B. Parker, Pharm. D. Used with permission.

9.7.4 An article by Tuzson et al. (A-9) in *Archives of Physical Medicine and Rehabilitation* reported the following data on peak knee velocity in walking (measured in degrees per second) at flexion and extension for 18 subjects with cerebral palsy.

Flexion (°/s)	Extension (°/s)
100	100
150	150
210	180
255	165
200	210
185	155
440	440
110	180
400	400
160	140
150	250
425	275
375	340
400	400
400	450
300	300
300	300
320	275

Source: Ann E. Tuzson, Kevin P. Granata, and Mark F. Abel, "Spastic Velocity Threshold Constrains Functional Performance in Cerebral Palsy," *Archives of Physical Medicine and Rehabilitation, 84* (2003), 1363–1368.

9.7.5 Amyotrophic lateral sclerosis (ALS) is characterized by a progressive decline of motor function. The degenerative process affects the respiratory system. Butz et al. (A-10) investigated the longitudinal impact of nocturnal noninvasive positive-pressure ventilation on patients with ALS. Prior to treatment, they measured partial pressure of arterial oxygen (Pao_2) and partial pressure of arterial carbon dioxide ($Paco_2$) in patients with the disease. The results were as follows:

$Paco_2$	Pao_2
40.0	101.0
47.0	69.0
34.0	132.0
42.0	65.0
54.0	72.0
48.0	76.0
53.6	67.2
56.9	70.9
58.0	73.0
45.0	66.0
54.5	80.0
54.0	72.0
43.0	105.0
44.3	113.0
53.9	69.2
41.8	66.7
33.0	67.0
43.1	77.5
52.4	65.1
37.9	71.0
34.5	86.5
40.1	74.7
33.0	94.0
59.9	60.4
62.6	52.5
54.1	76.9
45.7	65.3
40.6	80.3
56.6	53.2
59.0	71.9

Source: M. Butz, K. H. Wollinsky, U. Widemuth-Catrinescu, A. Sperfeld, S. Winter, H. H. Mehrkens, A. C. Ludolph, and H. Schreiber, "Longitudinal Effects of Noninvasive Positive-Pressure Ventilation in Patients with Amyotrophic Lateral Sclerosis," *American Journal of Medical Rehabilitation, 82* (2003) 597–604.

9.7.6 A simple random sample of 15 apparently healthy children between the ages of 6 months and 15 years yielded the following data on age, X, and liver volume per unit of body weight (ml/kg), Y:

X	Y	X	Y
.5	41	10.0	26
.7	55	10.1	35
2.5	41	10.9	25
4.1	39	11.5	31

(Continued)

X	Y	X	Y
5.9	50	12.1	31
6.1	32	14.1	29
7.0	41	15.0	23
8.2	42		

9.8 SOME PRECAUTIONS

Regression and correlation analysis are powerful statistical tools when properly employed. Their inappropriate use, however, can lead only to meaningless results. To aid in the proper use of these techniques, we make the following suggestions:

1. The assumptions underlying regression and correlation analysis should be reviewed carefully before the data are collected. Although it is rare to find that assumptions are met to perfection, practitioners should have some idea about the magnitude of the gap that exists between the data to be analyzed and the assumptions of the proposed model, so that they may decide whether they should choose another model; proceed with the analysis, but use caution in the interpretation of the results; or use the chosen model with confidence.
2. In simple linear regression and correlation analysis, the two variables of interest are measured on the same entity, called the *unit of association.* If we are interested in the relationship between height and weight, for example, these two measurements are taken on the same individual. It usually does not make sense to speak of the correlation, say, between the heights of one group of individuals and the weights of another group.
3. No matter how strong is the indication of a relationship between two variables, it should not be interpreted as one of cause and effect. If, for example, a significant sample correlation coefficient between two variables X and Y is observed, it can mean one of several things:

    a. X causes Y.

    b. Y causes X.

    c. Some third factor, either directly or indirectly, causes both X and Y.

    d. An unlikely event has occurred and a large sample correlation coefficient has been generated by chance from a population in which X and Y are, in fact, not correlated.

    e. The correlation is purely nonsensical, a situation that may arise when measurements of X and Y are not taken on a common unit of association.

4. The sample regression equation should not be used to predict or estimate outside the range of values of the independent variable represented in the sample. As illustrated in Section 9.5, this practice, called extrapolation, is risky. The true relationship between two variables, although linear over an interval of the independent variable, sometimes may be described at best as a curve outside this interval. If

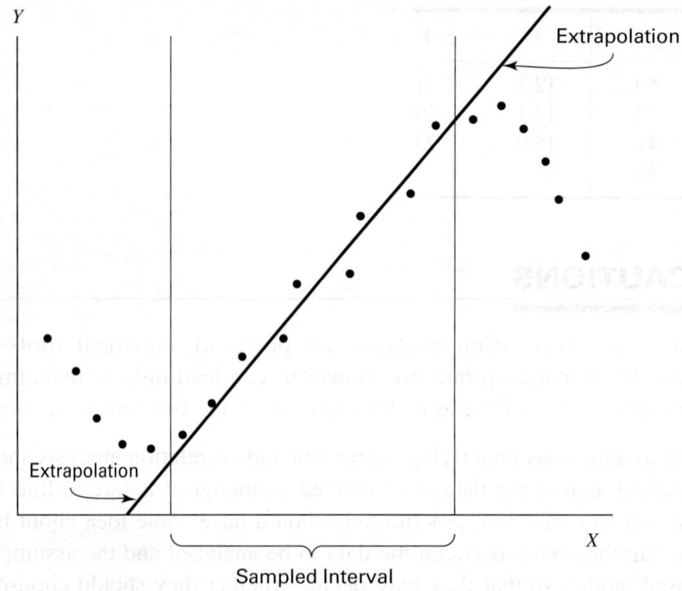

FIGURE 9.8.1 Example of extrapolation.

our sample by chance is drawn only from the interval where the relationship is linear, we have only a limited representation of the population, and to project the sample results beyond the interval represented by the sample may lead to false conclusions. Figure 9.8.1 illustrates the possible pitfalls of extrapolation.

9.9 SUMMARY

In this chapter, two important tools of statistical analysis, simple linear regression and correlation, are examined. The following outline for the application of these techniques has been suggested.

1. **Identify the model.** Practitioners must know whether the regression model or the correlation model is the appropriate one for answering their questions.
2. **Review assumptions.** It has been pointed out several times that the validity of the conclusions depends on how well the analyzed data fit the chosen model.
3. **Obtain the regression equation.** We have seen how the regression equation is obtained by the method of least squares. Although the computations, when done by hand, are rather lengthy, involved, and subject to error, this is not the problem today that it has been in the past. Computers are now in such widespread use that the researcher or statistician without access to one is the exception rather than the rule. No apology for lengthy computations is necessary to the researcher who has a computer available.
4. **Evaluate the equation.** We have seen that the usefulness of the regression equation for estimating and predicting purposes is determined by means of the analysis

of variance, which tests the significance of the regression mean square. The strength of the relationship between two variables under the correlation model is assessed by testing the null hypothesis that there is no correlation in the population. If this hypothesis can be rejected we may conclude, at the chosen level of significance, that the two variables are correlated.

5. **Use the equation.** Once it has been determined that it is likely that the regression equation provides a good description of the relationship between two variables, X and Y, it may be used for one of two purposes:

   a. To predict what value Y is likely to assume, given a particular value of X, or

   b. To estimate the mean of the subpopulation of Y values for a particular value of X.

This necessarily abridged treatment of simple linear regression and correlation may have raised more questions than it has answered. It may have occurred to the reader, for example, that a dependent variable can be more precisely predicted using two or more independent variables rather than one. Or, perhaps, he or she may feel that knowledge of the strength of the relationship among several variables might be of more interest than knowledge of the relationship between only two variables. The exploration of these possibilities is the subject of the next chapter, and the reader's curiosity along these lines should be at least partially relieved.

For those who would like to pursue further the topic of regression analysis a number of excellent references are available, including those by Dielman (7), Hocking (8), Mendenhall and Sincich (9), and Neter et al. (10).

SUMMARY OF FORMULAS FOR CHAPTER 9

Formula Number	Name	Formula
9.2.1	Assumption of linearity	$\mu_{y\|x} = \beta_0 + \beta_1 x$
9.2.2	Simple linear regression model	$y = \beta_0 + \beta_1 x + \epsilon$
9.2.3	Error (residual) term	$\epsilon = y - (\beta_0 + \beta_1 x) = y - \mu_{y\|x}$
9.3.1	Algebraic representation of a straight line	$y = a + bx$
9.3.2	Least square estimate of the slope of a regression line	$\hat{\beta}_1 = \dfrac{\sum_{i=1}^{n}(x_i - \bar{x})(y_i - \bar{y})}{\sum_{i=1}^{n}(x_i - \bar{x})^2}$

(*Continued*)

9.3.3	Least square estimate of the intercept of a regression line	$\hat{\beta}_0 = \bar{y} - \hat{\beta}_1\bar{x}$
9.4.1	Deviation equation	$(y_i - \bar{y}) = (\hat{y}_i - \bar{y}) + (y_i - \hat{y}_i)$
9.4.2	Sum-of-squares equation	$\sum(y_i - \bar{y})^2 = \sum(\hat{y}_i - \bar{y})^2 + \sum(y_i - \hat{y}_i)^2$
9.4.3	Estimated population coefficient of determination	$\bar{r}^2 = 1 - \dfrac{\sum(y_i - \hat{y}_i)^2/(n-2)}{\sum(y_i - \bar{y})^2/(n-1)}$
9.4.4–9.4.7	Means and variances of point estimators *a* and *b*	$\mu\hat{\beta}_0 = \beta_0$ $\sigma^2_{\hat{\beta}_0} = \dfrac{\sigma^2_{y\|x}\sum x_i^2}{n\sum_{i=1}^{n}(x_i - \bar{x})^2}$ $\mu_{\hat{\beta}_1} = \beta_1$ $\sigma^2_{\hat{\beta}_1} = \dfrac{\sigma^2_{y/x}}{\sum_{i=1}^{n}(x - \bar{x})^2}$
9.4.8	*z* statistic for testing hypotheses about β	$z = \dfrac{\hat{\beta}_1 - (\beta_1)_0}{\sigma_{\hat{\beta}_0}}$
9.4.9	*t* statistic for testing hypotheses about β	$t = \dfrac{\hat{\beta}_1 - (\beta_1)_0}{s_{\hat{\beta}_0}}$
9.5.1 9.5.2	Prediction interval for *Y* for a given *X* Confidence Interval for the mean of Y for a given X	$\hat{y} \pm t_{(1-a/2)} s_{y/x}\sqrt{1 + \dfrac{1}{n} + \dfrac{(x_p - \bar{x})^2}{\sum(x_i - \bar{x})^2}}$ $\hat{y} \pm t_{(1-a/2)} s_{y\|x}\sqrt{\dfrac{1}{n} + \dfrac{(x_p - \bar{x})}{\sum(x_i - \bar{x})}}$
9.7.1–9.7.2	Correlation coefficient	$r = \sqrt{\dfrac{\hat{\beta}_1^2[\sum x_i^2 - (\sum x_i)^2/n]}{\sum y_i^2 - (\sum y_i)^2/n}}$ $= \dfrac{n\sum x_i y_i - (\sum x_i)(\sum y_i)}{\sqrt{n\sum x_i^2 - (\sum x_i)^2}\sqrt{n\sum y_i^2 - (\sum y_i)^2}}$

9.7.3	t statistic for correlation coefficient	$t = r\sqrt{\frac{n-2}{1-r^2}}$
9.7.4	z statistic for correlation coefficient	$z_r = \frac{1}{2}\ln\frac{1+r}{1-r}$
9.7.5	Estimated standard deviation for z statistic	$\sigma_{z_p} = \frac{1}{\sqrt{n-3}}$
9.7.6	Z statistic for correlation coefficient	$Z = \frac{z_r - z_p}{1/\sqrt{n-3}}$
9.7.7	Z statistic for correlation coefficient when $n < 25$	$z^* = z_r - \frac{3z_r + r}{4n}$
9.7.8	Standard Deviation for z^*	$\sigma_{z^*} = \frac{1}{\sqrt{n-3}}$
9.7.9	Z^* statistic for correlation coefficient	$Z^* = \frac{z^* - \xi^*}{1/\sqrt{n-1}} = (z^* - \xi^*)\sqrt{n-1}$, where $\xi^* = Z_p - \frac{(3z_p + \rho)}{4n}$
9.7.10	Confidence interval for ρ	$Z_r = Z\left(1/\sqrt{n-3}\right)$
Symbol Key		• β_0 = regression intercept term • $\hat{\beta}_0$ = estimated regression intercept • α = probability of type I error or regression intercept • $\hat{\beta}_1$ = estimated regression slope • β_1 = regression slope • ϵ = error term • μ_x = population mean of statistic/variable x • n = sample size • σ_x^2 = population variance of statistic/variable x • ρ = population correlation coefficient • r = sample correlation coefficient • r^2 = sample coefficient of determination • t = t statistic • x_i = value of independent variable at i • $\bar{x}$ = sample mean of independent variable • y_i = value of dependent variable at i • $\bar{y}$ = sample mean of dependent variable • $\hat{y}$ = estimated y • z = z statistic

REVIEW QUESTIONS AND EXERCISES

1. What are the assumptions underlying simple linear regression analysis when one of the objectives is to make inferences about the population from which the sample data were drawn?
2. Why is the regression equation called the least-squares equation?
3. Explain the meaning of $\hat{\beta}_0$ in the sample regression equation.
4. Explain the meaning of $\hat{\beta}_1$ in the sample regression equation.
5. Explain the following terms:
   (a) Total sum of squares
   (b) Explained sum of squares
   (c) Unexplained sum of squares
6. Explain the meaning of and the method of computing the coefficient of determination.
7. What is the function of the analysis of variance in regression analysis?
8. Describe three ways in which one may test the null hypothesis that $\beta_1 = 0$.
9. For what two purposes can a regression equation be used?
10. What are the assumptions underlying simple correlation analysis when inference is an objective?
11. What is meant by the unit of association in regression and correlation analysis?
12. What are the possible explanations for a significant sample correlation coefficient?
13. Explain why it is risky to use a sample regression equation to predict or to estimate outside the range of values of the independent variable represented in the sample.
14. Describe a situation in your particular area of interest where simple regression analysis would be useful. Use real or realistic data and do a complete regression analysis.
15. Describe a situation in your particular area of interest where simple correlation analysis would be useful. Use real or realistic data and do a complete correlation analysis.

In each of the following exercises, carry out the required analysis and test hypotheses at the indicated significance levels. Compute the p value for each test.

16. A study by Scrogin et al. (A-11) was designed to assess the effects of concurrent manipulations of dietary NaCl and calcium on blood pressure as well as blood pressure and catecholamine responses to stress. Subjects were salt-sensitive, spontaneously hypertensive male rats. Among the analyses performed by the investigators was a correlation between baseline blood pressure and plasma epinephrine concentration (E). The following data on these two variables were collected. Let $\alpha = .01$.

BP	PlasmaE	BP	PlasmaE
163.90	248.00	143.20	179.00
195.15	339.20	166.00	160.40
170.20	193.20	160.40	263.50
171.10	307.20	170.90	184.70

(Continued)

BP	PlasmaE	BP	PlasmaE
148.60	80.80	150.90	227.50
195.70	550.00	159.60	92.35
151.00	70.00	141.60	139.35
166.20	66.00	160.10	173.80
177.80	120.00	166.40	224.80
165.10	281.60	162.00	183.60
174.70	296.70	214.20	441.60
164.30	217.30	179.70	612.80
152.50	88.00	178.10	401.60
202.30	268.00	198.30	132.00
171.70	265.50		

Source: Karie E. Scrogin. Used with permission.

17. Dean Parmalee (A-12) wished to know if the year-end grades assigned to Wright State University Medical School students are predictive of their second-year board scores. The following table shows, for 89 students, the year-end score (AVG, in percent of 100) and the score on the second-year medical board examination (BOARD).

AVG	BOARD	AVG	BOARD	AVG	BOARD
95.73	257	85.91	208	82.01	196
94.03	256	85.81	210	81.86	179
91.51	242	85.35	212	81.70	207
91.49	223	85.30	225	81.65	202
91.13	241	85.27	203	81.51	230
90.88	234	85.05	214	81.07	200
90.83	226	84.58	176	80.95	200
90.60	236	84.51	196	80.92	160
90.30	250	84.51	207	80.84	205
90.29	226	84.42	207	80.77	194
89.93	233	84.34	211	80.72	196
89.83	241	84.34	202	80.69	171
89.65	234	84.13	229	80.58	201
89.47	231	84.13	202	80.57	177
88.87	228	84.09	184	80.10	192
88.80	229	83.98	206	79.38	187
88.66	235	83.93	202	78.75	161
88.55	216	83.92	176	78.32	172
88.43	207	83.73	204	78.17	163
88.34	224	83.47	208	77.39	166
87.95	237	83.27	211	76.30	170
87.79	213	83.13	196	75.85	159
87.01	215	83.05	203	75.60	154
86.86	187	83.02	188	75.16	169
86.85	204	82.82	169	74.85	159
86.84	219	82.78	205	74.66	167

(Continued)

AVG	BOARD	AVG	BOARD	AVG	BOARD
86.30	228	82.57	183	74.58	154
86.13	210	82.56	181	74.16	148
86.10	216	82.45	173	70.34	159
85.92	212	82.24	185		

Source: Dean Parmalee, M.D. and the Wright State University Statistical Consulting Center. Used with permission.

Perform a complete regression analysis with AVG as the independent variable. Let $\alpha = .05$ for all tests.

18. Maria Mathias (A-13) conducted a study of hyperactive children. She measured the children's attitude, hyperactivity, and social behavior before and after treatment. The following table shows for 31 subjects the age and improvement scores from pre-treatment to post-treatment for attitude (ATT), social behavior (SOC), and hyperactivity (HYP). A negative score for HYP indicates an improvement in hyperactivity; a positive score in ATT or SOC indicates improvement. Perform an analysis to determine if there is evidence to indicate that age (years) is correlated with any of the three outcome variables. Let $\alpha = .05$ for all tests.

Subject No.	AGE	ATT	HYP	SOC
1	9	−1.2	−1.2	0.0
2	9	0.0	0.0	1.0
3	13	−0.4	0.0	0.2
4	6	−0.4	−0.2	1.2
5	9	1.0	−0.8	0.2
6	8	0.8	0.2	0.4
7	8	−0.6	−0.2	0.6
8	9	−1.2	−0.8	−0.6
9	7	0.0	0.2	0.8
10	12	0.4	−0.8	0.4
11	9	−0.8	0.8	−0.2
12	10	1.0	−0.8	1.2
13	12	1.4	−1.6	0.6
14	9	1.0	−0.2	−0.2
15	12	0.8	−0.8	1.0
16	9	1.0	0.4	0.4
17	10	0.4	−0.2	0.6
18	7	0.0	−0.4	0.6
19	12	1.1	−0.6	0.8
20	9	0.2	−0.4	0.2
21	7	0.4	−0.2	0.6
22	6	0.0	−3.2	1.0
23	11	0.6	−0.4	0.0
24	11	0.4	−0.4	0.0
25	11	1.0	−0.7	−0.6
26	11	0.8	−0.8	0.0

(Continued)

Subject No.	AGE	ATT	HYP	SOC
27	11	1.2	0.6	1.0
28	11	0.2	0.0	−0.2
29	11	0.8	−1.2	0.3
30	8	0.0	0.0	−0.4
31	9	0.4	−0.2	0.2

Source: Maria Mathias, M.D. and the Wright State University Statistical Consulting Center. Used with permission.

19. A study by Triller et al. (A-14) examined the length of time required for home health-care nurses to repackage a patient's medications into various medication organizers (i.e., pill boxes). For the 19 patients in the study, researchers recorded the time required for repackaging of medications. They also recorded the number of problems encountered in the repackaging session.

Patient No.	No. of Problems	Repackaging Time (Minutes)	Patient No.	No. of Problems	Repackaging Time (Minutes)
1	9	38	11	1	10
2	2	25	12	2	15
3	0	5	13	1	17
4	6	18	14	0	18
5	5	15	15	0	23
6	3	25	16	10	29
7	3	10	17	0	5
8	1	5	18	1	22
9	2	10	19	1	20
10	0	15			

Source: Darren M. Triller, Pharm.D. Used with permission.

Perform a complete regression analysis of these data using the number of problems to predict the time it took to complete a repackaging session. Let $\alpha = .05$ for all tests. What conclusions can be drawn from your analysis? How might your results be used by health-care providers?

20. The following are the pulmonary blood flow (PBF) and pulmonary blood volume (PBV) values recorded for 16 infants and children with congenital heart disease:

Y PBV (ml/sqM)	*X* PBF (L/min/sqM)
168	4.31
280	3.40
391	6.20
420	17.30
303	12.30
429	13.99
605	8.73
522	8.90
224	5.87
291	5.00

(*Continued*)

Y PBV (ml/sqM)	*X* PBF (L/min/sqM)
233	3.51
370	4.24
531	19.41
516	16.61
211	7.21
439	11.60

Find the regression equation describing the linear relationship between the two variables, compute r^2, and test $H_0: \beta_1 = 0$ by both the F test and the t test. Let $\alpha = .05$.

21. Fifteen specimens of human sera were tested comparatively for tuberculin antibody by two methods. The logarithms of the titers obtained by the two methods were as follows:

Method	
A (*X*)	*B* (*Y*)
3.31	4.09
2.41	3.84
2.72	3.65
2.41	3.20
2.11	2.97
2.11	3.22
3.01	3.96
2.13	2.76
2.41	3.42
2.10	3.38
2.41	3.28
2.09	2.93
3.00	3.54
2.08	3.14
2.11	2.76

Find the regression equation describing the relationship between the two variables, compute r^2, and test $H_0: \beta_1 = 0$ by both the F test and the t test.

22. The following table shows the methyl mercury intake and whole blood mercury values in 12 subjects exposed to methyl mercury through consumption of contaminated fish:

X Methyl Mercury Intake (μg Hg/day)	*Y* Mercury in Whole Blood (ng/g)
180	90
200	120
230	125
410	290

(*Continued*)

X Methyl Mercury Intake (μg Hg/day)	Y Mercury in Whole Blood (ng/g)
600	310
550	290
275	170
580	375
105	70
250	105
460	205
650	480

Find the regression equation describing the linear relationship between the two variables, compute r^2, and test $H_0: \beta_1 = 0$ by both the F and t tests.

23. The following are the weights (kg) and blood glucose levels (mg/100 ml) of 16 apparently healthy adult males:

Weight (*X*)	Glucose (*Y*)
64.0	108
75.3	109
73.0	104
82.1	102
76.2	105
95.7	121
59.4	79
93.4	107
82.1	101
78.9	85
76.7	99
82.1	100
83.9	108
73.0	104
64.4	102
77.6	87

Find the simple linear regression equation and test $H_0: \beta_1 = 0$ using both ANOVA and the t test. Test $H_0: \rho = 0$ and construct a 95 percent confidence interval for ρ. What is the predicted glucose level for a man who weighs 95 kg? Construct the 95 percent prediction interval for his weight. Let $\alpha = .05$ for all tests.

24. The following are the ages (years) and systolic blood pressures of 20 apparently healthy adults:

Age (*X*)	BP (*Y*)	Age (*X*)	BP (*Y*)
20	120	46	128
43	128	53	136

(*Continued*)

Age (*X*)	BP (*Y*)	Age (*X*)	BP (*Y*)
63	141	70	146
26	126	20	124
53	134	63	143
31	128	43	130
58	136	26	124
46	132	19	121
58	140	31	126
70	144	23	123

Find the simple linear regression equation and test $H_0: \beta_1 = 0$ using both ANOVA and the t test. Test $H_0: \rho = 0$ and construct a 95 percent confidence interval for ρ. Find the 95 percent prediction interval for the systolic blood pressure of a person who is 25 years old. Let $\alpha = .05$ for all tests.

25. The following data were collected during an experiment in which laboratory animals were inoculated with a pathogen. The variables are time in hours after inoculation and temperature in degrees Celsius.

Time	Temperature	Time	Temperature
24	38.8	44	41.1
28	39.5	48	41.4
32	40.3	52	41.6
36	40.7	56	41.8
40	41.0	60	41.9

Find the simple linear regression equation and test $H_0: \beta_1 = 0$ using both ANOVA and the t test. Test $H_0: \rho = 0$ and construct a 95 percent confidence interval for ρ. Construct the 95 percent prediction interval for the temperature at 50 hours after inoculation. Let $\alpha = .05$ for all tests.

For each of the studies described in Exercises 26 through 28, answer as many of the following questions as possible.

(a) Which is more relevant, regression analysis or correlation analysis, or are both techniques equally relevant?

(b) Which is the independent variable?

(c) Which is the dependent variable?

(d) What are the appropriate null and alternative hypotheses?

(e) Do you think the null hypothesis was rejected? Explain why or why not.

(f) Which is the more relevant objective, prediction or estimation, or are the two equally relevant?

(g) What is the sampled population?

(h) What is the target population?

(i) Are the variables directly or inversely related?

26. Lamarre-Cliche et al. (A-15) state, "The QT interval corrected for heart rate (QTc) is believed to reflect sympathovagal balance. It has also been established that β-blockers influence the autonomic nervous system." The researchers performed correlation analysis to measure the association between QTc interval, heart rate, heart rate change, and therapeutic blood pressure response for 73 hypertensive subjects taking β-blockers. The researchers found that QTc interval length, pre-treatment heart rate, and heart rate change with therapy were not good predictors of blood pressure response to β1-selective β-blockers in hypertensive subjects.

27. Skinner et al. (A-16) conducted a cross-sectional telephone survey to obtain 24-hour dietary recall of infants' and toddlers' food intakes, as reported by mothers or other primary caregivers. One finding of interest was that among 561 toddlers ages 15–24 months, the age in weeks of the child was negatively related to vitamin C density $\hat{\beta}_1 = -.43, p = .01$). When predicting calcium density, age in weeks of the child produced a slope coefficient of -1.47 with a p of .09.

28. Park et al. (A-17) studied 29 male subjects with clinically confirmed cirrhosis. Among other variables, they measured whole blood manganese levels (MnB), plasma manganese (MnP), urinary manganese (MnU), and pallidal index (PI), a measure of signal intensity in T1 weighted magnetic resonance imaging (MRI). They found a correlation coefficient of .559, $p < .01$, between MnB and PI. However, there were no significant correlations between MnP and Pi or MnU and Pi($r = .353, p > .05, r = .252, p > .05$, respectively).

For the studies described in Exercises 29 through 46, do the following:

(a) Perform a statistical analysis of the data (including hypothesis testing and confidence interval construction) that you think would yield useful information for the researchers.

(b) Construct graphs that you think would be helpful in illustrating the relationships among variables.

(c) Where you think appropriate, use techniques learned in other chapters, such as analysis of variance and hypothesis testing and interval estimation regarding means and proportions.

(d) Determine p values for each computed test statistic.

(e) State all assumptions that are necessary to validate your analysis.

(f) Describe the population(s) about which you think inferences based on your analysis would be applicable.

(g) If available, consult the cited reference and compare your analyses and results with those of the authors.

29. Moerloose et al. (A-18) conducted a study to evaluate the clinical usefulness of a new laboratory technique (method A) for use in the diagnosis of pulmonary embolism (PE). The performance of the new technique was compared with that of a standard technique (method B). Subjects consisted of patients with clinically suspected PE who were admitted to the emergency ward of a European university hospital. The following are the measurements obtained by the two techniques for 85 patients. The researchers performed two analyses: (1) on all 85 pairs of measurements and (2) on those pairs of measurements for which the value for method B was less than 1000.

B	A	B	A	B	A
9	119	703	599	2526	1830
84	115	725	610	2600	1880
86	108	727	3900	2770	2100
190	182	745	4050	3100	1780
208	294	752	785	3270	1870

(Continued)

B	A	B	A	B	A
218	226	884	914	3280	2480
251	311	920	1520	3410	1440
252	250	966	972	3530	2190
256	312	985	913	3900	2340
264	403	994	556	4260	3490
282	296	1050	1330	4300	4960
294	296	1110	1410	4560	7180
296	303	1170	484	4610	1390
311	336	1190	867	4810	1600
344	333	1250	1350	5070	3770
371	257	1280	1560	5470	2780
407	424	1330	1290	5576	2730
418	265	1340	1540	6230	1260
422	347	1400	1710	6260	2870
459	412	1530	1333	6370	2210
468	389	1560	1250	6430	2210
481	414	1840	764	6500	2380
529	667	1870	1680	7120	5220
540	486	2070	1310	7430	2650
562	720	2120	1360	7800	4910
574	343	2170	1770	8890	4080
646	518	2270	2240	9930	3840
664	801	2490	1910		
670	760	2520	2110		

Source: Dr. Philippe de Moerloose. Used with permission.

30. Research by Huhtaniemi et al. (A-19) focused on the quality of serum luteinizing hormone (LH) during pubertal maturation in boys. Subjects, consisting of healthy boys entering puberty (ages 11 years 5 months to 12 years), were studied over a period of 18 months. The following are the concentrations (IU/L) of bioactive LH (B-LH) and immunoreactive LH (I-LH) in serum samples taken from the subjects. Only observations in which the subjects' B/I ratio was greater than 3.5 are reported here.

I-LH	B-LH	I-LH	B-LH
.104	.37	.97	3.63
.041	.28	.49	2.26
.124	.64	1	4.55
.808	2.32	1.17	5.06
.403	1.28	1.46	4.81
.27	.9	1.97	8.18
.49	2.45	.88	2.48
.66	2.8	1.24	4.8
.82	2.6	1.54	3.12

(*Continued*)

I-LH	B-LH	I-LH	B-LH
1.09	4.5	1.71	8.4
1.05	3.2	1.11	6
.83	3.65	1.35	7.2
.89	5.25	1.59	7.6
.75	2.9		

Source: Dr. Ilpo T. Huhtaniemi. Used with permission.

31. Tsau et al. (A-20) studied urinary epidermal growth factor (EGF) excretion in normal children and those with acute renal failure (ARF). Random urine samples followed by 24-hour urine collection were obtained from 25 children. Subjects ranged in age from 1 month to 15 years. Urinary EGF excretion was expressed as a ratio of urinary EGF to urinary creatinine concentration (EGF/Cr). The authors conclude from their research results that it is reasonable to use random urine tests for monitoring EGF excretion. Following are the random (spot) and 24-hour urinary EGF/Cr concentrations (pmol/mmol) for the 25 subjects:

Subject	24-h Urine EGF/Cr (x)	Spot Urine EGF/Cr (y)	Subject	24-h Urine EGF/Cr (x)	Spot Urine EGF/Cr (y)
1	772	720	14	254	333
2	223	271	15[a]	93	84
3	494	314	16	303	512
4	432	350	17	408	277
5[a]	79	79	18	711	443
6[a]	155	118	19	209	309
7	305	387	20	131	280
8	318	432	21	165	189
9[a]	174	97	22	151	101
10	1318	1309	23	165	221
11	482	406	24	125	228
12	436	426	25	232	157
13	527	595			

[a] Subjects with ARF.

Source: Dr. Yong-Kwei Tsau. Used with permission.

32. One of the reasons for a study by Usaj and Starc (A-21) was an interest in the behavior of pH kinetics during conditions of long-term endurance and short-term endurance among healthy runners. The nine subjects participating in the study were marathon runners aged 26 ± 5 years. The authors report that they obtained a good correlation between pH kinetics and both short-term and long-term endurance. The following are the short- (V_{SE}) and long-term (V_{LE}) speeds and blood pH measurements for the participating subjects.

V_{LE}	V_{SE}	pH Range
5.4	5.6	.083
4.75	5.1	.1
4.6	4.6	.021

(Continued)

V_{LE}	V_{SE}	pH Range
4.6	5	.065
4.55	4.9	.056
4.4	4.6	.01
4.4	4.9	.058
4.2	4.4	.013
4.2	4.5	.03

Source: Anton Usaj, Ph.D.
Used with permission.

33. Bean et al. (A-22) conducted a study to assess the performance of the isoelectric focusing/immunoblotting/laser densitometry (IEF/IB/LD) procedure to evaluate carbohydrate-deficient transferrin (CDT) derived from dry blood spots. The investigators evaluated paired serum (S) and dry blood spot (DBS) specimens simultaneously for CDT. Assessment of CDT serves as a marker for alcohol abuse. The use of dry blood spots as a source of CDT for analysis by IEF/IB/LD results in simplified sampling, storage, and transportation of specimens. The following are the IEF/IB/LD values in densitometry units (DU) of CDT from 25 serum and dry blood spot specimens:

Specimen No.	S	DBS	Specimen No.	S	DBS
1	64	23	14	9	13
2	74	38	15	10	8
3	75	37	16	17	7
4	103	53	17	38	14
5	10	9	18	9	9
6	22	18	19	15	9
7	33	20	20	70	31
8	10	5	21	61	26
9	31	14	22	42	14
10	30	15	23	20	10
11	28	12	24	58	26
12	16	9	25	31	12
13	13	7			

Source: Dr. Pamela Bean.
Used with permission.

34. Kato et al. (A-23) measured the plasma concentration of adrenomedullin (AM) in patients with chronic congestive heart failure due to various cardiac diseases. AM is a hypotensive peptide, which, on the basis of other studies, the authors say, has an implied role as a circulating hormone in regulation of the cardiovascular system. Other data collected from the subjects included plasma concentrations of hormones known to affect the cardiovascular system. Following are the plasma AM (fmol/ml) and plasma renin activity (PRA) (ng/L · · · s) values for 19 heart failure patients:

Patient No.	Sex (1 = M, 2 = F)	Age (Years)	AM (fmol/ml)	PRA (ng/L · · · s)
1	1	70	12.11	.480594
2	1	44	7.306	.63894
3	1	72	6.906	1.219542

(Continued)

Patient No.	Sex (1 = M, 2 = F)	Age (Years)	AM (fmol/ml)	PRA (ng/L···s)
4	1	62	7.056	.450036
5	2	52	9.026	.19446
6	2	65	10.864	1.966824
7	2	64	7.324	.29169
8	1	71	9.316	1.775142
9	2	61	17.144	9.33408
10	1	68	6.954	.31947
11	1	63	7.488	1.594572
12	2	59	10.366	.963966
13	2	55	10.334	2.191842
14	2	57	13	3.97254
15	2	68	6.66	.52782
16	2	51	8.906	.350028
17	1	69	8.952	1.73625
18	1	71	8.034	.102786
19	1	46	13.41	1.13898

Source: Dr. Johji Kato. Used with permission.

35. In a study reported on in *Archives of Disease in Childhood,* Golden et al. (A-24) tested the hypothesis that plasma calprotectin (PCal) (a neutrophil cytosolic protein released during neutrophil activation or death) concentration is an early and sensitive indicator of inflammation associated with bacterial infection in cystic fibrosis (CF). Subjects were children with confirmed CF and a control group of age- and sex-matched children without the disease. Among the data collected were the following plasma calprotectin (μg/L) and plasma copper (PCu) (μmol/L) measurements. Plasma copper is an index of acute phase response in cystic fibrosis. The authors reported a correlation coefficient of .48 between plasma calprotectin ($\log_{10}$) and plasma copper.

CF Subject No.	PCal	PCu	CF Subject No.	PCal	PCu	CF Subject No.	PCal	PCu
1	452	17.46	12	1548	15.31	22	674	18.11
2	590	14.84	13	708	17.00	23	3529	17.42
3	1958	27.42	14	8050	20.00	24	1467	17.42
4	2015	18.51	15	9942	25.00	25	1116	16.73
5	417	15.89	16	791	13.10	26	611	18.11
6	2884	17.99	17	6227	23.00	27	1083	21.56
7	1862	21.66	18	1473	16.70	28	1432	21.56
8	10471	19.03	19	8697	18.11	29	4422	22.60
9	25850	16.41	20	621	18.80	30	3198	18.91
10	5011	18.51	21	1832	17.08	31	544	14.37
11	5128	22.70						

(*Continued*)

Control Subject No.	PCal	PCu	Control Subject No.	PCal	PCu
1	674	16.73	17	368	16.73
2	368	16.73	18	674	16.73
3	321	16.39	19	815	19.82
4	1592	14.32	20	598	16.1
5	518	16.39	21	684	13.63
6	815	19.82	22	684	13.63
7	684	17.96	23	674	16.73
8	870	19.82	24	368	16.73
9	781	18.11	25	1148	24.15
10	727	18.11	26	1077	22.30
11	727	18.11	27	518	9.49
12	781	18.11	28	1657	16.10
13	674	16.73	29	815	19.82
14	1173	20.53	30	368	16.73
15	815	19.82	31	1077	22.30
16	727	18.11			

Source: Dr. Barbara E. Golden. Used with permission.

36. Gelb et al. (A-25) conducted a study in which they explored the relationship between moderate to severe expiratory airflow limitation and the presence and extent of morphologic and CT scored emphysema in consecutively seen outpatients with chronic obstructive pulmonary disease. Among the data collected were the following measures of lung CT and pathology (PATH) for emphysema scoring:

CT Score	PATH	CT Score	PATH
5	15	45	50
90	70	45	40
50	20	85	75
10	25	7	0
12	25	80	85
35	10	15	5
40	35	45	40
45	30	37	35
5	5	75	45
25	50	5	5
60	60	5	20
70	60		

Source: Dr. Arthur F. Gelb. Used with permission.

37. The objective of a study by Witteman et al. (A-26) was to investigate skin reactivity with purified major allergens and to assess the relation with serum levels of immunoglobulin E (IgE) antibodies and to determine which additional factors contribute to the skin test result. Subjects consisted of patients with allergic rhinitis, allergic asthma, or both, who were seen in a European medical center. As part of their study, the researchers collected, from 23 subjects, the following measurements on specific IgE

(IU/ml) and skin test (ng/ml) in the presence of Lol p 5, a purified allergen from grass pollen. We wish to know the nature and strength of the relationship between the two variables. (*Note:* The authors converted the measurements to natural logarithms before investigating this relationship.)

IgE	Skin Test
24.87	.055
12.90	.041034
9.87	.050909
8.74	.046
6.88	.039032
5.90	.050909
4.85	.042142
3.53	.055
2.25	4.333333
2.14	.55
1.94	.050909
1.29	.446153
.94	.4
.91	.475
.55	4.461538
.30	4.103448
.14	7.428571
.11	4.461538
.10	6.625
.10	49.13043
.10	36.47058
.10	52.85714
.10	47.5

Source: Dr. Jaring S. van der Zee. Used with permission.

38. Garland et al. (A-27) conducted a series of experiments to delineate the complex maternal-fetal pharmacokinetics and the effects of zidovudine (AZT) in the chronically instrumented maternal and fetal baboon (*Papio* species) during both steady-state intravenous infusion and oral bolus dosage regimens. Among the data collected were the following measurements on dosage (mg/kg/h) and steady-state maternal plasma AZT concentration (ng/ml):

Dosage	AZT Concentration	Dosage	AZT Concentration
2.5	832	2.0	771
2.5	672	1.8	757
2.5	904	0.9	213
2.5	554	0.6	394
2.5	996	0.9	391
1.9	878	1.3	430
2.1	815	1.1	440
1.9	805	1.4	352

(*Continued*)

Dosage	AZT Concentration	Dosage	AZT Concentration
1.9	592	1.1	337
0.9	391	0.8	181
1.5	710	0.7	174
1.4	591	1.0	470
1.4	660	1.1	426
1.5	694	0.8	170
1.8	668	1.0	360
1.8	601	0.9	320

Source: Dr. Marianne Garland. Used with permission.

39. The purpose of a study by Halligan et al. (A-28) was to evaluate diurnal variation in blood pressure (BP) in women who were normotensive and those with pre-eclampsia. The subjects were similar in age, weight, and mean duration of gestation (35 weeks). The researchers collected the following BP readings. As part of their analysis they studied the relationship between mean day and night measurements and day/night differences for both diastolic and systolic BP in each group.

C1	C2	C3	C4	C5	C1	C2	C3	C4	C5
0	75	56	127	101	1	94	78	137	119
0	68	57	113	104	1	90	86	139	138
0	72	58	115	105	1	85	69	138	117
0	71	51	111	94	1	80	75	133	126
0	81	61	130	110	1	81	60	127	112
0	68	56	111	101	1	89	79	137	126
0	78	60	113	102	1	107	110	161	161
0	71	55	120	99	1	98	88	152	141
0	65	51	106	96	1	78	74	134	132
0	78	61	120	109	1	80	80	121	121
0	74	60	121	104	1	96	83	143	129
0	75	52	121	102	1	85	76	137	131
0	68	50	109	91	1	79	74	135	120
0	63	49	108	99	1	91	95	139	135
0	77	47	132	115	1	87	67	137	115
0	73	51	112	90	1	83	64	143	119
0	73	52	118	97	1	94	85	127	123
0	64	62	122	114	1	85	70	142	124
0	64	54	108	94	1	78	61	119	110
0	66	54	106	88	1	80	59	129	114
0	72	49	116	101	1	98	102	156	163
0	83	60	127	103	1	100	100	149	149
0	69	50	121	104	1	89	84	141	135
0	72	52	108	95	1	98	91	148	139

C1 = group (0 = normotensive, 1 = pre-eclamptic); C2 = day diastolic; C3 = night diastolic; C4 = day systolic; C5 = night systolic.

Source: Dr. Aidan Halligan. Used with permission.

40. Marks et al. (A-29) conducted a study to determine the effects of rapid weight loss on contraction of the gallbladder and to evaluate the effects of ursodiol and ibuprofen on saturation, nucleation and growth, and contraction. Subjects were obese patients randomly assigned to receive ursodiol, ibuprofen, or placebo. Among the data collected were the following cholesterol saturation index values (CSI) and nucleation times (NT) in days of 13 (six male, seven female) placebo-treated subjects at the end of 6 weeks:

CSI	NT
1.20	4.00
1.42	6.00
1.18	14.00
.88	21.00
1.05	21.00
1.00	18.00
1.39	6.00
1.31	10.00
1.17	9.00
1.36	14.00
1.06	21.00
1.30	8.00
1.71	2.00

Source: Dr. Jay W. Marks. Used with permission.

41. The objective of a study by Peacock et al. (A-30) was to investigate whether spinal osteoarthritis is responsible for the fact that lumbar spine bone mineral density (BMD) is greater when measured in the anteroposterior plane than when measured in the lateral plane. Lateral spine radiographs were studied from women (age range 34 to 87 years) who attended a hospital outpatient department for bone density measurement and underwent lumbar spine radiography. Among the data collected were the following measurements on anteroposterior (A) and lateral (L) BMD (g/cm^2):

ABMD	LBMD	ABMD	LBMD	ABMD	LBMD
.879	.577	1.098	.534	1.091	.836
.824	.622	.882	.570	.746	.433
.974	.643	.816	.558	1.127	.732
.909	.664	1.017	.675	1.411	.766
.872	.559	.669	.590	.751	.397
.930	.663	.857	.666	.786	.515
.912	.710	.571	.474	1.031	.574
.758	.592	1.134	.711	.622	.506
1.072	.702	.705	.492	.848	.657
.847	.655	.775	.348	.778	.537
1.000	.518	.968	.579	.784	.419
.565	.354	.963	.665	.659	.429
1.036	.839	.933	.626	.948	.485
.811	.572	.704	.194	.634	.544
.901	.612	.624	.429	.946	.550

(*Continued*)

ABMD	LBMD	ABMD	LBMD	ABMD	LBMD
1.052	.663	1.119	.707	1.107	.458
.731	.376	.686	.508	1.583	.975
.637	.488	.741	.484	1.026	.550
.951	.747	1.028	.787		
.822	.610	.649	.469		
.951	.710	1.166	.796		
1.026	.694	.954	.548		
1.022	.580	.666	.545		
		1.047	.706		
		.737	.526		

Source: Dr. Cyrus Cooper. Used with permission.

42. Sloan et al. (A-31) note that cardiac sympathetic activation and parasympathetic withdrawal result in heart rate increases during psychological stress. As indicators of cardiac adrenergic activity, plasma epinephrine (E) and norepinephrine (NE) generally increase in response to psychological challenge. Power spectral analysis of heart period variability also provides estimates of cardiac autonomic nervous system activity. The authors conducted a study to determine the relationship between neurohumoral and two different spectral estimates of cardiac sympathetic nervous system activity during a quiet resting baseline and in response to a psychologically challenging arithmetic task. Subjects were healthy, medication-free male and female volunteers with a mean age of 37.8 years. None had a history of cardiac, respiratory, or vascular disease. Among the data collected were the following measurements on E, NE, low-frequency (LF) and very-low-frequency (VLF) power spectral indices, and low-frequency/high frequency ratios (LH/HF). Measurements are given for three periods: baseline (B), a mental arithmetic task (MA), and change from baseline to task (DELTA).

Patient No.	E	NE	LF/HF	LF	Period	VLF
5	3.55535	6.28040	0.66706	7.71886	B	7.74600
5	0.05557	0.13960	−0.48115	−0.99826	DELTA	−2.23823
5	3.61092	6.41999	0.18591	6.72059	MA	5.50777
6	3.55535	6.24611	2.48308	7.33729	B	6.64353
6	0.10821	−0.05374	−2.03738	−0.77109	DELTA	−1.27196
6	3.66356	6.19236	0.44569	6.56620	MA	5.37157
7	3.29584	4.91998	−0.15473	7.86663	B	7.99450
7	0.59598	0.53106	0.14086	−0.81345	DELTA	−2.86401
7	3.89182	5.45104	−0.01387	7.05319	MA	5.13049
8	4.00733	5.97635	1.58951	8.18005	B	5.97126
8	0.29673	0.11947	−0.11771	−1.16584	DELTA	−0.39078
8	4.30407	6.09582	1.47180	7.01421	MA	5.58048
12	3.87120	5.35659	0.47942	6.56488	B	5.94960
12	*	*	0.19379	0.03415	DELTA	0.50134
12	*	*	0.67321	6.59903	MA	6.45094
13	3.97029	5.85507	0.13687	6.27444	B	5.58500
13	−0.20909	0.10851	1.05965	−0.49619	DELTA	−1.68911
13	3.76120	5.96358	1.19652	5.77825	MA	3.89589

(Continued)

Patient No.	E	NE	LF/HF	LF	Period	VLF
14	3.63759	5.62040	0.88389	6.08877	B	6.12490
14	0.31366	0.07333	1.06100	1.37098	DELTA	−1.07633
14	3.95124	5.69373	1.94489	7.45975	MA	5.04857
18	4.44265	5.88053	0.99200	7.52268	B	7.19376
18	0.35314	0.62824	−0.10297	−0.57142	DELTA	−2.06150
18	4.79579	6.50877	0.88903	6.95126	MA	5.13226
19	*	5.03044	0.62446	6.90677	B	7.39854
19	*	0.69966	0.09578	0.94413	DELTA	−0.88309
19	2.94444	5.73010	0.72024	7.85090	MA	6.51545
20	3.91202	5.86363	1.11825	8.26341	B	6.89497
20	−0.02020	0.21401	−0.60117	−1.13100	DELTA	−1.12073
20	3.89182	6.07764	0.51708	7.13241	MA	5.77424
21	3.55535	6.21860	0.78632	8.74397	B	8.26111
21	0.31585	−0.52487	−1.92114	−2.38726	DELTA	−2.08151
21	3.87120	5.69373	−1.13483	6.35671	MA	6.17960
22	4.18965	5.76832	−0.02785	8.66907	B	7.51529
22	0.16705	−0.05459	0.93349	−0.89157	DELTA	−1.00414
22	4.35671	5.71373	0.90563	7.77751	MA	6.51115
23	3.95124	5.52545	−0.24196	6.75330	B	6.93020
23	0.26826	0.16491	−0.00661	0.18354	DELTA	−1.18912
23	4.21951	5.69036	−0.24856	6.93684	MA	5.74108
24	3.78419	5.59842	−0.67478	6.26453	B	6.45268
24	0.32668	−0.17347	1.44970	0.52169	DELTA	0.39277
24	4.11087	5.42495	0.77493	6.78622	MA	6.84545
1	3.36730	6.13123	0.19077	6.75395	B	6.13708
1	0.54473	0.08538	0.79284	0.34637	DELTA	−0.56569
1	3.91202	6.21661	0.98361	7.10031	MA	5.57139
3	2.83321	5.92158	1.89472	7.92524	B	6.30664
3	1.15577	0.64930	−0.75686	−1.58481	DELTA	−1.95636
3	3.98898	6.57088	1.13786	6.34042	MA	4.35028
4	4.29046	5.73657	1.81816	7.02734	B	7.02882
4	0.14036	0.47000	−0.26089	−1.08028	DELTA	−1.43858
4	4.43082	6.20658	1.55727	5.94705	MA	5.59024
5	3.93183	5.62762	1.70262	6.76859	B	6.11102
5	0.80437	0.67865	−0.26531	−0.29394	DELTA	−0.94910
5	4.73620	6.30628	1.43731	6.47465	MA	5.16192
6	3.29584	5.47227	0.18852	6.49054	B	6.84279
6	−0.16034	0.27073	−0.16485	−1.12558	DELTA	−1.84288
6	3.13549	5.74300	0.02367	5.36496	MA	4.99991
8	3.25810	5.37064	−0.09631	7.23131	B	7.16371
8	0.40547	−0.13953	0.97906	−0.62894	DELTA	−2.15108
8	3.66356	5.23111	0.88274	6.60237	MA	5.01263
9	3.78419	5.94542	0.77839	5.86126	B	6.22910
9	0.64663	0.05847	−0.42774	−0.53530	DELTA	−2.18430
9	4.43082	6.00389	0.35066	5.32595	MA	4.04480
10	4.07754	5.87493	2.32137	6.71736	B	6.59769
10	0.23995	−0.00563	−0.25309	−0.00873	DELTA	−0.75357

(*Continued*)

Patient No.	E	NE	LF/HF	LF	Period	VLF
10	4.31749	5.86930	2.06827	6.70863	MA	5.84412
11	4.33073	5.84064	2.89058	7.22570	B	5.76079
11	−3.63759	−0.01464	−1.22533	−1.33514	DELTA	−0.55240
11	0.69315	5.82600	1.66525	5.89056	MA	5.20839
12	3.55535	6.04501	1.92977	8.50684	B	7.15797
12	0.13353	0.12041	−0.15464	−0.84735	DELTA	0.13525
12	3.68888	6.16542	1.77513	7.65949	MA	7.29322
13	3.33220	4.63473	−0.11940	6.35464	B	6.76285
13	1.16761	1.05563	0.85621	0.63251	DELTA	−0.52121
13	4.49981	5.69036	0.73681	6.98716	MA	6.24164
14	3.25810	5.96358	1.10456	7.01270	B	7.49426
14	*	*	0.26353	−1.20066	DELTA	−3.15046
14	*	*	1.36809	5.81204	MA	4.34381
15	5.42935	6.34564	2.76361	9.48594	B	7.05730
15	*	*	−1.14662	−1.58468	DELTA	−0.08901
15	*	*	1.61699	7.90126	MA	6.96829
16	4.11087	6.59441	−0.23319	6.68269	B	6.76872
16	−0.06782	−0.54941	0.34755	−0.29398	DELTA	−1.80868
16	4.04305	6.04501	0.11437	6.38871	MA	4.96004
17	*	6.28040	1.40992	6.09671	B	4.82671
17	*	−0.12766	−0.17490	−0.05945	DELTA	0.69993
17	*	6.15273	1.23501	6.03726	MA	5.52665
18	2.39790	6.03548	0.23183	6.39707	B	6.60421
18	1.06784	0.11299	0.27977	−0.38297	DELTA	−1.92672
18	3.46574	6.14847	0.51160	6.01410	MA	4.67749
19	4.21951	6.35784	1.08183	5.54214	B	5.69070
19	0.21131	−0.00347	0.12485	−0.54440	DELTA	−1.49802
19	4.43082	6.35437	1.20669	4.99774	MA	4.19268
20	4.14313	5.73334	0.89483	7.35045	B	6.93974
20	−0.11778	0.00000	0.17129	−0.58013	DELTA	−1.72916
20	4.02535	5.73334	1.06612	6.77032	MA	5.21058
21	3.66356	6.06843	−0.87315	5.09848	B	6.02972
21	0.20764	−0.10485	0.41178	−0.33378	DELTA	−2.00974
21	3.87120	5.96358	−0.46137	4.76470	MA	4.01998
22	3.29584	5.95324	2.38399	7.62877	B	7.54359
22	0.36772	0.68139	−0.75014	−0.89992	DELTA	−1.25555
22	3.66356	6.63463	1.63384	6.72884	MA	6.28804

* = missing data.

Source: Dr. Richard P. Sloan. Used with permission.

43. The purpose of a study by Chati et al. (A-32) was to ascertain the role of physical deconditioning in skeletal muscle metabolic abnormalities in patients with chronic heart failure (CHF). Subjects included ambulatory CHF patients (12 males, two females) ages 35 to 74 years. Among the data collected were the following measurements, during exercise, of workload (WL) under controlled conditions, peak oxygen consumption (V_{O_2}), anaerobic ventilatory threshold (AT), both measured in ml/kg/min, and exercise total time (ET) in seconds.

WL	Vo_2	AT	ET	WL	Vo_2	AT	ET
7.557	32.800	13.280	933.000	3.930	22.500	18.500	720.000
3.973	8.170	6.770	255.000	3.195	17.020	8.520	375.000
5.311	16.530	11.200	480.000	2.418	15.040	12.250	480.000
5.355	15.500	10.000	420.000	0.864	7.800	4.200	240.000
6.909	24.470	11.550	960.000	2.703	12.170	8.900	513.000
1.382	7.390	5.240	346.000	1.727	15.110	6.300	540.000
8.636	19.000	10.400	600.000	7.773	21.100	12.500	1200.000

Source: Dr. Zukaï Chati. Used with permission.

44. Czader et al. (A-33) investigated certain prognostic factors in patients with centroblastic-centrocytic non-Hodgkin's lymphomas (CB/CC NHL). Subjects consisted of men and women between the ages of 20 and 84 years at time of diagnosis. Among the data collected were the following measurements on two relevant factors, A and B. The authors reported a significant correlation between the two.

A	B	A	B	A	B
20.00	.154	22.34	.147	48.66	.569
36.00	.221	18.00	.132	20.00	.227
6.97	.129	18.00	.085	17.66	.125
13.67	.064	22.66	.577	14.34	.089
36.34	.402	45.34	.134	16.33	.051
39.66	.256	20.33	.246	18.34	.100
14.66	.188	16.00	.175	26.49	.202
27.00	.138	15.66	.105	13.33	.077
2.66	.078	23.00	.145	6.00	.206
22.00	.142	27.33	.129	15.67	.153
11.00	.086	6.27	.062	32.33	.549
20.00	.170			24.34	.147
22.66	.198			22.33	.769
7.34	.092			11.33	.130
29.67	.227			6.67	.099
11.66	.159				
8.05	.223				
22.66	.065				

Source: Dr. Magdalena Czader and Dr. Anna Porwit-MacDonald. Used with permission.

45. Fleroxacin, a fluoroquinolone derivative with a broad antibacterial spectrum and potent activity in vitro against gram-negative and many gram-positive bacteria, was the subject of a study by Reigner and Welker (A-34). The objectives of their study were to estimate the typical values of clearance over systemic availability (CL/F) and the volume of distribution over systemic availability (V/F) after the administration of therapeutic doses of fleroxacin and to identify factors that influence the disposition of fleroxacin and to quantify the degree to which they do so. Subjects were 172 healthy male and female volunteers and uninfected patients representing a wide age range. Among the data analyzed were the following measurements (ml/min) of CL/F and creatinine clearance (CLcr). According to the authors, previous studies have shown that there is a correlation between the two variables.

CL/F	CLer	CL/F	CLer	CL/F	CLer	CL/F	CLer
137.000	96.000	77.000	67.700	152.000	109.000	132.000	111.000
106.000	83.000	57.000	51.500	100.000	82.000	94.000	118.000
165.000	100.000	69.000	52.400	86.000	88.000	90.000	111.000
127.000	101.000	69.000	65.900	69.000	67.000	87.000	124.000
139.000	116.000	76.000	60.900	108.000	68.700	48.000	10.600
102.000	78.000	77.000	93.800	77.000	83.200	26.000	9.280
72.000	84.000	66.000	73.800	85.000	72.800	54.000	12.500
86.000	81.000	53.000	99.100	89.000	82.300	36.000	9.860
85.000	77.000	26.000	110.000	105.000	71.100	26.000	4.740
122.000	102.000	89.000	99.900	66.000	56.000	39.000	7.020
76.000	80.000	44.000	73.800	73.000	61.000	27.000	6.570
57.000	67.000	27.000	65.800	64.000	79.500	36.000	13.600
62.000	41.000	96.000	109.000	26.000	9.120	15.000	7.600
90.000	93.000	102.000	76.800	29.000	8.540	138.000	100.000
165.000	88.000	159.000	125.000	39.100	93.700	127.000	108.000
132.000	64.000	115.000	112.000	75.500	65.600	203.000	121.000
159.000	92.000	82.000	91.600	86.000	102.000	198.000	143.000
148.000	114.000	96.000	83.100	106.000	105.000	151.000	126.000
116.000	59.000	121.000	88.800	77.500	67.300	113.000	111.000
124.000	67.000	99.000	94.000	87.800	96.200	139.000	109.000
76.000	56.000	120.000	91.500	25.700	6.830	135.000	102.000
40.000	61.000	101.000	83.800	89.700	74.800	116.000	110.000
23.000	35.000	118.000	97.800	108.000	84.000	148.000	94.000
27.000	38.000	116.000	100.000	58.600	79.000	221.000	110.000
64.000	79.000	116.000	67.500	91.700	68.500	115.000	101.000
44.000	64.000	87.000	97.500	48.900	20.600	150.000	110.000
59.000	94.000	59.000	45.000	53.500	10.300	135.000	143.000
47.000	96.000	96.000	53.500	41.400	11.800	201.000	115.000
17.000	25.000	163.000	84.800	24.400	7.940	164.000	103.000
67.000	122.000	39.000	73.700	42.300	3.960	130.000	103.000
25.000	43.000	73.000	87.300	34.100	12.700	162.000	169.000
24.000	22.000	45.000	74.800	28.300	7.170	107.000	140.000
65.000	55.000	94.000	100.000	47.000	6.180	78.000	87.100
69.000	42.500	74.000	73.700	30.500	9.470	87.500	134.000
55.000	71.000	70.000	64.800	38.700	13.700	108.000	108.000
39.000	34.800	129.000	119.000	60.900	17.000	126.000	118.000
58.000	50.300	34.000	30.000	51.300	6.810	131.000	109.000
37.000	38.000	42.000	65.900	46.100	24.800	94.400	60.000
32.000	32.000	48.000	34.900	25.000	7.200	87.700	82.900
66.000	53.500	58.000	55.900	29.000	7.900	94.000	99.600
49.000	60.700	30.000	40.100	25.000	6.600	157.000	123.000
40.000	66.500	47.000	48.200	40.000	8.600		
34.000	22.600	35.000	14.800	28.000	5.500		
87.000	61.800	20.000	14.400				

Source: Dr. Bruno Reigner. Used with permission.

46. Yasu et al. (A-35) used noninvasive magnetic resonance spectroscopy to determine the short- and long-term effects of percutaneous transvenous mitral commissurotomy (PTMC) on exercise capacity and metabolic responses of skeletal muscles during exercise. Data were collected on 11 patients (2 males, 9 females) with symptomatic mitral stenosis. Their mean age was 52 years with a standard deviation of 11. Among the data collected were the following measurements on changes in mitral valve area (d-MVA) and peak oxygen consumption (d-Vo_2) 3, 30, and 90 days post-PTMC:

Subject	Days Post-PTMC	d-MVA (cm^2)	d-Vo_2 (ml/kg/min)
1	3	0.64	0.3
2	3	0.76	−0.9
3	3	0.3	1.9
4	3	0.6	−3.1
5	3	0.3	−0.5
6	3	0.4	−2.7
7	3	0.7	1.5
8	3	0.9	1.1
9	3	0.6	−7.4
10	3	0.4	−0.4
11	3	0.65	3.8
1	30	0.53	1.6
2	30	0.6	3.3
3	30	0.4	2.6
4	30	0.5	*
5	30	0.3	3.6
6	30	0.3	0.2
7	30	0.67	4.2
8	30	0.75	3
9	30	0.7	2
10	30	0.4	0.8
11	30	0.55	4.2
1	90	0.6	1.9
2	90	0.6	5.9
3	90	0.4	3.3
4	90	0.6	5
5	90	0.25	0.6
6	90	0.3	2.5
7	90	0.7	4.6
8	90	0.8	4
9	90	0.7	1
10	90	0.38	1.1
11	90	0.53	*

* = Missing data.

Source: Dr. Takanori Yasu. Used with permission.

Exercises for Use with Large Data Sets Available on the Following Website: www.wiley.com/college/daniel

1. Refer to the data for 1050 subjects with cerebral edema (CEREBRAL). Cerebral edema with consequent increased intracranial pressure frequently accompanies lesions resulting from head injury and other conditions that adversely affect the integrity of the brain. Available treatments for cerebral edema vary in effectiveness and undesirable side effects. One such treatment is glycerol, administered either orally or intravenously. Of interest to clinicians is the relationship between intracranial pressure and glycerol plasma concentration. Suppose you are a statistical consultant with a research team investigating the relationship between these two variables. Select a simple random sample from the population and perform the analysis that you think would be useful to the researchers. Present your findings and conclusions in narrative form and illustrate with graphs where appropriate. Compare your results with those of your classmates.

2. Refer to the data for 1050 subjects with essential hypertension (HYPERTEN). Suppose you are a statistical consultant to a medical research team interested in essential hypertension. Select a simple random sample from the population and perform the analyses that you think would be useful to the researchers. Present your findings and conclusions in narrative form and illustrate with graphs where appropriate. Compare your results with those of your classmates. Consult with your instructor regarding the size of sample you should select.

3. Refer to the data for 1200 patients with rheumatoid arthritis (CALCIUM). One hundred patients received the medicine at each dose level. Suppose you are a medical researchers wishing to gain insight into the nature of the relationship between dose level of prednisolone and total body calcium. Select a simple random sample of three patients from each dose level group and do the following.

   (a) Use the total number of pairs of observations to obtain the least-squares equation describing the relationship between dose level (the independent variable) and total body calcium.

   (b) Draw a scatter diagram of the data and plot the equation.

   (c) Compute r and test for significance at the .05 level. Find the p value.

   (d) Compare your results with those of your classmates.

REFERENCES

Methodology References

1. FREDERICK HARTWIG with BRAIN E. DEARING, *Exploratory Data Analysis,* Sage Publications, Beverly Hills, 1979.
2. IAIN M. JOHNSTONE and PAUL F. VELLEMAN, "The Resistant Line and Related Regression Methods," *Journal of the American Statistical Association, 80* (1985), 1041–1054.
3. DONALD R. MCNEIL, *Interactive Data Analysis: A Practical Primer,* Wiley, New York, 1977.
4. PAUL F. VELLEMAN and DAVID C. HOAGLIN, *Applications, Basics, and Computing of Exploratory Data Analysis,* Duxbury, Belmont, CA, 1981.
5. R. A. FISHER, "On the Probable Error of a Coefficient of Correlation Deduced from a Small Sample," *Metron, 1* (1921), 3–21.
6. H. HOTELLING, "New Light on the Correlation Coefficient and Its Transforms," *Journal of the Royal Statistical Society, Series B, 15* (1953), 193–232.
7. TERRY E. DIELMAN, *Applied Regression Analysis for Business and Economics,* Second Edition, Duxbury, Belmont, CA, 1996.
8. RONALD R. HOCKING, *Methods and Applications of Linear Models: Regression and the Analysis of Variance,* Wiley, New York, 1996.

9. William Mendenhall and Terry Sincich, *A Second Course in Statistics: Regression Analysis,* Fifth Edition, Prentice Hall, Upper Saddle River, NJ, 1996.
10. John Neter, Michael H. Kutner, Christopher J. Nachtsheim, and William Wasserman, *Applied Linear Regression Models,* Third Edition, Irwin, Chicago, 1996.

Applications References

A-1. Jean-Pierre Després, Denis Prud'homme, Marie-Christine Pouliot, Angelo Tremblay, and Claude Bouchard, "Estimation of Deep Abdominal Adipose-Tissue Accumulation from Simple Anthropometric Measurements in Men," *American Journal of Clinical Nutrition, 54* (1991), 471–477.

A-2. Mori J. Krantz, Ilana B. Kutinsky, Alastair D. Roberston, and Philip S. Mehler, "Dose-Related Effects of Methadone on QT Prolongation in a Series of Patients with Torsade de Pointes," *Pharmacotherapy, 23* (2003), 802–805.

A-3. Robert A. Reiss, Curtis E. Haas, Deborah L. Griffis, Bernadette Porter, and Mary Ann Tara, "Point-of-Care Versus Laboratory Monitoring of Patients Receiving Different Anticoagulant Therapies," *Pharmacotherapy, 22* (2002), 677–685.

A-4. Robert B. Parker, Ryan Yates, Judith E. Soberman, and Casey Laizure, "Effects of Grapefruit Juice on Intestial P-glycoprotein: Evaluation Using Digoxin in Humans," *Pharmacotherapy, 23* (2003), 979–987.

A-5. R. B. Evans, W. Gordon, and M. Conzemius, "Effect of Velocity on Ground Reaction Forces in Dogs with Lameness Attributable to Tearing of the Cranial Cruciate Ligament," *American Journal of Veterinary Research, 64* (2003), 1479–1481.

A-6. David Krieser, Andrew Rosenberg, and Gad Kainer, "The Relationship Between Serum Creatinine, Serum Cystatin C and Glomerular Filtration Rate in Pediatric Renal Transplant Recipients: A Pilot Study," *Pediatric Transplantation, 6* (2002), 392–395.

A-7. Olga Kwast-Rabben, Rolf Libelius, and Hannu Heikkilä, "Somatosensory Evoked Potentials Following Stimulation of Digital Nerves," *Muscle and Nerve, 26* (2002), 533–538.

A-8. Geri R. Brown and Kim Persley, "Hepatitis A Epidemic in the Elderly," *Southern Medical Journal, 95* (2002), 826–833.

A-9. Ann E. Tuzson, Kevin P. Granata, and Mark F. Abel, "Spastic Velocity Threshold Constrains Functional Performance in Cerebral Palsy," *Archives of Physical Medicine and Rehabilitation, 84* (2003), 1363–1368.

A-10. M. Butz, K. H. Wollinsky, U. Widemuth-Catrinescu, A. Sperfeld, S. Winter, H. H. Mehrkens, A. C. Ludolph, and H. Schreiber, "Longitudinal Effects of Noninvasive Positive-Pressure Ventilation in Patients with Amyotrophic Lateral Sclerosis," *American Journal of Medical Rehabilitation, 82* (2003), 597–604.

A-11. Karie E. Scrogin, Daniel C. Hatton, and David A. McCarron, "The Interactive Effects of Dietary Sodium Chloride and Calcium on Cardiovascular Stress Responses," *American Journal of Physiology (Regulatory Integrative Comp. Physiol. 30), 261* (1991), R945–R949.

A-12. Dean Parmalee, Data analyzed by the Wright State University Statistical Consulting Center, Wright State University, Dayton, OH (2003).

A-13. Maria Mathias, Data analyzed by the Wright State University Statistical Consulting Center, Wright State University, Dayton, OH (2001).

A-14. Darren M. Triller, Steven L. Clause, and Christopher Domarew, "Analysis of Medication Management Activities for Home-Dwelling Patients," *American Journal of Health-System Pharmacy, 59* (2002), 2356–2359.

A-15. Maxime Lamarre-Cliche, Yves Lacourcière, Jacques de Champlain, Luc Poirier, and Pierre Larochelle, "Does QTc Interval Predict the Response to Beta-Blockers and Calcium Channel Blockers in Hypertensives?" *Heart and Disease, 5* (2003), 244–252.

A-16. Jean D. Skinner, Paula Ziegler, and Michael Ponza, "Transitions in Infants' and Toddlers' Beverage Patterns," *Journal of the American Dietetic Association, Supplement, 104* (2004), 45–50.

A-17. Neung Hwa Park, Ji Kang Park, Younghee Choi, Cheol-In Yoo, Choong Ryeol Lee, Hun Lee, Hyo Kyung Kim, Sung-Ryul Kim, Tae-Heum Jeong, Jungsun Park, Chung Sik Yoon, and Yangho Kim, "Whole Blood Manganese Correlates with High Signal Intensities on T-Weighted MRI in Patients with Liver Cirrhosis," *NeuroToxicology, 24* (2003), 909–915.

A-18. Philippe de Moerloose, Sylvie Desmarais, Henri Bounameaux, Guido Reber, Arnaud Perrier, Georges Dupuy, and Jean-Louis Pittet, "Contribution of a New, Rapid, Individual and Quantitative Automated D-Dimer ELISA to Exclude Pulmonary Embolism," *Thrombosis and Haemostasis, 75* (1996), 11–13.

A-19. ILPO T. HUHTANIEMI, ANNE-MAARIT HAAVISTO, RAIJA ANTTILA, MARTTI A. SIIMES, and LEO DUNKEL, "Sensitive Immunoassay and *In Vitro* Bioassay Demonstrate Constant Bioactive/Immunoreactive Ratio of Luteinizing Hormone in Healthy Boys During the Pubertal Maturation," *Pediatric Research, 39* (1996), 180–184.

A-20. YONG-KWEI TSAU, JI-NAN SHEU, CHIUNG-HUI CHEN, RU-JENG TENG, and HUI-CHI CHEN, "Decreased Urinary Epidermal Growth Factor in Children with Acute Renal Failure: Epidermal Growth Factor/Creatinine Ratio Not a Reliable Parameter for Urinary Epidermal Growth Factor Excretion," *Pediatric Research, 39* (1996), 20–24.

A-21. A. USAJ and V. STARC, "Blood pH and Lactate Kinetics in the Assessment of Running Endurance," *International Journal of Sports Medicine, 17* (1996), 34–40.

A-22. PAMELA BEAN, MARY SUSAN SUTPHIN, PATRICIA NECESSARY, MELKON S. AGOPIAN, KARSTEN LIEGMANN, CARL LUDVIGSEN, and JAMES B. PETER, "Carbohydrate-Deficient Transferrin Evaluation in Dry Blood Spots," *Alcoholism: Clinical and Experimental Research, 20* (1996), 56–60.

A-23. JOHJI KATO, KOHJI KOBAYASHI, TAKUMA ETOH, MIHO TANAKA, KAZUO KITAMURA, TAKUROH IMAMURA, YASUSHI KOIWAYA, KENJI KANGAWA, and TANENAO ETO, "Plasma Adrenomedullin Concentration in Patients with Heart Failure," *Journal of Clinical Endocrinology and Metabolism, 81* (1996), 180–183.

A-24. B. E. GOLDEN, P. A. CLOHESSY, G. RUSSELL, and M. K. FAGERHOL, "Calprotectin as a Marker of Inflammation in Cystic Fibrosis," *Archives of Disease in Childhood, 74* (1996), 136–139.

A-25. ARTHUR F. GELB, JAMES C. HOGG, NESTOR L. MÜLLER, MARK J. SCHEIN, JOSEPH KUEI, DONALD P. TASHKIN, JOEL D. EPSTEIN, JOZEF KOLLIN, ROBERT H. GREEN, NOE ZAMEL, W. MARK ELLIOTT, and LIDA HADJIAGHAI, "Contribution of Emphysema and Small Airways in COPD," *Chest, 109* (1996), 353–359.

A-26. AGNES M. WITTEMAN, STEVEN O. STAPEL, GERRARD J. PERDOK, DEMAN H. S. SJAMSOEDIN, HENK M. JANSEN, ROB C. AALBERSE, and JARING S. VAN DER ZEE, "The Relationship Between RAST and Skin Test Results in Patients with Asthma or Rhinitis: A Quantitative Study with Purified Major Allergens," *Journal of Allergy and Clinical Immunology, 97* (1996), 16–25.

A-27. MARIANNE GARLAND, HAZEL H. SZETO, SALHA S. DANIEL, PAMELA J. TROPPER, MICHAEL M. MYERS, and RAYMOND I. STARK, "Zidovudine Kinetics in the Pregnant Baboon," *Journal of Acquired Immune Deficiency Syndromes and Human Retrovirology, 11* (1996), 117–127.

A-28. A. HALLIGAN, A. SHENNAN, P. C. LAMBERT, M. DE SWIET, and D. J. TAYLOR, "Diurnal Blood Pressure Difference in the Assessment of Preeclampsia," *Obstetrics & Gynecology, 87* (1996), 205–208.

A-29. Jay W. Marks, George G. Bonorris, and Leslie J. Schoenfield, "Effects of Ursodiol or Ibuprofen on Contraction of Gallbladder and Bile Among Obese Patients During Weight Loss," *Digestive Diseases and Sciences, 41* (1996), 242–249.

A-30. D. J. PEACOCK, P. EGGER, P. TAYLOR, M. I. D. CAWLEY, and C. COOPER, "Lateral Bone Density Measurements in Osteoarthritis of the Lumbar Spine," *Annals of the Rheumatic Diseases, 55* (1996), 196–198.

A-31. R. P. SLOAN, P. A. SHAPIRE, E. BAGIELLA, J. T. BIGGER, E. S. LO, and J. M. GORMAN, "Relationships Between Circulating Catecholamines and Low Frequency Heart Period Variability as Indices of Cardiac Sympathetic Activity During Mental Stress," *Psychosomatic Medicine, 58* (1996) 25–31.

A-32. ZUKAÏ CHATI, FAÏEZ ZANNAD, CLAUDE JEANDEL, BRIGITTE LHERBIER, JEAN-MARIE ESCANYE, JACQUES ROBERT, and ETIENNE ALIOT, "Physical Deconditioning May Be a Mechanism for the Skeletal Muscle Energy Phosphate Metabolism Abnormalities in Chronic Heart Failure," *American Heart Journal, 131* (1996), 560–566.

A-33. MAGDALENA CZADER, JOANNA MAZUR, MIKAEL PETTERSSON, JAN LILIEMARK, MATS STRÖMBERG, BIRGER CHRISTENSSON, BERNARD TRIBUKAIT, GERT AUER, ÅKE ÖST, and ANNA PORWIT, "Prognostic Significance of Proliferative and Apoptotic Fractions in Low Grade Follicle Center Cell-Derived Non-Hodgkin's Lymphomas," *Cancer, 77* (1996), 1180–1188.

A-34. B. G. REIGNER and H. A. WELKER, "Factors Influencing Elimination and Distribution of Fleroxacin: Metaanalysis of Individual Data from 10 Pharmacokinetic Studies," *Antimicrobial Agents and Chemotherapy, 40* (1996), 575–580.

A-35. TAKANORI YASU, TAKA'AKI KATSUKI, NOBUHIRO OHMURA, IKUKO NAKADA, MAFUMI OWA, MIKIHISA FUJII, AKIRA SAKAGUCHI, and MUNEYASU SAITO, "Delayed Improvement in Skeletal Muscle Metabolism and Exercise Capacity in Patients with Mitral Stenosis Following Immediate Hemodynamic Amelioration by Percutaneous Transvenous Mitral Commissurotomy," *American Journal of Cardiology, 77* (1996), 492–497.

CHAPTER **10**

STATISTICAL INFERENCE AND THE RELATIONSHIPS AMONG THREE OR MORE VARIABLES

CHAPTER OVERVIEW

This chapter provides extensions of the simple linear regression and bivariate correlation models discussed in Chapter 9. The concepts and techniques discussed here are useful when the researcher wishes to consider simultaneously the relationships among more than two variables. Although the concepts, computations, and interpretations associated with analysis of multiple-variable data may seem complex, they are natural extensions of material explored in previous chapters.

TOPICS

LEARNING OUTCOMES

After studying this chapter, the student will

1. understand how to include more than one independent variable in a regression equation.
2. be able to obtain a multiple regression model and use it to make predictions.
3. be able to evaluate the multiple regression coefficients and the suitability of the regression model.
4. understand how to calculate and interpret multiple, bivariate, and partial correlation coefficients.

10.1 INTRODUCTION

In Chapter 9 we explored the concepts and techniques for analyzing and making use of the linear relationship between two variables. We saw that this analysis may lead to an equation that can be used to predict the value of some dependent variable given the value of an associated independent variable.

Intuition tells us that, in general, we ought to be able to improve our predicting ability by including more independent variables in such an equation. For example, a researcher may find that intelligence scores of individuals may be predicted from physical factors such as birth order, birth weight, and length of gestation along with certain hereditary and external environmental factors. Length of stay in a chronic disease hospital may be related to the patient's age, marital status, sex, and income, not to mention the obvious factor of diagnosis. The response of an experimental animal to some drug may depend on the size of the dose and the age and weight of the animal. A nursing supervisor may be interested in the strength of the relationship between a nurse's performance on the job, score on the state board examination, scholastic record, and score on some achievement or aptitude test. Or a hospital administrator studying admissions from various communities served by the hospital may be interested in determining what factors seem to be responsible for differences in admission rates.

The concepts and techniques for analyzing the associations among several variables are natural extensions of those explored in the previous chapters. The computations, as one would expect, are more complex and tedious. However, as is pointed out in Chapter 9, this presents no real problem when a computer is available. It is not unusual to find researchers investigating the relationships among a dozen or more variables. For those who have access to a computer, the decision as to how many variables to include in an analysis is based not on the complexity and length of the computations but on such considerations as their meaningfulness, the cost of their inclusion, and the importance of their contribution.

In this chapter we follow closely the sequence of the previous chapter. The regression model is considered first, followed by a discussion of the correlation model. In considering the regression model, the following points are covered: a description of the model, methods for obtaining the regression equation, evaluation of the equation, and the uses that may be made of the equation. In both models the possible inferential procedures and their underlying assumptions are discussed.

10.2 THE MULTIPLE LINEAR REGRESSION MODEL

In the multiple regression model we assume that a linear relationship exists between some variable Y, which we call the dependent variable, and k independent variables, $X_1, X_2, \ldots, X_k$. The independent variables are sometimes referred to as *explanatory variables,* because of their use in explaining the variation in Y. They are also called *predictor variables,* because of their use in predicting Y.

Assumptions The assumptions underlying multiple regression analysis are as follows.

1. The X_i are nonrandom (fixed) variables. This assumption distinguishes the multiple regression model from the multiple correlation model, which will be presented in Section 10.6. This condition indicates that any inferences that are drawn from sample data apply only to the set of X values observed and not to some larger collection of X's. Under the regression model, correlation analysis is not meaningful. Under the correlation model to be presented later, the regression techniques that follow may be applied.
2. For each set of X_i values there is a subpopulation of Y values. To construct certain confidence intervals and test hypotheses, it must be known, or the researcher must be willing to assume, that these subpopulations of Y values are normally distributed. Since we will want to demonstrate these inferential procedures, the assumption of normality will be made in the examples and exercises in this chapter.
3. The variances of the subpopulations of Y are all equal.
4. The Y values are independent. That is, the values of Y selected for one set of X values do not depend on the values of Y selected at another set of X values.

The Model Equation The assumptions for multiple regression analysis may be stated in more compact fashion as

$$y_j = \beta_0 + \beta_1 x_{1j} + \beta_2 x_{2j} + \cdots + \beta_k x_{kj} + \epsilon_j \tag{10.2.1}$$

where y_j is a typical value from one of the subpopulations of Y values; the β_i are called the regression coefficients; $x_{1j}, x_{2j}, \ldots, x_{kj}$ are, respectively, particular values of the independent variables $X_1, X_2, \ldots X_k$; and ϵ_j is a random variable with mean 0 and variance σ^2, the common variance of the subpopulations of Y values. To construct confidence intervals for and test hypotheses about the regression coefficients, we assume that the ϵ_j are normally and independently distributed. The statements regarding ϵ_j are a consequence of the assumptions regarding the distributions of Y values. We will refer to Equation 10.2.1 as the *multiple linear regression model.*

When Equation 10.2.1 consists of one dependent variable and two independent variables, that is, when the model is written

$$y_j = \beta_0 + \beta_1 x_{1j} + \beta_2 x_{2j} + \epsilon_j \tag{10.2.2}$$

a *plane* in three-dimensional space may be fitted to the data points as illustrated in Figure 10.2.1. When the model contains more than two independent variables, it is described geometrically as a *hyperplane.*

In Figure 10.2.1 the observer should visualize some of the points as being located above the plane and some as being located below the plane. The deviation of a point from the plane is represented by

$$\epsilon_j = y_j - \beta_0 - \beta_1 x_{1j} - \beta_2 x_{2j} \tag{10.2.3}$$

In Equation 10.2.2, β_0 represents the point where the plane cuts the Y-axis; that is, it represents the Y-intercept of the plane. β_1 measures the average change in Y for a

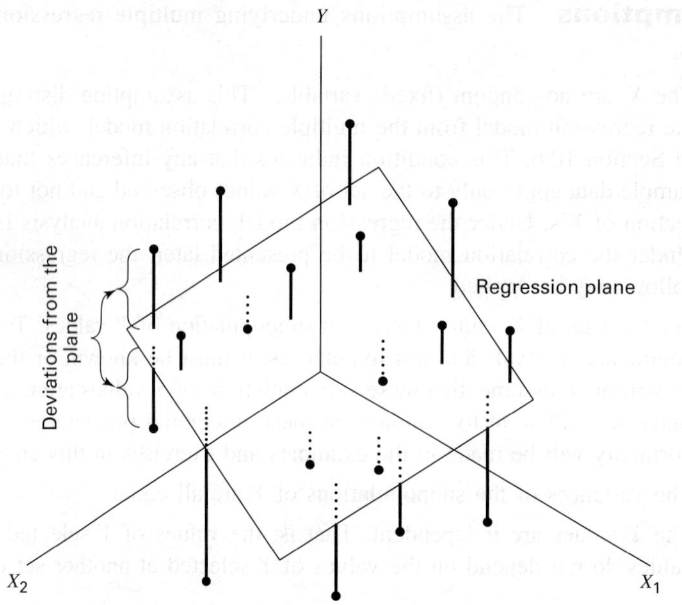

FIGURE 10.2.1 Multiple regression plane and scatter of points.

unit change in X_1 when X_2 remains unchanged, and β_2 measures the average change in Y for a unit change in X_2 when X_1 remains unchanged. For this reason β_1 and β_2 are referred to as *partial regression coefficients.*

10.3 OBTAINING THE MULTIPLE REGRESSION EQUATION

Unbiased estimates of the parameters $\beta_0, \beta_1, \ldots, \beta_k$ of the model specified in Equation 10.2.1 are obtained by the method of least squares. This means that the sum of the squared deviations of the observed values of Y from the resulting regression surface is minimized. In the three-variable case, as illustrated in Figure 10.2.1, the sum of the squared deviations of the observations from the plane are a minimum when β_0, β_1, and β_2 are estimated by the method of least squares. In other words, by the method of least squares, sample estimates of $\beta_0, \beta_1, \ldots, \beta_k$ are selected in such a way that the quantity

$$\sum \epsilon_j^2 = \sum (y_j - \beta_0 - \beta_1 x_{1j} - \beta_2 x_{2j} - \cdots - \beta_k x_{kj})^2$$

is minimized. This quantity, referred to as the sum of squares of the residuals, may also be written as

$$\sum \epsilon_j^2 = \sum (y_j - \hat{y}_j)^2 \tag{10.3.1}$$

indicating the fact that the sum of squares of deviations of the observed values of Y from the values of Y calculated from the estimated equation is minimized.

Estimates of the multiple regression parameters may be obtained by means of arithmetic calculations performed on a handheld calculator. This method of obtaining the estimates is tedious, time-consuming, subject to errors, and a waste of time when a computer is available. Those interested in examining or using the arithmetic approach may consult earlier editions of this text or those by Snedecor and Cochran (1) and Steel and Torrie (2), who give numerical examples for four variables, and Anderson and Bancroft (3), who illustrate the calculations involved when there are five variables. In the following example we use SPSS software to illustrate an interesting graphical summary of sample data collected on three variables. We then use MINITAB to illustrate the application of multiple regression analysis.

EXAMPLE 10.3.1

Researchers Jansen and Keller (A-1) used age and education level to predict the capacity to direct attention (CDA) in elderly subjects. CDA refers to neural inhibitory mechanisms that focus the mind on what is meaningful while blocking out distractions. The study collected information on 71 community-dwelling older women with normal mental status. The CDA measurement was calculated from results on standard visual and auditory measures requiring the inhibition of competing and distracting stimuli. In this study, CDA scores ranged from −7.65 to 9.61 with higher scores corresponding with better attentional functioning. The measurements on CDA, age in years, and education level (years of schooling) for 71 subjects are shown in Table 10.3.1. We wish to obtain the sample multiple regression equation.

TABLE 10.3.1 CDA Scores, Age, and Education Level for 71 Subjects Described in Example 10.3.1

Age	Ed-Level	CDA	Age	Ed-Level	CDA
72	20	4.57	79	12	3.17
68	12	−3.04	87	12	−1.19
65	13	1.39	71	14	0.99
85	14	−3.55	81	16	−2.94
84	13	−2.56	66	16	−2.21
90	15	−4.66	81	16	−0.75
79	12	−2.70	80	13	5.07
74	10	0.30	82	12	−5.86
69	12	−4.46	65	13	5.00
87	15	−6.29	73	16	0.63
84	12	−4.43	85	16	2.62
79	12	0.18	83	17	1.77
71	12	−1.37	83	8	−3.79
76	14	3.26	76	20	1.44
73	14	−1.12	77	12	−5.77
86	12	−0.77	83	12	−5.77
69	17	3.73	79	14	−4.62
66	11	−5.92	69	12	−2.03
65	16	5.74	66	14	−2.22

(*Continued*)

Age	Ed-Level	CDA	Age	Ed-Level	CDA
71	14	2.83	75	12	0.80
80	18	−2.40	77	16	−0.75
81	11	−0.29	78	12	−4.60
66	14	4.44	83	20	2.68
76	17	3.35	85	10	−3.69
70	12	−3.13	76	18	4.85
76	12	−2.14	75	14	−0.08
67	12	9.61	70	16	0.63
72	20	7.57	79	16	5.92
68	18	2.21	75	18	3.63
102	12	−2.30	94	8	−7.07
67	12	1.73	76	18	6.39
66	14	6.03	84	18	−0.08
75	18	−0.02	79	17	1.07
91	13	−7.65	78	16	5.31
74	15	4.17	79	12	0.30
90	15	−0.68			

Source: Debra A. Jansen, Ph.D., R.N. Used with permission.

Prior to analyzing the data using multiple regression techniques, it is useful to construct plots of the relationships among the variables. This is accomplished by making separate plots of each pair of variables, ($X1$, $X2$), ($X1$, Y), and ($X2$, Y). A software package such as SPSS displays each combination simultaneously in a matrix format as shown in Figure 10.3.1. From this figure it is apparent that we should expect a negative

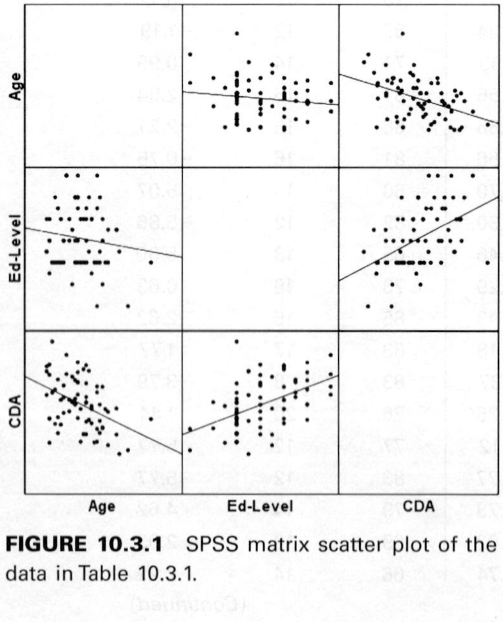

FIGURE 10.3.1 SPSS matrix scatter plot of the data in Table 10.3.1.

Dialog box:

Stat ➤ Regression ➤ Regression
Type *Y* in **Response** and *X1 X2*
in **Predictors.**
Check **Residuals.**
Check **Standard resids.**
Check **OK.**

Session command:

```
MTB > Name C4 = `SRES1'
      C5 = `FITS1' C6 = `RESI1'
MTB > Regress `y' 2 `x1' `x2';
SUBC> SResiduals `SRES1';
SUBC> Fits `FITS1';
SUBC> Constant;
SUBC> Residuals `RESI1'.
```

Output:

Regression Analysis: Y versus X1, X2

```
The regression equation is
Y = 5.49 - 0.184 X1 + 0.611 X2

Predictor        Coef    SE Coef       T       P
Constant        5.494      4.443    1.24   0.220
X1           -0.18412    0.04851   -3.80   0.000
X2             0.6108     0.1357    4.50   0.000

S = 3.134    R-Sq = 37.1%    R-Sq (adj) = 35.2%

Analysis of Variance

Source             DF        SS        MS       F       P
Regression          2    393.39    196.69   20.02   0.000
Residual Error     68    667.97      9.82
Total              70   1061.36

Source             DF    Seq SS
X1                  1    194.24
X2                  1    199.15

Unusual Observations
Obs    X1        Y      Fit   SE Fit   Residual   St Resid
 28    67    9.610    0.487    0.707      9.123      2.99R
 31   102   -2.300   -5.957    1.268      3.657      1.28X
 44    80    5.070   -1.296    0.425      6.366      2.05R
 67    94   -7.070   -6.927    1.159     -0.143     -0.05X

R denotes an observation with a large standardized residual.
X denotes an observation whose X value gives it large influence.
```

FIGURE 10.3.2 MINITAB procedure and output for Example 10.3.1.

```
                    The REG Procedure

                      Model: MODEL1
                 Dependent Variable: CDA

                  Analysis of Variance

                            Sum of           Mean
Source              DF     Squares         Square    F Value    Pr > F

Model                2   393.38832      196.69416      20.02    <.0001
Error               68   667.97084        9.82310
Corrected Total     70  1061.35915

       Root MSE               3.13418    R-Square     0.3706
       Dependent Mean         0.00676    Adj R-Sq     0.3521
       Coeff Var                46360

                   Parameter Estimates

                     Parameter     Standard
Variable      DF      Estimate        Error    t Value    Pr > |t|

Intercept      1       5.49407      4.44297       1.24      0.2205
AGE            1      -0.18412      0.04851      -3.80      0.0003
EDUC           1       0.61078      0.13565       4.50      <.0001
```

FIGURE 10.3.3 SAS® output for Example 10.3.1.

relationship between CDA and Age and a positive relationship between CDA and Ed-Level. We shall see that this is indeed the case when we use MINITAB to analyze the data.

Solution: We enter the observations on age, education level, and CDA in c1 through c3 and name them $X1$, $X2$, and Y, respectively. The MINITAB dialog box and session command, as well as the output, are shown in Figure 10.3.2. We see from the output that the sample multiple regression equation, in the notation of Section 10.2, is

$$\hat{y}_j = 5.49 - .184x_{1j} + .611x_{2j}$$

Other output entries will be discussed in the sections that follow.

The SAS output for Example 10.3.1 is shown in Figure 10.3.3. ■

After the multiple regression equation has been obtained, the next step involves its evaluation and interpretation. We cover this facet of the analysis in the next section.

EXERCISES

Obtain the regression equation for each of the following data sets.

10.3.1 Machiel Naeije (A-2) studied the relationship between maximum mouth opening and measurements of the lower jaw (mandible). He measured the dependent variable, maximum mouth opening (MMO, measured in mm), as well as predictor variables, mandibular length (ML, measured in mm) and angle of rotation of the mandible (RA, measured in degrees) of 35 subjects.

MMO (Y)	ML (X_1)	RA (X_2)	MMO (Y)	ML (X_1)	RA (X_2)
52.34	100.85	32.08	50.82	90.65	38.33
51.90	93.08	39.21	40.48	92.99	25.93
52.80	98.43	33.74	59.68	108.97	36.78
50.29	102.95	34.19	54.35	91.85	42.02
57.79	108.24	35.13	47.00	104.30	27.20
49.41	98.34	30.92	47.23	93.16	31.37
53.28	95.57	37.71	41.19	94.18	27.87
59.71	98.85	44.71	42.76	89.56	28.69
53.32	98.32	33.17	51.88	105.85	31.04
48.53	92.70	31.74	42.77	89.29	32.78
51.59	88.89	37.07	52.34	92.58	37.82
58.52	104.06	38.71	50.45	98.64	33.36
62.93	98.18	43.89	43.18	83.70	31.93
57.62	91.01	41.06	41.99	88.46	28.32
65.64	96.98	41.92	39.45	94.93	24.82
52.85	97.85	35.25	38.91	96.81	23.88
64.43	96.89	45.11	49.10	93.13	36.17
57.25	98.35	39.44			

Source: M. Naeije, D.D.S. Used with permission.

10.3.2 Family caregiving of older adults is more common in Korea than in the United States. Son et al. (A-3) studied 100 caregivers of older adults with dementia in Seoul, South Korea. The dependent variable was caregiver burden as measured by the Korean Burden Inventory (KBI). Scores ranged from 28 to 140, with higher scores indicating higher burden. Explanatory variables were indexes that measured the following:

ADL: total activities of daily living (low scores indicate that the elderly perform activities independently).

MEM: memory and behavioral problems (higher scores indicate more problems).

COG: cognitive impairment (lower scores indicate a greater degree of cognitive impairment).

The reported data are as follows:

KBI (Y)	ADL (X_1)	MEM (X_2)	COG (X_3)	KBI (Y)	ADL (X_1)	MEM (X_2)	COG (X_3)
28	39	4	18	88	76	50	5
68	52	33	9	54	79	44	11
59	89	17	3	73	48	57	9
91	57	31	7	87	90	33	6

(*Continued*)

KBI (Y)	ADL (X_1)	MEM (X_2)	COG (X_3)	KBI (Y)	ADL (X_1)	MEM (X_2)	COG (X_3)
70	28	35	19	47	55	11	20
38	34	3	25	60	83	24	11
46	42	16	17	65	50	21	25
57	52	6	26	57	44	31	18
89	88	41	13	85	79	30	20
48	90	24	3	28	24	5	22
74	38	22	13	40	40	20	17
78	83	41	11	87	35	15	27
43	30	9	24	80	55	9	21
76	45	33	14	49	45	28	17
72	47	36	18	57	46	19	17
61	90	17	0	32	37	4	21
63	63	14	16	52	47	29	3
77	34	35	22	42	28	23	21
85	76	33	23	49	61	8	7
31	26	13	18	63	35	31	26
79	68	34	26	89	68	65	6
92	85	28	10	67	80	29	10
76	22	12	16	43	43	8	13
91	82	57	3	47	53	14	18
78	80	51	3	70	60	30	16
103	80	20	18	99	63	22	18
99	81	20	1	53	28	9	27
73	30	7	17	78	35	18	14
88	27	27	27	112	37	33	17
64	72	9	0	52	82	25	13
52	46	15	22	68	88	16	0
71	63	52	13	63	52	15	0
41	45	26	18	49	30	16	18
85	77	57	0	42	69	49	12
52	42	10	19	56	52	17	20
68	60	34	11	46	59	38	17
57	33	14	14	72	53	22	21
84	49	30	15	95	65	56	2
91	89	64	0	57	90	12	0
83	72	31	3	88	88	42	6
73	45	24	19	81	66	12	23
57	73	13	3	104	60	21	7
69	58	16	15	88	48	14	13
81	33	17	21	115	82	41	13
71	34	13	18	66	88	24	14
91	90	42	6	92	63	49	5
48	48	7	23	97	79	34	3
94	47	17	18	69	71	38	17
57	32	13	15	112	66	48	13
49	63	32	15	88	81	66	1

Source: Gwi-Ryung Son, R.N., Ph.D. Used with permission.

10.3.3 In a study of factors thought to be related to patterns of admission to a large general hospital, an administrator obtained these data on 10 communities in the hospital's catchment area:

Community	Persons per 1000 Population Admitted During Study Period (Y)	Index of Availability of Other Health Services (X_1)	Index of Indigency (X_2)
1	61.6	6.0	6.3
2	53.2	4.4	5.5
3	65.5	9.1	3.6
4	64.9	8.1	5.8
5	72.7	9.7	6.8
6	52.2	4.8	7.9
7	50.2	7.6	4.2
8	44.0	4.4	6.0
9	53.8	9.1	2.8
10	53.5	6.7	6.7
Total	571.6	69.9	55.6

10.3.4 The administrator of a general hospital obtained the following data on 20 surgery patients during a study to determine what factors appear to be related to length of stay:

Postoperative Length of Stay in Days (Y)	Number of Current Medical Problems (X_1)	Preoperative Length of Stay in Days (X_2)
6	1	1
6	2	1
11	2	2
9	1	3
16	3	3
16	1	5
4	1	1
8	3	1
11	2	2
13	3	2
13	1	4
9	1	2
17	3	3
17	2	4
12	4	1
6	1	1
5	1	1

(Continued)

Postoperative Length of Stay in Days (Y)	Number of Current Medical Problems (X_1)	Preoperative Length of Stay in Days (X_2)
12	3	2
8	1	2
9	2	2
Total 208	38	43

10.3.5 A random sample of 25 nurses selected from a state registry yielded the following information on each nurse's score on the state board examination and his or her final score in school. Both scores relate to the nurse's area of affiliation. Additional information on the score made by each nurse on an aptitude test, taken at the time of entering nursing school, was made available to the researcher. The complete data are as follows:

State Board Score (Y)	Final Score (X_1)	Aptitude Test Score (X_2)
440	87	92
480	87	79
535	87	99
460	88	91
525	88	84
480	89	71
510	89	78
530	89	78
545	89	71
600	89	76
495	90	89
545	90	90
575	90	73
525	91	71
575	91	81
600	91	84
490	92	70
510	92	85
575	92	71
540	93	76
595	93	90
525	94	94
545	94	94
600	94	93
625	94	73
Total 13,425	2263	2053

10.3.6 The following data were collected on a simple random sample of 20 patients with hypertension. The variables are

Y = mean arterial blood pressure (mm Hg)
X_1 = age (years)
X_2 = weight (kg)
X_3 = body surface area (sq m)
X_4 = duration of hypertension (years)
X_5 = basal pulse (beatsthn/min)
X_6 = measure of stress

Patient	Y	X_1	X_2	X_3	X_4	X_5	X_6
1	105	47	85.4	1.75	5.1	63	33
2	115	49	94.2	2.10	3.8	70	14
3	116	49	95.3	1.98	8.2	72	10
4	117	50	94.7	2.01	5.8	73	99
5	112	51	89.4	1.89	7.0	72	95
6	121	48	99.5	2.25	9.3	71	10
7	121	49	99.8	2.25	2.5	69	42
8	110	47	90.9	1.90	6.2	66	8
9	110	49	89.2	1.83	7.1	69	62
10	114	48	92.7	2.07	5.6	64	35
11	114	47	94.4	2.07	5.3	74	90
12	115	49	94.1	1.98	5.6	71	21
13	114	50	91.6	2.05	10.2	68	47
14	106	45	87.1	1.92	5.6	67	80
15	125	52	101.3	2.19	10.0	76	98
16	114	46	94.5	1.98	7.4	69	95
17	106	46	87.0	1.87	3.6	62	18
18	113	46	94.5	1.90	4.3	70	12
19	110	48	90.5	1.88	9.0	71	99
20	122	56	95.7	2.09	7.0	75	99

10.4 EVALUATING THE MULTIPLE REGRESSION EQUATION

Before one uses a multiple regression equation to predict and estimate, it is desirable to determine first whether it is, in fact, worth using. In our study of simple linear regression we have learned that the usefulness of a regression equation may be evaluated by a consideration of the sample coefficient of determination and estimated slope. In evaluating a multiple regression equation we focus our attention on the *coefficient of multiple determination* and the partial regression coefficients.

The Coefficient of Multiple Determination In Chapter 9 the coefficient of determination is discussed in considerable detail. The concept extends logically to the multiple regression case. The total variation present in the Y values may be partitioned into two components—the explained variation, which measures the amount of the total variation that is explained by the fitted regression surface, and the unexplained variation, which is that part of the total variation not explained by fitting the regression surface. The measure of variation in each case is a sum of squared deviations. The total variation is the sum of squared deviations of each observation of Y from the mean of the observations and is designated by $\Sigma(y_j - \bar{y})^2$ or SST. The explained variation, designated $\Sigma(\hat{y}_j - \bar{y})^2$, is the sum of squared deviations of the calculated values from the mean of the observed Y values. This sum of squared deviations is called the *sum of squares due to regression* (SSR). The unexplained variation, written as $\Sigma(y_j - \hat{y}_j)^2$, is the sum of squared deviations of the original observations from the calculated values. This quantity is referred to as the *sum of squares about regression* or the *error sum of squares* (SSE). We may summarize the relationship among the three sums of squares with the following equation:

$$\Sigma(y_j - \bar{y})^2 = \Sigma(\hat{y}_j - \bar{y})^2 + \Sigma(y_j - \hat{y}_j)^2 \qquad (10.4.1)$$

$$SST = SSR + SSE$$

$$\text{total sum of squares} = \text{explained (regression) sum of squares} + \text{unexplained (error) sum of squares}$$

The coefficient of multiple determination, $R^2_{y.12\ldots k}$ is obtained by dividing the explained sum of squares by the total sum of squares. That is,

$$R^2_{y.12\ldots k} = \frac{\Sigma(\hat{y}_j - \bar{y})^2}{\Sigma(y_j - \bar{y})^2} = \frac{SSR}{SST} \qquad (10.4.2)$$

The subscript $y.12\ldots k$ indicates that in the analysis Y is treated as the dependent variable and the X variables from X_1 through X_k are treated as the independent variables. The value of $R^2_{y.12\ldots k}$ indicates what proportion of the total variation in the observed Y values is explained by the regression of Y on $X_1, X_2, \ldots, X_k$. In other words, we may say that $R^2_{y.12\ldots k}$ is a measure of the goodness of fit of the regression surface. This quantity is analogous to r^2, which was computed in Chapter 9.

EXAMPLE 10.4.1

Refer to Example 10.3.1. Compute $R^2_{y.12}$.

Solution: For our illustrative example we have in Figure 10.3.1

$$SST = 1061.36$$

$$SSR = 393.39$$

$$SSE = 667.97$$

$$R^2_{y.12} = \frac{393.39}{1061.36} = .3706 \approx .371$$

We say that about 37.1 percent of the total variation in the Y values is explained by the fitted regression plane, that is, by the linear relationship with age and education level. ■

Testing the Regression Hypothesis To determine whether the overall regression is significant (that is, to determine whether $R^2_{y.12}$ is significant), we may perform a hypothesis test as follows.

1. **Data.** The research situation and the data generated by the research are examined to determine if multiple regression is an appropriate technique for analysis.
2. **Assumptions.** We assume that the multiple regression model and its underlying assumptions as presented in Section 10.2 are applicable.
3. **Hypotheses.** In general, the null hypothesis is H_0: $\beta_1 = \beta_2 = \beta_3 = \cdots = \beta_k = 0$ and the alternative is H_A: not all $\beta_i = 0$. In words, the null hypothesis states that all the independent variables are of no value in explaining the variation in the Y values.
4. **Test statistic.** The appropriate test statistic is V.R., which is computed as part of an analysis of variance. The general ANOVA table is shown as Table 10.4.1. In Table 10.4.1, *MSR* stands for mean square due to regression and *MSE* stands for mean square about regression or, as it is sometimes called, the error mean square.
5. **Distribution of test statistic.** When H_0 is true and the assumptions are met, V.R. is distributed as F with k and $n - k - 1$ degrees of freedom.
6. **Decision rule.** Reject H_0 if the computed value of V.R. is equal to or greater than the critical value of F.
7. **Calculation of test statistic.** See Table 10.4.1.
8. **Statistical decision.** Reject or fail to reject H_0 in accordance with the decision rule.
9. **Conclusion.** If we reject H_0, we conclude that, in the population from which the sample was drawn, the dependent variable is linearly related to the independent variables as a group. If we fail to reject H_0, we conclude that, in the population from which our sample was drawn, there may be no linear relationship between the dependent variable and the independent variables as a group.
10. ***p* value.** We obtain the p value from the table of the F distribution.

We illustrate the hypothesis testing procedure by means of the following example.

TABLE 10.4.1 ANOVA Table for Multiple Regression

Source	SS	d.f.	MS	V.R.
Due to regression	SSR	k	$MSR = SSR/k$	MSR/MSE
About regression	SSE	$n - k - 1$	$MSE = SSE/(n - k - 1)$	
Total	SST	$n - 1$		

EXAMPLE 10.4.2

We wish to test the null hypothesis of no linear relationship among the three variables discussed in Example 10.3.1: CDA score, age, and education level.

Solution:

1. **Data.** See the description of the data given in Example 10.3.1.
2. **Assumptions.** We assume that the assumptions discussed in Section 10.2 are met.
3. **Hypotheses.**

$$H_0: = \beta_1 = \beta_2 = 0$$

$$H_A: = \text{not all } \beta_i = 0$$

4. **Test statistic.** The test statistic is V.R.
5. **Distribution of test statistic.** If H_0 is true and the assumptions are met, the test statistic is distributed as F with 2 numerator and 68 denominator degrees of freedom.
6. **Decision rule.** Let us use a significance level of $\alpha = .01$. The decision rule, then, is reject H_0 if the computed value of V.R. is equal to or greater than 4.95 (obtained by interpolation).
7. **Calculation of test statistic.** The ANOVA for the example is shown in Figure 10.3.1, where we see that the computed value of V.R. is 20.02.
8. **Statistical decision.** Since 20.02 is greater than 4.95, we reject H_0.
9. **Conclusion.** We conclude that, in the population from which the sample came, there is a linear relationship among the three variables.
10. ***p* value.** Since 20.02 is greater than 5.76, the p value for the test is less than .005. ■

Inferences Regarding Individual β's Frequently, we wish to evaluate the strength of the linear relationship between Y and the independent variables individually. That is, we may want to test the null hypothesis that $\beta_i = 0$ against the alternative $\beta_i \neq 0$ $(i = 1, 2, \ldots, k)$. The validity of this procedure rests on the assumptions stated earlier: that for each combination of X_i values there is a normally distributed subpopulation of Y values with variance σ^2.

Hypothesis Tests for the β_i To test the null hypothesis that β_i is equal to some particular value, say, β_{i0}, the following t statistic may be computed:

$$t = \frac{\hat{\beta}_i - \beta_{i0}}{s_{\hat{\beta}_i}} \tag{10.4.3}$$

where the degrees of freedom are equal to $n - k - 1$, and $s_{\hat{\beta}_i}$ is the standard deviation of the $\hat{\beta}_i$.

The standard deviations of the $\hat{\beta}_i$ are given as part of the output from most computer software packages that do regression analysis.

EXAMPLE 10.4.3

Let us refer to Example 10.3.1 and test the null hypothesis that age (years) is irrelevant in predicting the capacity to direct attention (CDA).

Solution:

1. **Data.** See Example 10.3.1.
2. **Assumptions.** See Section 10.2.
3. **Hypotheses.**

$$H_0: \beta_1 = 0$$
$$H_A: \beta_1 \neq 0$$
$$\text{Let } \alpha = .05$$

4. **Test statistic.** See Equation 10.4.3.
5. **Distribution of test statistic.** When H_0 is true and the assumptions are met, the test statistic is distributed as Student's t with 68 degrees of freedom.
6. **Decision rule.** Reject H_0 if the computed t is either greater than or equal to 1.9957 (obtained by interpolation) or less than or equal to -1.9957.
7. **Calculation of test statistic.** By Equation 10.4.3 and data from Figure 10.3.1 we compute

$$t = \frac{\hat{\beta}_1 - 0}{s_{\hat{\beta}_1}} = \frac{-.18412}{.04851} = -3.80$$

8. **Statistical decision.** The null hypothesis is rejected since the computed value of t, -3.80, is less than -1.9957.
9. **Conclusion.** We conclude, then, that there is a linear relationship between age and CDA in the presence of education level.
10. ***p* value.** For this test, $p < 2(.005) = .01$ because $-3.80 < -2.6505$ (obtained by interpolation). ■

Now, let us perform a similar test for the second partial regression coefficient, β_2:

$$H_0: \beta_2 = 0$$
$$H_A: \beta_2 \neq 0$$
$$\alpha = .05$$
$$t = \frac{\hat{\beta}_2 - 0}{s_{\hat{\beta}_2}} = \frac{.6108}{.1357} = 4.50$$

In this case also the null hypothesis is rejected, since 4.50 is greater than 1.9957. We conclude that there is a linear relationship between education level and CDA in the presence age, and that education level, used in this manner, is a useful variable for predicting CDA. [For this test, $p < 2(.005) = .01$.]

Confidence Intervals for the β_i When the researcher has been led to conclude that a partial regression coefficient is not 0, he or she may be interested in obtaining a confidence interval for this β_i. Confidence intervals for the β_i may be constructed in the usual way by using a value from the t distribution for the reliability factor and standard errors given above.

A $100(1-\alpha)$ percent confidence interval for β_i is given by

$$\hat{\beta}_i \pm t_{1-(\alpha/2),n-k-1} s_{\hat{\beta}_i}$$

For our illustrative example we may compute the following 95 percent confidence intervals for β_1 and β_2.

The 95 percent confidence interval for β_1 is

$$-.18412 \pm 1.9957(.04851)$$
$$-.18412 \pm .0968$$
$$-.28092, -.08732$$

The 95 percent confidence interval for β_2 is

$$.6108 \pm (1.9957)(.1357)$$
$$.6108 \pm .2708$$
$$.3400, .8816$$

We may give these intervals the usual probabilistic and practical interpretations. We are 95 percent confident, for example, that β_2 is contained in the interval from .3400 to .8816 since, in repeated sampling, 95 percent of the intervals that may be constructed in this manner will include the true parameter.

Some Precautions One should be aware of the problems involved in carrying out multiple hypothesis tests and constructing multiple confidence intervals from the same sample data. The effect on α of performing multiple hypothesis tests from the same data is discussed in Section 8.2. A similar problem arises when one wishes to construct confidence intervals for two or more partial regression coefficients. The intervals will not be independent, so that the tabulated confidence coefficient does not, in general, apply. In other words, all such intervals would not be $100(1-\alpha)$ percent confidence intervals.

In order to maintain approximate $100(1-\alpha)$ confidence intervals for partial regression coefficients, adjustments must be made to the calculation of errors in the previous equations. These adjustments are sometimes called *family-wise error rates*, and can be found in many computer software packages. The topic is discussed in detail by Kutner, et al. (4).

Another problem sometimes encountered in the application of multiple regression is an apparent incompatibility in the results of the various tests of significance that one may perform. In a given problem for a given level of significance, one or the other of the following situations may be observed.

1. R^2 and all $\hat{\beta}_i$ significant
2. R^2 and some but not all $\hat{\beta}_i$ significant
3. R^2 significant but none of the $\hat{\beta}_i$ significant
4. All $\hat{\beta}_i$ significant but not R^2
5. Some $\hat{\beta}_i$ significant, but not all nor R^2
6. Neither R^2 nor any $\hat{\beta}_i$ significant

Notice that situation 1 exists in our illustrative example, where we have a significant R^2 and two significant regression coefficients. This situation does not occur in all cases. In fact, situation 2 is very common, especially when a large number of independent variables have been included in the regression equation.

EXERCISES

10.4.1 Refer to Exercise 10.3.1. (a) Calculate the coefficient of multiple determination; (b) perform an analysis of variance; (c) test the significance of each $\hat{\beta}_i (i > 0)$. Let $\alpha = .05$ for all tests of significance and determine the p value for all tests; (d) construct a 95 percent confidence interval for each significant sample slope.

10.4.2 Refer to Exercise 10.3.2. Do the analysis suggested in Exercise 10.4.1.

10.4.3 Refer to Exercise 10.3.3. Do the analysis suggested in Exercise 10.4.1.

10.4.4 Refer to Exercise 10.3.4. Do the analysis suggested in Exercise 10.4.1.

10.4.5 Refer to Exercise 10.3.5. Do the analysis suggested in Exercise 10.4.1.

10.4.6 Refer to Exercise 10.3.6. Do the analysis suggested in Exercise 10.4.1.

10.5 USING THE MULTIPLE REGRESSION EQUATION

As we learned in the previous chapter, a regression equation may be used to obtain a computed value of Y, $\hat{y}$ when a particular value of X is given. Similarly, we may use our multiple regression equation to obtain a $\hat{y}$ value when we are given particular values of the two or more X variables present in the equation.

Just as was the case in simple linear regression, we may, in multiple regression, interpret a $\hat{y}$ value in one of two ways. First we may interpret $\hat{y}$ as an estimate of the mean of the subpopulation of Y values assumed to exist for particular combinations of X_i values. Under this interpretation $\hat{y}$ is called an *estimate,* and when it is used for this purpose, the equation is thought of as an *estimating equation.* The second interpretation of

$\hat{y}$ is that it is the value Y is most likely to assume for given values of the X_i. In this case $\hat{y}$ is called the *predicted value* of Y, and the equation is called a *prediction equation*. In both cases, intervals may be constructed about the $\hat{y}$ value when the normality assumption of Section 10.2 holds true. When $\hat{y}$ is interpreted as an estimate of a population mean, the interval is called a *confidence interval,* and when $\hat{y}$ is interpreted as a predicted value of Y, the interval is called a *prediction interval*. Now let us see how each of these intervals is constructed.

The Confidence Interval for the Mean of a Subpopulation of *Y* Values Given Particular Values of the X_i We have seen that a $100(1 - \alpha)$ percent confidence interval for a parameter may be constructed by the general procedure of adding to and subtracting from the estimator a quantity equal to the reliability factor corresponding to $1 - \alpha$ multiplied by the standard error of the estimator. We have also seen that in multiple regression the estimator is

$$\hat{y}_j = \hat{\beta}_0 + \hat{\beta}_1 x_{1j} + \hat{\beta}_2 x_{2j} + \cdots + \hat{\beta}_k x_{k_j} \tag{10.5.1}$$

If we designate the standard error of this estimator by $s_{\hat{y}}$, the $100(1 - \alpha)$ percent confidence interval for the mean of Y, given specified X_i is as follows:

$$\hat{y}_j \pm t_{(1-\alpha/2),n-k-1} s_{\hat{y}_j} \tag{10.5.2}$$

The Prediction Interval for a Particular Value of *Y* Given Particular Values of the X_i When we interpret $\hat{y}$ as the value Y is most likely to assume when particular values of the X_i are observed, we may construct a prediction interval in the same way in which the confidence interval was constructed. The only difference in the two is the standard error. The standard error of the prediction is slightly larger than the standard error of the estimate, which causes the prediction interval to be wider than the confidence interval.

If we designate the standard error of the prediction by $s'_{\hat{y}}$, the $100(1 - \alpha)$ percent prediction interval is

$$\hat{y}_j \pm t_{(1-\alpha/2),n-k-1} s'_{\hat{y}_j} \tag{10.5.3}$$

The calculations of $s_{\hat{y}_j}$ and $s'_{\hat{y}_j}$ in the multiple regression case are complicated and will not be covered in this text. The reader who wishes to see how these statistics are calculated may consult the book by Anderson and Bancroft (3), other references listed at the end of this chapter and Chapter 9, and previous editions of this text. The following example illustrates how MINITAB may be used to obtain confidence intervals for the mean of Y and prediction intervals for a particular value of Y.

EXAMPLE 10.5.1

We refer to Example 10.3.1. First, we wish to construct a 95 percent confidence interval for the mean CDA score (Y) in a population of 68-year-old subjects (X_1) who completed 12 years of education (X_2). Second, suppose we have a subject who is 68 years of age and has an education level of 12 years. What do we predict to be this subject's CDA score?

Solution: The point estimate of the mean CDA score is

$$\hat{y} = 5.494 - .18412(68) + .6108(12) = .3034$$

The point prediction, which is the same as the point estimate obtained previously, also is

$$\hat{y} = 5.494 - .18412(68) + .6108(12) = .3034$$

To obtain the confidence interval and the prediction interval for the parameters for which we have just computed a point estimate and a point prediction, we use MINITAB as follows. After entering the information for a regression analysis of our data as shown in Figure 10.3.1, we click on Options in the dialog box. In the box labeled "Prediction intervals for new observations," we type 68 and 12 and click OK twice. In addition to the regression analysis, we obtain the following output:

```
New Obs      Fit    SE Fit          95.0% CI               95.0% PI
      1    0.303     0.672    (-1.038,  1.644)   (-6.093,  6.699)
```

We interpret these intervals in the usual ways. We look first at the confidence interval. We are 95 percent confident that the interval from −1.038 to 1.644 includes the mean of the subpopulation of Y values for the specified combination of X_i values, since this parameter would be included in about 95 percent of the intervals that can be constructed in the manner shown.

Now consider the subject who is 68 years old and has 12 years of education. We are 95 percent confident that this subject would have a CDA score somewhere between −6.093 and 6.699. The fact that the P.I. is wider than the C.I. should not be surprising. After all, it is easier to estimate the mean response than it is estimate an individual observation. ■

EXERCISES

For each of the following exercises compute the y value and construct (a) 95 percent confidence and (b) 95 percent prediction intervals for the specified values of X_i.

10.5.1 Refer to Exercise 10.3.1 and let $x_{1j} = 95$ and $x_{2j} = 35$.

10.5.2 Refer to Exercise 10.3.2 and let $x_{1j} = 50, x_{2j} = 20$, and $x_{3j} = 22$.

10.5.3 Refer to Exercise 10.3.3 and let $x_{1j} = 5$ and $x_{2j} = 6$.

10.5.4 Refer to Exercise 10.3.4 and let $x_{1j} = 1$ and $x_{2j} = 2$.

10.5.5 Refer to Exercise 10.3.5 and let $x_{1j} = 90$ and $x_{2j} = 80$.

10.5.6 Refer to Exercise 10.3.6 and let $x_{1j} = 50, x_{2j} = 95.0,\ x_{3j} = 2.00,\ x_{4j} = 6.00, x_{5j} = 75$, and $x_{6j} = 70$.

10.6 THE MULTIPLE CORRELATION MODEL

We pointed out in the preceding chapter that while regression analysis is concerned with the form of the relationship between variables, the objective of correlation analysis is to gain insight into the strength of the relationship. This is also true in the multivariable case, and in this section we investigate methods for measuring the strength of the relationship among several variables. First, however, let us define the model and assumptions on which our analysis rests.

The Model Equation We may write the correlation model as

$$y_j = \beta_0 + \beta_1 x_{1j} + \beta_2 x_{2j} + \cdots + \beta_k x_{kj} + \epsilon_j \tag{10.6.1}$$

where y_j is a typical value from the population of values of the variable Y, the β's are the regression coefficients defined in Section 10.2, and the x_{ij} are particular (known) values of the random variables X_i. This model is similar to the multiple regression model, but there is one important distinction. In the multiple regression model, given in Equation 10.2.1, the X_i are nonrandom variables, but in the multiple correlation model the X_i are random variables. In other words, in the correlation model there is a joint distribution of Y and the X_i that we call a *multivariate distribution*. Under this model, the variables are no longer thought of as being dependent or independent, since logically they are interchangeable and either of the X_i may play the role of Y.

Typically, random samples of units of association are drawn from a population of interest, and measurements of Y and the X_i are made.

A least-squares plane or hyperplane is fitted to the sample data by methods described in Section 10.3, and the same uses may be made of the resulting equation. Inferences may be made about the population from which the sample was drawn if it can be assumed that the underlying distribution is normal, that is, if it can be assumed that the joint distribution of Y and X_i is a *multivariate normal distribution*. In addition, sample measures of the degree of the relationship among the variables may be computed and, under the assumption that sampling is from a multivariate normal distribution, the corresponding parameters may be estimated by means of confidence intervals, and hypothesis tests may be carried out. Specifically, we may compute an estimate of the *multiple correlation coefficient* that measures the dependence between Y and the X_i. This is a straightforward extension of the concept of correlation between two variables that we discuss in Chapter 9. We may also compute *partial correlation coefficients* that measure the intensity of the relationship between any two variables when the influence of all other variables has been removed.

The Multiple Correlation Coefficient As a first step in analyzing the relationships among the variables, we look at the multiple correlation coefficient.

The multiple correlation coefficient is the square root of the coefficient of multiple determination and, consequently, the sample value may be computed by taking the square root of Equation 10.4.2. That is,

$$R_{y.12\ldots k} = \sqrt{R^2_{y.12\ldots k}} = \sqrt{\frac{\Sigma(\hat{y}_j - \bar{y})^2}{\Sigma(y_j - \bar{y})^2}} = \sqrt{\frac{SSR}{SST}} \tag{10.6.2}$$

To illustrate the concepts and techniques of multiple correlation analysis, let us consider an example.

EXAMPLE 10.6.1

Wang et al. (A-4), using cadaveric human femurs from subjects ages 16 to 19 years, investigated toughness properties of the bone and measures of the collagen network within the bone. Two variables measuring the collagen network are porosity (P, expressed as a percent) and a measure of collagen network tensile strength (S). The measure of toughness (W, Newtons), is the force required for bone fracture. The 29 cadaveric femurs used in the study were free from bone-related pathologies. We wish to analyze the nature and strength of the relationship among the three variables. The measurements are shown in the following table.

TABLE 10.6.1 Bone Toughness and Collagen Network Properties for 29 Femurs

W	P	S
193.6	6.24	30.1
137.5	8.03	22.2
145.4	11.62	25.7
117.0	7.68	28.9
105.4	10.72	27.3
99.9	9.28	33.4
74.0	6.23	26.4
74.4	8.67	17.2
112.8	6.91	15.9
125.4	7.51	12.2
126.5	10.01	30.0
115.9	8.70	24.0
98.8	5.87	22.6
94.3	7.96	18.2
99.9	12.27	11.5
83.3	7.33	23.9
72.8	11.17	11.2
83.5	6.03	15.6
59.0	7.90	10.6

(Continued)

W	P	S
87.2	8.27	24.7
84.4	11.05	25.6
78.1	7.61	18.4
51.9	6.21	13.5
57.1	7.24	12.2
54.7	8.11	14.8
78.6	10.05	8.9
53.7	8.79	14.9
96.0	10.40	10.3
89.0	11.72	15.4

Source: Xiaodu Wang, Ph.D. Used with permission.

Solution: We use MINITAB to perform the analysis of our data. Readers interested in the derivation of the underlying formulas and the arithmetic procedures involved may consult the texts listed at the end of this chapter and Chapter 9, as well as previous editions of this text. If a least-squares prediction equation and multiple correlation coefficient are desired as part of the analysis, we may obtain them by using the previously described MINITAB multiple regression procedure. When we do this with the sample values of Y, X_1, and X_2, stored in Columns 1 through 3, respectively, we obtain the output shown in Figure 10.6.1.

The least-squares equation, then, is

$$\hat{y}_j = 35.61 + 1.451x_{1j} + 2.3960x_{2j}$$

```
The regression equation is
Y = 35.6 + 1.45 X1 + 2.40 X2

Predictor         Coef    SE Coef       T        P
Constant         35.61      29.13    1.22    0.232
X1               1.451      2.763    0.53    0.604
X2              2.3960     0.7301    3.28    0.003

S = 27.42     R-Sq = 29.4%     R-Sq(adj) = 24.0%

Analysis of Variance

Source            DF         SS        MS       F        P
Regression         2     8151.1    4075.6    5.42    0.011
Residual Error    26    19553.5     752.1
Total             28    27704.6
```

FIGURE 10.6.1 Output from MINITAB multiple regression procedure for the data in Table 10.6.1.

This equation may be used for estimation and prediction purposes and may be evaluated by the methods discussed in Section 10.4.

As we see in Figure 10.6.1, the multiple regression output also gives us the coefficient of multiple determination, which, in our present example, is

$$R^2_{y.12} = .294$$

The multiple correlation coefficient, therefore, is

$$R_{y.12} = \sqrt{.294} = .542$$

Interpretation of $R_{y.12}$

We interpret $R_{y.12}$ as a measure of the correlation among the variables force required to fracture, porosity, and collagen network strength in the sample of 29 femur bones from subjects ages 16 to 19. If our data constitute a random sample from the population of such persons, we may use $R_{y.12}$ as an estimate of $\rho_{y.12}$, the true population multiple correlation coefficient. We may also interpret $R_{y.12}$ as the simple correlation coefficient between y_j and $\hat{y}$, the observed and calculated values, respectively, of the "dependent" variable. Perfect correspondence between the observed and calculated values of Y will result in a correlation coefficient of 1, while a complete lack of a linear relationship between observed and calculated values yields a correlation coefficient of 0. The multiple correlation coefficient is always given a positive sign.

We may test the null hypothesis that $\rho_{y.12\ldots k} = 0$ by computing

$$F = \frac{R^2_{y.12\ldots k}}{1 - R^2_{y.12\ldots k}} \cdot \frac{n - k - 1}{k} \tag{10.6.3}$$

The numerical value obtained from Equation 10.6.3 is compared with the tabulated value of F with k and $n - k - 1$ degrees of freedom. The reader will recall that this is identical to the test of H_0: $\beta_1 = \beta_2 = \cdots = \beta_k = 0$ described in Section 10.4.

For our present example let us test the null hypothesis that $\rho_{y.12} = 0$ against the alternative that $\rho_{y.12} \neq 0$. We compute

$$F = \frac{.294}{1 - .294} \cdot \frac{29 - 2 - 1}{2} = 5.41$$

Since 5.41 is greater than 4.27, $p < .025$, so that we may reject the null hypothesis at the .025 level of significance and conclude that the force required for fracture is correlated with porosity and the measure of collagen network strength in the sampled population.

The computed value of F for testing H_0 that the population multiple correlation coefficient is equal to zero is given in the analysis of variance table in Figure 10.6.1 and is 5.42. The two computed values of F differ as a result of differences in rounding in the intermediate calculations. ■

Partial Correlation The researcher may wish to have a measure of the strength of the linear relationship between two variables when the effect of the remaining variables has been removed. Such a measure is provided by the *partial correlation* coefficient. For example, the partial sample correlation coefficient $r_{y.12}$ is a measure of the correlation between Y and X_1 after controlling for the effect of X_2.

The partial correlation coefficients may be computed from the *simple correlation coefficients*. The simple correlation coefficients measure the correlation between two variables when no effort has been made to control other variables. In other words, they are the coefficients for any pair of variables that would be obtained by the methods of simple correlation discussed in Chapter 9.

Suppose we have three variables, Y, X_1, and X_2. The sample partial correlation coefficient measuring the correlation between Y and X_1 after controlling for X_2, for example, is written $r_{y1.2}$. In the subscript, the symbol to the right of the decimal point indicates the variable whose effect is being controlled, while the two symbols to the left of the decimal point indicate which variables are being correlated. For the three-variable case, there are two other sample partial correlation coefficients that we may compute. They are $r_{y2.1}$ and $r_{12.y}$.

The Coefficient of Partial Determination The square of the partial correlation coefficient is called the coefficient of partial determination. It provides useful information about the interrelationships among variables. Consider $r_{y1.2}$, for example. Its square, $r^2_{y1.2}$ tells us what proportion of the remaining variability in Y is explained by X_1 after X_2 has explained as much of the total variability in Y as it can.

Calculating the Partial Correlation Coefficients For three variables the following simple correlation coefficients may be calculated:

r_{y1}, the simple correlation between Y and X_1

r_{y2}, the simple correlation between Y and X_2

r_{12}, the simple correlation between X_1 and X_2

The MINITAB correlation procedure may be used to compute these simple correlation coefficients as shown in Figure 10.6.2. As noted earlier, the sample observations are stored in Columns 1 through 3. From the output in Figure 10.6.2 we see that $r_{12} = -.08$, $r_{y1} = .043$, and $r_{y2} = .535$.

The sample partial correlation coefficients that may be computed from the simple correlation coefficients in the three-variable case are:

1. The partial correlation between Y and X_1 after controlling for the effect of X_2:

$$r_{y1.2} = (r_{y1} - r_{y2}r_{12})/\sqrt{(1 - r^2_{y2})(1 - r^2_{12})} \tag{10.6.4}$$

2. The partial correlation between Y and X_2 after controlling for the effect of X_1:

$$r_{y2.1} = (r_{y2} - r_{y1}r_{12})/\sqrt{(1 - r^2_{y1})(1 - r^2_{12})} \tag{10.6.5}$$

3. The partial correlation between X_1 and X_2 after controlling for the effect of Y:

$$r_{12.y} = (r_{12} - r_{y1}r_{y2})/\sqrt{(1 - r^2_{y1})(1 - r^2_{y2})} \tag{10.6.6}$$

Dialog box: **Session Command:**

Stat ➤ Basic Statistics ➤ Correlation `MTB> CORRELATION C1-C3`

Type *C1-C3* in **Variables.** Click **OK.**

Output:

```
          Y         X1
X1     0.043
       0.823

X2     0.535     -0.080
       0.003      0.679

Cell Contents: Pearson correlation
               P-Value
```

FIGURE 10.6.2 MINITAB procedure for calculating the simple correlation coefficients for the data in Table 10.6.1.

EXAMPLE 10.6.2

To illustrate the calculation of sample partial correlation coefficients, let us refer to Example 10.6.1, and calculate the partial correlation coefficients among the variables force to fracture (Y), porosity (X_1), and collagen network strength (X_2).

Solution: Instead of computing the partial correlation coefficients from the simple correlation coefficients by Equations 10.6.4 through 10.6.6, we use MINITAB to obtain them.

The MINITAB procedure for computing partial correlation coefficients is based on the fact that a given partial correlation coefficient is itself the simple correlation between two sets of residuals. A set of residuals is obtained as follows. Suppose we have measurements on two variables, X (independent) and Y (dependent). We obtain the least-squares prediction equation, $\hat{y} = \hat{\beta}_0 + \hat{\beta}_x$. For each value of X we compute a residual, which is equal to $(y_i - \hat{y}_i)$, the difference between the observed value of Y and the predicted value of Y associated with the X.

Now, suppose we have three variables, X_1, X_2, and Y. We want to compute the partial correlation coefficient between X_1 and Y while holding X_2 constant. We regress X_1 on X_2 and compute the residuals, which we may call residual set A. We regress Y on X_2 and compute the residuals, which we may call residual set B. The simple correlation coefficient measuring the strength of the relationship between residual set A and residual set B is the partial correlation coefficient between X_1 and Y after controlling for the effect of X_2.

When using MINITAB we store each set of residuals in a different column for future use in calculating the simple correlation coefficients between them.

We use session commands rather than a dialog box to calculate the partial correlation coefficients when we use MINITAB. With the observations on X_1, X_2, and Y stored in Columns 1 through 3, respectively, the procedure for the data of Table 10.6.1 is shown in Figure 10.6.3. The output shows that $r_{y1.2} = .102$, $r_{12.y} = -.122$, and $r_{y2.1} = .541$.

Partial correlations can be calculated directly using SPSS software as seen in Figure 10.6.5. This software displays, in a succinct table, both the partial correlation coefficient and the p value associated with each partial correlation. ■

Testing Hypotheses About Partial Correlation Coefficients

We may test the null hypothesis that any one of the population partial correlation coefficients is 0 by means of the t test. For example, to test H_0: $\rho_{y1.2\ldots k} = 0$, we compute

$$t = r_{y1.2\ldots k}\sqrt{\frac{n-k-1}{1-r^2_{y1.2\ldots k}}} \tag{10.6.7}$$

which is distributed as Student's t with $n - k - 1$ degrees of freedom.

Let us illustrate the procedure for our current example by testing H_0: $\rho_{y1.2} = 0$ against the alternative, H_A: $\rho_{y1.2} \neq 0$. The computed t is

$$t = .102\sqrt{\frac{29-2-1}{1-(.102)^2}} = .523$$

Since the computed t of .523 is smaller than the tabulated t of 2.0555 for 26 degrees of freedom and $\alpha = .05$ (two-sided test), we fail to reject H_0 at the .05 level of significance and conclude that there may be no correlation between force required for fracture and porosity after controlling for the effect of collagen network strength. Significance tests for the other two partial correlation coefficients will be left as an exercise for the reader. Note that p values for these tests are calculated by MINITAB as shown in Figure 10.6.3.

The SPSS statistical software package for the PC provides a convenient procedure for obtaining partial correlation coefficients. To use this feature choose "Analyze" from the menu bar, then "Correlate," and, finally, "Partial." Following this sequence of choices the Partial Correlations dialog box appears on the screen. In the box labeled "Variables:," enter the names of the variables for which partial correlations are desired. In the box labeled "Controlling for:" enter the names of the variable(s) for which you wish to control. Select either a two-tailed or one-tailed level of significance. Unless the option is deselected, actual significance levels will be displayed. For Example 10.6.2, Figure 10.6.4 shows the SPSS computed partial correlation coefficients between the other two variables when controlling, successively, for X_1 (porosity), X_2 (collagen network strength), and Y (force required for fracture).

```
MTB > regress C1 1 C2;
SUBC> residuals C4.

MTB > regress C3 1 C2;
SUBC> residuals C5.

MTB > regress C1 1 C3;
SUBC> residuals C6.

MTB > regress C2 1 C3;
SUBC> residuals C7.

MTB > regress C2 1 C1;
SUBC> residuals C8.

MTB > regress C3 1 C1;
SUBC> residuals C9.

MTB > corr C4 C5

Correlations: C4, C5

Pearson correlation of C4 and C5 = 0.102
P-Value = 0.597

MTB > corr C6 C7

Correlations: C6, C7

Pearson correlation of C6 and C7 = -0.122
P-Value = 0.527

MTB > corr C8 C9

Correlations: C8, C9

Pearson correlation of C8 and C9 = 0.541
P-Value = 0.002
```

FIGURE 10.6.3 MINITAB procedure for computing partial correlation coefficients from the data of Table 10.6.1.

```
Controlling for:  X1

               X2             Y

X2          1.0000          .5412
          (     0)        (    26)
          P=  .           P=  .003

Y            .5412         1.0000
          (    26)        (     0)
          P=  .003        P=  .

Controlling for:  X2

               Y              X1

Y           1.0000          .1024
          (     0)         (  26)
          P=  .           P=  .604

X1           .1024         1.0000
          (    26)        (     0)
          P=  .604        P=  .

Controlling for:  Y

               X1             X2

X1          1.0000         -.1225
          (     0)        (  26)
          P=  .           P=  .535

X2          -.1225         1.0000
          (    26)        (     0)
          P=  .535        P=  .

(Coefficient / (D.F.) / 2-tailed Significance)
"." is printed if a coefficient cannot be computed
```

FIGURE 10.6.4 Partial coefficients obtained with SPSS for Windows, Example 10.6.2.

Correlations

Control Variables			Force to Fracture (Y)	Porosity (X1)
Tensile Strength (X2)	Force to Fracture (Y)	Correlation	1.000	.102
		Significance (2-tailed)	.	.604
		df	0	26
	Porosity (X1)	Correlation	.102	1.000
		Significance (2-tailed)	.604	.
		df	26	0

(a)

Correlations

Control Variables			Porosity (X1)	Tensile Strength (X2)
Force to Fracture (Y)	Porocity (X1)	Correlation	1.000	−.122
		Significance (2-tailed)	.	.535
		df	0	26
	Tensile Strength (X2)	Correlation	−.122	1.000
		Significance (2-tailed)	.535	.
		df	26	0

(b)

Correlations

Control Variables			Tensile Strength (X2)	Force to Fracture (Y)
Porosity (X1)	Tensile Strength (X2)	Correlation	1.000	.541
		Significance (2-tailed)	.	.003
		df	0	26
	Force to Fracture (Y)	Correlation	.541	1.000
		Significance (2-tailed)	.003	.
		df	26	0

(c)

FIGURE 10.6.5 Partial correlation coefficients for the data in Example 10.6.1. (*a*) $r_{y1.2}$, (*b*) $r_{12.y}$, and (*c*) $r_{y2.1}$.

Although our illustration of correlation analysis is limited to the three-variable case, the concepts and techniques extend logically to the case of four or more variables. The number and complexity of the calculations increase rapidly as the number of variables increases.

EXERCISES

10.6.1 The objective of a study by Anton et al. (A-5) was to investigate the correlation structure of multiple measures of HIV burden in blood and tissue samples. They measured HIV burden four ways. Two measurements were derived from blood samples, and two measurements were made on rectal tissue. The two blood measures were based on HIV DNA assays and a second co-culture assay that was a modification of the first measure. The third and fourth measurements were quantitations of HIV-1 DNA and RNA from rectal biopsy tissue. The table below gives data on HIV levels from these measurements for 34 subjects.

HIV DNA Blood (Y)	HIV Co-Culture Blood (X_1)	HIV DNA Rectal Tissue (X_2)	HIV RNA Rectal Tissue (X_3)
115	.38	899	56
86	1.65	167	158
19	.16	73	152
6	.08	146	35
23	.02	82	60
147	1.98	2483	1993
27	.15	404	30
140	.25	2438	72
345	.55	780	12
92	.22	517	5
85	.09	346	5
24	.17	82	12
109	.41	1285	5
5	.02	380	5
95	.84	628	32
46	.02	451	5
25	.64	159	5
187	.20	1335	121
5	.04	30	5
47	.02	13	30
118	.24	5	5
112	.72	625	83
79	.45	719	70
52	.23	309	167
52	.06	27	29
7	.37	199	5
13	.13	510	42
80	.24	271	15
86	.96	273	45
26	.29	534	71
53	.25	473	264
185	.28	2932	108
30	.19	658	33
9	.03	103	5
76	.21	2339	5

(*Continued*)

HIV DNA Blood (Y)	HIV Co-Culture Blood (X_1)	HIV DNA Rectal Tissue (X_2)	HIV RNA Rectal Tissue (X_3)
51	.09	31	36
73	.06	158	5
47	.08	773	5
48	.12	545	67
16	.03	5	5

Source: Peter A. Anton, M.D. Used with permission.

(a) Compute the simple correlation coefficients between all possible pairs of variables.

(b) Compute the multiple correlation coefficient among the four variables. Test the overall correlation for significance.

(c) Calculate the partial correlations between HIV DNA blood and each one of the other variables while controlling for the other two. (These are called second-order partial correlation coefficients.)

(d) Calculate the partial correlation between HIV co-culture blood and HIV DNA, controlling for the other two variables.

(e) Calculate the partial correlation between HIV co-culture blood and HIV RNA, controlling for the other two variables.

(f) Calculate the partial correlations between HIV DNA and HIV RNA, controlling for the other two variables.

10.6.2 The following data were obtained on 12 males between the ages of 12 and 18 years (all measurements are in centimeters):

Height (Y)	Radius Length (X_1)	Femur Length (X_2)
149.0	21.00	42.50
152.0	21.79	43.70
155.7	22.40	44.75
159.0	23.00	46.00
163.3	23.70	47.00
166.0	24.30	47.90
169.0	24.92	48.95
172.0	25.50	49.90
174.5	25.80	50.30
176.1	26.01	50.90
176.5	26.15	50.85
179.0	26.30	51.10
Total 1992.1	290.87	573.85

(a) Find the sample multiple correlation coefficient and test the null hypothesis that $\rho_{y.12} = 0$.

(b) Find each of the partial correlation coefficients and test each for significance. Let $\alpha = .05$ for all tests.

(c) Determine the p value for each test.

(d) State your conclusions.

10.6.3 The following data were collected on 15 obese girls:

Weight in Kilograms (Y)	Lean Body Weight (X_1)	Mean Daily Caloric Intake (X_2)
79.2	54.3	2670
64.0	44.3	820
67.0	47.8	1210
78.4	53.9	2678
66.0	47.5	1205
63.0	43.0	815
65.9	47.1	1200
63.1	44.0	1180
73.2	44.1	1850
66.5	48.3	1260
61.9	43.5	1170
72.5	43.3	1852
101.1	66.4	1790
66.2	47.5	1250
99.9	66.1	1789
Total 1087.9	741.1	22739

(a) Find the multiple correlation coefficient and test it for significance.

(b) Find each of the partial correlation coefficients and test each for significance. Let $\alpha = .05$ for all tests.

(c) Determine the p value for each test.

(d) State your conclusions.

10.6.4 A research project was conducted to study the relationships among intelligence, aphasia, and apraxia. The subjects were patients with focal left hemisphere damage. Scores on the following variables were obtained through the application of standard tests.

Y = intelligence
X_1 = ideomotor apraxia
X_2 = constructive apraxia
X_3 = lesion volume (pixels)
X_4 = severity of aphasia

The results are shown in the following table. Find the multiple correlation coefficient and test for significance. Let $\alpha = .05$ and find the p value.

Subject	Y	X_1	X_2	X_3	X_4
1	66	7.6	7.4	2296.87	2
2	78	13.2	11.9	2975.82	8
3	79	13.0	12.4	2839.38	11

(Continued)

Subject	Y	X_1	X_2	X_3	X_4
4	84	14.2	13.3	3136.58	15
5	77	11.4	11.2	2470.50	5
6	82	14.4	13.1	3136.58	9
7	82	13.3	12.8	2799.55	8
8	75	12.4	11.9	2565.50	6
9	81	10.7	11.5	2429.49	11
10	71	7.6	7.8	2369.37	6
11	77	11.2	10.8	2644.62	7
12	74	9.7	9.7	2647.45	9
13	77	10.2	10.0	2672.92	7
14	74	10.1	9.7	2640.25	8
15	68	6.1	7.2	1926.60	5

10.7 SUMMARY

In this chapter we examine how the concepts and techniques of simple linear regression and correlation analysis are extended to the multiple-variable case. The least-squares method of obtaining the regression equation is presented and illustrated. This chapter also is concerned with the calculation of descriptive measures, tests of significance, and the uses to be made of the multiple regression equation. In addition, the methods and concepts of correlation analysis, including partial correlation, are discussed.

When the assumptions underlying the methods of regression and correlation presented in this and the previous chapter are not met, the researcher must resort to alternative techniques such as those discussed in Chapter 13.

SUMMARY OF FORMULAS FOR CHAPTER 10

Formula Number	Name	Formula
10.2.1	Representation of the multiple linear regression equation	$y_j = \beta_0 + \beta_1 x_{1j} + \beta_2 x_{2j} + \cdots + \beta_k x_{kj} + \epsilon_j$
10.2.2	Representation of the multiple linear regression equation with two independent variables	$y_j = \beta_0 + \beta_1 x_{1j} + \beta_2 x_{2j} + \epsilon_j$

(Continued)

10.2.3	Random deviation of a point from a plane when there are two independent variables	$\epsilon_j = y_j - \beta_0 - \beta_1 x_{1j} - \beta_2 x_{2j}$
10.3.1	Sum-of-squared residuals	$\sum \epsilon_j^2 = \sum (y_j - \hat{y}_j)^2$
10.4.1	Sum-of-squares equation	$\sum (y_j - \bar{y})^2 = \sum (\hat{y}_j - \bar{y})^2 + \sum (y_j - \hat{y}_j)^2$ SST = SSR + SSE
10.4.2	Coefficient of multiple determination	$R^2_{y.12...k} = \dfrac{\Sigma(y_j - \hat{y}_j)^2}{\Sigma(y_j - \bar{y})^2} = \dfrac{SSR}{SST}$
10.4.3	t statistic for testing hypotheses about β_i	$t = \dfrac{\hat{\beta}_j - \beta_{jo}}{s_{\hat{\beta}_i}}$
10.5.1	Estimation equation for multiple linear regression	$\hat{y}_j = \hat{\beta}_0 + \hat{\beta}_1 x_{1j} + \hat{\beta}_2 x_{2j} + \cdots + \hat{\beta}_k x_{kj}$
10.5.2	Confidence interval for the mean of Y for a given X	$\hat{y}_j \pm t_{(1-\alpha/2),n-k-1} s_{\hat{y}_j}$
10.5.3	Prediction interval for Y for a given X	$\hat{y}_j \pm t_{(1-\alpha/2),n-k-1} s'_{\hat{y}_j}$
10.6.1	Multiple correlation model	$y_j = \beta_0 + \beta_1 x_{1j} + \beta_2 x_{2j} + \cdots + \beta_k x_{kj} + \epsilon_j$
10.6.2	Multiple correlation coefficient	$R_{y.12...k} = \sqrt{R^2_{y.12...k}} = \sqrt{\dfrac{\Sigma(y_i - \hat{y}_j)^2}{\Sigma(y_i - \bar{y})^2}} = \sqrt{\dfrac{SSR}{SST}}$
10.6.3	F statistic for testing the multiple correlation coefficient	$F = \dfrac{R^2_{y.12...k}}{1 - R^2_{y.12...k}} \cdot \dfrac{n-k-1}{k}$
10.6.4–10.6.6	Partial correlation between two variables (1 and 2) after controlling for a third (3)	$r_{12.3} = (r_{12} - r_{13}r_{23})/\sqrt{(1 - r_{13}^2)(1 - r_{23}^2)}$

10.6.7	t statistic for testing hypotheses about partial correlation coefficients	$t = r_{y1.2...k}\sqrt{\dfrac{n-k-1}{1-r_{y1.2..k}^2}}$
Symbol Key	• $\hat{\beta}_x$ = estimated regression/correlation coefficient x • β_x = regression/correlation coefficient • ϵ = model error term • k = number of independent variables • n = sample size • $r_{12.3}$ = sample partial correlation coefficient between 1 and 2 after controlling for 3 • R = sample correlation coefficient • R^2 = multiple coefficient of determination • t = t statistic • x_i = value of independent variable at i • $\bar{x}$ = sample mean of independent variable • y_i = value of dependent variable at i • $\bar{y}$ = sample mean of dependent variable • $\hat{y}$ = estimated y • z = z statistic	

REVIEW QUESTIONS AND EXERCISES

1. What are the assumptions underlying multiple regression analysis when one wishes to infer about the population from which the sample data have been drawn?

2. What are the assumptions underlying the correlation model when inference is an objective?

3. Explain fully the following terms:

(a) Coefficient of multiple determination
(b) Multiple correlation coefficient
(c) Simple correlation coefficient
(d) Partial correlation coefficient

4. Describe a situation in your particular area of interest where multiple regression analysis would be useful. Use real or realistic data and do a complete regression analysis.

5. Describe a situation in your particular area of interest where multiple correlation analysis would be useful. Use real or realistic data and do a complete correlation analysis.

In Exercises 6 through 11 carry out the indicated analysis and test hypotheses at the indicated significance levels. Compute the p value for each test.

6. We learned in Example 9.7.1 that the purpose of a study by Kwast-Rabben et al. (A-6) was to analyze somatosensory evoked potentials (SEPs) and their interrelations following stimulation of digits I, III, and V in the hand. Healthy volunteers were recruited for the study. Researchers applied stimulation below-pain-level intensity to the fingers. Recordings of spinal responses were made with electrodes fixed by adhesive electrode cream to the subject's skin. Results are shown in the following table for 114 subjects. Use multiple regression to see how well you can predict the peak spinal latency (Cv) of the SEP for digit I when age (years) and arm length (cm) are the predictor variables. Evaluate the usefulness of your prediction equation.

Age	Arm Length	Cv Dig.I	Age	Arm Length	Cv Dig.I	Age	Arm Length	Cv Dig.I
35.07	76.5	13.50	32.00	82.0	16.30	42.08	94.0	17.70
35.07	76.5	13.50	32.00	82.0	15.40	40.09	94.0	17.70
21.01	77.0	13.00	38.09	86.5	16.60	40.09	94.0	17.40
21.01	77.0	13.60	38.09	86.5	16.00	42.09	92.5	18.40
47.06	75.5	14.30	58.07	85.0	17.00	20.08	95.0	19.00
47.06	75.5	14.90	58.07	85.0	16.40	50.08	94.5	19.10
26.00	80.0	15.40	54.02	88.0	17.60	50.08	94.5	19.20
26.00	80.0	14.70	48.10	92.0	16.80	47.11	97.5	17.80
53.04	82.0	15.70	48.10	92.0	17.00	47.11	97.5	19.30
53.04	82.0	15.80	54.02	88.0	17.60	26.05	96.0	17.50
43.07	79.0	15.20	45.03	91.5	17.30	26.05	96.0	18.00
39.08	83.5	16.50	45.03	91.5	16.80	43.02	98.0	18.00
39.08	83.5	17.00	35.11	94.0	17.00	43.02	98.0	18.80
43.07	79.0	14.70	26.04	88.0	15.60	32.06	98.5	18.30
29.06	81.0	16.00	51.07	87.0	16.80	32.06	98.5	18.60
29.06	81.0	15.80	51.07	87.0	17.40	33.09	97.0	18.80
50.02	86.0	15.10	26.04	88.0	16.50	33.09	97.0	19.20
25.07	81.5	14.60	35.11	94.0	16.60	35.02	100.0	18.50
25.07	81.5	15.60	52.00	88.5	18.00	35.02	100.0	18.50
25.10	82.5	14.60	44.02	90.0	17.40	26.05	96.0	19.00
47.04	86.0	17.00	44.02	90.0	17.30	26.05	96.0	18.50
47.04	86.0	16.30	24.05	91.0	16.40	25.08	100.5	19.80
37.00	83.0	16.00	24.00	87.0	16.10	25.06	100.0	18.80
37.00	83.0	16.00	24.00	87.0	16.10	25.06	100.0	18.40
34.10	84.0	16.30	24.00	87.0	16.00	25.08	100.5	19.00
47.01	87.5	17.40	24.00	87.0	16.00	30.05	101.0	18.00
47.01	87.5	17.00	53.05	90.0	17.50	30.05	101.0	18.20
30.04	81.0	14.10	53.05	90.0	17.50	36.07	104.5	18.90
23.06	81.5	14.20	52.06	90.0	18.00	36.07	104.5	19.20
23.06	81.5	14.70	52.06	90.0	17.90	35.09	102.0	21.00
30.04	81.0	13.90	53.04	93.0	18.40	35.09	102.0	19.20
78.00	81.0	17.20	22.04	90.0	16.40	21.01	101.5	18.60
41.02	83.5	16.70	22.04	90.0	15.80	21.01	101.5	18.60
41.02	83.5	16.50	46.07	95.5	18.80	40.00	95.5	20.00
28.07	78.0	14.80	46.07	95.5	18.60	42.09	92.5	18.40
28.07	78.0	15.00	47.00	93.5	18.00	42.08	94.0	18.50
36.05	88.0	17.30	47.00	93.5	17.90	35.04	86.0	16.00
35.04	86.0	15.30	39.05	94.5	17.40	36.05	88.0	16.60

Source: Olga Kwast-Rabben, Ph.D. Used with permission.

7. The following table shows the weight and total cholesterol and triglyceride levels in 15 patients with primary type II hyperlipoproteinemia just prior to initiation of treatment:

Y Weight (kg)	X_1 Total Cholesterol (mg/100 ml)	X_2 Triglyceride (mg/100 ml)
76	302	139
97	336	101
83	220	57
52	300	56
70	382	113
67	379	42
75	331	84
78	332	186
70	426	164
99	399	205
75	279	230
78	332	186
70	410	160
77	389	153
76	302	139

Compute the multiple correlation coefficient and test for significance at the .05 level.

8. In a study of the relationship between creatinine excretion, height, and weight, the data shown in the following table were collected on 20 infant males:

Infant	Creatinine Excretion (mg/day) Y	Weight (kg) X_1	Height (cm) X_2
1	100	9	72
2	115	10	76
3	52	6	59
4	85	8	68
5	135	10	60
6	58	5	58
7	90	8	70
8	60	7	65
9	45	4	54
10	125	11	83
11	86	7	64
12	80	7	66
13	65	6	61
14	95	8	66
15	25	5	57
16	125	11	81
17	40	5	59

(*Continued*)

Infant	Creatinine Excretion (mg/day) Y	Weight (kg) X_1	Height (cm) X_2
18	95	9	71
19	70	6	62
20	120	10	75

(a) Find the multiple regression equation describing the relationship among these variables.

(b) Compute R^2 and do an analysis of variance.

(c) Let $X_1 = 10$ and $X_2 = 60$ and find the predicted value of Y.

9. A study was conducted to examine those variables thought to be related to the job satisfaction of nonprofessional hospital employees. A random sample of 15 employees gave the following results:

Score on Job Satisfaction Test (Y)	Coded Intelligence Score (X_1)	Index of Personal Adjustment (X_2)
54	15	8
37	13	1
30	15	1
48	15	7
37	10	4
37	14	2
31	8	3
49	12	7
43	1	9
12	3	1
30	15	1
37	14	2
61	14	10
31	9	1
31	4	5

(a) Find the multiple regression equation describing the relationship among these variables.

(b) Compute the coefficient of multiple determination and do an analysis of variance.

(c) Let $X_1 = 10$ and $X_2 = 5$ and find the predicted value of Y.

10. A medical research team obtained the index of adiposity, basal insulin, and basal glucose values on 21 normal subjects. The results are shown in the following table. The researchers wished to investigate the strength of the association among these variables.

Index of Adiposity Y	Basal Insulin (μU/ml) X_1	Basal Glucose (mg/100 ml) X_2
90	12	98
112	10	103
127	14	101
137	11	102
103	10	90
140	38	108
105	9	100
92	6	101
92	8	92
96	6	91
114	9	95
108	9	95
160	41	117
91	7	101
115	9	86
167	40	106
108	9	84
156	43	117
167	17	99
165	40	104
168	22	85

Compute the multiple correlation coefficient and test for significance at the .05 level.

11. As part of a study to investigate the relationship between stress and certain other variables, the following data were collected on a simple random sample of 15 corporate executives.

(a) Find the least-squares regression equation for these data.

(b) Construct the analysis of variance table and test the null hypothesis of no relationship among the five variables.

(c) Test the null hypothesis that each slope in the regression model is equal to zero.

(d) Find the multiple coefficient of determination and the multiple correlation coefficient. Let $\alpha = .05$ and find the p value for each test.

Measure of Stress (Y)	Measure of Firm Size (X_1)	Number of Years in Present Position (X_2)	Annual Salary ($\times 1000$) (X_3)	Age (X_4)
101	812	15	\$30	38
60	334	8	20	52
10	377	5	20	27
27	303	10	54	36

(Continued)

Measure of Stress (Y)	Measure of Firm Size (X_1)	Number of Years in Present Position (X_2)	Annual Salary ($\times 1000$) (X_3)	Age (X_4)
89	505	13	52	34
60	401	4	27	45
16	177	6	26	50
184	598	9	52	60
34	412	16	34	44
17	127	2	28	39
78	601	8	42	41
141	297	11	84	58
11	205	4	31	51
104	603	5	38	63
76	484	8	41	30

For each of the studies described in Exercises 12 through 16, answer as many of the following questions as possible:

(a) Which is more relevant, regression analysis or correlation analysis, or are both techniques equally relevant?

(b) Which is the dependent variable?

(c) What are the independent variables?

(d) What are the appropriate null and alternative hypotheses?

(e) Which null hypotheses do you think were rejected? Why?

(f) Which is the more relevant objective, prediction or estimation, or are the two equally relevant? Explain your answer.

(g) What is the sampled population?

(h) What is the target population?

(i) Which variables are related to which other variables? Are the relationships direct or inverse?

(j) Write out the regression equation using appropriate numbers for parameter estimates.

(k) What is the numerical value of the coefficient of multiple determination?

(l) Give numerical values for any correlation coefficients that you can.

12. Hashimoto et al. (A-7) developed a multiple regression model to predict the number of visits to emergency rooms at Jikei University hospitals in Tokyo for children having an asthma attack. The researchers found that the number of visits per night increased significantly when climate conditions showed a rapid decrease from higher barometric pressure, from higher air temperature, and from higher humidity, as well as lower wind speed. The final model demonstrated that 22 percent of the variation in the number of visits was explained by variation in the predictor variables mentioned above with eight other significant climate variables.

13. Correlation was one of many procedures discussed in a study reported by Stenvinkel et al. (A-8). In a cohort of 204 subjects with end-stage renal disease, they found no significant correlations between log plasma adiponectin levels and age and no significant correlation between log plasma adiponectin and glomerular filtration rate.

14. Van Schuylenbergh et al. (A-9) used physiological and anthropometric measurements as independent variables to predict triathlon performance (expressed in minutes). Ten triathletes underwent extensive physiological testing in swimming, cycling, and running. Within 2 weeks after the last laboratory test, all subjects competed in the National University Triathlon Championship. The final regression model was

$$TP = 130 - 9.2MLSSR - 25.9MLSSS + 1.4BLCR$$

in which TP = triathlon performance in minutes, MLSSR = the running speed at MLSS (m/s), MLSSS = the swimming speed at MLSS, and BLCR = blood lactate concentration at running MLSS (mmol/L). MLSS refers to maximal lactate steady state and is generally acknowledged to be a good marker of functional aerobic power during prolonged exercise. It also differs for each physical activity. For the above model $R^2 = .98$.

15. Maximal static inspiratory (P_{Imax}) mouth pressure is a simple measurement of respiratory muscle strength. A study by Tomalak et al. (A-10) examined correlations among the variables with P_{Imax} (measured sitting), forced expiratory volume (FEV), peak expiratory flow (PEF), and maximal inspiratory flow (PIF) in 144 boys and 152 girls ages 7–14. The researchers found P_{Imax} was correlated with FEV, PEF, and PIF in boys ($p = .001, p = .0055$, and $p = .002$, respectively) and for girls the correlations were also significant ($p < .001, p < .001$, and $p < .001$, respectively).

16. Di Monaco et al. (A-11) used multiple regression to predict bone mineral density of the femoral neck (among other locations). Among 124 Caucasian, healthy postmenopausal women, they found that weight ($p < .001$), age ($p < .01$), and total lymphocyte count ($p < .001$) were each useful in predicting bone mineral density. In addition, $R^2 = .40$.

For each of the data sets given in Exercises 17 through 19, do as many of the following as you think appropriate:

(a) Obtain the least-squares multiple regression equation.

(b) Compute the sample coefficient of multiple determination.

(c) Compute the sample coefficient of multiple correlation.

(d) Compute simple coefficients of determination and correlation.

(e) Compute partial correlation coefficients.

(f) Construct graphs.

(g) Formulate relevant hypotheses, perform the appropriate tests, and find p values.

(h) State the statistical decisions and clinical conclusions that the results of your hypothesis tests justify.

(i) Use your regression equations to make predictions and estimates about the dependent variable for your selected values of the independent variables.

(j) Construct confidence intervals for relevant population parameters.

(k) Describe the population(s) to which you think your inferences are applicable.

17. Pellegrino et al. (A-12) hypothesized that maximal bronchoconstriction can be predicted from the bronchomotor effect of deep inhalation and the degree of airway sensitivity to methacholine (MCh). One group of participants consisted of 26 healthy or mildly asthmatic subjects (22 males, 4 females) who had limited bronchoconstriction to inhaled MCh. The mean age of the patients was 31 years with a standard deviation of 8. There was one smoker in the group. Among the data collected on each subject were the following observations on various lung function measurement variables:

(X_1) FEV_1	(X_2) FEV_1, % Pred	(X_3) FEV_1/FVC, %	(X_4) $\dot{V}m_{50}$	(X_5) $\dot{V}p_{50}$	(X_6) M/P Ratio	(X_7) MP Slope	(X_8) $PD_{15}FEV_1$ (ln mg)	(X_9) $PD_{40}\dot{V}m_{50}$ (ln mg)	(X_{10}) $PD_{40}\dot{V}p_{50}$ (ln mg)	(X_{11}) FEV_1 Max decr (%)	(X_{12}) $\dot{V}m_{50}$ Max decr (%)	(X_{13}) $\dot{V}p_{50}$ Max decr (%)
5.22	108.75	83.92	5.30	3.90	1.36	0.75	8.44	8.24	6.34	21.40	55.40	74.40
5.38	123.96	78.54	6.00	3.70	1.62	0.56	7.76	7.00	6.18	15.80	50.80	85.14
3.62	111.04	86.19	3.10	2.85	1.10	0.69	6.92	6.61	5.56	30.40	54.36	83.07
3.94	94.26	85.28	4.10	2.70	1.52	0.44	6.79	8.52	6.38	16.40	29.10	58.50
4.48	104.43	76.58	3.21	3.00	1.07	0.63	8.79	9.74	6.68	27.80	46.30	76.70
5.28	117.33	81.99	5.65	5.55	1.02	0.83	8.98	8.97	8.19	32.60	70.80	90.00
3.80	93.37	76.61	3.75	4.70	0.80	0.50	10.52	10.60	10.04	15.80	35.30	64.90
3.14	104.67	82.63	3.20	3.20	1.00	0.70	6.18	6.58	6.02	37.60	64.10	87.50
5.26	120.09	84.84	6.30	7.40	0.89	0.55	11.85	11.85	11.85	11.70	29.10	41.20
4.87	121.14	89.69	5.50	5.50	1.00	0.56	11.85	11.85	11.85	10.30	16.40	29.70
5.35	124.71	84.65	5.60	7.00	0.80	0.40	11.98	11.98	11.29	0.00	18.00	47.20
4.30	95.98	80.37	5.78	4.90	1.18	0.59	6.48	6.19	5.11	17.00	48.20	79.60
3.75	87.82	65.79	2.26	1.65	1.37	0.53	6.25	7.02	5.03	27.10	39.53	81.80
4.41	112.21	69.78	3.19	2.95	1.08	0.57	7.66	8.08	5.51	24.70	48.80	85.90
4.66	108.37	78.72	5.00	5.90	0.85	0.49	7.79	9.77	6.10	15.00	35.00	70.30
5.19	99.05	73.62	4.20	1.50	2.80	0.63	5.15	5.78	4.72	31.40	61.90	86.70
4.32	122.38	75.13	4.39	3.30	1.33	0.74	6.20	6.34	5.10	28.25	60.30	78.00
4.05	95.97	84.38	3.40	2.50	1.30	0.59	5.64	8.52	5.61	18.20	29.50	46.00
3.23	88.25	87.30	4.00	4.00	1.00	0.71	3.47	3.43	2.77	21.60	64.50	86.00
3.99	105.56	86.74	5.30	2.70	1.96	0.76	6.40	5.20	6.17	22.50	63.00	77.80
4.37	102.34	80.18	3.20	1.80	1.77	0.85	5.05	4.97	5.42	35.30	57.00	78.00
2.67	68.11	65.12	1.70	1.30	1.38	0.91	3.97	3.95	4.11	32.40	58.80	82.40
4.75	103.71	73.08	4.60	3.60	1.21	0.71	6.34	5.29	6.04	18.85	47.50	72.20
3.19	88.12	85.07	3.20	1.80	1.77	0.76	5.08	4.85	5.16	36.20	83.40	93.00
3.29	102.17	92.68	3.80	2.40	1.58	0.50	8.21	6.90	10.60	21.60	28.10	66.70
2.87	95.03	95.67	3.00	3.00	1.00	0.75	6.24	5.99	7.50	27.00	46.70	68.30

$\dot{V}m_{50}$ and $\dot{V}p_{50}$ = maximal and partial forced expiratory flows at 50 percent of control FVC; M/P ratio = ratio of $\dot{V}m_{50}$ to $\dot{V}p_{50}$ at control; MP slope = slope of the regression of percent decrements of Vm_{50} and $\dot{V}p_{50}$ recorded during the MCh inhalation challenge; $PD_{15}FEV_1$ = dose of MCh that decreased FEV_1 by 15 percent of control; PD_{40} Vm_{50} and $PD_{40}\dot{V}p_{50}$ = doses of MCh that decreased $\dot{V}m_{50}$ and Vp_{50} by 40 percent of control respectively; % max decr = percent maximal decrement at plateau.

Source: Dr. Riccardo Pellegrino. Used with permission.

18. The purpose of a study by O'Brien et al. (A-13) was to assess hypothalamic-pituitary-adrenal (HPA) axis function (known to be altered in depression) in patients with Alzheimer's disease (AD) by means of the adrenocorticotrophic hormone (ACTH) test, which assesses adrenal function by measuring cortisol production by the adrenal gland in response to an injection of ACTH. AD subjects (mean age 69.9 years with standard deviation of 9.8) were recruited from referrals to a hospital memory clinic. Normal control subjects consisted of spouses of patients and residents of a retirement hostel (mean age 73.8 with standard deviation of 11.6). There were eight males and eight females in the AD group and 10 males and eight females in the control group. Among the data collected were the following observations on age (C1), age at onset for AD subjects (C2), length of history of disease in months (C3), cognitive examination score (C4), peak cortisol level (C5), and total hormone response (C6):

Alzheimer's Disease Subjects						Controls					
C1	C2	C3	C4	C5	C6	C1	C2	C3	C4	C5	C6
73	69	48	75	400.00	44610	70	•	•	97	419.00	53175
87	83	48	39	565.00	63855	81	•	•	93	470.00	54285
60	54	72	67	307.00	31110	82	•	•	93	417.00	47160
62	57	60	64	335.00	36000	57	•	•	101	215.00	27120
75	70	48	51	352.00	44760	87	•	•	91	244.00	23895
63	60	24	79	426.00	47250	88	•	•	88	355.00	33565
81	77	48	51	413.00	51825	87	•	•	91	392.00	42810
66	64	24	61	402.00	41745	70	•	•	100	354.00	45105
78	73	60	32	518.00	66030	63	•	•	103	457.00	48765
72	64	72	61	505.00	49905	87	•	•	81	323.00	39360
69	65	48	73	427.00	55350	73	•	•	94	386.00	48150
76	73	36	63	409.00	51960	87	•	•	91	244.00	25830
46	41	60	73	333.00	33030	58	•	•	103	353.00	42060
77	75	18	63	591.00	73125	85	•	•	93	335.00	37425
64	61	36	59	559.00	60750	58	•	•	99	470.00	55140
72	69	30	47	511.00	54945	67	•	•	100	346.00	50745
						68	•	•	100	262.00	28440
						62	•	•	93	271.00	23595

• = Not applicable.

Source: Dr. John T. O'Brien. Used with permission.

19. Johnson et al. (A-14) note that the ability to identify the source of remembered information is a fundamental cognitive function. They conducted an experiment to explore the relative contribution of perceptual cues and cognitive operations information to age-related deficits in discriminating memories from different external sources (external source monitoring). Subjects for the experiment included 96 graduate and undergraduate students (41 males and 55 females) ranging in ages from 18 to 27 years. Among the data collected were the following performance recognition scores on source monitoring conditions (C1, C2, C3) and scores on the Benton Facial Recognition Test (C4), the Wechsler Adult Intelligence Scale—Revised (WAIS-R), WAIS-R Block Design subscale (C5), WAIS-R Vocabulary subscale (C6), the Benton Verbal Fluency Test (C7), and the Wisconsin Card Sorting Test (C8):

C1	C2	C3	C4	C5	C6	C7	C8
0.783	2.63	0.808	25	38	62	67	6
0.909	3.36	0.846	*	*	50	*	*
0.920	2.14	0.616	23	25	53	47	6
0.727	3.36	0.846	25	40	49	58	6
0.737	2.93	0.731	*	*	59	*	*
0.600	4.07	0.962	19	50	51	35	6
0.840	3.15	0.885	*	*	57	*	*
0.850	3.06	0.769	*	*	55	*	*

(*Continued*)

C1	C2	C3	C4	C5	C6	C7	C8
0.875	3.72	0.923	24	23	52	35	6
0.792	3.15	0.884	*	*	50	*	*
0.680	4.07	0.962	*	*	56	*	*
0.731	4.64	1.000	23	30	59	47	3
0.826	1.84	0.616	*	*	52	*	*
0.609	2.98	0.846	*	*	56	*	*
0.923	4.64	1.000	*	*	53	*	*
0.773	3.36	0.846	*	*	60	*	*
0.714	1.62	0.577	23	43	53	42	6
0.667	3.72	0.923	20	32	59	28	6
0.769	1.40	0.423	*	*	51	*	*
0.565	3.55	0.885	*	*	45	*	*
0.824	1.78	0.577	*	*	45	*	*
0.458	1.90	0.615	21	46	50	47	6
0.840	4.07	0.962	*	*	59	*	*
0.720	4.07	0.962	*	*	53	*	*
0.917	3.72	0.923	24	31	43	37	6
0.560	4.07	0.926	*	*	62	*	*
0.840	4.07	0.962	26	22	50	40	6
0.720	4.07	0.962	*	*	52	*	*
0.783	1.74	0.577	*	*	54	*	*
0.696	1.62	0.539	*	*	57	*	*
0.625	3.72	0.923	22	37	55	40	6
0.737	1.12	0.423	*	*	47	*	*
0.900	1.92	0.654	22	40	46	42	6
0.565	3.55	0.885	22	43	56	64	6
0.680	4.07	0.962	*	*	54	*	*
0.760	4.07	0.962	*	*	58	*	*
0.958	1.90	0.615	24	36	46	43	6
0.652	2.98	0.846	*	*	54	*	*
0.560	4.07	0.962	*	*	56	*	*
0.500	1.92	0.654	24	42	45	46	6
0.826	2.63	0.808	*	*	60	*	*
0.783	2.58	0.808	*	*	60	*	*
0.783	2.63	0.808	*	*	49	*	*
0.750	2.14	0.692	22	37	62	58	6
0.913	2.11	0.693	*	*	46	*	*
0.952	1.49	0.539	26	32	48	36	6
0.800	4.07	0.962	*	*	59	*	*
0.870	3.55	0.885	*	*	48	*	*
0.652	1.97	0.654	*	*	59	*	*
0.640	4.07	0.962	25	36	56	54	6
0.692	4.64	1.000	23	23	58	25	6
0.917	3.72	0.923	*	*	55	*	*
0.760	4.07	0.962	22	35	52	33	6
0.739	3.55	0.885	24	43	58	43	6

(Continued)

C1	C2	C3	C4	C5	C6	C7	C8
0.857	3.20	0.808	*	*	59	*	*
0.727	3.36	0.846	*	*	61	*	*
0.833	2.80	0.846	*	*	56	*	*
0.840	4.07	0.962	21	11	49	58	3
0.478	2.27	0.731	*	*	60	*	*
0.920	4.07	0.962	24	40	64	50	6
0.731	4.64	1.000	20	40	51	50	6
0.920	4.07	0.962	23	50	61	53	6
0.720	4.07	0.962	*	*	57	*	*
1.000	2.79	0.807	25	47	56	30	6
0.708	3.72	0.923	24	16	57	42	6
1.000	4.64	1.000	25	48	55	54	6
0.739	3.55	0.885	23	27	57	38	6
0.600	4.20	0.962	22	38	57	33	6
0.962	4.64	1.000	25	37	63	31	6
0.772	2.22	0.731	24	48	51	41	6
0.800	2.92	0.847	24	28	47	45	6
0.923	4.64	1.000	25	45	54	48	6
0.870	3.50	0.885	24	44	54	48	5
0.808	4.64	1.000	24	43	57	58	6
1.000	4.07	0.962	25	30	59	49	6
0.870	3.55	0.885	26	44	61	35	6
0.923	4.64	1.000	*	*	52	*	*
0.958	2.58	0.808	27	32	52	33	6
0.826	3.50	0.885	21	31	61	44	6
0.962	3.72	0.923	23	31	57	38	6
0.783	3.50	0.885	23	46	60	36	6
0.905	3.20	0.808	23	34	55	37	4
1.000	4.64	1.000	23	33	57	33	6
0.875	3.72	0.923	21	34	55	29	6
0.885	4.07	0.962	*	*	52	*	*
0.913	2.92	0.846	23	44	57	47	6
0.962	4.07	0.961	24	36	54	43	6
0.682	3.36	0.846	20	41	61	34	1
0.810	2.63	0.769	20	40	57	43	6
0.720	2.79	0.808	25	23	64	43	3
0.875	2.80	0.846	24	43	59	43	2
0.923	3.72	0.924	25	40	58	33	6
0.909	3.36	0.846	24	43	56	41	6
0.920	4.07	0.962	24	50	52	28	6
1.000	3.72	0.923	21	45	64	46	6
0.609	3.50	0.885	22	25	49	35	6

* = Missing data.

Source: Dr. Doreen M. De Leonardis. Used with permission.

Exercises for Use with the Large Data Sets Available on the Following Website: www.wiley.com/college/daniel

1. Winters et al. (A-15) conducted a study involving 248 high-school students enrolled in introductory physical education courses. The researchers wanted to know if social cognitive theory constructs were correlated with discretionary, "leisure-time" physical exercise. The main outcome variable is STREN, which is the number of days in a week that a high-school student engaged in strenuous physical activity (operationally defined as exercise that results in sweating, labored breathing, and rapid heart rate). Students in the study filled out lengthy questionnaires from which the following variables were derived:

   SELFR100—measures personal regulation of goal-directed behavior (higher values indicate more goal oriented).

   SS100—measures social support, social encouragement, and social expectation that are provided by friends and family for physical exercise (higher values indicate more support).

   SSE100—measures perceived ability to overcome barriers to exercise (higher values indicate higher ability).

   OEVNORM—measures outcome expectations and their associated expectancies for physical exercise (higher values indicate stronger perceived links to desired outcomes from exercise).

   With these data (LTEXER),

   (a) Calculate the bivariate correlation for each pair of variables and interpret the meaning of each.

   (b) Using STREN as the dependent variable, compute the multiple correlation coefficient.

   (c) Using STREN as the dependent variable, calculate the partial correlation coefficient for STREN and SELFR100 after controlling for SS100.

   (d) Using STREN as the dependent variable, calculate the partial correlation coefficient for STREN and SSE100 after controlling for OEVNORM.

   Note that there many missing values in this data set.

2. With data obtained from a national database on childbirth, Matulavich et al. (A-16) examined the number of courses of prescribed steroids a mother took during pregnancy (STEROIDS). The size of the baby was measured by length (cm), weight (grams), and head circumference (cm). Calculate the correlation of the number of courses of steroids with each of the three outcome variables. What are the hypotheses for your tests? What are the p-values? What are your conclusions? (The name of the data set is STERLENGTH.)

3. Refer to the data on cardiovascular risk factors (RISKFACT). The subjects are 1000 males engaged in sedentary occupations. You wish to study the relationships among risk factors in this population. The variables are

   Y = oxygen consumption
   X_1 = systolic blood pressure (mm Hg)
   X_2 = total cholesterol (mg/dl)
   X_3 = HDL cholesterol (mg/dl)
   X_4 = triglycerides (mg/dl)

   Select a simple random sample from this population and carry out an appropriate statistical analysis. Prepare a narrative report of your findings and compare them with those of your classmates. Consult with your instructor regarding the size of the sample.

4. Refer to the data on 500 patients who have sought treatment for the relief of respiratory disease symptoms (RESPDIS). A medical research team is conducting a study to determine what factors may be related to respiratory disease. The dependent variable Y is a measure of the severity of the disease. A larger value indicates a more serious condition. The independent variables are as follows:

X_1 = education (highest grade completed)
X_2 = measure of crowding of living quarters
X_3 = measure of air quality at place of residence (a larger number indicates poorer quality)
X_4 = nutritional status (a large number indicates a higher level of nutrition)
X_5 = smoking status (0 = smoker, 1 = nonsmoker)

Select a simple random sample of subjects from this population and conduct a statistical analysis that you think would be of value to the research team. Prepare a narrative report of your results and conclusions. Use graphic illustrations where appropriate. Compare your results with those of your classmates. Consult your instructor regarding the size of sample you should select.

REFERENCES

Methodology References

1. GEORGE W. SNEDECOR and WILLIAM G. COCHRAN, *Statistical Methods,* Sixth Edition, Iowa State University Press, Ames, 1967.
2. ROBERT G. D. STEEL and JAMES H. TORRIE, *Principles and Procedures of Statistics,* McGraw-Hill, New York, 1960.
3. R. L. ANDERSON and T. A. BANCROFT, *Statistical Theory in Research,* McGraw-Hill, New York, 1952.
4. M. H. KUTNER, C. J. NACHTSHEIM, J. NETER, and W. LI, *Applied Linear Statistical Models*, Fifth Edition, McGraw-Hill/Irwin, New York, 2005.

Applications References

A-1. DEBRA A. JANSEN and MARY L. KELLER, "Cognitive Function in Community-Dwelling Elderly Women," *Journal of Gerontological Nursing, 29* (2003), 34–43.

A-2. M. NAEIJE, "Local Kinematic and Anthropometric Factors Related to the Maximum Mouth Opening in Healthy Individuals," *Journal of Oral Rehabilitation, 29* (2002), 534–539.

A-3. GWI-RYUNG SON, MAY L. WYKLE, and JACLENE A. ZAUSZNIEWSKI, "Korean Adult Child Caregivers of Older Adults with Dementia," *Journal of Gerontological Nursing, 29* (2003), 19–28.

A-4. X. WANG, X. SHEN, X. LI, and C. MAULI AGRAWAL, "Age-Related Changes in the Collagen Network and Toughness of Bone," *Bone, 31* (2002), 1–7.

A-5. PETER A. ANTON, RONALD T. MITSUYASU, STEVEN G. DEEKS, DAVID T. SCADDEN, BRIDGET WAGNER, CHRISTINE HUANG, CATHERINE MACKEN, DOUGLAS D. RICHMAN, CINDY CHRISTOPHERSON, FLAVIA BORELLINI, RICHARD LAZAR, and KRISTEN M. HEGE, "Multiple Measures of HIV Burden in Blood and Tissue Are Correlated with Each Other but Not with Clinical Parameter in Aviremic Subjects," *AIDS, 17* (2003), 53–63.

A-6. OLGA KWAST-RABBEN, ROLF LIBELIUS, and HANNU HEIKKILÄ, "Somatosensory Evoked Potentials Following Stimulation of Digital Nerves," *Muscle and Nerve, 26* (2002), 533–538.

A-7. MITSUO HASHIMOTO, TAIKI FUKUDA, TETSUYA SHIMIZU, SHOU WATANABE, SATOSHI WATANUKI, YOSHIKATSU ETO, and MITSUYOSHI URASHIMA, "Influence of Climate Factors on Emergency Visits for Childhood Asthma Attack," *Pediatrics International, 46* (2004), 48–52.

A-8. PETER STENVINKEL, ALICIA MARCHELEWSKA, ROBERTO PECOITS-FILHO, OLOF HEIMBÜRGER, ZHENGZHONG ZHANG, CATHERINE HOFF, CLIFF HOLMES, JOANAS AXELSSON, SIVONNE ARVIDSSON, MARTIN SCHALLING, PETER BARANY, BENGT LINKHOLM, and LOUISE NORDFORS, "Adiponectin in Renal Disease: Relationship to

Phenotype and Genetic Variation in the Genetic Encoding Adiponectin," *Kidney International, 65* (2004), 274–281.

A-9. R. VAN SCHUYLENBERGH, B. VANDEN EYNDE, and P. HESPEL, "Prediction of Sprint Triathlon Performance from Laboratory Tests," *European Journal of Applied Physiology, 91* (2004), 94–99.

A-10. WALDEMAR TOMALAK, ANDRZEJ POGORZELSKI, and JAROSLAW PRUSAK, "Normal Values for Maximal Static Inspiratory and Expiratory Pressures in Healthy Children," *Pediatric Pulmonology, 34* (2002), 42–46.

A-11. MARCO DI MONACO, FULVIA VALLERO, ROBERTO DI MONACO, FULVIO MAUTINO, and ALBERTO CAVANNA, "Total Lymphocyte Count and Femoral Bone Mineral Density in Postmenopausal Women," *Journal of Bone Mineral Metabolism, 22* (2004), 58–63.

A-12. RICCARDO PELLEGRINO, BENEDETTO VIOLANTE, and VITO BRUSASCO, "Maximal Bronchoconstriction in Humans: Relationship to Deep Inhalation and Airway Sensitivity," *American Journal of Respiratory and Critical Care Medicine, 153* (1996), 115–121.

A-13. J. T. O'BRIEN, D. AMES, I. SCHWEITZER, M. MASTWYK, and P. COLMAN, "Enhanced Adrenal Sensitivity to Adrenocorticotrophic Hormone (ACTH) Is Evidence of HPA Axis Hyperactivity in Alzheimer's Disease," *Psychological Medicine, 26* (1996), 7–14.

A-14. MARCIA K. JOHNSON, DOREEN M. DE LEONARDIS, SHAHIN HARSHTROUDI, and SUSAN A. FERGUSON, "Aging and Single Versus Multiple Cues in Source Monitoring," *Psychology and Aging, 10* (1995), 507–517.

A-15. ERIC R. WINTERS, RICK L. PETOSA, and THOMAS E. CHARLETON, "Using Social Cognitive Theory to Explain Discretionary, 'Leisure-Time' Physical Exercise Among High School Students," *Journal of Adolescent Health, 32* (2003), 436–442.

A-16. ANGELA MATULAVICH, DONNA MILES-CURRY, BARBARA WARNER, BOBBE GRAY, and the National Institute of Child and Health Development. Data analyzed at the Wright State University Statistical Consulting Center.

CHAPTER **11**

ADDITIONAL TECHNIQUES FOR THE ANALYSIS OF RELATIONSHIPS AMONG VARIABLES

CHAPTER OVERVIEW

This chapter discusses some additional tools and concepts that are useful in regression analysis. The presentation includes expansions of the basic ideas and techniques of regression analysis that were introduced in Chapters 9 and 10.

TOPICS

LEARNING OUTCOMES

After studying this chapter, the student will

1. understand how to include qualitative variables in a regression analysis.
2. understand how to use automated variable selection procedures to develop regression models.
3. be able to perform logistic regression for dichotomous and polytomous dependent variables.

11.1 INTRODUCTION

The basic concepts and methodology of regression analysis are covered in Chapters 9 and 10. In Chapter 9 we discuss the situation in which the objective is to obtain an equation that can be used to make predictions and estimates about some dependent variable from knowledge of some other single variable that we call the independent, predictor, or explanatory variable. In Chapter 10 the ideas and techniques learned in Chapter 9 are expanded to cover the situation in which it is believed that the inclusion of information on two or more independent variables will yield a better equation for use in making predictions and estimations. Regression analysis is a complex and powerful statistical tool that is widely employed in health sciences research. To do the subject justice requires more space than is available in an introductory statistics textbook. However, for the benefit of those who wish additional coverage of regression analysis, we present in this chapter some additional topics that should prove helpful to the student and practitioner of statistics.

Regression Assumptions Revisited As we learned in Chapters 9 and 10, there are several assumptions underlying the appropriate use of regression procedures. Often there are certain measurements that strongly influence the shape of a distribution or impact the magnitude of the variance of a measured variable. Other times, certain independent variables that are being used to develop a model are highly correlated, leading to the development of a model that may not be unique or correct.

Non-Normal Data Many times the data that are used to build a regression model are not normally distributed. One may wish to explore the possibility that some of the observed data points are outliers or that they disproportionately affect the distribution of the data. Such an investigation may be accomplished informally by constructing a scatter plot and looking for observations that do not seem to fit with the others. Alternatively, many computer packages produce formal tests to evaluate potential outlying observations in either the dependent variable or the independent variables. It is always up to the researcher, however, to justify which observations are to be removed from the data set prior to analysis.

Often one may wish to attempt a transformation of the data. Mathematical transformations are useful because they do not affect the underlying relationships among variables. Since hypothesis tests for the regression coefficients are based on normal distribution statistics, data transformations can sometimes normalize the data to the extent necessary to perform such tests. Simple transformations, such as taking the square root of measurements or taking the logarithm of measurements, are quite common.

EXAMPLE 11.1.1

Researchers were interested in blood concentrations of delta-9-tetrahydrocannabinol (Δ-9-THC), the active psychotropic component in marijuana, from 25 research subjects. These data are presented in Table 11.1.1, as are these same data after using a $\log_{10}$ transformation.

TABLE 11.1.1 Data from a Random Sample of 25 Research Subjects Tested for Δ-9-THC, Example 11.1.1

Case No.	Concentration (μg/ml)	Log_{10} Concentration (μg/ml)
1	.30	−.52
2	2.75	.44
3	2.27	.36
4	2.37	.37
5	1.12	.05
6	.60	−.22
7	.61	−.21
8	.89	−.05
9	.33	−.48
10	.85	−.07
11	2.18	.34
12	3.59	.56
13	.28	−.55
14	1.90	.28
15	1.71	.23
16	.85	−.07
17	1.53	.18
18	2.25	.35
19	.88	−.05
20	.49	−.31
21	4.35	.64
22	.67	−.17
23	2.74	.44
24	.79	−.10
25	6.94	.84

Box-and-whisker plots from SPSS software for these data are shown in Figure 11.1.1. The raw data are clearly skewed, and an outlier is identified (observation 25). A $\log_{10}$ transformation, which is often useful for such skewed data, removes the magnitude of the outlier and results in a distribution that is much more nearly symmetric about the median. Therefore, the transformed data could be used in lieu of the raw data for constructing the regression model. Though symmetric data do not, necessarily, imply that the data are normal, they do result in a more appropriate model. Formal tests of normality, as previously mentioned, should always be carried out prior to analysis. ■

Unequal Error Variances When the variances of the error terms are not equal, we may obtain a satisfactory equation for the model, but, because the assumption that the error variances are equal is violated, we will not be able to perform appropriate hypothesis tests on the model coefficients. Just as was the case in overcoming the non-normality problem, transformations of the regression variables may reduce the impact of unequal error variances.

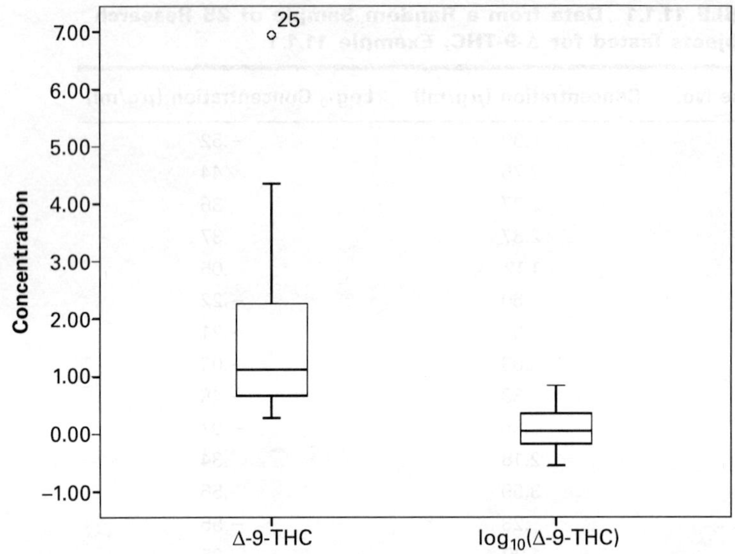

FIGURE 11.1.1 Box-and-whisker plots of data from Example 11.1.1.

Correlated Independent Variables *Multicollinearity* is a common problem that arises when one attempts to build a model using many independent variables. Multicollinearity occurs when there is a high degree of correlation among the independent variables. For example, imagine that we want to find an equation relating height and weight to blood pressure. A common variable that is derived from height and weight is called the body mass index (BMI). If we attempt to find an equation relating height, weight, and BMI to blood pressure, we can expect to run into analytical problems because BMI, by definition, is highly correlated with both height and weight.

The problem arises mathematically when the solutions for the regression coefficients are derived. Since the data are correlated, solutions may not be found that are unique to a given model. The least complex solution to multicollinearity is to calculate correlations among all of the independent variables and to retain only those variables that are not highly correlated. A conservative rule of thumb to remove redundancy in the data set is to eliminate variables that are related to others with a significant correlation coefficient above 0.7.

EXAMPLE 11.1.2

A study of obesity and metabolic syndrome used data collected from 15 students, and included systolic blood pressure (SBP), weight, and BMI. These data are presented in Table 11.1.2.

Correlations for the three variables are shown in Figure 11.1.2. The very large and significant correlation between the variables weight and BMI suggests that including both of these variables in the model is inappropriate because of the high level of redundancy in the information provided by these variables. This makes logical sense since BMI is a function of weight. The researcher is now faced with the task of deciding which of the variables to retain for constructing the regression model.

TABLE 11.1.2 Data from a Random Sample of 15 Students

Case No.	SBP	Weight (lbs.)	BMI
1	126	125	24.41
2	129	130	23.77
3	126	132	20.07
4	123	200	27.12
5	124	321	39.07
6	125	100	20.90
7	127	138	22.96
8	125	138	24.44
9	123	149	23.33
10	119	180	25.82
11	127	184	26.40
12	126	251	31.37
13	122	197	26.72
14	126	107	20.22
15	125	125	23.62

Correlations: SBP, Weight, BMI

```
            SBP      Weight

Weight   -0.289
p-value   0.296

BMI      -0.213      0.962
p-value   0.447      0.000
```

FIGURE 11.1.2 Correlations calculated in MINITAB software for the data in Example 11.1.2. ■

11.2 QUALITATIVE INDEPENDENT VARIABLES

The independent variables considered in the discussion in Chapter 10 were all quantitative; that is, they yielded numerical values that were either counts or measurements in the usual sense of the word. For example, some of the independent variables used in our examples and exercises were age, education level, collagen porosity, and collagen tensile strength. Frequently, however, it is desirable to use one or more qualitative variables as independent variables in the regression model. Qualitative variables, it will be recalled, are those variables whose "values" are categories and that convey the concept of attribute rather than amount or quantity. The variable marital status, for example, is a qualitative variable whose categories are "single," "married," "widowed," and "divorced." Other examples of qualitative variables include sex (male or female), diagnosis, race, occupation, and immunity status to some disease. In certain situations an investigator may suspect that including one or more variables such as these in the regression equation would contribute significantly to the reduction of the error sum of squares and thereby provide more precise estimates of the parameters of interest.

Suppose, for example, that we are studying the relationship between the dependent variable systolic blood pressure and the independent variables weight and age. We might also want to include the qualitative variable sex as one of the independent variables. Or suppose we wish to gain insight into the nature of the relationship between lung capacity and other relevant variables. Candidates for inclusion in the model might consist of such quantitative variables as height, weight, and age, as well as qualitative variables such as sex, area of residence (urban, suburban, rural), and smoking status (current smoker, ex-smoker, never smoked).

Dummy Variables In order to incorporate a qualitative independent variable in the multiple regression model, it must be quantified in some manner. This may be accomplished through the use of what are known as *dummy variables*.

DEFINITION

A *dummy variable* is a variable that assumes only a finite number of values (such as 0 or 1) for the purpose of identifying the different categories of a qualitative variable.

The term "dummy" is used to indicate the fact that the numerical values (such as 0 and 1) assumed by the variable have no quantitative meaning but are used merely to identify different categories of the qualitative variable under consideration. Qualitative variables are sometimes called *indicator* variables, and when there are only two categories, they are sometimes called *dichotomous* variables.

The following are some examples of qualitative variables and the dummy variables used to quantify them:

Qualitative Variable	Dummy Variable
Sex (male, female):	$x_1 = \begin{cases} 1 \text{ for male} \\ 0 \text{ for female} \end{cases}$
Place of residence (urban, rural, suburban):	$x_1 = \begin{cases} 1 \text{ for urban} \\ 0 \text{ for rural and suburban} \end{cases}$
	$x_2 = \begin{cases} 1 \text{ for rural} \\ 0 \text{ for urban and suburban} \end{cases}$
Smoking status [current smoker, ex-smoker (has not smoked for 5 years or less), ex-smoker (has not smoked for more than 5 years), never smoked]:	$x_1 = \begin{cases} 1 \text{ for current smoker} \\ 0 \text{ for otherwise} \end{cases}$
	$x_2 = \begin{cases} 1 \text{ for ex-smoker } (\leq 5 \text{ years}) \\ 0 \text{ otherwise} \end{cases}$
	$x_3 = \begin{cases} 1 \text{ for ex-smoker } (> 5 \text{ years}) \\ 0 \text{ otherwise} \end{cases}$

Note in these examples that when the qualitative variable has k categories, $k - 1$ dummy variables must be defined for all the categories to be properly coded. This rule is applicable for any multiple regression containing an intercept constant. The variable sex, with two categories, can be quantified by the use of only one dummy variable, while three dummy variables are required to quantify the variable smoking status, which has four categories.

The following examples illustrate some of the uses of qualitative variables in multiple regression. In the first example we assume that there is no interaction between the independent variables. Since the assumption of no interaction is not realistic in many instances, we illustrate, in the second example, the analysis that is appropriate when interaction between variables is accounted for.

EXAMPLE 11.2.1

In a study of factors thought to be associated with birth weight, a simple random sample of 100 birth records was selected from the North Carolina 2001 Birth Registry (A-1). Table 11.2.1 shows, for three variables, the data extracted from each record. There are two independent variables: length of gestation (weeks), which is quantitative, and smoking status of mother (smoke), a qualitative variable. The dependent variable is birth weight (grams).

TABLE 11.2.1 Data from a Simple Random Sample of 100 Births from the North Carolina Birth Registry, Example 11.2.1

Case No.	Grams	Weeks	Smoke	Case No.	Grams	Weeks	Smoke
1	3147	40	0	51	3232	38	0
2	2977	41	0	52	3317	40	0
3	3119	38	0	53	2863	37	0
4	3487	38	0	54	3175	37	0
5	4111	39	0	55	3317	40	0
6	3572	41	0	56	3714	34	0
7	3487	40	0	57	2240	36	0
8	3147	41	0	58	3345	39	0
9	3345	38	1	59	3119	39	0
10	2665	34	0	60	2920	37	0
11	1559	34	0	61	3430	41	0
12	3799	38	0	62	3232	35	0
13	2750	38	0	63	3430	38	0
14	3487	40	0	64	4139	39	0
15	3317	38	0	65	3714	39	0
16	3544	43	1	66	1446	28	1
17	3459	45	0	67	3147	39	1
18	2807	37	0	68	2580	31	0
19	3856	40	0	69	3374	37	0
20	3260	40	0	70	3941	40	0
21	2183	42	1	71	2070	37	0
22	3204	38	0	72	3345	40	0

(*Continued*)

Case No.	Grams	Weeks	Smoke	Case No.	Grams	Weeks	Smoke
23	3005	36	0	73	3600	40	0
24	3090	40	1	74	3232	41	0
25	3430	39	0	75	3657	38	1
26	3119	40	0	76	3487	39	0
27	3912	39	0	77	2948	38	0
28	3572	40	0	78	2722	40	0
29	3884	41	0	79	3771	40	0
30	3090	38	0	80	3799	45	0
31	2977	42	0	81	1871	33	0
32	3799	37	0	82	3260	39	0
33	4054	40	0	83	3969	38	0
34	3430	38	1	84	3771	40	0
35	3459	41	0	85	3600	40	0
36	3827	39	0	86	2693	35	1
37	3147	44	1	87	3062	45	0
38	3289	38	0	88	2693	36	0
39	3629	36	0	89	3033	41	0
40	3657	36	0	90	3856	42	0
41	3175	41	1	91	4111	40	0
42	3232	43	1	92	3799	39	0
43	3175	36	0	93	3147	38	0
44	3657	40	1	94	2920	36	0
45	3600	39	0	95	4054	40	0
46	3572	40	0	96	2296	36	0
47	709	25	0	97	3402	38	0
48	624	25	0	98	1871	33	1
49	2778	36	0	99	4167	41	0
50	3572	35	0	100	3402	37	1

Source: John P. Holcomb, sampled and coded from North Carolina Birth Registry data found at *www.irss.unc.edu/ncvital/bfd1down.html.*

Solution: For the analysis, we quantify smoking status by means of a dummy variable that is coded 1 if the mother is a smoker and 0 if she is a nonsmoker. The data in Table 11.2.1 are plotted as a scatter diagram in Figure 11.2.1. The scatter diagram suggests that, in general, longer periods of gestation are associated with larger birth weights.

To obtain additional insight into the nature of these data, we may enter them into a computer and employ an appropriate program to perform further analyses. For example, we enter the observations $y_1 = 3147$, $x_{11} = 40$, $x_{21} = 0$, for the first case; $y_2 = 2977$, $x_{12} = 41$, $x_{22} = 0$ for the second case; and so on. Figure 11.2.2 shows the computer output obtained with the use of the MINITAB multiple regression program.

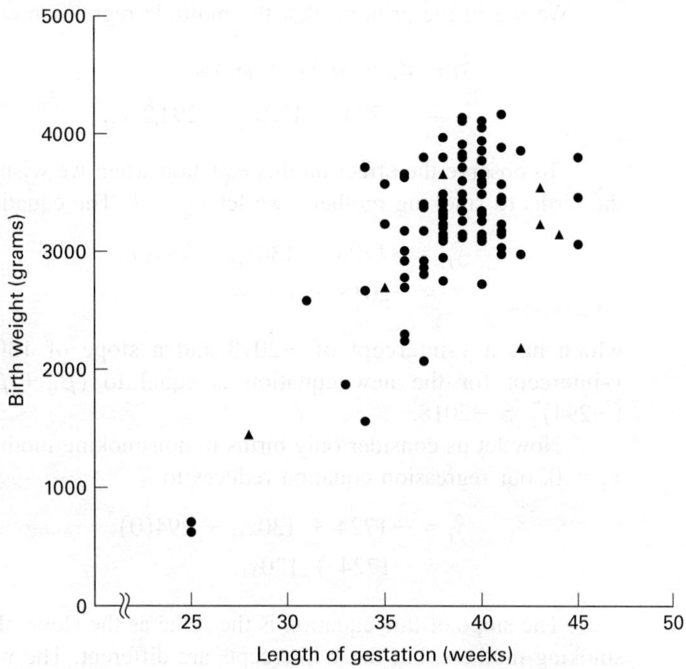

FIGURE 11.2.1 Birth weights and lengths of gestation for 100 births: (▲) smoking and (●) nonsmoking mothers.

The regression equation is

grams = −1724 + 130 x1 − 294 x2

Predictor	Coef	SE Coef	T	P
Constant	−1724.4	558.8	−3.09	0.003
weeks (x1)	130.05	14.52	8.96	0.000
smoke (x2)	−294.4	135.8	−2.17	0.033

S = 484.6   R-Sq = 46.4%   R-Sq(adj) = 45.3%

Analysis of Variance

SOURCE	DF	SS	MS	F	P
Regression	2	19689185	9844593	41.92	0.000
Residual Error	97	22781681	234863		
Total	99	42470867			

SOURCE	DF	Seq SS
x1	1	18585166
x2	1	1104020

FIGURE 11.2.2 Partial computer printout, MINITAB multiple regression analysis. Example 11.2.1.

We see in the printout that the multiple regression equation is

$$\hat{y}_j = \hat{\beta}_0 + \hat{\beta}_1 x_{1j} + \hat{\beta}_2 x_{2j}$$
$$\hat{y}_j = -1724 + 130x_{1j} - 294\hat{\beta}_2 x_{2j} \tag{11.2.1}$$

To observe the effect on this equation when we wish to consider only the births to smoking mothers, we let $x_{2j} = 1$. The equation then becomes

$$\begin{aligned}\hat{y}_j &= -1724 + 130x_{1j} - 294(1) \\ &= -2018 + 130x_{1j}\end{aligned} \tag{11.2.2}$$

which has a y-intercept of -2018 and a slope of 130. Note that the y-intercept for the new equation is equal to $(\hat{\beta}_0 + \hat{\beta}_1) = [-1724 + (-294)] = -2018$.

Now let us consider only births to nonsmoking mothers. When we let $x_2 = 0$, our regression equation reduces to

$$\begin{aligned}\hat{y}_j &= -1724 + 130x_{1j} - 294(0) \\ &= -1724 + 130x_{1j}\end{aligned} \tag{11.2.3}$$

The slope of this equation is the same as the slope of the equation for smoking mothers, but the y-intercepts are different. The y-intercept for the equation associated with nonsmoking mothers is larger than the one for the

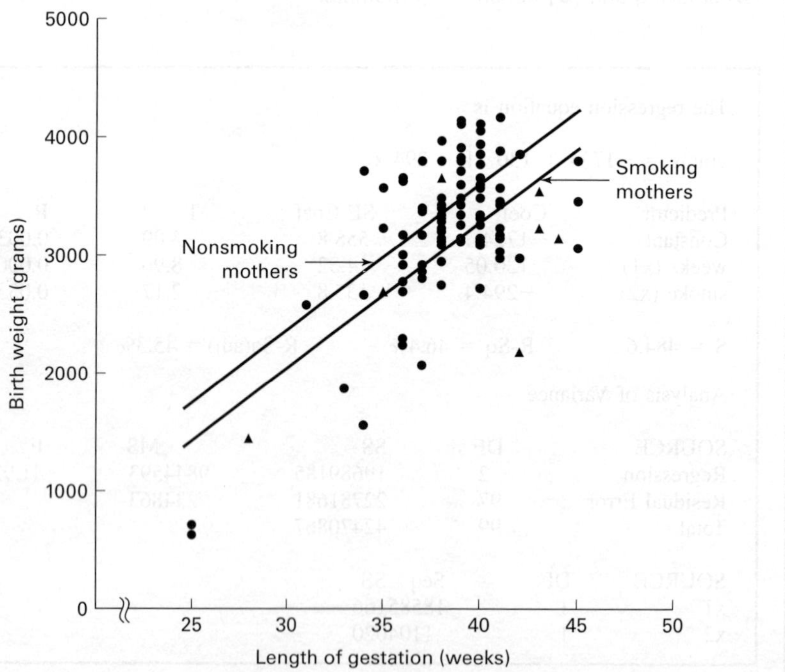

FIGURE 11.2.3 Birth weights and lengths of gestation for 100 births and the fitted regression lines: (▲) smoking and (●) nonsmoking mothers.

smoking mothers. These results show that for this sample, babies born to mothers who do not smoke weighed, on the average, more than babies born to mothers who do smoke, when length of gestation is taken into account. The amount of the difference, on the average, is 294 grams. Stated another way, we can say that for this sample, babies born to mothers who smoke weighed, on the average, 294 grams less than the babies born to mothers who do not smoke, when length of gestation is taken into account. Figure 11.2.3 shows the scatter diagram of the original data along with a plot of the two regression lines (Equations 11.2.2 and 11.2.3). ■

EXAMPLE 11.2.2

At this point a question arises regarding what inferences we can make about the sampled population on the basis of the sample results obtained in Example 11.2.1. First of all, we wish to know if the sample difference of 294 grams is significant. In other words, does smoking have an effect on birth weight? We may answer this question through the following hypothesis testing procedure.

Solution:

1. **Data.** The data are as given in Example 11.2.1.
2. **Assumptions.** We presume that the assumptions underlying multiple regression analysis are met.
3. **Hypotheses.** $H_0: \beta_2 = 0$; $H_A: \beta_2 \neq 0$. Suppose we let $\alpha = .05$.
4. **Test statistic.** The test statistic is $t = (\hat{\beta}_2 - 0)/s_{\hat{\beta}_2}$.
5. **Distribution of test statistic.** When the assumptions are met and H_0 is true the test statistic is distributed as Student's t with 97 degrees of freedom.
6. **Decision rule.** We reject H_0 if the computed t is either greater than or equal to 1.9848 or less than or equal to -1.9848 (obtained by interpolation).
7. **Calculation of test statistic.** The calculated value of the test statistic appears in Figure 11.2.2 as the t ratio for the coefficient associated with the variable appearing in Column 3 of Table 11.2.1. This coefficient, of course, is $\hat{\beta}_2$. We see that the computed t is -2.17.
8. **Statistical decision.** Since $-2.17 < -1.9848$, we reject H_0.
9. **Conclusion.** We conclude that, in the sampled population, whether the mothers smoke is associated with a reduction in the birth weights of their babies.
10. ***p* value.** For this test we have $p = .033$ from Figure 11.2.2. ■

A Confidence Interval for β_2 Given that we are able to conclude that in the sampled population the smoking status of the mothers does have an effect on the birth weights of their babies, we may now inquire as to the magnitude of the effect. Our best point estimate of the average difference in birth weights, when length of gestation is taken into account, is 294 grams in favor of babies born to mothers who do not smoke. We may obtain an interval estimate of the mean amount of the difference by using information from the computer printout by means of the following expression:

$$\hat{\beta}_2 \pm ts_{\hat{\beta}_2}$$

For a 95 percent confidence interval, we have

$$-294.4 \pm 1.9848(135.8)$$

$$-563.9, -24.9$$

Thus, we are 95 percent confident that the difference is somewhere between about 564 grams and 25 grams.

Advantages of Dummy Variables The reader may have correctly surmised that an alternative analysis of the data of Example 11.2.1 would consist of fitting two separate regression equations: one to the subsample of mothers who smoke and another to the subsample of those who do not. Such an approach, however, lacks some of the advantages of the dummy variable technique and is a less desirable procedure when the latter procedure is valid. If we can justify the assumption that the two separate regression lines have the same slope, we can get a better estimate of this common slope through the use of dummy variables, which entails pooling the data from the two subsamples. In Example 11.2.1 the estimate using a dummy variable is based on a total sample size of 100 observations, whereas separate estimates would be based on a sample of 85 smokers and only 15 nonsmokers. The dummy variables approach also yields more precise inferences regarding other parameters since more degrees of freedom are available for the calculation of the error mean square.

Use of Dummy Variables: Interaction Present Now let us consider the situation in which interaction between the variables is assumed to be present. Suppose, for example, that we have two independent variables: one quantitative variable X_1 and one qualitative variable with three response levels yielding the two dummy variables X_2 and X_3. The model, then, would be

$$y_j = \beta_0 + \beta_1 X_{1j} + \beta_2 X_{2j} + \beta_3 X_{3j} + \beta_4 X_{1j} X_{2j} + \beta_5 X_{1j} X_{3j} + \epsilon_j \qquad (11.2.4)$$

in which $\beta_4 X_{1j} X_{2j}$ and $\beta_5 X_{1j} X_{3j}$ are called *interaction terms* and represent the interaction between the quantitative and the qualitative independent variables. Note that there is no need to include in the model the term containing $X_{2j} X_{3j}$; it will always be zero because when $X_2 = 1, X_3 = 0$, and when $X_3 = 1, X_2 = 0$. The model of Equation 11.2.4 allows for a different slope and Y-intercept for each level of the qualitative variable.

Suppose we use dummy variable coding to quantify the qualitative variable as follows:

$$X_2 = \begin{cases} 1 \text{ for level 1} \\ 0 \text{ otherwise} \end{cases}$$

$$X_3 = \begin{cases} 1 \text{ for level 2} \\ 0 \text{ otherwise} \end{cases}$$

The three sample regression equations for the three levels of the qualitative variable, then, are as follows:

Level 1 ($X_2 = 1, X_3 = 0$)

$$\begin{aligned}\hat{y}_j &= \hat{\beta}_0 + \hat{\beta}_1 x_{1j} + \hat{\beta}_2(1) + \hat{\beta}_3(0) + \hat{\beta}_4 x_{1j}(1) + \hat{\beta}_5 x_{1j}(0)\\ &= \hat{\beta}_0 + \hat{\beta}_1 x_{1j} + \hat{\beta}_2 + \hat{\beta}_4 x_{1j}\\ &= (\hat{\beta}_0 + \hat{\beta}_2) + (\hat{\beta}_1 + \hat{\beta}_4)x_{1j}\end{aligned} \tag{11.2.5}$$

Level 2 ($X_2 = 0, X_3 = 1$)

$$\begin{aligned}\hat{y}_j &= \hat{\beta}_0 + \hat{\beta}_1 x_{1j} + \hat{\beta}_2(0) + \hat{\beta}_3(1) + \hat{\beta}_4 x_{1j}(0) + \hat{\beta}_5 x_{1j}(1)\\ &= \hat{\beta}_0 + \hat{\beta}_1 x_{1j} + \hat{\beta}_3 + \hat{\beta}_5 x_{1j}\\ &= (\hat{\beta}_0 + \hat{\beta}_3) + (\hat{\beta}_1 + \hat{\beta}_5)x_{1j}\end{aligned} \tag{11.2.6}$$

Level 3 ($X_2 = 0, X_3 = 0$)

$$\begin{aligned}\hat{y}_j &= \hat{\beta}_0 + \hat{\beta}_1 x_{1j} + \hat{\beta}_2(0) + \hat{\beta}_3(0) + \hat{\beta}_4 x_{1j}(0) + \hat{\beta}_5 x_{1j}(0)\\ &= \hat{\beta}_0 + \hat{\beta}_1 x_{1j}\end{aligned} \tag{11.2.7}$$

Let us illustrate these results by means of an example.

EXAMPLE 11.2.3

A team of mental health researchers wishes to compare three methods (A, B, and C) of treating severe depression. They would also like to study the relationship between age and treatment effectiveness as well as the interaction (if any) between age and treatment. Each member of a simple random sample of 36 patients, comparable with respect to diagnosis and severity of depression, was randomly assigned to receive treatment A, B, or C. The results are shown in Table 11.2.2. The dependent variable Y is treatment effectiveness, the quantitative independent variable X_1 is patient's age at nearest birthday, and the independent variable type of treatment is a qualitative variable that occurs at three levels. The following dummy variable coding is used to quantify the qualitative variable:

$$X_2 = \begin{cases} 1 \text{ for treatment A} \\ 0 \text{ otherwise} \end{cases}$$

$$X_3 = \begin{cases} 1 \text{ for treatment B} \\ 0 \text{ otherwise} \end{cases}$$

The scatter diagram for these data is shown in Figure 11.2.4. Table 11.2.3 shows the data as they were entered into a computer for analysis. Figure 11.2.5 contains the printout of the analysis using the MINITAB multiple regression program.

Solution: Now let us examine the printout to see what it provides in the way of insight into the nature of the relationships among the variables. The least-squares equation is

$$\hat{y}_j = 6.21 + 1.03x_{1j} + 41.3x_{2j} + 22.7x_{3j} - .703x_{1j}x_{2j} - .510x_{1j}x_{3j}$$

TABLE 11.2.2 Data for Example 11.2.3

Measure of Effectiveness	Age	Method of Treatment
56	21	A
41	23	B
40	30	B
28	19	C
55	28	A
25	23	C
46	33	B
71	67	C
48	42	B
63	33	A
52	33	A
62	56	C
50	45	C
45	43	B
58	38	A
46	37	C
58	43	B
34	27	C
65	43	A
55	45	B
57	48	B
59	47	C
64	48	A
61	53	A
62	58	B
36	29	C
69	53	A
47	29	B
73	58	A
64	66	B
60	67	B
62	63	A
71	59	C
62	51	C
70	67	A
71	63	C

The three regression equations for the three treatments are as follows:

Treatment A (Equation 11.2.5)

$$\hat{y}_j = (6.21 + 41.3) + (1.03 - .703)x_{1j}$$
$$= 47.51 + .327x_{1j}$$

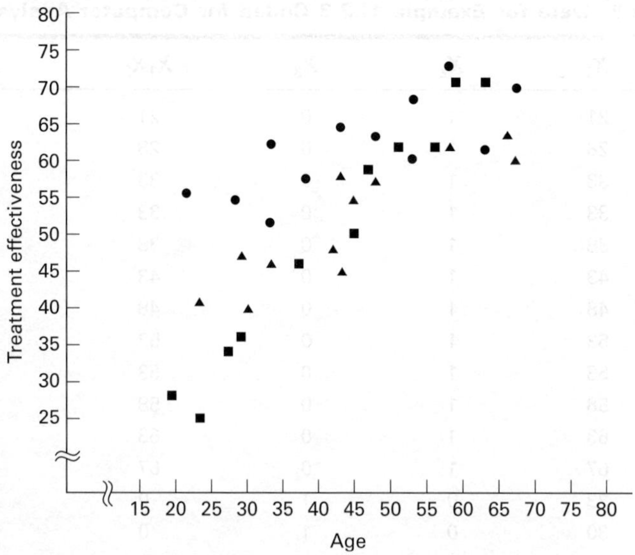

FIGURE 11.2.4 Scatter diagram of data for Example 11.2.3: (●) treatment A, (▲) treatment B, (■) treatment C.

Treatment B (Equation 11.2.6)

$$\hat{y}_j = (6.21 + 22.7) + (1.03 - .510)x_{1j}$$
$$= 28.91 + .520x_{1j}$$

Treatment C (Equation 11.2.7)

$$\hat{y}_j = 6.21 + 1.03x_{1j}$$

Figure 11.2.6 contains the scatter diagram of the original data along with the regression equations for the three treatments. Visual inspection of Figure 11.2.6 suggests that treatments A and B do not differ greatly with respect to their slopes, but their y-intercepts are considerably different. The graph suggests that treatment A is better than treatment B for younger patients, but the difference is less dramatic with older patients. Treatment C appears to be decidedly less desirable than both treatments A and B for younger patients but is about as effective as treatment B for older patients. These subjective impressions are compatible with the contention that there is interaction between treatments and age.

Inference Procedures

The relationships we see in Figure 11.2.6, however, are sample results. What can we conclude about the population from which the sample was drawn?

For an answer let us look at the t ratios on the computer printout in Figure 11.2.5. Each of these is the test statistic

$$t = \frac{\hat{\beta}_i - 0}{s_{\hat{\beta}_i}}$$

TABLE 11.2.3 Data for Example 11.2.3 Coded for Computer Analysis

Y	X_1	X_2	X_3	X_1X_2	X_1X_3
56	21	1	0	21	0
55	28	1	0	28	0
63	33	1	0	33	0
52	33	1	0	33	0
58	38	1	0	38	0
65	43	1	0	43	0
64	48	1	0	48	0
61	53	1	0	53	0
69	53	1	0	53	0
73	58	1	0	58	0
62	63	1	0	63	0
70	67	1	0	67	0
41	23	0	1	0	23
40	30	0	1	0	30
46	33	0	1	0	33
48	42	0	1	0	42
45	43	0	1	0	43
58	43	0	1	0	43
55	45	0	1	0	45
57	48	0	1	0	48
62	58	0	1	0	58
47	29	0	1	0	29
64	66	0	1	0	66
60	67	0	1	0	67
28	19	0	0	0	0
25	23	0	0	0	0
71	67	0	0	0	0
62	56	0	0	0	0
50	45	0	0	0	0
46	37	0	0	0	0
34	27	0	0	0	0
59	47	0	0	0	0
36	29	0	0	0	0
71	59	0	0	0	0
62	51	0	0	0	0
71	63	0	0	0	0

for testing $H_0: \beta_i = 0$. We see by Equation 11.2.5 that the y-intercept of the regression line for treatment A is equal to $\hat{\beta}_0 + \hat{\beta}_2$. Since the t ratio of 8.12 for testing $H_0: \beta_2 = 0$ is greater than the critical t of 2.0423 (for $\alpha = .05$), we can reject H_0 that $\beta_2 = 0$ and conclude that the y-intercept of the population regression line for treatment A is different from the y-intercept of the population regression line for treatment C, which has a

```
The regression equation is
y = 6.21 + 1.03 x1 + 41.3 x2 + 22.7 x3 − 0.703 x4 − 0.510 x5

Predictor     Coef        Stdev       t-ratio     p
Constant      6.211       3.350       1.85        0.074
x1            1.03339     0.07233     14.29       0.000
x2            41.304      5.085       8.12        0.000
x3            22.707      5.091       4.46        0.000
x4            −0.7029     0.1090      −6.45       0.000
x5            −0.5097     0.1104      −4.62       0.000

s = 3.925        R-sq = 91.4%        R-sq(adj) = 90.0%

Analysis of Variance

SOURCE        DF      SS          MS          F           p
Regression    5       4932.85     986.57      64.04       0.000
Error         30      462.15      15.40
Total         35      5395.00

SOURCE        DF      SEQ SS
x1            1       3424.43
x2            1       803.80
x3            1       1.19
x4            1       375.00
x5            1       328.42
```

FIGURE 11.2.5 Computer printout, MINITAB multiple regression analysis, Example 11.2.3.

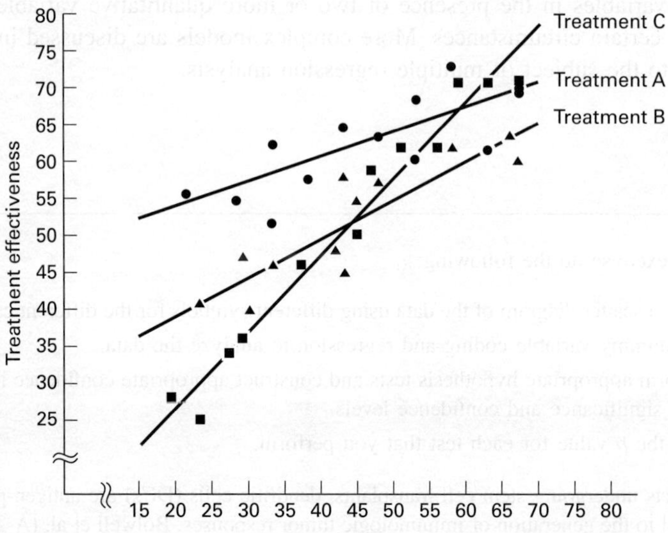

FIGURE 11.2.6 Scatter diagram of data for Example 11.2.3 with the fitted regression lines: (●) treatment A, (▲) treatment B, (■) treatment C.

y-intercept of β_0. Similarly, since the t ratio of 4.46 for testing H_0: $\beta_3 = 0$ is also greater than the critical t of 2.0423, we can conclude (at the .05 level of significance) that the y-intercept of the population regression line for treatment B is also different from the y-intercept of the population regression line for treatment C. (See the y-intercept of Equation 11.2.6.)

Now let us consider the slopes. We see by Equation 11.2.5 that the slope of the regression line for treatment A is equal to $\hat{\beta}_1$ (the slope of the line for treatment C) + $\hat{\beta}_4$. Since the t ratio of -6.45 for testing H_0: $\beta_4 = 0$ is less than the critical t of -2.0423, we can conclude (for $\alpha = .05$) that the slopes of the population regression lines for treatments A and C are different. Similarly, since the computed t ratio for testing H_0: $\beta_5 = 0$ is also less than -2.0423, we conclude (for $\alpha = .05$) that the population regression lines for treatments B and C have different slopes (see the slope of Equation 11.2.6). Thus we conclude that there is interaction between age and type of treatment. This is reflected by a lack of parallelism among the regression lines in Figure 11.2.6. ■

Another question of interest is this: Is the slope of the population regression line for treatment A different from the slope of the population regression line for treatment B? To answer this question requires computational techniques beyond the scope of this text. The interested reader is referred to books devoted specifically to regression analysis.

In Section 10.4 the reader was warned that there are problems involved in making multiple inferences from the same sample data. Again, books on regression analysis are available that may be consulted for procedures to be followed when multiple inferences, such as those discussed in this section, are desired.

We have discussed only two situations in which the use of dummy variables is appropriate. More complex models involving the use of one or more qualitative independent variables in the presence of two or more quantitative variables may be appropriate in certain circumstances. More complex models are discussed in the many books devoted to the subject of multiple regression analysis.

EXERCISES

For each exercise do the following:

(a) Draw a scatter diagram of the data using different symbols for the different categorical variables.

(b) Use dummy variable coding and regression to analyze the data.

(c) Perform appropriate hypothesis tests and construct appropriate confidence intervals using your choice of significance and confidence levels.

(d) Find the p value for each test that you perform.

11.2.1 For subjects undergoing stem cell transplants, dendritic cells (DCs) are antigen-presenting cells that are critical to the generation of immunologic tumor responses. Bolwell et al. (A-2) studied lymphoid DCs in 44 subjects who underwent autologous stem cell transplantation. The outcome variable is the concentration of DC2 cells as measured by flow cytometry. One of the independent variables is the age of the subject (years), and the second independent variable is the mobilization method. During chemotherapy, 11 subjects received granulocyte colony-stimulating factor (G-CSF) mobilizer (μg/kg/day) and 33 received etoposide (2 g/m^2). The mobilizer is a kind of blood progenitor cell that triggers the formation of the DC cells. The results were as follows:

G-CSF				Etoposide			
DC	Age	DC	Age	DC	Age	DC	Age
6.16	65	3.18	70	4.24	60	4.09	36
6.14	55	2.58	64	4.86	40	2.86	51
5.66	57	1.69	65	4.05	48	2.25	54
8.28	47	2.16	55	5.07	50	0.70	50
2.99	66	3.26	51	4.26	23	0.23	62
8.99	24	1.61	53	11.95	26	1.31	56
4.04	59	6.34	24	1.88	59	1.06	31
6.02	60	2.43	53	6.10	24	3.14	48
10.14	66	2.86	37	0.64	52	1.87	69
27.25	63	7.74	65	2.21	54	8.21	62
8.86	69	11.33	19	6.26	43	1.44	60

Source: Lisa Rybicki, M.S. Used with permission.

11.2.2 According to Pandey et al. (A-3) carcinoma of the gallbladder is not infrequent. One of the primary risk factors for gallbladder cancer is cholelithiasis, the asymptomatic presence of stones in the gallbladder. The researchers performed a case-control study of 50 subjects with gallbladder cancer and 50 subjects with cholelithiasis. Of interest was the concentration of lipid peroxidation products in gallbladder bile, a condition that may give rise to gallbladder cancer. The lipid peroxidation product melonaldehyde (MDA, μg/mg) was used to measure lipid peroxidation. One of the independent variables considered was the cytochrome P-450 concentration (CYTO, nmol/mg). Researchers used disease status (gallbladder cancer vs. cholelithiasis) and cytochrome P-450 concentration to predict MDA. The following data were collected.

Cholelithiasis				Gallbladder Cancer			
MDA	CYTO	MDA	CYTO	MDA	CYTO	MDA	CYTO
0.68	12.60	11.62	4.83	1.60	22.74	9.20	8.99
0.16	4.72	2.71	3.25	4.00	4.63	0.69	5.86
0.34	3.08	3.39	7.03	4.50	9.83	10.20	28.32
3.86	5.23	6.10	9.64	0.77	8.03	3.80	4.76
0.98	4.29	1.95	9.02	2.79	9.11	1.90	8.09
3.31	21.46	3.80	7.76	8.78	7.50	2.00	21.05
1.11	10.07	1.72	3.68	2.69	18.05	7.80	20.22
4.46	5.03	9.31	11.56	0.80	3.92	16.10	9.06
1.16	11.60	3.25	10.33	3.43	22.20	0.98	35.07
1.27	9.00	0.62	5.72	2.73	11.68	2.85	29.50

(Continued)

Cholelithiasis				Gallbladder Cancer			
MDA	CYTO	MDA	CYTO	MDA	CYTO	MDA	CYTO
1.38	6.13	2.46	4.01	1.41	19.10	3.50	45.06
3.83	6.06	7.63	6.09	6.08	36.70	4.80	8.99
0.16	6.45	4.60	4.53	5.44	48.30	1.89	48.15
0.56	4.78	12.21	19.01	4.25	4.47	2.90	10.12
1.95	34.76	1.03	9.62	1.76	8.83	0.87	17.98
0.08	15.53	1.25	7.59	8.39	5.49	4.25	37.18
2.17	12.23	2.13	12.33	2.82	3.48	1.43	19.09
0.00	0.93	0.98	5.26	5.03	7.98	6.75	6.05
1.35	3.81	1.53	5.69	7.30	27.04	4.30	17.05
3.22	6.39	3.91	7.72	4.97	16.02	0.59	7.79
1.69	14.15	2.25	7.61	1.11	6.14	5.30	6.78
4.90	5.67	1.67	4.32	13.27	13.31	1.80	16.03
1.33	8.49	5.23	17.79	7.73	10.03	3.50	5.07
0.64	2.27	2.79	15.51	3.69	17.23	4.98	16.60
5.21	12.35	1.43	12.43	9.26	9.29	6.98	19.89

Source: Manoj Pandey, M.D. Used with permission.

11.2.3 The purpose of a study by Krantz et al. (A-4) was to investigate dose-related effects of methadone in subjects with *torsades de pointes,* a polymorphic ventricular tachycardia. In the study of 17 subjects, 10 were men (sex = 0) and seven were women (sex = 1). The outcome variable, is the QTc interval, a measure of arrhythmia risk. The other independent variable, in addition to sex, was methadone dose (mg/day). Measurements on these variables for the 17 subjects were as follows.

Sex	Dose (mg/day)	QTc (msec)
0	1000	600
0	550	625
0	97	560
1	90	585
1	85	590
1	126	500
0	300	700
0	110	570
1	65	540
1	650	785
1	600	765
1	660	611
1	270	600
1	680	625
0	540	650
0	600	635
1	330	522

Source: Mori J. Krantz, M.D. Used with permission.

11.2.4 Refer to Exercise 9.7.2, which describes research by Reiss et al. (A-5), who collected samples from 90 patients and measured partial thromboplastin time (aPTT) using two different methods: the CoaguChek point-of-care assay and standard laboratory hospital assay. The subjects were also classified by their medication status: 30 receiving heparin alone, 30 receiving heparin with warfarin, and 30 receiving warfarin and enoxaparin. The data are as follows.

Heparin		Warfarin		Warfarin and Enoxaparin	
CoaguChek aPTT	Hospital aPTT	CoaguChek aPTT	Hospital aPTT	CoaguChek aPTT	Hospital aPTT
49.3	71.4	18.0	77.0	56.5	46.5
57.9	86.4	31.2	62.2	50.7	34.9
59.0	75.6	58.7	53.2	37.3	28.0
77.3	54.5	75.2	53.0	64.8	52.3
42.3	57.7	18.0	45.7	41.2	37.5
44.3	59.5	82.6	81.1	90.1	47.1
90.0	77.2	29.6	40.9	23.1	27.1
55.4	63.3	82.9	75.4	53.2	40.6
20.3	27.6	58.7	55.7	27.3	37.8
28.7	52.6	64.8	54.0	67.5	50.4
64.3	101.6	37.9	79.4	33.6	34.2
90.4	89.4	81.2	62.5	45.1	34.8
64.3	66.2	18.0	36.5	56.2	44.2
89.8	69.8	38.8	32.8	26.0	28.2
74.7	91.3	95.4	68.9	67.8	46.3
150.0	118.8	53.7	71.3	40.7	41.0
32.4	30.9	128.3	111.1	36.2	35.7
20.9	65.2	60.5	80.5	60.8	47.2
89.5	77.9	150.0	150.0	30.2	39.7
44.7	91.5	38.5	46.5	18.0	31.3
61.0	90.5	58.9	89.1	55.6	53.0
36.4	33.6	112.8	66.7	18.0	27.4
52.9	88.0	26.7	29.5	18.0	35.7
57.5	69.9	49.7	47.8	78.3	62.0
39.1	41.0	85.6	63.3	75.3	36.7
74.8	81.7	68.8	43.5	73.2	85.3
32.5	33.3	18.0	54.0	42.0	38.3
125.7	142.9	92.6	100.5	49.3	39.8
77.1	98.2	46.2	52.4	22.8	42.3
143.8	108.3	60.5	93.7	35.8	36.0

Source: Curtis E. Haas, Pharm.D. Used with permission.

Use the multiple regression to predict the hospital aPTT from the CoaguCheck aPTT level as well as the medication received. Is knowledge of medication useful in the prediction? Let $\alpha = .05$ for all tests.

11.3 VARIABLE SELECTION PROCEDURES

Health sciences researchers contemplating the use of multiple regression analysis to solve problems usually find that they have a large number of variables from which to select the independent variables to be employed as predictors of the dependent variable. Such investigators will want to include in their model as many variables as possible in order to maximize the model's predictive ability. The investigator must realize, however, that adding another independent variable to a set of independent variables always increases the coefficient of determination R^2. Therefore, independent variables should not be added to the model indiscriminately, but only for good reason. In most situations, for example, some potential predictor variables are more expensive than others in terms of data-collection costs. The cost-conscious investigator, therefore, will not want to include an expensive variable in a model unless there is evidence that it makes a worthwhile contribution to the predictive ability of the model.

The investigator who wishes to use multiple regression analysis most effectively must be able to employ some strategy for making intelligent selections from among those potential predictor variables that are available. Many such strategies are in current use, and each has its proponents. The strategies vary in terms of complexity and the tedium involved in their employment. Unfortunately, the strategies do not always lead to the same solution when applied to the same problem.

Stepwise Regression Perhaps the most widely used strategy for selecting independent variables for a multiple regression model is the stepwise procedure. The procedure consists of a series of steps. At each step of the procedure each variable then in the model is evaluated to see if, according to specified criteria, it should remain in the model.

Suppose, for example, that we wish to perform stepwise regression for a model containing k predictor variables. The criterion measure is computed for each variable. Of all the variables that do not satisfy the criterion for inclusion in the model, the one that least satisfies the criterion is removed from the model. If a variable is removed in this step, the regression equation for the smaller model is calculated and the criterion measure is computed for each variable now in the model. If any of these variables fail to satisfy the criterion for inclusion in the model, the one that least satisfies the criterion is removed. If a variable is removed at this step, the variable that was removed in the first step is reentered into the model, and the evaluation procedure is continued. This process continues until no more variables can be entered or removed.

The nature of the stepwise procedure is such that, although a variable may be deleted from the model in one step, it is evaluated for possible reentry into the model in subsequent steps.

MINITAB's STEPWISE procedure, for example, uses the associated F statistic as the evaluative criterion for deciding whether a variable should be deleted or added to the model. Unless otherwise specified, the cutoff value is $F = 4$. The printout of the STEPWISE results contains t statistics (the square root of F) rather than F statistics. At each step MINITAB calculates an F statistic for each variable then in the model. If the F statistic for any of these variables is less than the specified cutoff value (4 if some other value is not specified), the variable with the smallest F is removed from the model. The regression equation is refitted for the reduced model, the results are printed, and the procedure goes to the next step. If no variable can be removed, the procedure tries to add a variable. An F statistic is calculated for each variable not then in the model. Of these variables, the one with the largest associated F statistic is added, provided its F statistic is larger than the specified cutoff value (4 if some other value is not specified). The regression equation is refitted for the new model, the results are printed, and the procedure goes on to the next step. The procedure stops when no variable can be added or deleted.

The following example illustrates the use of the stepwise procedure for selecting variables for a multiple regression model.

EXAMPLE 11.3.1

A nursing director would like to use nurses' personal characteristics to develop a regression model for predicting the job performance (JOBPER). The following variables are available from which to choose the independent variables to include in the model:

$$X_1 = \text{assertiveness (ASRV)}$$
$$X_2 = \text{enthusiasm (ENTH)}$$
$$X_3 = \text{ambition (AMB)}$$
$$X_4 = \text{communication skills (COMM)}$$
$$X_5 = \text{problem-solving skills (PROB)}$$
$$X_6 = \text{initiative (INIT)}$$

We wish to use the stepwise procedure for selecting independent variables from those available in the table to construct a multiple regression model for predicting job performance.

Solution: Table 11.3.1 shows the measurements taken on the dependent variable, JOBPER, and each of the six independent variables for a sample of 30 nurses.

TABLE 11.3.1 Measurements on Seven Variables for Examples 11.3.1

Y	X_1	X_2	X_3	X_4	X_5	X_6
45	74	29	40	66	93	47
65	65	50	64	68	74	49
73	71	67	79	81	87	33
63	64	44	57	59	85	37
83	79	55	76	76	84	33
45	56	48	54	59	50	42
60	68	41	66	71	69	37
73	76	49	65	75	67	43

(Continued)

Y	X_1	X_2	X_3	X_4	X_5	X_6
74	83	71	77	76	84	33
69	62	44	57	67	81	43
66	54	52	67	63	68	36
69	61	46	66	64	75	43
71	63	56	67	60	64	35
70	84	82	68	64	78	37
79	78	53	82	84	78	39
83	65	49	82	65	55	38
75	86	63	79	84	80	41
67	61	64	75	60	81	45
67	71	45	67	80	86	48
52	59	67	64	69	79	54
52	71	32	44	48	65	43
66	62	51	72	71	81	43
55	67	51	60	68	81	39
42	65	41	45	55	58	51
65	55	41	58	71	76	35
68	78	65	73	93	77	42
80	76	57	84	85	79	35
50	58	43	55	56	84	40
87	86	70	81	82	75	30
84	83	38	83	69	79	41

We use MINITAB to obtain a useful model by the stepwise procedure. Observations on the dependent variable job performance (JOBPER) and the six candidate independent variables are stored in MINITAB Columns 1 through 7, respectively. Figure 11.3.1 shows the appropriate MINITAB procedure and the printout of the results.

To obtain the results in Figure 11.3.1, the values of F to enter and F to remove both were set automatically at 4. In step 1 there are no variables to be considered for deletion from the model. The variable AMB (Column 4) has the largest associated F statistic, which is $F = (9.74)^2 = 94.8676$. Since 94.8676 is greater than 4, AMB is added to the model. In step 2 the variable INIT (Column 7) qualifies for addition to the model since its associated F of $(-2.2)^2 = 4.84$ is greater than 4 and it is the variable with the largest associated F statistic. It is added to the model. After step 2 no other variable could be added or deleted, and the procedure stopped. We see, then, that the model chosen by the stepwise procedure is a two-independent-variable model with AMB and INIT as the independent variables. The estimated regression equation is

$$\hat{y} = 31.96 + .787x_3 - .45x_6$$

■

Dialog box:

Stat ➤ Regression ➤ Stepwise

Type *C1* in **Response** and *C2–C7* in **Predictors.**

Session command:

```
MTB > Stepwise C1 C2-C7;
SUBC> FEnter 4.0;
SUBC> FRemove 4.0.
```

Stepwise Regression

```
F-to-Enter:    4.00 F-to-Remove:    4.00

Response is C1 on 6 predictors, with N = 30

Step          1        2
Constant  7.226    31.955

C4         0.888    0.787
T-Ratio    9.74     8.13

C7                  -0.45
T-Ratio             -2.20

S          5.90     5.53
R-Sq      77.21    80.68
```

FIGURE 11.3.1 MINITAB stepwise procedure and output for the data of Table 11.3.1.

To change the criterion for allowing a variable to enter the model from 4 to some other value *K*, click on Options, then type the desired value of *K* in the Enter box. The new criterion *F* statistic, then, is *K* rather than 4. To change the criterion for deleting a variable from the model from 4 to some other value *K*, click on Options, then type the desired value of *K* in the Remove box. We must choose *K* to enter to be greater than or equal to *K* to remove.

Though the stepwise selection procedure is a common technique employed by researchers, other methods are available. Following is a brief discussion of two such tools. The final model obtained by each of these procedures is the same model that was found by using the stepwise procedure in Example 11.3.1.

Forward Selection This strategy is closely related to the stepwise regression procedure. This method builds a model using correlations. Variables are retained that meet the criteria for inclusion, as in stepwise selection. The first variable entered into the model is the one with the highest correlation with the dependent variable. If this variable meets the inclusion criterion, it is retained. The next variable to be considered for inclusion is the one with the highest partial correlation with the dependent variable. If it meets the inclusion criteria, it is retained. This procedure continues until all of the independent variables have been considered. The final model contains all of the independent variables that meet the inclusion criteria.

Backward Elimination This model-building procedure begins with all of the variables in the model. This strategy also builds a model using correlations and a predetermined inclusion criterion based on the F statistic. The first variable considered for removal from the model is the one with the smallest partial correlation coefficient. If this variable does not meet the criterion for inclusion, it is eliminated from the model. The next variable to be considered for elimination is the one with the next lowest partial correlation. It will be eliminated if it fails to meet the criterion for inclusion. This procedure continues until all variables have been considered for elimination. The final model contains all of the independent variables that meet the inclusion criteria.

EXERCISES

11.3.1 Refer to the data of Exercise 10.3.2 reported by Son et al. (A-6), who studied family caregiving in Korea of older adults with dementia. The outcome variable, caregiver burden (BURDEN), was measured by the Korean Burden Inventory (KBI) where scores ranged from 28 to 140 with higher scores indicating higher burden. Perform a stepwise regression analysis on the following independent variables reported by the researchers:

CGAGE: caregiver age (years)

CGINCOME: caregiver income (Won-Korean currency)

CGDUR: caregiver-duration of caregiving (month)

ADL: total activities of daily living where low scores indicate the elderly perform activities independently.

MEM: memory and behavioral problems with higher scores indicating more problems.

COG: cognitive impairment with lower scores indicating a greater degree of cognitive impairment.

SOCIALSU: total score of perceived social support (25–175, higher values indicating more support). The reported data are as follows.

CGAGE	CGINCOME	CGDUR	ADL	MEM	COG	SOCIALSU	BURDEN
41	200	12	39	4	18	119	28
30	120	36	52	33	9	131	68
41	300	60	89	17	3	141	59
35	350	2	57	31	7	150	91
37	600	48	28	35	19	142	70
42	90	4	34	3	25	148	38
49	300	26	42	16	17	172	46
39	500	16	52	6	26	147	57
49	309	30	88	41	13	98	89

(*Continued*)

CGAGE	CGINCOME	CGDUR	ADL	MEM	COG	SOCIALSU	BURDEN
40	250	60	90	24	3	147	48
40	300	36	38	22	13	146	74
70	60	10	83	41	11	97	78
49	450	24	30	9	24	139	43
55	300	18	45	33	14	127	76
27	309	30	47	36	18	132	72
39	250	10	90	17	0	142	61
39	260	12	63	14	16	131	63
44	250	32	34	35	22	141	77
33	200	48	76	33	23	106	85
42	200	12	26	13	18	144	31
52	200	24	68	34	26	119	79
48	300	36	85	28	10	122	92
53	300	12	22	12	16	110	76
40	300	11	82	57	3	121	91
35	200	8	80	51	3	142	78
47	150	60	80	20	18	101	103
33	180	19	81	20	1	117	99
41	200	48	30	7	17	129	73
43	300	36	27	27	27	142	88
25	309	24	72	9	0	137	64
35	250	12	46	15	22	148	52
35	200	6	63	52	13	135	71
45	200	7	45	26	18	144	41
36	300	24	77	57	0	128	85
52	600	60	42	10	19	148	52
41	230	6	60	34	11	141	68
40	200	36	33	14	14	151	57
45	400	96	49	30	15	124	84
48	75	6	89	64	0	105	91
50	200	30	72	31	3	117	83
31	250	30	45	24	19	111	73
33	300	2	73	13	3	146	57
30	200	30	58	16	15	99	69
36	250	6	33	17	21	115	81
45	500	12	34	13	18	119	71
32	300	60	90	42	6	134	91
55	200	24	48	7	23	165	48
50	309	20	47	17	18	101	94
37	250	30	32	13	15	148	57
40	1000	21	63	32	15	132	49
40	300	12	76	50	5	120	88
49	300	18	79	44	11	129	54
37	309	18	48	57	9	133	73
47	250	38	90	33	6	121	87
41	200	60	55	11	20	117	47

(*Continued*)

CGAGE	CGINCOME	CGDUR	ADL	MEM	COG	SOCIALSU	BURDEN
33	1000	18	83	24	11	140	60
28	309	12	50	21	25	117	65
33	400	120	44	31	18	138	57
34	330	18	79	30	20	163	85
40	200	18	24	5	22	157	28
54	200	12	40	20	17	143	40
32	300	32	35	15	27	125	87
44	280	66	55	9	21	161	80
44	350	40	45	28	17	142	49
42	280	24	46	19	17	135	57
44	500	14	37	4	21	137	32
25	600	24	47	29	3	133	52
41	250	84	28	23	21	131	42
28	1000	30	61	8	7	144	49
24	200	12	35	31	26	136	63
65	450	120	68	65	6	169	89
50	200	12	80	29	10	127	67
40	309	12	43	8	13	110	43
47	1000	12	53	14	18	120	47
44	300	24	60	30	16	115	70
37	309	54	63	22	18	101	99
36	300	12	28	9	27	139	53
55	200	12	35	18	14	153	78
45	2000	12	37	33	17	111	112
45	400	14	82	25	13	131	52
23	200	36	88	16	0	139	68
42	1000	12	52	15	0	132	63
38	200	36	30	16	18	147	49
41	230	36	69	49	12	171	42
25	200	30	52	17	20	145	56
47	200	12	59	38	17	140	46
35	100	12	53	22	21	139	72
59	150	60	65	56	2	133	95
49	300	60	90	12	0	145	57
51	200	48	88	42	6	122	88
54	250	6	66	12	23	133	81
53	30	24	60	21	7	107	104
49	100	36	48	14	13	118	88
44	300	48	82	41	13	95	115
36	200	18	88	24	14	100	66
64	200	48	63	49	5	125	92
51	120	2	79	34	3	116	97
43	200	66	71	38	17	124	69
54	150	96	66	48	13	132	112
29	309	19	81	66	1	152	88

Source: Gwi-Ryung Son, R.N., Ph.D. Used with permission.

11.3.2 Machiel Naeije (A-7) identifies variables useful in predicting maximum mouth opening (MMO, millimeters) for 35 healthy volunteers. The variables examined were:

AGE:	years
DOWN_CON:	downward condylar translation, mm
FORW_CON:	forward condylar translation, mm
Gender:	0 = Female, 1 = Male
MAN_LENG:	mandibular length, mm
MAN_WIDT:	mandibular width, mm

Use the following reported measurements to perform a stepwise regression.

AGE	DOWN_CON	FORW_CON	Gender	MAN_LENG	MAN_WIDT	MMO
21.00	4.39	14.18	1	100.86	121.00	52.34
26.00	1.39	20.23	0	93.08	118.29	51.90
30.00	2.42	13.45	1	98.43	130.56	52.80
28.00	−.18	19.66	1	102.95	125.34	50.29
21.00	4.10	22.71	1	108.24	125.19	57.79
20.00	4.49	13.94	0	98.34	113.84	49.41
21.00	2.07	19.35	0	95.57	115.41	53.28
19.00	−.77	25.65	1	98.86	118.30	59.71
24.00	7.88	18.51	1	98.32	119.20	53.32
18.00	6.06	21.72	0	92.70	111.21	48.53
22.00	9.37	23.21	0	88.89	119.07	51.59
21.00	3.77	23.02	1	104.06	127.34	58.52
20.00	1.10	19.59	0	98.18	111.24	62.93
22.00	2.52	16.64	0	91.01	113.81	57.62
24.00	5.99	17.38	1	96.98	114.94	65.64
22.00	5.28	22.57	0	97.86	111.58	52.85
22.00	1.25	20.89	0	96.89	115.16	64.43
22.00	6.02	20.38	1	98.35	122.52	57.25
19.00	1.59	21.63	0	90.65	118.71	50.82
26.00	6.05	10.59	0	92.99	119.10	40.48
22.00	−1.51	20.03	1	108.97	129.00	59.68
24.00	−.41	24.55	0	91.85	100.77	54.35
21.00	6.75	14.67	1	104.30	127.15	47.00
22.00	4.87	17.91	1	93.16	123.10	47.23
22.00	.64	17.60	1	94.18	113.86	41.19
29.00	7.18	15.19	0	89.56	110.56	42.76
25.00	6.57	17.25	1	105.85	140.03	51.88
20.00	1.51	18.01	0	89.29	121.70	42.77
27.00	4.64	19.36	0	92.58	128.01	52.34
26.00	3.58	16.57	1	98.64	129.00	50.45
23.00	6.64	12.47	0	83.70	130.98	43.18
25.00	7.61	18.52	0	88.46	124.97	41.99
22.00	5.39	11.66	1	94.93	129.99	39.45
31.00	5.47	12.85	1	96.81	132.97	38.91
23.00	2.60	19.29	0	93.13	121.03	49.10

Source: Machiel Naeije, D.D.S. Used with permission.

11.3.3 One purpose of a study by Connor et al. (A-8) was to examine reactive aggression among children and adolescents referred to a residential treatment center. The researchers used the Proactive/Reactive Rating Scale, obtained by presenting three statements to clinicians who examined the subjects. The respondents answered, using a scale from 1 to 5, with 5 indicating that the statement almost always applied to the child. An example of a reactive aggression statement is, "When this child has been teased or threatened, he or she gets angry easily and strikes back." The reactive score was the average response to three statements of this type. With this variable as the outcome variable, researchers also examined the following: AGE (years), VERBALIQ (verbal IQ), STIM (stimulant use), AGEABUSE (age when first abused), CTQ (a measure of hyperactivity in which higher scores indicate higher hyperactivity), TOTALHOS (total hostility as measured by an evaluator, with higher numbers indicating higher hostility). Perform stepwise regression to find the variables most useful in predicting reactive aggression in the following sample of 68 subjects.

REACTIVE	AGE	VERBALIQ	STIM	AGEABUSE	CTQ	TOTALHOS
4.0	17	91	0	0	0	8
3.7	12	94	0	1	29	10
2.3	14	105	0	1	12	10
5.0	16	97	0	1	9	11
2.0	15	97	0	2	17	10
2.7	8	91	0	0	6	4
2.0	10	111	0	0	6	6
3.3	12	105	0	0	28	7
2.0	17	101	1	0	12	9
4.3	13	102	1	1	8	11
4.7	15	83	0	0	9	9
4.3	15	66	0	1	5	8
2.0	15	90	0	2	3	8
4.0	13	88	0	1	28	8
2.7	13	98	0	1	17	10
2.7	9	135	0	0	30	11
2.7	18	72	0	0	10	9
2.0	13	93	0	2	20	8
3.0	14	94	0	2	10	11
2.7	13	93	0	1	4	8
3.7	16	73	0	0	11	11
2.7	12	74	0	1	10	7
2.3	14	97	0	2	3	11
4.0	13	91	1	1	21	11
4.0	12	88	0	1	14	9
4.3	13	90	0	0	15	2
3.7	14	104	1	1	10	10
3.0	18	82	0	0	1	7
4.3	14	79	1	3	6	7
1.0	16	93	0	0	5	8
4.3	16	99	0	1	21	11

(*Continued*)

REACTIVE	AGE	VERBALIQ	STIM	AGEABUSE	CTQ	TOTALHOS
2.3	14	73	0	2	8	9
3.0	12	112	0	0	15	9
1.3	15	102	0	1	1	5
3.0	16	78	1	1	26	8
2.3	9	95	1	0	23	10
1.0	15	124	0	3	0	11
3.0	17	73	0	1	1	10
3.3	11	105	0	0	23	5
4.0	11	89	0	0	27	8
1.7	9	88	0	1	2	8
2.3	16	96	0	1	5	7
4.7	15	76	1	1	17	9
1.7	16	87	0	2	0	4
1.7	15	90	0	1	10	12
4.0	12	76	0	0	22	10
5.0	12	83	1	1	19	7
4.3	10	88	1	0	10	5
5.0	9	98	1	0	8	9
3.7	12	100	0	0	6	4
3.3	14	80	0	1	3	10
2.3	16	84	0	1	3	9
1.0	17	117	0	2	1	9
1.7	12	145	1	0	0	5
3.7	12	123	0	0	1	3
2.0	16	94	0	2	6	6
3.7	17	70	0	1	11	13
4.3	14	113	0	0	8	8
2.0	12	123	1	0	2	8
3.0	7	107	0	0	11	9
3.7	12	78	1	0	15	11
4.3	14	73	0	1	2	8
2.3	18	91	0	3	8	10
4.7	12	91	0	0	6	9
3.7	15	111	0	0	2	9
1.3	15	71	0	1	20	10
3.7	7	102	0	0	14	9
1.7	9	89	0	0	24	6

Source: Daniel F. Connor, M.D. and Lang Lin. Used with permission.

11.4 LOGISTIC REGRESSION

Up to now our discussion of regression analysis has been limited to those situations in which the dependent variable is a continuous variable such as weight, blood pressure, or plasma levels of some hormone. Much research in the health sciences field is motivated by a desire to describe, understand, and make use of the relationship between independent variables and a dependent (or outcome) variable that is discrete. Particularly plentiful are circumstances in which the outcome variable is dichotomous. A dichotomous variable, we recall, is a variable that can assume only one of two mutually exclusive values. These values are usually coded $Y = 1$ for a success and $Y = 0$ for a nonsuccess, or failure. Dichotomous variables include those whose two possible values are such categories as died–did not die; cured–not cured; disease occurred–disease did not occur; and smoker–nonsmoker. The health sciences professional who either engages in research or needs to understand the results of research conducted by others will find it advantageous to have, at least, a basic understanding of *logistic regression,* the type of regression analysis that is usually employed when the dependent variable is dichotomous. The purpose of the present discussion is to provide the reader with this level of understanding. We shall limit our presentation to the case in which there is only one independent variable that may be either continuous or dichotomous.

The Logistic Regression Model Recall that in Chapter 9 we referred to regression analysis involving only two variables as simple linear regression analysis. The simple linear regression model was expressed by the equation

$$y = \beta_0 + \beta_1 x + \epsilon \tag{11.4.1}$$

in which y is an arbitrary observed value of the continuous dependent variable. When the observed value of Y is $\mu_{y|x}$, the mean of a subpopulation of Y values for a given value of X, the quantity ϵ, the difference between the observed Y and the regression line (see Figure 9.2.1) is zero, and we may write Equation 11.4.1 as

$$\mu_{y|x} = \beta_0 + \beta_1 x \tag{11.4.2}$$

which may also be written as

$$E_{(y|x)} = \beta_0 + \beta_1 x \tag{11.4.3}$$

Generally the right-hand side of Equations 11.4.1 through 11.4.3 may assume any value between minus infinity and plus infinity.

Even though only two variables are involved, the simple linear regression model is not appropriate when Y is a dichotomous variable because the expected value (or mean) of Y is the probability that $Y = 1$ and, therefore, is limited to the range 0 through 1, inclusive. Equations 11.4.1 through 11.4.3, then, are incompatible with the reality of the situation.

If we let $p = P(Y = 1)$, then the ratio $p/(1 - p)$ can take on values between 0 and plus infinity. Furthermore, the natural logarithm (ln) of $p/(1 - p)$ can take on values between minus infinity and plus infinity just as can the right-hand side of Equations 11.4.1 through 11.4.3. Therefore, we may write

$$\ln\left[\frac{p}{1-p}\right] = \beta_0 + \beta_1 x \tag{11.4.4}$$

Equation 11.4.4 is called the *logistic regression model* because the transformation of $\mu_{y|x}$ (that is, p) to $\ln[p/(1-p)]$ is called the *logit transformation*. Equation 11.4.4 may also be written as

$$p = \frac{\exp(\beta_0 + \beta_1 x)}{1 + \exp(\beta_0 + \beta_1 x)} \tag{11.4.5}$$

in which exp is the inverse of the natural logarithm.

The logistic regression model is widely used in health sciences research. For example, the model is frequently used by epidemiologists as a model for the probability (interpreted as the risk) that an individual will acquire a disease during some specified time period during which he or she is exposed to a condition (called a risk factor) known to be or suspected of being associated with the disease.

Logistic Regression: Dichotomous Independent Variable The simplest situation in which logistic regression is applicable is one in which both the dependent and the independent variables are dichotomous. The values of the dependent (or outcome) variable usually indicate whether or not a subject acquired a disease or whether or not the subject died. The values of the independent variable indicate the status of the subject relative to the presence or absence of some risk factor. In the discussion that follows we assume that the dichotomies of the two variables are coded 0 and 1. When this is the case the variables may be cross-classified in a table, such as Table 11.4.1, that contains two rows and two columns. The cells of the table contain the frequencies of occurrence of all possible pairs of values of the two variables: (1, 1), (1, 0), (0, 1), and (0, 0).

An objective of the analysis of data that meet these criteria is a statistic known as the *odds ratio*. To understand the concept of the odds ratio, we must understand the term

TABLE 11.4.1 Two Cross-Classified Dichotomous Variables Whose Values Are Coded 1 and 0

	Independent Variable (X)	
Dependent Variable (Y)	**1**	**0**
1	$n_{1,1}$	$n_{1,0}$
2	$n_{0,1}$	$n_{0,0}$

odds, which is frequently used by those who place bets on the outcomes of sporting events or participate in other types of gambling activities. Using probability terminology, we may define odds as follows.

DEFINITION

The odds for success are the ratio of the probability of success to the probability of failure.

The odds ratio is a measure of how much greater (or less) the odds are for subjects possessing the risk factor to experience a particular outcome. This conclusion assumes that the outcome is a rare event. For example, when the outcome is the contracting of a disease, the interpretation of the odds ratio assumes that the disease is rare.

Suppose, for example, that the outcome variable is the acquisition or nonacquisition of skin cancer and the independent variable (or risk factor) is high levels of exposure to the sun. Analysis of such data collected on a sample of subjects might yield an odds ratio of 2, indicating that the odds of skin cancer are two times higher among subjects with high levels of exposure to the sun than among subjects without high levels of exposure.

Computer software packages that perform logistic regression frequently provide as part of their output estimates of β_0 and β_1 and the numerical value of the odds ratio. As it turns out the odds ratio is equal to $\exp(\beta_1)$.

EXAMPLE 11.4.1

LaMont et al. (A-9) tested for obstructive coronary artery disease (OCAD) among 113 men and 35 women who complained of chest pain or possible equivalent to their primary care physician. Table 11.4.2 shows the cross-classification of OCAD with gender. We wish to use logistic regression analysis to determine how much greater the odds are of finding OCAD among men than among women.

Solution: We may use the SAS® software package to analyze these data. The independent variable is gender and the dependent variable is status with respect to having obstructive coronary artery disease (OCAD). Use of the SAS® command PROC LOGIST yields, as part of the resulting output, the statistics shown in Figure 11.4.1.

TABLE 11.4.2 Cases of Obstructive Coronary Artery Disease (OCAD) Classified by Sex

Disease	Males	Females	Total
OCAD present	92	15	107
OCAD not present	21	20	41
Total	113	35	148

Source: Matthew J. Budoff, M.D. Used with permission.

```
                  The LOGISTIC Procedure

        Analysis of Maximum Likelihood Estimates

Parameter  DF  Estimate  Standard       Wald
                            Error  Chi-Square  Pr > ChiSq

Intercept   1   -1.4773    0.2418     37.3118      <.0001
sex         1    1.7649    0.4185     17.7844      <.0001
```

FIGURE 11.4.1 Partial output from use of SAS® command PROC LOGISTIC with the data of Table 11.4.2.

We see that the estimate of α is -1.4773 and the estimate of β_1 is 1.7649. The estimated odds ratio, then, is $\widehat{OR} = \exp(1.7649) = 5.84$. Thus we estimate that the odds of finding a case of obstructive coronary artery disease to be almost six times higher among men than women. ■

Logistic Regression: Continuous Independent Variable Now let us consider the situation in which we have a dichotomous dependent variable and a continuous independent variable. We shall assume that a computer is available to perform the calculations. Our discussion, consequently, will focus on an evaluation of the adequacy of the model as a representation of the data at hand, interpretation of key elements of the computer printout, and the use of the results to answer relevant questions about the relationship between the two variables.

EXAMPLE 11.4.2

According to Gallagher et al. (A-10), cardiac rehabilitation programs offer "information, support, and monitoring for return to activities, symptom management, and risk factor modification." The researchers conducted a study to identify among women factors that are associated with participation in such programs. The data in Table 11.4.3 are the ages of 185 women discharged from a hospital in Australia who met eligibility criteria involving discharge for myocardial infarction, artery bypass surgery, angioplasty, or stent. We wish to use these data to obtain information regarding the relationship between age (years) and participation in a cardiac rehabilitation program ($ATT = 1$, if participated, and $ATT = 0$, if not). We wish also to know if we may use the results of our analysis to predict the likelihood of participation by a woman if we know her age.

Solution: The independent variable is the continuous variable age (AGE), and the dependent or response variable is status with respect to attendance in a cardiac rehabilitation program. The dependent variable is a dichotomous variable that can assume one of two values: $0 =$ did not attend, and $1 =$ did

TABLE 11.4.3 Ages of Women Participating and Not Participating in a Cardiac Rehabilitation Program

Nonparticipating (ATT = 0)				Participating (ATT = 1)	
50	73	46	74	74	62
59	75	57	59	50	74
42	71	53	81	55	61
50	69	40	74	66	69
34	78	73	77	49	76
49	69	68	59	55	71
67	74	72	75	73	61
44	86	59	68	41	46
53	49	64	81	64	69
45	63	78	74	46	66
79	63	68	65	65	57
46	72	67	81	50	60
62	64	55	62	61	63
58	72	71	85	64	63
70	79	80	84	59	56
60	75	75	39	73	70
67	70	69	52	73	70
64	73	80	67	65	63
62	66	79	82	67	63
50	75	71	84	60	65
61	73	69	79	69	67
69	71	78	81	61	68
74	72	75	74	79	84
65	69	71	85	66	69
80	76	69	92	68	78
69	60	77	69	61	69
77	79	81	83	63	79
61	78	78	82	70	83
72	62	76	85	68	67
67	73	84	82	59	47
			80	64	57
					66

Source: Robyn Gallagher, R.N., Ph.D. Used with permission.

attend. We use the SAS® software package to analyze the data. The SAS® command is PROC LOGISTIC, but if we wish to predict attendance in the cardiac program, we need to use the "descending" option with PROC LOGISTIC. (When you wish to predict the outcome labeled "1" of the dependent variable, use the "descending option" in SAS®. Consult SAS®

Parameter	DF	Estimate	Standard Error	Wald Chi-Square	Pr > ChiSq
Intercept	1	1.8744	0.9809	3.6518	0.0560
age	1	−0.0379	0.0146	6.7083	0.0096

FIGURE 11.4.2 Partial SAS® printout of the logistic regression analysis of the data in Table 11.4.3.

documentation for further details.) A partial printout of the analysis is shown in Figure 11.4.2.

The slope of our regression is −.0379, and the intercept is 1.8744. The regression equation, then, is

$$\hat{y}_i = 1.8744 - .0379x_i$$

where $\hat{y}_i = \ln[\hat{p}_i/(1 - \hat{p}_i)]$ and $\hat{p}_i$ is the predicted probability of attending cardiac rehabilitation for a woman aged x_i.

Test of H_0 that $\beta_1 = 0$

We reach a conclusion about the adequacy of the logistic model by testing the null hypothesis that the slope of the regression line is zero. The test statistic is $z = \hat{\beta}_1/s_{\hat{\beta}_1}$ where z is the standard normal statistic, $\hat{\beta}_1$ is the sample slope (−.0379), and $s_{\hat{\beta}_1}$ is its standard error (.0146) as shown in Figure 11.4.2. From these numbers we compute $z = -.0379/.0146 = -2.5959$, which has an associated two-sided p value of .0094. We conclude, therefore, that the logistic model is adequate. The square of z is chi-square with 1 degree of freedom, a statistic that is shown in Figure 11.4.2.

Using the Logistic Regression to Estimate p

We may use Equation 11.4.5 and the results of our analysis to estimate p, the probability that a woman of a given age (within the range of ages represented by the data) will attend a cardiac rehabilitation program. Suppose, for example, that we wish to estimate the probability that a woman who is 50 years of age will participate in a rehabilitation program. Substituting 50 and the results shown in Figure 11.4.2 into Equation 11.4.5 gives

$$\hat{p} = \frac{\exp[1.8744 - (.0379)(50)]}{1 + \exp[1.8744 - (.0379)(50)]} = .49485$$

SAS® calculates the estimated probabilities for the given values of X. We can see the estimated probabilities of attending cardiac rehabilitation programs for the age range of the subjects enrolled in the study in Figure 11.4.3. Since the slope was negative, we see a decreasing probability of attending a cardiac rehabilitation program for older women.

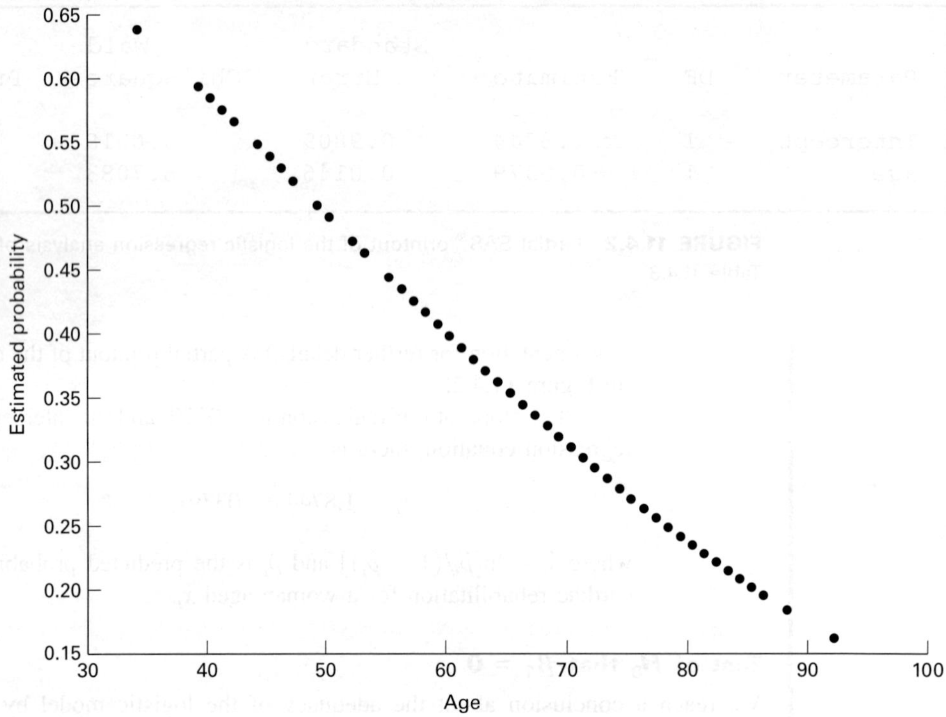

FIGURE 11.4.3 Estimated probabilities of attendance for ages within the study for Example 11.4.2. ■

Multiple Logistic Regression Practitioners often are interested in the relationships of several independent variables to a response variable. These independent variables may be either continuous or discrete or a combination of the two.

Multiple logistic models are constructed by expanding Equations 11.4.1 to 11.4.4. If we begin with Equation 11.4.4, multiple logistic regression can be represented as

$$\ln\left[\frac{p}{1-p}\right] = \beta_0 + \beta_1 x_{1j} + \beta_2 x_{2j} + \cdots + \beta_k x_{kj} \tag{11.4.6}$$

Using the logit transformation, we now have

$$p = \frac{\exp(\beta_0 + \beta_1 x_{1j} + \beta_2 x_{2j} + \cdots + \beta_k x_{kj})}{1 + \exp(\beta_0 + \beta_1 x_{1j} + \beta_2 x_{2j} + \cdots + \beta_k x_{kj})} \tag{11.4.7}$$

EXAMPLE 11.4.3

Consider the data presented in Review Exercise 24. In this study by Fils-Aime et al. (A-21), data were gathered and classified with regard to alcohol use. Subjects were classified as having either early (< 25 years) or late (> 25 years) onset of excessive alcohol use.

Parameter	B	S.E.	Wald	Df	Sig.	Exp(B)
5-HIAA	−.013	.006	5.878	1	.015	.987
TRYPT	.000	.000	.000	1	.983	1.000
Constant	2.076	1.049	3.918	1	.048	7.970

FIGURE 11.4.3 SPSS output for the data in Example 11.4.3.

Levels of cerebrospinal fluid (CSF) tryptophan (TRYPT) and 5-hydroxyindoleacetic acid (5-HIAA) concentrations were also obtained.

Solution: The independent variables are the concentrations of TRYPT and 5-HIAA, and the dependent variable is the dichotomous response for onset of excessive alcohol use. We use SPSS software to analyze the data. The output is presented in Figure 11.4.3.

The equation can be written as

$$\hat{y}_i = 2.076 - .013x_{1j} + 0x_{2j}$$

Note that the coefficient for TRYPT is 0, and therefore it is not playing a role in the model.

Test of H0 that $\beta_1 = 0$

Tests for significance of the regression coefficients can be obtained directly from Figure 11.4.3. Note that both the constant (intercept) and the 5-HIAA variables are significant in the model (both have p values, noted as "Sig." in the table, $< .05$); however, TRYPT is not significant and therefore need not be in the model, suggesting that it is not useful for identifying those study participants with early or late alcoholism onset.

As above, probabilities can be easily obtained by using equation 11.4.7 and substituting the values obtained from the analysis. ■

Polytomous Logistic Regression Thus far we have limited our discussion to situations in which there is a dichotomous response variable (e.g., successful or unsuccessful). Often we have a situation in which multiple categories make up the response. We may, for example, have subjects that are classified as positive, negative, and undetermined for a given disease (a standard polytomous response). There may also be times when we have a response variable that is ordered. We may, for example, classify our subjects by BMI as underweight, ideal weight, overweight, or obese (an ordinal polytomous response). The modeling process is slightly more complex and requires the use of a computer program. For those interested in exploring these valuable methods further, we recommend the book by Hosmer and Lemeshow (1).

Further Reading We have discussed only the basic concepts and applications of logistic regression. The technique has much wider application. Stepwise regression analysis may be used with logistic regression. There are also techniques available for constructing confidence intervals for odds ratios. The reader who wishes to learn more about logistic regression may consult the books by Hosmer and Lemeshow (1) and Kleinbaum (2).

EXERCISES

11.4.1 In a study of violent victimization of women and men, Porcerelli et al. (A-11) collected information from 679 women and 345 men ages 18 to 64 years at several family-practice centers in the metropolitan Detroit area. Patients filled out a health history questionnaire that included a question about victimization. The following table shows the sample subjects cross-classified by gender and whether the subject self-identified as being "hit, kicked, punched, or otherwise hurt by someone within the past year." Subjects answering yes to that question are classified "violently victimized." Use logistic regression analysis to find the regression coefficients and the estimate of the odds ratio. Write an interpretation of your results.

Victimization	Women	Men	Total
No victimization	611	308	919
Violently victimized	68	37	105
Total	679	345	1024

Source: John H. Porcerelli, Rosemary Cogan, Patricia P. West, Edward A. Rose, Dawn Lambrecht, Karen E. Wilson, Richard K. Severson, and Dunia Karana, "Violent Victimization of Women and Men: Physical and Psychiatric Symptoms," *Journal of the American Board of Family Practice, 16* (2003), 32–39.

11.4.2 Refer to the research of Gallagher et al. (A-10) discussed in Example 11.4.2. Another covariate of interest was a score using the Hospital Anxiety and Depression Index. A higher value for this score indicates a higher level of anxiety and depression. Use the following data to predict whether a woman in the study participated in a cardiac rehabilitation program.

Hospital Anxiety and Depression Index Scores for Nonparticipating Women				Hospital Anxiety and Depression Index Scores for Participating Women	
17	14	19	16	23	25
7	21	6	9	3	6
19	13	8	22	24	29
16	15	13	17	13	22
23	21	4	14	26	11
27	12	15	14	19	12
23	9	23	5	25	20
18	29	19	5	15	18
21	4	14	14	22	24
27	18	19	20	13	18
14	22	17	21	21	8

(*Continued*)

Hospital Anxiety and Depression Index Scores for Nonparticipating Women				Hospital Anxiety and Depression Index Scores for Participating Women	
25	5	13	17	15	10
19	27	14	17	12	17
23	16	14	10	25	14
6	11	17	13	29	21
8	19	26	10	17	25
15	23	15	20	21	25
30	22	19	3	8	16
18	25	16	18	19	23
10	11	10	9	16	19
29	20	15	10	24	24
8	11	22	5	17	11
12	28	8	15	26	17
27	12	15	13	12	19
12	19	20	16	19	20
9	18	12		13	17
16	13	2		23	31
6	12	6		11	0
22	7	14		17	18
10	12	19		29	18
9	14	14		6	15
11	13	19		20	

Source: Robyn Gallagher, R.N., Ph.D. Used with permission.

11.5 SUMMARY

This chapter is included for the benefit of those who wish to extend their understanding of regression analysis and their ability to apply techniques to models that are more complex than those covered in Chapters 9 and 10. In this chapter we present some additional topics from regression analysis. We discuss the analysis that is appropriate when one or more of the independent variables is dichotomous. In this discussion the concept of dummy variable coding is presented. A second topic that we discuss is how to select the most useful independent variables when we have a long list of potential candidates. The technique we illustrate for the purpose is stepwise regression analysis. Finally, we present the basic concepts and procedures that are involved in logistic regression analysis. We cover two situations: the case in which the independent variable is dichotomous, and the case in which the independent variable is continuous.

Since the calculations involved in obtaining useful results from data that are appropriate for analysis by means of the techniques presented in this chapter are complicated and time-consuming when attempted by hand, it is recommended that a computer be used to work the exercises.

SUMMARY OF FORMULAS FOR CHAPTER 11

Formula Number	Name	Formula
11.4.1–11.4.3	Representations of the simple linear regression model	$y = \beta_0 + \beta_1 x + \epsilon$ $\mu_{y\|x} = \beta_0 + \beta_1 x$ $E_{(y\|x)} = \beta_0 + \beta_1 x$
11.4.4	Simple logistic regression model	$\ln\left[\frac{p}{1-p}\right] = \beta_0 + \beta_1 x$
11.4.5	Alternative representation of the simple logistic regression model	$p = \frac{\exp(\beta_0 + \beta_1 x)}{1 + \exp(\beta_0 + \beta_1 x)}$
11.4.6	Alternative representation of the multiple logistic regression model	$\ln\left[\frac{p}{1-p}\right] = \beta_0 + \beta_1 x_{1j} + \beta_2 x_{2j} + \cdots + \beta_k x_{kj}$
11.4.7	Alternative representation of the multiple logistic regression model	$p = \frac{\exp(\beta_0 + \beta_1 x_{1j} + \beta_2 x_{2j} + \cdots + \beta_k x_{kj})}{1 + \exp(\beta_0 + \beta_1 x_{1j} + \beta_2 x_{2j} + \cdots + \beta_k x_{kj})}$
Symbol Key	• β_0 = regression intercept • β_i = regression coefficient • ϵ = regression model error term • $E_{(y\|x)}$ = expected value of y at x • $\ln\left[\frac{p}{1-p}\right]$ = logit transformation • $\mu_{y\|x}$ = mean of y at x • x_i = value of independent variable at i	

REVIEW QUESTIONS AND EXERCISES

1. What is a qualitative variable?
2. What is a dummy variable?
3. Explain and illustrate the technique of dummy variable coding.
4. Why is a knowledge of variable selection techniques important to the health sciences researcher?
5. What is stepwise regression?
6. Explain the basic concept involved in stepwise regression.

7. When is logistic regression used?

8. Write out and explain the components of the logistic regression model.

9. Define the word *odds*.

10. What is an odds ratio?

11. Give an example in your field in which logistic regression analysis would be appropriate when the independent variable is dichotomous.

12. Give an example in your field in which logistic regression analysis would be appropriate when the independent variable is continuous.

13. Find a published article in the health sciences field in which each of the following techniques is employed:
   (a) Dummy variable coding
   (b) Stepwise regression
   (c) Logistic regression

   Write a report on the article in which you identify the variables involved, the reason for the choice of the technique, and the conclusions that the authors reach on the basis of their analysis.

14. In Example 10.3.1, we saw that the purpose of a study by Jansen and Keller (A-12) was to predict the capacity to direct attention (CDA) in elderly subjects. The study collected information on 71 community-dwelling older women with normal mental status. Higher CDA scores indicate better attentional functioning. In addition to the variables age and education level, the researchers performed stepwise regression with two additional variables: IADL, a measure of activities of daily living (higher values indicate greater number of daily activities), and ADS, a measure of attentional demands (higher values indicate more attentional demands). Perform stepwise regression with the data in the following table and report your final model, p values, and conclusions.

CDA	Age	Edyrs	IADL	ADS	CDA	Age	Edyrs	IADL	ADS
4.57	72	20	28	27	3.17	79	12	28	18
−3.04	68	12	27	96	−1.19	87	12	21	61
1.39	65	13	24	97	0.99	71	14	28	55
−3.55	85	14	27	48	−2.94	81	16	27	124
−2.56	84	13	28	50	−2.21	66	16	28	42
−4.66	90	15	27	47	−0.75	81	16	28	64
−2.70	79	12	28	71	5.07	80	13	28	26
0.30	74	10	24	48	−5.86	82	12	28	84
−4.46	69	12	28	67	5.00	65	13	28	43
−6.29	87	15	21	81	0.63	73	16	26	70
−4.43	84	12	27	44	2.62	85	16	28	20
0.18	79	12	28	39	1.77	83	17	23	80
−1.37	71	12	28	124	−3.79	83	8	27	21
3.26	76	14	29	43	1.44	76	20	28	26
−1.12	73	14	29	30	−5.77	77	12	28	53
−0.77	86	12	26	44	−5.77	83	12	22	69
3.73	69	17	28	47	−4.62	79	14	27	82

(*Continued*)

CDA	Age	Edyrs	IADL	ADS	CDA	Age	Edyrs	IADL	ADS
−5.92	66	11	28	49	−2.03	69	12	28	77
5.74	65	16	28	48	−2.22	66	14	28	38
2.83	71	14	28	46	0.80	75	12	28	28
−2.40	80	18	28	25	−0.75	77	16	27	85
−0.29	81	11	28	27	4.60	78	12	22	82
4.44	66	14	29	54	2.68	83	20	28	34
3.35	76	17	29	26	−3.69	85	10	20	72
−3.13	70	12	25	100	4.85	76	18	28	24
−2.14	76	12	27	38	−0.08	75	14	29	49
9.61	67	12	26	84	0.63	70	16	28	29
7.57	72	20	29	44	5.92	79	16	27	83
2.21	68	18	28	52	3.63	75	18	28	32
−2.30	102	12	26	18	−7.07	94	8	24	80
1.73	67	12	27	80	6.39	76	18	28	41
6.03	66	14	28	54	−0.08	84	18	27	75
−0.02	75	18	26	67	1.07	79	17	27	21
−7.65	91	13	21	101	5.31	78	16	28	18
4.17	74	15	28	90	0.30	79	12	28	38

Source: Debra Jansen, Ph.D., R.N. Used with permission.

15. In the following table are the cardiac output (L/min) and oxygen consumption (V_{O_2}) values for a sample of adults (A) and children (C), who participated in a study designed to investigate the relationship among these variables. Measurements were taken both at rest and during exercise. Treat cardiac output as the dependent variable and use dummy variable coding and analyze the data by regression techniques. Explain the results. Plot the original data and the fitted regression equations.

Cardiac Output (L/min)	V_{O_2} (L/min)	Age Group	Cardiac Output (L/min)	V_{O_2} (L/min)	Age Group
4.0	.21	A	4.0	.25	C
7.5	.91	C	6.1	.22	A
3.0	.22	C	6.2	.61	C
8.9	.60	A	4.9	.45	C
5.1	.59	C	14.0	1.55	A
5.8	.50	A	12.9	1.11	A
9.1	.99	A	11.3	1.45	A
3.5	.23	C	5.7	.50	C
7.2	.51	A	15.0	1.61	A
5.1	.48	C	7.1	.83	C
6.0	.74	C	8.0	.61	A
5.7	.70	C	8.1	.82	A
14.2	1.60	A	9.0	1.15	C
4.1	.30	C	6.1	.39	A

16. A simple random sample of normal subjects between the ages of 6 and 18 yielded the data on total body potassium (mEq) and total body water (liters) shown in the following table. Let total potassium be the dependent variable and use dummy variable coding to quantify the qualitative variable. Analyze the data using regression techniques. Explain the results. Plot the original data and the fitted regression equations.

Total Body Potassium	Total Body Water	Sex	Total Body Potassium	Total Body Water	Sex
795	13	M	950	12	F
1590	16	F	2400	26	M
1250	15	M	1600	24	F
1680	21	M	2400	30	M
800	10	F	1695	26	F
2100	26	M	1510	21	F
1700	15	F	2000	27	F
1260	16	M	3200	33	M
1370	18	F	1050	14	F
1000	11	F	2600	31	M
1100	14	M	3000	37	M
1500	20	F	1900	25	F
1450	19	M	2200	30	F
1100	14	M			

17. The data shown in the following table were collected as part of a study in which the subjects were preterm infants with low birth weights born in three different hospitals. Use dummy variable coding and multiple regression techniques to analyze these data. May we conclude that the three sample hospital populations differ with respect to mean birth weight when gestational age is taken into account? May we conclude that there is interaction between hospital of birth and gestational age? Plot the original data and the fitted regression equations.

Birth Weight (kg)	Gestation Age (weeks)	Hospital of Birth	Birth Weight (kg)	Gestation Age (weeks)	Hospital of Birth
1.4	30	A	1.0	29	C
.9	27	B	1.4	33	C
1.2	33	A	.9	28	A
1.1	29	C	1.0	28	C
1.3	35	A	1.9	36	B
.8	27	B	1.3	29	B
1.0	32	A	1.7	35	C
.7	26	A	1.0	30	A
1.2	30	C	.9	28	A
.8	28	A	1.0	31	A
1.5	32	B	1.6	31	B
1.3	31	A	1.6	33	B
1.4	32	C	1.7	34	B
1.5	33	B	1.6	35	C
1.0	27	A	1.2	28	A
1.8	35	B	1.5	30	B

(Continued)

Birth Weight (kg)	Gestation Age (weeks)	Hospital of Birth	Birth Weight (kg)	Gestation Age (weeks)	Hospital of Birth
1.4	36	C	1.8	34	B
1.2	34	A	1.5	34	C
1.1	28	B	1.2	30	A
1.2	30	B	1.2	32	C

18. Refer to Chapter 9, Review Exercise 18. In the study cited in that exercise, Maria Mathias (A-13) investigated the relationship between ages (AGE) of boys and improvement in measures of hyperactivity, attitude, and social behavior. In the study, subjects were randomly assigned to two different treatments. The control group (TREAT = 0) received standard therapy for hyperactivity, and the treatment group (TREAT = 1) received standard therapy plus pet therapy. The results are shown in the following table. Create a scatter plot with age as the independent variable and ATT (change in attitude with positive numbers indicating positive change in attitude) as the dependent variable. Use different symbols for the two different treatment groups. Use multiple regression techniques to determine whether age, treatment, or the interaction are useful in predicting ATT. Report your results.

Subject	TREAT	AGE	ATT	Subject	TREAT	AGE	ATT
1	1	9	−1.2	17	0	10	0.4
2	1	9	0.0	18	0	7	0.0
3	1	13	−0.4	19	0	12	1.1
4	1	6	−0.4	20	0	9	0.2
5	1	9	1.0	21	0	7	0.4
6	1	8	0.8	22	0	6	0.0
7	1	8	−0.6	23	1	11	0.6
8	1	9	−1.2	24	1	11	0.4
9	0	7	0.0	25	1	11	1.0
10	0	12	0.4	26	1	11	0.8
11	0	9	−0.8	27	1	11	1.2
12	0	10	1.0	28	1	11	0.2
13	0	12	1.4	29	1	11	0.8
14	0	9	1.0	30	1	8	0.0
15	0	12	0.8	31	1	9	0.4
16	0	9	1.0				

Source: Maria Mathias, M.D. and the Wright State University Statistical Consulting Center. Used with permission.

For each study described in Exercises 19 through 21, answer as many of the following questions as possible:

(a) Which is the dependent variable?

(b) What are the independent variables?

(c) What are the appropriate null and alternative hypotheses?

(d) Which null hypotheses do you think were rejected? Why?

(e) Which is the more relevant objective, prediction or estimation, or are the two equally relevant? Explain your answer.

(f) What is the sampled population?

(g) What is the target population?

(h) Which variables are related to which other variables? Are the relationships direct or inverse?

(i) Write out the regression equation using appropriate numbers for parameter estimates.

(j) Give numerical values for any other statistics that you can.

(k) Identify each variable as to whether it is quantitative or qualitative.

(l) Explain the meaning of any statistics for which numerical values are given.

19. Golfinopoulos and Arhonditsis (A-14) used a multiple regression model in a study of trihalomethanes (THMs) in drinking water in Athens, Greece. THMs are of concern since they have been related to cancer and reproductive outcomes. The researchers found the following regression model useful in predicting THM:

$$\begin{aligned} THM = & -.26chla + 1.57\,pH + 28.74Br - 66.72Br^2 \\ & -43.63S + 1.13Sp + 2.62T \times S - .72T \times CL \end{aligned}$$

The variables were as follows: $chla$ = chlorophyll concentration, pH = acid/base scale, Br = bromide concentration, S = dummy variable for summer, Sp = dummy variable for spring, T = temperature, and CL = chlorine concentration. The researchers reported $R = .52$, $p < .001$.

20. In a study by Takata et al. (A-15), investigators evaluated the relationship between chewing ability and teeth number and measures of physical fitness in a sample of subjects ages 80 or higher in Japan. One of the outcome variables that measured physical fitness was leg extensor strength. To measure the ability to chew foods, subjects were asked about their ability to chew 15 foods (peanuts, vinegared octopus, and French bread, among others). Consideration of such variables as height, body weight, gender, systolic blood pressure, serum albumin, fasting glucose concentration, back pain, smoking, alcohol consumption, marital status, regular medical treatment, and regular exercise revealed that the number of chewable foods was significant in predicting leg extensor strength ($\hat{\beta}_1 = .075, p = .0366$). However, in the presence of the other variables, number of teeth was not a significant predictor ($\hat{\beta}_1 = .003, p = .9373$).

21. Varela et al. (A-16) examined 515 patients who underwent lung resection for bronchogenic carcinoma. The outcome variable was the occurrence of cardiorespiratory morbidity after surgery. Any of the following postoperative events indicated morbidity: pulmonary atelectasis or pneumonia, respiratory or ventilatory insufficiency at discharge, need for mechanical ventilation at any time after extubation in the operating room, pulmonary thromboembolism, arrhythmia, myocardial ischemia or infarct, and clinical cardiac insufficiency. Performing a stepwise logistic regression, the researchers found that age ($p < .001$) and postoperative forced expiratory volume ($p = .003$) were statistically significant in predicting the occurrence of cardiorespiratory morbidity.

For each of the data sets given in Exercises 22 through 29, do as many of the following as you think appropriate:

(a) Apply one or more of the techniques discussed in this chapter.

(b) Apply one or more of the techniques discussed in previous chapters.

(c) Construct graphs.

(d) Formulate relevant hypotheses, perform the appropriate tests, and find p values.

(e) State the statistical decisions and clinical conclusions that the results of your hypothesis tests justify.

(f) Describe the population(s) to which you think your inferences are applicable.

22. A study by Davies et al. (A-17) was motivated by the fact that, in previous studies of contractile responses to β-adrenoceptor agonists in single myocytes from failing and nonfailing human hearts, they had observed an age-related decline in maximum response to isoproterenol, at frequencies where the maximum response to high Ca^{2+} in the same cell was unchanged. For the present study, the investigators computed the isoproterenol/Ca^{2+} ratio (ISO/CA) from measurements taken on myocytes from patients ranging in age from 7 to 70 years. Subjects were classified as older (>50 years) and younger. The following are the (ISO/CA) values, age, and myocyte source of subjects in the study. Myocyte sources were reported as donor and biopsy.

Age	ISO/CA	Myocyte Source
7	1.37	Donor
21	1.39	Donor
28	1.17	Donor
35	0.71	Donor
38	1.14	Donor
50	0.95	Donor
51	0.86	Biopsy
52	0.72	Biopsy
55	0.53	Biopsy
56	0.81	Biopsy
61	0.86	Biopsy
70	0.77	Biopsy

Source: Dr. Sian E. Harding. Used with permission.

23. Hayton et al. (A-18) investigated the pharmacokinetics and bioavailability of cefetamet and cefetamet pivoxil in infants between the ages of 3.5 and 17.3 months who had received the antibiotic during and after urological surgery. Among the pharmacokinetic data collected were the following measurements of the steady-state apparent volume of distribution (V). Also shown are previously collected data on children ages 3 to 12 years (A-19) and adults (A-20). Weights (W) of subjects are also shown.

Infants		Children		Adults	
W (kg)	**V (liters)**	**W (kg)**	**V (liters)**	**W (kg)**	**V (liters)**
6.2	2.936	13	4.72	61	19.7
7.5	3.616	14	5.23	80	23.7
7.0	1.735	14	5.85	96	20.0
7.1	2.557	15	4.17	75	19.5
7.8	2.883	16	5.01	60	19.6
8.2	2.318	17	5.81	68	21.5
8.3	3.689	17	7.03	72.2	21.9
8.5	4.133	17.5	6.62	87	30.9
8.6	2.989	17	4.98	66.5	20.4
8.8	3.500	17.5	6.45		
10.0	4.235	20	7.73		
10.0	4.804	23	7.67		
10.2	2.833	25	9.82		

(Continued)

Infants		Children		Adults	
W (kg)	V (liters)	W (kg)	V (liters)	W (kg)	V (liters)
10.3	4.068	37	14.40		
10.6	3.640	28	10.90		
10.7	4.067	47	15.40		
10.8	8.366	29	9.86		
11.0	4.614	37	14.40		
12.5	3.168				
13.1	4.158				

Source: Dr. Klaus Stoeckel. Used with permission.

24. According to Fils-Aime et al. (A-21), epidemiologic surveys have found that alcoholism is the most common mental or substance abuse disorder among men in the United States. Fils-Aime and associates investigated the interrelationships of age at onset of excessive alcohol consumption, family history of alcoholism, psychiatric comorbidity, and cerebrospinal fluid (CSF) monoamine metabolite concentrations in abstinent, treatment-seeking alcoholics. Subjects were mostly white males classified as experiencing early (25 years or younger) or late (older than 25 years) onset of excessive alcohol consumption. Among the data collected were the following measurements on CSF tryptophan (TRYPT) and 5-hydroxyindoleacetic acid (5-HIAA) concentrations (pmol/ml).

5-HIAA	TRYPT	Onset 1 = Early 0 = Late	5-HIAA	TRYPT	Onset 1 = Early 0 = Late
57	3315	1	102	3181	1
116	2599	0	51	2513	1
81	3334	1	92	2764	1
78	2505	0	104	3098	1
206	3269	0	50	2900	1
64	3543	1	93	4125	1
123	3374	0	146	6081	1
147	2345	1	96	2972	1
102	2855	1	112	3962	0
93	2972	1	23	4894	1
128	3904	0	109	3543	1
69	2564	1	80	2622	1
20	8832	1	111	3012	1
66	4894	0	85	2685	1
90	6017	1	131	3059	0
103	3143	0	58	3946	1
68	3729	0	110	3356	0
81	3150	1	80	3671	1
143	3955	1	42	4155	1
121	4288	1	80	1923	1
149	3404	0	91	3589	1
82	2547	1	102	3839	0

(*Continued*)

5-HIAA	TRYPT	Onset 1 = Early 0 = Late	5-HIAA	TRYPT	Onset 1 = Early 0 = Late
100	3633	1	93	2627	0
117	3309	1	98	3181	0
41	3315	1	78	4428	0
223	3418	0	152	3303	0
96	2295	1	108	5386	1
87	3232	0	102	3282	1
96	3496	1	122	2754	1
34	2656	1	81	4321	1
98	4318	1	81	3386	1
86	3510	0	99	3344	1
118	3613	1	73	3789	1
84	3117	1	163	2131	1
99	3496	1	109	3030	0
114	4612	1	90	4731	1
140	3051	1	110	4581	1
74	3067	1	48	3292	0
45	2782	1	77	4494	0
51	5034	1	67	3453	1
99	2564	1	92	3373	1
54	4335	1	86	3787	0
93	2596	1	101	3842	1
50	2960	1	88	2882	1
118	3916	0	38	2949	1
96	2797	0	75	2248	0
49	3699	1	35	3203	0
133	2394	0	53	3248	1
105	2495	0	77	3455	0
61	2496	1	179	4521	1
197	2123	1	151	3240	1
87	3320	0	57	3905	1
50	3117	1	45	3642	1
109	3308	0	76	5233	0
59	3280	1	46	4150	1
107	3151	1	98	2579	1
85	3955	0	84	3249	1
156	3126	0	119	3381	0
110	2913	0	41	4020	1
81	3786	1	40	4569	1
53	3616	1	149	3781	1
64	3277	1	116	2346	1
57	2656	1	76	3901	1
29	4953	0	96	3822	1
34	4340	1			

Source: Dr. Markku Linnoila. Used with permission.

25. The objective of a study by Abrahamsson et al. (A-22) was to investigate the anti-thrombotic effects of an inhibitor of the plasminogen activator inhibitor-1 (PAI-1) in rats given endotoxin. Experimental subjects were male Sprague–Dawley rats weighing between 300 and 400 grams. Among the data collected were the following measurements on PAI-1 activity and the lung ^{125}I-concentration in anesthetized rats given three drugs:

Drugs	Plasma PAI-1 Activity (U/ml)	^{125}I-Fibrin in the Lungs (% of Ref. Sample)
Endotoxin	127	158
	175	154
	161	118
	137	77
	219	172
	260	277
	203	216
	195	169
	414	272
	244	192
Endotoxin + PRAP-1 low dose	107	49
	103	28
	248	187
	164	109
	176	96
	230	126
	184	148
	276	17
	201	97
	158	86
Endotoxin + PRAP-1 high dose	132	86
	130	24
	75	17
	140	41
	166	114
	194	110
	121	26
	111	53
	208	71
	211	90

Source: Dr. Tommy Abrahamsson. Used with permission.

26. Pearse and Sylvester (A-23) conducted a study to determine the separate contributions of ischemia and extracorporeal perfusion to vascular injury occurring in isolated sheep lungs and to determine the oxygen dependence of this injury. Lungs were subjected to ischemia alone, extracorporeal perfusion alone, and both ischemia and extracorporeal perfusion. Among the data collected were the following observations on change in pulmonary arterial pressure (mm Hg) and pulmonary vascular permeability assessed by estimation of the reflection coefficient for albumin in perfused lungs with and without preceding ischemia:

Ischemic–Perfused Lungs		Perfused Lungs	
Change in Pulmonary Pressure	Reflection Coefficient	Change in Pulmonary Pressure	Reflection Coefficient
8.0	0.220	34.0	0.693
3.0	0.560	31.0	0.470
10.0	0.550	4.0	0.651
23.0	0.806	48.0	0.999
15.0	0.472	32.0	0.719
43.0	0.759	27.0	0.902
18.0	0.489	25.0	0.736
27.0	0.546	25.0	0.718
13.0	0.548		
0.0	0.467		

Source: Dr. David B. Pearse. Used with permission.

27. The purpose of a study by Balzamo et al. (A-24) was to investigate, in anesthetized rabbits, the effects of mechanical ventilation on the concentration of substance P (SP) measured by radioimmunoassay in nerves and muscles associated with ventilation and participating in the sensory innervation of the respiratory apparatus and heart. SP is a neurotransmitter located in primary sensory neurons in the central and autonomic nervous systems. Among the data collected were the following measures of SP concentration in cervical vagus nerves (X) and corresponding nodose ganglia (NG), right and left sides:

SPXright	SPNGright	SPXleft	SPNGleft
0.6500	9.6300	3.3000	1.9300
2.5600	3.7800	0.6200	2.8700
1.1300	7.3900	0.9600	1.3100
1.5500	3.2800	2.7000	5.6400
35.9000	22.0000	4.5000	9.1000
19.0000	22.8000	8.6000	8.0000
13.6000	2.3000	7.0000	8.3000
8.0000	15.8000	4.1000	4.7000
7.4000	1.6000	5.5000	2.5000
3.3000	11.6000	9.7000	8.0000
19.8000	18.0000	13.8000	8.0000
8.5000	6.2000	11.0000	17.2000
5.4000	7.8000	11.9000	5.3000
11.9000	16.9000	8.2000	10.6000
47.7000	35.9000	3.9000	3.3000
14.2000	10.2000	3.2000	1.9000
2.9000	1.6000	2.7000	3.5000
6.6000	3.7000	2.8000	2.5000
3.7000	1.3000		

Source: Dr. Yves Jammes. Used with permission.

28. Scheeringa and Zeanah (A-25) examined the presence of posttraumatic stress disorder (PTSD), the severity of posttraumatic symptomatology, and the pattern of expression of symptom clusters in relation to six independent variables that may be salient to the development of a posttraumatic disorder in children under 48 months of age. The following data were collected during the course of the study.

Predictor Variables						Response Variables			
Gender	Age	Acute/ Rept.	Injury	Wit./ Exper.	Threat to Caregiver	Reexp	Numb	Arous	FrAgg
0	1	0	1	1	1	3	0	0	1
0	1	0	0	0	1	2	2	1	1
1	1	0	0	0	1	3	1	1	1
0	1	0	0	0	1	3	1	0	4
1	0	1	1	1	0	1	3	1	1
1	1	0	1	1	0	3	1	0	1
0	1	0	1	1	0	4	2	0	1
0	1	0	0	1	0	5	2	0	4
1	1	0	0	0	1	2	1	3	2
1	1	1	1	1	0	4	1	0	0
0	0	1	1	1	0	1	3	0	1
1	0	1	0	1	0	1	3	0	2
1	0	1	1	1	0	0	3	0	0
1	1	0	1	1	0	4	1	2	1
1	0	0	1	1	1	3	2	1	3
1	0	0	1	1	1	3	1	2	1
0	1	0	1	1	1	3	1	2	2
0	1	0	0	0	1	5	2	1	1
0	1	0	0	0	1	1	2	2	2
0	1	0	1	1	0	4	4	0	3
1	0	1	1	1	0	2	1	2	3
1	0	0	1	1	1	1	1	2	1
1	1	0	0	0	1	4	1	1	1
0	1	0	0	0	1	3	2	1	0
0	1	0	0	0	1	3	1	2	4
0	1	0	0	0	1	3	1	2	4
0	1	0	0	1	0	2	2	0	0
1	1	0	0	0	1	2	0	3	0
1	1	0	0	0	1	2	0	1	2
0	1	0	1	0	1	2	3	1	3
1	1	1	0	1	0	1	2	1	1
1	1	0	1	1	1	3	2	0	4
1	1	0	0	0	0	2	4	2	0
0	1	0	0	0	1	1	1	0	2
0	0	1	0	0	1	2	3	2	3
0	0	1	0	0	1	3	1	4	3
0	0	1	0	0	1	3	1	2	3

(Continued)

Predictor Variables						Response Variables			
Gender	Age	Acute/ Rept.	Injury	Wit./ Exper.	Threat to Caregiver	Reexp	Numb	Arous	FrAgg
0	0	0	0	1	0	1	1	0	0
1	0	0	0	0	1	4	3	2	3
1	0	0	1	1	0	4	2	3	2
0	0	1	1	1	0	1	2	2	1

Key: **Gender** 0 = male
1 = female
Age 0 = younger than 18 months at time of trauma
1 = older than 18 months
Acute/rept. 0 = trauma was acute, single blow
1 = trauma was repeated or chronic
Injury 0 = subject was not injured in the trauma
1 = subject was physically injured in the trauma
Wit./exper. 0 = subject witnessed but did not directly experience trauma
1 = subject directly experienced the trauma
Threat to caregiver 0 = caregiver was not threatened in the trauma
1 = caregiver was threatened in the trauma
Reexp = Reexperiencing cluster symptom count
Numb = Numbing of responsiveness/avoidance cluster symptom count
Arous = Hyperarousal cluster symptom count
FrAgg = New fears/aggression cluster symptom count

Source: Dr. Michael S. Scheeringa. Used with permission.

29. One of the objectives of a study by Mulloy and McNicholas (A-26) was to compare ventilation and gas exchange during sleep and exercise in chronic obstructive pulmonary disease (COPD). The investigators wished also to determine whether exercise studies could aid in the prediction of nocturnal desaturation in COPD. Subjects (13 male, 6 female) were ambulatory patients attending an outpatient respiratory clinic. The mean age of the patients, all of whom had severe, stable COPD, was 64.8 years with a standard deviation of 5.2. Among the data collected were measurements on the following variables:

Age (years)	BMI	Pa_{O_2} (mm Hg)	Pa_{CO_2} (mm Hg)	FEV_1 (% Predicted)	Lowest Ex. $Sa_{O_2}^a$	Mean Sleep $Sa_{O_2}^a$	Lowest Sleep $Sa_{O_2}^a$	Fall Sleep $Sa_{O_2}^a$
67	23.46	52.5	54	22	74	70.6	56	29.6
62	25.31	57.75	49.575	19	82	85.49	76	11.66
68	23.11	72	43.8	41	95	88.72	82	11.1
61	25.15	72	47.4	38	88	91.11	76	18.45
70	24.54	78	40.05	40	88	92.86	92	0.8
71	25.47	63.75	45.375	31	85	88.95	80	13
60	19.49	80.25	42.15	28	91	94.78	90	4
57	21.37	84.75	40.2	20	91	93.72	89	5.8

(*Continued*)

Age (years)	BMI	Pa_{O_2} (mm Hg)	Pa_{CO_2} (mm Hg)	FEV_1 (% Predicted)	Lowest Ex. Sa_{O_2}[a]	Mean Sleep Sa_{O_2}[a]	Lowest Sleep Sa_{O_2}[a]	Fall Sleep Sa_{O_2}[a]
69	25.78	68.25	43.8	32	85	90.91	79	13
57	22.13	83.25	43.725	20	88	94.39	86	9.5
74	26.74	57.75	51	33	75	89.89	80	14.11
63	19.07	78	44.175	36	81	93.95	82	13
64	19.61	90.75	40.35	27	90	95.07	92	4
73	30.30	69.75	38.85	53	87	90	76	18
63	26.12	51.75	46.8	39	67	69.31	46	34.9
62	21.71	72	41.1	27	88	87.95	72	22
67	24.75	84.75	40.575	45	87	92.95	90	2.17
57	25.98	84.75	40.05	35	94	93.4	86	8.45
66	32.00	51.75	53.175	30	83	80.17	71	16

[a] Treated as dependent variable in the authors' analyses. BMI = body mass index; Pa_{O_2} = arterial oxygen tension; Pa_{CO_2} = arterial carbon dioxide pressure; FEV_1 = forced expiratory volume in 1 second; Sa_{O_2} = arterial oxygen saturation.
Source: Dr. Eithne Mulloy. Used with permission.

Exercises for Use with the Large Data Sets Available on the Following Website: www.wiley.com/college/daniel

1. The goal of a study by Gyurcsik et al. (A-27) was to examine the usefulness of aquatic exercise-related goals, task self-efficacy, and scheduling self-efficacy for predicting aquatic exercise attendance by individuals with arthritis. The researchers collected data on 142 subjects participating in Arthritis Foundation Aquatics Programs. The outcome variable was the percentage of sessions attended over an 8-week period (ATTEND). The following predictor variables are all centered values. Thus, for each participant, the mean for all participants is subtracted from the individual score. The variables are:

   GOALDIFF—higher values indicate setting goals of higher participation.

   GOALSPEC—higher values indicate higher specificity of goals related to aquatic exercise.

   INTER—interaction of GOALDIFF and GOALSPEC.

   TSE—higher values indicate participants' confidence in their abilities to attend aquatic classes.

   SSE—higher values indicate participants' confidence in their abilities to perform eight tasks related to scheduling exercise into their daily routine for 8 weeks.

   MONTHS—months of participation in aquatic exercise prior to start of study.

   With the data set AQUATICS, perform a multiple regression to predict ATTEND with each of the above variables. What is the multiple correlation coefficient? What variables are significant in predicting ATTEND? What are your conclusions?

2. Rodehorst (A-28) conducted a prospective study of 212 rural elementary school teachers. The main outcome variable was the teachers' intent to manage children demonstrating symptoms of asthma in their classrooms. This variable was measured with a single-item question that used a seven-point Likert scale (INTENT, with possible responses of 1 = extremely probable to

7 = extremely improbable). Rodehorst used the following variables as independent variables to predict INTENT:

SS = Social Support. Scores range from 7 to 49, with higher scores indicating higher perceived social support for managing children with asthma in a school setting.

ATT = Attitude. Scores range from 15 to 90, with higher scores indicating more favorable attitudes toward asthma.

KNOW = Knowledge. Scores range from 0 to 24, with higher scores indicating higher general knowledge about asthma.

CHILD = Number of children with asthma the teacher has had in his or her class during his or her entire teaching career.

SE = Self-efficacy. Scores range from 12 to 60, with higher scores indicating higher self-efficacy for managing children with asthma in the school setting.

YRS = Years of teaching experience.

With the data TEACHERS, use stepwise regression analysis to select the most useful variables to include in a model for predicting INTENT.

3. Refer to the weight loss data on 588 cancer patients and 600 healthy controls (WGTLOSS). Weight loss among cancer patients is a well-known phenomenon. Of interest to clinicians is the role played in the process by metabolic abnormalities. One investigation into the relationships among these variables yielded data on whole-body protein turnover (Y) and percentage of ideal body weight for height (X). Subjects were lung cancer patients and healthy controls of the same age. Select a simple random sample of size 15 from each group and do the following:

(a) Draw a scatter diagram of the sample data using different symbols for each of the two groups.

(b) Use dummy variable coding to analyze these data.

(c) Plot the two regression lines on the scatter diagram. May one conclude that the two sampled populations differ with respect to mean protein turnover when percentage of ideal weight is taken into account?

May one conclude that there is interaction between health status and percentage of ideal body weight? Prepare a verbal interpretation of the results of your analysis and compare your results with those of your classmates.

REFERENCES

Methodology References

1. David W. Hosmer and Stanley Lemeshow, *Applied Logistic Regression*, 2nd ed. Wiley, New York, 2000.
2. David G. Kleinbaum, *Logistic Regression: A Self-Learning Text*, New York, Springer, 1994.

Applications References

A-1. North Carolina State Center for Health Statistics and Howard W. Odum Institute for Research in Social Science at the University of North Carolina at Chapel Hill, Birth Data set for 2001 found at www.irss.unc.edu/ncvital/bfd1down.html. All sampling and coding performed by John Holcomb and do not represent the findings of the Center or Institute.

A-2. B. Bolwell, R. Sobecks, B. Pohlman, S. Andresen, K. Theil, S. Serafino, L. Rybicki, and M. Kalaycio, "Etoposide (VP-16) Plus G-CSF Mobilizes Different Dendritic Cell Subsets than Does G-CSF Alone," *Bone Marrow Transplantation, 31* (2003), 95–98.

A-3. Manoj Pandey, Lal B. Sharma, and Vijay K. Shukla, "Cytochrome P-450 Expression and Lipid Peroxidation in Gallbladder Cancer," *Journal of Surgical Oncology, 82* (2003), 180–183.

A-4. Mori J. Krantz, Ilana B. Kutinsky, Alastair D. Robertson, and Philip S. Mehler, "Dose-Related Effects of Methadone on QT Prolongation in a Series of Patients with Torsade de Pointes," *Pharmacotherapy, 23* (2003), 802–805.

A-5. Robert A. Reiss, Curtis E. Haas, Deborah L. Griffis, Bernadette Porter, and Mary Ann Tara, "Point-of-Care versus Laboratory Monitoring of Patients Receiving Different Anticoagulant Therapies," *Pharmacotherapy 22,* (2002), 677–685.

A-6. Gwi-Ryung Son, May L. Wykle, and Jaclene A. Zauszniewski, "Korean Adult Child Caregivers of Older Adults with Dementia," *Journal of Gerontological Nursing, 29* (2003), 19–28.

A-7. M. Naeije, "Local Kinematic and Anthropometric Factors Related to the Maximum Mouth Opening in Healthy Individuals," *Journal of Oral Rehabilitation, 29* (2002), 534–539.

A-8. Daniel F. Connor, Ronald J. Steingard, Jennifer J. Anderson, and Richard H. Melloni, "Gender Differences in Reactive and Proactive Aggression," *Child Psychiatry and Human Development, 33* (2003), 279–294.

A-9. Daniel H. LaMont, Matthew J. Budoff, David M. Shavelle, Robert Shavelle, Bruce H. Brundage, and James M. Hagar, "Coronary Calcium Scanning Adds Incremental Value to Patients with Positive Stress Tests," *American Heart Journal, 143* (2002), 861–867.

A-10. Robyn Gallagher, Sharon McKinley, and Kathleen Dracup, "Predictors of Women's Attendance at Cardiac Rehabilitation Programs," *Progress in Cardiovascular Nursing, 18* (2003), 121–126.

A-11. John H. Porcerelli, Rosemary Cogan, Patricia P. West, Edward A. Rose, Dawn Lambrecht, Karen E. Wilson, Richard K. Severson, and Dunia Karana, "Violent Victimization of Women and Men: Physical and Psychiatric Symptoms," *Journal of the American Board of Family Practice, 16* (2003), 32–39.

A-12. Debra A. Jansen and Mary L. Keller, "Cognitive Function in Community-Dwelling Elderly Women," *Journal of Gerontological Nursing, 29* (2003), 34–43.

A-13. Maria Mathias and the Wright State University Statistical Consulting Center.

A-14. Spyros K. Golfinopoulos and George B. Arhonditsis, "Multiple Regression Models: A Methodology for Evaluating Trihalomethane Concentrations in Drinking Water from Raw Water Characteristics," *Chemosphere, 47* (2002), 1007–1018.

A-15. Y. Takata, T. Ansai, S. Awano, T. Hamasaki, Y. Yoshitake, Y. Kimura, K. Sonoki, M. Wakisaka, M. Fukuhara, and T. Takehara, "Relationship of Physical Fitness to Chewing in an 80-Year-Old Population," *Oral Diseases, 10* (2004), 44–49.

A-16. G. Varela, N. Novoa, M. F. Jiménez, and G. Santos, "Applicability of Logistic Regression (LR) Risk Modeling to Decision Making in Lung Cancer Resection," *Interactive Cardiovascular and Thoracic Surgery, 2* (2003), 12–15.

A-17. C. H. Davies, N. Ferrara, and S. E. Harding, "β-Adrenoceptor Function Changes with Age of Subject in Myocytes from Non-Failing Human Ventricle," *Cardiovascular Research, 31* (1996), 152–156.

A-18. William L. Hayton, Johannes Kneer, Ronald de Groot, and Klaus Stoeckel, "Influence of Maturation and Growth on Cefetamet Pivoxil Pharmacokinetics: Rational Dosing for Infants," *Antimicrobial Agents and Chemotherapy, 40* (1996), 567–574.

A-19. W. L. Hayton, R. A. Walstad, E. Thurmann-Nielsen, T. Kufaas, J. Kneer, R. J. Ambros, H. E. Rugstad, E. Monn, E. Bodd, and K. Stoeckel,"Pharmacokinetics of Intravenous Cefetamet and Oral Cefetamet Pivoxil in Children," *Antimicrobial Agents and Chemotherapy, 35* (1991), 720–725. Erratum, *36* (1992), 2575.

A-20. M. P. Ducharme, D. J. Edwards, P. J. McNamara, and K. Stoeckel, "Bioavailability of Syrup and Tablet Formulations of Cefetamet Pivoxil," *Antimicrobial Agents and Chemotherapy, 37* (1993), 2706–2709.

A-21. Marie-Lourdes Fils-Aime, Michael J. Eckardt, David T. George, Gerald L. Brown, Ivan Mefford, and Markku Linnoila, "Early-Onset Alcoholics Have Lower Cerebrospinal Fluid 5-Hydroxyindoleacetic Acid Levels than Late-Onset Alcoholics," *Archives of General Psychiatry, 53* (1996), 211–216.

A-22. T. Abrahamsson, V. Nerme, M. Strömqvist, B. Åkerblom, A. Legnehed, K. Pettersson, and A. Westin Eriksson, "Anti-thrombotic Effect of PAI-1 Inhibitor in Rats Given Endotoxin," *Thrombosis and Haemostasis, 75* (1996), 118–126.

A-23. David B. Pearse and J. T. Sylvester, "Vascular Injury in Isolated Sheep Lungs: Role of Ischemia, Extracorporeal Perfusion, and Oxygen," *American Journal of Respiratory and Critical Care Medicine, 153* (1996), 196–202.

A-24. EMMANUEL BALZAMO, PIERRE JOANNY, JEAN GUILLAUME STEINBERG, CHARLES OLIVER, and YVES JAMMES, "Mechanical Ventilation Increases Substance P Concentration in the Vagus, Sympathetic, and Phrenic Nerves," *American Journal of Respiratory and Critical Care Medicine, 153* (1996), 153–157.

A-25. MICHAEL S. SCHEERINGA and CHARLES H. ZEANAH, "Symptom Expression and Trauma Variables in Children Under 48 Months of Age," *Infant Mental Health Journal, 16* (1995), 259–270.

A-26. EITHNE MULLOY and WALTER T. MCNICHOLAS, "Ventilation and Gas Exchange During Sleep and Exercise in Severe COPD," *Chest, 109* (1996), 387–394.

A-27. NANCY C. GYURCSIK, PAUL A. ESTABROOKS, and MELISSA J. FRAHM-TEMPLAR, "Exercise-Related Goals and Self-Efficacy as Correlates of Aquatic Exercise in Individuals with Arthritis," *Arthritis Care and Research, 49* (2003), 306–313.

A-28. T. KIM RODEHORST, "Rural Elementary School Teachers' Intent to Manage Children with Asthma Symptoms," *Pediatric Nursing, 29* (2003), 184–194.

CHAPTER **12**

ANALYSIS OF FREQUENCY DATA: AN INTRODUCTION TO THE CHI-SQUARE DISTRIBUTION

CHAPTER OVERVIEW

This chapter explores techniques that are commonly used in the analysis of count or frequency data. Uses of the chi-square distribution, which was mentioned briefly in Chapter 6, are discussed and illustrated in greater detail. Additionally, statistical techniques often used in epidemiological studies are introduced and demonstrated by means of examples.

TOPICS

12.1 INTRODUCTION

12.2 THE MATHEMATICAL PROPERTIES OF THE CHI-SQUARE DISTRIBUTION

12.3 TESTS OF GOODNESS-OF-FIT

12.4 TESTS OF INDEPENDENCE

12.5 TESTS OF HOMOGENEITY

12.6 THE FISHER EXACT TEST

12.7 RELATIVE RISK, ODDS RATIO, AND THE MANTEL-HAENSZEL STATISTIC

12.8 SURVIVAL ANALYSIS

12.9 SUMMARY

LEARNING OUTCOMES

After studying this chapter, the student will

1. understand the mathematical properties of the chi-square distribution.
2. be able to use the chi-square distribution for goodness-of-fit tests.
3. be able to construct and use contingency tables to test independence and homogeneity.

4. be able to apply Fisher's exact test for 2 × 2 tables.
5. understand how to calculate and interpret the epidemiological concepts of relative risk, odds ratios, and the Mantel-Haenszel statistic.

12.1 INTRODUCTION

In the chapters on estimation and hypothesis testing, brief mention is made of the chi-square distribution in the construction of confidence intervals for, and the testing of, hypotheses concerning a population variance. This distribution, which is one of the most widely used distributions in statistical applications, has many other uses. Some of the more common ones are presented in this chapter along with a more complete description of the distribution itself, which follows in the next section.

The chi-square distribution is the most frequently employed statistical technique for the analysis of count or frequency data. For example, we may know for a sample of hospitalized patients how many are male and how many are female. For the same sample we may also know how many have private insurance coverage, how many have Medicare insurance, and how many are on Medicaid assistance. We may wish to know, for the population from which the sample was drawn, if the type of insurance coverage differs according to gender. For another sample of patients, we may have frequencies for each diagnostic category represented and for each geographic area represented. We might want to know if, in the population from which the same was drawn, there is a relationship between area of residence and diagnosis. We will learn how to use chi-square analysis to answer these types of questions.

There are other statistical techniques that may be used to analyze frequency data in an effort to answer other types of questions. In this chapter we will also learn about these techniques.

12.2 THE MATHEMATICAL PROPERTIES OF THE CHI-SQUARE DISTRIBUTION

The chi-square distribution may be derived from normal distributions. Suppose that from a normally distributed random variable Y with mean μ and variance σ^2 we randomly and independently select samples of size $n = 1$. Each value selected may be transformed to the standard normal variable z by the familiar formula

$$z_i = \frac{y_i - \mu}{\sigma} \tag{12.2.1}$$

Each value of z may be squared to obtain z^2. When we investigate the sampling distribution of z^2, we find that it follows a chi-square distribution with 1 degree of freedom. That is,

$$\chi^2_{(1)} = \left(\frac{y - \mu}{\sigma}\right)^2 = z^2$$

Now suppose that we randomly and independently select samples of size $n = 2$ from the normally distributed population of Y values. Within each sample we may transform each value of y to the standard normal variable z and square as before. If the resulting values of z^2 for each sample are added, we may designate this sum by

$$\chi^2_{(2)} = \left(\frac{y_1 - \mu}{\sigma}\right)^2 + \left(\frac{y_2 - \mu}{\sigma}\right)^2 = z_1^2 + z_2^2$$

since it follows the chi-square distribution with 2 degrees of freedom, the number of independent squared terms that are added together.

The procedure may be repeated for any sample size n. The sum of the resulting z^2 values in each case will be distributed as chi-square with n degrees of freedom. In general, then,

$$\chi^2_{(n)} = z_1^2 + z_2^2 + \cdots + z_n^2 \tag{12.2.2}$$

follows the chi-square distribution with n degrees of freedom. The mathematical form of the chi-square distribution is as follows:

$$f(u) = \frac{1}{\left(\frac{k}{2} - 1\right)!} \frac{1}{2^{k/2}} u^{(k/2)-1} e^{-(u/2)}, \quad u > 0 \tag{12.2.3}$$

where e is the irrational number 2.71828 $\cdots$ and k is the number of degrees of freedom. The variate u is usually designated by the Greek letter chi (χ) and, hence, the distribution is called the chi-square distribution. As we pointed out in Chapter 6, the chi-square distribution has been tabulated in Appendix Table F. Further use of the table is demonstrated as the need arises in succeeding sections.

The mean and variance of the chi-square distribution are k and $2k$, respectively. The modal value of the distribution is $k - 2$ for values of k greater than or equal to 2 and is zero for $k = 1$.

The shapes of the chi-square distributions for several values of k are shown in Figure 6.9.1. We observe in this figure that the shapes for $k = 1$ and $k = 2$ are quite different from the general shape of the distribution for $k > 2$. We also see from this figure that chi-square assumes values between 0 and infinity. It cannot take on negative values, since it is the sum of values that have been squared. A final characteristic of the chi-square distribution worth noting is that the sum of two or more independent chi-square variables also follows a chi-square distribution.

Types of Chi-Square Tests As already noted, we make use of the chi-square distribution in this chapter in testing hypotheses where the data available for analysis are in the form of frequencies. These hypothesis testing procedures are discussed under the topics of *tests of goodness-of-fit, tests of independence,* and *tests of homogeneity.* We will discover that, in a sense, all of the chi-square tests that we employ may be thought of as goodness-of-fit tests, in that they test the goodness-of-fit of observed frequencies to frequencies that one would expect if the data were generated under some particular theory or hypothesis. We, however, reserve the phrase "goodness-of-fit" for use in a more restricted sense. We use it to refer to a comparison of a sample distribution to some theoretical distribution that it is assumed describes the population from which the sample came. The justification of our use of the distribution in these situations is due to Karl Pearson (1), who showed that the chi-square distribution may be used as a test of the agreement between observation and hypothesis whenever the data are in the form of frequencies. An extensive treatment of the chi-square distribution is to be found in the book by Lancaster (2). Nikulin and Greenwood (3) offer practical advice for conducting chi-square tests.

Observed Versus Expected Frequencies The chi-square statistic is most appropriate for use with categorical variables, such as marital status, whose values are the categories married, single, widowed, and divorced. The quantitative data used in the computation of the test statistic are the frequencies associated with each category of the one or more variables under study. There are two sets of frequencies with which we are concerned, *observed frequencies* and *expected frequencies.* The observed frequencies are the number of subjects or objects in our sample that fall into the various categories of the variable of interest. For example, if we have a sample of 100 hospital patients, we may observe that 50 are married, 30 are single, 15 are widowed, and 5 are divorced. Expected frequencies are the number of subjects or objects in our sample that we would expect to observe if some null hypothesis about the variable is true. For example, our null hypothesis might be that the four categories of marital status are equally represented in the population from which we drew our sample. In that case we would expect our sample to contain 25 married, 25 single, 25 widowed, and 25 divorced patients.

The Chi-Square Test Statistic The test statistic for the chi-square tests we discuss in this chapter is

$$X^2 = \sum\left[\frac{(O_i - E_i)^2}{E_i}\right] \tag{12.2.4}$$

When the null hypothesis is true, X^2 is distributed approximately as χ^2 with $k - r$ degrees of freedom. In determining the degrees of freedom, k is equal to the number of groups for which observed and expected frequencies are available, and r is the number of restrictions or constraints imposed on the given comparison. A restriction is imposed when we force the sum of the expected frequencies to equal the sum of the observed frequencies, and an additional restriction is imposed for each parameter that is estimated from the sample.

In Equation 12.2.4, O_i is the observed frequency for the ith category of the variable of interest, and E_i is the expected frequency (given that H_0 is true) for the ith category.

The quantity X^2 is a measure of the extent to which, in a given situation, pairs of observed and expected frequencies agree. As we will see, the nature of X^2 is such that when there is close agreement between observed and expected frequencies it is small, and when the agreement is poor it is large. Consequently, only a sufficiently large value of X^2 will cause rejection of the null hypothesis.

If there is perfect agreement between the observed frequencies and the frequencies that one would expect, given that H_0 is true, the term $O_i - E_i$ in Equation 12.2.4 will be equal to zero for each pair of observed and expected frequencies. Such a result would yield a value of X^2 equal to zero, and we would be unable to reject H_0.

When there is disagreement between observed frequencies and the frequencies one would expect given that H_0 is true, at least one of the $O_i - E_i$ terms in Equation 12.2.4 will be a nonzero number. In general, the poorer the agreement between the O_i and the E_i, the greater or the more frequent will be these nonzero values. As noted previously, if the agreement between the O_i and the E_i is sufficiently poor (resulting in a sufficiently large X^2 value,) we will be able to reject H_0.

When there is disagreement between a pair of observed and expected frequencies, the difference may be either positive or negative, depending on which of the two frequencies is the larger. Since the measure of agreement, X^2, is a sum of component quantities whose magnitudes depend on the difference $O_i - E_i$, positive and negative differences must be given equal weight. This is achieved by squaring each $O_i - E_i$ difference. Dividing the squared differences by the appropriate expected frequency converts the quantity to a term that is measured in original units. Adding these individual $(O_i - E_i)^2/E_i$ terms yields X^2, a summary statistic that reflects the extent of the overall agreement between observed and expected frequencies.

The Decision Rule The quantity $\Sigma[(O_i - E_i)^2/E_i]$ will be small if the observed and expected frequencies are close together and will be large if the differences are large.

The computed value of X^2 is compared with the tabulated value of χ^2 with $k - r$ degrees of freedom. The decision rule, then, is: Reject H_0 if X^2 is greater than or equal to the tabulated χ^2 for the chosen value of α.

Small Expected Frequencies Frequently in applications of the chi-square test the expected frequency for one or more categories will be small, perhaps much less than 1. In the literature the point is frequently made that the approximation of X^2 to χ^2 is not strictly valid when some of the expected frequencies are small. There is disagreement among writers, however, over what size expected frequencies are allowable before making some adjustment or abandoning χ^2 in favor of some alternative test. Some writers, especially the earlier ones, suggest lower limits of 10, whereas others suggest that all expected frequencies should be no less than 5. Cochran (4, 5), suggests that for goodness-of-fit tests of unimodal distributions (such as the normal), the minimum expected frequency can be as low as 1. If, in practice, one encounters one or more expected frequencies less than 1, adjacent categories may be combined to achieve the suggested minimum. Combining reduces the number of categories and, therefore, the number of degrees of freedom. Cochran's suggestions appear to have been followed extensively by practitioners in recent years.

12.3 TESTS OF GOODNESS-OF-FIT

As we have pointed out, a goodness-of-fit test is appropriate when one wishes to decide if an observed distribution of frequencies is incompatible with some preconceived or hypothesized distribution.

We may, for example, wish to determine whether or not a sample of observed values of some random variable is compatible with the hypothesis that it was drawn from a population of values that is normally distributed. The procedure for reaching a decision consists of placing the values into mutually exclusive categories or class intervals and noting the frequency of occurrence of values in each category. We then make use of our knowledge of normal distributions to determine the frequencies for each category that one could expect if the sample had come from a normal distribution. If the discrepancy is of such magnitude that it could have come about due to chance, we conclude that the sample may have come from a normal distribution. In a similar manner, tests of goodness-of-fit may be carried out in cases where the hypothesized distribution is the binomial, the Poisson, or any other distribution. Let us illustrate in more detail with some examples of tests of hypotheses of goodness-of-fit.

EXAMPLE 12.3.1 *The Normal Distribution*

Cranor and Christensen (A-1) conducted a study to assess short-term clinical, economic, and humanistic outcomes of pharmaceutical care services for patients with diabetes in community pharmacies. For 47 of the subjects in the study, cholesterol levels are summarized in Table 12.3.1.

We wish to know whether these data provide sufficient evidence to indicate that the sample did not come from a normally distributed population. Let $\alpha = .05$

Solution:

1. **Data.** See Table 12.3.1.
2. **Assumptions.** We assume that the sample available for analysis is a simple random sample.

TABLE 12.3.1 Cholesterol Levels as Described in Example 12.3.1

Cholesterol Level (mg/dl)	Number of Subjects
100.0–124.9	1
125.0–149.9	3
150.0–174.9	8
175.0–199.9	18
200.0–224.9	6
225.0–249.9	4
250.0–274.9	4
275.0–299.9	3

Source: Carole W. Cranor, and Dale B. Christensen, "The Asheville Project: Short-Term Outcomes of a Community Pharmacy Diabetes Care Program," *Journal of the American Pharmaceutical Association*, *43* (2003), 149–159. Used with permission.

3. Hypotheses.

H_0: In the population from which the sample was drawn, cholesterol levels are normally distributed.

H_A: The sampled population is not normally distributed.

4. Test statistic. The test statistic is

$$X^2 = \sum_{i=1}^{k}\left[\frac{(O_i - E_i)^2}{E_i}\right]$$

5. Distribution of test statistic. If H_0 is true, the test statistic is distributed approximately as chi-square with $k - r$ degrees of freedom. The values of k and r will be determined later.

6. Decision rule. We will reject H_0 if the computed value of X^2 is equal to or greater than the critical value of chi-square.

7. Calculation of test statistic. Since the mean and variance of the hypothesized distribution are not specified, the sample data must be used to estimate them. These parameters, or their estimates, will be needed to compute the frequency that would be expected in each class interval when the null hypothesis is true. The mean and standard deviation computed from the grouped data of Table 12.3.1 are

$$\bar{x} = 198.67$$
$$s = 41.31$$

As the next step in the analysis, we must obtain for each class interval the frequency of occurrence of values that we would expect when the null hypothesis is true, that is, if the sample were, in fact, drawn from a normally distributed population of values. To do this, we first determine the expected relative frequency of occurrence of values for each class interval and then multiply these expected relative frequencies by the total number of values to obtain the expected number of values for each interval.

The Expected Relative Frequencies

It will be recalled from our study of the normal distribution that the relative frequency of occurrence of values equal to or less than some specified value, say, x_0, of the normally distributed random variable X is equivalent to the area under the curve and to the left of x_0 as represented by the shaded area in Figure 12.3.1. We obtain the numerical value of this area by converting x_0 to a standard normal deviation by the formula $z_0 = (x_0 - \mu)/\sigma$ and finding the appropriate value in Appendix Table D. We use this procedure to obtain the expected relative frequencies corresponding to each of the class intervals in Table 12.3.1. We estimate μ and σ with $\bar{x}$ and s as computed from the grouped sample data. The first step consists of obtaining z values corresponding to the lower limit of each class interval. The area between two successive z values will give the expected relative frequency of occurrence of values for the corresponding class interval.

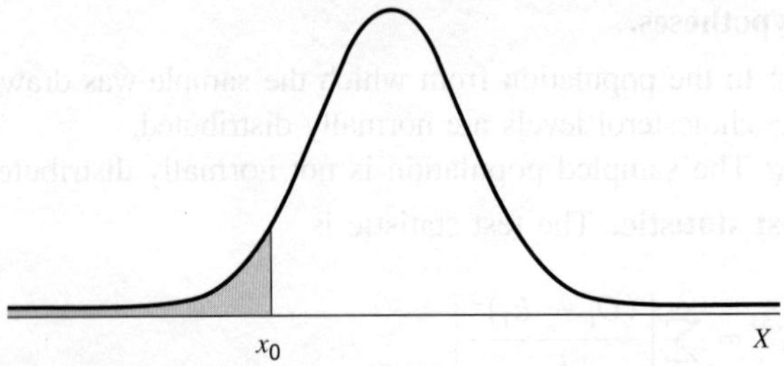

FIGURE 12.3.1 A normal distribution showing the relative frequency of occurrence of values less than or equal to x_0. The shaded area represents the relative frequency of occurrence of values equal to or less than x_0.

For example, to obtain the expected relative frequency of occurrence of values in the interval 100.0 to 124.9 we proceed as follows:

$$\text{The } z \text{ value corresponding to } X = 100.0 \text{ is } z = \frac{100.0 - 198.67}{41.31} = -2.39$$

$$\text{The } z \text{ value corresponding to } X = 125.0 \text{ is } z = \frac{125.0 - 198.67}{41.31} = -1.78$$

In Appendix Table D we find that the area to the left of -2.39 is .0084, and the area to the left of -1.78 is .0375. The area between -1.78 and -2.39 is equal to $.0375 - .0084 = .0291$, which is equal to the expected relative frequency of occurrence of cholesterol levels within the interval 100.0 to 124.9. This tells us that if the null hypothesis is true, that is, if the cholesterol levels are normally distributed, we should expect 2.91 percent of the values in our sample to be between 100.0 and 124.9. When we multiply our total sample size, 47, by .0291 we find the expected frequency for the interval to be 1.4. Similar calculations will give the expected frequencies for the other intervals as shown in Table 12.3.2.

TABLE 12.3.2 Class Intervals and Expected Frequencies for Example 12.3.1

Class Interval	$z = (x_i - \bar{x})/s$ At Lower Limit of Interval	Expected Relative Frequency	Expected Frequency
< 100		.0084	.4 } 1.8
100.0–124.9	−2.39	.0291	1.4 }
125.0–149.9	−1.78	.0815	3.8
150.0–174.9	−1.18	.1653	7.8
175.0–199.9	−.57	.2277	10.7
200.0–224.9	.03	.2269	10.7
225.0–249.9	.64	.1536	7.2
250.0–274.9	1.24	.0753	3.5
275.0–299.9	1.85	.0251	1.2 } 1.5
300.0 and greater	2.45	.0071	.3 }

Comparing Observed and Expected Frequencies

We are now interested in examining the magnitudes of the discrepancies between the observed frequencies and the expected frequencies, since we note that the two sets of frequencies do not agree. We know that even if our sample were drawn from a normal distribution of values, sampling variability alone would make it highly unlikely that the observed and expected frequencies would agree perfectly. We wonder, then, if the discrepancies between the observed and expected frequencies are small enough that we feel it reasonable that they could have occurred by chance alone, when the null hypothesis is true. If they are of this magnitude, we will be unwilling to reject the null hypothesis that the sample came from a normally distributed population.

If the discrepancies are so large that it does not seem reasonable that they could have occurred by chance alone when the null hypothesis is true, we will want to reject the null hypothesis. The criterion against which we judge whether the discrepancies are "large" or "small" is provided by the chi-square distribution.

The observed and expected frequencies along with each value of $(O_i - E_i)^2/E_i$ are shown in Table 12.3.3. The first entry in the last column, for example, is computed from $(1 - 1.8)^2/1.8 = .356$. The other values of $(O_i - E_i)^2/E_i$ are computed in a similar manner.

From Table 12.3.3 we see that $X^2 = \sum[(O_i - E_i)^2/E_i] = 10.566$. The appropriate degrees of freedom are 8 (the number of groups or class intervals) − 3 (for the three restrictions: making $\sum E_i = \sum O_i$, and estimating μ and σ from the sample data) = 5.

8. Statistical decision. When we compare $X^2 = 10.566$ with values of χ^2 in Appendix Table F, we see that it is less than $\chi^2_{.95} = 11.070$, so that, at the .05 level of significance, we cannot reject the null hypothesis that the sample came from a normally distributed population.

TABLE 12.3.3 Observed and Expected Frequencies and $(O_i - E_i)^2/E_i$ for Example 12.3.1

Class Interval	Observed Frequency (O_i)	Expected Frequency (E_i)	$(O_i - E_i)^2/E_i$
< 100	0	.4 } 1.8	.356
100.0–124.9	1	1.4 }	
125.0–149.9	3	3.8	.168
150.0–174.9	8	7.8	.005
175.0–199.9	18	10.7	4.980
200.0–224.9	6	10.7	2.064
225.0–249.9	4	7.2	1.422
250.0–274.9	4	3.5	.071
275.0–299.9	3	1.2 } 1.5	1.500
300.0 and greater	0	.3 }	
Total	47	47	10.566

9. Conclusion. We conclude that in the sampled population, cholesterol levels may follow a normal distribution.

10. p value. Since $11.070 > 10.566 > 9.236$, $.05 < p < .10$. In other words, the probability of obtaining a value of X^2 as large as 10.566, when the null hypothesis is true, is between .05 and .10. Thus we conclude that such an event is not sufficiently rare to reject the null hypothesis that the data come from a normal distribution. ■

Sometimes the parameters are specified in the null hypothesis. It should be noted that had the mean and variance of the population been specified as part of the null hypothesis in Example 12.3.1, we would not have had to estimate them from the sample and our degrees of freedom would have been $8 - 1 = 7$.

Alternatives Although one frequently encounters in the literature the use of chi-square to test for normality, it is not the most appropriate test to use when the hypothesized distribution is continuous. The Kolmogorov–Smirnov test, described in Chapter 13, was especially designed for goodness-of-fit tests involving continuous distributions.

EXAMPLE 12.3.2 *The Binomial Distribution*

In a study designed to determine patient acceptance of a new pain reliever, 100 physicians each selected a sample of 25 patients to participate in the study. Each patient, after trying the new pain reliever for a specified period of time, was asked whether it was preferable to the pain reliever used regularly in the past.

The results of the study are shown in Table 12.3.4.

TABLE 12.3.4 Results of Study Described in Example 12.3.2

Number of Patients Out of 25 Preferring New Pain Reliever	Number of Doctors Reporting this Number	Total Number of Patients Preferring New Pain Reliever by Doctor
0	5	0
1	6	6
2	8	16
3	10	30
4	10	40
5	15	75
6	17	102
7	10	70
8	10	80
9	9	81
10 or more	0	0
Total	100	500

We are interested in determining whether or not these data are compatible with the hypothesis that they were drawn from a population that follows a binomial distribution. Again, we employ a chi-square goodness-of-fit test.

Solution: Since the binomial parameter, p, is not specified, it must be estimated from the sample data. A total of 500 patients out of the 2500 patients participating in the study said they preferred the new pain reliever, so that our point estimate of p is $\hat{p} = 500/2500 = .20$. The expected relative frequencies can be obtained by evaluating the binomial function

$$f(x) = \binom{25}{x} .2^x .8^{25-x}$$

for $x = 0, 1, \ldots, 25$. For example, to find the probability that out of a sample of 25 patients none would prefer the new pain reliever, when in the total population the true proportion preferring the new pain reliever is .2, we would evaluate

$$f(0) = \binom{25}{0} .2^0 .8^{25-0}$$

This can be done most easily by consulting Appendix Table B, where we see that $P(X = 0) = .0038$. The relative frequency of occurrence of samples of size 25 in which no patients prefer the new pain reliever is .0038. To obtain the corresponding expected frequency, we multiply .0038 by 100 to get .38. Similar calculations yield the remaining expected frequencies, which, along with the observed frequencies, are shown in Table 12.3.5. We

TABLE 12.3.5 Calculations for Example 12.3.2

Number of Patients Out of 25 Preferring New Pain Reliever	Number of Doctors Reporting This Number (Observed Frequency, O_i)	Expected Relative Frequency	Expected Frequency E_i
0	5 } 11	.0038	.38 } 2.74
1	6	.0236	2.36
2	8	.0708	7.08
3	10	.1358	13.58
4	10	.1867	18.67
5	15	.1960	19.60
6	17	.1633	16.33
7	10	.1109	11.09
8	10	.0623	6.23
9	9	.0295	2.95
10 or more	0	.0173	1.73
Total	100	1.0000	100.00

see in this table that the first expected frequency is less than 1, so that we follow Cochran's suggestion and combine this group with the second group. When we do this, all the expected frequencies are greater than 1.

From the data, we compute

$$X^2 = \frac{(11 - 2.74)^2}{2.74} + \frac{(8 - 7.08)^2}{7.08} + \cdots + \frac{(0 - 1.73)^2}{1.73} = 47.624$$

The appropriate degrees of freedom are 10 (the number of groups left after combining the first two) less 2, or 8. One degree of freedom is lost because we force the total of the expected frequencies to equal the total observed frequencies, and one degree of freedom is sacrificed because we estimated p from the sample data.

We compare our computed X^2 with the tabulated χ^2 with 8 degrees of freedom and find that it is significant at the .005 level of significance; that is, $p < .005$. We reject the null hypothesis that the data came from a binomial distribution. ■

EXAMPLE 12.3.3 *The Poisson Distribution*

A hospital administrator wishes to test the null hypothesis that emergency admissions follow a Poisson distribution with $\lambda = 3$. Suppose that over a period of 90 days the numbers of emergency admissions were as shown in Table 12.3.6.

TABLE 12.3.6 Number of Emergency Admissions to a Hospital During a 90-Day Period

Day	Emergency Admissions	Day	Emergency Admissions	Day	Emergency Admissions	Day	Emergency Admissions
1	2	24	5	47	4	70	3
2	3	25	3	48	2	71	5
3	4	26	2	49	2	72	4
4	5	27	4	50	3	73	1
5	3	28	4	51	4	74	1
6	2	29	3	52	2	75	6
7	3	30	5	53	3	76	3
8	0	31	1	54	1	77	3
9	1	32	3	55	2	78	5
10	0	33	2	56	3	79	2
11	1	34	4	57	2	80	1
12	0	35	2	58	5	81	7
13	6	36	5	59	2	82	7
14	4	37	0	60	7	83	1
15	4	38	6	61	8	84	5
16	4	39	4	62	3	85	1

(*Continued*)

Day	Emergency Admissions	Day	Emergency Admissions	Day	Emergency Admissions	Day	Emergency Admissions
17	3	40	4	63	1	86	4
18	4	41	5	64	3	87	4
19	3	42	1	65	1	88	9
20	3	43	3	66	0	89	2
21	3	44	1	67	3	90	3
22	4	45	2	68	2		
23	3	46	3	69	1		

The data of Table 12.3.6 are summarized in Table 12.3.7.

Solution: To obtain the expected frequencies we first obtain the expected relative frequencies by evaluating the Poisson function given by Equation 4.4.1 for each entry in the left-hand column of Table 12.3.7. For example, the first expected relative frequency is obtained by evaluating

$$f(0) = \frac{e^{-3}3^0}{0!}$$

We may use Appendix Table C to find this and all the other expected relative frequencies that we need. Each of the expected relative frequencies

TABLE 12.3.7 Summary of Data Presented in Table 12.3.6

Number of Emergency Admissions in a Day	Number of Days This Number of Emergency Admissions Occurred
0	5
1	14
2	15
3	23
4	16
5	9
6	3
7	3
8	1
9	1
10 or more	0
Total	90

TABLE 12.3.8 Observed and Expected Frequencies and Components of X^2 for Example 12.3.3

Number of Emergency Admissions	Number of Days this Number Occurred, O_i	Expected Relative Frequency	Expected Frequency	$\frac{(O_i - E_i)^2}{E_i}$
0	5	.050	4.50	.056
1	14	.149	13.41	.026
2	15	.224	20.16	1.321
3	23	.224	20.16	.400
4	16	.168	15.12	.051
5	9	.101	9.09	.001
6	3	.050	4.50	.500
7	3	.022	1.98	.525
8	1 }	.008	.72 }	
9	1 } 2	.003	.27 } 1.08	.784
10 or more	0 }	.001	.09 }	
Total	90	1.000	90.00	3.664

is multiplied by 90 to obtain the corresponding expected frequencies. These values along with the observed and expected frequencies and the components of X^2, $(O_i - E_i)^2/E_i$, are displayed in Table 12.3.8, in which we see that

$$X^2 = \sum\left[\frac{(O_i - E_i)^2}{E_i}\right] = \frac{(5 - 4.50)^2}{4.50} + \cdots + \frac{(2 - 1.08)^2}{1.08} = 3.664$$

We also note that the last three expected frequencies are less than 1, so that they must be combined to avoid having any expected frequencies less than 1. This means that we have only nine effective categories for computing degrees of freedom. Since the parameter, λ, was specified in the null hypothesis, we do not lose a degree of freedom for reasons of estimation, so that the appropriate degrees of freedom are $9 - 1 = 8$. By consulting Appendix Table F, we find that the critical value of χ^2 for 8 degrees of freedom and $\alpha = .05$ is 15.507, so that we cannot reject the null hypothesis at the .05 level, or for that matter any reasonable level, of significance ($p > .10$). We conclude, therefore, that emergency admissions at this hospital may follow a Poisson distribution with $\lambda = 3$. At least the observed data do not cast any doubt on that hypothesis.

If the parameter λ has to be estimated from sample data, the estimate is obtained by multiplying each value x by its frequency, summing these products, and dividing the total by the sum of the frequencies. ■

EXAMPLE 12.3.4 *The Uniform Distribution*

The flu season in southern Nevada for 2005–2006 ran from December to April, the coldest months of the year. The Southern Nevada Health District reported the numbers of vaccine-preventable influenza cases shown in Table 12.3.9. We are interested in knowing whether the numbers of flu cases in the district are equally distributed among the five flu season months. That is, we wish to know if flu cases follow a uniform distribution.

Solution:

1. **Data.** See Table 12.3.9.
2. **Assumptions.** We assume that the reported cases of flu constitute a simple random sample of cases of flu that occurred in the district.
3. **Hypotheses.**

   H_0: Flu cases in southern Nevada are uniformly distributed over the five flu season months.

   H_A: Flu cases in southern Nevada are not uniformly distributed over the five flu season months.

   Let $\alpha = .01$.
4. **Test statistic.** The test statistic is

$$X^2 = \sum \frac{(O_i - E_i)^2}{E_i}$$

5. **Distribution of test statistic.** If H_0 is true, X^2 is distributed approximately as χ^2 with $(5 - 1) = 4$ degrees of freedom.
6. **Decision rule.** Reject H_0 if the computed value of X^2 is equal to or greater than 13.277.

TABLE 12.3.9 Reported Vaccine-Preventable Influenza Cases from Southern Nevada, December 2005–April 2006

Month	Number of Reported Cases of Influenza
December 2005	62
January 2006	84
February 2006	17
March 2006	16
April 2006	21
Total	200

Source: http://www.southernnevadahealthdistrict.org/epidemiology/disease_statistics.htm.

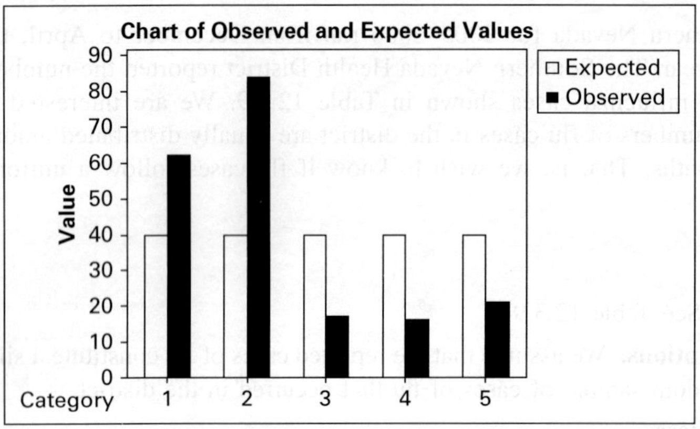

Chi-Square Goodness-of-Fit Test for Observed Counts in Variable: C1

Category	Observed	Test Proportion	Expected	Contribution to Chi-Sq
1	62	0.2	40	12.100
2	84	0.2	40	48.400
3	17	0.2	40	13.225
4	16	0.2	40	14.400
5	21	0.2	40	9.025

N	DF	Chi-Sq	P-Value
200	4	97.15	0.000

FIGURE 12.3.2 MINITAB output for Example 12.3.4.

7. **Calculation of test statistic.** If the null hypothesis is true, we would expect to observe $200/5 = 40$ cases per month. Figure 12.3.2 shows the computer printout obtained from MINITAB. The bar graph shows the observed and expected frequencies per month. The chi-square table provides the observed frequencies, the expected frequencies based on a uniform distribution, and the individual chi-square contribution for each test value.

8. **Statistical decision.** Since 97.15, the computed value of X^2, is greater than 13.277, we reject, based on these data, the null hypothesis of a

uniform distribution of flu cases during the flu season in southern Nevada.

9. **Conclusion.** We conclude that the occurrence of flu cases does not follow a uniform distribution.

10. ***p* value.** From the MINITAB output we see that $p = .000$ (i.e., $< .001$). ■

EXAMPLE 12.3.5

A certain human trait is thought to be inherited according to the ratio 1:2:1 for homozygous dominant, heterozygous, and homozygous recessive. An examination of a simple random sample of 200 individuals yielded the following distribution of the trait: dominant, 43; heterozygous, 125; and recessive, 32. We wish to know if these data provide sufficient evidence to cast doubt on the belief about the distribution of the trait.

Solution:

1. **Data.** See statement of the example.

2. **Assumptions.** We assume that the data meet the requirements for the application of the chi-square goodness-of-fit test.

3. **Hypotheses.**

   H_0: The trait is distributed according to the ratio 1:2:1 for homozygous dominant, heterozygous, and homozygous recessive.

   H_A: The trait is not distributed according to the ratio 1:2:1.

4. **Test statistic.** The test statistic is

$$X^2 = \sum\left[\frac{(O - E)^2}{E}\right]$$

5. **Distribution of test statistic.** If H_0 is true, X^2 is distributed as chi-square with 2 degrees of freedom.

6. **Decision rule.** Suppose we let the probability of committing a type I error be .05. Reject H_0 if the computed value of X^2 is equal to or greater than 5.991.

7. **Calculation of test statistic.** If H_0 is true, the expected frequencies for the three manifestations of the trait are 50, 100, and 50 for dominant, heterozygous, and recessive, respectively. Consequently,

$$X^2 = (43 - 50)^2/50 + (125 - 100)2/100 + (32 - 50)^2/50 = 13.71$$

8. **Statistical decision.** Since $13.71 > 5.991$, we reject H_0.

9. **Conclusion.** We conclude that the trait is not distributed according to the ratio 1:2:1.

10. ***p* value.** Since $13.71 > 10.597$, the p value for the test is $p < .005$. ■

EXERCISES

12.3.1 The following table shows the distribution of uric acid determinations taken on 250 patients. Test the goodness-of-fit of these data to a normal distribution with $\mu = 5.74$ and $\sigma = 2.01$. Let $\alpha = .01$.

Uric Acid Determination	Observed Frequency	Uric Acid Determination	Observed Frequency
<1	1	6 to 6.99	45
1 to 1.99	5	7 to 7.99	30
2 to 2.99	15	8 to 8.99	22
3 to 3.99	24	9 to 9.99	10
4 to 4.99	43	10 or higher	5
5 to 5.99	50		
Total			250

12.3.2 The following data were collected on 300 eight-year-old girls. Test, at the .05 level of significance, the null hypothesis that the data are drawn from a normally distributed population. The sample mean and standard deviation computed from grouped data are 127.02 and 5.08.

Height in Centimeters	Observed Frequency	Height in Centimeters	Observed Frequency
114 to 115.9	5	128 to 129.9	43
116 to 117.9	10	130 to 131.9	42
118 to 119.9	14	132 to 133.9	30
120 to 121.9	21	134 to 135.9	11
122 to 123.9	30	136 to 137.9	5
124 to 125.9	40	138 to 139.9	4
126 to 127.9	45		
Total			300

12.3.3 The face sheet of patients' records maintained in a local health department contains 10 entries. A sample of 100 records revealed the following distribution of erroneous entries:

Number of Erroneous Entries Out of 10	Number of Records
0	8
1	25
2	32
3	24
4	10
5 or more	1
Total	100

Test the goodness-of-fit of these data to the binomial distribution with $p = .20$. Find the p value for this test.

12.3.4 In a study conducted by Byers et al. (A-2), researchers tested a Poisson model for the distribution of activities of daily living (ADL) scores after a 7-month prehabilitation program designed to prevent functional decline among physically frail, community-living older persons. ADL measured the ability of individuals to perform essential tasks, including walking inside the house, bathing, upper and lower body dressing, transferring from a chair, toileting, feeding, and grooming. The scoring method used in this study assigned a value of 0 for no (personal) help and no difficulty, 1 for difficulty but no help, and 2 for help regardless of difficulty. Scores were summed to produce an overall score ranging from 0 to 16 (for eight tasks). There were 181 subjects who completed the study. Suppose we use the authors' scoring method to assess the status of another group of 181 subjects relative to their activities of daily living. Let us assume that the following results were obtained.

X	Observed Frequency X	Expected Frequency	X	Observed Frequency X	Expected Frequency
0	74	11.01	7	4	2.95
1	27	30.82	8	3	1.03
2	14	43.15	9	2	0.32
3	14	40.27	10	3	0.09
4	11	28.19	11	4	0.02
5	7	15.79	12 or more	13	0.01
6	5	7.37			

Source: Hypothetical data based on procedure reported by Amy L. Byers, Heather Allore, Thomas M. Gill, and Peter N. Peduzzi, "Application of Negative Binomial Modeling for Discrete Outcomes: A Case Study in Aging Research," *Journal of Clinical Epidemiology*, *56* (2003), 559–564.

Test the null hypothesis that these data were drawn from a Poisson distribution with $\lambda = 2.8$. Let $\alpha = .01$.

12.3.5 The following are the numbers of a particular organism found in 100 samples of water from a pond:

Number of Organisms per Sample	Frequency	Number of Organisms per Sample	Frequency
0	15	4	5
1	30	5	4
2	25	6	1
3	20	7	0
Total			100

Test the null hypothesis that these data were drawn from a Poisson distribution. Determine the p value for this test.

12.3.6 A research team conducted a survey in which the subjects were adult smokers. Each subject in a sample of 200 was asked to indicate the extent to which he or she agreed with the statement: "I would like to quit smoking." The results were as follows:

Response:	Strongly agree	Agree	Disagree	Strongly Disagree
Number Responding:	102	30	60	8

Can one conclude on the basis of these data that, in the sampled population, opinions are not equally distributed over the four levels of agreement? Let the probability of committing a type I error be .05 and find the p value.

12.4 TESTS OF INDEPENDENCE

Another, and perhaps the most frequent, use of the chi-square distribution is to test the null hypothesis that two criteria of classification, when applied to the same set of entities, are independent. We say that two criteria of classification are independent if the distribution of one criterion is the same no matter what the distribution of the other criterion. For example, if socioeconomic status and area of residence of the inhabitants of a certain city are independent, we would expect to find the same proportion of families in the low, medium, and high socioeconomic groups in all areas of the city.

The Contingency Table The classification, according to two criteria, of a set of entities, say, people, can be shown by a table in which the r rows represent the various levels of one criterion of classification and the c columns represent the various levels of the second criterion. Such a table is generally called a *contingency table*. The classification according to two criteria of a finite population of entities is shown in Table 12.4.1.

We will be interested in testing the null hypothesis that in the population the two criteria of classification are independent. If the hypothesis is rejected, we will conclude

TABLE 12.4.1 Two-Way Classification of a Finite Population of Entities

Second Criterion of Classification	**First Criterion of Classification Level**					
Level	**1**	**2**	**3**	**. . .**	***c***	**Total**
1	N_{11}	N_{12}	N_{13}	...	N_{1c}	$N_{1.}$
2	N_{21}	N_{22}	N_{23}	...	N_{2c}	$N_{2.}$
3	N_{31}	N_{32}	N_{33}	...	N_{3c}	$N_{3.}$
⋮	⋮	⋮	⋮	⋮	⋮	⋮
r	N_{r1}	N_{r2}	N_{r3}	...	N_{rc}	$N_{r.}$
Total	$N_{.1}$	$N_{.2}$	$N_{.3}$	...	$N_{.c}$	N

TABLE 12.4.2 Two-Way Classification of a Sample of Entities

Second Criterion of Classification Level	First Criterion of Classification Level					
	1	2	3	⋯	c	Total
1	n_{11}	n_{12}	n_{13}	⋯	n_{1c}	$n_{1.}$
2	n_{21}	n_{22}	n_{23}	⋯	n_{2c}	$n_{2.}$
3	n_{31}	n_{32}	n_{33}	⋯	n_{3c}	$n_{3.}$
⋮	⋮	⋮	⋮		⋮	⋮
r	n_{r1}	n_{r2}	n_{r3}	⋯	n_{rc}	$n_{r.}$
Total	$n_{.1}$	$n_{.2}$	$n_{.3}$	⋯	$n_{.c}$	n

that the two criteria of classification are not independent. A sample of size n will be drawn from the population of entities, and the frequency of occurrence of entities in the sample corresponding to the cells formed by the intersections of the rows and columns of Table 12.4.1 along with the marginal totals will be displayed in a table such as Table 12.4.2.

Calculating the Expected Frequencies The expected frequency, under the null hypothesis that the two criteria of classification are independent, is calculated for each cell.

We learned in Chapter 3 (see Equation 3.4.4) that if two events are independent, the probability of their joint occurrence is equal to the product of their individual probabilities. Under the assumption of independence, for example, we compute the probability that one of the n subjects represented in Table 12.4.2 will be counted in Row 1 and Column 1 of the table (that is, in Cell 11) by multiplying the probability that the subject will be counted in Row 1 by the probability that the subject will be counted in Column 1. In the notation of the table, the desired calculation is

$$\left(\frac{n_{1.}}{n}\right)\left(\frac{n_{.1}}{n}\right)$$

To obtain the expected frequency for Cell 11, we multiply this probability by the total number of subjects, n. That is, the expected frequency for Cell 11 is given by

$$\left(\frac{n_{1.}}{n}\right)\left(\frac{n_{.1}}{n}\right)(n)$$

Since the n in one of the denominators cancels into numerator n, this expression reduces to

$$\frac{(n_{1.})(n_{.1})}{n}$$

In general, then, we see that to obtain the expected frequency for a given cell, we multiply the total of the row in which the cell is located by the total of the column in which the cell is located and divide the product by the grand total.

Observed Versus Expected Frequencies The expected frequencies and observed frequencies are compared. If the discrepancy is sufficiently small, the null hypothesis is tenable. If the discrepancy is sufficiently large, the null hypothesis is rejected, and we conclude that the two criteria of classification are not independent. The decision as to whether the discrepancy between observed and expected frequencies is sufficiently large to cause rejection of H_0 will be made on the basis of the size of the quantity computed when we use Equation 12.2.4, where O_i and E_i refer, respectively, to the observed and expected frequencies in the cells of Table 12.4.2. It would be more logical to designate the observed and expected frequencies in these cells by O_{ij} and E_{ij}, but to keep the notation simple and to avoid the introduction of another formula, we have elected to use the simpler notation. It will be helpful to think of the cells as being numbered from 1 to k, where 1 refers to Cell 11 and k refers to Cell rc. It can be shown that X^2 as defined in this manner is distributed approximately as χ^2 with $(r - 1)(c - 1)$ degrees of freedom when the null hypothesis is true. If the computed value of X^2 is equal to or larger than the tabulated value of χ^2 for some α, the null hypothesis is rejected at the α level of significance. The hypothesis testing procedure is illustrated with the following example.

EXAMPLE 12.4.1

In 1992, the U.S. Public Health Service and the Centers for Disease Control and Prevention recommended that all women of childbearing age consume 400 mg of folic acid daily to reduce the risk of having a pregnancy that is affected by a neural tube defect such as spina bifida or anencephaly. In a study by Stepanuk et al. (A-3), 693 pregnant women called a teratology information service about their use of folic acid supplementation. The researchers wished to determine if preconceptional use of folic acid and race are independent. The data appear in Table 12.4.3.

Solution:

1. **Data.** See Table 12.4.3.
2. **Assumptions.** We assume that the sample available for analysis is equivalent to a simple random sample drawn from the population of interest.

TABLE 12.4.3 Race of Pregnant Caller and Use of Folic Acid

	Preconceptional Use of Folic Acid		
	Yes	No	Total
White	260	299	559
Black	15	41	56
Other	7	14	21
Total	282	354	636

Source: Kathleen M. Stepanuk, Jorge E. Tolosa, Dawneete Lewis, Victoria Meyers, Cynthia Royds, Juan Carlos Saogal, and Ron Librizzi, "Folic Acid Supplementation Use Among Women Who Contact a Teratology Information Service," *American Journal of Obstetrics and Gynecology, 187* (2002), 964–967.

3. **Hypotheses.**

H_0: Race and preconceptional use of folic acid are independent.

H_A: The two variables are not independent.

Let $\alpha = .05$.

4. **Test statistic.** The test statistic is

$$X^2 = \sum_{i=1}^{k} \left[\frac{(O_i - E_i)^2}{E_i} \right]$$

5. **Distribution of test statistic.** When H_0 is true, X^2 is distributed approximately as χ^2 with $(r - 1)(c - 1) = (3 - 1)(2 - 1) = (2)(1) = 2$ degrees of freedom.

6. **Decision rule.** Reject H_0 if the computed value of X^2 is equal to or greater than 5.991.

7. **Calculation of test statistic.** The expected frequency for the first cell is $(559 \times 282)/636 = 247.86$. The other expected frequencies are calculated in a similar manner. Observed and expected frequencies are displayed in Table 12.4.4. From the observed and expected frequencies we may compute

$$X^2 = \sum \left[\frac{(O_i - E_i)^2}{E_i} \right]$$

$$= \frac{(260 - 247.86)^2}{247.86} + \frac{(299 - 311.14)^2}{311.14} + \cdots + \frac{(14 - 11.69)^2}{11.69}$$

$$= .59461 + .47368 + \cdots + .45647 = 9.08960$$

8. **Statistical decision.** We reject H_0 since $9.08960 > 5.991$.

9. **Conclusion.** We conclude that H_0 is false, and that there is a relationship between race and preconceptional use of folic acid.

10. ***p* value.** Since $7.378 < 9.08960 < 9.210$, $.01 < p < .025$.

TABLE 12.4.4 Observed and Expected Frequencies for Example 12.4.1

	Preconceptional Use of Folic Acid		
	Yes	No	Total
White	260 (247.86)	299 (311.14)	559
Black	15 (24.83)	41 (31.17)	56
Other	7 (9.31)	14 (11.69)	21
Total	282	354	636

■

Computer Analysis The computer may be used to advantage in calculating X^2 for tests of independence and tests of homogeneity. Figure 12.4.1 shows the procedure and printout for Example 12.4.1 when the MINITAB program for computing X^2 from contingency tables is used. The data were entered into MINITAB Columns 1 and 2, corresponding to the columns of Table 12.4.3.

We may use SAS® to obtain an analysis and printout of contingency table data by using the PROC FREQ statement. Figure 12.4.2 shows a partial SAS® printout reflecting the analysis of the data of Example 12.4.1.

Data:

```
C1: 260 15  7
C2: 299 41 14
```

Dialog Box:

Stat ➤ Tables ➤ Chi-square Test

Type *C1-C2* in Columns containing the table.
Click **OK**.

Session command:

```
MTB  >  CHISQUARE  C1-C3
```

Output:

Chi-Square Test: C1, C2

```
Expected counts are printed below observed counts

              C1          C2     Total
    1        260         299       559
          247.86      311.14

    2         15          41        56
           24.83       31.17

    3          7          14        21
            9.31       11.69

Total        282         354       636

Chi-Sq  =   0.595  + 0.474  +
            3.892  + 3.100  +
            0.574  + 0.457  =  9.091
DF  =  2,  P-Value  =  0.011
```

FIGURE 12.4.1 MINITAB procedure and output for chi-square analysis of data in Table 12.4.3.

```
The SAS System

The FREQ Procedure

                Table of race by folic

         race        |    folic
         Frequency   |
         Percent     |
         Row Pct     |
         Col Pct     |    No    |   Yes   |   Total
         ------------+----------+---------+
         Black       |     41   |    15   |     56
                     |   6.45   |  2.36   |   8.81
                     |  73.21   | 26.79   |
                     |  11.58   |  5.32   |
         ------------+----------+---------+
         Other       |     14   |     7   |     21
                     |   2.20   |  1.10   |   3.30
                     |  66.67   | 33.33   |
                     |   3.95   |  2.48   |
         ------------+----------+---------+
         White       |    299   |   260   |    559
                     |  47.01   | 40.88   |  87.89
                     |  53.49   | 46.51   |
                     |  84.46   | 92.20   |
         ------------+----------+---------+
         Total            354       282        636
                        55.66     44.34     100.00

         Statistics for Table of race by folic

Statistic                       DF       Value        Prob
-----------------------------------------------------------
Chi-Square                       2      9.0913      0.0106
Likelihood Ratio Chi-Square      2      9.4808      0.0087
Mantel-Haenszel Chi-Square       1      8.9923      0.0027
Phi Coefficient                         0.1196
Contingency Coefficient                 0.1187
Cramer's V                              0.1196

Sample Size = 636
```

FIGURE 12.4.2 Partial SAS® printout for the chi-square analysis of the data from Example 12.4.1.

Note that the SAS® printout shows, in each cell, the percentage that cell frequency is of its row total, its column total, and the grand total. Also shown, for each row and column total, is the percentage that the total is of the grand total. In addition to the X^2 statistic, SAS® gives the value of several other statistics that may be computed from contingency table data. One of these, the Mantel–Haenszel chi-square statistic, will be discussed in a later section of this chapter.

Small Expected Frequencies The problem of small expected frequencies discussed in the previous section may be encountered when analyzing the data of contingency tables. Although there is a lack of consensus on how to handle this problem, many authors currently follow the rule given by Cochran (5). He suggests that for contingency tables with more than 1 degree of freedom a minimum expectation of 1 is allowable if no more than 20 percent of the cells have expected frequencies of less than 5. To meet this rule, adjacent rows and/or adjacent columns may be combined when to do so is logical in light of other considerations. If X^2 is based on less than 30 degrees of freedom, expected frequencies as small as 2 can be tolerated. We did not experience the problem of small expected frequencies in Example 12.4.1, since they were all greater than 5.

The 2 × 2 Contingency Table Sometimes each of two criteria of classification may be broken down into only two categories, or levels. When data are cross-classified in this manner, the result is a contingency table consisting of two rows and two columns. Such a table is commonly referred to as a 2×2 table. The value of X^2 may be computed by first calculating the expected cell frequencies in the manner discussed above. In the case of a 2×2 contingency table, however, X^2 may be calculated by the following shortcut formula:

$$X^2 = \frac{n(ad - bc)^2}{(a + c)(b + d)(a + b)(c + d)} \tag{12.4.1}$$

where a, b, c, and d are the observed cell frequencies as shown in Table 12.4.5. When we apply the $(r - 1)(c - 1)$ rule for finding degrees of freedom to a 2×2 table, the result is 1 degree of freedom. Let us illustrate this with an example.

TABLE 12.4.5 A 2 × 2 Contingency Table

Second Criterion of Classification	First Criterion of Classification 1	2	Total
1	a	b	$a + b$
2	c	d	$c + d$
Total	$a + c$	$b + d$	n

EXAMPLE 12.4.2

According to Silver and Aiello (A-4), falls are of major concern among polio survivors. Researchers wanted to determine the impact of a fall on lifestyle changes. Table 12.4.6 shows the results of a study of 233 polio survivors on whether fear of falling resulted in lifestyle changes.

Solution:

1. **Data.** From the information given we may construct the 2×2 contingency table displayed as Table 12.5.6.
2. **Assumptions.** We assume that the sample is equivalent to a simple random sample.
3. **Hypotheses.**
   H_0: Fall status and lifestyle change because of fear of falling are independent.
   H_1: The two variables are not independent.
   Let $\alpha = .05$.
4. **Test statistic.** The test statistic is

$$X^2 = \sum_{i=1}^{k} \left[\frac{(O_i - E_i)^2}{E_i} \right]$$

5. **Distribution of test statistic.** When H_0 is true, X^2 is distributed approximately as χ^2 with $(r - 1)(c - 1) = (2 - 1)(2 - 1) = (1)(1) = 1$ degree of freedom.
6. **Decision rule.** Reject H_0 if the computed value of X^2 is equal to or greater than 3.841.
7. **Calculation of test statistic.** By Equation 12.4.1 we compute

$$X^2 = \frac{233[(131)(36) - (52)(14)]^2}{(145)(88)(183)(50)} = 31.7391$$

8. **Statistical decision.** We reject H_0 since $31.7391 > 3.841$.

TABLE 12.4.6 Contingency Table for the Data of Example 12.4.2

	Made Lifestyle Changes Because of Fear of Falling		
	Yes	**No**	**Total**
Fallers	131	52	183
Nonfallers	14	36	50
Total	145	88	233

Source: J. K. Silver and D. D. Aiello, "Polio Survivors: Falls and Subsequent Injuries," *American Journal of Physical Medicine and Rehabilitation, 81* (2002), 567–570.

9. Conclusion. We conclude that H_0 is false, and that there is a relationship between experiencing a fall and changing one's lifestyle because of fear of falling.

10. *p* value. Since $31.7391 > 7.879, p < .005$. ■

Small Expected Frequencies The problems of how to handle small expected frequencies and small total sample sizes may arise in the analysis of 2 × 2 contingency tables. Cochran (5) suggests that the χ^2 test should not be used if $n < 20$ or if $20 < n < 40$ and any expected frequency is less than 5. When $n = 40$, an expected cell frequency as small as 1 can be tolerated.

Yates's Correction The observed frequencies in a contingency table are discrete and thereby give rise to a discrete statistic, X^2, which is approximated by the χ^2 distribution, which is continuous. Yates (6) in 1934 proposed a procedure for correcting for this in the case of 2 × 2 tables. The correction, as shown in Equation 12.4.2, consists of subtracting half the total number of observations from the absolute value of the quantity $ad - bc$ before squaring. That is,

$$X^2_{\text{corrected}} = \frac{n(|ad - bc| - .5n)^2}{(a + c)(b + d)(a + b)(c + d)} \tag{12.4.2}$$

It is generally agreed that no correction is necessary for larger contingency tables. Although Yates's correction for 2 × 2 tables has been used extensively in the past, more recent investigators have questioned its use. As a result, some practitioners recommend against its use.

We may, as a matter of interest, apply the correction to our current example. Using Equation 12.4.2 and the data from Table 12.4.6, we may compute

$$X^2 = \frac{233[|(131)(36) - (52)(14)| - .5(233)]^2}{(145)(88)(183)(50)} = 29.9118$$

As might be expected, with a sample this large, the difference in the two results is not dramatic.

Tests of Independence: Characteristics The characteristics of a chi-square test of independence that distinguish it from other chi-square tests are as follows:

1. A single sample is selected from a population of interest, and the subjects or objects are cross-classified on the basis of the two variables of interest.
2. The rationale for calculating expected cell frequencies is based on the probability law, which states that if two events (here the two criteria of classification) are independent, the probability of their joint occurrence is equal to the product of their individual probabilities.
3. The hypotheses and conclusions are stated in terms of the independence (or lack of independence) of two variables.

EXERCISES

In the exercises that follow perform the test at the indicated level of significance and determine the p value.

12.4.1 In the study by Silver and Aiello (A-4) cited in Example 12.4.2, a secondary objective was to determine if the frequency of falls was independent of wheelchair use. The following table gives the data for falls and wheelchair use among the subjects of the study.

	Wheelchair Use	
	Yes	**No**
Fallers	62	121
Nonfallers	18	32

Source: J. K. Silver and D. D. Aiello, "Polio Survivors: Falls and Subsequent Injuries," *American Journal of Physical Medicine and Rehabilitation, 81* (2002), 567–570.

Do these data provide sufficient evidence to warrant the conclusion that wheelchair use and falling are related? Let $\alpha = .05$.

12.4.2 Sternal surgical site infection (SSI) after coronary artery bypass graft surgery is a complication that increases patient morbidity and costs for patients, payers, and the health care system. Segal and Anderson (A-5) performed a study that examined two types of preoperative skin preparation before performing open heart surgery. These two preparations used aqueous iodine and insoluble iodine with the following results.

Comparison of Aqueous and Insoluble Preps

Prep Group	Infected	Not Infected
Aqueous iodine	14	94
Insoluble iodine	4	97

Source: Cynthia G. Segal and Jacqueline J. Anderson, "Preoperative Skin Preparation of Cardiac Patients," *AORN Journal, 76* (2002), 8231–827.

Do these data provide sufficient evidence at the $\alpha = .05$ level to justify the conclusion that the type of skin preparation and infection are related?

12.4.3 The side effects of nonsteroidal antiinflammatory drugs (NSAIDs) include problems involving peptic ulceration, renal function, and liver disease. In 1996, the American College of Rheumatology issued and disseminated guidelines recommending baseline tests (CBC, hepatic panel, and renal tests) when prescribing NSAIDs. A study was conducted by Rothenberg and Holcomb (A-6) to determine if physicians taking part in a national database of computerized medical records performed the recommended baseline tests when prescribing NSAIDs. The researchers classified physicians in the study into four categories—those practicing in internal medicine, family practice, academic family practice, and multispeciality groups. The data appear in the following table.

	Performed Baseline Tests	
Practice Type	**Yes**	**No**
Internal medicine	294	921
Family practice	98	2862
Academic family practice	50	3064
Multispecialty groups	203	2652

Source: Ralph Rothenberg and John P. Holcomb, "Guidelines for Monitoring of NSAIDs: Who Listened?," *Journal of Clinical Rheumatology, 6* (2000), 258–265.

Do the data above provide sufficient evidence for us to conclude that type of practice and performance of baseline tests are related? Use $\alpha = .01$.

12.4.4 Boles and Johnson (A-7) examined the beliefs held by adolescents regarding smoking and weight. Respondents characterized their weight into three categories: underweight, overweight, or appropriate. Smoking status was categorized according to the answer to the question, "Do you currently smoke, meaning one or more cigarettes per day?" The following table shows the results of a telephone study of adolescents in the age group 12–17.

	Smoking	
	Yes	**No**
Underweight	17	97
Overweight	25	142
Appropriate	96	816

Source: Sharon M. Boles and Patrick B. Johnson, "Gender, Weight Concerns, and Adolescent Smoking," *Journal of Addictive Diseases, 20* (2001), 5–14.

Do the data provide sufficient evidence to suggest that weight perception and smoking status are related in adolescents? $\alpha = .05$.

12.4.5 A sample of 500 college students participated in a study designed to evaluate the level of college students' knowledge of a certain group of common diseases. The following table shows the students classified by major field of study and level of knowledge of the group of diseases:

	Knowledge of Diseases		
Major	**Good**	**Poor**	**Total**
Premedical	31	91	122
Other	19	359	378
Total	50	450	500

Do these data suggest that there is a relationship between knowledge of the group of diseases and major field of study of the college students from which the present sample was drawn? Let $\alpha = .05$.

12.4.6 The following table shows the results of a survey in which the subjects were a sample of 300 adults residing in a certain metropolitan area. Each subject was asked to indicate which of three policies they favored with respect to smoking in public places.

	Policy Favored				
Highest Education Level	**No Restrictions on Smoking**	**Smoking Allowed in Designated Areas Only**	**No Smoking at All**	**No Opinion**	**Total**
College graduate	5	44	23	3	75
High-school graduate	15	100	30	5	150
Grade-school graduate	15	40	10	10	75
Total	35	184	63	18	300

Can one conclude from these data that, in the sampled population, there is a relationship between level of education and attitude toward smoking in public places? Let $\alpha = .05$.

12.5 TESTS OF HOMOGENEITY

A characteristic of the examples and exercises presented in the last section is that, in each case, the total sample was assumed to have been drawn before the entities were classified according to the two criteria of classification. That is, the observed number of entities falling into each cell was determined after the sample was drawn. As a result, the row and column totals are chance quantities not under the control of the investigator. We think of the sample drawn under these conditions as a single sample drawn from a single population. On occasion, however, either row or column totals may be under the control of the investigator; that is, the investigator may specify that independent samples be drawn from each of several populations. In this case, one set of marginal totals is said to be *fixed,* while the other set, corresponding to the criterion of classification applied to the samples, is *random.* The former procedure, as we have seen, leads to a chi-square test of independence. The latter situation leads to a chi-square *test of homogeneity*. The two situations not only involve different sampling procedures; they lead to different questions and null hypotheses. The test of independence is concerned with the question: Are the two criteria of classification independent? The homogeneity test is concerned with the question: Are the samples drawn from populations that are homogeneous with respect to some criterion of classification? In the latter case the null hypothesis states that the samples are drawn from the same population. Despite these differences in concept and sampling procedure, the two tests are mathematically identical, as we see when we consider the following example.

Calculating Expected Frequencies Either the row categories or the column categories may represent the different populations from which the samples are drawn. If, for example, three populations are sampled, they may be designated as populations 1, 2, and 3, in which case these labels may serve as either row or column headings. If the variable of interest has three categories, say, A, B, and C, these labels may serve as headings for rows or columns, whichever is not used for the populations. If we use notation similar to that adopted for Table 12.4.2, the contingency table for this situation, with columns used to represent the populations, is shown as Table 12.5.1. Before computing our test statistic we

TABLE 12.5.1 A Contingency Table for Data for a Chi-Square Test of Homogeneity

	Population			
Variable Category	**1**	**2**	**3**	**Total**
A	n_{A1}	n_{A2}	n_{A3}	$n_{A.}$
B	n_{B1}	n_{B2}	n_{B3}	$n_{B.}$
C	n_{C1}	n_{C2}	n_{C3}	$n_{C.}$
Total	$n_{.1}$	$n_{.2}$	$n_{.3}$	n

need expected frequencies for each of the cells in Table 12.5.1. If the populations are indeed homogeneous, or, equivalently, if the samples are all drawn from the same population, with respect to the categories A, B, and C, our best estimate of the proportion in the combined population who belong to category A is $n_{A.}/n$. By the same token, if the three populations are homogeneous, we interpret this probability as applying to each of the populations individually. For example, under the null hypothesis, $n_{A.}$ is our best estimate of the probability that a subject picked at random from the combined population will belong to category A. We would expect, then, to find $n_{.1}(n_{A.}/n)$ of those in the sample from population 1 to belong to category A, $n_{.2}(n_{A.}/n)$ of those in the sample from population 2 to belong to category A, and $n_{.3}(n_{A.}/n)$ of those in the sample from population 3 to belong to category A. These calculations yield the expected frequencies for the first row of Table 12.5.1. Similar reasoning and calculations yield the expected frequencies for the other two rows.

We see again that the shortcut procedure of multiplying appropriate marginal totals and dividing by the grand total yields the expected frequencies for the cells.

From the data in Table 12.5.1 we compute the following test statistic:

$$X^2 = \sum_{i=1}^{k} \left[\frac{(O_i - E_i)^2}{E_i} \right]$$

EXAMPLE 12.5.1

Narcolepsy is a disease involving disturbances of the sleep–wake cycle. Members of the German Migraine and Headache Society (A-8) studied the relationship between migraine headaches in 96 subjects diagnosed with narcolepsy and 96 healthy controls. The results are shown in Table 12.5.2. We wish to know if we may conclude, on the basis of these

TABLE 12.5.2 Frequency of Migraine Headaches by Narcolepsy Status

	Reported Migraine Headaches		
	Yes	**No**	**Total**
Narcoleptic subjects	21	75	96
Healthy controls	19	77	96
Total	40	152	192

Source: The DMG Study Group, "Migraine and Idiopathic Narcolepsy—A Case-Control Study," *Cephalagia, 23* (2003), 786–789.

data, that the narcolepsy population and healthy populations represented by the samples are not homogeneous with respect to migraine frequency.

Solution:

1. **Data.** See Table 12.5.2.
2. **Assumptions.** We assume that we have a simple random sample from each of the two populations of interest.
3. **Hypotheses.**

   H_0: The two populations are homogeneous with respect to migraine frequency.

   H_A: The two populations are not homogeneous with respect to migraine frequency.

   Let $\alpha = .05$.
4. **Test statistic.** The test statistic is

$$X^2 = \sum[(O_i - E_i)^2/E_i]$$

5. **Distribution of test statistic.** If H_0 is true, X^2 is distributed approximately as χ^2 with $(2-1)(2-1) = (1)(1) = 1$ degree of freedom.
6. **Decision rule.** Reject H_0 if the computed value of X^2 is equal to or greater than 3.841.
7. **Calculation of test statistic.** The MINITAB output is shown in Figure 12.5.1.

```
Chi-Square Test

Expected counts are printed below observed counts

Rows: Narcolepsy    Columns: Migraine

              No        Yes        All

 No           77         19         96
           76.00      20.00      96.00

Yes           75         21         96
           76.00      20.00      96.00

All          152         40        192
          152.00      40.00     192.00

Chi-Square = 0.126, DF = 1, P-Value = 0.722
```

FIGURE 12.5.1 MINITAB output for Example 12.5.1.

8. **Statistical decision.** Since .126 is less than the critical value of 3.841, we are unable to reject the null hypothesis.

9. **Conclusion.** We conclude that the two populations may be homogeneous with respect to migraine frequency.

10. ***p* value.** From the MINITAB output we see that $p = .722$. ■

Small Expected Frequencies The rules for small expected frequencies given in the previous section are applicable when carrying out a test of homogeneity.

In summary, the chi-square test of homogeneity has the following characteristics:

1. Two or more populations are identified in advance, and an independent sample is drawn from each.
2. Sample subjects or objects are placed in appropriate categories of the variable of interest.
3. The calculation of expected cell frequencies is based on the rationale that if the populations are homogeneous as stated in the null hypothesis, the best estimate of the probability that a subject or object will fall into a particular category of the variable of interest can be obtained by pooling the sample data.
4. The hypotheses and conclusions are stated in terms of homogeneity (with respect to the variable of interest) of populations.

Test of Homogeneity and H_0: $p_1 = p_2$ The chi-square test of homogeneity for the two-sample case provides an alternative method for testing the null hypothesis that two population proportions are equal. In Section 7.6, it will be recalled, we learned to test H_0: $p_1 = p_2$ against H_A: $p_1 \neq p_2$ by means of the statistic

$$z = \frac{(\hat{p}_1 - \hat{p}_2) - (p_1 - p_2)_0}{\sqrt{\dfrac{\overline{p}(1-\overline{p})}{n_1} + \dfrac{\overline{p}(1-\overline{p})}{n_2}}}$$

where $\overline{p}$ is obtained by pooling the data of the two independent samples available for analysis.

Suppose, for example, that in a test of H_0: $p_1 = p_2$ against H_A: $p_1 \neq p_2$, the sample data were as follows: $n_1 = 100$, $\hat{p}_1 = .60$ $n_2 = 120$, $\hat{p}_2 = .40$. When we pool the sample data we have

$$\overline{p} = \frac{.60(100) + .40(120)}{100 + 120} = \frac{108}{220} = .4909$$

and

$$z = \frac{.60 - .40}{\sqrt{\dfrac{(.4909)(.5091)}{100} + \dfrac{(.4909)(.5091)}{120}}} = 2.95469$$

which is significant at the .05 level since it is greater than the critical value of 1.96.

If we wish to test the same hypothesis using the chi-square approach, our contingency table will be

	Characteristic Present		
Sample	Yes	No	Total
1	60	40	100
2	48	72	120
Total	108	112	220

By Equation 12.4.1 we compute

$$X^2 = \frac{220[(60)(72) - (40)(48)]^2}{(108)(112)(100)(120)} = 8.7302$$

which is significant at the .05 level because it is greater than the critical value of 3.841. We see, therefore, that we reach the same conclusion by both methods. This is not surprising because, as explained in Section 12.2, $\chi^2_{(1)} = z^2$. We note that $8.7302 = (2.95469)^2$ and that $3.841 = (1.96)^2$.

EXERCISES

In the exercises that follow perform the test at the indicated level of significance and determine the p value.

12.5.1 Refer to the study by Carter et al. (A-9), who investigated the effect of age at onset of bipolar disorder on the course of the illness. One of the variables studied was subjects' family history. Table 3.4.1 shows the frequency of a family history of mood disorders in the two groups of interest: early age at onset (18 years or younger) and later age at onset (later than 18 years).

Family History of Mood Disorders	Early $\leq 18(E)$	Later $> 18(L)$	Total
Negative (A)	28	35	63
Bipolar disorder (B)	19	38	57
Unipolar (C)	41	44	85
Unipolar and bipolar (D)	53	60	113
Total	141	177	318

Source: Tasha D. Carter, Emanuela Mundo, Sagar V. Parkh, and James L. Kennedy, "Early Age at Onset as a Risk Factor for Poor Outcome of Bipolar Disorder," *Journal of Psychiatric Research, 37* (2003), 297–303.

Can we conclude on the basis of these data that subjects 18 or younger differ from subjects older than 18 with respect to family histories of mood disorders? Let $\alpha = .05$.

12.5.2 Coughlin et al. (A-10) examined breast and cervical screening practices of Hispanic and non-Hispanic women in counties that approximate the U.S. southern border region. The study used data from the Behavioral Risk Factor Surveillance System surveys of adults ages 18 years or older conducted in 1999 and 2000. The following table shows the number of observations of Hispanic and non-Hispanic women who had received a mammogram in the past 2 years cross-classified by marital status.

Marital Status	Hispanic	Non-Hispanic	Total
Currently married	319	738	1057
Divorced or separated	130	329	459
Widowed	88	402	490
Never married or living as an unmarried couple	41	95	136
Total	578	1564	2142

Source: Steven S. Coughlin, Robert J. Uhler, Thomas Richards, and Katherine M. Wilson, "Breast and Cervical Cancer Screening Practices Among Hispanic and Non-Hispanic Women Residing Near the United States–Mexico Border, 1999–2000," *Family and Community Health, 26*, (2003), 130–139.

We wish to know if we may conclude on the basis of these data that marital status and ethnicity (Hispanic and non-Hispanic) in border counties of the southern United States are not homogeneous. Let $\alpha = .05$.

12.5.3 Swor et al. (A-11) examined the effectiveness of cardiopulmonary resuscitation (CPR) training in people over 55 years of age. They compared the skill retention rates of subjects in this age group who completed a course in traditional CPR instruction with those who received chest-compression–only cardiopulmonary resuscitation (CC-CPR). Independent groups were tested 3 months after training. Among the 27 subjects receiving traditional CPR, 12 were rated as competent. In the CC-CPR group, 15 out of 29 were rated competent. Do these data provide sufficient evidence for us to conclude that the two populations are not homogeneous with respect to competency rating 3 months after training? Let $\alpha = .05$.

12.5.4 In an air pollution study, a random sample of 200 households was selected from each of two communities. A respondent in each household was asked whether or not anyone in the household was bothered by air pollution. The responses were as follows:

	Any Member of Household Bothered by Air Pollution?		
Community	**Yes**	**No**	**Total**
I	43	157	200
II	81	119	200
Total	124	276	400

Can the researchers conclude that the two communities differ with respect to the variable of interest? Let $\alpha = .05$.

12.5.5 In a simple random sample of 250 industrial workers with cancer, researchers found that 102 had worked at jobs classified as "high exposure" with respect to suspected cancer-causing agents. Of the remainder, 84 had worked at "moderate exposure" jobs, and 64 had experienced no known exposure because of their jobs. In an independent simple random sample of 250 industrial workers from the same area who had no history of cancer, 31 worked in "high exposure" jobs, 60 worked in "moderate exposure" jobs, and 159 worked in jobs involving no known exposure to suspected cancer-causing agents. Does it appear from these data that persons working in jobs that expose them to suspected cancer-causing agents have an increased risk of contracting cancer? Let $\alpha = .05$.

12.6 THE FISHER EXACT TEST

Sometimes we have data that can be summarized in a 2×2 contingency table, but these data are derived from very small samples. The chi-square test is not an appropriate method of analysis if minimum expected frequency requirements are not met. If, for example, n is less than 20 or if n is between 20 and 40 and one of the expected frequencies is less than 5, the chi-square test should be avoided.

A test that may be used when the size requirements of the chi-square test are not met was proposed in the mid-1930s almost simultaneously by Fisher (7, 8), Irwin (9), and Yates (10). The test has come to be known as the *Fisher exact test.* It is called exact because, if desired, it permits us to calculate the exact probability of obtaining the observed results or results that are more extreme.

Data Arrangement When we use the Fisher exact test, we arrange the data in the form of a 2×2 contingency table like Table 12.6.1. We arrange the frequencies in such a way that $A > B$ and choose the characteristic of interest so that $a/A > b/B$.

Some theorists believe that Fisher's exact test is appropriate only when both marginal totals of Table 12.6.1 are fixed by the experiment. This specific model does not appear to arise very frequently in practice. Many experimenters, therefore, use the test when both marginal totals are not fixed.

Assumptions The following are the assumptions for the Fisher exact test.

1. The data consist of A sample observations from population 1 and B sample observations from population 2.
2. The samples are random and independent.
3. Each observation can be categorized as one of two mutually exclusive types.

TABLE 12.6.1 A 2 × 2 Contingency Table for the Fisher Exact Test

Sample	With Characteristic	Without Characteristic	Total
1	a	$A - a$	A
2	b	$B - b$	B
Total	$a + b$	$A + B - a - b$	$A + B$

Hypotheses The following are the null hypotheses that may be tested and their alternatives.

1. (Two-sided)

   H_0: The proportion with the characteristic of interest is the same in both populations; that is, $p_1 = p_2$.
   H_A: The proportion with the characteristic of interest is not the same in both populations; $p_1 \neq p_2$.

2. (One-sided)

   H_0: The proportion with the characteristic of interest in population 1 is less than or the same as the proportion in population 2; $p_1 \leq p_2$.
   H_A: The proportion with the characteristic of interest is greater in population 1 than in population 2; $p_1 > p_2$.

Test Statistic The test statistic is b, the number in sample 2 with the characteristic of interest.

Decision Rule Finney (11) has prepared critical values of b for $A \leq 15$. Latscha (12) has extended Finney's tables to accommodate values of A up to 20. Appendix Table J gives these critical values of b for A between 3 and 20, inclusive. Significance levels of .05, .025, .01, and .005 are included. The specific decision rules are as follows:

1. *Two-sided test* Enter Table J with A, B, and a. If the observed value of b is equal to or less than the integer in a given column, reject H_0 at a level of significance equal to twice the significance level shown at the top of that column. For example, suppose $A = 8$, $B = 7$, $a = 7$, and the observed value of b is 1. We can reject the null hypothesis at the $2(.05) = .10$, the $2(.025) = .05$, and the $2(.01) = .02$ levels of significance, but not at the $2(.005) = .01$ level.
2. *One-sided test* Enter Table J with A, B, and a. If the observed value of b is less than or equal to the integer in a given column, reject H_0 at the level of significance shown at the top of that column. For example, suppose that $A = 16$, $B = 8$, $a = 4$, and the observed value of b is 3. We can reject the null hypothesis at the .05 and .025 levels of significance, but not at the .01 or .005 levels.

Large-Sample Approximation For sufficiently large samples we can test the null hypothesis of the equality of two population proportions by using the normal approximation. Compute

$$z = \frac{(a/A) - (b/B)}{\sqrt{\hat{p}(1 - \hat{p})(1/A + 1/B)}} \tag{12.6.1}$$

where

$$\hat{p} = (a + b)/(A + B) \tag{12.6.2}$$

and compare it for significance with appropriate critical values of the standard normal distribution. The use of the normal approximation is generally considered satisfactory if a,

b, $A - a$, and $B - b$ are all greater than or equal to 5. Alternatively, when sample sizes are sufficiently large, we may test the null hypothesis by means of the chi-square test.

Further Reading The Fisher exact test has been the subject of some controversy among statisticians. Some feel that the assumption of fixed marginal totals is unrealistic in most practical applications. The controversy then centers around whether the test is appropriate when both marginal totals are not fixed. For further discussion of this and other points, see the articles by Barnard (13, 14, 15), Fisher (16), and Pearson (17).

Sweetland (18) compared the results of using the chi-square test with those obtained using the Fisher exact test for samples of size $A + B = 3$ to $A + B = 69$. He found close agreement when A and B were close in size and the test was one-sided.

Carr (19) presents an extension of the Fisher exact test to more than two samples of equal size and gives an example to demonstrate the calculations. Neave (20) presents the Fisher exact test in a new format; the test is treated as one of independence rather than of homogeneity. He has prepared extensive tables for use with his approach.

The sensitivity of Fisher's exact test to minor perturbations in 2×2 contingency tables is discussed by Dupont (21).

EXAMPLE 12.6.1

The purpose of a study by Justesen et al. (A-12) was to evaluate the long-term efficacy of taking indinavir/ritonavir twice a day in combination with two nucleoside reverse transcriptase inhibitors among HIV-positive subjects who were divided into two groups. Group 1 consisted of patients who had no history of taking protease inhibitors (PI Naïve). Group 2 consisted of patients who had a previous history taking a protease inhibitor (PI Experienced). Table 12.6.2 shows whether these subjects remained on the regimen for the 120 weeks of follow-up. We wish to know if we may conclude that patients classified as group 1 have a lower probability than subjects in group 2 of remaining on the regimen for 120 weeks.

TABLE 12.6.2 Regimen Status at 120 Weeks for PI Naïve and PI Experienced Subjects Taking Indinavir/Ritonavir as Described in Example 12.6.1

		Remained in the Regimen for 120 Weeks	
	Total	Yes	No
1 (PI Naïve)	9	2	7
2 (PA Experienced)	12	8	4
Total	21	10	11

Source: U.S. Justesen, A. M. Lervfing, A. Thomsen, J. A. Lindberg, C. Pedersen, and P. Tauris, "Low-Dose Indinavir in Combination with Low-Dose Ritonavir: Steady-State Pharmacokinetics and Long-Term Clinical Outcome Follow-Up," *HIV Medicine*, *4* (2003), 250–254.

TABLE 12.6.3 Data of Table 12.6.2 Rearranged to Conform to the Layout of Table 12.6.1

	Remained in Regimen for 120 Weeks		
	Yes	No	Total
2 (PI Experienced)	$8 = a$	$4 = A - a$	$12 = A$
1 (PI Naïve)	$2 = b$	$7 = B - b$	$9 = B$
Total	$10 = a + b$	$11 = A + B - a - b$	$21 = A + B$

Solution:

1. **Data.** The data as reported are shown in Table 12.6.2. Table 12.6.3 shows the data rearranged to conform to the layout of Table 12.6.1. Remaining on the regimen is the characteristic of interest.
2. **Assumptions.** We presume that the assumptions for application of the Fisher exact test are met.
3. **Hypotheses.**

   H_0: The proportion of subjects remaining 120 weeks on the regimen in a population of patients classified as group 2 is the same as or less than the proportion of subjects remaining on the regimen 120 weeks in a population classified as group 1.

   H_A: Group 2 patients have a higher rate than group 1 patients of remaining on the regimen for 120 weeks.
4. **Test statistic.** The test statistic is the observed value of b as shown in Table 12.6.3.
5. **Distribution of test statistic.** We determine the significance of b by consulting Appendix Table J.
6. **Decision rule.** Suppose we let $\alpha = .05$. The decision rule, then, is to reject H_0 if the observed value of b is equal to or less than 1, the value of b in Table J for $A = 12$, $B = 9$, $a = 8$, and $\alpha = .05$.
7. **Calculation of test statistic.** The observed value of b, as shown in Table 12.6.3, is 2.
8. **Statistical decision.** Since $2 > 1$, we fail to reject H_0.
9. **Conclusion.** Since we fail to reject H_0, we conclude that the null hypothesis may be true. That is, it may be true that the rate of remaining on the regimen for 120 weeks is the same or less for the PI experienced group compared to the PI naïve group.
10. ***p* value.** We see in Table J that when $A = 12$, $B = 9$, $a = 8$, the value of $b = 2$ has an exact probability of occurring by chance alone, when H_0 is true, greater than .05. ■

PI * Remained Cross-Tabulation

Count

		Remained		
		Yes	No	Total
PI	Experienced	8	4	12
	Naive	2	7	9
Total		10	11	21

Chi-Square Tests

	Value	df	Asymp. Sig. (2-sided)	Exact Sig. (2-sided)	Exact Sig. (1-sided)
Pearson Chi-Square	4.073[b]	1	.044		
Continuity Correction[a]	2.486	1	.115		
Likelihood Ratio	4.253	1	.039		
Fisher's Exact Test				.080	.056
Linear-by-Linear Association	3.879	1	.049		
N of Valid Cases	21				

a. Computed only for a 2×2 table

b. 2 cells (50.0%) have expected count less than 5. The minimum expected count is 4.29.

FIGURE 12.6.1 SPSS output for Example 12.6.1.

Various statistical software programs perform the calculations for the Fisher exact test. Figure 12.6.1 shows the results of Example 12.6.1 as computed by SPSS. The exact p value is provided for both a one-sided and a two-sided test. Based on these results, we fail to reject H_0 (p value $> .05$), just as we did using the statistical tables in the Appendix. Note that in addition to the Fisher exact test several alternative tests are provided. The reader should be aware that these alternative tests are not appropriate if the assumptions underlying them have been violated.

EXERCISES

12.6.1 The goal of a study by Tahmassebi and Curzon (A-13) was to determine if drooling in children with cerebral palsy is due to hypersalivation. One of the procedures toward that end was to examine the salivary buffering capacity of cerebral palsied children and controls. The following table gives the results.

	Buffering Capacity	
Group	**Medium**	**High**
Cerebral palsy	2	8
Control	3	7

Source: J. F. Tahmassebi and M. E. J. Curzon, "The Cause of Drooling in Children with Cerebral Palsy—Hypersalivation or Swallowing Defect?" *International Journal of Paediatric Dentistry, 13* (2003), 106–111.

Test for a significant difference between cerebral palsied children and controls with respect to high or low buffering capacity. Let $\alpha = .05$ and find the p value.

12.6.2 In a study by Xiao and Shi (A-14), researchers studied the effect of cranberry juice in the treatment and prevention of *Helicobacter pylori* infection in mice. The eradication of *Helicobacter pylori* results in the healing of peptic ulcers. Researchers compared treatment with cranberry juice to "triple therapy (amoxicillin, bismuth subcitrate, and metronidazole) in mice infected with *Helicobacter pylori*. After 4 weeks, they examined the mice to determine the frequency of eradication of the bacterium in the two treatment groups. The following table shows the results.

	No. of Mice with *Helicobacter pylori* Eradicated	
	Yes	**No**
Triple therapy	8	2
Cranberry juice	2	8

Source: Shu Dong Xiao and Tong Shi, "Is Cranberry Juice Effective in the Treatment and Prevention of *Helicobacter Pylori* Infection of Mice," *Chinese Journal of Digestive Diseases, 4* (2003), 136–139.

May we conclude, on the basis of these data, that triple therapy is more effective than cranberry juice at eradication of the bacterium? Let $\alpha = .05$ and find the p value.

12.6.3 In a study by Shaked et al. (A-15), researchers studied 26 children with blunt pancreatic injuries. These injuries occurred from a direct blow to the abdomen, bicycle handlebars, fall from height, or car accident. Nineteen of the patients were classified as having minor injuries, and seven were classified as having major injuries. Pseudocyst formation was suspected when signs of clinical deterioration developed, such as increased abdominal pain, epigastric fullness, fever, and increased pancreatic enzyme levels. In the major injury group, six of the seven children developed pseudocysts while in the minor injury group, three of the 19 children developed pseudocysts. Is this sufficient evidence to allow us to conclude that the proportion of children developing pseudocysts is higher in the major injury group than in the minor injury group? Let $\alpha = .01$.

12.7 RELATIVE RISK, ODDS RATIO, AND THE MANTEL–HAENSZEL STATISTIC

In Chapter 8 we learned to use analysis of variance techniques to analyze data that arise from designed experiments, investigations in which at least one variable is manipulated in some way. Designed experiments, of course, are not the only sources of data that are of interest to clinicians and other health sciences professionals. Another important class of scientific investigation that is widely used is the *observational study.*

DEFINITION

An *observational study* is a scientific investigation in which neither the subjects under study nor any of the variables of interest are manipulated in any way.

An observational study, in other words, may be defined simply as an investigation that is not an experiment. The simplest form of observational study is one in which there are only two variables of interest. One of the variables is called the *risk factor,* or independent variable, and the other variable is referred to as the *outcome,* or dependent variable.

DEFINITION

The term *risk factor* is used to designate a variable that is thought to be related to some outcome variable. The risk factor may be a suspected cause of some specific state of the outcome variable.

In a particular investigation, for example, the outcome variable might be subjects' status relative to cancer and the risk factor might be their status with respect to cigarette smoking. The model is further simplified if the variables are categorical with only two categories per variable. For the outcome variable the categories might be cancer present and cancer absent. With respect to the risk factor subjects might be categorized as smokers and nonsmokers.

When the variables in observational studies are categorical, the data pertaining to them may be displayed in a contingency table, and hence the inclusion of the topic in the present chapter. We shall limit our discussion to the situation in which the outcome variable and the risk factor are both dichotomous variables.

Types of Observational Studies There are two basic types of observational studies, *prospective studies* and *retrospective studies.*

DEFINITION

A *prospective study* is an observational study in which two random samples of subjects are selected. One sample consists of subjects who possess the risk factor, and the other sample consists of subjects who do not possess the risk factor. The subjects are followed into the future (that is, they are followed prospectively), and a record is kept on the number of subjects in each sample who, at some point in time, are classifiable into each of the categories of the outcome variable.

The data resulting from a prospective study involving two dichotomous variables can be displayed in a 2×2 contingency table that usually provides information regarding the number of subjects with and without the risk factor and the number who did and did not

TABLE 12.7.1 Classification of a Sample of Subjects with Respect to Disease Status and Risk Factor

	Disease Status		
Risk Factor	**Present**	**Absent**	**Total at Risk**
Present	a	b	$a+b$
Absent	c	d	$c+d$
Total	$a+c$	$b+d$	n

succumb to the disease of interest as well as the frequencies for each combination of categories of the two variables.

DEFINITION

A *retrospective study* is the reverse of a prospective study. The samples are selected from those falling into the categories of the outcome variable. The investigator then looks back (that is, takes a retrospective look) at the subjects and determines which ones have (or had) and which ones do not have (or did not have) the risk factor.

From the data of a retrospective study we may construct a contingency table with frequencies similar to those that are possible for the data of a prospective study.

In general, the prospective study is more expensive to conduct than the retrospective study. The prospective study, however, more closely resembles an experiment.

Relative Risk The data resulting from a prospective study in which the dependent variable and the risk factor are both dichotomous may be displayed in a 2×2 contingency table such as Table 12.7.1. The risk of the development of the disease among the subjects with the risk factor is $a/(a+b)$. The risk of the development of the disease among the subjects without the risk factor is $c/(c+d)$. We define relative risk as follows.

DEFINITION

***Relative risk* is the ratio of the risk of developing a disease among subjects with the risk factor to the risk of developing the disease among subjects without the risk factor.**

We represent the relative risk from a prospective study symbolically as

$$\widehat{RR} = \frac{a/(a+b)}{c/(c+d)} \tag{12.7.1}$$

where a, b, c, and d are as defined in Table 12.7.1, and $\widehat{RR}$ indicates that the relative risk is computed from a sample to be used as an estimate of the relative risk, RR, for the population from which the sample was drawn.

We may construct a confidence interval for RR

$$100(1-\alpha)\%\text{CI} = \widehat{RR}^{1\pm(z_\alpha/\sqrt{X^2})} \tag{12.7.2}$$

where z_α is the two-sided z value corresponding to the chosen confidence coefficient and X^2 is computed by Equation 12.4.1.

Interpretation of *RR* The value of RR may range anywhere between zero and infinity. A value of 1 indicates that there is no association between the status of the risk factor and the status of the dependent variable. In most cases the two possible states of the dependent variable are disease present and disease absent. We interpret an RR of 1 to mean that the risk of acquiring the disease is the same for those subjects with the risk factor and those without the risk factor. A value of RR greater than 1 indicates that the risk of acquiring the disease is greater among subjects with the risk factor than among subjects without the risk factor. An RR value that is less than 1 indicates less risk of acquiring the disease among subjects with the risk factor than among subjects without the risk factor. For example, a risk factor of 2 is taken to mean that those subjects with the risk factor are twice as likely to acquire the disease as compared to subjects without the risk factor.

We illustrate the calculation of relative risk by means of the following example.

EXAMPLE 12.7.1

In a prospective study of pregnant women, Magann et al. (A-16) collected extensive information on exercise level of low-risk pregnant working women. A group of 217 women did no voluntary or mandatory exercise during the pregnancy, while a group of 238 women exercised extensively. One outcome variable of interest was experiencing preterm labor. The results are summarized in Table 12.7.2.

We wish to estimate the relative risk of preterm labor when pregnant women exercise extensively.

Solution: By Equation 12.7.1 we compute

$$\widehat{RR} = \frac{22/238}{18/217} = \frac{.0924}{.0829} = 1.1$$

TABLE 12.7.2 Subjects with and without the Risk Factor Who Became Cases of Preterm Labor

Risk Factor	Cases of Preterm Labor	Noncases of Preterm Labor	Total
Extreme exercising	22	216	238
Not exercising	18	199	217
Total	40	415	455

Source: Everett F. Magann, Sharon F. Evans, Beth Weitz, and John Newnham, "Antepartum, Intrapartum, and Neonatal Significance of Exercise on Healthy Low-Risk Pregnant Working Women," *Obstetrics and Gynecology, 99* (2002), 466–472.

Odds Ratio and Relative Risk Section

Parameter	Common Odds Ratio	Original Odds Ratio	Iterated Odds Ratio	Log Odds Ratio	Relative Risk
Upper 95% C.L.		2.1350	2.2683	0.7585	2.1192
Estimate	1.1260	1.1207	1.1207	0.1140	1.1144
Lower 95% C.L.		0.5883	0.5606	−0.5305	0.5896

FIGURE 12.7.1 NCSS output for the data in Example 12.7.1.

These data indicate that the risk of experiencing preterm labor when a woman exercises heavily is 1.1 times as great as it is among women who do not exercise at all.

We compute the 95 percent confidence interval for *RR* as follows. By Equation 12.4.1, we compute from the data in Table 12.7.2:

$$X^2 = \frac{455[(22)(199) - (216)(18)]^2}{(40)(415)(238)(217)} = .1274$$

By Equation 12.7.2, the lower and upper confidence limits are, respectively, $1.1^{1-1.96/\sqrt{.1274}} = .65$ and $1.1^{1+1.96/\sqrt{.1274}} = 1.86$. Since the interval includes 1, we conclude, at the .05 level of significance, that the population risk may be 1. In other words, we conclude that, in the population, there may not be an increased risk of experiencing preterm labor when a pregnant woman exercises extensively.

The data were processed by NCSS. The results are shown in Figure 12.7.1. The relative risk calculation is shown in the column at the far right of the output, along with the 95% confidence limits. Because of rounding errors, these values differ slightly from those given in the example. ■

Odds Ratio When the data to be analyzed come from a retrospective study, relative risk is not a meaningful measure for comparing two groups. As we have seen, a retrospective study is based on a sample of subjects with the disease (cases) and a separate sample of subjects without the disease (controls or noncases). We then retrospectively determine the distribution of the risk factor among the cases and controls. Given the results of a retrospective study involving two samples of subjects, cases, and controls, we may display the data in a 2 × 2 table such as Table 12.7.3, in which subjects are dichotomized with respect to the presence and absence of the risk factor. Note that the column headings in Table 12.7.3 differ from those in Table 12.7.1 to emphasize the fact that the data are from a retrospective study and that the subjects were selected because they were either cases or controls. When the data from a retrospective study are displayed as in Table 12.7.3, the ratio $a/(a + b)$, for example, is not an estimate of the risk of disease for subjects with the risk factor. The appropriate measure for comparing cases and controls in a retrospective study is the *odds ratio*. As noted in Chapter 11, in order to understand the

TABLE 12.7.3 Subjects of a Retrospective Study Classified According to Status Relativeto a Risk Factor and Whether They Are Cases or Controls

	Sample		
Risk Factor	**Cases**	**Controls**	**Total**
Present	a	b	$a + b$
Absent	c	d	$c + d$
Total	$a + c$	$b + d$	n

concept of the odds ratio, we must understand the term *odds,* which is frequently used by those who place bets on the outcomes of sporting events or participate in other types of gambling activities.

DEFINITION

The odds for success are the ratio of the probability of success to the probability of failure.

We use this definition of odds to define two odds that we can calculate from data displayed as in Table 12.7.3:

1. The odds of being a case (having the disease) to being a control (not having the disease) among subjects with the risk factor is $[a/(a+b)]/[b/(a+b)] = a/b$.
2. The odds of being a case (having the disease) to being a control (not having the disease) among subjects without the risk factor is $[c/(c+d)]/[d/(c+d)] = c/d$.

We now define the odds ratio that we may compute from the data of a retrospective study. We use the symbol $\widehat{OR}$ to indicate that the measure is computed from sample data and used as an estimate of the population odds ratio, OR.

DEFINITION

The estimate of the population odds ratio is

$$\widehat{OR} = \frac{a/b}{c/d} = \frac{ab}{bc} \qquad (12.7.3)$$

where *a, b, c,* and *d* are as defined in Table 12.7.3.

We may construct a confidence interval for OR by the following method:

$$100(1 - \alpha)\%\text{CI} = \widehat{OR}^{1 \pm (z_\alpha/\sqrt{X^2})} \qquad (12.7.4)$$

where z_α is the two-sided z value corresponding to the chosen confidence coefficient and X^2 is computed by Equation 12.4.1.

Interpretation of the Odds Ratio In the case of a rare disease, the population odds ratio provides a good approximation to the population relative risk. Consequently, the sample odds ratio, being an estimate of the population odds ratio, provides an indirect estimate of the population relative risk in the case of a rare disease.

The odds ratio can assume values between zero and ∞. A value of 1 indicates no association between the risk factor and disease status. A value less than 1 indicates reduced odds of the disease among subjects with the risk factor. A value greater than 1 indicates increased odds of having the disease among subjects in whom the risk factor is present.

EXAMPLE 12.7.2

Toschke et al. (A-17) collected data on obesity status of children ages 5–6 years and the smoking status of the mother during the pregnancy. Table 12.7.4 shows 3970 subjects classified as cases or noncases of obesity and also classified according to smoking status of the mother during pregnancy (the risk factor). We wish to compare the odds of obesity at ages 5–6 among those whose mother smoked throughout the pregnancy with the odds of obesity at age 5–6 among those whose mother did not smoke during pregnancy.

Solution: The odds ratio is the appropriate measure for answering the question posed. By Equation 12.7.3 we compute

$$\widehat{OR} = \frac{(64)(3496)}{(342)(68)} = 9.62$$

We see that obese children (cases) are 9.62 times as likely as nonobese children (noncases) to have had a mother who smoked throughout the pregnancy.

We compute the 95 percent confidence interval for OR as follows. By Equation 12.4.1 we compute from the data in Table 12.7.4

$$X^2 = \frac{3970[(64)(3496) - (342)(68)]^2}{(132)(3838)(406)(3564)} = 217.6831$$

TABLE 12.7.4 Subjects Classified According to Obesity Status and Mother's Smoking Status during Pregnancy

	Obesity Status		
Smoking Status During Pregnancy	Cases	Noncases	Total
Smoked throughout	64	342	406
Never smoked	68	3496	3564
Total	132	3838	3970

Source: A. M. Toschke, S. M. Montgomery, U. Pfeiffer, and R. von Kries, "Early Intrauterine Exposure to Tobacco-Inhaled Products and Obesity," *American Journal of Epidemiology, 158* (2003), 1068–1074.

Smoking_status * Obsesity_status Cross-Tabulation

Count

		Obesity status		
		Cases	Noncases	Total
Smoking_status	Smoked throughout	64	342	406
	Never smoked	68	3496	3564
Total		132	3838	3970

Risk Estimate

	Value	95% Confidence Interval	
		Lower	Upper
Odds Ratio for Smoking_status (Smoked throughout /Never smoked)	9.621	6.719	13.775
For cohort Obesity_ status = Cases	8.262	5.966	11.441
For cohort Obesity_ status = Noncases	.859	.823	.896
N of Valid Cases	3970		

FIGURE 12.7.2 SPSS output for Example 12.7.2.

The lower and upper confidence limits for the population *OR*, respectively, are $9.62^{1-1.96/\sqrt{217.6831}} = 7.12$ and $9.62^{1+1.96/\sqrt{217.6831}} = 13.00$. We conclude with 95 percent confidence that the population *OR* is somewhere between 7.12 and 13.00. Because the interval does not include 1, we conclude that, in the population, obese children (cases) are more likely than nonobese children (noncases) to have had a mother who smoked throughout the pregnancy.

The data from Example 12.7.2 were processed using SPSS. The results are shown in Figure 12.7.2. The odds ratio calculation, along with the 95% confidence limits, are shown in the top line of the Risk Estimate box. These values differ slightly from those in the example because of rounding error. ■

The Mantel–Haenszel Statistic Frequently when we are studying the relationship between the status of some disease and the status of some risk factor, we are aware of another variable that may be associated with the disease, with the risk factor, or with both in such a way that the true relationship between the disease status and the risk factor is masked. Such a variable is called a *confounding variable*. For example, experience might indicate the possibility that the relationship between some disease and a suspected risk factor differs among different ethnic groups. We would then treat ethnic membership as a confounding variable. When they can be identified, it is desirable to control for confounding variables so that an unambiguous measure of the relationship between disease status and risk factor may be calculated. A technique for accomplishing this objective is the Mantel–Haenszel (22) procedure, so called in recognition of the two men who developed it. The procedure allows us to test the null hypothesis that there is no association between status with respect to disease and risk factor status. Initially used only with data from retrospective studies, the Mantel–Haenszel procedure is also appropriate for use with data from prospective studies, as discussed by Mantel (23).

In the application of the Mantel–Haenszel procedure, case and control subjects are assigned to strata corresponding to different values of the confounding variable. The data are then analyzed within individual strata as well as across all strata. The discussion that follows assumes that the data under analysis are from a retrospective or a prospective study with case and noncase subjects classified according to whether they have or do not have the suspected risk factor. The confounding variable is categorical, with the different categories defining the strata. If the confounding variable is continuous it must be categorized. For example, if the suspected confounding variable is age, we might group subjects into mutually exclusive age categories. The data before stratification may be displayed as shown in Table 12.7.3.

Application of the Mantel–Haenszel procedure consists of the following steps.

1. Form k strata corresponding to the k categories of the confounding variable. Table 12.7.5 shows the data display for the ith stratum.
2. For each stratum compute the expected frequency e_i of the upper left-hand cell of Table 12.7.5 as follows:

$$e_i = \frac{(a_i + b_i)(a_i + c_i)}{n_i} \qquad (12.7.5)$$

TABLE 12.7.5 Subjects in the *i*th Stratum of a Confounding Variable Classified According to Status Relative to a Risk Factor and Whether They Are Cases or Controls

	Sample		
Risk Factor	**Cases**	**Controls**	**Total**
Present	a_i	b_i	$a_i + b_i$
Absent	c_i	d_i	$c_i + d_i$
Total	$a_i + c_i$	$b_i + d_i$	n_i

3. For each stratum compute

$$v_i = \frac{(a_i + b_i)(c_i + d_i)(a_i + c_i)(b_i + d_i)}{n_i^2(n_i - 1)} \tag{12.7.6}$$

4. Compute the Mantel–Haenszel test statistic, χ^2_{MH} as follows:

$$\chi^2_{\text{MH}} = \frac{\left(\sum_{i=1}^{k} a_i - \sum_{i=1}^{k} e_i\right)^2}{\sum_{i=1}^{k} v_i} \tag{12.7.7}$$

5. Reject the null hypothesis of no association between disease status and suspected risk factor status in the population if the computed value of χ^2_{MH} is equal to or greater than the critical value of the test statistic, which is the tabulated chi-square value for 1 degree of freedom and the chosen level of significance.

Mantel–Haenszel Estimator of the Common Odds Ratio When we have k strata of data, each of which may be displayed in a table like Table 12.7.5, we may compute the Mantel–Haenszel estimator of the common odds ratio, $\widehat{OR}_{\text{MH}}$ as follows:

$$\widehat{OR}_{\text{MH}} = \frac{\sum_{i=1}^{k} (a_i d_i / n_i)}{\sum_{i=1}^{k} (b_i c_i / n_i)} \tag{12.7.8}$$

When we use the Mantel–Haenszel estimator given by Equation 12.7.4, we assume that, in the population, the odds ratio is the same for each stratum.

We illustrate the use of the Mantel–Haenszel statistics with the following examples.

EXAMPLE 12.7.3

In a study by LaMont et al. (A-18), researchers collected data on obstructive coronary artery disease (OCAD), hypertension, and age among subjects identified by a treadmill stress test as being at risk. In Table 12.7.6, counts on subjects in two age strata are presented with hypertension as the risk factor and the presence of OCAD as the case/noncase variable.

Solution:

1. **Data.** See Table 12.7.6.
2. **Assumptions.** We assume that the assumptions discussed earlier for the valid use of the Mantel–Haenszel statistic are met.

TABLE 12.7.6 Patients Stratified by Age and Classified by Status Relative to Hypertension (the Risk Factor) and OCAD (Case/Noncase Variable)

Stratum 1 (55 and under)			
Risk Factor (Hypertension)	**Cases (OCAD)**	**Noncases**	**Total**
Present	21	11	32
Absent	16	6	22
Total	37	17	54
Stratum 2 (over 55)			
Risk Factor (Hypertension)	**Cases (OCAD)**	**Noncases**	**Total**
Present	50	14	64
Absent	18	6	24
Total	68	20	88

Source: Matthew J. Budoff, MD. Used with permission.

3. **Hypotheses.**

   H_0: There is no association between the presence of hypertension and occurrence of OCAD in subjects 55 and under and subjects over 55.

   H_A: There is a relationship between the two variables.

4. **Test statistic.**

$$\chi^2_{\mathrm{MH}} = \frac{\left(\sum_{i=1}^{k} a_i - \sum_{i=1}^{k} e_i\right)^2}{\sum_{i=1}^{k} v_i}$$

   as given in Equation 12.7.7.

5. **Distribution of test statistic.** Chi-square with 1 degree of freedom.

6. **Decision rule.** Suppose we let $\alpha = .05$. Reject H_0 if the computed value of the test statistic is greater than or equal to 3.841.

7. **Calculation of test statistic.** By Equation 12.7.5 we compute the following expected frequencies:

$$e_1 = (21 + 11)(21 + 16)/54 = (32)(37)/54 = 21.93$$
$$e_2 = (50 + 14)(50 + 18)/88 = (64)(68)/88 = 49.45$$

By Equation 12.7.6 we compute

$$v_1 = (32)(22)(37)(17)/(2916)(54 - 1) = 2.87$$
$$v_2 = (64)(24)(68)(20)/(7744)(88 - 1) = 3.10$$

Finally, by Equation 12.7.7 we compute

$$\chi^2_{\text{MH}} = \frac{[(21 + 50) - (21.93 + 49.45)]^2}{2.87 + 3.10} = .0242$$

8. **Statistical decision.** Since $.0242 < 3.841$, we fail to reject H_0.
9. **Conclusion.** We conclude that there may not be an association between hypertension and the occurrence of OCAD.
10. ***p* value.** Since $.0242 < 2.706$, the p value for this test is $p > .10$.

We now illustrate the calculation of the Mantel–Haenszel estimator of the common odds ratio. ■

EXAMPLE 12.7.4

Let us refer to the data in Table 12.7.6 and compute the common odds ratio.

Solution: From the stratified data in Table 12.7.6 we compute the numerator of the ratio as follows:

$$(a_1 d_1/n_1) + (a_2 d_2/n_2) = [(21)(6)/54] + [(50)(6)/88]$$
$$= 5.7424$$

The denominator of the ratio is

$$(b_1 c_1/n_1) + (b_2 c_2/n_2) = [(11)(16)/54] + [(14)(18)/88]$$
$$= 6.1229$$

Now, by Equation 12.7.7, we compute the common odds ratio:

$$\widehat{OR}_{\text{MH}} = \frac{5.7424}{6.1229} = .94$$

From these results we estimate that, regardless of age, patients who have hypertension are less likely to have OCAD than patients who do not have hypertension. ■

Hand calculation of the Mantel-Haenszel test statistics can prove to be a cumbersome task. Fortunately, the researcher can find relief in one of several statistical software packages that are available. To illustrate, results from the use of SPSS to process the data of Example 12.7.3 are shown in Figure 12.7.3. These results differ from those given in the example because of rounding error.

Smoking_status * Obsesity_status * Stratum Cross-Tabulation

Count

Stratum			Obesity status: Cases	Obesity status: Noncases	Total
55 and under	Smoking_status	Smoked throughout	21	11	32
		Never smoked	16	6	22
	Total		37	17	54
Over 55	Smoking_status	Smoked throughout	50	14	64
		Never smoked	18	6	24
	Total		68	20	88

Tests of Conditional Independence

	Chi-Squared	df	Asymp. Sig. (2-sided)
Cochran's	.025	1	.875
Mantel-Haenszel	.002	1	.961

Mantel-Haenszel Common Odds Ratio Estimate

Estimate			.938
ln(Estimate)			−.064
Std. Error of ln(Estimate)			.412
Asymp. Sig. (2-sided)			.876
Asymp. 95% confidence Interval	Common Odds Ratio	Lower Bound	.418
		Upper Bound	2.102
	ln(Common Odds Ratio)	Lower Bound	−.871
		Upper Bound	.743

FIGURE 12.7.3 SPSS output for Example 12.7.3.

EXERCISES

12.7.1 Davy et al. (A-19) reported the results of a study involving survival from cervical cancer. The researchers found that among subjects younger than age 50, 16 of 371 subjects had not survived for 1 year after diagnosis. In subjects age 50 or older, 219 of 376 had not survived for 1 year after diagnosis. Compute the relative risk of death among subjects age 50 or older. Does it appear from these data that older subjects diagnosed as having cervical cancer are prone to higher mortality rates?

12.7.2 The objective of a prospective study by Stenestrand et al. (A-20) was to compare the mortality rate following an acute myocardial infarction (AMI) among subjects receiving early revascularization to the mortality rate among subjects receiving conservative treatments. Among 2554 patients receiving revascularization within 14 days of AMI, 84 died in the year following the AMI. In the conservative treatment group (risk factor present), 1751 of 19,358 patients died within a year of AMI. Compute the relative risk of mortality in the conservative treatment group as compared to the revascularization group in patients experiencing AMI.

12.7.3 Refer to Example 12.7.2. Toschke et al. (A-17), who collected data on obesity status of children ages 5–6 years and the smoking status of the mother during the pregnancy, also reported on another outcome variable: whether the child was born premature (37 weeks or fewer of gestation). The following table summarizes the results of this aspect of the study. The same risk factor (smoking during pregnancy) is considered, but a case is now defined as a mother who gave birth prematurely.

Smoking Status During Pregnancy	**Premature Birth Status** Cases	Noncases	Total
Smoked throughout	36	370	406
Never smoked	168	3396	3564
Total	204	3766	3970

Source: A. M. Toschke, S. M. Montgomery, U. Pfeiffer, and R. von Kries, "Early Intrauterine Exposure to Tobacco-Inhaled Products and Obesity," *American Journal of Epidemiology*, *158* (2003), 1068–1074.

Compute the odds ratio to determine if smoking throughout pregnancy is related to premature birth. Use the chi-square test of independence to determine if one may conclude that there is an association between smoking throughout pregnancy and premature birth. Let $\alpha = .05$.

12.7.4 Sugiyama et al. (A-21) examined risk factors for allergic diseases among 13- and 14-year-old schoolchildren in Japan. One risk factor of interest was a family history of eating an unbalanced diet. The following table shows the cases and noncases of children exhibiting symptoms of rhinitis in the presence and absence of the risk factor.

Family History	**Rhinitis** Cases	Noncases	Total
Unbalanced diet	656	1451	2107
Balanced diet	677	1662	2339
Total	1333	3113	4446

Source: Takako Sugiyama, Kumiya Sugiyama, Masao Toda, Tastuo Yukawa, Sohei Makino, and Takeshi Fukuda, "Risk Factors for Asthma and Allergic Diseases Among 13–14-Year-Old Schoolchildren in Japan," *Allergology International*, *51* (2002), 139–150.

What is the estimated odds ratio of having rhinitis among subjects with a family history of an unbalanced diet compared to those eating a balanced diet? Compute the 95 percent confidence interval for the odds ratio.

12.7.5 According to Holben et al. (A-22), "Food insecurity implies a limited access to or availability of food or a limited/uncertain ability to acquire food in socially acceptable ways." These researchers collected data on 297 families with a child in the Head Start nursery program in a rural area of Ohio near Appalachia. The main outcome variable of the study was household status relative to food security. Households that were not food secure are considered to be cases. The risk factor of interest was the absence of a garden from which a household was able to supplement its food supply. In the following table, the data are stratified by the head of household's employment status outside the home.

Stratum 1 (Employed Outside the Home)

Risk Factor	Cases	Noncases	Total
No garden	40	37	77
Garden	13	38	51
Total	53	75	128

Stratum 2 (Not Employed Outside the Home)

Risk Factor	Cases	Noncases	Total
No garden	75	38	113
Garden	15	33	48
Total	90	71	161

Source: David H. Holben, Ph.D. and John P. Holcomb, Jr., Ph.D. Used with permission.

Compute the Mantel–Haenszel common odds ratio with stratification by employment status. Use the Mantel–Haenszel chi-square test statistic to determine if we can conclude that there is an association between the risk factor and food insecurity. Let $\alpha = .05$.

12.8 SURVIVAL ANALYSIS

In many clinical studies, an investigator may wish to monitor the progress of patients from some point in time, such as the time a surgical procedure is performed or a treatment regimen is initiated, until the occurrence of some well-defined event such as death or cessation of symptoms.

Suppose, for example, that patients who have experienced their first heart attack are enrolled in a study to assess the effectiveness of two competing medications for the prevention of a second myocardial infarction. The investigation begins when the first patient, following his or her first heart attack, is enrolled in the study. The study continues until each patient in the study experiences one or another of three events: (1) a myocardial infarction (the *event of interest*), (2) *loss to follow-up* for some reason such as death from a cause other than a heart attack or having moved to another locality, or

(3) the condition of being alive at the time the investigator decides to terminate the study.

For each patient in the study, the investigator records the amount of time (in months, days, years, or some other measures of time) elapsing between the point at which the patient entered the study and the point at which he or she experienced one of the terminating events. The time elapsing between enrollment in the study and the experiencing of one of the events is referred to as the patient's *survival time.* The set of such survival times recorded during the course of a study is referred to as *survival data.*

Suppose we have the following information on three of the patients in the study involving heart-attack patients. Patient A entered the study on January 1, 2002, and had a myocardial infarction on December 31, 2003. Patient A's survival time is 24 months. Patient B entered the study on July 1, 2002, and moved out of state on December 31, 2002. Patient B's survival time is 6 months. Patient C entered the study on August 1, 2002, and was still alive when the study was terminated on December 31, 2004. Patient C's survival time is 29 months. The survival time for patient B is called a *censored survival time,* since the terminating event was loss to follow-up rather than the event of interest. Similarly, since the terminating event for patient C was being alive at the end of the study, this patient's survival time is also called a censored survival time. The survival times for patient B and patient C are called *censored data.* The experiences of these three patients may be represented graphically as shown in Figure 12.8.1.

Censored data can arise in a variety of ways. In *singly censored data*, a fixed number of subjects enter into a study at the same time and remain in the study until one of two things occur. First, at the conclusion of the study, not all of the subjects may have experienced a given condition or endpoint of interest, and therefore their survival times are not known exactly. This is called *type I censoring*. Second, the study may be ended after a certain proportion of subjects have experienced a given condition. Those that did not experience the condition would not have a known survival time. This is called *type II censoring*. A third type of censoring leads to *progressively censored data*. In this type of censoring, the period of the study is fixed, but subjects may enter the experiment at different times. (For example, in a clinical study, patients can enter the study at any time after diagnosis.) Patients may then experience, or not experience, a condition by the end of the study. Those who did not experience the condition do not have known survival times. This is called *type III censoring*. Clearly the details surrounding censored data are complex and

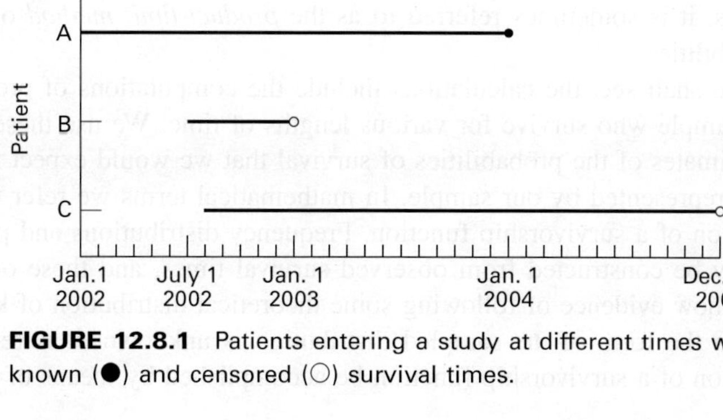

FIGURE 12.8.1 Patients entering a study at different times with known (●) and censored (○) survival times.

require much more detailed analysis than can be covered in this text. For those interested in further reading, we suggest the books by Lee (28) or Hosmer and Lemeshow (34).

Typically, for purposes of analysis, a dichotomous, or indicator, variable is used to distinguish survival times of those patients who experienced the event of interest from the censored times of those who did not experience the event of interest because of loss to follow-up or being alive at the termination of the study.

In studies involving the comparison of two treatments, we are interested in three items of information for each patient: (1) Which treatment, A or B, was given to the patient? (2) For what length of time was the patient observed? (3) Did the patient experience the event of interest during the study or was he or she either lost to follow-up or alive at the end of the study? (That is, is the observed time an event time or a censored time?) In studies that are not concerned with the comparison of treatments or other characteristics of patients, only the last two items of data are relevant.

Armed with these three items of information, we are able, in studies like our myocardial infarction example, to estimate the median survival time of the group of patients who received treatment A and compare it with the estimated median survival time of the group receiving treatment B. Comparison of the two medians allows us to answer the following question: Based on the information from our study, which treatment do we conclude delays for a longer period of time, on the average, the occurrence of the event of interest? In the case of our example, we may answer the question: Which treatment do we conclude delays for a longer period of time, on the average, the occurrence of a second myocardial infarction? The data collected in follow-up studies such as we have described may also be used to answer another question of considerable interest to the clinician: What is the estimated probability that a patient will survive for a specified length of time? The clinician involved in our myocardial infarction study, for example, might ask, "What is the estimated probability that, following a first heart attack, a patient receiving treatment A will survive for more than three years?" The methods employed to answer these questions by using the information collected during a follow-up study are known as *survival analysis* methods.

The Kaplan–Meier Procedure Now let us show how we may use the data usually collected in follow-up studies of the type we have been discussing to estimate the probability of surviving for a specified length of time. The method we use was introduced by Kaplan and Meier (24) and for that reason is called the *Kaplan–Meier procedure.* Since the procedure involves the successive multiplication of individual estimated probabilities, it is sometimes referred to as the *product-limit method* of estimating survival probabilities.

As we shall see, the calculations include the computations of proportions of subjects in a sample who survive for various lengths of time. We use these sample proportions as estimates of the probabilities of survival that we would expect to observe in the population represented by our sample. In mathematical terms we refer to the process as the estimation of a survivorship function. Frequency distributions and probability distributions may be constructed from observed survival times, and these observed distributions may show evidence of following some theoretical distribution of known functional form. When the form of the sampled distribution is unknown, it is recommended that the estimation of a survivorship function be accomplished by means of a *nonparametric technique,* of which the Kaplan–Meier procedure is one. Nonparametric techniques are defined and discussed in detail in Chapter 13.

Calculations for the Kaplan–Meier Procedure We let

n = the number of subjects whose survival times are available

p_1 = the proportion of subjects surviving at least the first time period (day, month, year, etc.)

p_2 = the proportion of subjects surviving the second time period after having survived the first time period

p_3 = the proportion of subjects surviving the third time period after having survived the second time period

⋮

p_k = the proportion of subjects surviving the kth time period after having survived the $(k - 1)$th time period

We use these proportions, which we may relabel $\hat{p}_1, \hat{p}_2, \hat{p}_3, \ldots, \hat{p}_k$ as estimates of the probability that a subject from the population represented by the sample will survive time periods $1, 2, 3, \ldots, k$, respectively.

For any time period, $t(1 \leq t \leq k)$, we estimate the probability of surviving the tth time period, p_t, as follows:

$$\hat{p}_t = \frac{\text{number of subjects surviving at least } (t-1) \text{ time periods who also survive the } t\text{th period}}{\text{number of subjects alive at end of time period } (t-1)} \qquad (12.8.1)$$

The probability of surviving to time t, $S(t)$, is estimated by

$$\hat{S}(t) = \hat{p}_1 \times \hat{p}_2 \times \cdots \times \hat{p}_t \qquad (12.8.2)$$

We illustrate the use of the Kaplan–Meier procedure with the following example.

EXAMPLE 12.8.1

To assess results and identify predictors of survival, Martini et al. (A-23) reviewed their total experience with primary malignant tumors of the sternum. They classified patients as having either low-grade (25 patients) or high-grade (14 patients) tumors. The event (status), time to event (months), and tumor grade for each patient are shown in Table 12.8.1. We wish to compare the 5-year survival experience of these two groups by means of the Kaplan–Meier procedure.

Solution: The data arrangement and necessary calculations are shown in Table 12.8.2. The entries for the table are obtained as follows.

1. We begin by listing the observed times in order from smallest to largest in Column 1.

TABLE 12.8.1 Survival Data, Subjects with Malignant Tumors of the Sternum

Subject	Time (Months)	Vital Status[a]	Tumor Grade[b]	Subject	Time (Months)	Vital Status[a]	Tumor Grade[b]
1	29	dod	L	21	155	ned	L
2	129	ned	L	22	102	dod	L
3	79	dod	L	23	34	ned	L
4	138	ned	L	24	109	ned	L
5	21	dod	L	25	15	dod	L
6	95	ned	L	26	122	ned	H
7	137	ned	L	27	27	dod	H
8	6	ned	L	28	6	dod	H
9	212	dod	L	29	7	dod	H
10	11	dod	L	30	2	dod	H
11	15	dod	L	31	9	dod	H
12	337	ned	L	32	17	dod	H
13	82	ned	L	33	16	dod	H
14	33	dod	L	34	23	dod	H
15	75	ned	L	35	9	dod	H
16	109	ned	L	36	12	dod	H
17	26	ned	L	37	4	dod	H
18	117	ned	L	38	0	dpo	H
19	8	ned	L	39	3	dod	H
20	127	ned	L				

[a] dod = dead of disease; ned = no evidence of disease; dpo = dead postoperation.
[b] L = low-grade; H = high-grade.
Source: Dr. Nael Martini. Used with permission.

2. Column 2 contains an indicator variable that shows vital status (1 = died, 0 = alive or censored).
3. In Column 3 we list the number of patients at risk for each time associated with the death of a patient. We need only be concerned about the times at which deaths occur because the survival rate does not change at censored times.
4. Column 4 contains the number of patients remaining alive just after one or more deaths.
5. Column 5 contains the estimated conditional probability of surviving, which is obtained by dividing Column 4 by Column 3. Note that, although there were two deaths at 15 months in the low-grade group and two deaths at 9 months in the high-grade group, we calculate only one survival proportion at these points. The calculations take the two deaths into account.
6. Column 6 contains the estimated cumulative probability of survival. We obtain the entries in this column by successive multiplication. Each entry

TABLE 12.8.2 Data Arrangement and Calculations for Kaplan–Meier Procedure, Example 12.8.1

1 Time (Months)	2 Vital Status 0 = Censored 1 = Dead	3 Patients at Risk	4 Patients Remaining Alive	5 Survival Proportion	6 Cumulative Survival Proportion
			Patients with Low-Grade Tumors		
6	0				
8	0				
11	1	23	22	22/23 = .956522	.956522
15	1				
15	1	22	20	20/22 = .909090	.869564
21	1	20	19	19/20 = .950000	.826086
26	0				
29	1	18	17	17/18 = .944444	.780192
33	1	17	16	16/17 = .941176	.734298
34	0				
75	0				
79	1	14	13	13/14 = .928571	.681847
82	0				
95	0				
102	1	11	10	10/11 = .909090	.619860
109	0				
109	0				
117	0				
127	0				
129	0				
137	0				
138	0				
155	0				
212	1	2	1	1/2 = .500000	.309930
337	0				

(*Continued*)

1	2	3	4	5	6
Time (Months)	Vital Status 0 = Censored 1 = Dead	Patients at Risk	Patients Remaining Alive	Survival Proportion	Cumulative Survival Proportion
			Patients with High-Grade Tumors		
0	1	14	13	13/14 = .928571	.928571
2	1	13	12	12/13 = .923077	.857142
3	1	12	11	11/12 = .916667	.785714
4	1	11	10	10/11 = .909090	.714285
6	1	10	9	9/10 = .900000	.642856
7	1	9	8	8/9 = .888889	.571428
9	1				
9	1	8	6	6/8 = .750000	.428572
12	1	6	5	5/6 = .833333	.357143
16	1	5	4	4/5 = .800000	.285714
17	1	4	3	3/4 = .750000	.214286
23	1	3	2	2/3 = .666667	.142857
27	1	2	1	1/2 = .500000	.071428
122	0	1	0		

after the first in Column 5 is multiplied by the cumulative product of all previous entries.

After the calculations are completed we examine Table 12.8.2 to determine what useful information it provides. From the table we note the following facts, which allow us to compare the survival experience of the two groups of subjects: those with low-grade tumors and those with high-grade tumors:

1. **Median survival time.** We can determine the median survival time by locating the time, in months, at which the cumulative survival proportion is equal to .5. None of the cumulative survival proportions are exactly .5, but we see that in the low-grade tumor group, the probability changes from .619860 to .309930 at 212 months; therefore, the median survival for this group is 212 months. In the high-grade tumor group, the

cumulative proportion changes from .571428 to .428572 at 9 months, which is the median survival for this group.

2. **Five-year survival rate.** We can determine the 5-year or 60-month survival rate for each group directly from the cumulative survival proportion at 60 months. For the low-grade tumor group, the 5-year survival rate is .734298 or 73 percent; for the high-grade tumor group, the 5-year survival rate is .071428 or 7 percent.

3. **Mean survival time.** We may compute for each group the mean of the survival times, which we will call $\overline{T}_L$ and $\overline{T}_H$ for the low-grade and high-grade groups, respectively. For the low-grade tumor group we compute $\overline{T}_L = 2201/25 = 88.04$, and for the high-grade tumor group we compute $\overline{T}_H = 257/14 = 18.35$. Since so many of the times in the low-grade group are censored, the true mean survival time for that group is, in reality, higher (perhaps, considerably so) than 88.04. The true mean survival time for the high-grade group is also likely higher than the computed 18.35, but with just one censored time we do not expect as great a difference between the calculated mean and the true mean. Thus, we see that we have still another indication that the survival experience of the low-grade tumor group is more favorable than the survival experience of the high-grade tumor group.

4. **Average hazard rate.** From the raw data of each group we may also calculate another descriptive statistic that can be used to compare the two survival experiences. This statistic is called the *average hazard rate*. It is a measure of nonsurvival potential rather than survival. A group with a higher average hazard rate will have a lower probability of surviving than a group with a lower average hazard rate. We compute the average hazard rate, designated $\overline{h}$ by dividing the number of subjects who do not survive by the sum of the observed survival times. For the low-grade tumor group, we compute $\overline{h}_L = 9/2201 = .004089$. For the high-grade tumor group we compute $\overline{h}_H = 13/257 = .05084$. We see that the average hazard rate for the high-grade group is higher than for the low-grade group, indicating a smaller chance of surviving for the high-grade group.

The cumulative survival proportion column of Table 12.8.2 may be portrayed visually in a survival curve graph in which the cumulative survival proportions are represented by the vertical axis and the time in months by the horizontal axis. We note that the graph resembles stairsteps with "steps" occurring at the times when deaths occurred. The graph also allows us to represent visually the median survival time and survival rates such as the 5-year survival rate. The graph for the cumulative survival data of Table 12.8.2 is shown in Figure 12.8.2.

These observations strongly suggest that the survival experience of patients with low-grade tumors is far more favorable than that of patients with high-grade tumors.

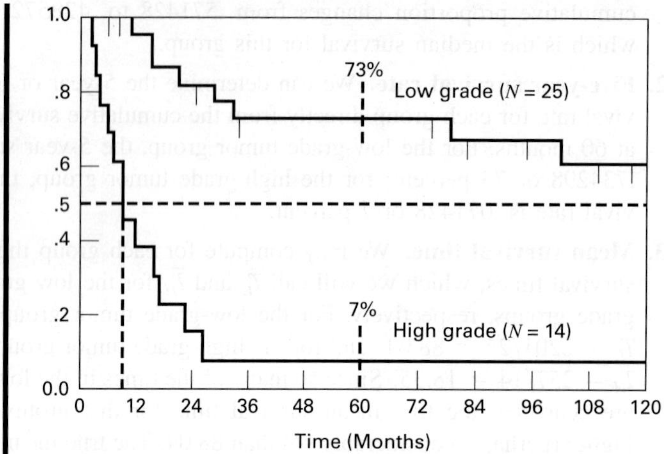

FIGURE 12.8.2 Kaplan–Meier survival curve, Example 12.8.1, showing median survival times and 5-year (60-month) survival rates. ■

The results of comparing the survival experiences of two groups will not always be as dramatic as those of our example. For an objective comparison of the survival experiences of two groups, it is desirable that we have an objective technique for determining whether they are statistically significantly different. We know also that the observed results apply strictly to the samples on which the analyses are based. Of much greater interest is a method for determining if we may conclude that there is a difference between survival experiences in the populations from which the samples were drawn. In other words, at this point, we desire a method for testing the null hypothesis that there is no difference in survival experience between two populations against the alternative that there is a difference. Such a test is provided by the *log-rank test*. The log-rank test is an application of the Mantel–Haenszel procedure discussed in Section 12.7. The extension of the procedure to survival data was proposed by Mantel (25). To calculate the log-rank statistic we proceed as follows:

1. Order the survival times until death for both groups combined, omitting censored times. Each time constitutes a stratum as defined in Section 12.7.
2. For each stratum or time, t_i, we construct a 2×2 table in which the first row contains the number of observed deaths, the second row contains the number of patients alive, the first column contains data for one group, say, group A, and the second column contains data for the other group, say, group B. Table 12.8.3 shows the table for time t_i.
3. For each stratum compute the expected frequency for the upper left-hand cell of its table by Equation 12.7.5.
4. For each stratum compute v_i by Equation 12.7.6.
5. Finally, compute the Mantel–Haenszel statistic (now called the log-rank statistic) by Equation 12.7.7.

We illustrate the calculation of the log-rank statistic with the following example.

TABLE 12.8.3 Contingency Table for Stratum (Time) t_i for Calculating the Log-Rank Test

	Group A	Group B	Total
Number of deaths observed	a_i	b_i	$a_i + b_i$
Number of patients alive	c_i	d_i	$c_i + d_i$
Number of patients "at risk"	$a_i + c_i$	$b_i + d_i$	$n_i = a_i + b_i + c_i + d_i$

EXAMPLE 12.8.2

Let us refer again to the data on primary malignant tumors of the sternum presented in Example 12.8.1. Examination of the data reveals that there are 20 time periods (strata). For each of these a 2×2 table following the pattern of Table 12.8.3 must be constructed. The first of these tables is shown as Table 12.8.4. By Equations 12.7.5 and 12.7.6 we compute e_i and v_i as follows:

$$e_i = \frac{(0 + 1)(0 + 25)}{39} = .641$$

$$v_i = \frac{(0 + 1)(25 + 13)(0 + 25)(1 + 13)}{39^2\,(38)} = .230$$

The data for Table 12.8.4 and similar data for the other 19 time periods are shown in Table 12.8.5. Using data from Table 12.8.5, we compute the log-rank statistic by Equation 12.7.7 as follows:

$$\chi^2_{\text{MH}} = \frac{(9 - 17.811)^2}{3.140} = 24.704$$

Reference to Appendix Table F reveals that since $24.704 > 7.879$, the p value for this test is $< .005$. We, therefore, reject the null hypothesis that the survival experience is the same for patients with low-grade tumors and high-grade tumors and conclude that they are different.

There are alternative procedures for testing the null hypothesis that two survival curves are identical. They include the Breslow test (also called the generalized Wilcoxon test) and the Tarone–Ware test. Both tests, as well as the log-rank test, are discussed in Parmar and Machin (26). Like the log-rank test, the Breslow test and the Tarone–Ware test are based on the weighted differences between actual and expected numbers of deaths at

TABLE 12.8.4 Contingency Table for First Stratum (Time Period) for Calculating the Log-Rank Test, Example 12.8.2

	Low-Grade	High-Grade	Total
Deaths	0	1	1
Patients alive	25	13	38
Patients at risk	25	13	39

TABLE 12.8.5 Intermediate Calculations for the Log-Rank Test, Example 12.8.2

Time, t_i	a_i	c_i	$a_i + c_i$	b_i	d_i	$b_i + d_i$	n_i	e_i	γ_i
0	0	25	25	1	13	14	39	0.641	0.230
2	0	25	25	1	12	13	38	0.658	0.225
3	0	25	25	1	11	12	37	0.676	0.219
4	0	25	25	1	10	11	36	0.694	0.212
6	0	25	25	1	9	10	35	0.714	0.204
7	0	24	24	1	8	9	33	0.727	0.198
9	0	23	23	2	6	8	31	1.484	0.370
11	1	22	23	0	6	6	29	0.793	0.164
12	0	22	22	1	5	6	28	0.786	0.168
15	2	20	22	0	5	5	27	1.630	0.290
16	0	20	20	1	4	5	25	0.800	0.160
17	0	20	20	1	3	4	24	0.833	0.139
21	1	19	20	0	3	3	23	0.870	0.113
23	0	19	19	1	2	3	22	0.864	0.118
27	0	18	18	1	1	2	20	0.900	0.090
29	1	17	18	0	1	1	19	0.947	0.050
33	1	16	17	0	1	1	18	0.944	0.052
79	1	13	14	0	1	1	15	0.933	0.062
102	1	10	11	0	1	1	12	0.917	0.076
212	1	1	2	0	0	0	2	1.000	0.000
Totals	9							17.811	3.140

the observed time points. Whereas the log-rank test ranks all deaths equally, the Breslow and Tarone–Ware tests give more weight to early deaths. For Example 12.8.1, SPSS computes a value of 24.93 ($p = .0000$) for the Breslow test and a value of 25.22 ($p = .0000$) for the Tarone–Ware test. Kleinbaum (27) discusses another test called the Peto test. Formulas for this test are found in Parmar and Machin (26). The Peto test also gives more weight to the early part of the survival curve, where we find the larger numbers of subjects at risk. When choosing a test, then, researchers who want to give more weight to the earlier part of the survival curve will select either the Breslow, the Tarone–Ware, or the Peto test. Otherwise, the log-rank test is appropriate.

We have covered only the basic concepts of survival analysis in this section. The reader wishing to pursue the subject in more detail may consult one or more of several books devoted to the topic, such as those by Kleinbaum (27), Lee (28), Marubini and Valsecchi (29), and Parmar and Machin (26).

Computer Analysis

Several of the available statistical software packages, such as SPSS, are capable of performing survival analysis and constructing supporting graphs as described in this section.

A standard SPSS analysis of the data discussed in Examples 12.8.1 and 12.8.2 is shown in Figure 12.8.3. ■

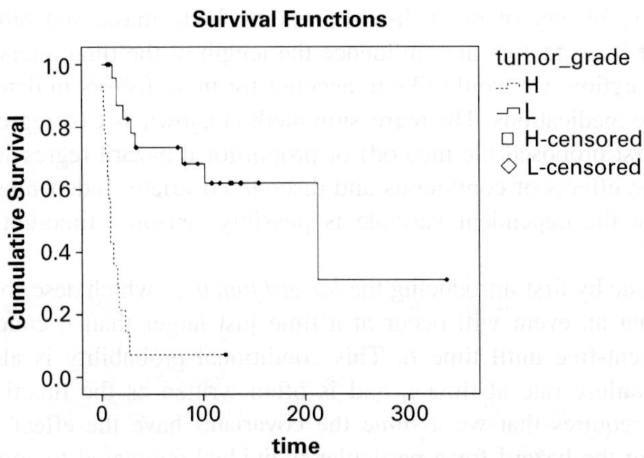

Means and Medians for Survival Time

tumor_grade	Mean[a]				Median			
			95% Confidence Interval				95% Confidence Interval	
	Estimate	Std. Error	Lower Bound	Upper Bound	Estimate	Std. Error	Lower Bound	Upper Bound
H	18.357	8.251	2.186	34.528	9.000	1.852	5.371	12.629
L	88.040	15.258	58.134	117.946	82.000	16.653	49.359	114.641
Overall	63.026	11.490	40.505	85.546	27.000	7.492	12.317	41.683

[a] Estimation is limited to the largest survival time if it is censored.

Overall Comparisons

	Chi-Square	df	Sig.
Log Rank (Mantel-Cox)	24.704	1	.000
Breslow (Generalized Wilcoxon)	24.927	1	.000
Tarone-Ware	25.217	1	.000

Test of equality of survival distributions for the different levels of tumor_grade.

FIGURE 12.8.3 SPSS output for Examples 12.8.1 and 12.8.2.

The Proportional Hazards or Cox Regression Model In previous chapters, we saw that regression models can be used for continuous outcome measures and for binary outcome measures (logistic regression). Additional regression techniques are available when the dependent measures may consist of a mixture of either time-until-event data or censored time observations. Returning to our example of a clinical trial of the effectiveness of two different medications to prevent a second myocardial infarction, we may wish to control for additional characteristics of the subjects enrolled in the study.

For example, we would expect subjects to be different in their baseline systolic blood pressure measurements, family history of heart disease, weight, body mass, and other characteristics. Because all of these factors may influence the length of the time interval until a second myocardial infarction, we would like to account for these factors in determining the effectiveness of the medications. The regression method known as Cox regression (after D. R. Cox, who first proposed the method) or proportional hazard regression can be used to account for the effects of continuous and discrete covariate (independent variable) measurements when the dependent variable is possibly censored time-until-event data.

We describe this technique by first introducing the *hazard function,* which describes the conditional probability that an event will occur at a time just larger than t_i conditional on having survived event-free until time t_i. This conditional probability is also known as the instantaneous failure rate at time t_i and is often written as the function $h(t_i)$. The regression model requires that we assume the covariates have the effect of either increasing or decreasing the hazard for a particular individual compared to some baseline value for the function. In our clinical trial example we might measure k covariates on each of the subjects where there are $I = 1, \ldots, n$ subjects and $h_0(t_i)$ is the baseline hazard function. We describe the regression model as

$$h(t_i) = h_0(t_i)\exp(\beta_1 z_{i1} + \beta_2 z_{i2} + \cdots + \beta_k z_{ik}) \tag{12.8.3}$$

The regression coefficients represent the change in the hazard that results from the risk factor, z_{ik}, that we have measured. Rearranging the above equation shows that the exponentiated coefficient represents the hazard ratio or the ratio of the conditional probabilities of an event. This is the basis for naming this method *proportional hazards regression.* You may recall that this is the same way we obtained the estimate of the odds ratio from the estimated coefficient when we discussed logistic regression in Chapter 11.

$$\frac{h(t_i)}{h_0(t_i)} = \exp(\beta_1 z_{i1} + \beta_2 z_{i2} + \cdots + \beta_k z_{ik}) \tag{12.8.4}$$

Estimating the covariate effects, $\hat{\beta}$ requires the use of a statistical software package because there is no straightforward single equation that will provide the estimates for this regression model. Computer output usually includes estimates of the regression coefficients, standard error estimates, hazard ratio estimates, and confidence intervals. In addition, computer output may also provide graphs of the hazard functions and survival functions for subjects with different covariate values that are useful to compare the effects of covariates on survival. In summary, Cox regression is a useful technique for determining the effects of covariates with survival data. Additional information can be found in the texts by Kleinbaum (27), Lee (28), Kalbfleisch and Prentice (30), Elandt-Johnson and Johnson (31), Cox and Oakes (32), and Fleming and Harrington (33).

EXERCISES

12.8.1 Fifty-three patients with medullary thyroid cancer (MTC) were the subjects of a study by Dottorini et al. (A-24), who evaluated the impact of different clinical and pathological factors and the type of treatment on their survival. Thirty-two of the patients were females, and the mean age of all patients was 46.11 years with a standard deviation of 14.04 (range 18–35 years). The following table shows the status of each patient at various periods of time following surgery. Analyze the data using the techniques discussed in this section.

Subject	Time[a] (Years)	Status[b]	Subject	Time[a] (Years)	Status[b]
1	0	doc	28	6	alive
2	1	mtc	29	6	alive
3	1	mtc	30	6	alive
4	1	mtc	31	6	alive
5	1	mtc	32	7	mtc
6	1	mtc	33	8	alive
7	1	mtc	34	8	alive
8	1	mtc	35	8	alive
9	1	alive	36	8	alive
10	2	mtc	37	8	alive
11	2	mtc	38	9	alive
12	2	mtc	39	10	alive
13	2	alive	40	11	mtc
14	2	alive	41	11	doc
15	3	mtc	42	12	mtc
16	3	mtc	43	12	doc
17	3	alive	44	13	mtc
18	4	mtc	45	14	alive
19	4	alive	46	15	alive
20	4	alive	47	16	mtc
21	4	alive	48	16	alive
22	5	alive	49	16	alive
23	5	alive	50	16	alive
24	5	alive	51	17	doc
25	5	alive	52	18	mtc
26	6	alive	53	19	alive
27	6	alive			

[a] Time is number of years after surgery.

[b] doc = dead of other causes; mtc = dead of medullary thyroid cancer.

Source: Dr. Massimo E. Dottorini. Used with permission.

12.8.2 Banerji et al. (A-25) followed non–insulin-dependent diabetes mellitus (NIDDM) patients from onset of their original hyperglycemia and the inception of their near–normoglycemic remission following treatment. Subjects were black men and women with a mean age of 45.4 years and a standard deviation of 10.4. The following table shows the relapse/remission experience of 62 subjects. Use the techniques covered in this section to analyze these data.

Total Duration of Remission (Months)	Remission Status[a]	Total Duration of Remission (Months)	Remission Status[a]	Total Duration of Remission (Months)	Remission Status[a]
3	1	8	2	26	1
3	2	9	2	27	1
3	1	10	1	28	2
3	1	10	1	29	1
3	1	11	2	31	2
4	1	13	1	31	1
4	1	16	1	33	2
4	1	16	2	39	2
5	1	17	2	41	1
5	1	18	2	44	1
5	1	20	1	46	1
5	1	22	1	46	2
5	1	22	2	48	1
5	1	22	2	48	2
5	1	23	1	48	1
6	1	24	2	49	1
6	1	25	2	50	1
6	1	25	2	53	1
7	1	26	1	70	2
8	2	26	1	94	1
8	1				
8	2				

[a] 1 = yes (the patient is still in remission); 2 = no (the patient has relapsed).

Source: Dr. Mary Ann Banerji. Used with permission.

12.8.3 If available in your library, read the article, "Impact of Obesity on Allogeneic Stem Cell Transplant Patients: A Matched Case-Controlled Study," by Donald R. Fleming et al. [*American Journal of Medicine, 102* (1997), 265–268] and answer the following questions:

(a) How was survival time determined?

(b) Why do you think the authors used the Wilcoxon test (Breslow test) for comparing the survival curves?

(c) Explain the meaning of the p values reported for Figures 1 through 4.

(d) What specific statistical results allow the authors to arrive at their stated conclusion?

12.8.4 If available in your library, read the article, "Improved Survival in Patients with Locally Advanced Prostate Cancer Treated with Radiotherapy and Goserelin," by Michel Bolla et al. [*New England Journal of Medicine, 337* (1997), 295–300], and answer the following questions:

(a) How was survival time determined?

(b) Why do you think the authors used the log-rank test for comparing the survival curves?

(c) Explain the meaning of the p values reported for Figures 1 and 2.
(d) What specific statistical results allow the authors to arrive at their stated conclusion?

12.8.5 Fifty subjects who completed a weight-reduction program at a fitness center were divided into two equal groups. Subjects in group 1 were immediately assigned to a support group that met weekly. Subjects in group 2 did not participate in support group activities. All subjects were followed for a period of 60 weeks. They reported weekly to the fitness center, where they were weighed and a determination was made as to whether they were within goal. Subjects were considered to be within goal if their weekly weight was within 5 pounds of their weight at time of completion of the weight-reduction program. Survival was measured from the date of completion of the weight-reduction program to the termination of follow-up or the point at which the subject exceeded goal. The following results were observed:

Subject	Time (Weeks)	Status (G = Within Goal G+ = Exceeded Goal L = Lost to Follow-Up)	Subject	Time (Weeks)	Status (G = Within Goal G+ = Exceeded Goal L = Lost to Follow-Up)
	Group 1			Group 2	
1	60	G	1	20	G+
2	32	L	2	26	G+
3	60	G	3	10	G+
4	22	L	4	2	G+
5	6	G+	5	36	G+
6	60	G	6	10	G+
7	60	G	7	20	G+
8	20	G+	8	18	L
9	32	G+	9	15	G+
10	60	G	10	22	G+
11	60	G	11	4	G+
12	8	G+	12	12	G+
13	60	G	13	24	G+
14	60	G	14	6	G+
15	60	G	15	18	G+
16	14	L	16	3	G+
17	16	G+	17	27	G+
18	24	L	18	22	G+
19	34	L	19	8	G+
20	60	G	20	10	L
21	40	L	21	32	G+
22	26	L	22	7	G+
23	60	G	23	8	G+
24	60	G	24	28	G+
25	52	L	25	7	G+

Analyze these data using the methods discussed in this section.

12.9 SUMMARY

In this chapter some uses of the versatile chi-square distribution are discussed. Chi-square goodness-of-fit tests applied to the normal, binomial, and Poisson distributions are presented. We see that the procedure consists of computing a statistic

$$X^2 = \sum\left[\frac{(O_i - E_i)^2}{E_i}\right]$$

that measures the discrepancy between the observed (O_i) and expected (E_i) frequencies of occurrence of values in certain discrete categories. When the appropriate null hypothesis is true, this quantity is distributed approximately as χ^2. When X^2 is greater than or equal to the tabulated value of χ^2 for some α, the null hypothesis is rejected at the α level of significance.

Tests of independence and tests of homogeneity are also discussed in this chapter. The tests are mathematically equivalent but conceptually different. Again, these tests essentially test the goodness-of-fit of observed data to expectation under hypotheses, respectively, of independence of two criteria of classifying the data and the homogeneity of proportions among two or more groups.

In addition, we discussed and illustrated in this chapter four other techniques for analyzing frequency data that can be presented in the form of a 2 × 2 contingency table: the Fisher exact test, the odds ratio, relative risk, and the Mantel–Haenszel procedure. Finally, we discussed the basic concepts of survival analysis and illustrated the computational procedures by means of two examples.

SUMMARY OF FORMULAS FOR CHAPTER 12

Formula Number	Name	Formula
12.2.1	Standard normal random variable	$z_i = \frac{y_i - \mu}{\sigma}$
12.2.2	Chi-square distribution with n degrees of freedom	$\chi^2_{(n)} = z_1^2 + z_2^2 + \cdots + z_n^2$
12.2.3	Chi-square probability density function	$f(u) = \frac{1}{\left(\frac{k}{2} - 1\right)!}\frac{1}{2^{k/2}}u^{(k/2)-1}e^{-(u/2)}$
12.2.4	Chi-square test statistic	$\chi^2 = \sum\left[\frac{(O_i - E_i)^2}{E_i}\right]$
12.4.1	Chi-square calculation formula for a 2 × 2 contingency table	$\chi^2 = \frac{n(ad - bc)^2}{(a + c)(b + d)(a + b)(c + d)}$

12.4.2	Yates's corrected chi-square calculation for a 2 × 2 contingency table	$\chi^2_{\text{corrected}} = \dfrac{n(\lvert ad - bc\rvert - .5n)^2}{(a+c)(b+d)(a+b)(c+d)}$
12.6.1–12.6.2	Large-sample approximation to the chi-square	$z = \dfrac{(a/A) - (b/B)}{\sqrt{\hat{p}(1-\hat{p})(1/A + 1/B)}}$ where $\hat{p} = (a+b)/(A+B)$
12.7.1	Relative risk estimate	$\widehat{RR} = \dfrac{a/(a+b)}{c/(c+d)}$
12.7.2	Confidence interval for the relative risk estimate	$100(1-\alpha)\%CI = \widehat{RR}^{1 \pm (z_\alpha/\sqrt{x^2})}$
12.7.3	Odds ratio estimate	$\widehat{OR} = \dfrac{a/b}{c/d} = \dfrac{ad}{bc}$
12.7.4	Confidence interval for the odds ratio estimate	$100(1-\alpha)\%CI = \widehat{OR}^{1 \pm (z_\alpha/\sqrt{x^2})}$
12.7.5	Expected frequency in the Mantel–Haenszel statistic	$e_i = \dfrac{(a_i + b_i)(a_i + c_i)}{n_i}$
12.7.6	Stratum expected frequency in the Mantel–Haenszel statistic	$v_i = \dfrac{(a_i + b_i)(c_i + d_i)(a_i + c_i)(b_i + d_i)}{n_i^2(n_i - 1)}$
12.7.7	Mantel–Haenszel test statistic	$\chi^2_{\text{MH}} = \dfrac{\left(\sum_{i=1}^{k} a_i - \sum_{i=1}^{k} e_i\right)^2}{\sum_{i=1}^{k} v_i}$
12.7.8	Mantel–Haenszel estimator of the common odds ratio	$\widehat{OR}_{\text{MH}} = \dfrac{\sum_{i=1}^{k}(a_i d_i/n_i)}{\sum_{i=1}^{k}(b_i c_i/n_i)}$

(Continued)

12.8.1	Survival probability	$\hat{p}_1 = \dfrac{\text{number of subjects surviving at least } (t-1) \text{ time period who also survive } t^{th} \textit{ period}}{\text{number of subjects alive at end of time period } (t-1)}$
12.8.2	Estimated survival function	$\hat{S}(t) = \hat{p}_1 \times \hat{p}_2 \times \cdots \times \hat{p}_t$
12.8.3	Hazard regression model	$h(t_i) = h_0(t_i)\exp(\beta_1 z_{i1} + \beta_2 z_{i2} + \cdots + \beta_k z_{ik})$
12.8.4	Proportional hazard model	$\dfrac{h(t_i)}{h_0(t_i)} = \exp(\beta_1 z_{i1} + \beta_2 z_{i2} + \cdots + \beta_k z_{ik})$
Symbol Key		• a, b, c, d = cell frequencies in a 2×2 contingency table • A, B = row totals in the 2×2 contingency table • β = regression coefficient • χ^2 (or X^2) = chi-square • e_i = expected frequency in the Mantel–Haenszel statistic • E_i = expected frequency • $E_{(y\|x)}$ = expected value of y at x • k = degrees of freedom in the chi-square distribution • μ = mean • O_i = observed frequency • $\widehat{OR}$ = odds ratio estimate • σ = standard deviation • $\widehat{RR}$ = relative risk estimate • v_i = stratum expected frequency in the Mantel–Haenszel statistic • y_i = data value at point i • z = normal variate

REVIEW QUESTIONS AND EXERCISES

1. Explain how the chi-square distribution may be derived.

2. What are the mean and variance of the chi-square distribution?

3. Explain how the degrees of freedom are computed for the chi-square goodness-of-fit tests.

4. State Cochran's rule for small expected frequencies in goodness-of-fit tests.

5. How does one adjust for small expected frequencies?

6. What is a contingency table?

7. How are the degrees of freedom computed when an X^2 value is computed from a contingency table?

8. Explain the rationale behind the method of computing the expected frequencies in a test of independence.

9. Explain the difference between a test of independence and a test of homogeneity.

10. Explain the rationale behind the method of computing the expected frequencies in a test of homogeneity.

11. When do researchers use the Fisher exact test rather than the chi-square test?

12. Define the following:

(a) Observational study **(b)** Risk factor
(c) Outcome **(d)** Retrospective study
(e) Prospective study **(f)** Relative risk
(g) Odds **(h)** Odds ratio
(i) Confounding variable

13. Under what conditions is the Mantel–Haenszel test appropriate?

14. Explain how researchers interpret the following measures:

(a) Relative risk
(b) Odds ratio
(c) Mantel–Haenszel common odds ratio

15. In a study of violent victimization of women and men, Porcerelli et al. (A-26) collected information from 679 women and 345 men ages 18 to 64 years at several family practice centers in the metropolitan Detroit area. Patients filled out a health history questionnaire that included a question about victimization. The following table shows the sample subjects cross-classified by gender and the type of violent victimization reported. The victimization categories are defined as no victimization, partner victimization (and not by others), victimization by a person other than a partner (friend, family member, or stranger), and those who reported multiple victimization.

Gender	No Victimization	Partner	Nonpartner	Multiple	Total
Women	611	34	16	18	679
Men	308	10	17	10	345
Total	919	44	33	28	1024

Source: John H. Porcerelli, Rosemary Cogan, Patricia P. West, Edward A. Rose, Dawn Lambrecht, Karen E. Wilson, Richard K. Severson, and Dunia Karana, "Violent Victimization of Women and Men: Physical and Psychiatric Symptoms," *Journal of the American Board of Family Practice, 16* (2003), 32–39.

Can we conclude on the basis of these data that victimization status and gender are not independent? Let $\alpha = .05$.

16. Refer to Exercise 15. The following table shows data reported by Porcerelli et al. for 644 African-American and Caucasian women. May we conclude on the basis of these data that for women, race and victimization status are not independent? Let $\alpha = .05$.

	No Victimization	Partner	Nonpartner	Multiple	Total
Caucasian	356	20	3	9	388
African-American	226	11	10	9	256
Total	582	31	13	18	644

Source: John H. Porcerelli, Rosemary Cogan, Patricia P. West, Edward A. Rose, Dawn Lambrecht, Karen E. Wilson, Richard K. Severson, and Dunia Karana, "Violent Victimization of Women and Men: Physical and Psychiatric Symptoms," *Journal of the American Board of Family Practice, 16* (2003), 32–39.

17. A sample of 150 chronic carriers of a certain antigen and a sample of 500 noncarriers revealed the following blood group distributions:

Blood Group	Carriers	Noncarriers	Total
0	72	230	302
A	54	192	246
B	16	63	79
AB	8	15	23
Total	150	500	650

Can one conclude from these data that the two populations from which the samples were drawn differ with respect to blood group distribution? Let $\alpha = .05$. What is the p value for the test?

18. The following table shows 200 males classified according to social class and headache status:

	Social Class			
Headache Group	A	B	C	Total
No headache (in previous year)	6	30	22	58
Simple headache	11	35	17	63
Unilateral headache (nonmigraine)	4	19	14	37
Migraine	5	25	12	42
Total	26	109	65	200

Do these data provide sufficient evidence to indicate that headache status and social class are related? Let $\alpha = .05$. What is the p value for this test?

19. The following is the frequency distribution of scores made on an aptitude test by 175 applicants to a physical therapy training facility ($\bar{x} = 39.71, s = 12.92$).

Score	Number of Applicants	Score	Number of Applicants
10–14	3	40–44	28
15–19	8	45–49	20

(Continued)

Score	Number of Applicants	Score	Number of Applicants
20–24	13	50–54	18
25–29	17	55–59	12
30–34	19	60–64	8
35–39	25	65–69	4
Total			175

Do these data provide sufficient evidence to indicate that the population of scores is not normally distributed? Let $\alpha = .05$. What is the p value for this test?

20. A local health department sponsored a venereal disease (VD) information program that was open to high-school juniors and seniors who ranged in age from 16 to 19 years. The program director believed that each age level was equally interested in knowing more about VD. Since each age level was about equally represented in the area served, she felt that equal interest in VD would be reflected by equal age-level attendance at the program. The age breakdown of those attending was as follows:

Age	Number Attending
16	26
17	50
18	44
19	40

Are these data incompatible with the program director's belief that students in the four age levels are equally interested in VD? Let $\alpha = .05$. What is the p value for this test?

21. A survey of children under 15 years of age residing in the inner-city area of a large city were classified according to ethnic group and hemoglobin level. The results were as follows:

	Hemoglobin Level (g/100 ml)			
Ethnic Group	**10.0 or Greater**	**9.0–9.9**	**<9.0**	**Total**
A	80	100	20	200
B	99	190	96	385
C	70	30	10	110
Total	249	320	126	695

Do these data provide sufficient evidence to indicate, at the .05 level of significance, that the two variables are related? What is the p value for this test?

22. A sample of reported cases of mumps in preschool children showed the following distribution by age:

Age (Years)	Number of Cases
Under 1	6
1	20
2	35
3	41
4	48
Total	150

Test the hypothesis that cases occur with equal frequency in the five age categories. Let $\alpha = .05$. What is the p value for this test?

23. Each of a sample of 250 men drawn from a population of suspected joint disease victims was asked which of three symptoms bother him most. The same question was asked of a sample of 300 suspected women joint disease victims. The results were as follows:

Most Bothersome Symptom	Men	Women
Morning stiffness	111	102
Nocturnal pain	59	73
Joint swelling	80	125
Total	250	300

Do these data provide sufficient evidence to indicate that the two populations are not homogeneous with respect to major symptoms? Let $\alpha = .05$. What is the p value for this test?

For each of the Exercises 24 through 34, indicate whether a null hypothesis of homogeneity or a null hypothesis of independence is appropriate.

24. A researcher wishes to compare the status of three communities with respect to immunity against polio in preschool children. A sample of preschool children was drawn from each of the three communities.

25. In a study of the relationship between smoking and respiratory illness, a random sample of adults were classified according to consumption of tobacco and extent of respiratory symptoms.

26. A physician who wished to know more about the relationship between smoking and birth defects studies the health records of a sample of mothers and their children, including stillbirths and spontaneously aborted fetuses where possible.

27. A health research team believes that the incidence of depression is higher among people with hypoglycemia than among people who do not suffer from this condition.

28. In a simple random sample of 200 patients undergoing therapy at a drug abuse treatment center, 60 percent belonged to ethnic group I. The remainder belonged to ethnic group II. In ethnic group I, 60 were being treated for alcohol abuse (A), 25 for marijuana abuse (B), and 20 for abuse of heroin, illegal methadone, or some other opioid (C). The remainder had abused barbiturates, cocaine, amphetamines, hallucinogens, or some other nonopioid besides marijuana (D). In ethnic group II the abused drug category and the numbers involved were as follows:

A(28) B(32) C(13) D(the remainder)

Can one conclude from these data that there is a relationship between ethnic group and choice of drug to abuse? Let $\alpha = .05$ and find the p value.

29. Solar keratoses are skin lesions commonly found on the scalp, face, backs of hands, forearms, ears, scalp, and neck. They are caused by long-term sun exposure, but they are not skin cancers. Chen et al. (A-27) studied 39 subjects randomly assigned (with a 3 to 1 ratio) to imiquimod cream and a control cream. The criterion for effectiveness was having 75 percent or more of the lesion area cleared after 14 weeks of treatment. There were 21 successes among 29 imiquimod-treated subjects and three successes among 10 subjects using the control cream. The researchers used Fisher's exact test and obtained a p value of .027. What are the variables involved? Are the variables quantitative or qualitative? What null and alternative hypotheses are appropriate? What are your conclusions?

30. Janardhan et al. (A-28) examined 125 patients who underwent surgical or endovascular treatment for intracranial aneurysms. At 30 days postprocedure, 17 subjects experienced transient/persistent neurological deficits. The researchers performed logistic regression and found that the 95 percent confidence interval for the odds ratio for aneurysm size was .09–.96. Aneurysm size was dichotomized as less than 13 mm and greater than or equal to 13 mm. The larger tumors indicated higher odds of deficits. Describe the variables as to whether they are continuous, discrete, quantitative, or qualitative. What conclusions may be drawn from the given information?

31. In a study of smoking cessation by Gold et al. (A-29), 189 subjects self-selected into three treatments: nicotine patch only (NTP), Bupropion SR only (B), and nicotine patch with Bupropion SR (NTP + B). Subjects were grouped by age into younger than 50 years old, between 50 and 64, and 65 and older. There were 15 subjects younger than 50 years old who chose NTP, 26 who chose B, and 16 who chose NTP + B. In the 50–64 years category, six chose NTP, 54 chose B, and 40 chose NTP + B. In the oldest age category, six chose NTP, 21 chose B, and five chose NTP + B. What statistical technique studied in this chapter would be appropriate for analyzing these data? Describe the variables involved as to whether they are continuous, discrete, quantitative, or qualitative. What null and alternative hypotheses are appropriate? If you think you have sufficient information, conduct a complete hypothesis test. What are your conclusions?

32. Kozinszky and Bártai (A-30) examined contraceptive use by teenage girls requesting abortion in Szeged, Hungary. Subjects were classified as younger than 20 years old or 20 years old or older. Of the younger than 20-year-old women, 146 requested an abortion. Of the older group, 1054 requested an abortion. A control group consisted of visitors to the family planning center who did not request an abortion or persons accompanying women who requested an abortion. In the control group, there were 147 women under 20 years of age and 1053 who were 20 years or older. One of the outcome variables of interest was knowledge of emergency contraception. The researchers report that, "Emergency contraception was significantly [(Mantel–Haenszel) $p < .001$] less well known among the would-be aborter teenagers as compared to the older women requesting artificial abortion (OR = .07) than the relevant knowledge of the teenage controls (OR = .10)." Explain the meaning of the reported statistics. What are your conclusions based on the given information?

33. The goal of a study by Crosignani et al. (A-31) was to assess the effect of road traffic exhaust on the risk of childhood leukemia. They studied 120 children in Northern Italy identified through a population-based cancer registry (cases). Four controls per case, matched by age and gender, were sampled from population files. The researchers used a diffusion model of benzene to estimate exposure to traffic exhaust. Compared to children whose homes were not exposed to road traffic emissions, the rate of childhood leukemia was significantly higher for heavily exposed children. Characterize this study as to whether it is observational, prospective, or retrospective. Describe the variables as to whether they are continuous, discrete, quantitative, qualitative, a risk factor, or a confounding variable. Explain the meaning of the reported results. What are your conclusions based on the given information?

34. Gallagher et al. (A-32) conducted a descriptive study to identify factors that influence women's attendance at cardiac rehabilitation programs following a cardiac event. One outcome variable of interest was actual attendance at such a program. The researchers enrolled women discharged from four metropolitan hospitals in Sydney, Australia. Of 183 women, only 57 women actually attended programs. The authors reported odds ratios and confidence intervals on the following variables that significantly affected outcome: age-squared (1.72; 1.10–2.70). Women over the age of 70 had the lowest odds, while women ages 55–70 years had the highest odds.), perceived control (.92; .85–1.00), employment (.20; .07–.58), diagnosis (6.82, 1.84–25.21, odds ratio was higher for women who experienced coronary artery bypass grafting vs. myocardial infarction), and stressful event (.21, .06–.73). Characterize this study as to whether it is observational, prospective, or retrospective. Describe the variables as to whether they are continuous, discrete, quantitative, qualitative, a risk factor, or a confounding variable. Explain the meaning of the reported odds ratios.

For each of the Exercises 35 through 54, do as many of the following as you think appropriate:

(a) Apply one or more of the techniques discussed in this chapter.
(b) Apply one or more of the techniques discussed in previous chapters.
(c) Construct graphs.
(d) Construct confidence intervals for population parameters.
(e) Formulate relevant hypotheses, perform the appropriate tests, and find p values.
(f) State the statistical decisions and clinical conclusions that the results of your hypothesis tests justify.
(g) Describe the population(s) to which you think your inferences are applicable.
(h) State the assumptions necessary for the validity of your analyses.

35. In a prospective, randomized, double-blind study, Stanley et al. (A-33) examined the relative efficacy and side effects of morphine and pethidine, drugs commonly used for patient-controlled analgesia (PCA). Subjects were 40 women, between the ages of 20 and 65 years, undergoing total abdominal hysterectomy. Patients were allocated randomly to receive morphine or pethidine by PCA. At the end of the study, subjects described their appreciation of nausea and vomiting, pain, and satisfaction by means of a three-point verbal scale. The results were as follows:

	Satisfaction			
Drug	**Unhappy/ Miserable**	**Moderately Happy**	**Happy/ Delighted**	**Total**
Pethidine	5	9	6	20
Morphine	9	9	2	20
Total	14	18	8	40

	Pain			
Drug	**Unbearable/ Severe**	**Moderate**	**Slight/ None**	**Total**
Pethidine	2	10	8	20
Morphine	2	8	10	20
Total	4	18	18	40

	Nausea			
Drug	**Unbearable/ Severe**	**Moderate**	**Slight/ None**	**Total**
Pethidine	5	9	6	20
Morphine	7	8	5	20
Total	12	17	11	40

Source: Dr. Balraj L. Appadu. Used with permission.

36. Screening data from a statewide lead poisoning prevention program between April 1990 and March 1991 were examined by Sargent et al. (A-34) in an effort to learn more about community risk factors for iron deficiency in young children. Study subjects ranged in age between 6 and 59 months. Among 1860 children with Hispanic surnames, 338 had iron deficiency. Four-hundred-fifty-seven of 1139 with Southeast Asian surnames and 1034 of 8814 children with other surnames had iron deficiency.

37. To increase understanding of HIV-infection risk among patients with severe mental illness, Horwath et al. (A-35) conducted a study to identify predictors of injection drug use among patients who did not have a primary substance use disorder. Of 192 patients recruited from inpatient and outpatient public psychiatric facilities, 123 were males. Twenty-nine of the males and nine of the females were found to have a history of illicit-drug injection.

38. Skinner et al. (A-36) conducted a clinical trial to determine whether treatment with melphalan, prednisone, and colchicine (MPC) is superior to colchicine (C) alone. Subjects consisted of 100 patients with primary amyloidosis. Fifty were treated with C and 50 with MPC. Eighteen months after the last person was admitted and 6 years after the trial began, 44 of those receiving C and 36 of those receiving MPC had died.

39. The purpose of a study by Miyajima et al. (A-37) was to evaluate the changes of tumor cell contamination in bone marrow (BM) and peripheral blood (PB) during the clinical course of patients with advanced neuroblastoma. Their procedure involved detecting tyrosine hydroxylase (TH) mRNA to clarify the appropriate source and time for harvesting hematopoietic stem cells for transplantation. The authors used Fisher's exact test in the analysis of their data. If available, read their article and decide if you agree that Fisher's exact text was the appropriate technique to use. If you agree, duplicate their procedure and see if you get the same results. If you disagree, explain why.

40. Cohen et al. (A-38) investigated the relationship between HIV seropositivity and bacterial vaginosis in a population at high risk for sexual acquisition of HIV. Subjects were 144 female commercial sex workers in Thailand of whom 62 were HIV-positive and 109 had a history of sexually transmitted diseases (STD). In the HIV-negative group, 51 had a history of STD.

41. The purpose of a study by Lipschitz et al. (A-39) was to examine, using a questionnaire, the rates and characteristics of childhood abuse and adult assaults in a large general outpatient population. Subjects consisted of 120 psychiatric outpatients (86 females, 34 males) in treatment at a large hospital-based clinic in an inner-city area. Forty-seven females and six males reported incidents of childhood sexual abuse.

42. Subjects of a study by O'Brien et al. (A-40) consisted of 100 low-risk patients having well-dated pregnancies. The investigators wished to evaluate the efficacy of a more gradual method for promoting cervical change and delivery. Half of the patients were randomly assigned to receive a placebo, and the remainder received 2 mg of intravaginal prostaglandin E_2 (PGE_2) for 5 consecutive days. One of the infants born to mothers in the experimental group and four born to those in the control group had macrosomia.

43. The purposes of a study by Adra et al. (A-41) were to assess the influence of route of delivery on neonatal outcome in fetuses with gastroschisis and to correlate ultrasonographic appearance of the fetal bowel with immediate postnatal outcome. Among 27 cases of prenatally diagnosed gastroschisis the ultrasonograph appearance of the fetal bowel was normal in 15. Postoperative complications were observed in two of the 15 and in seven of the cases in which the ultrasonographic appearance was not normal.

44. Liu et al. (A-42) conducted household surveys in areas of Alabama under tornado warnings. In one of the surveys (survey 2) the mean age of the 193 interviewees was 54 years. Of these 56.0 percent were women, 88.6 percent were white, and 83.4 percent had a high-school education or higher. Among the information collected were data on shelter-seeking activity and understanding of the term "tornado warning." One-hundred-twenty-eight respondents indicated that they usually seek shelter when made aware of a tornado warning. Of these, 118 understood the meaning of tornado warning. Forty-six of those who said they didn't usually seek shelter understood the meaning of the term.

45. The purposes of a study by Patel et al. (A-43) were to investigate the incidence of acute angle-closure glaucoma secondary to pupillary dilation and to identify screening methods for detecting angles at risk of occlusion. Of 5308 subjects studied, 1287 were 70 years of age or older. Seventeen of the older subjects and 21 of the younger subjects (40 through 69 years of age) were identified as having potentially occludable angles.

46. Voskuyl et al. (A-44) investigated those characteristics (including male gender) of patients with rheumatoid arthritis (RA) that are associated with the development of rheumatoid vasculitis (RV). Subjects consisted of 69 patients who had been diagnosed as having RV and 138 patients with RA who were not suspected to have vasculitis. There were 32 males in the RV group and 38 among the RA patients.

47. Harris et al. (A-45) conducted a study to compare the efficacy of anterior colporrhaphy and retropubic urethropexy performed for genuine stress urinary incontinence. The subjects were 76 women who had undergone one or the other surgery. Subjects in each group were comparable in age, social status, race, parity, and weight. In 22 of the 41 cases reported as cured the surgery had been performed by attending staff. In 10 of the failures, surgery had been performed by attending staff. All other surgeries had been performed by resident surgeons.

48. Kohashi et al. (A-46) conducted a study in which the subjects were patients with scoliosis. As part of the study, 21 patients treated with braces were divided into two groups, group $A(n_A = 12)$ and group $B(n_B = 9)$, on the basis of certain scoliosis progression factors. Two patients in group A and eight in group B exhibited evidence of progressive deformity, while the others did not.

49. In a study of patients with cervical intraepithelial neoplasia, Burger et al. (A-47) compared those who were human papillomavirus (HPV)-positive and those who were HPV-negative with respect to risk factors for HPV infection. Among their findings were 60 out of 91 nonsmokers with HPV infection and 44 HPV-positive patients out of 50 who smoked 21 or more cigarettes per day.

50. Thomas et al. (A-48) conducted a study to determine the correlates of compliance with follow-up appointments and prescription filling after an emergency department visit. Among 235 respondents, 158 kept their appointments. Of these, 98 were females. Of those who missed their appointments, 31 were males.

51. The subjects of a study conducted by O'Keefe and Lavan (A-49) were 60 patients with cognitive impairment who required parenteral fluids for at least 48 hours. The patients were randomly assigned to receive either intravenous (IV) or subcutaneous (SC) fluids. The mean age of the 30 patients in the SC group was 81 years with a standard deviation of 6. Fifty-seven percent were females. The mean age of the IV group was 84 years with a standard deviation of 7. Agitation related to the cannula or drip was observed in 11 of the SC patients and 24 of the IV patients.

52. The objective of a study by Lee et al. (A-50) was to improve understanding of the biologic behavior of gastric epithelioid stromal tumors. They studied the clinical features, histologic findings, and DNA ploidy of a series of the tumors to identify factors that might distinguish between benign and malignant variants of these tumors and have relevance for prognosis. Fifty-five patients with tumors were classified on the basis of whether their tumors were high-grade malignant (grade 2), low-grade malignant (grade 1), or benign (grade 0). Among the data collected were the following:

Patient	Tumor Grade	Outcome (1 = Death from Disease)	Number of Days to Last Follow-Up or Death	Patient	Tumor Grade	Outcome (1 = Death from Disease)	Number of Days to Last Follow-Up or Death
1	0	0	87	8	0	0	1616
2	0	0	775	9	0	0	1982
3	0	0	881	10	0	0	2035
4	0	0	914	11	0	0	2191
5	0	0	1155	12	0	0	2472
6	0	0	1162	13	0	0	2527
7	0	0	1271	14	0	0	2782
15	0	0	3108	36	0	0	7318
16	0	0	3158	37	0	0	7447
17	0	0	3609	38	0	0	9525
18	0	0	3772	39	0	0	9938
19	0	0	3799	40	0	0	10429

(*Continued*)

Patient	Tumor Grade	Outcome (1 = Death from Disease)	Number of Days to Last Follow-Up or Death	Patient	Tumor Grade	Outcome (1 = Death from Disease)	Number of Days to Last Follow-Up or Death
20	0	0	3819	41	1	1	450
21	0	0	4586	42	1	1	556
22	0	0	4680	43	1	1	2102
23	0	0	4989	44	1	0	2756
24	0	0	5675	45	1	0	3496
25	0	0	5936	46	1	1	3990
26	0	0	5985	47	1	0	5686
27	0	0	6175	48	1	0	6290
28	0	0	6177	49	1	0	8490
29	0	0	6214	50	2	1	106
30	0	0	6225	51	2	1	169
31	0	0	6449	52	2	1	306
32	0	0	6669	53	2	1	348
33	0	0	6685	54	2	1	549
34	0	0	6873	55	2	1	973
35	0	0	6951				

Source: Dr. Michael B. Farnell. Used with permission.

53. Girard et al. (A-51) conducted a study to identify prognostic factors of improved survival after resection of isolated pulmonary metastases (PM) from colorectal cancer. Among the data collected were the following regarding number of resected PM, survival, and outcome for 77 patients who underwent a complete resection at the first thoracic operation:

Patient	Number of Resected PM	Survival (Months)	Status	Patient	Number of Resected PM	Survival (Months)	Status
1	1	24	Alive	8	1	15	Dead
2	1	67	Alive	9	1	10	Dead
3	1	42	Alive	10	1	41	Dead
4	>1	28	Dead	11	>1	41	Dead
5	1	37	Dead	12	1	27	Dead
6	1	133	Alive	13	1	93	Alive
7	1	33	Dead	14	>1	0	Dead
15	1	60	Dead	47	1	54	Dead
16	1	43	Dead	48	>1	57	Alive
17	>1	73	Alive	49	>1	16	Dead
18	1	55	Alive	50	1	29	Dead
19	1	46	Dead	51	1	14	Dead
20	1	66	Alive	52	>1	29	Dead
21	1	10	Dead	53	>1	99	Dead
22	>1	3	Dead	54	>1	23	Dead

(Continued)

Patient	Number of Resected PM	Survival (Months)	Status	Patient	Number of Resected PM	Survival (Months)	Status
23	>1	7	Dead	55	1	74	Alive
24	>1	129	Alive	56	1	169	Alive
25	1	19	Alive	57	>1	24	Dead
26	>1	15	Dead	58	>1	9	Dead
27	1	39	Alive	59	1	43	Dead
28	1	15	Dead	60	1	3	Alive
29	>1	30	Dead	61	>1	20	Dead
30	1	35	Alive	62	1	2	Dead
31	>1	18	Dead	63	>1	41	Dead
32	1	27	Dead	64	>1	27	Dead
33	1	121	Alive	65	1	45	Alive
34	>1	8	Dead	66	1	26	Dead
35	1	24	Alive	67	>1	10	Dead
36	1	127	Alive	68	1	143	Alive
37	1	26	Dead	69	1	16	Dead
38	>1	7	Dead	70	1	29	Alive
39	>1	26	Dead	71	1	17	Dead
40	>1	17	Dead	72	>1	20	Dead
41	1	18	Dead	73	1	92	Alive
42	1	17	Dead	74	>1	15	Dead
43	>1	10	Dead	75	1	5	Dead
44	>1	33	Dead	76	>1	73	Alive
45	>1	42	Alive	77	1	19	Dead
46	1	40	Alive				

Source: Dr. Philippe Girard. Used with permission.

Exercises for Use with the Large Data Sets Available on the Following Website: www.wiley.com/college/daniel

1. Refer to the data on smoking, alcohol consumption, blood pressure, and respiratory disease among 1200 adults (SMOKING). The variables are as follows:

Sex (A): 1 = male, 0 = female
Smoking status (B): 0 = nonsmoker, 1 = smoker
Drinking level (C): 0 = nondrinker
1 = light to moderate drinker
2 = heavy drinker
Symptoms of respiratory disease (D): 1 = present, 0 = absent
High blood pressure status (E): 1 = present, 0 = absent

Select a simple random sample of size 100 from this population and carry out an analysis to see if you can conclude that there is a relationship between smoking status and symptoms of respiratory disease. Let $\alpha = .05$ and determine the p value for your test. Compare your results with those of your classmates.

2. Refer to Exercise 1. Select a simple random sample of size 100 from the population and carry out a test to see if you can conclude that there is a relationship between drinking status and high blood pressure status in the population. Let $\alpha = .05$ and determine the p value. Compare your results with those of your classmates.

3. Refer to Exercise 1. Select a simple random sample of size 100 from the population and carry out a test to see if you can conclude that there is a relationship between gender and smoking status in the population. Let $\alpha = .05$ and determine the p value. Compare your results with those of your classmates.

4. Refer to Exercise 1. Select a simple random sample of size 100 from the population and carry out a test to see if you can conclude that there is a relationship between gender and drinking level in the population. Let $\alpha = .05$ and find the p value. Compare your results with those of your classmates.

REFERENCES

Methodology References

1. KARL PEARSON, "On the Criterion that a Given System of Deviations from the Probable in the Case of a Correlated System of Variables Is Such that It Can Be Reasonably Supposed to Have Arisen from Random Sampling," *The London, Edinburgh and Dublin Philosophical Magazine and Journal of Science,* Fifth Series, *50* (1900), 157–175. Reprinted in *Karl Pearson's Early Statistical Papers,* Cambridge University Press, 1948.
2. H. O. LANCASTER, *The Chi-Squared Distribution,* Wiley, New York, 1969.
3. MIKHAIL S. NIKULIN and PRISCILLA E. GREENWOOD, *A Guide to Chi-Squared Testing,* Wiley, New York, 1996.
4. WILLIAM G. COCHRAN, "The χ^2 Test of Goodness of Fit," *Annals of Mathematical Statistics, 23* (1952), 315–345.
5. WILLIAM G. COCHRAN, "Some Methods for Strengthening the Common χ^2 Tests," *Biometrics, 10* (1954), 417–451.
6. F. YATES, "Contingency Tables Involving Small Numbers and the χ^2 Tests," *Journal of the Royal Statistical Society, Supplement, 1,* 1934 (Series B), 217–235.
7. R. A. FISHER, *Statistical Methods for Research Workers,* Fifth Edition, Oliver and Boyd, Edinburgh, 1934.
8. R. A. FISHER, "The Logic of Inductive Inference," *Journal of the Royal Statistical Society Series A, 98* (1935), 39–54.
9. J. O. IRWIN, "Tests of Significance for Differences Between Percentages Based on Small Numbers," *Metron, 12* (1935), 83–94.
10. F. YATES, "Contingency Tables Involving Small Numbers and the χ^2 Test," *Journal of the Royal Statistical Society, Supplement, 1* (1934), 217–235.
11. D. J. FINNEY, "The Fisher-Yates Test of Significance in 2 × 2 Contingency Tables," *Biometrika, 35* (1948), 145–156.
12. R. LATSCHA, "Tests of Significance in a 2 × 2 Contingency Table: Extension of Finney's Table," *Biometrika, 40* (1955), 74–86.
13. G. A. BARNARD, "A New Test for 2 × 2 Tables," *Nature, 156* (1945), 117.
14. G. A. BARNARD, "A New Test for 2 × 2 Tables," *Nature, 156* (1945), 783–784.
15. G. A. BARNARD, "Significance Tests for 2 × 2 Tables," *Biometrika, 34* (1947), 123–138.
16. R. A. FISHER, "A New Test for 2 × 2 Tables," *Nature, 156* (1945), 388.
17. E. S. PEARSON, "The Choice of Statistical Tests Illustrated on the Interpretation of Data Classed in a 2 × 2 Table," *Biometrika, 34* (1947), 139–167.
18. A. SWEETLAND, "A Comparison of the Chi-Square Test for 1 df and the Fisher Exact Test," Rand Corporation, Santa Monica, CA, 1972.

19. WENDELL E. CARR, "Fisher's Exact Text Extended to More than Two Samples of Equal Size," *Technometrics, 22* (1980), 269–270.
20. HENRY R. NEAVE, "A New Look at an Old Test," *Bulletin of Applied Statistics, 9* (1982), 165–178.
21. WILLIAM D. DUPONT, "Sensitivity of Fisher's Exact Text to Minor Perturbations in 2 × 2 Contingency Tables," *Statistics in Medicine, 5* (1986), 629–635.
22. N. MANTEL and W. HAENSZEL, "Statistical Aspects of the Analysis of Data from Retrospective Studies of Disease," *Journal of the National Cancer Institute, 22* (1959), 719–748.
23. N. MANTEL, "Chi-Square Tests with One Degree of Freedom: Extensions of the Mantel-Haenszel Procedure," *Journal of the American Statistical Association, 58* (1963), 690–700.
24. E. L. KAPLAN and P. MEIER, "Nonparametric Estimation from Incomplete Observations," *Journal of the American Statistical Association, 53* (1958), 457–481.
25. NATHAN MANTEL, "Evaluation of Survival Data and Two New Rank Order Statistics Arising in Its Consideration," *Cancer Chemotherapy Reports, 50* (March 1966), 163–170.
26. MAHESH K. B. PARMAR and DAVID MACHIN, *Survival Analysis: A Practical Approach,* Wiley, New York, 1995.
27. DAVID G. KLEINBAUM, *Survival Analysis: A Self-Learning Text,* Springer, New York, 1996.
28. ELISA T. LEE, *Statistical Methods for Survival Data Analysis,* Lifetime Learning Publications, Belmont, CA, 1980.
29. ETTORE MARUBINI and MARIA GRAZIA VALSECCHI, *Analysing Survival Data from Clinical Trials and Observational Studies,* Wiley, New York, 1995.
30. J. D. KALBFLEISCH and R. L. PRENTICE, *The Statistical Analysis of Failure Time Data,* Wiley, New York, 1980.
31. REGINA C. ELANDT-JOHNSON and NORMAN L. JOHNSON, *Survival Models and Data Analysis,* Wiley, New York, 1980.
32. D. R. COX and D. OAKES, *Analysis of Survival Data,* Chapman and Hall, London, 1984.
33. THOMAS R. FLEMING and DAVID P. HARRINGTON, *Counting Processes and Survival Analysis,* Wiley, New York, 1991.
34. DAVID W. HOSMER, JR. and STANLEY LEMESHOW, *Applied Survival Analysis: Regression Modeling of Time to Event Data,* Wiley, New York, 1999.

Applications References

A-1. CAROLE W. CRANOR and DALE B. CHRISTENSEN, "The Asheville Project: Short-Term Outcomes of a Community Pharmacy Diabetes Care Program," *Journal of the American Pharmaceutical Association, 43* (2003), 149–159.
A-2. AMY L. BYERS, HEATHER ALLORE, THOMAS M. GILL, and PETER N. PEDUZZI, "Application of Negative Binomial Modeling for Discrete Outcomes: A Case Study in Aging Research," *Journal of Clinical Epidemiology, 56* (2003), 559–564.
A-3. KATHLEEN M. STEPANUK, JORGE E. TOLOSA, DAWNEETE LEWIS, VICTORIA MEYERS, CYNTHIA ROYDS, JUAN CARLOS SAOGAL, and RON LIBRIZZI, "Folic Acid Supplementation Use Among Women Who Contact a Teratology Information Service," *American Journal of Obstetrics and Gynecology, 187* (2002), 964–967.
A-4. J. K. SILVER and D. D. AIELLO, "Polio Survivors: Falls and Subsequent Injuries," *American Journal of Physical Medicine and Rehabilitation, 81* (2002), 567–570.
A-5. CYNTHIA G. SEGAL and JACQUELINE J. ANDERSON, "Preoperative Skin Preparation of Cardiac Patients," *AORN Journal, 76* (2002), 821–827.
A-6. RALPH ROTHENBERG and JOHN P. HOLCOMB, "Guidelines for Monitoring of NSAIDs: Who Listened?," *Journal of Clinical Rheumatology, 6* (2000), 258–265.
A-7. SHARON M. BOLES and PATRICK B. JOHNSON, "Gender, Weight Concerns, and Adolescent Smoking," *Journal of Addictive Diseases, 20* (2001), 5–14.
A-8. The DMG Study Group, "Migraine and Idiopathic Narcolepsy—A Case-Control Study," *Cephalagia, 23* (2003), 786–789.
A-9. TASHA D. CARTER, EMANUELA MUNDO, SAGAR V. PARKH, and JAMES L. KENNEDY, "Early Age at Onset as a Risk Factor for Poor Outcome of Bipolar Disorder," *Journal of Psychiatric Research, 37* (2003), 297–303.

A-10. Steven S. Coughlin, Robert J. Uhler, Thomas Richards, and Katherine M. Wilson, "Breast and Cervical Cancer Screening Practices Among Hispanic and Non-Hispanic Women Residing Near the United States–Mexico Border, 1999–2000," *Family and Community Health, 26* (2003), 130–139.

A-11. Robert Swor, Scott Compton, Fern Vining, Lynn Ososky Farr, Sue Kokko, Rebecca Pascual, and Raymond E. Jackson, "A Randomized Controlled Trial of Chest Compression Only CPR for Older Adults: A Pilot Study," *Resuscitation, 58* (2003), 177–185.

A-12. U. S. Justesen, A. M. Lervfing, A. Thomsen, J. A. Lindberg, C. Pedersen, and P. Tauris, "Low-Dose Indinavir in Combination with Low-Dose Ritonavir: Steady-State Pharmacokinetics and Long-Term Clinical Outcome Follow-Up," *HIV Medicine, 4* (2003), 250–254.

A-13. J. F. Tahmassebi and M. E. J. Curzon, "The Cause of Drooling in Children with Cerebral Palsy—Hypersalivation or Swallowing Defect?" *International Journal of Paediatric Dentistry, 13* (2003), 106–111.

A-14. Shu Dong Xiao and Tong Shi, "Is Cranberry Juice Effective in the Treatment and Prevention of *Helicobacter Pylori* Infection of Mice?," *Chinese Journal of Digestive Diseases, 4* (2003), 136–139.

A-15. Gad Shaked, Oleg Kleiner, Robert Finally, Jacob Mordechai, Nitza Newman, and Zahavi Cohen, "Management of Blunt Pancreatic Injuries in Children," *European Journal of Trauma, 29* (2003), 151–155.

A-16. Everett F. Magann, Sharon F. Evans, Beth Weitz, and John Newnham, "Antepartum, Intrapartum, and Neonatal Significance of Exercise on Healthy Low-Risk Pregnant Working Women," *Obstetrics and Gynecology, 99* (2002), 466–472.

A-17. A. M. Toschke, S. M. Montgomery, U. Pfeiffer, and R. von Kries, "Early Intrauterine Exposure to Tobacco-Inhaled Products and Obesity," *American Journal of Epidemiology, 158* (2003), 1068–1074.

A-18. Daniel H. LaMont, Matthew J. Budoff, David M. Shavelle, Robert Shavelle, Bruce H. Brundage, and James M. Hagar, "Coronary Calcium Scanning Adds Incremental Value to Patients with Positive Stress Tests," *American Heart Journal, 143* (2002), 861–867.

A-19. Margaret L. J. Davy, Tom J. Dodd, Colin G. Luke, and David M. Roder, "Cervical Cancer: Effect of Glandular Cell Type on Prognosis, Treatment, and Survival," *Obstetrics and Gynecology, 101* (2003), 38–45.

A-20. U. Stenestrand and L. Wallentin, "Early Revascularization and 1-Year Survival in 14-Day Survivors of Acute Myocardial Infarction," *Lancet, 359* (2002), 1805–1811.

A-21. Takako Sugiyama, Kumiya Sugiyama, Masao Toda, Tastuo Yukawa, Sohei Makino, and Takeshi Fukuda, "Risk Factors for Asthma and Allergic Diseases Among 13–14-Year-Old Schoolchildren in Japan," *Allergology International, 51* (2002), 139–150.

A-22. D. Holben, M. C. McClincy, J. P. Holcomb, and K. L. Dean, "Food Security Status of Households in Appalachian Ohio with Children in Head Start," *Journal of American Dietetic Association, 104* (2004), 238–241.

A-23. Nael Martini, Andrew G. Huvos, Michael E. Burt, Robert T. Heelan, Manjit S. Bains, Patricia M. McCormack, Valerie W. Rusch, Michael Weber, Robert J. Downey, and Robert J. Ginsberg, "Predictions of Survival in Malignant Tumors of the Sternum," *Journal of Thoracic and Cardiovascular Surgery, 111* (1996), 96–106.

A-24. Massimo E. Dottorini, Agnese Assi, Maria Sironi, Gabriele Sangalli, Gianluigi Spreafico, and Luigia Colombo, "Multivariate Analysis of Patients with Medullary Thyroid Carcinoma," *Cancer, 77* (1996), 1556–1565.

A-25. Mary Ann Banerji, Rochelle L. Chaiken, and Harold E. Lebovitz, "Long-Term Normoglycemic Remission in Black Newly Diagnosed NIDDM Subjects," *Diabetes, 45* (1996), 337–341.

A-26. John H. Porcerelli, Rosemary Cogan, Patricia P. West, Edward A. Rose, Dawn Lambrecht, Karen E. Wilson, Richard K. Severson, and Dunia Karana, "Violent Victimization of Women and Men: Physical and Psychiatric Symptoms," *Journal of the American Board of Family Practice, 16* (2003), 32–39.

A-27. Keng Chen, Lee Mei Yap, Robin Marks, and Stephen Shumack, "Short-Course Therapy with Imiquimod 5% Cream for Solar Keratoses: A Randomized Controlled Trial," *Australasian Journal of Dermatology, 44* (2003), 250–255.

A-28. Vallabh Janardhan, Robert Friedlander, Howard Riina, and Philip Edwin Stieg, "Identifying Patients at Risk for Postprocedural Morbidity After Treatment of Incidental Intracranial Aneurysms: The Role of Aneurysm Size and Location," *Neurosurgical Focus, 13* (2002), 1–8.

A-29. Paul B. Gold, Robert N. Rubey, and Richard T. Harvey, "Naturalistic, Self-Assignment Comparative Trial of Bupropion SR, a Nicotine Patch, or Both for Smoking Cessation Treatment in Primary Care," *American Journal on Addictions, 11* (2002), 315–331.

A-30. Zoltán Kozinszky and György Bártai, "Contraceptive Behavior of Teenagers Requesting Abortion," *European Journal of Obstetrics and Gynecology and Reproductive Biology, 112* (2004), 80–83.

A-31. Paolo Crosignani, Andrea Tittarelli, Alessandro Borgini, Tiziana Codazzi, Adriano Rovelli, Emma Porro, Paolo Contiero, Nadia Bianchi, Giovanna Tagliabue, Rosaria Fissi, Francesco Rossitto, and Franco Berrino, "Childhood Leukemia and Road Traffic: A Population-Based Case-Control Study," *International Journal of Cancer, 108* (2004), 596–599.

A-32. Robyn Gallagher, Sharon McKinley, and Kathleen Dracup, "Predictors of Women's Attendance at Cardiac Rehabilitation Programs," *Progress in Cardiovascular Nursing, 18* (2003), 121–126.

A-33. G. Stanley, B. Appadu, M. Mead, and D. J. Rowbotham, "Dose Requirements, Efficacy and Side Effects of Morphine and Pethidine Delivered by Patient-Controlled Analgesia After Gynaecological Surgery," *British Journal of Anaesthesia, 76* (1996), 484–486.

A-34. James D. Sargent, Therese A. Stukel, Madeline A. Dalton, Jean L. Freeman, and Mary Jean Brown, "Iron Deficiency in Massachusetts Communities: Socioeconomic and Demographic Risk Factors Among Children," *American Journal of Public Health, 86* (1996), 544–550.

A-35. Ewald Horwath, Francine Cournos, Karen McKinnon, Jeannine R. Guido, and Richard Herman, "Illicit-Drug Injection Among Psychiatric Patients Without a Primary Substance Use Disorder," *Psychiatric Services, 47* (1996), 181–185.

A-36. Martha Skinner, Jennifer J. Anderson, Robert Simms, Rodney Falk, Ming Wang, Caryn A. Libbey, Lee Anna Jones, and Alan S. Cohen, "Treatment of 100 Patients with Primary Amyloidosis: A Randomized Trial of Melphalan, Prednisone, and Colchicine Versus Colchicine Only," *American Journal of Medicine, 100* (1996), 290–298.

A-37. Yuji Miyajima, Keizo Horibe, Minoru Fukuda, Kimikazu Matsumoto, Shin-ichiro Numata, Hiroshi Mori, and Koji Kato, "Sequential Detection of Tumor Cells in the Peripheral Blood and Bone Marrow of Patients with Stage IV Neuroblastoma by the Reverse Transcription-Polymerase Chain Reaction for Tyrosine Hydroxylase mRNA," *Cancer, 77* (1996), 1214–1219.

A-38. Craig R. Cohen, Ann Duerr, Niwat Pruithithada, Sungwal Rugpao, Sharon Hillier, Patricia Garcia, and Kenrad Nelson, "Bacterial Vaginosis and HIV Seroprevalence Among Female Commercial Sex Workers in Chiang Mai, Thailand," *AIDS, 9* (1995), 1093–1097.

A-39. Deborah S. Lipschitz, Margaret L. Kaplan, Jodie B. Sorkenn, Gianni L. Faedda, Peter Chorney, and Gregory M. Asnis, "Prevalence and Characteristics of Physical and Sexual Abuse Among Psychiatric Outpatients," *Psychiatric Services, 47* (1996), 189–191.

A-40. John M. O'Brien, Brian M. Mercer, Nancy T. Cleary, and Baha M. Sibai, "Efficacy of Outpatient Induction with Low-Dose Intravaginal Prostaglandin E_2: A Randomized, Double-Blind, Placebo-Controlled Trial," *American Journal of Obstetrics and Gynecology, 173* (1995), 1855–1859.

A-41. Abdallah M. Adra, Helain J. Landy, Jaime Nahmias, and Orlando Gómez-Marín, "The Fetus with Gastroschisis: Impact of Route of Delivery and Prenatal Ultrasonography," *American Journal of Obstetrics and Gynecology, 174* (1996), 540–546.

A-42. Simin Liu, Lynn E. Quenemoen, Josephine Malilay, Eric Noji, Thomas Sinks, and James Mendlein, "Assessment of a Severe-Weather Warning System and Disaster Preparedness, Calhoun Country, Alabama, 1994," *American Journal of Public Health, 86* (1996), 87–89.

A-43. Ketan H. Patel, Jonathan C. Javitt, James M. Tielsch, Debra A. Street, Joanne Katz, Harry A. Quigley, and Alfred Sommer, "Incidence of Acute Angle-Closure Glaucoma After Pharmacologic Mydriasis," *American Journal of Ophthalmology, 120* (1995), 709–717.

A-44. Alexandre E. Voskuyl, Aeilko H. Zwinderman, Marie Louise Westedt, Jan P. Vandenbroucke, Ferdinand C. Breedveld, and Johanna M. W. Hazes, "Factors Associated with the Development of Vasculitis in Rheumatoid Arthritis: Results of a Case-Control Study," *Annals of the Rheumatic Diseases, 55* (1996), 190–192.

A-45. Robert L. Harris, Christopher A. Yancey, Winfred L. Wiser, John C. Morrison, and G. Rodney Meeks, "Comparison of Anterior Colporrhaphy and Retropubic Urethropexy for Patients with Genuine Stress Urinary Incontinence," *American Journal of Obstetrics and Gynecology, 173* (1995), 1671–1675.

A-46. Yoshihiro Kohashi, Masayoshi Oga, and Yoichi Sugioka, "A New Method Using Top Views of the Spine to Predict the Progression of Curves in Idiopathic Scoliosis During Growth," *Spine, 21* (1996), 212–217.

A-47. M. P. M. Burger, H. Hollema, W. J. L. M. Pieters, F. P. Schröder, and W. G. V. Quint, "Epidemiological Evidence of Cervical Intraepithelial Neoplasia Without the Presence of Human Papillomavirus," *British Journal of Cancer, 73* (1996), 831–836.

A-48. Eric J. Thomas, Helen R. Burstin, Anne C. O'Neil, E. John Orav, and Troyen A. Brennan, "Patient Noncompliance with Medical Advice After the Emergency Department Visit," *Annals of Emergency Medicine, 27* (1996), 49–55.

A-49. S. T. O'Keefe and J. N. Lavan, "Subcutaneous Fluids in Elderly Hospital Patients with Cognitive Impairment," *Gerontology, 42* (1996), 36–39.

A-50. Joy S. Y. Lee, Antonio G. Nascimento, Michael B. Farnell, J. Aidan Carney, William S. Harmsen, and Duane M. Ilstrup, "Epithelioid Gastric Stromal Tumors (Leiomyoblastomas): A Study of Fifty-five Cases," *Surgery, 118* (1995), 653–661.

A-51. Philippe Girard, Michel Ducreux, Pierre Baldeyrou, Philippe Lasser, Brice Gayet, Pierre Ruffié, and Dominique Grunenwald, "Surgery for Lung Metastases from Colorectal Cancer: Analysis of Prognostic Factors," *Journal of Clinical Oncology, 14* (1996), 2047–2053.

CHAPTER **13**

SPECIAL TECHNIQUES FOR USE WHEN POPULATION PARAMETERS AND/OR POPULATION DISTRIBUTIONS ARE UNKNOWN

CHAPTER OVERVIEW

This chapter explores a wide variety of techniques that are useful when the underlying assumptions of traditional hypothesis tests are violated or one wishes to perform a test without making assumptions about the sampled population.

TOPICS

LEARNING OUTCOMES

After studying this chapter, the student will

1. understand the rank transformation and how nonparametric procedures can be used for weak measurement scales.
2. be able to calculate and interpret a wide variety of nonparametric tests commonly used in practice.
3. understand which nonparametric tests may be used in place of traditional parametric statistical tests when various test assumptions are violated.

13.1 INTRODUCTION

Most of the statistical inference procedures we have discussed up to this point are classified as *parametric statistics*. One exception is our use of chi-square—as a test of goodness-of-fit and as a test of independence. These uses of chi-square come under the heading of *nonparametric statistics*.

The obvious question now is, "What is the difference?" In answer, let us recall the nature of the inferential procedures that we have categorized as *parametric*. In each case, our interest was focused on estimating or testing a hypothesis about one or more population parameters. Furthermore, central to these procedures was a knowledge of the functional form of the population from which were drawn the samples providing the basis for the inference.

An example of a parametric statistical test is the widely used t test. The most common uses of this test are for testing a hypothesis about a single population mean or the difference between two population means. One of the assumptions underlying the valid use of this test is that the sampled population or populations are at least approximately normally distributed.

As we will learn, the procedures that we discuss in this chapter either are not concerned with population parameters or do not depend on knowledge of the sampled population. Strictly speaking, only those procedures that test hypotheses that are not statements about population parameters are classified as *nonparametric*, while those that make no assumption about the sampled population are called *distribution-free* procedures. Despite this distinction, it is customary to use the terms *nonparametric* and *distribution-free* interchangeably and to discuss the various procedures of both types under the heading *nonparametric statistics*. We will follow this convention.

The above discussion implies the following four advantages of nonparametric statistics.

1. They allow for the testing of hypotheses that are not statements about population parameter values. Some of the chi-square tests of goodness-of-fit and the tests of independence are examples of tests possessing this advantage.
2. Nonparametric tests may be used when the form of the sampled population is unknown.
3. Nonparametric procedures tend to be computationally easier and consequently more quickly applied than parametric procedures. This can be a desirable feature in certain cases, but when time is not at a premium, it merits a low priority as a criterion for choosing a nonparametric test. Indeed, most statistical software packages now

include a wide variety of nonparametric analysis options, making considerations about computation speed unnecessary.

4. Nonparametric procedures may be applied when the data being analyzed consist merely of rankings or classifications. That is, the data may not be based on a measurement scale strong enough to allow the arithmetic operations necessary for carrying out parametric procedures. The subject of measurement scales is discussed in more detail in the next section.

Although nonparametric statistics enjoy a number of advantages, their disadvantages must also be recognized.

1. The use of nonparametric procedures with data that can be handled with a parametric procedure results in a waste of data.
2. The application of some of the nonparametric tests may be laborious for large samples.

13.2 MEASUREMENT SCALES

As was pointed out in the previous section, one of the advantages of nonparametric statistical procedures is that they can be used with data that are based on a weak measurement scale. To understand fully the meaning of this statement, it is necessary to know and understand the meaning of measurement and the various measurement scales most frequently used. At this point the reader may wish to refer to the discussion of measurement scales in Chapter 1.

Many authorities are of the opinion that different statistical tests require different measurement scales. Although this idea appears to be followed in practice, there are alternative points of view.

Data based on ranks, as will be discussed in this chapter, are commonly encountered in statistics. We may, for example, simply note the order in which a sample of subjects complete an event instead of the actual time taken to complete it. More often, however, we use a *rank transformation* on the data by replacing, prior to analysis, the original data by their ranks. Although we usually lose some information by employing this procedure (for example, the ability to calculate the mean and variance), the transformed measurement scale allows the computation of most nonparametric statistical procedures. In fact, most of the commonly used nonparametric procedures, including most of those presented in this chapter, can be obtained by first applying the rank transformation and then using the standard parametric procedure on the transformed data instead of on the original data. For example, if we wish to determine whether two independent samples differ, we may employ the independent samples t test if the data are approximately normally distributed. If we cannot make the assumption of normal distributions, we may, as we shall see in the sections that follow, employ an appropriate nonparametric test. In lieu of these procedures, we could first apply the rank transformation on the data and then use the independent samples t test on the ranks. This will provide an equivalent test to the nonparametric test, and is a useful tool to employ if a desired nonparametric test is not available in your available statistical software package.

Readers should also keep in mind that other transformations (e.g., taking the logarithm of the original data) may sufficiently normalize the data such that standard parametric procedures can be used on the transformed data in lieu of using nonparametric methods.

13.3 THE SIGN TEST

The familiar t test is not strictly valid for testing (1) the null hypothesis that a population mean is equal to some particular value, or (2) the null hypothesis that the mean of a population of differences between pairs of measurements is equal to zero unless the relevant populations are at least approximately normally distributed. Case 2 will be recognized as a situation that was analyzed by the paired comparisons test in Chapter 7. When the normality assumptions cannot be made or when the data at hand are ranks rather than measurements on an interval or ratio scale, the investigator may wish for an optional procedure. Although the t test is known to be rather insensitive to violations of the normality assumption, there are times when an alternative test is desirable.

A frequently used nonparametric test that does not depend on the assumptions of the t test is the *sign test*. This test focuses on the median rather than the mean as a measure of central tendency or location. The median and mean will be equal in symmetric distributions. The only assumption underlying the test is that the distribution of the variable of interest is continuous. This assumption rules out the use of nominal data.

The sign test gets its name from the fact that pluses and minuses, rather than numerical values, provide the raw data used in the calculations. We illustrate the use of the sign test, first in the case of a single sample, and then by an example involving paired samples.

EXAMPLE 13.3.1

Researchers wished to know if instruction in personal care and grooming would improve the appearance of mentally retarded girls. In a school for the mentally retarded, 10 girls selected at random received special instruction in personal care and grooming. Two weeks after completion of the course of instruction the girls were interviewed by a nurse and a social worker who assigned each girl a score based on her general appearance. The investigators believed that the scores achieved the level of an ordinal scale. They felt that although a score of, say, 8 represented a better appearance than a score of 6, they were unwilling to say that the difference between scores of 6 and 8 was equal to the difference between, say, scores of 8 and 10; or that the difference between scores of 6 and 8 represented twice as much improvement as the difference between scores of 5 and 6. The scores are shown in Table 13.3.1. We wish to know if we can conclude that the median score of the population from which we assume this sample to have been drawn is different from 5.

Solution:

1. **Data.** See problem statement.
2. **Assumptions.** We assume that the measurements are taken on a continuous variable.

TABLE 13.3.1 General Appearance Scores of 10 Mentally Retarded Girls

Girl	Score	Girl	Score
1	4	6	6
2	5	7	10
3	8	8	7
4	8	9	6
5	9	10	6

3. **Hypotheses.**

   H_0: The population median is 5.

   H_A: The population median is not 5.

   Let $\alpha = .05$.

4. **Test statistic.** The test statistic for the sign test is either the observed number of plus signs or the observed number of minus signs. The nature of the alternative hypothesis determines which of these test statistics is appropriate. In a given test, any one of the following alternative hypotheses is possible:

   H_A: $P(+) > P(-)$ one-sided alternative

   H_A: $P(+) < P(-)$ one-sided alternative

   H_A: $P(+) \neq P(-)$ two-sided alternative

   If the alternative hypothesis is

   $$H_A: P(+) > P(-)$$

   a sufficiently small number of minus signs causes rejection of H_0. The test statistic is the number of minus signs. Similarly, if the alternative hypothesis is

   $$H_A: P(+) < P(-)$$

   a sufficiently small number of plus signs causes rejection of H_0. The test statistic is the number of plus signs. If the alternative hypothesis is

   $$H_A: P(+) \neq P(-)$$

   either a sufficiently small number of plus signs or a sufficiently small number of minus signs causes rejection of the null hypothesis. We may take as the test statistic the less frequently occurring sign.

5. **Distribution of test statistic.** As a first step in determining the nature of the test statistic, let us examine the data in Table 13.3.1 to determine which scores lie above and which ones lie below the hypothesized median of 5. If we assign a plus sign to those scores that lie above the

TABLE 13.3.2 Scores Above (+) and Below (−) the Hypothesized Median Based on Data of Example 13.3.1

Girl	1	2	3	4	5	6	7	8	9	10
Score relative to hypothesized median	−	0	+	+	+	+	+	+	+	+

hypothesized median and a minus to those that fall below, we have the results shown in Table 13.3.2.

If the null hypothesis were true, that is, if the median were, in fact, 5, we would expect the numbers of scores falling above and below 5 to be approximately equal. This line of reasoning suggests an alternative way in which we could have stated the null hypothesis, namely, that the probability of a plus is equal to the probability of a minus, and these probabilities are equal to .5. Stated symbolically, the hypothesis would be

$$H_0: P(+) = P(-) = .5$$

In other words, we would expect about the same number of plus signs as minus signs in Table 13.3.2 when H_0 is true. A look at Table 13.3.2 reveals a preponderance of pluses; specifically, we observe eight pluses, one minus, and one zero, which was assigned to the score that fell exactly on the median. The usual procedure for handling zeros is to eliminate them from the analysis and reduce n, the sample size, accordingly. If we follow this procedure, our problem reduces to one consisting of nine observations of which eight are plus and one is minus.

Since the number of pluses and minuses is not the same, we wonder if the distribution of signs is sufficiently disproportionate to cast doubt on our hypothesis. Stated another way, we wonder if this small a number of minuses could have come about by chance alone when the null hypothesis is true, or if the number is so small that something other than chance (that is, a false null hypothesis) is responsible for the results.

Based on what we learned in Chapter 4, it seems reasonable to conclude that the observations in Table 13.3.2 constitute a set of n independent random variables from the Bernoulli population with parameter p. If we let k = the test statistic, the sampling distribution of k is the binomial probability distribution with parameter $p = .5$ if the null hypothesis is true.

6. Decision rule. The decision rule depends on the alternative hypothesis.

For $H_A: P(+) > P(-)$, reject H_0 if, when H_0 is true, the probability of observing k or fewer minus signs is less than or equal to α.

For $H_A: P(+) < P(-)$, reject H_0 if the probability of observing, when H_0 is true, k or fewer plus signs is equal to or less than α.

For $H_A: P(+) \neq P(-)$, reject H_0 if (given that H_0 is true) the probability of obtaining a value of k as extreme as or more extreme than was actually computed is equal to or less than $\alpha/2$.

For this example the decision rule is: Reject H_0 if the p value for the computed test statistic is less than or equal to .05.

7. **Calculation of test statistic.** We may determine the probability of observing x or fewer minus signs when given a sample of size n and parameter p by evaluating the following expression:

$$P(k \leq x \mid n, p) = \sum_{k=0}^{x} {}_nC_k p^k q^{n-k} \tag{13.3.1}$$

For our example we would compute

$$_9C_0(.5)^0(.5)^{9-0} + {}_9C_1(.5)^1(.5)^{9-1} = .00195 + .01758 = .0195$$

8. **Statistical decision.** In Appendix Table B we find

$$P(k \leq 1 \mid 9, .5) = .0195$$

With a two-sided test either a sufficiently small number of minuses or a sufficiently small number of pluses would cause rejection of the null hypothesis. Since, in our example, there are fewer minuses, we focus our attention on minuses rather than pluses. By setting α equal to .05, we are saying that if the number of minuses is so small that the probability of observing this few or fewer is less than .025 (half of α), we will reject the null hypothesis. The probability we have computed, .0195, is less than .025. We, therefore, reject the null hypothesis.

9. **Conclusion.** We conclude that the median score is not 5.

10. **p value.** The p value for this test is $2(.0195) = .0390$. ■

Sign Test: Paired Data When the data to be analyzed consist of observations in matched pairs and the assumptions underlying the t test are not met, or the measurement scale is weak, the sign test may be employed to test the null hypothesis that the median difference is 0. An alternative way of stating the null hypothesis is

$$P(X_i > Y_i) = P(X_i < Y_i) = .5$$

One of the matched scores, say, Y_i, is subtracted from the other score, X_i. If Y_i is less than X_i, the sign of the difference is $+$, and if Y_i is greater than X_i, the sign of the difference is $-$. If the median difference is 0, we would expect a pair picked at random to be just as likely to yield a $+$ as a $-$ when the subtraction is performed. We may state the null hypothesis, then, as

$$H_0: P(+) = P(-) = .5$$

In a random sample of matched pairs, we would expect the number of $+$'s and $-$'s to be about equal. If there are more $+$'s or more $-$'s than can be accounted for by chance alone when the null hypothesis is true, we will entertain some doubt about the truth of our null hypothesis. By means of the sign test, we can decide how many of one sign constitutes more than can be accounted for by chance alone.

EXAMPLE 13.3.2

A dental research team wished to know if teaching people how to brush their teeth would be beneficial. Twelve pairs of patients seen in a dental clinic were obtained by carefully matching on such factors as age, sex, intelligence, and initial oral hygiene scores. One member of each pair received instruction on how to brush his or her teeth and on other oral hygiene matters. Six months later all 24 subjects were examined and assigned an oral hygiene score by a dental hygienist unaware of which subjects had received the instruction. A low score indicates a high level of oral hygiene. The results are shown in Table 13.3.3.

Solution:

1. **Data.** See problem statement.
2. **Assumptions.** We assume that the population of differences between pairs of scores is a continuous variable.
3. **Hypotheses.** If the instruction produces a beneficial effect, this fact would be reflected in the scores assigned to the members of each pair. If we take the differences $X_i - Y_i$, we would expect to observe more $-$'s than $+$'s if instruction had been beneficial, since a low score indicates a higher level of oral hygiene. If, in fact, instruction is beneficial, the median of the hypothetical population of all such differences would be less than 0, that is, negative. If, on the other hand, instruction has no effect, the median of this population would be zero. The null and alternate hypotheses, then, are:

TABLE 13.3.3 Oral Hygiene Scores of 12 Subjects Receiving Oral Hygiene Instruction (X_i) and 12 Subjects Not Receiving Instruction (Y_i)

	Score	
Pair Number	Instructed (X_i)	Not Instructed (Y_i)
1	1.5	2.0
2	2.0	2.0
3	3.5	4.0
4	3.0	2.5
5	3.5	4.0
6	2.5	3.0
7	2.0	3.5
8	1.5	3.0
9	1.5	2.5
10	2.0	2.5
11	3.0	2.5
12	2.0	2.5

TABLE 13.3.4 Signs of Differences ($X_i - Y_i$) in Oral Hygiene Scores of 12 Subjects Instructed (X_i) and 12 Matched Subjects Not Instructed (Y_i)

Pair	1	2	3	4	5	6	7	8	9	10	11	12
Sign of score differences	−	0	−	+	−	−	−	−	−	−	+	−

H_0: The median of the differences is zero $[P(+) = P(-)]$.

H_A: The median of the differences is negative $[P(+) < P(-)]$.

Let α be .05.

4. **Test statistic.** The test statistic is the number of plus signs.
5. **Distribution of test statistic.** The sampling distribution of k is the binomial distribution with parameters n and .5 if H_0 is true.
6. **Decision rule.** Reject H_0 if $P(k \leq 2 \mid 11, .5) \leq .05$.
7. **Calculation of test statistic.** As will be seen, the procedure here is identical to the single sample procedure once the score differences have been obtained for each pair. Performing the subtractions and observing signs yields the results shown in Table 13.3.4.

   The nature of the hypothesis indicates a one-sided test so that all of $\alpha = .05$ is associated with the rejection region, which consists of all values of k (where k is equal to the number of + signs) for which the probability of obtaining that many or fewer pluses due to chance alone when H_0 is true is equal to or less than .05. We see in Table 13.3.4 that the experiment yielded one zero, two pluses, and nine minuses. When we eliminate the zero, the effective sample size is $n = 11$ with two pluses and nine minuses. In other words, since a "small" number of plus signs will cause rejection of the null hypothesis, the value of our test statistic is $k = 2$.
8. **Statistical decision.** We want to know the probability of obtaining no more than two pluses out of 11 tries when the null hypothesis is true. As we have seen, the answer is obtained by evaluating the appropriate binomial expression. In this example we find

$$P(k \leq 2 \mid 11, .5) = \sum_{k=0}^{2} {}_{11}C_k(.5)^k(.5)^{11-k}$$

   By consulting Appendix Table B, we find this probability to be .0327. Since .0327 is less than .05, we must reject H_0.
9. **Conclusion.** We conclude that the median difference is negative. That is, we conclude that the instruction was beneficial.
10. **p value.** For this test, $p = .0327$. ■

Sign Test with "Greater Than" Tables As has been demonstrated, the sign test may be used with a single sample or with two samples in which each member of one sample is matched with a member of the other sample to form a sample of matched pairs. We have also seen that the alternative hypothesis may lead to either a one-sided or a two-sided test. In either case we concentrate on the less frequently occurring sign and calculate the probability of obtaining that few or fewer of that sign.

We use the least frequently occurring sign as our test statistic because the binomial probabilities in Appendix Table B are "less than or equal to" probabilities. By using the least frequently occurring sign, we can obtain the probability we need directly from Table B without having to do any subtracting. If the probabilities in Table B were "greater than or equal to" probabilities, which are often found in tables of the binomial distribution, we would use the more frequently occurring sign as our test statistic in order to take advantage of the convenience of obtaining the desired probability directly from the table without having to do any subtracting. In fact, we could, in our present examples, use the more frequently occurring sign as our test statistic, but because Table B contains "less than or equal to" probabilities we would have to perform a subtraction operation to obtain the desired probability. As an illustration, consider the last example. If we use as our test statistic the most frequently occurring sign, it is 9, the number of minuses. The desired probability, then, is the probability of nine or more minuses, when $n = 11$ and $p = .5$. That is, we want

$$P(k = 9 \mid 11, .5)$$

However, since Table B contains "less than or equal to" probabilities, we must obtain this probability by subtraction. That is,

$$\begin{aligned} P(k \geq 9 \mid 11, .5) &= 1 - P(k \leq 8 \mid 11, .5) \\ &= 1 - .9673 \\ &= .0327 \end{aligned}$$

which is the result obtained previously.

Sample Size We saw in Chapter 5 that when the sample size is large and when p is close to .5, the binomial distribution may be approximated by the normal distribution. The rule of thumb used was that the normal approximation is appropriate when both np and nq are greater than 5. When $p = .5$, as was hypothesized in our two examples, a sample of size 12 would satisfy the rule of thumb. Following this guideline, one could use the normal approximation when the sign test is used to test the null hypothesis that the median or median difference is 0 and n is equal to or greater than 12. Since the procedure involves approximating a continuous distribution by a discrete distribution, the continuity correction of .5 is generally used. The test statistic then is

$$z = \frac{(k \pm .5) - .5n}{.5\sqrt{n}} \tag{13.3.2}$$

which is compared with the value of z from the standard normal distribution corresponding to the chosen level of significance. In Equation 13.3.2, $k + .5$ is used when $k < n/2$ and $k - .5$ is used when $k \geq n/2$.

Data:

```
C1: 4 5 8 8 9 6 10 7 6 6
```

Dialog box:

Stat ➤ Nonparametrics ➤ 1-Sample Sign

Type *C1* in **Variables.** Choose **Test median** and type *5* in the text box. Click **OK.**

Session command:

```
MTB > STest 5 C1;
SUBC>   Alternative 0.
```

Output:

Sign Test for Median: C1

```
Sign test of median  =  5.00 versus N.E. 5.000
                 N  BELOW  EQUAL  ABOVE  P-VALUE  MEDIAN
C1              10      1      1      8   0.0391   6.500
```

FIGURE 13.3.1 MINITAB procedure and output for Example 13.3.1.

Computer Analysis Many statistics software packages will perform the sign test. For example, if we use MINITAB to perform the test for Example 13.3.1 in which the data are stored in Column 1, the procedure and output would be as shown in Figure 13.3.1.

EXERCISES

13.3.1 A random sample of 15 student nurses was given a test to measure their level of authoritarianism with the following results:

Student Number	Authoritarianism Score	Student Number	Authoritarianism Score
1	75	9	82
2	90	10	104
3	85	11	88
4	110	12	124
5	115	13	110
6	95	14	76
7	132	15	98
8	74		

Test at the .05 level of significance, the null hypothesis that the median score for the sampled population is 100. Determine the p value.

13.3.2 Determining the effects of grapefruit juice on pharmacokinetics of oral digoxin (a drug often prescribed for heart ailments) was the goal of a study by Parker et al. (A-1). Seven healthy nonsmoking volunteers participated in the study. Subjects took digoxin with water for 2 weeks, no digoxin for 2 weeks, and digoxin with grapefruit juice for 2 weeks. The average peak plasma digoxin concentration (Cmax) when subjects took digoxin with water is given in the first column of the following table. The second column gives the Cmax concentration when subjects took digoxin with grapefruit juice. May we conclude on the basis of these data that the Cmax concentration is higher when digoxin is taken with grapefruit juice? Let $\alpha = .05$.

	Cmax	
Subject	**H_2O**	**GFJ**
1	2.34	3.03
2	2.46	3.46
3	1.87	1.97
4	3.09	3.81
5	5.59	3.07
6	4.05	2.62
7	6.21	3.44

Source: Robert B. Parker, Pharm.D. Used with permission.

13.3.3 A sample of 15 patients suffering from asthma participated in an experiment to study the effect of a new treatment on pulmonary function. Among the various measurements recorded were those of forced expiratory volume (liters) in 1 second (FEV_1) before and after application of the treatment. The results were as follows:

Subject	Before	After	Subject	Before	After
1	1.69	1.69	9	2.58	2.44
2	2.77	2.22	10	1.84	4.17
3	1.00	3.07	11	1.89	2.42
4	1.66	3.35	12	1.91	2.94
5	3.00	3.00	13	1.75	3.04
6	.85	2.74	14	2.46	4.62
7	1.42	3.61	15	2.35	4.42
8	2.82	5.14			

On the basis of these data, can one conclude that the treatment is effective in increasing the FEV_1 level? Let $\alpha = .05$ and find the p value.

13.4 THE WILCOXON SIGNED-RANK TEST FOR LOCATION

Sometimes we wish to test a null hypothesis about a population mean, but for some reason neither z nor t is an appropriate test statistic. If we have a small sample ($n < 30$) from a population that is known to be grossly nonnormally distributed, and the central limit theorem is not applicable, the z statistic is ruled out. The t statistic is not appropriate because the sampled population does not sufficiently approximate a normal distribution. When confronted with such a situation we usually look for an appropriate nonparametric statistical procedure. As we have seen, the sign test may be used when our data consist of a single sample or when we have paired data. If, however, the data for analysis are measured on at least an interval scale, the sign test may be undesirable because it would not make full use of the information contained in the data. A more appropriate procedure might be the Wilcoxon (1) signed-rank test, which makes use of the magnitudes of the differences between measurements and a hypothesized location parameter rather than just the signs of the differences.

Assumptions The Wilcoxon test for location is based on the following assumptions about the data.

1. The sample is random.
2. The variable is continuous.
3. The population is symmetrically distributed about its mean μ.
4. The measurement scale is at least interval.

Hypotheses The following are the null hypotheses (along with their alternatives) that may be tested about some unknown population mean μ_0.

(a) $H_0: \mu = \mu_0$ **(b)** $H_0: \mu \geq \mu_0$ **(c)** $H_0: \mu \leq \mu_0$

$H_A: \mu \neq \mu_0$ $H_A: \mu < \mu_0$ $H_A: \mu > \mu_0$

When we use the Wilcoxon procedure, we perform the following calculations.

1. Subtract the hypothesized mean μ_0 from each observation x_i, to obtain

$$d_i = x_i - \mu_0$$

If any x_i is equal to the mean, so that $d_i = 0$, eliminate that d_i from the calculations and reduce n accordingly.

2. Rank the usable d_i from the smallest to the largest without regard to the sign of d_i. That is, consider only the absolute value of the d_i, designated $|d_i|$, when ranking them. If two or more of the $|d_i|$ are equal, assign each tied value the mean of the rank positions the tied values occupy. If, for example, the three smallest $|d_i|$ are all equal, place them in rank positions 1, 2, and 3, but assign each a rank of $(1 + 2 + 3)/3 = 2$.
3. Assign each rank the sign of the d_i that yields that rank.
4. Find T_+, the sum of the ranks with positive signs, and T_-, the sum of the ranks with negative signs.

The Test Statistic The Wilcoxon test statistic is either T_+ or T_-, depending on the nature of the alternative hypothesis. If the null hypothesis is true, that is, if the true population mean is equal to the hypothesized mean, and if the assumptions are met, the probability of observing a positive difference $d_i = x_i - \mu_0$ of a given magnitude is equal to the probability of observing a negative difference of the same magnitude. Then, in repeated sampling, when the null hypothesis is true and the assumptions are met, the expected value of T_+ is equal to the expected value of T_-. We do not expect T_+ and T_- computed from a given sample to be equal. However, when H_0 is true, we do not expect a large difference in their values. Consequently, a sufficiently small value of T_+ or a sufficiently small value of T_- will cause rejection of H_0.

When the alternative hypothesis is two-sided ($\mu \neq \mu_0$), either a sufficiently small value of T_+ or a sufficiently small value of T_- will cause us to reject H_0: $\mu = \mu_0$. The test statistic, then, is T_+ or T_-, whichever is smaller. To simplify notation, we call the smaller of the two T.

When H_0: $\mu \geq \mu_0$ is true, we expect our sample to yield a large value of T_+. Therefore, when the one-sided alternative hypothesis states that the true population mean is less than the hypothesized mean ($\mu < \mu_0$), a sufficiently small value of T_+ will cause rejection of H_0, and T_+ is the test statistic.

When H_0: $\mu \leq \mu_0$ is true, we expect our sample to yield a large value of T_-. Therefore, for the one-sided alternative H_A: $\mu > \mu_0$, a sufficiently small value of T_- will cause rejection of H_0 and T_- is the test statistic.

Critical Values Critical values of the Wilcoxon test statistic are given in Appendix Table K. Exact probability levels (P) are given to four decimal places for all possible rank totals (T) that yield a different probability level at the fourth decimal place from .0001 up through .5000. The rank totals (T) are tabulated for all sample sizes from $n = 5$ through $n = 30$. The following are the decision rules for the three possible alternative hypotheses:

(a) H_A: $\mu \neq \mu_0$. Reject H_0 at the α level of significance if the calculated T is smaller than or equal to the tabulated T for n and preselected $\alpha/2$. Alternatively, we may enter Table K with n and our calculated value of T to see whether the tabulated P associated with the calculated T is less than or equal to our stated level of significance. If so, we may reject H_0.

(b) H_A: $\mu < \mu_0$. Reject H_0 at the α level of significance if T_+ is less than or equal to the tabulated T for n and preselected α.

(c) H_A: $\mu > \mu_0$. Reject H_0 at the α level of significance if T_- is less than or equal to the tabulated T for n and preselected α.

EXAMPLE 13.4.1

Cardiac output (liters/minute) was measured by thermodilution in a simple random sample of 15 postcardiac surgical patients in the left lateral position. The results were as follows:

4.91	4.10	6.74	7.27	7.42	7.50	6.56	4.64
5.98	3.14	3.23	5.80	6.17	5.39	5.77	

We wish to know if we can conclude on the basis of these data that the population mean is different from 5.05.

Solution:

1. **Data.** See statement of example.
2. **Assumptions.** We assume that the requirements for the application of the Wilcoxon signed-ranks test are met.
3. **Hypotheses.**

   H_0: $\mu = 5.05$

   H_A: $\mu \neq 5.05$

   Let $\alpha = 0.05$.
4. **Test statistic.** The test statistic will be T_+ or T_-, whichever is smaller. We will call the test statistic T.
5. **Distribution of test statistic.** Critical values of the test statistic are given in Table K of the Appendix.
6. **Decision rule.** We will reject H_0 if the computed value of T is less than or equal to 25, the critical value for $n = 15$, and $\alpha/2 = .0240$, the closest value to .0250 in Table K.
7. **Calculation of test statistic.** The calculation of the test statistic is shown in Table 13.4.1.
8. **Statistical decision.** Since 34 is greater than 25, we are unable to reject H_0.
9. **Conclusion.** We conclude that the population mean may be 5.05.
10. ***p* value.** From Table K we see that $p = 2(.0757) = .1514$.

TABLE 13.4.1 Calculation of the Test Statistic for Example 13.4.1

Cardiac Output	$d_i = x_i - 5.05$	Rank of $\lvert d_i \rvert$	Signed Rank of $\lvert d_i \rvert$
4.91	−.14	1	−1
4.10	−.95	7	−7
6.74	+1.69	10	+10
7.27	+2.22	13	+13
7.42	+2.37	14	+14
7.50	+2.45	15	+15
6.56	+1.51	9	+9
4.64	−.41	3	−3
5.98	+.93	6	+6
3.14	−1.91	12	−12
3.23	−1.82	11	−11
5.80	+.75	5	+5
6.17	+1.12	8	+8
5.39	+.34	2	+2
5.77	+.72	4	+4
			$T_+ = 86$, $T_- = 34$, $T = 34$

■

Dialog box:	**Session command:**
Stat ➤ Nonparametrics ➤ 1-Sample Wilcoxon	`MTB > WTEST 5.05 C1;` `SUBC> Alternative 0.`
Type *C1* in **Variables.** Choose **Test median.** Type 5.05 in the text box. Click **OK.**	

Output:

Wilcoxon Signed Rank Test: C1

```
TEST OF MEDIAN = 5.050 VERSUS MEDIAN N.E. 5.050

                   N FOR WILCOXON              ESTIMATED
           N    TEST  STATISTIC  P-VALUE       MEDIAN
C1        15      15       86.0    0.148        5.747
```

FIGURE 13.4.1 MINITAB procedure and output for Example 13.4.1.

Wilcoxon Matched-Pairs Signed-Ranks Test The Wilcoxon test may be used with paired data under circumstances in which it is not appropriate to use the paired-comparisons *t* test described in Chapter 7. In such cases obtain each of the n d_i values, the difference between each of the n pairs of measurements. If we let μ_D = the mean of a population of such differences, we may follow the procedure described above to test any one of the following null hypotheses: H_0: $\mu_D = 0$, H_0: $\mu_D \geq 0$, and H_0: $\mu_D \geq 0$.

Computer Analysis Many statistics software packages will perform the Wilcoxon signed-rank test. If, for example, the data of Example 13.4.1 are stored in Column 1, we could use MINITAB to perform the test as shown in Figure 13.4.1.

EXERCISES

13.4.1 Sixteen laboratory animals were fed a special diet from birth through age 12 weeks. Their weight gains (in grams) were as follows:

63 68 79 65 64 63 65 64 76 74 66 66 67 73 69 76

Can we conclude from these data that the diet results in a mean weight gain of less than 70 grams? Let $\alpha = .05$, and find the p value.

13.4.2 Amateur and professional singers were the subjects of a study by Grape et al. (A-2). The researchers investigated the possible beneficial effects of singing on well-being during a single singing lesson. One of the variables of interest was the change in cortisol as a result of the signing lesson. Use the data in the following table to determine if, in general, cortisol (nmol/L) increases after a singing lesson. Let $\alpha = .05$. Find the p value.

Subject	1	2	3	4	5	6	7	8
Before	214	362	202	158	403	219	307	331
After	232	276	224	412	562	203	340	313

Source: Christina Grape, M.P.H., Licensed Nurse. Used with permission.

13.4.3 In a study by Zuckerman and Heneghan (A-3), hemodynamic stresses were measured on subjects undergoing laparoscopic cholecystectomy. An outcome variable of interest was the ventricular end diastolic volume (LVEDV) measured in milliliters. A portion of the data appear in the following table. Baseline refers to a measurement taken 5 minutes after induction of anesthesia, and the term "5 minutes" refers to a measurement taken 5 minutes after baseline.

	LVEDV (ml)	
Subject	**Baseline**	**5 Minutes**
1	51.7	49.3
2	79.0	72.0
3	78.7	87.3
4	80.3	88.3
5	72.0	103.3
6	85.0	94.0
7	69.7	94.7
8	71.3	46.3
9	55.7	71.7
10	56.3	72.3

Source: R. S. Zuckerman, MD. Used with permission.

May we conclude, on the basis of these data, that among subjects undergoing laparoscopic cholecystectomy, the average LVEDV levels change? Let $\alpha = .01$.

13.5 THE MEDIAN TEST

A nonparametric procedure that may be used to test the null hypothesis that two independent samples have been drawn from populations with equal medians is the median test. The test, attributed mainly to Mood (2) and Westenberg (3), is also discussed by Brown and Mood (4).

We illustrate the procedure by means of an example.

EXAMPLE 13.5.1

Do urban and rural male junior high school students differ with respect to their level of mental health?

Solution:

1. Data. Members of a random sample of 12 male students from a rural junior high school and an independent random sample of 16 male

TABLE 13.5.1 Level of Mental Health Scores of Junior High Boys

School			
Urban	**Rural**	**Urban**	**Rural**
35	29	25	50
26	50	27	37
27	43	45	34
21	22	46	31
27	42	33	
38	47	26	
23	42	46	
25	32	41	

students from an urban junior high school were given a test to measure their level of mental health. The results are shown in Table 13.5.1.

To determine if we can conclude that there is a difference, we perform a hypothesis test that makes use of the median test. Suppose we choose a .05 level of significance.

2. **Assumptions.** The assumptions underlying the test are (a) the samples are selected independently and at random from their respective populations; (b) the populations are of the same form, differing only in location; and (c) the variable of interest is continuous. The level of measurement must be, at least, ordinal. The two samples do not have to be of equal size.
3. **Hypotheses.**

$$H_0: M_U = M_R$$
$$H_A: M_U \neq M_R$$

M_U is the median score of the sampled population of urban students, and M_R is the median score of the sampled population of rural students. Let $\alpha = .05$.

4. **Test statistic.** As will be shown in the discussion that follows, the test statistic is X^2 as computed, for example, by Equation 12.4.1 for a 2×2 contingency table.
5. **Distribution of test statistic.** When H_0 is true and the assumptions are met, X^2 is distributed approximately as χ^2 with 1 degree of freedom.
6. **Decision rule.** Reject H_0 if the computed value of X^2 is ≥ 3.841 (since $\alpha = .05$).
7. **Calculation of test statistic.** The first step in calculating the test statistic is to compute the common median of the two samples combined. This is done by arranging the observations in ascending order and, because

TABLE 13.5.2 Level of Mental Health Scores of Junior High School Boys

	Urban	Rural	Total
Number of scores above median	6	8	14
Number of scores below median	10	4	14
Total	16	12	28

the total number of observations is even, obtaining the mean of the two middle numbers. For our example the median is $(33 + 34)/2 = 33.5$.

We now determine for each group the number of observations falling above and below the common median. The resulting frequencies are arranged in a 2×2 table. For the present example we construct Table 13.5.2.

If the two samples are, in fact, from populations with the same median, we would expect about one-half the scores in each sample to be above the combined median and about one-half to be below. If the conditions relative to sample size and expected frequencies for a 2×2 contingency table as discussed in Chapter 12 are met, the chi-square test with 1 degree of freedom may be used to test the null hypothesis of equal population medians. For our examples we have, by Formula 12.4.1,

$$X^2 = \frac{28[(6)(4) - (8)(10)]^2}{(16)(12)(14)(14)} = 2.33$$

8. **Statistical decision.** Since $2.33 < 3.841$, the critical value of χ^2 with $\alpha = .05$ and 1 degree of freedom, we are unable to reject the null hypothesis on the basis of these data.

9. **Conclusion.** We conclude that the two samples may have been drawn from populations with equal medians.

10. **p value.** Since $2.33 < 2.706$, we have $p > .10$. ■

Handling Values Equal to the Median Sometimes one or more observed values will be exactly equal to the common median and, hence, will fall neither above nor below it. We note that if $n_1 + n_2$ is odd, at least one value will always be exactly equal to the median. This raises the question of what to do with observations of this kind. One solution is to drop them from the analysis if $n_1 + n_2$ is large and there are only a few values that fall at the combined median. Or we may dichotomize the scores into those that exceed the median and those that do not, in which case the observations that equal the median will be counted in the second category.

Median Test Extension The median test extends logically to the case where it is desired to test the null hypothesis that $k \geq 3$ samples are from populations with equal medians. For this test a $2 \times k$ contingency table may be constructed by using the

Dialog box: **Session command:**

Stat ➤ Nonparametrics ➤ Mood's Median Test `MTB > Mood C1 C2.`

Type *C1* in **Response** and *C2* in **Factor.** Click **OK.**

Output:

Mood Median Test: C1 versus C2

```
Mood median test of C1

Chisquare = 2.33 df = 1 p = 0.127

                                       Individual 95.0% CIs
 C2 N<= N> Median  Q3-Q1  --------+---------+---------+--------
  1  10  6   27.0   15.0    (-+-------------------)
  2   4  8   39.5   14.8             (------------+----------)
                          --------+---------+---------+--------
                                30.0      36.0      42.0
Overall median = 33.5

A 95.0% C.I. for median (1) - median(2): (-17.1,3.1)
```

FIGURE 13.5.1 MINITAB procedure and output for Example 13.5.1.

frequencies that fall above and below the median computed from combined samples. If conditions as to sample size and expected frequencies are met, X^2 may be computed and compared with the critical χ^2 with $k - 1$ degrees of freedom.

Computer Analysis The median test calculations may be carried out using MINITAB. To illustrate using the data of Example 13.5.1 we first store the measurements in MINITAB Column 1. In MINITAB Column 2 we store codes that identify the observations as to whether they are for an urban (1) or rural (2) subject. The MINITAB procedure and output are shown in Figure 13.5.1.

EXERCISES

13.5.1 Fifteen patient records from each of two hospitals were reviewed and assigned a score designed to measure level of care. The scores were as follows:

Hospital A: 99, 85, 73, 98, 83, 88, 99, 80, 74, 91, 80, 94, 94, 98, 80

Hospital B: 78, 74, 69, 79, 57, 78, 79, 68, 59, 91, 89, 55, 60, 55, 79

Would you conclude, at the .05 level of significance, that the two population medians are different? Determine the p value.

13.5.2 The following serum albumin values were obtained from 17 normal and 13 hospitalized subjects:

Serum Albumin (g/100 ml)		Serum Albumin (g/100 ml)	
Normal Subjects	**Hospitalized Subjects**	**Normal Subjects**	**Hospitalized Subjects**
2.4 3.0	1.5 3.1	3.4 4.0	3.8 1.5
3.5 3.2	2.0 1.3	4.5 3.5	3.5
3.1 3.5	3.4 1.5	5.0 3.6	
4.0 3.8	1.7 1.8	2.9	
4.2 3.9	2.0 2.0		

Would you conclude at the .05 level of significance that the medians of the two populations sampled are different? Determine the p value.

13.6 THE MANN–WHITNEY TEST

The median test discussed in the preceding section does not make full use of all the information present in the two samples when the variable of interest is measured on at least an ordinal scale. Reducing an observation's information content to merely that of whether or not it falls above or below the common median is a waste of information. If, for testing the desired hypothesis, there is available a procedure that makes use of more of the information inherent in the data, that procedure should be used if possible. Such a nonparametric procedure that can often be used instead of the median test is the Mann–Whitney test (5), sometimes called the Mann–Whitney–Wilcoxon test. Since this test is based on the ranks of the observations, it utilizes more information than does the median test.

Assumptions The assumptions underlying the Mann–Whitney test are as follows:

1. The two samples, of size n and m, respectively, available for analysis have been independently and randomly drawn from their respective populations.
2. The measurement scale is at least ordinal.
3. The variable of interest is continuous.
4. If the populations differ at all, they differ only with respect to their medians.

Hypotheses When these assumptions are met we may test the null hypothesis that the two populations have equal medians against either of the three possible alternatives: (1) the populations do not have equal medians (two-sided test), (2) the median of population 1 is larger than the median of population 2 (one-sided test), or (3) the median of population 1 is smaller than the median of population 2 (one-sided test). If the two populations are symmetric, so that within each population the mean and median are the same, the conclusions we reach regarding the two population medians will also apply to the two population means. The following example illustrates the use of the Mann–Whitney test.

EXAMPLE 13.6.1

A researcher designed an experiment to assess the effects of prolonged inhalation of cadmium oxide. Fifteen laboratory animals served as experimental subjects, while 10 similar animals served as controls. The variable of interest was hemoglobin level following the experiment. The results are shown in Table 13.6.1. We wish to know if we can conclude that prolonged inhalation of cadmium oxide reduces hemoglobin level.

Solution:

1. **Data.** See Table 13.6.1.
2. **Assumptions.** We assume that the assumptions of the Mann– Whitney test are met.
3. **Hypotheses.** The null and alternative hypotheses are as follows:

$$H_0: M_X \geq M_Y$$
$$H_A: M_X < M_Y$$

where M_X is the median of a population of animals exposed to cadmium oxide and M_Y is the median of a population of animals not exposed to the substance. Suppose we let $\alpha = .05$.

4. **Test statistic.** To compute the test statistic we combine the two samples and rank all observations from smallest to largest while keeping track of the sample to which each observation belongs. Tied observations are assigned a rank equal to the mean of the rank positions for which they are tied. The results of this step are shown in Table 13.6.2.

TABLE 13.6.1 Hemoglobin Determinations (grams) for 25 Laboratory Animals

Exposed Animals (X)	Unexposed Animals (Y)
14.4	17.4
14.2	16.2
13.8	17.1
16.5	17.5
14.1	15.0
16.6	16.0
15.9	16.9
15.6	15.0
14.1	16.3
15.3	16.8
15.7	
16.7	
13.7	
15.3	
14.0	

TABLE 13.6.2 Original Data and Ranks, Example 13.6.1

X	Rank	Y	Rank
13.7	1		
13.8	2		
14.0	3		
14.1	4.5		
14.1	4.5		
14.2	6		
14.4	7		
		15.0	8.5
		15.0	8.5
15.3	10.5		
15.3	10.5		
15.6	12		
15.7	13		
15.9	14		
		16.0	15
		16.2	16
		16.3	17
16.5	18		
16.6	19		
16.7	20		
		16.8	21
		16.9	22
		17.1	23
		17.4	24
		17.5	25
Total	145		

The test statistic is

$$T = S - \frac{n(n+1)}{2} \tag{13.6.1}$$

where n is the number of sample X observations and S is the sum of the ranks assigned to the sample observations from the population of X values. The choice of which sample's values we label X is arbitrary.

5. **Distribution of test statistic.** Critical values from the distribution of the test statistic are given in Appendix Table L for various levels of α.
6. **Decision rule.** If the median of the X population is, in fact, smaller than the median of the Y population, as specified in the alternative hypothesis, we would expect (for equal sample sizes) the sum of the ranks

assigned to the observations from the X population to be smaller than the sum of the ranks assigned to the observations from the Y population. The test statistic is based on this rationale in such a way that a sufficiently small value of T will cause rejection of H_0: $M_X \geq M_Y$. In general, for one-sided tests of the type illustrated here the decision rule is:

Reject H_0: $M_X = M_Y$ if the computed T is less than w_α, where w_α is the critical value of T obtained by entering Appendix Table L with n, the number of X observations; m, the number of Y observations; and α, the chosen level of significance.

If we use the Mann–Whitney procedure to test

$$H_0: M_X \leq M_Y$$

against

$$H_A: M_X > M_Y$$

sufficiently large values of T will cause rejection so that the decision rule is:

Reject H_0: $M_X \leq M_Y$ if computed T is greater than $w_{1-\alpha}$, where $w_{1-\alpha} = nm - w_\alpha$.

For the two-sided test situation with

$$H_0: M_X = M_Y$$
$$H_A: M_X \neq M_Y$$

computed values of T that are either sufficiently large or sufficiently small will cause rejection of H_0. The decision rule for this case, then, is:

Reject H_0: $M_X = M_Y$ if the computed value of T is either less than $w_{\alpha/2}$ or greater than $w_{1-(\alpha/2)}$, where $w_{\alpha/2}$ is the critical value of T for n, m, and $\alpha/2$ given in Appendix Table L, and $w_{1-(\alpha/2)} = nm - w_{\alpha/2}$.

For this example the decision rule is:

Reject H_0 if the computed value of T is smaller than 45, the critical value of the test statistic for $n = 15$, $m = 10$, and $\alpha = .05$ found in Table L.

The rejection regions for each set of hypotheses are shown in Figure 13.6.1.

7. Calculation of test statistic. For our present example we have, as shown in Table 13.6.2, $S = 145$, so that

$$T = 145 - \frac{15(15+1)}{2} = 25$$

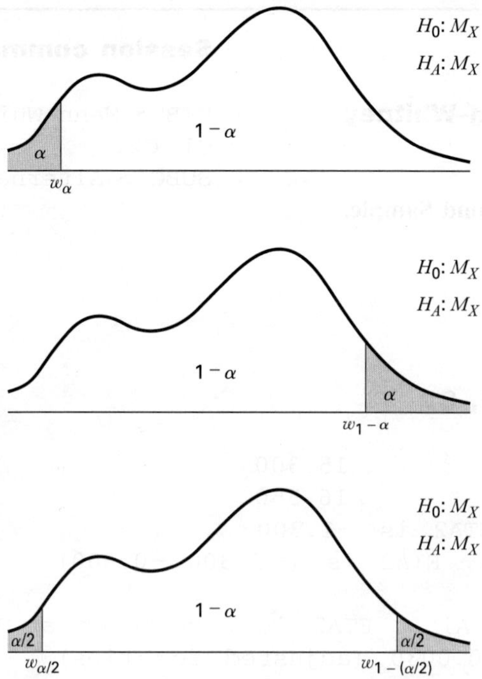

FIGURE 13.6.1 Mann–Whitney test rejection regions for three sets of hypotheses.

8. **Statistical decision.** When we enter Table L with $n = 15$, $m = 10$, and $\alpha = .05$, we find the critical value of w_α to be 45. Since $25 < 45$, we reject H_0.
9. **Conclusion.** We conclude that M_X is smaller than M_Y. This leads to the conclusion that prolonged inhalation of cadmium oxide does reduce the hemoglobin level.
10. ***p* value.** Since $22 < 25 < 30$, we have for this test $.005 > p > .001$. ■

Large-Sample Approximation When either n or m is greater than 20 we cannot use Appendix Table L to obtain critical values for the Mann–Whitney test. When this is the case we may compute

$$z = \frac{T - mn/2}{\sqrt{nm(n + m + 1)/12}} \tag{13.6.2}$$

and compare the result, for significance, with critical values of the standard normal distribution.

Mann–Whitney Statistic and the Wilcoxon Statistic As was noted at the beginning of this section, the Mann–Whitney test is sometimes referred to as the

Dialog box:

Stat ➤ Nonparametrics ➤ Mann–Whitney

Type *C1* in **First Sample** and *C2* in **Second Sample.**
At **Alternative** choose less than.
Click **OK.**

Session command:

```
MTB > Mann-Whitney 95.0
C1 C2;
SUBC > Alternative -1.
```

Output:

Mann–Whitney Test and CI: C1, C2

```
C1     N = 15     Median =        15.300
C2     N = 10     Median =        16.550
Point estimate for ETA1 - ETA2 is -1.300
95.1 Percent C.I. for ETA1 - ETA2 is (-2.300,-0.600)
W = 145.0
Test of ETA1 = ETA2 vs. ETA1 < ETA2 is significant at 0.0030
The test is significant at 0.0030 (adjusted for ties)
```

FIGURE 13.6.2 MINITAB procedure and output for Example 13.6.1.

Ranks

	y	N	Mean Rank	Sum of Rank
x	1.000000	15	9.67	146.00
	2.000000	10	18.00	180.00
	Total	25		

Test Statistic[b]

	x
Mann-Whitney U	25.000
Wilcoxon W	145.000
Z	−2.775
Asymp. Sig. (2-tailed)	.006
Exact Sig. [2*(1-tailed Sig.)]	.004[a]

a. Not corrected for ties
b. Grouping Variable: y

FIGURE 13.6.3 SPSS output for Example 13.6.1.

Mann–Whitney-Wilcoxon test. Indeed, many computer packages give the test value of both the Mann–Whitney test (U) and the Wilcoxon test (W). These two tests are algebraically equivalent tests, and are related by the following equality when there are no ties in the data:

$$U + W = \frac{m(m + 2n + 1)}{2} \tag{13.6.3}$$

Computer Analysis Many statistics software packages will perform the Mann–Whitney test. With the data of two samples stored in Columns 1 and 2, for example, MINITAB will perform a one-sided or two-sided test. The MINITAB procedure and output for Example 13.6.1 are shown in Figure 13.6.2.

The SPSS output for Example 13.6.1 is shown in Figure 13.6.3. As we see this output provides the Mann–Whitney test, the Wilcoxon test, and large-sample z approximation.

EXERCISES

13.6.1 Cranor and Christensen (A-4) studied diabetics insured by two employers. Group 1 subjects were employed by the City of Asheville, North Carolina, and group 2 subjects were employed by Mission–St. Joseph's Health System. At the start of the study, the researchers performed the Mann–Whitney test to determine if a significant difference in weight existed between the two study groups. The data are displayed in the following table.

Weight (Pounds)					
Group 1			Group 2		
252	215	240	185	195	220
240	190	302	310	210	295
205	270	312	212	190	202
200	159	126	238	172	268
170	204	268	184	190	220
170	215	215	136	140	311
320	254	183	200	280	164
148	164	287	270	264	206
214	288	210	200	270	170
270	138	225	212	210	190
265	240	258	182	192	
203	217	221	225	126	

Source: Carole W. Carnor, Ph.D. Used with permission.

May we conclude, on the basis of these data, that patients in the two groups differ significantly with respect to weight? Let $\alpha = .05$.

13.6.2 One of the purposes of a study by Liu et al. (A-5) was to determine the effects of MRZ 2/579 (a receptor antagonist shown to provide neuroprotective activity in vivo and in vitro) on neurological deficit in Sprague–Dawley rats. In the study, 10 rats were to receive MRZ 2/579 and nine rats were to receive regular saline. Prior to treatment, researchers studied the blood gas levels in the two groups of rats. The following table shows the pO_2 levels for the two groups.

Saline (mmHg)	MRZ 2/579 (mmHg)
112.5	133.3
106.3	106.4
99.5	113.1
98.3	117.2
103.4	126.4
109.4	98.1
108.9	113.4
107.4	116.8
	116.5

Source: Ludmila Belayev, M.D. Used with permission.

May we conclude, on the basis of these data, that, in general, subjects on saline have, on average, lower pO_2 levels at baseline? Let $\alpha = .01$.

13.6.3 The purpose of a study by researchers at the Cleveland (Ohio) Clinic (A-6) was to determine if the use of Flomax® reduced the urinary side effects commonly experienced by patients following brachytherapy (permanent radioactive seed implant) treatment for prostate cancer. The following table shows the American Urological Association (AUA) symptom index scores for two groups of subjects after 8 weeks of treatment. The higher the AUA index, the more severe the urinary obstruction and irritation.

AUA Index (Flomax®)			AUA Index (Placebo)		
1	5	11	1	6	12
1	5	11	1	6	12
2	6	11	2	6	13
2	6	11	2	6	14
2	7	12	2	6	17
2	7	12	3	7	18
3	7	13	3	8	19
3	7	14	3	8	20
3	8	16	3	9	23
4	8	16	4	9	23
4	8	18	4	10	
4	8	21	4	10	
4	9	31	5	11	
4	9		5	11	
4	10		5	12	

Source: Chandana Reddy, M.S. Used with permission.

May we conclude, on the basis of these data, that the median AUA index in the Flomax® group differs significantly from the median AUA index of the placebo group? Let $\alpha = .05$.

13.7 THE KOLMOGOROV–SMIRNOV GOODNESS-OF-FIT TEST

When one wishes to know how well the distribution of sample data conforms to some theoretical distribution, a test known as the Kolmogorov–Smirnov goodness-of-fit test provides an alternative to the chi-square goodness-of-fit test discussed in Chapter 12. The test gets its name from A. Kolmogorov and N. V. Smirnov, two Russian mathematicians who introduced two closely related tests in the 1930s.

Kolmogorov's work (6) is concerned with the one-sample case as discussed here. Smirnov's work (7) deals with the case involving two samples in which interest centers on testing the hypothesis that the distributions of the two-parent populations are identical. The test for the first situation is frequently referred to as the Kolmogorov–Smirnov one-sample test. The test for the two-sample case, commonly referred to as the Kolmogorov–Smirnov two-sample test, will not be discussed here.

The Test Statistic In using the Kolmogorov–Smirnov goodness-of-fit test, a comparison is made between some theoretical cumulative distribution function, $F_T(x)$, and a sample cumulative distribution function, $F_S(x)$. The sample is a random sample from a population with unknown cumulative distribution function $F(x)$. It will be recalled (Section 4.2) that a cumulative distribution function gives the probability that X is equal to or less than a particular value, x. That is, by means of the sample cumulative distribution function, $F_S(x)$, we may estimate $P(X \leq x)$. If there is close agreement between the theoretical and sample cumulative distributions, the hypothesis that the sample was drawn from the population with the specified cumulative distribution function, $F_T(x)$, is supported. If, however, there is a discrepancy between the theoretical and observed cumulative distribution functions too great to be attributed to chance alone, when H_0 is true, the hypothesis is rejected.

The difference between the theoretical cumulative distribution function, $F_T(x)$, and the sample cumulative distribution function, $F_S(x)$, is measured by the statistic D, which is the greatest vertical distance between $F_S(x)$ and $F_T(x)$. When a two-sided test is appropriate, that is, when the hypotheses are

$$H_0: F(x) = F_T(x) \quad \text{for all } x \quad \text{from } -\infty \text{ to } +\infty$$
$$H_A: F(x) \neq F_T(x) \quad \text{for at least one } x$$

the test statistic is

$$D = \sup_x |F_S(x) - F_T(x)| \tag{13.7.1}$$

which is read, "D equals the supremum (greatest), over all x, of the absolute value of the difference $F_S(X)$ minus $F_T(X)$."

The null hypothesis is rejected at the α level of significance if the computed value of D exceeds the value shown in Appendix Table M for $1 - \alpha$ (two-sided) and the sample size n.

Assumptions The assumptions underlying the Kolmogorov–Smirnov test include the following:

1. The sample is a random sample.
2. The hypothesized distribution $F_T(x)$ is continuous.

When values of D are based on a discrete theoretical distribution, the test is conservative. When the test is used with discrete data, then, the investigator should bear in mind that the true probability of committing a type I error is at most equal to α, the stated level of significance. The test is also conservative if one or more parameters have to be estimated from sample data.

EXAMPLE 13.7.1

Fasting blood glucose determinations made on 36 nonobese, apparently healthy, adult males are shown in Table 13.7.1. We wish to know if we may conclude that these data are not from a normally distributed population with a mean of 80 and a standard deviation of 6.

Solution:

1. **Data.** See Table 13.7.1.
2. **Assumptions.** The sample available is a simple random sample from a continuous population distribution.
3. **Hypotheses.** The appropriate hypotheses are

$$H_0: F(x) = F_T(x) \quad \text{for all } x \text{ from } -\infty \text{ to } +\infty$$
$$H_A: F(x) \neq F_T(x) \quad \text{for at least one } x$$

   Let $\alpha = .05$.

4. **Test statistic.** See Equation 13.7.1.
5. **Distribution of test statistic.** Critical values of the test statistic for selected values of α are given in Appendix Table M.
6. **Decision rule.** Reject H_0 if the computed value of D exceeds .221, the critical value of D for $n = 36$ and $\alpha = .05$.
7. **Calculation of test statistic.** Our first step is to compute values of $F_S(x)$ as shown in Table 13.7.2.

TABLE 13.7.1 Fasting Blood Glucose Values (mg/100 ml) for 36 Nonobese, Apparently Healthy, Adult Males

75	92	80	80	84	72
84	77	81	77	75	81
80	92	72	77	78	76
77	86	77	92	80	78
68	78	92	68	80	81
87	76	80	87	77	86

TABLE 13.7.2 Values of $F_S(x)$ for Example 13.7.1

x	Frequency	Cumulative Frequency	$F_S(x)$
68	2	2	.0556
72	2	4	.1111
75	2	6	.1667
76	2	8	.2222
77	6	14	.3889
78	3	17	.4722
80	6	23	.6389
81	3	26	.7222
84	2	28	.7778
86	2	30	.8333
87	2	32	.8889
92	4	36	1.0000
	36		

Each value of $F_S(x)$ is obtained by dividing the corresponding cumulative frequency by the sample size. For example, the first value of $F_S(x) = 2/36 = .0556$.

We obtain values of $F_T(x)$ by first converting each observed value of x to a value of the standard normal variable, z. From Appendix Table D we then find the area between $-\infty$ and z. From these areas we are able to compute values of $F_T(x)$. The procedure, which is similar to that used to obtain expected relative frequencies in the chi-square goodness-of-fit test, is summarized in Table 13.7.3.

TABLE 13.7.3 Steps in Calculation of $F_T(x)$ for Example 13.7.1

x	$z = (x - 80)/6$	$F_T(x)$
68	−2.00	.0228
72	−1.33	.0918
75	−.83	.2033
76	−.67	.2514
77	−.50	.3085
78	−.33	.3707
80	.00	.5000
81	.17	.5675
84	.67	.7486
86	1.00	.8413
87	1.17	.8790
92	2.00	.9772

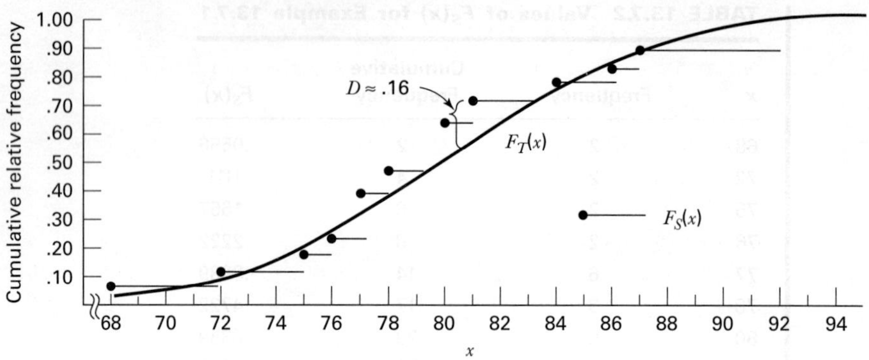

FIGURE 13.7.1 $F_S(x)$ and $F_T(x)$ for Example 13.7.1.

The test statistic D may be computed algebraically, or it may be determined graphically by actually measuring the largest vertical distance between the curves of $F_S(x)$ and $F_T(x)$ on a graph. The graphs of the two distributions are shown in Figure 13.7.1.

Examination of the graphs of $F_S(x)$ and $F_T(x)$ reveals that $D \approx .16 = (.72 - .56)$. Now let us compute the value of D algebraically. The possible values of $|F_S(x) - F_T(x)|$ are shown in Table 13.7.4. This table shows that the exact value of D is .1547.

8. **Statistical decision.** Reference to Table M reveals that a computed D of .1547 is not significant at any reasonable level. Therefore, we are not willing to reject H_0.

9. **Conclusion.** The sample may have come from the specified distribution.

10. ***p* value.** Since we have a two-sided test, and since $.1547 < .174$, we have $p > .20$.

TABLE 13.7.4 Calculation of $|F_S(x) - F_T(x)|$ for Example 13.7.1

x	$F_S(x)$	$F_T(x)$	$\lvert F_S(x) - F_T(x)\rvert$
68	.0556	.0228	.0328
72	.1111	.0918	.0193
75	.1667	.2033	.0366
76	.2222	.2514	.0292
77	.3889	.3085	.0804
78	.4722	.3707	.1015
80	.6389	.5000	.1389
81	.7222	.5675	.1547
84	.7778	.7486	.0292
86	.8333	.8413	.0080
87	.8889	.8790	.0099
92	1.0000	.9772	.0228

■

Kolmogorov–Smirnov One-Sample Test

kolmogorov (response = glucose, method = asymp, di = no (mean = 80, stddev = 6), time_limit = none);

Data File:

Column Variable: Glucose

Sample Size: 36

Summary of the Test Statistic:

	Type	Mean	Std. Dev
Hypothesized distribution F(X)	Normal	80	6

Let S(X) be the empirical distribution.

Inference:

	Statistic		
Item	**Sup{|S(X) - F(X)|}**	**Sup{S(X) - F(X)}**	**Sup{F(X) - S(X)}**
Observed Statistic	0.156	0.156	0.09122
Stand. Statistic	0.9362	0.9362	0.5473
Asymptotic p-value	0.3447	0.1732	0.5493

FIGURE 13.7.2 StatXact output for Example 13.7.1

StatXact is often used for nonparametric statistical analysis. This particular software program has a nonparametric module that contains nearly all of the commonly used nonparametric tests, and many less common, but useful, procedures as well. Computer analysis using StatXact for the data in Example 13.7.1 is shown in Figure 13.7.2. Note that it provides the test statistic of D = 0.156 and the exact two-sided p value of .3447.

A Precaution The reader should be aware that in determining the value of D, *it is not always sufficient to compute and choose from the possible values of* $|F_S(x) - F_T(x)|$. *The largest vertical distance between* $F_S(x)$ and $F_T(x)$ *may not occur at an observed value, x, but at some other value of X.* Such a situation is illustrated in Figure 13.7.3. We see that if only values of $|F_S(x) - F_T(x)|$ at the left endpoints of the horizontal bars are considered, we would incorrectly compute D as $|.2 - .4| = .2$. One can see by examining the graph, however, that the largest vertical distance between $F_S(x)$ and $F_T(x)$ occurs at the right endpoint of the horizontal bar originating at the point corresponding to $x = .4$, and the correct value of D is $|.5 - .2| = .3$.

One can determine the correct value of D algebraically by computing, in addition to the differences $|F_S(x) - F_T(x)|$, the differences $|F_S(x_{i-1}) - F_T(x_i)|$ for all values of $i = 1, 2, \ldots, r + 1$, where $r =$ the number of different values of x and $F_S(x_0) = 0$. The correct value of the test statistic will then be

$$D = \underset{1 \leq i \leq r}{\text{maximum}} \left\{ \text{maximum}\left[|F_S(x_i) - F_T(x_i)|, |F_S(x_{i-1}) - F_T(x_i)| \right] \right\} \quad (13.7.2)$$

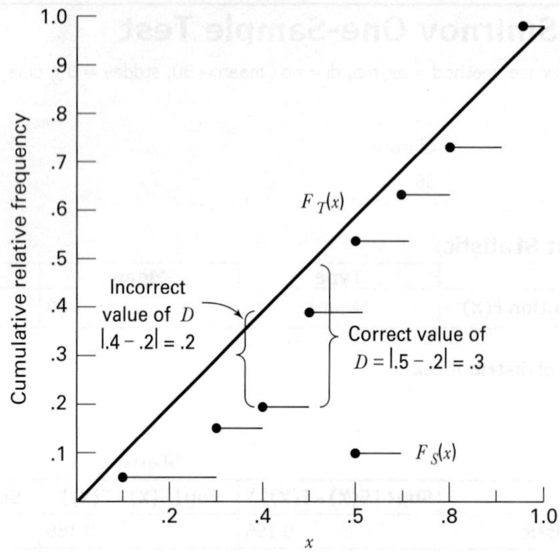

FIGURE 13.7.3 Graph of fictitious data showing correct calculation of *D*.

Advantages and Disadvantages The following are some important points of comparison between the Kolmogorov–Smirnov and the chi-square goodness-of-fit tests.

1. The Kolmogorov–Smirnov test does not require that the observations be grouped as is the case with the chi-square test. The consequence of this difference is that the Kolmogorov–Smirnov test makes use of all the information present in a set of data.
2. The Kolmogorov–Smirnov test can be used with any size sample. It will be recalled that certain minimum sample sizes are required for the use of the chi-square test.
3. As has been noted, the Kolmogorov–Smirnov test is not applicable when parameters have to be estimated from the sample. The chi-square test may be used in these situations by reducing the degrees of freedom by 1 for each parameter estimated.
4. The problem of the assumption of a continuous theoretical distribution has already been mentioned.

EXERCISES

13.7.1 The weights at autopsy of the brains of 25 adults suffering from a certain disease were as follows:

Weight of Brain (grams)

859	1073	1041	1166	1117
962	1051	1064	1141	1202
973	1001	1016	1168	1255
904	1012	1002	1146	1233
920	1039	1086	1140	1348

Can one conclude from these data that the sampled population is not normally distributed with a mean of 1050 and a standard deviation of 50? Determine the p value for this test.

13.7.2 IQs of a sample of 30 adolescents arrested for drug abuse in a certain metropolitan jurisdiction were as follows:

IQ					
95	100	91	106	109	110
98	104	97	100	107	119
92	106	103	106	105	112
101	91	105	102	101	110
101	95	102	104	107	118

Do these data provide sufficient evidence that the sampled population of IQ scores is not normally distributed with a mean of 105 and a standard deviation of 10? Determine the p value.

13.7.3 For a sample of apparently normal subjects who served as controls in an experiment, the following systolic blood pressure readings were recorded at the beginning of the experiment:

162	177	151	167
130	154	179	146
147	157	141	157
153	157	134	143
141	137	151	161

Can one conclude on the basis of these data that the population of blood pressures from which the sample was drawn is not normally distributed with $\mu = 150$ and $\sigma = 12$? Determine the p value.

13.8 THE KRUSKAL–WALLIS ONE-WAY ANALYSIS OF VARIANCE BY RANKS

In Chapter 8 we discuss how one-way analysis of variance may be used to test the null hypothesis that several population means are equal. When the assumptions underlying this technique are not met, that is, when the populations from which the samples are drawn are not normally distributed with equal variances, or when the data for analysis consist only of ranks, a nonparametric alternative to the one way analysis of variance may be used to test the hypothesis of equal location parameters. As was pointed out in Section 13.5, the median test may be extended to accommodate the situation involving more than two samples. A deficiency of this test, however, is the fact that it uses only a small amount of the information available. The test uses only information as to whether or not the observations are above or below a single number, the median of the combined samples. The test does not directly use measurements of known quantity. Several nonparametric analogs to analysis of variance are available that use more information by taking into account the magnitude of each observation relative to the magnitude of every other observation. Perhaps the best known of these procedures is the Kruskal–Wallis one-way analysis of variance by ranks (8).

The Kruskal–Wallis Procedure The application of the test involves the following steps.

1. The $n_1, n_2, \ldots, n_k$ observations from the k samples are combined into a single series of size n and arranged in order of magnitude from smallest to largest. The observations are then replaced by ranks from 1, which is assigned to the smallest observation, to n, which is assigned to the largest observation. When two or more observations have the same value, each observation is given the mean of the ranks for which it is tied.
2. The ranks assigned to observations in each of the k groups are added separately to give k rank sums.
3. The test statistic

$$H = \frac{12}{n(n+1)} \sum_{j=1}^{k} \frac{R_j^2}{n_j} - 3(n+1) \tag{13.8.1}$$

is computed. In Equation 13.8.1,

k = the number of samples

n_j = the number of observations in the jth sample

n = the number of observations in all samples combined

R_j = the sum of the ranks in the jth sample

4. When there are three samples and five or fewer observations in each sample, the significance of the computed H is determined by consulting Appendix Table N. When there are more than five observations in one or more of the samples, H is compared with tabulated values of χ^2 with $k - 1$ degrees of freedom.

EXAMPLE 13.8.1

In a study of pulmonary effects on guinea pigs, Lacroix et al. (A-7) exposed ovalbumin (OA)-sensitized guinea pigs to regular air, benzaldehyde, or acetaldehyde. At the end of exposure, the guinea pigs were anesthetized and allergic responses were assessed in bronchoalveolar lavage (BAL). One of the outcome variables examined was the count of eosinophil cells, a type of white blood cell that can increase with allergies. Table 13.8.1 gives the eosinophil cell count ($\times 10^6$) for the three treatment groups.

Can we conclude that the three populations represented by the three samples differ with respect to eosinophil cell count? We can so conclude if we can reject the null hypothesis that the three populations do not differ in eosinophil cell count.

TABLE 13.8.1 Eosinophil Count for Ovalbumin-Sensitized Guinea Pigs

Eosinophil Cell Count ($\times 10^6$)		
Air	**Benzaldehyde**	**Acetaldehyde**
12.22	3.68	54.36
28.44	4.05	27.87
28.13	6.47	66.81
38.69	21.12	46.27
54.91	3.33	30.19

Source: G. Lacroix. Used with permission.

Solution:

1. **Data.** See Table 13.8.1.
2. **Assumptions.** The samples are independent random samples from their respective populations. The measurement scale employed is at least ordinal. The distributions of the values in the sampled populations are identical except for the possibility that one or more of the populations are composed of values that tend to be larger than those of the other populations.
3. **Hypotheses.**

   H_0: The population centers are all equal.

   H_A: At least one of the populations tends to exhibit larger values than at least one of the other populations.

   Let $\alpha = .01$.
4. **Test statistic.** See Equation 13.8.1.
5. **Distribution of test statistic.** Critical values of H for various sample sizes and α levels are given in Appendix Table N.
6. **Decision rule.** The null hypothesis will be rejected if the computed value of H is so large that the probability of obtaining a value that large or larger when H_0 is true is equal to or less than the chosen significance level, α.
7. **Calculation of test statistic.** When the three samples are combined into a single series and ranked, the table of ranks shown in Tablc 13.8.2 may be constructed.

   The null hypothesis implies that the observations in the three samples constitute a single sample of size 15 from a single population. If this is true, we would expect the ranks to be well distributed among the three groups. Consequently, we would expect the total sum of ranks to be divided among the three groups in proportion to group size. Departures

TABLE 13.8.2 The Data of Table 13.8.1 Replaced by Ranks

Air	Benzaldehyde	Acetaldehyde
5	2	13
9	3	7
8	4	15
11	6	12
14	1	10
$R_1 = 47$	$R_2 = 16$	$R_3 = 57$

from these conditions are reflected in the magnitude of the test statistics H.

From the data in Table 13.8.2 and Equation 13.8.1, we obtain

$$H = \frac{12}{15(16)}\left[\frac{(47)^2}{5} + \frac{(16)^2}{5} + \frac{(57)^2}{5}\right] - 3(15 + 1) = 9.14$$

8. Statistical decision. Table N shows that when the n_j are 5, 5, and 5, the probability of obtaining a value of $H = 9.14$ is less than .009. The null hypothesis can be rejected at the .01 level of significance.

9. Conclusion. We conclude that there is a difference in the average eosinophil cell count among the three populations.

10. *p* value. For this test, $p < .009$. ■

Ties When ties occur among the observations, we may adjust the value of H by dividing it by

$$1 - \frac{\sum T}{n^3 - n} \tag{13.8.2}$$

where $T = t^3 - t$. The letter t is used to designate the number of tied observations in a group of tied values. In our example there are no groups of tied values but, in general, there may be several groups of tied values resulting in several values of T.

The effect of the adjustment for ties is usually negligible. Note also that the effect of the adjustment is to increase H, so that if the unadjusted H is significant at the chosen level, there is no need to apply the adjustment.

More than Three Samples/Large Samples Now let us illustrate the procedure when there are more than three samples and at least one of the n_j is greater than 5.

TABLE 13.8.3 Net Book Value of Equipment per Bed by Hospital Type

Type Hospital				
A	B	C	D	E
\$1735(11)	\$5260(35)	\$2790(20)	\$3475(26)	\$6090(40)
1520(2)	4455(28)	2400(12)	3115(22)	6000(38)
1476(1)	4480(29)	2655(16)	3050(21)	5894(37)
1688(7)	4325(27)	2500(13)	3125(23)	5705(36)
1702(10)	5075(32)	2755(19)	3275(24)	6050(39)
2667(17)	5225(34)	2592(14)	3300(25)	6150(41)
1575(4)	4613(30)	2601(15)	2730(18)	5110(33)
1602(5)	4887(31)	1648(6)		
1530(3)		1700(9)		
1698(8)				
$R_1 = 68$	$R_2 = 246$	$R_3 = 124$	$R_4 = 159$	$R_5 = 264$

EXAMPLE 13.8.2

Table 13.8.3 shows the net book value of equipment capital per bed for a sample of hospitals from each of five types of hospitals. We wish to determine, by means of the Kruskal–Wallis test, if we can conclude that the average net book value of equipment capital per bed differs among the five types of hospitals. The ranks of the 41 values, along with the sum of ranks for each sample, are shown in the table.

Solution: From the sums of the ranks we compute

$$H = \frac{12}{41(41+1)}\left[\frac{(68)^2}{10} + \frac{(246)^2}{8} + \frac{(124)^2}{9} + \frac{(159)^2}{7} + \frac{(264)^2}{7}\right] - 3(41+1)$$
$$= 36.39$$

Reference to Appendix Table F with $k - 1 = 4$ degrees of freedom indicates that the probability of obtaining a value of H as large as or larger than 36.39, due to chance alone, when there is no difference among the populations, is less than .005. We conclude, then, that there is a difference among the five populations with respect to the average value of the variable of interest. ■

Computer Analysis The MINITAB software package computes the Kruskal–Wallis test statistic and provides additional information. After we enter the eosinophil counts in Table 13.8.1 into Column 1 and the group codes into Column 2, the MINITAB procedure and output are as shown in Figure 13.8.1.

Data:

```
C1: 12.22 28.44 28.13 38.69 54.91 3.68 4.05 6.47 21.12 3.33 54.36 27.87 66.81 46.27 30.19
C2: 1 1 1 1 1 2 2 2 2 2 3 3 3 3 3
```

Dialog box:

Stat ➤ Nonparametrics ➤ Kruskal–Wallis
Type *C1* in **Response** and *C2* in **Factor.** Click **OK.**

Session command:

```
MTB > Kruskal-Wallis C1 C2.
```

Output:

Kruskal–Wallis Test: C1 versus C2

```
Kruskal-Wallis Test on C1

C2         N    Median    Ave Rank       Z
1          5    28.440         9.4    0.86
2          5     4.050         3.2   -2.94
3          5    46.270        11.4    2.08
Overall   15                   8.0

H = 9.14  DF = 2  P = 0.010
```

FIGURE 13.8.1 MINITAB procedure and output, Kruskal–Wallis test of eosinophil count data in Table 13.8.1.

EXERCISES

For the following exercises, perform the test at the indicated level of significance and determine the *p* value.

13.8.1 In a study of healthy subjects grouped by age (Younger: 19–50 years, Seniors: 65–75 years, and Longeval: 85–102 years), Herrmann et al. (A-8) measured their vitamin B-12 levels (ng/L). All elderly subjects were living at home and able to carry out normal day-to-day activities. The following table shows vitamin B-12 levels for 50 subjects in the young group, 92 seniors, and 90 subjects in the longeval group.

Young (19–50 Years)		Senior (65–75 Years)				Longeval (85–102 Years)			
230	241	319	371	566	170	148	149	631	198
477	442	190	460	290	542	1941	409	305	321
561	491	461	440	271	282	128	229	393	2772
347	279	163	520	308	194	145	183	282	428
566	334	377	256	440	445	174	193	273	259
260	247	190	335	238	921	495	161	157	111
300	314	375	137	525	1192	460	400	1270	262

(Continued)

Young (19–50 Years)		Senior (65–75 Years)				Longeval (85–102 Years)			
230	254	229	452	298	748	548	348	252	161
215	419	193	437	153	187	198	175	262	1113
260	335	294	236	323	350	165	540	381	409
349	455	740	432	205	1365	226	293	162	378
315	297	194	411	248	232	557	196	340	203
257	456	780	268	371	509	166	632	370	221
536	668	245	703	668	357	218	438	483	917
582	240	258	282	197	201	186	368	222	244
293	320	419	290	260	177	346	262	277	
569	562	372	286	198	872	239	190	226	
325	360	413	143	336		240	241	203	
275	357	685	310	421		136	195	369	
172	609	136	352	712		359	220	162	
2000	740	441	262	461		715	164	95	
240	430	423	404	631		252	279	178	
235	645	617	380	1247		414	297	530	
284	395	985	322	1033		372	474	334	
883	302	170	340	285		236	375	521	

Source: W. Herrmann and H. Schorr. Used with permission.

May we conclude, on the basis of these data, that the populations represented by these samples differ with respect to vitamin B-12 levels? Let $\alpha = .01$.

13.8.2 The following are outpatient charges $(-\$100)$ made to patients for a certain surgical procedure by samples of hospitals located in three different areas of the country:

Area		
I	II	III
\$80.75	\$58.63	\$84.21
78.15	72.70	101.76
85.40	64.20	107.74
71.94	62.50	115.30
82.05	63.24	126.15

Can we conclude at the .05 level of significance that the three areas differ with respect to the charges?

13.8.3 A study of young children by Flexer et al. (A-9) published in the *Hearing Journal* examines the effectiveness of an FM sound field when teaching phonics to children. In the study, children in a classroom with no phonological or phonemic awareness training (control) were compared to a class with phonological and phonemic awareness (PPA) and to a class that utilized phonological and phonemic awareness training and the FM sound field (PPA/FM). A total of 53 students from three separate preschool classrooms participated in this study. Students were given a measure of phonemic awareness in preschool and then at the end of the first semester of kindergarten. The improvement scores are listed in the following table as measured by the Yopp–Singer Test of Phonemic Segmentation.

Improvement (Control)		Improvement PPA	Improvement PPA/FM	
0	1	2	1	19
−1	1	3	3	20
0	2	15	7	21
1	2	18	9	21
4	3	19	11	22
5	6	20	17	22
9	7	5	17	15
9	8		17	17
13	9		18	17
18	18		18	19
0	20		19	22
0			19	

Source: John P. Holcomb, Jr., Ph.D. Used with permission.

Test for a significant difference among the three groups. Let $\alpha = .05$.

13.8.4 Refer to Example 13.8.1. Another variable of interest to Lacroix et al. (A-7) was the number of alveolar cells in three groups of subjects exposed to air, benzaldehyde, or acetaldehyde. The following table gives the information for six guinea pigs in each of the three treatment groups.

Number of Alveolar Cells ($\times 10^6$)		
Air	**Benzaldehyde**	**Acetaldehyde**
0.55	0.81	0.65
0.48	0.56	13.69
7.8	1.11	17.11
8.72	0.74	7.43
0.65	0.77	5.48
1.51	0.83	0.99
0.55	0.81	0.65

Source: G. Lacroix. Used with permission.

May we conclude, on the basis of these data, that the number of alveolar cells in ovalbumin-sensitized guinea pigs differs with type of exposure? Let $\alpha = .05$.

13.8.5 The following table shows the pesticide residue levels (ppb) in blood samples from four populations of human subjects. Use the Kruskal–Wallis test to test at the .05 level of significance the null hypothesis that there is no difference among the populations with respect to average level of pesticide residue.

Population				Population			
A	**B**	**C**	**D**	**A**	**B**	**C**	**D**
10	4	15	7	44	11	9	4
37	35	5	11	12	7	11	5
12	32	10	10	15	32	9	2

(*Continued*)

Population				Population			
A	B	C	D	A	B	C	D
31	19	12	8	42	17	14	6
11	33	6	2	23	8	15	3
9	18	6	5				

13.8.6 Hepatic γ-glutamyl transpeptidase (GGTP) activity was measured in 22 patients undergoing percutaneous liver biopsy. The results were as follows:

Subject	Diagnosis	Hepatic GGTP Level
1	Normal liver	27.7
2	Primary biliary cirrhosis	45.9
3	Alcoholic liver disease	85.3
4	Primary biliary cirrhosis	39.0
5	Normal liver	25.8
6	Persistent hepatitis	39.6
7	Chronic active hepatitis	41.8
8	Alcoholic liver disease	64.1
9	Persistent hepatitis	41.1
10	Persistent hepatitis	35.3
11	Alcoholic liver disease	71.5
12	Primary biliary cirrhosis	40.9
13	Normal liver	38.1
14	Primary biliary cirrhosis	40.4
15	Primary biliary cirrhosis	34.0
16	Alcoholic liver disease	74.4
17	Alcoholic liver disease	78.2
18	Persistent hepatitis	32.6
19	Chronic active hepatitis	46.3
20	Normal liver	39.6
21	Chronic active hepatitis	52.7
22	Chronic active hepatitis	57.2

Can we conclude from these sample data that the average population GGTP level differs among the five diagnostic groups? Let $\alpha = .05$ and find the p value.

13.9 THE FRIEDMAN TWO-WAY ANALYSIS OF VARIANCE BY RANKS

Just as we may on occasion have need of a nonparametric analog to the parametric one-way analysis of variance, we may also find it necessary to analyze the data in a two-way classification by nonparametric methods analogous to the two-way analysis of variance. Such a need may arise because the assumptions necessary for parametric analysis of variance are not met, because the measurement scale employed is weak, or because results are needed in a hurry. A test frequently employed under these circumstances is the Friedman two-way analysis of variance by ranks (9, 10). This test is appropriate whenever the data are measured on, at least, an ordinal scale and can be meaningfully arranged in a two-way classification as is given for the randomized block experiment discussed in Chapter 8. The following example illustrates this procedure.

EXAMPLE 13.9.1

A physical therapist conducted a study to compare three models of low-volt electrical stimulators. Nine other physical therapists were asked to rank the stimulators in order of preference. A rank of 1 indicates first preference. The results are shown in Table 13.9.1. We wish to know if we can conclude that the models are not preferred equally.

Solution:

1. **Data.** See Table 13.9.1.
2. **Assumptions.** The observations appearing in a given block are independent of the observations appearing in each of the other blocks, and within each block measurement on at least an ordinal scale is achieved.
3. **Hypothesis.** In general, the hypotheses are:

   H_0: The treatments all have identical effects.

   H_A: At least one treatment tends to yield larger observations than at least one of the other treatments.

   For our present example we state the hypotheses as follows:

   H_0: The three models are equally preferred.

   H_A: The three models are not equally preferred.

   Let $\alpha = .05$.

TABLE 13.9.1 Physical Therapists' Rankings of Three Models of Low-Volt Electrical Stimulators

	Model		
Therapist	A	B	C
1	2	3	1
2	2	3	1
3	2	3	1
4	1	3	2
5	3	2	1
6	1	2	3
7	2	3	1
8	1	3	2
9	1	3	2
R_j	15	25	14

4. **Test statistic.** By means of the Friedman test we will be able to determine if it is reasonable to assume that the columns of ranks have been drawn from the same population. If the null hypothesis is true we would expect the observed distribution of ranks within any column to be the result of chance factors and, hence, we would expect the numbers 1, 2, and 3 to occur with approximately the same frequency in each column. If, on the other hand, the null hypothesis is false (that is, the models are not equally preferred), we would expect a preponderance of relatively high (or low) ranks in at least one column. This condition would be reflected in the sums of the ranks. The Friedman test will tell us whether or not the observed sums of ranks are so discrepant that it is not likely they are a result of chance when H_0 is true.

   Since the data already consist of rankings within blocks (rows), our first step is to sum the ranks within each column (treatment). These sums are the R_j shown in Table 13.9.1. A test statistic, denoted by Friedman as χ_r^2, is computed as follows:

$$\chi_r^2 = \frac{12}{nk(k+1)} \sum_{j=1}^{k} (R_j)^2 - 3n(k+1) \tag{13.9.1}$$

   where n = the number of rows (blocks) and k = the number of columns (treatments).

5. **Distribution of test statistic.** Critical values for various values of n and k are given in Appendix Table O.

6. **Decision rule.** Reject H_0 if the probability of obtaining (when H_0 is true) a value of χ_r^2 as large as or larger than actually computed is less than or equal to α.

7. **Calculation of test statistic.** Using the data in Table 13.9.1 and Equations 13.9.1, we compute

$$\chi_r^2 = \frac{12}{9(3)(3+1)} [(15)^2 + (25)^2 + (14)^2] - 3(9)(3+1) = 8.222$$

8. **Statistical decision.** When we consult Appendix Table Oa, we find that the probability of obtaining a value of χ_r^2 as large as 8.222 due to chance alone, when the null hypothesis is true, is .016. We are able, therefore, to reject the null hypothesis.

9. **Conclusion.** We conclude that the three models of low-volt electrical stimulator are not equally preferred.

10. ***p* value.** For this test, $p = .016$. ■

Ties When the original data consist of measurements on an interval or a ratio scale instead of ranks, the measurements are assigned ranks based on their relative magnitudes within blocks. If ties occur, each value is assigned the mean of the ranks for which it is tied.

Large Samples When the values of k and/or n exceed those given in Table O, the critical value of χ_r^2 is obtained by consulting the χ^2 table (Table F) with the chosen α and $k - 1$ degrees of freedom.

EXAMPLE 13.9.2

Table 13.9.2 shows the responses, in percent decrease in salivary flow, of 16 experimental animals following different dose levels of atropine. The ranks (in parentheses) and the sum of the ranks are also given in the table. We wish to see if we may conclude that the different dose levels produce different responses. That is, we wish to test the null hypothesis of no difference in response among the four dose levels.

Solution: From the data, we compute

$$\chi_r^2 = \frac{12}{16(4)(4+1)}\left[(20)^2 + (36.5)^2 + (44)^2 + (59.5)^2\right] - 3(16)(4+1) = 30.32$$

Reference to Table F indicates that with $k - 1 = 3$ degrees of freedom the probability of getting a value of χ_r^2 as large as 30.32 due to chance alone is, when H_0 is true, less than .005. We reject the null hypothesis and conclude that the different dose levels do produce different responses.

TABLE 13.9.2 Percent Decrease in Salivary Flow of Experimental Animals Following Different Dose Levels of Atropine

Animal Number	Dose Level A	B	C	D
1	29(1)	48(2)	75(3)	100(4)
2	72(2)	30(1)	100(3.5)	100(3.5)
3	70(1)	100(4)	86(2)	96(3)
4	54(2)	35(1)	90(3)	99(4)
5	5(1)	43(3)	32(2)	81(4)
6	17(1)	40(2)	76(3)	81(4)
7	74(1)	100(3)	100(3)	100(3)
8	6(1)	34(2)	60(3)	81(4)
9	16(1)	39(2)	73(3)	79(4)
10	52(2)	34(1)	88(3)	96(4)
11	8(1)	42(3)	31(2)	79(4)
12	29(1)	47(2)	72(3)	99(4)
13	71(1)	100(3.5)	97(2)	100(3.5)
14	7(1)	33(2)	58(3)	79(4)
15	68(1)	99(4)	84(2)	93(3)
16	70(2)	30(1)	99(3.5)	99(3.5)
R_j	20	36.5	44	59.5

■

Dialog box:

Stat ➤ Nonparametrics ➤ Friedman

Type *C3* in **Response,** *C1* in **Treatment** and *C2* in **Blocks.** Click **OK.**

Session command:

```
MTB > FRIEDMAN C3 C1 C2
```

Output:

Friedman Test: C3 versus C1 blocked by C2

```
S = 8.22 d.f. = 2 p = 0.017

                        Est.     Sum of
   C1      N          Median     RANKS
    1      9          2.0000      15.0
    2      9          2.6667      25.0
    3      9          1.3333      14.0

Grand median = 2.0000
```

FIGURE 13.9.1 MINITAB procedure and output for Example 13.9.1.

Computer Analysis Many statistics software packages, including MINITAB, will perform the Friedman test. To use MINITAB we form three columns of data. We may, for example, set up the columns so that Column 1 contains numbers that indicate the treatment to which the observations belong, Column 2 contains numbers indicating the blocks to which the observations belong, and Column 3 contains the observations. If we do this for Example 13.9.1, the MINITAB procedure and output are as shown in Figure 13.9.1.

EXERCISES

For the following exercises perform the test at the indicated level of significance and determine the *p* value.

13.9.1 The following table shows the scores made by nine randomly selected student nurses on final examinations in three subject areas:

	Subject Area		
Student Number	**Fundamentals**	**Physiology**	**Anatomy**
1	98	95	77
2	95	71	79

(*Continued*)

Student Number	Subject Area: Fundamentals	Physiology	Anatomy
3	76	80	91
4	95	81	84
5	83	77	80
6	99	70	93
7	82	80	87
8	75	72	81
9	88	81	83

Test the null hypothesis that student nurses constituting the population from which the above sample was drawn perform equally well in all three subject areas against the alternative hypothesis that they perform better in, at least, one area. Let $\alpha = .05$.

13.9.2 Fifteen randomly selected physical therapy students were given the following instructions: "Assume that you will marry a person with one of the following handicaps (the handicaps were listed and designated by the letters A to J). Rank these handicaps from 1 to 10 according to your first, second, third (and so on) choice of a handicap for your marriage partner." The results are shown in the following table.

Student Number	Handicap: A	B	C	D	E	F	G	H	I	J
1	1	3	5	9	8	2	4	6	7	10
2	1	4	5	7	8	2	3	6	9	10
3	2	3	7	8	9	1	4	6	5	10
4	1	4	7	8	9	2	3	6	5	10
5	1	4	7	8	10	2	3	6	5	9
6	2	3	7	9	8	1	4	5	6	10
7	2	4	6	9	8	1	3	7	5	10
8	1	5	7	9	10	2	3	4	6	8
9	1	4	5	7	8	2	3	6	9	10
10	2	3	6	8	9	1	4	7	5	10
11	2	4	5	8	9	1	3	7	6	10
12	2	3	6	8	10	1	4	5	7	9
13	3	2	6	9	8	1	4	7	5	10
14	2	5	7	8	9	1	3	4	6	10
15	2	3	6	7	8	1	5	4	9	10

Test the null hypothesis of no preference for handicaps against the alternative that some handicaps are preferred over others. Let $\alpha = .05$.

13.9.3 Ten subjects with exercise-induced asthma participated in an experiment to compare the protective effect of a drug administered in four dose levels. Saline was used as a control. The variable of interest was change in FEV_1 after administration of the drug or saline. The results were as follows:

		Dose Level of Drug (mg/ml)			
Subject	Saline	2	10	20	40
1	−.68	−.32	−.14	−.21	−.32
2	−1.55	−.56	−.31	−.21	−.16
3	−1.41	−.28	−.11	−.08	−.83
4	−.76	−.56	−.24	−.41	−.08
5	−.48	−.25	−.17	−.04	−.18
6	−3.12	−1.99	−1.22	−.55	−.75
7	−1.16	−.88	−.87	−.54	−.84
8	−1.15	−.31	−.18	−.07	−.09
9	−.78	−.24	−.39	−.11	−.51
10	−2.12	−.35	−.28	+.11	−.41

Can one conclude on the basis of these data that different dose levels have different effects? Let $\alpha = .05$ and find the p value.

13.10 THE SPEARMAN RANK CORRELATION COEFFICIENT

Several nonparametric measures of correlation are available to the researcher. Of these a frequently used procedure that is attractive because of the simplicity of the calculations involved is due to Spearman (11). The measure of correlation computed by this method is called the Spearman rank correlation coefficient and is designated by r_s. This procedure makes use of the two sets of ranks that may be assigned to the sample values of X and Y, the independent and continuous variables of a bivariate distribution.

Hypotheses The usually tested hypotheses and their alternatives are as follows:

(a) H_0: X and Y are mutually independent.
H_A: X and Y are not mutually independent.

(b) H_0: X and Y are mutually independent.
H_A: There is a tendency for large values of X and large values of Y to be paired together.

(c) H_0: X and Y are mutually independent.
H_A: There is a tendency for large values of X to be paired with small values of Y.

The hypotheses specified in (a) lead to a two-sided test and are used when it is desired to detect any departure from independence. The one-sided tests indicated by (b) and (c) are used, respectively, when investigators wish to know if they can conclude that the variables are directly or inversely correlated.

The Procedure The hypothesis-testing procedure involves the following steps.

1. Rank the values of X from 1 to n (numbers of pairs of values of X and Y in the sample). Rank the values of Y from 1 to n.

2. Compute d_i for each pair of observations by subtracting the rank of Y_i from the rank of X_i.
3. Square each d_i and compute $\sum d_i^2$, the sum of the squared values.
4. Compute

$$r_s = 1 - \frac{6\sum d_i^2}{n(n^2 - 1)} \tag{13.10.1}$$

5. If n is between 4 and 30, compare the computed value of r_s with the critical values, r_s^*, of Appendix Table P. For the two-sided test, H_0 is rejected at the α significance level if r_s is greater than r_s^* or less than $-r_s^*$, where r_s^* is at the intersection of the column headed $\alpha/2$ and the row corresponding to n. For the one-sided test with H_A specifying direct correlation, H_0 is rejected at the α significance level if r_s is greater than r_s^* for α and n. The null hypothesis is rejected at the α significance level in the other one-sided test if r_s is less than $-r_s^*$ for α and n.
6. If n is greater than 30, one may compute

$$z = r_s\sqrt{n - 1} \tag{13.10.2}$$

   and use Appendix Table D to obtain critical values.
7. Tied observations present a problem. The use of Table P is strictly valid only when the data do not contain any ties (unless some random procedure for breaking ties is employed). In practice, however, the table is frequently used after some other method for handling ties has been employed. If the number of ties is large, the following correction for ties may be employed:

$$T = \frac{t^3 - t}{12} \tag{13.10.3}$$

   where t = the number of observations that are tied for some particular rank. When this correction factor is used, r_s is computed from

$$r_s = \frac{\sum x^2 + \sum y^2 - \sum d_i^2}{2\sqrt{\sum x^2 \sum y^2}} \tag{13.10.4}$$

   instead of from Equation 13.10.1.

   In Equation 13.10.4,

$$\sum x^2 = \frac{n^3 - n}{12} - \sum T_x$$

$$\sum y^2 = \frac{n^3 - n}{12} - \sum T_y$$

   T_x = the sum of the values of T for the various tied ranks in X

   T_y = the sum of the values of T for the various tied ranks in Y

   Most authorities agree that unless the number of ties is excessive, the correction makes very little difference in the value of r_s. When the number of ties is small, we can follow the usual procedure of assigning the tied observations the mean of the ranks for which they are tied and proceed with steps 2 to 6.

EXAMPLE 13.10.1

In a study of the relationship between age and the EEG, data were collected on 20 subjects between ages 20 and 60 years. Table 13.10.1 shows the age and a particular EEG output value for each of the 20 subjects. The investigator wishes to know if it can be concluded that this particular EEG output is inversely correlated with age.

Solution:

1. **Data.** See Table 13.10.1.
2. **Assumptions.** We assume that the sample available for analysis is a simple random sample and that both X and Y are measured on at least the ordinal scale.
3. **Hypotheses.**

   H_0: This EEG output and age are mutually independent.

   H_A: There is a tendency for this EEG output to decrease with age.

   Suppose we let $\alpha = .05$.

TABLE 13.10.1 Age and EEG Output Value for 20 Subjects

Subject Number	Age (X)	EEG Output Value (Y)
1	20	98
2	21	75
3	22	95
4	24	100
5	27	99
6	30	65
7	31	64
8	33	70
9	35	85
10	38	74
11	40	68
12	42	66
13	44	71
14	46	62
15	48	69
16	51	54
17	53	63
18	55	52
19	58	67
20	60	55

4. **Test statistic.** See Equation 13.10.1.
5. **Distribution of test statistic.** Critical values of the test statistic are given in Appendix Table P.
6. **Decision rule.** For the present test we will reject H_0 if the computed value of r_s is less than $-.3789$.
7. **Calculation of test statistic.** When the X and Y values are ranked, we have the results shown in Table 13.10.2. The d_i, d_i^2, and $\sum d_i^2$ are shown in the same table.

   Substitution of the data from Table 13.10.2 into Equation 13.10.1 gives

$$r_s = 1 - \frac{6(2340)}{20[(20)^2 - 1]} = -.76$$

8. **Statistical decision.** Since our computed $r_s = -.76$ is less than the critical r_s^*, we reject the null hypothesis.
9. **Conclusion.** We conclude that the two variables are inversely related.
10. ***p* value.** Since $-.76 < -0.6586$, we have for this test $p < .001$.

TABLE 13.10.2 Ranks for Data of Example 13.10.1

Subject Number	Rank (*X*)	Rank (*Y*)	d_i	d_i^2
1	1	18	−17	289
2	2	15	−13	169
3	3	17	−14	196
4	4	20	−16	256
5	5	19	−14	196
6	6	7	−1	1
7	7	6	1	1
8	8	12	−4	16
9	9	16	−7	49
10	10	14	−4	16
11	11	10	1	1
12	12	8	4	16
13	13	13	0	0
14	14	4	10	100
15	15	11	4	16
16	16	2	14	196
17	17	5	12	144
18	18	1	17	289
19	19	9	10	100
20	20	3	17	289
				$\sum d_i^2 = 2340$

■

Let us now illustrate the procedure for a sample with $n > 30$ and some tied observations.

EXAMPLE 13.10.2

In Table 13.10.3 are shown the ages and concentrations (ppm) of a certain mineral in the tissue of 35 subjects on whom autopsies were performed as part of a large research project.

The ranks, d_i, d_i^2, and $\sum d_i^2$ are shown in Table 13.10.4. Let us test, at the .05 level of significance, the null hypothesis that X and Y are mutually independent against the two-sided alternative that they are not mutually independent.

Solution: From the data in Table 13.10.4 we compute

$$r_s = 1 - \frac{6(1788.5)}{35[(35)^2 - 1]} = .75$$

To test the significance of r_s we compute

$$z = .75\sqrt{35 - 1} = 4.37$$

TABLE 13.10.3 Age and Mineral Concentration (ppm) in Tissue of 35 Subjects

Subject Number	Age (X)	Mineral Concentration (Y)	Subject Number	Age (X)	Mineral Concentration (Y)
1	82	169.62	19	50	4.48
2	85	48.94	20	71	46.93
3	83	41.16	21	54	30.91
4	64	63.95	22	62	34.27
5	82	21.09	23	47	41.44
6	53	5.40	24	66	109.88
7	26	6.33	25	34	2.78
8	47	4.26	26	46	4.17
9	37	3.62	27	27	6.57
10	49	4.82	28	54	61.73
11	65	108.22	29	72	47.59
12	40	10.20	30	41	10.46
13	32	2.69	31	35	3.06
14	50	6.16	32	75	49.57
15	62	23.87	33	50	5.55
16	33	2.70	34	76	50.23
17	36	3.15	35	28	6.81
18	53	60.59			

TABLE 13.10.4 Ranks for Data of Example 13.10.2

Subject Number	Rank (X)	Rank (Y)	d_i	d_i^2	Subject Number	Rank (X)	Rank (Y)	d_i	d_i^2
1	32.5	35	−2.5	6.25	19	17	9	8	64.00
2	35	27	8	64.00	20	28	25	3	9.00
3	34	23	11	121.00	21	21.5	21	.5	.25
4	25	32	−7	49.00	22	23.5	22	1.5	2.25
5	32.5	19	13.5	182.25	23	13.5	24	−10.5	110.25
6	19.5	11	8.5	72.25	24	27	34	−7	49.00
7	1	14	−13	169.00	25	6	3	3	9.00
8	13.5	8	5.5	30.25	26	12	7	5	25.00
9	9	6	3	9.00	27	2	15	−13	169.00
10	15	10	5	25.00	28	21.5	31	−9.5	90.25
11	26	33	−7	49.00	29	29	26	3	9.00
12	10	17	−7	49.00	30	11	18	−7	49.00
13	4	1	3	9.00	31	7	4	3	9.00
14	17	13	4	16.00	32	30	28	2	4.00
15	23.5	20	3.5	12.25	33	17	12	5	25.00
16	5	2	3	9.00	34	31	29	2	4.00
17	8	5	3	9.00	35	3	16	−13	169.00
18	19.5	30	−10.5	110.25					
									$\sum d_i^2 = 1788.5$

Since 4.37 is greater than $z = 3.89$, $p < 2(.0001) = .0002$, and we reject H_0 and conclude that the two variables under study are not mutually independent.

For comparative purposes let us correct for ties using Equation 13.10.3 and then compute r_s by Equation 13.10.4.

In the rankings of X we had six groups of ties that were broken by assigning the values 13.5, 17, 19.5, 21.5, 23.5, and 32.5. In five of the groups two observations tied, and in one group three observations tied. We, therefore, compute five values of

$$T_x = \frac{2^3 - 2}{12} = \frac{6}{12} = .5$$

and one value of

$$T_x = \frac{3^3 - 3}{12} = \frac{24}{12} = 2$$

From these computations, we have $\sum T_x = 5(.5) + 2 = 4.5$, so that

$$\sum x^2 = \frac{35^2 - 35}{12} - 4.5 = 3565.5$$

Dialog box:

Stat ➤ Basic Statistics ➤ Correlation

Type *C3–C4* in **Variables.** Click **OK.**

Session command:

```
MTB > CORRELATION C3 C4
```

Output:

Correlations (Pearson)

```
Correlation of (X)Rank and (Y)Rank = -0.759
```

FIGURE 13.10.1 MINITAB procedure and output for computing Spearman rank correlation coefficient, Example 13.10.1.

Since no ties occurred in the Y rankings, we have $\sum T_y = 0$ and

$$\sum y^2 = \frac{35^3 - 35}{12} - 0 = 3570.0$$

From Table 13.10.4 we have $\sum d_i^2 = 1788.5$. From these data we may now compute by Equation 13.10.4

$$r_s = \frac{3565.5 + 3570.0 - 1788.5}{2\sqrt{(3565.5)(3570)}} = .75$$

We see that in this case the correction for ties does not make any difference in the value of r_s. ■

Computer Analysis We may use MINITAB, as well as many other statistical software packages, to compute the Spearman correlation coefficient. To use MINITAB, we must first have MINITAB rank the observations and store the ranks in separate columns, one for the X ranks and one for the Y ranks. If we rank the X and Y values of Example 13.10.1 and store them in Columns 3 and 4, we may obtain the Spearman rank correlation coefficient with the procedure shown in Figure 13.10.1. Other software packages such as SAS® and SPSS, for example, automatically rank the measurements before computing the coefficient, thereby eliminating an extra step in the procedure.

EXERCISES

For the following exercises perform the test at the indicated level of significance and determine the p value.

13.10.1 The following table shows 15 randomly selected geographic areas ranked by population density and age-adjusted death rate. Can we conclude at the .05 level of significance that population density and age-adjusted death rate are not mutually independent?

	Rank by			Rank by	
Area	**Population Density (X)**	**Age-Adjusted Death Rate (Y)**	**Area**	**Population Density (X)**	**Age-Adjusted Death Rate (Y)**
1	8	10	9	6	8
2	2	14	10	14	5
3	12	4	11	7	6
4	4	15	12	1	2
5	9	11	13	13	9
6	3	1	14	15	3
7	10	12	15	11	13
8	5	7			

13.10.2 The following table shows 10 communities ranked by decayed, missing, or filled (DMF) teeth per 100 children and fluoride concentration in ppm in the public water supply:

	Rank by			Rank by	
Community	**DMF Teeth per 100 Children (X)**	**Fluoride Concentration (Y)**	**Community**	**DMF Teeth per 100 Children (X)**	**Fluoride Concentration (Y)**
1	8	1	6	4	7
2	9	3	7	1	10
3	7	4	8	5	6
4	3	9	9	6	5
5	2	8	10	10	2

Do these data provide sufficient evidence to indicate that the number of DMF teeth per 100 children tends to decrease as fluoride concentration increases? Let $\alpha = .05$.

13.10.3 The purpose of a study by Nozawa et al. (A-10) was to evaluate the outcome of surgical repair of pars interarticularis defect by segmental wire fixation in young adults with lumbar spondylolysis. The authors cite literature indicating that segmental wire fixation has been successful in the treatment of nonathletes with spondylolysis and point out that no information existed on the results of this type of surgery in athletes. In a retrospective study of subjects having surgery between 1993 and 2000, the authors found 20 subjects who had undergone the surgery. The following table shows the age (years) at surgery and duration (months) of follow-up care for these subjects.

Duration of Follow-Up (Months)	Age (Years)	Duration of Follow-Up (Months)	Age (Years)
103	37	38	27
68	27	36	31
62	12	34	24
60	18	30	23
60	18	19	14

(*Continued*)

Duration of Follow-Up (Months)	Age (Years)	Duration of Follow-Up (Months)	Age (Years)
54	28	19	23
49	25	19	18
44	20	19	29
42	18	17	24
41	30	16	27

Source: Satoshi Nozawa, Katsuji Shimizu, Kei Miyamoto, and Mizuo Tanaka, "Repair of Pars Interarticularis Defect by Segmental Wire Fixation in Young Athletes with Spondylolysis," *American Journal of Sports Medicine, 31* (2003), pp. 359–364.

May we conclude, on the basis of these data, that in a population of similar subjects there is an association between age and duration of follow-up? Let $\alpha = .05$.

13.10.4 Refer to Exercise 13.10.3. Nozawa et al. (A-10) also calculated the Japanese Orthopaedic Association score for measuring back pain (JOA). The results for the 20 subjects along with the duration of follow-up are shown in the following table. The higher the number, the lesser the degree of pain.

Duration of Follow-Up (Months)	JOA	Duration of Follow-Up (Months)	JOA
103	21	38	13
68	14	36	24
62	26	34	21
60	24	30	22
60	13	19	25
54	24	19	23
49	22	19	20
44	23	19	21
42	18	17	25
41	24	16	21

Source: Satoshi Nozawa, Katsuji Shimizu, Kei Miyamoto, and Mizuo Tanaka, "Repair of Pars Interarticularis Defect by Segmental Wire Fixation in Young Athletes with Spondylolysis," *American Journal of Sports Medicine, 31* (2003), pp. 359–364.

Can we conclude from these data that in general there is a relationship between length of follow-up and JOA score at the time of the operation? Let $\alpha = .05$.

13.10.5 Butz et al. (A-11) studied the use of noninvasive positive-pressure ventilation by patients with amyotrophic lateral sclerosis. They evaluated the benefit of the procedure on patients' symptoms, quality of life, and survival. Two variables of interest are $PaCO_2$, partial pressure of arterial carbon dioxide, and PaO_2, partial pressure of arterial oxygen. The following table shows, for 30 subjects, values of these variables (mm Hg) obtained from baseline arterial blood gas analyses.

$PaCO_2$	PaO_2	$PaCO_2$	PaO_2	$PaCO_2$	PaO_2
40	101	54.5	80	34.5	86.5
47	69	54	72	40.1	74.7

(Continued)

$PaCO_2$	PaO_2	$PaCO_2$	PaO_2	$PaCO_2$	PaO_2
34	132	43	105	33	94
42	65	44.3	113	59.9	60.4
54	72	53.9	69.2	62.6	52.5
48	76	41.8	66.7	54.1	76.9
53.6	67.2	33	67	45.7	65.3
56.9	70.9	43.1	77.5	40.6	80.3
58	73	52.4	65.1	56.6	53.2
45	66	37.9	71	59	71.9

Source: M. Butz, K. H. Wollinsky, U. Widemuth-Catrinescu, A. Sperfeld, S. Winter, H. H. Mehrkens, A. C. Ludolph, and H. Schreiber, "Longitudinal Effects of Noninvasive Positive-Pressure Ventilation in Patients with Amyotrophic Lateral Sclerosis," *American Journal of Medical Rehabilitation, 82* (2003) 597–604.

On the basis of these data may we conclude that there is an association between $PaCO_2$ and PaO_2 values? Let $\alpha = .05$.

13.10.6 Seventeen patients with a history of congestive heart failure participated in a study to assess the effects of exercise on various bodily functions. During a period of exercise the following data were collected on the percent change in plasma norepinephrine (Y) and the percent change in oxygen consumption (X):

Subject	X	Y	Subject	X	Y
1	500	525	10	50	60
2	475	130	11	175	105
3	390	325	12	130	148
4	325	190	13	76	75
5	325	90	14	200	250
6	205	295	15	174	102
7	200	180	16	201	151
8	75	74	17	125	130
9	230	420			

On the basis of these data can one conclude that there is an association between the two variables? Let $\alpha = .05$.

13.11 NONPARAMETRIC REGRESSION ANALYSIS

When the assumptions underlying simple linear regression analysis as discussed in Chapter 9 are not met, we may employ nonparametric procedures. In this section we present estimators of the slope and intercept that are easy-to-calculate alternatives to the least-squares estimators described in Chapter 9.

Theil's Slope Estimator Theil (12) proposes a method for obtaining a point estimate of the slope coefficient β. We assume that the data conform to the classic regression model

$$y_i = \beta_0 + \beta_1 x_1 + \epsilon_i, \qquad i = 1, \ldots, n$$

where the x_i are known constants, β_0 and β_1 are unknown parameters, and y_i is an observed value of the continuous random variable Y at x_i. For each value of x_i, we assume a subpopulation of Y values, and the ϵ_i are mutually independent. The x_i are all distinct (no ties), and we take $x_1 < x_2 < \cdots < x_n$.

The data consist of n pairs of sample observations, $(x_1, y_1), (x_2, y_2), \ldots, (x_n, y_n)$, where the ith pair represents measurements taken on the ith unit of association.

To obtain Theil's estimator of β_1 we first form all possible sample slopes $S_{ij} = (y_j - y_i)/(x_j - x_i)$, where $i < j$. There will be $N = {}_nC_2$ values of S_{ij}. The estimator of β_1 which we designate by $\hat{\beta}_1$ is the median of S_{ij} values. That is,

$$\hat{\beta}_1 = \text{median}\{S_{ij}\} \tag{13.11.1}$$

The following example illustrates the calculation of $\hat{\beta}_1$.

EXAMPLE 13.11.1

In Table 13.11.1 are the plasma testosterone (ng/ml) levels (Y) and seminal citric acid (mg/ml) levels in a sample of eight adult males. We wish to compute the estimate of the population regression slope coefficient by Theil's method.

Solution: The $N = {}_8C_2 = 28$ ordered values of S_{ij} are shown in Table 13.11.2.

If we let $i = 1$ and $j = 2$, the indicators of the first and second values of Y and X in Table 13.11.1, we may compute S_{12} as follows:

$$S_{12} = (175 - 230)/(278 - 421) = -.3846$$

When all the slopes are computed in a similar manner and ordered as in Table 13.11.2, $-.3846$ winds up as the tenth value in the ordered array.

The median of the S_{ij} values is .4878. Consequently, our estimate of the population slope coefficient $\hat{\beta}_1 = .4878$.

TABLE 13.11.1 Plasma Testosterone and Seminal Citric Acid Levels in Adult Males

Testosterone:	230	175	315	290	275	150	360	425
Citric acid:	421	278	618	482	465	105	550	750

TABLE 13.11.2 Ordered Values of S_{ij} for Example 13.11.1

−.6618	.5037
.1445	.5263
.1838	.5297
.2532	.5348
.2614	.5637
.3216	.5927
.3250	.6801
.3472	.8333
.3714	.8824
.3846	.9836
.4118	1.0000
.4264	1.0078
.4315	1.0227
.4719	1.0294

■

An Estimator of the Intercept Coefficient Dietz (13) recommends two intercept estimators. The first, designated $(\hat{\beta}_0)_{1,M}$ is the median of the n terms $y_i - \hat{\beta}_1 x_i$ in which $\hat{\beta}_1$ is the Theil estimator. It is recommended when the researcher is not willing to assume that the error terms are symmetric about 0. If the researcher is willing to assume a symmetric distribution of error terms, Dietz recommends the estimator $(\hat{\beta}_0)_{2,M}$ which is the median of the $n(n+1)/2$ pairwise averages of the $y_i - \hat{\beta}_1 x_i$ terms. We illustrate the calculation of each in the following example.

EXAMPLE 13.11.2

Refer to Example 13.11.1. Let us compute $\hat{\alpha}_{1,M}$ and $\hat{\alpha}_{2,M}$ from the data on testosterone and citric acid levels.

Solution: The ordered $y_i - .4878x_i$ terms are: 13.5396, 24.6362, 39.3916, 48.1730, 54.8804, 59.1500, 91.7100, and 98.7810. The median, 51.5267, is the estimator $(\hat{\beta}_0)_{1,M}$.

The $8(8+1)/2 = 36$ ordered pairwise averages of the $y_i - .4878x_i$ are

13.5396	49.2708	75.43
19.0879	51.5267	76.8307
24.6362	52.6248	78.9655
26.4656	53.6615	91.71
30.8563	54.8804	95.2455
32.0139	56.1603	98.781
34.21	57.0152	
36.3448	58.1731	

(*Continued*)

36.4046	59.15
39.3916	61.7086
39.7583	65.5508
41.8931	69.0863
43.7823	69.9415
47.136	73.2952
48.173	73.477

The median of these averages, 53.1432, is the estimator $\hat{\alpha}_{2,M}$. The estimating equation, then, is $y_i = 53.1432 + .4878x_i$ if we are willing to assume that the distribution of error terms is symmetric about 0. If we are not willing to make the assumption of symmetry, the estimating equation is $y_i = 51.5267 + .4878x_i$. ■

EXERCISES

13.11.1 The following are the heart rates (HR: beats/minute) and oxygen consumption values (VO_2: cal/kg/24 h) for nine infants with chronic congestive heart failure:

HR(X):	163	164	156	151	152	167	165	153	155
$VO_2(Y)$:	53.9	57.4	41.0	40.0	42.0	64.4	59.1	49.9	43.2

Compute $\hat{\beta}_1$, $(\hat{\beta}_0)_{1,M}$, and $(\hat{\beta}_0)_{2,M}$

13.11.2 The following are the body weights (grams) and total surface area (cm^2) of nine laboratory animals:

Body weight (X):	660.2	706.0	924.0	936.0	992.1	888.9	999.4	890.3	841.2
Surface area (Y):	781.7	888.7	1038.1	1040.0	1120.0	1071.5	1134.5	965.3	925.0

Compute the slope estimator and two intercept estimators.

13.12 SUMMARY

This chapter is concerned with nonparametric statistical tests. These tests may be used either when the assumptions underlying the parametric tests are not realized or when the data to be analyzed are measured on a scale too weak for the arithmetic procedures necessary for the parametric tests.

Nine nonparametric tests are described and illustrated. Except for the Kolmogorov–Smirnov goodness-of-fit test, each test provides a nonparametric alternative to a well-known parametric test. There are a number of other nonparametric tests available. The interested reader is referred to the many books devoted to nonparametric methods, including those by Gibbons (14) and Pett (15).

SUMMARY OF FORMULAS FOR CHAPTER 13

Formula Number	Name	Formula
13.3.1	Sign test statistic	$P(k \leq x \mid n, p) = \sum_{k=0}^{x} {}_nC_k p^k q^{n-k}$
13.3.2	Large-sample approximation of the sign test	$z = \dfrac{(k + 0.5) + 0.5n}{0.5\sqrt{n}}$, if $k < \dfrac{n}{2}$ $z = \dfrac{(k - 0.5) - 0.5n}{0.5\sqrt{n}}$, if $k \geq \dfrac{n}{2}$
13.6.1	Mann–Whitney test statistic	$T = S - \dfrac{n(n+1)}{2}$
13.6.2	Large-sample approximation of the Mann–Whitney test	$Z = \dfrac{T - mn/2}{\sqrt{nm(n+m+1)/12}}$
13.6.3	Equivalence of the Mann–Whitney and Wilcoxon two-sample statistics	$U + W = \dfrac{m(m + 2n + 1)}{2}$
13.7.1–13.7.2	Kolmogorov–Smirnov test statistic	$D = \sup_{x} \lvert F_s(x) - F_T(x) \rvert$ $= \max_{1 \leq i \leq r} \left\{ \max \left[\lvert F_s(x_i) - F_T(x_i) \rvert, \lvert F_s(x_{i-1}) - F_T(x_{i-1}) \rvert \right] \right\}$
13.8.1	Kruskal–Wallis test statistic	$H = \dfrac{12}{n(n+1)} \sum_{j=1}^{k} \dfrac{R_j^2}{n_j} - 3(n+1)$
13.8.2	Kruskal–Wallis test statistic adjustment for ties	$1 - \dfrac{\sum T}{n^3 - n}$
13.9.2	Friedman test statistic	$\chi_r^2 = \dfrac{12}{nk(k+1)} \sum_{j=1}^{k} (R_j)^2 - 3n(k+1)$
13.10.1	Spearman rank correlation test statistic	$r_s = 1 - \dfrac{6 \sum d_i^2}{n(n^2 - 1)}$

13.10.2	Large-sample approximation of the Spearman rank correlation	$z = r_s\sqrt{n-1}$
13.10.3–13.10.4	Correction for tied observations in the Spearman rank correlation	$T = \frac{t^3 - t}{12}$ with $r_s = \frac{\sum x^2 + \sum y^2 - \sum d_i^2}{2\sqrt{\sum x^2 \sum y^2}}$
13.11.1	Theil's estimator of β	$\hat{\beta} = median\{S_i\}$
Symbol Key	• $\hat{\beta}$ = Theil's estimator of β • χ^2(or X^2) = chi-square • D = Kolmogorov–Smirnov test statistic • $F_i(x)$ = distribution function of i • H = Friedman test statistic • k = sign test statistic and the number of columns in the Friedman test • m = sample size of the smaller of two samples • n = sample size of the larger of two samples • p = probability of success • $q = 1 - p$ = probability of failure • R = rank • r_s = Spearman rank correlation coefficient • S = sum of ranks • S_{ij} = slope between point i and j • sup = supremum (greatest) • t = number of tied observations • T = correction for tied observations • x and y = data value for variables x and y • U = Mann–Whitney test statistic • W = Wilcoxon test statistic • z = normal variate	

REVIEW QUESTIONS AND EXERCISES

1. Define nonparametric statistics.

2. What is meant by the term *distribution-free statistical tests*?

3. What are some of the advantages of using nonparametric statistical tests?

4. What are some of the disadvantages of the nonparametric tests?

5. Describe a situation in your particular area of interest where each of the following tests could be used. Use real or realistic data and test an appropriate hypothesis using each test.

(a) The sign test
(b) The median test
(c) The Wilcoxon test
(d) The Mann–Whitney test
(e) The Kolmogorov–Smirnov goodness-of-fit test
(f) The Kruskal–Wallis one-way analysis of variance by ranks
(g) The Friedman two-way analysis of variance by ranks
(h) The Spearman rank correlation coefficient
(i) Nonparametric regression analysis

6. The following are the ranks of the ages (X) of 20 surgical patients and the dose (Y) of an analgesic agent required to block one spinal segment.

Rank of Age in Years (X)	Rank of Dose Requirement (Y)	Rank of Age in Years (X)	Rank of Dose Requirement (Y)
1	1	11	13
2	7	12	5
3	2	13	11
4	4	14	16
5	6	15	20
6	8	16	18
7	3	17	19
8	15	18	17
9	9	19	10
10	12	20	14

Compute r_s and test (two-sided) for significance. Let $\alpha = .05$. Determine the p value for this test.

7. Otani and Kishi (A-12) studied seven subjects with diabetic macular edema. They measured the foveal thickness (μm) in seven eyes pre- and post-unilateral vitrectomy surgery. The results are shown in the following table:

Subject	Pre-op Foveal Thickness (μm)	Post-op Foveal Thickness (μm)
1	690	200
2	840	280
3	470	230
4	690	200
5	730	560
6	500	210
7	440	200

Source: Tomohiro Otani, M.D. Used with permission.

Use the Wilcoxon signed-rank test to determine whether one should conclude that the surgery is effective in reducing foveal thickness. Let $\alpha = .05$. What is the p value?

8. The subjects of a study by J. Jose and S. R. Ell (A-13) were 303 healthy volunteers who self-assessed their own nasal flow status by indicating whether their nasal airway was (1) totally clear, (2) not very clear, (3) very blocked, or (4) totally blocked. Following the self-assessment, an In-Check meter was used to measure peak inspiratory nasal flow rate (PINFR, L/min). Data on 175 subjects in three of the self-assessment categories are displayed in the following table. The authors performed a Kruskal–Wallis test to determine if these data provide sufficient evidence to indicate a difference in population centers of PINFR among these three response groups. Let $\alpha = .01$. What is the test statistic value for this test?

Peak Inspiratory Nasal Flow Rate (L/min)							
Totally Clear				**Not Very Clear**			**Partially Blocked**
180	105	150	120	160	190	130	100
150	150	110	95	200	95	110	100
200	240	130	140	70	130	110	100
130	120	100	135	75	240	130	105
200	90	170	100	150	180	125	95
120	135	80	130	80	140	100	85
150	110	125	180	130	150	230	50
150	155	115	155	160	130	110	105
160	105	140	130	180	90	270	200
150	140	140	140	90	115	180	
110	200	95	120	180	130	130	
190	170	110	290	140	210	125	
150	150	160	170	230	190	90	
120	120	90	280	220	135	210	
180	170	135	150	130	130	140	
140	200	110	185	180	210	125	
130	160	130	150	140	90	210	
230	180	170	150	140	125	120	
200	170	130	170	120	140	115	
140	160	115	210	140	160	100	
150	150	145	140	150	230	130	
170	100	130	140	190	100	130	
180	100	170	160	210	120	110	
160	180	160	120	130	120	150	
200	130	90	230	190	150	110	
90	200	110	100	220	110	90	
130	120	130	190	160	150	120	
140	145	130	90	105	130	115	
200	130	120	100	120	150	140	
220	100	130	125	140	130	130	
200	130	180	180	130	145	160	
120	160	140	200	115	160	110	
310	125	175	160	115	120	165	
160	100	185	170	100	220	120	
115	140	190	85	150	145	150	

(*Continued*)

Peak Inspiratory Nasal Flow Rate (L/min)							
Totally Clear				Not Very Clear			Partially Blocked
170	185	130	150	130	150	170	
130	180	160	280	130	120	110	
220	115	160	140	170	155	120	
250	260	130	100	130	100	85	
160	160	135	140	145	140		
130	170	130	90				
130	115	120	190				
150	150	190	130				
160	130	170					

Source: J. Jose, MS, FRCS. Used with permission.

9. Ten subjects with bronchial asthma participated in an experiment to evaluate the relative effectiveness of three drugs. The following table shows the change in FEV_1 (forced expired volume in 1 second) values (expressed as liters) 2 hours after drug administration:

	Drug				Drug		
Subject	**A**	**B**	**C**	**Subject**	**A**	**B**	**C**
1	.00	.13	.26	6	.03	.18	.25
2	.04	.17	.23	7	.05	.21	.32
3	.02	.20	.21	8	.02	.23	.38
4	.02	.27	.19	9	.00	.24	.30
5	.04	.11	.36	10	.12	.08	.30

Are these data sufficient to indicate a difference in drug effectiveness? Let $\alpha = .05$. What is the p value for this test?

10. One facet of the nursing curriculum at Wright State University requires that students use mathematics to perform appropriate dosage calculations. In a study by Wendy Gantt (A-14), undergraduate nursing students were given a standardized mathematics test to determine their mathematical aptitude (scale: 0–100). The students were divided into two groups: traditional college age (18–24 years, 26 observations) and nontraditional (25+, eight observations). Scores on the mathematics test appear in the following table:

Traditional Students' Scores			Nontraditional Students' Scores
70	6	88	77
57	79	68	72
85	14	88	54
55	82	92	87
87	45	85	85

(*Continued*)

Traditional Students' Scores			Nontraditional Students' Scores
84	57	56	62
56	91	31	77
68	76	80	86
94	60		

Source: Wendy Gantt and the Wright State University Statistical Consulting Center. Used with permission.

Do these data provide sufficient evidence to indicate a difference in population medians? Let $\alpha = .05$. What is the p value for this test? Use both the median test and the Mann–Whitney test and compare the results.

11. The following are the $PaCO_2$ (mm Hg) values in 16 patients with bronchopulmonary disease:

39, 40, 45, 48, 49, 56, 60, 75, 42, 48, 32, 37, 32, 33, 33, 36

Use the Kolmogorov–Smirnov test to test the null hypothesis that $PaCO_2$ values in the sampled population are normally distributed with $\mu = 44$ and $\sigma = 12$.

12. The following table shows the caloric intake (cal/day/kg) and oxygen consumption VO_2 (ml/min/kg) in 10 infants:

Calorie Intake (X)	$VO_2(Y)$	Calorie Intake (X)	$VO_2(Y)$
50	7.0	100	10.8
70	8.0	150	12.0
90	10.5	110	10.0
120	11.0	75	9.5
40	9.0	160	11.9

Test the null hypothesis that the two variables are mutually independent against the alternative that they are directly related. Let $\alpha = .05$. What is the p value for this test?

13. Mary White (A-15) surveyed physicians to measure their opinions regarding the importance of ethics in medical practice. The measurement tool utilized a scale from 1 to 5 in which a higher value indicated higher opinion of the importance of ethics. The ages and scores of the study subjects are shown in the following table. Can one conclude on the basis of these results that age and ethics score are directly related? Let the probability of committing a type I error be .05. What is the p value?

Age	Ethics	Age	Ethics	Age	Ethics
25	4.00	26	4.50	26	4.50
34	4.00	29	4.75	27	5.00
30	4.25	30	4.25	22	3.75
31	3.50	26	4.50	22	4.25
25	4.75	30	4.25	24	4.50

(Continued)

Age	Ethics	Age	Ethics	Age	Ethics
25	3.75	25	3.75	22	4.25
25	4.75	24	4.75	24	3.75
29	4.50	24	4.00	38	4.50
29	4.50	25	4.50	22	4.50
26	3.75	25	4.00	22	4.50
25	3.25	26	4.75	25	4.00
29	4.50	34	3.25	23	3.75
27	3.75	23	4.50	22	4.25
29	4.25	26	3.25	23	4.00
25	3.75	23	5.00	22	4.25
25	4.50	24	4.25	25	3.50
25	4.00	45	3.25	26	4.25
26	4.25	23	3.75	25	4.25
26	4.00	25	3.75	27	4.75
24	4.00	25	3.75	23	3.75
25	4.00	23	3.75	22	4.00
22	3.75	23	4.75	26	4.75
26	4.50	26	4.00	22	4.25
				23	4.00

Source: Mary White, Ph.D. and Wright State University Statistical Consulting Center. Used with permission.

14. Dominic Sprott (A-16) conducted an experiment with rabbits in which the outcome variable was the fatty infiltration in the shoulder mass (PFI, measured as a percent). At baseline, 15 rabbits had a randomly chosen shoulder muscle detached. The shoulder was then reattached. Six weeks later, five randomly chosen rabbits were sacrificed and the differences in the PFI between the reattached shoulder and the nondetached shoulder were recorded (group A). Six months later, the 10 remaining rabbits were sacrificed and again the differences in the PFI between the reattached shoulder and the nondetached shoulder were recorded (group B).

Percent Fatty Infiltration Difference (Nondetached–Reattached)

Group A	Group B	
2.55	1.04	1.38
0.9	3.29	0.75
0.2	0.99	0.36
−0.29	1.79	0.74
1.11	−0.85	0.3

Source: Dominic Sprott, M.D. and the Wright State University Statistical Consulting Center. Used with permission.

Can we conclude, at the .05 level of significance, that the treatments have a differential effect on PFI between the two shoulder muscles? What is the p value for the test?

In each of the Exercises 15 through 29, do one or more of the following that you think are appropriate:

(a) Apply one or more of the techniques discussed in this chapter.

(b) Apply one or more of the techniques discussed in previous chapters.

(c) Formulate relevant hypotheses, perform the appropriate tests, and find p values.

(d) State the statistical decisions and clinical conclusions that the results of your hypothesis tests justify.

(e) Describe the population(s) to which you think your inferences are applicable.

(f) State the assumptions necessary for the validity of your analyses.

15. The purpose of a study by Damm et al. (A-17) was to investigate insulin sensitivity and insulin secretion in women with previous gestational diabetes (GDM). Subjects were 12 normal-weight glucose-tolerant women (mean age, 36.6 years; standard deviation, 4.16) with previous gestational diabetes and 11 controls (mean age, 35 years; standard deviation, 3.3). Among the data collected were the following fasting plasma insulin values (mmol/L). Use the Mann–Whitney test to determine if you can conclude on the basis of these data that the two populations represented differ with respect to average fasting plasma insulin level.

Controls	Previous GDM	Controls	Previous GDM
46.25	30.00	40.00	31.25
40.00	41.25	30.00	56.25
31.25	56.25	51.25	61.25
38.75	45.00	32.50	50.00
41.25	46.25	43.75	53.75
38.75	46.25		62.50

Source: Dr. Peter Damm. Used with permission.

16. Gutin et al. (A-18) compared three measures of body composition, including dual-energy x-ray absorptiometry (DXA). Subjects were apparently healthy children (21 boys and 22 girls) between the ages of 9 and 11 years. Among the data collected were the following measurements of body-composition compartments by DXA. The investigators were interested in the correlation between all possible pairs of these variables.

Percent Fat	Fat Mass	Fat-Free Mass	Bone Mineral Content	Fat-Free Soft Tissue
11.35	3.8314	29.9440	1.19745	28.7465
22.90	6.4398	21.6805	0.79250	20.8880
12.70	4.0072	27.6290	0.95620	26.6728
42.20	24.0329	32.9164	1.45740	31.4590
24.85	9.4303	28.5009	1.32505	27.1758
26.25	9.4292	26.4344	1.17412	25.2603
23.80	8.4171	26.9938	1.11230	25.8815
37.40	20.2313	33.8573	1.40790	32.4494
14.00	3.9892	24.4939	0.95505	23.5388
19.35	7.2981	30.3707	1.45545	28.9153
29.35	11.1863	26.8933	1.17775	25.7156

(*Continued*)

Percent Fat	Fat Mass	Fat-Free Mass	Bone Mineral Content	Fat-Free Soft Tissue
18.05	5.8449	26.5341	1.13820	25.3959
13.95	4.6777	28.9144	1.23730	27.6771
32.85	13.2474	27.0849	1.17515	25.9097
11.40	3.7912	29.5245	1.42780	28.0967
9.60	3.2831	30.8228	1.14840	29.6744
20.90	7.2277	27.3302	1.24890	26.0813
44.70	25.7246	31.8461	1.51800	30.3281
17.10	5.1219	24.8233	0.84985	23.9734
16.50	5.0749	25.7040	1.09240	24.6116
14.35	5.0341	30.0228	1.40080	28.6220
15.45	4.8695	26.6403	1.07285	25.5674
28.15	10.6715	27.2746	1.24320	26.0314
18.35	5.3847	23.9875	0.94965	23.0379
15.10	5.6724	31.9637	1.32300	30.6407
37.75	25.8342	42.6004	1.88340	40.7170
39.05	19.6950	30.7579	1.50540	29.2525
22.25	7.2755	25.4560	0.88025	24.5757
15.50	4.4964	24.4888	0.96500	23.5238
14.10	4.3088	26.2401	1.17000	25.0701
26.65	11.3263	31.2088	1.48685	29.7219
20.25	8.0265	31.5657	1.50715	30.0586
23.55	10.1197	32.8385	1.34090	31.4976
46.65	24.7954	28.3651	1.22575	27.1394
30.55	10.0462	22.8647	1.01055	21.8541
26.80	9.5499	26.0645	1.05615	25.0083
28.10	9.4096	24.1042	0.97540	23.1288
24.55	14.5113	44.6181	2.17690	42.4412
17.85	6.6987	30.8043	1.23525	29.5690
20.90	6.5967	24.9693	0.97875	23.9905
33.00	12.3689	25.1049	0.96725	24.1377
44.00	26.1997	33.3471	1.42985	31.9172
19.00	5.0785	21.6926	0.78090	20.9117

Source: Dr. Mark Litaker. Used with permission.

17. The concern of a study by Crim et al. (A-19) was the potential role of flow cytometric analysis of bronchoalveolar lavage fluid (BALF) in diagnosing acute lung rejection. The investigators note that previous studies suggested an association of acute lung rejection with increases in CD8+ lymphocytes, and increased expression of human lymphocyte antigen (HLA)-DR antigen and interleukin-2 receptor (IL-2R). Subjects consisted of lung transplant (LT) recipients who had no histologic evidence of rejection or infection, normal human volunteers (NORM), healthy heart transplant (HT) recipient volunteers, and lung transplant recipients who were experiencing acute lung rejection (AR). Among the data collected were the following percentages of BALF CD8+ lymphocytes that also express IL-2R observed in the four groups of subjects.

Norm	HT	LT	AR	
0	0	1	6	12
2	0	0	6	0
1	5	5	8	9
0	4	0	16	7
0	6	0	24	2
2	0	5	5	6
3	0	18	3	14
0	4	2	22	10
0	8	2	10	3
1	8	8	0	0
		0	8	0
		7	3	1
		2	4	1
		5	4	0
		1	18	0
			0	4

Source: Dr. Courtney Crim.
Used with permission.

18. Ichinose et al. (A-20) studied the involvement of endogenous tachykinins in exercise-induced airway narrowing in patients with asthma by means of a selective neurokinin 1-receptor antagonist, FK-888. Nine subjects (eight male, one female) ages 18 to 43 years with at least a 40 percent fall in the specific airway conductance participated in the study. The following are the oxygen consumption (ml/min) data for the subjects at rest and during exercise while under treatment with a placebo and FK-888:

Placebo		FK-888	
At Rest	**Exercise**	**At Rest**	**Exercise**
303	2578	255	2406
288	2452	348	2214
285	2768	383	3134
280	2356	328	2536
295	2112	321	1942
270	2716	234	2652
274	2614	387	2824
185	1524	198	1448
364	2538	312	2454

Source: Dr. Kunio Shirato.
Used with permission.

19. Transforming growth factor α (TGFα), according to Tomiya and Fujiwara (A-21), is alleged to play a role in malignant progression as well as normal cell growth in an autocrine manner, and its serum levels have been reported to increase during this progression. The present investigators have developed an enzyme-linked immunosorbent assay (ELISA) for measuring serum TGFα levels in the diagnosis of hepatocellular carcinoma (HCC) complicating cirrhosis. In a study in which they evaluated the significance of serum TGFα levels for diagnostic purposes, they collected the following measurements on the liver function tests, TGFα (pg/ml), and serum α-fetoprotein (AFP) (ng/ml) from HCC patients:

TGFα	AFP	TGFα	AFP	TGFα	AFP	TGFα	AFP
32.0	12866	44.0	23077	100.0	479	15.0	921
65.9	9	75.0	371	12.0	47	34.0	118
25.0	124.3	36.0	291	32.0	177	100.0	6.2
30.0	9	65.0	700	98.0	9	26.0	19
22.0	610	44.0	40	20.0	1063	53.0	594
40.0	238	56.0	9538	20.0	21	140.0	10
52.0	153	34.0	19	9.0	206	24.0	292
28.0	23	300.0	11	58.0	32	20.0	11
11.0	28	39.0	42246	39.0	628	35.0	37
45.0	240	82.0	12571			52.0	35
29.0	66	85.0	20			50.0	742
45.0	83	24.0	29			95.0	10
21.0	4	40.0	310			18.0	291
38.0	214	9.0	19				

Source: Dr. Kenji Fujiwara. Used with permission.

20. The objective of a study by Sakhaee et al. (A-22) was to ascertain body content of aluminum (A1) noninvasively using the increment in serum and urinary Al following the intravenous administration of deferoxamine (DFO) in patients with kidney stones and osteoporotic women undergoing long-term treatment with potassium citrate (K_3Cit) or tricalcium dicitrate (Ca_3Cit_2), respectively. Subjects consisted of 10 patients with calcium nephrolithiasis and five patients with osteoporosis who were maintained on potassium citrate or calcium citrate for 2–8 years, respectively, plus 16 normal volunteers without a history of regular aluminum-containing antacid use. Among the data collected were the following 24-hour urinary aluminum excretion measurements (μg/day) before (PRE) and after (POST) 2-hour infusion of DFO.

Group	PRE	POST	Group	PRE	POST
Control	41.04	135.00	Control	9.39	12.32
Control	70.00	95.20	Control	10.72	13.42
Control	42.60	74.00	Control	16.48	17.40
Control	15.48	42.24	Control	10.20	14.20
Control	26.90	104.30	Control	11.40	20.32
Control	16.32	66.90	Control	8.16	12.80
Control	12.80	10.68	Control	14.80	62.00
Control	68.88	46.48	Patient	15.20	27.15
Control	25.50	73.80	Patient	8.70	38.72
Patient	0.00	14.16	Patient	5.52	7.84
Patient	2.00	20.72	Patient	13.28	31.70
Patient	4.89	15.72	Patient	3.26	17.04
Patient	25.90	52.40	Patient	29.92	151.36
Patient	19.35	35.70	Patient	15.00	61.38

(Continued)

Group	PRE	POST	Group	PRE	POST
Patient	4.88	70.20	Patient	36.80	142.45
Patient	42.75	86.25			

Source: Dr. Khashayar Sakhaee. Used with permission.

21. The purpose of a study by Dubuis et al. (A-23) was to determine whether neuropsychological deficit of children with the severe form of congenital hypothyroidism can be avoided by earlier onset of therapy and higher doses of levothyroxine. Subjects consisted of 10 infants (ages 3 to 24 days) with severe and 35 infants (ages 2 to 10 days) with moderate congenital hypothyroidism. Among the data collected were the following measurements on plasma T_4 (nmol/L) levels at screening:

Severe Cases		Moderate Cases			
Sex	T_4 (nmol/L)	Sex	T_4 (nmol/L)	Sex	T_4 (nmol/L)
M	16	F	20	F	62
M	57	F	34	M	50
M	40	F	188	F	40
F	50	F	69	F	116
F	57	F	162	F	80
F	38	F	148	F	97
F	51	F	108	F	51
F	38	F	54	F	84
M	*	F	96	F	51
F	60	M	76	F	94
		M	122	M	158
		M	43	F	*
		F	40	M	47
		F	29	M	143
		F	83	M	128
		F	62	M	112
				M	111
				F	84
				M	55

* = Missing data.

Source: Dr. Guy Van Vliet. Used with permission.

22. Kuna et al. (A-24) conducted a study concerned with chemokines in seasonal allergic rhinitis. Subjects included 18 atopic individuals with seasonal allergic rhinitis caused by ragweed pollen. Among the data collected on these subjects were the following eosinophil cationic protein (ECP) and histamine measurements:

ECP (ng/ml)	Histamine (ng/ml)	ECP (ng/ml)	Histamine (ng/ml)
511.0	31.2	25.3	5.6
388.0	106.0	31.1	62.7
14.1	37.0	325.0	138.0
314.0	90.0	437.0	116.0

(Continued)

ECP (ng/ml)	Histamine (ng/ml)	ECP (ng/ml)	Histamine (ng/ml)
74.1	29.0	277.0	70.6
8.8	87.0	602.0	184.0
144.0	45.0	33.0	8.6
56.0	151.8	661.0	264.0
205.0	86.0	162.0	92.0

Source: Dr. Allen P. Kaplan. Used with permission.

23. The purpose of a study by Kim et al. (A-25) was to investigate the serial changes in Lp(a) lipoprotein levels with the loss of female sex hormones by surgical menopause and with estrogen replacement therapy in the same women. Subjects were 44 premenopausal women who underwent a transabdominal hysterectomy (TAH). Thirty-one of the women had a TAH and unilateral salpingo-oophorectomy (USO), and 13 had a TAH and bilateral salpingo-oophorectomy (BSO). The women ranged in age from 30 to 53 years. Subjects in the BSO group received .625 mg of conjugated equine estrogen daily 2 months after the operation. The following were the subjects' total cholesterol levels before (TC0), 2 months after (TC2), and 4 months after (TC4) the surgical procedure and hormone replacement therapy.

USO				USO			
Subject	TC0	TC2	TC4	Subject	TC0	TC2	TC4
1	202	203	196	25	134	131	135
2	204	183	203	26	163	190	185
3	206	199	192	27	196	183	192
4	166	180	176	28	181	194	208
5	150	171	154	29	160	162	181
6	137	134	129	30	188	200	181
7	164	168	171	31	172	188	189
8	207	249	223				
9	126	121	140	BSO			
10	131	141	167	Subject	TC0	TC2	TC4
11	133	159	149				
12	142	152	140	32	224	218	239
13	225	193	180	33	202	196	231
14	158	182	179	34	181	182	208
15	184	177	182	35	191	230	208
16	223	244	234	36	248	284	279
17	154	178	187	37	224	228	199
18	176	137	162	38	229	318	272
19	205	253	288	39	147	199	194
20	167	156	136	40	248	258	302
21	164	176	191	41	160	218	229
22	177	168	185	42	175	187	166
23	140	175	167	43	262	260	247
24	167	186	195	44	189	199	181

Source: Dr. Chee Jeong Kim. Used with permission.

24. Velthuis et al. (A-26) conducted a study to evaluate whether the combination of passively immobilized heparin-coating and standard heparization can reduce complement activation in patients undergoing cardiac surgical intervention. The investigators note that heparin-coated extracorporeal circuits reduce complement activation during cardiac operations, but that little in vivo information is available on the reduction in alternative and classic pathway activation. Complement activation initiates a systemic inflammatory response during and after cardiac operations and is associated with pathophysiologic events such as postoperative cardiac depression, pulmonary capillary leakage, and hemolysis. Subjects were 20 patients undergoing elective cardiopulmonary bypass (CPB) grafting randomly allocated to be treated with either heparin-coated extracorporeal circuits (H) or uncoated circuits (U). Among the data collected were the following plasma terminal complement complex (SC5b-9) concentrations at baseline, 10 minutes after start of CPB, at cessation of CPB, and after the administration of protamine sulfate:

Patient	Treatment	Baseline	10 min CPB	End CPB	Protamine
1	U	0.37	0.81	1.88	2.12
2	U	0.48	0.73	3.28	3.31
3	U	0.48	0.42	2.94	1.46
4	H	0.37	0.44	1.28	3.82
5	H	0.38	0.31	0.50	0.68
6	U	0.38	0.43	1.39	5.04
7	H	0.46	0.57	1.03	1.29
8	H	0.32	0.35	0.75	1.10
9	U	0.41	0.94	1.57	2.53
10	U	0.37	0.38	2.07	1.69
11	H	0.48	0.33	1.12	1.04
12	H	0.39	0.39	1.69	1.62
13	U	0.27	0.41	1.28	2.26
14	H	0.51	0.27	1.17	1.05
15	H	0.97	0.75	1.82	1.31
16	U	0.53	1.57	4.49	2.15
17	U	0.41	0.47	1.60	1.87
18	U	0.46	0.65	1.49	1.24
19	H	0.75	0.78	1.49	1.57
20	H	0.64	0.52	2.11	2.44

Source: Dr. Henk te Velthuis. Used with permission.

25. Heijdra et al. (A-27) state that many patients with severe chronic obstructive pulmonary disease (COPD) have low arterial oxygen saturation during the night. These investigators conducted a study to determine whether there is a causal relationship between respiratory muscle dysfunction and nocturnal saturation. Subjects were 20 (five females, 15 males) patients with COPD randomly assigned to receive either target-flow inspiratory muscle training (TF-IMT) at 60 percent of their maximal inspiratory mouth pressure (PI_{max}) or sham TF-IMT at 10 percent of PI_{max}. Among the data collected were the following endurance times (Time, s) for each subject at the beginning of training and 10 weeks later:

Time (s) TF-IMT 60% PI_{max}		Time (s) TF-IMT 10% PI_{max}	
Week 0	Week 10	Week 0	Week 10
330	544	430	476
400	590	400	320
720	624	900	650
249	330	420	330
144	369	679	486
440	789	522	369
440	459	116	110
289	529	450	474
819	1099	570	700
540	930	199	259

Source: Dr. Yvonne F. Heijdra. Used with permission.

26. The three objectives of a study by Wolkin et al. (A-28) were to determine (a) the effects of chronic haloperidol treatment on cerebral metabolism in schizophrenic patients, (b) the relation between negative symptoms and haloperidol-induced regional changes in cerebral glucose utilization, and (c) the relation between metabolic change and clinical antipsychotic effect. Subjects were 18 male veterans' hospital inpatients (10 black, five white, and three Hispanic) with either acute or chronic decompensation of schizophrenia. Subjects ranged in age from 26 to 44 years, and their duration of illness ranged from 7 to 27 years. Among the data collected were the following pretreatment scores on the digit-symbol substitution subtest of the WAIS-R (DSY1RW) and haloperidol-induced change in absolute left dorsolateral prefrontal cortex (DLLA3V1) and absolute right dorsolateral prefrontal cortex (DLRA3V1) measured in units of μmol glucose/100 g tissue/min:

DSY1RW	DLLA3V1	DLRA3V1	DSY1RW	DLLA3V1	DLRA3V1
47	−7.97	−17.17	18	−4.91	−9.58
16	−8.08	−9.59	0	−1.71	.40
31	−10.15	−11.58	29	−4.62	−4.57
34	−5.46	−2.16	17	9.48	11.31
22	−17.12	−12.95	38	−6.59	−6.47
70	−12.12	−13.01	64	−12.19	−13.61
59	−9.70	−12.61	52	−15.13	−11.81
41	−9.02	−7.48	50	−10.82	−9.45
0	4.67	7.26	62	−4.92	−1.87

Source: Dr. Adam Wolkin. Used with permission.

27. The purpose of a study by Maltais et al. (A-29) was to compare and correlate the increase in arterial lactic acid (La) during exercise and the oxidative capacity of the skeletal muscle in patients with chronic obstructive pulmonary disease (COPD) and control subjects (C). There were nine subjects in each group. The mean age of the patients was 62 years with a standard deviation of 5. Control subjects had a mean age of 54 years with a standard deviation of 3. Among the data collected were the following values for the activity of phosphofructokinase (PFK), hexokinase (HK), and lactate dehydrogenase (LDH) for the two groups:

PFK		HK		LDH	
C	COPD	C	COPD	C	COPD
106.8	49.3	2.0	2.3	241.5	124.3
19.6	107.1	3.2	1.4	216.8	269.6
27.3	62.9	2.5	1.0	105.6	247.8
51.6	53.2	2.6	3.6	133.9	200.7
73.2	105.7	2.4	1.3	336.4	540.5
89.6	61.3	2.4	2.9	131.1	431.1
47.7	28.2	3.5	2.2	241.4	65.3
113.5	68.5	2.2	1.5	297.1	204.7
46.4	40.8	2.4	1.6	156.6	137.6

Source: Dr. François Maltais. Used with permission.

28. Torre et al. (A-30) conducted a study to determine serum levels of nitrite in pediatric patients with human immunodeficiency virus type 1 (HIV-1) infection. Subjects included 10 healthy control children (six boys and four girls) with a mean age of 9.7 years and a standard deviation of 3.3. The remainder of the subjects were 21 children born to HIV-1-infected mothers. Of these, seven (three boys and four girls) were affected by AIDS. They had a mean age of 6 years with a standard deviation of 2.8. The remaining 14 children (seven boys and seven girls) became seronegative for HIV-1 during the first year of life. Their mean age was 3.3 years with a standard deviation of 2.3 years. Among the data collected were the following serum levels of nitrite (μmol/L):

Controls $n = 10$	Seronegativized Children $n = 14$	HIV-1-Positive Patients $n = 7$
0.301	0.335	0.503
0.167	0.986	0.268
0.201	0.846	0.335
0.234	1.006	0.946
0.268	2.234	0.846
0.268	1.006	0.268
0.201	0.803	0.268
0.234	0.301	
0.268	0.936	
0.301	0.268	
	0.134	
	0.335	
	0.167	
	0.234	

Source: Dr. Donato Torre. Used with permission.

29. Seghaye et al. (A-31) analyzed the influence of low-dose aprotinin on complement activation, leukocyte stimulation, cytokine production, and the acute-phase response in children undergoing cardiac operations. Inclusion criterion for the study was a noncyanotic congenital cardiac defect requiring a relatively simple primary surgical procedure associated with a low postoperative risk. Among the data collected were the following measurements on interleukin-6 (IL-6) and C-reactive protein (CRP) obtained 4 and 24 hours postoperatively, respectively:

IL-6	CRP	IL-6	CRP	IL-6	CRP
122	32	467	53	215	50
203	39	421	29	415	41
458	63	421	44	66	12
78	7	227	24	58	14
239	62	265	31	213	9
165	22	97	12		

Source: Dr. Marie-Christine Seghaye. Used with permission.

Exercises for Use with Large Data Sets Available on the Following Website: www.wiley.com/college/daniel

1. California State Assembly Bill 2071 (AB 2071) mandated that patients at methadone clinics be required to undergo a minimum of 50 minutes of counseling per month. Evan Kletter (A-32) collected data on 168 subjects who were continuously active in treatment through the Bay Area Addiction Research and Treatment (BAART) centers for 1 year prior to, and 2 years after AB 2071's implementation. Prior to AB 2071, BAART center counselors spent two sessions of at least 15 minutes per session per month with each client. The subjects in the study were also identified as cocaine abusers. The observations in KLETTER are the percentages of failing a cocaine drug test for each of the subjects pre- and post-AB 2071. For example, a pre-value of 60 implies that the patient failed a cocaine test 60 percent of the time prior to adoption of AB 2071. Dr. Kletter performed a Wilcoxon rank sum test to determine if the percentage of failed tests decreased significantly after the passage of AB 2071. Use the data to determine what conclusion he was able to reach. Report the test statistic and p value.

REFERENCES

Methodology References

1. FRANK WILCOXON, "Individual Comparisons by Ranking Methods," *Biometrics, 1* (1945), 80–83.
2. A. M. MOOD, *Introduction to the Theory of Statistics,* McGraw-Hill, New York, 1950.
3. J. WESTENBERG, "Significance Test for Median and Interquartile Range in Samples from Continuous Populations of Any Form," *Proceedings Koninklijke Nederlandse Akademie Van Wetenschappen, 51* (1948), 252–261.
4. G. W. BROWN and A. M. MOOD, "On Median Tests for Linear Hypotheses," *Proceedings of the Second Berkeley Symposium on Mathematical Statistics and Probability,* University of California Press, Berkeley, 1951, 159–166.
5. H. B. MANN and D. R. WHITNEY, "On a Test of Whether One of Two Random Variables Is Stochastically Larger than the Other," *Annals of Mathematical Statistics, 18* (1947), 50–60.
6. A. N. KOLMOGOROV, "Sulla Determinazione Empirial di una Legge di Distribuizione," *Giornale dell' Institute Italiano degli Altuari, 4* (1933), 83–91.
7. N. V. SMIRNOV, "Estimate of Deviation Between Empirical Distribution Functions in Two Independent Samples" (in Russian), *Bulletin Moscow University, 2* (1939), 3–16.

8. W. H. Kruskal and W. A. Wallis, "Use of Ranks in One-Criterion Analysis of Variance," *Journal of the American Statistical Association, 47* (1952), 583–621; errata, *ibid., 48* (1953), 907–911.
9. M. Friedman, "The Use of Ranks to Avoid the Assumption of Normality Implicit in the Analysis of Variance," *Journal of the American Statistical Association, 32* (1937), 675–701.
10. M. Friedman, "A Comparison of Alternative Tests of Significance for the Problem of *m* Rankings," *Annals of Mathematical Statistics, II* (1940), 86–92.
11. C. Spearman, "The Proof and Measurement of Association Between Two Things," *American Journal of Psychology, 15* (1904), 72–101.
12. H. Theil, "A Rank-Invariant Method of Linear and Polynomial Regression Analysis. III," *Koninklijke Nederlandse Akademie Van Wetenschappen, Proceedings, Series A, 53* (1950), 1397–1412.
13. E. Jacquelin Dietz, "Teaching Regression in a Nonparametric Statistic Course," *American Statistician, 43* (1989), 35–40.
14. Jean D. Gibbons, *Nonparametric Methods for Quantitative Analysis,* Third Edition, American Sciences Press, Syracuse, NY, 1996.
15. Marjorie A. Pett, *Nonparametric Statistics for Health Care Research,* Sage Publications, Thousand Oaks, CA, 1997.

Applications References

A-1. Robert B. Parker, Ryan Yates, Judith E. Soberman, and Casey Laizure, "Effects of Grapefruit Juice on Intestial P-glycoprotein: Evaluation Using Digoxin in Humans," *Pharmacotherapy, 23* (2003), 979–987.

A-2. Christina Grape, Maria Sandgren, Lars-Olaf Hansson, Mats Ericson, and Töres Theorell, "Does Singing Promote Well-Being?: An Empirical Study of Professional and Amateur Singers During a Singing Lesson," *Integrative Physiological and Behavioral Science, 38* (2003), 65–74.

A-3. R. S. Zuckerman and S. Heneghan, "The Duration of Hemodynamic Depression During Laparoscopic Cholecystectomy," *Surgical Endoscopy, 16* (2002), 1233–1236.

A-4. Carole W. Cranor and Dale B. Christensen, "The Asheville Project: Short-Term Outcomes of a Community Pharmacy Diabetes Care Program," *Journal of the American Pharmaceutical Association, 43* (2003), 149–159.

A-5. Yitao Liu, Ludmila Belayev, Weizhao Zhao, Raul Busto, and Myron D. Ginsburg, "MRZ 2/579, a Novel Uncompetitive *N*-Methyl-D-Aspartate Antagonist, Reduces Infarct Volume and Brain Swelling and Improves Neurological Deficit After Focal Cerebral Ischemia in Rats," *Brain Research, 862* (2000), 111–119.

A-6. Mohamed A. Elshaikh, James C. Ulchaker, Chandana A. Reddy, Kenneth Angermeier, Eric A. Klein, and Jay P. Ciezki, "Prophylactic Tamsulosin (Flomax®) in Patients Undergoing Prostate ^{125}I Brachytherapy for Prostate Carcinoma. Final Report of a Double-Blind Placebo-Controlled Randomized Study," *International Journal of Radiation Oncology, Biology, Physics* (forthcoming).

A-7. G. Lacroix, S. Tissot, F. Rogerieux, R. Beaulieu, L. Cornu, C. Gillet, F. Robidel, J. P. Lefèvre, and F. Y. Bois, "Decrease in Ovalbumin-Induced Pulmonary Allergic Response by Benzaldehyde but Not Acetaldehyde Exposure in a Guinea Pig Model," *Journal of Toxicology and Environmental Health, Part A, 65* (2002), 995–1012.

A-8. W. Herrmann, H. Schorr, J. P. Knapp, A. Müller, G. Stein, and J. Geisel, "Role of Homocysteine, Cystathionine and Methylmalonic Acid Measurement for Diagnosis of Vitamin Deficiency in High-Aged Subjects," *European Journal of Clinical Investigation, 30* (2000), 1083–1089.

A-9. Carol Flexer, Kate Kemp Biley, Alyssa Hinkley, Cheryl Harkema, and John. Holcomb, "Using Sound-Field to Teach Phonemic Awareness to Pre-schoolers," *Hearing Journal,* 55(3) (2002), 38–44.

A-10. Satoshi Nozawa, Katsuji Shimizu, Kei Miyamoto, and Mizuo Tanaka, "Repair of Pars Interarticularis Defect by Segmental Wire Fixation in Young Athletes with Spondylolysis," *American Journal of Sports Medicine, 31* (2003), 359–364.

A-11. M. Butz, K. H. Wollinsky, U. Widemuth-Catrinescu, A. Sperfeld, S. Winter, H. H. Mehrkens, A. C. Ludolph, and H. Schreiber, "Longitudinal Effects of Noninvasive Positive-Pressure Ventilation in Patients with Amyotrophic Lateral Sclerosis," *American Journal of Medical Rehabilitation, 82* (2003), 597–604.

A-12. Tomohiro Otani and Shoji Kishi, "A Controlled Study of Vitrectomy for Diabetic Macular Edema," *American Journal of Ophthalmology, 134* (2002), 214–219.

A-13. J. Jose and S. R. Ell, "The Association of Subjective Nasal Patency with Peak Inspiratory Nasal Flow in a Large Healthy Population," *Clinical Otolaryngology, 28* (2003), 352–354.

A-14. WENDY GANTT and the Wright State University Statistical Consulting Center (2002).

A-15. MARY WHITE and the Wright State University Statistical Consulting Center (2001).

A-16. DOMINIC SPROTT and the WRIGHT State University Statistical Consulting Center (2003).

A-17. PETER DAMM, HENRIK VESTERGAARD, CLAUS KÜHL, and OLUF PEDERSEN, "Impaired Insulin-Stimulated Nonoxidative Glucose Metabolism in Glucose-Tolerant Women with Previous Gestational Diabetes," *American Journal of Obstetrics and Gynecology, 174* (1996), 722–729.

A-18. BERNARD GUTIN, MARK LITAKER, SYED ISLAM, TINA MANOS, CLAYTON SMITH, and FRANK TREIBER, "Body-Composition Measurement in 9–11-yr-Old Children by Dual-Energy X-Ray Absorptiometry, Skinfold-Thickness Measurements, and Bioimpedance Analysis," *American Journal of Clinical Nutrition, 63* (1996), 287–292.

A-19. COURTNEY CRIM, CESAR A. KELLER, CHERIE H. DUNPHY, HORACIO M. MALUF, and JILL A. OHAR, "Flow Cytometric Analysis of Lung Lymphocytes in Lung Transplant Recipients," *American Journal of Respiratory and Critical Care Medicine, 153* (1996), 1041–1046.

A-20. MASAKAZU ICHINOSE, MOTOHIKO MIURA, HIDEYUKI YAMAUCHI, NATSUKO KAGEYAMA, MASAFUMI TOMAKI, TATSUYA OYAKE, YUZURU OHUCHI, WATARU HIDA, HIROSHI MIKI, GEN TAMURA, and KUNIO SHIRATO, "A Neurokinin 1-Receptor Antagonist Improves Exercise-Induced Airway Narrowing in Asthmatic Patients," *American Journal of Respiratory and Critical Care Medicine, 153* (1996), 936–941.

A-21. TOMOAKI TOMIYA and KENJI FUJIWARA, "Serum Transforming Growth Factor *a* Level as a Marker of Hepatocellular Carcinoma Complicating Cirrhosis," *Cancer, 77* (1996), 1056–1060.

A-22. KHASHAYAR SAKHAEE, LISA RUML, PAULETTE PADALINO, SHARON HAYNES, and CHARLES Y. C. PAK, "The Lack of Influence of Long-Term Potassium Citrate and Calcium Citrate Treatment in Total Body Aluminum Burden in Patients with Functioning Kidneys," *Journal of the American College of Nutrition, 15* (1996), 102–106.

A-23. JEAN-MICHEL DUBUIS, JACQUELINE GLORIEUX, FAISCA RICHER, CHERI L. DEAL, JEAN H. DUSSAULT, and GUY VAN VLIET, "Outcome of Severe Congenital Hypothyroidism: Closing the Developmental Gap with Early High Dose Levothyroxine Treatment," *Journal of Clinical Endocrinology and Metabolism, 81* (1996), 222–227.

A-24. PIOTR KUNA, MARK LAZAROVICH, and ALLEN P. KAPLAN, "Chemokines in Seasonal Allergic Rhinitis," *Journal of Allergy and Clinical Immunology, 97* (1996), 104–112.

A-25. CHEE JEONG KIM, WANG SEONG RYU, JU WON KWAK, CHONG TAIK PARK, and UN HO RYOO, "Changes in Lp(a) Lipoprotein and Lipid Levels After Cessation of Female Sex Hormone Production and Estrogen Replacement Therapy," *Archives of Internal Medicine, 156* (1996), 500–504.

A-26. HENK TE VELTHUIS, PIET G. M. JANSEN, C. ERIK HACK, LEÓN EIJSMAN, and CHARLES R. H. WILDEVUUR, "Specific Complement Inhibition with Heparin-Coated Extracorporeal Circuits," *Annals of Thoracic Surgery, 61* (1996), 1153–1157.

A-27. YVONNE F. HEIJDRA, P. N. RICHARD DEKHUIJZEN, CEES L. A. VAN HERWAARDEN, and HANS TH. M. FOLGERING, "Nocturnal Saturation Improves by Target-Flow Inspiratory Muscle Training in Patients with COPD," *American Journal of Respiratory and Critical Care Medicine, 153* (1996), 260–265.

A-28. ADAM WOLKIN, MICHAEL SANFILIPO, ERICA DUNCAN, BURTON ANGRIST, ALFRED P. WOLF, THOMAS B. COOPER, JONATHAN D. BRODIE, EUGENE LASKA, and JOHN P. ROSTROSEN, "Blunted Change in Cerebral Glucose Utilization After Haloperidol Treatment in Schizophrenic Patients with Prominent Negative Symptoms," *American Journal of Psychiatry, 153* (1996), 346–354.

A-29. FRANÇOIS MALTAIS, ANDRÉE-ANNE SIMARD, CLERMONT SIMARD, JEAN JOBIN, PIERRE DESGAGNÉS, and PIERRE LEBLANC, "Oxidative Capacity of the Skeletal Muscle and Lactic Acid Kinetics During Exercise in Normal Subjects and in Patients with COPD," *American Journal of Respiratory and Critical Care Medicine, 153* (1996), 288–293.

A-30. DONATO TORRE, GIULIO FERRARIO, FILIPPO SPERANZA, ROBERTO MARTEGANI, and CLAUDIA ZEROLI, "Increased Levels of Nitrite in the Sera of Children Infected with Human Immunodeficiency Virus Type 1," *Clinical Infectious Diseases*, 22 (1996), 650–653.

A-31. MARIE-CHRISTINE SEGHAYE, JEAN DUCHATEAU, RALPH G. GRABITZ, KARSTEN JABLONKA, TOBIAS WENZL, CHRISTIANE MARCUS, BRUNO J. MESSMER, and GOETZ von BERNUTH, "Influence of Low-Dose Aprotinin on the Inflammatory Reaction Due to Cardiopulmonary Bypass in Children," *Annals of Thoracic Surgery, 61* (1996), 1205–1211.

A-32. EVAN KLETTER, "Counseling as an Intervention for the Cocaine-Abusing Methadone Maintenance Patient," *Journal of Psychoactive Drugs, 35* (2003), 271–277.

CHAPTER **14**

EVALUATING THE HEALTH OF HUMAN GROUPS: VITAL STATISTICS

CHAPTER OVERVIEW

The focus of this chapter is on measurements and other tools that are used in the evaluation of the health of various human groups of interest, such as local, state, and national populations. The techniques employed are concerned with certain aspects of death, disease, and fertility.

TOPICS

LEARNING OUTCOMES

After studying this chapter, the student will

1. understand the basic concepts and methodologies of epidemiology and vital statistics and how they are used by professionals in the health sciences field.
2. be able to describe the differences between a rate and a ratio, and to explain when the use of each is appropriate.
3. understand the concepts of mortality, morbidity, and fertility.
4. know how to calculate useful measures from mortality, morbidity, and fertility data.

14.1 INTRODUCTION

The private physician arrives at a diagnosis and treatment plan for an individual patient by means of a case history, a physical examination, and various laboratory tests. The community may be thought of as a living complex organism for which the public health team is the physician. To carry out this role satisfactorily the public health team must also make use of appropriate tools and techniques for evaluating the health status of the community. Traditionally, these tools and techniques have consisted of the community's vital statistics, which include the counts of births, deaths, illnesses, and the various rates and ratios that may be computed from them.

The idea of community-based, public health medicine as just described is most often studied using the concepts and tools of *epidemiology*. Epidemiologists study the mechanisms by which diseases and other health-related conditions arise and how they are distributed among populations. While private physicians diagnose and treat individual patients who have a medical condition, epidemiologists and public health professionals are additionally interested in studying those members of a population who are well, and how those who have an illness differ from those who are free of the illness. To that end, the use of vital statistics and epidemiological tools are employed in determining the *prevalence* of a given condition at a point in time and *incidence* at which new conditions arise in the population. Standard epidemiological measures such as those of mortality, morbidity, and fertility are explored in this chapter.

In succeeding sections we give some of the more useful and widely used rates and ratios. Before proceeding, however, let us distinguish between the terms *rate* and *ratio* by defining each as follows.

1. **Rate.** Although there are some exceptions, the term *rate* usually is reserved to refer to those calculations that involve the frequency of the occurrence of some event. A rate is expressed in the form

$$\left(\frac{a}{a+b}\right)k \qquad (14.1.1)$$

where

a = the frequency with which an event has occurred during some specified period of time

$a + b$ = the number of persons exposed to the risk of the event during the same period of time

k = some number such as 10, 100, 1000, 10,000, or 100,000

As indicated by Expression 14.1.1, the numerator of a rate is a component part of the denominator. The purpose of the multiplier, k, called the base, is to avoid results involving the very small numbers that may arise in the calculation of rates and to facilitate comprehension of the rate. The value chosen for k will depend on the magnitudes of the numerator and denominator.

2. **Ratio.** A ratio is a fraction of the form

$$\left(\frac{c}{d}\right)k \tag{14.1.2}$$

where k is some base as already defined and both c and d refer to the frequency of occurrence of some event or item. In the case of a ratio, as opposed to a rate, the numerator is not a component part of the denominator. We can speak, for example, of the person–doctor ratio or the person–hospital-bed ratio of a certain geographic area. The values of k most frequently used in ratios are 1 and 100.

14.2 DEATH RATES AND RATIOS

The rates and ratios discussed in this section are concerned with the occurrence of death. Death rates express the relative frequency of the occurrence of death within some specified interval of time in a specific population. The denominator of a death rate is referred to as the population at risk. The numerator represents only those deaths that occurred in the population specified by the denominator.

1. **Annual crude death rate.** The annual crude death rate is defined as

$$\frac{\text{total number of deaths during year (January 1 to December 31)}}{\text{total population as of July 1}} \cdot k$$

where the value of k is usually chosen as 1000. This is the most widely used rate for measuring the overall health of a community. To compare the crude death rates of two communities is hazardous, unless it is known that the communities are comparable with respect to the many characteristics, other than health conditions, that influence the death rate. Variables that enter into the picture include age, race, sex, and socioeconomic status. When two populations must be compared on the basis of death rates, adjustments may be made to reconcile the population differences with respect to these variables. The same precautions should be exercised when comparing the annual death rates for the same community for 2 different years.

2. **Annual specific death rates.** It is usually more meaningful and enlightening to observe the death rates of small, well-defined subgroups of the total population. Rates of this type are called *specific death rates* and are defined as

$$\frac{\text{total number of deaths in a specific subgroup during a year}}{\text{total population in the specific subgroup as of July 1}} \cdot k$$

where k is usually equal to 1000. Subgroups for which specific death rates may be computed include those groups that may be distinguished on the basis of sex, race, and age. Specific rates may be computed for two or more characteristics simultaneously. For example, we may compute the death rate for white males, thus obtaining a race-sex specific rate. Cause-specific death rates may also be computed by including in the numerator only those deaths due to a particular cause of death, say, cancer, heart disease, or accidents. Because of the small fraction that results, the base, k, for a cause-specific rate is usually 100,000 or 1,000,000.

3. **Adjusted or standardized death rates.** As we have already pointed out, the usefulness of the crude death rate is restricted by the fact that it does not reflect the composition of the population with respect to certain characteristics by which it is influenced. We have seen that by means of specific death rates various segments of the population may be investigated individually. If, however, we attempt to obtain an overall impression of the health of a population by looking at individual specific death rates, we are soon overwhelmed by their great number.

What is wanted is a single figure that measures the forces of mortality in a population while holding constant one or more of the compositional factors such as age, race, or sex. Such a figure, called an *adjusted death rate,* is available. It is most commonly obtained by what is known as the *direct method* of adjustment. The method consists essentially of applying to a *standard population* specific rates observed in the population of interest. From the resulting expected numbers we may compute an overall rate that tells us what the rate for the population of interest would be if that population had the same composition as the standard population. This method is not restricted to the computation of adjusted death rates only, but may be used to obtain other adjusted rates, for example, an adjusted birth rate. If two or more populations are adjusted in this manner, they are then directly comparable on the basis of the adjustment factors. Opinions differ as to what population should be used as the standard. The population of the United States as of the last decennial census is frequently used. For adjustment calculations a population of 1,000,000, reflecting the composition of the standard population and called the *standard million,* is usually used. In the following example we illustrate the direct method of adjustment to obtain an age-adjusted death rate.

EXAMPLE 14.2.1

The 2000 crude death rate for Georgia was 7.8 deaths per 1000 population (A-1). Let us obtain an age-adjusted death rate for Georgia by using the 2000 United States census as the standard population. In other words, we want a death rate that could have been expected in Georgia if the age composition of the Georgia population had been the same as that of the United States in 2000.

Solution: The data necessary for the calculations are shown in Table 14.2.1.

The procedure for calculating an age-adjusted death rate by the direct method consists of the following steps.

1. The population of interest is listed (Column 2) according to age group (Column 1).
2. The deaths in the population of interest are listed (Column 3) by age group.
3. The age-specific death rates (Column 4) for each age group are calculated by dividing Column 3 by Column 2 and multiplying by 100,000.

TABLE 14.2.1 Calculations of Age-Adjusted Death Rate for Georgia, 2000, by Direct Method

1	2	3	4	5	6
Age (Years)	Population[a]	Deaths[a]	Age-Specific Death Rates (per 100,000)	Standard Population Based on U.S. Population 2000[b]	Number of Expected Deaths in Standard Population
0–4	595,150	1,299	218.3	68,139	149
5–9	615,584	101	16.4	73,020	12
10–14	607,759	136	22.4	72,944	16
15–19	596,277	447	75.0	71,849	54
20–44	3,244,960	5,185	159.8	369,567	591
45–64	1,741,448	13,092	751.8	220,141	1,655
65 and over	785,275	43,397	5526.3	124,339	6,871
Total	8,186,453	63,657		1,000,000[c]	9,348

Sources: [a] *Georgia Vital and Morbidity Statistics 2000,* Georgia Division of Public Health, Atlanta (A-1).
[b] *Profile of General Demographic Characteristics: 2000,* U.S. Census Bureau DP-1 (A-2).
[c] Total does not reflect actual sum because of rounding to the nearest person.

4. The standard population (Column 5) is listed by age group. We obtain the standard population as follows. The 2000 U.S. population by age group is shown in Table 14.2.2. We divide the total for each age group by the grand total and multiply by 1,000,000. For example, to obtain the standard population count for the 0–4 age group, we divide 19,175,798 by 281,421,906 and multiply the result by 1,000,000. That is, 1,000,000(19,175,798/281,421,906) = 68,139. Similar calculations yield the standard population counts for the other age groups as shown in Table 14.2.1.

TABLE 14.2.2 Population of the United States, 2000

Age (Years)	Population
0–4	19,175,798
5–9	20,549,505
10–14	20,528,072
15–19	20,219,890
20–44	104,004,252
45–64	61,952,636
65 and over	34,991,753
Total	281,421,906

Source: *Profile of General Demographic Characteristics: 2000,* U.S. Census Bureau DP-1 (A-2).

5. The expected number of deaths in the standard population for each group (Column 6) is computed by multiplying Column 4 by Column 5 and dividing by 100,000. The entries in Column 6 are the deaths that would be expected in the standard population if the persons in this population had been exposed to the same risk of death experienced by the population being adjusted.
6. The entries in Column 6 are summed to obtain the total number of expected deaths in the standard population.
7. The age-adjusted death rate is computed in the same manner as a crude death rate. That is, the age-adjusted death rate is equal to

$$\frac{\text{total number of expected deaths}}{\text{total standard population}} \cdot 1000$$

In the present example we have an age-adjusted death rate of

$$\frac{9348}{1{,}000{,}000} \cdot 1000 = 9.3$$

We see, then, that by adjusting the 2000 population of Georgia to the age distribution of the standard population, we obtain an adjusted death rate that is 1.5 per 1000 greater than the crude death rate $(9.3 - 7.8)$. This increase in the death rate following adjustment reflects the fact that in 2000 the population of Georgia was slightly younger than the population of the United States as a whole. For example, only 9.6 percent of the Georgia population was 65 years of age or older, whereas 12.4 percent of the U.S. population was in that age group. ■

4. **Maternal mortality rate.** This rate is defined as

$$\frac{\text{deaths from all puerperal causes during a year}}{\text{total live births during the year}} \cdot k$$

where k is taken as 1000 or 100,000. The preferred denominator for this rate is the number of women who were pregnant during the year. This denominator, however, is impossible to determine.

A death from a puerperal cause is a death that can be ascribed to some phase of childbearing. Because of the decline in the maternal mortality rate in the United States, it is more convenient to use $k = 100{,}000$. In some countries, however, $k = 1000$ results in a more convenient rate. The decline in the maternal mortality rate in this country also has had the effect of reducing its usefulness as a discriminator among communities with varying qualities of medical care and health facilities.

Some limitations of the maternal mortality rate include the following:

(a) Fetal deaths are not included in the denominator. This results in an inflated rate, since a mother can die from a puerperal cause without producing a live birth.

(b) A maternal death can be counted only once, although twins or larger multiple births may have occurred. Such cases cause the denominator to be too large and, hence, there is a too small rate.

(c) Under-registration of live births, which result in a too small denominator, causes the rate to be too large.

(d) A maternal death may occur in a year later than the year in which the birth occurred. Although there are exceptions, in most cases the transfer of maternal deaths will balance out in a given year.

5. **Infant mortality rate.** This rate is defined as

$$\frac{\text{number of deaths under 1 year of age during a year}}{\text{total number of live births during the year}} \cdot k$$

where k is generally taken as 1000. Use and interpretation of this rate must be made in light of its limitations, which are similar to those that characterize the maternal mortality rate. Many of the infants who die in a given calendar year were born during the previous year; and, similarly, many children born in a given calendar year will die during the following year. In populations with a stable birth rate this does not pose a serious problem. In periods of rapid change, however, some adjustment should be made. One way to make an adjustment is to allocate the infant deaths to the calendar year in which the infants were born before computing the rate.

6. **Neonatal mortality rate.** In an effort to better understand the nature of infant deaths, rates for ages less than a year are frequently computed. Of these, the one most frequently computed is the *neonatal mortality rate,* which is defined as

$$\frac{\text{number of deaths under 28 days of age during a year}}{\text{total number of live births during the year}} \cdot k$$

where $k = 1000$.

7. **Fetal death rate.** This rate is defined as

$$\frac{\text{total number of fetal deaths during a year}}{\text{total deliveries during the year}} \cdot k$$

where k is usually taken to be 1000. A fetal death is defined as a product of conception that shows no sign of life after complete birth. There are several problems associated with the use and interpretation of this rate. There is variation among reporting areas with respect to the duration of gestation. Some areas report all fetal deaths regardless of length of gestation, while others have a minimum gestation period that must be reached before reporting is required. Another objection to the fetal death rate is that it does not take into account the extent to which a community is trying to reproduce. The ratio to be considered next has been proposed to overcome this objection.

8. **Fetal death ratio.** This ratio is defined as

$$\frac{\text{total number of fetal deaths during a year}}{\text{total number of live births during the year}} \cdot k$$

where k is taken as 100 or 1000.

Some authorities have suggested that the number of fetal deaths as well as live births be included in the denominator in an attempt to include all pregnancies in the computation of the ratio. The objection to this suggestion rests on the incompleteness of fetal death reporting.

9. **Perinatal mortality rate.** Since fetal deaths occurring late in pregnancy and neonatal deaths frequently have the same underlying causes, it has been suggested that the two be combined to obtain what is known as the *perinatal mortality rate.* This rate is computed as

$$\frac{(\text{number of fetal deaths of 28 weeks or more}) + (\text{infant deaths under 7 days})}{(\text{number of fetal deaths of 28 weeks or more}) + (\text{number of live births})} \cdot k$$

where $k = 1000$.

10. **Cause-of-death ratio.** This ratio is defined as

$$\frac{\text{number of deaths due to a specific disease during a year}}{\text{total number of deaths due to all causes during the year}} \cdot k$$

where $k = 100$. This index is used to measure the relative importance of a given cause of death. It should be used with caution in comparing one community with another. A higher cause-of-death ratio in one community than that in another may be because the first community has a low mortality from other causes.

11. **Proportional mortality ratio.** This index has been suggested as a single measure for comparing the overall health conditions of different communities. It is defined as

$$\frac{\text{number of deaths in a particular subgroup}}{\text{total number of deaths}} \cdot k$$

where $k = 100$. The specified class is usually an age group such as 50 years and over, or a cause of death category, such as accidents.

EXERCISES

14.2.1 The following 2000 data were reported for the Clayton County (Georgia) Health District—Morrow:

		Number	
	Total	White	Black[a]
Estimated population as of July 1, 2000	236,517	89,741	121,927
Total live births	4,350	1,629	2,494
Fetal deaths	41	8	32
Deaths			
Total deaths all ages	1366	898	446
Under 1 year	41	13	28
Under 28 days	24	6	18
Between 28 and 364 days	17	7	10
Cause of death			
Malignant neoplasms	303	212	85
Major cardiovascular diseases	471	329	136

Source: *2000 Mortality Vital Statistics Reports,* Georgia Office of Health Information and Policy (A-3).

[a] The Black category does not include "other" races. The total includes white, black, and "other."

From these data compute the following rates and ratios: (a) crude death rate, (b) race-specific death rates for white and black, (c) infant mortality rate, (d) neonatal mortality rate, (e) fetal death ratio, and (f) cause of death ratios for malignant neoplasms and major cardiovascular diseases.

14.2.2 The following table shows the deaths and estimated population by age for the state of Georgia for 1999. Use these data to compute the age-adjusted death rate for Georgia, 1999. Use the same standard population that was used in Example 14.2.1.

Age (Years)	Population	Deaths
0–4	580,222	1,197
5–9	577,151	126
10–14	563,293	153
15–19	570,547	460
20–44	3,083,521	4,989
45–64	1,653,515	12,633
65 and over	761,457	42,076
Total	7,789,706	61,634

Source: *1999 Mortality Vital Statistics Reports,* Georgia Office of Health Information and Policy (A-4).

14.2.3 The following table shows, by age group, deaths in North Carolina during 2000, the population of North Carolina for 2000, and the population of the United States for 2000. Use these data to compute the 2000 age-adjusted death rate for North Carolina. Use the 2000 U.S. population as the standard population.

Age (Years)	Population[a]	Deaths[b]	U.S. Population[c]
0–4	539,509	1,178	19,175,798
5–14	1,113,920	224	41,077,577
15–24	1,117,439	954	39,183,891
25–34	1,213,415	1,384	39,891,724
35–44	1,287,120	2,823	45,148,527
45–54	1,085,150	5,271	37,677,952
55–64	723,712	8,035	24,274,684
65 and over	969,048	51,863	34,991,753
Total	8,049,313	71,732	281,421,906

Sources: [a] *Profile of General Demographic Characteristics: 2000,* U.S. Census Bureau, DP-1 for North Carolina (A-5).

[b] *Selected Vital Statistics for 2000 and 1996–2000,* North Carolina State Center for Health Statistics, (A-6).

[c] *Profile of General Demographic Characteristics: 2000,* U.S. Census Bureau DP-1 (A-2).

14.2.4 The following data are for North Carolina for year 2000. Use these data to compute rates and/or ratios of interest.

	Total	White	Minority[a]
Live births	120,247	86,341	33,906
Low birth weight	10,572	6,150	4,422
C-section	28,198	19,895	8,303
Perinatal deaths	1,683	913	770
Fetal deaths	937	516	421
Neonatal (under 28 days)	746	397	349
Postneonatal (28 days–1 year)	288	148	140
Infant deaths (under 1 year)	1,034	545	489
Death: males (excluding fetals)	71,732	55,241	16,491
Death: females (excluding fetals)	35,227	26,823	8,404
Cause of death: heart disease	19,649	15,419	4230
Cause of death: cancer	15,747	12,375	3372

Source: *Selected Vital Statistics for 2000 and 1996–2000* and *Leading Causes of Death by Age Group,* North Carolina State Center for Health Statistics (A-6).

[a] The minority category includes all nonwhite races.

14.3 MEASURES OF FERTILITY

The term *fertility* as used by American demographers refers to the actual bearing of children as opposed to the capacity to bear children, for which phenomenon the term *fecundity* is used. Knowledge of the "rate" of childbearing in a community is important to the health worker in planning services and facilities for mothers, infants, and children. The following are the six basic measures of fertility.

1. **Crude birth rate.** This rate is the most widely used of the fertility measures. It is obtained from

$$\frac{\text{total number of live births during a year}}{\text{total population as of July 1}} \cdot k$$

where $k = 1000$. For an illustration of the computation of this and the other five rates, see Table 14.3.1.

2. **General fertility rate.** This rate is defined as

$$\frac{\text{number of live births during a year}}{\text{total number of women of childbearing age}} \cdot k$$

TABLE 14.3.1 Illustration of Procedures for Computing Six Basic Measures of Fertility, for Georgia, 2000

1 Age of Woman (Years)	2 Number of Women in Population[a]	3 Number of Births to Women of Specified Age[b]	4 Age-Specific Birth Rate per 1000 Women	5 U.S. Population for Year 2000[c]	6 Standard Population Based on U.S. Population 2000	7 Expected Births	8 Cumulative Fertility Rate
10–14	296,114	396	1.3	20,528,072	112,524	146	6.7
15–19	286,463	17,915	62.5	20,219,890	110,835	6,927	319.4
20–24	285,733	36,512	127.8	18,964,001	103,951	13,285	958.3
25–29	316,000	35,206	111.4	19,381,336	106,239	11,835	1,515.4
30–34	326,709	27,168	83.2	20,512,388	112,438	9,355	1,931.1
35–39	350,943	12,685	36.1	22,706,664	124,466	4,493	2,111.9
40–54	887,104	2,404[d]	2.7	60,119,815	329,546	890	2,152.5
Total	2,749,066	132,286		182,432,166	1,000,000	46,931	

Computation of six basic rates:

(1) Crude birth rate = total births divided by total population
= (132,286/8,186,453)(1000) = 16.2.

(2) General fertility rates = (132,286/2,749,066)(1000) = 48.1.

(3) Age-specific fertility rates = entries in Column 3 divided by entries in Column 2 multiplied by 1000 for each group. Results appear in Column 4.

(4) Expected births = entries in Column 4 multiplied by entries in Column 6 divided by 1000 for each group. Results appear in Column 7.

(5) Total fertility rate = the sum of each age-specific rate multiplied by the age interval width = 1.3(5) + 62.5(5) + (127.8)(5) + 111.4(5) + 83.2(5) + 36.1(5) + 2.7(15) = 2,152.5.

(6) Cumulative fertility rate = age-specific birth rate multiplied by age interval width cumulated by age. See Column 8.

(7) Standardized general fertility rate = (46,943)/(1,000,000)(1000) = 46.9.

Sources: [a] Obtained from Georgia Division of Public Health Online Analytical Statistical Information System (OASIS) (A-7).

[b] *Georgia Natality Vital Statistics Reports, 2000,* Georgia Division of Public Health, Atlanta (A-8).

[c] *Age Groups and Sex: 2000,* U.S. Census Bureau 2000 Summary file QT-P1 for United States (A-9).

[d] May include births to women age 55 years.

where $k = 1000$ and the childbearing age is usually defined as ages 15 through 44 or ages 15 through 49. The attractive feature of this rate, when compared to the crude birth rate, is the fact that the denominator approximates the number of persons actually exposed to the risk of bearing a child.

3. **Age-specific fertility rate.** Since the rate of childbearing is not uniform throughout the childbearing ages, a rate that permits the analysis of fertility rates for shorter maternal age intervals is desirable. The rate used is the age-specific fertility rate, which is defined as

$$\frac{\text{number of births to women of a certain age in a year}}{\text{total number of women of the specified age}} \cdot k$$

where $k = 1000$. Age-specific rates may be computed for single years of age or any age interval. Rates for 5-year age groups are the ones most frequently computed. Specific fertility rates may be computed also for other population subgroups such as those defined by race, socioeconomic status, and various demographic characteristics.

4. **Total fertility rate.** If the age-specific fertility rates for all ages are added and multiplied by the interval into which the ages were grouped, the result is called the *total fertility rate.* The resulting figure is an estimate of the number of children a cohort of 1000 women would have if, during their reproductive years, they reproduced at the rates represented by the age-specific fertility rates from which the total fertility rate is computed.

5. **Cumulative fertility rate.** The cumulative fertility rate is computed in the same manner as the total fertility rate except that the adding process can terminate at the end of any desired age group. The numbers in Column 7 of Table 14.3.1 are the cumulative fertility rates through the ages indicated in Column 1. The final entry in the cumulative fertility rate column is the total fertility rate.

6. **Standardized fertility rate.** Just as the crude death rate may be standardized or adjusted, so may we standardize the general fertility rate. The procedure is identical to that discussed in Section 14.2 for adjusting the crude death rate. The necessary computations for computing the age-standardized fertility rate are shown in Table 14.3.1. Note the total population figure for Georgia in 2000 is 8,186,453 (see Table 14.2.1).

EXERCISES

14.3.1 The data in the following table are for the state of Georgia for 2001:

Age of Woman (Years)	Number of Women in Population[a]	Number of Births to Women of Specified Age[b]
10–14	307,496	391
15–19	287,916	17,249

(Continued)

Age of Woman (Years)	Number of Women in Population[a]	Number of Births to Women of Specified Age[b]
25–29	309,759	34,884
30–34	339,018	28,334
35–39	353,387	12,911
40–54	923,315	2,413[c]
Total	2,815,251	133,468

Sources: [a] Obtained from Georgia Division of Public Health Online Analytical Statistical Information System (OASIS) (A-7).

[b] *Georgia Natality Vital Statistics Reports, 2001,* Georgia Division of Public Health, Atlanta (A-10).

[c] May include births to women age 55 years.

From the above data compute the following rates:

(a) Age-specific fertility rates for each age group

(b) Total fertility rate

(c) Cumulative fertility rate through each age group

(d) General fertility rate standardized by age

Use the standard population shown in Table 14.3.1.

14.3.2 There was a total of 133,468 live births in Georgia in 2001 (A-10). The estimated total population as of July 1, 2001, was 8,186,453 (A-11), and the number of women between the ages of 10 and 54 was 2,815,251 (A-10). Use these data to compute:

(a) The crude birth rate **(b)** The general fertility rate

14.3.3 The following data are for the state of North Carolina for 2000:

Age of Woman (Years)	Number of Births to Women of Specified Age[a]	Number of Women in Population[b]	U.S. Population for Year 2000[c]
10–14	335	276,837	20,528,072
15–19	15,343	262,292	20,219,890
20–24	33,030	274,738	18,964,001
25–29	32,975	290,046	19,381,336
30–34	25,529	303,023	20,512,388
35–39	11,032	326,552	22,706,664
40–44	1,917	320,657	22,441,863
45–54	82[d]	559,312	37,677,952
Total	120,243	2,613,457	182,432,166

Sources: [a] *North Carolina Resident Births for 2000 by Age of Mother and Birth Order for All Women,* North Carolina Department of Health and Human Services, State Center for Health Statistics (A-12).

[b] *Age Groups and Sex: 2000,* U.S. Census Bureau 2000 Summary file QT-P1 for North Carolina (A-13).

[c] *Age Groups and Sex: 2000,* U.S. Census Bureau 2000 Summary file QT-P1 for the United States (A-9).

[d] May contain births to women older than 54.

From these data, compute the rates specified in Exercise 14.3.1. To the extent possible, compare the results with those obtained in Exercise 14.3.1 and comment on the comparison.

14.4 MEASURES OF MORBIDITY

Another area that concerns the health worker who is analyzing the health of a community is *morbidity*. The word "morbidity" refers to the community's status with respect to disease. Data for the study of the morbidity of a community are not, as a rule, as readily available and complete as are the data on births and deaths because of incompleteness of reporting and differences among states with regard to laws requiring the reporting of diseases. The two rates most frequently used in the study of diseases in a community are the *incidence rate* and the *prevalence rate.*

1. **Incidence rate.** This rate is defined as

$$\frac{\text{total number of new cases of a specific disease during a year}}{\text{total population as of July 1}} \cdot k$$

where the value of k depends on the magnitude of the numerator. A base of 1000 is used when convenient, but 100 can be used for the more common diseases, and 10,000 or 100,000 can be used for those less common or rare. This rate, which measures the degree to which new cases are occurring in the community, is useful in helping determine the need for initiation of preventive measures. It is a meaningful measure for both chronic and acute diseases.

2. **Prevalence rate.** Although it is referred to as a rate, the *prevalence rate* is really a ratio, since it is computed from

$$\frac{\text{total number of cases, new or old, existing at a point in time}}{\text{total population at that point in time}} \cdot k$$

where the value of k is selected by the same criteria as for the incidence rate. This rate is especially useful in the study of chronic diseases, but it may also be computed for acute diseases.

3. **Case-fatality ratio.** This ratio is useful in determining how well the treatment program for a certain disease is succeeding. It is defined as

$$\frac{\text{total number of deaths due to a disease}}{\text{total number of cases due to the disease}} \cdot k$$

where $k = 100$. The period of time covered is arbitrary, depending on the nature of the disease, and it may cover several years for an endemic disease. Note that this ratio can be interpreted as the probability of dying following contraction of the disease in question and, as such, reveals the seriousness of the disease.

4. **Immaturity ratio.** This ratio is defined as

$$\frac{\text{number of live births under 2500 grams during a year}}{\text{total number of live births during the year}} \cdot k$$

where $k = 100$.

5. **Secondary attack rate.** This rate measures the occurrence of a contagious disease among susceptible persons who have been exposed to a primary case and is defined as

$$\frac{\text{number of additional cases among contacts of a primary case within the maximum incubation period}}{\text{total number of susceptible contacts}} \cdot k$$

where $k = 100$. This rate is used to measure the spread of infection and is usually applied to closed groups such as a household or classroom, where it can reasonably be assumed that all members were, indeed, contacts.

EXERCISES

14.4.1 Use the following facts to compute and label appropriate measures of morbidity.

(a) In 1997, in one of Georgia's state districts, 504 of 6927 babies born weighed less than 2500 grams at time of birth. In 2001, in the same district, 675 of 8336 babies born weighed less than 2500 grams. [SOURCE: *Georgia Vital Statistics Report, 1997* (A-14), and *Georgia Natality Vital Statistics Report, 2001* (A-10).]

(b) In 1999, the estimated number of cigarette smokers in Nevada was 430,000, and in the same year, the estimated population of the state was 1,935,000. In the same year, 57,296,000 Americans were current users of cigarettes, and the estimated population as of July 1, 1999 was 279,040,000. [SOURCE: U.S. Substance Abuse and Mental Health Services Administration, *National Household Survey on Drug Abuse, 1999,* as reported in Statistical Abstracts of the United States: 2002 (118th edition), U.S. Bureau of the Census, Washington, DC, 2002, Table No. 183; U.S. Census Bureau State Population Estimates, as reported in Statistical Abstract of the United States: 2002 (118th edition), U.S. Bureau of the Census, Washington, DC, 2002, Table No. 18 (A-15).]

(c) In 2000, in the United States, there were 40,758 reported new cases of acquired immunodeficiency syndrome (AIDS). The estimated population of the United States as of July 1, 2000, was 281,421,906. [SOURCE: U.S. Centers for Disease Control and Prevention, Atlanta, GA, Summary of Notifiable Diseases, United States, 2000, Morbidity and Mortality Weekly Report, Vol. 49, No. 53, June 14, 2002, and Statistical Abstract of the United States: 2002 (118th edition), U.S. Bureau of the Census, Washington, DC, 2002, Table No. 174 (A-16).]

14.5 SUMMARY

This chapter is concerned with the computation and interpretation of various rates and ratios that are useful in studying the health of a community. More specifically, we discuss the more important rates and ratios relating to births, deaths, and morbidity. Individuals who wish to continue their reading in this area may find the books by Kittleson (1) and Smith (2) of interest.

SUMMARY OF FORMULAS FOR CHAPTER 14

Formula Number	Name	Formula
14.1.1	Rate	$\left(\frac{a}{a+b}\right)k$
14.1.2	Ratio	$\left(\frac{c}{d}\right)k$
14.2—death rates and ratios	Annual crude death rate	$\frac{\text{total number of deaths during year}}{\text{total population as of July 1}} \cdot k$
	Annual specific death rate	$\frac{\text{total number of deaths in a specific subgroup during a year}}{\text{total population in the specific subgroup as of July 1}} \cdot k$
	Age-adjusted death rate	$\frac{\text{total number of expected deaths}}{\text{total standard population}} \cdot k$
	Maternal mortality rate	$\frac{\text{deaths from all puerperal causes during a year}}{\text{total live births during the year}} \cdot k$
	Infant mortality rate	$\frac{\text{number of deaths under 1 year of age during a year}}{\text{total number of live births during the year}} \cdot k$
	Neonatal mortality rate	$\frac{\text{number of deaths under 28 days of age during a year}}{\text{total number of live births during the year}} \cdot k$
	Fetal death rate	$\frac{\text{total number of fetal deaths during a year}}{\text{total number of deliveries during the year}} \cdot k$
	Fetal death ratio	$\frac{\text{total number of fetal deaths during a year}}{\text{total number of live births during the year}} \cdot k$
	Perinatal mortality	$\frac{\text{(number of fetal deaths of 28 weeks or more)} + \text{(infant deaths under 7 days)}}{\text{(number of fetal deaths of 28 weeks or more)} + \text{(number of live births)}} \cdot k$
	Cause-of-death ratio	$\frac{\text{number of deaths due to a specific disease during a year}}{\text{total number of deaths due to all causes during the year}} \cdot k$
	Proportional mortality ratio	$\frac{\text{number of deaths in a particular subgroup}}{\text{total number of deaths}} \cdot k$
14.3—measures of fertility	Crude birth rate	$\frac{\text{total number of live births during a year}}{\text{total population as of July 1}} \cdot k$

	General fertility rate	$\frac{\text{total number of live births during a year}}{\text{total number of women of childbearing age}} \cdot k$
	Age-specific fertility rate	$\frac{\text{number of births to women of a certain age in a year}}{\text{total number of women of the specified age}} \cdot k$
14.4—measures of morbidity	Incidence rate	$\frac{\text{total number of new cases of a specific disease during a year}}{\text{total population as of July 1}} \cdot k$
	Prevalence rate	$\frac{\text{total number of all diseases cases at a given point in time}}{\text{total population at a given point in time}} \cdot k$
	Case-fatality ratio	$\frac{\text{total number of deaths due to a disease}}{\text{total number of cases due to the disease}} \cdot k$
	Immaturity ratio	$\frac{\text{total number of live births under 2500 grams during a year}}{\text{total number of live births during the year}} \cdot k$
	Secondary attack rate	$\frac{\text{number of additional cases among contacts of a primary case within the maximum incubation period}}{\text{total number of susceptible contacts}} \cdot k$
Symbol Key	• a, b, c, d = frequencies associated with the definition of rates and ratios • k = multiplier for rates and ratios	

REVIEW QUESTIONS AND EXERCISES

1. Explain the difference between a rate and a ratio.
2. Explain the difference between a crude death rate and a specific death rate.
3. Why is it inadvisable to compare the crude death rates of two geographic areas?
4. What is an adjusted or standardized death rate? What is the advantage of an adjusted death rate over a crude death rate?
5. Define and discuss each of the following:
   - (a) Maternal mortality rate
   - (b) Infant mortality rate
   - (c) Fetal death rate
   - (d) Fetal death ratio
   - (e) Perinatal mortality rate
   - (f) Cause-of-death ratio
   - (g) Proportional mortality ratio
   - (h) Crude birth rate
   - (i) General fertility rate
   - (j) Age-specific fertility rate
   - (k) Total fertility rate
   - (l) Cumulative fertility rate
   - (m) Standardized fertility rate
6. What is morbidity?

7. Define and discuss the following morbidity measures:
   (a) Incidence rate
   (b) Prevalence rate
   (c) Case-fatality ratio
   (d) Immaturity ratio
   (e) Secondary attack rate

8. For one or more geographical areas (for example, the United States, your state of residence, and your county of residence) collect data and compute as many of the rates and ratios discussed in this chapter as possible.

9. The following table shows, by age group, deaths in North Carolina during 2001, the estimated population of North Carolina for 2001, and the population of the United States for 2000. Use these data to compute the 2001 age-adjusted death rate for North Carolina. Use the 2000 U.S. population as the standard population.

Age (Years)	Population[a]	Deaths[b]	U.S. Population[c]
0–4	566,740	1,178	19,175,798
5–14	1,140,934	224	41,077,577
15–24	1,130,671	954	39,183,891
25–34	1,210,066	1,384	39,891,724
35–44	1,293,741	2,823	45,148,527
45–54	1,124,305	5,271	37,677,952
55–64	753,533	8,035	24,274,684
65 and over	986,115	51,863	34,991,753
Total	8,206,105	71,732	281,421,906

Sources: [a] Table ST-EST2002-ASRO-02-37—*State Characteristic Estimates,* Population Division, U.S. Census Bureau (A-17).

[b] *Selected Vital Statistics for 2001 and 1997–2001,* North Carolina State Center for Health Statistics (A-18).

[c] *Profile of General Demographic Characteristics: 2000,* U.S. Census Bureau, DP-1 for the United States (A-2).

10. The following data (numbers of specified events) are for a specific county in North Carolina for the year 2001. Use these data to compute as many of the rates and ratios mentioned in this chapter as possible.

		Number	
	Total	White	Minority
Estimated population as of July 1, 2001	301,098	176,961	124,137
Total live births	5,315	3,093	2,222
Fetal deaths	45	21	24
Deaths			
Total deaths all ages (excluding fetals)	1,989	1,222	767
Under 1 year	56	17	39
Under 28 days	40	9	31
Between 28 and 364 days	16	8	8

(Continued)

	Total	Number	
		White	Minority
Low birth weight	480	201	279
Mother smoked	667	472	195
C-section	1,243	694	549

Source: *Selected Vital Statistics for 2001 and 1997–2001,* North Carolina State Center for Health Statistics (A-18).

11. The following death data (number of deaths per category) are for the state of New York for 2000. Use these data to compute rates and/or ratios of interest. To the extent possible, compare your results with those of Exercises 14.2.4 and comment on the comparison.

	Total	White	Minority[a]
Live births	285,455	182,285	103,170
Low birth weight	20,020	12,235	7,785
C-section	64,424	45,577	18,847
Infant deaths (under 1 year)	1,632	957	675
Death: males	74,636	61,873	12,763
Death: females	82,776	69,890	12,886
Cause of death: heart disease	57,921	49,624	8,297
Cause of death: cancer	37,277	31,351	5,926
Cause of death: AIDS	2,299	1,048	1,251

Source: *Vital Statistics of New York State 2000 Tables,* New York State Department of Health (A-19).

[a] The minority category includes all nonwhite and not stated races.

12. The following data are for the state of Georgia for 1994:

Age of Woman (Years)	Number of Women in Population[b]	Number of Births to Women of Specified Age[b]	U.S. Population[c]
10–14	246,026	599	17,114,249
15–19	251,197	17,329	17,754,015
20–24	259,413	31,333	19,020,312
25–29	287,951	28,940	21,313,045
30–34	300,358	22,588	21,862,887
35–39	298,578	8,777	19,963,117
40 44	278,679	1,329	17,615,786
45–49	229,377	43[a]	13,872,573
Unknown	—	46	—

[a] May include some births to women over 49 years of age.

Sources: [b] *Georgia Vital Statistics Report 1994,* Georgia Department of Human Resources, Atlanta, October 1995.

[c] *1990 Census of Population, United States, CP-1-1,* Bureau of Census, Washington DC, November 1992, Table 14.

Use these data to compute the rates specified in Exercise 14.3.1. Compare the results with those obtained in Exercises 14.3.1 and 14.3.3 where appropriate and comment on the comparison.

13. There was a total of 110,984 live births in Georgia in 1994. The estimated total population as of July 1, 1994, was 6,965,539. See Review Exercise 12 for the number of women of childbearing age. From these data compute the crude birth rate and the general fertility rate. Compare the results with those of Exercises 14.3.2 and 14.3.4 and comment on the comparison.

14. Use the following facts to compute and label appropriate measures of morbidity.

(a) In 2000, in the United States, there were 1560 cases of Malaria. The estimated population of the United States as of July 1, 2000, was 281,422,000. [SOURCE: U.S. Centers for Disease Control and Prevention, Atlanta, GA, *Summary of Notifiable Diseases, United States, 2000. Morbidity and Mortality Weekly Report,* Vol. 49, No. 53, June 14, 2002, and *Statistical Abstract of the United States: 2002* (118th edition), U.S. Bureau of the Census, Washington, DC, 2002, Table Nos. 18 and 174 (A-20).]

(b) In 2000, in the United States, there were 17,730 reported cases of Lyme disease. The estimated population of the United States as of July 1, 2000, was 281,422,000. [SOURCE: U.S. Centers for Disease Control and Prevention, Atlanta, GA, *Summary of Notifiable Diseases, United States, 2000. Morbidity and Mortality Weekly Report,* Vol. 49, No. 53, June 14, 2002, and *Statistical Abstract of the United States: 2002* (118th edition), U.S. Bureau of the Census, Washington, DC, 2002, Table Nos. 18 and 174 (A-20).]

(c) In 1999, in Ohio, there were 436,000 estimated current users of marijuana. The estimated population of Ohio as of July 1, 2000, was 11,335,000. [SOURCE: U.S. Substance Abuse and Mental Health Services Administration, *National Household Survey on Drug Abuse, 1999,* and *Statistical Abstract of the United States: 2002* (118th edition), U.S. Bureau of the Census, Washington, DC, 2002, Table Nos. 18 and 183 (A-21).]

REFERENCES

Methodology References

1. MARK J. KITTLESON, *Vital Statistics for the Public Health Educator,* Southern Illinois University Press, Carbondale, 1996.
2. DAVID P. SMITH, *Formal Demography,* Plenum Press, New York, 1992.

Applications References

A-1. *Georgia Vital and Morbidity Statistics 2000,* Georgia Division of Public Health, Atlanta, http://www.ph.dhr.state.ga.us/programs/ohip/birthdeath.shtml.

A-2. *Profile of General Demographic Characteristics: 2000,* U.S. Census Bureau DP-1, http://www.census.gov/Press-Release/www/2001/tables/dp_us_2000.PDF.

A-3. *2000 Mortality Vital Statistics Reports,* Georgia Office of Health Information and Policy, http://www.ph.dhr.state.ga.us/pdfs/ohip/vsr/mortality.00.pdf, p. 25.

A-4. *1999 Mortality Vital Statistics Reports,* Georgia Office of Health Information and Policy, http://www.ph.dhr.state.ga.us/programs/ohip/pdfs/mortality1999rev.pdf, p. 14.

A-5. *Profile of General Demographic Characteristics: 2000,* U.S. Census Bureau, DP-1 for North Carolina, http://www.census.gov/census2000/states/nc.html.

A-6. *Selected Vital Statistics for 2000 and 1996–2000,* North Carolina State Center for Health Statistics, http://www.schs.state.nc.us/SCHS/vitalstats/volume1/2000/north_carolina.html.

A-7. Georgia Division of Public Health Online Analytical Statistical Information System (OASIS), http://oasis.state.ga.us/webquery/population.html.

A-8. *Georgia Natality Vital Statistics Reports, 2000,* Georgia Division of Public Health, Atlanta, http://www.ph.dhr.state.ga.us/pdfs/ohip/vsr/natality.00.pdf, p. 23.

A-9. *Age Groups and Sex: 2000,* U.S. Census Bureau 2000 Summary file QT-P1 for United States. http://factfinder.census.gov/servlet/QTTable?_bm=y&-geo_id=D&-qr_name=DEC_2000_SFI_U_QTP1&-ds_name=D&-_lang=en&-redoLog=false.

A-10. *Georgia Natality Vital Statistics Reports, 2001*, Georgia Division of Public Health, Atlanta, http://www.ph.dhr.state.ga.us/pdfs/ohip/vsr/natality.01.pdf, p. 23.

A-11. *Georgia Vital Statistics Reports, Births and Deaths, 2001*, Georgia Division of Public Health, Atlanta, http://www.ph.dhr.state.ga.us/pdfs/ohip/vsr/population.01.pdf, p. 23.

A-12. *North Carolina Resident Births for 2000 by Age of Mother and Birth Order for All Women,* North Carolina Department of Health and Human Services, State Center for Health Statistics, http://www.schs.state.nc.us/SCHS/births/babybook/2000/counties/northcarolina.pdf.

A-13. *Age Groups and Sex: 2000,* U.S. Census Bureau 2000 Summary file QT-P1 for North Carolina, http://factfinder.census.gov/servlet/DTGeoSearchByListServlet?ds_name=DEC_2000_SF1_U&_lang=en&_ts=103465481955.

A-14. *Georgia Vital Statistics Report, 1997,* Georgia Division of Public Health, Atlanta, http://www.ph.dhr.state.ga.us/programs/ohip/pdfs/birthdeath97/statebirthdeath.pdf.

A-15. Statistical Abstract of the United States: 2002 (118th edition), U.S. Bureau of the Census, Washington, DC, 2002, Table No. 18, Table No. 183. http://www.census.gov/prod/2003pubs/02statab/pop.pdf [for population data]/health.pdf [for smoking data].

A-16. Statistical Abstract of the United States: 2002 (118th edition), U.S. Bureau of the Census, Washington, DC, 2002, Table No. 174, http://www.census.gov/prod/www/statistical-abstract-02.html.

A-17. *State Characteristic Estimates,* Table ST-EST2002-ASRO-02-37, Population Division, U.S. Census Bureau, http://eire.census.gov/popest/.

A-18. *Selected Vital Statistics for 2001 and 1997–2001,* North Carolina State Center for Health Statistics, http://www.schs.state.nc.us/SCHS/vitalstats/volume1/2001/nc.html.

A-19. *Vital Statistics of New York State 2000 Tables,* New York State Department of Health, http://www.health.state.ny.us/nysdoh/vital_statistics/2000/toc.htm.

A-20. *Statistical Abstract of the United States: 2002* (118th edition), U.S. Bureau of the Census, Washington, DC, 2002, Table Nos. 18 and 174, http://www.census.gov/prod/www/statistical-abstract-02.html.

A-21. *Statistical Abstract of the United States: 2002* (118th edition), U.S. Bureau of the Census, Washington, DC, 2002, Table Nos. 18 and 183, http://www.census.gov/prod/www/statistical-abstract-02.html.

APPENDIX

STATISTICAL TABLES

List of Tables

TABLE A Random Digits

	00000 12345	**00001 67890**	**11111 12345**	**11112 67890**	**22222 12345**	**22223 67890**	**33333 12345**	**33334 67890**	**44444 12345**	**44445 67890**
01	85967	73152	14511	85285	36009	95892	36962	67835	63314	50162
02	07483	51453	11649	86348	76431	81594	95848	36738	25014	15460
03	96283	01898	61414	83525	04231	13604	75339	11730	85423	60698
04	49174	12074	98551	37895	93547	24769	09404	76548	05393	96770
05	97366	39941	21225	93629	19574	71565	33413	56087	40875	13351
06	90474	41469	16812	81542	81652	45554	27931	93994	22375	00953
07	28599	64109	09497	76235	41383	31555	12639	00619	22909	29563
08	25254	16210	89717	65997	82667	74624	36348	44018	64732	93589
09	28785	02760	24359	99410	77319	73408	58993	61098	04393	48245
10	84725	86576	86944	93296	10081	82454	76810	52975	10324	15457
11	41059	66456	47679	66810	15941	84602	14493	65515	19251	41642
12	67434	41045	82830	47617	36932	46728	71183	36345	41404	81110
13	72766	68816	37643	19959	57550	49620	98480	25640	67257	18671
14	92079	46784	66125	94932	64451	29275	57669	66658	30818	58353
15	29187	40350	62533	73603	34075	16451	42885	03448	37390	96328
16	74220	17612	65522	80607	19184	64164	66962	82310	18163	63495
17	03786	02407	06098	92917	40434	60602	82175	04470	78754	90775
18	75085	55558	15520	27038	25471	76107	90832	10819	56797	33751
19	09161	33015	19155	11715	00551	24909	31894	37774	37953	78837
20	75707	48992	64998	87080	39333	00767	45637	12538	67439	94914
21	21333	48660	31288	00086	79889	75532	28704	62844	92337	99695
22	65626	50061	42539	14812	48895	11196	34335	60492	70650	51108
23	84380	07389	87891	76255	89604	41372	10837	66992	93183	56920
24	46479	32072	80083	63868	70930	89654	05359	47196	12452	38234
25	59847	97197	55147	76639	76971	55928	36441	95141	42333	67483
26	31416	11231	27904	57383	31852	69137	96667	14315	01007	31929
27	82066	83436	67914	21465	99605	83114	97885	74440	99622	87912
28	01850	42782	39202	18582	46214	99228	79541	78298	75404	63648
29	32315	89276	89582	87138	16165	15984	21466	63830	30475	74729
30	59388	42703	55198	80380	67067	97155	34160	85019	03527	78140
31	58089	27632	50987	91373	07736	20436	96130	73483	85332	24384
32	61705	57285	30392	23660	75841	21931	04295	00875	09114	32101
33	18914	98982	60199	99275	41967	35208	30357	76772	92656	62318
34	11965	94089	34803	48941	69709	16784	44642	89761	66864	62803
35	85251	48111	80936	81781	93248	67877	16498	31924	51315	79921
36	66121	96986	84844	93873	46352	92183	51152	85878	30490	15974
37	53972	96642	24199	58080	35450	03482	66953	49521	63719	57615
38	14509	16594	78883	43222	23093	58645	60257	89250	63266	90858
39	37700	07688	65533	72126	23611	93993	01848	03910	38552	17472
40	85466	59392	72722	15473	73295	49759	56157	60477	83284	56367
41	52969	55863	42312	67842	05673	91878	82738	36563	79540	61935
42	42744	68315	17514	02878	97291	74851	42725	57894	81434	62041
43	26140	13336	67726	61876	29971	99294	96664	52817	90039	53211
44	95589	56319	14563	24071	06916	59555	18195	32280	79357	04224
45	39113	13217	59999	49952	83021	47709	53105	19295	88318	41626
46	41392	17622	18994	98283	07249	52289	24209	91139	30715	06604
47	54684	53645	79246	70183	87731	19185	08541	33519	07223	97413
48	89442	61001	36658	57444	95388	36682	38052	46719	09428	94012
49	36751	16778	54888	15357	68003	43564	90976	58904	40512	07725
50	98159	02564	21416	74944	53049	88749	02865	25772	89853	88714

TABLE B Cumulative Binomial Probability Distribution

$$P(X \le x \mid n, p) = \sum_{X=0}^{x} \binom{n}{x} p^x q^{n-x}$$

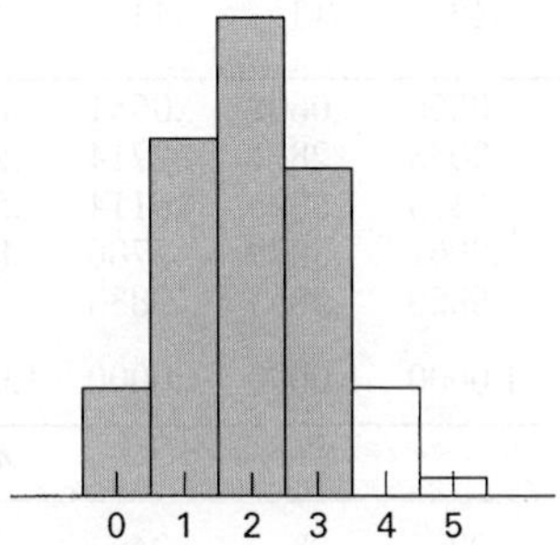

$p(x \le (3 \mid 5, .40) = .9130$

$n = 5$

x \ p	.01	.02	.03	.04	.05	.06	.07	.08	.09	.10
0	.9510	.9039	.8587	.8154	.7738	.7339	.6957	.6591	.6240	.5905
1	.9990	.9962	.9915	.9852	.9774	.9681	.9575	.9456	.9326	.9185
2	1.0000	.9999	.9997	.9994	.9988	.9980	.9969	.9955	.9937	.9914
3	1.0000	1.0000	1.0000	1.0000	1.0000	.9999	.9999	.9998	.9997	.9995
4	1.0000	1.0000	1.0000	1.0000	1.0000	1.0000	1.0000	1.0000	1.0000	1.0000

x \ p	.11	.12	.13	.14	.15	.16	.17	.18	.19	.20
0	.5584	.5277	.4984	.4704	.4437	.4182	.3939	.3707	.3487	.3277
1	.9035	.8875	.8708	.8533	.8352	.8165	.7973	.7776	.7576	.7373
2	.9888	.9857	.9821	.9780	.9734	.9682	.9625	.9563	.9495	.9421
3	.9993	.9991	.9987	.9983	.9978	.9971	.9964	.9955	.9945	.9933
4	1.0000	1.0000	1.0000	.9999	.9999	.9999	.9999	.9998	.9998	.9997
5	1.0000	1.0000	1.0000	1.0000	1.0000	1.0000	1.0000	1.0000	1.0000	1.0000

x \ p	.21	.22	.23	.24	.25	.26	.27	.28	.29	.30
0	.3077	.2887	.2707	.2536	.2373	.2219	.2073	.1935	.1804	.1681
1	.7167	.6959	.6749	.6539	.6328	.6117	.5907	.5697	.5489	.5282
2	.9341	.9256	.9164	.9067	.8965	.8857	.8743	.8624	.8499	.8369
3	.9919	.9903	.9886	.9866	.9844	.9819	.9792	.9762	.9728	.9692
4	.9996	.9995	.9994	.9992	.9990	.9988	.9986	.9983	.9979	.9976
5	1.0000	1.0000	1.0000	1.0000	1.0000	1.0000	1.0000	1.0000	1.0000	1.0000

x \ p	.31	.32	.33	.34	.35	.36	.37	.38	.39	.40
0	.1564	.1454	.1350	.1252	.1160	.1074	.0992	.0916	.0845	.0778
1	.5077	.4875	.4675	.4478	.4284	.4094	.3907	.3724	.3545	.3370
2	.8234	.8095	.7950	.7801	.7648	.7491	.7330	.7165	.6997	.6826
3	.9653	.9610	.9564	.9514	.9460	.9402	.9340	.9274	.9204	.9130
4	.9971	.9966	.9961	.9955	.9947	.9940	.9931	.9921	.9910	.9898
5	1.0000	1.0000	1.0000	1.0000	1.0000	1.0000	1.0000	1.0000	1.0000	1.0000

TABLE B (continued)

n = 5 (*continued*)

x \ p	.41	.42	.43	.44	.45	.46	.47	.48	.49	.50
0	.0715	.0656	.0602	.0551	.0503	.0459	.0418	.0380	.0345	.0312
1	.3199	.3033	.2871	.2714	.2562	.2415	.2272	.2135	.2002	.1875
2	.6651	.6475	.6295	.6114	.5931	.5747	.5561	.5375	.5187	.5000
3	.9051	.8967	.8879	.8786	.8688	.8585	.8478	.8365	.8247	.8125
4	.9884	.9869	.9853	.9835	.9815	.9794	.9771	.9745	.9718	.9688
5	1.0000	1.0000	1.0000	1.0000	1.0000	1.0000	1.0000	1.0000	1.0000	1.0000

n = 6

x \ p	.01	.02	.03	.04	.05	.06	.07	.08	.09	.10
0	.9415	.8858	.8330	.7828	.7351	.6899	.6470	.6064	.5679	.5314
1	.9985	.9943	.9875	.9784	.9672	.9541	.9392	.9227	.9048	.8857
2	1.0000	.9998	.9995	.9988	.9978	.9962	.9942	.9915	.9882	.9841
3	1.0000	1.0000	1.0000	1.0000	.9999	.9998	.9997	.9995	.9992	.9987
4	1.0000	1.0000	1.0000	1.0000	1.0000	1.0000	1.0000	1.0000	1.0000	.9999
5	1.0000	1.0000	1.0000	1.0000	1.0000	1.0000	1.0000	1.0000	1.0000	1.0000

x \ p	.11	.12	.13	.14	.15	.16	.17	.18	.19	.20
0	.4970	.4644	.4336	.4046	.3771	.3513	.3269	.3040	.2824	.2621
1	.8655	.8444	.8224	.7997	.7765	.7528	.7287	.7044	.6799	.6554
2	.9794	.9739	.9676	.9605	.9527	.9440	.9345	.9241	.9130	.9011
3	.9982	.9975	.9966	.9955	.9941	.9925	.9906	.9884	.9859	.9830
4	.9999	.9999	.9998	.9997	.9996	.9995	.9993	.9990	.9987	.9984
5	1.0000	1.0000	1.0000	1.0000	1.0000	1.0000	1.0000	1.0000	1.0000	.9999
6	1.0000	1.0000	1.0000	1.0000	1.0000	1.0000	1.0000	1.0000	1.0000	1.0000

x \ p	.21	.22	.23	.24	.25	.26	.27	.28	.29	.30
0	.2431	.2252	.2084	.1927	.1780	.1642	.1513	.1393	.1281	.1176
1	.6308	.6063	.5820	.5578	.5339	.5104	.4872	.4644	.4420	.4202
2	.8885	.8750	.8609	.8461	.8306	.8144	.7977	.7804	.7626	.7443
3	.9798	.9761	.9720	.9674	.9624	.9569	.9508	.9443	.9372	.9295
4	.9980	.9975	.9969	.9962	.9954	.9944	.9933	.9921	.9907	.9891
5	.9999	.9999	.9999	.9998	.9998	.9997	.9996	.9995	.9994	.9993
6	1.0000	1.0000	1.0000	1.0000	1.0000	1.0000	1.0000	1.0000	1.0000	1.0000

x \ p	.31	.32	.33	.34	.35	.36	.37	.38	.39	.40
0	.1079	.0989	.0905	.0827	.0754	.0687	.0625	.0568	.0515	.0467
1	.3988	.3780	.3578	.3381	.3191	.3006	.2828	.2657	.2492	.2333
2	.7256	.7064	.6870	.6672	.6471	.6268	.6063	.5857	.5650	.5443
3	.9213	.9125	.9031	.8931	.8826	.8714	.8596	.8473	.8343	.8208
4	.9873	.9852	.9830	.9805	.9777	.9746	.9712	.9675	.9635	.9590
5	.9991	.9989	.9987	.9985	.9982	.9978	.9974	.9970	.9965	.9959
6	1.0000	1.0000	1.0000	1.0000	1.0000	1.0000	1.0000	1.0000	1.0000	1.0000

TABLE B (*continued*)

n = 6 (*continued*)

x \ p	.41	.42	.43	.44	.45	.46	.47	.48	.49	.50
0	.0422	.0381	.0343	.0308	.0277	.0248	.0222	.0198	.0176	.0156
1	.2181	.2035	.1895	.1762	.1636	.1515	.1401	.1293	.1190	.1094
2	.5236	.5029	.4823	.4618	.4415	.4214	.4015	.3820	.3627	.3437
3	.8067	.7920	.7768	.7610	.7447	.7280	.7107	.6930	.6748	.6562
4	.9542	.9490	.9434	.9373	.9308	.9238	.9163	.9083	.8997	.8906
5	.9952	.9945	.9937	.9927	.9917	.9905	.9892	.9878	.9862	.9844
6	1.0000	1.0000	1.0000	1.0000	1.0000	1.0000	1.0000	1.0000	1.0000	1.0000

n = 7

x \ p	.01	.02	.03	.04	.05	.06	.07	.08	.09	.10
0	.9321	.8681	.8080	.7514	.6983	.6485	.6017	.5578	.5168	.4783
1	.9980	.9921	.9829	.9706	.9556	.9382	.9187	.8974	.8745	.8503
2	1.0000	.9997	.9991	.9980	.9962	.9937	.9903	.9860	.9807	.9743
3	1.0000	1.0000	1.0000	.9999	.9998	.9996	.9993	.9988	.9982	.9973
4	1.0000	1.0000	1.0000	1.0000	1.0000	1.0000	1.0000	.9999	.9999	.9998
5	1.0000	1.0000	1.0000	1.0000	1.0000	1.0000	1.0000	1.0000	1.0000	1.0000

x \ p	.11	.12	.13	.14	.15	.16	.17	.18	.19	.20
0	.4423	.4087	.3773	.3479	.3206	.2951	.2714	.2493	.2288	.2097
1	.8250	.7988	.7719	.7444	.7166	.6885	.6604	.6323	.6044	.5767
2	.9669	.9584	.9487	.9380	.9262	.9134	.8995	.8846	.8687	.8520
3	.9961	.9946	.9928	.9906	.9879	.9847	.9811	.9769	.9721	.9667
4	.9997	.9996	.9994	.9991	.9988	.9983	.9978	.9971	.9963	.9953
5	1.0000	1.0000	1.0000	1.0000	.9999	.9999	.9999	.9998	.9997	.9996
6	1.0000	1.0000	1.0000	1.0000	1.0000	1.0000	1.0000	1.0000	1.0000	1.0000

x \ p	.21	.22	.23	.24	.25	.26	.27	.28	.29	.30
0	.1920	.1757	.1605	.1465	.1335	.1215	.1105	.1003	.0910	.0824
1	.5494	.5225	.4960	.4702	.4449	.4204	.3965	.3734	.3510	.3294
2	.8343	.8159	.7967	.7769	.7564	.7354	.7139	.6919	.6696	.6471
3	.9606	.9539	.9464	.9383	.9294	.9198	.9095	.8984	.8866	.8740
4	.9942	.9928	.9912	.9893	.9871	.9847	.9819	.9787	.9752	.9712
5	.9995	.9994	.9992	.9989	.9987	.9983	.9979	.9974	.9969	.9962
6	1.0000	1.0000	1.0000	1.0000	.9999	.9999	.9999	.9999	.9998	.9998
7	1.0000	1.0000	1.0000	1.0000	1.0000	1.0000	1.0000	1.0000	1.0000	1.0000

TABLE B (continued)

n = 7 (*continued*)

x \ p	.31	.32	.33	.34	.35	.36	.37	.38	.39	.40
0	.0745	.0672	.0606	.0546	.0490	.0440	.0394	.0352	.0314	.0280
1	.3086	.2887	.2696	.2513	.2338	.2172	.2013	.1863	.1721	.1586
2	.6243	.6013	.5783	.5553	.5323	.5094	.4866	.4641	.4419	.4199
3	.8606	.8466	.8318	.8163	.8002	.7833	.7659	.7479	.7293	.7102
4	.9668	.9620	.9566	.9508	.9444	.9375	.9299	.9218	.9131	.9037
5	.9954	.9945	.9935	.9923	.9910	.9895	.9877	.9858	.9836	.9812
6	.9997	.9997	.9996	.9995	.9994	.9992	.9991	.9989	.9986	.9984
7	1.0000	1.0000	1.0000	1.0000	1.0000	1.0000	1.0000	1.0000	1.0000	1.0000

x \ p	.41	.42	.43	.44	.45	.46	.47	.48	.49	.50
0	.0249	.0221	.0195	.0173	.0152	.0134	.0117	.0103	.0090	.0078
1	.1459	.1340	.1228	.1123	.1024	.0932	.0847	.0767	.0693	.0625
2	.3983	.3771	.3564	.3362	.3164	.2973	.2787	.2607	.2433	.2266
3	.6906	.6706	.6502	.6294	.6083	.5869	.5654	.5437	.5219	.5000
4	.8937	.8831	.8718	.8598	.8471	.8337	.8197	.8049	.7895	.7734
5	.9784	.9754	.9721	.9684	.9643	.9598	.9549	.9496	.9438	.9375
6	.9981	.9977	.9973	.9968	.9963	.9956	.9949	.9941	.9932	.9922
7	1.0000	1.0000	1.0000	1.0000	1.0000	1.0000	1.0000	1.0000	1.0000	1.0000

n = 8

x \ p	.01	.02	.03	.04	.05	.06	.07	.08	.09	.10
0	.9227	.8508	.7837	.7214	.6634	.6096	.5596	.5132	.4703	.4305
1	.9973	.9897	.9777	.9619	.9428	.9208	.8965	.8702	.8423	.8131
2	.9999	.9996	.9987	.9969	.9942	.9904	.9853	.9789	.9711	.9619
3	1.0000	1.0000	.9999	.9998	.9996	.9993	.9987	.9978	.9966	.9950
4	1.0000	1.0000	1.0000	1.0000	1.0000	1.0000	.9999	.9999	.9997	.9996
5	1.0000	1.0000	1.0000	1.0000	1.0000	1.0000	1.0000	1.0000	1.0000	1.0000

x \ p	.11	.12	.13	.14	.15	.16	.17	.18	.19	.20
0	.3937	.3596	.3282	.2992	.2725	.2479	.2252	.2044	.1853	.1678
1	.7829	.7520	.7206	.6889	.6572	.6256	.5943	.5634	.5330	.5033
2	.9513	.9392	.9257	.9109	.8948	.8774	.8588	.8392	.8185	.7969
3	.9929	.9903	.9871	.9832	.9786	.9733	.9672	.9603	.9524	.9437
4	.9993	.9990	.9985	.9979	.9971	.9962	.9950	.9935	.9917	.9896
5	1.0000	.9999	.9999	.9998	.9998	.9997	.9995	.9993	.9991	.9988
6	1.0000	1.0000	1.0000	1.0000	1.0000	1.0000	1.0000	1.0000	.9999	.9999
7	1.0000	1.0000	1.0000	1.0000	1.0000	1.0000	1.0000	1.0000	1.0000	1.0000

TABLE B (*continued*)

n = 8 (*continued*)

x \ p	.21	.22	.23	.24	.25	.26	.27	.28	.29	.30
0	.1517	.1370	.1236	.1113	.1001	.0899	.0806	.0722	.0646	.0576
1	.4743	.4462	.4189	.3925	.3671	.3427	.3193	.2969	.2756	.2553
2	.7745	.7514	.7276	.7033	.6785	.6535	.6282	.6027	.5772	.5518
3	.9341	.9235	.9120	.8996	.8862	.8719	.8567	.8406	.8237	.8059
4	.9871	.9842	.9809	.9770	.9727	.9678	.9623	.9562	.9495	.9420
5	.9984	.9979	.9973	.9966	.9958	.9948	.9936	.9922	.9906	.9887
6	.9999	.9998	.9998	.9997	.9996	.9995	.9994	.9992	.9990	.9987
7	1.0000	1.0000	1.0000	1.0000	1.0000	1.0000	1.0000	1.0000	.9999	.9999
8	1.0000	1.0000	1.0000	1.0000	1.0000	1.0000	1.0000	1.0000	1.0000	1.0000

x \ p	.31	.32	.33	.34	.35	.36	.37	.38	.39	.40
0	.0514	.0457	.0406	.0360	.0319	.0281	.0248	.0218	.0192	.0168
1	.2360	.2178	.2006	.1844	.1691	.1548	.1414	.1289	.1172	.1064
2	.5264	.5013	.4764	.4519	.4278	.4042	.3811	.3585	.3366	.3154
3	.7874	.7681	.7481	.7276	.7064	.6847	.6626	.6401	.6172	.5941
4	.9339	.9250	.9154	.9051	.8939	.8820	.8693	.8557	.8414	.8263
5	.9866	.9841	.9813	.9782	.9747	.9707	.9664	.9615	.9561	.9502
6	.9984	.9980	.9976	.9970	.9964	.9957	.9949	.9939	.9928	.9915
7	.9999	.9999	.9999	.9998	.9998	.9997	.9996	.9996	.9995	.9993
8	1.0000	1.0000	1.0000	1.0000	1.0000	1.0000	1.0000	1.0000	1.0000	1.0000

x \ p	.41	.42	.43	.44	.45	.46	.47	.48	.49	.50
0	.0147	.0128	.0111	.0097	.0084	.0072	.0062	.0053	.0046	.0039
1	.0963	.0870	.0784	.0705	.0632	.0565	.0504	.0448	.0398	.0352
2	.2948	.2750	.2560	.2376	.2201	.2034	.1875	.1724	.1581	.1445
3	.5708	.5473	.5238	.5004	.4770	.4537	.4306	.4078	.3854	.3633
4	.8105	.7938	.7765	.7584	.7396	.7202	.7001	.6795	.6584	.6367
5	.9437	.9366	.9289	.9206	.9115	.9018	.8914	.8802	.8682	.8555
6	.9900	.9883	.9864	.9843	.9819	.9792	.9761	.9728	.9690	.9648
7	.9992	.9990	.9988	.9986	.9983	.9980	.9976	.9972	.9967	.9961
8	1.0000	1.0000	1.0000	1.0000	1.0000	1.0000	1.0000	1.0000	1.0000	1.0000

n = 9

x \ p	.01	.02	.03	.04	.05	.06	.07	.08	.09	.10
0	.9135	.8337	.7602	.6925	.6302	.5730	.5204	.4722	.4279	.3874
1	.9966	.9869	.9718	.9522	.9288	.9022	.8729	.8417	.8088	.7748
2	.9999	.9994	.9980	.9955	.9916	.9862	.9791	.9702	.9595	.9470
3	1.0000	1.0000	.9999	.9997	.9994	.9987	.9977	.9963	.9943	.9917
4	1.0000	1.0000	1.0000	1.0000	1.0000	.9999	.9998	.9997	.9995	.9991
5	1.0000	1.0000	1.0000	1.0000	1.0000	1.0000	1.0000	1.0000	1.0000	.9999
6	1.0000	1.0000	1.0000	1.0000	1.0000	1.0000	1.0000	1.0000	1.0000	1.0000

TABLE B (continued)

n = 9 (*continued*)

x \ p	.11	.12	.13	.14	.15	.16	.17	.18	.19	.20
0	.3504	.3165	.2855	.2573	.2316	.2082	.1869	.1676	.1501	.1342
1	.7401	.7049	.6696	.6343	.5995	.5652	.5315	.4988	.4670	.4362
2	.9327	.9167	.8991	.8798	.8591	.8371	.8139	.7895	.7643	.7382
3	.9883	.9842	.9791	.9731	.9661	.9580	.9488	.9385	.9270	.9144
4	.9986	.9979	.9970	.9959	.9944	.9925	.9902	.9875	.9842	.9804
5	.9999	.9998	.9997	.9996	.9994	.9991	.9987	.9983	.9977	.9969
6	1.0000	1.0000	1.0000	1.0000	1.0000	.9999	.9999	.9998	.9998	.9997
7	1.0000	1.0000	1.0000	1.0000	1.0000	1.0000	1.0000	1.0000	1.0000	1.0000

x \ p	.21	.22	.23	.24	.25	.26	.27	.28	.29	.30
0	.1199	.1069	.0952	.0846	.0751	.0665	.0589	.0520	.0458	.0404
1	.4066	.3782	.3509	.3250	.3003	.2770	.2548	.2340	.2144	.1960
2	.7115	.6842	.6566	.6287	.6007	.5727	.5448	.5171	.4898	.4628
3	.9006	.8856	.8696	.8525	.8343	.8151	.7950	.7740	.7522	.7297
4	.9760	.9709	.9650	.9584	.9511	.9429	.9338	.9238	.9130	.9012
5	.9960	.9949	.9935	.9919	.9900	.9878	.9851	.9821	.9787	.9747
6	.9996	.9994	.9992	.9990	.9987	.9983	.9978	.9972	.9965	.9957
7	1.0000	1.0000	.9999	.9999	.9999	.9999	.9998	.9997	.9997	.9996
8	1.0000	1.0000	1.0000	1.0000	1.0000	1.0000	1.0000	1.0000	1.0000	1.0000

x \ p	.31	.32	.33	.34	.35	.36	.37	.38	.39	.40
0	.0355	.0311	.0272	.0238	.0207	.0180	.0156	.0135	.0117	.0101
1	.1788	.1628	.1478	.1339	.1211	.1092	.0983	.0882	.0790	.0705
2	.4364	.4106	.3854	.3610	.3373	.3144	.2924	.2713	.2511	.2318
3	.7065	.6827	.6585	.6338	.6089	.5837	.5584	.5331	.5078	.4826
4	.8885	.8748	.8602	.8447	.8283	.8110	.7928	.7738	.7540	.7334
5	.9702	.9652	.9596	.9533	.9464	.9388	.9304	.9213	.9114	.9006
6	.9947	.9936	.9922	.9906	.9888	.9867	.9843	.9816	.9785	.9750
7	.9994	.9993	.9991	.9989	.9986	.9983	.9979	.9974	.9969	.9962
8	1.0000	1.0000	1.0000	.9999	.9999	.9999	.9999	.9998	.9998	.9997
9	1.0000	1.0000	1.0000	1.0000	1.0000	1.0000	1.0000	1.0000	1.0000	1.0000

x \ p	.41	.42	.43	.44	.45	.46	.47	.48	.49	.50
0	.0087	.0074	.0064	.0054	.0046	.0039	.0033	.0028	.0023	.0020
1	.0628	.0558	.0495	.0437	.0385	.0338	.0296	.0259	.0225	.0195
2	.2134	.1961	.1796	.1641	.1495	.1358	.1231	.1111	.1001	.0898
3	.4576	.4330	.4087	.3848	.3614	.3386	.3164	.2948	.2740	.2539
4	.7122	.6903	.6678	.6449	.6214	.5976	.5735	.5491	.5246	.5000
5	.8891	.8767	.8634	.8492	.8342	.8183	.8015	.7839	.7654	.7461
6	.9710	.9666	.9617	.9563	.9502	.9436	.9363	.9283	.9196	.9102
7	.9954	.9945	.9935	.9923	.9909	.9893	.9875	.9855	.9831	.9805
8	.9997	.9996	.9995	.9994	.9992	.9991	.9989	.9986	.9984	.9980
9	1.0000	1.0000	1.0000	1.0000	1.0000	1.0000	1.0000	1.0000	1.0000	1.0000

TABLE B ***(continued)***

$n = 10$

x \ p	.01	.02	.03	.04	.05	.06	.07	.08	.09	.10
0	.9044	.8171	.7374	.6648	.5987	.5386	.4840	.4344	.3894	.3487
1	.9957	.9838	.9655	.9418	.9139	.8824	.8483	.8121	.7746	.7361
2	.9999	.9991	.9972	.9938	.9885	.9812	.9717	.9599	.9460	.9298
3	1.0000	1.0000	.9999	.9996	.9990	.9980	.9964	.9942	.9912	.9872
4	1.0000	1.0000	1.0000	1.0000	.9999	.9998	.9997	.9994	.9990	.9984
5	1.0000	1.0000	1.0000	1.0000	1.0000	1.0000	1.0000	1.0000	.9999	.9999
6	1.0000	1.0000	1.0000	1.0000	1.0000	1.0000	1.0000	1.0000	1.0000	1.0000

x \ p	.11	.12	.13	.14	.15	.16	.17	.18	.19	.20
0	.3118	.2785	.2484	.2213	.1969	.1749	.1552	.1374	.1216	.1074
1	.6972	.6583	.6196	.5816	.5443	.5080	.4730	.4392	.4068	.3758
2	.9116	.8913	.8692	.8455	.8202	.7936	.7659	.7372	.7078	.6778
3	.9822	.9761	.9687	.9600	.9500	.9386	.9259	.9117	.8961	.8791
4	.9975	.9963	.9947	.9927	.9901	.9870	.9832	.9787	.9734	.9672
5	.9997	.9996	.9994	.9990	.9986	.9980	.9973	.9963	.9951	.9936
6	1.0000	1.0000	.9999	.9999	.9999	.9998	.9997	.9996	.9994	.9991
7	1.0000	1.0000	1.0000	1.0000	1.0000	1.0000	1.0000	1.0000	.9999	.9999
8	1.0000	1.0000	1.0000	1.0000	1.0000	1.0000	1.0000	1.0000	1.0000	1.0000

x \ p	.21	.22	.23	.24	.25	.26	.27	.28	.29	.30
0	.0947	.0834	.0733	.0643	.0563	.0492	.0430	.0374	.0326	.0282
1	.3464	.3185	.2921	.2673	.2440	.2222	.2019	.1830	.1655	.1493
2	.6474	.6169	.5863	.5558	.5256	.4958	.4665	.4378	.4099	.3828
3	.8609	.8413	.8206	.7988	.7759	.7521	.7274	.7021	.6761	.6496
4	.9601	.9521	.9431	.9330	.9219	.9096	.8963	.8819	.8663	.8497
5	.9918	.9896	.9870	.9839	.9803	.9761	.9713	.9658	.9596	.9527
6	.9988	.9984	.9979	.9973	.9965	.9955	.9944	.9930	.9913	.9894
7	.9999	.9998	.9998	.9997	.9996	.9994	.9993	.9990	.9988	.9984
8	1.0000	1.0000	1.0000	1.0000	1.0000	1.0000	.9999	.9999	.9999	.9999
9	1.0000	1.0000	1.0000	1.0000	1.0000	1.0000	1.0000	1.0000	1.0000	1.0000

x \ p	.31	.32	.33	.34	.35	.36	.37	.38	.39	.40
0	.0245	.0211	.0182	.0157	.0135	.0115	.0098	.0084	.0071	.0060
1	.1344	.1206	.1080	.0965	.0860	.0764	.0677	.0598	.0527	.0464
2	.3566	.3313	.3070	.2838	.2616	.2405	.2206	.2017	.1840	.1673
3	.6228	.5956	.5684	.5411	.5138	.4868	.4600	.4336	.4077	.3823
4	.8321	.8133	.7936	.7730	.7515	.7292	.7061	.6823	.6580	.6331
5	.9449	.9363	.9268	.9164	.9051	.8928	.8795	.8652	.8500	.8338
6	.9871	.9845	.9815	.9780	.9740	.9695	.9644	.9587	.9523	.9452
7	.9980	.9975	.9968	.9961	.9952	.9941	.9929	.9914	.9897	.9877
8	.9998	.9997	.9997	.9996	.9995	.9993	.9991	.9989	.9986	.9983
9	1.0000	1.0000	1.0000	1.0000	1.0000	1.0000	1.0000	.9999	.9999	.9999
10	1.0000	1.0000	1.0000	1.0000	1.0000	1.0000	1.0000	1.0000	1.0000	1.0000

TABLE B (*continued*)

n = 10 (*continued*)

x \ p	.41	.42	.43	.44	.45	.46	.47	.48	.49	.50
0	.0051	.0043	.0036	.0030	.0025	.0021	.0017	.0014	.0012	.0010
1	.0406	.0355	.0309	.0269	.0233	.0201	.0173	.0148	.0126	.0107
2	.1517	.1372	.1236	.1111	.0996	.0889	.0791	.0702	.0621	.0547
3	.3575	.3335	.3102	.2877	.2660	.2453	.2255	.2067	.1888	.1719
4	.6078	.5822	.5564	.5304	.5044	.4784	.4526	.4270	.4018	.3770
5	.8166	.7984	.7793	.7593	.7384	.7168	.6943	.6712	.6474	.6230
6	.9374	.9288	.9194	.9092	.8980	.8859	.8729	.8590	.8440	.8281
7	.9854	.9828	.9798	.9764	.9726	.9683	.9634	.9580	.9520	.9453
8	.9979	.9975	.9969	.9963	.9955	.9946	.9935	.9923	.9909	.9893
9	.9999	.9998	.9998	.9997	.9997	.9996	.9995	.9994	.9992	.9990
10	1.0000	1.0000	1.0000	1.0000	1.0000	1.0000	1.0000	1.0000	1.0000	1.0000

n = 11

x \ p	.01	.02	.03	.04	.05	.06	.07	.08	.09	.10
0	.8953	.8007	.7153	.6382	.5688	.5063	.4501	.3996	.3544	.3138
1	.9948	.9805	.9587	.9308	.8981	.8618	.8228	.7819	.7399	.6974
2	.9998	.9988	.9963	.9917	.9848	.9752	.9630	.9481	.9305	.9104
3	1.0000	1.0000	.9998	.9993	.9984	.9970	.9947	.9915	.9871	.9815
4	1.0000	1.0000	1.0000	1.0000	.9999	.9997	.9995	.9990	.9983	.9972
5	1.0000	1.0000	1.0000	1.0000	1.0000	1.0000	1.0000	.9999	.9998	.9997
6	1.0000	1.0000	1.0000	1.0000	1.0000	1.0000	1.0000	1.0000	1.0000	1.0000

x \ p	.11	.12	.13	.14	.15	.16	.17	.18	.19	.20
0	.2775	.2451	.2161	.1903	.1673	.1469	.1288	.1127	.0985	.0859
1	.6548	.6127	.5714	.5311	.4922	.4547	.4189	.3849	.3526	.3221
2	.8880	.8634	.8368	.8085	.7788	.7479	.7161	.6836	.6506	.6174
3	.9744	.9659	.9558	.9440	.9306	.9154	.8987	.8803	.8603	.8389
4	.9958	.9939	.9913	.9881	.9841	.9793	.9734	.9666	.9587	.9496
5	.9995	.9992	.9988	.9982	.9973	.9963	.9949	.9932	.9910	.9883
6	1.0000	.9999	.9999	.9998	.9997	.9995	.9993	.9990	.9986	.9980
7	1.0000	1.0000	1.0000	1.0000	1.0000	1.0000	.9999	.9999	.9998	.9998
8	1.0000	1.0000	1.0000	1.0000	1.0000	1.0000	1.0000	1.0000	1.0000	1.0000

x \ p	.21	.22	.23	.24	.25	.26	.27	.28	.29	.30
0	.0748	.0650	.0564	.0489	.0422	.0364	.0314	.0270	.0231	.0198
1	.2935	.2667	.2418	.2186	.1971	.1773	.1590	.1423	.1270	.1130
2	.5842	.5512	.5186	.4866	.4552	.4247	.3951	.3665	.3390	.3127
3	.8160	.7919	.7667	.7404	.7133	.6854	.6570	.6281	.5989	.5696
4	.9393	.9277	.9149	.9008	.8854	.8687	.8507	.8315	.8112	.7897
5	.9852	.9814	.9769	.9717	.9657	.9588	.9510	.9423	.9326	.9218
6	.9973	.9965	.9954	.9941	.9924	.9905	.9881	.9854	.9821	.9784
7	.9997	.9995	.9993	.9991	.9988	.9984	.9979	.9973	.9966	.9957
8	1.0000	1.0000	.9999	.9999	.9999	.9998	.9998	.9997	.9996	.9994
9	1.0000	1.0000	1.0000	1.0000	1.0000	1.0000	1.0000	1.0000	1.0000	1.0000

TABLE B (*continued*)

$n = 11$ (*continued*)

x \ p	.31	.32	.33	.34	.35	.36	.37	.38	.39	.40
0	.0169	.0144	.0122	.0104	.0088	.0074	.0062	.0052	.0044	.0036
1	.1003	.0888	.0784	.0690	.0606	.0530	.0463	.0403	.0350	.0302
2	.2877	.2639	.2413	.2201	.2001	.1814	.1640	.1478	.1328	.1189
3	.5402	.5110	.4821	.4536	.4256	.3981	.3714	.3455	.3204	.2963
4	.7672	.7437	.7193	.6941	.6683	.6419	.6150	.5878	.5603	.5328
5	.9099	.8969	.8829	.8676	.8513	.8339	.8153	.7957	.7751	.7535
6	.9740	.9691	.9634	.9570	.9499	.9419	.9330	.9232	.9124	.9006
7	.9946	.9933	.9918	.9899	.9878	.9852	.9823	.9790	.9751	.9707
8	.9992	.9990	.9987	.9984	.9980	.9974	.9968	.9961	.9952	.9941
9	.9999	.9999	.9999	.9998	.9998	.9997	.9996	.9995	.9994	.9993
10	1.0000	1.0000	1.0000	1.0000	1.0000	1.0000	1.0000	1.0000	1.0000	10000

x \ p	.41	.42	.43	.44	.45	.46	.47	.48	.49	.50
0	.0030	.0025	.0021	.0017	.0014	.0011	.0009	.0008	.0006	.0005
1	.0261	.0224	.0192	.0164	.0139	.0118	.0100	.0084	.0070	.0059
2	.1062	.0945	.0838	.0740	.0652	.0572	.0501	.0436	.0378	.0327
3	.2731	.2510	.2300	.2100	.1911	.1734	.1567	.1412	.1267	.1133
4	.5052	.4777	.4505	.4236	.3971	.3712	.3459	.3213	.2974	.2744
5	.7310	.7076	.6834	.6586	.6331	.6071	.5807	.5540	.5271	.5000
6	.8879	.8740	.8592	.8432	.8262	.8081	.7890	.7688	.7477	.7256
7	.9657	.9601	.9539	.9468	.9390	.9304	.9209	.9105	.8991	.8867
8	.9928	.9913	.9896	.9875	.9852	.9825	.9794	.9759	.9718	.9673
9	.9991	.9988	.9986	.9982	.9978	.9973	.9967	.9960	.9951	.9941
10	.9999	.9999	.9999	.9999	.9998	.9998	.9998	.9997	.9996	.9995
11	1.0000	1.0000	1.0000	1.0000	1.0000	1.0000	1.0000	1.0000	1.0000	1.0000

$n = 12$

x \ p	.01	.02	.03	.04	.05	.06	.07	.08	.09	.10
0	.8864	.7847	.6938	.6127	.5404	.4759	.4186	.3677	.3225	.2824
1	.9938	.9769	.9514	.9191	.8816	.8405	.7967	.7513	.7052	.6590
2	.9998	.9985	.9952	.9893	.9804	.9684	.9532	.9348	.9134	.8891
3	1.0000	.9999	.9997	.9990	.9978	.9957	.9925	.9880	.9820	.9744
4	1.0000	1.0000	1.0000	.9999	.9998	.9996	.9991	.9984	.9973	.9957
5	1.0000	1.0000	1.0000	1.0000	1.0000	1.0000	.9999	.9998	.9997	.9995
6	1.0000	1.0000	1.0000	1.0000	1.0000	1.0000	1.0000	1.0000	1.0000	.9999
7	1.0000	1.0000	1.0000	1.0000	1.0000	1.0000	1.0000	1.0000	1.0000	1.0000

TABLE B (continued)

n = 12 (*continued*)

x \ p	.11	.12	.13	.14	.15	.16	.17	.18	.19	.20
0	.2470	.2157	.1880	.1637	.1422	.1234	.1069	.0924	.0798	.0687
1	.6133	.5686	.5252	.4834	.4435	.4055	.3696	.3359	.3043	.2749
2	.8623	.8333	.8023	.7697	.7358	.7010	.6656	.6298	.5940	.5583
3	.9649	.9536	.9403	.9250	.9078	.8886	.8676	.8448	.8205	.7946
4	.9935	.9905	.9867	.9819	.9761	.9690	.9607	.9511	.9400	.9274
5	.9991	.9986	.9978	.9967	.9954	.9935	.9912	.9884	.9849	.9806
6	.9999	.9998	.9997	.9996	.9993	.9990	.9985	.9979	.9971	.9961
7	1.0000	1.0000	1.0000	1.0000	.9999	.9999	.9998	.9997	.9996	.9994
8	1.0000	1.0000	1.0000	1.0000	1.0000	1.0000	1.0000	1.0000	1.0000	.9999
9	1.0000	1.0000	1.0000	1.0000	1.0000	1.0000	1.0000	1.0000	1.0000	1.0000

x \ p	.21	.22	.23	.24	.25	.26	.27	.28	.29	.30
0	.0591	.0507	.0434	.0371	.0317	.0270	.0229	.0194	.0164	.0138
1	.2476	.2224	.1991	.1778	.1584	.1406	.1245	.1100	.0968	.0850
2	.5232	.4886	.4550	.4222	.3907	.3603	.3313	.3037	.2775	.2528
3	.7674	.7390	.7096	.6795	.6488	.6176	.5863	.5548	.5235	.4925
4	.9134	.8979	.8808	.8623	.8424	.8210	.7984	.7746	.7496	.7237
5	.9755	.9696	.9626	.9547	.9456	.9354	.9240	.9113	.8974	.8822
6	.9948	.9932	.9911	.9887	.9857	.9822	.9781	.9733	.9678	.9614
7	.9992	.9989	.9984	.9979	.9972	.9964	.9953	.9940	.9924	.9905
8	.9999	.9999	.9998	.9997	.9996	.9995	.9993	.9990	.9987	.9983
9	1.0000	1.0000	1.0000	1.0000	1.0000	.9999	.9999	.9999	.9998	.9998
10	1.0000	1.0000	1.0000	1.0000	1.0000	1.0000	1.0000	1.0000	1.0000	1.0000

x \ p	.31	.32	.33	.34	.35	.36	.37	.38	.39	.40
0	.0116	.0098	.0082	.0068	.0057	.0047	.0039	.0032	.0027	.0022
1	.0744	.0650	.0565	.0491	.0424	.0366	.0315	.0270	.0230	.0196
2	.2296	.2078	.1876	.1687	.1513	.1352	.1205	.1069	.0946	.0834
3	.4619	.4319	.4027	.3742	.3467	.3201	.2947	.2704	.2472	.2253
4	.6968	.6692	.6410	.6124	.5833	.5541	.5249	.4957	.4668	.4382
5	.8657	.8479	.8289	.8087	.7873	.7648	.7412	.7167	.6913	.6652
6	.9542	.9460	.9368	.9266	.9154	.9030	.8894	.8747	.8589	.8418
7	.9882	.9856	.9824	.9787	.9745	.9696	.9641	.9578	.9507	.9427
8	.9978	.9972	.9964	.9955	.9944	.9930	.9915	.9896	.9873	.9847
9	.9997	.9996	.9995	.9993	.9992	.9989	.9986	.9982	.9978	.9972
10	1.0000	1.0000	1.0000	.9999	.9999	.9999	.9999	.9998	.9998	.9997
11	1.0000	1.0000	1.0000	1.0000	1.0000	1.0000	1.0000	1.0000	1.0000	1.0000

TABLE B (continued)

n = 12 (*continued*)

x \ p	.41	.42	.43	.44	.45	.46	.47	.48	.49	.50
0	.0018	.0014	.0012	.0010	.0008	.0006	.0005	.0004	.0003	.0002
1	.0166	.0140	.0118	.0099	.0083	.0069	.0057	.0047	.0039	.0032
2	.0733	.0642	.0560	.0487	.0421	.0363	.0312	.0267	.0227	.0193
3	.2047	.1853	.1671	.1502	.1345	.1199	.1066	.0943	.0832	.0730
4	.4101	.3825	.3557	.3296	.3044	.2802	.2570	.2348	.2138	.1938
5	.6384	.6111	.5833	.5552	.5269	.4986	.4703	.4423	.4145	.3872
6	.8235	.8041	.7836	.7620	.7393	.7157	.6911	.6657	.6396	.6128
7	.9338	.9240	.9131	.9012	.8883	.8742	.8589	.8425	.8249	.8062
8	.9817	.9782	.9742	.9696	.9644	.9585	.9519	.9445	.9362	.9270
9	.9965	.9957	.9947	.9935	.9921	.9905	.9886	.9863	.9837	.9807
10	.9996	.9995	.9993	.9991	.9989	.9986	.9983	.9979	.9974	.9968
11	1.0000	1.0000	1.0000	.9999	.9999	.9999	.9999	.9999	.9998	.9998
12	1.0000	1.0000	1.0000	1.0000	1.0000	1.0000	1.0000	1.0000	1.0000	1.0000

n = 13

x \ p	.01	.02	.03	.04	.05	.06	.07	.08	.09	.10
0	.8775	.7690	.6730	.5882	.5133	.4474	.3893	.3383	.2935	.2542
1	.9928	.9730	.9436	.9068	.8646	.8186	.7702	.7206	.6707	.6213
2	.9997	.9980	.9938	.9865	.9755	.9608	.9422	.9201	.8946	.8661
3	1.0000	.9999	.9995	.9986	.9969	.9940	.9897	.9837	.9758	.9658
4	1.0000	1.0000	1.0000	.9999	.9997	.9993	.9987	.9976	.9959	.9935
5	1.0000	1.0000	1.0000	1.0000	1.0000	.9999	.9999	.9997	.9995	.9991
6	1.0000	1.0000	1.0000	1.0000	1.0000	1.0000	1.0000	1.0000	.9999	.9999
7	1.0000	1.0000	1.0000	1.0000	1.0000	1.0000	1.0000	1.0000	1.0000	1.0000

x \ p	.11	.12	.13	.14	.15	.16	.17	.18	.19	.20
0	.2198	.1898	.1636	.1408	.1209	.1037	.0887	.0758	.0646	.0550
1	.5730	.5262	.4814	.4386	.3983	.3604	.3249	.2920	.2616	.2336
2	.8349	.8015	.7663	.7296	.6920	.6537	.6152	.5769	.5389	.5017
3	.9536	.9391	.9224	.9033	.8820	.8586	.8333	.8061	.7774	.7473
4	.9903	.9861	.9807	.9740	.9658	.9562	.9449	.9319	.9173	.9009
5	.9985	.9976	.9964	.9947	.9925	.9896	.9861	.9817	.9763	.9700
6	.9998	.9997	.9995	.9992	.9987	.9981	.9973	.9962	.9948	.9930
7	1.0000	1.0000	.9999	.9999	.9998	.9997	.9996	.9994	.9991	.9988
8	1.0000	1.0000	1.0000	1.0000	1.0000	1.0000	1.0000	.9999	.9999	.9998
9	1.0000	1.0000	1.0000	1.0000	1.0000	1.0000	1.0000	1.0000	1.0000	1.0000

TABLE B (continued)

n = 13 (*continued*)

x \ p	.21	.22	.23	.24	.25	.26	.27	.28	.29	.30
0	.0467	.0396	.0334	.0282	.0238	.0200	.0167	.0140	.0117	.0097
1	.2080	.1846	.1633	.1441	.1267	.1111	.0971	.0846	.0735	.0637
2	.4653	.4301	.3961	.3636	.3326	.3032	.2755	.2495	.2251	.2025
3	.7161	.6839	.6511	.6178	.5843	.5507	.5174	.4845	.4522	.4206
4	.8827	.8629	.8415	.8184	.7940	.7681	.7411	.7130	.6840	.6543
5	.9625	.9538	.9438	.9325	.9198	.9056	.8901	.8730	.8545	.8346
6	.9907	.9880	.9846	.9805	.9757	.9701	.9635	.9560	.9473	.9376
7	.9983	.9976	.9968	.9957	.9944	.9927	.9907	.9882	.9853	.9818
8	.9998	.9996	.9995	.9993	.9990	.9987	.9982	.9976	.9969	.9960
9	1.0000	1.0000	.9999	.9999	.9999	.9998	.9997	.9996	.9995	.9993
10	1.0000	1.0000	1.0000	1.0000	1.0000	1.0000	1.0000	1.0000	.9999	.9999
11	1.0000	1.0000	1.0000	1.0000	1.0000	1.0000	1.0000	1.0000	1.0000	1.0000

x \ p	.31	.32	.33	.34	.35	.36	.37	.38	.39	.40
0	.0080	.0066	.0055	.0045	.0037	.0030	.0025	.0020	.0016	.0013
1	.0550	.0473	.0406	.0347	.0296	.0251	.0213	.0179	.0151	.0126
2	.1815	.1621	.1443	.1280	.1132	.0997	.0875	.0765	.0667	.0579
3	.3899	.3602	.3317	.3043	.2783	.2536	.2302	.2083	.1877	.1686
4	.6240	.5933	.5624	.5314	.5005	.4699	.4397	.4101	.3812	.3530
5	.8133	.7907	.7669	.7419	.7159	.6889	.6612	.6327	.6038	.5744
6	.9267	.9146	.9012	.8865	.8705	.8532	.8346	.8147	.7935	.7712
7	.9777	.9729	.9674	.9610	.9538	.9456	.9365	.9262	.9149	.9023
8	.9948	.9935	.9918	.9898	.9874	.9846	.9813	.9775	.9730	.9679
9	.9991	.9988	.9985	.9980	.9975	.9968	.9960	.9949	.9937	.9922
10	.9999	.9999	.9998	.9997	.9997	.9995	.9994	.9992	.9990	.9987
11	1.0000	1.0000	1.0000	1.0000	1.0000	1.0000	.9999	.9999	.9999	.9999
12	1.0000	1.0000	1.0000	1.0000	1.0000	1.0000	1.0000	1.0000	1.0000	1.0000

x \ p	.41	.42	.43	.44	.45	.46	.47	.48	.49	.50
0	.0010	.0008	.0007	.0005	.0004	.0003	.0003	.0002	.0002	.0001
1	.0105	.0088	.0072	.0060	.0049	.0040	.0033	.0026	.0021	.0017
2	.0501	.0431	.0370	.0316	.0269	.0228	.0192	.0162	.0135	.0112
3	.1508	.1344	.1193	.1055	.0929	.0815	.0712	.0619	.0536	.0461
4	.3258	.2997	.2746	.2507	.2279	.2065	.1863	.1674	.1498	.1334
5	.5448	.5151	.4854	.4559	.4268	.3981	.3701	.3427	.3162	.2905
6	.7476	.7230	.6975	.6710	.6437	.6158	.5873	.5585	.5293	.5000
7	.8886	.8736	.8574	.8400	.8212	.8012	.7800	.7576	.7341	.7095
8	.9621	.9554	.9480	.9395	.9302	.9197	.9082	.8955	.8817	.8666
9	.9904	.9883	.9859	.9830	.9797	.9758	.9713	.9662	.9604	.9539
10	.9983	.9979	.9973	.9967	.9959	.9949	.9937	.9923	.9907	.9888
11	.9998	.9998	.9997	.9996	.9995	.9993	.9991	.9989	.9986	.9983
12	1.0000	1.0000	1.0000	1.0000	1.0000	1.0000	.9999	.9999	.9999	.9999
13	1.0000	1.0000	1.0000	1.0000	1.0000	1.0000	1.0000	1.0000	1.0000	1.0000

TABLE B ***(continued)***

$n = 14$

x \ p	.01	.02	.03	.04	.05	.06	.07	.08	.09	.10
0	.8687	.7536	.6528	.5647	.4877	.4205	.3620	.3112	.2670	.2288
1	.9916	.9690	.9355	.8941	.8470	.7963	.7436	.6900	.6368	.5846
2	.9997	.9975	.9923	.9833	.9699	.9522	.9302	.9042	.8745	.8416
3	1.0000	.9999	.9994	.9981	.9958	.9920	.9864	.9786	.9685	.9559
4	1.0000	1.0000	1.0000	.9998	.9996	.9990	.9980	.9965	.9941	.9908
5	1.0000	1.0000	1.0000	1.0000	1.0000	.9999	.9998	.9996	.9992	.9985
6	1.0000	1.0000	1.0000	1.0000	1.0000	1.0000	1.0000	1.0000	.9999	.9998
7	1.0000	1.0000	1.0000	1.0000	1.0000	1.0000	1.0000	1.0000	1.0000	1.0000
x \ p	**.11**	**.12**	**.13**	**.14**	**.15**	**.16**	**.17**	**.18**	**.19**	**.20**
0	.1956	.1670	.1423	.1211	.1028	.0871	.0736	.0621	.0523	.0440
1	.5342	.4859	.4401	.3969	.3567	.3193	.2848	.2531	.2242	.1979
2	.8061	.7685	.7292	.6889	.6479	.6068	.5659	.5256	.4862	.4481
3	.9406	.9226	.9021	.8790	.8535	.8258	.7962	.7649	.7321	.6982
4	.9863	.9804	.9731	.9641	.9533	.9406	.9259	.9093	.8907	.8702
5	.9976	.9962	.9943	.9918	.9885	.9843	.9791	.9727	.9651	.9561
6	.9997	.9994	.9991	.9985	.9978	.9968	.9954	.9936	.9913	.9884
7	1.0000	.9999	.9999	.9998	.9997	.9995	.9992	.9988	.9983	.9976
8	1.0000	1.0000	1.0000	1.0000	1.0000	.9999	.9999	.9998	.9997	.9996
9	1.0000	1.0000	1.0000	1.0000	1.0000	1.0000	1.0000	1.0000	1.0000	1.0000
x \ p	**.21**	**.22**	**.23**	**.24**	**.25**	**.26**	**.27**	**.28**	**.29**	**.30**
0	.0369	.0309	.0258	.0214	.0178	.0148	.0122	.0101	.0083	.0068
1	.1741	.1527	.1335	.1163	.1010	.0874	.0754	.0648	.0556	.0475
2	.4113	.3761	.3426	.3109	.2811	.2533	.2273	.2033	.1812	.1608
3	.6634	.6281	.5924	.5568	.5213	.4864	.4521	.4187	.3863	.3552
4	.8477	.8235	.7977	.7703	.7415	.7116	.6807	.6490	.6168	.5842
5	.9457	.9338	.9203	.9051	.8883	.8699	.8498	.8282	.8051	.7805
6	.9848	.9804	.9752	.9690	.9617	.9533	.9437	.9327	.9204	.9067
7	.9967	.9955	.9940	.9921	.9897	.9868	.9833	.9792	.9743	.9685
8	.9994	.9992	.9989	.9984	.9978	.9971	.9962	.9950	.9935	.9917
9	.9999	.9999	.9998	.9998	.9997	.9995	.9993	.9991	.9988	.9983
10	1.0000	1.0000	1.0000	1.0000	1.0000	.9999	.9999	.9999	.9998	.9998
11	1.0000	1.0000	1.0000	1.0000	1.0000	1.0000	1.0000	1.0000	1.0000	1.0000

TABLE B (*continued*)

$n = 14$ (*continued*)

x \ p	.31	.32	.33	.34	.35	.36	.37	.38	.39	.40
0	.0055	.0045	.0037	.0030	.0024	.0019	.0016	.0012	.0010	.0008
1	.0404	.0343	.0290	.0244	.0205	.0172	.0143	.0119	.0098	.0081
2	.1423	.1254	.1101	.0963	.0839	.0729	.0630	.0543	.0466	.0398
3	.3253	.2968	.2699	.2444	.2205	.1982	.1774	.1582	.1405	.1243
4	.5514	.5187	.4862	.4542	.4227	.3920	.3622	.3334	.3057	.2793
5	.7546	.7276	.6994	.6703	.6405	.6101	.5792	.5481	.5169	.4859
6	.8916	.8750	.8569	.8374	.8164	.7941	.7704	.7455	.7195	.6925
7	.9619	.9542	.9455	.9357	.9247	.9124	.8988	.8838	.8675	.8499
8	.9895	.9869	.9837	.9800	.9757	.9706	.9647	.9580	.9503	.9417
9	.9978	.9971	.9963	.9952	.9940	.9924	.9905	.9883	.9856	.9825
10	.9997	.9995	.9994	.9992	.9989	.9986	.9981	.9976	.9969	.9961
11	1.0000	.9999	.9999	.9999	.9999	.9998	.9997	.9997	.9995	.9994
12	1.0000	1.0000	1.0000	1.0000	1.0000	1.0000	1.0000	1.0000	1.0000	.9999
13	1.0000	1.0000	1.0000	1.0000	1.0000	1.0000	1.0000	1.0000	1.0000	1.0000

x \ p	.41	.42	.43	.44	.45	.46	.47	.48	.49	.50
0	.0006	.0005	.0004	.0003	.0002	.0002	.0001	.0001	.0001	.0001
1	.0066	.0054	.0044	.0036	.0029	.0023	.0019	.0015	.0012	.0009
2	.0339	.0287	.0242	.0203	.0170	.0142	.0117	.0097	.0079	.0065
3	.1095	.0961	.0839	.0730	.0632	.0545	.0468	.0399	.0339	.0287
4	.2541	.2303	.2078	.1868	.1672	.1490	.1322	.1167	.1026	.0898
5	.4550	.4246	.3948	.3656	.3373	.3100	.2837	.2585	.2346	.2120
6	.6645	.6357	.6063	.5764	.5461	.5157	.4852	.4549	.4249	.3953
7	.8308	.8104	.7887	.7656	.7414	.7160	.6895	.6620	.6337	.6047
8	.9320	.9211	.9090	.8957	.8811	.8652	.8480	.8293	.8094	.7880
9	.9788	.9745	.9696	.9639	.9574	.9500	.9417	.9323	.9218	.9102
10	.9951	.9939	.9924	.9907	.9886	.9861	.9832	.9798	.9759	.9713
11	.9992	.9990	.9987	.9983	.9978	.9973	.9966	.9958	.9947	.9935
12	.9999	.9999	.9999	.9998	.9997	.9997	.9996	.9994	.9993	.9991
13	1.0000	1.0000	1.0000	1.0000	1.0000	1.0000	1.0000	1.0000	1.0000	.9999
14	1.0000	1.0000	1.0000	1.0000	1.0000	1.0000	1.0000	1.0000	1.0000	1.0000

$n = 15$

x \ p	.01	.02	.03	.04	.05	.06	.07	.08	.09	.10
0	.8601	.7386	.6333	.5421	.4633	.3953	.3367	.2863	.2430	.2059
1	.9904	.9647	.9270	.8809	.8290	.7738	.7168	.6597	.6035	.5490
2	.9996	.9970	.9906	.9797	.9638	.9429	.9171	.8870	.8531	.8159
3	1.0000	.9998	.9992	.9976	.9945	.9896	.9825	.9727	.9601	.9444
4	1.0000	1.0000	.9999	.9998	.9994	.9986	.9972	.9950	.9918	.9873
5	1.0000	1.0000	1.0000	1.0000	.9999	.9999	.9997	.9993	.9987	.9978
6	1.0000	1.0000	1.0000	1.0000	1.0000	1.0000	1.0000	.9999	.9998	.9997
7	1.0000	1.0000	1.0000	1.0000	1.0000	1.0000	1.0000	1.0000	1.0000	1.0000

TABLE B (*continued*)

$n = 15$ (*continued*)

x \ p	.11	.12	.13	.14	.15	.16	.17	.18	.19	.20
0	.1741	.1470	.1238	.1041	.0874	.0731	.0611	.0510	.0424	.0352
1	.4969	.4476	.4013	.3583	.3186	.2821	.2489	.2187	.1915	.1671
2	.7762	.7346	.6916	.6480	.6042	.5608	.5181	.4766	.4365	.3980
3	.9258	.9041	.8796	.8524	.8227	.7908	.7571	.7218	.6854	.6482
4	.9813	.9735	.9639	.9522	.9383	.9222	.9039	.8833	.8606	.8358
5	.9963	.9943	.9916	.9879	.9832	.9773	.9700	.9613	.9510	.9389
6	.9994	.9990	.9985	.9976	.9964	.9948	.9926	.9898	.9863	.9819
7	.9999	.9999	.9998	.9996	.9994	.9990	.9986	.9979	.9970	.9958
8	1.0000	1.0000	1.0000	1.0000	.9999	.9999	.9998	.9997	.9995	.9992
9	1.0000	1.0000	1.0000	1.0000	1.0000	1.0000	1.0000	1.0000	.9999	.9999
10	1.0000	1.0000	1.0000	1.0000	1.0000	1.0000	1.0000	1.0000	1.0000	1.0000

x \ p	.21	.22	.23	.24	.25	.26	.27	.28	.29	.30
0	.0291	.0241	.0198	.0163	.0134	.0109	.0089	.0072	.0059	.0047
1	.1453	.1259	.1087	.0935	.0802	.0685	.0583	.0495	.0419	.0353
2	.3615	.3269	.2945	.2642	.2361	.2101	.1863	.1645	.1447	.1268
3	.6105	.5726	.5350	.4978	.4613	.4258	.3914	.3584	.3268	.2969
4	.8090	.7805	.7505	.7190	.6865	.6531	.6190	.5846	.5500	.5155
5	.9252	.9095	.8921	.8728	.8516	.8287	.8042	.7780	.7505	.7216
6	.9766	.9702	.9626	.9537	.9434	.9316	.9183	.9035	.8870	.8689
7	.9942	.9922	.9896	.9865	.9827	.9781	.9726	.9662	.9587	.9500
8	.9989	.9984	.9977	.9969	.9958	.9944	.9927	.9906	.9879	.9848
9	.9998	.9997	.9996	.9994	.9992	.9989	.9985	.9979	.9972	.9963
10	1.0000	1.0000	.9999	.9999	.9999	.9998	.9998	.9997	.9995	.9993
11	1.0000	1.0000	1.0000	1.0000	1.0000	1.0000	1.0000	1.0000	.9999	.9999
12	1.0000	1.0000	1.0000	1.0000	1.0000	1.0000	1.0000	1.0000	1.0000	1.0000

x \ p	.31	.32	.33	.34	.35	.36	.37	.38	.39	.40
0	.0038	.0031	.0025	.0020	.0016	.0012	.0010	.0008	.0006	.0005
1	.0296	.0248	.0206	.0171	.0142	.0117	.0096	.0078	.0064	.0052
2	.1107	.0962	.0833	.0719	.0617	.0528	.0450	.0382	.0322	.0271
3	.2686	.2420	.2171	.1940	.1727	.1531	.1351	.1187	.1039	.0905
4	.4813	.4477	.4148	.3829	.3519	.3222	.2938	.2668	.2413	.2173
5	.6916	.6607	.6291	.5968	.5643	.5316	.4989	.4665	.4346	.4032
6	.8491	.8278	.8049	.7806	.7548	.7278	.6997	.6705	.6405	.6098
7	.9401	.9289	.9163	.9023	.8868	.8698	.8513	.8313	.8098	.7869
8	.9810	.9764	.9711	.9649	.9578	.9496	.9403	.9298	.9180	.9050
9	.9952	.9938	.9921	.9901	.9876	.9846	.9810	.9768	.9719	.9662
10	.9991	.9988	.9984	.9978	.9972	.9963	.9953	.9941	.9925	.9907
11	.9999	.9998	.9997	.9996	.9995	.9994	.9991	.9989	.9985	.9981
12	1.0000	1.0000	1.0000	1.0000	.9999	.9999	.9999	.9998	.9998	.9997
13	1.0000	1.0000	1.0000	1.0000	1.0000	1.0000	1.0000	1.0000	1.0000	1.0000

TABLE B (continued)

n = 15 (*continued*)

x \ p	.41	.42	.43	.44	.45	.46	.47	.48	.49	.50
0	.0004	.0003	.0002	.0002	.0001	.0001	.0001	.0001	.0000	.0000
1	.0042	.0034	.0027	.0021	.0017	.0013	.0010	.0008	.0006	.0005
2	.0227	.0189	.0157	.0130	.0107	.0087	.0071	.0057	.0046	.0037
3	.0785	.0678	.0583	.0498	.0424	.0359	.0303	.0254	.0212	.0176
4	.1948	.1739	.1546	.1367	.1204	.1055	.0920	.0799	.0690	.0592
5	.3726	.3430	.3144	.2869	.2608	.2359	.2125	.1905	.1699	.1509
6	.5786	.5470	.5153	.4836	.4522	.4211	.3905	.3606	.3316	.3036
7	.7626	.7370	.7102	.6824	.6535	.6238	.5935	.5626	.5314	.5000
8	.8905	.8746	.8573	.8385	.8182	.7966	.7735	.7490	.7233	.6964
9	.9596	.9521	.9435	.9339	.9231	.9110	.8976	.8829	.8667	.8491
10	.9884	.9857	.9826	.9789	.9745	.9695	.9637	.9570	.9494	.9408
11	.9975	.9968	.9960	.9949	.9937	.9921	.9903	.9881	.9855	.9824
12	.9996	.9995	.9993	.9991	.9989	.9986	.9982	.9977	.9971	.9963
13	1.0000	1.0000	.9999	.9999	.9999	.9998	.9998	.9997	.9996	.9995
14	1.0000	1.0000	1.0000	1.0000	1.0000	1.0000	1.0000	1.0000	1.0000	1.0000

n = 16

x \ p	.01	.02	.03	.04	.05	.06	.07	.08	.09	.10
0	.8515	.7238	.6143	.5204	.4401	.3716	.3131	.2634	.2211	.1853
1	.9891	.9601	.9182	.8673	.8108	.7511	.6902	.6299	.5711	.5147
2	.9995	.9963	.9887	.9758	.9571	.9327	.9031	.8688	.8306	.7892
3	1.0000	.9998	.9989	.9968	.9930	.9868	.9779	.9658	.9504	.9316
4	1.0000	1.0000	.9999	.9997	.9991	.9981	.9962	.9932	.9889	.9830
5	1.0000	1.0000	1.0000	1.0000	.9999	.9998	.9995	.9990	.9981	.9967
6	1.0000	1.0000	1.0000	1.0000	1.0000	1.0000	.9999	.9999	.9997	.9995
7	1.0000	1.0000	1.0000	1.0000	1.0000	1.0000	1.0000	1.0000	1.0000	.9999
8	1.0000	1.0000	1.0000	1.0000	1.0000	1.0000	1.0000	1.0000	1.0000	1.0000

x \ p	.11	.12	.13	.14	.15	.16	.17	.18	.19	.20
0	.1550	.1293	.1077	.0895	.0743	.0614	.0507	.0418	.0343	.0281
1	.4614	.4115	.3653	.3227	.2839	.2487	.2170	.1885	.1632	.1407
2	.7455	.7001	.6539	.6074	.5614	.5162	.4723	.4302	.3899	.3518
3	.9093	.8838	.8552	.8237	.7899	.7540	.7164	.6777	.6381	.5981
4	.9752	.9652	.9529	.9382	.9209	.9012	.8789	.8542	.8273	.7982
5	.9947	.9918	.9880	.9829	.9765	.9685	.9588	.9473	.9338	.9183
6	.9991	.9985	.9976	.9962	.9944	.9920	.9888	.9847	.9796	.9733
7	.9999	.9998	.9996	.9993	.9989	.9984	.9976	.9964	.9949	.9930
8	1.0000	1.0000	.9999	.9999	.9998	.9997	.9996	.9993	.9990	.9985
9	1.0000	1.0000	1.0000	1.0000	1.0000	1.0000	.9999	.9999	.9998	.9998
10	1.0000	1.0000	1.0000	1.0000	1.0000	1.0000	1.0000	1.0000	1.0000	1.0000

TABLE B (*continued*)

	n = 16 (*continued*)									
x \ *p*	.21	.22	.23	.24	.25	.26	.27	.28	.29	.30
0	.0230	.0188	.0153	.0124	.0100	.0081	.0065	.0052	.0042	.0033
1	.1209	.1035	.0883	.0750	.0635	.0535	.0450	.0377	.0314	.0261
2	.3161	.2827	.2517	.2232	.1971	.1733	.1518	.1323	.1149	.0994
3	.5582	.5186	.4797	.4417	.4050	.3697	.3360	.3041	.2740	.2459
4	.7673	.7348	.7009	.6659	.6302	.5940	.5575	.5212	.4853	.4499
5	.9008	.8812	.8595	.8359	.8103	.7831	.7542	.7239	.6923	.6598
6	.9658	.9568	.9464	.9342	.9204	.9049	.8875	.8683	.8474	.8247
7	.9905	.9873	.9834	.9786	.9729	.9660	.9580	.9486	.9379	.9256
8	.9979	.9970	.9959	.9944	.9925	.9902	.9873	.9837	.9794	.9743
9	.9996	.9994	.9992	.9988	.9984	.9977	.9969	.9959	.9945	.9929
10	.9999	.9999	.9999	.9998	.9997	.9996	.9994	.9992	.9989	.9984
11	1.0000	1.0000	1.0000	1.0000	1.0000	.9999	.9999	.9999	.9998	.9997
12	1.0000	1.0000	1.0000	1.0000	1.0000	1.0000	1.0000	1.0000	1.0000	1.0000
x \ *p*	.31	.32	.33	.34	.35	.36	.37	.38	.39	.40
0	.0026	.0021	.0016	.0013	.0010	.0008	.0006	.0005	.0004	.0003
1	.0216	.0178	.0146	.0120	.0098	.0079	.0064	.0052	.0041	.0033
2	.0856	.0734	.0626	.0533	.0451	.0380	.0319	.0266	.0222	.0183
3	.2196	.1953	.1730	.1525	.1339	.1170	.1018	.0881	.0759	.0651
4	.4154	.3819	.3496	.3187	.2892	.2613	.2351	.2105	.1877	.1666
5	.6264	.5926	.5584	.5241	.4900	.4562	.4230	.3906	.3592	.3288
6	.8003	.7743	.7469	.7181	.6881	.6572	.6254	.5930	.5602	.5272
7	.9119	.8965	.8795	.8609	.8406	.8187	.7952	.7702	.7438	.7161
8	.9683	.9612	.9530	.9436	.9329	.9209	.9074	.8924	.8758	.8577
9	.9908	.9883	.9852	.9815	.9771	.9720	.9659	.9589	.9509	.9417
10	.9979	.9972	.9963	.9952	.9938	.9921	.9900	.9875	.9845	.9809
11	.9996	.9995	.9993	.9990	.9987	.9983	.9977	.9970	.9962	.9951
12	1.0000	.9999	.9999	.9999	.9998	.9997	.9996	.9995	.9993	.9991
13	1.0000	1.0000	1.0000	1.0000	1.0000	1.0000	1.0000	.9999	.9999	.9999
14	1.0000	1.0000	1.0000	1.0000	1.0000	1.0000	1.0000	1.0000	1.0000	1.0000

TABLE B (continued)

$n = 16$ (*continued*)

x \ p	.41	.42	.43	.44	.45	.46	.47	.48	.49	.50
0	.0002	.0002	.0001	.0001	.0001	.0001	.0000	.0000	.0000	.0000
1	.0026	.0021	.0016	.0013	.0010	.0008	.0006	.0005	.0003	.0003
2	.0151	.0124	.0101	.0082	.0066	.0053	.0042	.0034	.0027	.0021
3	.0556	.0473	.0400	.0336	.0281	.0234	.0194	.0160	.0131	.0106
4	.1471	.1293	.1131	.0985	.0853	.0735	.0630	.0537	.0456	.0384
5	.2997	.2720	.2457	.2208	.1976	.1759	.1559	.1374	.1205	.1051
6	.4942	.4613	.4289	.3971	.3660	.3359	.3068	.2790	.2524	.2272
7	.6872	.6572	.6264	.5949	.5629	.5306	.4981	.4657	.4335	.4018
8	.8381	.8168	.7940	.7698	.7441	.7171	.6889	.6596	.6293	.5982
9	.9313	.9195	.9064	.8919	.8759	.8584	.8393	.8186	.7964	.7728
10	.9766	.9716	.9658	.9591	.9514	.9426	.9326	.9214	.9089	.8949
11	.9938	.9922	.9902	.9879	.9851	.9817	.9778	.9732	.9678	.9616
12	.9988	.9984	.9979	.9973	.9965	.9956	.9945	.9931	.9914	.9894
13	.9998	.9998	.9997	.9996	.9994	.9993	.9990	.9987	.9984	.9979
14	1.0000	1.0000	1.0000	1.0000	.9999	.9999	.9999	.9999	.9998	.9997
15	1.0000	1.0000	1.0000	1.0000	1.0000	1.0000	1.0000	1.0000	1.0000	1.0000

$n = 17$

x \ p	.01	.02	.03	.04	.05	.06	.07	.08	.09	.10
0	.8429	.7093	.5958	.4996	.4181	.3493	.2912	.2423	.2012	.1668
1	.9877	.9554	.9091	.8535	.7922	.7283	.6638	.6005	.5396	.4818
2	.9994	.9956	.9866	.9714	.9497	.9218	.8882	.8497	.8073	.7618
3	1.0000	.9997	.9986	.9960	.9912	.9836	.9727	.9581	.9397	.9174
4	1.0000	1.0000	.9999	.9996	.9988	.9974	.9949	.9911	.9855	.9779
5	1.0000	1.0000	1.0000	1.0000	.9999	.9997	.9993	.9985	.9973	.9953
6	1.0000	1.0000	1.0000	1.0000	1.0000	1.0000	.9999	.9998	.9996	.9992
7	1.0000	1.0000	1.0000	1.0000	1.0000	1.0000	1.0000	1.0000	1.0000	.9999
8	1.0000	1.0000	1.0000	1.0000	1.0000	1.0000	1.0000	1.0000	1.0000	1.0000

x \ p	.11	.12	.13	.14	.15	.16	.17	.18	.19	.20
0	.1379	.1138	.0937	.0770	.0631	.0516	.0421	.0343	.0278	.0225
1	.4277	.3777	.3318	.2901	.2525	.2187	.1887	.1621	.1387	.1182
2	.7142	.6655	.6164	.5676	.5198	.4734	.4289	.3867	.3468	.3096
3	.8913	.8617	.8290	.7935	.7556	.7159	.6749	.6331	.5909	.5489
4	.9679	.9554	.9402	.9222	.9013	.8776	.8513	.8225	.7913	.7582
5	.9925	.9886	.9834	.9766	.9681	.9577	.9452	.9305	.9136	.8943
6	.9986	.9977	.9963	.9944	.9917	.9882	.9837	.9780	.9709	.9623
7	.9998	.9996	.9993	.9989	.9983	.9973	.9961	.9943	.9920	.9891
8	1.0000	.9999	.9999	.9998	.9997	.9995	.9992	.9988	.9982	.9974
9	1.0000	1.0000	1.0000	1.0000	1.0000	.9999	.9999	.9998	.9997	.9995
10	1.0000	1.0000	1.0000	1.0000	1.0000	1.0000	1.0000	1.0000	1.0000	.9999
11	1.0000	1.0000	1.0000	1.0000	1.0000	1.0000	1.0000	1.0000	1.0000	1.0000

TABLE B (*continued*)

n = 17 (*continued*)

x \ p	.21	.22	.23	.24	.25	.26	.27	.28	.29	.30
0	.0182	.0146	.0118	.0094	.0075	.0060	.0047	.0038	.0030	.0023
1	.1004	.0849	.0715	.0600	.0501	.0417	.0346	.0286	.0235	.0193
2	.2751	.2433	.2141	.1877	.1637	.1422	.1229	.1058	.0907	.0774
3	.5073	.4667	.4272	.3893	.3530	.3186	.2863	.2560	.2279	.2019
4	.7234	.6872	.6500	.6121	.5739	.5357	.4977	.4604	.4240	.3887
5	.8727	.8490	.8230	.7951	.7653	.7339	.7011	.6671	.6323	.5968
6	.9521	.9402	.9264	.9106	.8929	.8732	.8515	.8279	.8024	.7752
7	.9853	.9806	.9749	.9680	.9598	.9501	.9389	.9261	.9116	.8954
8	.9963	.9949	.9930	.9906	.9876	.9839	.9794	.9739	.9674	.9597
9	.9993	.9989	.9984	.9978	.9969	.9958	.9943	.9925	.9902	.9873
10	.9999	.9998	.9997	.9996	.9994	.9991	.9987	.9982	.9976	.9968
11	1.0000	1.0000	1.0000	.9999	.9999	.9998	.9998	.9997	.9995	.9993
12	1.0000	1.0000	1.0000	1.0000	1.0000	1.0000	1.0000	1.0000	.9999	.9999
13	1.0000	1.0000	1.0000	1.0000	1.0000	1.0000	1.0000	1.0000	1.0000	1.0000

x \ p	.31	.32	.33	.34	.35	.36	.37	.38	.39	.40
0	.0018	.0014	.0011	.0009	.0007	.0005	.0004	.0003	.0002	.0002
1	.0157	.0128	.0104	.0083	.0067	.0054	.0043	.0034	.0027	.0021
2	.0657	.0556	.0468	.0392	.0327	.0272	.0225	.0185	.0151	.0123
3	.1781	.1563	.1366	.1188	.1028	.0885	.0759	.0648	.0550	.0464
4	.3547	.3222	.2913	.2622	.2348	.2094	.1858	.1640	.1441	.1260
5	.5610	.5251	.4895	.4542	.4197	.3861	.3535	.3222	.2923	.2639
6	.7464	.7162	.6847	.6521	.6188	.5848	.5505	.5161	.4818	.4478
7	.8773	.8574	.8358	.8123	.7872	.7605	.7324	.7029	.6722	.6405
8	.9508	.9405	.9288	.9155	.9006	.8841	.8659	.8459	.8243	.8011
9	.9838	.9796	.9746	.9686	.9617	.9536	.9443	.9336	.9216	.9081
10	.9957	.9943	.9926	.9905	.9880	.9849	.9811	.9766	.9714	.9652
11	.9991	.9987	.9983	.9977	.9970	.9960	.9949	.9934	.9916	.9894
12	.9998	.9998	.9997	.9996	.9994	.9992	.9989	.9985	.9981	.9975
13	1.0000	1.0000	1.0000	.9999	.9999	.9999	.9998	.9998	.9997	.9995
14	1.0000	1.0000	1.0000	1.0000	1.0000	1.0000	1.0000	1.0000	1.0000	.9999
15	1.0000	1.0000	1.0000	1.0000	1.0000	1.0000	1.0000	1.0000	1.0000	1.0000

TABLE B (continued)

$n = 17$ (*continued*)

x \ p	.41	.42	.43	.44	.45	.46	.47	.48	.49	.50
0	.0001	.0001	.0001	.0001	.0000	.0000	.0000	.0000	.0000	.0000
1	.0016	.0013	.0010	.0008	.0006	.0004	.0003	.0002	.0002	.0001
2	.0100	.0080	.0065	.0052	.0041	.0032	.0025	.0020	.0015	.0012
3	.0390	.0326	.0271	.0224	.0184	.0151	.0123	.0099	.0080	.0064
4	.1096	.0949	.0817	.0699	.0596	.0505	.0425	.0356	.0296	.0245
5	.2372	.2121	.1887	.1670	.1471	.1288	.1122	.0972	.0838	.0717
6	.4144	.3818	.3501	.3195	.2902	.2623	.2359	.2110	.1878	.1662
7	.6080	.5750	.5415	.5079	.4743	.4410	.4082	.3761	.3448	.3145
8	.7762	.7498	.7220	.6928	.6626	.6313	.5992	.5665	.5333	.5000
9	.8930	.8764	.8581	.8382	.8166	.7934	.7686	.7423	.7145	.6855
10	.9580	.9497	.9403	.9295	.9174	.9038	.8888	.8721	.8538	.8338
11	.9867	.9835	.9797	.9752	.9699	.9637	.9566	.9483	.9389	.9283
12	.9967	.9958	.9946	.9931	.9914	.9892	.9866	.9835	.9798	.9755
13	.9994	.9992	.9989	.9986	.9981	.9976	.9969	.9960	.9950	.9936
14	.9999	.9999	.9998	.9998	.9997	.9996	.9995	.9993	.9991	.9988
15	1.0000	1.0000	1.0000	1.0000	1.0000	1.0000	.9999	.9999	.9999	.9999
16	1.0000	1.0000	1.0000	1.0000	1.0000	1.0000	1.0000	1.0000	1.0000	1.0000

$n = 18$

x \ p	.01	.02	.03	.04	.05	.06	.07	.08	.09	.10
0	.8345	.6951	.5780	.4796	.3972	.3283	.2708	.2229	.1831	.1501
1	.9862	.9505	.8997	.8393	.7735	.7055	.6378	.5719	.5091	.4503
2	.9993	.9948	.9843	.9667	.9419	.9102	.8725	.8298	.7832	.7338
3	1.0000	.9996	.9982	.9950	.9891	.9799	.9667	.9494	.9277	.9018
4	1.0000	1.0000	.9998	.9994	.9985	.9966	.9933	.9884	.9814	.9718
5	1.0000	1.0000	1.0000	.9999	.9998	.9995	.9990	.9979	.9962	.9936
6	1.0000	1.0000	1.0000	1.0000	1.0000	1.0000	.9999	.9997	.9994	.9988
7	1.0000	1.0000	1.0000	1.0000	1.0000	1.0000	1.0000	1.0000	.9999	.9998
8	1.0000	1.0000	1.0000	1.0000	1.0000	1.0000	1.0000	1.0000	1.0000	1.0000

x \ p	.11	.12	.13	.14	.15	.16	.17	.18	.19	.20
0	.1227	.1002	.0815	.0662	.0536	.0434	.0349	.0281	.0225	.0180
1	.3958	.3460	.3008	.2602	.2241	.1920	.1638	.1391	.1176	.0991
2	.6827	.6310	.5794	.5287	.4797	.4327	.3881	.3462	.3073	.2713
3	.8718	.8382	.8014	.7618	.7202	.6771	.6331	.5888	.5446	.5010
4	.9595	.9442	.9257	.9041	.8794	.8518	.8213	.7884	.7533	.7164
5	.9898	.9846	.9778	.9690	.9581	.9449	.9292	.9111	.8903	.8671
6	.9979	.9966	.9946	.9919	.9882	.9833	.9771	.9694	.9600	.9487
7	.9997	.9994	.9989	.9983	.9973	.9959	.9940	.9914	.9880	.9837
8	1.0000	.9999	.9998	.9997	.9995	.9992	.9987	.9980	.9971	.9957
9	1.0000	1.0000	1.0000	1.0000	.9999	.9999	.9998	.9996	.9994	.9991
10	1.0000	1.0000	1.0000	1.0000	1.0000	1.0000	1.0000	.9999	.9999	.9998
11	1.0000	1.0000	1.0000	1.0000	1.0000	1.0000	1.0000	1.0000	1.0000	1.0000

TABLE B (*continued*)

$n = 18$ (*continued*)

x \ p	.21	.22	.23	.24	.25	.26	.27	.28	.29	.30
0	.0144	.0114	.0091	.0072	.0056	.0044	.0035	.0027	.0021	.0016
1	.0831	.0694	.0577	.0478	.0395	.0324	.0265	.0216	.0176	.0142
2	.2384	.2084	.1813	.1570	.1353	.1161	.0991	.0842	.0712	.0600
3	.4586	.4175	.3782	.3409	.3057	.2728	.2422	.2140	.1881	.1646
4	.6780	.6387	.5988	.5586	.5187	.4792	.4406	.4032	.3671	.3327
5	.8414	.8134	.7832	.7512	.7174	.6824	.6462	.6093	.5719	.5344
6	.9355	.9201	.9026	.8829	.8610	.8370	.8109	.7829	.7531	.7217
7	.9783	.9717	.9637	.9542	.9431	.9301	.9153	.8986	.8800	.8593
8	.9940	.9917	.9888	.9852	.9807	.9751	.9684	.9605	.9512	.9404
9	.9986	.9980	.9972	.9961	.9946	.9927	.9903	.9873	.9836	.9790
10	.9997	.9996	.9994	.9991	.9988	.9982	.9975	.9966	.9954	.9939
11	1.0000	.9999	.9999	.9998	.9998	.9997	.9995	.9993	.9990	.9986
12	1.0000	1.0000	1.0000	1.0000	1.0000	.9999	.9999	.9999	.9998	.9997
13	1.0000	1.0000	1.0000	1.0000	1.0000	1.0000	1.0000	1.0000	1.0000	1.0000

x \ p	.31	.32	.33	.34	.35	.36	.37	.38	.39	.40
0	.0013	.0010	.0007	.0006	.0004	.0003	.0002	.0002	.0001	.0001
1	.0114	.0092	.0073	.0058	.0046	.0036	.0028	.0022	.0017	.0013
2	.0502	.0419	.0348	.0287	.0236	.0193	.0157	.0127	.0103	.0082
3	.1432	.1241	.1069	.0917	.0783	.0665	.0561	.0472	.0394	.0328
4	.2999	.2691	.2402	.2134	.1886	.1659	.1451	.1263	.1093	.0942
5	.4971	.4602	.4241	.3889	.3550	.3224	.2914	.2621	.2345	.2088
6	.6889	.6550	.6202	.5849	.5491	.5133	.4776	.4424	.4079	.3743
7	.8367	.8122	.7859	.7579	.7283	.6973	.6651	.6319	.5979	.5634
8	.9280	.9139	.8981	.8804	.8609	.8396	.8165	.7916	.7650	.7368
9	.9736	.9671	.9595	.9506	.9403	.9286	.9153	.9003	.8837	.8653
10	.9920	.9896	.9867	.9831	.9788	.9736	.9675	.9603	.9520	.9424
11	.9980	.9973	.9964	.9953	.9938	.9920	.9898	.9870	.9837	.9797
12	.9996	.9995	.9992	.9989	.9986	.9981	.9974	.9966	.9956	.9942
13	.9999	.9999	.9999	.9998	.9997	.9996	.9995	.9993	.9990	.9987
14	1.0000	1.0000	1.0000	1.0000	1.0000	.9999	.9999	.9999	.9998	.9998
15	1.0000	1.0000	1.0000	1.0000	1.0000	1.0000	1.0000	1.0000	1.0000	1.0000

TABLE B (continued)

$n = 18$ (*continued*)										
x \ p	.41	.42	.43	.44	.45	.46	.47	.48	.49	.50
0	.0001	.0001	.0000	.0000	.0000	.0000	.0000	.0000	.0000	.0000
1	.0010	.0008	.0006	.0004	.0003	.0002	.0002	.0001	.0001	.0001
2	.0066	.0052	.0041	.0032	.0025	.0019	.0015	.0011	.0009	.0007
3	.0271	.0223	.0182	.0148	.0120	.0096	.0077	.0061	.0048	.0038
4	.0807	.0687	.0582	.0490	.0411	.0342	.0283	.0233	.0190	.0154
5	.1849	.1628	.1427	.1243	.1077	.0928	.0795	.0676	.0572	.0481
6	.3418	.3105	.2807	.2524	.2258	.2009	.1778	.1564	.1368	.1189
7	.5287	.4938	.4592	.4250	.3915	.3588	.3272	.2968	.2678	.2403
8	.7072	.6764	.6444	.6115	.5778	.5438	.5094	.4751	.4409	.4073
9	.8451	.8232	.7996	.7742	.7473	.7188	.6890	.6579	.6258	.5927
10	.9314	.9189	.9049	.8893	.8720	.8530	.8323	.8098	.7856	.7597
11	.9750	.9693	.9628	.9551	.9463	.9362	.9247	.9117	.8972	.8811
12	.9926	.9906	.9882	.9853	.9817	.9775	.9725	.9666	.9598	.9519
13	.9983	.9978	.9971	.9962	.9951	.9937	.9921	.9900	.9875	.9846
14	.9997	.9996	.9994	.9993	.9990	.9987	.9983	.9977	.9971	.9962
15	1.0000	.9999	.9999	.9999	.9999	.9998	.9997	.9996	.9995	.9993
16	1.0000	1.0000	1.0000	1.0000	1.0000	1.0000	1.0000	1.0000	.9999	.9999
17	1.0000	1.0000	1.0000	1.0000	1.0000	1.0000	1.0000	1.0000	1.0000	1.0000

$n = 19$										
x \ p	.01	.02	.03	.04	.05	.06	.07	.08	.09	.10
0	.8262	.6812	.5606	.4604	.3774	.3086	.2519	.2051	.1666	.1351
1	.9847	.9454	.8900	.8249	.7547	.6829	.6121	.5440	.4798	.4203
2	.9991	.9939	.9817	.9616	.9335	.8979	.8561	.8092	.7585	.7054
3	1.0000	.9995	.9978	.9939	.9868	.9757	.9602	.9398	.9147	.8850
4	1.0000	1.0000	.9998	.9993	.9980	.9956	.9915	.9853	.9765	.9648
5	1.0000	1.0000	1.0000	.9999	.9998	.9994	.9986	.9971	.9949	.9914
6	1.0000	1.0000	1.0000	1.0000	1.0000	.9999	.9998	.9996	.9991	.9983
7	1.0000	1.0000	1.0000	1.0000	1.0000	1.0000	1.0000	.9999	.9999	.9997
8	1.0000	1.0000	1.0000	1.0000	1.0000	1.0000	1.0000	1.0000	1.0000	1.0000
x \ p	.11	.12	.13	.14	.15	.16	.17	.18	.19	.20
0	.1092	.0881	.0709	.0569	.0456	.0364	.0290	.0230	.0182	.0144
1	.3658	.3165	.2723	.2331	.1985	.1682	.1419	.1191	.0996	.0829
2	.6512	.5968	.5432	.4911	.4413	.3941	.3500	.3090	.2713	.2369
3	.8510	.8133	.7725	.7292	.6841	.6380	.5915	.5451	.4995	.4551
4	.9498	.9315	.9096	.8842	.8556	.8238	.7893	.7524	.7136	.6733
5	.9865	.9798	.9710	.9599	.9463	.9300	.9109	.8890	.8643	.8369
6	.9970	.9952	.9924	.9887	.9837	.9772	.9690	.9589	.9468	.9324
7	.9995	.9991	.9984	.9974	.9959	.9939	.9911	.9874	.9827	.9767
8	.9999	.9998	.9997	.9995	.9992	.9986	.9979	.9968	.9953	.9933
9	1.0000	1.0000	1.0000	.9999	.9999	.9998	.9996	.9993	.9990	.9984
10	1.0000	1.0000	1.0000	1.0000	1.0000	1.0000	.9999	.9999	.9998	.9997
11	1.0000	1.0000	1.0000	1.0000	1.0000	1.0000	1.0000	1.0000	1.0000	1.0000

TABLE B (*continued*)

n = 19 (*continued*)

x \ p	.21	.22	.23	.24	.25	.26	.27	.28	.29	.30
0	.0113	.0089	.0070	.0054	.0042	.0033	.0025	.0019	.0015	.0011
1	.0687	.0566	.0465	.0381	.0310	.0251	.0203	.0163	.0131	.0104
2	.2058	.1778	.1529	.1308	.1113	.0943	.0795	.0667	.0557	.0462
3	.4123	.3715	.3329	.2968	.2631	.2320	.2035	.1776	.1542	.1332
4	.6319	.5900	.5480	.5064	.4654	.4256	.3871	.3502	.3152	.2822
5	.8071	.7749	.7408	.7050	.6677	.6295	.5907	.5516	.5125	.4739
6	.9157	.8966	.8751	.8513	.8251	.7968	.7664	.7343	.7005	.6655
7	.9693	.9604	.9497	.9371	.9225	.9059	.8871	.8662	.8432	.8180
8	.9907	.9873	.9831	.9778	.9713	.9634	.9541	.9432	.9306	.9161
9	.9977	.9966	.9953	.9934	.9911	.9881	.9844	.9798	.9742	.9674
10	.9995	.9993	.9989	.9984	.9977	.9968	.9956	.9940	.9920	.9895
11	.9999	.9999	.9998	.9997	.9995	.9993	.9990	.9985	.9980	.9972
12	1.0000	1.0000	1.0000	.9999	.9999	.9999	.9998	.9997	.9996	.9994
13	1.0000	1.0000	1.0000	1.0000	1.0000	1.0000	1.0000	1.0000	.9999	.9999
14	1.0000	1.0000	1.0000	1.0000	1.0000	1.0000	1.0000	1.0000	1.0000	1.0000

x \ p	.31	.32	.33	.34	.35	.36	.37	.38	.39	.40
0	.0009	.0007	.0005	.0004	.0003	.0002	.0002	.0001	.0001	.0001
1	.0083	.0065	.0051	.0040	.0031	.0024	.0019	.0014	.0011	.0008
2	.0382	.0314	.0257	.0209	.0170	.0137	.0110	.0087	.0069	.0055
3	.1144	.0978	.0831	.0703	.0591	.0495	.0412	.0341	.0281	.0230
4	.2514	.2227	.1963	.1720	.1500	.1301	.1122	.0962	.0821	.0696
5	.4359	.3990	.3634	.3293	.2968	.2661	.2373	.2105	.1857	.1629
6	.6294	.5927	.5555	.5182	.4812	.4446	.4087	.3739	.3403	.3081
7	.7909	.7619	.7312	.6990	.6656	.6310	.5957	.5599	.5238	.4878
8	.8997	.8814	.8611	.8388	.8145	.7884	.7605	.7309	.6998	.6675
9	.9595	.9501	.9392	.9267	.9125	.8965	.8787	.8590	.8374	.8139
10	.9863	.9824	.9777	.9720	.9653	.9574	.9482	.9375	.9253	.9115
11	.9962	.9949	.9932	.9911	.9886	.9854	.9815	.9769	.9713	.9648
12	.9991	.9988	.9983	.9977	.9969	.9959	.9946	.9930	.9909	.9884
13	.9998	.9998	.9997	.9995	.9993	.9991	.9987	.9983	.9977	.9969
14	1.0000	1.0000	.9999	.9999	.9999	.9998	.9998	.9997	.9995	.9994
15	1.0000	1.0000	1.0000	1.0000	1.0000	1.0000	1.0000	1.0000	.9999	.9999
16	1.0000	1.0000	1.0000	1.0000	1.0000	1.0000	1.0000	1.0000	1.0000	1.0000

TABLE B (*continued*)

n = 19 (*continued*)

x \ p	.41	.42	.43	.44	.45	.46	.47	.48	.49	.50
0	.0000	.0000	.0000	.0000	.0000	.0000	.0000	.0000	.0000	.0000
1	.0006	.0005	.0004	.0003	.0002	.0001	.0001	.0001	.0001	.0000
2	.0043	.0033	.0026	.0020	.0015	.0012	.0009	.0007	.0005	.0004
3	.0187	.0151	.0122	.0097	.0077	.0061	.0048	.0037	.0029	.0022
4	.0587	.0492	.0410	.0340	.0280	.0229	.0186	.0150	.0121	.0096
5	.1421	.1233	.1063	.0912	.0777	.0658	.0554	.0463	.0385	.0318
6	.2774	.2485	.2213	.1961	.1727	.1512	.1316	.1138	.0978	.0835
7	.4520	.4168	.3824	.3491	.3169	.2862	.2570	.2294	.2036	.1796
8	.6340	.5997	.5647	.5294	.4940	.4587	.4238	.3895	.3561	.3238
9	.7886	.7615	.7328	.7026	.6710	.6383	.6046	.5701	.5352	.5000
10	.8960	.8787	.8596	.8387	.8159	.7913	.7649	.7369	.7073	.6762
11	.9571	.9482	.9379	.9262	.9129	.8979	.8813	.8628	.8425	.8204
12	.9854	.9817	.9773	.9720	.9658	.9585	.9500	.9403	.9291	.9165
13	.9960	.9948	.9933	.9914	.9891	.9863	.9829	.9788	.9739	.9682
14	.9991	.9988	.9984	.9979	.9972	.9964	.9954	.9940	.9924	.9904
15	.9999	.9998	.9997	.9996	.9995	.9993	.9990	.9987	.9983	.9978
16	1.0000	1.0000	1.0000	.9999	.9999	.9999	.9999	.9998	.9997	.9996
17	1.0000	1.0000	1.0000	1.0000	1.0000	1.0000	1.0000	1.0000	1.0000	1.0000

n = 20

x \ p	.01	.02	.03	.04	.05	.06	.07	.08	.09	.10
0	.8179	.6676	.5438	.4420	.3585	.2901	.2342	.1887	.1516	.1216
1	.9831	.9401	.8802	.8103	.7358	.6605	.5869	.5169	.4516	.3917
2	.9990	.9929	.9790	.9561	.9245	.8850	.8390	.7879	.7334	.6769
3	1.0000	.9994	.9973	.9926	.9841	.9710	.9529	.9294	.9007	.8670
4	1.0000	1.0000	.9997	.9990	.9974	.9944	.9893	.9817	.9710	.9568
5	1.0000	1.0000	1.0000	.9999	.9997	.9991	.9981	.9962	.9932	.9887
6	1.0000	1.0000	1.0000	1.0000	1.0000	.9999	.9997	.9994	.9987	.9976
7	1.0000	1.0000	1.0000	1.0000	1.0000	1.0000	1.0000	.9999	.9998	.9996
8	1.0000	1.0000	1.0000	1.0000	1.0000	1.0000	1.0000	1.0000	1.0000	.9999
9	1.0000	1.0000	1.0000	1.0000	1.0000	1.0000	1.0000	1.0000	1.0000	1.0000

TABLE B ***(continued)***

n = 20 (*continued*)

x \ p	.11	.12	.13	.14	.15	.16	.17	.18	.19	.20
0	.0972	.0776	.0617	.0490	.0388	.0306	.0241	.0189	.0148	.0115
1	.3376	.2891	.2461	.2084	.1756	.1471	.1227	.1018	.0841	.0692
2	.6198	.5631	.5080	.4550	.4049	.3580	.3146	.2748	.2386	.2061
3	.8290	.7873	.7427	.6959	.6477	.5990	.5504	.5026	.4561	.4114
4	.9390	.9173	.8917	.8625	.8298	.7941	.7557	.7151	.6729	.6296
5	.9825	.9740	.9630	.9493	.9327	.9130	.8902	.8644	.8357	.8042
6	.9959	.9933	.9897	.9847	.9781	.9696	.9591	.9463	.9311	.9133
7	.9992	.9986	.9976	.9962	.9941	.9912	.9873	.9823	.9759	.9679
8	.9999	.9998	.9995	.9992	.9987	.9979	.9967	.9951	.9929	.9900
9	1.0000	1.0000	.9999	.9999	.9998	.9996	.9993	.9989	.9983	.9974
10	1.0000	1.0000	1.0000	1.0000	1.0000	.9999	.9999	.9998	.9996	.9994
11	1.0000	1.0000	1.0000	1.0000	1.0000	1.0000	1.0000	1.0000	.9999	.9999
12	1.0000	1.0000	1.0000	1.0000	1.0000	1.0000	1.0000	1.0000	1.0000	1.0000

x \ p	.21	.22	.23	.24	.25	.26	.27	.28	.29	.30
0	.0090	.0069	.0054	.0041	.0032	.0024	.0018	.0014	.0011	.0008
1	.0566	.0461	.0374	.0302	.0243	.0195	.0155	.0123	.0097	.0076
2	.1770	.1512	.1284	.1085	.0913	.0763	.0635	.0526	.0433	.0355
3	.3690	.3289	.2915	.2569	.2252	.1962	.1700	.1466	.1256	.1071
4	.5858	.5420	.4986	.4561	.4148	.3752	.3375	.3019	.2685	.2375
5	.7703	.7343	.6965	.6573	.6172	.5765	.5357	.4952	.4553	.4164
6	.8929	.8699	.8442	.8162	.7858	.7533	.7190	.6831	.6460	.6080
7	.9581	.9464	.9325	.9165	.8982	.8775	.8545	.8293	.8018	.7723
8	.9862	.9814	.9754	.9680	.9591	.9485	.9360	.9216	.9052	.8867
9	.9962	.9946	.9925	.9897	.9861	.9817	.9762	.9695	.9615	.9520
10	.9991	.9987	.9981	.9972	.9961	.9945	.9926	.9900	.9868	.9829
11	.9998	.9997	.9996	.9994	.9991	.9986	.9981	.9973	.9962	.9949
12	1.0000	1.0000	.9999	.9999	.9998	.9997	.9996	.9994	.9991	.9987
13	1.0000	1.0000	1.0000	1.0000	1.0000	1.0000	.9999	.9999	.9998	.9997
14	1.0000	1.0000	1.0000	1.0000	1.0000	1.0000	1.0000	1.0000	1.0000	1.0000

TABLE B (continued)

n = 20 (*continued*)

x \ p	.31	.32	.33	.34	.35	.36	.37	.38	.39	.40
0	.0006	.0004	.0003	.0002	.0002	.0001	.0001	.0001	.0001	.0000
1	.0060	.0047	.0036	.0028	.0021	.0016	.0012	.0009	.0007	.0005
2	.0289	.0235	.0189	.0152	.0121	.0096	.0076	.0060	.0047	.0036
3	.0908	.0765	.0642	.0535	.0444	.0366	.0300	.0245	.0198	.0160
4	.2089	.1827	.1589	.1374	.1182	.1011	.0859	.0726	.0610	.0510
5	.3787	.3426	.3082	.2758	.2454	.2171	.1910	.1671	.1453	.1256
6	.5695	.5307	.4921	.4540	.4166	.3803	.3453	.3118	.2800	.2500
7	.7409	.7078	.6732	.6376	.6010	.5639	.5265	.4892	.4522	.4159
8	.8660	.8432	.8182	.7913	.7624	.7317	.6995	.6659	.6312	.5956
9	.9409	.9281	.9134	.8968	.8782	.8576	.8350	.8103	.7837	.7553
10	.9780	.9721	.9650	.9566	.9468	.9355	.9225	.9077	.8910	.8725
11	.9931	.9909	.9881	.9846	.9804	.9753	.9692	.9619	.9534	.9435
12	.9982	.9975	.9966	.9955	.9940	.9921	.9898	.9868	.9833	.9790
13	.9996	.9994	.9992	.9989	.9985	.9979	.9972	.9963	.9951	.9935
14	.9999	.9999	.9999	.9998	.9997	.9996	.9994	.9991	.9988	.9984
15	1.0000	1.0000	1.0000	1.0000	1.0000	.9999	.9999	.9998	.9998	.9997
16	1.0000	1.0000	1.0000	1.0000	1.0000	1.0000	1.0000	1.0000	1.0000	1.0000

x \ p	.41	.42	.43	.44	.45	.46	.47	.48	.49	.50
0	.0000	.0000	.0000	.0000	.0000	.0000	.0000	.0000	.0000	.0000
1	.0004	.0003	.0002	.0002	.0001	.0001	.0001	.0000	.0000	.0000
2	.0028	.0021	.0016	.0012	.0009	.0007	.0005	.0004	.0003	.0002
3	.0128	.0102	.0080	.0063	.0049	.0038	.0029	.0023	.0017	.0013
4	.0423	.0349	.0286	.0233	.0189	.0152	.0121	.0096	.0076	.0059
5	.1079	.0922	.0783	.0660	.0553	.0461	.0381	.0313	.0255	.0207
6	.2220	.1959	.1719	.1499	.1299	.1119	.0958	.0814	.0688	.0577
7	.3804	.3461	.3132	.2817	.2520	.2241	.1980	.1739	.1518	.1316
8	.5594	.5229	.4864	.4501	.4143	.3793	.3454	.3127	.2814	.2517
9	.7252	.6936	.6606	.6264	.5914	.5557	.5196	.4834	.4474	.4119
10	.8520	.8295	.8051	.7788	.7507	.7209	.6896	.6568	.6229	.5881
11	.9321	.9190	.9042	.8877	.8692	.8489	.8266	.8024	.7762	.7483
12	.9738	.9676	.9603	.9518	.9420	.9306	.9177	.9031	.8867	.8684
13	.9916	.9893	.9864	.9828	.9786	.9735	.9674	.9603	.9520	.9423
14	.9978	.9971	.9962	.9950	.9936	.9917	.9895	.9867	.9834	.9793
15	.9996	.9994	.9992	.9989	.9985	.9980	.9973	.9965	.9954	.9941
16	.9999	.9999	.9999	.9998	.9997	.9996	.9995	.9993	.9990	.9987
17	1.0000	1.0000	1.0000	1.0000	1.0000	.9999	.9999	.9999	.9999	.9998
18	1.0000	1.0000	1.0000	1.0000	1.0000	1.0000	1.0000	1.0000	1.0000	1.0000

TABLE B (*continued*)

$n = 25$

x \ p	.01	.02	.03	.04	.05	.06	.07	.08	.09	.10
0	.7778	.6035	.4670	.3604	.2774	.2129	.1630	.1244	.0946	.0718
1	.9742	.9114	.8280	.7358	.6424	.5527	.4696	.3947	.3286	.2712
2	.9980	.9868	.9620	.9235	.8729	.8129	.7466	.6768	.6063	.5371
3	.9999	.9986	.9938	.9835	.9659	.9402	.9064	.8649	.8169	.7636
4	1.0000	.9999	.9992	.9972	.9928	.9850	.9726	.9549	.9314	.9020
5	1.0000	1.0000	.9999	.9996	.9988	.9969	.9935	.9877	.9790	.9666
6	1.0000	1.0000	1.0000	1.0000	.9998	.9995	.9987	.9972	.9946	.9905
7	1.0000	1.0000	1.0000	1.0000	1.0000	.9999	.9998	.9995	.9989	.9977
8	1.0000	1.0000	1.0000	1.0000	1.0000	1.0000	1.0000	.9999	.9998	.9995
9	1.0000	1.0000	1.0000	1.0000	1.0000	1.0000	1.0000	1.0000	1.0000	.9999
10	1.0000	1.0000	1.0000	1.0000	1.0000	1.0000	1.0000	1.0000	1.0000	1.0000

x \ p	.11	.12	.13	.14	.15	.16	.17	.18	.19	.20
0	.0543	.0409	.0308	.0230	.0172	.0128	.0095	.0070	.0052	.0038
1	.2221	.1805	.1457	.1168	.0931	.0737	.0580	.0454	.0354	.0274
2	.4709	.4088	.3517	.3000	.2537	.2130	.1774	.1467	.1204	.0982
3	.7066	.6475	.5877	.5286	.4711	.4163	.3648	.3171	.2734	.2340
4	.8669	.8266	.7817	.7332	.6821	.6293	.5759	.5228	.4708	.4207
5	.9501	.9291	.9035	.8732	.8385	.7998	.7575	.7125	.6653	.6167
6	.9844	.9757	.9641	.9491	.9305	.9080	.8815	.8512	.8173	.7800
7	.9959	.9930	.9887	.9827	.9745	.9639	.9505	.9339	.9141	.8909
8	.9991	.9983	.9970	.9950	.9920	.9879	.9822	.9748	.9652	.9532
9	.9998	.9996	.9993	.9987	.9979	.9965	.9945	.9917	.9878	.9827
10	1.0000	.9999	.9999	.9997	.9995	.9991	.9985	.9976	.9963	.9944
11	1.0000	1.0000	1.0000	1.0000	.9999	.9998	.9997	.9994	.9990	.9985
12	1.0000	1.0000	1.0000	1.0000	1.0000	1.0000	.9999	.9999	.9998	.9996
13	1.0000	1.0000	1.0000	1.0000	1.0000	1.0000	1.0000	1.0000	1.0000	.9999
14	1.0000	1.0000	1.0000	1.0000	1.0000	1.0000	1.0000	1.0000	1.0000	1.0000

TABLE B (continued)

n = 25 (*continued*)

x \ p	.21	.22	.23	.24	.25	.26	.27	.28	.29	.30
0	.0028	.0020	.0015	.0010	.0008	.0005	.0004	.0003	.0002	.0001
1	.0211	.0162	.0123	.0093	.0070	.0053	.0039	.0029	.0021	.0016
2	.0796	.0640	.0512	.0407	.0321	.0252	.0196	.0152	.0117	.0090
3	.1987	.1676	.1403	.1166	.0962	.0789	.0642	.0519	.0417	.0332
4	.3730	.3282	.2866	.2484	.2137	.1826	.1548	.1304	.1090	.0905
5	.5675	.5184	.4701	.4233	.3783	.3356	.2956	.2585	.2245	.1935
6	.7399	.6973	.6529	.6073	.5611	.5149	.4692	.4247	.3817	.3407
7	8642	.8342	.8011	.7651	.7265	.6858	.6435	.6001	.5560	.5118
8	.9386	.9212	.9007	.8772	.8506	.8210	.7885	.7535	.7162	.6769
9	.9760	.9675	.9569	.9440	.9287	.9107	.8899	.8662	.8398	.8106
10	.9918	.9883	.9837	.9778	.9703	.9611	.9498	.9364	.9205	.9022
11	.9976	.9964	.9947	.9924	.9893	.9852	.9801	.9736	.9655	.9558
12	.9994	.9990	.9985	.9977	.9966	.9951	.9931	.9904	.9870	.9825
13	.9999	.9998	.9996	.9994	.9991	.9986	.9979	.9970	.9957	.9940
14	1.0000	1.0000	.9999	.9999	.9998	.9997	.9995	.9992	.9988	.9982
15	1.0000	1.0000	1.0000	1.0000	1.0000	.9999	.9999	.9998	.9997	.9995
16	1.0000	1.0000	1.0000	1.0000	1.0000	1.0000	1.0000	1.0000	.9999	.9999
17	1.0000	1.0000	1.0000	1.0000	1.0000	1.0000	1.0000	1.0000	1.0000	1.0000

x \ p	.31	.32	.33	.34	.35	.36	.37	.38	.39	.40
0	.0001	.0001	.0000	.0000	.0000	.0000	.0000	.0000	.0000	.0000
1	.0011	.0008	.0006	.0004	.0003	.0002	.0002	.0001	.0001	.0001
2	.0068	.0051	.0039	.0029	.0021	.0016	.0011	.0008	.0006	.0004
3	.0263	.0207	.0162	.0126	.0097	.0074	.0056	.0043	.0032	.0024
4	.0746	.0610	.0496	.0400	.0320	.0255	.0201	.0158	.0123	.0095
5	.1656	.1407	.1187	.0994	.0826	.0682	.0559	.0454	.0367	.0294
6	.3019	.2657	.2321	.2013	.1734	.1483	.1258	.1060	.0886	.0736
7	.4681	.4253	.3837	.3439	.3061	.2705	.2374	.2068	.1789	.1536
8	.6361	.5943	.5518	.5092	.4668	.4252	.3848	.3458	.3086	.2735
9	.7787	.7445	.7081	.6700	.6303	.5896	.5483	.5067	.4653	.4246
10	.8812	.8576	.8314	.8025	.7712	.7375	.7019	.6645	.6257	.5858
11	.9440	.9302	.9141	.8956	.8746	.8510	.8249	.7964	.7654	.7323
12	.9770	.9701	.9617	.9515	.9396	.9255	.9093	.8907	.8697	.8462
13	.9917	.9888	.9851	.9804	.9745	.9674	.9588	.9485	.9363	.9222
14	.9974	.9964	.9950	.9931	.9907	.9876	.9837	.9788	.9729	.9656
15	.9993	.9990	.9985	.9979	.9971	.9959	.9944	.9925	.9900	.9868
16	.9998	.9998	.9996	.9995	.9992	.9989	.9984	.9977	.9968	.9957
17	1.0000	1.0000	.9999	.9999	.9998	.9997	.9996	.9994	.9992	.9988
18	1.0000	1.0000	1.0000	1.0000	1.0000	.9999	.9999	.9999	.9998	.9997
19	1.0000	1.0000	1.0000	1.0000	1.0000	1.0000	1.0000	1.0000	1.0000	.9999
20	1.0000	1.0000	1.0000	1.0000	1.0000	1.0000	1.0000	1.0000	1.0000	1.0000

TABLE B ***(continued)***

$n = 25$ (*continued*)

x \ p	.41	.42	.43	.44	.45	.46	.47	.48	.49	.50
0	.0000	.0000	.0000	.0000	.0000	.0000	.0000	.0000	.0000	.0000
1	.0000	.0000	.0000	.0000	.0000	.0000	.0000	.0000	.0000	.0000
2	.0003	.0002	.0002	.0001	.0001	.0000	.0000	.0000	.0000	.0000
3	.0017	.0013	.0009	.0007	.0005	.0003	.0002	.0002	.0001	.0001
4	.0073	.0055	.0042	.0031	.0023	.0017	.0012	.0009	.0006	.0005
5	.0233	.0184	.0144	.0112	.0086	.0066	.0050	.0037	.0028	.0020
6	.0606	.0495	.0401	.0323	.0258	.0204	.0160	.0124	.0096	.0073
7	.1308	.1106	.0929	.0773	.0639	.0523	.0425	.0342	.0273	.0216
8	.2407	.2103	.1823	.1569	.1340	.1135	.0954	.0795	.0657	.0539
9	.3849	.3465	.3098	.2750	.2424	.2120	.1840	.1585	.1354	.1148
10	.5452	.5044	.4637	.4235	.3843	.3462	.3098	.2751	.2426	.2122
11	.6971	.6603	.6220	.5826	.5426	.5022	.4618	.4220	.3829	.3450
12	.8203	.7920	.7613	.7285	.6937	.6571	.6192	.5801	.5402	.5000
13	.9059	.8873	.8664	.8431	.8173	.7891	.7587	.7260	.6914	.6550
14	.9569	.9465	.9344	.9203	.9040	.8855	.8647	.8415	.8159	.7878
15	.9829	.9780	.9720	.9647	.9560	.9457	.9337	.9197	.9036	.8852
16	.9942	.9922	.9897	.9866	.9826	.9778	.9719	.9648	.9562	.9461
17	.9983	.9977	.9968	.9956	.9942	.9923	.9898	.9868	.9830	.9784
18	.9996	.9994	.9992	.9988	.9984	.9977	.9969	.9959	.9945	.9927
19	.9999	.9999	.9998	.9997	.9996	.9995	.9992	.9989	.9985	.9980
20	1.0000	1.0000	1.0000	1.0000	.9999	.9999	.9998	.9998	.9997	.9995
21	1.0000	1.0000	1.0000	1.0000	1.0000	1.0000	1.0000	1.0000	.9999	.9999
22	1.0000	1.0000	1.0000	1.0000	1.0000	1.0000	1.0000	1.0000	1.0000	1.0000

TABLE C Cumulative Poisson Distribution $P(X \leq X|\lambda)$. 1000 Times the Probability of X or Fewer Occurrences of Event That Has Average Number of Occurrences Equal to λ

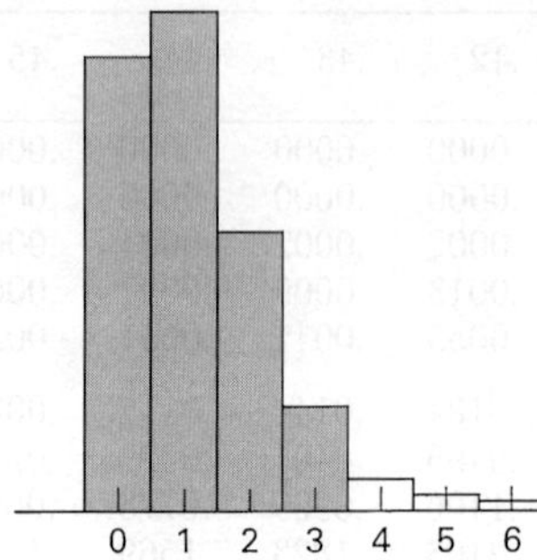

$p(x \leq 2 \mid \lambda = 1.00) = .920$

x \ λ	.02	.04	.06	.08	.10	.15	.20	.25
0	980	961	942	923	905	861	819	779
1	1000	999	998	997	995	990	982	974
2		1000	1000	1000	1000	999	999	998
3						1000	1000	1000

x \ λ	.30	.35	.40	.45	.50	.55	.60	.65
0	741	705	670	638	607	577	549	522
1	963	951	938	925	910	894	878	861
2	996	994	992	989	986	982	977	972
3	1000	1000	999	999	998	998	997	998
4			1000	1000	1000	1000	1000	999
5								1000

x \ λ	.70	.75	.80	.85	.90	.95	1.0	1.1
0	497	472	449	427	407	387	368	333
1	844	827	809	791	772	754	736	699
2	966	959	953	945	937	929	920	900
3	994	993	991	989	987	984	981	974
4	999	999	999	998	998	997	996	995
5	1000	1000	1000	1000	1000	1000	999	999
6							1000	1000

TABLE C ***(continued)***

x \ λ	1.2	1.3	1.4	1.5	1.6	1.7	1.8	1.9
0	301	273	247	223	202	183	165	150
1	663	627	592	558	525	493	463	434
2	879	857	833	809	783	757	731	704
3	966	957	946	934	921	907	891	875
4	992	989	986	981	976	970	964	956
5	998	998	997	996	994	992	990	987
6	1000	1000	999	999	999	998	997	997
7			1000	1000	1000	1000	999	999
8							1000	1000

x \ λ	2.0	2.2	2.4	2.6	2.8	3.0	3.2	3.4
0	135	111	091	074	061	050	041	033
1	406	355	308	267	231	199	171	147
2	677	623	570	518	469	423	380	340
3	857	819	779	736	692	647	603	558
4	947	928	904	877	848	815	781	744
5	983	975	964	951	935	916	895	871
6	995	993	988	983	976	966	955	942
7	999	998	997	995	992	988	983	977
8	1000	1000	999	999	998	997	994	992
9			1000	1000	999	999	998	997
10					1000	1000	1000	999
11								1000

x \ λ	3.6	3.8	4.0	4.2	4.4	4.6	4.8	5.0
0	027	022	018	015	012	010	008	007
1	126	107	092	078	066	056	048	040
2	303	269	238	210	185	163	143	125
3	515	473	433	395	359	326	294	265
4	706	668	629	590	551	513	476	440
5	844	816	785	753	720	686	651	616
6	927	909	889	867	844	818	791	762
7	969	960	949	936	921	905	887	867
8	988	984	979	972	964	955	944	932
9	996	994	992	989	985	980	975	968
10	999	998	997	996	994	992	990	986
11	1000	999	999	999	998	997	996	995
12		1000	1000	1000	999	999	999	998
13					1000	1000	1000	999
14								1000

TABLE C (continued)

x \ λ	5.2	5.4	5.6	5.8	6.0	6.2	6.4	6.6
0	006	005	004	003	002	002	002	001
1	034	029	024	021	017	015	012	010
2	109	095	082	072	062	054	046	040
3	238	213	191	170	151	134	119	105
4	406	373	342	313	285	259	235	213
5	581	546	512	478	446	414	384	355
6	732	702	670	638	606	574	542	511
7	845	822	797	771	744	716	687	658
8	918	903	886	867	847	826	803	780
9	960	951	941	929	916	902	886	869
10	982	977	972	965	957	949	939	927
11	993	990	988	984	980	975	969	963
12	997	996	995	993	991	989	986	982
13	999	999	998	997	996	995	994	992
14	1000	999	999	999	999	998	997	997
15		1000	1000	1000	999	999	999	999
16					1000	1000	1000	999
17								1000

x \ λ	6.8	7.0	7.2	7.4	7.6	7.8	8.0	8.5
0	001	001	001	001	001	000	000	000
1	009	007	006	005	004	004	003	002
2	034	030	025	022	019	016	014	009
3	093	082	072	063	055	048	042	030
4	192	173	156	140	125	112	100	074
5	327	301	276	253	231	210	191	150
6	480	450	420	392	365	338	313	256
7	628	599	569	539	510	481	453	386
8	755	729	703	676	648	620	593	523
9	850	830	810	788	765	741	717	653
10	915	901	887	871	854	835	816	763
11	955	947	937	926	915	902	888	849
12	978	973	967	961	954	945	936	909
13	990	987	984	980	976	971	966	949
14	996	994	993	991	989	986	983	973
15	998	998	997	996	995	993	992	986
16	999	999	999	998	998	997	996	993
17	1000	1000	999	999	999	999	998	997
18			1000	1000	1000	1000	999	999
19							1000	999
20								1000

TABLE C ***(continued)***

x \ λ	9.0	9.5	10.0	10.5	11.0	11.5	12.0	12.5
1	001	001	000	000	000	000	000	000
2	006	004	003	002	001	001	001	000
3	021	015	010	007	005	003	002	002
4	055	040	029	021	015	011	008	005
5	116	089	067	050	038	028	020	015
6	207	165	130	102	079	060	046	035
7	324	269	220	179	143	114	090	070
8	456	392	333	279	232	191	155	125
9	587	522	458	397	341	289	242	201
10	706	645	583	521	460	402	347	297
11	803	752	697	639	579	520	462	406
12	876	836	792	742	689	633	576	519
13	926	898	864	825	781	733	682	628
14	959	940	917	888	854	815	772	725
15	978	967	951	932	907	878	844	806
16	989	982	973	960	944	924	899	869
17	995	991	986	978	968	954	937	916
18	998	996	993	988	982	974	963	948
19	999	998	997	994	991	986	979	969
20	1000	999	998	997	995	992	988	983
21		1000	999	999	998	996	994	991
22			1000	999	999	998	997	995
23				1000	1000	999	999	998
24						1000	999	999
25							1000	999
26								1000

TABLE C (*continued*)

x \ λ	13.0	13.5	14.0	14.5	15	16	17	18
3	001	001	000	000	000	000	000	000
4	004	003	002	001	001	000	000	000
5	011	008	006	004	003	001	001	000
6	026	019	014	010	008	004	002	001
7	054	041	032	024	018	010	005	003
8	100	079	062	048	037	022	013	007
9	166	135	109	088	070	043	026	015
10	252	211	176	145	118	077	049	030
11	353	304	260	220	185	127	085	055
12	463	409	358	311	268	193	135	092
13	573	518	464	413	363	275	201	143
14	675	623	570	518	466	368	281	208
15	764	718	669	619	568	467	371	287
16	835	798	756	711	664	566	468	375
17	890	861	827	790	749	659	564	469
18	930	908	883	853	819	742	655	562
19	957	942	923	901	875	812	736	651
20	975	965	952	936	917	868	805	731
21	986	980	971	960	947	911	861	799
22	992	989	983	976	967	942	905	855
23	996	994	991	986	981	963	937	899
24	998	997	995	992	989	978	959	932
25	999	998	997	996	994	987	975	955
26	1000	999	999	998	997	993	985	972
27		1000	999	999	998	996	991	983
28			1000	999	999	998	995	990
29				1000	1000	999	997	994
30						999	999	997
31						1000	999	998
32							1000	999
33								1000

TABLE C ***(continued)***

x \ λ	19	20	21	22	23	24	25
6	001	000	000	000	000	000	000
7	002	001	000	000	000	000	000
8	004	002	001	001	000	000	000
9	009	005	003	002	001	000	000
10	018	011	006	004	002	001	001
11	035	021	013	008	004	003	001
12	061	039	025	015	009	005	003
13	098	066	043	028	017	011	006
14	150	105	072	048	031	020	012
15	215	157	111	077	052	034	022
16	292	221	163	117	082	056	038
17	378	297	227	169	123	087	060
18	469	381	302	232	175	128	092
19	561	470	384	306	238	180	134
20	647	559	471	387	310	243	185
21	725	644	558	472	389	314	247
22	793	721	640	556	472	392	318
23	849	787	716	637	555	473	394
24	893	843	782	712	635	554	473
25	927	888	838	777	708	632	553
26	951	922	883	832	772	704	629
27	969	948	917	877	827	768	700
28	980	966	944	913	873	823	763
29	988	978	963	940	908	868	818
30	993	987	976	959	936	904	863
31	996	992	985	973	956	932	900
32	998	995	991	983	971	953	929
33	999	997	994	989	981	969	950
34	999	999	997	994	988	979	966
35	1000	999	998	996	993	987	978
36		1000	999	998	996	992	985
37			999	999	997	995	991
38			1000	999	999	997	994
39				1000	999	998	997
40					1000	999	998
41						999	999
42						1000	999
43							1000

TABLE D Normal Curve Areas $P(z \leq z_0)$. Entries in the Body of the Table Are Areas Between $-\infty$ and z

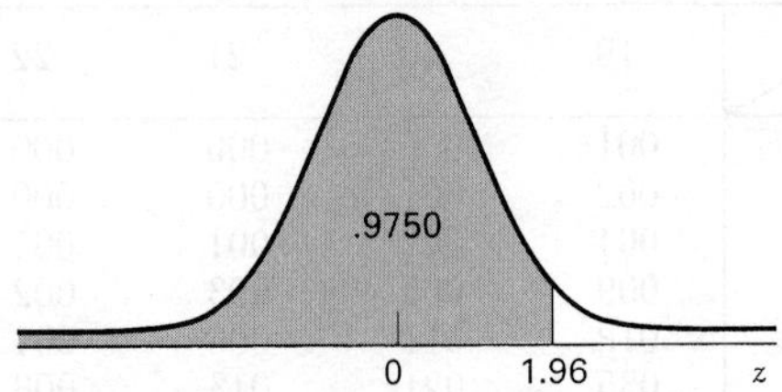

z	−0.09	−0.08	−0.07	−0.06	−0.05	−0.04	−0.03	−0.02	−0.01	0.00	z
−3.80	.0001	.0001	.0001	.0001	.0001	.0001	.0001	.0001	.0001	.0001	−3.80
−3.70	.0001	.0001	.0001	.0001	.0001	.0001	.0001	.0001	.0001	.0001	−3.70
−3.60	.0001	.0001	.0001	.0001	.0001	.0001	.0001	.0001	.0002	.0002	−3.60
−3.50	.0002	.0002	.0002	.0002	.0002	.0002	.0002	.0002	.0002	.0002	−3.50
−3.40	.0002	.0003	.0003	.0003	.0003	.0003	.0003	.0003	.0003	.0003	−3.40
−3.30	.0003	.0004	.0004	.0004	.0004	.0004	.0004	.0005	.0005	.0005	−3.30
−3.20	.0005	.0005	.0005	.0006	.0006	.0006	.0006	.0006	.0007	.0007	−3.20
−3.10	.0007	.0007	.0008	.0008	.0008	.0008	.0009	.0009	.0009	.0010	−3.10
−3.00	.0010	.0010	.0011	.0011	.0011	.0012	.0012	.0013	.0013	.0013	−3.00
−2.90	.0014	.0014	.0015	.0015	.0016	.0016	.0017	.0018	.0018	.0019	−2.90
−2.80	.0019	.0020	.0021	.0021	.0022	.0023	.0023	.0024	.0025	.0026	−2.80
−2.70	.0026	.0027	.0028	.0029	.0030	.0031	.0032	.0033	.0034	.0035	−2.70
−2.60	.0036	.0037	.0038	.0039	.0040	.0041	.0043	.0044	.0045	.0047	−2.60
−2.50	.0048	.0049	.0051	.0052	.0054	.0055	.0057	.0059	.0060	.0062	−2.50
−2.40	.0064	.0066	.0068	.0069	.0071	.0073	.0075	.0078	.0080	.0082	−2.40
−2.30	.0084	.0087	.0089	.0091	.0094	.0096	.0099	.0102	.0104	.0107	−2.30
−2.20	.0110	.0113	.0116	.0119	.0122	.0125	.0129	.0132	.0136	.0139	−2.20
−2.10	.0143	.0146	.0150	.0154	.0158	.0162	.0166	.0170	.0174	.0179	−2.10
−2.00	.0183	.0188	.0192	.0197	.0202	.0207	.0212	.0217	.0222	.0228	−2.00
−1.90	.0233	.0239	.0244	.0250	.0256	.0262	.0268	.0274	.0281	.0287	−1.90
−1.80	.0294	.0301	.0307	.0314	.0322	.0329	.0336	.0344	.0351	.0359	−1.80
−1.70	.0367	.0375	.0384	.0392	.0401	.0409	.0418	.0427	.0436	.0446	−1.70
−1.60	.0455	.0465	.0475	.0485	.0495	.0505	.0516	.0526	.0537	.0548	−1.60
−1.50	.0559	.0571	.0582	.0594	.0606	.0618	.0630	.0643	.0655	.0668	−1.50
−1.40	.0681	.0694	.0708	.0721	.0735	.0749	.0764	.0778	.0793	.0808	−1.40
−1.30	.0823	.0838	.0853	.0869	.0885	.0901	.0918	.0934	.0951	.0968	−1.30
−1.20	.0985	.1003	.1020	.1038	.1056	.1075	.1093	.1112	.1131	.1151	−1.20
−1.10	.1170	.1190	.1210	.1230	.1251	.1271	.1292	.1314	.1335	.1357	−1.10
−1.00	.1379	.1401	.1423	.1446	.1469	.1492	.1515	.1539	.1562	.1587	−1.00
−0.90	.1611	.1635	.1660	.1685	.1711	.1736	.1762	.1788	.1814	.1841	−0.90
−0.80	.1867	.1894	.1922	.1949	.1977	.2005	.2033	.2061	.2090	.2119	−0.80
−0.70	.2148	.2177	.2206	.2236	.2266	.2296	.2327	.2358	.2389	.2420	−0.70
−0.60	.2451	.2483	.2514	.2546	.2578	.2611	.2643	.2676	.2709	.2743	−0.60
−0.50	.2776	.2810	.2843	.2877	.2912	.2946	.2981	.3015	.3050	.3085	−0.50
−0.40	.3121	.3156	.3192	.3228	.3264	.3300	.3336	.3372	.3409	.3446	−0.40
−0.30	.3483	.3520	.3557	.3594	.3632	.3669	.3707	.3745	.3783	.3821	−0.30
−0.20	.3859	.3897	.3936	.3974	.4013	.4052	.4090	.4129	.4168	.4207	−0.20
−0.10	.4247	.4286	.4325	.4364	.4404	.4443	.4483	.4522	.4562	.4602	−0.10
0.00	.4641	.4681	.4721	.4761	.4801	.4840	.4880	.4920	.4960	.5000	0.00

TABLE D (*continued*)

z	0.00	0.01	0.02	0.03	0.04	0.05	0.06	0.07	0.08	0.09	z
0.00	.5000	.5040	.5080	.5120	.5160	.5199	.5239	.5279	.5319	.5359	0.00
0.10	.5398	.5438	.5478	.5517	.5557	.5596	.5636	.5675	.5714	.5753	0.10
0.20	.5793	.5832	.5871	.5910	.5948	.5987	.6026	.6064	.6103	.6141	0.20
0.30	.6179	.6217	.6255	.6293	.6331	.6368	.6406	.6443	.6480	.6517	0.30
0.40	.6554	.6591	.6628	.6664	.6700	.6736	.6772	.6808	.6844	.6879	0.40
0.50	.6915	.6950	.6985	.7019	.7054	.7088	.7123	.7157	.7190	.7224	0.50
0.60	.7257	.7291	.7324	.7357	.7389	.7422	.7454	.7486	.7517	.7549	0.60
0.70	.7580	.7611	.7642	.7673	.7704	.7734	.7764	.7794	.7823	.7852	0.70
0.80	.7881	.7910	.7939	.7967	.7995	.8023	.8051	.8078	.8106	.8133	0.80
0.90	.8159	.8186	.8212	.8238	.8264	.8289	.8315	.8340	.8365	.8389	0.90
1.00	.8413	.8438	.8461	.8485	.8508	.8531	.8554	.8577	.8599	.8621	1.00
1.10	.8643	.8665	.8686	.8708	.8729	.8749	.8770	.8790	.8810	.8830	1.10
1.20	.8849	.8869	.8888	.8907	.8925	.8944	.8962	.8980	.8997	.9015	1.20
1.30	.9032	.9049	.9066	.9082	.9099	.9115	.9131	.9147	.9162	.9177	1.30
1.40	.9192	.9207	.9222	.9236	.9251	.9265	.9279	.9292	.9306	.9319	1.40
1.50	.9332	.9345	.9357	.9370	.9382	.9394	.9406	.9418	.9429	.9441	1.50
1.60	.9452	.9463	.9474	.9484	.9495	.9505	.9515	.9525	.9535	.9545	1.60
1.70	.9554	.9564	.9573	.9582	.9591	.9599	.9608	.9616	.9625	.9633	1.70
1.80	.9641	.9649	.9656	.9664	.9671	.9678	.9686	.9693	.9699	.9706	1.80
1.90	.9713	.9719	.9726	.9732	.9738	.9744	.9750	.9756	.9761	.9767	1.90
2.00	.9772	.9778	.9783	.9788	.9793	.9798	.9803	.9808	.9812	.9817	2.00
2.10	.9821	.9826	.9830	.9834	.9838	.9842	.9846	.9850	.9854	.9857	2.10
2.20	.9861	.9864	.9868	.9871	.9875	.9878	.9881	.9884	.9887	.9890	2.20
2.30	.9893	.9896	.9898	.9901	.9904	.9906	.9909	.9911	.9913	.9916	2.30
2.40	.9918	.9920	.9922	.9925	.9927	.9929	.9931	.9932	.9934	.9936	2.40
2.50	.9938	.9940	.9941	.9943	.9945	.9946	.9948	.9949	.9951	.9952	2.50
2.60	.9953	.9955	.9956	.9957	.9959	.9960	.9961	.9962	.9963	.9964	2.60
2.70	.9965	.9966	.9967	.9968	.9969	.9970	.9971	.9972	.9973	.9974	2.70
2.80	.9974	.9975	.9976	.9977	.9977	.9978	.9979	.9979	.9980	.9981	2.80
2.90	.9981	.9982	.9982	.9983	.9984	.9984	.9985	.9985	.9986	.9986	2.90
3.00	.9987	.9987	.9987	.9988	.9988	.9989	.9989	.9989	.9990	.9990	3.00
3.10	.9990	.9991	.9991	.9991	.9992	.9992	.9992	.9992	.9993	.9993	3.10
3.20	.9993	.9993	.9994	.9994	.9994	.9994	.9994	.9995	.9995	.9995	3.20
3.30	.9995	.9995	.9995	.9996	.9996	.9996	.9996	.9996	.9996	.9997	3.30
3.40	.9997	.9997	.9997	.9997	.9997	.9997	.9997	.9997	.9997	.9998	3.40
3.50	.9998	.9998	.9998	.9998	.9998	.9998	.9998	.9998	.9998	.9998	3.50
3.60	.9998	.9998	.9999	.9999	.9999	.9999	.9999	.9999	.9999	.9999	3.60
3.70	.9999	.9999	.9999	.9999	.9999	.9999	.9999	.9999	.9999	.9999	3.70
3.80	.9999	.9999	.9999	.9999	.9999	.9999	.9999	.9999	.9999	.9999	3.80

TABLE E Percentiles of the *t* Distribution

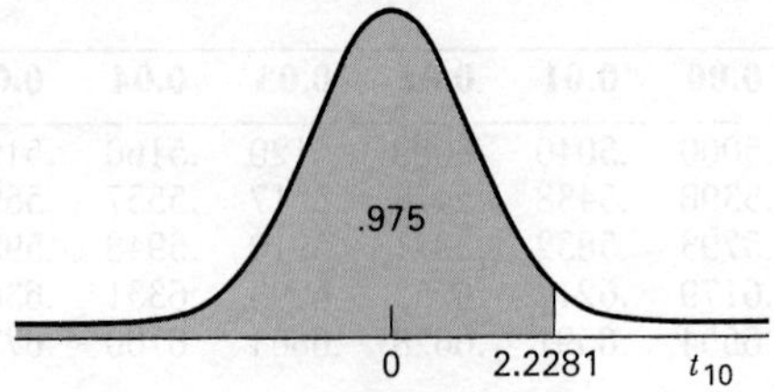

$p(t_{10} \leq 2.2281) = .975$

d.f.	$t_{.90}$	$t_{.95}$	$t_{.975}$	$t_{.99}$	$t_{.995}$
1	3.078	6.3138	12.706	31.821	63.657
2	1.886	2.9200	4.3027	6.965	9.9248
3	1.638	2.3534	3.1825	4.541	5.8409
4	1.533	2.1318	2.7764	3.747	4.6041
5	1.476	2.0150	2.5706	3.365	4.0321
6	1.440	1.9432	2.4469	3.143	3.7074
7	1.415	1.8946	2.3646	2.998	3.4995
8	1.397	1.8595	2.3060	2.896	3.3554
9	1.383	1.8331	2.2622	2.821	3.2498
10	1.372	1.8125	2.2281	2.764	3.1693
11	1.363	1.7959	2.2010	2.718	3.1058
12	1.356	1.7823	2.1788	2.681	3.0545
13	1.350	1.7709	2.1604	2.650	3.0123
14	1.345	1.7613	2.1448	2.624	2.9768
15	1.341	1.7530	2.1315	2.602	2.9467
16	1.337	1.7459	2.1199	2.583	2.9208
17	1.333	1.7396	2.1098	2.567	2.8982
18	1.330	1.7341	2.1009	2.552	2.8784
19	1.328	1.7291	2.0930	2.539	2.8609
20	1.325	1.7247	2.0860	2.528	2.8453
21	1.323	1.7207	2.0796	2.518	2.8314
22	1.321	1.7171	2.0739	2.508	2.8188
23	1.319	1.7139	2.0687	2.500	2.8073
24	1.318	1.7109	2.0639	2.492	2.7969
25	1.316	1.7081	2.0595	2.485	2.7874
26	1.315	1.7056	2.0555	2.479	2.7787
27	1.314	1.7033	2.0518	2.473	2.7707
28	1.313	1.7011	2.0484	2.467	2.7633
29	1.311	1.6991	2.0452	2.462	2.7564
30	1.310	1.6973	2.0423	2.457	2.7500
35	1.3062	1.6896	2.0301	2.438	2.7239
40	1.3031	1.6839	2.0211	2.423	2.7045
45	1.3007	1.6794	2.0141	2.412	2.6896
50	1.2987	1.6759	2.0086	2.403	2.6778
60	1.2959	1.6707	2.0003	2.390	2.6603
70	1.2938	1.6669	1.9945	2.381	2.6480
80	1.2922	1.6641	1.9901	2.374	2.6388
90	1.2910	1.6620	1.9867	2.368	2.6316
100	1.2901	1.6602	1.9840	2.364	2.6260
120	1.2887	1.6577	1.9799	2.358	2.6175
140	1.2876	1.6558	1.9771	2.353	2.6114
160	1.2869	1.6545	1.9749	2.350	2.6070
180	1.2863	1.6534	1.9733	2.347	2.6035
200	1.2858	1.6525	1.9719	2.345	2.6006
∞	1.282	1.645	1.96	2.326	2.576

TABLE F Percentiles of the Chi-Square Distribution

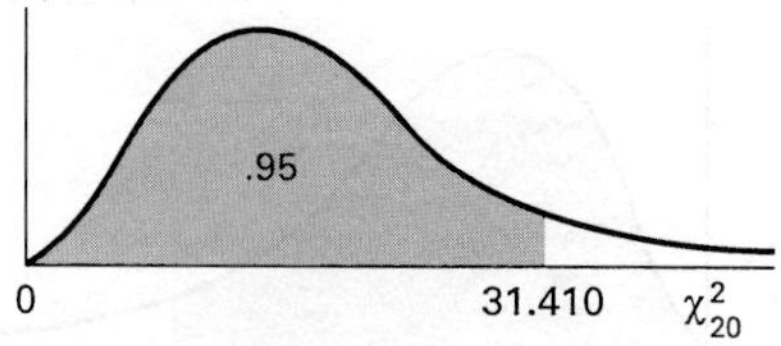

$p(\chi^2_{20} \le 31.410) = .95$

d.f.	$\chi^2_{.005}$	$\chi^2_{.025}$	$\chi^2_{.05}$	$\chi^2_{.90}$	$\chi^2_{.95}$	$\chi^2_{.975}$	$\chi^2_{.99}$	$\chi^2_{.995}$
1	.0000393	.000982	.00393	2.706	3.841	5.024	6.635	7.879
2	.0100	.0506	.103	4.605	5.991	7.378	9.210	10.597
3	.0717	.216	.352	6.251	7.815	9.348	11.345	12.838
4	.207	.484	.711	7.779	9.488	11.143	13.277	14.860
5	.412	.831	1.145	9.236	11.070	12.832	15.086	16.750
6	.676	1.237	1.635	10.645	12.592	14.449	16.812	18.548
7	.989	1.690	2.167	12.017	14.067	16.013	18.475	20.278
8	1.344	2.180	2.733	13.362	15.507	17.535	20.090	21.955
9	1.735	2.700	3.325	14.684	16.919	19.023	21.666	23.589
10	2.156	3.247	3.940	15.987	18.307	20.483	23.209	25.188
11	2.603	3.816	4.575	17.275	19.675	21.920	24.725	26.757
12	3.074	4.404	5.226	18.549	21.026	23.336	26.217	28.300
13	3.565	5.009	5.892	19.812	22.362	24.736	27.688	29.819
14	4.075	5.629	6.571	21.064	23.685	26.119	29.141	31.319
15	4.601	6.262	7.261	22.307	24.996	27.488	30.578	32.801
16	5.142	6.908	7.962	23.542	26.296	28.845	32.000	34.267
17	5.697	7.564	8.672	24.769	27.587	30.191	33.409	35.718
18	6.265	8.231	9.390	25.989	28.869	31.526	34.805	37.156
19	6.844	8.907	10.117	27.204	30.144	32.852	36.191	38.582
20	7.434	9.591	10.851	28.412	31.410	34.170	37.566	39.997
21	8.034	10.283	11.591	29.615	32.671	35.479	38.932	41.401
22	8.643	10.982	12.338	30.813	33.924	36.781	40.289	42.796
23	9.260	11.688	13.091	32.007	35.172	38.076	41.638	44.181
24	9.886	12.401	13.848	33.196	36.415	39.364	42.980	45.558
25	10.520	13.120	14.611	34.382	37.652	40.646	44.314	46.928
26	11.160	13.844	15.379	35.563	38.885	41.923	45.642	48.290
27	11.808	14.573	16.151	36.741	40.113	43.194	46.963	49.645
28	12.461	15.308	16.928	37.916	41.337	44.461	48.278	50.993
29	13.121	16.047	17.708	39.087	42.557	45.722	49.588	52.336
30	13.787	16.791	18.493	40.256	43.773	46.979	50.892	53.672
35	17.192	20.569	22.465	46.059	49.802	53.203	57.342	60.275
40	20.707	24.433	26.509	51.805	55.758	59.342	63.691	66.766
45	24.311	28.366	30.612	57.505	61.656	65.410	69.957	73.166
50	27.991	32.357	34.764	63.167	67.505	71.420	76.154	79.490
60	35.535	40.482	43.188	74.397	79.082	83.298	88.379	91.952
70	43.275	48.758	51.739	85.527	90.531	95.023	100.425	104.215
80	51.172	57.153	60.391	96.578	101.879	106.629	112.329	116.321
90	59.196	65.647	69.126	107.565	113.145	118.136	124.116	128.299
100	67.328	74.222	77.929	118.498	124.342	129.561	135.807	140.169

TABLE G Percentiles of the *F* Distribution

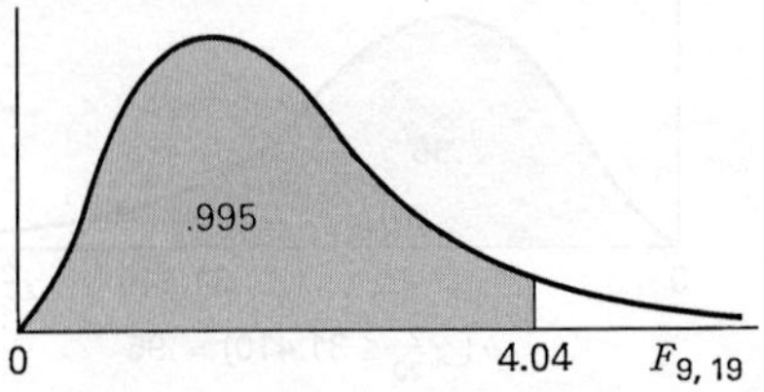

$p(F_{9,19} \leq 4.04) = .995$

$F_{.995}$

Denominator Degrees of Freedom	Numerator Degrees of Freedom								
	1	2	3	4	5	6	7	8	9
1	16211	20000	21615	22500	23056	23437	23715	23925	24091
2	198.5	199.0	199.2	199.2	199.3	199.3	199.4	199.4	199.4
3	55.55	49.80	47.47	46.19	45.39	44.84	44.43	44.13	43.88
4	31.33	26.28	24.26	23.15	22.46	21.97	21.62	21.35	21.14
5	22.78	18.31	16.53	15.56	14.94	14.51	14.20	13.96	13.77
6	18.63	14.54	12.92	12.03	11.46	11.07	10.79	10.57	10.39
7	16.24	12.40	10.88	10.05	9.52	9.16	8.89	8.68	8.51
8	14.69	11.04	9.60	8.81	8.30	7.95	7.69	7.50	7.34
9	13.61	10.11	8.72	7.96	7.47	7.13	6.88	6.69	6.54
10	12.83	9.43	8.08	7.34	6.87	6.54	6.30	6.12	5.97
11	12.23	8.91	7.60	6.88	6.42	6.10	5.86	5.68	5.54
12	11.75	8.51	7.23	6.52	6.07	5.76	5.52	5.35	5.20
13	11.37	8.19	6.93	6.23	5.79	5.48	5.25	5.08	4.94
14	11.06	7.92	6.68	6.00	5.56	5.26	5.03	4.86	4.72
15	10.80	7.70	6.48	5.80	5.37	5.07	4.85	4.67	4.54
16	10.58	7.51	6.30	5.64	5.21	4.91	4.69	4.52	4.38
17	10.38	7.35	6.16	5.50	5.07	4.78	4.56	4.39	4.25
18	10.22	7.21	6.03	5.37	4.96	4.66	4.44	4.28	4.14
19	10.07	7.09	5.92	5.27	4.85	4.56	4.34	4.18	4.04
20	9.94	6.99	5.82	5.17	4.76	4.47	4.26	4.09	3.96
21	9.83	6.89	5.73	5.09	4.68	4.39	4.18	4.01	3.88
22	9.73	6.81	5.65	5.02	4.61	4.32	4.11	3.94	3.81
23	9.63	6.73	5.58	4.95	4.54	4.26	4.05	3.88	3.75
24	9.55	6.66	5.52	4.89	4.49	4.20	3.99	3.83	3.69
25	9.48	6.60	5.46	4.84	4.43	4.15	3.94	3.78	3.64
26	9.41	6.54	5.41	4.79	4.38	4.10	3.89	3.73	3.60
27	9.34	6.49	5.36	4.74	4.34	4.06	3.85	3.69	3.56
28	9.28	6.44	5.32	4.70	4.30	4.02	3.81	3.65	3.52
29	9.23	6.40	5.28	4.66	4.26	3.98	3.77	3.61	3.48
30	9.18	6.35	5.24	4.62	4.23	3.95	3.74	3.58	3.45
40	8.83	6.07	4.98	4.37	3.99	3.71	3.51	3.35	3.22
60	8.49	5.79	4.73	4.14	3.76	3.49	3.29	3.13	3.01
120	8.18	5.54	4.50	3.92	3.55	3.28	3.09	2.93	2.81
∞	7.88	5.30	4.28	3.72	3.35	3.09	2.90	2.74	2.62

TABLE G ***(continued)***

Denominator Degrees of Freedom	Numerator Degrees of Freedom									
	10	12	15	20	24	30	40	60	120	∞
1	24224	24426	24630	24836	24940	25044	25148	25253	25359	25465
2	199.4	199.4	199.4	199.4	199.5	199.5	199.5	199.5	199.5	199.5
3	43.69	43.39	43.08	42.78	42.62	42.47	42.31	42.15	41.99	41.83
4	20.97	20.70	20.44	20.17	20.03	19.89	19.75	19.61	19.47	19.32
5	13.62	13.38	13.15	12.90	12.78	12.66	12.53	12.40	12.27	12.14
6	10.25	10.03	9.81	9.59	9.47	9.36	9.24	9.12	9.00	8.88
7	8.38	8.18	7.97	7.75	7.65	7.53	7.42	7.31	7.19	7.08
8	7.21	7.01	6.81	6.61	6.50	6.40	6.29	6.18	6.06	5.95
9	6.42	6.23	6.03	5.83	5.73	5.62	5.52	5.41	5.30	5.19
10	5.85	5.66	5.47	5.27	5.17	5.07	4.97	4.86	4.75	4.64
11	5.42	5.24	5.05	4.86	4.76	4.65	4.55	4.44	4.34	4.23
12	5.09	4.91	4.72	4.53	4.43	4.33	4.23	4.12	4.01	3.90
13	4.82	4.64	4.46	4.27	4.17	4.07	3.97	3.87	3.76	3.65
14	4.60	4.43	4.25	4.06	3.96	3.86	3.76	3.66	3.55	3.44
15	4.42	4.25	4.07	3.88	3.79	3.69	3.58	3.48	3.37	3.26
16	4.27	4.10	3.92	3.73	3.64	3.54	3.44	3.33	3.22	3.11
17	4.14	3.97	3.79	3.61	3.51	3.41	3.31	3.21	3.10	2.98
18	4.03	3.86	3.68	3.50	3.40	3.30	3.20	3.10	2.99	2.87
19	3.93	3.76	3.59	3.40	3.31	3.21	3.11	3.00	2.89	2.78
20	3.85	3.68	3.50	3.32	3.22	3.12	3.02	2.92	2.81	2.69
21	3.77	3.60	3.43	3.24	3.15	3.05	2.95	2.84	2.73	2.61
22	3.70	3.54	3.36	3.18	3.08	2.98	2.88	2.77	2.66	2.55
23	3.64	3.47	3.30	3.12	3.02	2.92	2.82	2.71	2.60	2.48
24	3.59	3.42	3.25	3.06	2.97	2.87	2.77	2.66	2.55	2.43
25	3.54	3.37	3.20	3.01	2.92	2.82	2.72	2.61	2.50	2.38
26	3.49	3.33	3.15	2.97	2.87	2.77	2.67	2.56	2.45	2.33
27	3.45	3.28	3.11	2.93	2.83	2.73	2.63	2.52	2.41	2.29
28	3.41	3.25	3.07	2.89	2.79	2.69	2.59	2.48	2.37	2.25
29	3.38	3.21	3.04	2.86	2.76	2.66	2.56	2.45	2.33	2.21
30	3.34	3.18	3.01	2.82	2.73	2.63	2.52	2.42	2.30	2.18
40	3.12	2.95	2.78	2.60	2.50	2.40	2.30	2.18	2.06	1.93
60	2.90	2.74	2.57	2.39	2.29	2.19	2.08	1.96	1.83	1.69
120	2.71	2.54	2.37	2.19	2.09	1.98	1.87	1.75	1.61	1.43
∞	2.52	2.36	2.19	2.00	1.90	1.79	1.67	1.53	1.36	1.00

TABLE G (*continued*)

$F_{.99}$

Denominator Degrees of Freedom	Numerator Degrees of Freedom								
	1	2	3	4	5	6	7	8	9
1	4052	4999.5	5403	5625	5764	5859	5928	5981	6022
2	98.50	99.00	99.17	99.25	99.30	99.33	99.36	99.37	99.39
3	34.12	30.82	29.46	28.71	28.24	27.91	27.67	27.49	27.35
4	21.20	18.00	16.69	15.98	15.52	15.21	14.98	14.80	14.66
5	16.26	13.27	12.06	11.39	10.97	10.67	10.46	10.29	10.16
6	13.75	10.92	9.78	9.15	8.75	8.47	8.26	8.10	7.98
7	12.25	9.55	8.45	7.85	7.46	7.19	6.99	6.84	6.72
8	11.26	8.65	7.59	7.01	6.63	6.37	6.18	6.03	5.91
9	10.56	8.02	6.99	6.42	6.06	5.80	5.61	5.47	5.35
10	10.04	7.56	6.55	5.99	5.64	5.39	5.20	5.06	4.94
11	9.65	7.21	6.22	5.67	5.32	5.07	4.89	4.74	4.63
12	9.33	6.93	5.95	5.41	5.06	4.82	4.64	4.50	4.39
13	9.07	6.70	5.74	5.21	4.86	4.62	4.44	4.30	4.19
14	8.86	6.51	5.56	5.04	4.69	4.46	4.28	4.14	4.03
15	8.68	6.36	5.42	4.89	4.56	4.32	4.14	4.00	3.89
16	8.53	6.23	5.29	4.77	4.44	4.20	4.03	3.89	3.78
17	8.40	6.11	5.18	4.67	4.34	4.10	3.93	3.79	3.68
18	8.29	6.01	5.09	4.58	4.25	4.01	3.84	3.71	3.60
19	8.18	5.93	5.01	4.50	4.17	3.94	3.77	3.63	3.52
20	8.10	5.85	4.94	4.43	4.10	3.87	3.70	3.56	3.46
21	8.02	5.78	4.87	4.37	4.04	3.81	3.64	3.51	3.40
22	7.95	5.72	4.82	4.31	3.99	3.76	3.59	3.45	3.35
23	7.88	5.66	4.76	4.26	3.94	3.71	3.54	3.41	3.30
24	7.82	5.61	4.72	4.22	3.90	3.67	3.50	3.36	3.26
25	7.77	5.57	4.68	4.18	3.85	3.63	3.46	3.32	3.22
26	7.72	5.53	4.64	4.14	3.82	3.59	3.42	3.29	3.18
27	7.68	5.49	4.60	4.11	3.78	3.56	3.39	3.26	3.15
28	7.64	5.45	4.57	4.07	3.75	3.53	3.36	3.23	3.12
29	7.60	5.42	4.54	4.04	3.73	3.50	3.33	3.20	3.09
30	7.56	5.39	4.51	4.02	3.70	3.47	3.30	3.17	3.07
40	7.31	5.18	4.31	3.83	3.51	3.29	3.12	2.99	2.89
60	7.08	4.98	4.13	3.65	3.34	3.12	2.95	2.82	2.72
120	6.85	4.79	3.95	3.48	3.17	2.96	2.79	2.66	2.56
∞	6.63	4.61	3.78	3.32	3.02	2.80	2.64	2.51	2.41

TABLE G (continued)

Denominator Degrees of Freedom	Numerator Degrees of Freedom									
	10	12	15	20	24	30	40	60	120	∞
1	6056	6106	6157	6209	6235	6261	6287	6313	6339	6366
2	99.40	99.42	99.43	99.45	99.46	99.47	99.47	99.48	99.49	99.50
3	27.23	27.05	26.87	26.69	26.60	26.50	26.41	26.32	26.22	26.13
4	14.55	14.37	14.20	14.02	13.93	13.84	13.75	13.65	13.56	13.46
5	10.05	9.89	9.72	9.55	9.47	9.38	9.29	9.20	9.11	9.02
6	7.87	7.72	7.56	7.40	7.31	7.23	7.14	7.06	6.97	6.88
7	6.62	6.47	6.31	6.16	6.07	5.99	5.91	5.82	5.74	5.65
8	5.81	5.67	5.52	5.36	5.28	5.20	5.12	5.03	4.95	4.86
9	5.26	5.11	4.96	4.81	4.73	4.65	4.57	4.48	4.40	4.31
10	4.85	4.71	4.56	4.41	4.33	4.25	4.17	4.08	4.00	3.91
11	4.54	4.40	4.25	4.10	4.02	3.94	3.86	3.78	3.69	3.60
12	4.30	4.16	4.01	3.86	3.78	3.70	3.62	3.54	3.45	3.36
13	4.10	3.96	3.82	3.66	3.59	3.51	3.43	3.34	3.25	3.17
14	3.94	3.80	3.66	3.51	3.43	3.35	3.27	3.18	3.09	3.00
15	3.80	3.67	3.52	3.37	3.29	3.21	3.13	3.05	2.96	2.87
16	3.69	3.55	3.41	3.26	3.18	3.10	3.02	2.93	2.84	2.75
17	3.59	3.46	3.31	3.16	3.08	3.00	2.92	2.83	2.75	2.65
18	3.51	3.37	3.23	3.08	3.00	2.92	2.84	2.75	2.66	2.57
19	3.43	3.30	3.15	3.00	2.92	2.84	2.76	2.67	2.58	2.49
20	3.37	3.23	3.09	2.94	2.86	2.78	2.69	2.61	2.52	2.42
21	3.31	3.17	3.03	2.88	2.80	2.72	2.64	2.55	2.46	2.36
22	3.26	3.12	2.98	2.83	2.75	2.67	2.58	2.50	2.40	2.31
23	3.21	3.07	2.93	2.78	2.70	2.62	2.54	2.45	2.35	2.26
24	3.17	3.03	2.89	2.74	2.66	2.58	2.49	2.40	2.31	2.21
25	3.13	2.99	2.85	2.70	2.62	2.54	2.45	2.36	2.27	2.17
26	3.09	2.96	2.81	2.66	2.58	2.50	2.42	2.33	2.23	2.13
27	3.06	2.93	2.78	2.63	2.55	2.47	2.38	2.29	2.20	2.10
28	3.03	2.90	2.75	2.60	2.52	2.44	2.35	2.26	2.17	2.06
29	3.00	2.87	2.73	2.57	2.49	2.41	2.33	2.23	2.14	2.03
30	2.98	2.84	2.70	2.55	2.47	2.39	2.30	2.21	2.11	2.01
40	2.80	2.66	2.52	2.37	2.29	2.20	2.11	2.02	1.92	1.80
60	2.63	2.50	2.35	2.20	2.12	2.03	1.94	1.84	1.73	1.60
120	2.47	2.34	2.19	2.03	1.95	1.86	1.76	1.66	1.53	1.38
∞	2.32	2.18	2.04	1.88	1.79	1.70	1.59	1.47	1.32	1.00

TABLE G (continued)

	$F_{.975}$								
Denominator Degrees of Freedom	**Numerator Degrees of Freedom**								
	1	**2**	**3**	**4**	**5**	**6**	**7**	**8**	**9**
1	647.8	799.5	864.2	899.6	921.8	937.1	948.2	956.7	963.3
2	38.51	39.00	39.17	39.25	39.30	39.33	39.36	39.37	39.39
3	17.44	16.04	15.44	15.10	14.88	14.73	14.62	14.54	14.47
4	12.22	10.65	9.98	9.60	9.36	9.20	9.07	8.98	8.90
5	10.01	8.43	7.76	7.39	7.15	6.98	6.85	6.76	6.68
6	8.81	7.26	6.60	6.23	5.99	5.82	5.70	5.60	5.52
7	8.07	6.54	5.89	5.52	5.29	5.12	4.99	4.90	4.82
8	7.57	6.06	5.42	5.05	4.82	4.65	4.53	4.43	4.36
9	7.21	5.71	5.08	4.72	4.48	4.32	4.20	4.10	4.03
10	6.94	5.46	4.83	4.47	4.24	4.07	3.95	3.85	3.78
11	6.72	5.26	4.63	4.28	4.04	3.88	3.76	3.66	3.59
12	6.55	5.10	4.47	4.12	3.89	3.73	3.61	3.51	3.44
13	6.41	4.97	4.35	4.00	3.77	3.60	3.48	3.39	3.31
14	6.30	4.86	4.24	3.89	3.66	3.50	3.38	3.29	3.21
15	6.20	4.77	4.15	3.80	3.58	3.41	3.29	3.20	3.12
16	6.12	4.69	4.08	3.73	3.50	3.34	3.22	3.12	3.05
17	6.04	4.62	4.01	3.66	3.44	3.28	3.16	3.06	2.98
18	5.98	4.56	3.95	3.61	3.38	3.22	3.10	3.01	2.93
19	5.92	4.51	3.90	3.56	3.33	3.17	3.05	2.96	2.88
20	5.87	4.46	3.86	3.51	3.29	3.13	3.01	2.91	2.84
21	5.83	4.42	3.82	3.48	3.25	3.09	2.97	2.87	2.80
22	5.79	4.38	3.78	3.44	3.22	3.05	2.93	2.84	2.76
23	5.75	4.35	3.75	3.41	3.18	3.02	2.90	2.81	2.73
24	5.72	4.32	3.72	3.38	3.15	2.99	2.87	2.78	2.70
25	5.69	4.29	3.69	3.35	3.13	2.97	2.85	2.75	2.68
26	5.66	4.27	3.67	3.33	3.10	2.94	2.82	2.73	2.65
27	5.63	4.24	3.65	3.31	3.08	2.92	2.80	2.71	2.63
28	5.61	4.22	3.63	3.29	3.06	2.90	2.78	2.69	2.61
29	5.59	4.20	3.61	3.27	3.04	2.88	2.76	2.67	2.59
30	5.57	4.18	3.59	3.25	3.03	2.87	2.75	2.65	2.57
40	5.42	4.05	3.46	3.13	2.90	2.74	2.62	2.53	2.45
60	5.29	3.93	3.34	3.01	2.79	2.63	2.51	2.41	2.33
120	5.15	3.80	3.23	2.89	2.67	2.52	2.39	2.30	2.22
∞	5.02	3.69	3.12	2.79	2.57	2.41	2.29	2.19	2.11

TABLE G (*continued*)

Denominator Degrees of Freedom	Numerator Degrees of Freedom									
	10	12	15	20	24	30	40	60	120	∞
1	968.6	976.7	984.9	993.1	997.2	1001	1006	1010	1014	1018
2	39.40	39.41	39.43	39.45	39.46	39.46	39.47	39.48	39.49	39.50
3	14.42	14.34	14.25	14.17	14.12	14.08	14.04	13.99	13.95	13.90
4	8.84	8.75	8.66	8.56	8.51	8.46	8.41	8.36	8.31	8.26
5	6.62	6.52	6.43	6.33	6.28	6.23	6.18	6.12	6.07	6.02
6	5.46	5.37	5.27	5.17	5.12	5.07	5.01	4.96	4.90	4.85
7	4.76	4.67	4.57	4.47	4.42	4.36	4.31	4.25	4.20	4.14
8	4.30	4.20	4.10	4.00	3.95	3.89	3.84	3.78	3.73	3.67
9	3.96	3.87	3.77	3.67	3.61	3.56	3.51	3.45	3.39	3.33
10	3.72	3.62	3.52	3.42	3.37	3.31	3.26	3.20	3.14	3.08
11	3.53	3.43	3.33	3.23	3.17	3.12	3.06	3.00	2.94	2.88
12	3.37	3.28	3.18	3.07	3.02	2.96	2.91	2.85	2.79	2.72
13	3.25	3.15	3.05	2.95	2.89	2.84	2.78	2.72	2.66	2.60
14	3.15	3.05	2.95	2.84	2.79	2.73	2.67	2.61	2.55	2.49
15	3.06	2.96	2.86	2.76	2.70	2.64	2.59	2.52	2.46	2.40
16	2.99	2.89	2.79	2.68	2.63	2.57	2.51	2.45	2.38	2.32
17	2.92	2.82	2.72	2.62	2.56	2.50	2.44	2.38	2.32	2.25
18	2.87	2.77	2.67	2.56	2.50	2.44	2.38	2.32	2.26	2.19
19	2.82	2.72	2.62	2.51	2.45	2.39	2.33	2.27	2.20	2.13
20	2.77	2.68	2.57	2.46	2.41	2.35	2.29	2.22	2.16	2.09
21	2.73	2.64	2.53	2.42	2.37	2.31	2.25	2.18	2.11	2.04
22	2.70	2.60	2.50	2.39	2.33	2.27	2.21	2.14	2.08	2.00
23	2.67	2.57	2.47	2.36	2.30	2.24	2.18	2.11	2.04	1.97
24	2.64	2.54	2.44	2.33	2.27	2.21	2.15	2.08	2.01	1.94
25	2.61	2.51	2.41	2.30	2.24	2.18	2.12	2.05	1.98	1.91
26	2.59	2.49	2.39	2.28	2.22	2.16	2.09	2.03	1.95	1.88
27	2.57	2.47	2.36	2.25	2.19	2.13	2.07	2.00	1.93	1.85
28	2.55	2.45	2.34	2.23	2.17	2.11	2.05	1.98	1.91	1.83
29	2.53	2.43	2.32	2.21	2.15	2.09	2.03	1.96	1.89	1.81
30	2.51	2.41	2.31	2.20	2.14	2.07	2.01	1.94	1.87	1.79
40	2.39	2.29	2.18	2.07	2.01	1.94	1.88	1.80	1.72	1.64
60	2.27	2.17	2.06	1.94	1.88	1.82	1.74	1.67	1.58	1.48
120	2.16	2.05	1.94	1.82	1.76	1.69	1.61	1.53	1.43	1.31
∞	2.05	1.94	1.83	1.71	1.64	1.57	1.48	1.39	1.27	1.00

TABLE G (*continued*)

$F_{.95}$

Denominator Degrees of Freedom	Numerator Degrees of Freedom								
	1	2	3	4	5	6	7	8	9
1	161.4	199.5	215.7	224.6	230.2	234.0	236.8	238.9	240.5
2	18.51	19.00	19.16	19.25	19.30	19.33	19.35	19.37	19.38
3	10.13	9.55	9.28	9.12	9.01	8.94	8.89	8.85	8.81
4	7.71	6.94	6.59	6.39	6.26	6.16	6.09	6.04	6.00
5	6.61	5.79	5.41	5.19	5.05	4.95	4.88	4.82	4.77
6	5.99	5.14	4.76	4.53	4.39	4.28	4.21	4.15	4.10
7	5.59	4.74	4.35	4.12	3.97	3.87	3.79	3.73	3.68
8	5.32	4.46	4.07	3.84	3.69	3.58	3.50	3.44	3.39
9	5.12	4.26	3.86	3.63	3.48	3.37	3.29	3.23	3.18
10	4.96	4.10	3.71	3.48	3.33	3.22	3.14	3.07	3.02
11	4.84	3.98	3.59	3.36	3.20	3.09	3.01	2.95	2.90
12	4.75	3.89	3.49	3.26	3.11	3.00	2.91	2.85	2.80
13	4.67	3.81	3.41	3.18	3.03	2.92	2.83	2.77	2.71
14	4.60	3.74	3.34	3.11	2.96	2.85	2.76	2.70	2.65
15	4.54	3.68	3.29	3.06	2.90	2.79	2.71	2.64	2.59
16	4.49	3.63	3.24	3.01	2.85	2.74	2.66	2.59	2.54
17	4.45	3.59	3.20	2.96	2.81	2.70	2.61	2.55	2.49
18	4.41	3.55	3.16	2.93	2.77	2.66	2.58	2.51	2.46
19	4.38	3.52	3.13	2.90	2.74	2.63	2.54	2.48	2.42
20	4.35	3.49	3.10	2.87	2.71	2.60	2.51	2.45	2.39
21	4.32	3.47	3.07	2.84	2.68	2.57	2.49	2.42	2.37
22	4.30	3.44	3.05	2.82	2.66	2.55	2.46	2.40	2.34
23	4.28	3.42	3.03	2.80	2.64	2.53	2.44	2.37	2.32
24	4.26	3.40	3.01	2.78	2.62	2.51	2.42	2.36	2.30
25	4.24	3.39	2.99	2.76	2.60	2.49	2.40	2.34	2.28
26	4.23	3.37	2.98	2.74	2.59	2.47	2.39	2.32	2.27
27	4.21	3.35	2.96	2.73	2.57	2.46	2.37	2.31	2.25
28	4.20	3.34	2.95	2.71	2.56	2.45	2.36	2.29	2.24
29	4.18	3.33	2.93	2.70	2.55	2.43	2.35	2.28	2.22
30	4.17	3.32	2.92	2.69	2.53	2.42	2.33	2.27	2.21
40	4.08	3.23	2.84	2.61	2.45	2.34	2.25	2.18	2.12
60	4.00	3.15	2.76	2.53	2.37	2.25	2.17	2.10	2.04
120	3.92	3.07	2.68	2.45	2.29	2.17	2.09	2.02	1.96
∞	3.84	3.00	2.60	2.37	2.21	2.10	2.01	1.94	1.88

TABLE G (*continued*)

Denominator Degrees of Freedom	Numerator Degrees of Freedom									
	10	12	15	20	24	30	40	60	120	∞
1	241.9	243.9	245.9	248.0	249.1	250.1	251.1	252.2	253.3	254.3
2	19.40	19.41	19.43	19.45	19.45	19.46	19.47	19.48	19.49	19.50
3	8.79	8.74	8.70	8.66	8.64	8.62	8.59	8.57	8.55	8.53
4	5.96	5.91	5.86	5.80	5.77	5.75	5.72	5.69	5.66	5.63
5	4.74	4.68	4.62	4.56	4.53	4.50	4.46	4.43	4.40	4.36
6	4.06	4.00	3.94	3.87	3.84	3.81	3.77	3.74	3.70	3.67
7	3.64	3.57	3.51	3.44	3.41	3.38	3.34	3.30	3.27	3.23
8	3.35	3.28	3.22	3.15	3.12	3.08	3.04	3.01	2.97	2.93
9	3.14	3.07	3.01	2.94	2.90	2.86	2.83	2.79	2.75	2.71
10	2.98	2.91	2.85	2.77	2.74	2.70	2.66	2.62	2.58	2.54
11	2.85	2.79	2.72	2.65	2.61	2.57	2.53	2.49	2.45	2.40
12	2.75	2.69	2.62	2.54	2.51	2.47	2.43	2.38	2.34	2.30
13	2.67	2.60	2.53	2.46	2.42	2.38	2.34	2.30	2.25	2.21
14	2.60	2.53	2.46	2.39	2.35	2.31	2.27	2.22	2.18	2.13
15	2.54	2.48	2.40	2.33	2.29	2.25	2.20	2.16	2.11	2.07
16	2.49	2.42	2.35	2.28	2.24	2.19	2.15	2.11	2.06	2.01
17	2.45	2.38	2.31	2.23	2.19	2.15	2.10	2.06	2.01	1.96
18	2.41	2.34	2.27	2.19	2.15	2.11	2.06	2.02	1.97	1.92
19	2.38	2.31	2.23	2.16	2.11	2.07	2.03	1.98	1.93	1.88
20	2.35	2.28	2.20	2.12	2.08	2.04	1.99	1.95	1.90	1.84
21	2.32	2.25	2.18	2.10	2.05	2.01	1.96	1.92	1.87	1.81
22	2.30	2.23	2.15	2.07	2.03	1.98	1.94	1.89	1.84	1.78
23	2.27	2.20	2.13	2.05	2.01	1.96	1.91	1.86	1.81	1.76
24	2.25	2.18	2.11	2.03	1.98	1.94	1.89	1.84	1.79	1.73
25	2.24	2.16	2.09	2.01	1.96	1.92	1.87	1.82	1.77	1.71
26	2.22	2.15	2.07	1.99	1.95	1.90	1.85	1.80	1.75	1.69
27	2.20	2.13	2.06	1.97	1.93	1.88	1.84	1.79	1.73	1.67
28	2.19	2.12	2.04	1.96	1.91	1.87	1.82	1.77	1.71	1.65
29	2.18	2.10	2.03	1.94	1.90	1.85	1.81	1.75	1.70	1.64
30	2.16	2.09	2.01	1.93	1.89	1.84	1.79	1.74	1.68	1.62
40	2.08	2.00	1.92	1.84	1.79	1.74	1.69	1.64	1.58	1.51
60	1.99	1.92	1.84	1.75	1.70	1.65	1.59	1.53	1.47	1.39
120	1.91	1.83	1.75	1.66	1.61	1.55	1.50	1.43	1.35	1.25
∞	1.83	1.75	1.67	1.57	1.52	1.46	1.39	1.32	1.22	1.00

TABLE G (continued)

$F_{.90}$

Denominator Degrees of Freedom	Numerator Degrees of Freedom								
	1	2	3	4	5	6	7	8	9
1	39.86	49.50	53.59	55.83	57.24	58.20	58.91	59.44	59.86
2	8.53	9.00	9.16	9.24	9.29	9.33	9.35	9.37	9.38
3	5.54	5.46	5.39	5.34	5.31	5.28	5.27	5.25	5.24
4	4.54	4.32	4.19	4.11	4.05	4.01	3.98	3.95	3.94
5	4.06	3.78	3.62	3.52	3.45	3.40	3.37	3.34	3.32
6	3.78	3.46	3.29	3.18	3.11	3.05	3.01	2.98	2.96
7	3.59	3.26	3.07	2.96	2.88	2.83	2.78	2.75	2.72
8	3.46	3.11	2.92	2.81	2.73	2.67	2.62	2.59	2.56
9	3.36	3.01	2.81	2.69	2.61	2.55	2.51	2.47	2.44
10	3.29	2.92	2.73	2.61	2.52	2.46	2.41	2.38	2.35
11	3.23	2.86	2.66	2.54	2.45	2.39	2.34	2.30	2.27
12	3.18	2.81	2.61	2.48	2.39	2.33	2.28	2.24	2.21
13	3.14	2.76	2.56	2.43	2.35	2.28	2.23	2.20	2.16
14	3.10	2.73	2.52	2.39	2.31	2.24	2.19	2.15	2.12
15	3.07	2.70	2.49	2.36	2.27	2.21	2.16	2.12	2.09
16	3.05	2.67	2.46	2.33	2.24	2.18	2.13	2.09	2.06
17	3.03	2.64	2.44	2.31	2.22	2.15	2.10	2.06	2.03
18	3.01	2.62	2.42	2.29	2.20	2.13	2.08	2.04	2.00
19	2.99	2.61	2.40	2.27	2.18	2.11	2.06	2.02	1.98
20	2.97	2.59	2.38	2.25	2.16	2.09	2.04	2.00	1.96
21	2.96	2.57	2.36	2.23	2.14	2.08	2.02	1.98	1.95
22	2.95	2.56	2.35	2.22	2.13	2.06	2.01	1.97	1.93
23	2.94	2.55	2.34	2.21	2.11	2.05	1.99	1.95	1.92
24	2.93	2.54	2.33	2.19	2.10	2.04	1.98	1.94	1.91
25	2.92	2.53	2.32	2.18	2.09	2.02	1.97	1.93	1.89
26	2.91	2.52	2.31	2.17	2.08	2.01	1.96	1.92	1.88
27	2.90	2.51	2.30	2.17	2.07	2.00	1.95	1.91	1.87
28	2.89	2.50	2.29	2.16	2.06	2.00	1.94	1.90	1.87
29	2.89	2.50	2.28	2.15	2.06	1.99	1.93	1.89	1.86
30	2.88	2.49	2.28	2.14	2.05	1.98	1.93	1.88	1.85
40	2.84	2.44	2.23	2.09	2.00	1.93	1.87	1.83	1.79
60	2.79	2.39	2.18	2.04	1.95	1.87	1.82	1.77	1.74
120	2.75	2.35	2.13	1.99	1.90	1.82	1.77	1.72	1.68
∞	2.71	2.30	2.08	1.94	1.85	1.77	1.72	1.67	1.63

TABLE G (continued)

Denominator Degrees of Freedom	Numerator Degrees of Freedom									
	10	12	15	20	24	30	40	60	120	∞
1	60.19	60.71	61.22	61.74	62.00	62.26	62.53	62.79	63.06	63.33
2	9.39	9.41	9.42	9.44	9.45	9.46	9.47	9.47	9.48	9.49
3	5.23	5.22	5.20	5.18	5.18	5.17	5.16	5.15	5.14	5.13
4	3.92	3.90	3.87	3.84	3.83	3.82	3.80	3.79	3.78	3.76
5	3.30	3.27	3.24	3.21	3.19	3.17	3.16	3.14	3.12	3.10
6	2.94	2.90	2.87	2.84	2.82	2.80	2.78	2.76	2.74	2.72
7	2.70	2.67	2.63	2.59	2.58	2.56	2.54	2.51	2.49	2.47
8	2.54	2.50	2.46	2.42	2.40	2.38	2.36	2.34	2.32	2.29
9	2.42	2.38	2.34	2.30	2.28	2.25	2.23	2.21	2.18	2.16
10	2.32	2.28	2.24	2.20	2.18	2.16	2.13	2.11	2.08	2.06
11	2.25	2.21	2.17	2.12	2.10	2.08	2.05	2.03	2.00	1.97
12	2.19	2.15	2.10	2.06	2.04	2.01	1.99	1.96	1.93	1.90
13	2.14	2.10	2.05	2.01	1.98	1.96	1.93	1.90	1.88	1.85
14	2.10	2.05	2.01	1.96	1.94	1.91	1.89	1.86	1.83	1.80
15	2.06	2.02	1.97	1.92	1.90	1.87	1.85	1.82	1.79	1.76
16	2.03	1.99	1.94	1.89	1.87	1.84	1.81	1.78	1.75	1.72
17	2.00	1.96	1.91	1.86	1.84	1.81	1.78	1.75	1.72	1.69
18	1.98	1.93	1.89	1.84	1.81	1.78	1.75	1.72	1.69	1.66
19	1.96	1.91	1.86	1.81	1.79	1.76	1.73	1.70	1.67	1.63
20	1.94	1.89	1.84	1.79	1.77	1.74	1.71	1.68	1.64	1.61
21	1.92	1.87	1.83	1.78	1.75	1.72	1.69	1.66	1.62	1.59
22	1.90	1.86	1.81	1.76	1.73	1.70	1.67	1.64	1.60	1.57
23	1.89	1.84	1.80	1.74	1.72	1.69	1.66	1.62	1.59	1.55
24	1.88	1.83	1.78	1.73	1.70	1.67	1.64	1.61	1.57	1.53
25	1.87	1.82	1.77	1.72	1.69	1.66	1.63	1.59	1.56	1.52
26	1.86	1.81	1.76	1.71	1.68	1.65	1.61	1.58	1.54	1.50
27	1.85	1.80	1.75	1.70	1.67	1.64	1.60	1.57	1.53	1.49
28	1.84	1.79	1.74	1.69	1.66	1.63	1.59	1.56	1.52	1.48
29	1.83	1.78	1.73	1.68	1.65	1.62	1.58	1.55	1.51	1.47
30	1.82	1.77	1.72	1.67	1.64	1.61	1.57	1.54	1.50	1.46
40	1.76	1.71	1.66	1.61	1.57	1.54	1.51	1.47	1.42	1.38
60	1.71	1.66	1.60	1.54	1.51	1.48	1.44	1.40	1.35	1.29
120	1.65	1.60	1.55	1.48	1.45	1.41	1.37	1.32	1.26	1.19
∞	1.60	1.55	1.49	1.42	1.38	1.34	1.30	1.24	1.17	1.00

TABLE H Percentage Points of the Studentized Range for 2 Through 20 Treatments Upper 5% Points

Error df	2	3	4	5	6	7	8	9	10
1	17.97	26.98	32.82	37.08	40.41	43.12	45.40	47.36	49.07
2	6.08	8.33	9.80	10.88	11.74	12.44	13.03	13.54	13.99
3	4.50	5.91	6.82	7.50	8.04	8.48	8.85	9.18	9.46
4	3.93	5.04	5.76	6.29	6.71	7.05	7.35	7.60	7.83
5	3.64	4.60	5.22	5.67	6.03	6.33	6.58	6.80	6.99
6	3.46	4.34	4.90	5.30	5.63	5.90	6.12	6.32	6.49
7	3.34	4.16	4.68	5.06	5.36	5.61	5.82	6.00	6.16
8	3.26	4.04	4.53	4.89	5.17	5.40	5.60	5.77	5.92
9	3.20	3.95	4.41	4.76	5.02	5.24	5.43	5.59	5.74
10	3.15	3.88	4.33	4.65	4.91	5.12	5.30	5.46	5.60
11	3.11	3.82	4.26	4.57	4.82	5.03	5.20	5.35	5.49
12	3.08	3.77	4.20	4.51	4.75	4.95	5.12	5.27	5.39
13	3.06	3.73	4.15	4.45	4.69	4.88	5.05	5.19	5.32
14	3.03	3.70	4.11	4.41	4.64	4.83	4.99	5.13	5.25
15	3.01	3.67	4.08	4.37	4.59	4.78	4.94	5.08	5.20
16	3.00	3.65	4.05	4.33	4.56	4.74	4.90	5.03	5.15
17	2.98	3.63	4.02	4.30	4.52	4.70	4.86	4.99	5.11
18	2.97	3.61	4.00	4.28	4.49	4.67	4.82	4.96	5.07
19	2.96	3.59	3.98	4.25	4.47	4.65	4.79	4.92	5.04
20	2.95	3.58	3.96	4.23	4.45	4.62	4.77	4.90	5.01
24	2.92	3.53	3.90	4.17	4.37	4.54	4.68	4.81	4.92
30	2.89	3.49	3.85	4.10	4.30	4.46	4.60	4.72	4.82
40	2.86	3.44	3.79	4.04	4.23	4.39	4.52	4.63	4.73
60	2.83	3.40	3.74	3.98	4.16	4.31	4.44	4.55	4.65
120	2.80	3.36	3.68	3.92	4.10	4.24	4.36	4.47	4.56
∞	2.77	3.31	3.63	3.86	4.03	4.17	4.29	4.39	4.47

Error df	11	12	13	14	15	16	17	18	19	20
1	50.59	51.96	53.20	54.33	55.36	56.32	57.22	58.04	58.83	59.56
2	14.39	14.75	15.08	15.38	15.65	15.91	16.14	16.37	16.57	16.77
3	9.72	9.95	10.15	10.35	10.52	10.69	10.84	10.98	11.11	11.24
4	8.03	8.21	8.37	8.52	8.66	8.79	8.91	9.03	9.13	9.23
5	7.17	7.32	7.47	7.60	7.72	7.83	7.93	8.03	8.12	8.21
6	6.65	6.79	6.92	7.03	7.14	7.24	7.34	7.43	7.51	7.59
7	6.30	6.43	6.55	6.66	6.76	6.85	6.94	7.02	7.10	7.17
8	6.05	6.18	6.29	6.39	6.48	6.57	6.65	6.73	6.80	6.87
9	5.87	5.98	6.09	6.19	6.28	6.36	6.44	6.51	6.58	6.64
10	5.72	5.83	5.93	6.03	6.11	6.19	6.27	6.34	6.40	6.47
11	5.61	5.71	5.81	5.90	5.98	6.06	6.13	6.20	6.27	6.33
12	5.51	5.61	5.71	5.80	5.88	5.95	6.02	6.09	6.15	6.21
13	5.43	5.53	5.63	5.71	5.79	5.86	5.93	5.99	6.05	6.11
14	5.36	5.46	5.55	5.64	5.71	5.79	5.85	5.91	5.97	6.03
15	5.31	5.40	5.49	5.57	5.65	5.72	5.78	5.85	5.90	5.96

TABLE H (*continued*)

Error df	11	12	13	14	15	16	17	18	19	20
16	5.26	5.35	5.44	5.52	5.59	5.66	5.73	5.79	5.84	5.90
17	5.21	5.31	5.39	5.47	5.54	5.61	5.67	5.73	5.79	5.84
18	5.17	5.27	5.35	5.43	5.50	5.57	5.63	5.69	5.74	5.79
19	5.14	5.23	5.31	5.39	5.46	5.53	5.59	5.65	5.70	5.75
20	5.11	5.20	5.28	5.36	5.43	5.49	5.55	5.61	5.66	5.71
24	5.01	5.10	5.18	5.25	5.32	5.38	5.44	5.49	5.55	5.59
30	4.92	5.00	5.08	5.15	5.21	5.27	5.33	5.38	5.43	5.47
40	4.82	4.90	4.98	5.04	5.11	5.16	5.22	5.27	5.31	5.36
60	4.73	4.81	4.88	4.94	5.00	5.06	5.11	5.15	5.20	5.24
120	4.64	4.71	4.78	4.84	4.90	4.95	5.00	5.04	5.09	5.13
∞	4.55	4.62	4.68	4.74	4.80	4.85	4.89	4.93	4.97	5.01

Upper 1% Points

Error df	2	3	4	5	6	7	8	9	10
1	90.03	135.0	164.3	185.6	202.2	215.8	227.2	237.0	245.6
2	14.04	19.02	22.29	24.72	26.63	28.20	29.53	30.68	31.69
3	8.26	10.62	12.17	13.33	14.24	15.00	15.64	16.20	16.69
4	6.51	8.12	9.17	9.96	10.58	11.10	11.55	11.93	12.27
5	5.70	6.98	7.80	8.42	8.91	9.32	9.67	9.97	10.24
6	5.24	6.33	7.03	7.56	7.97	8.32	8.61	8.87	9.10
7	4.95	5.92	6.54	7.01	7.37	7.68	7.94	8.17	8.37
8	4.75	5.64	6.20	6.62	6.96	7.24	7.47	7.68	7.86
9	4.60	5.43	5.96	6.35	6.66	6.91	7.13	7.33	7.49
10	4.48	5.27	5.77	6.14	6.43	6.67	6.87	7.05	7.21
11	4.39	5.15	5.62	5.97	6.25	6.48	6.67	6.84	6.99
12	4.32	5.05	5.50	5.84	6.10	6.32	6.51	6.67	6.81
13	4.26	4.96	5.40	5.73	5.98	6.19	6.37	6.53	6.67
14	4.21	4.89	5.32	5.63	5.88	6.08	6.26	6.41	6.54
15	4.17	4.84	5.25	5.56	5.80	5.99	6.16	6.31	6.44
16	4.13	4.79	5.19	5.49	5.72	5.92	6.08	6.22	6.35
17	4.10	4.74	5.14	5.43	5.66	5.85	6.01	6.15	6.27
18	4.07	4.70	5.09	5.38	5.60	5.79	5.94	6.08	6.20
19	4.05	4.67	5.05	5.33	5.55	5.73	5.89	6.02	6.14
20	4.02	4.64	5.02	5.29	5.51	5.69	5.84	5.97	6.09
24	3.96	4.55	4.91	5.17	5.37	5.54	5.69	5.81	5.92
30	3.89	4.45	4.80	5.05	5.24	5.40	5.54	5.65	5.76
40	3.82	4.37	4.70	4.93	5.11	5.26	5.39	5.50	5.60
60	3.76	4.28	4.59	4.82	4.99	5.13	5.25	5.36	5.45
120	3.70	4.20	4.50	4.71	4.87	5.01	5.12	5.21	5.30
∞	3.64	4.12	4.40	4.60	4.76	4.88	4.99	5.08	5.16

TABLE H (continued)

Error df	11	12	13	14	15	16	17	18	19	20
1	253.2	260.0	266.2	271.8	277.0	281.8	286.3	290.4	294.3	298.0
2	32.59	33.40	34.13	34.81	35.43	36.00	36.53	37.03	37.50	37.95
3	17.13	17.53	17.89	18.22	18.52	18.81	19.07	19.32	19.55	19.77
4	12.57	12.84	13.09	13.32	13.53	13.73	13.91	14.08	14.24	14.40
5	10.48	10.70	10.89	11.08	11.24	11.40	11.55	11.68	11.81	11.93
6	9.30	9.48	9.65	9.81	9.95	10.08	10.21	10.32	10.43	10.54
7	8.55	8.71	8.86	9.00	9.12	9.24	9.35	9.46	9.55	9.65
8	8.03	8.18	8.31	8.44	8.55	8.66	8.76	8.85	8.94	9.03
9	7.65	7.78	7.91	8.03	8.13	8.23	8.33	8.41	8.49	8.57
10	7.36	7.49	7.60	7.71	7.81	7.91	7.99	8.08	8.15	8.23
11	7.13	7.25	7.36	7.46	7.56	7.65	7.73	7.81	7.88	7.95
12	6.94	7.06	7.17	7.26	7.36	7.44	7.52	7.59	7.66	7.73
13	6.79	6.90	7.01	7.10	7.19	7.27	7.35	7.42	7.48	7.55
14	6.66	6.77	6.87	6.96	7.05	7.13	7.20	7.27	7.33	7.39
15	6.55	6.66	6.76	6.84	6.93	7.00	7.07	7.14	7.20	7.26
16	6.46	6.56	6.66	6.74	6.82	6.90	6.97	7.03	7.09	7.15
17	6.38	6.48	6.57	6.66	6.73	6.81	6.87	6.94	7.00	7.05
18	6.31	6.41	6.50	6.58	6.65	6.73	6.79	6.85	6.91	6.97
19	6.25	6.34	6.43	6.51	6.58	6.65	6.72	6.78	6.84	6.89
20	6.19	6.28	6.37	6.45	6.52	6.59	6.65	6.71	6.77	6.82
24	6.02	6.11	6.19	6.26	6.33	6.39	6.45	6.51	6.56	6.61
30	5.85	5.93	6.01	6.08	6.14	6.20	6.26	6.31	6.36	6.41
40	5.69	5.76	5.83	5.90	5.96	6.02	6.07	6.12	6.16	6.21
60	5.53	5.60	5.67	5.73	5.78	5.84	5.89	5.93	5.97	6.01
120	5.37	5.44	5.50	5.56	5.61	5.66	5.71	5.75	5.79	5.83
∞	5.23	5.29	5.35	5.40	5.45	5.49	5.54	5.57	5.61	5.65

TABLE I Transformation of r to z (the Body of the Table Contains Values of $z = .5[\ln(1 + r)/(1 - r)] = \tanh^{-1} r$ for Corresponding Values of r, the Correlation Coefficient)

r	.00	.01	.02	.03	.04	.05	.06	.07	.08	.09
.0	.00000	.01000	.02000	.03001	.04002	.05004	.06007	.07012	.08017	.09024
.1	.10034	.11045	.12058	.13074	.14093	.15114	.16139	.17167	.18198	.19234
.2	.20273	.21317	.22366	.23419	.24477	.25541	.26611	.27686	.28768	.29857
.3	.30952	.32055	.33165	.34283	.35409	.36544	.37689	.38842	.40006	.41180
.4	.42365	.43561	.44769	.45990	.47223	.48470	.49731	.51007	.52298	.53606
.5	.54931	.56273	.57634	.59014	.60415	.61838	.63283	.64752	.66246	.67767
.6	.69315	.70892	.72500	.74142	.75817	.77530	.79281	.81074	.82911	.84795
.7	.86730	.88718	.90764	.92873	.95048	.97295	.99621	1.02033	1.04537	1.07143
.8	1.09861	1.12703	1.15682	1.18813	1.22117	1.25615	1.29334	1.33308	1.37577	1.42192
.9	1.47222	1.52752	1.58902	1.65839	1.73805	1.83178	1.94591	2.09229	2.29756	2.64665

TABLE J Significance Tests in a 2 × 2 Contingency Table[a]

	a	Probability 0.05	0.025	0.01	0.005
$A = 3$ $B = 3$	3	$\mathbf{0}_{.050}$	—	—	—
$A = 4$ $B = 4$	4	$\mathbf{0}_{.014}$	$\mathbf{0}_{.014}$	—	—
3	4	$\mathbf{0}_{.029}$	—	—	—
$A = 5$ $B = 5$	5	$\mathbf{1}_{.024}$	$\mathbf{1}_{.024}$	$\mathbf{0}_{.004}$	$\mathbf{0}_{.004}$
	4	$\mathbf{0}_{.024}$	$\mathbf{0}_{.024}$	—	—
4	5	$\mathbf{1}_{.048}$	$\mathbf{0}_{.008}$	$\mathbf{0}_{.008}$	—
	4	$\mathbf{0}_{.040}$	—	—	—
3	5	$\mathbf{0}_{.018}$	$\mathbf{0}_{.018}$	—	—
2	5	$\mathbf{0}_{.048}$	—	—	—
$A = 6$ $B = 6$	6	$\mathbf{2}_{.030}$	$\mathbf{1}_{.008}$	$\mathbf{1}_{.008}$	$\mathbf{0}_{.001}$
	5	$\mathbf{1}_{.040}$	$\mathbf{0}_{.008}$	$\mathbf{0}_{.008}$	—
	4	$\mathbf{0}_{.030}$	—	—	—
5	6	$\mathbf{1}_{.015+}$	$\mathbf{1}_{.015+}$	$\mathbf{0}_{.002}$	$\mathbf{0}_{.002}$
	5	$\mathbf{0}_{.013}$	$\mathbf{0}_{.013}$	—	—
	4	$\mathbf{0}_{.045+}$	—	—	—
4	6	$\mathbf{1}_{.033}$	$\mathbf{0}_{.005-}$	$\mathbf{0}_{.005-}$	$\mathbf{0}_{.005-}$
	5	$\mathbf{0}_{.024}$	$\mathbf{0}_{.024}$	—	—
3	6	$\mathbf{0}_{.012}$	$\mathbf{0}_{.012}$	—	—
	5	$\mathbf{0}_{.048}$	—	—	—
2	6	$\mathbf{0}_{.036}$	—	—	—
$A = 7$ $B = 7$	7	$\mathbf{3}_{.035-}$	$\mathbf{2}_{.010+}$	$\mathbf{1}_{.002}$	$\mathbf{1}_{.002}$
	6	$\mathbf{1}_{.015-}$	$\mathbf{1}_{.015-}$	$\mathbf{0}_{.002}$	$\mathbf{0}_{.002}$
	5	$\mathbf{0}_{.010+}$	$\mathbf{0}_{.010+}$	—	—
	4	$\mathbf{0}_{.035-}$	—	—	—
6	7	$\mathbf{2}_{.021}$	$\mathbf{2}_{.021}$	$\mathbf{1}_{.005-}$	$\mathbf{1}_{.005-}$
	6	$\mathbf{1}_{.025+}$	$\mathbf{0}_{.004}$	$\mathbf{0}_{.004}$	$\mathbf{0}_{.004}$
	5	$\mathbf{0}_{.016}$	$\mathbf{0}_{.016}$	—	—
	4	$\mathbf{0}_{.049}$	—	—	—
5	7	$\mathbf{2}_{.045+}$	$\mathbf{1}_{.010+}$	$\mathbf{0}_{.001}$	$\mathbf{0}_{.001}$
	6	$\mathbf{1}_{.045+}$	$\mathbf{0}_{.008}$	$\mathbf{0}_{.008}$	—
	5	$\mathbf{0}_{.027}$	—	—	—
4	7	$\mathbf{1}_{.024}$	$\mathbf{1}_{.024}$	$\mathbf{0}_{.003}$	$\mathbf{0}_{.003}$
	6	$\mathbf{0}_{.015+}$	$\mathbf{0}_{.015+}$	—	—
	5	$\mathbf{0}_{.045+}$	—	—	—
3	7	$\mathbf{0}_{.008}$	$\mathbf{0}_{.008}$	$\mathbf{0}_{.008}$	—
	6	$\mathbf{0}_{.033}$	—	—	—
2	7	$\mathbf{0}_{.028}$	—	—	—
$A = 8$ $B = 8$	8	$\mathbf{4}_{.038}$	$\mathbf{3}_{.013}$	$\mathbf{2}_{.003}$	$\mathbf{2}_{.003}$
	7	$\mathbf{2}_{.020}$	$\mathbf{2}_{.020}$	$\mathbf{1}_{.005+}$	$\mathbf{0}_{.001}$
	6	$\mathbf{1}_{.020}$	$\mathbf{1}_{.020}$	$\mathbf{0}_{.003}$	$\mathbf{0}_{.003}$
	5	$\mathbf{0}_{.013}$	$\mathbf{0}_{.013}$	—	—
	4	$\mathbf{0}_{.038}$	—	—	—
$A = 8$ $B = 7$	8	$\mathbf{3}_{.026}$	$\mathbf{2}_{.007}$	$\mathbf{2}_{.007}$	$\mathbf{1}_{.001}$
	7	$\mathbf{2}_{.035-}$	$\mathbf{1}_{.009}$	$\mathbf{1}_{.009}$	$\mathbf{0}_{.001}$
	6	$\mathbf{1}_{.032}$	$\mathbf{0}_{.006}$	$\mathbf{0}_{.006}$	—
	5	$\mathbf{0}_{.019}$	$\mathbf{0}_{.019}$	—	—
6	8	$\mathbf{2}_{.015-}$	$\mathbf{2}_{.015-}$	$\mathbf{1}_{.003}$	$\mathbf{1}_{.003}$

[a]Bold type, for given a, A, and B, shows the value of b ($< a$), which is just significant at the probability level quoted (single-tail test). Small type, for given A, B, and $r = a + b$, shows the exact probability (if there is independence) that b is equal to or less than the integer shown in bold type.

TABLE J ***(continued)***

		Probability			
	a	**0.05**	**0.025**	**0.01**	**0.005**
	7	$\mathbf{1}_{.016}$	$\mathbf{1}_{.016}$	$\mathbf{0}_{.002}$	$\mathbf{0}_{.002}$
	6	$\mathbf{0}_{.009}$	$\mathbf{0}_{.009}$	$\mathbf{0}_{.009}$	—
	5	$\mathbf{0}_{.028}$	—	—	—
5	8	$\mathbf{2}_{.035^-}$	$\mathbf{1}_{.007}$	$\mathbf{1}_{.007}$	$\mathbf{0}_{.001}$
	7	$\mathbf{1}_{.032}$	$\mathbf{0}_{.005^-}$	$\mathbf{0}_{.005^-}$	$\mathbf{0}_{.005^-}$
	6	$\mathbf{0}_{.016}$	$\mathbf{0}_{.016}$	—	—
	5	$\mathbf{0}_{.044}$	—	—	—
4	8	$\mathbf{1}_{.018}$	$\mathbf{1}_{.018}$	$\mathbf{0}_{.002}$	$\mathbf{0}_{.002}$
	7	$\mathbf{0}_{.010^+}$	$\mathbf{0}_{.010^+}$	—	—
	6	$\mathbf{0}_{.030}$	—	—	—
3	8	$\mathbf{0}_{.006}$	$\mathbf{0}_{.006}$	$\mathbf{0}_{.006}$	—
	7	$\mathbf{0}_{.024}$	$\mathbf{0}_{.024}$	—	—
2	8	$\mathbf{0}_{.022}$	$\mathbf{0}_{.022}$	—	—
$A = 9$ $B = 9$	9	$\mathbf{5}_{.041}$	$\mathbf{4}_{.015^-}$	$\mathbf{3}_{.005^-}$	$\mathbf{3}_{.005^-}$
	8	$\mathbf{3}_{.025^-}$	$\mathbf{3}_{.025^-}$	$\mathbf{2}_{.008}$	$\mathbf{1}_{.002}$
	7	$\mathbf{2}_{.028}$	$\mathbf{1}_{.008}$	$\mathbf{1}_{.008}$	$\mathbf{0}_{.001}$
	6	$\mathbf{1}_{.025^-}$	$\mathbf{1}_{.025^-}$	$\mathbf{0}_{.005^-}$	$\mathbf{0}_{.005^-}$
	5	$\mathbf{0}_{.015^-}$	$\mathbf{0}_{.015^-}$	—	—
	4	$\mathbf{0}_{.041}$	—	—	—
8	9	$\mathbf{4}_{.029}$	$\mathbf{3}_{.009}$	$\mathbf{3}_{.009}$	$\mathbf{2}_{.002}$
	8	$\mathbf{3}_{.043}$	$\mathbf{2}_{.013}$	$\mathbf{1}_{.003}$	$\mathbf{1}_{.003}$
	7	$\mathbf{2}_{.044}$	$\mathbf{1}_{.012}$	$\mathbf{0}_{.002}$	$\mathbf{0}_{.002}$
	6	$\mathbf{1}_{.036}$	$\mathbf{0}_{.007}$	$\mathbf{0}_{.007}$	—
	5	$\mathbf{0}_{.020}$	$\mathbf{0}_{.020}$	—	—
7	9	$\mathbf{3}_{.019}$	$\mathbf{3}_{.019}$	$\mathbf{2}_{.005^-}$	$\mathbf{2}_{.005^-}$
	8	$\mathbf{2}_{.024}$	$\mathbf{2}_{.024}$	$\mathbf{1}_{.006}$	$\mathbf{0}_{.001}$
	7	$\mathbf{1}_{.020}$	$\mathbf{1}_{.020}$	$\mathbf{0}_{.003}$	$\mathbf{0}_{.003}$
	6	$\mathbf{0}_{.010^+}$	$\mathbf{0}_{.010^+}$	—	—
	5	$\mathbf{0}_{.029}$	—	—	—
6	9	$\mathbf{3}_{.044}$	$\mathbf{2}_{.011}$	$\mathbf{1}_{.002}$	$\mathbf{1}_{.002}$
	8	$\mathbf{2}_{.047}$	$\mathbf{1}_{.011}$	$\mathbf{0}_{.001}$	$\mathbf{0}_{.001}$
	7	$\mathbf{1}_{.035^-}$	$\mathbf{0}_{.006}$	$\mathbf{0}_{.006}$	—
	6	$\mathbf{0}_{.017}$	$\mathbf{0}_{.017}$	—	—
	5	$\mathbf{0}_{.042}$	—	—	—
5	9	$\mathbf{2}_{.027}$	$\mathbf{1}_{.005^-}$	$\mathbf{1}_{.005^-}$	$\mathbf{1}_{.005^-}$
	8	$\mathbf{1}_{.023}$	$\mathbf{1}_{.023}$	$\mathbf{0}_{.003}$	$\mathbf{0}_{.003}$
	7	$\mathbf{0}_{.010^+}$	$\mathbf{0}_{.010^+}$	—	—
	6	$\mathbf{0}_{.028}$	—	—	—
4	9	$\mathbf{1}_{.014}$	$\mathbf{1}_{.014}$	$\mathbf{0}_{.001}$	$\mathbf{0}_{.001}$
	8	$\mathbf{0}_{.007}$	$\mathbf{0}_{.007}$	$\mathbf{0}_{.007}$	—
	7	$\mathbf{0}_{.021}$	$\mathbf{0}_{.021}$	—	—
	6	$\mathbf{0}_{.049}$	—	—	—
3	9	$\mathbf{1}_{.045^+}$	$\mathbf{0}_{.005^-}$	$\mathbf{0}_{.005^-}$	$\mathbf{0}_{.005^-}$
	8	$\mathbf{0}_{.018}$	$\mathbf{0}_{.018}$	—	—
	7	$\mathbf{0}_{.045^+}$	—	—	—
2	9	$\mathbf{0}_{.018}$	$\mathbf{0}_{.018}$	—	—
$A = 10$ $B = 10$	10	$\mathbf{6}_{.043}$	$\mathbf{5}_{.016}$	$\mathbf{4}_{.005^+}$	$\mathbf{3}_{.002}$
	9	$\mathbf{4}_{.029}$	$\mathbf{3}_{.010^-}$	$\mathbf{3}_{.010^-}$	$\mathbf{2}_{.003}$
	8	$\mathbf{3}_{.035^-}$	$\mathbf{2}_{.012}$	$\mathbf{1}_{.003}$	$\mathbf{1}_{.003}$
	7	$\mathbf{2}_{.035^-}$	$\mathbf{1}_{.010^-}$	$\mathbf{1}_{.010^-}$	$\mathbf{0}_{.002}$

TABLE J ***(continued)***

		Probability			
	a	0.05	0.025	0.01	0.005
	6	$\mathbf{1}_{.029}$	$\mathbf{0}_{.005+}$	$\mathbf{0}_{.005+}$	—
	5	$\mathbf{0}_{.016}$	$\mathbf{0}_{.016}$	—	—
	4	$\mathbf{0}_{.043}$	—	—	—
$A = 10$ $B = 9$	10	$\mathbf{5}_{.033}$	$\mathbf{4}_{.011}$	$\mathbf{3}_{.003}$	$\mathbf{3}_{.003}$
	9	$\mathbf{4}_{.050-}$	$\mathbf{3}_{.017}$	$\mathbf{2}_{.005-}$	$\mathbf{2}_{.005-}$
	8	$\mathbf{2}_{.019}$	$\mathbf{2}_{.019}$	$\mathbf{1}_{.004}$	$\mathbf{1}_{.004}$
	7	$\mathbf{1}_{.015-}$	$\mathbf{1}_{.015-}$	$\mathbf{0}_{.002}$	$\mathbf{0}_{.002}$
	6	$\mathbf{1}_{.040}$	$\mathbf{0}_{.008}$	$\mathbf{0}_{.008}$	—
	5	$\mathbf{0}_{.022}$	$\mathbf{0}_{.022}$	—	—
8	10	$\mathbf{4}_{.023}$	$\mathbf{4}_{.023}$	$\mathbf{3}_{.007}$	$\mathbf{2}_{.002}$
	9	$\mathbf{3}_{.032}$	$\mathbf{2}_{.009}$	$\mathbf{2}_{.009}$	$\mathbf{1}_{.002}$
	8	$\mathbf{2}_{.031}$	$\mathbf{1}_{.008}$	$\mathbf{1}_{.008}$	$\mathbf{0}_{.001}$
	7	$\mathbf{1}_{.023}$	$\mathbf{1}_{.023}$	$\mathbf{0}_{.004}$	$\mathbf{0}_{.004}$
	6	$\mathbf{0}_{.011}$	$\mathbf{0}_{.011}$	—	—
	5	$\mathbf{0}_{.029}$	—	—	—
7	10	$\mathbf{3}_{.015-}$	$\mathbf{3}_{.015-}$	$\mathbf{2}_{.003}$	$\mathbf{2}_{.003}$
	9	$\mathbf{2}_{.018}$	$\mathbf{2}_{.018}$	$\mathbf{1}_{.004}$	$\mathbf{1}_{.004}$
	8	$\mathbf{1}_{.013}$	$\mathbf{1}_{.013}$	$\mathbf{0}_{.002}$	$\mathbf{0}_{.002}$
	7	$\mathbf{1}_{.036}$	$\mathbf{0}_{.006}$	$\mathbf{0}_{.006}$	—
	6	$\mathbf{0}_{.017}$	$\mathbf{0}_{.017}$	—	—
	5	$\mathbf{0}_{.041}$	—	—	—
6	10	$\mathbf{3}_{.036}$	$\mathbf{2}_{.008}$	$\mathbf{2}_{.008}$	$\mathbf{1}_{.001}$
	9	$\mathbf{2}_{.036}$	$\mathbf{1}_{.008}$	$\mathbf{1}_{.008}$	$\mathbf{0}_{.001}$
	8	$\mathbf{1}_{.024}$	$\mathbf{1}_{.024}$	$\mathbf{0}_{.003}$	$\mathbf{0}_{.003}$
	7	$\mathbf{0}_{.010+}$	$\mathbf{0}_{.010+}$	—	—
	6	$\mathbf{0}_{.026}$	—	—	—
5	10	$\mathbf{2}_{.022}$	$\mathbf{2}_{.022}$	$\mathbf{1}_{.004}$	$\mathbf{1}_{.004}$
	9	$\mathbf{1}_{.017}$	$\mathbf{1}_{.017}$	$\mathbf{0}_{.002}$	$\mathbf{0}_{.002}$
	8	$\mathbf{1}_{.047}$	$\mathbf{0}_{.007}$	$\mathbf{0}_{.007}$	—
	7	$\mathbf{0}_{.019}$	$\mathbf{0}_{.019}$	—	—
	6	$\mathbf{0}_{.042}$	—	—	—
4	10	$\mathbf{1}_{.011}$	$\mathbf{1}_{.011}$	$\mathbf{0}_{.001}$	$\mathbf{0}_{.001}$
	9	$\mathbf{1}_{.041}$	$\mathbf{0}_{.005-}$	$\mathbf{0}_{.005-}$	$\mathbf{0}_{.005-}$
	8	$\mathbf{0}_{.015-}$	$\mathbf{0}_{.015-}$	—	—
	7	$\mathbf{0}_{.035-}$	—	—	—
3	10	$\mathbf{1}_{.038}$	$\mathbf{0}_{.003}$	$\mathbf{0}_{.003}$	$\mathbf{0}_{.003}$
	9	$\mathbf{0}_{.014}$	$\mathbf{0}_{.014}$	—	—
	8	$\mathbf{0}_{.035-}$	—	—	—
2	10	$\mathbf{0}_{.015+}$	$\mathbf{0}_{.015+}$	—	—
	9	$\mathbf{0}_{.045+}$	—	—	—
$A = 11$ $B = 11$	11	$\mathbf{7}_{.045+}$	$\mathbf{6}_{.018}$	$\mathbf{5}_{.006}$	$\mathbf{4}_{.002}$
	10	$\mathbf{5}_{.032}$	$\mathbf{4}_{.012}$	$\mathbf{3}_{.004}$	$\mathbf{3}_{.004}$
	9	$\mathbf{4}_{.040}$	$\mathbf{3}_{.015-}$	$\mathbf{2}_{.004}$	$\mathbf{2}_{.004}$
	8	$\mathbf{3}_{.043}$	$\mathbf{2}_{.015-}$	$\mathbf{1}_{.004}$	$\mathbf{1}_{.004}$
	7	$\mathbf{2}_{.040}$	$\mathbf{1}_{.012}$	$\mathbf{0}_{.002}$	$\mathbf{0}_{.002}$
	6	$\mathbf{1}_{.032}$	$\mathbf{0}_{.006}$	$\mathbf{0}_{.006}$	—
	5	$\mathbf{0}_{.018}$	$\mathbf{0}_{.018}$	—	—
	4	$\mathbf{0}_{.045+}$	—	—	—
10	11	$\mathbf{6}_{.035+}$	$\mathbf{5}_{.012}$	$\mathbf{4}_{.004}$	$\mathbf{4}_{.004}$
	10	$\mathbf{4}_{.021}$	$\mathbf{4}_{.021}$	$\mathbf{3}_{.007}$	$\mathbf{2}_{.002}$

TABLE J (*continued*)

		Probability			
	a	**0.05**	**0.025**	**0.01**	**0.005**
	9	$3_{.024}$	$3_{.024}$	$2_{.007}$	$1_{.002}$
	8	$2_{.023}$	$2_{.023}$	$1_{.006}$	$0_{.001}$
	7	$1_{.017}$	$1_{.017}$	$0_{.003}$	$0_{.003}$
	6	$1_{.043}$	$0_{.009}$	$0_{.009}$	—
	5	$0_{.023}$	$0_{.023}$	—	—
9	11	$5_{.026}$	$4_{.008}$	$4_{.008}$	$3_{.002}$
	10	$4_{.038}$	$3_{.012}$	$2_{.003}$	$2_{.003}$
	9	$3_{.040}$	$2_{.012}$	$1_{.003}$	$1_{.003}$
	8	$2_{.035-}$	$1_{.009}$	$1_{.009}$	$0_{.001}$
	7	$1_{.025-}$	$1_{.025-}$	$0_{.004}$	$0_{.004}$
	6	$0_{.012}$	$0_{.012}$	—	—
	5	$0_{.030}$	—	—	—
***A* = 11 *B* = 8**	11	$4_{.018}$	$4_{.018}$	$3_{.005-}$	$3_{.005-}$
	10	$3_{.024}$	$3_{.024}$	$2_{.006}$	$1_{.001}$
	9	$2_{.022}$	$2_{.022}$	$1_{.005-}$	$1_{.005-}$
	8	$1_{.015-}$	$1_{.015-}$	$0_{.002}$	$0_{.002}$
	7	$1_{.037}$	$0_{.007}$	$0_{.007}$	—
	6	$0_{.017}$	$0_{.017}$	—	—
	5	$0_{.040}$	—	—	—
7	11	$4_{.043}$	$3_{.011}$	$2_{.002}$	$2_{.002}$
	10	$3_{.047}$	$2_{.013}$	$1_{.002}$	$1_{.002}$
	9	$2_{.039}$	$1_{.009}$	$1_{.009}$	$0_{.001}$
	8	$1_{.025-}$	$1_{.025-}$	$0_{.004}$	$0_{.004}$
	7	$0_{.010+}$	$0_{.010+}$	—	—
	6	$0_{.025-}$	$0_{.025-}$	—	—
6	11	$3_{.029}$	$2_{.006}$	$2_{.006}$	$1_{.001}$
	10	$2_{.028}$	$1_{.005+}$	$1_{.005+}$	$0_{.001}$
	9	$1_{.018}$	$1_{.018}$	$0_{.002}$	$0_{.002}$
	8	$1_{.043}$	$0_{.007}$	$0_{.007}$	—
	7	$0_{.017}$	$0_{.017}$	—	—
	6	$0_{.037}$	—	—	—
5	11	$2_{.018}$	$2_{.018}$	$1_{.003}$	$1_{.003}$
	10	$1_{.013}$	$1_{.013}$	$0_{.001}$	$0_{.001}$
	9	$1_{.036}$	$0_{.005-}$	$0_{.005-}$	$0_{.005-}$
	8	$0_{.013}$	$0_{.013}$	—	—
	7	$0_{.029}$	—	—	—
4	11	$1_{.009}$	$1_{.009}$	$1_{.009}$	$0_{.001}$
	10	$1_{.033}$	$0_{.004}$	$0_{.004}$	$0_{.004}$
	9	$0_{.011}$	$0_{.011}$	—	—
	8	$0_{.026}$	—	—	—
3	11	$1_{.033}$	$0_{.003}$	$0_{.003}$	$0_{.003}$
	10	$0_{.011}$	$0_{.011}$	—	—
	9	$0_{.027}$	—	—	—
2	11	$0_{.013}$	$0_{.013}$	—	—
	10	$0_{.038}$	—	—	—
***A* = 12 *B* = 12**	12	$8_{.047}$	$7_{.019}$	$6_{.007}$	$5_{.002}$
	11	$6_{.034}$	$5_{.014}$	$4_{.005-}$	$4_{.005-}$
	10	$5_{.045-}$	$4_{.018}$	$3_{.006}$	$2_{.002}$
	9	$4_{.050-}$	$3_{.020}$	$2_{.006}$	$1_{.001}$
	8	$3_{.050-}$	$2_{.018}$	$1_{.005-}$	$1_{.005-}$

TABLE J (*continued*)

		Probability			
	a	**0.05**	**0.025**	**0.01**	**0.005**
	7	$2_{.045-}$	$1_{.014}$	$0_{.002}$	$0_{.002}$
	6	$1_{.034}$	$0_{.007}$	$0_{.007}$	—
	5	$0_{.019}$	$0_{.019}$	—	—
	4	$0_{.047}$	—	—	—
11	12	$7_{.037}$	$6_{.014}$	$5_{.005-}$	$5_{.005-}$
	11	$5_{.024}$	$5_{.024}$	$4_{.008}$	$3_{.002}$
	10	$4_{.029}$	$3_{.010+}$	$2_{.003}$	$2_{.003}$
	9	$3_{.030}$	$2_{.009}$	$2_{.009}$	$1_{.002}$
	8	$2_{.026}$	$1_{.007}$	$1_{.007}$	$0_{.001}$
	7	$1_{.019}$	$1_{.019}$	$0_{.003}$	$0_{.003}$
	6	$1_{.045-}$	$0_{.009}$	$0_{.009}$	—
	5	$0_{.024}$	$0_{.024}$	—	—
10	12	$6_{.029}$	$5_{.010-}$	$5_{.010-}$	$4_{.003}$
	11	$5_{.043}$	$4_{.015+}$	$3_{.005-}$	$3_{.005-}$
	10	$4_{.048}$	$3_{.017}$	$2_{.005-}$	$2_{.005-}$
	9	$3_{.046}$	$2_{.015-}$	$1_{.004}$	$1_{.004}$
	8	$2_{.038}$	$1_{.010+}$	$0_{.002}$	$0_{.002}$
	7	$1_{.026}$	$0_{.005-}$	$0_{.005-}$	$0_{.005-}$
	6	$0_{.012}$	$0_{.012}$	—	—
	5	$0_{.030}$	—	—	—
***A* = 12 *B* = 9**	12	$5_{.021}$	$5_{.021}$	$4_{.006}$	$3_{.002}$
	11	$4_{.029}$	$3_{.009}$	$3_{.009}$	$2_{.002}$
	10	$3_{.029}$	$2_{.008}$	$2_{.008}$	$1_{.002}$
	9	$2_{.024}$	$2_{.024}$	$1_{.006}$	$0_{.001}$
	8	$1_{.016}$	$1_{.016}$	$0_{.002}$	$0_{.002}$
	7	$1_{.037}$	$0_{.007}$	$0_{.007}$	—
	6	$0_{.017}$	$0_{.017}$	—	—
	5	$0_{.039}$	—	—	—
8	12	$5_{.049}$	$4_{.014}$	$3_{.004}$	$3_{.004}$
	11	$3_{.018}$	$3_{.018}$	$2_{.004}$	$2_{.004}$
	10	$2_{.015+}$	$2_{.015+}$	$1_{.003}$	$1_{.003}$
	9	$2_{.040}$	$1_{.010-}$	$1_{.010-}$	$0_{.001}$
	8	$1_{.025-}$	$1_{.025-}$	$0_{.004}$	$0_{.004}$
	7	$0_{.010+}$	$0_{.010+}$	—	—
	6	$0_{.024}$	$0_{.024}$	—	—
7	12	$4_{.036}$	$3_{.009}$	$3_{.009}$	$2_{.002}$
	11	$3_{.038}$	$2_{.010-}$	$2_{.010-}$	$1_{.002}$
	10	$2_{.029}$	$1_{.006}$	$1_{.006}$	$0_{.001}$
	9	$1_{.017}$	$1_{.017}$	$0_{.002}$	$0_{.002}$
	8	$1_{.040}$	$0_{.007}$	$0_{.007}$	—
	7	$0_{.016}$	$0_{.016}$	—	—
	6	$0_{.034}$	—	—	—
6	12	$3_{.025-}$	$3_{.025-}$	$2_{.005-}$	$2_{.005-}$
	11	$2_{.022}$	$2_{.022}$	$1_{.004}$	$1_{.004}$
	10	$1_{.013}$	$1_{.013}$	$0_{.002}$	$0_{.002}$
	9	$1_{.032}$	$0_{.005-}$	$0_{.005-}$	$0_{.005-}$
	8	$0_{.011}$	$0_{.011}$	—	—
	7	$0_{.025-}$	$0_{.025-}$	—	—
	6	$0_{.050-}$	—	—	—
5	12	$2_{.015-}$	$2_{.015-}$	$1_{.002}$	$1_{.002}$

TABLE J ***(continued)***

			Probability			
		a	**0.05**	**0.025**	**0.01**	**0.005**
		11	$\mathbf{1}_{.010^-}$	$\mathbf{1}_{.010^-}$	$\mathbf{1}_{.010^-}$	$\mathbf{0}_{.001}$
		10	$\mathbf{1}_{.028}$	$\mathbf{0}_{.003}$	$\mathbf{0}_{.003}$	$\mathbf{0}_{.003}$
		9	$\mathbf{0}_{.009}$	$\mathbf{0}_{.009}$	$\mathbf{0}_{.009}$	—
		8	$\mathbf{0}_{.020}$	$\mathbf{0}_{.020}$	—	—
		7	$\mathbf{0}_{.041}$	—	—	—
	4	12	$\mathbf{2}_{.050}$	$\mathbf{1}_{.007}$	$\mathbf{1}_{.007}$	$\mathbf{0}_{.001}$
		11	$\mathbf{1}_{.027}$	$\mathbf{0}_{.003}$	$\mathbf{0}_{.003}$	$\mathbf{0}_{.003}$
		10	$\mathbf{0}_{.008}$	$\mathbf{0}_{.008}$	$\mathbf{0}_{.008}$	—
		9	$\mathbf{0}_{.019}$	$\mathbf{0}_{.019}$	—	—
		8	$\mathbf{0}_{.038}$	—	—	—
	3	12	$\mathbf{1}_{.029}$	$\mathbf{0}_{.002}$	$\mathbf{0}_{.002}$	$\mathbf{0}_{.002}$
		11	$\mathbf{0}_{.009}$	$\mathbf{0}_{.009}$	$\mathbf{0}_{.009}$	—
		10	$\mathbf{0}_{.022}$	$\mathbf{0}_{.022}$	—	—
		9	$\mathbf{0}_{.044}$	—	—	—
	2	12	$\mathbf{0}_{.011}$	$\mathbf{0}_{.011}$	—	—
		11	$\mathbf{0}_{.033}$	—	—	—
A **= 13** ***B*** **= 13**		13	$\mathbf{9}_{.048}$	$\mathbf{8}_{.020}$	$\mathbf{7}_{.007}$	$\mathbf{6}_{.003}$
		12	$\mathbf{7}_{.037}$	$\mathbf{6}_{.015^+}$	$\mathbf{5}_{.006}$	$\mathbf{4}_{.002}$
		11	$\mathbf{6}_{.048}$	$\mathbf{5}_{.021}$	$\mathbf{4}_{.008}$	$\mathbf{3}_{.002}$
		10	$\mathbf{4}_{.024}$	$\mathbf{4}_{.024}$	$\mathbf{3}_{.008}$	$\mathbf{2}_{.002}$
		9	$\mathbf{3}_{.024}$	$\mathbf{3}_{.024}$	$\mathbf{2}_{.008}$	$\mathbf{1}_{.002}$
		8	$\mathbf{2}_{.021}$	$\mathbf{2}_{.021}$	$\mathbf{1}_{.006}$	$\mathbf{0}_{.001}$
		7	$\mathbf{2}_{.048}$	$\mathbf{1}_{.015^+}$	$\mathbf{0}_{.003}$	$\mathbf{0}_{.003}$
		6	$\mathbf{1}_{.037}$	$\mathbf{0}_{.007}$	$\mathbf{0}_{.007}$	—
		5	$\mathbf{0}_{.020}$	$\mathbf{0}_{.020}$	—	—
		4	$\mathbf{0}_{.048}$	—	—	—
	12	13	$\mathbf{8}_{.039}$	$\mathbf{7}_{.015^-}$	$\mathbf{6}_{.005^+}$	$\mathbf{5}_{.002}$
		12	$\mathbf{6}_{.027}$	$\mathbf{5}_{.010^-}$	$\mathbf{5}_{.010^-}$	$\mathbf{4}_{.003}$
		11	$\mathbf{5}_{.033}$	$\mathbf{4}_{.013}$	$\mathbf{3}_{.004}$	$\mathbf{3}_{.004}$
		10	$\mathbf{4}_{.036}$	$\mathbf{3}_{.013}$	$\mathbf{2}_{.004}$	$\mathbf{2}_{.004}$
A **= 13** ***B*** **= 12**		9	$\mathbf{3}_{.034}$	$\mathbf{2}_{.011}$	$\mathbf{1}_{.003}$	$\mathbf{1}_{.003}$
		8	$\mathbf{2}_{.029}$	$\mathbf{1}_{.008}$	$\mathbf{1}_{.008}$	$\mathbf{0}_{.001}$
		7	$\mathbf{1}_{.020}$	$\mathbf{1}_{.020}$	$\mathbf{0}_{.004}$	$\mathbf{0}_{.004}$
		6	$\mathbf{1}_{.046}$	$\mathbf{0}_{.010^-}$	$\mathbf{0}_{.010^-}$	—
		5	$\mathbf{0}_{.024}$	$\mathbf{0}_{.024}$	—	—
	11	13	$\mathbf{7}_{.031}$	$\mathbf{6}_{.011}$	$\mathbf{5}_{.003}$	$\mathbf{5}_{.003}$
		12	$\mathbf{6}_{.048}$	$\mathbf{5}_{.018}$	$\mathbf{4}_{.006}$	$\mathbf{3}_{.002}$
		11	$\mathbf{4}_{.021}$	$\mathbf{4}_{.021}$	$\mathbf{3}_{.007}$	$\mathbf{2}_{.002}$
		10	$\mathbf{3}_{.021}$	$\mathbf{3}_{.021}$	$\mathbf{2}_{.006}$	$\mathbf{1}_{.001}$
		9	$\mathbf{3}_{.050^-}$	$\mathbf{2}_{.017}$	$\mathbf{1}_{.004}$	$\mathbf{1}_{.004}$
		8	$\mathbf{2}_{.040}$	$\mathbf{1}_{.011}$	$\mathbf{0}_{.002}$	$\mathbf{0}_{.002}$
		7	$\mathbf{1}_{.027}$	$\mathbf{0}_{.005^-}$	$\mathbf{0}_{.005^-}$	$\mathbf{0}_{.005^-}$
		6	$\mathbf{0}_{.013}$	$\mathbf{0}_{.013}$	—	—
		5	$\mathbf{0}_{.030}$	—	—	—
	10	13	$\mathbf{6}_{.024}$	$\mathbf{6}_{.024}$	$\mathbf{5}_{.007}$	$\mathbf{4}_{.002}$
		12	$\mathbf{5}_{.035^-}$	$\mathbf{4}_{.012}$	$\mathbf{3}_{.003}$	$\mathbf{3}_{.003}$
		11	$\mathbf{4}_{.037}$	$\mathbf{3}_{.012}$	$\mathbf{2}_{.003}$	$\mathbf{2}_{.003}$
		10	$\mathbf{3}_{.033}$	$\mathbf{2}_{.010^+}$	$\mathbf{1}_{.002}$	$\mathbf{1}_{.002}$
		9	$\mathbf{2}_{.026}$	$\mathbf{1}_{.006}$	$\mathbf{1}_{.006}$	$\mathbf{0}_{.001}$
		8	$\mathbf{1}_{.017}$	$\mathbf{1}_{.017}$	$\mathbf{0}_{.003}$	$\mathbf{0}_{.003}$

TABLE J ***(continued)***

		Probability			
	a	0.05	0.025	0.01	0.005
	7	$1_{.038}$	$0_{.007}$	$0_{.007}$	—
	6	$0_{.017}$	$0_{.017}$	—	—
	5	$0_{.038}$	—	—	—
9	13	$5_{.017}$	$5_{.017}$	$4_{.005-}$	$4_{.005-}$
	12	$4_{.023}$	$4_{.023}$	$3_{.007}$	$2_{.001}$
	11	$3_{.022}$	$3_{.022}$	$2_{.006}$	$1_{.001}$
	10	$2_{.017}$	$2_{.017}$	$1_{.004}$	$1_{.004}$
	9	$2_{.040}$	$1_{.010+}$	$0_{.001}$	$0_{.001}$
	8	$1_{.025-}$	$1_{.025-}$	$0_{.004}$	$0_{.004}$
	7	$0_{.010+}$	$0_{.010+}$	—	—
	6	$0_{.023}$	$0_{.023}$	—	—
	5	$0_{.049}$	—	—	—
8	13	$5_{.042}$	$4_{.012}$	$3_{.003}$	$3_{.003}$
	12	$4_{.047}$	$3_{.014}$	$2_{.003}$	$2_{.003}$
	11	$3_{.041}$	$2_{.011}$	$1_{.002}$	$1_{.002}$
	10	$2_{.029}$	$1_{.007}$	$1_{.007}$	$0_{.001}$
	9	$1_{.017}$	$1_{.017}$	$0_{.002}$	$0_{.002}$
	8	$1_{.037}$	$0_{.006}$	$0_{.006}$	—
	7	$0_{.015-}$	$0_{.015-}$	—	—
	6	$0_{.032}$	—	—	—
7	13	$4_{.031}$	$3_{.007}$	$3_{.007}$	$2_{.001}$
	12	$3_{.031}$	$2_{.007}$	$2_{.007}$	$1_{.001}$
	11	$2_{.022}$	$2_{.022}$	$1_{.004}$	$1_{.004}$
	10	$1_{.012}$	$1_{.012}$	$0_{.002}$	$0_{.002}$
	9	$1_{.029}$	$0_{.004}$	$0_{.004}$	$0_{.004}$
	8	$0_{.010+}$	$0_{.010+}$	—	—
	7	$0_{.022}$	$0_{.022}$	—	—
	6	$0_{.044}$	—	—	—
6	13	$3_{.021}$	$3_{.021}$	$2_{.004}$	$2_{.004}$
	12	$2_{.017}$	$2_{.017}$	$1_{.003}$	$1_{.003}$
	11	$2_{.046}$	$1_{.010-}$	$1_{.010-}$	$0_{.001}$
	10	$1_{.024}$	$1_{.024}$	$0_{.003}$	$0_{.003}$
	9	$1_{.050-}$	$0_{.008}$	$0_{.008}$	—
	8	$0_{.017}$	$0_{.017}$	—	—
	7	$0_{.034}$	—	—	—
5	13	$2_{.012}$	$2_{.012}$	$1_{.002}$	$1_{.002}$
	12	$2_{.044}$	$1_{.008}$	$1_{.008}$	$0_{.001}$
	11	$1_{.022}$	$1_{.022}$	$0_{.002}$	$0_{.002}$
	10	$1_{.047}$	$0_{.007}$	$0_{.007}$	—
	9	$0_{.015-}$	$0_{.015-}$	—	—
	8	$0_{.029}$	—	—	—
$A = 13$ $B = 4$	13	$2_{.044}$	$1_{.006}$	$1_{.006}$	$0_{.000}$
	12	$1_{.022}$	$1_{.022}$	$0_{.002}$	$0_{.002}$
	11	$0_{.006}$	$0_{.006}$	$0_{.006}$	—
	10	$0_{.015-}$	$0_{.015-}$	—	—
	9	$0_{.029}$	—	—	—
3	13	$1_{.025}$	$1_{.025}$	$0_{.002}$	$0_{.002}$
	12	$0_{.007}$	$0_{.007}$	$0_{.007}$	—
	11	$0_{.018}$	$0_{.018}$	—	—
	10	$0_{.036}$	—	—	—

TABLE J *(continued)*

			Probability			
		a	**0.05**	**0.025**	**0.01**	**0.005**
	2	13	$0_{.010-}$	$0_{.010-}$	$0_{.010-}$	—
		12	$0_{.029}$	—	—	—
***A* = 14 *B* = 14**		14	$10_{.049}$	$9_{.020}$	$8_{.008}$	$7_{.003}$
		13	$8_{.038}$	$7_{.016}$	$6_{.006}$	$5_{.002}$
		12	$6_{.023}$	$6_{.023}$	$5_{.009}$	$4_{.003}$
		11	$5_{.027}$	$4_{.011}$	$3_{.004}$	$3_{.004}$
		10	$4_{.028}$	$3_{.011}$	$2_{.003}$	$2_{.003}$
		9	$3_{.027}$	$2_{.009}$	$2_{.009}$	$1_{.002}$
		8	$2_{.023}$	$2_{.023}$	$1_{.006}$	$0_{.001}$
		7	$1_{.016}$	$1_{.016}$	$0_{.003}$	$0_{.003}$
		6	$1_{.038}$	$0_{.008}$	$0_{.008}$	—
		5	$0_{.020}$	$0_{.020}$	—	—
		4	$0_{.049}$	—	—	—
	13	14	$9_{.041}$	$8_{.016}$	$7_{.006}$	$6_{.002}$
		13	$7_{.029}$	$6_{.011}$	$5_{.004}$	$5_{.004}$
		12	$6_{.037}$	$5_{.015+}$	$4_{.005+}$	$3_{.002}$
		11	$5_{.041}$	$4_{.017}$	$3_{.006}$	$2_{.001}$
		10	$4_{.041}$	$3_{.016}$	$2_{.005-}$	$2_{.005-}$
		9	$3_{.038}$	$2_{.013}$	$1_{.003}$	$1_{.003}$
		8	$2_{.031}$	$1_{.009}$	$1_{.009}$	$0_{.001}$
		7	$1_{.021}$	$1_{.021}$	$0_{.004}$	$0_{.004}$
		6	$1_{.048}$	$0_{.010+}$	—	—
		5	$0_{.025-}$	$0_{.025-}$	—	—
	12	14	$8_{.033}$	$7_{.012}$	$6_{.004}$	$6_{.004}$
		13	$6_{.021}$	$6_{.021}$	$5_{.007}$	$4_{.002}$
		12	$5_{.025+}$	$4_{.009}$	$4_{.009}$	$3_{.003}$
		11	$4_{.026}$	$3_{.009}$	$3_{.009}$	$2_{.002}$
		10	$3_{.024}$	$3_{.024}$	$2_{.007}$	$1_{.002}$
		9	$2_{.019}$	$2_{.019}$	$1_{.005-}$	$1_{.005-}$
		8	$2_{.042}$	$1_{.012}$	$0_{.002}$	$0_{.002}$
		7	$1_{.028}$	$0_{.005+}$	$0_{.005+}$	—
		6	$0_{.013}$	$0_{.013}$	—	—
		5	$0_{.030}$	—	—	—
	11	14	$7_{.026}$	$6_{.009}$	$6_{.009}$	$5_{.003}$
		13	$6_{.039}$	$5_{.014}$	$4_{.004}$	$4_{.004}$
		12	$5_{.043}$	$4_{.016}$	$3_{.005-}$	$3_{.005-}$
		11	$4_{.042}$	$3_{.015-}$	$2_{.004}$	$2_{.004}$
		10	$3_{.036}$	$2_{.011}$	$1_{.003}$	$1_{.003}$
		9	$2_{.027}$	$1_{.007}$	$1_{.007}$	$0_{.001}$
		8	$1_{.017}$	$1_{.017}$	$0_{.003}$	$0_{.003}$
		7	$1_{.038}$	$0_{.007}$	$0_{.007}$	—
		6	$0_{.017}$	$0_{.017}$	—	—
		5	$0_{.038}$	—	—	—
	10	14	$6_{.020}$	$6_{.020}$	$5_{.006}$	$4_{.002}$
		13	$5_{.028}$	$4_{.009}$	$4_{.009}$	$3_{.002}$
		12	$4_{.028}$	$3_{.009}$	$3_{.009}$	$2_{.002}$
		11	$3_{.024}$	$3_{.024}$	$2_{.007}$	$1_{.001}$
		10	$2_{.018}$	$2_{.018}$	$1_{.004}$	$1_{.004}$
		9	$2_{.040}$	$1_{.011}$	$0_{.002}$	$0_{.002}$
		8	$1_{.024}$	$1_{.024}$	$0_{.004}$	$0_{.004}$

TABLE J ***(continued)***

		Probability			
	a	**0.05**	**0.025**	**0.01**	**0.005**
	7	$0_{.010^{-}}$	$0_{.010^{-}}$	$0_{.010^{-}}$	—
$A = 14$ $B = 10$	6	$0_{.022}$	$0_{.022}$	—	—
	5	$0_{.047}$	—	—	—
9	14	$6_{.047}$	$5_{.014}$	$4_{.004}$	$4_{.004}$
	13	$4_{.018}$	$4_{.018}$	$3_{.005^{-}}$	$3_{.005^{-}}$
	12	$3_{.017}$	$3_{.017}$	$2_{.004}$	$2_{.004}$
	11	$3_{.042}$	$2_{.012}$	$1_{.002}$	$1_{.002}$
	10	$2_{.029}$	$1_{.007}$	$1_{.007}$	$0_{.001}$
	9	$1_{.017}$	$1_{.017}$	$0_{.002}$	$0_{.002}$
	8	$1_{.036}$	$0_{.006}$	$0_{.006}$	—
	7	$0_{.014}$	$0_{.014}$	—	—
	6	$0_{.030}$	—	—	—
8	14	$5_{.036}$	$4_{.010^{-}}$	$4_{.010^{-}}$	$3_{.002}$
	13	$4_{.039}$	$3_{.011}$	$2_{.002}$	$2_{.002}$
	12	$3_{.032}$	$2_{.008}$	$2_{.008}$	$1_{.001}$
	11	$2_{.022}$	$2_{.022}$	$1_{.005^{-}}$	$1_{.005^{-}}$
	10	$2_{.048}$	$1_{.012}$	$0_{.002}$	$0_{.002}$
	9	$1_{.026}$	$0_{.004}$	$0_{.004}$	$0_{.004}$
	8	$0_{.009}$	$0_{.009}$	$0_{.009}$	—
	7	$0_{.020}$	$0_{.020}$	—	—
	6	$0_{.040}$	—	—	—
7	14	$4_{.026}$	$3_{.006}$	$3_{.006}$	$2_{.001}$
	13	$3_{.025}$	$2_{.006}$	$2_{.006}$	$1_{.001}$
	12	$2_{.017}$	$2_{.017}$	$1_{.003}$	$1_{.003}$
	11	$2_{.041}$	$1_{.009}$	$1_{.009}$	$0_{.001}$
	10	$1_{.021}$	$1_{.021}$	$0_{.003}$	$0_{.003}$
	9	$1_{.043}$	$0_{.007}$	$0_{.007}$	—
	8	$0_{.015^{-}}$	$0_{.015^{-}}$	—	—
	7	$0_{.030}$	—	—	—
6	14	$3_{.018}$	$3_{.018}$	$2_{.003}$	$2_{.003}$
	13	$2_{.014}$	$2_{.014}$	$1_{.002}$	$1_{.002}$
	12	$2_{.037}$	$1_{.007}$	$1_{.007}$	$0_{.001}$
	11	$1_{.018}$	$1_{.018}$	$0_{.002}$	$0_{.002}$
	10	$1_{.038}$	$0_{.005^{+}}$	$0_{.005^{+}}$	—
	9	$0_{.012}$	$0_{.012}$	—	—
	8	$0_{.024}$	$0_{.024}$	—	—
	7	$0_{.044}$	—	—	—
5	14	$2_{.010^{+}}$	$2_{.010^{+}}$	$1_{.001}$	$1_{.001}$
	13	$2_{.037}$	$1_{.006}$	$1_{.006}$	$0_{.001}$
	12	$1_{.017}$	$1_{.017}$	$0_{.002}$	$0_{.002}$
	11	$1_{.038}$	$0_{.005^{-}}$	$0_{.005^{-}}$	$0_{.005^{-}}$
	10	$0_{.011}$	$0_{.011}$	—	—
	9	$0_{.022}$	$0_{.022}$	—	—
	8	$0_{.040}$	—	—	—
4	14	$2_{.039}$	$1_{.005^{-}}$	$1_{.005^{-}}$	$1_{.005^{-}}$
	13	$1_{.019}$	$1_{.019}$	$0_{.002}$	$0_{.002}$
	12	$1_{.044}$	$0_{.005^{-}}$	$0_{.005^{-}}$	$0_{.005^{-}}$
	11	$0_{.011}$	$0_{.011}$	—	—
	10	$0_{.023}$	$0_{.023}$	—	—
	9	$0_{.041}$	—	—	—

TABLE J ***(continued)***

		Probability			
	a	0.05	0.025	0.01	0.005
3	14	$1_{.022}$	$1_{.022}$	$0_{.001}$	$0_{.001}$
	13	$0_{.006}$	$0_{.006}$	$0_{.006}$	—
	12	$0_{.015-}$	$0_{.015-}$	—	—
	11	$0_{.029}$	—	—	—
2	14	$0_{.008}$	$0_{.008}$	$0_{.008}$	—
	13	$0_{.025}$	$0_{.025}$	—	—
	12	$0_{.050}$	—	—	—
$A = 15$ $B = 15$	15	$11_{.050-}$	$10_{.021}$	$9_{.008}$	$8_{.003}$
	14	$9_{.040}$	$8_{.018}$	$7_{.007}$	$6_{.003}$
	13	$7_{.025+}$	$6_{.010+}$	$5_{.004}$	$5_{.004}$
	12	$6_{.030}$	$5_{.013}$	$4_{.005-}$	$4_{.005-}$
$A = 15$ $B = 15$	11	$5_{.033}$	$4_{.013}$	$3_{.005-}$	$3_{.005-}$
	10	$4_{.033}$	$3_{.013}$	$2_{.004}$	$2_{.004}$
	9	$3_{.030}$	$2_{.010+}$	$1_{.003}$	$1_{.003}$
	8	$2_{.025+}$	$1_{.007}$	$1_{.007}$	$0_{.001}$
	7	$1_{.018}$	$1_{.018}$	$0_{.003}$	$0_{.003}$
	6	$1_{.040}$	$0_{.008}$	$0_{.008}$	—
	5	$0_{.021}$	$0_{.021}$	—	—
	4	$0_{.050-}$	—	—	—
14	15	$10_{.042}$	$9_{.017}$	$8_{.006}$	$7_{.002}$
	14	$8_{.031}$	$7_{.013}$	$6_{.005-}$	$6_{.005-}$
	13	$7_{.041}$	$6_{.017}$	$5_{.007}$	$4_{.002}$
	12	$6_{.046}$	$5_{.020}$	$4_{.007}$	$3_{.002}$
	11	$5_{.048}$	$4_{.020}$	$3_{.007}$	$2_{.002}$
	10	$4_{.046}$	$3_{.018}$	$2_{.006}$	$1_{.001}$
	9	$3_{.041}$	$2_{.014}$	$1_{.004}$	$1_{.004}$
	8	$2_{.033}$	$1_{.009}$	$1_{.009}$	$0_{.001}$
	7	$1_{.022}$	$1_{.022}$	$0_{.004}$	$0_{.004}$
	6	$1_{.049}$	$0_{.011}$	—	—
	5	$0_{.025+}$	—	—	—
13	15	$9_{.035-}$	$8_{.013}$	$7_{.005-}$	$7_{.005-}$
	14	$7_{.023}$	$7_{.023}$	$6_{.009}$	$5_{.003}$
	13	$6_{.029}$	$5_{.011}$	$4_{.004}$	$4_{.004}$
	12	$5_{.031}$	$4_{.012}$	$3_{.004}$	$3_{.004}$
	11	$4_{.030}$	$3_{.011}$	$2_{.003}$	$2_{.003}$
	10	$3_{.026}$	$2_{.008}$	$2_{.008}$	$1_{.002}$
	9	$2_{.020}$	$2_{.020}$	$1_{.005+}$	$0_{.001}$
	8	$2_{.043}$	$1_{.013}$	$0_{.002}$	$0_{.002}$
	7	$1_{.029}$	$0_{.005+}$	$0_{.005+}$	—
	6	$0_{.013}$	$0_{.013}$	—	—
	5	$0_{.031}$	—	—	—
12	15	$8_{.028}$	$7_{.010-}$	$7_{.010-}$	$6_{.003}$
	14	$7_{.043}$	$6_{.016}$	$5_{.006}$	$4_{.002}$
	13	$6_{.049}$	$5_{.019}$	$4_{.007}$	$3_{.002}$
	12	$5_{.049}$	$4_{.019}$	$3_{.006}$	$2_{.002}$
	11	$4_{.045+}$	$3_{.017}$	$2_{.005-}$	$2_{.005-}$
	10	$3_{.038}$	$2_{.012}$	$1_{.003}$	$1_{.003}$
	9	$2_{.028}$	$1_{.007}$	$1_{.007}$	$0_{.001}$
	8	$1_{.018}$	$1_{.018}$	$0_{.003}$	$0_{.003}$
	7	$1_{.038}$	$0_{.007}$	$0_{.007}$	—

TABLE J ***(continued)***

		Probability			
	a	**0.05**	**0.025**	**0.01**	**0.005**
	6	$0_{.017}$	$0_{.017}$	—	—
	5	$0_{.037}$	—	—	—
11	15	$7_{.022}$	$7_{.022}$	$6_{.007}$	$5_{.002}$
	14	$6_{.032}$	$5_{.011}$	$4_{.003}$	$4_{.003}$
	13	$5_{.034}$	$4_{.012}$	$3_{.003}$	$3_{.003}$
	12	$4_{.032}$	$3_{.010+}$	$2_{.003}$	$2_{.003}$
	11	$3_{.026}$	$2_{.008}$	$2_{.008}$	$1_{.002}$
	10	$2_{.019}$	$2_{.019}$	$1_{.004}$	$1_{.004}$
	9	$2_{.040}$	$1_{.011}$	$0_{.002}$	$0_{.002}$
	8	$1_{.024}$	$1_{.024}$	$0_{.004}$	$0_{.004}$
	7	$1_{.049}$	$0_{.010-}$	$0_{.010-}$	—
	6	$0_{.022}$	$0_{.022}$	—	—
	5	$0_{.046}$	—	—	—
10	15	$6_{.017}$	$6_{.017}$	$5_{.005-}$	$5_{.005-}$
	14	$5_{.023}$	$5_{.023}$	$4_{.007}$	$3_{.002}$
	13	$4_{.022}$	$4_{.022}$	$3_{.007}$	$2_{.001}$
	12	$3_{.018}$	$3_{.018}$	$2_{.005-}$	$2_{.005-}$
	11	$3_{.042}$	$2_{.013}$	$1_{.003}$	$1_{.003}$
	10	$2_{.029}$	$1_{.007}$	$1_{.007}$	$0_{.001}$
	9	$1_{.016}$	$1_{.016}$	$0_{.002}$	$0_{.002}$
	8	$1_{.034}$	$0_{.006}$	$0_{.006}$	—
$A = 15$ $B = 10$	7	$0_{.013}$	$0_{.013}$	—	—
	6	$0_{.028}$	—	—	—
9	15	$6_{.042}$	$5_{.012}$	$4_{.003}$	$4_{.003}$
	14	$5_{.047}$	$4_{.015-}$	$3_{.004}$	$3_{.004}$
	13	$4_{.042}$	$3_{.013}$	$2_{.003}$	$2_{.003}$
	12	$3_{.032}$	$2_{.009}$	$2_{.009}$	$1_{.002}$
	11	$2_{.021}$	$2_{.021}$	$1_{.005-}$	$1_{.005-}$
	10	$2_{.045-}$	$1_{.011}$	$0_{.002}$	$0_{.002}$
	9	$1_{.024}$	$1_{.024}$	$0_{.004}$	$0_{.004}$
	8	$1_{.048}$	$0_{.009}$	$0_{.009}$	—
	7	$0_{.019}$	$0_{.019}$	—	—
	6	$0_{.037}$	—	—	—
8	15	$5_{.032}$	$4_{.008}$	$4_{.008}$	$3_{.002}$
	14	$4_{.033}$	$3_{.009}$	$3_{.009}$	$2_{.002}$
	13	$3_{.026}$	$2_{.006}$	$2_{.006}$	$1_{.001}$
	12	$2_{.017}$	$2_{.017}$	$1_{.003}$	$1_{.003}$
	11	$2_{.037}$	$1_{.008}$	$1_{.008}$	$0_{.001}$
	10	$1_{.019}$	$1_{.019}$	$0_{.003}$	$0_{.003}$
	9	$1_{.038}$	$0_{.006}$	$0_{.006}$	—
	8	$0_{.013}$	$0_{.013}$	—	—
	7	$0_{.026}$	—	—	—
	6	$0_{.050-}$	—	—	—
7	15	$4_{.023}$	$4_{.023}$	$3_{.005-}$	$3_{.005-}$
	14	$3_{.021}$	$3_{.021}$	$2_{.004}$	$2_{.004}$
	13	$2_{.014}$	$2_{.014}$	$1_{.002}$	$1_{.002}$
	12	$2_{.032}$	$1_{.007}$	$1_{.007}$	$0_{.001}$
	11	$1_{.015+}$	$1_{.015+}$	$0_{.002}$	$0_{.002}$
	10	$1_{.032}$	$0_{.005-}$	$0_{.005-}$	$0_{.005-}$
	9	$0_{.010+}$	$0_{.010+}$	—	—

TABLE J ***(continued)***

		Probability			
	a	**0.05**	**0.025**	**0.01**	**0.005**
	8	$\mathbf{0}_{.020}$	$\mathbf{0}_{.020}$	—	—
	7	$\mathbf{0}_{.038}$	—	—	—
6	15	$\mathbf{3}_{.015+}$	$\mathbf{3}_{.015+}$	$\mathbf{2}_{.003}$	$\mathbf{2}_{.003}$
	14	$\mathbf{2}_{.011}$	$\mathbf{2}_{.011}$	$\mathbf{1}_{.002}$	$\mathbf{1}_{.002}$
	13	$\mathbf{2}_{.031}$	$\mathbf{1}_{.006}$	$\mathbf{1}_{.006}$	$\mathbf{0}_{.001}$
	12	$\mathbf{1}_{.014}$	$\mathbf{1}_{.014}$	$\mathbf{0}_{.002}$	$\mathbf{0}_{.002}$
	11	$\mathbf{1}_{.029}$	$\mathbf{0}_{.004}$	$\mathbf{0}_{.004}$	$\mathbf{0}_{.004}$
	10	$\mathbf{0}_{.009}$	$\mathbf{0}_{.009}$	$\mathbf{0}_{.009}$	—
	9	$\mathbf{0}_{.017}$	$\mathbf{0}_{.017}$	—	—
	8	$\mathbf{0}_{.032}$	—	—	—
5	15	$\mathbf{2}_{.009}$	$\mathbf{2}_{.009}$	$\mathbf{2}_{.009}$	$\mathbf{1}_{.001}$
	14	$\mathbf{2}_{.032}$	$\mathbf{1}_{.005-}$	$\mathbf{1}_{.005-}$	$\mathbf{1}_{.005-}$
	13	$\mathbf{1}_{.014}$	$\mathbf{1}_{.014}$	$\mathbf{0}_{.001}$	$\mathbf{0}_{.001}$
	12	$\mathbf{1}_{.031}$	$\mathbf{0}_{.004}$	$\mathbf{0}_{.004}$	$\mathbf{0}_{.004}$
	11	$\mathbf{0}_{.008}$	$\mathbf{0}_{.008}$	$\mathbf{0}_{.008}$	—
	10	$\mathbf{0}_{.016}$	$\mathbf{0}_{.016}$	—	—
	9	$\mathbf{0}_{.030}$	—	—	—
4	15	$\mathbf{2}_{.035+}$	$\mathbf{1}_{.004}$	$\mathbf{1}_{.004}$	$\mathbf{1}_{.004}$
	14	$\mathbf{1}_{.016}$	$\mathbf{1}_{.016}$	$\mathbf{0}_{.001}$	$\mathbf{0}_{.001}$
	13	$\mathbf{1}_{.037}$	$\mathbf{0}_{.004}$	$\mathbf{0}_{.004}$	$\mathbf{0}_{.004}$
	12	$\mathbf{0}_{.009}$	$\mathbf{0}_{.009}$	$\mathbf{0}_{.009}$	—
	11	$\mathbf{0}_{.018}$	$\mathbf{0}_{.018}$	—	—
	10	$\mathbf{0}_{.033}$	—	—	—
3	15	$\mathbf{1}_{.020}$	$\mathbf{1}_{.020}$	$\mathbf{0}_{.001}$	$\mathbf{0}_{.001}$
	14	$\mathbf{0}_{.005-}$	$\mathbf{0}_{.005-}$	$\mathbf{0}_{.005-}$	$\mathbf{0}_{.005-}$
	13	$\mathbf{0}_{.012}$	$\mathbf{0}_{.012}$	—	—
	12	$\mathbf{0}_{.025-}$	$\mathbf{0}_{.025-}$	—	—
	11	$\mathbf{0}_{.043}$	—	—	—
2	15	$\mathbf{0}_{.007}$	$\mathbf{0}_{.007}$	$\mathbf{0}_{.007}$	—
	14	$\mathbf{0}_{.022}$	$\mathbf{0}_{.022}$	—	—
	13	$\mathbf{0}_{.044}$	—	—	—
$A = 16$ $B = 16$	16	$\mathbf{11}_{.022}$	$\mathbf{11}_{.022}$	$\mathbf{10}_{.009}$	$\mathbf{9}_{.003}$
	15	$\mathbf{10}_{.041}$	$\mathbf{9}_{.019}$	$\mathbf{8}_{.008}$	$\mathbf{7}_{.003}$
	14	$\mathbf{8}_{.027}$	$\mathbf{7}_{.012}$	$\mathbf{6}_{.005-}$	$\mathbf{6}_{.005-}$
	13	$\mathbf{7}_{.033}$	$\mathbf{6}_{.015-}$	$\mathbf{5}_{.006}$	$\mathbf{4}_{.002}$
	12	$\mathbf{6}_{.037}$	$\mathbf{5}_{.016}$	$\mathbf{4}_{.006}$	$\mathbf{3}_{.002}$
	11	$\mathbf{5}_{.038}$	$\mathbf{4}_{.016}$	$\mathbf{3}_{.006}$	$\mathbf{2}_{.002}$
	10	$\mathbf{4}_{.037}$	$\mathbf{3}_{.015-}$	$\mathbf{2}_{.005-}$	$\mathbf{2}_{.005-}$
	9	$\mathbf{3}_{.033}$	$\mathbf{2}_{.012}$	$\mathbf{1}_{.003}$	$\mathbf{1}_{.003}$
	8	$\mathbf{2}_{.027}$	$\mathbf{1}_{.008}$	$\mathbf{1}_{.008}$	$\mathbf{0}_{.001}$
	7	$\mathbf{1}_{.019}$	$\mathbf{1}_{.019}$	$\mathbf{0}_{.003}$	$\mathbf{0}_{.003}$
	6	$\mathbf{1}_{.041}$	$\mathbf{0}_{.009}$	$\mathbf{0}_{.009}$	—
	5	$\mathbf{0}_{.022}$	$\mathbf{0}_{.022}$	—	—
15	16	$\mathbf{11}_{.043}$	$\mathbf{10}_{.018}$	$\mathbf{9}_{.007}$	$\mathbf{8}_{.002}$
	15	$\mathbf{9}_{.033}$	$\mathbf{8}_{.014}$	$\mathbf{7}_{.005+}$	$\mathbf{6}_{.002}$
	14	$\mathbf{8}_{.044}$	$\mathbf{7}_{.019}$	$\mathbf{6}_{.008}$	$\mathbf{5}_{.003}$
	13	$\mathbf{6}_{.023}$	$\mathbf{6}_{.023}$	$\mathbf{5}_{.009}$	$\mathbf{4}_{.003}$
	12	$\mathbf{5}_{.024}$	$\mathbf{5}_{.024}$	$\mathbf{4}_{.009}$	$\mathbf{3}_{.003}$
	11	$\mathbf{4}_{.023}$	$\mathbf{4}_{.023}$	$\mathbf{3}_{.008}$	$\mathbf{2}_{.002}$
	10	$\mathbf{4}_{.049}$	$\mathbf{3}_{.020}$	$\mathbf{2}_{.006}$	$\mathbf{1}_{.001}$

TABLE J (*continued*)

		Probability			
	a	**0.05**	**0.025**	**0.01**	**0.005**
	9	$3_{.043}$	$2_{.016}$	$1_{.004}$	$1_{.004}$
	8	$2_{.035-}$	$1_{.010+}$	$0_{.002}$	$0_{.002}$
	7	$1_{.023}$	$1_{.023}$	$0_{.004}$	$0_{.004}$
	6	$0_{.011}$	$0_{.011}$	—	—
	5	$0_{.026}$	—	—	—
14	16	$10_{.037}$	$9_{.014}$	$8_{.005+}$	$7_{.002}$
	15	$8_{.025+}$	$7_{.010-}$	$7_{.010-}$	$6_{.003}$
	14	$7_{.032}$	$6_{.013}$	$5_{.005-}$	$5_{.005-}$
	13	$6_{.035+}$	$5_{.014}$	$4_{.005+}$	$3_{.001}$
	12	$5_{.035+}$	$4_{.014}$	$3_{.005-}$	$3_{.005-}$
	11	$4_{.033}$	$3_{.012}$	$2_{.004}$	$2_{.004}$
	10	$3_{.028}$	$2_{.009}$	$2_{.009}$	$1_{.002}$
	9	$2_{.021}$	$2_{.021}$	$1_{.006}$	$0_{.001}$
	8	$2_{.045-}$	$1_{.013}$	$0_{.002}$	$0_{.002}$
	7	$1_{.030}$	$0_{.006}$	$0_{.006}$	—
	6	$0_{.013}$	$0_{.013}$	—	—
	5	$0_{.031}$	—	—	—
13	16	$9_{.030}$	$8_{.011}$	$7_{.004}$	$7_{.004}$
	15	$8_{.047}$	$7_{.019}$	$6_{.007}$	$5_{.002}$
	14	$6_{.023}$	$6_{.023}$	$5_{.008}$	$4_{.003}$
	13	$5_{.023}$	$5_{.023}$	$4_{.008}$	$3_{.003}$
	12	$4_{.022}$	$4_{.022}$	$3_{.007}$	$2_{.002}$
	11	$4_{.048}$	$3_{.018}$	$2_{.005+}$	$1_{.001}$
	10	$3_{.039}$	$2_{.013}$	$1_{.003}$	$1_{.003}$
	9	$2_{.029}$	$1_{.008}$	$1_{.008}$	$0_{.001}$
	8	$1_{.018}$	$1_{.018}$	$0_{.003}$	$0_{.003}$
	7	$1_{.038}$	$0_{.007}$	$0_{.007}$	—
	6	$0_{.017}$	$0_{.017}$	—	—
	5	$0_{.037}$	—	—	—
12	16	$8_{.024}$	$8_{.024}$	$7_{.008}$	$6_{.002}$
	15	$7_{.036}$	$6_{.013}$	$5_{.004}$	$5_{.004}$
	14	$6_{.040}$	$5_{.015-}$	$4_{.005-}$	$4_{.005-}$
	13	$5_{.039}$	$4_{.014}$	$3_{.004}$	$3_{.004}$
	12	$4_{.034}$	$3_{.012}$	$2_{.003}$	$2_{.003}$
	11	$3_{.027}$	$2_{.008}$	$2_{.008}$	$1_{.002}$
	10	$2_{.019}$	$2_{.019}$	$1_{.005-}$	$1_{.005-}$
	9	$2_{.040}$	$1_{.011}$	$0_{.002}$	$0_{.002}$
	8	$1_{.024}$	$1_{.024}$	$0_{.004}$	$0_{.004}$
	7	$1_{.048}$	$0_{.010-}$	$0_{.010-}$	—
	6	$0_{.021}$	$0_{.021}$	—	—
	5	$0_{.044}$	—	—	—
***A* = 16 *B* = 11**	**16**	$7_{.019}$	$7_{.019}$	$6_{.006}$	$5_{.002}$
	15	$6_{.027}$	$5_{.009}$	$5_{.009}$	$4_{.002}$
	14	$5_{.027}$	$4_{.009}$	$4_{.009}$	$3_{.002}$
	13	$4_{.024}$	$4_{.024}$	$3_{.008}$	$2_{.002}$
	12	$3_{.019}$	$3_{.019}$	$2_{.005+}$	$1_{.001}$
	11	$3_{.041}$	$2_{.013}$	$1_{.003}$	$1_{.003}$
	10	$2_{.028}$	$1_{.007}$	$1_{.007}$	$0_{.001}$
	9	$1_{.016}$	$1_{.016}$	$0_{.002}$	$0_{.002}$
	8	$1_{.033}$	$0_{.006}$	$0_{.006}$	—

TABLE J ***(continued)***

		Probability			
	a	**0.05**	**0.025**	**0.01**	**0.005**
	7	$\mathbf{0}_{.013}$	$\mathbf{0}_{.013}$	—	—
	6	$\mathbf{0}_{.027}$	—	—	—
10	16	$\mathbf{7}_{.046}$	$\mathbf{6}_{.014}$	$\mathbf{5}_{.004}$	$\mathbf{5}_{.004}$
	15	$\mathbf{5}_{.018}$	$\mathbf{5}_{.018}$	$\mathbf{4}_{.005^{+}}$	$\mathbf{3}_{.001}$
	14	$\mathbf{4}_{.018}$	$\mathbf{4}_{.018}$	$\mathbf{3}_{.005^{-}}$	$\mathbf{3}_{.005^{-}}$
	13	$\mathbf{4}_{.042}$	$\mathbf{3}_{.014}$	$\mathbf{2}_{.003}$	$\mathbf{2}_{.003}$
	12	$\mathbf{3}_{.032}$	$\mathbf{2}_{.009}$	$\mathbf{2}_{.009}$	$\mathbf{1}_{.002}$
	11	$\mathbf{2}_{.021}$	$\mathbf{2}_{.021}$	$\mathbf{1}_{.005^{-}}$	$\mathbf{1}_{.005^{-}}$
	10	$\mathbf{2}_{.042}$	$\mathbf{1}_{.011}$	$\mathbf{0}_{.002}$	$\mathbf{0}_{.002}$
	9	$\mathbf{1}_{.023}$	$\mathbf{1}_{.023}$	$\mathbf{0}_{.004}$	$\mathbf{0}_{.004}$
	8	$\mathbf{1}_{.045^{-}}$	$\mathbf{0}_{.008}$	$\mathbf{0}_{.008}$	—
	7	$\mathbf{0}_{.017}$	$\mathbf{0}_{.017}$	—	—
	6	$\mathbf{0}_{.035^{-}}$	—	—	—
9	16	$\mathbf{6}_{.037}$	$\mathbf{5}_{.010^{-}}$	$\mathbf{5}_{.010^{-}}$	$\mathbf{4}_{.002}$
	15	$\mathbf{5}_{.040}$	$\mathbf{4}_{.012}$	$\mathbf{3}_{.003}$	$\mathbf{3}_{.003}$
	14	$\mathbf{4}_{.034}$	$\mathbf{3}_{.010^{-}}$	$\mathbf{3}_{.010^{-}}$	$\mathbf{2}_{.002}$
	13	$\mathbf{3}_{.025^{+}}$	$\mathbf{2}_{.007}$	$\mathbf{2}_{.007}$	$\mathbf{1}_{.001}$
	12	$\mathbf{2}_{.016}$	$\mathbf{2}_{.016}$	$\mathbf{1}_{.003}$	$\mathbf{1}_{.003}$
	11	$\mathbf{2}_{.033}$	$\mathbf{1}_{.008}$	$\mathbf{1}_{.008}$	$\mathbf{0}_{.001}$
	10	$\mathbf{1}_{.017}$	$\mathbf{1}_{.017}$	$\mathbf{0}_{.002}$	$\mathbf{0}_{.002}$
	9	$\mathbf{1}_{.034}$	$\mathbf{0}_{.006}$	$\mathbf{0}_{.006}$	—
	8	$\mathbf{0}_{.012}$	$\mathbf{0}_{.012}$	—	—
	7	$\mathbf{0}_{.024}$	$\mathbf{0}_{.024}$	—	—
	6	$\mathbf{0}_{.045^{+}}$	—	—	—
8	16	$\mathbf{5}_{.028}$	$\mathbf{4}_{.007}$	$\mathbf{4}_{.007}$	$\mathbf{3}_{.001}$
	15	$\mathbf{4}_{.028}$	$\mathbf{3}_{.007}$	$\mathbf{3}_{.007}$	$\mathbf{2}_{.001}$
	14	$\mathbf{3}_{.021}$	$\mathbf{3}_{.021}$	$\mathbf{2}_{.005^{-}}$	$\mathbf{2}_{.005^{-}}$
	13	$\mathbf{3}_{.047}$	$\mathbf{2}_{.013}$	$\mathbf{1}_{.002}$	$\mathbf{1}_{.002}$
	12	$\mathbf{2}_{.028}$	$\mathbf{1}_{.006}$	$\mathbf{1}_{.006}$	$\mathbf{0}_{.001}$
	11	$\mathbf{1}_{.014}$	$\mathbf{1}_{.014}$	$\mathbf{0}_{.002}$	$\mathbf{0}_{.002}$
	10	$\mathbf{1}_{.027}$	$\mathbf{0}_{.004}$	$\mathbf{0}_{.004}$	$\mathbf{0}_{.004}$
	9	$\mathbf{0}_{.009}$	$\mathbf{0}_{.009}$	$\mathbf{0}_{.009}$	—
	8	$\mathbf{0}_{.017}$	$\mathbf{0}_{.017}$	—	—
	7	$\mathbf{0}_{.033}$	—	—	—
7	16	$\mathbf{4}_{.020}$	$\mathbf{4}_{.020}$	$\mathbf{3}_{.004}$	$\mathbf{3}_{.004}$
	15	$\mathbf{3}_{.017}$	$\mathbf{3}_{.017}$	$\mathbf{2}_{.003}$	$\mathbf{2}_{.003}$
	14	$\mathbf{3}_{.045^{+}}$	$\mathbf{2}_{.011}$	$\mathbf{1}_{.002}$	$\mathbf{1}_{.002}$
	13	$\mathbf{2}_{.026}$	$\mathbf{1}_{.005^{-}}$	$\mathbf{1}_{.005^{-}}$	$\mathbf{1}_{.005^{-}}$
	12	$\mathbf{1}_{.012}$	$\mathbf{1}_{.012}$	$\mathbf{0}_{.001}$	$\mathbf{0}_{.001}$
	11	$\mathbf{1}_{.024}$	$\mathbf{1}_{.024}$	$\mathbf{0}_{.003}$	$\mathbf{0}_{.003}$
	10	$\mathbf{1}_{.045^{-}}$	$\mathbf{0}_{.007}$	$\mathbf{0}_{.007}$	—
	9	$\mathbf{0}_{.014}$	$\mathbf{0}_{.014}$	—	—
	8	$\mathbf{0}_{.026}$	—	—	—
	7	$\mathbf{0}_{.047}$	—	—	—
6	16	$\mathbf{3}_{.013}$	$\mathbf{3}_{.013}$	$\mathbf{2}_{.002}$	$\mathbf{2}_{.002}$
	15	$\mathbf{3}_{.046}$	$\mathbf{2}_{.009}$	$\mathbf{2}_{.009}$	$\mathbf{1}_{.001}$
	14	$\mathbf{2}_{.025^{+}}$	$\mathbf{1}_{.004}$	$\mathbf{1}_{.004}$	$\mathbf{1}_{.004}$
	13	$\mathbf{1}_{.011}$	$\mathbf{1}_{.011}$	$\mathbf{0}_{.001}$	$\mathbf{0}_{.001}$
	12	$\mathbf{1}_{.023}$	$\mathbf{1}_{.023}$	$\mathbf{0}_{.003}$	$\mathbf{0}_{.003}$
	11	$\mathbf{1}_{.043}$	$\mathbf{0}_{.006}$	$\mathbf{0}_{.006}$	—

TABLE J (*continued*)

		Probability			
	a	**0.05**	**0.025**	**0.01**	**0.005**
	10	$\mathbf{0}_{.012}$	$\mathbf{0}_{.012}$	—	—
***A* = 16 *B* = 6**	9	$\mathbf{0}_{.023}$	$\mathbf{0}_{.023}$	—	—
	8	$\mathbf{0}_{.040}$	—	—	—
5	16	$\mathbf{3}_{.048}$	$\mathbf{2}_{.008}$	$\mathbf{2}_{.008}$	$\mathbf{1}_{.001}$
	15	$\mathbf{2}_{.028}$	$\mathbf{1}_{.004}$	$\mathbf{1}_{.004}$	$\mathbf{1}_{.004}$
	14	$\mathbf{1}_{.011}$	$\mathbf{1}_{.011}$	$\mathbf{0}_{.001}$	$\mathbf{0}_{.001}$
	13	$\mathbf{1}_{.025+}$	$\mathbf{0}_{.003}$	$\mathbf{0}_{.003}$	$\mathbf{0}_{.003}$
	12	$\mathbf{1}_{.047}$	$\mathbf{0}_{.006}$	$\mathbf{0}_{.006}$	—
	11	$\mathbf{0}_{.012}$	$\mathbf{0}_{.012}$	—	—
	10	$\mathbf{0}_{.023}$	$\mathbf{0}_{.023}$	—	—
	9	$\mathbf{0}_{.039}$	—	—	—
4	16	$\mathbf{2}_{.032}$	$\mathbf{1}_{.004}$	$\mathbf{1}_{.004}$	$\mathbf{1}_{.004}$
	15	$\mathbf{1}_{.013}$	$\mathbf{1}_{.013}$	$\mathbf{0}_{.001}$	$\mathbf{1}_{.001}$
	14	$\mathbf{1}_{.032}$	$\mathbf{0}_{.003}$	$\mathbf{0}_{.003}$	$\mathbf{0}_{.003}$
	13	$\mathbf{0}_{.007}$	$\mathbf{0}_{.007}$	$\mathbf{0}_{.007}$	—
	12	$\mathbf{0}_{.014}$	$\mathbf{0}_{.014}$	—	—
	11	$\mathbf{0}_{.026}$	—	—	—
	10	$\mathbf{0}_{.043}$	—	—	—
3	16	$\mathbf{1}_{.018}$	$\mathbf{1}_{.018}$	$\mathbf{0}_{.001}$	$\mathbf{0}_{.001}$
	15	$\mathbf{0}_{.004}$	$\mathbf{0}_{.004}$	$\mathbf{0}_{.004}$	$\mathbf{0}_{.004}$
	14	$\mathbf{0}_{.010+}$	$\mathbf{0}_{.010+}$	—	—
	13	$\mathbf{0}_{.021}$	$\mathbf{0}_{.021}$	—	—
	12	$\mathbf{0}_{.036}$	—	—	—
2	16	$\mathbf{0}_{.007}$	$\mathbf{0}_{.007}$	$\mathbf{0}_{.007}$	—
	15	$\mathbf{0}_{.020}$	$\mathbf{0}_{.020}$	—	—
	14	$\mathbf{0}_{.039}$	—	—	—
***A* = 17 *B* = 17**	17	$\mathbf{12}_{.022}$	$\mathbf{12}_{.022}$	$\mathbf{11}_{.009}$	$\mathbf{10}_{.004}$
	16	$\mathbf{11}_{.043}$	$\mathbf{10}_{.020}$	$\mathbf{9}_{.008}$	$\mathbf{8}_{.003}$
	15	$\mathbf{9}_{.029}$	$\mathbf{8}_{.013}$	$\mathbf{7}_{.005+}$	$\mathbf{6}_{.002}$
	14	$\mathbf{8}_{.035+}$	$\mathbf{7}_{.016}$	$\mathbf{6}_{.007}$	$\mathbf{5}_{.002}$
	13	$\mathbf{7}_{.040}$	$\mathbf{6}_{.018}$	$\mathbf{5}_{.007}$	$\mathbf{4}_{.003}$
	12	$\mathbf{6}_{.042}$	$\mathbf{5}_{.019}$	$\mathbf{4}_{.007}$	$\mathbf{3}_{.002}$
	11	$\mathbf{5}_{.042}$	$\mathbf{4}_{.018}$	$\mathbf{3}_{.007}$	$\mathbf{2}_{.002}$
	10	$\mathbf{4}_{.040}$	$\mathbf{3}_{.016}$	$\mathbf{2}_{.005+}$	$\mathbf{1}_{.001}$
	9	$\mathbf{3}_{.035+}$	$\mathbf{2}_{.013}$	$\mathbf{1}_{.003}$	$\mathbf{1}_{.003}$
	8	$\mathbf{2}_{.029}$	$\mathbf{1}_{.008}$	$\mathbf{1}_{.008}$	$\mathbf{0}_{.001}$
	7	$\mathbf{1}_{.020}$	$\mathbf{1}_{.020}$	$\mathbf{0}_{.004}$	$\mathbf{0}_{.004}$
	6	$\mathbf{1}_{.043}$	$\mathbf{0}_{.009}$	$\mathbf{0}_{.009}$	—
	5	$\mathbf{0}_{.022}$	$\mathbf{0}_{.022}$	—	—
16	17	$\mathbf{12}_{.044}$	$\mathbf{11}_{.018}$	$\mathbf{10}_{.007}$	$\mathbf{9}_{.003}$
	16	$\mathbf{10}_{.035-}$	$\mathbf{9}_{.015-}$	$\mathbf{8}_{.006}$	$\mathbf{7}_{.002}$
	15	$\mathbf{9}_{.046}$	$\mathbf{8}_{.021}$	$\mathbf{7}_{.009}$	$\mathbf{6}_{.003}$
	14	$\mathbf{7}_{.025+}$	$\mathbf{6}_{.011}$	$\mathbf{5}_{.004}$	$\mathbf{5}_{.004}$
	13	$\mathbf{6}_{.027}$	$\mathbf{5}_{.011}$	$\mathbf{4}_{.004}$	$\mathbf{4}_{.004}$
	12	$\mathbf{5}_{.027}$	$\mathbf{4}_{.011}$	$\mathbf{3}_{.004}$	$\mathbf{3}_{.004}$
	11	$\mathbf{4}_{.025+}$	$\mathbf{3}_{.009}$	$\mathbf{3}_{.009}$	$\mathbf{2}_{.003}$
	10	$\mathbf{3}_{.022}$	$\mathbf{3}_{.022}$	$\mathbf{2}_{.007}$	$\mathbf{1}_{.002}$
	9	$\mathbf{3}_{.046}$	$\mathbf{2}_{.017}$	$\mathbf{1}_{.004}$	$\mathbf{1}_{.004}$
	8	$\mathbf{2}_{.036}$	$\mathbf{1}_{.011}$	$\mathbf{0}_{.002}$	$\mathbf{0}_{.002}$
	7	$\mathbf{1}_{.024}$	$\mathbf{1}_{.024}$	$\mathbf{0}_{.005-}$	$\mathbf{0}_{.005-}$

TABLE J ***(continued)***

			Probability			
		a	**0.05**	**0.025**	**0.01**	**0.005**
		6	$0_{.011}$	$0_{.011}$	—	—
		5	$0_{.026}$	—	—	—
	15	17	$11_{.038}$	$10_{.015^-}$	$9_{.006}$	$8_{.002}$
		16	$9_{.027}$	$8_{.011}$	$7_{.004}$	$7_{.004}$
		15	$8_{.035^+}$	$7_{.015^-}$	$6_{.006}$	$5_{.002}$
		14	$7_{.040}$	$6_{.017}$	$5_{.006}$	$4_{.002}$
		13	$6_{.041}$	$5_{.017}$	$4_{.006}$	$3_{.002}$
		12	$5_{.039}$	$4_{.016}$	$3_{.005^+}$	$2_{.001}$
		11	$4_{.035^+}$	$3_{.013}$	$2_{.004}$	$2_{.004}$
		10	$3_{.029}$	$2_{.010^-}$	$2_{.010^-}$	$1_{.002}$
		9	$2_{.022}$	$2_{.022}$	$1_{.006}$	$0_{.001}$
$A = 17$ $B = 15$		8	$2_{.046}$	$1_{.014}$	$0_{.002}$	$0_{.002}$
		7	$1_{.030}$	$0_{.006}$	$0_{.006}$	—
		6	$0_{.014}$	$0_{.014}$	—	—
		5	$0_{.031}$	—	—	—
	14	17	$10_{.032}$	$9_{.012}$	$8_{.004}$	$8_{.004}$
		16	$8_{.021}$	$8_{.021}$	$7_{.008}$	$6_{.003}$
		15	$7_{.026}$	$6_{.010^-}$	$6_{.010^-}$	$5_{.003}$
		14	$6_{.028}$	$5_{.011}$	$4_{.004}$	$4_{.004}$
		13	$5_{.027}$	$4_{.010^-}$	$4_{.010^-}$	$3_{.003}$
		12	$4_{.024}$	$4_{.024}$	$3_{.008}$	$2_{.002}$
		11	$4_{.049}$	$3_{.019}$	$2_{.006}$	$1_{.001}$
		10	$3_{.040}$	$2_{.014}$	$1_{.003}$	$1_{.003}$
		9	$2_{.029}$	$1_{.008}$	$1_{.008}$	$0_{.001}$
		8	$1_{.018}$	$1_{.018}$	$0_{.003}$	$0_{.003}$
		7	$1_{.038}$	$0_{.007}$	$0_{.007}$	—
		6	$0_{.017}$	$0_{.017}$	—	—
		5	$0_{.036}$	—	—	—
	13	17	$9_{.026}$	$8_{.009}$	$8_{.009}$	$7_{.003}$
		16	$8_{.040}$	$7_{.015^+}$	$6_{.005^+}$	$5_{.002}$
		15	$7_{.045^+}$	$6_{.018}$	$5_{.006}$	$4_{.002}$
		14	$6_{.045^+}$	$5_{.018}$	$4_{.006}$	$3_{.002}$
		13	$5_{.042}$	$4_{.016}$	$3_{.005^+}$	$2_{.001}$
		12	$4_{.035^+}$	$3_{.013}$	$2_{.004}$	$2_{.004}$
		11	$3_{.028}$	$2_{.009}$	$2_{.009}$	$1_{.002}$
		10	$2_{.019}$	$2_{.019}$	$1_{.005^-}$	$1_{.005^-}$
		9	$2_{.040}$	$1_{.011}$	$0_{.002}$	$0_{.002}$
		8	$1_{.024}$	$1_{.024}$	$0_{.004}$	$0_{.004}$
		7	$1_{.047}$	$0_{.010^-}$	$0_{.010^-}$	—
		6	$0_{.021}$	$0_{.021}$	—	—
		5	$0_{.043}$	—	—	—
	12	17	$8_{.021}$	$8_{.021}$	$7_{.007}$	$6_{.002}$
		16	$7_{.030}$	$6_{.011}$	$5_{.003}$	$5_{.003}$
		15	$6_{.033}$	$5_{.012}$	$4_{.004}$	$4_{.004}$
		14	$5_{.030}$	$4_{.011}$	$3_{.003}$	$3_{.003}$
		13	$4_{.026}$	$3_{.008}$	$3_{.008}$	$2_{.002}$
		12	$3_{.020}$	$3_{.020}$	$2_{.006}$	$1_{.001}$
		11	$3_{.041}$	$2_{.013}$	$1_{.003}$	$1_{.003}$
		10	$2_{.028}$	$1_{.007}$	$1_{.007}$	$0_{.001}$
		9	$1_{.016}$	$1_{.016}$	$0_{.002}$	$0_{.002}$

TABLE J (continued)

		Probability			
	a	0.05	0.025	0.01	0.005
	8	$1_{.032}$	$0_{.006}$	$0_{.006}$	—
	7	$0_{.012}$	$0_{.012}$	—	—
	6	$0_{.026}$	—	—	—
11	17	$7_{.016}$	$7_{.016}$	$6_{.005-}$	$6_{.005-}$
	16	$6_{.022}$	$6_{.022}$	$5_{.007}$	$4_{.002}$
	15	$5_{.022}$	$5_{.022}$	$4_{.007}$	$3_{.002}$
	14	$4_{.019}$	$4_{.019}$	$3_{.006}$	$2_{.001}$
	13	$4_{.042}$	$3_{.014}$	$2_{.004}$	$2_{.004}$
	12	$3_{.031}$	$2_{.009}$	$2_{.009}$	$1_{.002}$
	11	$2_{.020}$	$2_{.020}$	$1_{.005-}$	$1_{.005-}$
	10	$2_{.040}$	$1_{.011}$	$0_{.001}$	$0_{.001}$
	9	$1_{.022}$	$1_{.022}$	$0_{.004}$	$0_{.004}$
	8	$1_{.042}$	$0_{.008}$	$0_{.008}$	—
	7	$0_{.016}$	$0_{.016}$	—	—
	6	$0_{.033}$	—	—	—
10	17	$7_{.041}$	$6_{.012}$	$5_{.003}$	$5_{.003}$
	16	$6_{.047}$	$5_{.015+}$	$4_{.004}$	$4_{.004}$
	15	$5_{.043}$	$4_{.014}$	$3_{.004}$	$3_{.004}$
	14	$4_{.034}$	$3_{.010+}$	$2_{.002}$	$2_{.002}$
	13	$3_{.024}$	$3_{.024}$	$2_{.007}$	$1_{.001}$
	12	$3_{.049}$	$2_{.015+}$	$1_{.003}$	$1_{.003}$
$A = 17$ $B = 10$	11	$2_{.031}$	$1_{.007}$	$1_{.007}$	$0_{.001}$
	10	$1_{.016}$	$1_{.016}$	$0_{.002}$	$0_{.002}$
	9	$1_{.031}$	$0_{.005+}$	$0_{.005+}$	—
	8	$0_{.011}$	$0_{.011}$	—	—
	7	$0_{.022}$	$0_{.022}$	—	—
	6	$0_{.042}$	—	—	—
9	17	$6_{.032}$	$5_{.008}$	$5_{.008}$	$4_{.002}$
	16	$5_{.034}$	$4_{.010-}$	$4_{.010-}$	$3_{.002}$
	15	$4_{.028}$	$3_{.008}$	$3_{.008}$	$2_{.002}$
	14	$3_{.020}$	$3_{.020}$	$2_{.005-}$	$2_{.005-}$
	13	$3_{.042}$	$2_{.012}$	$1_{.002}$	$1_{.002}$
	12	$2_{.025+}$	$1_{.006}$	$1_{.006}$	$0_{.001}$
	11	$2_{.048}$	$1_{.012}$	$0_{.002}$	$0_{.002}$
	10	$1_{.024}$	$1_{.024}$	$0_{.004}$	$0_{.004}$
	9	$1_{.045-}$	$0_{.008}$	$0_{.008}$	—
	8	$0_{.016}$	$0_{.016}$	—	—
	7	$0_{.030}$	—	—	—
8	17	$5_{.024}$	$5_{.024}$	$4_{.006}$	$3_{.001}$
	16	$4_{.023}$	$4_{.023}$	$3_{.006}$	$2_{.001}$
	15	$3_{.017}$	$3_{.017}$	$2_{.004}$	$2_{.004}$
	14	$3_{.039}$	$2_{.010-}$	$2_{.010-}$	$1_{.002}$
	13	$2_{.022}$	$2_{.022}$	$1_{.004}$	$1_{.004}$
	12	$2_{.043}$	$1_{.010-}$	$1_{.010-}$	$0_{.001}$
	11	$1_{.020}$	$1_{.020}$	$0_{.003}$	$0_{.003}$
	10	$1_{.038}$	$0_{.006}$	$0_{.006}$	—
	9	$0_{.012}$	$0_{.012}$	—	—
	8	$0_{.022}$	$0_{.022}$	—	—
	7	$0_{.040}$	—	—	—
7	17	$4_{.017}$	$4_{.017}$	$3_{.003}$	$3_{.003}$

TABLE J (*continued*)

		Probability			
	a	**0.05**	**0.025**	**0.01**	**0.005**
	16	$\mathbf{3}_{.014}$	$\mathbf{3}_{.014}$	$\mathbf{2}_{.003}$	$\mathbf{2}_{.003}$
	15	$\mathbf{3}_{.038}$	$\mathbf{2}_{.009}$	$\mathbf{2}_{.009}$	$\mathbf{1}_{.001}$
	14	$\mathbf{2}_{.021}$	$\mathbf{2}_{.021}$	$\mathbf{1}_{.004}$	$\mathbf{1}_{.004}$
	13	$\mathbf{2}_{.042}$	$\mathbf{1}_{.009}$	$\mathbf{1}_{.009}$	$\mathbf{0}_{.001}$
	12	$\mathbf{1}_{.018}$	$\mathbf{1}_{.018}$	$\mathbf{0}_{.002}$	$\mathbf{0}_{.002}$
	11	$\mathbf{1}_{.034}$	$\mathbf{0}_{.005^-}$	$\mathbf{0}_{.005^-}$	$\mathbf{0}_{.005^-}$
	10	$\mathbf{0}_{.010^-}$	$\mathbf{0}_{.010^-}$	$\mathbf{0}_{.010^-}$	—
	9	$\mathbf{0}_{.019}$	$\mathbf{0}_{.019}$	—	—
	8	$\mathbf{0}_{.033}$	—	—	—
6	17	$\mathbf{3}_{.011}$	$\mathbf{3}_{.011}$	$\mathbf{2}_{.002}$	$\mathbf{2}_{.002}$
	16	$\mathbf{3}_{.040}$	$\mathbf{2}_{.008}$	$\mathbf{2}_{.008}$	$\mathbf{1}_{.001}$
	15	$\mathbf{2}_{.021}$	$\mathbf{2}_{.021}$	$\mathbf{1}_{.003}$	$\mathbf{1}_{.003}$
	14	$\mathbf{2}_{.045^+}$	$\mathbf{1}_{.009}$	$\mathbf{1}_{.009}$	$\mathbf{0}_{.001}$
	13	$\mathbf{1}_{.018}$	$\mathbf{1}_{.018}$	$\mathbf{0}_{.002}$	$\mathbf{0}_{.002}$
	12	$\mathbf{1}_{.035^-}$	$\mathbf{0}_{.005^-}$	$\mathbf{0}_{.005^-}$	$\mathbf{0}_{.005^-}$
	11	$\mathbf{0}_{.009}$	$\mathbf{0}_{.009}$	$\mathbf{0}_{.009}$	—
	10	$\mathbf{0}_{.017}$	$\mathbf{0}_{.017}$	—	—
	9	$\mathbf{0}_{.030}$	—	—	—
	8	$\mathbf{0}_{.050^-}$	—	—	—
5	17	$\mathbf{3}_{.043}$	$\mathbf{2}_{.006}$	$\mathbf{2}_{.006}$	$\mathbf{1}_{.001}$
	16	$\mathbf{2}_{.024}$	$\mathbf{2}_{.024}$	$\mathbf{1}_{.003}$	$\mathbf{1}_{.003}$
	15	$\mathbf{1}_{.009}$	$\mathbf{1}_{.009}$	$\mathbf{1}_{.009}$	$\mathbf{0}_{.001}$
	14	$\mathbf{1}_{.021}$	$\mathbf{1}_{.021}$	$\mathbf{0}_{.002}$	$\mathbf{0}_{.002}$
	13	$\mathbf{1}_{.039}$	$\mathbf{0}_{.005^-}$	$\mathbf{0}_{.005^-}$	$\mathbf{0}_{.005^-}$
	12	$\mathbf{0}_{.010^-}$	$\mathbf{0}_{.010^-}$	$\mathbf{0}_{.010^-}$	—
	11	$\mathbf{0}_{.018}$	$\mathbf{0}_{.018}$	—	—
	10	$\mathbf{0}_{.030}$	—	—	—
	9	$\mathbf{0}_{.049}$	—	—	—
4	17	$\mathbf{2}_{.029}$	$\mathbf{1}_{.003}$	$\mathbf{1}_{.003}$	$\mathbf{1}_{.003}$
	16	$\mathbf{1}_{.011}$	$\mathbf{1}_{.011}$	$\mathbf{0}_{.001}$	$\mathbf{0}_{.001}$
	15	$\mathbf{1}_{.028}$	$\mathbf{0}_{.003}$	$\mathbf{0}_{.003}$	$\mathbf{0}_{.003}$
	14	$\mathbf{0}_{.006}$	$\mathbf{0}_{.006}$	$\mathbf{0}_{.006}$	—
$A = 17$ $B = 4$	13	$\mathbf{0}_{.012}$	$\mathbf{0}_{.012}$	—	—
	12	$\mathbf{0}_{.021}$	$\mathbf{0}_{.021}$	—	—
	11	$\mathbf{0}_{.035^+}$	—	—	—
3	17	$\mathbf{1}_{.016}$	$\mathbf{1}_{.016}$	$\mathbf{0}_{.001}$	$\mathbf{0}_{.001}$
	16	$\mathbf{1}_{.046}$	$\mathbf{0}_{.004}$	$\mathbf{0}_{.004}$	$\mathbf{0}_{.004}$
	15	$\mathbf{0}_{.009}$	$\mathbf{0}_{.009}$	$\mathbf{0}_{.009}$	—
	14	$\mathbf{0}_{.018}$	$\mathbf{0}_{.018}$	—	—
	13	$\mathbf{0}_{.031}$	—	—	—
	12	$\mathbf{0}_{.049}$	—	—	—
2	17	$\mathbf{0}_{.006}$	$\mathbf{0}_{.006}$	$\mathbf{0}_{.006}$	—
	16	$\mathbf{0}_{.018}$	$\mathbf{0}_{.018}$	—	—
	15	$\mathbf{0}_{.035^+}$	—	—	—
$A = 18$ $B = 18$	18	$\mathbf{13}_{.023}$	$\mathbf{13}_{.023}$	$\mathbf{12}_{.010^-}$	$\mathbf{11}_{.004}$
	17	$\mathbf{12}_{.044}$	$\mathbf{11}_{.020}$	$\mathbf{10}_{.009}$	$\mathbf{9}_{.004}$
	16	$\mathbf{10}_{.030}$	$\mathbf{9}_{.014}$	$\mathbf{8}_{.006}$	$\mathbf{7}_{.002}$
	15	$\mathbf{9}_{.038}$	$\mathbf{8}_{.018}$	$\mathbf{7}_{.008}$	$\mathbf{6}_{.003}$
	14	$\mathbf{8}_{.043}$	$\mathbf{7}_{.020}$	$\mathbf{6}_{.009}$	$\mathbf{5}_{.003}$
	13	$\mathbf{7}_{.046}$	$\mathbf{6}_{.022}$	$\mathbf{5}_{.009}$	$\mathbf{4}_{.003}$

TABLE J ***(continued)***

		Probability			
	a	**0.05**	**0.025**	**0.01**	**0.005**
	12	$\mathbf{6}_{.047}$	$\mathbf{5}_{.022}$	$\mathbf{4}_{.009}$	$\mathbf{3}_{.003}$
	11	$\mathbf{5}_{.046}$	$\mathbf{4}_{.020}$	$\mathbf{3}_{.008}$	$\mathbf{2}_{.002}$
	10	$\mathbf{4}_{.043}$	$\mathbf{3}_{.018}$	$\mathbf{2}_{.006}$	$\mathbf{1}_{.001}$
	9	$\mathbf{3}_{.038}$	$\mathbf{2}_{.014}$	$\mathbf{1}_{.004}$	$\mathbf{1}_{.004}$
	8	$\mathbf{2}_{.030}$	$\mathbf{1}_{.009}$	$\mathbf{1}_{.009}$	$\mathbf{0}_{.001}$
	7	$\mathbf{1}_{.020}$	$\mathbf{1}_{.020}$	$\mathbf{0}_{.004}$	$\mathbf{0}_{.004}$
	6	$\mathbf{1}_{.044}$	$\mathbf{0}_{.010-}$	$\mathbf{0}_{.010-}$	—
	5	$\mathbf{0}_{.023}$	$\mathbf{0}_{.023}$	—	—
17	18	$\mathbf{13}_{.045+}$	$\mathbf{12}_{.019}$	$\mathbf{11}_{.008}$	$\mathbf{10}_{.003}$
	17	$\mathbf{11}_{.036}$	$\mathbf{10}_{.016}$	$\mathbf{9}_{.007}$	$\mathbf{8}_{.002}$
	16	$\mathbf{10}_{.049}$	$\mathbf{9}_{.023}$	$\mathbf{8}_{.010-}$	$\mathbf{7}_{.004}$
	15	$\mathbf{8}_{.028}$	$\mathbf{7}_{.012}$	$\mathbf{6}_{.005-}$	$\mathbf{6}_{.005-}$
	14	$\mathbf{7}_{.030}$	$\mathbf{6}_{.013}$	$\mathbf{5}_{.005+}$	$\mathbf{4}_{.002}$
	13	$\mathbf{6}_{.031}$	$\mathbf{5}_{.013}$	$\mathbf{4}_{.005-}$	$\mathbf{4}_{.005-}$
	12	$\mathbf{5}_{.030}$	$\mathbf{4}_{.012}$	$\mathbf{3}_{.004}$	$\mathbf{3}_{.004}$
	11	$\mathbf{4}_{.028}$	$\mathbf{3}_{.010+}$	$\mathbf{2}_{.003}$	$\mathbf{2}_{.003}$
	10	$\mathbf{3}_{.023}$	$\mathbf{3}_{.023}$	$\mathbf{2}_{.008}$	$\mathbf{1}_{.002}$
	9	$\mathbf{3}_{.047}$	$\mathbf{2}_{.018}$	$\mathbf{1}_{.005-}$	$\mathbf{1}_{.005-}$
	8	$\mathbf{2}_{.037}$	$\mathbf{1}_{.011}$	$\mathbf{0}_{.002}$	$\mathbf{0}_{.002}$
	7	$\mathbf{1}_{.025-}$	$\mathbf{1}_{.025-}$	$\mathbf{0}_{.005-}$	$\mathbf{0}_{.005-}$
	6	$\mathbf{0}_{.011}$	$\mathbf{0}_{.011}$	—	—
	5	$\mathbf{0}_{.026}$	—	—	—
16	18	$\mathbf{12}_{.039}$	$\mathbf{11}_{.016}$	$\mathbf{10}_{.006}$	$\mathbf{9}_{.002}$
	17	$\mathbf{10}_{.029}$	$\mathbf{9}_{.012}$	$\mathbf{8}_{.005-}$	$\mathbf{8}_{.005-}$
	16	$\mathbf{9}_{.038}$	$\mathbf{8}_{.017}$	$\mathbf{7}_{.007}$	$\mathbf{6}_{.002}$
	15	$\mathbf{8}_{.043}$	$\mathbf{7}_{.019}$	$\mathbf{6}_{.008}$	$\mathbf{5}_{.003}$
	14	$\mathbf{7}_{.046}$	$\mathbf{6}_{.020}$	$\mathbf{5}_{.008}$	$\mathbf{4}_{.003}$
	13	$\mathbf{6}_{.045+}$	$\mathbf{5}_{.020}$	$\mathbf{4}_{.007}$	$\mathbf{3}_{.002}$
	12	$\mathbf{5}_{.042}$	$\mathbf{4}_{.018}$	$\mathbf{3}_{.006}$	$\mathbf{2}_{.002}$
	11	$\mathbf{4}_{.037}$	$\mathbf{3}_{.015-}$	$\mathbf{2}_{.004}$	$\mathbf{2}_{.004}$
	10	$\mathbf{3}_{.031}$	$\mathbf{2}_{.011}$	$\mathbf{1}_{.003}$	$\mathbf{1}_{.003}$
	9	$\mathbf{2}_{.023}$	$\mathbf{2}_{.023}$	$\mathbf{1}_{.006}$	$\mathbf{0}_{.001}$
	8	$\mathbf{2}_{.046}$	$\mathbf{1}_{.014}$	$\mathbf{0}_{.002}$	$\mathbf{0}_{.002}$
	7	$\mathbf{1}_{.030}$	$\mathbf{0}_{.006}$	$\mathbf{0}_{.006}$	—
	6	$\mathbf{0}_{.014}$	$\mathbf{0}_{.014}$	—	—
	5	$\mathbf{0}_{.031}$	—	—	—
15	18	$\mathbf{11}_{.033}$	$\mathbf{10}_{.013}$	$\mathbf{9}_{.005-}$	$\mathbf{9}_{.005-}$
	17	$\mathbf{9}_{.023}$	$\mathbf{9}_{.023}$	$\mathbf{8}_{.009}$	$\mathbf{7}_{.003}$
	16	$\mathbf{8}_{.029}$	$\mathbf{7}_{.012}$	$\mathbf{6}_{.004}$	$\mathbf{6}_{.004}$
	15	$\mathbf{7}_{.031}$	$\mathbf{6}_{.013}$	$\mathbf{5}_{.005-}$	$\mathbf{5}_{.005-}$
	14	$\mathbf{6}_{.031}$	$\mathbf{5}_{.013}$	$\mathbf{4}_{.004}$	$\mathbf{4}_{.004}$
	13	$\mathbf{5}_{.029}$	$\mathbf{4}_{.011}$	$\mathbf{3}_{.004}$	$\mathbf{3}_{.004}$
***A* = 18 *B* = 15**	12	$\mathbf{4}_{.025+}$	$\mathbf{3}_{.009}$	$\mathbf{3}_{.009}$	$\mathbf{2}_{.003}$
	11	$\mathbf{3}_{.020}$	$\mathbf{3}_{.020}$	$\mathbf{2}_{.006}$	$\mathbf{1}_{.001}$
	10	$\mathbf{3}_{.041}$	$\mathbf{2}_{.014}$	$\mathbf{1}_{.004}$	$\mathbf{1}_{.004}$
	9	$\mathbf{2}_{.030}$	$\mathbf{1}_{.008}$	$\mathbf{1}_{.008}$	$\mathbf{0}_{.001}$
	8	$\mathbf{1}_{.018}$	$\mathbf{1}_{.018}$	$\mathbf{0}_{.003}$	$\mathbf{0}_{.003}$
	7	$\mathbf{1}_{.038}$	$\mathbf{0}_{.007}$	$\mathbf{0}_{.007}$	—
	6	$\mathbf{0}_{.017}$	$\mathbf{0}_{.017}$	—	—
	5	$\mathbf{0}_{.036}$	—	—	—

TABLE J ***(continued)***

		Probability			
	a	**0.05**	**0.025**	**0.01**	**0.005**
14	18	$\mathbf{10}_{.028}$	$\mathbf{9}_{.010-}$	$\mathbf{9}_{.010-}$	$\mathbf{8}_{.003}$
	17	$\mathbf{9}_{.043}$	$\mathbf{8}_{.017}$	$\mathbf{7}_{.006}$	$\mathbf{6}_{.002}$
	16	$\mathbf{8}_{.050-}$	$\mathbf{7}_{.021}$	$\mathbf{6}_{.008}$	$\mathbf{5}_{.003}$
	15	$\mathbf{6}_{.022}$	$\mathbf{6}_{.022}$	$\mathbf{5}_{.008}$	$\mathbf{4}_{.003}$
	14	$\mathbf{6}_{.049}$	$\mathbf{5}_{.020}$	$\mathbf{4}_{.007}$	$\mathbf{3}_{.002}$
	13	$\mathbf{5}_{.044}$	$\mathbf{4}_{.017}$	$\mathbf{3}_{.006}$	$\mathbf{2}_{.001}$
	12	$\mathbf{4}_{.037}$	$\mathbf{3}_{.013}$	$\mathbf{2}_{.004}$	$\mathbf{2}_{.004}$
	11	$\mathbf{3}_{.028}$	$\mathbf{2}_{.009}$	$\mathbf{2}_{.009}$	$\mathbf{1}_{.002}$
	10	$\mathbf{2}_{.020}$	$\mathbf{2}_{.020}$	$\mathbf{1}_{.005-}$	$\mathbf{1}_{.005-}$
	9	$\mathbf{2}_{.039}$	$\mathbf{1}_{.011}$	$\mathbf{0}_{.002}$	$\mathbf{0}_{.002}$
	8	$\mathbf{1}_{.024}$	$\mathbf{1}_{.024}$	$\mathbf{0}_{.004}$	$\mathbf{0}_{.004}$
	7	$\mathbf{1}_{.047}$	$\mathbf{0}_{.009}$	$\mathbf{0}_{.009}$	—
	6	$\mathbf{0}_{.020}$	$\mathbf{0}_{.020}$	—	—
	5	$\mathbf{0}_{.043}$	—	—	—
13	18	$\mathbf{9}_{.023}$	$\mathbf{9}_{.023}$	$\mathbf{8}_{.008}$	$\mathbf{7}_{.002}$
	17	$\mathbf{8}_{.034}$	$\mathbf{7}_{.012}$	$\mathbf{6}_{.004}$	$\mathbf{6}_{.004}$
	16	$\mathbf{7}_{.037}$	$\mathbf{6}_{.014}$	$\mathbf{5}_{.005-}$	$\mathbf{5}_{.005-}$
	15	$\mathbf{6}_{.036}$	$\mathbf{5}_{.014}$	$\mathbf{4}_{.004}$	$\mathbf{4}_{.004}$
	14	$\mathbf{5}_{.032}$	$\mathbf{4}_{.012}$	$\mathbf{3}_{.004}$	$\mathbf{3}_{.004}$
	13	$\mathbf{4}_{.027}$	$\mathbf{3}_{.009}$	$\mathbf{3}_{.009}$	$\mathbf{2}_{.002}$
	12	$\mathbf{3}_{.020}$	$\mathbf{3}_{.020}$	$\mathbf{2}_{.006}$	$\mathbf{1}_{.001}$
	11	$\mathbf{3}_{.040}$	$\mathbf{2}_{.013}$	$\mathbf{1}_{.003}$	$\mathbf{1}_{.003}$
	10	$\mathbf{2}_{.027}$	$\mathbf{1}_{.007}$	$\mathbf{1}_{.007}$	$\mathbf{0}_{.001}$
	9	$\mathbf{1}_{.015+}$	$\mathbf{1}_{.015+}$	$\mathbf{0}_{.002}$	$\mathbf{0}_{.002}$
	8	$\mathbf{1}_{.031}$	$\mathbf{0}_{.006}$	$\mathbf{0}_{.006}$	—
	7	$\mathbf{0}_{.012}$	$\mathbf{0}_{.012}$	—	—
	6	$\mathbf{0}_{.025+}$	—	—	—
12	18	$\mathbf{8}_{.018}$	$\mathbf{8}_{.018}$	$\mathbf{7}_{.006}$	$\mathbf{6}_{.002}$
	17	$\mathbf{7}_{.026}$	$\mathbf{6}_{.009}$	$\mathbf{6}_{.009}$	$\mathbf{5}_{.003}$
	16	$\mathbf{6}_{.027}$	$\mathbf{5}_{.009}$	$\mathbf{5}_{.009}$	$\mathbf{4}_{.003}$
	15	$\mathbf{5}_{.024}$	$\mathbf{5}_{.024}$	$\mathbf{4}_{.008}$	$\mathbf{3}_{.002}$
	14	$\mathbf{4}_{.020}$	$\mathbf{4}_{.020}$	$\mathbf{3}_{.006}$	$\mathbf{2}_{.001}$
	13	$\mathbf{4}_{.042}$	$\mathbf{3}_{.014}$	$\mathbf{2}_{.004}$	$\mathbf{2}_{.004}$
	12	$\mathbf{3}_{.030}$	$\mathbf{2}_{.009}$	$\mathbf{2}_{.009}$	$\mathbf{1}_{.002}$
	11	$\mathbf{2}_{.019}$	$\mathbf{2}_{.019}$	$\mathbf{1}_{.005-}$	$\mathbf{1}_{.005-}$
	10	$\mathbf{2}_{.038}$	$\mathbf{1}_{.010+}$	$\mathbf{0}_{.001}$	$\mathbf{0}_{.001}$
	9	$\mathbf{1}_{.021}$	$\mathbf{1}_{.021}$	$\mathbf{0}_{.003}$	$\mathbf{0}_{.003}$
	8	$\mathbf{1}_{.040}$	$\mathbf{0}_{.007}$	$\mathbf{0}_{.007}$	—
	7	$\mathbf{0}_{.016}$	$\mathbf{0}_{.016}$	—	—
	6	$\mathbf{0}_{.031}$	—	—	—
11	18	$\mathbf{8}_{.045+}$	$\mathbf{7}_{.014}$	$\mathbf{6}_{.004}$	$\mathbf{6}_{.004}$
	17	$\mathbf{6}_{.018}$	$\mathbf{6}_{.018}$	$\mathbf{5}_{.006}$	$\mathbf{4}_{.001}$
	16	$\mathbf{5}_{.018}$	$\mathbf{5}_{.018}$	$\mathbf{4}_{.005+}$	$\mathbf{3}_{.001}$
	15	$\mathbf{5}_{.043}$	$\mathbf{4}_{.015-}$	$\mathbf{3}_{.004}$	$\mathbf{3}_{.004}$
	14	$\mathbf{4}_{.033}$	$\mathbf{3}_{.011}$	$\mathbf{2}_{.003}$	$\mathbf{2}_{.003}$
	13	$\mathbf{3}_{.023}$	$\mathbf{3}_{.023}$	$\mathbf{2}_{.007}$	$\mathbf{1}_{.001}$
	12	$\mathbf{3}_{.046}$	$\mathbf{2}_{.014}$	$\mathbf{1}_{.003}$	$\mathbf{1}_{.003}$
	11	$\mathbf{2}_{.029}$	$\mathbf{1}_{.007}$	$\mathbf{1}_{.007}$	$\mathbf{0}_{.001}$
	10	$\mathbf{1}_{.015-}$	$\mathbf{1}_{.015-}$	$\mathbf{0}_{.002}$	$\mathbf{0}_{.002}$
	9	$\mathbf{1}_{.029}$	$\mathbf{0}_{.005-}$	$\mathbf{0}_{.005-}$	$\mathbf{0}_{.005-}$

TABLE J ***(continued)***

		Probability			
	a	**0.05**	**0.025**	**0.01**	**0.005**
	8	$0_{.010^+}$	$0_{.010^+}$	—	—
	7	$0_{.020}$	$0_{.020}$	—	—
	6	$0_{.039}$	—	—	—
$A = 18$ $B = 10$	18	$7_{.037}$	$6_{.010^+}$	$5_{.003}$	$5_{.003}$
	17	$6_{.041}$	$5_{.013}$	$4_{.003}$	$4_{.003}$
	16	$5_{.036}$	$4_{.011}$	$3_{.003}$	$3_{.003}$
	15	$4_{.028}$	$3_{.008}$	$3_{.008}$	$2_{.002}$
	14	$3_{.019}$	$3_{.019}$	$2_{.005^-}$	$2_{.005^-}$
	13	$3_{.039}$	$2_{.011}$	$1_{.002}$	$1_{.002}$
	12	$2_{.023}$	$2_{.023}$	$1_{.005^+}$	$0_{.001}$
	11	$2_{.043}$	$1_{.011}$	$0_{.001}$	$0_{.001}$
	10	$1_{.022}$	$1_{.022}$	$0_{.003}$	$0_{.003}$
	9	$1_{.040}$	$0_{.007}$	$0_{.007}$	—
	8	$0_{.014}$	$0_{.014}$	—	—
	7	$0_{.027}$	—	—	—
	6	$0_{.049}$	—	—	—
9	18	$6_{.029}$	$5_{.007}$	$5_{.007}$	$4_{.002}$
	17	$5_{.030}$	$4_{.008}$	$4_{.008}$	$3_{.002}$
	16	$4_{.023}$	$4_{.023}$	$3_{.006}$	$2_{.001}$
	15	$3_{.016}$	$3_{.016}$	$2_{.004}$	$2_{.004}$
	14	$3_{.034}$	$2_{.009}$	$2_{.009}$	$1_{.002}$
	13	$2_{.019}$	$2_{.019}$	$1_{.004}$	$1_{.004}$
	12	$2_{.037}$	$1_{.009}$	$1_{.009}$	$0_{.001}$
	11	$1_{.018}$	$1_{.018}$	$0_{.002}$	$0_{.002}$
	10	$1_{.033}$	$0_{.005^+}$	$0_{.005^+}$	—
	9	$0_{.010^+}$	$0_{.010^+}$	—	—
	8	$0_{.020}$	$0_{.020}$	—	—
	7	$0_{.036}$	—	—	—
8	18	$5_{.022}$	$5_{.022}$	$4_{.005^-}$	$4_{.005^-}$
	17	$4_{.020}$	$4_{.020}$	$3_{.004}$	$3_{.004}$
	16	$3_{.014}$	$3_{.014}$	$2_{.003}$	$2_{.003}$
	15	$3_{.032}$	$2_{.008}$	$2_{.008}$	$1_{.001}$
	14	$2_{.017}$	$2_{.017}$	$1_{.003}$	$1_{.003}$
	13	$2_{.034}$	$1_{.007}$	$1_{.007}$	$0_{.001}$
	12	$1_{.015^+}$	$1_{.015^+}$	$0_{.002}$	$0_{.002}$
	11	$1_{.028}$	$0_{.004}$	$0_{.004}$	$0_{.004}$
	10	$1_{.049}$	$0_{.008}$	$0_{.008}$	—
	9	$0_{.016}$	$0_{.016}$	—	—
	8	$0_{.028}$	—	—	—
	7	$0_{.048}$	—	—	—
7	18	$4_{.015^+}$	$4_{.015^+}$	$3_{.003}$	$3_{.003}$
	17	$3_{.012}$	$3_{.012}$	$2_{.002}$	$2_{.002}$
	16	$3_{.032}$	$2_{.007}$	$2_{.007}$	$1_{.001}$
	15	$2_{.017}$	$2_{.017}$	$1_{.003}$	$1_{.003}$
	14	$2_{.034}$	$1_{.007}$	$1_{.007}$	$0_{.001}$
	13	$1_{.014}$	$1_{.014}$	$0_{.002}$	$0_{.002}$
	12	$1_{.027}$	$0_{.004}$	$0_{.004}$	$0_{.004}$
	11	$1_{.046}$	$0_{.007}$	$0_{.007}$	—
	10	$0_{.013}$	$0_{.013}$	—	—
	9	$0_{.024}$	$0_{.024}$	—	—

TABLE J (continued)

		Probability			
	a	0.05	0.025	0.01	0.005
	8	$\mathbf{0}_{.040}$	—	—	—
6	18	$\mathbf{3}_{.010-}$	$\mathbf{3}_{.010-}$	$\mathbf{3}_{.010-}$	$\mathbf{2}_{.001}$
	17	$\mathbf{3}_{.035+}$	$\mathbf{2}_{.006}$	$\mathbf{2}_{.006}$	$\mathbf{1}_{.001}$
	16	$\mathbf{2}_{.018}$	$\mathbf{2}_{.018}$	$\mathbf{1}_{.003}$	$\mathbf{1}_{.003}$
	15	$\mathbf{2}_{.038}$	$\mathbf{1}_{.007}$	$\mathbf{1}_{.007}$	$\mathbf{0}_{.001}$
	14	$\mathbf{1}_{.015-}$	$\mathbf{1}_{.015-}$	$\mathbf{0}_{.002}$	$\mathbf{0}_{.002}$
	13	$\mathbf{1}_{.028}$	$\mathbf{0}_{.003}$	$\mathbf{0}_{.003}$	$\mathbf{0}_{.003}$
	12	$\mathbf{1}_{.048}$	$\mathbf{0}_{.007}$	$\mathbf{0}_{.007}$	—
	11	$\mathbf{0}_{.013}$	$\mathbf{0}_{.013}$	—	—
	10	$\mathbf{0}_{.022}$	$\mathbf{0}_{.022}$	—	—
	9	$\mathbf{0}_{.037}$	—	—	—
5	18	$\mathbf{3}_{.040}$	$\mathbf{2}_{.006}$	$\mathbf{2}_{.006}$	$\mathbf{1}_{.001}$
	17	$\mathbf{2}_{.021}$	$\mathbf{2}_{.021}$	$\mathbf{1}_{.003}$	$\mathbf{1}_{.003}$
	16	$\mathbf{2}_{.048}$	$\mathbf{1}_{.008}$	$\mathbf{1}_{.008}$	$\mathbf{0}_{.001}$
$A = 18$ $B = 5$	15	$\mathbf{1}_{.017}$	$\mathbf{1}_{.017}$	$\mathbf{0}_{.002}$	$\mathbf{0}_{.002}$
	14	$\mathbf{1}_{.033}$	$\mathbf{0}_{.004}$	$\mathbf{0}_{.004}$	$\mathbf{0}_{.004}$
	13	$\mathbf{0}_{.007}$	$\mathbf{0}_{.007}$	$\mathbf{0}_{.007}$	—
	12	$\mathbf{0}_{.014}$	$\mathbf{0}_{.014}$	—	—
	11	$\mathbf{0}_{.024}$	$\mathbf{0}_{.024}$	—	—
	10	$\mathbf{0}_{.038}$	—	—	—
4	18	$\mathbf{2}_{.026}$	$\mathbf{1}_{.003}$	$\mathbf{1}_{.003}$	$\mathbf{1}_{.003}$
	17	$\mathbf{1}_{.010-}$	$\mathbf{1}_{.010-}$	$\mathbf{1}_{.010-}$	$\mathbf{0}_{.001}$
	16	$\mathbf{1}_{.024}$	$\mathbf{1}_{.024}$	$\mathbf{0}_{.002}$	$\mathbf{0}_{.002}$
	15	$\mathbf{1}_{.046}$	$\mathbf{0}_{.005-}$	$\mathbf{0}_{.005-}$	$\mathbf{0}_{.005-}$
	14	$\mathbf{0}_{.010-}$	$\mathbf{0}_{.010-}$	$\mathbf{0}_{.010-}$	—
	13	$\mathbf{0}_{.017}$	$\mathbf{0}_{.017}$	—	—
	12	$\mathbf{0}_{.029}$	—	—	—
	11	$\mathbf{0}_{.045+}$	—	—	—
3	18	$\mathbf{1}_{.014}$	$\mathbf{1}_{.014}$	$\mathbf{0}_{.001}$	$\mathbf{0}_{.001}$
	17	$\mathbf{1}_{.041}$	$\mathbf{0}_{.003}$	$\mathbf{0}_{.003}$	$\mathbf{0}_{.003}$
	16	$\mathbf{0}_{.008}$	$\mathbf{0}_{.008}$	$\mathbf{0}_{.008}$	—
	15	$\mathbf{0}_{.015+}$	$\mathbf{0}_{.015+}$	—	—
	14	$\mathbf{0}_{.026}$	—	—	—
	13	$\mathbf{0}_{.042}$	—	—	—
2	18	$\mathbf{0}_{.005+}$	$\mathbf{0}_{.005+}$	$\mathbf{0}_{.005+}$	—
	17	$\mathbf{0}_{.016}$	$\mathbf{0}_{.016}$	—	—
	16	$\mathbf{0}_{.032}$	—	—	—
$A = 19$ $B = 19$	19	$\mathbf{14}_{.023}$	$\mathbf{14}_{.023}$	$\mathbf{13}_{.010-}$	$\mathbf{12}_{.004}$
	18	$\mathbf{13}_{.045-}$	$\mathbf{12}_{.021}$	$\mathbf{11}_{.009}$	$\mathbf{10}_{.004}$
	17	$\mathbf{11}_{.031}$	$\mathbf{10}_{.015-}$	$\mathbf{9}_{.006}$	$\mathbf{8}_{.003}$
	16	$\mathbf{10}_{.039}$	$\mathbf{9}_{.019}$	$\mathbf{8}_{.009}$	$\mathbf{7}_{.003}$
	15	$\mathbf{9}_{.046}$	$\mathbf{8}_{.022}$	$\mathbf{6}_{.004}$	$\mathbf{6}_{.004}$
	14	$\mathbf{8}_{.050-}$	$\mathbf{7}_{.024}$	$\mathbf{5}_{.004}$	$\mathbf{5}_{.004}$
	13	$\mathbf{6}_{.025+}$	$\mathbf{5}_{.011}$	$\mathbf{4}_{.004}$	$\mathbf{4}_{.004}$
	12	$\mathbf{5}_{.024}$	$\mathbf{5}_{.024}$	$\mathbf{3}_{.003}$	$\mathbf{3}_{.003}$
	11	$\mathbf{5}_{.050-}$	$\mathbf{4}_{.022}$	$\mathbf{3}_{.009}$	$\mathbf{2}_{.003}$
	10	$\mathbf{4}_{.046}$	$\mathbf{3}_{.019}$	$\mathbf{2}_{.006}$	$\mathbf{1}_{.002}$
	9	$\mathbf{3}_{.039}$	$\mathbf{2}_{.015-}$	$\mathbf{1}_{.004}$	$\mathbf{1}_{.004}$
	8	$\mathbf{2}_{.031}$	$\mathbf{1}_{.009}$	$\mathbf{1}_{.009}$	$\mathbf{0}_{.002}$
	7	$\mathbf{1}_{.021}$	$\mathbf{1}_{.021}$	$\mathbf{0}_{.004}$	$\mathbf{0}_{.004}$

TABLE J ***(continued)***

		Probability			
	a	0.05	0.025	0.01	0.005
	6	$\mathbf{1}_{.045-}$	$\mathbf{0}_{.010-}$	$\mathbf{0}_{.010-}$	—
	5	$\mathbf{0}_{.023}$	$\mathbf{0}_{.023}$	—	—
18	19	$\mathbf{14}_{.046}$	$\mathbf{13}_{.020}$	$\mathbf{12}_{.008}$	$\mathbf{11}_{.003}$
	18	$\mathbf{12}_{.037}$	$\mathbf{11}_{.017}$	$\mathbf{10}_{.007}$	$\mathbf{9}_{.003}$
	17	$\mathbf{10}_{.024}$	$\mathbf{10}_{.024}$	$\mathbf{8}_{.004}$	$\mathbf{8}_{.004}$
	16	$\mathbf{9}_{.030}$	$\mathbf{8}_{.014}$	$\mathbf{7}_{.006}$	$\mathbf{6}_{.002}$
	15	$\mathbf{8}_{.033}$	$\mathbf{7}_{.015+}$	$\mathbf{6}_{.006}$	$\mathbf{5}_{.002}$
	14	$\mathbf{7}_{.035+}$	$\mathbf{6}_{.016}$	$\mathbf{5}_{.006}$	$\mathbf{4}_{.002}$
	13	$\mathbf{6}_{.035-}$	$\mathbf{5}_{.015+}$	$\mathbf{4}_{.006}$	$\mathbf{3}_{.002}$
	12	$\mathbf{5}_{.033}$	$\mathbf{4}_{.014}$	$\mathbf{3}_{.005-}$	$\mathbf{3}_{.005-}$
	11	$\mathbf{4}_{.030}$	$\mathbf{3}_{.011}$	$\mathbf{2}_{.004}$	$\mathbf{2}_{.004}$
	10	$\mathbf{3}_{.025-}$	$\mathbf{3}_{.025-}$	$\mathbf{2}_{.008}$	$\mathbf{1}_{.002}$
	9	$\mathbf{3}_{.049}$	$\mathbf{2}_{.019}$	$\mathbf{1}_{.005+}$	$\mathbf{0}_{.001}$
	8	$\mathbf{2}_{.038}$	$\mathbf{1}_{.012}$	$\mathbf{0}_{.002}$	$\mathbf{0}_{.002}$
	7	$\mathbf{1}_{.025+}$	$\mathbf{0}_{.005-}$	$\mathbf{0}_{.005-}$	$\mathbf{0}_{.005-}$
	6	$\mathbf{0}_{.012}$	$\mathbf{0}_{.012}$	—	—
	5	$\mathbf{0}_{.027}$	—	—	—
17	19	$\mathbf{13}_{.040}$	$\mathbf{12}_{.016}$	$\mathbf{11}_{.006}$	$\mathbf{10}_{.002}$
	18	$\mathbf{11}_{.030}$	$\mathbf{10}_{.013}$	$\mathbf{9}_{.005+}$	$\mathbf{8}_{.002}$
	17	$\mathbf{10}_{.040}$	$\mathbf{9}_{.018}$	$\mathbf{8}_{.008}$	$\mathbf{7}_{.003}$
	16	$\mathbf{9}_{.047}$	$\mathbf{8}_{.022}$	$\mathbf{7}_{.009}$	$\mathbf{6}_{.003}$
	15	$\mathbf{8}_{.050-}$	$\mathbf{7}_{.023}$	$\mathbf{6}_{.010-}$	$\mathbf{5}_{.004}$
	14	$\mathbf{6}_{.023}$	$\mathbf{6}_{.023}$	$\mathbf{5}_{.010-}$	$\mathbf{4}_{.003}$
	13	$\mathbf{6}_{.049}$	$\mathbf{5}_{.022}$	$\mathbf{4}_{.008}$	$\mathbf{3}_{.003}$
$\boldsymbol{A = 19\ B = 17}$	12	$\mathbf{5}_{.045-}$	$\mathbf{4}_{.019}$	$\mathbf{3}_{.007}$	$\mathbf{2}_{.002}$
	11	$\mathbf{4}_{.039}$	$\mathbf{3}_{.015+}$	$\mathbf{2}_{.005-}$	$\mathbf{2}_{.005-}$
	10	$\mathbf{3}_{.032}$	$\mathbf{2}_{.011}$	$\mathbf{1}_{.003}$	$\mathbf{1}_{.003}$
	9	$\mathbf{2}_{.024}$	$\mathbf{2}_{.024}$	$\mathbf{1}_{.007}$	$\mathbf{0}_{.001}$
	8	$\mathbf{2}_{.047}$	$\mathbf{1}_{.015-}$	$\mathbf{0}_{.002}$	$\mathbf{0}_{.002}$
	7	$\mathbf{1}_{.031}$	$\mathbf{0}_{.006}$	$\mathbf{0}_{.006}$	—
	6	$\mathbf{0}_{.014}$	$\mathbf{0}_{.014}$	—	—
	5	$\mathbf{0}_{.031}$	—	—	—
16	19	$\mathbf{12}_{.035-}$	$\mathbf{11}_{.013}$	$\mathbf{10}_{.005-}$	$\mathbf{10}_{.005-}$
	18	$\mathbf{10}_{.024}$	$\mathbf{10}_{.024}$	$\mathbf{9}_{.010-}$	$\mathbf{8}_{.004}$
	17	$\mathbf{9}_{.031}$	$\mathbf{8}_{.013}$	$\mathbf{7}_{.005+}$	$\mathbf{6}_{.002}$
	16	$\mathbf{8}_{.035-}$	$\mathbf{7}_{.015+}$	$\mathbf{6}_{.006}$	$\mathbf{5}_{.002}$
	15	$\mathbf{7}_{.036}$	$\mathbf{6}_{.015+}$	$\mathbf{5}_{.006}$	$\mathbf{4}_{.002}$
	14	$\mathbf{6}_{.034}$	$\mathbf{5}_{.014}$	$\mathbf{4}_{.005+}$	$\mathbf{3}_{.002}$
	13	$\mathbf{5}_{.031}$	$\mathbf{4}_{.013}$	$\mathbf{3}_{.004}$	$\mathbf{3}_{.004}$
	12	$\mathbf{4}_{.027}$	$\mathbf{3}_{.010-}$	$\mathbf{3}_{.010-}$	$\mathbf{2}_{.003}$
	11	$\mathbf{3}_{.021}$	$\mathbf{3}_{.021}$	$\mathbf{2}_{.007}$	$\mathbf{1}_{.002}$
	10	$\mathbf{3}_{.042}$	$\mathbf{2}_{.015-}$	$\mathbf{1}_{.004}$	$\mathbf{1}_{.004}$
	9	$\mathbf{2}_{.030}$	$\mathbf{1}_{.009}$	$\mathbf{1}_{.009}$	$\mathbf{0}_{.001}$
	8	$\mathbf{1}_{.018}$	$\mathbf{1}_{.018}$	$\mathbf{0}_{.003}$	$\mathbf{0}_{.003}$
	7	$\mathbf{1}_{.037}$	$\mathbf{0}_{.007}$	$\mathbf{0}_{.007}$	—
	6	$\mathbf{0}_{.017}$	$\mathbf{0}_{.017}$	—	—
	5	$\mathbf{0}_{.036}$	—	—	—
15	19	$\mathbf{11}_{.029}$	$\mathbf{10}_{.011}$	$\mathbf{9}_{.004}$	$\mathbf{9}_{.004}$
	18	$\mathbf{10}_{.046}$	$\mathbf{9}_{.019}$	$\mathbf{8}_{.007}$	$\mathbf{7}_{.002}$
	17	$\mathbf{8}_{.023}$	$\mathbf{8}_{.023}$	$\mathbf{7}_{.009}$	$\mathbf{6}_{.003}$

TABLE J (continued)

		Probability			
	a	**0.05**	**0.025**	**0.01**	**0.005**
	16	$\mathbf{7}_{.025-}$	$\mathbf{7}_{.025-}$	$\mathbf{6}_{.010-}$	$\mathbf{5}_{.003}$
	15	$\mathbf{6}_{.024}$	$\mathbf{6}_{.024}$	$\mathbf{5}_{.009}$	$\mathbf{4}_{.003}$
	14	$\mathbf{5}_{.022}$	$\mathbf{5}_{.022}$	$\mathbf{4}_{.008}$	$\mathbf{3}_{.002}$
	13	$\mathbf{5}_{.045+}$	$\mathbf{4}_{.018}$	$\mathbf{3}_{.006}$	$\mathbf{2}_{.002}$
	12	$\mathbf{4}_{.037}$	$\mathbf{3}_{.014}$	$\mathbf{2}_{.004}$	$\mathbf{2}_{.004}$
	11	$\mathbf{3}_{.029}$	$\mathbf{2}_{.009}$	$\mathbf{2}_{.009}$	$\mathbf{1}_{.002}$
	10	$\mathbf{2}_{.020}$	$\mathbf{2}_{.020}$	$\mathbf{1}_{.005+}$	$\mathbf{0}_{.001}$
	9	$\mathbf{2}_{.039}$	$\mathbf{1}_{.011}$	$\mathbf{0}_{.002}$	$\mathbf{0}_{.002}$
	8	$\mathbf{1}_{.023}$	$\mathbf{1}_{.023}$	$\mathbf{0}_{.004}$	$\mathbf{0}_{.004}$
	7	$\mathbf{1}_{.046}$	$\mathbf{0}_{.009}$	$\mathbf{0}_{.009}$	—
	6	$\mathbf{0}_{.020}$	$\mathbf{0}_{.020}$	—	—
	5	$\mathbf{0}_{.042}$	—	—	—
14	19	$\mathbf{10}_{.024}$	$\mathbf{10}_{.024}$	$\mathbf{9}_{.008}$	$\mathbf{8}_{.003}$
	18	$\mathbf{9}_{.037}$	$\mathbf{8}_{.014}$	$\mathbf{7}_{.005-}$	$\mathbf{7}_{.005-}$
	17	$\mathbf{8}_{.042}$	$\mathbf{7}_{.017}$	$\mathbf{6}_{.006}$	$\mathbf{5}_{.002}$
	16	$\mathbf{7}_{.042}$	$\mathbf{6}_{.017}$	$\mathbf{5}_{.006}$	$\mathbf{4}_{.002}$
	15	$\mathbf{6}_{.039}$	$\mathbf{5}_{.015+}$	$\mathbf{4}_{.005+}$	$\mathbf{3}_{.001}$
	14	$\mathbf{5}_{.034}$	$\mathbf{4}_{.013}$	$\mathbf{3}_{.004}$	$\mathbf{3}_{.004}$
	13	$\mathbf{4}_{.027}$	$\mathbf{3}_{.009}$	$\mathbf{3}_{.009}$	$\mathbf{2}_{.003}$
	12	$\mathbf{3}_{.020}$	$\mathbf{3}_{.020}$	$\mathbf{2}_{.006}$	$\mathbf{1}_{.001}$
	11	$\mathbf{3}_{.040}$	$\mathbf{2}_{.013}$	$\mathbf{1}_{.003}$	$\mathbf{1}_{.003}$
	10	$\mathbf{2}_{.027}$	$\mathbf{1}_{.007}$	$\mathbf{1}_{.007}$	$\mathbf{0}_{.001}$
	9	$\mathbf{1}_{.015-}$	$\mathbf{1}_{.015-}$	$\mathbf{0}_{.002}$	$\mathbf{0}_{.002}$
	8	$\mathbf{1}_{.030}$	$\mathbf{0}_{.005+}$	$\mathbf{0}_{.005+}$	—
	7	$\mathbf{0}_{.012}$	$\mathbf{0}_{.012}$	—	—
	6	$\mathbf{0}_{.024}$	$\mathbf{0}_{.024}$	—	—
	5	$\mathbf{0}_{.049}$	—	—	—
13	19	$\mathbf{9}_{.020}$	$\mathbf{9}_{.020}$	$\mathbf{8}_{.006}$	$\mathbf{7}_{.002}$
	18	$\mathbf{8}_{.029}$	$\mathbf{7}_{.010+}$	$\mathbf{6}_{.003}$	$\mathbf{6}_{.003}$
	17	$\mathbf{7}_{.031}$	$\mathbf{6}_{.011}$	$\mathbf{5}_{.004}$	$\mathbf{5}_{.004}$
	16	$\mathbf{6}_{.029}$	$\mathbf{5}_{.011}$	$\mathbf{4}_{.003}$	$\mathbf{4}_{.003}$
	15	$\mathbf{5}_{.025+}$	$\mathbf{4}_{.009}$	$\mathbf{4}_{.009}$	$\mathbf{3}_{.003}$
	14	$\mathbf{4}_{.020}$	$\mathbf{4}_{.020}$	$\mathbf{3}_{.006}$	$\mathbf{2}_{.002}$
	13	$\mathbf{4}_{.041}$	$\mathbf{3}_{.015-}$	$\mathbf{2}_{.004}$	$\mathbf{2}_{.004}$
***A* = 19 *B* = 13**	12	$\mathbf{3}_{.029}$	$\mathbf{2}_{.009}$	$\mathbf{2}_{.009}$	$\mathbf{1}_{.002}$
	11	$\mathbf{2}_{.019}$	$\mathbf{2}_{.019}$	$\mathbf{1}_{.005-}$	$\mathbf{1}_{.005-}$
	10	$\mathbf{2}_{.036}$	$\mathbf{1}_{.010-}$	$\mathbf{1}_{.010-}$	$\mathbf{0}_{.001}$
	9	$\mathbf{1}_{.020}$	$\mathbf{1}_{.020}$	$\mathbf{0}_{.003}$	$\mathbf{0}_{.003}$
	8	$\mathbf{1}_{.038}$	$\mathbf{0}_{.007}$	$\mathbf{0}_{.007}$	—
	7	$\mathbf{0}_{.015-}$	$\mathbf{0}_{.015-}$	—	—
	6	$\mathbf{0}_{.030}$	—	—	—
12	19	$\mathbf{9}_{.049}$	$\mathbf{8}_{.016}$	$\mathbf{7}_{.005-}$	$\mathbf{7}_{.005-}$
	18	$\mathbf{7}_{.022}$	$\mathbf{7}_{.022}$	$\mathbf{6}_{.007}$	$\mathbf{5}_{.002}$
	17	$\mathbf{6}_{.022}$	$\mathbf{6}_{.022}$	$\mathbf{5}_{.007}$	$\mathbf{4}_{.002}$
	16	$\mathbf{5}_{.019}$	$\mathbf{5}_{.019}$	$\mathbf{4}_{.006}$	$\mathbf{3}_{.002}$
	15	$\mathbf{5}_{.042}$	$\mathbf{4}_{.015+}$	$\mathbf{3}_{.004}$	$\mathbf{3}_{.004}$
	14	$\mathbf{4}_{.032}$	$\mathbf{3}_{.011}$	$\mathbf{2}_{.003}$	$\mathbf{2}_{.003}$
	13	$\mathbf{3}_{.023}$	$\mathbf{3}_{.023}$	$\mathbf{2}_{.006}$	$\mathbf{1}_{.001}$
	12	$\mathbf{3}_{.043}$	$\mathbf{2}_{.014}$	$\mathbf{1}_{.003}$	$\mathbf{1}_{.003}$
	11	$\mathbf{2}_{.027}$	$\mathbf{1}_{.007}$	$\mathbf{1}_{.007}$	$\mathbf{0}_{.001}$
	10	$\mathbf{2}_{.050-}$	$\mathbf{1}_{.014}$	$\mathbf{0}_{.002}$	$\mathbf{0}_{.002}$

TABLE J ***(continued)***

		Probability			
	a	**0.05**	**0.025**	**0.01**	**0.005**
	9	$\mathbf{1}_{.027}$	$\mathbf{0}_{.005-}$	$\mathbf{0}_{.005-}$	$\mathbf{0}_{.005-}$
	8	$\mathbf{1}_{.050-}$	$\mathbf{0}_{.010-}$	$\mathbf{0}_{.010-}$	—
	7	$\mathbf{0}_{.019}$	$\mathbf{0}_{.019}$	—	—
	6	$\mathbf{0}_{.037}$	—	—	—
11	19	$\mathbf{8}_{.041}$	$\mathbf{7}_{.012}$	$\mathbf{6}_{.003}$	$\mathbf{6}_{.003}$
	18	$\mathbf{7}_{.047}$	$\mathbf{6}_{.016}$	$\mathbf{5}_{.004}$	$\mathbf{5}_{.004}$
	17	$\mathbf{6}_{.043}$	$\mathbf{5}_{.015-}$	$\mathbf{4}_{.004}$	$\mathbf{4}_{.004}$
	16	$\mathbf{5}_{.035+}$	$\mathbf{4}_{.012}$	$\mathbf{3}_{.003}$	$\mathbf{3}_{.003}$
	15	$\mathbf{4}_{.027}$	$\mathbf{3}_{.008}$	$\mathbf{3}_{.008}$	$\mathbf{2}_{.002}$
	14	$\mathbf{3}_{.018}$	$\mathbf{3}_{.018}$	$\mathbf{2}_{.005-}$	$\mathbf{2}_{.005-}$
	13	$\mathbf{3}_{.035+}$	$\mathbf{2}_{.010+}$	$\mathbf{1}_{.002}$	$\mathbf{1}_{.002}$
	12	$\mathbf{2}_{.021}$	$\mathbf{2}_{.021}$	$\mathbf{1}_{.005-}$	$\mathbf{1}_{.005-}$
	11	$\mathbf{2}_{.040}$	$\mathbf{1}_{.010+}$	$\mathbf{0}_{.001}$	$\mathbf{0}_{.001}$
	10	$\mathbf{1}_{.020}$	$\mathbf{1}_{.020}$	$\mathbf{0}_{.003}$	$\mathbf{0}_{.003}$
	9	$\mathbf{1}_{.037}$	$\mathbf{0}_{.006}$	$\mathbf{0}_{.006}$	—
	8	$\mathbf{0}_{.013}$	$\mathbf{0}_{.013}$	—	—
	7	$\mathbf{0}_{.025-}$	$\mathbf{0}_{.025-}$	—	—
	6	$\mathbf{0}_{.046}$	—	—	—
10	19	$\mathbf{7}_{.033}$	$\mathbf{6}_{.009}$	$\mathbf{6}_{.009}$	$\mathbf{5}_{.002}$
	18	$\mathbf{6}_{.036}$	$\mathbf{5}_{.011}$	$\mathbf{4}_{.003}$	$\mathbf{4}_{.003}$
	17	$\mathbf{5}_{.030}$	$\mathbf{4}_{.009}$	$\mathbf{4}_{.009}$	$\mathbf{3}_{.002}$
	16	$\mathbf{4}_{.022}$	$\mathbf{4}_{.022}$	$\mathbf{3}_{.006}$	$\mathbf{2}_{.001}$
	15	$\mathbf{4}_{.047}$	$\mathbf{3}_{.015-}$	$\mathbf{2}_{.004}$	$\mathbf{2}_{.004}$
	14	$\mathbf{3}_{.030}$	$\mathbf{2}_{.008}$	$\mathbf{2}_{.008}$	$\mathbf{1}_{.002}$
	13	$\mathbf{2}_{.017}$	$\mathbf{2}_{.017}$	$\mathbf{1}_{.004}$	$\mathbf{1}_{.004}$
	12	$\mathbf{2}_{.033}$	$\mathbf{1}_{.008}$	$\mathbf{1}_{.008}$	$\mathbf{0}_{.001}$
	11	$\mathbf{1}_{.016}$	$\mathbf{1}_{.016}$	$\mathbf{0}_{.002}$	$\mathbf{0}_{.002}$
	10	$\mathbf{1}_{.029}$	$\mathbf{0}_{.005-}$	$\mathbf{0}_{.005-}$	$\mathbf{0}_{.005-}$
	9	$\mathbf{0}_{.009}$	$\mathbf{0}_{.009}$	$\mathbf{0}_{.009}$	—
	8	$\mathbf{0}_{.018}$	$\mathbf{0}_{.018}$	—	—
	7	$\mathbf{0}_{.032}$	—	—	—
9	19	$\mathbf{6}_{.026}$	$\mathbf{5}_{.006}$	$\mathbf{5}_{.006}$	$\mathbf{4}_{.001}$
	18	$\mathbf{5}_{.026}$	$\mathbf{4}_{.007}$	$\mathbf{4}_{.007}$	$\mathbf{3}_{.001}$
	17	$\mathbf{4}_{.020}$	$\mathbf{4}_{.020}$	$\mathbf{3}_{.005-}$	$\mathbf{3}_{.005-}$
	16	$\mathbf{4}_{.044}$	$\mathbf{3}_{.013}$	$\mathbf{2}_{.003}$	$\mathbf{2}_{.003}$
	15	$\mathbf{3}_{.028}$	$\mathbf{2}_{.007}$	$\mathbf{2}_{.007}$	$\mathbf{1}_{.001}$
	14	$\mathbf{2}_{.015-}$	$\mathbf{2}_{.015-}$	$\mathbf{1}_{.003}$	$\mathbf{1}_{.003}$
	13	$\mathbf{2}_{.029}$	$\mathbf{1}_{.006}$	$\mathbf{1}_{.006}$	$\mathbf{0}_{.001}$
	12	$\mathbf{1}_{.013}$	$\mathbf{1}_{.013}$	$\mathbf{0}_{.002}$	$\mathbf{0}_{.002}$
	11	$\mathbf{1}_{.024}$	$\mathbf{1}_{.024}$	$\mathbf{0}_{.004}$	$\mathbf{0}_{.004}$
	10	$\mathbf{1}_{.042}$	$\mathbf{0}_{.007}$	$\mathbf{0}_{.007}$	—
	9	$\mathbf{0}_{.013}$	$\mathbf{0}_{.013}$	—	—
	8	$\mathbf{0}_{.024}$	$\mathbf{0}_{.024}$	—	—
	7	$\mathbf{0}_{.043}$	—	—	—
$A = 19$ $B = 8$	19	$\mathbf{5}_{.019}$	$\mathbf{5}_{.019}$	$\mathbf{4}_{.004}$	$\mathbf{4}_{.004}$
	18	$\mathbf{4}_{.017}$	$\mathbf{4}_{.017}$	$\mathbf{3}_{.004}$	$\mathbf{3}_{.004}$
	17	$\mathbf{4}_{.044}$	$\mathbf{3}_{.011}$	$\mathbf{2}_{.002}$	$\mathbf{2}_{.002}$
	16	$\mathbf{3}_{.027}$	$\mathbf{2}_{.006}$	$\mathbf{2}_{.006}$	$\mathbf{1}_{.001}$
	15	$\mathbf{2}_{.013}$	$\mathbf{2}_{.013}$	$\mathbf{1}_{.002}$	$\mathbf{1}_{.002}$
	14	$\mathbf{2}_{.027}$	$\mathbf{1}_{.006}$	$\mathbf{1}_{.006}$	$\mathbf{0}_{.001}$
	13	$\mathbf{2}_{.049}$	$\mathbf{1}_{.011}$	$\mathbf{0}_{.001}$	$\mathbf{0}_{.001}$

TABLE J ***(continued)***

		Probability			
	a	**0.05**	**0.025**	**0.01**	**0.005**
	12	$1_{.021}$	$1_{.021}$	$0_{.003}$	$0_{.003}$
	11	$1_{.038}$	$0_{.006}$	$0_{.006}$	—
	10	$0_{.011}$	$0_{.011}$	—	—
	9	$0_{.020}$	$0_{.020}$	—	—
	8	$0_{.034}$	—	—	—
7	19	$4_{.013}$	$4_{.013}$	$3_{.002}$	$3_{.002}$
	18	$4_{.047}$	$3_{.010+}$	$2_{.002}$	$2_{.002}$
	17	$3_{.028}$	$2_{.006}$	$2_{.006}$	$1_{.001}$
	16	$2_{.014}$	$2_{.014}$	$1_{.002}$	$1_{.002}$
	15	$2_{.028}$	$1_{.005+}$	$1_{.005+}$	$0_{.001}$
	14	$1_{.011}$	$1_{.011}$	$0_{.001}$	$0_{.001}$
	13	$1_{.021}$	$1_{.021}$	$0_{.003}$	$0_{.003}$
	12	$1_{.037}$	$0_{.005+}$	$0_{.005+}$	—
	11	$0_{.010-}$	$0_{.010-}$	$0_{.010-}$	—
	10	$0_{.017}$	$0_{.017}$	—	—
	9	$0_{.030}$	—	—	—
	8	$0_{.048}$	—	—	—
6	19	$4_{.050-}$	$3_{.009}$	$3_{.009}$	$2_{.001}$
	18	$3_{.031}$	$2_{.005+}$	$2_{.005+}$	$1_{.001}$
	17	$2_{.015+}$	$2_{.015+}$	$1_{.002}$	$1_{.002}$
	16	$2_{.032}$	$1_{.006}$	$1_{.006}$	$0_{.000}$
	15	$1_{.012}$	$1_{.012}$	$0_{.001}$	$0_{.001}$
	14	$1_{.023}$	$1_{.023}$	$0_{.003}$	$0_{.003}$
	13	$1_{.039}$	$0_{.005+}$	$0_{.005+}$	—
	12	$0_{.010-}$	$0_{.010-}$	$0_{.010-}$	—
	11	$0_{.017}$	$0_{.017}$	—	—
	10	$0_{.028}$	—	—	—
	9	$0_{.045+}$	—	—	—
5	19	$3_{.036}$	$2_{.005-}$	$2_{.005-}$	$2_{.005-}$
	18	$2_{.018}$	$2_{.018}$	$1_{.002}$	$1_{.002}$
	17	$2_{.042}$	$1_{.006}$	$1_{.006}$	$0_{.000}$
	16	$1_{.014}$	$1_{.014}$	$0_{.001}$	$0_{.001}$
	15	$1_{.028}$	$0_{.003}$	$0_{.003}$	$0_{.003}$
	14	$1_{.047}$	$0_{.006}$	$0_{.006}$	—
	13	$0_{.011}$	$0_{.011}$	—	—
	12	$0_{.019}$	$0_{.019}$	—	—
	11	$0_{.030}$	—	—	—
	10	$0_{.047}$	—	—	—
4	19	$2_{.024}$	$2_{.024}$	$1_{.002}$	$1_{.002}$
	18	$1_{.009}$	$1_{.009}$	$1_{.009}$	$0_{.001}$
	17	$1_{.021}$	$1_{.021}$	$0_{.002}$	$0_{.002}$
	16	$1_{.040}$	$0_{.004}$	$0_{.004}$	$0_{.004}$
	15	$0_{.008}$	$0_{.008}$	$0_{.008}$	—
	14	$0_{.014}$	$0_{.014}$	—	—
	13	$0_{.024}$	$0_{.024}$	—	—
	12	$0_{.037}$	—	—	—
3	19	$1_{.013}$	$1_{.013}$	$0_{.001}$	$0_{.001}$
	18	$1_{.038}$	$0_{.003}$	$0_{.003}$	$0_{.003}$
	17	$0_{.006}$	$0_{.006}$	$0_{.006}$	—
	16	$0_{.013}$	$0_{.013}$	—	—
	15	$0_{.023}$	$0_{.023}$	—	—

TABLE J (*continued*)

		Probability			
	a	**0.05**	**0.025**	**0.01**	**0.005**
	14	$\mathbf{0}_{.036}$	—	—	—
2	19	$\mathbf{0}_{.005-}$	$\mathbf{0}_{.005-}$	$\mathbf{0}_{.005-}$	$\mathbf{0}_{.005-}$
	18	$\mathbf{0}_{.014}$	$\mathbf{0}_{.014}$	—	—
***A* = 19 *B* = 2**	17	$\mathbf{0}_{.029}$	—	—	—
	16	$\mathbf{0}_{.048}$	—	—	—
***A* = 20 *B* = 20**	20	$\mathbf{15}_{.024}$	$\mathbf{15}_{.024}$	$\mathbf{13}_{.004}$	$\mathbf{13}_{.004}$
	19	$\mathbf{14}_{.046}$	$\mathbf{13}_{.022}$	$\mathbf{12}_{.010-}$	$\mathbf{11}_{.004}$
	18	$\mathbf{12}_{.032}$	$\mathbf{11}_{.015+}$	$\mathbf{10}_{.007}$	$\mathbf{9}_{.003}$
	17	$\mathbf{11}_{.041}$	$\mathbf{10}_{.020}$	$\mathbf{9}_{.009}$	$\mathbf{8}_{.004}$
	16	$\mathbf{10}_{.048}$	$\mathbf{9}_{.024}$	$\mathbf{7}_{.005-}$	$\mathbf{7}_{.005-}$
	15	$\mathbf{8}_{.027}$	$\mathbf{7}_{.012}$	$\mathbf{6}_{.005+}$	$\mathbf{5}_{.002}$
	14	$\mathbf{7}_{.028}$	$\mathbf{6}_{.013}$	$\mathbf{5}_{.005+}$	$\mathbf{4}_{.002}$
	13	$\mathbf{6}_{.028}$	$\mathbf{5}_{.012}$	$\mathbf{4}_{.005-}$	$\mathbf{4}_{.005-}$
	12	$\mathbf{5}_{.027}$	$\mathbf{4}_{.011}$	$\mathbf{3}_{.004}$	$\mathbf{3}_{.004}$
	11	$\mathbf{4}_{.024}$	$\mathbf{4}_{.024}$	$\mathbf{3}_{.009}$	$\mathbf{2}_{.003}$
	10	$\mathbf{4}_{.048}$	$\mathbf{3}_{.020}$	$\mathbf{2}_{.007}$	$\mathbf{1}_{.002}$
	9	$\mathbf{3}_{.041}$	$\mathbf{2}_{.015+}$	$\mathbf{1}_{.004}$	$\mathbf{1}_{.004}$
	8	$\mathbf{2}_{.032}$	$\mathbf{1}_{.010-}$	$\mathbf{1}_{.010-}$	$\mathbf{0}_{.002}$
	7	$\mathbf{1}_{.022}$	$\mathbf{1}_{.022}$	$\mathbf{0}_{.004}$	$\mathbf{0}_{.004}$
	6	$\mathbf{1}_{.046}$	$\mathbf{0}_{.010+}$	—	—
	5	$\mathbf{0}_{.024}$	$\mathbf{0}_{.024}$	—	—
19	20	$\mathbf{15}_{.047}$	$\mathbf{14}_{.020}$	$\mathbf{13}_{.008}$	$\mathbf{12}_{.003}$
	19	$\mathbf{13}_{.039}$	$\mathbf{12}_{.018}$	$\mathbf{11}_{.008}$	$\mathbf{10}_{.003}$
	18	$\mathbf{11}_{.026}$	$\mathbf{10}_{.012}$	$\mathbf{9}_{.005-}$	$\mathbf{9}_{.005-}$
	17	$\mathbf{10}_{.032}$	$\mathbf{9}_{.015-}$	$\mathbf{8}_{.006}$	$\mathbf{7}_{.002}$
	16	$\mathbf{9}_{.036}$	$\mathbf{8}_{.017}$	$\mathbf{7}_{.007}$	$\mathbf{6}_{.003}$
	15	$\mathbf{8}_{.038}$	$\mathbf{7}_{.018}$	$\mathbf{6}_{.008}$	$\mathbf{5}_{.003}$
	14	$\mathbf{7}_{.039}$	$\mathbf{6}_{.018}$	$\mathbf{5}_{.007}$	$\mathbf{4}_{.003}$
	13	$\mathbf{6}_{.038}$	$\mathbf{5}_{.017}$	$\mathbf{4}_{.007}$	$\mathbf{3}_{.002}$
	12	$\mathbf{5}_{.035+}$	$\mathbf{4}_{.015+}$	$\mathbf{3}_{.005+}$	$\mathbf{2}_{.002}$
	11	$\mathbf{4}_{.031}$	$\mathbf{3}_{.012}$	$\mathbf{2}_{.004}$	$\mathbf{2}_{.004}$
	10	$\mathbf{3}_{.026}$	$\mathbf{2}_{.009}$	$\mathbf{2}_{.009}$	$\mathbf{1}_{.002}$
	9	$\mathbf{2}_{.019}$	$\mathbf{2}_{.019}$	$\mathbf{1}_{.005+}$	$\mathbf{0}_{.001}$
	8	$\mathbf{2}_{.039}$	$\mathbf{1}_{.012}$	$\mathbf{0}_{.002}$	$\mathbf{0}_{.002}$
	7	$\mathbf{1}_{.026}$	$\mathbf{0}_{.005+}$	$\mathbf{0}_{.005+}$	—
	6	$\mathbf{0}_{.012}$	$\mathbf{0}_{.012}$	—	—
	5	$\mathbf{0}_{.027}$	—	—	—
18	20	$\mathbf{14}_{.041}$	$\mathbf{13}_{.017}$	$\mathbf{12}_{.007}$	$\mathbf{11}_{.003}$
	19	$\mathbf{12}_{.032}$	$\mathbf{11}_{.014}$	$\mathbf{10}_{.006}$	$\mathbf{9}_{.002}$
	18	$\mathbf{11}_{.043}$	$\mathbf{10}_{.020}$	$\mathbf{9}_{.008}$	$\mathbf{8}_{.003}$
	17	$\mathbf{10}_{.050-}$	$\mathbf{9}_{.024}$	$\mathbf{7}_{.004}$	$\mathbf{7}_{.004}$
	16	$\mathbf{8}_{.026}$	$\mathbf{7}_{.011}$	$\mathbf{6}_{.005-}$	$\mathbf{6}_{.005-}$
	15	$\mathbf{7}_{.027}$	$\mathbf{6}_{.012}$	$\mathbf{5}_{.004}$	$\mathbf{5}_{.004}$
	14	$\mathbf{6}_{.026}$	$\mathbf{5}_{.011}$	$\mathbf{4}_{.004}$	$\mathbf{4}_{.004}$
	13	$\mathbf{5}_{.024}$	$\mathbf{5}_{.024}$	$\mathbf{4}_{.009}$	$\mathbf{3}_{.003}$
	12	$\mathbf{5}_{.047}$	$\mathbf{4}_{.020}$	$\mathbf{3}_{.007}$	$\mathbf{2}_{.002}$
	11	$\mathbf{4}_{.041}$	$\mathbf{3}_{.016}$	$\mathbf{2}_{.005+}$	$\mathbf{1}_{.001}$
	10	$\mathbf{3}_{.033}$	$\mathbf{2}_{.012}$	$\mathbf{1}_{.003}$	$\mathbf{1}_{.003}$
	9	$\mathbf{2}_{.024}$	$\mathbf{2}_{.024}$	$\mathbf{1}_{.007}$	$\mathbf{0}_{.001}$
	8	$\mathbf{2}_{.048}$	$\mathbf{1}_{.015-}$	$\mathbf{0}_{.003}$	$\mathbf{0}_{.003}$
	7	$\mathbf{1}_{.031}$	$\mathbf{0}_{.006}$	$\mathbf{0}_{.006}$	—

TABLE J (continued)

			Probability			
		a	**0.05**	**0.025**	**0.01**	**0.005**
		6	$0_{.014}$	$0_{.014}$	—	—
		5	$0_{.031}$	—	—	—
	17	20	$13_{.036}$	$12_{.014}$	$11_{.005+}$	$10_{.002}$
		19	$11_{.026}$	$10_{.011}$	$9_{.004}$	$9_{.004}$
		18	$10_{.034}$	$9_{.015-}$	$8_{.006}$	$7_{.002}$
		17	$9_{.038}$	$8_{.017}$	$7_{.007}$	$6_{.003}$
		16	$8_{.040}$	$7_{.018}$	$6_{.007}$	$5_{.003}$
		15	$7_{.039}$	$6_{.017}$	$5_{.007}$	$4_{.002}$
		14	$6_{.037}$	$5_{.016}$	$4_{.006}$	$3_{.002}$
		13	$5_{.033}$	$4_{.013}$	$3_{.005-}$	$3_{.005-}$
		12	$4_{.028}$	$3_{.010+}$	$2_{.003}$	$2_{.003}$
		11	$3_{.022}$	$3_{.022}$	$2_{.007}$	$1_{.002}$
A **= 20 *B* = 17**		10	$3_{.042}$	$2_{.015+}$	$1_{.004}$	$1_{.004}$
		9	$2_{.031}$	$1_{.009}$	$1_{.009}$	$0_{.001}$
		8	$1_{.019}$	$1_{.019}$	$0_{.003}$	$0_{.003}$
		7	$1_{.037}$	$0_{.008}$	$0_{.008}$	—
		6	$0_{.017}$	$0_{.017}$	—	—
		5	$0_{.036}$	—	—	—
	16	20	$12_{.031}$	$11_{.012}$	$10_{.004}$	$10_{.004}$
		19	$11_{.049}$	$10_{.021}$	$9_{.008}$	$8_{.003}$
		18	$9_{.026}$	$8_{.011}$	$7_{.004}$	$7_{.004}$
		17	$8_{.028}$	$7_{.012}$	$6_{.004}$	$6_{.004}$
		16	$7_{.028}$	$6_{.012}$	$5_{.004}$	$5_{.004}$
		15	$6_{.026}$	$5_{.011}$	$4_{.004}$	$4_{.004}$
		14	$5_{.023}$	$5_{.023}$	$4_{.009}$	$3_{.003}$
		13	$5_{.046}$	$4_{.019}$	$3_{.007}$	$2_{.002}$
		12	$4_{.038}$	$3_{.014}$	$2_{.004}$	$2_{.004}$
		11	$3_{.029}$	$2_{.010-}$	$2_{.010-}$	$1_{.002}$
		10	$2_{.020}$	$2_{.020}$	$1_{.005+}$	$0_{.001}$
		9	$2_{.039}$	$1_{.011}$	$0_{.002}$	$0_{.002}$
		8	$1_{.023}$	$1_{.023}$	$0_{.004}$	$0_{.004}$
		7	$1_{.045+}$	$0_{.009}$	$0_{.009}$	—
		6	$0_{.020}$	$0_{.020}$	—	—
		5	$0_{.041}$	—	—	—
	15	20	$11_{.026}$	$10_{.009}$	$10_{.009}$	$9_{.003}$
		19	$10_{.040}$	$9_{.016}$	$8_{.006}$	$7_{.002}$
		18	$9_{.046}$	$8_{.019}$	$7_{.007}$	$6_{.002}$
		17	$8_{.047}$	$7_{.020}$	$6_{.008}$	$5_{.002}$
		16	$7_{.045-}$	$6_{.019}$	$5_{.007}$	$4_{.002}$
		15	$6_{.040}$	$5_{.017}$	$4_{.006}$	$3_{.002}$
		14	$5_{.034}$	$4_{.013}$	$3_{.004}$	$3_{.004}$
		13	$4_{.028}$	$3_{.010-}$	$3_{.010-}$	$2_{.003}$
		12	$3_{.020}$	$3_{.020}$	$2_{.006}$	$1_{.001}$
		11	$3_{.039}$	$2_{.013}$	$1_{.003}$	$1_{.003}$
		10	$2_{.026}$	$1_{.007}$	$1_{.007}$	$0_{.001}$
		9	$2_{.049}$	$1_{.015-}$	$0_{.002}$	$0_{.002}$
		8	$1_{.029}$	$0_{.005+}$	$0_{.005+}$	—
		7	$0_{.012}$	$0_{.012}$	—	—
		6	$0_{.024}$	$0_{.024}$	—	—
		5	$0_{.048}$	—	—	—
	14	20	$10_{.022}$	$10_{.022}$	$9_{.007}$	$8_{.002}$

TABLE J (*continued*)

		Probability			
	a	**0.05**	**0.025**	**0.01**	**0.005**
	19	$9_{.032}$	$8_{.012}$	$7_{.004}$	$7_{.004}$
	18	$8_{.035+}$	$7_{.014}$	$6_{.005-}$	$6_{.005-}$
	17	$7_{.035-}$	$6_{.013}$	$5_{.005-}$	$5_{.005-}$
	16	$6_{.031}$	$5_{.012}$	$4_{.004}$	$4_{.004}$
	15	$5_{.026}$	$4_{.009}$	$4_{.009}$	$3_{.003}$
	14	$4_{.020}$	$4_{.020}$	$3_{.007}$	$2_{.002}$
	13	$4_{.040}$	$3_{.015-}$	$2_{.004}$	$2_{.004}$
	12	$3_{.029}$	$2_{.009}$	$2_{.009}$	$1_{.002}$
	11	$2_{.018}$	$2_{.018}$	$1_{.005-}$	$1_{.005-}$
	10	$2_{.035+}$	$1_{.010-}$	$1_{.010-}$	$0_{.001}$
	9	$1_{.019}$	$1_{.019}$	$0_{.003}$	$0_{.003}$
	8	$1_{.037}$	$0_{.007}$	$0_{.007}$	—
	7	$0_{.014}$	$0_{.014}$	—	—
	6	$0_{.029}$	—	—	—
13	20	$9_{.017}$	$9_{.017}$	$8_{.005+}$	$7_{.002}$
	19	$8_{.025-}$	$8_{.025-}$	$7_{.008}$	$6_{.003}$
	18	$7_{.026}$	$6_{.009}$	$6_{.009}$	$5_{.003}$
	17	$6_{.024}$	$6_{.024}$	$5_{.008}$	$4_{.002}$
	16	$5_{.020}$	$5_{.020}$	$4_{.007}$	$3_{.002}$
	15	$5_{.041}$	$4_{.015+}$	$3_{.005-}$	$3_{.005-}$
	14	$4_{.031}$	$3_{.011}$	$2_{.003}$	$2_{.003}$
***A* = 20 *B* = 13**	13	$3_{.022}$	$3_{.022}$	$2_{.006}$	$1_{.001}$
	12	$3_{.041}$	$2_{.013}$	$1_{.003}$	$1_{.003}$
	11	$2_{.026}$	$1_{.007}$	$1_{.007}$	$0_{.001}$
	10	$2_{.047}$	$1_{.013}$	$0_{.002}$	$0_{.002}$
	9	$1_{.026}$	$0_{.004}$	$0_{.004}$	$0_{.004}$
	8	$1_{.047}$	$0_{.009}$	$0_{.009}$	—
	7	$0_{.018}$	$0_{.018}$	—	—
	6	$0_{.035-}$	—	—	—
12	20	$9_{.044}$	$8_{.014}$	$7_{.004}$	$7_{.004}$
	19	$7_{.019}$	$7_{.019}$	$6_{.006}$	$5_{.002}$
	18	$6_{.018}$	$6_{.018}$	$5_{.006}$	$4_{.002}$
	17	$6_{.043}$	$5_{.016}$	$4_{.005-}$	$4_{.005-}$
	16	$5_{.034}$	$4_{.012}$	$3_{.003}$	$3_{.003}$
	15	$4_{.025+}$	$3_{.008}$	$3_{.008}$	$2_{.002}$
	14	$4_{.049}$	$3_{.017}$	$2_{.005-}$	$2_{.005-}$
	13	$3_{.033}$	$2_{.010-}$	$2_{.010-}$	$1_{.002}$
	12	$2_{.020}$	$2_{.020}$	$1_{.005-}$	$1_{.005-}$
	11	$2_{.036}$	$1_{.009}$	$1_{.009}$	$0_{.001}$
	10	$1_{.018}$	$1_{.018}$	$0_{.003}$	$0_{.003}$
	9	$1_{.034}$	$0_{.006}$	$0_{.006}$	—
	8	$0_{.012}$	$0_{.012}$	—	—
	7	$0_{.023}$	$0_{.023}$	—	—
	6	$0_{.043}$	—	—	—
11	20	$8_{.037}$	$7_{.010+}$	$6_{.003}$	$6_{.003}$
	19	$7_{.042}$	$6_{.013}$	$5_{.004}$	$5_{.004}$
	18	$6_{.037}$	$5_{.012}$	$4_{.003}$	$4_{.003}$
	17	$5_{.029}$	$4_{.009}$	$4_{.009}$	$3_{.002}$
	16	$4_{.021}$	$4_{.021}$	$3_{.006}$	$2_{.001}$
	15	$4_{.042}$	$3_{.014}$	$2_{.003}$	$2_{.003}$
	14	$3_{.028}$	$2_{.008}$	$2_{.008}$	$1_{.001}$

TABLE J ***(continued)***

		Probability			
	a	0.05	0.025	0.01	0.005
	13	$2_{.016}$	$2_{.016}$	$1_{.003}$	$1_{.003}$
	12	$2_{.029}$	$1_{.007}$	$1_{.007}$	$0_{.001}$
	11	$1_{.014}$	$1_{.014}$	$0_{.002}$	$0_{.002}$
	10	$1_{.026}$	$0_{.004}$	$0_{.004}$	$0_{.004}$
	9	$1_{.046}$	$0_{.008}$	$0_{.008}$	—
	8	$0_{.016}$	$0_{.016}$	—	—
	7	$0_{.029}$	—	—	—
10	20	$7_{.030}$	$6_{.008}$	$6_{.008}$	$5_{.002}$
	19	$6_{.031}$	$5_{.009}$	$5_{.009}$	$4_{.002}$
	18	$5_{.026}$	$4_{.007}$	$4_{.007}$	$3_{.002}$
	17	$4_{.018}$	$4_{.018}$	$3_{.005-}$	$3_{.005-}$
	16	$4_{.039}$	$3_{.012}$	$2_{.003}$	$2_{.003}$
	15	$3_{.024}$	$3_{.024}$	$2_{.006}$	$1_{.001}$
	14	$3_{.045+}$	$2_{.013}$	$1_{.003}$	$1_{.003}$
	13	$2_{.025+}$	$1_{.006}$	$1_{.006}$	$0_{.001}$
	12	$2_{.045-}$	$1_{.011}$	$0_{.001}$	$0_{.001}$
	11	$1_{.021}$	$1_{.021}$	$0_{.003}$	$0_{.003}$
	10	$1_{.037}$	$0_{.006}$	$0_{.006}$	—
	9	$0_{.012}$	$0_{.012}$	—	—
	8	$0_{.022}$	$0_{.022}$	—	—
	7	$0_{.038}$	—	—	—
9	20	$6_{.023}$	$6_{.023}$	$5_{.005+}$	$4_{.001}$
	19	$5_{.022}$	$5_{.022}$	$4_{.005+}$	$3_{.001}$
	18	$4_{.016}$	$4_{.016}$	$3_{.004}$	$3_{.004}$
	17	$4_{.037}$	$3_{.010+}$	$2_{.002}$	$2_{.002}$
	16	$3_{.022}$	$3_{.022}$	$2_{.005+}$	$1_{.001}$
	15	$3_{.043}$	$2_{.012}$	$1_{.002}$	$1_{.002}$
	14	$2_{.023}$	$2_{.023}$	$1_{.005-}$	$1_{.005-}$
	13	$2_{.041}$	$1_{.009}$	$1_{.009}$	$0_{.001}$
	12	$1_{.018}$	$1_{.018}$	$0_{.002}$	$0_{.002}$
	11	$1_{.032}$	$0_{.005-}$	$0_{.005-}$	$0_{.005-}$
***A* = 20 *B* = 9**	10	$0_{.009}$	$0_{.009}$	$0_{.009}$	—
	9	$0_{.017}$	$0_{.017}$	—	—
	8	$0_{.029}$	—	—	—
	7	$0_{.050-}$	—	—	—
8	20	$5_{.017}$	$5_{.017}$	$4_{.003}$	$4_{.003}$
	19	$4_{.015-}$	$4_{.015-}$	$3_{.003}$	$3_{.003}$
	18	$4_{.038}$	$3_{.009}$	$3_{.009}$	$2_{.002}$
	17	$3_{.022}$	$3_{.022}$	$2_{.005-}$	$2_{.005-}$
	16	$3_{.044}$	$2_{.011}$	$1_{.002}$	$1_{.002}$
	15	$2_{.022}$	$2_{.022}$	$1_{.004}$	$1_{.004}$
	14	$2_{.040}$	$1_{.009}$	$1_{.009}$	$0_{.001}$
	13	$1_{.016}$	$1_{.016}$	$0_{.002}$	$0_{.002}$
	12	$1_{.029}$	$0_{.004}$	$0_{.004}$	$0_{.004}$
	11	$1_{.048}$	$0_{.008}$	$0_{.008}$	—
	10	$0_{.014}$	$0_{.014}$	—	—
	9	$0_{.024}$	$0_{.024}$	—	—
	8	$0_{.041}$	—	—	—
7	20	$4_{.012}$	$4_{.012}$	$3_{.002}$	$3_{.002}$
	19	$4_{.042}$	$3_{.009}$	$3_{.009}$	$2_{.001}$
	18	$3_{.024}$	$3_{.024}$	$2_{.005-}$	$2_{.005-}$

TABLE J ***(continued)***

		Probability			
	a	**0.05**	**0.025**	**0.01**	**0.005**
	17	$3_{.050^-}$	$2_{.011}$	$1_{.002}$	$1_{.002}$
	16	$2_{.023}$	$2_{.023}$	$1_{.004}$	$1_{.004}$
	15	$2_{.043}$	$1_{.009}$	$1_{.009}$	$0_{.001}$
	14	$1_{.016}$	$1_{.016}$	$0_{.002}$	$0_{.002}$
	13	$1_{.029}$	$0_{.004}$	$0_{.004}$	$0_{.004}$
	12	$1_{.048}$	$0_{.007}$	$0_{.007}$	—
	11	$0_{.013}$	$0_{.013}$	—	—
	10	$0_{.022}$	$0_{.022}$	—	—
	9	$0_{.036}$	—	—	—
6	20	$4_{.046}$	$3_{.008}$	$3_{.008}$	$2_{.001}$
	19	$3_{.028}$	$2_{.005^-}$	$2_{.005^-}$	$2_{.005^-}$
	18	$2_{.013}$	$2_{.013}$	$1_{.002}$	$1_{.002}$
	17	$2_{.028}$	$1_{.004}$	$1_{.004}$	$1_{.004}$
	16	$1_{.010^-}$	$1_{.010^-}$	$1_{.010^-}$	$0_{.001}$
	15	$1_{.018}$	$1_{.018}$	$0_{.002}$	$0_{.002}$
	14	$1_{.032}$	$0_{.004}$	$0_{.004}$	$0_{.004}$
	13	$0_{.007}$	$0_{.007}$	$0_{.007}$	—
	12	$0_{.013}$	$0_{.013}$	—	—
	11	$0_{.022}$	$0_{.022}$	—	—
	10	$0_{.035^-}$	—	—	—
5	20	$3_{.033}$	$2_{.004}$	$2_{.004}$	$2_{.004}$
	19	$2_{.016}$	$2_{.016}$	$1_{.002}$	$1_{.002}$
	18	$2_{.038}$	$1_{.005^+}$	$1_{.005^+}$	$0_{.000}$
	17	$1_{.012}$	$1_{.012}$	$0_{.001}$	$0_{.001}$
	16	$1_{.023}$	$1_{.023}$	$0_{.002}$	$0_{.002}$
	15	$1_{.040}$	$0_{.005^-}$	$0_{.005^-}$	$0_{.005^-}$
	14	$0_{.009}$	$0_{.009}$	$0_{.009}$	—
	13	$0_{.015^-}$	$0_{.015^-}$	—	—
	12	$0_{.024}$	$0_{.024}$	—	—
	11	$0_{.038}$	—	—	—
4	20	$2_{.022}$	$2_{.022}$	$1_{.002}$	$1_{.002}$
	19	$1_{.008}$	$1_{.008}$	$1_{.008}$	$0_{.000}$
	18	$1_{.018}$	$1_{.018}$	$0_{.001}$	$0_{.001}$
	17	$1_{.035^+}$	$0_{.003}$	$0_{.003}$	$0_{.003}$
	16	$0_{.007}$	$0_{.007}$	$0_{.007}$	—
	15	$0_{.012}$	$0_{.012}$	—	—
	14	$0_{.020}$	$0_{.020}$	—	—
	13	$0_{.031}$	—	—	—
	12	$0_{.047}$	—	—	—
3	20	$1_{.012}$	$1_{.012}$	$0_{.001}$	$0_{.001}$
	19	$1_{.034}$	$0_{.002}$	$0_{.002}$	$0_{.002}$
***A* = 20 *B* = 3**	18	$0_{.006}$	$0_{.006}$	$0_{.006}$	—
	17	$0_{.011}$	$0_{.011}$	—	—
	16	$0_{.020}$	$0_{.020}$	—	—
	15	$0_{.032}$	—	—	—
	14	$0_{.047}$	—	—	—
2	20	$0_{.004}$	$0_{.004}$	$0_{.004}$	$0_{.004}$
	19	$0_{.013}$	$0_{.013}$	—	—
	18	$0_{.026}$	—	—	—
	17	$0_{.043}$	—	—	—
1	20	$0_{.048}$	—	—	—

TABLE K Probability Levels for the Wilcoxon Signed-Rank Test

$n = 5$

T	P
[a]0	.0313
1	.0625
2	.0938
3	.1563
4	.2188
5	.3125
6	.4063
7	.5000

$n = 6$

T	P
0	.0156
1	.0313
[a]2	.0469
3	.0781
4	.1094
5	.1563
6	.2188
7	.2813
8	.3438
9	.4219
10	.5000

$n = 7$

T	P
0	.0078
1	.0156
2	.0234
[a]3	.0391
4	.0547
5	.0781
6	.1094
7	.1484
8	.1875
9	.2344
10	.2891
11	.3438
12	.4063

$n = 8$

T	P
0	.0039
1	.0078
2	.0117
3	.0195
4	.0273
[a]5	.0391
6	.0547
7	.0742
8	.0977
9	.1250
10	.1563
11	.1914
12	.2305
13	.2734
14	.3203
15	.3711
16	.4219
17	.4727
18	.5273

$n = 9$

T	P
0	.0020
1	.0039
2	.0059
3	.0098
4	.0137
5	.0195
6	.0273
7	.0371
[a]8	.0488
9	.0645
10	.0820
11	.1016
12	.1250
13	.1504
14	.1797
15	.2129

$n = 10$

T	P
0	.0010
1	.0020
2	.0029
3	.0049
4	.0068
5	.0098
6	.0137
7	.0186
8	.0244
9	.0322
[a]10	.0420
11	.0527
12	.0654
13	.0801
14	.0967
15	.1162
16	.1377
17	.1611
18	.1875
19	.2158
20	.2461
21	.2783
22	.3125
23	.3477
24	.3848
25	.4229
26	.4609
27	.5000

$n = 11$

T	P
0	.0005
1	.0010
2	.0015
3	.0024
4	.0034
5	.0049
6	.0068
7	.0093
8	.0122
9	.0161
10	.0210
11	.0269
12	.0337
[a]13	.0415
14	.0508
15	.0615
16	.0737
17	.0874
18	.1030
19	.1201
20	.1392
21	.1602
22	.1826
23	.2065
24	.2324
25	.2598
26	.2886
27	.3188
28	.3501
29	.3823
30	.4155
31	.4492
32	.4829
33	.5171

$n = 12$

T	P
0	.0002
1	.0005
2	.0007
3	.0012
4	.0017
5	.0024
6	.0034
7	.0046
8	.0061
9	.0081
10	.0105
11	.0134
12	.0171
13	.0212
14	.0261
15	.0320
16	.0386
[a]17	.0461
18	.0549
19	.0647
20	.0757
21	.0881
22	.1018
23	.1167
24	.1331
25	.1506
26	.1697
27	.1902
28	.2119
29	.2349
30	.2593
31	.2847
32	.3110
33	.3386
34	.3667
35	.3955

$n = 13$

T	P
0	.0001
1	.0002
2	.0004
3	.0006
4	.0009
5	.0012
6	.0017
7	.0023
8	.0031
9	.0040
10	.0052
11	.0067
12	.0085
13	.0107
14	.0133
15	.0164
16	.0199
17	.0239
18	.0287
19	.0341
20	.0402
[a]21	.0471
22	.0549
23	.0636
24	.0732
25	.0839
26	.0955
27	.1082
28	.1219
29	.1367
30	.1527
31	.1698
32	.1879
33	.2072
34	.2274
35	.2487

[a]For given n, the smallest rank total for which the probability level is equal to or less than 0.0500.

TABLE K (*continued*)

n = 7		*n* = 9		*n* = 12		*n* = 13	
T	P	T	P	T	P	T	P
13	.4688	16	.2480	36	.4250	36	.2709
14	.5313	17	.2852	37	.4548	37	.2939
		18	.3262	38	.4849	38	.3177
		19	.3672	39	.5151	39	.3424
		20	.4102			40	.3677
		21	.4551			41	.3934
		22	.5000			42	.4197
						43	.4463
						44	.4730
						45	.5000

n = 14		*n* = 14		*n* = 15		*n* = 16		*n* = 17		*n* = 17	
T	P	T	P	T	P	T	P	T	P	T	P
0	.0001	50	.4516	47	.2444	39	.0719	25	.0064	74	.4633
2	.0002	51	.4758	48	.2622	40	.0795	26	.0075	75	.4816
3	.0003	52	.5000	49	.2807	41	.0877	27	.0087	76	.5000
4	.0004			50	.2997	42	.0964	28	.0101		
5	.0006	*n* = 15		51	.3193	43	.1057	29	.0116	*n* = 18	
6	.0009	1	.0001	52	.3394	44	.1156	30	.0133	6	.0001
7	.0012	3	.0002	53	.3599	45	.1261	31	.0153	10	.0002
8	.0015	5	.0003	54	.3808	46	.1372	32	.0174	12	.0003
9	.0020	6	.0004	55	.4020	47	.1489	33	.0198	14	.0004
10	.0026	7	.0006	56	.4235	48	.1613	34	.0224	15	.0005
11	.0034	8	.0008	57	.4452	49	.1742	35	.0253	16	.0006
12	.0043	9	.0010	58	.4670	50	.1877	36	.0284	17	.0008
13	.0054	10	.0013	59	.4890	51	.2019	37	.0319	18	.0010
14	.0067	11	.0017	60	.5110	52	.2166	38	.0357	19	.0012
15	.0083	12	.0021	*n* = 16		53	.2319	39	.0398	20	.0014
16	.0101	13	.0027	3	.0001	54	.2477	40	.0443	21	.0017
17	.0123	14	.0034	5	.0002	55	.2641	[a]41	.0492	22	.0020
18	.0148	15	.0042	7	.0003	56	.2809	42	.0544	23	.0024
19	.0176	16	.0051	8	.0004	57	.2983	43	.0601	24	.0028
20	.0209	17	.0062	9	.0005	58	.3161	44	.0662	25	.0033
21	.0247	18	.0075	10	.0007	59	.3343	45	.0727	26	.0038
22	.0290	19	.0090	11	.0008	60	.3529	46	.0797	27	.0045
23	.0338	20	.0108	12	.0011	61	.3718	47	.0871	28	.0052
24	.0392	21	.0128	13	.0013	62	.3910	48	.0950	29	.0060

TABLE K (continued)

$n = 14$		$n = 15$		$n = 16$		$n = 16$		$n = 17$		$n = 18$	
T	P	T	P	T	P	T	P	T	P	T	P
[a]25	.0453	22	.0151	14	.0017	63	.4104	49	.1034	30	.0069
26	.0520	23	.0177	15	.0021	64	.4301	50	.1123	31	.0080
27	.0594	24	.0206	16	.0026	65	.4500	51	.1218	32	.0091
28	.0676	25	.0240	17	.0031	66	.4699	52	.1317	33	.0104
29	.0765	26	.0277	18	.0038	67	.4900	53	.1421	34	.0118
30	.0863	27	.0319	19	.0046	68	.5100	54	.1530	35	.0134
31	.0969	28	.0365	20	.0055			55	.1645	36	.0152
32	.1083	29	.0416	21	.0065	$n = 17$		56	.1764	37	.0171
33	.1206	[a]30	.0473	22	.0078	4	.0001	57	.1889	38	.0192
34	.1338	31	.0535	23	.0091	8	.0002	58	.2019	39	.0216
35	.1479	32	.0603	24	.0107	9	.0003	59	.2153	40	.0241
36	.1629	33	.0677	25	.0125	11	.0004	60	.2293	41	.0269
37	.1788	34	.0757	26	.0145	12	.0005	61	.2437	42	.0300
38	.1955	35	.0844	27	.0168	13	.0007	62	.2585	43	.0333
39	.2131	36	.0938	28	.0193	14	.0008	63	.2738	44	.0368
40	.2316	37	.1039	29	.0222	15	.0010	64	.2895	45	.0407
41	.2508	38	.1147	30	.0253	16	.0013	65	.3056	46	.0449
42	.2708	39	.1262	31	.0288	17	.0016	66	.3221	[a]47	.0494
43	.2915	40	.1384	32	.0327	18	.0019	67	.3389	48	.0542
44	.3129	41	.1514	33	.0370	19	.0023	68	.3559	49	.0594
45	.3349	42	.1651	34	.0416	20	.0028	69	.3733	50	.0649
46	.3574	43	.1796	[a]35	.0467	21	.0033	70	.3910	51	.0708
47	.3804	44	.1947	36	.0523	22	.0040	70	.4088	52	.0770
48	.4039	45	.2106	37	.0583	23	.0047	72	.4268	53	.0837
49	.4276	46	.2271	38	.0649	24	.0055	73	.4450	54	.0907

$n = 18$		$n = 19$		$n = 19$		$n = 20$		$n = 20$		$n = 21$	
T	P	T	P	T	P	T	P	T	P	T	P
55	.0982	30	.0036	79	.2706	48	.0164	97	.3921	61	.0298
56	.1061	31	.0041	80	.2839	49	.0181	98	.4062	62	.0323
57	.1144	32	.0047	81	.2974	50	.0200	99	.4204	63	.0351
58	.1231	33	.0054	82	.3113	51	.0220	100	.4347	64	.0380
59	.1323	34	.0062	83	.3254	52	.0242	101	.4492	65	.0411
60	.1419	35	.0070	84	.3397	53	.0266	102	.4636	66	.0444
61	.1519	36	.0080	85	.3543	54	.0291	103	.4782	[a]67	.0479
62	.1624	37	.0090	86	.3690	55	.0319	104	.4927	68	.0516
63	.1733	38	.0102	87	.3840	56	.0348	105	.5073	69	.0555
64	.1846	39	.0115	88	.3991	57	.0379	$n = 21$		70	.0597
65	.1964	40	.0129	89	.4144	58	.0413	14	.0001	71	.0640
66	.2086	41	.0145	90	.4298	59	.0448	20	.0002	72	.0686

TABLE K (*continued*)

n = 18		*n* = 19		*n* = 19		*n* = 20		*n* = 21		*n* = 21	
T	P	T	P	T	P	T	P	T	P	T	P
67	.2211	42	.0162	91	.4453	[a]60	.0487	22	.0003	73	.0735
68	.2341	43	.0180	92	.4609	61	.0527	24	.0004	74	.0786
69	.2475	44	.0201	93	.4765	62	.0570	26	.0005	75	.0839
70	.2613	45	.0223	94	.4922	63	.0615	27	.0006	76	.0895
71	.2754	46	.0247	95	.5078	64	.0664	28	.0007	77	.0953
72	.2899	47	.0273			65	.0715	29	.0008	78	.1015
73	.3047	48	.0301	*n* = 20		66	.0768	30	.0009	79	.1078
74	.3198	49	.0331	11	.0001	67	.0825	31	.0011	80	.1145
75	.3353	50	.0364	16	.0002	68	.0884	32	.0012	81	.1214
76	.3509	51	.0399	19	.0003	69	.0947	33	.0014	82	.1286
77	.3669	52	.0437	20	.0004	70	.1012	34	.0016	83	.1361
78	.3830	[a]53	.0478	22	.0005	71	.1081	35	.0019	84	.1439
79	.3994	54	.0521	23	.0006	72	.1153	36	.0021	85	.1519
80	.4159	55	.0567	24	.0007	73	.1227	37	.0024	86	.1602
81	.4325	56	.0616	25	.0008	74	.1305	38	.0028	87	.1688
82	.4493	57	.0668	26	.0010	75	.1387	39	.0031	88	.1777
83	.4661	58	.0723	27	.0012	76	.1471	40	.0036	89	.1869
84	.4831	59	.0782	28	.0014	77	.1559	41	.0040	90	.1963
85	.5000	60	.0844	29	.0016	78	.1650	42	.0045	91	.2060
		61	.0909	30	.0018	79	.1744	43	.0051	92	.2160
n = 19		62	.0978	31	.0021	80	.1841	44	.0057	93	.2262
9	.0001	63	.1051	32	.0024	81	.1942	45	.0063	94	.2367
13	.0002	64	.1127	33	.0028	82	.2045	46	.0071	95	.2474
15	.0003	65	.1206	34	.0032	83	.2152	47	.0079	96	.2584
17	.0004	66	.1290	35	.0036	84	.2262	48	.0088	97	.2696
18	.0005	67	.1377	36	.0042	85	.2375	49	.0097	98	.2810
19	.0006	68	.1467	37	.0047	86	.2490	50	.0108	99	.2927
20	.0007	69	.1562	38	.0053	87	.2608	51	.0119	100	.3046
21	.0008	70	.1660	39	.0060	88	.2729	52	.0132	101	.3166
22	.0010	71	.1762	40	.0068	89	.2853	53	.0145	102	.3289
23	.0012	72	.1868	41	.0077	90	.2979	54	.0160	103	.3414
24	.0014	73	.1977	42	.0086	91	.3108	55	.0175	104	.3540
25	.0017	74	.2090	43	.0096	92	.3238	56	.0192	105	.3667
26	.0020	75	.2207	44	.0107	93	.3371	57	.0210	106	.3796
27	.0023	76	.2327	45	.0120	94	.3506	58	.0230	107	.3927
28	.0027	77	.2450	46	.0133	95	.3643	59	.0251	108	.4058
29	.0031	78	.2576	47	.0148	96	.3781	60	.0273	109	.4191

TABLE K ***(continued)***

n = 21		*n* = 22		*n* = 22		*n* = 23		*n* = 23		*n* = 24	
T	P	T	P	T	P	T	P	T	P	T	P
110	.4324	67	.0271	116	.3751	68	.0163	117	.2700	62	.0053
111	.4459	68	.0293	117	.3873	69	.0177	118	.2800	63	.0058
112	.4593	69	.0317	118	.3995	70	.0192	119	.2902	64	.0063
113	.4729	70	.0342	119	.4119	71	.0208	120	.3005	65	.0069
114	.4864	71	.0369	120	.4243	72	.0224	121	.3110	66	.0075
115	.5000	72	.0397	121	.4368	73	.0242	122	.3217	67	.0082
		73	.0427	122	.4494	74	.0261	123	.3325	68	.0089
		74	.0459	123	.4620	75	.0281	124	.3434	69	.0097
n = 22		[a]75	.0492	124	.4746	76	.0303	125	.3545	70	.0106
18	.0001	76	.0527	125	.4873	77	.0325	126	.3657	71	.0115
23	.0002	77	.0564	126	.5000	78	.0349	127	.3770	72	.0124
26	.0003	78	.0603			79	.0374	128	.3884	73	.0135
29	.0004	79	.0644	*n* = 23		80	.0401	129	.3999	74	.0146
30	.0005	80	.0687	21	.0001	81	.0429	130	.4115	75	.0157
32	.0006	81	.0733	28	.0002	82	.0459	131	.4231	76	.0170
33	.0007	82	.0780	31	.0003	[a]83	.0490	132	.4348	77	.0183
34	.0008	83	.0829	33	.0004	84	.0523	133	.4466	78	.0197
35	.0010	84	.0881	35	.0005	85	.0557	134	.4584	79	.0212
36	.0011	85	.0935	36	.0006	86	.0593	135	.4703	80	.0228
37	.0013	86	.0991	38	.0007	87	.0631	136	.4822	81	.0245
38	.0014	87	.1050	39	.0008	88	.0671	137	.4941	82	.0263
39	.0016	88	.1111	40	.0009	89	.0712	138	.5060	83	.0282
40	.0018	89	.1174	41	.0011	90	.0755			84	.0302
41	.0021	90	.1240	42	.0012	91	.0801	*n* = 24		85	.0323
42	.0023	91	.1308	43	.0014	92	.0848	25	.0001	86	.0346
43	.0026	92	.1378	44	.0015	93	.0897	32	.0002	87	.0369
44	.0030	93	.1451	45	.0017	94	.0948	36	.0003	88	.0394
45	.0033	94	.1527	46	.0019	95	.1001	38	.0004	89	.0420
46	.0037	95	.1604	47	.0022	96	.1056	40	.0005	90	.0447
47	.0042	96	.1685	48	.0024	97	.1113	42	.0006	[a]91	.0475
48	.0046	97	.1767	49	.0027	98	.1172	43	.0007	92	.0505
49	.0052	98	.1853	50	.0030	99	.1234	44	.0008	93	.0537
50	.0057	99	.1940	51	.0034	100	.1297	45	.0009	94	.0570
51	.0064	100	.2030	52	.0037	101	.1363	46	.0010	95	.0604
52	.0070	101	.2122	53	.0041	102	.1431	47	.0011	96	.0640
53	.0078	102	.2217	54	.0046	103	.1501	48	.0013	97	.0678
54	.0086	103	.2314	55	.0051	104	.1573	49	.0014	98	.0717

TABLE K (*continued*)

n = 22		*n* = 22		*n* = 23		*n* = 23		*n* = 24		*n* = 24	
T	P	T	P	T	P	T	P	T	P	T	P
55	.0095	104	.2413	56	.0056	105	.1647	50	.0016	99	.0758
56	.0104	105	.2514	57	.0061	106	.1723	51	.0018	100	.0800
57	.0115	106	.2618	58	.0068	107	.1802	52	.0020	101	.0844
58	.0126	107	.2723	59	.0074	108	.1883	53	.0022	102	.0890
59	.0138	108	.2830	60	.0082	109	.1965	54	.0024	103	.0938
60	.0151	109	.2940	61	.0089	110	.2050	55	.0027	104	.0987
61	.0164	110	.3051	62	.0098	111	.2137	56	.0029	105	.1038
62	.0179	111	.3164	63	.0107	112	.2226	57	.0033	106	.1091
63	.0195	112	.3278	64	.0117	113	.2317	58	.0036	107	.1146
64	.0212	113	.3394	65	.0127	114	.2410	59	.0040	108	.1203
65	.0231	114	.3512	66	.0138	115	.2505	60	.0044	109	.1261
66	.0250	115	.3631	67	.0150	116	.2601	61	.0048	110	.1322

n = 24		*n* = 25		*n* = 25		*n* = 25		*n* = 26		*n* = 26	
T	P	T	P	T	P	T	P	T	P	T	P
111	.1384	50	.0008	99	.0452	148	.3556	81	.0076	130	.1289
112	.1448	51	.0009	[a]100	.0479	149	.3655	82	.0082	131	.1344
113	.1515	52	.0010	101	.0507	150	.3755	83	.0088	132	.1399
114	.1583	53	.0011	102	.0537	151	.3856	84	.0095	133	.1457
115	.1653	54	.0013	103	.0567	152	.3957	85	.0102	134	.1516
116	.1724	55	.0014	104	.0600	153	.4060	86	.0110	135	.1576
117	.1798	56	.0015	105	.0633	154	.4163	87	.0118	136	.1638
118	.1874	57	.0017	106	.0668	155	.4266	88	.0127	137	.1702
119	.1951	58	.0019	107	.0705	156	.4370	89	.0136	138	.1767
120	.2031	59	.0021	108	.0742	157	.4474	90	.0146	139	.1833
121	.2112	60	.0023	109	.0782	158	.4579	91	.0156	140	.1901
122	.2195	61	.0025	110	.0822	159	.4684	92	.0167	141	.1970
123	.2279	62	.0028	111	.0865	160	.4789	93	.0179	142	.2041
124	.2366	63	.0031	112	.0909	161	.4895	94	.0191	143	.2114
125	.2454	64	.0034	113	.0954	162	.5000	95	.0204	144	.2187
126	.2544	65	.0037	114	.1001			96	.0217	145	.2262
127	.2635	66	.0040	115	.1050	*n* = 26		97	.0232	146	.2339
128	.2728	67	.0044	116	.1100	34	.0001	98	.0247	147	.2417
129	.2823	68	.0048	117	.1152	42	.0002	99	.0263	148	.2496
130	.2919	69	.0053	118	.1205	46	.0003	100	.0279	149	.2577
131	.3017	70	.0057	119	.1261	49	.0004	101	.0297	150	.2658
132	.3115	71	.0062	120	.1317	51	.0005	102	.0315	151	.2741
133	.3216	72	.0068	121	.1376	53	.0006	103	.0334	152	.2826
134	.3317	73	.0074	122	.1436	55	.0007	104	.0355	153	.2911

TABLE K (*continued*)

n = 24		*n* = 25		*n* = 25		*n* = 26		*n* = 26		*n* = 26	
T	P	T	P	T	P	T	P	T	P	T	P
135	.3420	74	.0080	123	.1498	56	.0008	105	.0376	154	.2998
136	.3524	75	.0087	124	.1562	57	.0009	106	.0398	155	.3085
137	.3629	76	.0094	125	.1627	58	.0010	107	.0421	156	.3174
138	.3735	77	.0101	126	.1694	59	.0011	108	.0455	157	.3264
139	.3841	78	.0110	127	.1763	60	.0012	109	.0470	158	.3355
140	.3949	79	.0118	128	.1833	61	.0013	[a]110	.0497	159	.3447
141	.4058	80	.0128	129	.1905	62	.0015	111	.0524	160	.3539
142	.4167	81	.0137	130	.1979	63	.0016	112	.0553	161	.3633
143	.4277	82	.0148	131	.2054	64	.0018	113	.0582	162	.3727
144	.4387	83	.0159	132	.2131	65	.0020	114	.0613	163	.3822
145	.4498	84	.0171	133	.2209	66	.0021	115	.0646	164	.3918
146	.4609	85	.0183	134	.2289	67	.0023	116	.0679	165	.4014
147	.4721	86	.0197	135	.2371	68	.0026	117	.0714	166	.4111
148	.4832	87	.0211	136	.2454	69	.0028	118	.0750	167	.4208
149	.4944	88	.0226	137	.2539	70	.0031	119	.0787	168	.4306
150	.5056	89	.0241	130	.2625	71	.0033	120	.0825	169	.4405
		90	.0258	139	.2712	72	.0036	121	.0865	170	.4503
n = 25		91	.0275	140	.2801	73	.0040	122	.0907	171	.4602
29	.0001	92	.0294	141	.2891	74	.0043	123	.0950	172	.4702
37	.0002	93	.0313	142	.2983	75	.0047	124	.0994	173	.4801
41	.0003	94	.0334	143	.3075	76	.0051	125	.1039	174	.4900
43	.0004	95	.0355	144	.3169	77	.0055	126	.1086	175	.5000
45	.0005	96	.0377	145	.3264	78	.0060	127	.1135		
47	.0006	97	.0401	146	.3360	79	.0065	128	.1185		
48	.0007	98	.0426	147	.3458	80	.0070	129	.1236		

n = 27		*n* = 27		*n* = 27		*n* = 28		*n* = 28		*n* = 28	
T	P	T	P	T	P	T	P	T	P	T	P
39	.0001	105	.0218	154	.2066	74	.0012	123	.0349	172	.2466
47	.0002	106	.0231	155	.2135	75	.0013	124	.0368	173	.2538
52	.0003	107	.0246	156	.2205	76	.0015	125	.0387	174	.2611
55	.0004	108	.0260	157	.2277	77	.0016	126	.0407	175	.2685
57	.0005	109	.0276	158	.2349	78	.0017	127	.0428	176	.2759
59	.0006	110	.0292	159	.2423	79	.0019	128	.0450	177	.2835
61	.0007	111	.0309	160	.2498	80	.0020	129	.0473	178	.2912
62	.0008	112	.0327	161	.2574	81	.0022	[a]130	.0496	179	.2990
64	.0009	113	.0346	162	.2652	82	.0024	131	.0521	180	.3068
65	.0010	114	.0366	163	.2730	83	.0026	132	.0546	181	.3148
66	.0011	115	.0386	164	.2810	84	.0028	133	.0573	182	.3228
67	.0012	116	.0407	165	.2890	85	.0030	134	.0600	183	.3309

TABLE K (*continued*)

n = 27		*n* = 27		*n* = 27		*n* = 28		*n* = 28		*n* = 28	
T	P	T	P	T	P	T	P	T	P	T	P
68	.0014	117	.0430	166	.2972	86	.0033	135	.0628	184	.3391
69	.0015	118	.0453	167	.3055	87	.0035	136	.0657	185	.3474
70	.0016	[a]119	.0477	168	.3138	88	.0038	137	.0688	186	.3557
71	.0018	120	.0502	169	.3223	89	.0041	138	.0719	187	.3641
72	.0019	121	.0528	170	.3308	90	.0044	139	.0751	188	.3725
73	.0021	122	.0555	171	.3395	91	.0048	140	.0785	189	.3811
74	.0023	123	.0583	172	.3482	92	.0051	141	.0819	190	.3896
75	.0025	124	.0613	173	.3570	93	.0055	142	.0855	191	.3983
76	.0027	125	.0643	174	.3659	94	.0059	143	.0891	192	.4070
77	.0030	126	.0674	175	.3748	95	.0064	144	.0929	193	.4157
78	.0032	127	.0707	176	.3838	96	.0068	145	.0968	194	.4245
79	.0035	128	.0741	177	.3929	97	.0073	146	.1008	195	.4333
80	.0038	129	.0776	178	.4020	98	.0078	147	.1049	196	.4421
81	.0041	130	.0812	179	.4112	99	.0084	148	.1091	197	.4510
82	.0044	131	.0849	180	.4204	100	.0089	149	.1135	198	.4598
83	.0048	132	.0888	181	.4297	101	.0096	150	.1180	199	.4687
84	.0052	133	.0927	182	.4390	102	.0102	151	.1225	200	4777
85	.0056	134	.0968	183	.4483	103	.0109	152	.1273	201	.4866
86	.0060	135	.1010	184	.4577	104	.0116	153	.1321	202	.4955
87	.0065	136	.1054	185	.4670	105	.0124	154	.1370	203	.5045
88	.0070	137	.1099	186	.4764	106	.0132	155	.1421		
89	.0075	138	.1145	187	.4859	107	.0140	156	.1473	*n* = 28	
90	.0081	139	.1193	188	.4953	108	.0149	157	.1526	50	.0001
91	.0087	140	.1242	189	.5047	109	.0159	158	.1580	59	.0002
92	.0093	141	.1292			110	.0168	159	.1636	65	.0003
93	.0100	142	.1343	*n* = 28		111	.0179	160	.1693	68	.0004
94	.0107	143	.1396	44	.0001	112	.0190	161	.1751	71	.0005
95	.0115	144	.1450	53	.0002	113	.0201	162	.1810	73	.0006
96	.0123	145	.1506	58	.0003	114	.0213	163	.1870	75	.0007
97	.0131	146	.1563	61	.0004	115	.0226	164	.1932	76	.0008
98	.0140	147	.1621	64	.0005	116	.0239	165	.1995	78	.0009
99	.0150	148	.1681	66	.0006	117	.0252	166	.2059	79	.0010
100	.0159	149	.1742	68	.0007	118	.0267	167	.2124	80	.0011
101	.0170	150	.1804	69	.0008	119	.0282	168	.2190	81	.0012
102	.0181	151	.1868	70	.0009	120	.0298	169	.2257	82	.0013
103	.0193	152	.1932	72	.0010	121	.0314	170	.2326	83	.0014
104	.0205	153	.1999	73	.0011	122	.0331	171	.2395	84	.0015

TABLE K ***(continued)***

$n = 29$		$n = 29$		$n = 29$		$n = 30$		$n = 30$		$n = 30$	
T	P	T	P	T	P	T	P	T	P	T	P
85	.0016	134	.0362	183	.2340	90	.0013	139	.0275	188	.1854
86	.0018	135	.0380	184	.2406	91	.0014	140	.0288	189	.1909
87	.0019	136	.0399	185	.2473	92	.0015	141	.0303	190	.1965
88	.0021	137	.0418	186	.2541	93	.0016	142	.0318	191	.2022
89	.0022	138	.0439	187	.2611	94	.0017	143	.0333	192	.2081
90	.0024	139	.0460	188	.2681	95	.0019	144	.0349	193	.2140
91	.0026	[a]140	.0482	189	.2752	96	.0020	145	.0366	194	.2200
92	.0028	141	.0504	190	.2824	97	.0022	146	.0384	195	.2261
93	.0030	142	.0528	191	.2896	98	.0023	147	.0402	196	.2323
94	.0032	143	.0552	192	.2970	99	.0025	148	.0420	197	.2386
95	.0035	144	.0577	193	.3044	100	.0027	149	.0440	198	.2449
96	.0037	145	.0603	194	.3120	101	.0029	150	.0460	119	.2514
97	.0040	146	.0630	195	.3196	102	.0031	[a]151	.0481	200	.2579
98	.0043	147	.0658	196	.3272	103	.0033	152	.0502	201	.2646
99	.0046	148	.0687	197	.3350	104	.0036	153	.0524	202	.2713
100	.0049	149	.0716	198	.3428	105	.0038	154	.0547	203	.2781
101	.0053	150	.0747	199	.3507	106	.0041	155	.0571	204	.2849
102	.0057	151	.0778	200	.3586	107	.0044	156	.0595	205	.2919
103	.0061	152	.0811	201	.3666	108	.0047	157	.0621	206	.2989
104	.0065	153	.0844	202	.3747	109	.0050	158	.0647	207	.3060
105	.0069	154	.0879	203	.3828	110	.0053	159	.0674	208	.3132
106	.0074	155	.0914	204	.3909	111	.0057	160	.0701	209	.3204
107	.0079	156	.0951	205	.3991	112	.0060	161	.0730	210	.3277
108	.0084	157	.0988	206	.4074	113	.0064	162	.0759	211	.3351
109	.0089	158	.1027	207	.4157	114	.0068	163	.0790	212	.3425
110	.0095	159	.1066	208	.4240	115	.0073	164	.0821	213	.3500
111	.0101	160	.1107	209	.4324	116	.0077	165	.0853	214	.3576
112	.0108	161	.1149	210	.4408	117	.0082	166	.0886	215	.3652
113	.0115	162	.1191	211	.4492	118	.0087	167	.0920	216	.3728
114	.0122	163	.1235	212	.4576	119	.0093	168	.0955	217	.3805
115	.0129	164	.1280	213	.4661	120	.0098	169	.0990	218	.3883
116	.0137	165	.1326	214	.4745	121	.0104	170	.1027	219	.3961
117	.0145	166	.1373	215	.4830	122	.0110	171	.1065	220	.4039
118	.0154	167	.1421	216	.4915	123	.0117	172	.1103	221	.4118
119	.0163	168	.1471	217	.5000	124	.0124	173	.1143	222	.4197
120	.0173	169	.1521			125	.0131	174	.1183	223	.4276

TABLE K (*continued*)

n = 29		*n* = 29		*n* = 30		*n* = 30		*n* = 30		*n* = 30	
T	P	T	P			T	P	T	P	T	P
121	.0183	170	.1572			126	.0139	175	.1225	224	.4356
122	.0193	171	.1625	55	.0001	127	.0147	176	.1267	225	.4436
123	.0204	172	.1679	66	.0002	128	.0155	177	.1311	226	.4516
124	.0216	173	.1733	71	.0003	129	.0164	178	.1355	227	.4596
125	.0228	174	.1789	75	.0004	130	.0173	179	.1400	228	.4677
126	.0240	175	.1846	78	.0005	131	.0182	180	.1447	229	.4758
127	.0253	176	.1904	80	.0006	132	.0192	181	.1494	230	.4838
128	.0267	177	.1963	82	.0007	133	.0202	182	.1543	231	.4919
129	.0281	178	.2023	84	.0008	134	.0213	183	.1592	232	.5000
130	.0296	179	.2085	85	.0009	135	.0225	184	.1642		
131	.0311	180	.2147	87	.0010	136	.0236	185	.1694		
132	.0328	181	.2210	88	.0011	137	.0249	186	.1746		
133	.0344	182	.2274	89	.0012	138	.0261	187	.1799		

TABLE L Quantiles of the Mann-Whitney Test Statistic

n	*p*	*m* = 2	3	4	5	6	7	8	9	10	11	12	13	14	15	16	17	18	19	20
	.001	0	0	0	0	0	0	0	0	0	0	0	0	0	0	0	0	0	0	0
	.005	0	0	0	0	0	0	0	0	0	0	0	0	0	0	0	0	0	1	1
2	.01	0	0	0	0	0	0	0	0	0	0	0	1	1	1	1	1	1	2	2
	.025	0	0	0	0	0	0	1	1	1	1	2	2	2	2	2	3	3	3	3
	.05	0	0	0	1	1	1	2	2	2	2	3	3	4	4	4	4	5	5	5
	.10	0	1	1	2	2	2	3	3	4	4	5	5	5	6	6	7	7	8	8
	.001	0	0	0	0	0	0	0	0	0	0	0	0	0	0	0	1	1	1	1
	.005	0	0	0	0	0	0	0	1	1	1	2	2	2	3	3	3	3	4	4
3	.01	0	0	0	0	0	1	1	2	2	2	3	3	3	4	4	5	5	5	6
	.025	0	0	0	1	2	2	3	3	4	4	5	5	6	6	7	7	8	8	9
	.05	0	1	1	2	3	3	4	5	5	6	6	7	8	8	9	10	10	11	12
	.10	1	2	2	3	4	5	6	6	7	8	9	10	11	11	12	13	14	15	16
	.001	0	0	0	0	0	0	0	0	1	1	1	2	2	2	3	3	4	4	4
	.005	0	0	0	0	1	1	2	2	3	3	4	4	5	6	6	7	7	8	9
4	.01	0	0	0	1	2	2	3	4	4	5	6	6	7	9	8	9	10	10	11
	.025	0	0	1	2	3	4	5	5	6	7	8	9	10	11	12	12	13	14	15
	.05	0	1	2	3	4	5	6	7	8	9	10	11	12	13	15	16	17	18	19
	.10	1	2	4	5	6	7	8	10	11	12	13	14	16	17	18	19	21	22	23
	.001	0	0	0	0	0	0	1	2	2	3	3	4	4	5	6	6	7	8	8
	.005	0	0	0	1	2	2	3	4	5	6	7	8	8	9	10	11	12	13	14
5	.01	0	0	1	2	3	4	5	6	7	8	9	10	11	12	13	14	15	16	17
	.025	0	1	2	3	4	6	7	8	9	10	12	13	14	15	16	18	19	20	21
	.05	1	2	3	5	6	7	9	10	12	13	14	16	17	19	20	21	23	24	26
	.10	2	3	5	6	8	9	11	13	14	16	18	19	21	23	24	26	28	29	31

TABLE L (continued)

n	*p*	*m* = 2	3	4	5	6	7	8	9	10	11	12	13	14	15	16	17	18	19	20
	.001	0	0	0	0	0	0	2	3	4	5	5	6	7	8	9	10	11	12	13
	.005	0	0	1	2	3	4	5	6	7	8	10	11	12	13	14	16	17	18	19
6	.01	0	0	2	3	4	5	7	8	9	10	12	13	14	16	17	19	20	21	23
	.025	0	2	3	4	6	7	9	11	12	14	15	17	18	20	22	23	25	26	28
	.05	1	3	4	6	8	9	11	13	15	17	18	20	22	24	26	27	29	31	33
	.10	2	4	6	8	10	12	14	16	18	20	22	24	26	28	30	32	35	37	39
	.001	0	0	0	0	1	2	3	4	6	7	8	9	10	11	12	14	15	16	17
	.005	0	0	1	2	4	5	7	8	10	11	13	14	16	17	19	20	22	23	25
7	.01	0	1	2	4	5	7	8	10	12	13	15	17	18	20	22	24	25	27	29
	.025	0	2	4	6	7	9	11	13	15	17	19	21	23	25	27	29	31	33	35
	.05	1	3	5	7	9	12	14	16	18	20	22	25	27	29	31	34	36	38	40
	.10	2	5	7	9	12	14	17	19	22	24	27	29	32	34	37	39	42	44	47
	.001	0	0	0	1	2	3	5	6	7	9	10	12	13	15	16	18	19	21	22
	.005	0	0	2	3	5	7	8	10	12	14	16	18	19	21	23	25	27	29	31
8	.01	0	1	3	5	7	8	10	12	14	16	18	21	23	25	27	29	31	33	35
	.025	1	3	5	7	9	11	14	16	18	20	23	25	27	30	32	35	37	39	42
	.05	2	4	6	9	11	14	16	19	21	24	27	29	32	34	37	40	42	45	48
	.10	3	6	8	11	14	17	20	23	25	28	31	34	37	40	43	46	49	52	55
	.001	0	0	0	2	3	4	6	8	9	11	13	15	16	18	20	22	24	26	27
	.005	0	1	2	4	6	8	10	12	14	17	19	21	23	25	28	30	32	34	37
9	.01	0	2	4	6	8	10	12	15	17	19	22	24	27	29	32	34	37	39	41
	.025	1	3	5	8	11	13	16	18	21	24	27	29	32	35	38	40	43	46	49
	.05	2	5	7	10	13	16	19	22	25	28	31	34	37	40	43	46	49	52	55
	.10	3	6	10	13	16	19	23	26	29	32	36	39	42	46	49	53	56	59	63
	.001	0	0	1	2	4	6	7	9	11	13	15	18	20	22	24	26	28	30	33
	.005	0	1	3	5	7	10	12	14	17	19	22	25	27	30	32	35	38	40	43
10	.01	0	2	4	7	9	12	14	17	20	23	25	28	31	34	37	39	42	45	48
	.025	1	4	6	9	12	15	18	21	24	27	30	34	37	40	43	46	49	53	46
	.05	2	5	8	12	15	18	21	25	28	32	35	38	42	45	49	52	56	59	63
	.10	4	7	11	14	18	22	25	29	33	37	40	44	48	52	55	59	63	67	71

TABLE L ***(continued)***

n	p	$m = 2$	3	4	5	6	7	8	9	10	11	12	13	14	15	16	17	18	19	20
	.001	0	0	1	3	5	7	9	11	13	16	18	21	23	25	28	30	33	35	38
	.005	0	1	3	6	8	11	14	17	19	22	25	28	31	34	37	40	43	46	49
11	.01	0	2	5	8	10	13	16	19	23	26	29	32	35	38	42	45	48	51	54
	.025	1	4	7	10	14	17	20	24	27	31	34	38	41	45	48	52	56	59	63
	.05	2	6	9	13	17	20	24	28	32	35	39	43	47	51	55	58	62	66	70
	.10	4	8	12	16	20	24	28	32	37	41	45	49	53	58	62	66	70	74	79
	.001	0	0	1	3	5	8	10	13	15	18	21	24	26	29	32	35	38	41	43
	.005	0	2	4	7	10	13	16	19	22	25	28	32	35	38	42	45	48	52	55
12	.01	0	3	6	9	12	15	18	22	25	29	32	36	39	43	47	50	54	57	61
	.025	2	5	8	12	15	19	23	27	30	34	38	42	46	50	54	58	62	66	70
	.05	3	6	10	14	18	22	27	31	35	39	43	48	52	56	61	65	69	73	78
	.10	5	9	13	18	22	27	31	36	40	45	50	54	59	64	68	73	78	82	87
	.001	0	0	2	4	6	9	12	15	18	21	24	27	30	33	36	39	43	46	49
	.005	0	2	4	8	11	14	18	21	25	28	32	35	39	43	46	50	54	58	61
13	.01	1	3	6	10	13	17	21	24	28	32	36	40	44	48	52	56	60	64	68
	.025	2	5	9	13	17	21	25	29	34	38	42	46	51	55	60	64	68	73	77
	.05	3	7	11	16	20	25	29	34	38	43	48	52	57	62	66	71	76	81	85
	.10	5	10	14	19	24	29	34	39	44	49	54	59	64	69	75	80	85	90	95
	.001	0	0	2	4	7	10	13	16	20	23	26	30	33	37	40	44	47	51	55
	.005	0	2	5	8	12	16	19	23	27	31	35	39	43	47	51	55	59	64	68
14	.01	1	3	7	11	14	18	23	27	31	35	39	44	48	52	57	61	66	70	74
	.025	2	6	10	14	18	23	27	32	37	41	46	51	56	60	65	70	75	79	84
	.05	4	8	12	17	22	27	32	37	42	47	52	57	62	67	72	78	83	88	93
	.10	5	11	16	21	26	32	37	42	48	53	59	64	70	75	81	86	92	98	103
	.001	0	0	2	5	8	11	15	18	22	25	29	33	37	41	44	48	52	56	60
	.005	0	3	6	9	13	17	21	25	30	34	38	43	47	52	56	61	65	70	74
15	.01	1	4	8	12	16	20	25	29	34	38	43	48	52	57	62	67	71	76	81
	.025	2	6	11	15	20	25	30	35	40	45	50	55	60	65	71	76	81	86	91
	.05	4	8	13	19	24	29	34	40	45	51	56	62	67	73	78	84	89	95	101
	.10	6	11	17	23	28	34	40	46	52	58	64	69	75	81	87	93	99	105	111

TABLE L (*continued*)

n	*p*	*m* = 2	3	4	5	6	7	8	9	10	11	12	13	14	15	16	17	18	19	20
	.001	0	0	3	6	9	12	16	20	24	28	32	36	40	44	49	53	57	61	66
	.005	0	3	6	10	14	19	23	28	32	37	42	46	51	56	61	66	71	75	80
16	.01	1	4	8	13	17	22	27	32	37	42	47	52	57	62	67	72	77	83	88
	.025	2	7	12	16	22	27	32	38	43	48	54	60	65	71	76	82	87	93	99
	.05	4	9	15	20	26	31	37	43	49	55	61	66	72	78	84	90	96	102	108
	.10	6	12	18	24	30	37	43	49	55	62	68	75	81	87	94	100	107	113	120
	.001	0	1	3	6	10	14	18	22	26	30	35	39	44	48	53	58	62	67	71
	.005	0	3	7	11	16	20	25	30	35	40	45	50	55	61	66	71	76	82	87
17	.01	1	5	9	14	19	24	29	34	39	45	50	56	61	67	72	78	83	89	94
	.025	3	7	12	18	23	29	35	40	46	52	58	64	70	76	82	88	94	100	106
	.05	4	10	16	21	27	34	40	46	52	58	65	71	78	84	90	97	103	110	116
	.10	7	13	19	26	32	39	46	53	59	66	73	80	86	93	100	107	114	121	128
	.001	0	1	4	7	11	15	19	24	28	33	38	43	47	52	57	62	67	72	77
	.005	0	3	7	12	17	22	27	32	38	43	48	54	59	65	71	76	82	88	93
18	.01	1	5	10	15	20	25	31	37	42	48	54	60	66	71	77	83	89	95	101
	.025	3	8	13	19	25	31	37	43	49	56	62	68	75	81	87	94	100	107	113
	.05	5	10	17	23	29	36	42	49	56	62	69	76	83	89	96	103	110	117	124
	.10	7	14	21	28	35	42	49	56	63	70	78	85	92	99	107	114	121	129	136
	.001	0	1	4	8	12	16	21	26	30	35	41	46	51	56	61	67	72	78	83
	.005	1	4	8	13	18	23	29	34	40	46	52	58	64	70	75	82	88	94	100
19	.01	2	5	10	16	21	27	33	39	45	51	57	64	70	76	83	89	95	102	108
	.025	3	8	14	20	26	33	39	46	53	59	66	73	79	86	93	100	107	114	120
	.05	5	11	18	24	31	38	45	52	59	66	73	81	88	95	102	110	117	124	131
	.10	8	15	22	29	37	44	52	59	67	74	82	90	98	105	113	121	129	136	144
	.001	0	1	4	8	13	17	22	27	33	38	43	49	55	60	66	71	77	83	89
	.005	1	4	9	14	19	25	31	37	43	49	55	61	68	74	80	87	93	100	106
20	.01	2	6	11	17	23	29	35	41	48	54	61	68	74	81	88	94	101	108	115
	.025	3	9	15	21	28	35	42	49	56	63	70	77	84	91	99	106	113	120	128
	.05	5	12	19	26	33	40	48	55	63	70	78	85	93	101	108	116	124	131	139
	.10	8	16	23	31	39	47	55	63	71	79	87	95	103	111	120	128	136	144	152

TABLE M Quantiles of the Kolmogorov Test Statistic

One-Sided Test p =	**.90**	**.95**	**.975**	**.99**	**.995**
Two-Sided Test p =	**.80**	**.90**	**.95**	**.98**	**.99**
n = 1	.900	.950	.975	.990	.995
2	.684	.776	.842	.900	.929
3	.565	.636	.708	.785	.829
4	.493	.565	.624	.689	.734
5	.447	.509	.563	.627	.669
6	.410	.468	.519	.577	.617
7	.381	.436	.483	.538	.576
8	.358	.410	.454	.507	.542
9	.339	.387	.430	.480	.513
10	.323	.369	.409	.457	.489
11	.308	.352	.391	.437	.468
12	.296	.338	.375	.419	.449
13	.285	.325	.361	.404	.432
14	.275	.314	.349	.390	.418
15	.266	.304	.338	.377	.404
16	.258	.295	.327	.366	.392
17	.250	.286	.318	.355	.381
18	.244	.279	.309	.346	.371
19	.237	.271	.301	.337	.361
20	.232	.265	.294	.329	.352
21	.226	.259	.287	.321	.344
22	.221	.253	.281	.314	.337
23	.216	.247	.275	.307	.330
24	.212	.242	.269	.301	.323
25	.208	.238	.264	.295	.317
26	.204	.233	.259	.290	.311
27	.200	.229	.254	.284	.305
28	.197	.225	.250	.279	.300
29	.193	.221	.246	.275	.295
30	.190	.218	.242	.270	.290
31	.187	.214	.238	.266	.285
32	.184	.211	.234	.262	.281
33	.182	.208	.231	.258	.277
34	.179	.205	.227	.254	.273
35	.177	.202	.224	.251	.269
36	.174	.199	.221	.247	.265
37	.172	.196	.218	.244	.262
38	.170	.194	.215	.241	.258
39	.168	.191	.213	.238	.255
40	.165	.189	.210	.235	.252
Approximation for $n > 40$	$\frac{1.07}{\sqrt{n}}$	$\frac{1.22}{\sqrt{n}}$	$\frac{1.36}{\sqrt{n}}$	$\frac{1.52}{\sqrt{n}}$	$\frac{1.63}{\sqrt{n}}$

TABLE N Critical Values of the Kruskal-Wallis Test Statistic

Sample Sizes			Critical Value	α	Sample Sizes			Critical Value	α
n_1	n_2	n_3			n_1	n_2	n_3		
2	1	1	2.7000	.500	4	3	2	6.4444	.008
2	2	1	3.6000	.200				6.3000	.011
2	2	2	4.5714	.067				5.4444	.046
			3.7143	.200				5.4000	.051
3	1	1	3.2000	.300				4.5111	.098
3	2	1	4.2857	.100				4.4444	.102
			3.8571	.133	4	3	3	6.7455	.010
3	2	2	5.3572	.029				6.7091	.013
			4.7143	.048				5.7909	.046
			4.5000	.067				5.7273	.050
			4.4643	.105				4.7091	.092
3	3	1	5.1429	.043				4.7000	.101
			4.5714	.100	4	4	1	6.6667	.010
			4.0000	.129				6.1667	.022
3	3	2	6.2500	.011				4.9667	.048
			5.3611	.032				4.8667	.054
			5.1389	.061				4.1667	.082
			4.5556	.100				4.0667	.102
			4.2500	.121	4	4	2	7.0364	.006
3	3	3	7.2000	.004				6.8727	.011
			6.4889	.011				5.4545	.046
			5.6889	.029				5.2364	.052
			5.6000	.050				4.5545	.098
			5.0667	.086				4.4455	.103
			4.6222	.100	4	4	3	7.1439	.010
4	1	1	3.5714	.200				7.1364	.011
4	2	1	4.8214	.057				5.5985	.049
			4.5000	.076				5.5758	.051
			4.0179	.114				4.5455	.099
4	2	2	6.0000	.014				4.4773	.102
			5.3333	.033	4	4	4	7.6538	.008
			5.1250	.052				7.5385	.011
			4.4583	.100				5.6923	.049
			4.1667	.105				5.6538	.054
4	3	1	5.8333	.021				4.6539	.097
			5.2083	.050				4.5001	.104
			5.0000	.057	5	1	1	3.8571	.143
			4.0556	.093	5	2	1	5.2500	.036
			3.8889	.129				5.0000	.048
								4.4500	.071
								4.2000	.095
								4.0500	.119

TABLE N ***(continued)***

Sample Sizes n_1	n_2	n_3	Critical Value	α
5	2	2	6.5333	.008
			6.1333	.013
			5.1600	.034
			5.0400	.056
			4.3733	.090
			4.2933	.122
5	3	1	6.4000	.012
			4.9600	.048
			4.8711	.052
			4.0178	.095
			3.8400	.123
5	3	2	6.9091	.009
			6.8218	.010
			5.2509	.049
			5.1055	.052
			4.6509	.091
			4.4945	.101
5	3	3	7.0788	.009
			6.9818	.011
			5.6485	.049
			5.5152	.051
			4.5333	.097
			4.4121	.109
5	4	1	6.9545	.008
			6.8400	.011
			4.9855	.044
			4.8600	.056
			3.9873	.098
			3.9600	.102
5	4	2	7.2045	.009
			7.1182	.010
			5.2727	.049
			5.2682	.050
			4.5409	.098
			4.5182	.101
5	4	3	7.4449	.010
			7.3949	.011
			5.6564	.049
			5.6308	.050
			4.5487	.099
			4.5231	.103
5	4	4	7.7604	.009
			7.7440	.011
			5.6571	.049
			5.6176	.050
			4.6187	.100
			4.5527	.102
5	5	1	7.3091	.009
			6.8364	.011
			5.1273	.046
			4.9091	.053
			4.1091	.086
			4.0364	.105
5	5	2	7.3385	.010
			7.2692	.010
			5.3385	.047
			5.2462	.051
			4.6231	.097
			4.5077	.100
5	5	3	7.5780	.010
			7.5429	.010
			5.7055	.046
			5.6264	.051
			4.5451	.100
			4.5363	.102
5	5	4	7.8229	.010
			7.7914	.010
			5.6657	.049
			5.6429	.050
			4.5229	.099
			4.5200	.101
5	5	5	8.0000	.009
			7.9800	.010
			5.7800	.049
			5.6600	.051
			4.5600	.100
			4.5000	.102

TABLE Oa Exact Distribution of χ_r^2 for Tables with from 2 to 9 Sets of Three Ranks (k = 3; n = 2, 3, 4, 5, 6, 7, 8, 9; P is the Probability of Obtaining a Value of χ_r^2 as Great as or Greater Than the Corresponding Value of χ_r^2)

$n = 2$		$n = 3$		$n = 4$		$n = 5$	
χ_r^2	P	χ_r^2	P	χ_r^2	P	χ_r^2	P
0	1.000	.000	1.000	.0	1.000	.0	1.000
1	.833	.667	.944	.5	.931	.4	.954
3	.500	2.000	.528	1.5	.653	1.2	.691
4	.167	2.667	.361	2.0	.431	1.6	.522
		4.667	1.94	3.5	.273	2.8	.367
		6.000	.028	4.5	.125	3.6	.182
				6.0	.069	4.8	.124
				6.5	.042	5.2	.093
				8.0	.0046	6.4	.039
						7.6	.024
						8.4	.0085
						10.0	.00077

$n = 6$		$n = 7$		$n = 8$		$n = 9$	
.00	1.000	.000	1.000	.00	1.000	.000	1.000
0.33	.956	.286	.964	.25	.967	.222	.971
1.00	.740	.857	.768	.75	.794	.667	.814
1.33	.570	1.143	.620	1.00	.654	.889	.865
2.33	.430	2.000	.486	1.75	.531	1.556	.569
3.00	.252	2.571	.305	2.25	.355	2.000	.398
4.00	.184	3.429	.237	3.00	.285	2.667	.328
4.33	.142	3.714	.192	3.25	.236	2.889	.278
5.33	.072	4.571	.112	4.00	.149	3.556	.187
6.33	.052	5.429	.085	4.75	.120	4.222	.154
7.00	.029	6.000	.052	5.25	.079	4.667	.107
8.33	.012	7.143	.027	6.25	.047	5.556	.069
9.00	.0081	7.714	.021	6.75	.038	6.000	.057
9.33	.0055	8.000	.016	7.00	.030	6.222	.048
10.33	.0017	8.857	.0084	7.75	.018	6.889	.031
12.00	.00013	10.286	.0036	9.00	.0099	8.000	.019
		10.571	.0027	9.25	.0080	8.222	.016
		11.143	.0012	9.75	.0048	8.667	.010
		12.286	.00032	10.75	.0024	9.556	.0060
		14.000	.000021	12.00	.0011	10.667	.0035
				12.25	.00086	10.889	.0029
				13.00	.00026	11.556	.0013
				14.25	.000061	12.667	.00066
				16.00	.0000036	13.556	.00035
						14.000	.00020
						14.222	.000097
						14.889	.000054
						16.222	.000011
						18.000	.0000006

TABLE Ob Exact Distribution of χ_r^2 for Tables with from 2 to 9 Sets of Three Ranks (k = 4; n = 2, 3, 4; P is the Probability of Obtaining a Value of χ_r^2 as Great as or Greater Than the Corresponding Value of χ_r^2)

$n = 2$		$n = 3$		$n = 4$			
χ_r^2	P	χ_r^2	P	χ_r^2	P	χ_r^2	P
.0	1.000	.2	1.000	.0	1.000	5.7	.141
.6	.958	.6	.958	.3	.992	6.0	.105
1.2	.834	1.0	.910	.6	.928	6.3	.094
1.8	.792	1.8	.727	.9	.900	6.6	.077
2.4	.625	2.2	.608	1.2	.800	6.9	.068
3.0	.542	2.6	.524	1.5	.754	7.2	.054
3.6	.458	3.4	.446	1.8	.677	7.5	.052
4.2	.375	3.8	.342	2.1	.649	7.8	.036
4.8	.208	4.2	.300	2.4	.524	8.1	.033
5.4	.167	5.0	.207	2.7	.508	8.4	.019
6.0	.042	5.4	.175	3.0	.432	8.7	.014
		5.8	.148	3.3	.389	9.3	.012
		6.6	.075	3.6	.355	9.6	.0069
		7.0	.054	3.9	.324	9.9	.0062
		7.4	.033	4.5	.242	10.2	.0027
		8.2	.017	4.8	.200	10.8	.0016
		9.0	.0017	5.1	.190	11.1	.00094
				5.4	.158	12.0	.000072

TABLE P Critical Values of the Spearman Test Statistic. Approximate Upper-Tail Critical Values r_s^*, Where $P(r > r_s^*) \leq \alpha$, $n = 4(1)30$ Significance Level, α

n	.001	.005	.010	.025	.050	.100
4	—	—	—	—	.8000	.8000
5	—	—	.9000	.9000	.8000	.7000
6	—	.9429	.8857	.8286	.7714	.6000
7	.9643	.8929	.8571	.7450	.6786	.5357
8	.9286	.8571	.8095	.7143	.6190	.5000
9	.9000	.8167	.7667	.6833	.5833	.4667
10	.8667	.7818	.7333	.6364	.5515	.4424
11	.8364	.7545	.7000	.6091	.5273	.4182
12	.8182	.7273	.6713	.5804	.4965	.3986
13	.7912	.6978	.6429	.5549	.4780	.3791
14	.7670	.6747	.6220	.5341	.4593	.3626
15	.7464	.6536	.6000	.5179	.4429	.3500
16	.7265	.6324	.5824	.5000	.4265	.3382
17	.7083	.6152	.5637	.4853	.4118	.3260
18	.6904	.5975	.5480	.4716	.3994	.3148
19	.6737	.5825	.5333	.4579	.3895	.3070
20	.6586	.5684	.5203	.4451	.3789	.2977
21	.6455	.5545	.5078	.4351	.3688	.2909
22	.6318	.5426	.4963	.4241	.3597	.2829
23	.6186	.5306	.4852	.4150	.3518	.2767
24	.6070	.5200	.4748	.4061	.3435	.2704
25	.5962	.5100	.4654	.3977	.3362	.2646
26	.5856	.5002	.4564	.3894	.3299	.2588
27	.5757	.4915	.4481	.3822	.3236	.2540
28	.5660	.4828	.4401	.3749	.3175	.2490
29	.5567	.4744	.4320	.3685	.3113	.2443
30	.5479	.4665	.4251	.3620	.3059	.2400

Note: The corresponding lower-tail critical value for r_s is $-r_s^*$.

ACKNOWLEDGMENTS FOR TABLES

A. From *A Million Random Digits with 100,000 Normal Deviates* by The Rand Corporation, the Free Press, Glencoe, IL, 1955. Reprinted by permission.

E. Reproduced from *Documenta Geigy, Scientific Tables,* Seventh Edition, 1970, courtesy of CIBA-Geigy Limited, Basel, Switzerland.

F. From A. Hald and S. A. Sinkbaek, "A Table of Percentage Points of the X^2 Distribution." *Skandinavisk Aktuarietidskrift, 33* (1950), 168–175. Used by permission.

G. From *Biometrika Tables for Statisticians,* Third Edition, Vol. I, Bentley House, London, 1970. Reprinted by permission.

H. From *Biometrika Tables for Statisticians,* Third Edition, Vol. I, Bentley House, London, 1970. Used by permission.

J. Entries for $A = 3, B = 3$ through $A = 15, B = 2$ from Table 38 of E. S. Pearson and H. O. Hartley, *Biometrika Tables for Statisticians,* Volume 1, Third Edition, London: The Syndics of the Cambridge University Press, 1966. Entries for $A = 16$, $B = 16$ through $A = 20, B = 1$ from R. Latscha, "Tests of Significance in a 2×2 Contingency Table: Extension of Finney's Table," *Biometrika, 40* (1953), 74–86; used by permission of the Biometrika Trustees.

K. From Frank Wilcoxon, S. K. Katti, and Roberta A. Wilcox, "Critical Values and Probability Levels for the Wilcoxon Rank Sum Test and the Wilcoxon Signed Rank Test," originally prepared and distributed by Lederle Laboratories Division, American Cyanamid Company, Pearl River, New York, in cooperation with the Department of Statistics, The Florida State University, Tallahassee, Florida. Revised October 1968. Copyright 1963 by the American Cyanamid Company and the Florida State University. Reproduced by permission of S. K. Katti.

L. Adapted from L. R. Verdooren, "Extended Tables of Critical Values for Wilcoxon's Test Statistic," *Biometrika, 50* (1963), 177–186. Used by permission of the author and E. S. Pearson on behalf of the Biometrika Trustees. The adaptation is due to W. J. Conover, *Practical Nonparametric Statistics,* Wiley, New York, 1971, 384–388.

M. From L. H. Miller, "Table of Percentage Points of Kolmogorov Statistics," *Journal of the American Statistical Association, 51* (1956), 111–121. Reprinted by permission of the American Statistical Association. The table as reprinted here follows the format found in W. J. Conover, *Practical Nonparametric Statistics,* © 1971, by John Wiley & Sons, Inc.

N. From W. H. Kruskal and W. A. Wallis, "Use of Ranks in One-Criterion Analysis of Variance," *Journal of the American Statistical Association, 47* (1952), 583–621; errata, *ibid., 48* (1953), 907–911. Reprinted by permission of the American Statistical Association.

O. From M. Friedman, "The Use of Ranks to Avoid the Assumption of Normality Implicit in the Analysis of Variance," *Journal of the American Statistical Association, 32* (1937), 675–701. Reprinted by permission.

P. From Gerald J. Glasser and Robert F. Winter, "Critical Values of the Coefficient of Rank Correlation for Testing the Hypothesis of Independence," *Biometrika, 48* (1961), 444–448. Used by permission. The table as reprinted here contains corrections given in W. J. Conover, *Practical Nonparametric Statistics,* © 1971, by John Wiley & Sons, Inc.

ANSWERS TO ODD-NUMBERED EXERCISES

Chapter 1

Review Exercises

7. Situation A.

(a) 300 households **(b)** all households in the small southern town **(c)** number of school-aged children present **(d)** all that reported one or more children **(e)** nominal (categories: 0 children, 1 child, and so on)

Situation B.

(a) 250 patients **(b)** all patients admitted to the hospital during the past year **(c)** distance patient lives from the hospital **(d)** 250 distances **(e)** ratio

Chapter 2

2.3.1. (a)

Class Interval	Frequency	Cumulative Frequency	Relative Frequency	Cumulative Relative Frequency
0–0.49	3	3	3.33	3.33
.5–0.99	3	6	3.33	6.67
1.0–1.49	15	21	16.67	23.33

(Continued)

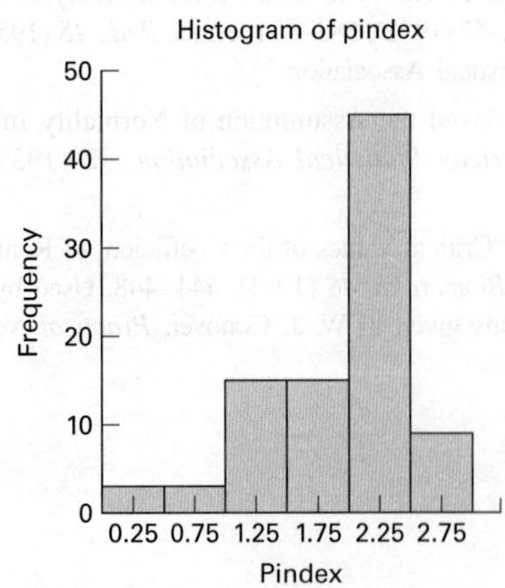

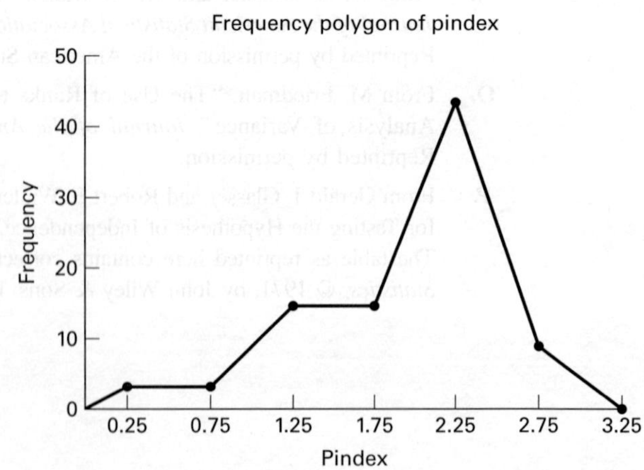

Class Interval	Frequency	Cumulative Frequency	Relative Frequency	Cumulative Relative Frequency
1.5–1.99	15	36	16.67	40.00
2.0–2.49	45	81	50.00	90.00
2.5–2.99	9	90	10.00	100.00

(b) 40.0% **(c)** .7667 **(d)** 16.67% **(e)** 9 **(f)** 16.67%
(g) 2.17, because it composes almost 25 percent of the data and is the most frequently occurring value in the data set. **(h)** Skewed to the left.

2.3.3. (a)

Class Interval	Frequency	Cumulative Frequency	Relative Frequency	Cumulative Relative Frequency
20–24.99	2	2	0.069	6.90
25–29.99	11	13	0.3793	44.83
30–34.99	6	19	0.2069	65.52
35–39.99	2	21	0.069	72.41
40–44.99	5	26	0.1724	89.66
45–49.99	2	28	0.069	96.55
50–54.99	1	29	0.0345	100.00

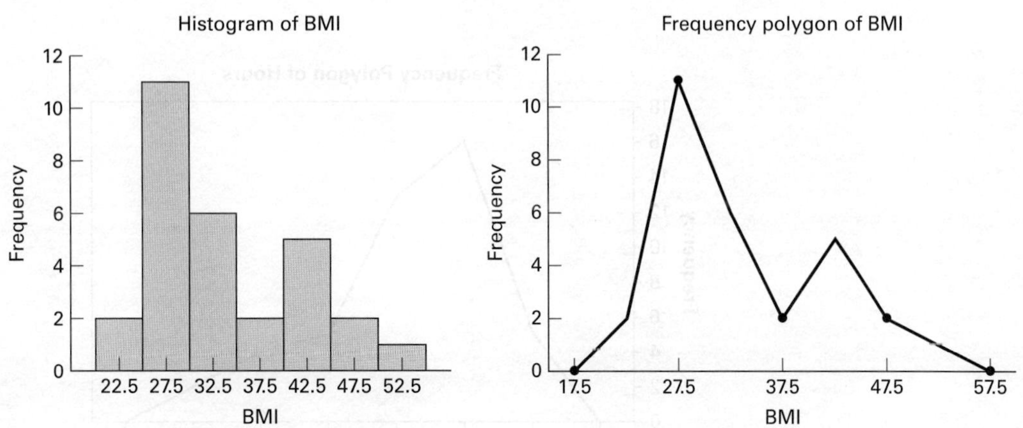

(b) 44.83% **(c)** 24.14% **(d)** 34.48% **(e)** The data are right skewed **(f)** 21

2.3.5. (a)

Class Interval	Frequency	Relative Frequency
0–2	5	0.1111
3–5	16	0.3556
6–8	13	0.2889
9–11	5	0.1111
12–14	4	0.0889
15–17	2	0.0444
	45	1.000

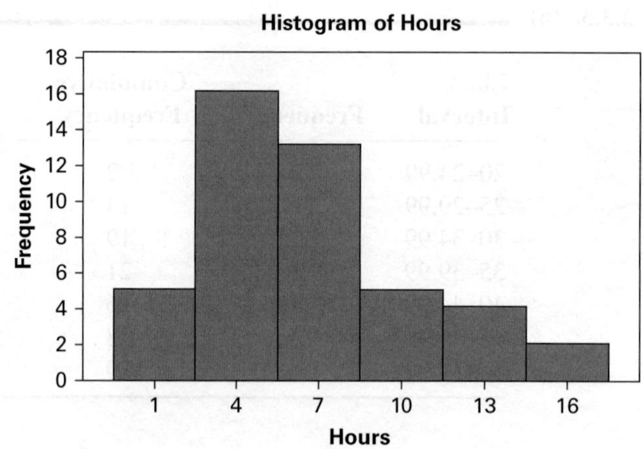

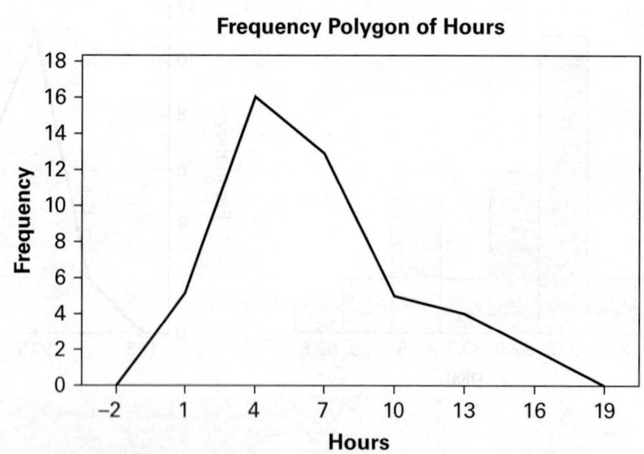

(b) Skewed right

2.3.7. (a)

Class Interval	Frequency	Cumulative Frequency	Relative Frequency	Cumulative Relative Frequency
110–139	8	8	0.0516	0.0516
140–169	16	24	0.1032	0.1548
170–199	46	70	0.2968	0.4516
200–229	49	119	0.3161	0.7677
230–259	26	145	0.1677	0.9354
260–289	9	154	0.0581	0.9935
290–319	1	155	0.0065	1.0000

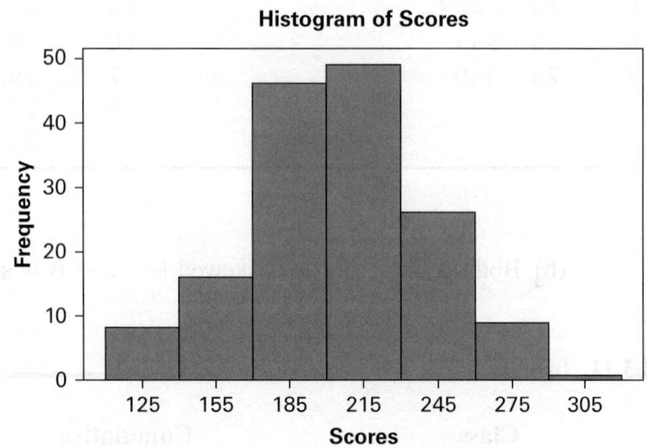

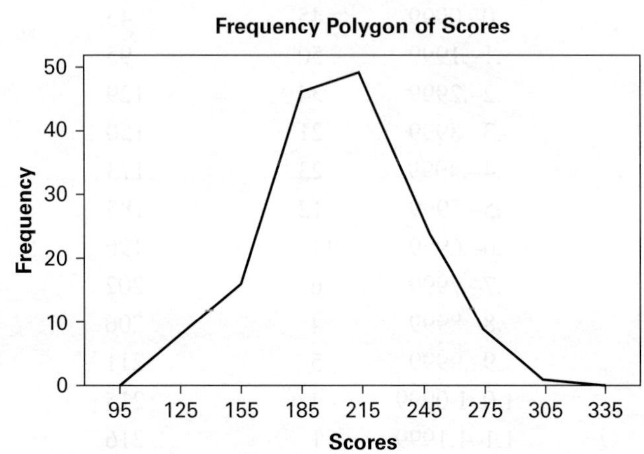

(b) Not greatly skewed

2.3.9. (a)

Stem-and-Leaf Display: Hospital A			Stem-and-Leaf Display: Hospital B		
Stem-and-leaf of C1 N = 25			Stem-and-leaf of C2 N = 25		
Leaf Unit = 1.0			Leaf Unit = 1.0		
1	17	1	1	12	5
2	18	4	2	13	5
4	19	15	4	14	35
9	20	11259	9	15	02445
(6)	21	233447	(4)	16	5678
10	22	2259	12	17	38
6	23	389	10	18	466
3	24	589	7	19	0059
			3	20	3
			2	21	24

(b) Both asymmetric: A is skewed left, and B is skewed right.

2.3.11. (a)

Class Interval	Frequency	Cumulative Frequency	Relative Frequency	Cumulative Relative Frequency
.0–.0999	45	45	20.83	20.83
.1–.1999	50	95	23.15	43.98
.2–.2999	34	129	15.74	59.72
.3–.3999	21	150	9.72	69.44
.4–.4999	23	173	10.65	80.09
.5–.5999	12	185	5.56	85.65
.6–.6999	11	196	5.09	90.74
.7–.7999	6	202	2.78	93.52
.8–.8999	4	206	1.85	95.37
.9–.9999	5	211	2.31	97.69
1.0–1.0999	4	215	1.85	99.54
1.1–1.1999	1	216	0.46	100.00

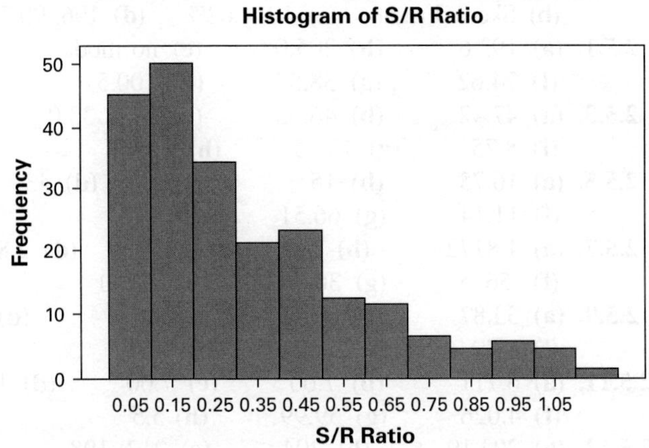

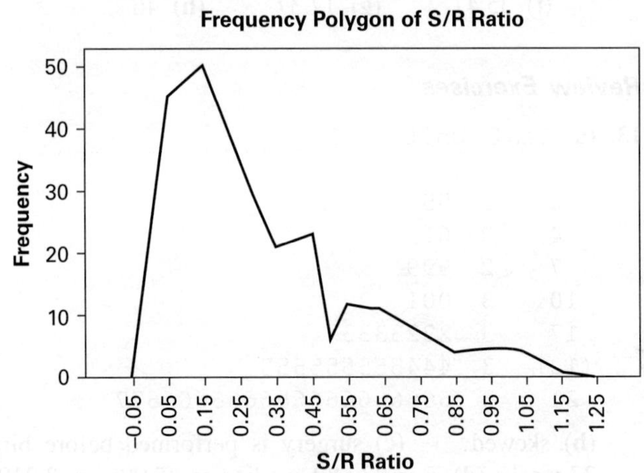


```
Stem-and-leaf of C1            N  =  216
Leaf Unit  =  0.010

 46     0  1245566777888899999999999999999999999999999
 96     1  00000000001122233333444455555566666667777778888999
(34)    2  0011111223444444445566666788889999
 86     3  001111244445556668999
 65     4  000011222233334445688 99
 42     5  00233444599
 30     6  02236788999
 19     7  012289
 13     8  0237
  9     9  05588
  4    10  236
  1    11  6
```

(b) Skewed right **(c)** 10, 4.62% **(d)** 196, 90.74%; 67, 31.02%, 143, 19.91%

2.5.1. **(a)** 193.6 **(b)** 205.0 **(c)** no mode **(d)** 255 **(e)** 5568.09 **(f)** 74.62 **(g)** 38.53 **(h)** 100.5

2.5.3. **(a)** 47.42 **(b)** 46.35 **(c)** 54.0, 33.0 **(d)** 29.6 **(e)** 76.54 **(f)** 8.75 **(g)** 18.45 **(h)** 13.73

2.5.5. **(a)** 16.75 **(b)** 15 **(c)** 15 **(d)** 43 **(e)** 124.02 **(f)** 11.14 **(g)** 66.51 **(h)** 8.25

2.5.7. **(a)** 1.8172 **(b)** 2 **(c)** 2.17 **(d)** 2.83 **(e)** .3165 **(f)** .5625 **(g)** 30.96 **(h)** .6700

2.5.9. **(a)** 33.87 **(b)** 30.49 **(c)** none **(d)** 29.84 **(e)** 64.00 **(f)** 8.00 **(g)** 23.62 **(h)** 13.4

2.5.11. **(a)** 6.711 **(b)** 7.00 **(c)** 7.00 **(d)** 16 **(e)** 16.21 **(f)** 4.026 **(g)** 59.99 **(h)** 5.5

2.5.13. **(a)** 204.19 **(b)** 204 **(c)** 212, 198 **(d)** 196 **(e)** 1258.12 **(f)** 35.47 **(g)** 17.37 **(h)** 46

Review Exercises

13. **(a)**
```
Leaf Unit = 1.0

   2   2  55
   4   2  67
   7   2  999
  10   3  001
  17   3  2223333
(12)   3  444555555555
  21   3  666666666666666666677
```
(b) skewed **(c)** surgery is performed before birth; birth is generally around 37 weeks **(d)** $\bar{x} = 33.680$, median $= 35.00$, $s = 3.210$, $s^2 = 10.304$

15. **(a)** $\bar{x} = 43.39$, median $= 42$, $s = 17.09$, C.V. $= 39.387$, $s^2 = 292.07$

(b)
```
Stem-and-leaf of GFR   N = 28
Leaf Unit = 1.0

   1   1  8
   6   2  11377
  12   3  022267
 (7)   4  1223388
   9   5  158
   6   6  02378
   1   7
   1   8  8
```
(c) See graph on following page (A-113)

(d) 67.9%, 96.55%, 100%

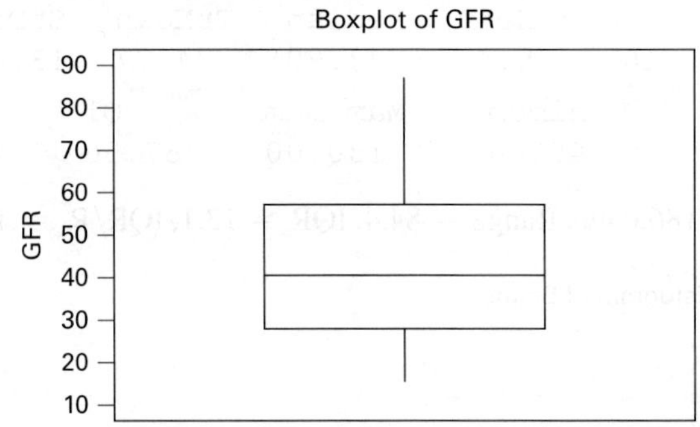

17. Some examples include difference, diversity, departure, discrepancy, deviation, and entropy

19. $\bar{x} = 3.95$, Median $= 3$, $s = 3.605$, $s^2 = 12.998$

21. Answers will vary: It is not uncommon for students to score higher on exams as a semester progresses; therefore, the exam scores are likely to be left skewed, making the median, which is less affected by skew, to be the better choice.

23. Answers will vary: Using Sturges's rule, where $w = \frac{R}{k}$, $k = 1 + 3.322(\log_{10}300) \simeq 9.23$. An estimate of sample standard deviation can be found by dividing the sample range by 4. Therefore, $s \approx \frac{R}{4}$ so that $R \approx 4s$. Using this formula, then $R = 160$ and $w = \frac{160}{9.23} = 17.33$ suggesting that (d) or (e) may be appropriate.

25. Answers will vary: Imagine you are examining protein intake among college students. In general most students are likely to consume the average daily protein intake, but among this population, there is likely to be a fair number of athletes who consume large amounts of protein owing to the demands of their sport. In that case, the data are likely to be positively skewed, and the median better represents the central tendency of the data.

27.

```
Variable       N      Mean    Median   TrMean    StDev   SE Mean
S/R          216    0.3197    0.2440   0.2959   0.2486    0.0169

Variable     Minimum     Maximum          Q1         Q3
S/R           0.0269      1.1600      0.1090     0.4367
```

IQR $= .3277$, Range $= 1.1331$, IQR/R $= .2892$

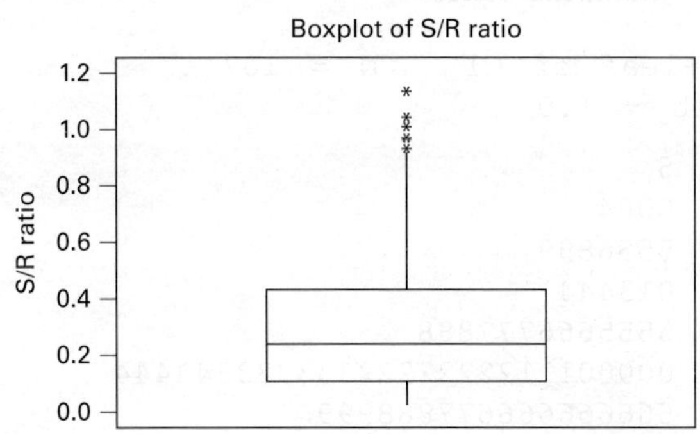

29. (a)

```
Variable      N    Mean   Median   TrMean   StDev   SE Mean
nutri       107   75.40    73.80    74.77   13.64      1.32

Variable     Minimum    Maximum       Q1       Q3
nutri          45.60     130.00    67.50    80.60
```

Variance = 186.0496, Range = 84.4, IQR = 13.1, IQR/R = .1552

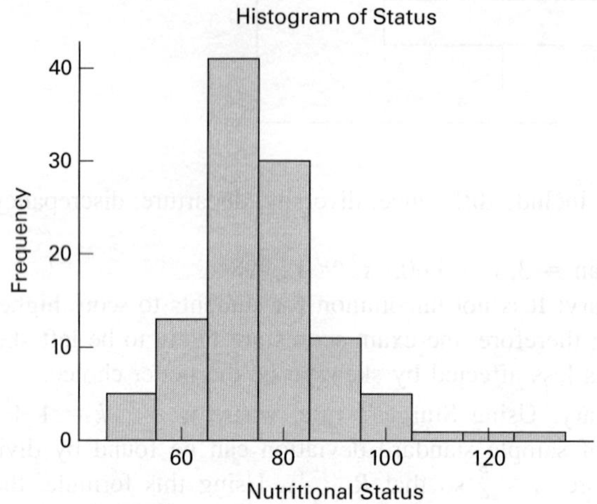

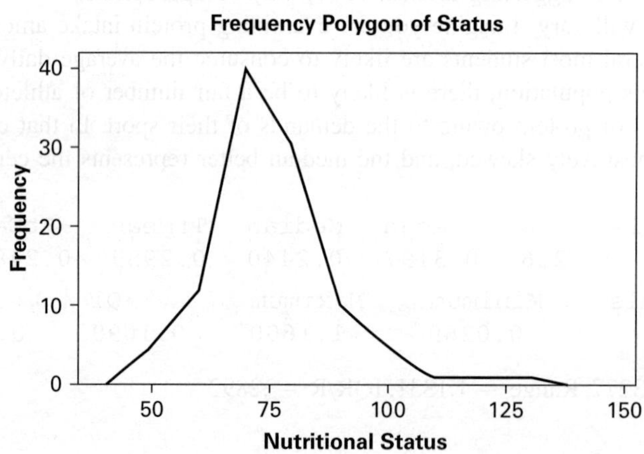


```
Stem-and-leaf of C1     N = 107
Leaf Unit = 1.0

   1    4  5
   5    5  0004
  12    5  5556899
  18    6  013444
  31    6  5555666777888
 (28)   7  0000011122222222333333344444
  48    7  66666666677888999
```

30	8	000002234444
18	8	56889
13	9	01223
8	9	679
5	10	00
3	10	9
2	11	
2	11	
2	12	3
1	12	
1	13	0

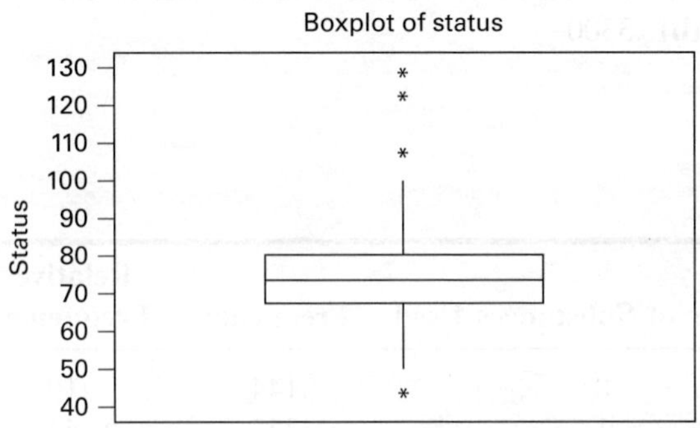

(d) 75.4 ± 13.64; 61.76, 89.04; $79/107 = .7383$; $75.4 \pm 2(13.64)$; 48.12, 102.68; $103/107 = .9626$; $75.4 \pm 3(13.64)$; 34.48, 116.32; $105/107 = .9813$ **(e)** $102/107 = .9533$ **(f)** $1/107 = .0093$

Chapter 3

3.4.1. **(a)** .6631 **(b)** marginal **(d)** .0332 **(e)** joint **(f)** .0493 **(g)** conditional **(h)** .3701 **(i)** addition rule

3.4.3. **(a)** male and split drugs, .3418 **(b)** male or split drugs or both, .8747 **(c)** male given split drugs, .6134 **(d)** male, .6592

3.4.5. .95

3.4.7. .301

3.5.1. **(a)** A subject having the symptom (S) and not having the disease. **(b)** A subject not having S but having the disease. **(c)** .96 **(d)** .9848 **(e)** .0595 **(f)** .99996 **(g)** .00628, .999996, .3895, .9996, .8753, .9955 **(h)** predictive value increases as the hypothetical disease rate increases

3.5.3. .9999977

Review Exercises

3. **(a)** .2143 **(b)** .5519 **(c)** .1536 **(d)** .4575 **(e)** .5065

5. **(a)** .1349 **(b)** .6111 **(c)** .3333 **(d)** .5873 **(e)** .3571 **(f)** .6667 **(g)** 0 **(h)** .3269

7. **(a)** **1.** .2200 **2.** .5000 **3.** .0555 **4.** .1100 **5.** .5900
(b) **1.** .3000 **2.** .3900 **3.** .3900 **4.** .1667 **5.** .0700 **6.** .6000

9. **(a)** .0432 **(b)** .0256 **(c)** .0247 **(d)** .9639 **(e)** .5713
(f) .9639 **(g)** .9810

11. .0060

13. .0625

15. mothers under the age of 24

17. null set, as events are mutually exclusive

19. **(a)** plasma lipoprotein between 10–15 or greater than or equal to 30.
(b) plasma lipoprotein between 10–15 and greater than or equal to 30.
(c) plasma lipoprotein between 10–15 and less than or equal to 20.
(d) plasma lipoprotein between 10–15 or less than or equal to 20.

21. **(a)** .7456 **(b)** .3300

23. .0125

Chapter 4

4.2.1 (a)

Number of Substances Used	Frequency	Relative Frequency	Cumulative Frequency
0	144	.19	.19
1	342	.44	.63
2	142	.18	.81
3	72	.09	.90
4	39	.05	.95
5	20	.03	.98
6	6	.01	.99
7	9	.01	1.00
8	2	.003	1.003
9	1	.001	1.004
Total	777	1.004	

(b)

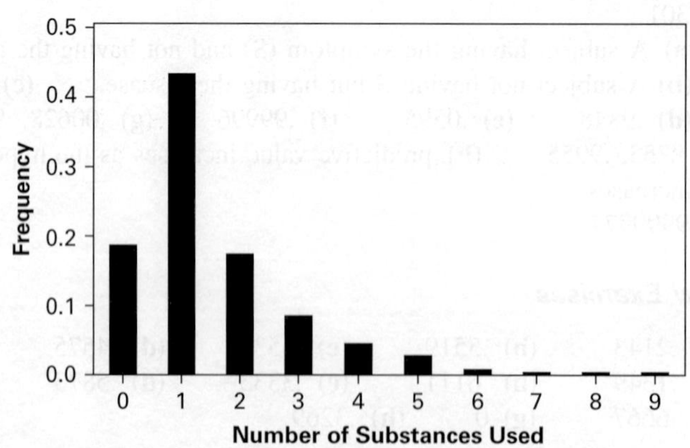

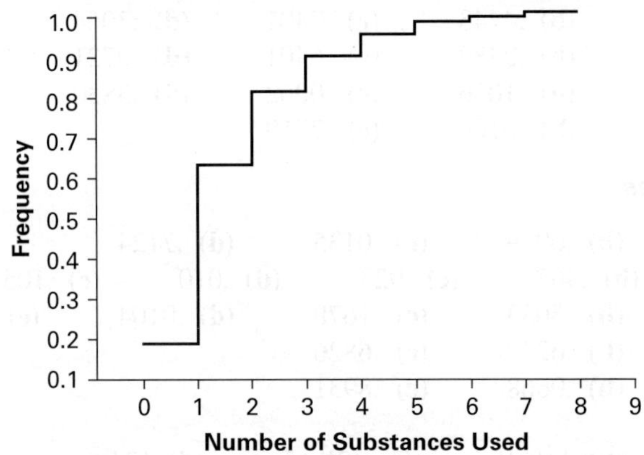

4.2.3. $\bar{x} = 1.58, s^2 = 2.15, s = 1.47$

4.3.1. **(a)** .1484 **(b)** .8915 **(c)** .1085 **(d)** .8080

4.3.3. **(a)** .2536, **(b)** .3461 **(c)** .7330 **(d)** .9067 **(e)** .0008

4.3.5. mean = 4.8, variance = 3.264

4.3.7. **(a)** .5314 **(b)** .3740 **(c)** .0946 **(d)** .9931 **(e)** .0946 **(f)** .0069

4.3.9.

Number of Successes, x	Probability, $f(x)$
0	$\frac{3!}{0!\,3!}(.2)^3(.8)^0 = .008$
1	$\frac{3!}{1!\,2!}(.2)^2(.8)^1 = .096$
2	$\frac{3!}{2!\,1!}(.2)^1(.8)^2 = .384$
3	$\frac{3!}{3!\,0!}(.2)^0(.8)^3 = .512$
Total	1

4.4.1. **(a)** .156 **(b)** .215 **(c)** .629 **(d)** .320

4.4.3. **(a)** .105 **(b)** .032 **(c)** .007 **(d)** .440

4.4.5. **(a)** .086 **(b)** .946 **(c)** .463 **(d)** .664 **(e)** .026

4.6.1. .4236

4.6.3. .2912

4.6.5. .0099

4.6.7. .95

4.6.9. .901

4.6.11. −2.54

4.6.13. 1.77

4.6.15. 1.32

4.7.1. **(a)** .6321 **(b)** .4443 **(c)** .0401 **(d)** .3064
4.7.3. **(a)** .1357 **(b)** .2389 **(c)** .6401 **(d)** .0721 **(e)** .1575
4.7.5. **(a)** .3413 **(b)** .1056 **(c)** .0062 **(d)** .3830
4.7.7. **(a)** .0630 **(b)** .0166 **(c)** .7719

Review Exercises

15. **(a)** .0212 **(b)** .0949 **(c)** .0135 **(d)** .7124
17. **(a)** .034 **(b)** .467 **(c)** .923 **(d)** .010 **(e)** .105
19. **(a)** .4967 **(b)** .5033 **(c)** .1678 **(d)** .0104 **(e)** .8218
21. **(a)** .0668 **(b)** .6247 **(c)** .6826
23. **(a)** .0013 **(b)** .0668 **(c)** .8931
25. 57.1
27. **(a)** 64.75 **(b)** 118.45 **(c)** 130.15 **(d)** 131.8
29. 14.90
31. 10.6
33. **(a)** Bernoulli assuming there is an equal probability of both genders
(b) Not Bernoulli—more than two possible outcomes
(c) Not Bernoulli—weight is not a binary variable

Chapter 5

5.3.1. 204, 6.2225
5.3.3. **(a)** .1841 **(b)** .7980 **(c)** .0668
5.3.5. **(a)** .0020 **(b)** .1736 **(c)** .9777 **(d)** .4041
5.3.7. **(a)** .9876 **(b)** .0668 **(c)** .0668 **(d)** .6902
5.3.9.

Sample	$\bar{x}$
6, 8, 10	8.00
6, 8, 12	8.67
6, 8, 14	9.33
6, 10, 12	9.33
6, 10, 14	10.00
6, 12, 14	10.67
8, 10, 12	10.00
8, 10, 14	10.67
8, 12, 14	11.33
10, 12, 14	12.00
$\mu_{\bar{x}} = 10, \sigma_{\bar{x}} = 1.333$	

5.4.1. .3897
5.4.3. .0038
5.4.5. .0139
5.5.1. .1131
5.5.3. .0808
5.5.5. **(a)** .1539 **(b)** .3409 **(c)** .5230
5.6.1. .1056
5.6.3. .7938

Review Exercises

11. .0003
13. .0262
15. .1335
17. .1093
19. .1292
21. 252
23. .53, .0476
25. At least approximately normally distributed
27. .9942
29. **(a)** No **(b)** Yes **(c)** No **(d)** Yes **(e)** Yes **(f)** Yes

Chapter 6

6.2.1. **(a)** 88, 92 **(b)** 87, 93 **(c)** 86, 94
6.2.3. **(a)** 7.63, 8.87 **(b)** 7.51, 8.99 **(c)** 7.28, 9.22
6.2.5. 1603.688, 1891.563; 1576.125, 1919.125; 1521.875, 1973.375
6.3.1. **(a)** 2.1448 **(b)** 2.8073 **(c)** 1.8946 **(d)** 2.0452
6.3.3. **(a)** 1.549, .387 **(b)** 2.64, 4.36; .49, .91 **(c)** Nitric oxide diffusion rates are normally distributed in the population from which the sample was drawn. **(f)** narrower **(g)** wider
6.3.5. 66.2, 76.8; 65.1, 77.9; 62.7, 80.3
6.4.1. −549.82, −340.17; −571.28, −318.72; −615.52, −274.48
6.4.3. −5.90, 17.70; −8.15, 19.95; −12.60, 24.40
6.4.5. 64.09, 479.91; 19.19, 524.81; −77.49, 621.49
6.4.7. 2.1, 4.5; 1.8, 4.8; 1.3, 5.3
6.4.9. −32.58, −25.42; −33.33, −24.67, −34.87, −23.13
6.5.1. .1028, .1642
6.5.3. .4415, .6615
6.6.1. .0268, .1076
6.6.3. −.0843, .2667
6.7.1. 27, 16
6.7.3. 19
6.8.1. 683, 1068
6.8.3. 385, 289
6.9.1. 6.334, 44.63
6.9.3. 793.92, 1370.41
6.9.5. 1.17, 2.09
6.9.7. $170.98503 \leq \sigma^2 \leq 630.65006$
6.10.1. .44, 17.37
6.10.3. .49, 2.95
6.10.5. .9, 3.52
6.10.7. 5.13, 60.30

Review Exercises

13. $\bar{x} = 79.87, s^2 = 28.1238, s = 5.3$; 76.93, 82.81
15. $\hat{p} = .30$; .19, .41

17. $\hat{p}_1 = .20; \hat{p}_2 = .54, .26, .42$
19. $\hat{p} = .90; .87, .93$
21. $\bar{x} = 19.23, s^2 = 20.2268; 16.01, 22.45$
23. −2.18, 2.82
25. 362.73, 507.27
27. .44, .74
29. −.416, .188
31. Level of confidence decreases. The interval would have no width. The level of confidence would be zero.
33. z, 8.1, 8.1
35. All drivers ages 55 and older. Drivers 55 and older participating in the vision study.
37. .2865, .3529 (Use z since $n > 30$)

Chapter 7

7.2.1. Reject H_0 because $-2.57 < -2.33, p = .0051 < .01$
7.2.3. Fail to reject H_0 because $.76 < 1.333. p > .10$
7.2.5. Yes, reject H_0, $z = -5.73 < -1.645. p < .0001$
7.2.7. No, fail to reject H_0. $t = -1.5 > -1.709. .05 < p < .10$
7.2.9. Yes, reject H_0, $z = 3.08. p = .0010$
7.2.11. $z = 4$, reject $H_0, p < .0001$
7.2.13. $t = .1271$, fail to reject H_0. $p > .2$
7.2.15. $z = -4.18$, reject H_0. $p < .0001$
7.2.17. $z = 1.67$, fail to reject H_0. $p = 2(.0475) = .095$
7.2.19. Reject H_0 since $z = -4.00. p < .0001$
7.3.1. Reject H_0 because $-10.9 < -2.388, p < .005$
7.3.3. Reject H_0 because $-9.60 < -2.6591, p < 2(.005)$
7.3.5. Reject H_0 because $z = 3.39 > 1.96. p = 2(.0003)$
7.3.7. $s_p^2 = 5421.25, t = -6.66$. Reject H_0. $p < 2(.005) = .010$
7.3.9. $z = 3.39$. Reject H_0. $p = 2(1 - .9997) = .0006$
7.3.11. $t = -3.3567$. Reject H_0. $p < .01$
7.4.1. Reject H_0 because $3.17 > 2.624, p < .005$
7.4.3. Reject H_0. $-3.1553 < -1.8125, .005 < p < .01$
7.4.5. Reject H_0 since $-4.4580 < -2.4469, p < .01$
7.5.1. Reject H_0 since $-1.62 > -1.645. p = .0526$
7.5.3. Reject H_0 because $-1.77 < -1.645, p = .0384$
7.5.5. Reject H_0 because $z = -2.21, p = .0136$
7.6.1. Reject H_0 because $-2.86 < -2.58, p = .0042$
7.6.3. Fail to reject H_0 because $z = 1.70 < 1.96, p = .088$
7.7.1. Do not reject H_0, since $5.142 < 20.723 < 34.267, p > .01$ (two-sided test).
7.7.3. $\chi^2 = 6.75$. Do not reject H_0. $p > .05$ (two-sided test)
7.7.5. $\chi^2 = 28.8$. Do not reject H_0. $p > .10$
7.7.7. $\chi^2 = 22.036, .10 > p > .05$
7.8.1. Fail to reject because V.R. $= 1.226 < 1.74 p > .10$
7.8.3. No, V.R. $= 1.83, p > .10$
7.8.5. Reject H_0. V.R. $= 4, .02 < p < .05$
7.8.7. V.R. $= 2.1417, p > .10$

7.9.1.

Alternative Value of μ	β	Value of Power Function $1 - \beta$
516	0.9500	0.0500
521	0.8461	0.1539
528	0.5596	0.4404
533	0.3156	0.6844
539	0.1093	0.8907
544	0.0314	0.9686
547	0.0129	0.9871

7.9.3.

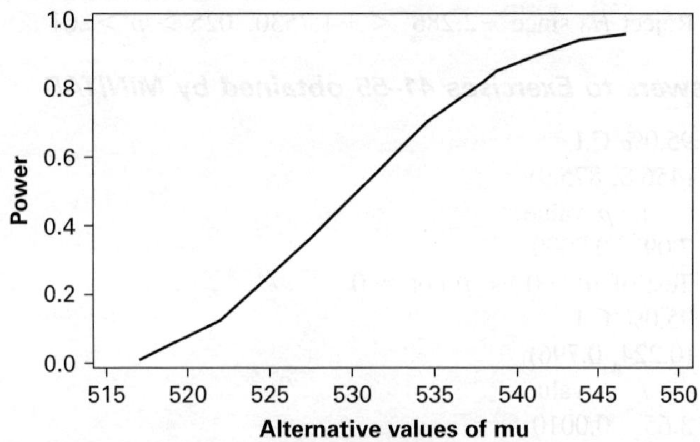

Alternative Value of μ	β	Value of Power Function $1 - \beta$
4.25	0.9900	0.0100
4.50	0.8599	0.1401
4.75	0.4325	0.5675
5.00	0.0778	0.9222
5.25	0.0038	0.9962

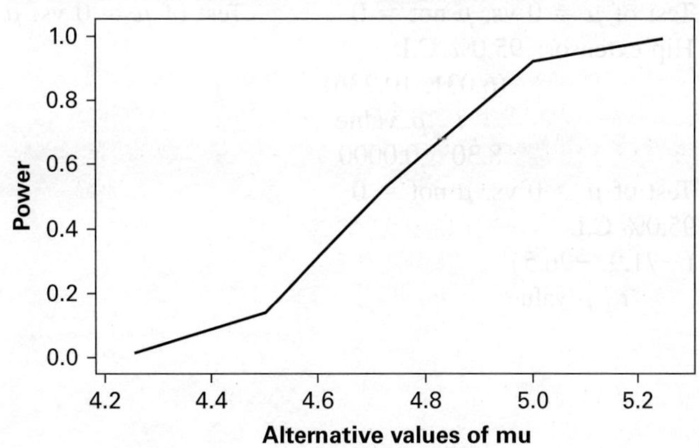

7.10.1. $n = 548$; $C = 518.25$. Select a sample of size 548 and compute $\bar{x}$. If $\bar{x} \geq 518.25$, reject H_0. If $\bar{x} < 518.25$ do not reject H_0.

7.10.3. $n = 103$; $C = 4.66$. Select a sample of size 103 and compute $\bar{x}$. If $\bar{x} \geq 4.66$, reject H_0. If $\bar{x} < 4.66$, do not reject H_0.

Review Exercises

19. Reject H_0 since $29.49 > 2.33$. $p < .0001$
21. Fail to reject the null because $z = 1.48 < 1.96$. $p = .1388$
23. Reject H_0 since $12.79 > 2.58$. $p < .0001$
25. Fail to reject H_0 because $1.10 < 1.645$, $p = .1357$
27. $t = 3.873$, $p < .005$
29. $\bar{d} = 11.49$, $s_d^2 = 256.679$, $s_d = 16.02$, $t = 2.485$, $.025 > p > .01$
31. Reject H_0 since $-2.286 < -1.7530$, $.025 > p > .01$

Answers to Exercises 41–55 obtained by MINITAB

41. 95.0% C.I.
(456.8, 875.9)
t p value
7.09 0.0000
Test of $\mu = 0$ vs. μ not $= 0$

43. 95.0% C.I.
(0.224, 0.796)
t p value
3.65 0.0010
Test of $\mu = 0$ vs. μ not $= 0$

45. Leg press: 95.0% C.I.
(32.22, 56.45)
t p value
7.85 0.0000
Test of $\mu = 0$ vs. μ not $= 0$

Hip flexor: 95.0% C.I.
(3.079, 6.388)
t p value
6.14 0.0000
Test of $\mu = 0$ vs. μ not $= 0$

Hip extensor: 95.0% C.I.
(6.031, 10.236)
t p value
8.30 0.0000
Test of $\mu = 0$ vs. μ not $= 0$

Arm abductor: 95.0% C.I.
(3.717, 7.217)
t p value
6.70 0.0000
Test of $\mu = 0$ vs. μ not $= 0$

Arm abductor: 95.0% C.I.
(4.597, 7.670)
t p value
8.56 0.0000
Test of $\mu = 0$ vs. μ not $= 0$

47. 95.0% C.I.
(−71.9, −26.5)
t p value

−4.34 0.0001
Test of $\mu = 0$ vs. μ not $= 0$

49. 95.0% C.I. for $\mu_1 - \mu_2$: $(-83.8, -20)$
t test $\mu_1 = \mu_2$ (vs. not =): $t = -3.30$ $p = .0021$ d.f. $= 42$
t test $\mu_1 = \mu_2$ (vs. <): $t = -3.30$ $p = .0011$ d.f. $= 42$

51. 95.0% C.I. for μ GROUP 1 $-$ μ GROUP 2 (.5, 26.4)
t test μ GROUP 1 $=$ μ GROUP 2 (vs. not =): $t = 2.88$, $p = .045$ d.f. $= 4$

53. 95.0% C.I. for $\mu_1 - \mu_2$: $(-3.00, 2.2)$
t test $\mu_1 = \mu_2$ (vs. not =): $t = -.29$ $p = .77$ d.f. $= 53$
Both use Pooled StDev $= 4.84$

55. 95.0% C.I. for $\mu_{PT} - \mu_C$: (7.6, 18.8)
t test $\mu_{PT} = \mu_C$ (vs. not =): $t = 4.78$ $p = 0.0000$ d.f. $= 31$

Chapter 8

Answers for 8.2.1–8.2.7 obtained by SAS®

8.2.1.
```
F  =  6.24
p  =  .0004
Alpha                                      0.05
Error Degrees of Freedom                    325
Error Mean Square                      1.068347
Critical Value of Studentized Range     3.65207

Comparisons significant at the 0.05 level are indicated
by ***.

                  Difference           Simultaneous
 Group              Between           95% Confidence
 Comparison          Means                Limits
  90  -  30         0.1911           -0.1831   0.5652
  90  -  120        0.6346            0.1320   1.1372  ***
  90  -  60         0.6386            0.1360   1.1413  ***
  30  -  90        -0.1911           -0.5652   0.1831
  30  -  120        0.4436           -0.1214   1.0085
  30  -  60         0.4476           -0.1173   1.0125
 120  -  90        -0.6346           -1.1372  -0.1320  ***
 120  -  30        -0.4436           -1.0085   0.1214
 120  -  60         0.0040           -0.6531   0.6611
  60  -  90        -0.6386           -1.1413  -0.1320  ***
  60  -  30        -0.4476           -1.0125   0.1173
  60  -  120       -0.0040           -0.6611   0.6531
```

8.2.3.
```
F  =  9.36
p  =  < .0001
```

Alpha	0.05
Error Degrees of Freedom	109
Error Mean Square	211252.3
Critical Value of Studentized Range	3.68984

Comparisons significant at the 0.05 level are indicated by ***.

Group Comparison	Difference Between Means	Simultaneous 95% Confidence Limits		
A - B	455.72	45.73	865.71	***
A - C	574.54	235.48	913.59	***
A - D	596.63	287.88	905.39	***
B - A	-455.72	-865.71	-45.73	***
B - C	118.82	-271.45	509.09	
B - D	140.91	-223.34	505.17	
C - A	-574.54	-913.59	-235.48	***
C - B	-118.82	-509.0	271.45	
C - D	22.10	-259.95	304.14	
D - A	-596.63	-905.39	-287.88	***
D - B	-140.91	-505.17	223.34	
D - C	-22.10	-304.14	259.95	

8.2.5. F = 9.26
p = .0009

Alpha	0.05
Error Degrees of Freedom	26
Error Mean Square	637.384
Critical Value of Studentized Range	3.51417

Comparisons significant at the 0.05 level are indicated by ***.

Group Comparison	Difference Between Means	Simultaneous 95% Confidence Limits		
Y - MA	18.16	-10.67	46.98	
Y - E	48.13	20.07	76.19	***
MA - Y	-18.16	-46.98	10.67	
MA - E	29.97	1.15	58.80	***
E - Y	-48.13	-76.19	-20.07	***
E - MA	-29.97	-58.80	-1.15	***

8.2.7. F = 4.94
p = .0026

Alpha	0.05
Error Degrees of Freedom	174
Error Mean Square	0.17783
Critical Value of Studentized Range	3.66853

```
Comparisons significant at the 0.05 level are indicated
by ***.
                  Difference              Simultaneous
 Group             Between             95% Confidence
 Comparison         Means                    Limits
 0 - 1            0.10165          -0.13866 0.34196
 0 - 2            0.17817          -0.2188  0.37822
                  Difference              Simultaneous
 Group             Between             95% Confidence
 Comparison         Means                    Limits
 0 - 3            0.35447           0.10511 0.60383  ***
 1 - 0           -0.10165          -0.34196 0.13866
 1 - 2            0.07652          -0.17257 0.32561
 1 - 3            0.25281          -0.03737 0.54300
 2 - 0           -0.17817          -0.37822 0.02188
 2 - 1           -0.07652          -0.32561 0.17257
 2 - 3            0.17630          -0.08154 0.43413
 3 - 0           -0.35447          -0.60383 -0.10511 ***
 3 - 1           -0.25281          -0.54300 0.03737
 3 - 2           -0.17630          -0.43413 0.08154
```

8.3.1. V.R. = 19.79, $p < .005$
8.3.3. V.R. = 30.22, $p < .005$
8.3.5. V.R. = 7.37, $.025 > p > .01$
8.3.7. Total *d.f.* = 41 Block (Dogs) *d.f.* = 5 Treatments (Times) *d.f.* = 6, Error *d.f.* = 30
8.4.1. V.R. = 48.78, $p < .005$
8.4.3. V.R. = 16.45, $p < .005$
8.4.5. Total *d.f.* = 29, Block (Subject) *d.f.* = 9, Treatment (Time) *d.f.* = 2, Error *d.f.* = 18
8.5.1. Ion V.R. = 6.18, $p = .023$; dose V.R. = 74.59, $p = .000$; inter V.R. = .89 $p = .427$
8.5.3. Migraine V.R. = 19.98, $p < .0001$; treat V.R. = 2.13, $p = .1522$; interaction V.R. = 1.42, $p = .2404$

Review Exercises

13. V.R. = 7.04, $p = .000$. The sample mean for the healthy subjects is significantly different from the means of categories B, C, and D. No other differences between sample means are significant.
15. V.R. = 1.35, $p = .274$. Do not reject H_0.
17. Smoking status V.R. = 3.16, $p = .052$, Vital Exhaustion Group V.R. = 6.84, $p = .003$, Interaction V.R. = 2.91, $p = .032$
19. V.R. = 4.23, $p < .001$
21. V.R. = 6.320, $p = .008$
23. V.R. = 3.1187, $p = .043$. The sample D mean is significantly different from the sample B mean. No other differences between sample means are significant.

25. V.R. (Age) = 29.38, $p < .001$; Occupation V.R. = 31.47, $p < .001$; Interaction V.R. = 7.12, $p < .001$

27. 499.5, 9, 166.5, 61.1667, 2.8889, 57.6346, <.005

29. **(a)** Completely randomized, **(b)** 3, **(c)** 30, **(d)** No, because 1.0438 < 3.35

31. V.R. = 26.06, $p < .001$. HSD = 2.4533. All differences significant except $\mu_{\text{Light}} - \mu_{\text{Moderate}}$

33. V.R. = 2.37, $p = .117$, Tukey HSD not necessary.

35. **(a)** One-way ANOVA
(b) Response: post-pre training score
(c) Factors: Groups of years of experience (with 4 levels)
(d) surgical experience and interest in obstetrics
(e) no carryover effects
(f) treatment is years of experience d.f. = 3, total d.f. = 29, error d.f. = 26.

37. **(a)** repeated measures
(b) Response: BMD
(c) Factors: time periods (with six levels)
(d) diet, exercise, and calcium intake
(e) no carryover effects
(f) time factor d.f. = 5, subject factor d.f. = 25, total, error d.f. = 125.

39.

Analysis of Variance for bilirubi

Source	DF	SS	MS	F	P
subject	17	2480.83	145.93	45.57	0.000
time	6	89.09	14.85	4.64	0.000
Error	102	326.65	3.20		
Total	125	2896.57			

41. CR = Compression Ratio

Analysis of Variance for C.R.

Source	DF	SS	MS	F	P
Group	4	9092	2273	8.12	0.001
Error	19	5319	280		
Total	23	14411			

Level	N	Mean	StDev
Control	6	79.96	5.46
I	4	78.69	21.44
II	4	47.84	23.74
III	5	43.51	10.43
IV	5	33.32	20.40

```
Individual 95% CIs For Mean
Based on Pooled StDev
---+---------+---------+---------+---
                      (-----*-----)
                    (------*------)
    (------*-------)
  (-----*-----)
(-----*-----)
---+---------+---------+---------+---
  25        50        75       100
```

Pooled StDev = 16.73

Tukey's pairwise comparisons

```
    Family error rate = 0.0500
Individual error rate = 0.00728

Critical Value = 4.25

Intervals for (column level mean) -   (row level mean)

          Control          I          II         III
I          -31.19
            33.72

II          -0.34      -4.70
            64.58      66.41

III          6.00       1.45     -29.41
            66.89      68.91      38.05

IV          16.19      11.64     -19.21      -21.61
            77.08      79.10      48.24       41.99
```

43. Two-way ANOVA: BC versus heat, chromo

Analysis of Variance for BC

Source	DF	SS	MS	F	P
heat	1	0.1602	0.1602	3.95	0.061
chromo	1	0.6717	0.6717	16.55	0.001
Interaction	1	0.0000	0.0000	0.00	0.994
Error	20	0.8119	0.0406		
Total	23	1.6438			

Analysis of Variance for AC

Source	DF	SS	MS	F	P
heat	1	0.0468	0.0468	1.99	0.174
chromo	1	0.4554	0.4554	19.34	0.000
Interaction	1	0.0039	0.0039	0.16	0.690
Error	20	0.4709	0.0235		
Total	23	0.9769			

Analysis of Variance for AC/BC

Source	DF	SS	MS	F	P
heat	1	0.04524	0.04524	15.62	0.001
chromo	1	0.00000	0.00000	0.00	1.000
Interaction	1	0.00385	0.00385	1.33	0.262
Error	20	0.05793	0.00290		
Total	23	0.10702			

45. C.A. = Congruence angle

Analysis of Variance for C.A.

Source	DF	SS	MS	F	P
Group	3	7598	2533	14.83	0.000
Error	86	14690	171		
Total	89	22288			

```
                                 Individual 95% CIs For Mean
                                 Based on Pooled StDev
Level      N     Mean   StDev   -----+---------+---------+---------+-
Lateral   27     6.78   15.10                                (----*----)
Medial    26   -10.81   10.80            (----*----)
Multi     17   -18.29   15.09   (------*-----)
Normal    20    -7.00   10.76               (-----*-----)
                                -----+---------+---------+---------+-
Pooled StDev = 13.07              -20       -10         0        10
Tukey's pairwise comparisons

   Family error rate = 0.0500
Individual error rate = 0.0103

Critical value = 3.71

Intervals for (column level mean) - (row level mean)
          Lateral   Medial    Multi
Medial       8.16
            27.01

Multi       14.46     -3.21
            35.69     18.18

Normal       3.66    -14.01   -22.60
            23.89      6.39     0.02
```

47.

```
Analysis of Variance for response
Source     DF      SS      MS      F      P
subject     5   25.78    5.16   4.72  0.018
temp        2   30.34   15.17  13.87  0.001
Error      10   10.93    1.09
Total      17   67.06
```

49. G.C. = glucose concentration

```
Analysis of Variance for G.C.
Source     DF       SS      MS      F      P
group       3    8.341   2.780  10.18  0.001
subject     5    8.774   1.755   6.43  0.002
Error      15    4.096   0.273
Total      23   21.210
```

51.

```
Analysis of Variance for T3
Source     DF      SS     MS      F      P
subject    11    8967    815   2.55  0.030
day         2   12466   6233  19.50  0.000
Error      22    7033    320
Total      35   28467
```

53. BBL = blood bilirubin levels

```
Analysis of Variance for BBL
Source    DF      SS      MS       F        P
Group      2    4077    2039    3.31    0.090
Error      8    4931     616
Total     10    9008

                                        Individual 95% CIs For Mean
                                        Based on Pooled StDev
Level       N     Mean    StDev    ---+---------+---------+---------+---
Control     4    63.50    28.25        (---------*---------)
Hypercar    4    50.00    22.69    (---------*---------)
Hyperosm    3    98.00    22.27                (-----------*----------)
                                   ---+---------+---------+---------+---
Pooled StDev =   24.83                30        60        90       120

Tukey's pairwise comparisons

   Family error rate = 0.0500
Individual error rate = 0.0213

Critical value = 4.04

Intervals for (column level mean) - (row level mean)

            Control    Hypercar

Hypercar      -36.7
               63.7

Hyperosm      -88.7     -102.2
               19.7        6.2
```

55.

```
Analysis of Variance for breathing scores
Source    DF        SS        MS        F        P
group      2    244.17    122.08    14.50    0.000
Error     38    319.88      8.42
Total     40    564.05

                                       Individual 95% CIs For Mean
                                       Based on Pooled StDev
Level      N      Mean    StDev    ----+---------+---------+---------+--
1         13    13.231    1.739    (------*-----)
2         14    13.786    2.833       (-----*-----)
3         14    18.643    3.713                       (------*-----)
                                   ----+---------+---------+---------+--
Pooled StDev = 2.901                  12.5      15.0      17.5      20.0

Tukey's pairwise comparisons

   Family error rate = 0.0500
```

```
Individual error rate = 0.0195
Critical value = 3.45

Intervals for (column level mean) - (row level mean)

              1          2

    2      -3.281
            2.171

    3      -8.138    -7.532
           -2.686    -2.182
```

57. PSWQ = PSWQ score

```
Analysis of Variance for PSWQ
Source       DF          SS         MS         F        P
Group         3     16654.9     5551.6     74.11    0.000
Error       115      8614.6       74.9
Total       118     25269.5

                                          Individual 95% CIs For Mean
                                          Based on Pooled StDev

Level       N       Mean      StDev       -----+---------+---------+---------+-
1          15     62.933      8.556                                (---*---)
2          30     38.333      7.494       (--*--)
3          19     64.158     10.259                                 (---*---)
4          55     66.536      8.678                                     (--*-)
                                          -----+---------+---------+---------+-
Pooled StDev = 8.655                          40        50        60        70

Tukey's pairwise comparisons

    Family error rate = 0.0500
Individual error rate = 0.0103

Critical value = 3.69

Intervals for (column level mean) - (row level mean)

                1            2           3

    2       17.459
            31.741

    3       -9.025      -32.446
             6.575      -19.203

    4      -10.181      -33.329      -8.388
             2.975      -23.077       3.631
```

59.

```
Analysis of Variance for Age
Source     DF         SS         MS         F        P
Group       2    16323.2     8161.6    139.79    0.000
Error     189    11034.7       58.4
Total     191    27357.9

                                   Individual 95% CIs For Mean
                                   Based on Pooled StDev
Level          N      Mean     StDev  -+---------+---------+---------+-----
Daughter      50    49.420     7.508  (--*-)
Husband       65    71.985     7.516                                  (-*-)
Wife          77    68.649     7.828                              (-*-)
                                      -+---------+---------+---------+-----
Pooled StDev =       7.641          48.0      56.0      64.0      72.0

Tukey's pairwise comparisons

    Family error rate = 0.0500
Individual error rate = 0.0192

Critical value = 3.34

Intervals for (column level mean) - (row level mean)

           Daughter    Husband

Husband     -25.959
            -19.170

Wife        -22.507      0.296
            -15.952      6.375
```

61. SAP = serum alkaline phosphatase level

```
Analysis of Variance for SAP
Source    DF        SS       MS       F       P
Grade      2     36181    18091    5.55   0.009
Error     29     94560     3261
Total     31    130742
```

```
                                   Individual 95% CIs For Mean
                                   Based on Pooled StDev
Level     N      Mean     StDev    -+---------+---------+---------+-----
I         9    118.00     61.85     (---------*--------)
II        8    143.63     55.90           (---------*---------)
III      15    194.80     54.82                        (-------*------)
                                   -+---------+---------+---------+-----
Pooled StDev = 57.10               80        120       160       200

Tukey's pairwise comparisons

    Family error rate = 0.0500
Individual error rate = 0.0197

Critical value = 3.49

Intervals for (column level mean) - (row level mean)

               I           II

II          -94.1
             42.8

III        -136.2       -112.9
            -17.4         10.5
```

63.

```
Analysis of Variance for Hematocrit
Source     DF        SS        MS        F        P
Group       2     817.5     408.8    20.26    0.000
Error      27     544.8      20.2
Total      29    1362.3

                                       Individual 95% CIs For Mean
                                       Based on Pooled StDev
Level          N     Mean    StDev     --+---------+---------+---------+----
Sham          10   38.200    2.573      (----*----)
Treated       15   40.200    5.348           (---*---)
Untreated      5   53.200    4.604                          (------*------)
                                       --+---------+---------+---------+----
Pooled StDev =      4.492              36.0      42.0      48.0      54.0

Tukey's pairwise comparisons

    Family error rate = 0.0500
Individual error rate = 0.0196
```

```
Critical value = 3.51

Intervals for (column level mean) - (row level mean)

                Sham     Treated

Treated        -6.551
                2.551
Untreated     -21.106    -18.757
               -8.894     -7.243
```

65. Both = rhIGF-I + rhGH

```
Analysis of Variance for Response
Source     DF        SS        MS        F        P
Group       3     4.148     1.383     1.39    0.282
Error      16    15.898     0.994
Total      19    20.046

                                   Individual 95% CIs For Mean
                                   Based on Pooled StDev
Level       N      Mean     StDev  ---------+---------+---------+-------
Both        5    11.520     0.653                (--------*---------)
rhGH        5    11.250     0.570             (--------*---------)
rhIGF-I     6    10.800     1.418        (---------*--------)
Saline      4    10.250     0.971   (---------*----------)
                                   ---------+---------+---------+-------
Pooled StDev =   0.997                     10.0      11.0      12.0

Tukey's pairwise comparisons

    Family error rate = 0.0500
Individual error rate = 0.0113

Critical value = 4.05

Intervals for (column level mean) - (row level mean)

              Both        rhGH     rhIGF-I

rhGH       -1.5354
            2.0754

rhIGF-I    -1.0086     -1.2786
            2.4486      2.1786

Saline     -0.6450     -0.9150     -1.2927
            3.1850      2.9150      2.3927
```

Chapter 9

9.3.1. **(a)** Direct, **(b)** Direct, **(c)** Inverse

9.3.3. $\hat{y} = 560 + 0.140x$

9.3.5. $\hat{y} = 68.6 - 19.5x$

9.3.7. $\hat{y} = 0.193 + 0.00628x$

9.4.1.

Predictor	Coef	SE Coef	T	P
Constant	559.90	29.13	19.22	0.000
Meth Dos	0.13989	0.06033	2.32	0.035

S = 68.28 R-Sq = 26.4% R-Sq(adj) = 21.5%

Analysis of Variance

Source	DF	SS	MS	F	P
Regression	1	25063	25063	5.38	0.035
Residual Error	15	69923	4662		
Total	16	94986			

Confidence interval for $\hat{\beta}_1$ .011, .268

9.4.3.

Predictor	Coef	SE Coef	T	P
Constant	68.64	16.68	4.12	0.006
Cmax w/	-19.529	4.375	-4.46	0.004

S = 18.87 R-Sq = 76.9% R-Sq(adj) = 73.0%

Analysis of Variance

Source	DF	SS	MS	F	P
Regression	1	7098.4	7098.4	19.93	0.004
Residual Error	6	2137.4	356.2		
Total	7	9235.9			

Confidence interval for $\hat{\beta}_1$ -30.23, -8.82

9.4.5.

Predictor	Coef	SE Coef	T	P
Constant	0.19290	0.04849	3.98	0.001
DTPA GFR	0.006279	0.001059	5.93	0.000

S = 0.09159 R-Sq = 58.5% R-Sq(adj) = 56.8%

Analysis of Variance

Source	DF	SS	MS	F	P
Regression	1	0.29509	0.29509	35.18	0.000
Residual Error	25	0.20972	0.00839		
Total	26	0.50481			

Confidence interval for $\hat{\beta}_1$ 0.0041, 0.0085

9.5.1. **(a)** 580.6, 651.2 **(b)** 466.1, 765.6
9.5.3. **(a)** −30.42, 5.22 **(b)** −62.11, 36.92
9.5.5. **(a)** 0.3727, 0.4526 **(b)** 0.2199, 0.6055
9.7.1. $r = .466$, $t = 2.23$, $p = .038$, $.030 < \rho < .775$
9.7.3. $r = -.812$, $t = -3.11$, $p = .027$, $-1 < \rho < -.152$
9.7.5. $r = -.531$, $t = -3.31$, $p = .003$, $-.770 < \rho < -.211$

Review Exercises

17. BOARD = −191 + 4.68 AVG, $r^2 = .772$, $t = 17.163$, $p < .001$

19.
```
y-hat = 12.6 + 1.80x

Predictor      Coef    SE Coef      T       P
Constant     12.641      2.133   5.93   0.000
no. of p     1.8045     0.5585   3.23   0.005

S = 7.081    R-Sq = 38.0%    R-Sq(adj) = 34.4%

Analysis of Variance

Source            DF       SS       MS       F       P
Regression         1   523.41   523.41   10.44   0.005
Residual Error    17   852.38    50.14
Total             18  1375.79
```

21.
```
The regression equation is
B = 1.28 + 0.851 A

Predictor      Coef    SE Coef      T       P
Constant     1.2763     0.3935   3.24   0.006
A            0.8513     0.1601   5.32   0.000

S = 0.2409    R-Sq = 68.5%    R-Sq(adj) = 66.1%

Analysis of Variance

Source            DF       SS       MS       F       P
Regression         1   1.6418   1.6418   28.29   0.000
Residual Error    13   0.7545   0.0580
Total             14   2.3963
```

23. $\hat{y} = 61.8819 + .509687x$; V.R. = 4.285; $.10 > p > .05$; $t = 2.07$; 95% C.I. for ρ: −.03, .79; 110.3022; 87.7773, 132.8271

25. $\hat{y} = 37.4564 + .0798x$; V.R. = 77.13; $p < .005$; $t = 8.78$; 95% C.I. for ρ: .91, 1; 40.63, 42.27.

29.
```
The regression equation is
A = 570 + 0.429 B

Predictor        Coef     SE Coef       T        P
Constant        569.8       141.2    4.03    0.000
B             0.42927     0.04353    9.86    0.000

S = 941.6    R-Sq = 54.0%    R-Sq(adj) = 53.4%

Pearson correlation of B and A = 0.735
P-Value = 0.000
```

31.
```
The regression equation is
y = 45.0 + 0.867 x

Predictor        Coef     SE Coef        T        P
Constant        44.99       33.54     1.34    0.193
x             0.86738     0.07644    11.35    0.000

S = 102.9    R-Sq = 84.8%    R-Sq(adj) = 84.2%

Pearson correlation of x and y = 0.921
P-Value = 0.000
```

33.
```
The regression equation is
S = -1.26 + 2.10 DBS

Predictor       Coef    SE Coef        T        P
Constant      -1.263      3.019    -0.42    0.680
DBS           2.0970     0.1435    14.62    0.000

S = 8.316    R-Sq = 90.3%    R-Sq(adj) = 89.9%

Pearson correlation of S and DBS = 0.950
P-Value = 0.000
```

35.
```
The regression equation is
log y = 2.06 + 0.0559 PCu

Predictor        Coef     SE Coef       T        P
Constant       2.0603      0.3007    6.85    0.000
PCu           0.05593     0.01631    3.43    0.001
```

```
S = 0.3873    R-Sq = 16.4%    R-Sq(adj) = 15.0%

Pearson correlation of PCu and log y = 0.405
P-Value = 0.001
```

37. The regression equation is

```
C6 = -0.141 - 1.33 C5

Predictor        Coef    SE Coef        T       P
Constant      -0.1413     0.2267    -0.62   0.540
C5            -1.3286     0.1242   -10.69   0.000

S = 1.086    R-Sq = 84.5%    R-Sq(adj) = 83.7%

Pearson correlation of IGELogE and SkinLogE = -0.919
P-Value = 0.000
```

39. Normotensive C6 = C4 − C5, C7 = (C4 + C5)/2, C8 = C2 − C3, C9 = (C2 + C3)/2

```
The regression equation is
C6 = 4.2 + 0.106 C7

Predictor       Coef    SE Coef      T       P
Constant        4.19      17.30   0.24   0.811
C7            0.1060     0.1590   0.67   0.512

S = 5.251    R-Sq = 2.0%    R-Sq(adj) = 0.0%

Pearson correlation of C6 and C7 = 0.141
P-Value = 0.512

The regression equation is
C8 = 0.2 + 0.268 C9

Predictor       Coef    SE Coef      T       P
Constant        0.25      18.53   0.01   0.989
C9            0.2682     0.2932   0.91   0.370

S = 5.736    R-Sq = 3.7%    R-Sq(adj) = 0.0%

Pearson correlation of C8 and C9 = 0.191
P-Value = 0.370
```

Preeclamptic

```
The regression equation is
C6 = 57.9 - 0.363 C7
```

```
Predictor      Coef     SE Coef      T        P
Constant      57.89      17.10      3.39    0.003
C7          -0.3625     0.1273     -2.85    0.009

S = 7.109    R-Sq = 26.9%    R-Sq(adj) = 23.6%

Pearson correlation of C6 and C7 = -0.519
P-Value = 0.009

The regression equation is
C8 = 54.4 - 0.540 C9

Predictor       Coef    SE Coef      T        P
Constant      54.377      9.771     5.56    0.000
C9           -0.5403     0.1154    -4.68    0.000

S = 5.787    R-Sq = 49.9%    R-Sq(adj) = 47.6%

Pearson correlation of C8 and C9 = -0.707
P-Value = 0.000
```

41.
```
The regression equation is
LBMD = 0.131 + 0.511 ABMD

Predictor       Coef    SE Coef      T        P
Constant     0.13097    0.05413    2.42    0.018
ABMD         0.51056    0.05935    8.60    0.000

S = 0.09188    R-Sq = 53.6%    R-Sq(adj) = 52.9%

Pearson correlation of ABMD and LBMD = 0.732
P-Value = 0.000
```

43. WL, VO_2

```
The regression equation is
WL = 0.01 + 0.262 VO2

Predictor       Coef    SE Coef      T        P
Constant       0.013      1.308    0.01    0.992
VO2          0.26237    0.07233    3.63    0.003

S = 1.835    R-Sq = 52.3%    R-Sq(adj) = 48.3%

Pearson correlation of WL and VO2 = 0.723
P-Value = 0.003
```

WL, AT

```
The regression equation is
WL = 0.75 + 0.367 AT

Predictor     Coef    SE Coef      T        P
Constant     0.752     1.761     0.43    0.677
AT          0.3668    0.1660     2.21    0.047

S = 2.241    R-Sq = 28.9%    R-Sq(adj) = 23.0%

Pearson correlation of WL and AT = 0.538
P-Value = 0.047
```

WL, ET

```
The regression equation is
WL = 0.74 + 0.00637 ET

Predictor        Coef      SE Coef      T        P
Constant        0.739       1.173     0.63    0.541
ET           0.006375    0.001840     3.46    0.005

S = 1.879        R-Sq = 50.0%        R-Sq(adj) = 45.8%

Pearson correlation of WL and ET = 0.707
P-Value = 0.005
```

45.

```
The regression equation is
CL/F = 19.4 + 0.893 CLer

Predictor        Coef     SE Coef       T        P
Constant       19.393       4.496     4.31    0.000
CLer          0.89250     0.05671    15.74    0.000

S = 28.20    R-Sq = 59.3%    R-Sq(adj) = 59.1%

Pearson correlation of CL/F and CLer = 0.770
P-Value = 0.000
```

Chapter 10

10.3.1. $\hat{y} = -31.4 + 0.473x_1 + 1.07x_2$

10.3.3. $\hat{y} = 13.45 + 4.02x_1 + 2.81x_2$

10.3.5. $\hat{y} = -422.00 + 11.17x_1 - .63x_2$

10.4.1.

Analysis of Variance

Source	DF	Sum of Squares	Mean Square	F value	Pr > F
Model	2	1576.99011	788.49506	185.13	<.0001
Error	32	136.29516	4.25922		
Corrected Total	34	1713.28527			

Root MSE	2.06379	R-Square	0.9204
Dependent Mean	51.25086	Adj R-Sq	0.9155
Coeff Var	4.02684		

Parameter Estimates

Variable	DF	Parameter Estimate	Standard Error	t Value	Pr > |t|	95% Confidence Limits	
Intercept	1	-31.42480	6.14747	-5.11	<.0001	-43.94678	-18.90282
X1	1	0.47317	0.06117	7.74	<.0001	0.34858	0.59776
X2	1	1.07117	0.06280	17.06	<.0001	0.94326	1.19909

(a) .9204 **(c)** X_1 p-value $< .0001$, X_2 p-value $< .0001$ **(d)** 95% C.I. for slope for X_1: $(0.34858 - 0.59776)$, 95% C.I. for slope for X_2: $(0.94326 - 1.19909)$

10.4.3.

Analysis of Variance

Source	DF	Sum of Squares	Mean Square	F Value	Pr > F
Model	2	452.56375	226.28188	7.05	0.0210
Error	7	224.70025	32.10004		
Corrected Total	9	677.26400			

Root MSE	5.66569	R-Square	0.6682
Dependent Mean	57.16000	Adj R-Sq	0.5734
Coeff Var	9.91198		

Parameter Estimates

Variable	DF	Parameter Estimate	Standard Error	t Value	Pr > |t|	95% Confidence Limits	
Intercept	1	13.44923	13.23156	1.02	0.3433	-17.83843	44.73689
X1	1	4.01680	1.07136	3.75	0.0072	1.48344	6.55016
X2	1	2.81175	1.37859	2.04	0.0808	-0.44809	6.07160

(a) .6682 **(c)** X_1 p-value $= .0072$, X_2 p-value $= .0808$ **(d)** 95% C.I. for slope for X_1: $(1.48344 - 6.55016)$

10.4.5.

Analysis of Variance

Source	DF	Sum of Squares	Mean Square	F Value	Pr > F
Model	2	17018	8508.89242	4.89	0.0175
Error	22	38282	1740.10069		
Corrected Total	24	55300			

Root MSE	41.71451	R-Square	0.3077
Dependent Mean	537.00000	Adj R-Sq	0.2448
Coeff Var	7.76807		

Parameter Estimates

Variable	DF	Parameter Estimate	Standard Error	t Value	Pr > |t|	95% Confidence Limits	
Intercept	1	-421.99671	339.76199	-1.24	0.2273	-1126.61995	282.62653
X1	1	11.16613	3.65523	3.05	0.0058	3.58564	18.74663
X2	1	-0.63032	0.93826	-0.67	0.5087	-2.57615	1.31552

(a) .3077 **(c)** X_1 p-value = .0058, X_2 p-value = .5807 **(d)** 95% C.I. for slope for X_1: (3.58564 − 18.74663)

10.5.1. C.I.: 50.289, 51.747. P.I.: 46.751, 55.284

10.5.3. C.I.: 44.22, 56.59; P.I.: 35.64, 65.17

10.5.5. C.I.: 514.31, 550.75; P.I.: 444.12, 620.94

10.6.1. **(a)** Pairwise correlations:

	DNA-Bloo	Co-cult	DNA-Rect
Co-cult	0.360		
DNA-Rect	0.532	0.303	
RNA-Rect	0.202	0.674	0.430

(b) $R = .370, F = 7.06, p = .001$ **(c)** $r_{y1.23} = .3472, r_{y2.13} = .5232, r_{y3.12} = -.2538$ **(d)** $r_{12.y3} = -.1660$ **(e)** $r_{13.y2} = .6615$ **(f)** $r_{23.y1} = .3969$

10.6.3. **(a)** $R = .9517, F = 57.638, p < .005$
(b), (c)
$r_{y1.2} = .9268, t = 8.549, p < .01; r_{y2.1} = .3785, t = 1.417, .20 > p > .10;$
$r_{12.y} = -.1789, t = -.630, p > .20$

Review Exercises

7. $R = .3496$ $F = .83$ $(p > .10)$

9. **(a)** $\hat{y} = 11.419 + 1.2598x_1 + 3.1067x_2$ **(b)** $R^2 = .92$

(c)

Source	SS	d.f.	MS	V.R.	p
Regression	1827.004659	2	913.50	69.048	$<.005$
Residual	158.728641	12	13.23		
	1985.7333	14			

(d) $\hat{y} = 11.419 + 1.2598(10) + 3.1067(5) = 39.55$

11. (a) $\hat{y} = -126.505 + .176x_1 - 1.563x_2 + 1.575x_3 + 1.6292x_4$

(b)

Source	SS	d.f.	MS	V.R.	p
Regression	30873.47	4	7718.367	13.37	$<.005$
Residual	5774.93	10	577.493		
	36648.40	14			

(c) $t_1 = 4.40; t_2 = -.78; t_3 = 3.53; t_4 = 2.59$

(d) $R^2_{y.1234} = .842423; R_{y.1234} = .91784$

13. (a) correlation

(b) log plasma adiponectin levels

(c) age and glomerular filtration rate

(d) both correlations were not significant

(h) subjects with end-stage renal disease

15. (a) correlation

(b) static inspiratory mouth pressure

(c) forced expiratory volume, peak expiratory flow, and maximal inspiratroy flow

(d) both correlations were not significant

(h) boys and girls ages 7–14

17.

	x1	x2	x3	x4	x5	x6	x7	x8	x9	x10	x11	x12	x13
X1	1												
X2	.737(**)	1											
X3	−.109	.244	1										
X4	.760(**)	.698(**)	.316	1									
X5	.556(**)	.608(**)	.273	.760(**)	1								
X6	.040	−.213	−.136	−.101	−.647(**)	1							
X7	−.291	−.289	−.093	−.293	−.412(*)	.231	1						
X8	.570(**)	.659(**)	.227	.568(**)	.763(**)	−.481(*)	−.555(**)	1					
X9	.555(**)	.566(**)	.146	.454(*)	.717(**)	−.503(**)	−.650(**)	.922(**)	1				
X10	.345	.508(**)	.419(*)	.455(*)	.640(**)	−.377	−.480(*)	.905(**)	.788(**)	1			
X11	−.467(*)	−.400(*)	−.224	−.621(**)	−.702(**)	.388(*)	.732(**)	−.652(**)	−.646(**)	−.582(**)	1		
X12	−.250	−.260	−.178	−.228	−.448(*)	.390(*)	.778(**)	−.641(**)	−.717(**)	−.667(**)	.796(**)	1	
X13	−.271	−.305	−.380	−.346	−.518(**)	.348	.524(**)	−.645(**)	−.707(**)	−.729(**)	.744(**)	.864(**)	1

** correlation is significant at the 0.01 level (2-tailed).

* correlation is significant at the 0.05 level (2-tailed).

19.

	v1	v2	v3	v4	v5	v6	v7	v8
v1	1							
v2	.123	1						
v3	.115	.963(**)	1					
v4	.417(**)	−.063	−.041	1				
v5	.005	−.102	−.103	−.059	1			
v6	.001	.270(**)	.295(**)	−.036	.137	1		
v7	−.113	−.074	−.076	.052	.134	.061	1	
v8	.077	−.002	−.023	.146	.165	−.202	−.032	1

** Correlation is significant at the 0.01 level (2-tailed).

Chapter 11

11.2.1. mobilizer: 0 G-CSF, 1-Etoposide

```
The regression equation is
conc = 12.9 - 0.0757 age - 5.48 mobilizer

Predictor        Coef     SE Coef       T        P
Constant       12.933       2.787    4.64    0.000
age          -0.07566     0.04388   -1.72    0.092
mobilize       -5.480       1.429   -3.83    0.000

S = 3.965     R-Sq = 27.1%     R-Sq(adj) = 23.6%

Analysis of Variance

Source            DF       SS        MS       F       P
Regression         2   240.02    120.01    7.63   0.002
Residual Error    41   644.60     15.72
Total             43   884.62
```

11.2.3.

```
The regression equation is
QTc = 23.0 + 39.4 sex + 0.825 dose

Predictor        Coef    SE Coef        T        P
Constant        22.98      46.92     0.49    0.632
sex             39.40      42.14     0.93    0.366
dose          0.82456    0.07556    10.91    0.000

S = 84.10     R-Sq = 89.6%       R-Sq(adj) = 88.1%
```

Analysis of Variance

Source	DF	SS	MS	F	P
Regression	2	850164	425082	60.10	0.000
Residual Error	14	99018	7073		
Total	16	949182			

11.3.1.

Step	1	2	3
Constant	51.93	116.07	115.54
MEM	0.66	0.60	0.57
T-Value	5.75	5.87	5.53
P-Value	0.000	0.000	0.000
SOCIALSU		-0.476	-0.492
T-Value		-5.28	-5.51
P-Value		0.000	0.000
CGDUR			0.122
T-Value			1.88
P-Value			0.064
S	17.4	15.4	15.2
R-Sq	25.20	41.92	43.97
R-Sq(adj)	24.44	40.72	42.22

11.3.3.

Alpha-to-Enter: 0.15 Alpha-to-Remove: 0.15

Response is REACTIVE on 6 predictors, with N = 68

Step	1	2	3
Constant	3.374	5.476	5.418
AGEABUSE	-0.38	-0.45	-0.42
T-Value	-2.49	-3.00	-2.91
P-Value	0.015	0.004	0.005
VERBALIQ		-0.0219	-0.0228
T-Value		-2.75	-2.93
P-Value		0.008	0.005
STIM			0.61
T-Value			2.05
P-Value			0.044

S	1.06	1.01	0.990
R-Sq	8.57	18.10	23.15
R-Sq(adj)	7.19	15.58	19.55
C-p	10.4	4.6	2.5

11.4.1.

Parameter	DF	Estimate	Standard Error	Wald Chi-Square	Pr > ChiSq
Intercept	1	2.1192	0.1740	148.3439	<.0001
sex	1	0.0764	0.2159	0.1252	0.7234

Odds Ratio Estimates

Effect	Point Estimate	95% Wald Confidence Limits	
sex	1.079	0.707	1.648

Review Exercises

15. $\hat{y} = 1.87 + 6.3772x_1 + 1.9251x_2$

Coefficient	Standard Error	t
1.867	.3182	5.87
6.3772	.3972	16.06
1.9251	.3387	5.68
$R^2 = .942$		

Source	SS	*d.f.*	MS	V.R.
Regression	284.6529	2	142.3265	202.36954
Residual	17.5813	25	.7033	
	302.2342	27		

17. $\hat{y} = -1.1361 + .07648x_1 + .7433x_2 - .8239x_3 - .02772x_1x_2 + .03204x_1x_3$

Coefficient	Standard Deviation	t	p^a
−1.1361	.4904	−2.32	$.05 > p > .02$
.07648	.01523	5.02	$< .01$
.7433	.6388	1.16	$> .20$
−.8239	.6298	−1.31	$.20 > p > .10$
−.02772	.02039	−1.36	$.20 > p > .10$
.03204	.01974	1.62	$.20 > p > .10$

[a] Approximate. Obtained by using 35 d.f.

$R^2 = .834$

Source	SS	*d.f.*	MS	V.R.
Regression	3.03754	5	.60751	34.04325
Residual	.60646	34	.01784	
	3.64400	39		

$$x_2 = \begin{cases} 1 \text{ if } A \\ 0 \text{ if otherwise} \end{cases} \qquad x_3 = \begin{cases} 1 \text{ if } B \\ 0 \text{ if otherwise} \end{cases}$$

For A: $\hat{y} = (-1.1361 + .7433) + (.07648 - .02772)x_1 = -.3928 + .04875x_1$
For B: $\hat{y} = (-1.1361 + .8239) + (.07648 + .03204)x_1 = -1.96 + .10852x_1$
For C: $\hat{y} = -1.1361 + .07648x_1$

23.

Response = V, Dummy1 = 1 if infant, 0 otherwise, Dummy2 = 1 if Child, 0 otherwise

```
The regression equation is
V = 11.7 + 0.137 W - 11.4 DUMMY1 -11.7 DUMMY2 + 0.226 INTER1 + 0.223
INTER2

Predictor         Coef      SE Coef         T        P
Constant        11.750        3.822      3.07    0.004
W              0.13738      0.05107      2.69    0.010
DUMMY1         -11.421        4.336     -2.63    0.012
DUMMY2         -11.731        3.966     -2.96    0.005
INTER1          0.2264       0.2208      1.03    0.311
INTER2         0.22332      0.06714      3.33    0.002

S = 1.73234     R=sq = 94.9%     R=sq(adj) = 94.3%

Analysis of Variance

Source            DF        SS        MS         F        P
Regression         5   2304.47    460.89    153.58    0.000
Residual Error    41    123.04      3.00
Total             46   2427.51

Source      DF    Seq SS
W            1   2265.07
DUMMY1       1      5.59
DUMMY2       1      0.00
INTER1       1      0.60
INTER2       1     33.20
```

```
Unusual Observations

Obs      W         V       Fit   SE Fit   Residual   St Resid
 17   10.8     8.366     4.257    0.496      4.109      2.48R
 36   47.0    15.400    16.971    1.145     -1.571     -1.21X
 41   96.0    20.000    24.938    1.265     -4.938    -4.17RX
 46   87.0    30.900    23.702    0.881      7.198      4.83R

R denotes an observation with a large standardized residual.
X denotes an observation whose X value gives it large influence.
```

Chapter 12

12.3.1. $X^2 = 2.072, p > .05$
12.3.3. $X^2 = 3.417, p > .10$
12.3.5. $X^2 = 2.21, p > .10$
12.4.1. $X^2 = .078, p > .10$
12.4.3. $X^2 = 816.410, p < .005$
12.4.5. $X^2 = 42.579, p < .005$
12.5.1. $X^2 = 3.622, d.f. = 3, p > .10$
12.5.3. $X^2 = .297, d.f. = 1, p > .10$
12.5.5. $X^2 = 82.373, d.f. = 2, p < .005$
12.6.1. Since $b = 7 > 3$ (for $A = 10$, $B = 10$, $a = 8$), $p > 2(.035) = .070$. Do not reject H_0.
12.6.3. Since $b = 1$ (for $A = 9$, $B = 7$, $a = 16$), $p = 2(.002) = .004$. Reject H_0.
12.7.1. $\widehat{RR} = 13.51$, 95% C.I. 9.7, 18.8
12.7.3. $X^2 = 12.898, p < .005, \widehat{OR} = 1.967$
12.7.5. $\widehat{OR}_{MH} = 3.733, X^2_{MH} = 25.095, p < .005$
12.8.1.

```
Number of Cases:  53  Censored:34  (64.15%)  Events:  19

        Survival Time    Standard Error    95% Confidence Interval

Mean:          12.57               1.10     (   10.40,      14.73  )
  (Limited to    19.00  )
Median:        16.00               1.80     (   12.47,      19.53  )

                       Percentiles
                     25.00   50.00   75.00

Value                18.00   16.00    4.00
Standard Error                1.80    3.76
```

12.8.5. Support group:

```
Number of Cases:  22  Censored:  0  (  .00%)  Events:  22
```

```
          Survival Time   Standard Error  95% Confidence  Interval

Mean:          45.09            3.98       (     37.29,     52.89 )
Median:        60.00             .00       (         . ,        . )

                       Percentiles

                     25.00          50.00     75.00
Value                60.00          60.00     26.00
Standard Error         .              .        6.96
```

Nonsupport group:

```
Number of Cases: 28  Censored: 0   ( .00%)  Events: 28

          Survival Time  Standard Error   95% Confidence  Interval

Mean:        16.04              1.86       (     12.39,     19.68 )
Median:      15.00              5.29       (      4.63,     25.37 )

                       Percentiles

                     25.00          50.00     75.00

Value                22.00          15.00      7.00
Standard Error        3.44           5.29       .92

Log Rank Statistic and (Significance): 29.22 ( .0000)

Breslow Statistic and (Significance): 23.42 ( .0000)

Tarone-Ware Statistic and (Significance): 26.28 ( .0000)
```

Review Exercises

15. $X^2 = 7.124$, $d.f. = 3$, $p > .05$, Fail to reject.
17. $X^2 = 2.40516, p > .10$
19. $X^2 = 5.1675, p > .10$
21. $X^2 = 67.8015\ p < .005$
23. $X^2 = 7.2577\ .05 > p > .025$
25. Independence
27. Homogeneity
35. Overall Satisfaction
$X^2 = 3.143$
d.f. $= 2$, $p = 0.208$
2 cells with expected counts less than 5.0
Pain
$X^2 = 0.444$
d.f. $= 2$, $p = 0.801$

2 cells with expected counts less than 5.0
Nausea and Vomiting
$X^2 = 0.483$
d.f. = 2, $p = 0.785$

37. $\widehat{OR} = 2.06$; 95% C.I.: .92, 4.61

41. $X^2 = 13.530$
d.f. = 1, $p = 0.000$

43. Test statistic = 2, $p = .019$ (one-sided test)

45. $X^2 = 8.749$
d.f. = 1, $p = 0.003$

47. $X^2 = 4.875$
d.f. = 1, $p = 0.027$

49. $\widehat{OR} = 3.79$; 95% C.I.: 1.52, 9.48OR

51. $X^2 = 11.589$
d.f. = 1, $p = 0.001$

53. No. of Resected PMs = 1

```
Number  of  Cases:  46   Censored:  0    (  .00%)  Events:  46

        Survival  Time    Standard Error   95% Confidence Interval

Mean:          45                      6       (        34,          57  )
Median:        33                      7       (        20,          46  )

                          Percentiles

                     25.00        50.00     75.00

Value                55.00        33.00     18.00
Standard  Error      12.15         6.78      1.79
```

No. of Resected PMs > 1

```
Number  of  Cases: 31  Censored: 0   (  .00%)  Events:  31

      Survival  Time   Standard  Error  95%  Confidence  Interval

Mean:          31                  5      (       20,            41  )
Median:        23                  4      (       14,            32  )

                          Percentiles

                     25.00       50.00      75.00

Value                41.00       23.00      10.00
Standard  Error       6.60        4.45       3.65
```

Log Rank Statistic and (Significance): 3.75 (.0527)

Breslow Statistic and (Significance): 4.11 (.0427)

Tarone–Ware Statistic and (Significance): 4.07 (.0437)

Chapter 13

13.3.1. $P = .3036, p = .6072$
13.3.3. $P(x \leq 2 \mid 13, .5) = .0112$. Since $.0112 < .05$, reject H_0. $p = .0112$
13.4.1. $T_+ = 48.5$. $.1613 < p < .174$
13.4.3. $T = 11.5, .1054 < p < .1308$
13.5.1. $X^2 = 16.13, p < .005$.
13.6.1. $T = 712.5, p = .2380$, Fail to reject H_0.
13.6.3. $S = 1772.5, p = .7566$, Fail to reject H_0.
13.7.1. $D = .3241, p < .01$
13.7.3. $D = .1319, p > .20$
13.8.1. $H = 11.38, p = .003, d.f. = 3$.
13.8.3. $H = 18.13, p < .0001, d.f. = 2$.
13.8.5. $H = 19.61$ (adjusted for ties), $p < .005$
13.9.1. $\chi_r^2 = 8.67, p = .01$
13.9.3. $\chi_r^2 = 29.38, p < .005$
13.10.1. $r_s = -0.07, p > .20$
13.10.3. $r_s = .018, n = 20, p > .05$
13.10.5. $r_s = -.43, n = 30, .01 < p < .02$
13.11.1. $\hat{\beta}_1 = 1.429$
$(\hat{\beta}_0)_{1,M} = -176.685$
$(\hat{\beta}_0)_{2,M} = -176.63$

Review Exercises

7. $T = 0, n = 7, p = .0078$
9. $\chi_r^2 = 16.2, p < .005$
11. $D = .1587, p > .20$
13. $r_s = .09, p = .4532$
15. $T = 29.5, p = 0.0263$, Reject H_0
17. $H = 9.02$, d.f. $= 3, p = 0.029$
$H = 9.30$, d.f. $= 3, p = 0.026$ (adjusted for ties)
19. $r_s = -.036, p = .802$
21. $T = 62.5, p = .0072$, Reject H_0
23. USO: $\chi_r^2 = 3.94, p = .140$
BSO: $\chi_r^2 = 4.77, p = .093$
25. $T = 89, p = .0046$, Reject H_0
27.
PFK: T $= 38, p = .8598$, Fail to reject H_0
HK: T $= 61.5, p = .0703$, Fail to reject H_0
LDH: T $= 37, p = .7911$, Fail to reject H_0
29. $r_s = .733, p = .001$

Chapter 14

14.2.1. **(a)** 5.8 **(b)** White: 10.0, Black: 3.7, **(c)** 9.43 **(d)** 5.5
(e) 9.43 **(f)** MN 22.2, MCD 34.5

14.2.3

Age (Years)	Population[a]	Deaths[b]	U.S. Population[c]	Age-Specific Death Rates (per 100,000)	Standard Population Based on U.S. Population 2000	Number of Expected Deaths in Standard Population
0–4	539,509	1,178	19,175,798	218.3	68139	149
5–14	1,113,920	224	41,077,577	20.1	145964	29
15–24	1,117,439	954	39,183,891	85.4	139235	119
25–34	1,213,415	1,384	39,891,724	114.1	141751	162
35–44	1,287,120	2,823	45,148,527	219.3	160430	352
45–54	1,085,150	5,271	37,677,952	485.7	133884	650
55–64	723,712	8,035	24,274,684	1110.2	86257	958
65 and Over	969,048	51,863	34,991,753	5352.0	124339	6655
Total	8,049,313	71,732	281,421,906	891.2	1000000	9073

Age-adjusted death rate = 9.1

14.3.1. **(a)** (10–14): 1.3, (15–19): 59.9, (20–24): 126.7, (25–29): 112.6, (30–34): 83.6, (35–39): 36.5, (40–over): 2.6; **(b)** 2142.1 **(c)** (10–14): 6.3, (15–19): 305.9, (20–24): 939.2, (25–29): 1502.3, (30–34): 1920.2, (35–39): 2102.9, (40–over): 2142.1 **(d)** 46.7

14.3.3. **(a)** (10–14): 1.2, (15–19): 58.5, (20–24): 120.2, (24–29): 113.7, (30–34): 84.2, (35–39): 33.8, (40–44): 6.0, (45 and over):.5; **(b)** 2089.6 **(c)** (10–14): 6.1, (15–19): 298.5, (20–24): 899.6, (25–29): 1468.1, (30–34): 1889.3, (35–39): 2058.2, (40–44): 2088.1, (45 and over): 2089.6 **(d)** 45.6

14.4.1. **(a)** immaturity ratio: 1997—7.3, 2001—8.1 **(b)** prevalence ratio: Nevada—22.2, United States—20.5 **(c)** incidence rate—14.5 per 100,000

Review Exercises

9. 8.9

11. Infant mortality: Total—5.7; white—5.3; nonwhite—6.5;
Cause of death: heart disease total—36.8; white 37.7; nonwhite 32.3
Cancer total—23.7; white—23.8; nonwhite—23.1
AIDS total—1.5; white .8; nonwhite 4.9
Immaturity ratio: total—7.0; white—6.7; nonwhite—7.5
Incident rate C-section: total—22.6; white 25.0; nonwhite—18.3

13. 15.9, 51.6

INDEX

Numbers preceded by A refer to Appendix pages.

H

I

J

K

L

M